T0214656

Lecture Notes in Computer Science 11477

Commenced Publication in 1973
Founding and Former Series Editors:
Gerhard Goos, Juris Hartmanis, and Jan van Leeuwen

More information about this series at http://www.springer.com/series/7410

Yuval Ishai · Vincent Rijmen (Eds.)

Advances in Cryptology – EUROCRYPT 2019

38th Annual International Conference on the Theory
and Applications of Cryptographic Techniques
Darmstadt, Germany, May 19–23, 2019
Proceedings, Part II

 Springer

Editors
Yuval Ishai
Technion
Haifa, Israel

Vincent Rijmen
COSIC Group
KU Leuven
Heverlee, Belgium

ISSN 0302-9743   ISSN 1611-3349   (electronic)
Lecture Notes in Computer Science
ISBN 978-3-030-17655-6   ISBN 978-3-030-17656-3   (eBook)
https://doi.org/10.1007/978-3-030-17656-3

LNCS Sublibrary: SL4 – Security and Cryptology

This Springer imprint is published by the registered company Springer Nature Switzerland AG
The registered company address is: Gewerbestrasse 11, 6330 Cham, Switzerland

Preface

Eurocrypt 2019, the 38th Annual International Conference on the Theory and Applications of Cryptographic Techniques, was held in Darmstadt, Germany, during May 19–23, 2019. The conference was sponsored by the International Association for Cryptologic Research (IACR). Marc Fischlin (Technische Universität Darmstadt, Germany) was responsible for the local organization. He was supported by a local organizing team consisting of Andrea Püchner, Felix Günther, Christian Janson, and the Cryptoplexity Group. We are deeply indebted to them for their support and smooth collaboration.

The conference program followed the now established parallel track system where the works of the authors were presented in two concurrently running tracks. The invited talks and the talks presenting the best paper/best young researcher spanned over both tracks.

We received a total of 327 submissions. Each submission was anonymized for the reviewing process and was assigned to at least three of the 58 Program Committee members. Committee members were allowed to submit at most one paper, or two if both were co-authored. Submissions by committee members were held to a higher standard than normal submissions. The reviewing process included a rebuttal round for all submissions. After extensive deliberations the Program Committee accepted 76 papers. The revised versions of these papers are included in these three volume proceedings, organized topically within their respective track.

The committee decided to give the Best Paper Award to the paper "Quantum Lightning Never Strikes the Same State Twice" by Mark Zhandry. The runner-up was the paper "Compact Adaptively Secure ABE for NC^1 from k Lin" by Lucas Kowalczyk and Hoeteck Wee. The Best Young Researcher Award went to the paper "Efficient Verifiable Delay Functions" by Benjamin Wesolowski. All three papers received invitations for the *Journal of Cryptology*.

The program also included an IACR Distinguished Lecture by Cynthia Dwork, titled "Differential Privacy and the People's Data," and invited talks by Daniele Micciancio, titled "Fully Homomorphic Encryption from the Ground Up," and François-Xavier Standaert, titled "Toward an Open Approach to Secure Cryptographic Implementations."

We would like to thank all the authors who submitted papers. We know that the Program Committee's decisions can be very disappointing, especially rejections of very good papers that did not find a slot in the sparse number of accepted papers. We sincerely hope that these works eventually get the attention they deserve.

We are also indebted to the members of the Program Committee and all external reviewers for their voluntary work. The committee's work is quite a workload. It has been an honor to work with everyone. The committee's work was tremendously simplified by Shai Halevi's submission software and his support, including running the service on IACR servers.

Finally, we thank everyone else—speakers, session chairs, and rump-session chairs—for their contribution to the program of Eurocrypt 2019. We would also like to thank the many sponsors for their generous support, including the Cryptography Research Fund that supported student speakers.

May 2019

Yuval Ishai
Vincent Rijmen

Eurocrypt 2019

**The 38th Annual International Conference
on the Theory and Applications of Cryptographic Techniques**

Sponsored by *the International Association for Cryptologic Research*

May 19–23, 2019
Darmstadt, Germany

General Chair

Marc Fischlin   Technische Universität Darmstadt, Germany

Program Co-chairs

Yuval Ishai   Technion, Israel
Vincent Rijmen   KU Leuven, Belgium and University of Bergen,
                 Norway

Program Committee

Michel Abdalla	CNRS and ENS Paris, France
Adi Akavia	University of Haifa, Israel
Martin Albrecht	Royal Holloway, UK
Elena Andreeva	KU Leuven, Belgium
Paulo S. L. M. Barreto	University of Washington Tacoma, USA
Amos Beimel	Ben-Gurion University, Israel
Alex Biryukov	University of Luxembourg, Luxembourg
Nir Bitansky	Tel Aviv University, Israel
Andrej Bogdanov	Chinese University of Hong Kong, SAR China
Christina Boura	University of Versailles and Inria, France
Xavier Boyen	QUT, Australia
David Cash	University of Chicago, USA
Melissa Chase	MSR Redmond, USA
Kai-Min Chung	Academia Sinica, Taiwan
Dana Dachman-Soled	University of Maryland, USA
Ivan Damgård	Aarhus University, Denmark
Itai Dinur	Ben-Gurion University, Israel
Stefan Dziembowski	University of Warsaw, Poland
Serge Fehr	Centrum Wiskunde & Informatica (CWI) and Leiden University, The Netherlands
Juan A. Garay	Texas A&M University, USA
Sanjam Garg	UC Berkeley, USA

Marshall Ball
James Bartusek
Balthazar Bauer
Carsten Baum
Christof Beierle
Fabrice Benhamouda
Iddo Bentov
Mario Berta
Ward Beullens
Ritam Bhaumik
Jean-François Biasse
Koen de Boer
Dan Boneh
Xavier Bonnetain
Charlotte Bonte
Carl Bootland
Jonathan Bootle
Joppe Bos
Adam Bouland
Florian Bourse
Benedikt Bünz
Wouter Castryck
Siu On Chan
Nishanth Chandran
Eshan Chattopadhyay
Yi-Hsiu Chen
Yilei Chen
Yu Long Chen
Jung-Hee Cheon
Mahdi Cheraghchi
Celine Chevalier
Nai-Hui Chia
Ilaria Chillotti
Chongwon Cho
Wutichai Chongchitmate
Michele Ciampi
Ran Cohen
Sandro Coretti
Ana Costache
Jan Czajkowski
Yuanxi Dai
Deepesh Data
Bernardo David
Alex Davidson
Thomas Debris-Alazard
Thomas De Cnudde

Thomas Decru
Luca De Feo
Akshay Degwekar
Cyprien Delpech de Saint Guilhem
Ioannis Demertzis
Ronald de Wolf
Giovanni Di Crescenzo
Christoph Dobraunig
Jack Doerner
Javad Doliskani
Leo Ducas
Yfke Dulek
Nico Döttling
Aner Ben Efraim
Maria Eichlseder
Naomi Ephraim
Daniel Escudero
Saba Eskandarian
Thomas Espitau
Pooya Farshim
Prastudy Fauzi
Rex Fernando
Houda Ferradi
Dario Fiore
Ben Fisch
Mathias Fitzi
Cody Freitag
Georg Fuchsbauer
Benjamin Fuller
Tommaso Gagliardoni
Steven Galbraith
Nicolas Gama
Chaya Ganesh
Sumegha Garg
Romain Gay
Peter Gazi
Craig Gentry
Marios Georgiou
Benedikt Gierlichs
Huijing Gong
Rishab Goyal
Lorenzo Grassi
Hannes Gross
Jens Groth
Paul Grubbs

Divya Gupta
Felix Günther
Helene Haagh
Björn Haase
Mohammad Hajiabadi
Carmit Hazay
Pavel Hubáček
Andreas Huelsing
Ilia Iliashenko
Muhammad Ishaq
Joseph Jaeger
Eli Jaffe
Aayush Jain
Abhishek Jain
Stacey Jeffery
Zhengfeng Ji
Yael Kalai
Daniel Kales
Chethan Kamath
Nathan Keller
Eike Kiltz
Miran Kim
Sam Kim
Taechan Kim
Karen Klein
Yash Kondi
Venkata Koppula
Mukul Kulkarni
Ashutosh Kumar
Ranjit Kumaresan
Rio LaVigne
Virginie Lallemand
Esteban Landerreche
Brandon Langenberg
Douglass Lee
Eysa Lee
François Le Gall
Chaoyun Li
Wei-Kai Lin
Qipeng Liu
Tianren Liu
Alex Lombardi
Julian Loss
Yun Lu
Vadim Lyubashevsky
Fermi Ma

Saeed Mahloujifar
Christian Majenz
Rusydi Makarim
Nikolaos Makriyannis
Nathan Manohar
Antonio Marcedone
Daniel Masny
Alexander May
Noam Mazor
Willi Meier
Rebekah Mercer
David Mestel
Peihan Miao
Brice Minaud
Matthias Minihold
Konstantinos Mitropoulos
Tarik Moataz
Hart Montgomery
Andrew Morgan
Pratyay Mukherjee
Luka Music
Michael Naehrig
Gregory Neven
Phong Nguyen
Jesper Buus Nielsen
Ryo Nishimaki
Daniel Noble
Adam O'Neill
Maciej Obremski
Sabine Oechsner
Michele Orrù
Emmanuela Orsini
Daniel Ospina
Giorgos Panagiotakos
Omer Paneth
Lorenz Panny
Anat Paskin-Cherniavsky
Alain Passelègue
Kenny Paterson
Chris Peikert
Geovandro Pereira
Léo Perrin
Edoardo Persichetti
Naty Peter

Rachel Player
Oxana Poburinnaya
Yuriy Polyakov
Antigoni Polychroniadou
Eamonn Postlethwaite
Willy Quach
Ahmadreza Rahimi
Sebastian Ramacher
Adrián Ranea
Peter Rasmussen
Shahram Rasoolzadeh
Ling Ren
Joao Ribeiro
Silas Richelson
Thomas Ricosset
Tom Ristenpart
Mike Rosulek
Dragos Rotaru
Yann Rotella
Lior Rotem
Yannis Rouselakis
Arnab Roy
Louis Salvail
Simona Samardziska
Or Sattath
Guillaume Scerri
John Schanck
Peter Scholl
André Schrottenloher
Sruthi Sekar
Srinath Setty
Brian Shaft
Ido Shahaf
Victor Shoup
Jad Silbak
Mark Simkin
Shashank Singh
Maciej Skórski
Caleb Smith
Fang Song
Pratik Soni
Katerina Sotiraki
Florian Speelman
Akshayaram Srinivasan

Uri Stemmer
Noah
    Stephens-Davidowitz
Alan Szepieniec
Gelo Noel Tabia
Aishwarya
    Thiruvengadam
Sergei Tikhomirov
Rotem Tsabary
Daniel Tschudy
Yiannis Tselekounis
Aleksei Udovenko
Dominique Unruh
Cédric Van Rompay
Prashant Vasudevan
Muthu
    Venkitasubramaniam
Daniele Venturi
Benoît Viguier
Fernando Virdia
Ivan Visconti
Giuseppe Vitto
Petros Wallden
Alexandre Wallet
Qingju Wang
Bogdan Warinschi
Gaven Watson
Hoeteck Wee
Friedrich Wiemer
Tim Wood
Keita Xagawa
Sophia Yakoubov
Takashi Yamakawa
Arkady Yerukhimovich
Eylon Yogev
Nengkun Yu
Yu Yu
Aaram Yun
Thomas Zacharias
Greg Zaverucha
Liu Zeyu
Mark Zhandry
Chen-Da Liu Zhang

Abstracts of Invited Talks

Differential Privacy and the People's Data

IACR DISTINGUISHED LECTURE

Cynthia Dwork[1]

Harvard University
dwork@seas.harvard.edu

Abstract. Differential Privacy will be the confidentiality protection method of the 2020 US Decennial Census. We explore the technical and social challenges to be faced as the technology moves from the realm of information specialists to the large community of consumers of census data.

Differential Privacy is a definition of privacy tailored to the statistical analysis of large datasets. Roughly speaking, differential privacy ensures that anything learnable about an individual could be learned independent of whether the individual opts in or opts out of the data set under analysis. The term has come to denote a field of study, inspired by cryptography and guided by theoretical lower bounds and impossibility results, comprising algorithms, complexity results, sample complexity, definitional relaxations, and uses of differential privacy when privacy is not itself a concern.

From its inception, a motivating scenario for differential privacy has been the US Census: data of the people, analyzed for the benefit of the people, to allocate the people's resources (hundreds of billions of dollars), with a legal mandate for privacy. Over the past 4–5 years, differential privacy has been adopted in a number of industrial settings by Google, Microsoft, Uber, and, with the most fanfare, by Apple. In 2020 it will be the confidentiality protection method for the US Decennial Census.

Census data are used throughout government and in thousands of research studies every year. This mainstreaming of differential privacy, the transition from the realm of technically sophisticated information specialists and analysts into much broader use, presents enormous technical and social challenges. The Fundamental Theorem of Information Reconstruction tells us that overly accurate estimates of too many statistics completely destroys privacy. Differential privacy provides a measure of privacy loss that permits the tracking and control of cumulative privacy loss as data are analyzed and re-analyzed. But provably no method can permit the data to be explored without bound. How will the privacy loss "budget" be allocated? Who will enforce limits?

More pressing for the scientific community are questions of how the multitudes of census data consumers will interact with the data moving forward. The Decennial Census is simple, and the tabulations can be handled well with existing technology. In contrast, the annual American Community Survey, which covers only a few million households yearly, is rich in personal details on subjects from internet access in the home to employment to ethnicity, relationships among persons in the home, and fertility. We are not (yet?) able to

---

[1] Supported in part by NSF Grant 1763665 and the Sloan Foundation.

offer differentially private algorithms for every kind of analysis carried out on these data. Historically, confidentiality has been handled by a combination of data summaries, restricted use access to the raw data, and the release of public-use microdata, a form of noisy individual records. Summary statistics are the bread and butter of differential privacy, but giving even trusted and trust-worthy researchers access to raw data is problematic, as their published findings are a vector for privacy loss: think of the researcher as an arbitrary non-differentially private algorithm that produces outputs in the form of published findings. The very *choice* of statistic to be published is inherently not privacy-preserving! At the same time, past microdata noising techniques can no longer be considered to provide adequate privacy, but generating synthetic public-use microdata while ensuring differential privacy is a computationally hard problem. Nonetheless, combinations of exciting new techniques give reason for optimism.

Towards an Open Approach to Secure Cryptographic Implementations

François-Xavier Standaert[1]

UCL Crypto Group, Université Catholique de Louvain, Belgium

Abstract. In this talk, I will discuss how recent advances in side-channel analysis and leakage-resilience could lead to both stronger security properties and improved confidence in cryptographic implementations. For this purpose, I will start by describing how side-channel attacks exploit physical leakages such as an implementation's power consumption or electromagnetic radiation. I will then discuss the definitional challenges that these attacks raise, and argue why heuristic hardware-level countermeasures are unlikely to solve the problem convincingly. Based on these premises, and focusing on the symmetric setting, securing cryptographic implementations can be viewed as a tradeoff between the design of modes of operation, underlying primitives and countermeasures.

Regarding modes of operation, I will describe a general design strategy for leakage-resilient authenticated encryption, propose models and assumptions on which security proofs can be based, and show how this design strategy encourages so-called leveled implementations, where only a part of the computation needs strong (hence expensive) protections against side-channel attacks.

Regarding underlying primitives and countermeasures, I will first emphasize the formal and practically-relevant guarantees that can be obtained thanks to masking (i.e., secret sharing at the circuit level), and how considering the implementation of such countermeasures as an algorithmic design goal (e.g., for block ciphers) can lead to improved performances. I will then describe how limiting the leakage of the less protected parts in a leveled implementations can be combined with excellent performances, for instance with respect to the energy cost.

I will conclude by putting forward the importance of sound evaluation practices in order to empirically validate (by lack of falsification) the assumptions needed both for leakage-resilient modes of operation and countermeasures like masking, and motivate the need of an open approach for this purpose. That is, by allowing adversaries and evaluators to know implementation details, we can expect to enable a better understanding of the fundamentals of physical security, therefore leading to improved security and efficiency in the long term.

---

[1] The author is a Senior Research Associate of the Belgian Fund for Scientific Research (FNRS-F.R.S.). This work has been funded in part by the ERC Project 724725.

Fully Homomorphic Encryption
from the Ground Up

Daniele Micciancio ⓘ

University of California, Mail Code 0404, La Jolla,
San Diego, CA, 92093, USA
daniele@cs.ucsd.edu
http://cseweb.ucsd.edu/~daniele/

Abstract. The development of fully homomorphic encryption (FHE), i.e., encryption schemes that allow to perform arbitrary computations on encrypted data, has been one of the main achievements of theoretical cryptography of the past 20 years, and probably the single application that brought most attention to lattice cryptography. While lattice cryptography, and fully homomorphic encryption in particular, are often regarded as a highly technical topic, essentially all constructions of FHE proposed so far are based on a small number of rather simple ideas. In this talk, I will try highlight the basic principles that make FHE possible, using lattices to build a simple private key encryption scheme that enjoys a small number of elementary, but very useful properties: a simple decryption algorithm (requiring, essentially, just the computation of a linear function), a basic form of circular security (i.e., the ability to securely encrypt its own key), and a very weak form of linear homomorphism (supporting only a bounded number of addition operations.)

All these properties are easily established using simple linear algebra and the hardness of the Learning With Errors (LWE) problem or standard worst-case complexity assumptions on lattices. Then, I will use this scheme (and its abstract properties) to build in a modular way a tower of increasingly more powerful encryption schemes supporting a wider range of operations: multiplication by arbitrary constants, multiplication between ciphertexts, and finally the evaluation of arithmetic circuits of arbitrary, but a-priory bounded depth. The final result is a *leveled*[1] FHE scheme based on standard lattice problems, i.e., a scheme supporting the evaluation of arbitrary circuits on encrypted data, as long as the depth of the circuit is provided at key generation time. Remarkably, lattices are used only in the construction (and security analysis) of the basic scheme: all the remaining steps in the construction do not make any direct use of lattices, and can be expressed in a simple, abstract way, and analyzed using solely the weakly homomorphic properties of the basic scheme.

Keywords: Lattice-based cryptography · Fully homomorphic encryption · Circular security · FHE bootstrapping

---

[1] The "leveled" restriction in the final FHE scheme can be lifted using "circular security" assumptions that have become relatively standard in the FHE literature, but that are still not well understood. Achieving (non-leveled) FHE from standard lattice assumptions is the main theoretical problem still open in the area.

Contents – Part II

Homomorphic Primitives

Homomorphic Secret Sharing from Lattices Without FHE 3
Elette Boyle, Lisa Kohl, and Peter Scholl

Improved Bootstrapping for Approximate Homomorphic Encryption 34
Hao Chen, Ilaria Chillotti, and Yongsoo Song

Minicrypt Primitives with Algebraic Structure and Applications 55
Navid Alamati, Hart Montgomery, Sikhar Patranabis, and Arnab Roy

Standards

Attacks only Get Better: How to Break FF3 on Large Domains 85
Viet Tung Hoang, David Miller, and Ni Trieu

Session Resumption Protocols and Efficient Forward Security
for TLS 1.3 0-RTT 117
Nimrod Aviram, Kai Gellert, and Tibor Jager

An Analysis of NIST SP 800-90A 151
Joanne Woodage and Dan Shumow

Searchable Encryption and ORAM

Computationally Volume-Hiding Structured Encryption 183
Seny Kamara and Tarik Moataz

Locality-Preserving Oblivious RAM 214
*Gilad Asharov, T.-H. Hubert Chan, Kartik Nayak, Rafael Pass,
Ling Ren, and Elaine Shi*

Private Anonymous Data Access. 244
Ariel Hamlin, Rafail Ostrovsky, Mor Weiss, and Daniel Wichs

Proofs of Work and Space

Reversible Proofs of Sequential Work 277
*Hamza Abusalah, Chethan Kamath, Karen Klein, Krzysztof Pietrzak,
and Michael Walter*

Incremental Proofs of Sequential Work 292
  Nico Döttling, Russell W. F. Lai, and Giulio Malavolta

Tight Proofs of Space and Replication........................... 324
  Ben Fisch

Secure Computation

Founding Secure Computation on Blockchains 351
  Arka Rai Choudhuri, Vipul Goyal, and Abhishek Jain

Uncovering Algebraic Structures in the MPC Landscape 381
  Navneet Agarwal, Sanat Anand, and Manoj Prabhakaran

Quantum I

Quantum Circuits for the CSIDH: Optimizing Quantum
Evaluation of Isogenies 409
  Daniel J. Bernstein, Tanja Lange, Chloe Martindale, and Lorenz Panny

A Quantum-Proof Non-malleable Extractor: With Application to Privacy
Amplification Against Active Quantum Adversaries 442
  Divesh Aggarwal, Kai-Min Chung, Han-Hsuan Lin, and Thomas Vidick

Secure Computation and NIZK

A Note on the Communication Complexity of Multiparty Computation
in the Correlated Randomness Model 473
  Geoffroy Couteau

Degree 2 is Complete for the Round-Complexity of Malicious MPC 504
  Benny Applebaum, Zvika Brakerski, and Rotem Tsabary

Two Round Information-Theoretic MPC with Malicious Security 532
  *Prabhanjan Ananth, Arka Rai Choudhuri, Aarushi Goel,
  and Abhishek Jain*

Designated-Verifier Pseudorandom Generators, and Their Applications 562
  Geoffroy Couteau and Dennis Hofheinz

Reusable Designated-Verifier NIZKs for all NP from CDH 593
  Willy Quach, Ron D. Rothblum, and Daniel Wichs

Designated Verifier/Prover and Preprocessing NIZKs
from Diffie-Hellman Assumptions.............................. 622
  *Shuichi Katsumata, Ryo Nishimaki, Shota Yamada,
  and Takashi Yamakawa*

Lattice-Based Cryptography

Building an Efficient Lattice Gadget Toolkit: Subgaussian
Sampling and More.   655
   Nicholas Genise, Daniele Micciancio, and Yuriy Polyakov

Approx-SVP in Ideal Lattices with Pre-processing.   685
   Alice Pellet-Mary, Guillaume Hanrot, and Damien Stehlé

The General Sieve Kernel and New Records in Lattice Reduction.   717
   Martin R. Albrecht, Léo Ducas, Gottfried Herold, Elena Kirshanova,
   Eamonn W. Postlethwaite, and Marc Stevens

Misuse Attacks on Post-quantum Cryptosystems.   747
   Ciprian Băetu, F. Betül Durak, Loïs Huguenin-Dumittan,
   Abdullah Talayhan, and Serge Vaudenay

Author Index   777

Homomorphic Primitives

Homomorphic Secret Sharing
from Lattices Without FHE

Elette Boyle[1($\boxtimes$)], Lisa Kohl[2], and Peter Scholl[3]

[1] IDC Herzliya, Herzliya, Israel
Elette.Boyle@idc.ac.il
[2] Karlsruhe Institute of Technology, Karlsruhe, Germany
Lisa.Kohl@kit.edu
[3] Aarhus University, Aarhus, Denmark
Peter.Scholl@cs.au.dk

Abstract. Homomorphic secret sharing (HSS) is an analog of somewhat- or fully homomorphic encryption (S/FHE) to the setting of secret sharing, with applications including succinct secure computation, private manipulation of remote databases, and more. While HSS can be viewed as a relaxation of S/FHE, the only constructions from lattice-based assumptions to date build *atop* specific forms of threshold or multi-key S/FHE. In this work, we present new techniques directly yielding efficient 2-party HSS for polynomial-size branching programs from a range of lattice-based encryption schemes, *without S/FHE*. More concretely, we avoid the costly *key-switching* and *modulus-reduction* steps used in S/FHE ciphertext multiplication, replacing them with a new *distributed decryption* procedure for performing "restricted" multiplications of an input with a partial computation value. Doing so requires new methods for handling the blowup of "noise" in ciphertexts in a distributed setting, and leverages several properties of lattice-based encryption schemes together with new tricks in share conversion.

The resulting schemes support a superpolynomial-size plaintext space and negligible correctness error, with share sizes comparable to SHE ciphertexts, but cost of homomorphic multiplication roughly one order of magnitude faster. Over certain rings, our HSS can further support some level of packed SIMD homomorphic operations. We demonstrate the practical efficiency of our schemes within two application settings, where we compare favorably with current best approaches: 2-server private database pattern-match queries, and secure 2-party computation of low-degree polynomials.

E. Boyle—Supported in part by ISF grant 1861/16, AFOSR Award FA9550-17-1-0069, and ERC grant 742754 (project NTSC).
L. Kohl—Supported by ERC Project PREP-CRYPTO (724307), by DFG grant HO 4534/2-2 and by a DAAD scholarship. This work was done in part while visiting the FACT Center at IDC Herzliya, Israel.
P. Scholl—Supported by the European Union's Horizon 2020 research and innovation programme under grant agreement No. 731583 (SODA), and the Danish Independent Research Council under Grant-ID DFF-6108-00169 (FoCC).

Y. Ishai and V. Rijmen (Eds.): EUROCRYPT 2019, LNCS 11477, pp. 3–33, 2019.
https://doi.org/10.1007/978-3-030-17656-3_1

1   Introduction

Homomorphic secret sharing (HSS) [7] is a form of secret sharing that supports a compact local evaluation on its shares. HSS can be viewed as the analog of fully (or somewhat-) homomorphic encryption (S/FHE) [26,38] to the setting of secret sharing: a relaxation where homomorphic evaluation can be distributed among two parties who do not interact with each other. Over the past years, there has been a wave of HSS constructions for rich function classes (e.g., [6,7,9,21,25]) as well as an expanding range of corresponding applications. HSS suffices for many scenarios in which S/FHE can be applied (and even some for which it *cannot*), including low-communication secure computation [6,7,10], private manipulation of remote databases [8,9,19,29,40], methods of succinctly generating correlated randomness [5,6], and more.

One of the appealing features of HSS compared to FHE is that allowing homomorphic evaluation to be distributed among two parties may constitute a simpler target to achieve. Indeed, forms of HSS for branching programs have been built from *discrete logarithm* type assumptions [7]; in contrast, obtaining encryption schemes from these structures that support comparable homomorphism on ciphertexts seems well beyond reach of current techniques. In regard to structures from which FHE does exist, the Learning With Errors (LWE) assumption [37] (and in turn its Ring LWE (RLWE) variant [36]) is known to imply strong versions of HSS [8,11,22].

However, in spite of its potential for comparative simplicity, all HSS constructions based on LWE or RLWE to date remain *at least* as complex as S/FHE. In particular, underlying each such HSS scheme is the common approach of beginning with and building atop some existing construction of FHE—relying on specific forms of threshold FHE, multi-key FHE, or even FHE-based "spooky encryption" [2,8,11,22]. In this sense, current lattice-based HSS constructions serve predominantly as statements of feasibility, and have not been explored as a competitive alternative for use within applications.

Given the rapidly expanding set of HSS applications, together with the demonstrated power and success of leveraging lattices as a tool for advanced cryptography, a natural question is whether this situation can be improved. In particular, can we construct HSS from LWE and RLWE *without* (in some sense) S/FHE?

1.1   Our Results

In this work we consider precisely this question. We present and leverage new approaches for directly obtaining 2-party HSS schemes from LWE and RLWE, bypassing the intermediate step of fully (or even somewhat-) homomorphic encryption (S/FHE).

More concretely, our techniques avoid the costly *key-switching* and repeated *modulus-reduction* steps typically required for homomorphic multiplication of ciphertexts in existing (R)LWE-based FHE schemes [14,27], and replace them instead with a new *distributed decryption* procedure for multiplying an encrypted value by a value in secret shared form, resulting in secret shares of the product. The cost of a homomorphic multiplication thus drops roughly to the cost of a decryption operation per party. This operation requires a new toolkit of methods for handling the blowup of "noise" from ciphertexts in a distributed setting, and leverages properties of lattice-based encryption schemes (such as key-dependent message security) in new ways.

Our construction takes inspiration from the HSS framework of [7], and yields a similar result: namely, HSS for the class of polynomial-size branching programs (capturing NC1 and logspace computations). However, as discussed below, our construction offers several strong advantages over existing DDH-based schemes [6,7,10,21], including *negligible* correctness error and *superpolynomial-size* plaintext space, as well as over S/FHE-based solutions for the same program class [2,22], including cheaper multiplication, simpler setup, and *no noise growth*. We showcase these advantages via two sample applications: (1) Generating correlated randomness for secure 2-party computation in the prepocessing model, and (2) 2-server Private Information Retrieval for various private database queries such as conjunctive keyword search and pattern matching.

We now proceed to describe our main results.

HSS from Nearly Linear Decryption. Our core approach leverages the *"nearly linear"* structure of ciphertext decryption common to a range of lattice-based encryption schemes:

Definition 1 (Informal - *Nearly Linear Decryption*). *Let $R = \mathbb{Z}[X]/ (X^N + 1)$ for N a power of 2. Let $p, q \in \mathbb{N}$ be moduli with $p|q$ and $1 \ll p \ll q$. We say that an encryption scheme supports* nearly linear decryption *for messages $m \in R_p := R/pR$ if the secret key is $\mathbf{s} \in R_q^d$, and for any ciphertext $\mathbf{c} \in R_q^d$ encrypting m,*

$$\langle \mathbf{s}, \mathbf{c} \rangle = (q/p) \cdot m + e \quad \mod q$$

for some "small" noise $e \in R$.

This captures, for example, the LWE-based schemes of Regev [37] and Applebaum *et al.* [1] (with $N = 1$, i.e. $R = \mathbb{Z}$, $d = \mathsf{poly}(\lambda)$), RLWE-based schemes of Lyubashevsky-Peikert-Regev [36] and Brakerski-Vaikutanathan [15] (with $N = \mathsf{poly}(\lambda), d = 2$), as well as various schemes based on Learning With Rounding [3] and Module-LWE [34]. For simplicity, we restrict ourselves to the decryption structure and polynomial rings R of the form specified above; however, our techniques extend to more general polynomial rings, as well as to encryption schemes which encode messages in low-order symbols (i.e., for which $\langle \mathbf{s}, \mathbf{c} \rangle = p \cdot e + m \mod q$).

We demonstrate how to exploit the near-linearity of decryption to support sequences of homomorphic *additions* over R, as well as homomorphic *"restricted" multiplications* over R of an evaluated value with an input. Ultimately, our scheme supports any sequence of these two homomorphic operations over R_r (for $r \in \mathbb{N}$ of choice) with negligible correctness error, subject to the requirement that the magnitude of computation values remain bounded by some chosen value B with $r \leq B \ll p$ (where the required size of p, and thus q, must grow with this bound). We remark that taking magnitude bound $B = 2$ suffices already to evaluate the class of polynomial-size branching programs with polynomial overhead [7]. Since efficiency is the primary focus of this work, however, we state our results as a function of these two operations directly, and applications can further exploit the ability to support large message spaces.

We achieve the stronger "public-key" variant of HSS [7], where the secret-sharing process can be split into a one-time setup phase (resulting in a public key pk and evaluation shares $(\mathsf{ek}_0, \mathsf{ek}_1)$) together with a separate "input encryption" phase, wherein any user can use pk to load his input x_i into secret-shared form (and where homomorphic evaluation can take place *across* parties' inputs). This variant of HSS facilitates applications of secure multi-party computation.

Theorem 1 (Informal - *Main HSS Construction*). *Given any encryption scheme with nearly linear decryption (as above) over ring R with parameters $p, q, d \in \mathbb{N}$, as well as magnitude bound $B \in \mathbb{N}$ for which $B \ll p \ll q/B$, and output modulus $r \leq B$, there exists 2-party public-key HSS for inputs in R_r with size:*

- *Public key $\mathsf{pk} = pk$ of the encryption scheme, Evaluation keys $\mathsf{ek}_b \in R_q^{d-1}$*
- *HSS shares of each input $x_i \in R_r$ consist of d ciphertexts.*

supporting any polynomial number of the following homomorphic operations over R_r (subject to the ℓ_∞ magnitude bound $\|y\|_\infty \leq B$ (in R) for all partial computation values y), with negligible correctness error, and the specified complexities:

- *Loading an input into memory $(y_i \leftarrow x_i)$:*   *d decryptions*
- *Addition of memory values $(y_k \leftarrow y_i + y_j)$:*   *1 addition over R_q^d*
- *Multiplication of input with memory value $(y_k \leftarrow x_i \cdot y_j)$:*   *$d$ decryptions*
  "Terminal" multiplication (s.t. y_k appears in no future mult):   *1 decryption*

where "decryption" is essentially one inner product over R_q^d.

Asymptotically, the modulus q is of size $O(\lambda + \log B)$ bits, giving a share size in $O(Nd(\lambda + \log B))$ for $R = \mathbb{Z}[X]/(X^N + 1)$. The cost of multiplication is $\widetilde{O}(Nd)$ operations in $\mathbb{Z}_q$.

Plugging in the "LPR" RLWE-based scheme of [36] (where ciphertexts consist of $d = 2$ ring elements), our HSS shares consist of 4 R_q-elements per R_r-element plaintext, and homomorphic multiplication of an input and memory value in R_r is dominated by 4 R_q-multiplications (with correctness if the resulting product y over R maintains $\|y\|_\infty \leq B$). For concrete parameters, the resulting HSS shares will be of comparable size to the analogous SHE-based approach, but will offer significantly cheaper homomorphic multiplication operations—faster by approximately one order of magnitude. (See "Comparison to SHE-based solutions" discussion below.)

We further explore extensions and optimizations to the core construction, within the following settings:

1. *Secret-key HSS*: For applications where all secrets in the system originate from a single party; e.g., the client in 2-server Private Information Retrieval.
2. *Degree-2 computations*: For the special case of homomorphic evaluation of degree-2 polynomials (and extension to low-degree polynomials), with applications to secure computation.
3. *SIMD computations*: Direct support for homomorphic evaluation of "packed" single-instruction multiple data (SIMD) parallel computations on data items encoded as vectors. This has useful application to parallel computations, and to PIR-type applications, where one wishes to perform several identical evaluations on different database items.

Comparison to Existing Approaches. We briefly discuss our resulting HSS in reference to existing approaches for comparable function classes.

Comparison to Group-Based HSS. Our core construction framework resembles the HSS schemes of Boyle, Gilboa, and Ishai [7] and successors [6,10,21], which rely on various flavors of discrete logarithm type assumptions in cryptographically hard Abelian groups (e.g., Decisional Diffie Hellman (DDH) or circular security of ElGamal). We refer to this line as "group-based" HSS.

Despite many algorithmic and heuristic advances, all works in this line are subject to a common computation barrier: In addition to their upper bounds, Dinur, Keller, and Klein [21] showed that (barring a breakthrough in discrete logarithm techniques[1]) performing a homomorphic multiplication via this general approach with plaintext space size B and correctness error δ *requires* runtime $T = \Omega(\sqrt{B/\delta})$. Of particular note, this inherently restricts support to plaintext spaces of polynomial size B, as well as inverse-polynomial error δ.

- *Superpolynomial-size plaintext space.* In contrast, our HSS scheme can directly support operations within superpolynomial plaintext spaces over the ring $\mathbb{Z}[X]/(X^N + 1)$, with complexity growing roughly as the logarithm of the maximum magnitude B. This circumvents blowups associated with artificial emulation of larger input spaces by breaking input elements into small pieces (e.g., bits) and operating piecewise. Such blowups manifest even when operating over small inputs, as encodings of the *secret key* as a plaintext are necessary in order to support homomorphism.
- *Negligible error.* Our HSS also enjoys negligible correctness error. Beyond a theoretical distinction, this greatly affects efficiency. Even to obtain constant failure probability in group-based approaches, homomorphic evaluation of S multiplications requires computation scaling as $S^{3/2}$ [21], since the error of each individual multiplication must be pushed down to $\sim \delta/S$ to reach overall error δ. The presence (or non-presence) of error may further *leak* information about the secret inputs; this adds layers of complexity and overhead to HSS-based applications, wherein effects of error must be sanitized before homomorphically evaluated output shares can be exchanged [6,7,10].

---

[1] Namely, solving the Discrete Logarithm in a Interval problem with interval length R in time $o(\sqrt{R})$.

– *Cheaper operations.* Overall, the resulting lattice-based schemes require cheaper operations than group-based alternatives, e.g. replacing cryptographic group exponentiations by simple polynomial ring multiplications (efficiently implementable using FFT). Most stark in contrast: the expensive "share-conversion" steps in group-based approaches—requiring large scales of repeated group multiplications and pattern matches, and dominating computation costs in homomorphic evaluation—are trivialized given our new techniques.

Comparison to SHE-based solutions. Two-server HSS can also be constructed from threshold variants of somewhat or fully homomorphic encryption, by replacing the (interactive) distributed decryption procedure with the non-interactive "rounding" technique from [22] to give additive shares of the output. Using FHE this can give HSS for circuits, although the computational overhead either grows with the depth of the circuit [14] or is independent of the circuit size but very large due to a costly bootstrapping step.[2]

SHE is reasonably practical for evaluating low-depth circuits, where the dominant cost is homomorphic multiplication. There are several different approaches to SHE multiplication, all of which have various complications due to the need for a "key-switching" procedure [14,27] to avoid ciphertext expansion, as well as either modulus switching [14] or scale-invariant operations [13,24] to reduce noise growth.[3]

Our approach avoids these complications, leading to a conceptually simpler and more efficient scheme with several advantages over SHE-based HSS.

– *Cheaper multiplication.* Our homomorphic multiplication procedure is much simpler and cheaper than in SHE, since we avoid costly ciphertext expansion or key/modulus-switching procedures which are inherent in most SHE schemes. Concretely, our multiplication procedure has roughly the cost of *2 decryption operations* in SHE, which we estimate improves our performance by around an order of magnitude based on recent implementation results [31]. While our supported multiplications are of a "restricted" form, requiring one of the multiplicands to be an original input value, this has a mild effect within low-degree computations, which are anyway the competitive regime for SHE.
– *Simpler setup.* Since we do not need key-switching, we also avoid the cost of setting up the key-switching material in a distributed manner, which is a source of additional complexity in threshold FHE [2] as it requires generating several "quasi-encryptions" of s^2, where s is the secret-shared private key.
– *No noise growth.* Unlike FHE, ciphertexts in our homomorphic evaluation procedure do not incur noise growth, which increases ciphertext size and

---

[2] Although the cost of bootstrapping has fallen dramatically in recent years [16,17, 23,32], the efficiency is still orders of magnitude worse than low-depth somewhat homomorphic encryption using SIMD operations.

[3] So-called "third generation" SHE schemes based on GSW [28] have simpler homomorphic multiplication, but much larger ciphertexts that grow with $\Omega(N \log^2 q)$ instead of $O(N \log q)$, for (R)LWE dimension N and modulus q.

limits the number of homomorphic operations. Instead, we are only limited in that the parameters must be chosen based on an upper bound on the maximum size of any plaintext value (without modular reduction) during the computation.

Sample applications. To illustrate the potential of our techniques, we consider two example use-cases of HSS for branching programs. Firstly, we look at secure two-party computation of low-degree polynomials, and its application to generating various forms of correlated randomness. Many MPC protocols use pre-processed, correlated randomness such as Beaver multiplication triples, matrix multiplication triples or truth-table correlations to achieve a very fast "online" protocol. Protocols such as SPDZ [20] often use SHE to generate this randomness, whereas using HSS (considered in [6]) has potential to greatly improve computational costs, and reduce the round complexity to just a single round, while paying a slight overhead with larger ciphertexts. Secondly, we look at 2-server Private Information Retrieval (PIR), which allows a client to perform private queries to a public database. Our HSS for branching programs allows a much richer set of queries than previous, practical schemes based on one-way functions [40], and in this case we can reduce the share size compared with using SHE, as well as the computation.

1.2   Technical Overview

Recall our HSS is with respect to an encryption scheme (Gen, Enc, Dec) with *nearly linear decryption* over ring $\mathbb{Z}[X]/(X^N + 1)$ (as discussed above), with moduli $r \leq B \ll p \ll q/B$ and parameter d. Ciphertexts and the secret key of the encryption scheme are elements $\mathbf{c}, \mathbf{s} \in R_q^d$ with $\mathbf{s} = (1, \widehat{\mathbf{s}}) \in R_q \times R_q^{d-1}$, the plaintext space of encryption is R_p, and we will support homomorphic operations over R_r for computations for which all intermediate computation values y (as performed over R) remain bounded by $\|y\|_\infty \leq B$. (We will denote $y \in [R]_B$ to highlight that arithmetic is *not* performed modulo B.)

The core of our HSS resembles the DDH-based framework of [7], translated to the setting of lattice-based encryption. The HSS public key pk is precisely the public key of the encryption scheme.[4] The evaluation keys $(\mathsf{ek}_0, \mathsf{ek}_1) \in R_q^d \times R_q^d$ are additive secret shares of the key $\mathbf{s} \in R_q^d$ over R_q.[5] Homomorphic evaluation maintains the invariant that for every intermediate computation value $y \in [R]_B$, Party 0 and Party 1 will hold *additive shares* $(\mathbf{t}_0^y, \mathbf{t}_1^y) \in R_q^d \times R_q^d$ of the product $y \cdot \mathbf{s} \in R_q^d$ over R_q. This directly admits homomorphic addition, by locally adding the corresponding secret shares.

As usual, the challenge comes in addressing multiplication. We support homomorphic "restricted" multiplications, between any intermediate computation

---

[4] Note that nearly linear decryption generically implies existence of a public-key encryption procedure.

[5] This can be decreased to $(d-1)$ R_q-elements communicated, as $s_1 = 1 \in R_q$.

value y and input value x. To aid this operation, the HSS sharing of input $x \in [R]_B$ will be a (componentwise) encryption of $x \cdot \mathbf{s}$: i.e., d ciphertexts $\mathbf{C}^x = (\mathbf{c}^{x \cdot s_1}, \ldots, \mathbf{c}^{x \cdot s_d}) \in (R_q^d)^d$. Interestingly, these encryptions can be generated given just pk of the encryption, leveraging a weak form of key-dependent message (KDM) security implied by nearly linear decryption—see "KDM Security" discussion below. Combining the HSS encoding $\mathbf{C}^x$ of x with the secret shares $(\mathbf{t}_0^y, \mathbf{t}_1^y)$ for y, nearly linear decryption then gives us:

$$\text{for every } i \in [d] : \quad \langle \mathbf{t}_0^y, \mathbf{c}^{x \cdot s_i} \rangle + \langle \mathbf{t}_1^y, \mathbf{c}^{x \cdot s_i} \rangle = \langle y \cdot \mathbf{s}, \mathbf{c}^{x \cdot s_i} \rangle \approx (q/p) \cdot xy \cdot s_i \quad \text{over } R_q.$$

Collectively, this *almost* yields the desired additive shares of $xy \cdot \mathbf{s} \in R_q^d$ to maintain the homomorphic evaluation invariant.

Rounding. Our first observation is that we can use the non-interactive rounding trick as in [22] to locally convert the approximate shares of $(q/p) \cdot xy \cdot s_i$ over R_q from above, to *exact* shares of $xy \cdot s_i$ over R_p. Concretely, each party simply scales his share by (p/q) and *locally* rounds to the nearest integer value. This operation heavily relies on the fact that there are 2 parties, and provides correct output shares over R_p with error probability pB/q, negligible for $p \ll q/B$.

However, this is not quite what we need: the resulting secret shares of $xy \cdot \mathbf{s}$ are over R_p, *not* R_q. This means we cannot use the shares again to "distributively decrypt" the original set of ciphertexts $\mathbf{C}^{x'}$ for any input x', a task whose operations must take place over R_q. (In fact, information about the $s_i \in R_q$ may even be *lost* when taken mod p.) Performing a second analogous multiplication would then necessitate a *second set* of ciphertexts, over a smaller modulus: namely with R_p playing the original role of R_q, and some $p_1 \ll p$ playing the previous role of p. In such fashion, one can devise a *leveled* HSS scheme operating via a sequence of decreasing moduli $q \gg p \gg p_1 \gg p_2 \gg \cdots p_{\deg}$, where each step must drop by a superpolynomial factor to guarantee negligible correctness error. The size and complexity of such HSS scheme, however, would grow significantly with the desired depth of homomorphic computation.

Lifting. We avoid the above conundrum by (quite literally) doing *nothing*. Our observation is as follows. In general, converting secret shares up to a higher modulus constitutes a problem: e.g., even from $\mathbb{Z}_2$ to $\mathbb{Z}_3$ we have $1 + 1 \equiv 0 \in \mathbb{Z}_2$ turning to $1 + 1 \equiv 2 \in \mathbb{Z}_3$. However, if we can guarantee that the secret shared payload is *very small* compared to the modulus, this wraparound problem drops to a negligible fraction of possible secret shares. Concretely, given shares $t_0 + t_1 \equiv t \mod p$ for $t_0, t_1, t \in (-\lfloor p/2 \rfloor, \ldots, \lfloor (p-1)/2 \rfloor]$, then $t_0 + t_1 = t$ with equality (over $\mathbb{Z}$) unless $t - t_0$ falls $\leq -\lfloor p/2 \rfloor$ or $> \lfloor (p-1)/2 \rfloor$. If t_0 is randomly chosen and $t \ll p$, then these corner cases occur with only negligible probability. Conditioned on this, conversion to shares modulo q is immediate: it *already* holds that $t_0 + t_1 = t \mod q$.

Recall that we wish to perform share conversion on payload values of the form $y \cdot s_i \in R_q$. To use this trick, we must thus adjust the construction to guarantee that the shares t_b are distributed randomly, and that any such payload value has low magnitude $\|y \cdot s_i\|_\infty \ll p$. Rerandomizing shares is accomplished by each party shifting his share by the same pseudorandom offset (determined via a

PRF). Ensuring low magnitude of $y \cdot s_i$ is done via two pieces. First, we leverage a result of Applebaum *et al.* [1], which allows us to replace a randomly sampled secret key $\mathbf{s} \in R_q^d$ of encryption with one sampled from the *low-magnitude* noise distribution, without loss of security. Second, we introduce an additional modulus level $B \ll p$, and address only computations which remain bounded in magnitude by $\|y\|_\infty \leq B$. Together, this will ensure each value $y \cdot s_i$ has small norm in comparison to p, and thus the shares of $y \cdot s_i$ over R_p can be *directly* interpreted as shares of $y \cdot s_i$ over R_q, successfully returning us to the desired homomorphic evaluation invariant. Finally, using the same low-magnitude-payload share conversion trick, in the final step of computation, the parties can convert their shares of $y \cdot \mathbf{s} = (y, y \cdot \widehat{\mathbf{s}})$ (recall $\mathbf{s} = (1, \widehat{\mathbf{s}})$) over R_q to shares of y over R_r for target output modulus r.

Ultimately, the HSS scheme uses three moduli levels: $B \ll p \ll q/B$. Correctness holds as long as the magnitude of computation values is bounded within $[R]_B$. Homomorphic evaluation maintains secret shares over R_q as its invariant. Each homomorphic multiplication drops down to R_p to remove effects of noise, then steps back up to shares over R_q to reinstate the invariant. An advantageous side effect of this structure is that, conditioned on remaining within magnitude bound B (e.g., Boolean computations), the size of our HSS shares is completely *independent* of the depth or size of the homomorphic computation.

To conclude, we highlight some of the additional ideas and techniques arising within our scheme and extensions.

KDM security. Our HSS reveals encryptions of the form $\mathsf{Enc}(x \cdot s_i)$, for input x and secret key $\mathbf{s}$ of the underlying encryption scheme. Key-dependent message (KDM) security of the encryption scheme with respect to this class of linear functions of $\mathbf{s}$ follows from its nearly linear decryption structure. As typical in KDM literature (following [1,15]), this is shown by demonstrating as an intermediate step that such encryptions can be efficiently generated from knowledge only of the public key and rerandomized to "look like" fresh encryptions.

In our construction, we *leverage* this efficient generation procedure. This enables pk of the HSS to consist purely of the public key of the encryption scheme, while still allowing parties to encode their respective inputs x as $\{\mathsf{Enc}(x \cdot s_i)\}_{i \in [d]}$. The corresponding encoding procedure is simpler and achieves better parameters than publishing $\{\mathsf{Enc}(s_i)\}_{i \in [d]}$ as part of pk (as was done in [7]), as this introduces extra ciphertexts as well as a second noise term that must be "drowned" by larger noise when scaling by x and rerandomizing.

Secret key as an input. For our RLWE-based HSS schemes with plaintext space $R_r = \mathbb{Z}_r[X]/(X^N + 1)$, and when sampling the secret key $\mathbf{s} = (1, \widehat{s}) \in R_q^2$ from the low-magnitude noise distribution, it holds that $\widehat{s}$ *itself* lies within the supported input space R_r of the HSS. This is implicitly exploited in our attained HSS efficiency, e.g. where our encryptions of $x \cdot \mathbf{s} = (x, x\widehat{s}) \in R_q^2$ can consist of just *two* ciphertexts (instead of λ).

However, this also opens *qualitatively* new approaches toward optimization. For example, suppose the evaluation keys $\mathsf{ek}_0, \mathsf{ek}_1$ are augmented with shares of $(\widehat{s})^2$, as well as shares of $\widehat{s}$. We can then view the shares of $(\widehat{s}, (\widehat{s})^2) = \widehat{s}(1, \widehat{s})$ as

HSS sharings of the computation value $\hat{s}$, and thus use them to homomorphically multiply an input x by $\hat{s}$. For degree 2, this allows us to save sending encryptions of $x \cdot \hat{s}$ for inputs x, since we can now generate these "for free" using $\mathsf{ek}_0, \mathsf{ek}_1$ and the encryptions of x, reducing the share size by a factor of two.

SIMD operations. If the underlying encryption scheme is over a ring R of the right form, then our basic HSS supports homomorphic evaluation of "single instruction, multiple data" (SIMD) operations. Namely, suppose $R = \mathbb{Z}[X]/(X^N+1)$ where X^N+1 splits over R_r (for some prime $r \geq 2$) into pairwise different irreducible polynomials of degree $k \in \mathbb{N}$: that is, $R_r \cong \mathbb{F}_{r^k} \times \cdots \times \mathbb{F}_{r^k}$ for N/k copies of $\mathbb{F}_{r^k}$. In such case, our homomorphic additions and multiplications over R_r directly translate to corresponding SIMD operations within the individual computation "slots."

A caveat of this correspondence is that the magnitude bound requirement B over R_r (in *coefficient* embedding) does not translate directly to a per-slot magnitude bound of B (in CRT embedding). Thus current SIMD support can effectively handle low-degree computations, but suffers performance degradation as the degree increases, even if the magnitude of the SIMD computations is bounded. An interesting goal for future investigation will be to devise new ways of packing to mitigate this disadvantage.

Beyond 2 parties. A major open problem is constructing an efficient HSS scheme with additive reconstruction for a reasonably expressive function class in the setting of 3 or more parties. Unfortunately, even for 3 parties, the rounding and lifting techniques we use would result in an error with *constant* probability for any choice of underlying parameters.

The authors of [22] manage to extend their construction of "spooky encryption" to the multiparty setting by using a GMW-like approach [30]. We can apply this to our scheme for degree-2 functions, by performing pairwise products on shares involving pairwise secret keys. However, it is not clear how to extend this further, as one would need to obtain shares of the pairwise outputs under all other secret keys.

2   Preliminaries

We begin this section by introducing some notation. For notation that we consider common knowledge we refer to the full version [12]. We denote our security parameter by λ. Throughout this paper we consider all parameters to implicitly depend on λ, e.g. by $\ell \in \mathbb{N}$ we actually consider ℓ to be a function $\ell \colon \mathbb{N} \to \mathbb{N}$, but simply write ℓ in order to refer to $\ell(\lambda)$.

For a real number $x \in \mathbb{R}$, by $\lfloor x \rceil \in \mathbb{Z}$ we denote the element closest to $x \in \mathbb{R}$, where we round up when the first decimal place of x is 5 or higher.

For $x \in \mathbb{Z}[X]/(X^N + 1)$ the maximum norm of x is defined as $\|x\|_\infty := \max |(x_1, \ldots, x_N)|$, where $x_i \in \mathbb{Z}$ such $x = \sum_{i=1}^{N} x_i X^{i-1} \mod X^N + 1$.

For $p \in \mathbb{N}$, by R_p we denote R/pR. Note that we consider R_p as elements for which all coefficients are in the interval $(-\lfloor p/2 \rfloor, \ldots, \lfloor (p-1)/2 \rceil]$. For $B \in \mathbb{N}$, we denote $[R]_B := \{x \in R \mid \|x\|_\infty \leq B\}$. More generally, for an interval $I \subseteq \mathbb{Z}$, we write $R|_I$ to denote all elements of R that have only coefficients in I.

We denote vectors by bold lower-case letters and matrices by bold upper-case letters. We interpret vectors as column-vectors. For a vector $\mathbf{x} \in R^\ell$, by x_i we refer to the i-th entry (for $i \in \{1, \ldots, \ell\}$).

We consider *public-key encryption schemes* that satisfy the security notion of *pseudorandomness of ciphertexts*. For a formal definition we refer to the full version.

2.1 Homomorphic Secret Sharing

We consider homomorphic secret sharing (HSS) as introduced in [7]. By default, in this work, the term HSS refers to a public-key variant of HSS. Unlike [7], we do not need to consider non-negligible δ error failure probability.

Definition 2 (Homomorphic Secret Sharing). *A (2-party, public-key) Homomorphic Secret Sharing (HSS) scheme for a class of programs $\mathcal{P}$ over a ring R with input space $\mathcal{I} \subseteq R$ consists of PPT algorithms* (HSS.Gen, HSS.Enc, HSS.Eval) *with the following syntax:*

- HSS.Gen(1^λ): *On input a security parameter 1^λ, the key generation algorithm outputs a public key* pk *and a pair of evaluation keys* $(\mathsf{ek}_0, \mathsf{ek}_1)$.
- HSS.Enc(pk, x): *Given public key* pk *and secret input value $x \in \mathcal{I}$, the encryption algorithm outputs a ciphertext* ct.
- HSS.Eval($b, \mathsf{ek}_b, (\mathsf{ct}^{(1)}, \ldots, \mathsf{ct}^{(\rho)}), P, r$): *On input party index $b \in \{0, 1\}$, evaluation key ek_b, vector of ρ ciphertexts, a program $P \in \mathcal{P}$ with ρ input values and an integer $r \geq 2$, the homomorphic evaluation algorithm outputs $y_b \in R_r$, constituting party b's share of an output $y \in R_r$.*

The algorithms (HSS.Gen, HSS.Enc, HSS.Eval) *should satisfy the following correctness and security requirements:*

- **Correctness:** *For all $\lambda \in \mathbb{N}$, for all $x^{(1)}, \ldots, x^{(\rho)} \in \mathcal{I}$, for all programs $P \in \mathcal{P}$ with size $|P| \leq \mathsf{poly}(\lambda)$ and $P(x^{(1)}, \ldots, x^{(\rho)}) \neq \perp$, for integer $r \geq 2$, for $(\mathsf{pk}, \mathsf{ek}_0, \mathsf{ek}_1) \leftarrow \mathsf{HSS.Gen}(1^\lambda)$ and for $\mathsf{ct}^{(i)} \leftarrow \mathsf{HSS.Enc}(1^\lambda, \mathsf{pk}, x^{(i)})$ we have*

$$\mathrm{Pr}^{\mathsf{cor}}_{\mathsf{HSS}, (x^{(i)})_i, P, r}(\lambda) := \mathrm{Pr}\left[y_0 + y_1 = P(x^{(1)}, \ldots, x^{(\rho)}) \mod r\right] \geq 1 - \lambda^{-\omega(1)},$$

where

$$y_b \leftarrow \mathsf{HSS.Eval}(b, \mathsf{ek}_b, (\mathsf{ct}^{(i)})_i, P, r)$$

for $b \in \{0, 1\}$ and where the probability is taken over the random coins of HSS.Gen, HSS.Enc *and* HSS.Eval.
- **Security:** *For all security parameters $\lambda \in \mathbb{N}$, for all PPT adversaries $\mathcal{A}$ that on input 1^λ output a bit $b \in \{0, 1\}$ (specifying which encryption key to*

corrupt), and input values $x_0, x_1 \in \mathcal{I}$, we require the following advantage to be negligible in λ:

$$\text{Adv}^{\text{sec}}_{\text{HSS},\mathcal{A}}(\lambda) := \left| \Pr \left[\mathcal{A}(\text{input}_b) = \beta \middle| \begin{array}{l} (b, x_0, x_1, \text{state}) \leftarrow \mathcal{A}(1^\lambda), \\ \beta \leftarrow \{0, 1\}, \\ (\text{pk}, (\text{ek}_0, \text{ek}_1)) \leftarrow \text{HSS.Gen}(1^\lambda), \\ \text{ct} \leftarrow \text{HSS.Enc}(\text{pk}, x_\beta), \\ \text{input}_b := (\text{state}, \text{pk}, \text{ek}_b, \text{ct}) \end{array} \right] - \frac{1}{2} \right|.$$

Remark 1. Within applications, we additionally consider a secret-key variant of HSS. For details we refer to the full version.

2.2   Computational Models

Our main HSS scheme naturally applies to programs P in a computational model known as *Restricted Multiplication Straight-line (RMS)* programs [7,18].

Definition 3 (RMS programs). *An RMS program consists of a magnitude bound B_{max} and an arbitrary sequence of the four following instructions, sorted according to a unique identifier $\text{id} \in \mathcal{S}_{\text{id}}$:*

- *Load an input into memory: $(\text{id}, \widehat{y}_j \leftarrow \widehat{x}_i)$.*
- *Add values in memory: $(\text{id}, \widehat{y}_k \leftarrow \widehat{y}_i + \widehat{y}_j)$.*
- *Add input values: $(\text{id}, \widehat{x}_k \leftarrow \widehat{x}_i + \widehat{x}_j)$.*
- *Multiply memory value by input: $(\text{id}, \widehat{y}_k \leftarrow \widehat{x}_i \cdot \widehat{y}_j)$.*
- *Output from memory, as R element: $(\text{id}, r, \widehat{O}_j \leftarrow \widehat{y}_i)$.*

If at any step of execution the size of a memory value exceeds the bound B_{max} (i.e. $\|\widehat{y}_j\|_\infty > B_{\text{max}}$), the output of the program on the corresponding input is defined to be $\bot$. Otherwise the output is the sequence of $\widehat{O}_j$ values, sorted by id. We define the size (resp., multiplicative size) of an RMS program P as the number of instructions (resp., multiplication and load input instructions). Note that we consider addition of input values merely for the purpose of efficiency. We denote the maximum number of additions on input values by $P_{\text{inp}+}$.

3   HSS from Encryption with Nearly Linear Decryption

As explained in the introduction, the core of our HSS construction is an encryption scheme with nearly linear decryption, where nearly linear means that for message $m \in R_p := R/pR$, secret key $\mathbf{s} \in R_q^d$, and ciphertext $\mathbf{c} \in R_q^d$ encrypting m, for some "small" noise $e \in R$ we have

$$\langle \mathbf{s}, \mathbf{c} \rangle = (q/p) \cdot m + e \mod q.$$

We begin in Sect. 3.1 by explaining our two main share conversion tricks, which allow two parties holding secret shares of $(q/p) \cdot m + e \mod q$ for small m to locally modify their values, such that in the end each party holds a secret share of the message m *modulo* q. In Sect. 3.2 we present our formal definition of nearly linear decryption, and prove two properties that it implies. Then, in Sect. 3.3 we give our HSS construction based on any such encryption scheme.

3.1 Computation on 2-Party Secret Shared Values

First, we present a local rounding trick as in [22] which allows to recover the shares of m (modulo p). The idea is that if q/p is large, the probability that the error term e leads to a rounding error is small. Note that it is crucial here that we are in the 2-party setting, where the secret shares of $(q/p) \cdot m + e \mod q$ have both approximately (that is, except for the error e) the same distance from some multiple of q/p. In fact, even for arbitrarily large gap between p and q, rounding for 3 or more parties fails with constant probability. For a proof of the rounding lemma we refer to the full version.

Lemma 1 (Rounding of noisy shares). *Let $p, q \in \mathbb{N}$ be modulus values with $q/p \geq \lambda^{\omega(1)}$. Let $R = \mathbb{Z}[X]/(X^N + 1)$ for N a power of 2 (i.e. $N = 2^n$ for $n \in \mathbb{N}_0$). Let $t_0, t_1 \in R_q$ random subject to*

$$t_0 + t_1 = (q/p) \cdot m + e \mod q$$

for some $m \in R_p, e \in R$ with $q/(p \cdot \|e\|_\infty) \geq \lambda^{\omega(1)}$. Then, for the the deterministic polynomial time procedure Round that on input $t_b \in R_q$ outputs

$$\lfloor (p/q) \cdot t_b \rceil \mod p \in R_p$$

it holds:

$$\mathsf{Round}(t_0) + \mathsf{Round}(t_1) = m \mod p$$

with probability at least $1 - N \cdot (\|e\|_\infty + 1) \cdot p/q \geq 1 - \lambda^{-\omega(1)}$ over the choice of the shares t_0, t_1.

The following simple observation constitutes a crucial step of our HSS construction, as it will allow to have several levels of multiplication without requiring a sequence of decreasing moduli. While in general the conversion of secret shares from one modulus to another constitutes a problem, we observe that whenever the secret shared value is *small* in comparison to the modulus, and we use the centered representation of R_p with coefficients in $(- \lfloor p/2 \rfloor , \ldots, \lfloor (p - 1)/2 \rfloor]$, then with high probability the secret sharing actually constitutes a secret sharing over R, so switching to an arbitrary modulus is trivial. Note that (as for rounding) this only holds true in the 2-party setting. For the proof we refer to the full version.

Lemma 2 (Lifting the modulus of shares). *Let $p \in \mathbb{N}$ be a modulus with $p \geq \lambda^{\omega(1)}$. Let $R = \mathbb{Z}[X]/(X^N + 1)$ for N a power of 2. Let $m \in R$ and $z_0, z_1 \in R_p$ be random, subject to*

$$z_0 + z_1 = m \mod p.$$

Then, we have

$$z_0 + z_1 = m \text{ over } R$$

with probability at least $1 - (N \cdot (\|m\|_\infty + 1)/p) \geq 1 - \lambda^{-\omega(1)}$ over the choice of the shares z_0, z_1.

3.2 Encryption with Nearly Linear Decryption

We now formally introduce encryption with nearly linear decryption. Basically, we require the following properties: First, there is a way to encrypt certain key-dependent messages without knowledge of the secret key. Second, it is possible to "distributively decrypt" a ciphertext. More precisely, given an encryption of message m and secret shares of some multiple x of the secret key $\mathbf{s}$, there is a way to obtain secret shares of $x \cdot m$ *over the same modulus as the original secret shares.* These properties together enable us to perform several stages of distributed decryption. That is, given an encryption of $x \cdot \mathbf{s}$ (for some value x) and a secret share of $x' \cdot \mathbf{s}$ modulo q, distributed decryption results in a secret share of $x \cdot x' \cdot \mathbf{s}$ modulo q, which can serve as input to another distributed decryption. One way to achieve both properties at once is to require nearly linear decryption.

Definition 4 (Encryption scheme with nearly linear decryption). *Let* PKE := (PKE.Gen, PKE.Enc, PKE.Dec) *be a public-key encryption scheme with pseudorandom ciphertexts. We say that* PKE *is a public-key encryption scheme with nearly linear decryption if it further satisfies the following properties:*

- **Parameters:** *The scheme is parametrized by modulus values $p, q \in \mathbb{N}$, dimension $d \in \mathbb{N}$, and bounds $B_{\mathsf{sk}}, B_{\mathsf{ct}} \in \mathbb{N}$, where $p | q$, $p \geq \lambda^{\omega(1)}$, $q/p \geq \lambda^{\omega(1)}$ and $d, B_{\mathsf{sk}}, B_{\mathsf{ct}} \leq \mathsf{poly}(\lambda)$, as well as a ring $R = \mathbb{Z}[X]/(X^N + 1)$, where $N \leq \mathsf{poly}(\lambda)$ is a power of 2.*[6]
- **Message space and secret key:** *The scheme has message space $\mathcal{M} := R_p := R/pR$ and ciphertext space $\mathcal{C} := R_q^d := (R/qR)^d$. The secret key $\mathbf{s}$ returned by* PKE.Gen *on input 1^λ is an element of R^d satisfying $\|\mathbf{s}\|_\infty \leq B_{\mathsf{sk}}$. Further, $\mathbf{s}$ is of the form $(1, \widehat{\mathbf{s}})$ for some $\widehat{\mathbf{s}} \in R_p^{d-1}$.*
- **Nearly linear decryption:** *For any $\lambda \in \mathbb{N}$, for any $(\mathsf{pk}, \mathbf{s})$ in the image of* Gen(1^λ), *for any message $m \in R_p$ and for any ciphertext $\mathbf{c} \in R_q^d$ in the image of* PKE.Enc(pk, m), *for some $e \in R$ with $\|e\|_\infty \leq B_{\mathsf{ct}}$ it holds*

$$\langle \mathbf{s}, \mathbf{c} \rangle = (q/p) \cdot m + e \mod q.$$

Notation. *For $(\mathsf{pk}, \mathbf{s}) \leftarrow$ Gen(1^λ) and $\mathbf{m} = (m_1, \ldots, m_d) \in R_p^d$, we denote by* PKE.**Enc**$(\mathsf{pk}, \mathbf{m})$ *the componentwise encryption $\mathbf{C} \leftarrow ($PKE.Enc $(\mathsf{pk}, m_1), \ldots,$ PKE.Enc$(\mathsf{pk}, m_d))$; we denote by* **Dec**$(\mathsf{sk}, \mathbf{C})$ *the decryption* (PKE.Dec$(\mathsf{sk}, \mathbf{c}_1), \ldots,$ PKE.Dec$(\mathsf{sk}, \mathbf{c}_d)) \in R_p^d$ *of the matrix of d ciphertexts $\mathbf{C} = (\mathbf{c}_1 | \ldots | \mathbf{c}_d) \in R_q^{d \times d}$.*

Remark 2. Encryption with nearly linear decryption can be instantiated based on LWE (e.g. with [1,37], where $d = \lambda$) and based on RLWE (e.g. with [15,36], where $d = 2$). Further, it can be instantiated with schemes based on assumptions like module-LWE [34] and LWR [3]. For more details on the instantiation from the RLWE-based encryption scheme of LPR [36], we refer to Sect. 4, and for the instantiation from the LWE-based encryption scheme of Regev [37] we refer to the full version.

---

[6] To simplify the analysis, we restrict the definition to 2-power cyclotomic rings. However, our construction can be generalized to arbitrary cyclotomics.

$\text{Exp}^{\text{kdm}-\text{ind}}_{\text{PKE.OKDM},\mathcal{A}}(\lambda):$	$\mathcal{O}_{\text{KDM}}(x,j):$
$(\text{pk},\text{sk}) \leftarrow \text{PKE.Gen}(1^\lambda)$	if $\beta = 0$
$\beta \leftarrow \{0,1\}$	$\quad$ return PKE.OKDM(pk,x,j)
$\beta' \leftarrow \mathcal{A}^{\mathcal{O}_{\text{KDM}}(\cdot,\cdot)}(1^\lambda, \text{pk})$	else
if $\beta = \beta'$ return 1	$\quad$ return PKE.Enc$(\text{pk},0)$
else return 0	

Fig. 1. Security challenge experiment for the KDM oracle.

We prove that our two desired properties are satisfied by any encryption scheme with nearly linear decryption. The first property allows anyone to compute an encryption of any linear function of the secret key without having access to the secret key itself, serving as a "KDM oracle." A similar notion, but for secret-key encryption schemes and with deterministic procedure, was introduced in [4]. For the proof of the following lemma we refer to the full version.

Lemma 3 (KDM oracle). *Let* PKE $:=$ (PKE.Gen, PKE.Enc, PKE.Dec) *be a public-key encryption scheme with nearly linear decryption and parameters* $(p, q, d, B_{\text{sk}}, B_{\text{ct}}, R)$.

Then, for the PPT procedure PKE.OKDM *that on input of a public key* pk, *a value* $x \in R$ *and an index* $j \in \{1, \ldots, d\}$ *computes an encryption* $\mathbf{c} \leftarrow$ PKE.Enc$(\text{pk}, 0)$ *and outputs*

$$\mathbf{c}_j := (q/p) \cdot x \cdot \mathbf{e}_j + \mathbf{c} \mod q,$$

where $\mathbf{e}_j \in R_q^d$ *is the* j-*th unit vector, the following properties are satisfied.*

- *Nearly linear decryption to the message* $x \cdot s_j$: *For any* $\lambda \in \mathbb{N}$, *for any* $(\text{pk}, \mathbf{s})$ *in the image of* Gen(1^λ), *and for any ciphertext* $\mathbf{c}_j \in R_q^d$ *in the image of* PKE.OKDM(pk, x, j), *it holds*

$$\langle \mathbf{s}, \mathbf{c}_j \rangle = (q/p) \cdot (x \cdot s_j) + e \mod q$$

  for some $e \in R$ *with* $\|e\|_\infty \leq B_{\text{ct}}$.
- *Security: For any* $\lambda \in \mathbb{N}$ *and any PPT adversary* $\mathcal{A}$ *we have that*

$$\text{Adv}^{\text{kdm}-\text{ind}}_{\text{PKE.OKDM},\mathcal{A}}(\lambda) := \left| \Pr\left[\text{Exp}^{\text{kdm}-\text{ind}}_{\text{PKE.OKDM},\mathcal{A}}(\lambda) = 1 \right] - 1/2 \right|$$

  is negligible in λ, *where* $\text{Exp}^{\text{kdm}-\text{ind}}_{\text{PKE.OKDM},\mathcal{A}}(\lambda)$ *is as defined in Fig. 1*

By PKE.**OKDM**(pk, x) *we denote the KDM oracle that returns a componentwise encryption of* $x \cdot \mathbf{s}$, *i.e. that outputs the matrix* (PKE.OKDM$(\text{pk}, x, 1)$, $\ldots$, PKE.OKDM(pk, x, d)) $\in R_q^{d \times d}$.

The following shows that any encryption with nearly linear decryption allows two parties to perform decryption *distributively*, employing their respective shares of the secret key to obtain a secret share of the corresponding message *modulo q*. Further, the scheme inherently supports homomorphic addition of ciphertexts, and the distributed decryption property holds accordingly for any sum of a bounded number of ciphertexts (generated from Enc or OKDM).

Lemma 4 (Distributed decryption of sums of ciphertexts). *Let* PKE :=
(PKE.Gen, PKE.Enc, PKE.Dec) *be a public-key encryption scheme with nearly
linear decryption and parameters* $(p, q, d, B_{\mathsf{sk}}, B_{\mathsf{ct}}, R)$, *where* R *has dimension
N. Let* PKE.OKDM *be the KDM oracle from Lemma 3. Let* $B_{\mathsf{add}} \in \mathbb{N}$ *with
$B_{\mathsf{add}} \leq \mathsf{poly}(\lambda)$. Then the deterministic polynomial time decryption procedure*
PKE.DDec *that on input* $b \in \{0, 1\}, \mathbf{t}_b \in R^d, \mathbf{c} \in R_q^d$ *outputs*

$$\mathsf{Round}(\langle \mathbf{t}_b, \mathbf{c} \rangle \mod q) \in R_q$$

(where Round *is as in Lemma 1) satisfies the following:*

For all $x \in R_p$ *with* $p/\|x\|_\infty \geq \lambda^{\omega(1)}$ *and* $q/(p \cdot \|x\|_\infty) \geq \lambda^{\omega(1)}$, *for all*
$(\mathsf{pk}, \mathbf{s}) \leftarrow \mathsf{Gen}(1^\lambda)$, *for all messages* $m_1 \ldots, m_{B_{\mathsf{add}}} \in R_p$, *for all encryptions* $\mathbf{c}_i$ *of*
m_i *that are either output of* PKE.Enc *or of* PKE.OKDM *(in that case we have*
$m_i = x_i \cdot s_j$ *for some value* $x_i \in R_p$ *and some index* $j \in \{1, \ldots, d\}$) *and for
shares* $\mathbf{t}_0, \mathbf{t}_1 \in R_q^d$ *random subject to*

$$\mathbf{t}_0 + \mathbf{t}_1 = x \cdot \mathbf{s} \mod q$$

for $\mathbf{c} := \sum_{i=1}^{B_{\mathsf{add}}} \mathbf{c}_i$ *and* $m := \sum_{i=1}^{B_{\mathsf{add}}} m_i$ *it holds*

$$\mathsf{PKE.DDec}(0, \mathbf{t}_0, \mathbf{c}) + \mathsf{PKE.DDec}(1, \mathbf{t}_1, \mathbf{c}) = x \cdot m \mod q$$

with probability over the random choice of the shares $\mathbf{t}_0, \mathbf{t}_1$ *of at least*

$$1 - N \cdot (N \cdot B_{\mathsf{add}} \cdot \|x\|_\infty \cdot B_{\mathsf{ct}} \cdot p/q + \|x \cdot m\|_\infty/p + p/q + 1/p) \geq 1 - \lambda^{-\omega(1)}.$$

For $\mathbf{C} = (\mathbf{c}_1 | \ldots | \mathbf{c}_d) \in R_p^{d \times d}$ *by* $\mathbf{m} \leftarrow \mathsf{PKE.\mathbf{DDec}}(b, \mathbf{t}_b, \mathbf{C})$ *we denote the
componentwise decryption* $\mathbf{m} \leftarrow (\mathsf{PKE.DDec}(b, \mathbf{t}_b, \mathbf{c}_1), \ldots, \mathsf{PKE.DDec}(b, \mathbf{t}_b, \mathbf{c}_d))$
$\in R_p^d$.

For a proof that PKE.DDec indeed satisfies the required we refer to the full
version. The idea is that nearly linear decryption allows (almost) homomorphic
addition of ciphertexts with linear growth in the error. As $q/(p \cdot \|x\|_\infty) \geq \lambda^{\omega(1)}$
and the vectors $\mathbf{t}_b$ are individually random, by Lemma 1 we can recover $x \cdot m$
mod p with overwhelming probability. Finally, as $p \geq \lambda^{\omega(1)}$, by Lemma 2 we can
lift the modulus q (as with overwhelming probability the shares constitute a
correct sharing of $x \cdot m$ over R and thus R_q).

Remark 3. Note that our techniques also extend to encryption schemes which
encrypt messages in low-order symbols, e.g. where $\langle \mathbf{s}, \mathbf{c} \rangle = m + p \cdot e \mod q$ for
p and q coprime. For more details we refer to the full version.

3.3   HSS from Encryption with Nearly Linear Decryption

We now present our construction of a public-key HSS from an encryption scheme
with nearly linear decryption. For various extensions that allow to improve the
efficiency in specific applications, we refer to Sect. 3.4.

HSS.Gen(1^λ) :

- Generate a key pair $(\mathsf{pk}, \mathbf{s}) \leftarrow \mathsf{PKE.Gen}(1^\lambda)$ for encryption and draw a PRF key $K \overset{\$}{\leftarrow} \mathcal{K}$.   //Recall $\mathbf{s} = (1, \widehat{\mathbf{s}}) \in R_q \times R_q^{d-1}$.
- **Secret share the secret key.** Choose $\mathbf{s}_0 \overset{\$}{\leftarrow} R_q^d$ at random. Define

$$\mathbf{s}_1 := \mathbf{s} - \mathbf{s}_0 \mod q.$$

- Output pk and $\mathsf{ek}_b \leftarrow (K, \mathbf{s}_b)$.

HSS.Enc$(1^\lambda, \mathsf{pk}, x)$:

- **Encrypt the input.** Compute and output $\mathbf{C}^x \leftarrow \mathsf{PKE.OKDM}(\mathsf{pk}, x)$.
  //This corresponds to $\mathbf{C}^x = \mathsf{PKE.Enc}(\mathsf{pk}, x \cdot \mathbf{s}) \in R_q^{d \times d}$.

HSS.Eval$(b, \mathsf{ek}_b, (\mathbf{C}^{x^{(1)}}, \ldots, \mathbf{C}^{x^{(\rho)}}), P, r)$:

Parse $(K, \mathbf{s}_b) =: \mathsf{ek}_b$, parse P as a sequence of RMS operations and proceed as follows.

- **Load an input into memory:** On instruction $(\mathsf{id}, \mathbf{C}^x)$ compute

$$\mathbf{t}_b^x := \mathsf{PKE.DDec}(b, \mathbf{s}_b, \mathbf{C}^x) + (1 - 2b) \cdot \mathsf{PRF}(K, \mathsf{id}) \mod q.$$

- **Add values in memory:** On instruction $(\mathsf{id}, \mathbf{t}_b^x, \mathbf{t}_b^{x'})$ compute

$$\mathbf{t}_b^{x+x'} \leftarrow \mathbf{t}_b^x + \mathbf{t}_b^{x'} + (1 - 2b) \cdot \mathsf{PRF}(K, \mathsf{id}) \mod q.$$

- **Add input values:** On instruction $(\mathsf{id}, \mathbf{C}^x, \mathbf{C}^{x'})$ compute

$$\mathbf{C}^{x+x'} \leftarrow \mathbf{C}^x + \mathbf{C}^{x'} \mod q.$$

- **Multiply memory value by input:** On instruction $(\mathsf{id}, \mathbf{t}_b^x, \mathbf{C}^{x'})$ compute

$$\mathbf{t}_b^{x \cdot x'} := \mathsf{PKE.DDec}(b, \mathbf{t}_b^x, \mathbf{C}^{x'}) + (1 - 2b) \cdot \mathsf{PRF}(K, \mathsf{id}) \mod q.$$

- **Output from memory, as element in R_r:** On instruction $(\mathsf{id}, \mathbf{t}_b^x)$ parse $\mathbf{t}_b^x =: (x_b, \widehat{\mathbf{t}_b^x})$ for some $x_b \in R_q, \widehat{\mathbf{t}_b^x} \in R_q^{d-1}$ and output

$$x_b \mod r.$$

Fig. 2. 2-party public-key homomorphic secret sharing scheme HSS for the class of RMS programs from encryption with nearly linear decryption. Here, $x \in R$ with $\|x\|_\infty \leq B_{\mathsf{inp}}$ is an input value. Throughout, *input values* $x \in R$ are represented by encryptions $\mathbf{C}^x$ of $x \cdot \mathbf{s}$ and *memory values* $x \in R$ are represented by shares $(\mathbf{t}_0^x, \mathbf{t}_1^x) \in R_q^d \times R_q^d$ with $\mathbf{t}_0^x + \mathbf{t}_1^x = x \cdot \mathbf{s} \mod q$.

$$
\begin{array}{|ll|}
\hline
\mathbf{G}^0_{\mathsf{HSS},\mathcal{A}}(\lambda): & \mathbf{G}^1_{\mathsf{HSS},\mathcal{A}}(\lambda): \\
\hline
(b, x_0, x_1, \text{state}) \leftarrow \mathcal{A}(1^\lambda) & (b, x_0, x_1, \text{state}) \leftarrow \mathcal{A}(1^\lambda) \\
\beta \leftarrow \{0, 1\} & \beta \leftarrow \{0, 1\} \\
(\mathsf{pk}, (\mathsf{ek}_0, \mathsf{ek}_1)) \leftarrow \mathsf{HSS.Gen}(1^\lambda) & (\mathsf{pk}, (\mathsf{ek}_0, \mathsf{ek}_1)) \leftarrow \mathsf{HSS.Gen}(1^\lambda) \\
//\text{Encrypt } x_\beta \cdot \mathsf{s}. & //\text{Encrypt } \mathbf{0} \in R^d. \\
\mathbf{C} \leftarrow \mathsf{PKE.OKDM}(\mathsf{pk}, x_\beta) & \mathbf{C} \leftarrow \mathsf{PKE.Enc}(\mathsf{pk}, \mathbf{0}) \\
\beta' \leftarrow \mathcal{A}(\text{state}, \mathsf{pk}, \mathsf{ek}_b, \mathbf{C}) & \beta' \leftarrow \mathcal{A}(\text{state}, \mathsf{pk}, \mathsf{ek}_b, \mathbf{C}) \\
\text{if } \beta' = \beta \text{ return } 1 & \text{if } \beta' = \beta \text{ return } 1 \\
\text{else return } 0 & \text{else return } 0 \\
\hline
\end{array}
$$

Fig. 3. Games $\mathbf{G}^0_{\mathsf{HSS},\mathcal{A}}(\lambda)$ and $\mathbf{G}^1_{\mathsf{HSS},\mathcal{A}}(\lambda)$ in the proof of Theorem 2 (Sec. of HSS).

Theorem 2 (HSS from encryption with nearly linear decryption). *Let* PKE $:=$ (PKE.Gen, PKE.Enc, PKE.Dec) *be a secure public-key encryption scheme with nearly linear decryption and parameters* $(p, q, d, B_{\mathsf{sk}}, B_{\mathsf{ct}}, R)$.

- *Let* $B_{\mathsf{inp}} \in \mathbb{N}$ *with* $p/B_{\mathsf{inp}} \geq \lambda^{\omega(1)}$ *and* $q/(B_{\mathsf{inp}} \cdot p) \geq \lambda^{\omega(1)}$.
- *Let* PKE.OKDM *be the KDM oracle from Lemma 3.*
- *Let* PKE.DDec *be the distributed decryption from Lemma 4.*
- *Let* PRF$: \mathcal{K} \times \mathcal{S}_{\mathsf{id}} \to R^d_q$ *be a pseudorandom function.*

Then, the scheme HSS = (HSS.Gen, HSS.Enc, HSS.Eval) *given in Fig. 2 is a 2-party public-key homomorphic secret sharing scheme with input space* $[R]_{B_{\mathsf{inp}}}$ *for the class of RMS programs with magnitude bound* B_{max}, *where* $p/B_{\mathsf{max}} \geq \lambda^{\omega(1)}$ *and* $q/(B_{\mathsf{max}} \cdot p) \geq \lambda^{\omega(1)}$. *More precisely,* HSS *satisfies the following (Fig. 3).*

- **Correctness:** *For any* $\lambda \in \mathbb{N}$, *for any* $x^{(1)}, \ldots, x^{(\rho)} \in [R]_{B_{\mathsf{inp}}}$, *for any polynomial-sized RMS program* P *with* $P(x^{(1)}, \ldots, x^{(\rho)}) \neq \bot$ *and magnitude bound* B_{max} *with* $p/B_{\mathsf{max}} \geq \lambda^{\omega(1)}$ *and* $q/(B_{\mathsf{max}} \cdot p) \geq \lambda^{\omega(1)}$, *and for any integer* $r \geq 2$, *there exist a PPT adversary* $\mathcal{B}$ *on the pseudorandomness of* PRF *such that*

$$
\Pr^{\mathsf{cor}}_{\mathsf{HSS}, (x^{(i)})_i, P, r}(\lambda) \geq 1 - \left(\mathsf{Adv}^{\mathsf{prf}}_{\mathsf{PRF}, \mathcal{B}}(\lambda) + \lambda^{-\omega(1)} \right).
$$

- **Security:** *For every PPT adversary* $\mathcal{A}$ *on the security of* HSS, *there exists an PPT adversary* $\mathcal{B}$ *on the security of* PKE.OKDM *such that*

$$
\mathsf{Adv}^{\mathsf{sec}}_{\mathsf{HSS}, \mathcal{A}}(\lambda) \leq \mathsf{Adv}^{\mathsf{kdm-ind}}_{\mathsf{PKE.OKDM}, \mathcal{B}}(\lambda).
$$

We prove correctness in the following lemma and refer to the full version for the proof of security.

Lemma 5 (Correctness of the HSS). *Let* HSS *be the HSS from Fig. 2 with underlying ring* $R = \mathbb{Z}[X]/(X^N + 1)$. *Then, for all* $\lambda \in \mathbb{N}$, *for all inputs* $x^{(1)}, \ldots, x^{(\rho)} \in [R]_{B_{\mathsf{inp}}}$, *for all RMS programs* P, *s.t.*

– P is of size $|P| \leq \mathsf{poly}(\lambda)$
– P has magnitude bound B_{max} with $p/B_{\mathsf{max}} \geq \lambda^{\omega(1)}$ and $q/(B_{\mathsf{max}} \cdot p) \geq \lambda^{\omega(1)}$,
– P has maximum number of input addition instructions $P_{\mathsf{inp}+}$

for $(\mathsf{pk}, \mathsf{ek}_0, \mathsf{ek}_1) \leftarrow \mathsf{HSS.Gen}(1^\lambda)$, for $\mathbf{C}^{x^{(i)}} \leftarrow \mathsf{HSS.Enc}(1^\lambda, \mathsf{pk}, x^{(i)})$, there exists an PPT adversary $\mathcal{B}$ on the pseudorandom function PRF with such that correctness holds with probability at least

$$\mathrm{Pr}^{\mathsf{cor}}_{\mathsf{HSS},(x^{(i)})_i,P}(\lambda) \geq 1 - \mathrm{Adv}^{\mathsf{prf}}_{\mathsf{PRF},\mathcal{B}}(\lambda) - N \cdot (B_{\mathsf{max}} + 1)/q$$
$$- |P| \cdot d \cdot N^2 \cdot P_{\mathsf{inp}+} \cdot B_{\mathsf{max}} \cdot (B_{\mathsf{ct}} \cdot p/q + B_{\mathsf{sk}}/p).$$
$$- |P| \cdot d \cdot N \cdot (p/q + 1/p).$$

Proof. We prove correctness via a hybrid argument. Let $\varepsilon_0 := \mathrm{Pr}^{\mathsf{cor}}_{\mathsf{HSS},(x^{(i)})_i,P,r}(\lambda)$. Recall that by ε^0 we denote the probability that homomorphic evaluation of a program P on input $(x^{(1)}, \ldots, x^{(\rho)}) \in [R]^\rho_{B_{\mathsf{inp}}}$ employing our HSS presented in Fig. 2 is successful (over the random choices of $\mathsf{HSS.Gen}, \mathsf{HSS.Enc}$). Our goal is to prove that for all $x^{(1)}, \ldots, x^{(\rho)} \in [R]_{B_{\mathsf{inp}}}$ and for all bounded RMS programs P the probability ε_0 is negligible in λ.

To this end, let $\varepsilon_1 := \mathrm{Pr}^1_{\mathsf{HSS},(x^{(i)})_i,P,r}(\lambda)$ denote the probability that evaluation yields the correct output, where we replace every evaluation of the PRF by inserting a value $\mathbf{r} \xleftarrow{\$} R_q^d$ chosen at random. We show that if the probabilities ε_0 and ε_1 differ significantly, then there exists an adversary $\mathcal{B}$ attacking the underlying PRF PRF. Namely, $\mathcal{B}$ homomorphically evaluates the program P on input $(x^{(1)}, \ldots, x^{(\rho)})$, but instead of evaluating $\mathsf{PRF}(K, \mathsf{id})$ the adversary $\mathcal{B}$ queries its PRF oracle. Finally, $\mathcal{B}$ returns *real* if homomorphic evaluation does not yield the correct result, and *random* otherwise. This yields

$$|\varepsilon_0 - \varepsilon_1| \leq \mathrm{Adv}^{\mathsf{prf}}_{\mathsf{PRF},\mathcal{B}}(\lambda).$$

It is left to give a lower bound for the probability ε_1. To that end, we prove that with overwhelming probability over the choice of $\mathbf{r} \leftarrow R_q^d$ (in place of the PRF evaluation) all shares $(\mathbf{t}_0^x, \mathbf{t}_1^x)$ computed during homomorphic evaluation of P satisfy

$$\mathbf{t}_0^x + \mathbf{t}_1^x = x \cdot \mathbf{s} = (x, x \cdot \widehat{\mathbf{s}}) \mod q \tag{1}$$

if the function evaluation of P at point $(\mathbf{t}_0^x, \mathbf{t}_1^x)$ corresponds to $x \in R$, where $\mathbf{s} = (1, \widehat{\mathbf{s}}) \in R \times R^{d-1}$ is the secret key returned by $\mathsf{PKE.Gen}$ on input 1^λ. Further, we have that $(\mathbf{t}_0^x, \mathbf{t}_1^x)$ are distributed uniformly at random conditioned on Eq. 1.

Assuming Eq. 1 is true, by Lemma 2 we have $x_0 + x_1 = x$ over R (and thus over R_r) with probability at least $1 - N \cdot (B_{\mathsf{max}} + 1)/q$.

It is left to prove that indeed Eq. 1 holds true during homomorphic evaluation of P except with negligible probability. Recall that $\mathsf{PKE.DDec}$ is the procedure for distributed decryption from Lemma 4. First, assume that distributed decryption is always successful. In this case we prove that any instruction preserves correctness. Note that we do not need to consider the addition of input values and the output of a memory value, as those do not affect the shares.

– **Load an input into memory:** Consider instruction $(\mathsf{id}, \mathbf{C}^x)$ for $b \in \{0,1\}$. Assuming correctness of distributed decryption it holds

$$\mathbf{t}_0^x + \mathbf{t}_1^x = \mathsf{PKE.DDec}(0, \mathbf{s}_0, \mathbf{C}^x) + \mathbf{r} + \mathsf{PKE.DDec}(1, \mathbf{s}_1, \mathbf{C}^x) - \mathbf{r} \mod q$$
$$= 1 \cdot (x \cdot \mathbf{s}) \mod q = x \cdot \mathbf{s} \mod q.$$

– **Add values in memory:** Assuming correctness holds for shares $(\mathbf{t}_0^x, \mathbf{t}_1^x)$ and $(\mathbf{t}_0^{x'}, \mathbf{t}_1^{x'})$ we have, as required,

$$\mathbf{t}_0^{x+x'} + \mathbf{t}_1^{x+x'} = \mathbf{t}_0^x + \mathbf{t}_0^{x'} + \mathbf{r} + \mathbf{t}_1^x + \mathbf{t}_1^{x'} - \mathbf{r} \mod q$$
$$= x \cdot \mathbf{s} + x' \cdot \mathbf{s} \mod q = (x + x') \cdot \mathbf{s} \mod q.$$

– **Multiply memory value by input:** Assuming correctness holds for the share $(\mathbf{t}_0^x, \mathbf{t}_1^x)$ and assuming correctness of distributed decryption it holds

$$\mathbf{t}_0^{x \cdot x'} + \mathbf{t}_1^{x \cdot x'} = \mathsf{PKE.DDec}(0, \mathbf{t}_0^x, \mathbf{C}^{x'}) + \mathsf{PKE.DDec}(1, \mathbf{t}_1^x, \mathbf{C}^{x'}) \mod q$$
$$= x \cdot (x' \cdot \mathbf{s}) \mod q = (x \cdot x') \cdot \mathbf{s} \mod q.$$

As $\mathbf{r}$ is chosen at random, the distribution of $(\mathbf{t}_0^y, \mathbf{t}_1^y) \in R_q^d$ for $y \in \{x, x + x', x \cdot x'\}$ is random conditioned on Eq. 1.

It is left to bound the probability that distributed decryption fails. As for all x computed throughout the evaluation of program P the distribution of $(\mathbf{t}_0^x, \mathbf{t}_1^x) \in R_q^d$ is random conditioned on Eq. 1, by Lemma 4 for all messages $m_1 \dots, m_{P_{\mathsf{inp}+}} \in R_p$ and for all encryptions $\mathbf{c}_i$ of m_i that are output of $\mathsf{PKE.OKDM}$ distributed decryption of $\sum_{i=1}^{P_{\mathsf{inp}+}} \mathbf{c}_i$ fails with probability at most

$$N^2 \cdot P_{\mathsf{inp}+} \cdot \|x\|_\infty \cdot B_{\mathsf{ct}} \cdot p/q + N \cdot \|x \cdot m\|_\infty / p + N \cdot (p/q + 1/p),$$

where $m := \sum_{i=1}^{P_{\mathsf{inp}+}} m_i$. Throughout the evaluation of P we are guaranteed $\|x\|_\infty \leq B_{\mathsf{max}}$ for all intermediary values $x \in R$. Further, for the messages $m_i = x_i \cdot s_{j_i}$ corresponding to outputs of $\mathsf{PKE.OKDM}$ we have

$$\left\| x \cdot \sum_{i=1}^{P_{\mathsf{inp}+}} x_i \cdot s_{j_i} \right\|_\infty \leq \sum_{i=1}^{P_{\mathsf{inp}+}} \|x \cdot x_i \cdot s_{j_i}\|_\infty \leq P_{\mathsf{inp}+} \cdot N \cdot B_{\mathsf{max}} \cdot B_{\mathsf{sk}}.$$

Finally, applying a union bound over all $|P| \cdot d$ decryptions yields

$$\varepsilon_1 \geq 1 - N \cdot (B_{\mathsf{max}} + 1)/q - |P| \cdot d \cdot N^2 \cdot P_{\mathsf{inp}+} \cdot B_{\mathsf{max}} \cdot (B_{\mathsf{ct}} \cdot p/q + B_{\mathsf{sk}}/p)$$
$$- |P| \cdot d \cdot N \cdot (p/q + 1/p).$$

3.4 Extensions

In the following we briefly describe some extensions which are tailored to special applications and improve the HSS construction introduced in the previous section in terms of efficiency. For a complete treatment, we refer the reader to the full version.

Secret-key HSS. For certain applications, where all secret inputs originate from a single party, it is sufficient to consider a *secret-key* HSS. This allows a more efficient instantiation for two reasons. First, the underlying encryption scheme is not required to support ciphertexts from a KDM oracle (but has to be KDM secure), which slightly saves in noise parameters. Further, we can save in terms of computations (at the cost of a larger share size), by replacing the DDec steps for loading an input x into memory, by instead sending the secret shares of $x \cdot \mathbf{s}$ as an additional part of the HSS share.

HSS for degree-2 polynomials. For the restricted class of degree-2 polynomials, we can achieve improved efficiency in both the secret-key and public-key setting, by leveraging the fact that our HSS need only support terminal multiplications.

For the secret-key case, as we do not need to load inputs, we actually only need one level of distributed decryption. This has two advantages: First, it suffices to encrypt $x \in R_p$ instead of $x \cdot \mathbf{s} \in R_p^d$, as the output is not required to allow another distributed encryption. Second, for the same reason, we do not need to lift the modulus of the output of the distributed decryption back to q. Thus, we can choose $p \leq \mathsf{poly}(\lambda)$ and $q \geq \lambda^{\omega(1)}$ (as we no longer must apply Lemma 2).

The idea of our public-key HSS is to change the way inputs are loaded into memory. The idea is to obtain the shares of $x \cdot \mathbf{s} = (x, x \cdot s_2, \ldots, x \cdot s_d) \in R^d$ by decrypting $\mathsf{PKE.Enc}(\mathsf{pk}, x)$ with $\mathbf{s}$ *and* with $s_2 \cdot \mathbf{s}, \ldots, s_d \cdot \mathbf{s}$. This strategy requires a quadratic number of secret shares (namely shares of $\mathbf{s} \cdot \mathbf{s}^\top$), but reduces the number of required encryption from d to 1 (as only encryptions of x are required). An additional advantage of this approach is that we only have to require the underlying encryption scheme to be IND-CPA secure (instead of satisfying pseudorandomness of ciphertexts).

HSS supporting SIMD operations. As first observed by [39], if the underlying ring R is of the right form, one can "pack" multiple plaintexts in one ciphertext. We show that our basic HSS supports "single instruction, multiple input" (SIMD) in this case. More precisely, we show that if $R = \mathbb{Z}[X]/(X^N + 1)$ for $N \in \mathbb{N}$, $N \leq \mathsf{poly}(\lambda)$ a power of 2, such that $X^N + 1$ splits over R_r (for some prime $r \geq 2$) into pairwise different irreducible polynomials of degree $k \in \mathbb{N}$ (i.e. $R_r \cong (\mathbb{F}_{r^k})^{N/k}$), one can evaluate a program P simultaneously on N/k inputs in $\mathbb{F}_{r^k}$. However, there are some caveats regarding magnitude growth with respect to the SIMD versus coefficient representations (see the full version for more details).

4   Instantiations and Efficiency Analysis

Our HSS schemes can be instantiated in a number of ways, using LWE or RLWE-based encryption schemes satisfying the nearly-linear decryption property from

---

LPR.Gen(1^λ) :

1. Sample $a \leftarrow R_q, \widehat{s} \leftarrow \mathcal{D}_{sk}, e \leftarrow \mathcal{D}_{err}$ and compute $b = a \cdot \widehat{s} + e$ in R_q.
2. Let $\mathbf{s} = (1, \widehat{s})$ and output $\mathsf{pk} = (a, b), \mathsf{sk} = \mathbf{s}$.

LPR.Enc(pk, m) :

1. To encrypt $m \in R_p$, first sample $v \leftarrow \mathcal{D}_{sk}, e_0, e_1 \leftarrow \mathcal{D}_{err}$.
2. Output the ciphertext $(c_0, c_1) \in R_q^2$, where $c_1 = -av + e_0$ and $c_0 = bv + e_1 + (q/p) \cdot m$.

LPR.OKDM(pk, m) :

1. Compute $\mathbf{c}^0 = \mathsf{LPR.Enc}(0)$ and $\mathbf{c}^m = \mathsf{LPR.Enc}(m)$.
2. Output the tuple $(\mathbf{c}^m, \mathbf{c}^0 + (0, (q/p) \cdot m))$ as encryptions of $m \cdot \mathbf{s}$.

LPR.DDec$(b, \mathbf{t}_b, \mathbf{c}^x)$:

1. Given $b \in \{0, 1\}$, a ciphertext $\mathbf{c}^x$ and a share $\mathbf{t}_b$ of $m \cdot \mathbf{s}$, first parse $\mathbf{c}^x = (c_0, c_1)$ and $\mathbf{t}_b = (t_{b,0}, t_{b,1})$.
2. Output $(d_0, d_1) := (\lfloor (p/q) \cdot (c_0 \cdot t_{b,0} + c_1 \cdot t_{b,1}) \rceil \mod p) \mod q$

Fig. 4. Ring-LWE based instantiation of PKE with approximately linear decryption, with procedures for HSS from Sect. 3

Definition 4. In this section we focus on a particularly efficient RLWE-based instantiation using a variant of the "LPR" encryption scheme [35] over 2-power cyclotomic rings. In the full version we also show how to use standard Regev encryption based on LWE [37], but this is less efficient in terms of share size.

4.1 Instantiation from Ring-LWE

Definition 5 (Decisional Ring Learning With Errors). *Let N be a power of 2, $q \geq 2$ be an integer, $R = \mathbb{Z}[X]/(X^N + 1)$ and $R_q = R/(qR)$. Let $\mathcal{D}_{err}$ be an error distribution over R and $\mathcal{D}_{sk}$ be a secret key distribution over R. Let $s \leftarrow \mathcal{D}_{sk}$. The $\mathsf{RLWE}_{N,q,\mathcal{D}_{err},\mathcal{D}_{sk}}$ problem is to distinguish the following two distributions over R_q^2:*

- *$\mathcal{O}_{\mathcal{D}_{err},s}$: Output (a, b) where $a \leftarrow R_q, e \leftarrow \mathcal{D}_{err}$ and $b = a \cdot s + e$*
- *U: Output $(a, u) \leftarrow R_q^2$*

Formally, for a PPT adversary $\mathcal{A}$ we define the advantage $\mathrm{Adv}_{N,q,\mathcal{D}_{err},\mathcal{D}_{sk}}^{\mathsf{rlwe}}(\lambda) = |\mathrm{Pr}_{s \leftarrow \mathcal{D}_{sk}}[\mathcal{A}^{\mathcal{O}_{\mathcal{D}_{err},s}}(\lambda) = 1] - \mathrm{Pr}_{s \leftarrow \mathcal{D}_{sk}}[\mathcal{A}^{U}(\lambda) = 1]|$.

In Fig. 4 we present the core algorithms for our RLWE-based instantiation using the LPR [35] public-key encryption scheme $\mathsf{LPR} = (\mathsf{LPR.Gen}, \mathsf{LPR.Enc})$, as well as the auxiliary algorithms LPR.OKDM and LPR.DDec used by our HSS constructions. We use an error distribution $\mathcal{D}_{\mathsf{err}}$ where each coefficient is a rounded Gaussian with parameter σ, which gives $B_{\mathsf{err}} = 8\sigma$ as a high-probability bound on the ℓ_∞ norm of samples from $\mathcal{D}_{\mathsf{sk}}$, with failure probability $\mathtt{erf}(8/\sqrt{2}) \approx 2^{-49}$. We choose the secret-key distribution such that each coefficient of s is uniform in $\{0, \pm 1\}$, subject to the constraint that only h_{sk} coefficients are non-zero.[7]

The following lemma (proven in the full version) shows that LPR satisfies the nearly-linear decryption property for our HSS scheme. Furthermore, notice that ciphertexts output by LPR.Enc are pseudorandom under the decisional ring-LWE assumption, by a standard hybrid argument [36]. Therefore, the correctness and security properties of the LPR.OKDM and LPR.DDec procedures follow from Lemmas 3 and 4.

Lemma 6. *Assuming hardness of* $\mathsf{RLWE}_{N,q,\mathcal{D}_{\mathsf{err}},\mathcal{D}_{\mathsf{sk}}}$, *the scheme* LPR *(Fig. 4) is a public-key encryption scheme with nearly-linear decryption over* $R = \mathbb{Z}[X]/(X^N + 1)$, *with ciphertext dimension* $d = 2$ *and bounds* B_{sk} *and* $B_{\mathsf{ct}} = B_{\mathsf{err}} \cdot (2h_{\mathsf{sk}} + 1)$.

4.2 Parameters and Efficiency Analysis

We now analyse the efficiency of our RLWE-based instantiation and compare it with using HSS constructed from somewhat homomorphic encryption, for various different settings of parameters.

For comparison with HSS based on DDH [7], we remark that for non-SIMD computations, DDH-based HSS shares can be smaller than both our approach and SHE. However, we estimate that homomorphic evaluation is around an order or magnitude faster than the times reported in [6] due to the expensive share conversion procedure, and when using SIMD both this and the share size can be dramatically improved.

Parameter estimation. We derived parameters for our HSS based on LPR using the bounds for correctness from Lemma 5, chosen to ensure that each RMS multiplication of a ring-element during evaluation is correct with probability $1 - 2^{-\kappa}$, where we chose $\kappa = 40$. To compare with constructing HSS from SHE, we estimated parameters for the "BFV" scheme based on RLWE [13,24], currently one of the leading candidate SHE schemes. To modify this to achieve HSS with additive output sharing, we need to increase the size of q by around 2κ bits. With both schemes we chose parameters estimated to have at least 80 bits of computational security, see the full version for more details.

Share size. Tables 1 and 2 show BFV ciphertext parameters for different multiplicative depths of circuit, and plaintext modulus 2 or $\approx 2^{128}$, respectively, to illustrate different kinds of Boolean and arithmetic computations. Table 3 gives

---

[7] Choosing a sparse secret like this does incur a small loss in security, and only gives us a small gain in parameters for the HSS. The main reason we choose s like this is to allow a fair comparison with SHE schemes, which typically have to use sparse secrets to obtain reasonable parameters.

our HSS parameters for various choices of $B_{\max}$, the maximum value any plaintext coefficient can hold during the computation. Note that in contrast to SHE, our parameters depend only on this bound and not the multiplicative depth, although we are more restricted in that we can only perform homomorphic multiplications where one value is an input.

This means that comparing parameters of the two schemes is very application-dependent. For instance, for Boolean computations where we can have $B_{\max} = 2$, our scheme has smaller parameters than SHE for all computations of depth > 3, so this can give a significant advantage for very high degree functions that can be expressed as an RMS program. However, if SIMD computations are required then $B_{\max}$ must be chosen to account for the worst-case coefficient growth, which is not directly related to the plaintexts, so our scheme would likely have larger ciphertexts than SHE in most cases. For operations on large integers, the parameters in both schemes quickly get very large, though our parameters grow slightly quicker due to the increase in $B_{\max}$.

Computational efficiency. The relative computational efficiency of the schemes is much clearer, and is the main advantage of our scheme over SHE. The cost of a homomorphic RMS multiplication with RLWE is roughly twice the cost of a decryption in any RLWE-based scheme (including BFV) with the same parameters. Recently, Halevi et al. [31] described an optimized implementation of BFV using CRT arithmetic, where according to their single-threaded runtimes, decryption costs between 20–30x less than multiplication (including key-switching) for the ranges of parameters we consider (cf. [31, Table 3]). This indicates a *10–15x improvement in performance* for homomorphic evaluation with our scheme compared with SHE, assuming similar parameters and numbers of multiplications. We remark that this comparison deserves some caution, since other SHE schemes such as BGV [14] may have different characteristics; we have not run experiments with BGV, but due to the complications in key-switching and modulus-switching we expect the improvement to still be around an order of magnitude.

5   Applications

In this section we highlight some applications of HSS for which our scheme seems well-suited. There are four primary approaches to compare: approaches not relying on HSS, using DDH-based or one-way function-based HSS, using HSS based on SHE, or using our new HSS. We remark that the concrete practicality of SHE-based HSS approaches has also not been considered before this work.

5.1   Secure 2-PC for Low-Degree Polynomials

Perhaps the most natural application of HSS is to achieve a very succinct form of multi-party computation. After a setup phase to create the key material $\mathsf{pk}, (\mathsf{ek}_0, \mathsf{ek}_1)$, each party publishes HSS-shares of its input, which can then be directly used to compute additive shares of the output. Even the simplest case of evaluating degree-2 polynomials has many interesting applications, and also

Table 1. BFV parameters with plaintext modulus 2

Depth	N	$\log q$	Security
1	4096	102	145.1
2	4096	118	122.6
3	4096	134	106.2
4	4096	150	93.73
5	4096	164	85.53
6	8192	186	157.5
7	8192	202	142.9
8	8192	220	129.8
9	8192	236	120.1
10	8192	252	111.9

Table 2. BFV parameters with plaintext modulus $\approx 2^{128}$

Depth	N	$\log q$	Security
1	16384	456	124.3
2	16384	602	92.44
3	32768	750	154.2

Table 3. RLWE based HSS parameters for RMS programs with maximum plaintext size B_{max}

B_{max}	N	$\log q$	Security
2	4096	137	103.3
2^{16}	4096	167	83.74
2^{32}	8192	203	142.0
2^{64}	8192	267	104.9
2^{128}	16384	399	143.9
2^{256}	16384	655	84.60

allows us to use our optimized HSS scheme from Sect. 3.4, where shares consist of *a single* RLWE ciphertext, instead of two. The main motivating example we look at is to MPC protocols in the *preprocessing model*, where correlated randomness is pre-generated ahead of time to help increase efficiency when the actual computation takes place. This correlated randomness can take many forms, but the most common are Beaver triples, namely additive shares of (a, b, c) where $c = a \cdot b$ and a, b are random elements of a (typically) large prime field. These can easily be generated using degree-2 HSS, where each party inputs two field elements, and are also highly amenable to SIMD processing.

Looking at Tables 2 and 3, for an example of degree 2 functions over a 128-bit message space, BFV with depth 1 requires a dimension $N = 16384$ and modulus $\log q = 456$, whereas our scheme would need to use $B_{\mathsf{max}} \approx 2^{256}$, giving the same dimension and a slightly larger modulus of around 655 bits. Therefore, our communication cost will be slightly larger than using SHE-based HSS, but we expect to gain from the lower computational costs that come with our multiplication.

Using DDH-type HSS [6], an m-bit triple can be created with $3712(5m/4 + 160)$ bits of communication, giving $148\,\mathrm{kB}$ for $m = 128$, meaning our communication is 20x higher for producing a single triple (at $2682\,\mathrm{kB}$), but orders of magnitude smaller (~900x) when amortized using SIMD (over $N = 16834$ triples). Computation requirements will greatly favor our approach.

We can also compare this with other approaches to Beaver triple generation. The SPDZ protocol [20] uses SHE (without HSS) to create triples; as well as the more complex homomorphic multiplication, this incurs extra costs in an interactive distributed decryption protocol, which adds a round of interaction that we can avoid using HSS with local rounding. The latest version of SPDZ [33] uses linearly-homomorphic encryption instead of SHE, and reports ciphertexts with $\log q$ as small as 327 bits, around half the size of ours. This would likely beat HSS in terms of communication and computation, but still has the undesirable feature of 2 rounds of interaction, whereas with HSS (and a small one-time setup), the triples are obtained after just one message from each party.

The recent work of Boyle et al. [6] considered an interesting alternative approach to triple generation using so-called "cryptographic capsules", where HSS evaluation of a local PRG is used to expand a small, initial amount of correlated randomness into many more triples. This allows communication complexity *sub-linear in the number triples*. They showed that this can be done with $O(\beta^2)$ Boolean RMS multiplications, where β is a parameter related to the locality of the PRG. With a DDH-like scheme, their protocol involves significant complications to ensure correct triples in the presence of a non-negligible failure probability for multiplication, making it quite impractical. However, using our HSS or SHE-based HSS with negligible error considerably simplifies this approach; it is not immediately clear of the best way to instantiate the parameters, but since it is a Boolean computation with relatively small degree it seems well-suited to our HSS scheme. We leave this exploration, as well as extending to other distributions, as an interesting direction of future research.

5.2   2-Server PIR

An attractive application of HSS is to obtain highly succinct Private Information Retrieval (PIR) protocols for $m \geq 2$ servers. Here, m servers hold a public database DB and allow clients to submit private queries to DB, such that both the query and response remain hidden to up to $m - 1$ colluding servers.[8] When using HSS, we can obtain a very simple, 1-round protocol where the client first sends an encryption of its query to both servers, who respond with an additive share of the result. Note that we only need the more efficient, secret-key version of HSS, such as our scheme from Sect. 3.4 with $m = 2$ servers.

Recent works on 2-server PIR have used HSS for point functions[9] to support basic queries including equality conditions, range queries and disjoint OR clauses, based on simple schemes using only one-way functions [9,40]. However these techniques degrade dramatically for more complex queries, due to the relatively weak homomorphic ability of the underlying HSS. With HSS for branching programs we can significantly increase the expressiveness of queries, at the cost of some overhead in ciphertext size and running time.

In a bit more detail, suppose that a client issues a simple $COUNT$ query,[10] which applies some predicate Q to each row $x_i \in$ DB, and returns $\sum_i Q(x_i)$, that is, the number of rows in DB that match Q. The general idea is that the client splits Q into HSS shares s^1, s^2, and sends s^j to server j. For each row $x_i \in$ DB,

---

[8] Using S/FHE alone instead of HSS allows for the stronger setting of single-server PIR. However, a major advantage of HSS with additive reconstruction is that shares across many rows can easily be combined, allowing more expressive queries with simpler computation.

[9] Actually, these works use *function secret-sharing* [8] for point functions, which in this case is equivalent to HSS for the same class of functions.

[10] Other queries such as returning the record identifier, or min/max and range queries can easily be supported with similar techniques, as previously shown in [6,40].

the servers then use homomorphic evaluation with the function $f_{x_i}(Q) := Q(x_i)$ on the shares, to obtain a shared 0/1 value indicating whether a match occurred. Given additive shares modulo r of the results $q_1, \ldots, q_D$ (where $D = |\mathsf{DB}|$), the servers can sum up the shares and send the result to the client, who reconstructs the result $q = \sum q_i$ (this assumes that $r < N$, so wraparound does not occur).

Below we analyse some useful classes of predicates that are much more expressive than function classes that can be handled using one-way function based approaches, and seem well-suited for our scheme supporting RMS programs.

Conjunctive keyword search. Suppose that each entry in DB is a document x with a list of keywords $W_x = \{w_1^x, \ldots, w_m^x\}$, and the query is a $COUNT$ query consisting of an arbitrary conjunction of keywords, each in $\{0,1\}^\ell$. That is, for a query $W = \{w_1, \ldots, w_k\}$ containing keywords shared bit-by-bit using the HSS, the servers will compute a sharing of

$$\#\{(x, W_x) \in \mathsf{DB} : W \subseteq W_x\}$$

To evaluate the query on a single entry of DB as an RMS program, we maintain the result f as a secret-shared memory value, which is initially set to 1. We then iterate over each query keyword $w_i \in W$, letting w_{ij} denote the j-th bit of w_i, and update f as

$$f := \sum_{w^x \in W_x} f \cdot \prod_{j=1}^m (1 \oplus w_j^x \oplus w_{ij})$$

Note that the i-th product evaluates to 1 iff $x^x = w_i$, and since all w^x are distinct, at most one of these will be 1. Multiplication by f applies a conjunction with the previous keyword, and must be performed inside the summation as f is a memory value. All other product terms are linear functions (over $\mathbb{Z}$) in the inputs w_i (via $a \oplus b = a + b - 2ab$), so each product can be evaluated left-to-right as an RMS program, for a total of $m \cdot \ell \cdot k$ RMS multiplications after iterating over all k query keywords.

Comparison to SHE-based HSS. When using SHE, the number of homomorphic multiplications is roughly the same as our case, and the multiplicative depth is $\log(m\ell k)$. For a concrete example, suppose that each document has $m = 10$ keywords of length $\ell = 128$ bits, and a client's query has $k = 4$ keywords. Using either our HSS scheme or HSS from SHE would need around 5120 multiplications per document, with a multiplicative depth of 13. This needs SHE parameters of $\log q \approx 300$ and dimension $N = 8192$ for the BFV scheme as above, whereas with our scheme we can use the best case of $B_{\mathsf{max}} = 2$, giving $\log q \approx 137$ and $N = 4096$. Using our secret-key HSS and LPR instantiation, the share size is $3N \log q$ bits $\approx 210\,\mathrm{kB}$, around 1/3 of the SHE ciphertext size using BFV. The communication cost for the whole query would be $107\,\mathrm{MB}$ for our HSS, and $314\,\mathrm{MB}$ with BFV, whilst we estimate the computational costs of homomorphic evaluation per document are around $2.5\,\mathrm{s}$ and $300\,\mathrm{s}$, respectively, so even with the relatively high communication cost, for matching several documents using our HSS would certainly give a significant performance improvement.

However, one drawback of our approach is that handling SIMD computations is more challenging, since the B_{max} bound must be chosen much larger to account for the coefficient growth of the plaintext polynomials, which may continue to grow even when the packed plaintext messsages themselves are only bits. If the number of documents in the database is large enough to warrant SIMD processing then it seems likely that SHE will be preferable, since $N = 8192$ documents could be searched at once without increasing the parameters.

Pattern-matching queries. Suppose here that the client wants to search for the occurrence of a pattern $p = (p_1, \ldots, p_m) \in \{0, 1\}^m$ in each row $x = (x_1, \ldots, x_N) \in \{0, 1\}^N$. An RMS program for computing the pattern-matching predicate, with public input x and private input p, can be done with $m \cdot N$ multiplications using a similar method to the previous example, modified slightly to compute the OR of matching p with every position in x.

Comparison to SHE-based HSS. When using SHE, this computation has depth $\log(nm)$, also requiring around $N \cdot m$ homomorphic multiplications. The comparison with our scheme is then similar to the keyword search example, depending on the parameters chosen. For another example, if we have a fairly large string of length $N = 10000$, and a pattern of size $m = 100$, then the SHE-based HSS must support depth 20, giving parameters $(N, \log q) = (16384, 434)$. Again, we can use our HSS with parameters for $B_{\mathsf{max}} = 2$, which lead to ciphertexts around 8.5x smaller than with SHE.

Acknowledgements. We would like to thank the anonymous reviewers of Eurocrypt 2019 for their thorough and generous comments.

References

1. Applebaum, B., Cash, D., Peikert, C., Sahai, A.: Fast cryptographic primitives and circular-secure encryption based on hard learning problems. In: Halevi, S. (ed.) CRYPTO 2009. LNCS, vol. 5677, pp. 595–618. Springer, Heidelberg (2009). https://doi.org/10.1007/978-3-642-03356-8_35
2. Asharov, G., Jain, A., López-Alt, A., Tromer, E., Vaikuntanathan, V., Wichs, D.: Multiparty computation with low communication, computation and interaction via threshold FHE. In: Pointcheval, D., Johansson, T. (eds.) EUROCRYPT 2012. LNCS, vol. 7237, pp. 483–501. Springer, Heidelberg (2012). https://doi.org/10.1007/978-3-642-29011-4_29
3. Banerjee, A., Peikert, C., Rosen, A.: Pseudorandom functions and lattices. In: Pointcheval, D., Johansson, T. (eds.) EUROCRYPT 2012. LNCS, vol. 7237, pp. 719–737. Springer, Heidelberg (2012). https://doi.org/10.1007/978-3-642-29011-4_42
4. Böhl, F., Davies, G.T., Hofheinz, D.: Encryption schemes secure under related-key and key-dependent message attacks. In: Krawczyk, H. (ed.) PKC 2014. LNCS, vol. 8383, pp. 483–500. Springer, Heidelberg (2014). https://doi.org/10.1007/978-3-642-54631-0_28

5. Boyle, E., Couteau, G., Gilboa, N., Ishai, Y.: Compressing vector OLE. In: Proceedings of the 2017 ACM SIGSAC Conference on Computer and Communications Security, CCS 2018 (2018)
6. Boyle, E., Couteau, G., Gilboa, N., Ishai, Y., Orrù, M.: Homomorphic secret sharing: optimizations and applications. In: ACM CCS 2017. ACM Press (2017)
7. Boyle, E., Gilboa, N., Ishai, Y.: Breaking the circuit size barrier for secure computation under DDH. In: Robshaw, M., Katz, J. (eds.) CRYPTO 2016, Part I. LNCS, vol. 9814, pp. 509–539. Springer, Heidelberg (2016). https://doi.org/10.1007/978-3-662-53018-4_19
8. Boyle, E., Gilboa, N., Ishai, Y.: Function secret sharing. In: Oswald, E., Fischlin, M. (eds.) EUROCRYPT 2015, Part II. LNCS, vol. 9057, pp. 337–367. Springer, Heidelberg (2015). https://doi.org/10.1007/978-3-662-46803-6_12
9. Boyle, E., Gilboa, N., Ishai, Y.: Function secret sharing: improvements and extensions. In: ACM CCS 2016. ACM Press, October 2016
10. Boyle, E., Gilboa, N., Ishai, Y.: Group-based secure computation: optimizing rounds, communication, and computation. In: Coron, J.-S., Nielsen, J.B. (eds.) EUROCRYPT 2017, Part II. LNCS, vol. 10211, pp. 163–193. Springer, Cham (2017). https://doi.org/10.1007/978-3-319-56614-6_6
11. Boyle, E., Gilboa, N., Ishai, Y., Lin, H., Tessaro, S.: Foundations of homomorphic secret sharing. In: ITCS 2018. LIPIcs, January 2018
12. Boyle, E., Kohl, L., Scholl, P.: Homomorphic secret sharing from lattices without FHE. Cryptology ePrint Archive, Report 2019/129. https://eprint.iacr.org/2019/129
13. Brakerski, Z.: Fully homomorphic encryption without modulus switching from classical GapSVP. In: Safavi-Naini, R., Canetti, R. (eds.) CRYPTO 2012. LNCS, vol. 7417, pp. 868–886. Springer, Heidelberg (2012). https://doi.org/10.1007/978-3-642-32009-5_50
14. Brakerski, Z., Gentry, C., Vaikuntanathan, V.: (Leveled) fully homomorphic encryption without bootstrapping. In: ITCS 2012. ACM, January 2012
15. Brakerski, Z., Vaikuntanathan, V.: Fully homomorphic encryption from ring-LWE and security for key dependent messages. In: Rogaway, P. (ed.) CRYPTO 2011. LNCS, vol. 6841, pp. 505–524. Springer, Heidelberg (2011). https://doi.org/10.1007/978-3-642-22792-9_29
16. Chillotti, I., Gama, N., Georgieva, M., Izabachène, M.: Faster fully homomorphic encryption: bootstrapping in less than 0.1 seconds. In: Cheon, J.H., Takagi, T. (eds.) ASIACRYPT 2016, Part I. LNCS, vol. 10031, pp. 3–33. Springer, Heidelberg (2016). https://doi.org/10.1007/978-3-662-53887-6_1
17. Chillotti, I., Gama, N., Georgieva, M., Izabachène, M.: Faster packed homomorphic operations and efficient circuit bootstrapping for TFHE. In: Takagi, T., Peyrin, T. (eds.) ASIACRYPT 2017, Part I. LNCS, vol. 10624, pp. 377–408. Springer, Cham (2017). https://doi.org/10.1007/978-3-319-70694-8_14
18. Cleve, R.: Towards optimal simulations of formulas by bounded-width programs. Comput. Complexity 1, 91–105 (1991). https://doi.org/10.1007/BF01200059
19. Corrigan-Gibbs, H., Boneh, D., Mazières, D.: Riposte: an anonymous messaging system handling millions of users. In: 2015 IEEE Symposium on Security and Privacy. IEEE Computer Society Press, May 2015
20. Damgård, I., Pastro, V., Smart, N.P., Zakarias, S.: Multiparty computation from somewhat homomorphic encryption. In: Safavi-Naini, R., Canetti, R. (eds.) CRYPTO 2012. LNCS, vol. 7417, pp. 643–662. Springer, Heidelberg (2012). https://doi.org/10.1007/978-3-642-32009-5_38

21. Dinur, I., Keller, N., Klein, O.: An optimal distributed discrete log protocol with applications to homomorphic secret sharing. In: Shacham, H., Boldyreva, A. (eds.) CRYPTO 2018, Part III. LNCS, vol. 10993, pp. 213–242. Springer, Cham (2018). https://doi.org/10.1007/978-3-319-96878-0_8

22. Dodis, Y., Halevi, S., Rothblum, R.D., Wichs, D.: Spooky encryption and its applications. In: Robshaw, M., Katz, J. (eds.) CRYPTO 2016, Part III. LNCS, vol. 9816, pp. 93–122. Springer, Heidelberg (2016). https://doi.org/10.1007/978-3-662-53015-3_4

23. Ducas, L., Micciancio, D.: FHEW: bootstrapping homomorphic encryption in less than a second. In: Oswald, E., Fischlin, M. (eds.) EUROCRYPT 2015, Part I. LNCS, vol. 9056, pp. 617–640. Springer, Heidelberg (2015). https://doi.org/10.1007/978-3-662-46800-5_24

24. Fan, J., Vercauteren, F.: Somewhat practical fully homomorphic encryption. Cryptology ePrint Archive, Report 2012/144 (2012). http://eprint.iacr.org/2012/144

25. Fazio, N., Gennaro, R., Jafarikhah, T., Skeith III, W.E.: Homomorphic secret sharing from Paillier encryption. In: Okamoto, T., Yu, Y., Au, M.H., Li, Y. (eds.) ProvSec 2017. LNCS, vol. 10592, pp. 381–399. Springer, Cham (2017). https://doi.org/10.1007/978-3-319-68637-0_23

26. Gentry, C.: Fully homomorphic encryption using ideal lattices. In: 41st ACM STOC. ACM Press (2009)

27. Gentry, C., Halevi, S., Smart, N.P.: Homomorphic evaluation of the AES circuit. In: Safavi-Naini, R., Canetti, R. (eds.) CRYPTO 2012. LNCS, vol. 7417, pp. 850–867. Springer, Heidelberg (2012). https://doi.org/10.1007/978-3-642-32009-5_49

28. Gentry, C., Sahai, A., Waters, B.: Homomorphic encryption from learning with errors: conceptually-simpler, asymptotically-faster, attribute-based. In: Canetti, R., Garay, J.A. (eds.) CRYPTO 2013, Part I. LNCS, vol. 8042, pp. 75–92. Springer, Heidelberg (2013). https://doi.org/10.1007/978-3-642-40041-4_5

29. Gilboa, N., Ishai, Y.: Distributed point functions and their applications. In: Nguyen, P.Q., Oswald, E. (eds.) EUROCRYPT 2014. LNCS, vol. 8441, pp. 640–658. Springer, Heidelberg (2014). https://doi.org/10.1007/978-3-642-55220-5_35

30. Goldreich, O., Micali, S., Wigderson, A.: How to play any mental game or a completeness theorem for protocols with honest majority. In: 19th ACM STOC. ACM Press, May 1987

31. Halevi, S., Polyakov, Y., Shoup, V.: An improved RNS variant of the BFV homomorphic encryption scheme. Cryptology ePrint Archive, Report 2018/117 (2018). https://eprint.iacr.org/2018/117

32. Halevi, S., Shoup, V.: Bootstrapping for HElib. In: Oswald, E., Fischlin, M. (eds.) EUROCRYPT 2015, Part I. LNCS, vol. 9056, pp. 641–670. Springer, Heidelberg (2015). https://doi.org/10.1007/978-3-662-46800-5_25

33. Keller, M., Pastro, V., Rotaru, D.: Overdrive: making SPDZ great again. In: Nielsen, J.B., Rijmen, V. (eds.) EUROCRYPT 2018, Part III. LNCS, vol. 10822, pp. 158–189. Springer, Cham (2018). https://doi.org/10.1007/978-3-319-78372-7_6

34. Langlois, A., Stehlé, D.: Worst-case to average-case reductions for module lattices. Des. Codes Cryptogr. **75**(3), 565–599 (2015)

35. Lyubashevsky, V., Peikert, C., Regev, O.: A toolkit for ring-LWE cryptography. In: Johansson, T., Nguyen, P.Q. (eds.) EUROCRYPT 2013. LNCS, vol. 7881, pp. 35–54. Springer, Heidelberg (2013). https://doi.org/10.1007/978-3-642-38348-9_3

36. Lyubashevsky, V., Peikert, C., Regev, O.: On ideal lattices and learning with errors over rings. In: Gilbert, H. (ed.) EUROCRYPT 2010. LNCS, vol. 6110, pp. 1–23. Springer, Heidelberg (2010). https://doi.org/10.1007/978-3-642-13190-5_1

37. Regev, O.: On lattices, learning with errors, random linear codes, and cryptography. In: 37th ACM STOC. ACM Press, May 2005
38. Rivest, R.L., Adleman, L., Dertouzos, M.L.: On data banks and privacy homomorphisms. In: Foundations of Secure Computation (Workshop, Georgia Institute of Technology, 1977), pp. 169–179. Academic, New York (1978)
39. Smart, N.P., Vercauteren, F.: Fully homomorphic SIMD operations. Des. Codes Cryptogr. **71**(1), 57–81 (2014). https://doi.org/10.1007/s10623-012-9720-4. ISSN 0925-1022
40. Wang, F., Yun, C., Goldwasser, S., Vaikuntanathan, V., Zaharia, M.: Splinter: practical private queries on public data. In: 14th USENIX Symposium on Networked Systems Design and Implementation, NSDI 2017, pp. 299–313 (2017)

Improved Bootstrapping for Approximate Homomorphic Encryption

Hao Chen[1(✉)], Ilaria Chillotti[2], and Yongsoo Song[1]

[1] Microsoft Research, Redmond, USA
{haoche,Yongsoo.Song}@microsoft.com
[2] imec-COSIC, KU Leuven, Leuven, Belgium
ilaria.chillotti@kuleuven.be

Abstract. Since Cheon et al. introduced a homomorphic encryption scheme for approximate arithmetic (Asiacrypt '17), it has been recognized as suitable for important real-life usecases of homomorphic encryption, including training of machine learning models over encrypted data. A follow up work by Cheon et al. (Eurocrypt '18) described an approximate bootstrapping procedure for the scheme. In this work, we improve upon the previous bootstrapping result. We improve the amortized bootstrapping time per plaintext slot by two orders of magnitude, from $\sim$1 s to $\sim$0.01 s. To achieve this result, we adopt a smart level-collapsing technique for evaluating DFT-like linear transforms on a ciphertext. Also, we replace the Taylor approximation of the sine function with a more accurate and numerically stable Chebyshev approximation, and design a modified version of the Paterson-Stockmeyer algorithm for fast evaluation of Chebyshev polynomials over encrypted data.

Keywords: Fully Homomorphic Encryption · Bootstrapping

1 Introduction

Homomorphic Encryption (HE) refers to a specific class of encryption schemes which allows computing directly on encrypted data without having to decrypt. Due to this special property, it has numerous potential applications in data-heavy industries, where one challenge is to gain meaningful insights from data while keeping the data itself private. Since the first construction of HE by Gentry [23], the field has witnessed a lot of growth: more efficient schemes (e.g. [7–9,17,18,20, 22]) have been proposed, and there has been various design and implementations (e.g. [4,6,10,12,19,25]) of confidential computing applications using HE.

However, the HE-based solutions have two issues when we apply them to applications which requires arithmetic on real numbers. The first issue is plaintext growth: the native plaintext in the HE schemes belong to a certain finite

---

Most of this work was done while the second author was an intern in the Cryptography Research group at Microsoft Research (Redmond, USA).

Y. Ishai and V. Rijmen (Eds.): EUROCRYPT 2019, LNCS 11477, pp. 34–54, 2019.
https://doi.org/10.1007/978-3-030-17656-3_2

space. In order to encrypt a real number, one needs to first scale it up to an integer, so that the fractional part becomes the less significant digits. The size of this scaled integer will grow as we perform homomorphic multiplications on the ciphertext. Since the plaintext space is finite, after a certain number of multiplications, it is impossible to recover the actual number. So, we need to perform "scale down and truncate" operation on encrypted data. However, it is an expensive operation since the available HE schemes only support addition and multiplication.

The second issue is ciphertext growth. Following Gentry's blueprint, a fresh encryption contains a small amount of "noise", and the noise level grows as operations are done on the ciphertext. It is necessary that the noise does not overwhelm the actual data within a ciphertext. To achieve this, one could use the Somewhat Homomorphic Encryption (SHE) approach, where the parameters are scaled up with the level of the circuit to be evaluated so that noise overflow is unlikely. Using SHE, both the ciphertext size and the performance overhead of HE will grow at least linearly with the circuit level, hence this approach has scaling issues. The other option is the Fully Homomorphic Encryption (FHE) approach, which uses Gentry's *bootstrapping* technique to refresh the noise in a ciphertext, so that circuit of arbitrary level can be evaluated on a fixed set of parameters. The FHE approach solves the ciphertext growth problem, with the caveat being that bootstrapping is expensive in practice despite the continuous effort in optimizations.

In 2017, Cheon et al. [16] proposed a HE scheme (denoted by CKKS scheme from now) which performs approximate arithmetic on encrypted data, by introducing a novel encoding technique and a fast "scaling down" operation, which effectively controls the growth of plaintext. Due to the nice properties, the CKKS scheme performs well at tasks such as training a logistic regression over encrypted data on a medium-sized data set with around 1000 samples [30]. Recently, a bootstrapping algorithm for the CKKS scheme was proposed in [14]. Using the bootstrapping procedure, one could train a logistic regression model over data sets with more than 4×10^5 samples in around 17 h [29].

1.1 Previous Bootstrapping Method for CKKS

In the CKKS scheme, the ciphertext modulus q decreases after each homomorphic multiplication, and decryption is correct if and only if the norm of the message is smaller than q. Hence, one can only perform a certain number of sequential multiplications before q gets too low for the next multiplication. Hence, bootstrapping amounts to the following function: giving a ciphertext ct with modulus q encrypted under the secret sk such that

$$[\langle \mathsf{ct}, \mathsf{sk} \rangle]_q = m,$$

bootstrapping outputs a ciphertext ct′ in a larger modulus $Q > q$ such that

$$[\langle \mathsf{ct}', \mathsf{sk} \rangle]_Q \approx m,$$

Note that we do not hope to have exact equality, due to the approximate nature of CKKS.

Given this goal, the bootstrapping method in previous work [14] starts by the following observation: if ct is a ciphertext with modulus q and message $m(X)$, then for a larger modulus $Q \gg q$, the same ciphertext decrypts to $t(X) = m(X) + q \cdot I(X)$ for a polynomial $I(X)$ with small coefficients. The next step approximately evaluates the modulo q function on coefficients to recover the coefficients $m_i = [t_i]_q$ of the input plaintext. It is done by first taking the d-th Taylor polynomial of the scaled exponential function $\exp(2\pi it/(2^r \cdot q))$, raising the polynomial to power 2^r through repeated squaring, and finally taking the imaginary part and scale by $q/(2\pi)$. In other words, we have an approximation polynomial indexed by d and r:

$$K_{d,r}(t) = \frac{q}{2\pi} \left[\sum_{k=0}^{d} \frac{1}{k!} \left(\frac{2\pi it}{2^r \cdot q} \right)^k \right]^{2^r},$$

whose imaginary part approximates values of $(q/2\pi) \cdot \sin(2\pi t/q) \approx [t]_q$, as desired. One issue remaining is that homomorphic operations are not performed on coefficients but on plaintext slots. Before and after the evaluation of an exponential function, we have to shift the coefficients into the plaintext slots, and vice versa. It can be done by evaluating the encoding and decoding algorithms which are linear transformations on plaintext vectors.

Why the previous method does not scale well. There have remained some efficiency issues in the previous work. First, the parameters of $K_{d,r}(t)$ were chosen by $d = O(1)$ and $r = O(\log q)$ to guarantee the accuracy of approximation. It requires only $O(\log q)$ homomorphic operations to evaluate the exponential function, but the depth $O(\log q)$ is somewhat large. Meanwhile, the linear transformations require only one level, but their complexity grows linearly with the number of plaintext slots. As a result, the previous solution was not scalable when a ciphertext is densely-packed, and it was not optimal with respect to the level consumption.

1.2   Our Contribution

In this paper, we suggest two improvements upon the bootstrapping algorithm in [14].

Linear transforms. To improve the linear transform step, we first observed that the linear transforms involved in the bootstrapping process admit FFT-like algorithms, which requires more levels but less operations. Then, in order to fully explore the trade-off between level consumption and number of operations, we adopted an idea from Halevi and Shoup in [27], which uses a dynamic programming approach to decide the optimal level collapsing strategy for a generic multi-leveled linear transforms. As a result, our linear transforms are faster while being able to operate on 2-128x more slots, resulting in a large increase on the bootstrapping throughput.

Sine approximation. Then, we used a Chebyshev interpolant to approximate the scaled sine function, which not only consumes less levels but also is more accurate than the original method. Our results indicate that in order to achieve the same level of approximation error, our method only requires $\max\{\log K + 2, \log\log q\}$ levels, whereas the previous solution requires $O(\log(Kq))$ levels. Here q is closely related to the plaintext size before bootstrapping, and $K = O(\lambda)$ is related to the security parameter and is small in practice.

In order to evaluate a Chebyshev interpolant of form $\sum_{k=0}^{n} c_k T_k(x)$ efficiently on encrypted inputs, we proposed a modified Paterson-Stockmeyer algorithm which works for polynomials represented in Chebyshev base. As a result, our approach requires $O(\sqrt{\max\{K, \log q\}})$ ciphertext multiplications to evaluate the sine approximation, compared to $O(\log(Kq))$ in the previous work.

1.3   Related Works

There has been a few works which focus on improving the performance of bootstrapping. In terms of throughput, the works [11, 24, 28] designed optimized bootstrapping algorithms for BGV/BFV schemes. In terms of latency, the line of work [17, 18, 20] designed a specific RLWE-based HE scheme suitable for bootstrapping, and through extensive optimizations brought the bootstrapping time down to 13 ms. However, the scheme encrypts every bit separately, and bootstrapping needs to be performed after every single binary gate. Hence the overhead is still quite large for it to be practical in large scale applications.

Our major point of comparison is [14], bootstrapping for the CKKS approximate homomorphic encryption scheme. It is based on a novel idea of using a scaled sine function $\frac{1}{2\pi}\sin(2\pi t/q)$ to approximate the modulus reduction function $[t]_q$.

1.4   Road Map

In Sect. 2, we recall the constructions and properties of the CKKS scheme and its bootstrapping algorithm. In Sect. 3, we describe our optimization of the linear transforms. In Sect. 4, we discuss our optimization of the sine evaluation step in CKKS bootstrapping using Chebyshev interpolants. We analyze our improved bootstrapping algorithm and present performance results in Sect. 5. Finally, we conclude in Sect. 6 with future research directions.

2   Background

2.1   The CKKS Scheme

We restate the CKKS scheme [16] below. For a power-of-two integer N, we denote $R = \mathbb{Z}[X]/(X^N + 1)$ be the ring of integers of the $(2N)$-th cyclotomic field. A single CKKS ciphertext can encrypt a complex vector with $\ell \leq (N/2)$ entries. To be precise, let $\zeta = \exp(\pi i/2\ell)$ be a (4ℓ)-th primitive root of unity for a power-of-two integer $1 \leq \ell \leq N/2$. The decoding algorithm takes as the input an element $m(Y)$ of the cyclotomic ring $\mathbb{R}[Y]/(Y^{2\ell} + 1)$ and returns a vector $\mathsf{Decode}(m) = (m(\zeta), m(\zeta^5), \ldots, m(\zeta^{4\ell-3}))$. Note that Decode is a ring isomorphism between $\mathbb{R}[Y]/(Y^{2\ell} + 1)$ and $\mathbb{C}^{\ell/2}$. If we identify $m(Y)$ with the vector $\boldsymbol{m} = (m_0, \ldots, m_{2\ell-1})$ of its coefficients, then the decoding algorithm can be viewed a linear transformation whose matrix representation is given by

$$
M_\ell = \begin{bmatrix}
1 & \zeta & \zeta^2 & \cdots & \zeta^{2\ell-1} \\
1 & \zeta^5 & \zeta^{5 \cdot 2} & \cdots & \zeta^{5(2\ell-1)} \\
\vdots & \vdots & \vdots & \ddots & \vdots \\
1 & \zeta^{4\ell-3} & \zeta^{(4\ell-3) \cdot 5} & \cdots & \zeta^{(4\ell-3)(2\ell-1)}
\end{bmatrix},
$$

i.e., $\mathsf{Decode}(m) = M_\ell \cdot \boldsymbol{m}$. The encoding algorithm is defined by its inverse. When we implement the decoding function, we first define the *special Fourier transformation* matrix

$$
\mathsf{SF}_\ell = \begin{bmatrix}
1 & \zeta & \cdots & \zeta^{\ell-1} \\
1 & \zeta^5 & \cdots & \zeta^{5(\ell-1)} \\
\vdots & \vdots & \ddots & \vdots \\
1 & \zeta^{4\ell-3} & \cdots & \zeta^{(4\ell-3)(\ell-1)}
\end{bmatrix},
$$

which is an $\ell \times \ell$ square matrix satisfying $M_\ell = [\mathsf{SF}_\ell | i \cdot \mathsf{SF}_\ell]$. Then, the decoding and encoding algorithms can be represented using the multiplication with SF_ℓ, its inverse and some conjugations.

We embed plaintext polynomials in $\mathbb{R}[Y]/(Y^{2\ell} + 1)$ into $\mathbb{R}[X]/(X^N + 1)$ by $Y \mapsto X^{N/2\ell}$. We say that a plaintext is fully packed (full-slot) when $\ell = N/2$. An encoded polynomial should be rounded to the closest integral polynomial in R to be encrypted.

- $\mathsf{Setup}(1^\lambda)$. Given the security parameter λ, choose a power-of-two integer N. Set the distributions $\chi_{\mathsf{key}}, \chi_{\mathsf{err}}, \chi_{\mathsf{enc}}$ on R for the secret, error, and encryption, respectively. For a base integer p and the number of levels L, set the chain of ciphertext moduli $q_\ell = p^\ell$ for $1 \leq \ell \leq L$. Choose an integer P.
- $\mathsf{Keygen}()$. Sample $s \leftarrow \chi_{\mathsf{key}}$ and set the secret key as $\mathsf{sk} \leftarrow (1, s)$. Sample $a \leftarrow U(R_{q_L})$ and $e \leftarrow \chi_{\mathsf{err}}$, and set the public key as $\mathsf{pk} \leftarrow (b, a) \in R_{q_L}^2$ for $b = -as + e \pmod{q_L}$. Sample $a' \leftarrow R_{P \cdot q_L}, e' \leftarrow \chi_{\mathsf{err}}$ and set evaluation key as $\mathsf{evk} \leftarrow (b', a') \in R_{P \cdot q_L}^2$ for $b' = -a's + e' + Ps^2 \pmod{P \cdot q_L}$.
- $\mathsf{Enc}_{\mathsf{pk}}(m)$. Sample $r \leftarrow \chi_{\mathsf{enc}}$ and $e_0, e_1 \leftarrow \chi_{\mathsf{err}}$. Output the ciphertext $\mathsf{ct} = r \cdot \mathsf{pk} + (m + e_0, e_1) \pmod{q_L}$. Note that $\langle \mathsf{ct}, \mathsf{sk} \rangle \pmod{q_L}$ is approximately equal to m.
- $\mathsf{Dec}_{\mathsf{sk}}(\mathsf{ct})$. For an input ciphertext of level ℓ, compute and output $m = \langle \mathsf{ct}, \mathsf{sk} \rangle \pmod{q_\ell}$.

We remark that the encryption procedure of CKKS introduces an error so its decrypted value is not exactly same as the input value. We describe homomorphic operations (addition, multiplication, scalar multiplication, and rescaling) as follows.

- Add(ct, ct'). For ciphertexts ct, ct' in the same level ℓ, output $\mathsf{ct_{add}} = \mathsf{ct} + \mathsf{ct}'$ (mod q_ℓ).
- CMult$_{\mathsf{evk}}(a, \mathsf{ct})$. For a constant $a \in R$ and a ciphertext ct of level ℓ, output $\mathsf{ct_{cmult}} = (d_0, d_1) + \lfloor P^{-1} \cdot d_2 \cdot \mathsf{evk} \rceil$ (mod q_ℓ).
- Mult$_{\mathsf{evk}}(\mathsf{ct}, \mathsf{ct}')$. For $\mathsf{ct} = (c_0, c_1), \mathsf{ct}' = (c_0', c_1') \in R_{q_\ell}^2$, let $(d_0, d_1, d_2) = (c_0 c_0', c_0 c_1' + c_0' c_1, c_1 c_1')$ (mod q_ℓ). Output $\mathsf{ct_{mult}} = (d_0, d_1) + \lfloor P^{-1} \cdot d_2 \cdot \mathsf{evk} \rceil$ (mod q_ℓ).
- Rescale$_{\ell \to \ell'}(\mathsf{ct})$. For an input ciphertext of level ℓ, output $\mathsf{ct}' = \lfloor p^{\ell' - \ell} \cdot \mathsf{ct} \rceil \in$ (mod $q_{\ell'}$).

We note that $\{1, 5, \ldots, 2\ell - 3\}$ is a cyclic subgroup of the multiplicative group $\mathbb{Z}_{2\ell}^\times$ generated by the integer 5. One can rotate or take the conjugate of an encrypted plaintext by evaluating the maps $Y \mapsto Y^5$ or $Y \mapsto Y^{-1}$ based on the key-switching technique. Certain evaluation keys should be published to perform these algorithms (see [14] for details).

- Rotate$_{\mathsf{rk}}(\mathsf{ct}; k)$. For an input encryption of $m(Y)$, return an encryption of $m(Y^{5^k})$ in the same level. The encrypted plaintext vector is shifted by k slots.
- Conjugate$_{\mathsf{ck}}(\mathsf{ct})$. For an input encryption of $m(Y)$, return an encryption of $m(Y^{-1})$ in the same level. It takes the conjugation of the encrypted plaintext.

In applications of CKKS, we usually multiply a *scaling factor* to plaintexts to maintain the precision of computations. The rescaling algorithm can divide an encrypted plaintext by a power of p and preserve the size of scaling factors during homomorphic arithmetic.

2.2 Previous Bootstrapping for CKKS

Cheon et al. [14] showed how to refresh a ciphertext of the CKKS scheme. In this section, we briefly explain the previous solution.

Suppose that we have a *low-level* ciphertext $\mathsf{ct} \in R_q^2$ encrypting $m(Y) \in \mathbb{Z}[Y]/(Y^{2\ell} + 1) \subseteq R$, i.e., $\langle \mathsf{ct}, \mathsf{sk} \rangle$ (mod q) $\approx m(Y)$. Recall that $m(Y)$ can be identified with an ℓ-dimensional complex vector $\mathbf{z} = \mathsf{Decode}(m)$. The goal of boostrapping is to generate a *high-level* ciphertext ct' satisfying $\langle \mathsf{ct}', \mathsf{sk} \rangle$ (mod Q) $\approx m(Y)$ by evaluating the decryption circuit homomorphically.

The first step raises up the modulus of an input ciphertext. We have that $[\langle \mathsf{ct}, \mathsf{sk} \rangle]_{Q_0} \approx q \cdot I(X) + m(Y)$ for some $Q_0 > q$ and $I(X) \in R$. The coefficients of $I(X)$ is bounded by a constant K which depends on the secret distribution χ_{key}. Then, we perform the *subsum* procedure which generates a ciphertext ct' such that $\langle \mathsf{ct}', \mathsf{sk} \rangle \approx (N/2\ell) \cdot t(Y)$ (mod Q_0) for $J(Y) = I_0 + I_{N/2\ell} \cdot Y + \cdots + I_{(2\ell-1)N/2\ell} \cdot Y^{N-1}$ and $t(Y) = q \cdot J(Y) + m(Y)$.[1] The constant $(N/2\ell)$ can be canceled by the rescaling process.

---

[1] The subsum algorithm can be understood as the evaluation of trace with respect to the field extension $\mathbb{Q}[X]/(X^N + 1) \geq \mathbb{Q}[Y]/(Y^{2\ell} + 1)$. It does nothing when $\ell = N/2$.

The *coefficients to slots* step, denoted by coeffToSlot, is to generate an encryption of the coefficients of $t(Y) = q \cdot J(Y) + m(Y)$, i.e., a ciphertext ct'' which satisfies that

$$[\langle ct'', sk \rangle]_{Q_1} \approx \mathsf{Encode}(t)$$

for some Q_1. This step can be done by homomorphically evaluating the encoding algorithm which is a variant of complex Fourier transformation. We point out that the resulting ciphertext should encrypt an (2ℓ)-dimensional vector $(t_0, \ldots, t_{2\ell-1})$ compared to the input ciphertext with ℓ plaintext slots, so we need to generate two ciphertexts encrypting halves of coefficients when the full-slot case $\ell = N/2$.

Now we have one or two ciphertexts which encrypt $t_i = q \cdot J_i + m_i$ for $0 \le i < 2\ell$ in their plaintext slots. The goal of next step (evalExp) is to homomorphically evaluate the reduction modulo q function and return ciphertexts encrypting $m_i = [t_i]_q$ in plaintext slots. Since the modulo reduction is not a polynomial function, the previous work used the following approximation by a trigonometric function which has a good accuracy under the condition that $|m| \ll q$:

$$[t]_q = m \approx \frac{q}{2\pi} \sin\left(\frac{2\pi t}{q}\right).$$

For the evaluation of this sine function, we first evaluate the polynomial

$$P_{-r}(t) = \sum_{k=0}^{d} \frac{1}{k!} \left(\frac{2\pi t}{2^r \cdot q}\right)^k \approx \exp\left(\frac{2\pi i t}{2^r \cdot q}\right)$$

for some integers r and d, which is the d-th Taylor polynomial of complex exponential function. Then, we can recursively perform the squaring r times $P_{i+1}(x) = P_i(x)^2$ to get an encryption of

$$P_0(t) = [P_{-r}(t)]^{2^r} \approx \exp(2\pi i t / q)$$

whose imaginary part is $\sin(2\pi i t/q)$ as desired. The output of evalExp is one or two ciphertexts which contains approximate values of $[t_i]_q = m_i$ in their plaintext slots.

During the evalExp step, one needs to multiply a scaling factor $\delta \cdot q$ to encrypted values for an appropriate constant δ to keep the precision of computation. A larger scaling factor will consume more ciphertext modulus while a small scaling factor makes the result less accurate. Table 1 summarizes the consumption of modulus bits and relative error from approximation based on the parameter Set-I which uses the initial polynomial of degree $d = 7$ and $r = 6$ iterations.

Finally, the *slots to coefficients* (slotToCoeff) stage is exactly the inverse of coeffToSlot. It homomorphically evaluates the decoding algorithm to get a ciphertext such that $[\langle ct''', sk \rangle]_{Q_2} \approx m(Y)$ for some Q_2. We stress again that ct''' has ℓ plaintext slots, the same as the input ciphertext ct. The slotToCoeff step merges two output ciphertexts of evalExp and returns a fully packed ciphertext in the full-slot case. Otherwise, the number of plaintext slots is reduced from 2ℓ to ℓ during the evaluation.

Table 1. Comparison of different log T and log I values

Params	$\log \delta$	Mod bit consumption	Relative error
	4	337	0.00083
Set-I	3	327	0.002
	2	317	0.003

3 Improved Linear Transforms from Level-Collapsing

In this section, we present a method to improve the performance of linear transformations coeffToSlot and slotToCoeff.

3.1 FFT-like Algorithms for coeffToSlot and slotToCoeff

The coeffToSlot and slotToCoeff steps in the original bootstrapping algorithm amounts to two linear transforms that are mutual inverses to each other. More precisely, slotToCoeff includes the computation $\mathbf{z} \mapsto \mathsf{SF}_\ell \cdot \mathbf{z}$ where SF_ℓ is the special Fourier transformation matrix defined in the previous section. Meanwhile, coeffToSlot is equivalent to computing the map SF_ℓ^{-1} on the plaintext vector. In order to evaluate these transforms on a ciphertext encrypting the vector $\mathbf{z}$, the previous work [14] adopted the diagonal method combined with a baby step-giant step trick.

We begin by noting that similar to the Cooley-Tukey butterfly algorithm for DFT, the linear transform SF_ℓ can be expressed as a sequence of "butterfly" operations. The following algorithm is taken from the HEAANBOOT library [13].

Algorithm 1: FFT-like algorithm for evaluating SF_ℓ

Input: $\ell > 1$ a power of 2 integer; $\mathbf{z} \in \mathbb{C}^\ell$, and a precomputed table Ψ of complex 4ℓ-roots of unities $\Psi[j] = \exp(\pi i j / 2\ell)$, $0 \le j < 4\ell$.

Output: $\mathbf{w} = \mathsf{SF}_\ell \cdot \mathbf{z}$

1 $\mathbf{w} = \mathbf{z}$
2 bitReverse$(\mathbf{w}, \ell)$
3 **for** ($m = 2$; $m \le \ell$; $m = 2m$) {
4   **for** ($i = 0$; $i < \ell$; $i = i + m$) {
5   **for** ($j = 0$; $j < m/2$; $j = j + 1$) {
6              $k = (5^j \mod 4m) \cdot \ell/m$
7              $U = \mathbf{w}[i + j]$
8              $V = \mathbf{w}[i + j + m/2]$
9              $V = V \cdot \Psi[k]$
10             $\mathbf{w}[i + j] = U + V$
11             $\mathbf{w}[i + j + m/2] = U - V$
12   }
13   }
14 }
15 **return w**

In the beginning of Algorithm 1, a bit-reversal is performed, which effectively permutes the input vector. Then, the algorithm performs $\log \ell$ layers of transforms. Similarly, we can invert the above algorithm to obtain an FFT-like algorithm to compute SF_ℓ^{-1}, which starts with ℓ levels of transforms, followed by a bit-reversal.

3.2 Our Solution

First, we observe that for the purpose of bootstrapping, the bit-reversal operations are not necessary in the linear transforms. This is because bit-reversal is a permutation of order 2, and the sine evaluation is a SIMD (single instruction multiple data) operation, i.e., the same operation is performed independently on each slot. Hence, bit-reversals right before and after the sine evaluation will cancel themselves out. Therefore, we only need to perform the butterfly transforms homomorphically. For ease of notations, we still use SF_ℓ to denote the linear transform in lines 3–15 of Algorithm 1.

Next, we note that each layer of Algorithm 1 can be implemented using two slot rotations and three SIMD plaintext multiplications. More precisely, the i-th iteration in Algorithm 1 can be represented as

$$\mathbf{w} := \mathbf{a}[i] \odot \mathbf{w} + \mathbf{b}[i] \odot (\mathbf{w} \ll 2^{i-1}) + \mathbf{c}[i] \odot (\mathbf{w} \gg 2^{i-1}),$$

where $\mathbf{w} \ll j$ (resp. $\mathbf{w} \gg j$) denotes rotating the vector $\mathbf{w}$ to the left (resp. right) by j slots, and $\odot$ denotes the component-wise multiplication between vectors. The vectors $\mathbf{a}[i], \mathbf{b}[i], \mathbf{c}[i] \in \mathbb{C}^\ell$ can be precomputed. This gives us a direct algorithm to evaluate linear transform SF_ℓ on an encrypted vector in CKKS scheme using $\log \ell$ levels and $O(\log \ell)$ operations. In contrast, the approach in [14] requires one level and $O(\ell)$ operations to evaluate SF_ℓ.

In practice, a hybrid approach might work better than the above two extremes. For example, we can trade operations for levels by "collapsing" some levels in the above algorithm. We will elaborate on this method below.

3.3 Optimal Level-Collapsing from Dynamical Programming

First we recall the idea of Halevi and Shoup [27]. The task is to apply a sequence of linear transforms $L_1 \circ \cdots \circ L_\ell$ on some input, and each evaluation consumes one "level". One is allowed to collapse some levels by merging some adjacent transforms into one. For example, for $n = 4$ we could merge into two levels by letting $M_1 = L_1 \circ L_2$ and $M_2 = L_3 \circ L_4$. Assuming there is a cost function associated to every linear transform, it is an optimization problem to find the best level collapsing strategy that minimizes the cost. More precisely, let $\mathsf{Cost}(a, b)$ denote the cost of evaluating $L_a \circ \cdots \circ L_{b-1}$ and let $\ell' \leq \ell$ be an upper bound on the level. Then we wish to solve the following optimization problem:

$$\min_{\substack{a_0=1<a_1<\ldots<a_k<a_{k+1}=\ell+1, \\ k+1\leq\ell'}} \sum_{i=0}^{k} \mathsf{Cost}(a_{i-1}, a_i).$$

To solve for an optimal solution, we recall the idea outlined in [27] as follows. Let $\mathsf{Opt}(d, \ell')$ be the optimal cost to evaluate the first d linear transforms using ℓ' levels. Then

$$\mathsf{Opt}(d, \ell') = \min_{1 \le d' \le d} \mathsf{Cost}(d - d', d + 1) + \mathsf{Opt}(d - d', \ell' - 1).$$

We can then use a dynamic programming algorithm to compute the optimal strategy as a list of splitting points $(a_1, \ldots, a_k)$. Given this optimal level collapsing strategy, we can generate the collapsed levels by merging the individual layers.

Applying level-collapsing to our case. First, we give an example of how levels can be merged. Recall that the i-th level of Algorithm 1

$$\mathbf{w} := \mathbf{a}[i] \odot \mathbf{w} + \mathbf{b}[i] \odot (\mathbf{w} \ll 2^{i-1}) + \mathbf{c}[i] \odot (\mathbf{w} \gg 2^{i-1}).$$

Suppose we merge the layers i and $i + 1$. Then the new linear transform is

$$
\begin{aligned}
\mathbf{w} := \ & \mathbf{a}[i + 1] \odot \left(\mathbf{a}[i] \odot \mathbf{w} + \mathbf{b}[i] \odot (\mathbf{w} \ll 2^{i-1}) + \mathbf{c}[i] \odot (\mathbf{w} \gg 2^{i-1})\right) \\
& + \mathbf{b}[i + 1] \odot \left(\mathbf{a}[i] \odot \mathbf{w} + \mathbf{b}[i] \odot (\mathbf{w} \ll 2^{i-1}) + \mathbf{c}[i] \odot (\mathbf{w} \gg 2^{i-1}))\right) \ll 2^i \\
& + \mathbf{c}[i + 1] \odot \left(\mathbf{a}[i] \odot \mathbf{w} + \mathbf{b}[i] \odot (\mathbf{w} \ll 2^{i-1}) + \mathbf{c}[i] \odot (\mathbf{w} \gg 2^{i-1}))\right) \gg 2^i \\
= \ & A \odot \mathbf{w} + B \odot (\mathbf{w} \ll 2^{i-1}) + C \odot (\mathbf{w} \gg 2^{i-1}) + D \odot (\mathbf{w} \ll 2^i) \\
& + E \odot (\mathbf{w} \gg 2^i) + F \odot (\mathbf{w} \ll 3 \cdot 2^{i-1}) + G \odot (\mathbf{w} \gg 3 \cdot 2^{i-1})
\end{aligned}
$$

for some vectors $A, B, \ldots, G$. Overall, this merged layer requires 6 rotations and 7 plaintext multiplications. In general, if we merge some layers together, then we end up with a merged layer which looks like

$$\mathbf{w} := \sum_{i=1}^{k} \mathbf{p}[i] \odot (\mathbf{w} \ll t_i)$$

for some precomputable vectors $\mathbf{p}[i]$ and integers t_i, and requires $(k-1)$ rotations and k plaintext multiplications to evaluate. To further reduce the complexity, we can utilize a baby step-giant step method to reduce the number of rotations to about $2\sqrt{k}$. Note that in a new version of the implementation of the CKKS scheme [1], plaintext multiplication takes much less time than rotation. Therefore, we define the cost of the merged layer as $2\sqrt{k}$. In the following Fig. 1, we present the optimal costs for different ℓ and level upper bounds.

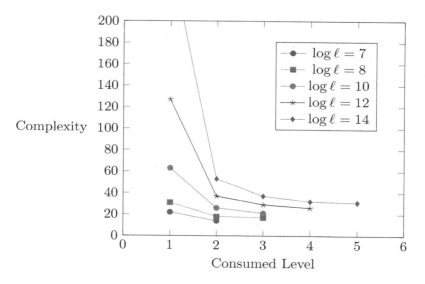

Fig. 1. Optimal complexity (number of rotations) of FFT-like algorithm with respect to the depth and number of slots

4   Improved Sine Evaluation from Chebyshev Approximations

4.1   Background: Chebyshev Polynomials and Chebyshev Interpolants

Recall that the *Chebyshev polynomials* is a family of polynomials $\{T_n(x)\}_{n\geq 0}$ defined by the recurrence relation:

$$\begin{aligned}
T_0(x) &= 1 \\
T_1(x) &= x \\
T_{2n}(x) &= 2T_n(x)^2 - 1 \\
T_{2n+1}(x) &= 2T_n(x) \cdot T_{n+1}(x) - x.
\end{aligned} \tag{1}$$

Given a Lipschitz continuous function f defined on the interval $[-1, 1]$, the n-th *Chebyshev interpolant* of f is defined as

$$p_n^{cheb}(x) = \sum_{k=0}^{n} c_k T_k(x)$$

where the coefficients c_k are uniquely determined such that $p_n^{cheb}(x_j) = f(x_j)$ for

$$x_j = \cos(j\pi/n), \text{ for } 0 \leq j \leq n.$$

Let p_n^* denote the *minimax* polynomial of degree $\leq n$ which minimizes the infinity norm $\|f - p_n^*\|_\infty$. It would be optimal to use p_n^* as a polynomial approximation to f. However, computing such polynomials is not trivial in practice.

On the other hand, Chebyshev interpolants are not only easy to compute, but also almost as good as the minimax approximation. More precisely, we have the following formula from [21]:

$$\|f - p_n^{cheb}\|_\infty \leq \left(\frac{2}{\pi} \log n + 2\right) \cdot \|f - p_n^*\|_\infty. \tag{2}$$

4.2 Chebyshev Interpolants of the Sine Function

Recall that in the bootstrapping procedure, we need to homomorphically evaluate

$$\frac{q}{2\pi} \sin\left(\frac{2\pi t}{q}\right).$$

with $t \in [-Kq, Kq]$. After a change of variables, we see that it suffices to evaluate

$$g(x) := \frac{1}{2\pi} \sin(2\pi K x)$$

with $x \in [-1, 1]$. Our goal is to find a polynomial $p(x)$ with small degree such that $\|g - p\|_\infty$ is small. How good can the approximation be? For the scaled sine function g, it has been shown (see e.g. [26]) that the minimax error $\epsilon_n = \|g - p_n^*\|_\infty$ satisfies

$$\limsup_{n \to \infty} n \epsilon_n^{1/n} = \frac{eK}{2}. \tag{3}$$

Therefore, ϵ_n decreases like $\left(\frac{eK}{2n}\right)^n$ as $n \to \infty$, i.e., the approximation error decreases super-exponentially as a function of the degree n. So, the $\log n$ loss factor from replacing the minimax approximation with Chebyshev interpolant is almost negligible compared to the decreasing speed of ϵ_n. Hence, Chebyshev interpolants provide a decent approximation of the sine function in our bootstrapping algorithm.

We compare the Chebyshev interpolant approach with the approach in [14]. Recall that [14] first uses a Taylor polynomial of $\exp(2\pi i K x/2^r)$ of degree d to approximate it. Then, it performs r repeated squaring operations to obtain an approximation of $\exp(2\pi i K x)$. Finally, $g(x)$ is equal to $1/(2\pi)$ times the imaginary part of $\exp(2\pi i K x)$. In Fig. 2 below, we present the log-log plot of approximation error versus polynomial degree for different values of d.

From the plot, we see that the Chebyshev interpolant achieves small error quickly for degree less than 128. On the other hand, the [14] approach requires a much larger degree to reach the same error when $d = 7$. For a larger $d = 55$, the difference between the approaches becomes smaller. However, since the Taylor coefficients of $\exp(2\pi K i x/2^r)$ decrease super-exponentially, evaluating such a large degree Taylor approximation is likely to result in large numerical errors. Therefore, we decided to use Chebyshev interpolants for approximating the sine function.

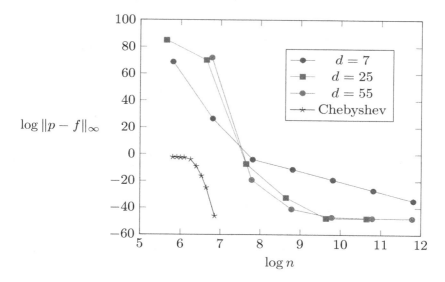

Fig. 2. Polynomial approximation errors to $\frac{1}{2\pi}\sin(2\pi K x)$ ($K = 12$).

4.3 Computing Chebyshev Polynomials in FHE

The Chebyshev coefficients c_k of the scaled sine function g can be precomputed and stored. Next, our task is to evaluate $\sum_{k=0}^{n} c_k T_k(x)$ homomorphically. There are several choices:

Since each $T_k(x)$ is a polynomial in x, we could rewrite $p_n^{cheb}(x)$ as $\sum_{k=0}^{n} c'_k x^k$, and use any existing method for homomorphic evaluation of polynomials in one variable in the literature. However, the transition matrix between the c_k and c'_k coefficients is ill-conditioned (actually, its conditional number grows exponentially as a function of n (see e.g. [3]), and the coefficients c'_k differ by many orders of magnitude. Therefore, the evaluation is likely to generate large numerical errors, even over unencrypted input.

A better method is to use the recurrence relation (1) to evaluate $T_k(x)$ for $0 \leq k \leq n$, and then compute $\sum c_k T_k(x)$ using scalar multiplications and additions. This method yields smaller numerical errors in practice. However, the efficiency is sub-optimal: we still need $O(n)$ homomorphic multiplications in order to evaluate a degree-n Chebyshev interpolant.

Paterson-Stockmeyer for Chebyshev. Our next idea is to use the Paterson-Stockmeyer algorithm [31], which requires only $O(\sqrt{n})$ non-scalar multiplications to evaluate a polynomial of degree n in x. However, we could not directly apply this algorithm since it requires the polynomial to be presented in the *power base* $1, x, \ldots, x^n$. Of course, one could rewrite the Chebyshev interpolant in power base first, and then execute the Paterson-Stockmeyer algorithm. But as we discussed above, such method is subject to large numerical errors, hence it is not a desirable solution.

Instead, we propose a new approach by modifying the Paterson-Stockmeyer algorithm to directly evaluate Chebyshev interpolants. As a result, we can evaluate a Chebyshev interpolant of degree n with $\sqrt{2n} + O(\log n)$ non-scalar multiplications. In order to describe our algorithm, we first recall the Paterson-Stockmeyer algorithm in [31]:

Algorithm 2: The original Paterson-Stockmeyer algorithm

Input: $(a_0, a_1, \ldots, a_n), u$
Output: $f(u) = \sum_i a_i u^i$

1   Find positive integers k, m such that $k \approx \sqrt{n/2}$ and $k(2^m - 1) > n$
2   $\tilde{f}(x) = f(x) + x^{k(2^m-1)}$
3   Compute powers $\mathbf{bs} = (u, u^2, \ldots, u^k)$, $\mathbf{gs} = (u^k, u^{2k}, u^{4k}, \ldots, u^{2^{m-1}k})$
4   Using long division, write $\tilde{f}(x) = x^{k2^{m-1}} q(x) + r(x)$
5   Set $\tilde{r}(x) = r(x) - x^{k(2^{m-1}-1)}$
6   Using long division, write $\tilde{r}(x) = c(x)q(x) + s(x)$
7   Evaluate $c(x)$ at u using the precomputed powers
8   Set $\tilde{s}(x) = s(x) + x^{k(2^{m-1}-1)}$
9   Recursively evaluate $q(x)$ and $\tilde{s}(x)$ at u (with lines 1-3 skipped)
10   Compute $\tilde{f}(u) = (u^{k(2^{m-1})} + c(u))q(u) + \tilde{s}(u)$.
11   Compute $u^{k(2^m-1)}$ by multiplying all values in $\mathbf{gs}$
12   return $f(u) = \tilde{f}(u) - u^{k(2^m-1)}$

Now suppose we wish to use the Chebyshev basis $\{T_k(x)\}_k$ instead of the power base in Algorithm 2. We can start by replacing every occurrence of x^i in the algorithm with $T_i(x)$. Line 3 requires computing certain $T_i(x)$ values, which can be done in $k+m$ operations using the recurrence formula (1). Thus we only need an algorithm for long division of polynomials in Chebyshev base. That is, given Chebyshev coefficients of polynomials f and g, output Chebyshev coefficients of the quotient and remainder polynomials q and r such that $\deg q = \deg f - \deg g$, $\deg r < \deg g$ and $f = qg + r$. A first attempt is to convert f and g to the power base, perform long division as usual, and convert the resulting q and r back to Chebyshev base. Again, this approach is likely to generate a lot of numerical errors since the transform matrices are ill-conditioned. To resolve this issue, we present a direct algorithm.

Long division for polynomials in Chebyshev base.

Lemma 1 (Long Division). *Given two polynomials f and g with positive degrees n and k given by their Chebyshev coefficients, there exists an algorithm with $O(k(n-k))$ operations to compute the Chebyshev coefficients of polynomials $q(x)$ and $r(x)$, such that $\deg q = \deg f - \deg g$, $\deg r < \deg g$, and*

$$f(x) = g(x)q(x) + r(x).$$

Proof. For simplicity, we assume both f and g are monic, meaning their highest Chebyshev coefficient is 1. We do it with induction on $n = \deg f$. If $n \leq \deg g$ then we are done. Now suppose $n > k = \deg g$ and $k \geq 1$. Let

$$r_0(x) = T_n(x) - 2g(x)T_{n-k}(x).$$

Using the formula

$$T_m(x) = 2T_i(x)T_{m-i}(x) - T_{|m-2i|}(x),$$

we see that $\deg(r_0) < n$, and we may compute the Chebyshev coefficients of $r_0(x)$. Now we could recursively perform the division r_0 by g to finish the algorithm. The correctness is easy to verify, and since computing r_0 requires $O(k)$ operations, the algorithm requires $O(k(n-k))$ operations. This finishes the proof.

Given the above lemma, we can modify Algorithm 2 to directly perform long division of polynomials in Chebyshev base. We omit the detailed description of the modified algorithm since it is straightforward. As a result, we have

Theorem 1. *There exists an algorithm to evaluate a polynomial of degree n given in Chebyshev base with $\sqrt{2n} + O(\log n)$ non-scalar multiplications and $O(n)$ scalar multiplications.*

5   Putting it Together

5.1   Asymptotic Analysis

Combining the optimization techniques in Sects. 3 and 4, we come up with a new bootstrapping algorithm for the CKKS scheme, whose complexity improves upon the algorithm in [14]. We make a detailed comparison below:

Linear transforms. The subSum step remains unchanged from [14], which requires $O(N/2\ell)$ rotations. For the two transforms coeffToSlot and slotToCoeff, recall that [14] takes $O(\sqrt{\ell})$ rotations and ℓ plaintext multiplications, whereas our algorithm provides a spectrum of trade-offs between level consumption and operation counts. For example, if we fix the level budget to be $\ell' = 2$, then both the coeffToSlot and slotToCoeff requires $O(\ell^{1/4})$ rotations and $O(\sqrt{\ell})$ plaintext multiplications.

Sine evaluation. The approach of [14] to evaluate the sine approximation requires a polynomial of degree $d \cdot 2^r$ and $O(d + r)$ ciphertext multiplications. They took $d = O(1)$ and $r = O(\log(Kq))$ in order to achieve an approximation error of $O(1)$ for the function $(q/2\pi)\sin(2\pi t/q)$. Thus, both the required level and the number of operations are $O(\log(Kq))$.

In our case, we used a Chebyshev interpolant to approximate the sine function. From the results in Sect. 4, we see that it suffices to take $n = \max\{4K, \log q\}$ to achieve $1/q$ approximation error from (2) and (3). Therefore, our approach consumes only $\log n \leq \max\{\log K + 2, \log \log q\}$ levels. In terms of the number of operations, by using the modified Paterson-Stockmeyer algorithm, we can evaluate the Chebyshev interpolant in $O(\sqrt{n})$ ciphertext multiplications.

5.2   Implementation and Performance

Recently, the authors of [16] published a improved version [1] of the implementation of the CKKS scheme with faster operations. We implemented our bootstrapping algorithm on top of the new version. In order to separate the causes of speedups, we also experimented with the original bootstrapping algorithm with the new library. We summarize our findings in Table 4.

Parameter choices. To benchmark the original bootstrapping algorithm, we used the same parameter sets (Table 2) from [14]. We modified these parameters slightly for our new bootstrapping algorithm. The modified parameters are presented in Table 3. We note that these modifications do not involve $\log N, \log Q$ or the initial noise in the ciphertexts, hence the security level remains the same as previous work.

Table 2. Parameter sets

Parameter	$\log N$	$\log Q_0$	$\log p$	$\log q$	r
Set-I	15	620	23	29	6
Set-II			27	37	7
Set-III	16	1240	31	41	7
Set-IV			39	54	9

Table 3. New parameter sets

Parameter	$\log p$	$\log q$	l_{ctos}	l_{stoc}
Set-I*	25	29	2	2
Set-II*	25	34	2	2
Set-II**	27	37	2	1
Set-III*	33	41	2	2
Set-III**	35	41	3	3
Set-IV*	43	54	3	3
Set-IV**	43	54	4	4

In Table 3, the columns labeled l_{ctos} and l_{stoc} denote the level consumption for coeffToSlot and slotToCoeff, respectively. Note that larger levels result in less operations. For the sine evaluation, we fixed $K = 12$ and a Chebyshev interpolant of degree $n = 119$ based on experimental results. All experiments are performed on a laptop with 2.8 GHz Intel Core i7 Processor and 16 GB memory, running on a single thread.

Table 4. Performance comparisons for bootstrapping: LT (linear transformations) timing is the sum of the timings for subSum, coeffToSlot and slotToCoeff. Precision is averaged among all slots. The "After Level" column shows how many levels of multiplications are allowed after each bootstrapping.

Params	logSlots	Method	LT	Sine Eval	Total Time (s)	Amortized Time (s)	Average Precision	After Level
Set-I	7	[13]	139.2	12.3	151.5	1.2	7.64	8
	7	[13] + [1]	36.1	5.26	41.36	0.32	7.64	8
Set-I*	10	This work	28.78	9.55	38.33	0.04	6.92	5
Set-II	7	[13]	127.3	12.5	139.8	1.1	9.9	1
	7	[13] + [1]	43.9	8.73	52.63	0.41	9.9	1
Set-II*	8	This work	16.87	9.18	26.05	0.04	10.03	2
Set-II**	10	This work	37.11	9.18	85.83	0.08	9.1	1
Set-III	7	[13]	528	63	591	4.6	13.2	19
	7	[13] + [1]	158.2	29.3	187.5	1.46	13.2	19
Set-III*	10	This work	154.28	47.7	201.98	0.2	13.7	17
Set-III**	12	This work	134.35	43.7	178.05	0.04	11.75	13
Set-IV	7	[13]	456	68	524	4.1	20.1	7
	7	[13] + [1]	224.2	80.7	304.9	2.38	20.1	7
Set-IV*	12	This work	127.49	40.38	167.87	0.04	20.86	6
Set-IV**	14	This work	119.76	38.56	158.32	0.01	18.63	3

5.3   Comparison

In order to make a meaningful comparison of the efficiency of the different bootstrapping methods/implementations, we need to provide a common measure, and one such measure is the number of slots times the number of levels allowed after bootstrapping, divided by the bootstrapping time. We argue that this definition makes sense, since in the process of evaluating a typical circuit homomorphically, the frequency of bootstrapping should be inverse proportional to the allowed multiplicative depth after bootstrapping. Also, since the complexity of bootstrapping depends on the bit precision of the output, we plot the utility versus precision in the following Fig. 3.

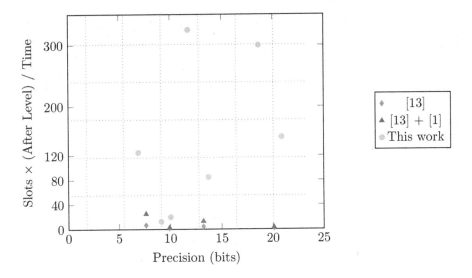

Fig. 3. Bootstrapping utility comparisons

From Fig. 3, we see that our new algorithms can improve the utility of bootstrapping by two orders of magnitude. For example, [14] could bootstrap numbers with around 20 bits of precision with a utility of 2.94 (Level × Slot/Second). With a slightly larger precision, we achieved a utility of 150, yielding a 50x improvement.

6 Conclusion and Future Work

In this work, we showed that algorithmic improvements to the linear transforms and sine evaluation steps could boost the efficiency of bootstrapping for the CKKS approximate homomorphic encryption scheme by two orders of magnitude.

Our results suggest that using Chebyshev interpolant together with the Paterson-Stockmeyer algorithm is a promising solution for approximately evaluating non-polynomial functions in FHE. For example, we could apply this idea to evaluate the sigmoid function or the RELU function, which is interesting from the point of view of doing machine learning over encrypted data. Also, this idea can be applied to the absolute value function, which may expedite evaluation of a sorting network over encrypted data.

The improved linear transform technique for the CKKS scheme can be used to provide a fast evaluation of discrete Fourier transform (DFT) over encrypted data, which might be of independent interest. Also, we could utilize our algorithm to provide an efficient implementation of the conversion between CKKS ciphertexts and ciphertexts from TFHE or BFV/BGV schemes, outlined in a recent work [5].

Recently, there is another variant of the CKKS scheme [15] based on the *Residue Number System* (RNS), following an idea of Bajard et al. [2]. The reported performance numbers of this new variant are up to 10x better than the original implementation. Thus, it would be interesting to implement our bootstrapping algorithm on this RNS variant to obtain even better performance.

References

1. HEAAN with Faster Multiplication (2018). https://github.com/snucrypto/HEAAN/releases/tag/2.1
2. Bajard, J.-C., Eynard, J., Hasan, M.A., Zucca, V.: A full RNS variant of FV like somewhat homomorphic encryption schemes. In: Avanzi, R., Heys, H. (eds.) SAC 2016. LNCS, vol. 10532, pp. 423–442. Springer, Cham (2017). https://doi.org/10.1007/978-3-319-69453-5_23
3. Beckermann, B.: On the numerical condition of polynomial bases: estimates for the condition number of Vandermonde, Krylov and Hankel matrices. Ph.D. thesis, Verlag nicht ermittelbar (1997)
4. Bonte, C., Bootland, C., Bos, J.W., Castryck, W., Iliashenko, I., Vercauteren, F.: Faster homomorphic function evaluation using non-integral base encoding. In: Fischer, W., Homma, N. (eds.) CHES 2017. LNCS, vol. 10529, pp. 579–600. Springer, Cham (2017). https://doi.org/10.1007/978-3-319-66787-4_28
5. Boura, C., Gama, N., Georgieva, M.: Chimera: a unified framework for B/FV, TFHE and HEAAN fully homomorphic encryption and predictions for deep learning. Cryptology ePrint Archive, Report 2018/758 (2018). https://eprint.iacr.org/2018/758
6. Bourse, F., Minelli, M., Minihold, M., Paillier, P.: Fast homomorphic evaluation of deep discretized neural networks. In: Shacham, H., Boldyreva, A. (eds.) CRYPTO 2018. LNCS, vol. 10993, pp. 483–512. Springer, Cham (2018). https://doi.org/10.1007/978-3-319-96878-0_17
7. Brakerski, Z., Gentry, C., Vaikuntanathan, V.: (Leveled) fully homomorphic encryption without bootstrapping. In: Proceedings of ITCS, pp. 309–325. ACM (2012)
8. Brakerski, Z., Vaikuntanathan, V.: Efficient fully homomorphic encryption from (standard) LWE. In: Proceedings of the 2011 IEEE 52nd Annual Symposium on Foundations of Computer Science, FOCS 2011, pp. 97–106. IEEE Computer Society (2011)
9. Brakerski, Z., Vaikuntanathan, V.: Fully homomorphic encryption from ring-LWE and security for key dependent messages. In: Rogaway, P. (ed.) CRYPTO 2011. LNCS, vol. 6841, pp. 505–524. Springer, Heidelberg (2011). https://doi.org/10.1007/978-3-642-22792-9_29
10. Chen, H., et al.: Logistic regression over encrypted data from fully homomorphic encryption. BMC Med. Genomics **11**(4), 81 (2018)
11. Chen, H., Han, K.: Homomorphic lower digits removal and improved FHE bootstrapping. In: Nielsen, J.B., Rijmen, V. (eds.) EUROCRYPT 2018. LNCS, vol. 10820, pp. 315–337. Springer, Cham (2018). https://doi.org/10.1007/978-3-319-78381-9_12
12. Chen, H., Laine, K., Rindal, P.: Fast private set intersection from homomorphic encryption. In: Proceedings of the 2017 ACM SIGSAC Conference on Computer and Communications Security, pp. 1243–1255. ACM (2017)

13. Cheon, J.H., Han, K., Kim, A., Kim, M., Song, Y.: Implementation of boostrapping for HEAAN (2017). https://github.com/kimandrik/HEAANBOOT

14. Cheon, J.H., Han, K., Kim, A., Kim, M., Song, Y.: Bootstrapping for approximate homomorphic encryption. In: Nielsen, J.B., Rijmen, V. (eds.) EUROCRYPT 2018. LNCS, vol. 10820, pp. 360–384. Springer, Cham (2018). https://doi.org/10.1007/978-3-319-78381-9_14

15. Cheon, J.H., Han, K., Kim, A., Kim, M., Song, Y.: A full RNS variant of approximate homomorphic encryption. In: Cid, C., Jacobson Jr., M. (eds.) SAC 2018. LNCS, vol. 11349, pp. 347–368. Springer, Cham (2018). https://doi.org/10.1007/978-3-030-10970-7_16

16. Cheon, J.H., Kim, A., Kim, M., Song, Y.: Homomorphic encryption for arithmetic of approximate numbers. In: Takagi, T., Peyrin, T. (eds.) ASIACRYPT 2017. LNCS, vol. 10624, pp. 409–437. Springer, Cham (2017). https://doi.org/10.1007/978-3-319-70694-8_15

17. Chillotti, I., Gama, N., Georgieva, M., Izabachène, M.: Faster fully homomorphic encryption: bootstrapping in less than 0.1 seconds. In: Cheon, J.H., Takagi, T. (eds.) ASIACRYPT 2016. LNCS, vol. 10031, pp. 3–33. Springer, Heidelberg (2016). https://doi.org/10.1007/978-3-662-53887-6_1

18. Chillotti, I., Gama, N., Georgieva, M., Izabachène, M.: Faster packed homomorphic operations and efficient circuit bootstrapping for TFHE. In: Takagi, T., Peyrin, T. (eds.) ASIACRYPT 2017. LNCS, vol. 10624, pp. 377–408. Springer, Cham (2017). https://doi.org/10.1007/978-3-319-70694-8_14

19. Crawford, J.L., Gentry, C., Halevi, S., Platt, D., Shoup, V.: Doing real work with FHE: the case of logistic regression. In: Proceedings of the 6th Workshop on Encrypted Computing & Applied Homomorphic Cryptography, pp. 1–12. ACM (2018)

20. Ducas, L., Micciancio, D.: FHEW: bootstrapping homomorphic encryption in less than a second. In: Oswald, E., Fischlin, M. (eds.) EUROCRYPT 2015. LNCS, vol. 9056, pp. 617–640. Springer, Heidelberg (2015). https://doi.org/10.1007/978-3-662-46800-5_24

21. Ehlich, H., Zeller, K.: Auswertung der normen von interpolationsoperatoren. Math. Ann. **164**(2), 105–112 (1966)

22. Fan, J., Vercauteren, F.: Somewhat practical fully homomorphic encryption. IACR Cryptology ePrint Archive 2012:144 (2012)

23. Gentry, C.: Fully homomorphic encryption using ideal lattices. In: Proceedings of the 41st Annual ACM Symposium on Theory of Computing, pp. 169–178. ACM (2009)

24. Gentry, C., Halevi, S., Smart, N.P.: Better bootstrapping in fully homomorphic encryption. In: Fischlin, M., Buchmann, J., Manulis, M. (eds.) PKC 2012. LNCS, vol. 7293, pp. 1–16. Springer, Heidelberg (2012). https://doi.org/10.1007/978-3-642-30057-8_1

25. Gilad-Bachrach, R., et al.: CryptoNets: applying neural networks to encrypted data with high throughput and accuracy. In: International Conference on Machine Learning, pp. 201–210 (2016)

26. Giroux, A.: Approximation of entire functions over bounded domains. J. Approx. Theory **28**(1), 45–53 (1980)

27. Halevi, S., Shoup, V.: Algorithms in HElib. In: Garay, J.A., Gennaro, R. (eds.) CRYPTO 2014. LNCS, vol. 8616, pp. 554–571. Springer, Heidelberg (2014). https://doi.org/10.1007/978-3-662-44371-2_31

28. Halevi, S., Shoup, V.: Bootstrapping for `HElib`. In: Oswald, E., Fischlin, M. (eds.) EUROCRYPT 2015. LNCS, vol. 9056, pp. 641–670. Springer, Heidelberg (2015). https://doi.org/10.1007/978-3-662-46800-5_25
29. Han, K., Hong, S., Cheon, J.H., Park, D.: Efficient logistic regression on large encrypted data. Cryptology ePrint Archive, Report 2018/662 (2018). https://eprint.iacr.org/2018/662
30. Kim, A., Song, Y., Kim, M., Lee, K., Cheon, J.H.: Logistic regression model training based on the approximate homomorphic encryption. BMC Med. Genomics **11**(4), 83 (2018)
31. Paterson, M.S., Stockmeyer, L.J.: On the number of nonscalar multiplications necessary to evaluate polynomials. SIAM J. Comput. **2**(1), 60–66 (1973)

Minicrypt Primitives with Algebraic Structure and Applications

Navid Alamati[1,2](✉), Hart Montgomery[2], Sikhar Patranabis[2,3],
and Arnab Roy[2]

[1] University of Michigan, Ann Arbor, USA
alamati@gmail.com
[2] Fujitsu Laboratories of America, Sunnyvale, USA
[3] IIT Kharagpur, Kharagpur, India

Abstract. Algebraic structure lies at the heart of Cryptomania as we know it. An interesting question is the following: instead of building (Cryptomania) primitives from concrete assumptions, can we build them from *simple* Minicrypt primitives endowed with some additional *algebraic* structure? In this work, we affirmatively answer this question by adding algebraic structure to the following Minicrypt primitives:

- One-Way Function (OWF)
- Weak Unpredictable Function (wUF)
- Weak Pseudorandom Function (wPRF)

The algebraic structure that we consider is group homomorphism over the input/output spaces of these primitives. We also consider a "bounded" notion of homomorphism where the primitive only supports an a priori bounded number of homomorphic operations in order to capture lattice-based and other "noisy" assumptions. We show that these structured primitives can be used to construct many cryptographic protocols. In particular, we prove that:

- (Bounded) *Homomorphic OWFs* (HOWFs) imply collision-resistant hash functions, Schnorr-style signatures and chameleon hash functions.
- (Bounded) *Input-Homomorphic weak UFs* (IHwUFs) imply CPA-secure PKE, non-interactive key exchange, trapdoor functions, blind batch encryption (which implies anonymous IBE, KDM-secure and leakage-resilient PKE), CCA2 deterministic PKE, and hinting PRGs (which in turn imply transformation of CPA to CCA security for ABE/1-sided PE).
- (Bounded) *Input-Homomorphic weak PRFs* (IHwPRFs) imply PIR, lossy trapdoor functions, OT and MPC (in the plain model).

In addition, we show how to realize any CDH/DDH-based protocol with certain properties in a generic manner using IHwUFs/IHwPRFs, and how to instantiate such a protocol from many concrete assumptions.

© International Association for Cryptologic Research 2019
Y. Ishai and V. Rijmen (Eds.): EUROCRYPT 2019, LNCS 11477, pp. 55–82, 2019.
https://doi.org/10.1007/978-3-030-17656-3_3

We also consider primitives with substantially richer structure, namely *Ring IHwPRFs* and *L-composable IHwPRFs*. In particular, we show the following:

- Ring IHwPRFs with certain properties imply FHE.
- 2-composable IHwPRFs imply (black-box) IBE, and *L*-composable IHwPRFs imply non-interactive $(L + 1)$-party key exchange.

Our framework allows us to categorize many cryptographic protocols based on which structured Minicrypt primitive implies them. In addition, it potentially makes showing the *existence* of many cryptosystems from novel assumptions substantially easier in the future.

1   Introduction

An important question in the theory of cryptography is also one of the simplest to state: what implies public-key cryptography? Ever since the (public) invention of public-key encryption [DH76, RSA78], people have debated this important question.

The history of symmetric-key cryptography goes back millenia–the Caesar cipher is a classic example of old cryptography–and it has continued to evolve through the centuries in different ways. There is a long list of ciphers, notably including the Viginère cipher, the Enigma machine, and even modern ciphers like AES, that can be thought of as the output of an enormous amount of human effort to build secure symmetric-key encryption.

On the other hand, public-key cryptography is a very recent development compared to symmetric-key cryptography. Many people thought that public-key cryptography was impossible before the seminal work by Diffie and Hellman [DH76]. Although we can build symmetric-key ciphers from many different assumptions, including some very simple ones, the known methods for realizing public-key cryptography require at least some kind of mathematical structure. This has led many to conjecture that public-key cryptography does, in fact, require some mathematical structure.

Barak ruminated on this question in his recent work "The Complexity of Public Key Cryptography" [Bar17]. As he puts it, "... it seems that you can't throw a rock without hitting a one-way function" but public-key cryptography is somehow "special". Barak implicitly argues that there is some mathematical structure inherent in public-key cryptography: "One way to phrase the question we are asking is to understand what type of structure is needed for public-key cryptography."

But many cryptosystems that interest people today are substantially more complicated than basic public-key encryption (PKE). In recent years, primitives like identity-based encryption [Sha84], fully homomorphic encryption [Gen09], and functional encryption [BSW11] have captivated cryptographers. It is natural to ask: is there any sort of mathematical structure that is inherent to these primitives as well? While there has been a substantial amount of work relating relatively similar primitives, to our knowledge no one has attempted to comprehensively examine the relationship between a broader collection of these higher-level primitives.

In a celebrated work, Impagliazzo [Imp95] proposed "five worlds" of relative complexity, which range from *Algorithmica*–where "efficient" algorithms for all (worst-case) problems in NP exist and cryptography is essentially nonexistent–to *Cryptomania*, a world in which public-key cryptography exists. Only two of these worlds allow for cryptography: *Minicrypt*, where symmetric cryptographic primitives exist but public-key cryptography does not, and the aforementioned *Cryptomania*.

It turns out that Minicrypt is a fairly simple world. A number of famous works have shown how to build the most commonly studied and used Minicrypt primitives from one-way functions in a generic manner. For instance, one-way functions imply pseudorandom generators [BM82, HILL99], which in turn can be used to build pseudorandom functions [GGM84]. From these primitives, it has long been known how to generically build symmetric-key encryption schemes and digital signature schemes [Rom90].

On the other hand, Cryptomania is a significantly more complicated class. It contains primitives that are very different, and it seems difficult to relate them in a generic manner. We cannot expect to, say, build FHE from PKE in a black-box manner, and there are many black-box separation results for cryptosystems in Cryptomania (we discuss this more in our related work section). In fact, recently it has even become popular to separate Cryptomania into two worlds: a world where indistinguishability obfuscation (iO) [BGI+01, GGH+13b] doesn't exist, and a world called Obfustopia [GPSZ17] where it does.

This, of course, raises a fundamental question in the complexity of public-key cryptography: can we construct classes of primitives within Cryptomania (i.e. "continents" of Cryptomania) that are tightly tied to each other through generic constructions? Ideally, we would want these "continents" to have strong relationships with a particular primitive (similar to the relationship between one-way functions and Minicrypt) where all of the cryptographic algorithms in the class could be built from the given primitive in a generic manner, and the given primitive would be conceptually the simplest function in the class.

The fact that most of the concrete assumptions that imply PKE (and also many other cryptographic primitives) have some algebraic structure seems to imply that perhaps we can classify cryptosystems by the algebraic structure necessary for them to function. This leads us to the following question:

Is it possible to construct Cryptomania primitives from simple Minicrypt primitives that are additionally equipped with some algebraic structure?

1.1 Our Contributions

In this work, we provide a constructive answer to the question of building PKE (and other primitives in Cryptomania) from Minicrypt primitives with algebraic structure. Let's start by considering the following Minicrypt primitives:

1. One-way Functions
2. Weak Unpredictable Functions
3. Weak Pseudorandom Functions

To add *algebraic structure* to the mentioned primitives, we assume that they are *(Input-)Homomorphic*: the input and output spaces of the primitive are groups, and the primitive is (bounded) homomorphic with respect to an efficiently computable group homomorphism. We use the following primitives and abbreviations throughout the paper:

- *Homomorphic One-way Functions* (HOWFs)[1]
- *Input-Homomorphic Weak Unpredictable Functions* (IHwUFs)
- *Input-Homomorphic Weak Pseudorandom Functions* (IHwPRFs)[2]

In the body of the paper we also consider "bounded" homomorphisms, where the number of allowed homomorphisms is bounded by some function $\gamma = \gamma(\lambda)$ where λ is the security parameter, which lets us work with lattice-based and other "noisy" cryptographic assumptions.

At this point we can informally state our main contribution: we present a framework for building cryptographic primitives from HOWFs/IHwUFs/IHwPRFs (see Fig. 1). This framework lets us categorize cryptographic primitives by the type of structured Minicrypt primitive that implies them. However, we need to be able to instantiate the above *general* primitives from *concrete* assumptions to have a useful framework. It turns out that we can instantiate our primitives (in most cases) from a wide variety of assumptions, typically including the assumptions that would be expected for such applications.

Instantiations from Concrete Assumptions. We show that "mainstream" cryptographic assumptions such as DDH and LWE naturally imply (bounded) HOWFs/IHwUFs/IHwPRFs. We also show that a (bounded) group-homomorphic PKE implies a (bounded) IHwPRF. This allows instantiating these primitives from any concrete assumption that implies a (bounded) homomorphic PKE (e.g. QR and DCR). Unfortunately, there is a caveat to this: the transformation from homomorphic PKE to IHwPRF comes with a disadvantage that the input space may *depend* on the key.[3] The reader may refer to Fig. 2 for an overview of instantiations from concrete assumptions.[4]

---

[1] When the function does not have a key (i.e. a one-way function) we will drop the "I" and refer to the function as simply homomorphic.

[2] In case of IHwUFs/IHwPRFs we do not assume any homomorphism on the key space.

[3] This property is necessary to realize certain cryptographic primitives from IHwUFs or IHwPRFs.

[4] Notice that search to decision reductions are mostly for Gaussian-like distributions, and there are certain distributions for which search to decision reduction is not available.

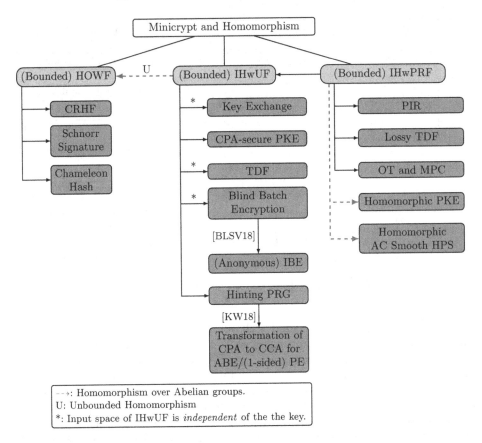

Fig. 1. Cryptographic primitives from Minicrypt and Homomorphism.

Building Cryptosystems from New Assumptions. One of the benefits of our work is the implications for new assumptions. Rather than manually building lots of different cryptosystems from a new assumption, researchers only need to build one (or more) of our simple structured primitives, and the existence of a whole host of cryptosystems immediately follows.

To illustrate how this might be useful, let's look at the history of lattice-based cryptography: Ajtai and Dwork [AD97] gave a lattice-based PKE (following Ajtai's worst-case to average-case reductions for lattice problems [Ajt96]), but lattice cryptography may have begun in earnest with Regev's LWE paper [Reg05] in 2005. This work, in addition to introducing the LWE problem, showed how to build a basic PKE scheme from LWE as well. However, it took a while for the cryptographic community to "catch up" to other group-based cryptosystems: for instance, the first private information retrieval scheme from lattices was presented in [AMG07], and the first identity-based encryption was given in [GPV08].

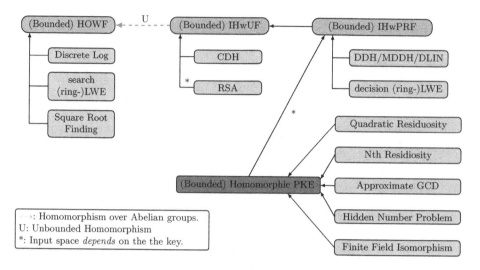

Fig. 2. Instantiations from Concrete Assumptions

These works used sophisticated techniques on lattices in order to extend the range of lattice-based cryptosystems. With our work, the existence of all of these types of cryptosystems based on the LWE assumption follows immediately from the simple observation that LWE implies a (bounded) IHwPRF. While the necessary tools for many of our constructions were not around in 2008 (particularly [DG17b] and the line of work following it), we do hope that this paper is useful for public-key cryptography assumptions in the future in terms of *feasibility* results. Ideally, it will be easy to show the existence of many types of cryptosystems for new assumptions using the tools from this paper.

More Primitives from Richer Structures. Although the main focus of this work is to construct many cryptographic primitives from IHwUFs/IHwPRFs, one might ask: what if we consider richer structures? For instance, what would happen if we have a *ring homomorphism* for an IHwPRF instead of just a group homomorphism? To partially answer this question, we consider two additional structures over wPRFs:

- *Ring Homomorphism:* We consider Ring IHwPRFs (RIHwPRFs) where the input and output spaces are rings, and the homomorphism is with respect to ring operations (instead of just group operations).
- *L-composability:* We consider L-composable IHwPRFs, where L levels of IHw-PRF operations compose with each other under certain conditions.

We summarize our results for these richly structured primitives in Fig. 3. We remark that "*" means the order of the output ring of RIHwPRF is polynomial in the security parameter.

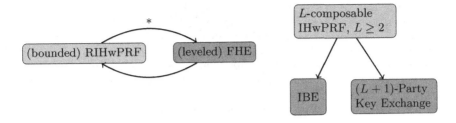

Fig. 3. Cryptographic primitives from richer structures.

While the structure of 2-composability appears similar to that of bilinear pairing groups, we partially explore a possible separation between the two. We argue that 2-composability suffices to achieve three-party non-interactive key exchange and simple black-box constructions of IBE. Subsequently, we also present a discussion on why this primitive does not naturally yield other cryptographic protocols implied by bilinear pairings. This leaves open the interesting question of whether there exists some concrete assumption that implies 2-composability but not bilinear pairings. The separation seemingly extends to the general L-composability setting, in the sense that the structure of L-composability appears to be weaker than that of a full-fledged multilinear map [GGH13a].

On the Categorization of Primitives. This work enables us to categorize different primitives based upon which structured Minicrypt primitive implies them. But it is also possible to ask whether a given cryptosystem may be constructed from some other structured Minicrypt primitive. For instance, is it possible to construct PKE from a HOWF? A positive answer would imply that one can base PKE on the discrete log problem, a long-standing (and potentially possible) goal in cryptography. We can build PKE from IHwUFs, but can we hope to do better? Our work gives rise to interesting questions like this for future work, and we discuss this more later in the paper.

It is easy to see that none of the three primitives HOWF/IHwUF/IHwPRF can be built from PKE in a *black-box* manner [HHRS07], as all of them imply collision-resistant hash functions. In addition to input homomorphisms, one may consider other structures on Minicrypt primitives.

One of the simplest structures is what we term *dual-computable*. This notion is certainly folklore, and some earlier works on PKE and key exchange implicitly constructed this primitive. A *dual-computable* primitive is a tuple of keyed functions (F_1, G_1, F_2, G_2) such that $G_1(k_1, F_2(k_2, x)) = G_2(k_2, F_1(k_1, x))$ where x represents the input and k_i represent keys. The reader may notice that this primitive is almost an abstraction of key exchange if the functions are unpredictable. It is not clear what kind of (minimal) structure over OWFs would imply dual-computable functions (Fig. 4).

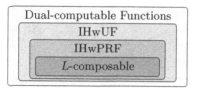

Fig. 4. Implication Landscape

1.2 Related Works

Realizing public-key cryptography via some form of structure and hardness has been studied seemingly since its invention. However, several recent works have discussed this relationship in more detail. For instance, [BDV17] examined the relationship of structure and hardness through obfuscation lens, while a recent work by Berman *et al.* showed that laconic zero-knowledge protocols imply PKE [BDRV18]. Pietrzak and Sjödin [PS08] showed that a certain input property of weak PRFs implies PKE. A recent survey [BR17] briefly discusses structure and PKE through the lens of (strengthened) PRFs.

A number of works have shown how to build certain cryptosystems from cryptographic primitives with algebraic structure. These include commitment schemes, CRHF, IND-CCA secure PKE, PIR, and key-dependent message (KDM) secure PKE [IKO05, HO12, KO97, HKS16]. Of particular relevance to us is the work of Hajiabadi *et al.* [HKS16] on using homomorphic weak PRFs to build KDM secure PKE.[5]

There are other related black-box constructions (or implications in a non-black-box way) between cryptographic primitives, some of which we utilize in our work. For instance, Ishai *et al.* showed how to construct secure computation protocols from enhanced trapdoor functions (or homomorphic PKE) [IKLP06]. Rothblum [Rot11] showed a transformation of a secret-key encryption (SKE) scheme with some special form of weak homomorphism into a PKE that has similar properties. Black-box constructions have been shown for resettable zero-knowledge arguments [OSV15] and cryptographic accumulators [DHS15]. Many cryptographic primitives have been realized in a black-box manner from lossy trapdoor functions [PW08, BHY09, GPR16]. Very recently, Friolo *et al.* [FMV18] showed how to build secure multi-party computation from what they call strongly uniform key agreement and Fischlin and Harasser [FH18] showed the equivalence of invisible sanitizable signatures and PKE.

Understanding the complexity of various public-key primitives also requires knowledge of black-box separations, which have been extensively studied in the literature. This (non-exhaustively) includes studies separating IBE from CRHFs (and thus FHE) [MM16], separating indistinguishability obfuscation (iO) from certain primitives (for instance, CRHFs) [AS15, MMN+16], separating succinct non-interactive arguments from falsifiable assumptions [GW11], and showing

---

[5] As mentioned earlier, we refer to this primitive as Input-Homomorphic weak PRF (IHwPRF) to emphasize that the homomorphism is on the input space and not on the key space.

that garbling of circuits having one-way function gates are not sufficient to realize PKE [GHMM18]. These separations (and related works) allow us to clearly see that some primitives are *not* equivalent, at least modulo certain assumptions. We refer the reader to [RTV04,Fis12,BBF13] for a survey on black-box reductions and separations.

2 Technical Overview

In this section, we aim to explain some of the intuition behind our results. We will start by focusing on one particular primitive–the input homomorphic weak PRF–and some of its applications. The results for other primitives are not exactly the same, but the general structure of how we build cryptosystems from these other primitives is relatively similar. We will discuss these other primitives later in this section.

2.1 PKE from IHwUFs/IHwPRFs

Let's start by considering the notion of a general input-homomorphic weak PRF, or, as we have been abbreviating, an IHwPRF, which we will define as a function $F : \mathcal{K} \times \mathcal{X} \to \mathcal{Y}$. Recall that, informally speaking, a weak PRF is a function that is indistinguishable from a random function *with respect to uniformly sampled inputs*. This "weakness" as compared to a regular PRF will be critical.

We will also endow our weak PRF F with a homomorphism over the input. Suppose our input space $\mathcal{X}$ and our output space $\mathcal{Y}$ are groups with group operations $\oplus$ and $\otimes$, respectively. Roughly speaking, an IHwPRF is just a regular weak PRF with the following property:

$$F(k, x_1 \oplus x_2) = F(k, x_1) \otimes F(k, x_2).$$

We also consider what we call γ-bounded IHwPRFs. These IHwPRFs have a homomorphism that can only be computed a maximum of γ times, where γ is a pre-determined parameter. This concept lets us consider noisy assumptions like LWE, which are only approximately homomorphic. The notion is very similar to definitions of the *almost* key-homomorphic PRFs of [BLMR13]. γ-bounded IHwPRFs work for almost all of the applications that we consider in almost the same way that unbounded IHwPRFs do. For the rest of this technical overview, though, we will assume we have an unbounded IHwPRF. Also, we occasionally refer to an Input-Homomorphic weak Unpredictable Function (IHwUF), which has the same properties as IHwPRF except for the fact that its output on a uniformly random input is just *unpredictable* and not necessarily *pseudorandom*.

DDH-Based Instantiation of IHwPRF. In general, it is simple to build IHwPRFs from assumptions that are widely used in cryptography. Here we show how to build an IHwPRF from the DDH assumption. Let $\mathbb{G}$ be a group of prime order q where the DDH problem is hard. For a uniformly sampled key $k \leftarrow \mathbb{Z}_q$ and an input $x \in \mathbb{G}$, consider the following function:

$$F(k, x) = x^k.$$

If we are only allowed to see the evaluation of F on random inputs x_i (as the weak PRF definition requires), then it is easy to see that F is a weak PRF based on the DDH assumption. Moreover, the homomorphism property is also satisfied:

$$x_1^k \cdot x_2^k = (x_1 \cdot x_2)^k .$$

Thus F is an IHwPRF. Building a *bounded* IHwPRF from LWE is similarly straightforward, but we defer this to later in the paper.

On the Input Space. It is useful to note that the "discrete logarithm problem" on the input space of an IHwPRF must be hard by its weak pseudorandomness property. Concretely, given two evaluations $(x_1, F(k, x_1))$ and $(x_2, F(k, x_2))$, an adversary can compute some value c such that $x_1^c = x_2$, then they can check if

$$F(k, x_1)^c = F(k, x_2)$$

and use this to break the (weak) pseudorandomness of F. In the context of (bounded) IHwPRFs over arbitrary groups, we note that there must exist an equivalent "discrete log" problem that allows us to capture the aforementioned property.[6] This property is crucial to the security of nearly all constructions presented in this paper.

PKE Construction. We now illustrate how to construct a CPA-secure PKE given an IHwPRF. To provide more intuition, we will present an instantiation of the encryption scheme using the DDH assumption in parallel. The construction from IHwPRF is highlighted for clarity.

Setup:
- **IHwPRF Construction:** Select an IHwPRF $F : \mathcal{K} \times \mathcal{X} \rightarrow \mathcal{Y}$ over groups $(\mathcal{X}, \oplus)$ and $(\mathcal{Y}, \otimes)$ with key space $\mathcal{K}$, input space $\mathcal{X}$, and output space $\mathcal{Y}$ and some integer $n > 3 \log(|\mathcal{X}|)$. Select a set X of $2n$ uniform "base elements" from $\mathcal{X}$ as

$$X = \{x_{j,b} \leftarrow \mathcal{X}\}_{j \in [n], b \in \{0,1\}} .$$

Select a random key $k \leftarrow \mathcal{K}$. Create a tuple Y of $2n$ elements from $\mathcal{Y}$ as

$$Y = \{y_{j,b}\}_{j \in [n], b \in \{0,1\}}$$

such that $y_{j,b} = F(k, x_{j,b})$. Output the secret key and public key as:[7]

$$\mathsf{sk} = k, \quad \mathsf{pk} = (X, Y) .$$

---

[6] For our LWE-based bounded IHwPRF, the "discrete log" problem equivalent is the ISIS problem.

[7] We implicitly assume that the description of IHwPRF is publicly available. This is similar to the assumption that in a DDH-based encryption scheme like ElGamal, the description of the cyclic group $\mathbb{G}$ is public.

- **DDH Instantiation:** Let $F : \mathbb{Z}_q \times \mathbb{G} \to \mathbb{G}$ be the function defined as $F(k \in \mathbb{Z}_q, g \in \mathbb{G}) = g^k$. Select a set G of $2n$ randomly sampled elements from $\mathbb{G}$ as

$$G = \{g_{j,b} \leftarrow \mathbb{G}\}_{j \in [n], b \in \{0,1\}} .$$

Select a random key $k \leftarrow \mathbb{Z}_q$. Create a tuple H of $2n$ elements from $\mathbb{G}$ as

$$H = \{h_{j,b}\}_{j \in [n], b \in \{0,1\}}$$

such that $h_{j,b} = g_{j,b}^k$. Output the secret key and the public key as

$$\mathsf{sk} = k, \quad \mathsf{pk} = (G, H).$$

Encrypt:
- **IHwPRF Construction:** On input a message $\mathsf{m} \in \mathcal{Y}$, sample a vector $\mathsf{s} = (s_1, \ldots, s_n) \leftarrow \{0,1\}^n$. Set

$$x^* = \bigoplus_{j \in [n]} x_{j,s_j}, \quad y^* = \bigotimes_{j \in [n]} y_{j,s_j} .$$

Output the ciphertext $\mathsf{ct} = (x^*, y^* \otimes \mathsf{m})$.
- **DDH Instantiation:** On input a message $\mathsf{m} \in \mathbb{G}$, sample a vector $\mathsf{s} = (s_1, \ldots, s_n) \leftarrow \{0,1\}^n$. Set

$$g^* = \prod_{j=1}^{n} g_{j,s_j}, \quad h^* = \prod_{j=1}^{n} h_{j,s_j} .$$

Output the ciphertext $\mathsf{ct} = (g^*, h^* \cdot \mathsf{m})$.

By the leftover hash lemma, our "subset sum" process gives us outputs that are statistically close to uniform for arbitrary groups. This may be viewed as a generalization of the "exponentiation" operation to arbitrary groups.

Decrypt:
- **IHwPRF Construction:** On input a ciphertext $\mathsf{ct} = (\mathsf{ct}_1, \mathsf{ct}_2) \in \mathcal{X} \times \mathcal{Y}$, output

$$\mathsf{m}' = [F(k, \mathsf{ct}_1)]^{-1} \otimes \mathsf{ct}_2.$$

If $(\mathsf{ct}_1, \mathsf{ct}_2) = (x^*, y^* \otimes \mathsf{m})$, we have

$$\mathsf{m}' = [F(k, \mathsf{ct}_1)]^{-1} \otimes \mathsf{ct}_2 = (y^*)^{-1} \otimes (y^* \otimes \mathsf{m}) = \mathsf{m}.$$

- **DDH Instantiation:** On input a ciphertext $\mathsf{ct} = (\mathsf{ct}_1, \mathsf{ct}_2) \in \mathbb{G} \times \mathbb{G}$, output

$$\mathsf{m}' = \left(\mathsf{ct}_1^k\right)^{-1} \cdot \mathsf{ct}_2.$$

If $(\mathsf{ct}_1, \mathsf{ct}_2) = (g^*, h^* \cdot \mathsf{m})$, we have

$$\mathsf{m}' = \left(\mathsf{ct}_1^k\right)^{-1} \cdot \mathsf{ct}_2 = (h^*)^{-1} \cdot (h^* \cdot \mathsf{m}) = \mathsf{m}.$$

Setup:

$$F : \mathcal{K} \times \mathcal{X} \to \mathcal{Y} \qquad\qquad\rightsquigarrow\qquad (\mathbb{G}, q)$$
$$X = \{x_{j,b} \leftarrow \mathcal{X}\}_{j\in[n],b\in\{0,1\}} \qquad\rightsquigarrow\qquad G = \{g_{j,b} \leftarrow \mathbb{G}\}_{j\in[n],b\in\{0,1\}}$$
$$k \leftarrow \mathcal{K} \qquad\qquad\qquad\rightsquigarrow\qquad k \leftarrow \mathbb{Z}_q$$
$$Y = \{y_{j,b} = F(k,x_{j,b})\}_{j\in[n],b\in\{0,1\}} \quad\rightsquigarrow\quad H = \{h_{j,b} = g_{j,b}^k\}_{j\in[n],b\in\{0,1\}}$$

$$\boxed{\mathsf{sk} = k, \quad \mathsf{pk} = (X, Y)} \qquad\rightsquigarrow\qquad \boxed{\mathsf{sk} = k, \quad \mathsf{pk} = (G, H)}$$

Encrypt:

$$m \in \mathcal{Y} \qquad\qquad\rightsquigarrow\qquad m \in \mathbb{G}$$
$$s \leftarrow \{0,1\}^n \qquad\qquad\rightsquigarrow\qquad s \leftarrow \{0,1\}^n$$
$$x^* = \bigoplus_{j\in[n]} x_{j,s_j} \qquad\rightsquigarrow\qquad g^* = \prod_{j=1}^n g_{j,s_j}$$
$$y^* = \bigotimes_{j\in[n]} y_{j,s_j} \qquad\rightsquigarrow\qquad h^* = \prod_{j=1}^n h_{j,s_j}$$

$$\boxed{\mathsf{ct} = (x^*, y^* \otimes m)} \qquad\rightsquigarrow\qquad \boxed{\mathsf{ct} = (g^*, h^* \cdot m)}$$

Decrypt:

$$\mathsf{ct} = (\mathsf{ct}_1, \mathsf{ct}_2) \qquad\rightsquigarrow\qquad \mathsf{ct} = (\mathsf{ct}_1, \mathsf{ct}_2)$$

$$\boxed{m' = [F(k,\mathsf{ct}_1)]^{-1} \otimes \mathsf{ct}_2} \quad\rightsquigarrow\quad \boxed{m' = \left(\mathsf{ct}_1^k\right)^{-1} \cdot \mathsf{ct}_2}$$

Fig. 5. PKE from IHwPRF and DDH Instantiation

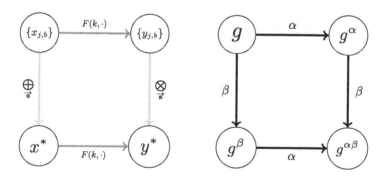

Fig. 6. Visualization of Non-Interactive Key Exchange from IHwPRF

Note that the decryption in the IHwPRF construction works even when $\mathcal{X}$ and $\mathcal{Y}$ are non-abelian groups.

We summarize the main steps in the construction of PKE from IHwPRF in Fig. 5, and compare it with the DDH-instantiation over cyclic groups of prime order. Observe that the DDH-based PKE described above is very similar to

ElGamal encryption [ElG84]. In fact, it can be viewed as a form of ElGamal encryption where we use a less efficient method to create the group elements (g, h) and (g^*, h^*): namely, in order to get a random element, we take a subset product of many public elements rather than just raising a single element to a random power.

This leads us to the following question: how far can we go if we take traditional DDH-based schemes and write them as IHwPRFs? For schemes that require two exponentiations, we could write the first exponentiation as a "subset sum", and then the second as a IHwPRF evaluation. This is essentially how our DDH-based instantiation of PKE from IHwPRF works. In what follows, we illustrate this comparison via a non-interactive key exchange protocol.

We show a non-interactive key exchange protocol from IHwPRFs in Fig. 6. For illustration, we compare it with the Diffie-Hellman key exchange protocol. In the IHwPRF setting, the (randomly sampled) "base elements" $\{x_{j,b}\}_{j \in [n], b \in \{0,1\}}$ are publicly available to both parties at the beginning of the protocol. Given the "base elements", there are two ways to arrive at the final secret y^*. The first way is to apply the IHwPRF on the "base elements", followed by applying a "subset product" in the output space of the IHwPRF. The second way is to first do a "subset sum" on the base elements, and then apply the IHwPRF. The two parties involved in the protocol each use one of these strategies. Security of the protocol follows from the weak pseudorandomness of F and one-wayness of "subset sums" in its input space, where the latter is also implied by the weak pseudorandomness of F.

Finally, the reader may observe that the protocol is secure even if the function F is an IHwUF instead of an IHwPRF, provided that both parties extract a "hardcore bit" from the secret y^* and use it as the key.[8] Similarly, one can construct a CPA-secure PKE from IHwUF by using the hardcore bit of the secret y^* to mask the message bit.

2.2 Extending the Scheme with a General Protocol

It turns out that we can do substantially more than just PKE, as an examination of the above protocol might suggest. It turns out we can take *any* one-round[9] CDH/DDH-based protocol and convert it into a (less efficient) protocol using a general IHwUF/IHwPRF. The basic idea is the following: visualize one-round CDH/DDH schemes as protocols played by two parties with the following four phases. Below is a rough description of this protocol:

- **Initialization:** Setting up the group and any random elements needed for the protocol.

---

[8] Note that the protocol assumes that the input space of the IHwUF/IHwPRF is independent of the choice of key.

[9] Informally, in our context this means a protocol that can be "played" by two parties with a simple out-and-back communication flow, along with any PPT computation the parties choose to do before, during, or after the communication.

- **Pre-evaluation:** The first party exponentiates some (or all) of the random elements from the initialization stage and sends some (or all) of these to the second player.
- **Evaluation:** The second party exponentiates some of the elements from the first player and potentially some of the elements from initialization as well. The second player potentially publishes some of these elements as well.
- **Post-evaluation:** Either party can multiply/invert/process the elements, and may publish some outputs of these.

It turns out that the vast majority of CDH/DDH-based cryptosystems fall into this archetype, and thus we can build them using an IHwUF/IHwPRF. Among other implications, this approach encompasses recent constructions such as (anonymous) IBE from CDH/DDH and a number of other works in the same vein [DG17b, DG17a, BLSV18, DGHM18, GH18, KW18, GGH18]. Although these works use many novel techniques, we show that the CDH/DDH-related portion of the constructions can be boiled down to something that fits within the above framework. The few protocols that cannot be handled involve at least three exponentiations (and cannot be rewritten as less efficient protocols with two or less exponentiations).

We can use our general protocol and the ideas around it to build many cryptosystems. In the following subsection, we outline some of the constructions that we consider interesting.

2.3 Batch Encryption from IHwUFs

In a recent work, Brakerski *et al.* [BLSV18] introduced and formalized a powerful cryptographic primitive called *batch encryption*. Roughly speaking, the basic idea of batch encryption is the following: a user encrypts a $2 \times N$ matrix of bits, and decryption *selectively* reveals only N of these bits–one in each column. For a given column, which bit is revealed depends on the value of the secret key used for decryption.

Brakerski *et al.* showed that batch encryption can be used in conjunction with garbled circuits to construct identity-based encryption (IBE).[10] In fact, when equipped with a stronger property called "blinding", batch encryption was shown to imply anonymous IBE, KDM-CPA secure PKE, and leakage resilient PKE [BLSV18]. The authors of [BLSV18] showed how to construct batch encryption from concrete assumptions, so it is natural to ask the following question: is there a generic primitive that implies batch encryption?

In this subsection, we answer this question in the affirmative by showing that IHwUFs are sufficient to construct blind batch encryption. This in turn implies that IHwUFs are sufficient to construct anonymous IBE, KDM-secure

---

[10] An equivalent cryptosystem, named as *hash encryption*, was introduced by Döttling *et al.* in [DGHM18].

PKE and leakage-resilient PKE as well.[11] We begin by defining blind batch encryption informally, and then illustrate how to construct the same from any IHwUF family. [12]

Batch Encryption. A batch encryption scheme is a public-key encryption scheme in which the key generation algorithm Gen "projects" a secret string $s \in \{0,1\}^n$ onto a corresponding hash value $h \in \{0,1\}^\ell$, such that $\ell < n$. Corresponding to this "projection" function, there should exist encryption and decryption algorithms such that:

- The encryption algorithm $\mathsf{Enc}(\mathsf{pp}, h, i, (\mathsf{m}_0, \mathsf{m}_1))$ takes as input the public parameter pp associated with the projection function, a hash $h \in \{0,1\}^\ell$, a position index $i \in [n]$ and a pair of message-bits $(\mathsf{m}_0, \mathsf{m}_1) \in \{0,1\}^2$, and outputs a ciphertext ct.
- The decryption algorithm $\mathsf{Dec}(\mathsf{pp}, s, i, \mathsf{ct})$ takes as input a ciphertext ct and a secret string s, and then recovers m_{s_i} where s_i is the value of the ith-bit of s, provided that ct was generated using $h = \mathsf{Gen}(\mathsf{pp}, s)$.

In other words, a decryptor can use the knowledge of the preimage s of a hash output string $h \in \{0,1\}^\ell$ to decrypt *exactly one* of the two encrypted messages, depending on the ith-bit of s. The security requirement is roughly that the distributions

$$\{\mathsf{pp}, s, \mathsf{Enc}(\mathsf{pp}, h = \mathsf{Gen}(\mathsf{pp}, s), i, (\mathsf{m}_{s_i}, \mathsf{m}_{1-s_i}))\}_{s \in \{0,1\}^n} \quad \text{and}$$

$$\{\mathsf{pp}, s, \mathsf{Enc}(\mathsf{pp}, h = \mathsf{Gen}(\mathsf{pp}, s), i, (\mathsf{m}_{s_i}, \mathsf{m}^*))\}_{s \in \{0,1\}^n, \mathsf{m}^* \leftarrow \{0,1\}}$$

are computationally indistinguishable. In fact, as Brakerski et al. pointed out in [BLSV18], a weaker *selective* notion of security suffices, where the adversary commits to a string $s \in \{0,1\}^n$ and an index $i \in [n]$ before the public parameter pp is published.

Note that the adaptive security guarantee implicitly requires the projection function to be collision-resistant; otherwise, a PPT adversary could distinguish an encryption of m_{1-s_i} from random with non-negligible probability simply by generating a different preimage s' of h such that $s_i' \neq s_i$.

An additional security requirement, called "blindness" was formalized with respect to batch encryption in [BLSV18]. Roughly, a batch encryption scheme is said to be blind if the ciphertext ct can be decomposed into parts $(\mathsf{ct}_1, \mathsf{ct}_2)$ such that the marginal distribution of ct_1 is independent of both the image string h and the message pair $(\mathsf{m}_0, \mathsf{m}_1)$, while the marginal distribution of ct_2 is uniform whenever the message pair $(\mathsf{m}_0, \mathsf{m}_1)$ is uniform in $\{0,1\}^2$.

---

[11] The construction of anonymous IBE requires an additional primitive - "blind garbled circuits" besides blind batch encryption. However, blind garbled circuits are implied by any one-way function, and are hence also implied by IHwUFs.

[12] We can analogously construct blind batch encryption from γ-bounded IHwUFs. For simplicity, we show the construction from an "unbounded" IHwUF here.

Projection Function from IHwUF. The first step in instantiating a batch encryption scheme is to realize the projection function. Given an IHwUF $F :$ $\mathcal{K} \times \mathcal{X} \to \mathcal{Y}$, we define $\mathsf{Gen}_{\mathrm{IHwUF}}(\mathsf{pp}, \mathbf{s})$ to output

$$h = \bigoplus_{j \in [n]} x_{j,s_j},$$

where $\{x_{j,b}\}_{j \in [n], b \in \{0,1\}}$ is a set of uniformly random elements in the input group of the IHwUF, published as part of the public parameter pp. We claim that this function is both one-way and collision resistant, provided that $n > 3 \log |\mathcal{X}|$.[13]

One-Wayness. To see that this function is one-way, consider a PPT adversary $\mathcal{A}$ that, given uniformly random group elements $\{x_{j,b}\}_{j \in [n], b \in \{0,1\}}$ and a "target" element x^*, outputs a vector $\mathbf{s} \in \{0,1\}^n$ such that

$$x^* = \bigoplus_{j \in [n]} x_{j,s_j}.$$

One can then construct a PPT algorithm $\mathcal{B}$ that on input $\{x_{j,b}, F(k, x_{j,b})\}_{j \in [n], b \in \{0,1\}}$ (where each $x_{j,b}$ is uniformly random) and a uniformly random target element x^*, invokes $\mathcal{A}$ as a subroutine on the tuple $\left\{x^*, \{x_{j,b}\}_{j \in [n], b \in \{0,1\}}\right\}$ to obtain $\mathbf{s} \in \{0,1\}^n$ and outputs

$$F(k, x^*) = \bigotimes_{j \in [n]} F(k, x_{j,s_j}),$$

which violates the weak unpredictability of the function F. We note that the reduction is valid because for $n > 3 \log |\mathcal{X}|$, the distribution of $\bigoplus_{j \in [n]} x_{j,s_j}$ is statistically indistinguishable from uniform by the leftover hash lemma [IZ89].

Collision-Resistance. To see that this function is collision-resistant, consider a PPT adversary $\mathcal{A}$ that, given uniformly random group elements $\{x_{j,b}\}_{j \in [n], b \in \{0,1\}}$, outputs $(\mathbf{s}, \mathbf{s}') \in \{0,1\}^n \times \{0,1\}^n$ such that $\mathbf{s} \neq \mathbf{s}'$ and

$$\bigoplus_{j \in [n]} x_{j,s_j} = \bigoplus_{j \in [n]} x_{j,s'_j}.$$

One can then construct a PPT algorithm $\mathcal{B}$ that on input $\{x_{j,b}, F(k, x_{j,b})\}_{j \in [n], b \in \{0,1\}}$ (where each $x_{j,b}$ is uniformly random) and a random target element x^*, uniformly guesses $i \leftarrow [n]$, resets $x_{i,0} := x^*$ and invokes $\mathcal{A}$ as a subroutine on the modified set $\{x_{j,b}\}_{j \in [n], b \in \{0,1\}}$ to obtain a collision $(\mathbf{s}, \mathbf{s}')$. If $s_i = s'_i$, it aborts. Otherwise, it exploits the homomorphism of the function F to output $F(k, x^*)$. Since the probability that $\mathbf{s}$ and $\mathbf{s}'$ differ in the ith bit is at least $1/n$, $\mathcal{B}$ breaks the weak unpredictability of F.

---

[13] We note that it is possible to use a smaller constant, but we use 3 through the whole paper for the sake of simplicity.

Encryption and Decryption. Corresponding to the projection function as described above, we realize our encryption procedure $\mathsf{Enc}_{\mathrm{IHwUF}}(\mathsf{pp}, h, i, (\mathsf{m}_0, \mathsf{m}_1))$ as follows: sample $k_0, k_1 \leftarrow \mathcal{K}$ and set the following

$$y_{j,0}^{(0)} = F(k_0, x_{j,b}), \quad y_{j,1}^{(1)} = F(k_1, x_{j,b}) \quad \text{for } j \in [n] \setminus \{i\}, b \in \{0,1\}$$
$$y_{i,0}^{(0)} = F(k_0, x_{i,0}), \quad y_{i,0}^{(1)} = \perp,$$
$$y_{i,1}^{(0)} = \perp, \quad y_{i,1}^{(1)} = F(k_1, x_{i,1}).$$

Next, mask the messages $(\mathsf{m}_0, \mathsf{m}_1) \in \{0,1\} \times \{0,1\}$ as follows:[14]

$$\mathsf{e}_0 = \mathrm{XOR}\left(\mathrm{HardCore}\left(F(k_0, h)\right), \mathsf{m}_0\right)$$
$$\mathsf{e}_1 = \mathrm{XOR}\left(\mathrm{HardCore}\left(F(k_1, h)\right), \mathsf{m}_1\right).$$

Output the ciphertext as

$$\mathsf{ct} = \left(\mathsf{ct}_1 = \left\{y_{j,b}^{(0)}, y_{j,b}^{(1)}\right\}_{j \in [n], b \in \{0,1\}}, \mathsf{ct}_2 = (\mathsf{e}_0, \mathsf{e}_1)\right).$$

Given a preimage string s, our decryption algorithm $\mathsf{Dec}_{\mathrm{IHwUF}}(\mathsf{pp}, \mathsf{s}, i, \mathsf{ct})$ now recovers m_{s_i} as

$$\mathsf{m}_{s_i} = \mathrm{XOR}\left(\mathrm{HardCore}\left(\bigotimes_{j \in [n]} y_{j,s_j}^{(s_i)}\right), \mathsf{e}_{s_i}\right).$$

Correctness follows from the homomorphic property of the function F. Observe that irrespective of the value of the bit s_i, m_{s_i} can always be recovered as the decryptor has access to $y_{i,b}^{(b)}$ for each $b \in \{0,1\}$. However, it cannot recover m_{1-s_i} since it does not have access to $y_{i,1-b}^{(b)}$ for either $b = 0$ or $b = 1$. In addition, we note that, unlike existing constructions, our construction does not require the groups $(\mathcal{X}, \oplus)$ and $(\mathcal{Y}, \otimes)$ to be abelian for correctness to hold.

Security. We now sketch our security proof. Suppose we are given an adversary $\mathcal{A}$ that breaks the security of this scheme. We construct a PPT algorithm $\mathcal{B}$ that breaks the weak unpredictability of the function F. We assume that $\mathcal{B}$ has oracle access to an IHwUF F with key k.

In our security game, $\mathcal{B}$ receives a uniformly random challenge element x^* and a bit $\mathsf{e}^* \in \{0,1\}$ such that $\mathsf{e}^* = \mathrm{HardCore}(F(k, x^*))$ (the "real" case) or e^* is a uniform bit (the "random" case). The goal of $\mathcal{B}$ is to output a bit b, such that

$$b = \begin{cases} 0 & \text{if } \mathsf{e}^* = \mathrm{HardCore}(F(k, x^*)) \\ 1 & \text{if } \mathsf{e}^* \leftarrow \{0,1\} \end{cases}$$

---

[14] We assume that each group element $y \in \mathcal{Y}$ has a deterministic hardcore bit, denoted as $\mathrm{HardCore}(y)$. If a deterministic hardcore bit is not known then we can use the Goldreich-Levin [GL89] construction.

In other words, $\mathcal{B}$ must distinguish the hardcore bit associated with the output of $F(k, x^*)$ from random (which is equivalent to constructing the entire output $F(k, x^*))$[15] using the adversary $\mathcal{A}$.

We note here that the exact value of n is typically chosen by the adversary $\mathcal{A}$ at the beginning of the game, subject to the restriction that $n > 3 \log |\mathcal{X}|$. For simplicity, we describe the interaction between $\mathcal{B}$ and $\mathcal{A}$ after the value of n has been chosen.

- The adversary $\mathcal{A}$ chooses an arbitrary preimage string $\mathbf{s} \in \{0, 1\}^n$ and an index $i \in [n]$, and provides $(\mathbf{s}, i)$ to $\mathcal{B}$.
- $\mathcal{B}$ queries the IHwUF F a total of $2n$ times, getting a tuple of the form

$$\{x_{j,b}, F(k, x_{j,b})\}_{j \in [n], b \in \{0,1\}} .$$

- $\mathcal{B}$ now resets

$$x_{i,s_i} := \left(\bigoplus_{j \in [i-1]} x_{j,s_j} \right)^{-1} \oplus x^* \oplus \left(\bigoplus_{j \in [i+1,n]} x_{j,s_j} \right)^{-1},$$

and provides $\mathsf{pp} = \{x_{j,b}\}_{j \in [n], b \in \{0,1\}}$ to $\mathcal{A}$. In other words, $\mathcal{B}$ fixes x^* to be the image of $\mathbf{s}$ under the projection function parameterized by pp.

- The adversary $\mathcal{A}$ generates $\mathbf{m}^{(0)} = \left(m_0^{(0)}, m_1^{(0)} \right)$ and $\mathbf{m}^{(1)} = \left(m_0^{(1)}, m_1^{(1)} \right)$ such that $m_{s_i}^{(0)} = m_{s_i}^{(1)}$, and sends them to $\mathcal{B}$.

- In response, $\mathcal{B}$ samples $k' \leftarrow \mathcal{K}$, and implicitly fixes $k_{s_i} := k'$ and $k_{1-s_i} := k$. It then sets the following

$$y_{j,s_j}^{(s_i)} = F(k', x_{j,s_j}), \quad y_{j,s_j}^{(1-s_i)} = F(k, x_{j,s_j}) \quad \text{for } j \in [n] \setminus \{i\}, b \in \{0, 1\},$$

$$y_{i,s_i}^{(s_i)} = F(k', x_{i,s_i}), \quad y_{i,s_i}^{(1-s_i)} = \bot,$$

$$y_{i,1-s_i}^{(s_i)} = \bot, \quad y_{i,1-s_i}^{(1-s_i)} = F(k, x_{i,1-s_i}).$$

To mask the messages, $\mathcal{B}$ sets the following

$$e_{s_i}^{(0)} = \mathrm{XOR}\left(\mathrm{HardCore}(F(k', x^*)), m_{s_i}^{(0)} \right), \quad e_{1-s_i}^{(0)} = \mathrm{XOR}\left(e^*, m_{1-s_i}^{(0)} \right),$$

$$e_{s_i}^{(1)} = \mathrm{XOR}\left(\mathrm{HardCore}(F(k', x^*)), m_{s_i}^{(1)} \right), \quad e_{1-s_i}^{(1)} = \mathrm{XOR}\left(e^*, m_{1-s_i}^{(1)} \right).$$

Finally, $\mathcal{B}$ samples $b^* \leftarrow \{0, 1\}$ and sends ct to $\mathcal{A}$ where

$$\mathsf{ct} = \left(\mathsf{ct}_1 = \left\{ y_{j,b}^{(0)}, y_{j,b}^{(1)} \right\}_{j \in [n], b \in \{0,1\}}, \mathsf{ct}_2 = \left(e_0^{(b^*)}, e_1^{(b^*)} \right) \right).$$

- $\mathcal{A}$ outputs a bit b'. If $b^* = b'$, $\mathcal{B}$ outputs 1. Otherwise it outputs 0.

Note that when $e^* = \mathrm{HardCore}(F(k, x^*))$, the challenge ciphertext is generated perfectly. On the other hand, when e^* is a uniform bit, the adversary $\mathcal{A}$ has no advantage since $m_{s_i}^{(0)} = m_{s_i}^{(1)}$ by definition. Hence, the advantage of $\mathcal{B}$ is negligibly different from the advantage of $\mathcal{A}$.

---

[15] By the Goldreich-Levin Theorem [GL89], this can be used to build an algorithm that constructs $F(k, x^*)$ with only polynomial loss in advantage.

Blindness. The aforementioned batch encryption scheme is additionally "blind". This follows from the fact that the ciphertext component ct_1 is independent of both the image string h and the message-pair (m_0, m_1). Additionally, if (m_0, m_1) is uniform in $\{0,1\}^2$, then the distribution of ct_2 is also uniform.

2.4   More Primitives

Recyclable OWFE. In a recent work, Garg and Hajiabadi [GH18] introduced a cryptographic primitive called recyclable *one-way function with encryption* (OWFE), and showed that recyclable OWFEs imply trapdoor functions (TDFs) with negligibly small inversion error. They also showed how to construct recyclable OWFE from the CDH assumption, which in turn gave the first TDF construction from the CDH assumption. In a more recent follow-up, Garg *et al.* [GGH18] introduced a strengthened version of recyclable OWFE called *smooth* recyclable OWFE, and showed how to realize the same from CDH assumption. They showed that this strengthened primitive implies TDFs with almost-perfect correctness and CCA2-secure deterministic encryption, where the CCA2-security holds with respect to plaintexts sampled from distributions with super-logarithmic min-entropy.

We show that IHwUFs imply smooth recyclable OWFE, thereby answering the question of whether this cryptosystem can be constructed from a generic primitive. This shows that IHwUFs also imply TDFs with almost-perfect correctness and CCA2-secure deterministic encryption for plaintexts sampled from distributions with super-logarithmic min-entropy. The techniques for this construction are similar to those presented for batch encryption.

Hinting PRG. A "hinting PRG" is a stronger variant of traditional PRGs introduced by Koppula and Waters in [KW18], who show that hinting PRGs can be used to generically transform any CPA-secure attribute-based encryption scheme or one-sided predicate encryption scheme into a CCA-secure counterpart. Informally, a hinting PRG takes n bits as input and outputs $n \cdot \ell$ output bits with the restriction that no PPT adversary can distinguish between $2n$ uniformly random strings and $2n$ strings such that half the strings are output by the PRG, and the remaining half are uniformly random, where the strings are arranged as a $2 \times n$ matrix as follows: in the ith column of this matrix, the top entry is pseudorandom and the bottom entry is random if the ith bit of the seed is 0; otherwise the bottom entry is pseudorandom and top entry is random.

Koppula and Waters [KW18] showed explicit constructions of hinting PRG families from the CDH and LWE assumptions. We show that any IHwUF family can be used to construct a hinting PRG, thereby answering the question of whether hinting PRGs can be constructed from a generic primitive. The techniques for our construction are also similar to those presented for batch encryption.

CRHF and More from HOWF. Informally, a HOWF is just a one-way function $f : \mathcal{X} \rightarrow \mathcal{Y}$ with the following additional properties: the input space $\mathcal{X}$ and the output space $\mathcal{Y}$ are groups with group operations $\oplus$ and $\otimes$, respectively, and

$$f(x_1 \oplus x_2) = f(x_1) \otimes f(x_2).$$

In this paper, we show that any HOWF can used to construct a collision-resistant hash function (CRHF) family that maps bit strings to elements in the output space of the HOWF. In addition, we show constructions of Schnorr signatures and chameleon hash functions from HOWFs.[16]

Richer Structures. As mentioned earlier, we can also consider richer structures than just a group homomorphism over a Minicrypt primitive. In this section, we provide more details for two of these more structured primitives, namely Ring IHwPRFs and L-composable IHwPRFs.

Ring IHwPRFs. We first informally define a Ring Input-Homomorphic weak PRF (RIHwPRF). Let $(R, +, \times)$ and $(\boxed{R}, \boxplus, \boxtimes)$ be two efficiently samplable rings such that the ring operations are efficiently computable. An RIHwPRF is a weak PRF

$$F : \mathcal{K} \times \boxed{R} \rightarrow R$$

(with input space $\boxed{R}$ and output space R) such that for every key $k \in \mathcal{K}$ the mapping $F(k, \cdot) : \boxed{R} \rightarrow R$ is a ring homomorphism from $\boxed{R}$ to R.[17]

We outline a simple construction of symmetric-key FHE from an RIHwPRF F provided that the size of output space of F is polynomial in the security parameter, i.e., $|R| \leq \text{poly}(\lambda)$. Using the generic transformation in [Rot11], one can obtain a public-key FHE from a symmetric-key FHE. The construction is as follows:

- Given an RIHwPRF $F : \mathcal{K} \times \boxed{R} \rightarrow R$, publish its description as the public parameters. To generate a secret key sample a key $k \leftarrow \mathcal{K}$.
- To encrypt a bit $\mathsf{m} \in \{0, 1\}$ under key k, sample a preimage $\mathsf{ct} \leftarrow \boxed{R}$ such that $F(k, \mathsf{ct}) = \mathsf{m}_R$ and publish ct as the ciphertext.[18] (Notice that 0_R and 1_R are the multiplicative and the additive identity elements of R, respectively.)

---

[16] Here we use "unbounded" HOWF for simplicity. We also consider "bounded" HOWFs for which only a bounded number of homomorphic operations is allowed. The notion of bounded HOWFs works for all of the applications that we consider in almost the same way that unbounded HOWFs do.

[17] It is also possible to define (bounded) RIHwPRFs similar to IHwPRFs, but we only consider unbounded homomorphism here for the sake of simplicity.

[18] Such a preimage can be efficiently sampled by weak pseudorandomness of F and the fact that the order of the ring is polynomial.

- To decrypt a ciphertext $\mathsf{ct} \in \boxed{R}$ under key k, output m' where

$$\mathsf{m}' = \begin{cases} 0 & \text{if } F(k, \boxed{r}) = 0_R \\ 1 & \text{if } F(k, \boxed{r}) = 1_R \\ \bot & \text{otherwise.} \end{cases}$$

- To evaluate a (homomorphic) $\mathsf{NAND}(\mathsf{ct}, \mathsf{ct}')$ operation, output $\boxed{1} \boxminus \mathsf{ct} \boxtimes \mathsf{ct}'$ where $\boxed{1}$ is the identity element of $\boxed{R}$ with respect to addition, and $\boxminus$ is the subtraction in the ring $\boxed{R}$.

The security of the scheme follows from a standard hybrid argument. Observe that by ring-homomorphism of F, if ct and ct' are valid ciphertexts encrypting m and m' respectively, decrypting $\boxed{1} \boxminus \mathsf{ct} \boxtimes \mathsf{ct}'$ gives $\mathsf{NAND}(\mathsf{m}, \mathsf{m}')$.

L-Composable IHwPRFs. We first describe 2-Composable IHwPRFs before generalizing to $L \geq 2$. Informally, a two-composable IHwPRF is a collection of two functions and two "composers"

$$F_1 : \mathcal{K} \times \mathcal{X}_1 \to \mathcal{Y}_1 , \quad F_2 : \mathcal{K} \times \mathcal{X}_2 \to \mathcal{Y}_2,$$
$$C_1 : \mathcal{Y}_1 \times \mathcal{X}_2 \to \mathcal{Z} , \quad C_2 : \mathcal{Y}_2 \times \mathcal{X}_1 \to \mathcal{Z}.$$

such that the functions are IHwPRFs and the composers are weak PRFs. Additionally, the following composition property holds: for every $k \in \mathcal{K}$ and for every $x_1, x_2 \in \mathcal{X}$, we have:

$$C_1\left(F_1\left(k, x_1\right), x_2\right) = C_2\left(F_2\left(k, x_2\right), x_1\right), \text{ both denoted } F_T(k, (x_1, x_2)).$$

This primitive gives us 3-party non-interactive key exchange (NIKE) in the following way: the public key includes vectors $\mathbf{x}^{(1)}$ and $\mathbf{x}^{(2)}$. Two of the parties generate secret subsets $\mathbf{s}_1$ and $\mathbf{s}_2$, and publish the group elements

$$\bigoplus_{j \in [n]} x^{(1)}_{j, s_{1,j}}, \quad \bigoplus_{j \in [n]} x^{(2)}_{j, s_{2,j}},$$

respectively. The 3rd party generates a secret key k and publishes $F_1(k, \mathbf{x}^{(1)})$ and $F_2(k, \mathbf{x}^{(2)})$. Each party computes the shared key:

$$F_T\left(k, \left(\bigoplus_{j \in [n]} x^{(1)}_{j, s_{1,j}}, \bigoplus_{j \in [n]} x^{(2)}_{j, s_{2,j}}\right)\right),$$

which can be computed from any party's secret and the other parties' outputs, using the composition property and input homomorphism of F_1 and F_2. Security follows by the weak PRF properties and LHL.

We argue that 2-composable IHwPRFs are seemingly much weaker than bilinear pairing groups. Specifically, we argue that the general abstraction of dual system groups (DSG [CGW15]) is hard to capture in the 2-Composable IHwPRF setting due to the following limitations:

1. DSG seems to require properties that translate to the requirement of key homomorphism in the 2-composable IHwPRF setting.
2. DSG also requires algebraic interaction on both of the coordinates. Realizing this in the IHwPRF setting forces both the coordinate domains $\mathcal{X}_1$ and $\mathcal{X}_2$ to be *ring homomorphic* on a single ring, where all the algebra can take place.

The currently known constructions of rich ABEs like fuzzy IBEs [SW05], spatial encryption [BH08] and monotone span program ABEs [GPSW06] from bilinear groups all require at least one of the properties just described. Since the only instantiation of 2-composable IHwPRFs we know of are bilinear groups, it seems difficult to achieve these rich ABEs without restricting 2-composable IHwPRFs to almost traditional bilinear groups.

Thus we see a seeming separation in the amount of structure that we need for 3-party NIKE and simple IBE (in RO) from that seemingly necessary for NIZKs (without RO) and rich ABEs. This poses a tantalizing question: *Can we construct a 3-party NIKE protocol from a weaker primitive than bilinear pairing groups?* In other words, can we achieve the structure of 2-composability from concrete assumptions, e.g., lattice-based assumptions, that do not naturally imply bilinear pairings?

Generalizing to L $\geq$ 2. In the general setting, we consider L inner IHwPRFs F_i and L different composers which satisfy an analogous composition property as the 2-composable setting. By a straightforward generalization, we get an $(L+1)$-party non-interactive key exchange from an L-Composable IHwPRF, which is not known from any $(<L)$-Composable IHwPRFs. We also do not know how to construct such a protocol from any hard $(<L)$-multilinear group. We still observe an analogous seeming separation in the amount of structure that we need for multi-party non-interactive key exchange from that seemingly necessary for circuit ABEs and iOs. The corresponding open question is whether we can build the former from weaker primitives that may lack the structure needed for the latter.

2.5   Conclusion and Future Work

In this paper, we presented a framework to build many cryptosystems from Minicrypt primitives with structure. Our framework allows us to categorize many cryptosystems based on which structured Minicrypt primitive implies them, and potentially makes showing the *existence* of many cryptosystems from novel assumptions substantially easier in the future. In addition, some of our constructions are novel in their own right. Although our framework yields new constructions from less studied assumptions, the main focus of this work is to investigate what kind of structure, when added to simple and natural Minicrypt primitives, implies advanced cryptosystems like IBE. Hence, we are not explicitly examining new constructions from a mainstream assumption. We believe that our work opens up a substantial number of questions, some of which we mention here.

Primitives from Weaker Assumptions. A pertinent open question is: can we build some of the Cryptomania primitives discussed in this paper from weaker Minicrypt primitives with structure. For instance, can we build PKE from HOWFs (which would imply PKE from discrete log)? Can we build PIR/lossy TDFs from IHwUFs (which would imply the first PIR/lossy TDFs from CDH)? Is it possible to build round-optimal OT and MPC in the plain model from IHwUFs/IHwPRFs?

More Primitives. While we constructed many popularly used Cryptomania primitives from our framework, we could not encompass many others. These (non-exhaustively) include primitives implied by bilinear pairings such as NIZK, unique signatures, VRFs, ABE and PE, and primitives known from specific assumptions such as worst-case smooth hash proof systems, KDM-CCA secure PKE and dual-mode cryptosystems. It is open to construct one or more of these primitives from simple Minicrypt primitives with structure.

New Assumptions. One of the nicest aspects of our work is the implications for new assumptions. If a new assumption implies one of the Minicrypt primitives with structure discussed in this paper, then it immediately implies a whole host of cryptographic primitives. We leave it open to build HOWFs/IHwUFs/IHwPRFs from new concrete assumptions, which in conjunction with our framework would allow building a large number of Cryptomania primitives from such assumptions.

"Continents" of Cryptomania. We leave it open to explore if there are even weaker forms of structure that, when endowed upon Minicrypt primitives, lead to interesting implications in Cryptomania. It is also interesting to explore non-trivial separations between these structured primitives, e.g., between HOWFs and IHwUFs. Such separations would potentially allow us to divide the world of Cryptomania into many "continents" of primitives, where each "continent" is entirely implied by some simple Minicrypt primitive with structure.

References

[AD97] Ajtai, M., Dwork, C.: A public-key cryptosystem with worst-case/average-case equivalence. In: 29th ACM STOC, pp. 284–293. ACM Press, May 1997

[Ajt96] Ajtai, M.: Generating hard instances of lattice problems (extended abstract). In: 28th ACM STOC, pp. 99–108. ACM Press, May 1996

[AMG07] Aguilar-Melchor, C., Gaborit, P.: A lattice-based computationally-efficient private information retrieval protocol. In: Western European Workshop on Research in Cryptology. Citeseer (2007)

[AS15] Asharov, G., Segev, G.: Limits on the power of indistinguishability obfuscation and functional encryption. In: Guruswami, V. (ed.) 56th FOCS, pp. 191–209. IEEE Computer Society Press, October 2015

[Bar17] Barak, B.: The complexity of public-key cryptography. Cryptology ePrint Archive, Report 2017/365 (2017). https://eprint.iacr.org/2017/365

[BBF13] Baecher, P., Brzuska, C., Fischlin, M.: Notions of black-box reductions, revisited. In: Sako, K., Sarkar, P. (eds.) ASIACRYPT 2013, Part I. LNCS, vol. 8269, pp. 296–315. Springer, Heidelberg (2013). https://doi.org/10.1007/978-3-642-42033-7_16

[BDRV18] Berman, I., Degwekar, A., Rothblum, R.D., Vasudevan, P.N.: From laconic zero-knowledge to public-key cryptography - extended abstract. In: Shacham, H., Boldyreva, A. (eds.) CRYPTO 2018, Part III. LNCS, vol. 10993, pp. 674–697. Springer, Cham (2018). https://doi.org/10.1007/978-3-319-96878-0_23

[BDV17] Bitansky, N., Degwekar, A., Vaikuntanathan, V.: Structure vs. hardness through the obfuscation lens. In: Katz, J., Shacham, H. (eds.) CRYPTO 2017, Part I. LNCS, vol. 10401, pp. 696–723. Springer, Cham (2017). https://doi.org/10.1007/978-3-319-63688-7_23

[BGI+01] Barak, B., Goldreich, O., Impagliazzo, R., Rudich, S., Sahai, A., Vadhan, S.P., Yang, K.: On the (im)possibility of obfuscating programs. In: Kilian, J. (ed.) CRYPTO 2001. LNCS, vol. 2139, pp. 1–18. Springer, Heidelberg (2001). https://doi.org/10.1007/3-540-44647-8_1

[BH08] Boneh, D., Hamburg, M.: Generalized identity based and broadcast encryption schemes. In: Pieprzyk, J. (ed.) ASIACRYPT 2008. LNCS, vol. 5350, pp. 455–470. Springer, Heidelberg (2008). https://doi.org/10.1007/978-3-540-89255-7_28

[BHY09] Bellare, M., Hofheinz, D., Yilek, S.: Possibility and impossibility results for encryption and commitment secure under selective opening. In: Joux, A. (ed.) EUROCRYPT 2009. LNCS, vol. 5479, pp. 1–35. Springer, Heidelberg (2009). https://doi.org/10.1007/978-3-642-01001-9_1

[BLMR13] Boneh, D., Lewi, K., Montgomery, H., Raghunathan, A.: Key homomorphic PRFs and their applications. In: Canetti, R., Garay, J.A. (eds.) CRYPTO 2013, Part I. LNCS, vol. 8042, pp. 410–428. Springer, Heidelberg (2013). https://doi.org/10.1007/978-3-642-40041-4_23

[BLSV18] Brakerski, Z., Lombardi, A., Segev, G., Vaikuntanathan, V.: Anonymous IBE, leakage resilience and circular security from new assumptions. In: Nielsen, J.B., Rijmen, V. (eds.) EUROCRYPT 2018, Part I. LNCS, vol. 10820, pp. 535–564. Springer, Cham (2018). https://doi.org/10.1007/978-3-319-78381-9_20

[BM82] Blum, M., Micali, S.: How to generate cryptographically strong sequences of pseudo random bits. In: 23rd FOCS, pp. 112–117. IEEE Computer Society Press, November 1982

[BR17] Bogdanov, A., Rosen, A.: Pseudorandom functions: three decades later. Cryptology ePrint Archive, Report 2017/652 (2017). https://eprint.iacr.org/2017/652

[BSW11] Boneh, D., Sahai, A., Waters, B.: Functional encryption: definitions and challenges. In: Ishai, Y. (ed.) TCC 2011. LNCS, vol. 6597, pp. 253–273. Springer, Heidelberg (2011). https://doi.org/10.1007/978-3-642-19571-6_16

[CGW15] Chen, J., Gay, R., Wee, H.: Improved dual system ABE in prime-order groups via predicate encodings. In: Oswald, E., Fischlin, M. (eds.) EUROCRYPT 2015, Part II. LNCS, vol. 9057, pp. 595–624. Springer, Heidelberg (2015). https://doi.org/10.1007/978-3-662-46803-6_20

[DG17a] Döttling, N., Garg, S.: From selective IBE to full IBE and selective HIBE. In: Kalai, Y., Reyzin, L. (eds.) TCC 2017, Part I. LNCS, vol. 10677, pp. 372–408. Springer, Cham (2017). https://doi.org/10.1007/978-3-319-70500-2_13

[DG17b] Döttling, N., Garg, S.: Identity-based encryption from the Diffie-Hellman assumption. In: Katz, J., Shacham, H. (eds.) CRYPTO 2017, Part I. LNCS, vol. 10401, pp. 537–569. Springer, Cham (2017). https://doi.org/10.1007/978-3-319-63688-7_18

[DGHM18] Döttling, N., Garg, S., Hajiabadi, M., Masny, D.: New constructions of identity-based and key-dependent message secure encryption schemes. In: Abdalla, M., Dahab, R. (eds.) PKC 2018, Part I. LNCS, vol. 10769, pp. 3–31. Springer, Cham (2018). https://doi.org/10.1007/978-3-319-76578-5_1

[DH76] Diffie, W., Hellman, M.E.: New directions in cryptography. IEEE Trans. Inf. Theor. **22**(6), 644–654 (1976)

[DHS15] Derler, D., Hanser, C., Slamanig, D.: Revisiting cryptographic accumulators, additional properties and relations to other primitives. In: Nyberg, K. (ed.) CT-RSA 2015. LNCS, vol. 9048, pp. 127–144. Springer, Cham (2015). https://doi.org/10.1007/978-3-319-16715-2_7

[ElG84] ElGamal, T.: A public key cryptosystem and a signature scheme based on discrete logarithms. In: Blakley, G.R., Chaum, D. (eds.) CRYPTO 1984. LNCS, vol. 196, pp. 10–18. Springer, Heidelberg (1985). https://doi.org/10.1007/3-540-39568-7_2

[FH18] Fischlin, M., Harasser, P.: Invisible sanitizable signatures and public-key encryption are equivalent. In: Preneel, B., Vercauteren, F. (eds.) ACNS 2018. LNCS, vol. 10892, pp. 202–220. Springer, Cham (2018). https://doi.org/10.1007/978-3-319-93387-0_11

[Fis12] Fischlin, M.: Black-box reductions and separations in cryptography. In: Mitrokotsa, A., Vaudenay, S. (eds.) AFRICACRYPT 2012. LNCS, vol. 7374, pp. 413–422. Springer, Heidelberg (2012). https://doi.org/10.1007/978-3-642-31410-0_26

[FMV18] Friolo, D., Masny, D., Venturi, D.: Secure multi-party computation from strongly uniform key agreement. Cryptology ePrint Archive, Report 2018/473 (2018). http://eprint.iacr.org/

[Gen09] Gentry, C.: Fully homomorphic encryption using ideal lattices. In: Mitzenmacher, M. (ed.) 41st ACM STOC, pp. 169–178. ACM Press, May/June (2009)

[GGH13a] Garg, S., Gentry, C., Halevi, S.: Candidate multilinear maps from ideal lattices. In: Johansson, T., Nguyen, P.Q. (eds.) EUROCRYPT 2013. LNCS, vol. 7881, pp. 1–17. Springer, Heidelberg (2013). https://doi.org/10.1007/978-3-642-38348-9_1

[GGH+13b] Garg, S., Gentry, C., Halevi, S., Raykova, M., Sahai, A., Waters, B.: Candidate indistinguishability obfuscation and functional encryption for all circuits. In: 54th FOCS, pp. 40–49. IEEE Computer Society Press, October 2013

[GGH18] Garg, S., Gay, R., Hajiabadi, M.: New techniques for efficient trapdoor functions and applications. Cryptology ePrint Archive, Report 2018/872 (2018). http://eprint.iacr.org/

[GGM84] Goldreich, O., Goldwasser, S., Micali, S.: How to construct random functions (extended abstract). In: 25th FOCS, pp. 464–479. IEEE Computer Society Press, October 1984

[GH18]   Garg, S., Hajiabadi, M.: Trapdoor Functions from the computational Diffie-Hellman assumption. In: Shacham, H., Boldyreva, A. (eds.) CRYPTO 2018, Part II. LNCS, vol. 10992, pp. 362–391. Springer, Cham (2018). https://doi.org/10.1007/978-3-319-96881-0_13

[GHMM18]   Garg, S., Hajiabadi, M., Mahmoody, M., Mohammed, A.: Limits on the power of garbling techniques for public-key encryption. In: Shacham, H., Boldyreva, A. (eds.) CRYPTO 2018, Part III. LNCS, vol. 10993, pp. 335–364. Springer, Cham (2018). https://doi.org/10.1007/978-3-319-96878-0_12

[GL89]   Goldreich, O., Levin, L.A.: A hard-core predicate for all one-way functions. In: 21st ACM STOC, pp. 25–32. ACM Press, May 1989

[GPR16]   Goyal, V., Pandey, O., Richelson, S.: Textbook non-malleable commitments. In: STOC (2016)

[GPSW06]   Goyal, V., Pandey, O., Sahai, A., Waters, B.: Attribute-based encryption for fine-grained access control of encrypted data. In: Juels, A., Wright, R.N., Vimercati, S. (ed.) ACM CCS 2006, pp. 89–98. ACM Press, October/November 2006. Available as Cryptology ePrint Archive Report 2006/309

[GPSZ17]   Garg, S., Pandey, O., Srinivasan, A., Zhandry, M.: Breaking the subexponential barrier in obfustopia. In: Coron, J.-S., Nielsen, J.B. (eds.) EUROCRYPT 2017, Part III. LNCS, vol. 10212, pp. 156–181. Springer, Cham (2017). https://doi.org/10.1007/978-3-319-56617-7_6

[GPV08]   Gentry, C., Peikert, C., Vaikuntanathan, V.: Trapdoors for hard lattices and new cryptographic constructions. In: Ladner, R.E., Dwork, C. (eds.) 40th ACM STOC, pp. 197–206. ACM Press, May 2008

[GW11]   Gentry, C., Wichs, D.: Separating succinct non-interactive arguments from all falsifiable assumptions. In: Fortnow, L., Vadhan, S.P. (eds.) 43rd ACM STOC, pp. 99–108. ACM Press, June 2011

[HHRS07]   Haitner, I., Hoch, J.J., Reingold, O., Segev, G.: Finding collisions in interactive protocols - a tight lower bound on the round complexity of statistically-hiding commitments. In: 48th FOCS, pp. 669–679. IEEE Computer Society Press, October 2007

[HILL99]   Håstad, J., Impagliazzo, R., Levin, L.A., Luby, M.: A pseudorandom generator from any one-way function. SIAM J. Comput. **28**(4), 1364–1396 (1999)

[HKS16]   Hajiabadi, M., Kapron, B.M., Srinivasan, V.: On generic constructions of circularly-secure, leakage-resilient public-key encryption schemes. In: Cheng, C.-M., Chung, K.-M., Persiano, G., Yang, B.-Y. (eds.) PKC 2016, Part II. LNCS, vol. 9615, pp. 129–158. Springer, Heidelberg (2016). https://doi.org/10.1007/978-3-662-49387-8_6

[HO12]   Hemenway, B., Ostrovsky, R.: On homomorphic encryption and chosen-ciphertext security. In: Fischlin, M., Buchmann, J., Manulis, M. (eds.) PKC 2012. LNCS, vol. 7293, pp. 52–65. Springer, Heidelberg (2012). https://doi.org/10.1007/978-3-642-30057-8_4

[IKLP06]   Ishai, Y., Kushilevitz, E., Lindell, Y., Petrank, E.: Black-box constructions for secure computation. In: Kleinberg, J.M. (ed.) 38th ACM STOC, pp. 99–108. ACM Press, May 2006

[IKO05]   Ishai, Y., Kushilevitz, E., Ostrovsky, R.: Sufficient conditions for collision-resistant hashing. In: Kilian, J. (ed.) TCC 2005. LNCS, vol. 3378, pp. 445–456. Springer, Heidelberg (2005). https://doi.org/10.1007/978-3-540-30576-7_24

[Imp95] Impagliazzo, R.: A personal view of average-case complexity. In: Tenth Annual IEEE Conference on Proceedings of Structure in Complexity Theory, pp. 134–147, June 1995. ISSN 1063-6870

[IZ89] Impagliazzo, R., Zuckerman, D.: How to recycle random bits. In: 30th FOCS, pp. 248–253. IEEE Computer Society Press, October/November 1989

[KO97] Kushilevitz, E., Ostrovsky, R.: Replication is NOT needed: SINGLE database, computationally-private information retrieval. In: FOCS, pp. 364–373 (1997)

[KW18] Koppula, V., Waters, B.: Realizing chosen ciphertext security generically in attribute-based encryption and predicate encryption. Cryptology ePrint Archive, Report 2018/847 (2018). http://eprint.iacr.org/

[MM16] Mahmoody, M., Mohammed, A.: On the power of hierarchical identity-based encryption. In: Fischlin, M., Coron, J.-S. (eds.) EUROCRYPT 2016, Part II. LNCS, vol. 9666, pp. 243–272. Springer, Heidelberg (2016). https://doi.org/10.1007/978-3-662-49896-5_9

[MMN+16] Mahmoody, M., Mohammed, A., Nematihaji, S., Pass, R., Shelat, A.: Lower bounds on assumptions behind indistinguishability obfuscation. In: Kushilevitz, E., Malkin, T. (eds.) TCC 2016, Part I. LNCS, vol. 9562, pp. 49–66. Springer, Heidelberg (2016). https://doi.org/10.1007/978-3-662-49096-9_3

[OSV15] Ostrovsky, R., Scafuro, A., Venkitasubramanian, M.: Resettably sound zero-knowledge arguments from OWFs - the (semi) black-box way. In: Dodis, Y., Nielsen, J.B. (eds.) TCC 2015, Part I. LNCS, vol. 9014, pp. 345–374. Springer, Heidelberg (2015). https://doi.org/10.1007/978-3-662-46494-6_15

[PS08] Pietrzak, K., Sjödin, J.: Weak pseudorandom functions in Minicrypt. In: Aceto, L., Damgård, I., Goldberg, L.A., Halldórsson, M.M., Ingólfsdóttir, A., Walukiewicz, I. (eds.) ICALP 2008, Part II. LNCS, vol. 5126, pp. 423–436. Springer, Heidelberg (2008). https://doi.org/10.1007/978-3-540-70583-3_35

[PW08] Peikert, C., Waters, B.: Lossy trapdoor functions and their applications. In: Ladner, R.E., Dwork, C. (eds.) 40th ACM STOC, pp. 187–196. ACM Press, May 2008

[Reg05] Regev, O.: On lattices, learning with errors, random linear codes, and cryptography. In: Gabow, H.N., Fagin, R. (eds.) 37th ACM STOC, pp. 84–93. ACM Press, May 2005

[Rom90] Rompel, J.: One-way functions are necessary and sufficient for secure signatures. In: 22nd ACM STOC, pp. 387–394. ACM Press, May 1990

[Rot11] Rothblum, R.: Homomorphic encryption: from private-key to public-key. In: Ishai, Y. (ed.) TCC 2011. LNCS, vol. 6597, pp. 219–234. Springer, Heidelberg (2011). https://doi.org/10.1007/978-3-642-19571-6_14

[RSA78] Rivest, R.L., Shamir, A., Adleman, L.: A method for obtaining digital signatures and public-key cryptosystems. Commun. ACM **21**(2), 120–126 (1978)

[RTV04] Reingold, O., Trevisan, L., Vadhan, S.: Notions of reducibility between cryptographic primitives. In: Naor, M. (ed.) TCC 2004. LNCS, vol. 2951, pp. 1–20. Springer, Heidelberg (2004). https://doi.org/10.1007/978-3-540-24638-1_1

[Sha84] Shamir, A.: Identity-based cryptosystems and signature schemes. In: Blakley, G.R., Chaum, D. (eds.) CRYPTO 1984. LNCS, vol. 196, pp. 47–53. Springer, Heidelberg (1985). https://doi.org/10.1007/3-540-39568-7_5

[SW05] Sahai, A., Waters, B.R.: Fuzzy identity-based encryption. In: Cramer, R. (ed.) EUROCRYPT 2005. LNCS, vol. 3494, pp. 457–473. Springer, Heidelberg (2005). https://doi.org/10.1007/11426639_27

Standards

Attacks only Get Better:
How to Break FF3 on Large Domains

Viet Tung Hoang[1(✉)], David Miller[1], and Ni Trieu[2]

[1] Department of Computer Science, Florida State University, Tallahassee, USA
tvhoang@cs.fsu.edu
[2] Department of Computer Science, Oregon State University, Corvallis, USA

Abstract. We improve the attack of Durak and Vaudenay (CRYPTO'17) on NIST Format-Preserving Encryption standard FF3, reducing the running time from $O(N^5)$ to $O(N^{17/6})$ for domain $\mathbb{Z}_N \times \mathbb{Z}_N$. Concretely, DV's attack needs about 2^{50} operations to recover encrypted 6-digit PINs, whereas ours only spends about 2^{30} operations. In realizing this goal, we provide a pedagogical example of how to use distinguishing attacks to speed up slide attacks. In addition, we improve the running time of DV's known-plaintext attack on 4-round Feistel of domain $\mathbb{Z}_N \times \mathbb{Z}_N$ from $O(N^3)$ time to just $O(N^{5/3})$ time. We also generalize our attacks to a general domain $\mathbb{Z}_M \times \mathbb{Z}_N$, allowing one to recover encrypted SSNs using about 2^{50} operations. Finally, we provide some proof-of-concept implementations to empirically validate our results.

Keywords: Format-Preserving Encryption · Attacks

1 Introduction

Format-Preserving Encryption (FPE) [6,12] is a form of deterministic symmetric encryption mechanism that preserves the *format* of plaintexts. For example, encrypting a 16-digit credit-card number under FPE would result in a 16-digit number, and encrypting a valid SSN would produce a ciphertext of nine decimal digits. FPE is widely used in practice by several companies, such as HPE Voltage, Verifone, Protegrity, Ingenico, to encrypt credit-card numbers and protect legacy databases. Recent research [4,15,20] however show that existing FPE standards FF1 and FF3 (NIST SP 800-38G, ANSI ASC X9.124) are somewhat vulnerable in small domains. The most damaging attack, due to Durak and Vaudenay (DV) [15], can recover the entire codebook of FF3 using $O(N^5)$ expected time, for domain $\mathbb{Z}_N \times \mathbb{Z}_N$.

Still, the attacks above are feasible only if the domain size is small; their cost becomes prohibitive for moderate and large domains. For example, for domain $\mathbb{Z}_{10}^6$ (namely encrypting 6-digit PINs), DV's attack would use about 2^{50} operations. In this paper, we improve DV's attack to break FF3 on large domains. Our attack can reduce the cost of breaking FF3 on domain $\mathbb{Z}_N \times \mathbb{Z}_N$ to $O(N^{17/6})$

© International Association for Cryptologic Research 2019
Y. Ishai and V. Rijmen (Eds.): EUROCRYPT 2019, LNCS 11477, pp. 85–116, 2019.
https://doi.org/10.1007/978-3-030-17656-3_4

Table 1. Our attack versus DV's. The first column indicates the values of N in the domain $\mathbb{Z}_N \times \mathbb{Z}_N$. The second column and third column show the number of queries in our attack and that of DV respectively; in both attacks, the queries are made over two tweaks. The fourth and fifth columns show our recovery rate and that of DV respectively, and the fifth and sixth columns show our time and DV's time respectively.

N	Our queries	DV's queries	Our rate	DV's rate	Our time	DV's time
128	16,384	17,388	39%	56.85%	2^{20}	2^{35}
256	52,012	55,176	50%	55.9%	2^{23}	2^{40}
512	165,140	175,164	33%	77.4%	2^{26}	2^{45}

expected time, meaning that it will need about 2^{30} operations to break FF3 of the domain $\mathbb{Z}_{10}^6$ above. Achieving this efficiency involves an elegant paradigm of combining distinguishing attacks with slide attacks [10,11], and improved cryptanalyses of 4-round Feistel. We give rigorous analyses to justify the advantage of our attack; proofs omitted due to lack of space appear in the full version of this paper. We also provide proof-of-concept implementations in Sect. 5 that empirically confirm our analyses.

We note that our attack essentially performs the same queries as DV's, and thus the two attacks have the same scenario and asymptotic data/space complexity $\Theta(N^{11/6})$ for domain $\mathbb{Z}_N \times \mathbb{Z}_N$. However, DV use more aggressive choices of the parameters, and thus our attack is concretely better in both data and space complexity, albeit at the cost of lower recovery rate. A concrete comparison of the two attacks are given in Table 1. Still, one can improve the recovery rate by relaunching our attack with different tweaks. For example, for domain $\mathbb{Z}_{128} \times \mathbb{Z}_{128}$, if one relaunches our attack another time, the recovery rate would become $1 - (1 - 0.39)^2 \approx 62\%$. See Sect. 3.3 for further details.

EXISTING CRYPTANALYSIS. Let us begin by reviewing prior attacks on the standards FF1 and FF3. Bellare, Hoang, and Tessaro (BHT) [4] give the first attack on these schemes, showing that one can fully recover a target message using $O(N^6 \log(N))$ pairs of plaintext/ciphertext, on domain $\mathbb{Z}_N \times \mathbb{Z}_N$. Their attack however requires that a designated, partially known message must have the same right half as the target, but it is unclear how one could mount such a correlation in practice. Hoang, Tessaro, and Trieu (HTT) [20] subsequently improve BHT's attack, requiring no correlation between the known messages and the target. Even better, they can reuse the known plaintext/ciphertext pairs to attack multiple targets, thus reduce the *amortized* cost to $O(N^5 \log^2(N))$ pairs per target. Both attacks above apply to a *generic* Feistel-based FPE, meaning that they break both FF1 and FF3, and the only way to thwart them is to increase the round count of the underlying Feistel networks.

In a different direction, Durak and Vaudenay (DV) [15] give a dedicated attack on FF3, exploiting a bug in its design of round functions. They show that on domain $\mathbb{Z}_N \times \mathbb{Z}_N$, one can recover the entire codebook of FF3 using $O(N^{11/6})$

pairs of chosen plaintext/ciphertext, within $O(N^5)$ expected running time. We stress that DV's attack does not apply to FF1, and it can be fixed without hurting performance by restricting the tweak space, as DV already suggested.

In response to DV's attack, NIST has temporarily suspended the use of FF3, whereas a draft update of the ANSI ASC X9.124 standard additionally recommends using double encryption on small domains to cope with the other attacks.

A BIRD'S-EYE VIEW ON DV'S ATTACK. We now briefly sketch a blueprint of DV's attack. Recall that in the balanced setting, the encryption scheme of FF3 is simply a tweakable blockcipher $\mathsf{F.E} : \mathsf{F.Keys} \times \mathsf{F.Twk} \times (\mathbb{Z}_N \times \mathbb{Z}_N) \to (\mathbb{Z}_N \times \mathbb{Z}_N)$ that is based on an 8-round balanced Feistel network. Due to a bug in the round functions of FF3, one can find two tweaks T and T^* such that $\mathsf{F.E}(K, T, \cdot)$ is the cascade $g(f(\cdot))$ of two 4-round Feistel networks f and g, whereas $\mathsf{F.E}(K, T^*, \cdot) = f(g(\cdot))$. Then, by mounting a slide attack using $O(N^{11/6})$ encryption queries, we obtain $O(N^2)$ instances, each of $O(N^{5/3})$ pairs of plaintext/ciphertext for f and also $O(N^{5/3})$ pairs for g. However, often just one of those instances provides the correct ciphertexts under f or g; in the remaining instances, the ciphertexts are random strings, independent of the plaintexts. DV resolve this by developing a codebook-recovery attack on 4-round Feistel networks using $O(N^3)$ expected running time. They then try this attack on every instance, using totally $O(N^5)$ expected time.

CONTRIBUTION: ELIMINATING FALSE INSTANCES. To improve the running time of DV's attack, we observe that it is an overkill to use an expensive codebook-recovery attack on false instances. A better solution is to find a cheap test to tell whether an instance is true or false, and then use the codebook-recovery attack on the true instances. A natural choice for such a test is a distinguishing attack on 4-round Feistel. However, the requirement here is a lot more stringent. To eliminate most of the random instances, our distinguishing attack should output 1 with probability about $1/N$ if it is given a false instance. To ensure that we will *not* incorrectly eliminate all true instances, the distinguishing attack should output 1 with high probability, say $1/2$, if it is given a true instance.

Our starting point is Patarin's distinguishing attack on 4-round Feistel [25],[1] which uses $O(\sqrt{N})$ pairs of plaintext/ciphertext. However, using this attack for our purpose runs into two obstacles. First, Patarin's asymptotic analysis is insufficient to pinpoint the hidden constant in the Big-Oh. Next, Patarin's attack fails to meet the requirement above, as given a false instance, the attack outputs 1 with constant probability.

Given the issues above, we instead design a new distinguishing attack, Left-Half Differential (LHD), such that (1) in the ideal world, it returns 1 with probability at most $\frac{1}{\sqrt{N}}$, and (2) in the real world, it returns 1 with probability at least $1 - \frac{1}{8\sqrt{N}} - \frac{10}{N} - \frac{1}{N^{3/4}}$. The LHD attack uses $O(N^{5/6})$ pairs of plaintext/ciphertext, and runs in $O(N^{5/6})$ time. Our analyses are generalized enough to include Patarin's

---

[1] While Patarin's attack is given for classic Feistel (meaning that $N = 2^n$, and the underlying operator is xor), generalizing it to cover FF3 setting is straightforward.

Table 2. A list of attacks on generic 4-round Feistel of domain $\mathbb{Z}_N \times \mathbb{Z}_N$. While the distinguishing attack was discovered by Patarin [24] and independently by Aiello and Venkatesan [1], the analyses in those papers are asymptotic. Our paper gives the first concrete treatment for this attack.

Type	Power	Source	Data	Time
Known-plaintext	Distinguishing	[1,24] Here	$O(\sqrt{N})$	$O(\sqrt{N})$
Known-plaintext	Full recovery	[15]	$O(N^{5/3})$	$O(N^3)$
Known-plaintext	Full recovery	Here	$O(N^{5/3})$	$O(N^{5/3})$
Chosen plaintext & ciphertext	Full recovery	[9]	$O(N^{3/2})$	$O(N^{3/2})$

attack as a special case. As a result, we can show that for $N \geq 2^{16}$, if one uses $\lceil 7 \cdot \sqrt{N} \rceil$ pairs of plaintext/ciphertext then Patarin's attack achieves advantage at least $1/2$.

In our test, we run LHD *twice*, first on the plaintext/ciphertext pairs of f, and then on those of g. Thus given a false instance, the chance that we fail to eliminate it is at most $\frac{1}{N}$, whereas given a true instance, the chance that we accept it is at least $\left(1 - \frac{1}{8\sqrt{N}} - \frac{10}{N} - \frac{1}{N^{3/4}}\right)^2$. Even better, our experiments indicate that in practice our test is nearly perfect, meaning that empirically, we never miss a true instance, and eliminate almost all false instances.

We note that while the idea of using distinguishing attacks to eliminate false instances in slide attacks was already known in the literature [2], to the best of our knowledge, nobody has ever explored this direction. Our analyses of FF3 thus provide a pedagogical example of this paradigm.

CONTRIBUTION: A BETTER ATTACK ON 4-ROUND FEISTEL. Thanks to the LHD tests above, we are now left with $O(N)$ false instances and a few true instances. If one uses DV's codebook recovery attack on 4-round Feistel, one would end up with $O(N^4)$ expected time, which is still very expensive. The core part of DV's attack needs to find all directed 3-cycles of zero weight in a (random) directed graph $\mathcal{G} = (V, E)$. DV's approach is to enumerate all directed 3-cycles via some sparse matrix multiplications, and then pick those of zero weight, spending $O(|V| \cdot |E|)$ time. We instead give an elementary algorithm that uses $O(|V| + |E|)$ expected time. In addition, DV's attack relies on a conjecture of Feistel networks. They however can only empirically verify this conjecture for $N \in \{2, 2^2, \ldots, 2^9\}$. In this work, we resolve this conjecture, solving an open problem posed by DV.

Our algorithm above leads to the best known-plaintext attack to 4-round Feistel in the literature, using $O(N^{5/3})$ data and time complexity. A prior work by Biryukov, Leurent, and Perrin [9] is slightly better, recovering the codebook within $O(N^{3/2})$ data and time, but this attack requires chosen plaintexts and ciphertexts. A comparison of the attacks on 4-round Feistel is listed in Table 2.

OTHER CONTRIBUTIONS. We also generalize our FF3 attack to unbalanced settings, for a general domain $\mathbb{Z}_M \times \mathbb{Z}_N$, with $M \geq N \geq 64$, so that we can recover, say encrypted SSNs. The asymmetry $M \geq N$ however requires some care in the extension of the attack on 4-round Feistel. In particular, due to the symmetry of 4-round Feistel, given plaintext/ciphertext pairs $(M_1, C_1), \ldots, (M_p, C_p)$, one can view $M_1, \ldots, M_p$ as the "ciphertexts" of $C_1, \ldots, C_p$ under an inverse 4-round Feistel, leading to a dual attack. The two attacks yield no difference in the balanced setting $M = N$, but if $M \gg N$, we find that the dual attack provides a superior recovery rate. We also introduce some tricks that substantially improve both the data complexity and the recovery rate.

On the other hand, there are often some gaps between the choices of the parameters according to DV's analyses, and what their experiments suggest. Even worse, the performance of their attacks is highly sensitive: in some experiments, if they triple the number of plaintext/ciphertext pairs, ironically, the recovery rate drops from 77% to 0%. DV thus have to calibrate concrete choices of the parameters via extensive experiments. In contrast, we choose to err on the conservative side in our analyses, and our estimates are consistent with the experiments. We also add some fail-safe to avoid the performance degradation when the number of plaintext/ciphertext pairs increases.

LIMITATION OF OUR ATTACK ON FF3. Our attack exploits the same bug of FF3 as DV's attack, and thus it can be thwarted without hurting performance by restricting the tweak space, as DV suggested. In addition, both of our attack and DV's requires that the adversary can *adaptively* make chosen plaintexts on $\Theta(N^2)$ queries for domain $\mathbb{Z}_N \times \mathbb{Z}_N$, but it is unclear how to mount this kind of attack, especially with that many queries, in practice.

ADDITIONAL RELATED WORK. There have been two separate lines of building FPE schemes. On the theoretical side, we have provably secure constructions that are based on card shuffling, such as Swap-or-Not [18], Mix-and-Cut [26], or Sometimes-Recurse [22] that are too slow for performance-hungry applications. On the practical side, in addition to FF1/FF3, there are other industry proposals, such as FNR from Cisco [14], or DTP from Protegrity [21], that have no theoretical justification. Hoang, Tessaro, and Trieu [20] however show that FNR is somewhat vulnerable in tiny domains, and DTP is completely broken even in large domains.

In a different direction, Bellare and Hoang [3] study the security of DFF, an FPE scheme currently proposed to NIST for standardization [28], and show that for appropriately large domains, DFF provides a way to localize and limit the damage from key exposure. However, as DFF is based on a 10-round Feistel network, it is still subject to prior attacks on generic Feistel-based FPE [5, 20] on tiny domains.

Very recently, Durak and Vaudenay [16] give some theoretical codebook-recovery attacks on generic balanced r-round Feistel, for $r \geq 5$. They conclude that on domain $\mathbb{Z}_N \times \mathbb{Z}_N$, FF1 cannot provide 128-bit security for $N \leq 11$, and FF3 for $N \leq 17$.

2 Preliminaries

NOTATION. If y is a string then let $|y|$ denote its length and let $y[i]$ denote its i-th bit for $1 \leq i \leq |y|$. We write $y[i : j]$ to denote the substring of y, from the ith bit to the j-th bit, inclusive. If X is a finite set, we let $x \leftarrow_s X$ denote picking an element of X uniformly at random and assigning it to x. We use the code based game playing framework of [7]. In particular, by $\Pr[G]$ we denote the probability that the execution of game G returns true.

FPE. An FPE scheme F is a pair of deterministic algorithms (F.E, F.D), where F.E : F.Keys × F.Twk × F.Dom → F.Dom is the encryption algorithm, F.D : F.Keys × F.Twk × F.Dom → F.Dom the decryption algorithm, F.Keys the key space, F.Twk the tweak space, and F.Dom the domain. For every key $K \in$ F.Keys and tweak $T \in \mathsf{T}$, the map F.E$(K, T, \cdot)$ is a permutation over F.Dom, and F.D$(K, T, \cdot)$ reverses F.E$(K, T, \cdot)$.

FEISTEL-BASED FPEs. Most existing FPE schemes, including FF3, are based on Feistel networks. Following BHT [5], we specify Feistel-based FPE in a general, parameterized way. This allows us to refer to both schemes of ideal round functions for the analysis, and schemes of some concrete round functions for realizing the standards.

We associate to parameters $r, M, N, \boxplus, \mathrm{PL}$ an FPE scheme F = **Feistel** $[r, M, N, \boxplus, \mathrm{PL}]$. Here $r \geq 2$ is an integer, the number of rounds, and $\boxplus$ is an operation for which $(\mathbb{Z}_M, \boxplus)$ and $(\mathbb{Z}_N, \boxplus)$ are Abelian groups. We let $\boxminus$ denote the inverse operator of $\boxplus$, meaning that $(X \boxplus Y) \boxminus Y = X$ for every X and Y. Integers $M, N \geq 1$ define the domain of F as F.Dom $= \mathbb{Z}_M \times \mathbb{Z}_N$. The parameter $\mathrm{PL} = (\mathcal{T}, \mathcal{K}, F_1, \ldots, F_r)$ specifies the set $\mathcal{T}$ of tweaks and a set $\mathcal{K}$ of keys, meaning F.Twk $= \mathcal{T}$ and F.Keys $= \mathcal{K}$, and the round functions $F_1, \ldots, F_r$ such that $F_i : \mathcal{K} \times \mathcal{T} \times \mathbb{Z}_N \to \mathbb{Z}_M$ if i is odd, and $F_i : \mathcal{K} \times \mathcal{T} \times \mathbb{Z}_M \to \mathbb{Z}_N$ if i is even. The code of F.E and F.D is shown in Fig. 1.

Classic Feistel schemes correspond to the boolean case, where $M = 2^m$ and $N = 2^n$ are powers of two, and $\boxplus$ is the bitwise xor operator $\oplus$. The scheme is balanced if $M = N$ and unbalanced otherwise. For $X = (L, R) \in \mathbb{Z}_M \times \mathbb{Z}_N$, we call L and R the *left segment* and *right segment* of X, respectively. We write LH(X) and RH(X) to refer to the left and right segments of X respectively. For simplicity, we assume that 0 is the zero element of the groups $(\mathbb{Z}_M, \boxplus)$ and $(\mathbb{Z}_N, \boxplus)$.

FEISTEL-BASED BLOCKCIPHERS. If the tweak space $\mathcal{T}$ is a singleton set then FPE degenerates into a blockcipher (of a general domain). For such a blockcipher F, we write F.E(K, M) and F.D(K, C) instead of F.E(K, T, M) and F.D(K, T, C) respectively.

In our analysis of Feistel-based blockciphers, the round functions are modeled as truly random. We write **Feistel**$[r, M, N, \boxplus]$ to denote **Feistel**$[r, M, N, \boxplus, \mathrm{PL}]$, for the ideal choice of $\mathrm{PL} = (\mathcal{T}, \mathcal{K}, F_1, \ldots, F_r)$ in which (i) $\mathcal{T} = \{\varepsilon\}$ where ε is the empty string, and (ii) $\mathcal{K}$ is the set **RF**(r, M, N) of all tuples of functions

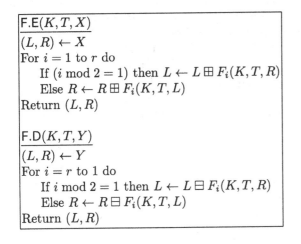
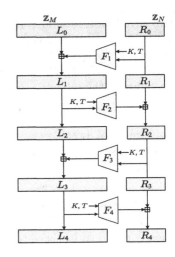

```
F.E(K, T, X)
(L, R) ← X
For i = 1 to r do
    If (i mod 2 = 1) then L ← L ⊞ F_i(K, T, R)
    Else R ← R ⊞ F_i(K, T, L)
Return (L, R)

F.D(K, T, Y)
(L, R) ← Y
For i = r to 1 do
    If i mod 2 = 1 then L ← L ⊟ F_i(K, T, R)
    Else R ← R ⊟ F_i(K, T, L)
Return (L, R)
```

Fig. 1. Left: The code for the encryption and decryption algorithms of $\mathsf{F} = \textbf{Feistel}$ $[r, M, N, \boxplus, \mathrm{PL}]$ where $\mathrm{PL} = (\mathcal{T}, \mathcal{K}, F_1, \ldots, F_r)$. Right: An illustration of encryption with $r = 4$ rounds.

$(G_1, \ldots, G_r)$ such that $G_i : \mathbb{Z}_N \to \mathbb{Z}_M$ if i is odd, and $G_i : \mathbb{Z}_M \to \mathbb{Z}_N$ if i is even, and (iii) for $1 \leq i \leq r$, the function $F_i(K, \cdot)$ is defined as $G_i(\cdot)$, where $(G_1, \ldots, G_r) \leftarrow K$.

3 Breaking FF3

In this section, we describe a chosen-plaintext codebook-recovery attack on FF3 that we call *Slide-then-Differential* (SD) attack.[2] This is based on Triangle-Finding (TF) attack, a known-plaintext codebook-recovery attack on 4-round Feistel that we will present in the next section. The running time of TF is $O(M^{5/3})$, and it actually recovers the round functions of the Feistel network, using

$$p = \max\left\{\lfloor 2^{1/3} M^{2/3} N \rfloor, \lceil M(\ln(M) + 5) \rceil\right\} \tag{1}$$

known plaintext/ciphertext pairs. We note that TF is used in a modular way; one does not need to know its technical details to understand SD.

THE FF3 SCHEME. FF3 is a Feistel-based FPE scheme $\mathsf{F} = \textbf{Feistel}[8, M, N,$ $\boxplus, \mathrm{PL}]$ of 8 rounds, where M and N are integers such that $M \geq N \geq 2$ and $MN \geq 100$.[3] The parameter PL specifies tweak space $\mathsf{F}.\mathsf{Twk} = \{0, 1\}^{2\tau}$, and

---

[2] While the notion of chosen-plaintext codebook-recovery attacks on blockciphers is folklore, one has to exercise some care in carrying this notion to FPE, because FPE domains can be tiny. In the full version we give a formal definition of chosen-plaintext codebook-recovery attacks on FPE.

[3] In NIST specification, the $\boxplus$ operation is the modular addition in $\mathbb{Z}_N$ and $\mathbb{Z}_M$, but here we will consider a generic group operator. Moreover, FF3 uses near-balanced Feistel, and thus the values of M and N are very close: if one wants to encrypt m characters in radix d, then $M = d^{\lceil m/2 \rceil}$ and $N = d^{\lfloor m/2 \rfloor}$.

two keyed hash functions $H_1 : \mathsf{F.Keys} \times \{0,1\}^\tau \times \mathbb{Z}_N \to \mathbb{Z}_M$, and $H_2 : \mathsf{F.Keys} \times \{0,1\}^\tau \times \mathbb{Z}_M \to \mathbb{Z}_N$. For each $i \leq 8$, if i is odd then the round function F_i is constructed via $F_i(K,T,X) = H_1(K, T[1:\tau] \oplus [i-1]_\tau, X)$, otherwise if i is even then $F_i(K,T,X) = H_2(K, T[\tau+1:2\tau] \oplus [i-1]_\tau, X)$, where $[j]_\tau$ is a τ-bit encoding of the integer j and $\oplus$ is the bitwise xor operator.

In analysis, the hash functions H_1 and H_2 are modeled as truly random. Formally, let $\mathcal{K}$ be the set $\mathbf{RF}(\tau, M, N)$ of all pairs of functions (G_1, G_2) such that $G_1 : \{0,1\}^{2\tau} \times \mathbb{Z}_N \to \mathbb{Z}_M$, and $G_2 : \{0,1\}^{2\tau} \times \mathbb{Z}_M \to \mathbb{Z}_N$. Then for each $j \leq 2$, define $H_j(K, \cdot, \cdot) = G_j(\cdot, \cdot)$, where $(G_1, G_2) \leftarrow K$, and we write $\mathbf{FF3}[M, N, \tau, \boxplus]$ to denote this ideal version of FF3.

In our attack to FF3, we will consider $M \geq N \geq 64$ and $MN \geq 2p$, where p is specified as in Eq. (1). While there are indeed applications of smaller values of M and N, they are already susceptible to prior attacks [4,15,20] whose running time is practical in those tiny domains. In addition, to simplify our asymptotic analysis, we will assume that $N = \Omega(\sqrt{M})$, which applies to the setting of the FF3 scheme, since FF3 uses near-balanced Feistel. Thus $p \in O(M^{2/3}N)$.

3.1   DV's Blueprint for Breaking FF3

Let $\mathsf{F} = \mathbf{FF3}[M, N, \tau, \boxplus]$. Let K and T be a key and tweak for F, respectively. Recall that $\mathsf{F.E}(K, T, \cdot)$ is an 8-round Feistel network. View $\mathsf{F.E}(K, T, \cdot)$ as the cascade of two 4-round Feistel networks f and g, meaning $\mathsf{F.E}(K, T, X) = g(f(X))$ for every $X \in \mathbb{Z}_M \times \mathbb{Z}_N$. DV [15] observe that $\mathsf{F.E}(K, T', \cdot)$ is the cascade of g and f—note that the ordering of f and g is now reversed—where $T' = T \oplus ([4]_\tau \,\|\, [4]_\tau)$. See Fig. 2 for an illustration.

A SKETCH OF DV'S ATTACK. From the observation above, one can launch a chosen-plaintext codebook-recovery attack on F as follows; this can also be viewed as a slide attack [10,11]. Let p be as specified in Eq. (1) and let $s = \left\lceil \sqrt{MN/2p} \right\rceil \geq 1.$[4] Sample s elements uniformly and independently from $\mathbb{Z}_M \times \mathbb{Z}_N$, and let S be the set of these elements. Repeat this process, and let S^* be the resulting set. Recall that the adversary is given an encryption oracle $\textsc{Enc}$ in this attack. Now, for each $U_0 \in S$, we iterate $U_i \leftarrow \textsc{Enc}(T, U_{i-1})$, for $i = 1, \ldots, 2p$, forming a U-chain $U_0 \to U_1 \to \cdots \to U_{2p}$. For each $V_0 \in S^*$, let $V_i \leftarrow \textsc{Enc}(T', V_{i-1})$ for $i = 1, \ldots, 2p$, forming a V-chain $V_0 \to V_1 \to \cdots \to V_{2p}$.

Consider a U-chain and a V-chain such that each chain has at least p distinct elements.[5] If there is some index $i < p$ such that $V_0 = f(U_i)$ then the pair (U_i, V_0) is called a *slid pair*, and $V_k = f(U_{i+k})$ and $U_{i+k+1} = g(V_k)$ for every $0 \leq k < p$. Likewise, if there is some index $j < p$ such that $U_0 = g(V_j)$ then the

---

[4] DV actually use different concrete choices of p and s to aggressively improve the recovery rate.

[5] To test if, say a U-chain $(U_0, \ldots, U_{2p})$ contains at least p distinct elements, we only need to check if $U_0 \notin \{U_1, \ldots, U_{p-1}\}$, since $|\{U_0, \ldots, U_{2p}\}| < p$ if and only if U_0 is within a cycle of length $k < p$ in the functional graph of the permutation $f(g(\cdot))$.

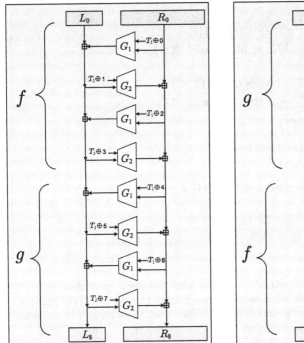 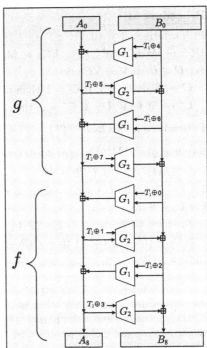

Fig. 2. Left: Encryption $\mathsf{F.E}(K, T, \cdot)$ as a cascade of 4-round Feistel networks f and g. Right: Slided encryption $\mathsf{F.E}(K, T', \cdot)$ as a cascade of g and f, with $T' = T \oplus ([4]_\tau \parallel [4]_\tau)$. Here T_1 and T_2 are the left half and right half of the tweak T, respectively. For simplicity, in the picture, instead of writing, say $T_1 \oplus [0]_\tau$, we simply write $T_1 \oplus 0$.

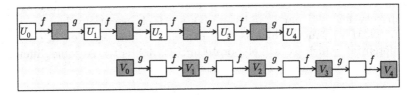

Fig. 3. Illustration of the slide attack. Here (U_1, V_0) is a slid pair.

pair (V_j, U_0) is also called a slid pair, and $U_k = g(V_{j+k})$ and $V_{j+k+1} = f(U_k)$ for every $0 \le k < p$. See Fig. 3 for an illustration.

Suppose that somehow we manage to find a slid pair. Then we get p input/output pairs for f, and can run TF to recover the codebook of f. Likewise, we can also recover the codebook of g. By composing the codebook of f and g, we finally recover the codebook of F on tweak T. We can also compose g and f to recover the codebook of F on tweak T'.

Procedure SD^{ENC}()

Pick arbitrary $T \in \{0,1\}^{2\tau}$; $T' \leftarrow T \oplus ([4]_\tau \parallel [4]_\tau)$

$UCh \leftarrow \mathsf{MakeChain}^{\text{ENC}}(T)$; $VCh \leftarrow \mathsf{MakeChain}^{\text{ENC}}(T')$

For $U \in UCh$, $V \in VCh$ do

    $C \leftarrow_\$ \mathsf{Slide}(U, V)$; If $C \neq \perp$ then return (T, C)

    $C' \leftarrow_\$ \mathsf{Slide}(V, U)$; If $C' \neq \perp$ then return (T', C')

Procedure MakeChain$^{\text{ENC}}(T)$

$p \leftarrow \max \left\{ \lfloor 2^{1/3} M^{2/3} N \rfloor, \lceil M(\ln(M) + 5) \rceil \right\}$

$s \leftarrow \lfloor \sqrt{MN/2p} \rfloor$; $S, UCh \leftarrow \emptyset$

For $i = 1$ to s do $U \leftarrow_\$ \mathbb{Z}_M \times \mathbb{Z}_N$; $S \leftarrow S \cup \{U\}$

For $U_0 \in S$ do

    For $i = 1$ to $2p$ do $U_i \leftarrow \mathsf{ENC}(T, U_{i-1})$

    If $U_0 \notin \{U_1, \ldots, U_{p-1}\}$ then $UCh \leftarrow UCh \cup \{(U_0, \ldots, U_{2p})\}$

Return UCh

Fig. 4. The blueprint of DV's attack, which is also the main procedure of the SD attack. Procedure $\mathsf{Slide}(U, V)$ takes as input two chains $U = (U_0, \ldots, U_{2p})$ and $V = (V_0, \ldots, V_{2p})$, tries to find a slid pair (U_i, V_0), and then uses TF to recover the codebook. Numbers M, N, τ are global parameters.

The code of the blueprint of DV's attack is given in Fig. 4, which is also the main procedure of our SD attack. The two attacks however differ in how they implement procedure Slide for finding a slid pair, among $2s^2 p \approx MN$ candidates. DV simply try every possible candidate, by running (a slow version of) the TF algorithm to recover the codebook of f and g. As we will show below, there are often very few slid pairs, and thus DV's attack essentially has to run TF for about $\Theta(MN)$ times, which is very expensive. The key idea in our Slide-then-Differential, which we will elaborate in Sect. 3.2, is to use some differential analysis to quickly eliminate false candidates.

THE NUMBER OF SLID PAIRS. Clearly, the attack above only works if there exists at least one slid pair. Let P be the random variable for the number of slid pairs. DV use a heuristic[6] to estimate that $\Pr[P \geq 1] \approx 1 - e^{-2s^2 p/MN} \approx 1 - 1/e$, under the model that the cascade of f and g is an ideal permutation. We instead give a rigorous lower bound of $\Pr[P \geq 1]$ for a *generic* value p in Lemma 1 in the same model; the proof is in the full version. For $s = 1$ (equivalently, $M < 1024$), we can compute the exact probability $\Pr[P \geq 1]$, but we stress that this result only holds in the model above. The experiments in Sect. 5 show that empirically, the event $P \geq 1$ happens with higher probability.

---

[6] While DV only consider balanced Feistel networks, their heuristic can be easily generalized to the general case. For completeness, in the proof of Lemma 1, we also describe this heuristic argument.

Lemma 1. *Let $M \geq N \geq 8$ and let $\mathsf{F} = \mathbf{FF3}[M, N, \tau, \boxplus]$. Let $p \geq 1$ be an integer such that $p \leq MN/2$ and let $s = \lfloor \sqrt{MN/2p} \rfloor$. Let f and g be as above, and let π be the cascade of f and g. Let P be the random variable for the number of slid pairs. Let $\delta = \frac{2p}{MN} - \frac{(2.5p^2 - 1.5p)}{(MN)^2}$. We will model π as an ideal random permutation on $\mathbb{Z}_M \times \mathbb{Z}_N$.*

(a) If $s = 1$ then $\Pr[P \geq 1] = \delta \approx \frac{3}{8}$.
(b) If $s \geq 2$ then $\Pr[P \geq 1] \geq \frac{s^2 \delta}{2} \approx \frac{1}{2}$.

Above, we show that it is quite likely that there are one or more slid pairs. However, often there will be very few of them. Lemma 2 below bounds the expected number of slid pairs for a *generic* value of p; the proof is in the full version. Combining this with Markov's inequality, one can show that with probability at least 0.8, there are at most 5 slid pairs.

Lemma 2. *Let $M \geq N \geq 64$ and let $\mathsf{F} = \mathbf{FF3}[M, N, \tau, \boxplus]$. Let $p \geq 1$ be an integer such that $p \leq MN/2$ and let $s = \lfloor \sqrt{MN/2p} \rfloor$. Let f and g be as above, and let π be the cascade of f and g. Let P be the random variable for the number of slid pairs. If we model π as an ideal random permutation on $\mathbb{Z}_M \times \mathbb{Z}_N$ then $\mathbf{E}[P] \leq \frac{2s^2 p}{MN} \leq 1$.*

3.2   Distinguishing Slid-Pair Candidates

As shown above, we often have very few slid pairs, among $2s^2 p \approx MN$ candidates, and using TF to find the actual slid pairs is an overkill. Note that each candidate gives us p plaintext/ciphertext pairs for f. If a candidate is indeed a slid pair, then the ciphertexts for f are indeed the images of the corresponding plaintexts under f, otherwise we can view them as produced from an ideal permutation on $\mathbb{Z}_M \times \mathbb{Z}_N$. The analogous claim also holds for g. A natural solution is to find a quick distinguishing attack for 4-round Feistel, so that we can tell the true candidates from the false ones.

OUR DISTINGUISHING ATTACK ON 4-ROUND FEISTEL. Below, we will give a distinguishing attack Left-Half Differential (LHD) of 4-round Feistel such that (1) in the ideal world, it returns 1 with probability at most $\frac{N^{5/6}}{M^{4/3}}$, and (2) in the real world, it returns 1 with probability at least $1 - \frac{\sqrt{N}}{8M} - \frac{9.7}{M} - \frac{0.88N^{3/4}}{M^{3/2}}$. For each slid-pair candidate, we run LHD on the plaintext/ciphertext pairs of f, and also on those of g. We will accept the candidate if LHD returns 1 in both cases. Then, for each false candidate, the chance that we incorrectly accept it is at most $N^{5/3}/M^{8/3}$. Since we have at most MN false candidates, on average we will have at most

$$MN \cdot \frac{N^{5/3}}{M^{8/3}} = \frac{N^{8/3}}{M^{5/3}} \leq N$$

false candidates that survived our test. In addition, for each true candidate, the chance that we incorrectly reject it is at most

$$1 - \left(1 - \frac{\sqrt{N}}{8M} - \frac{9.7}{M} - \frac{0.88N^{3/4}}{M^{3/2}}\right)^2 \leq 0.37$$

for $M, N \geq 64$. We note that our bounds are very conservative, since we obtain them via Chebyshev's inequality, which is loose. In fact, our empirical results, presented in Sect. 5, significantly outperform the theoretical estimates. In particular, on average just one (possibly false) candidate survives our test, and we never incorrectly reject a true candidate.

Proceeding into details, the LHD algorithm is based on the following Lemma 3, which is a generalization of a result by Patarin for balanced, boolean Feistel [24]. A more general version of Lemma 3 appears in [5] for a Feistel network of an even number of rounds, but this result only provides a (very tight) approximation of the bound, instead of an exact one. The proof is in the full version

Lemma 3. *Let $M, N \geq 8$ be integers and $\overline{\mathsf{F}} = \mathbf{Feistel}[4, M, N, \boxplus]$. Let X and X' be two distinct messages in $\mathbb{Z}_M \times \mathbb{Z}_N$ such that $\mathsf{RH}(X) = \mathsf{RH}(X')$. Let C and C' be the ciphertexts of X and X' under $\overline{\mathsf{F}}$ with a uniformly random key. Then*

$$\Pr[\mathsf{LH}(C) \boxminus \mathsf{LH}(C') = \mathsf{LH}(X) \boxminus \mathsf{LH}(X')] = \frac{M + N - 1}{MN} \ .$$

Lemma 3 above shows that if we encrypt two messages X and X' of the same right segment under a 4-round Feistel network, then there will be some bias in the distribution of the ciphertexts C and C': (1) the chance that $\mathsf{LH}(C) \boxminus \mathsf{LH}(C') = \mathsf{LH}(X) \boxminus \mathsf{LH}(X')$ is $\frac{M+N-1}{MN}$, (2) had we instead sampled C and C' uniformly without replacement from $\mathbb{Z}_M \times \mathbb{Z}_N$, this probability would have been just $\frac{N}{MN-1}$. Our distinguishing attack LHD will amplify this bias, by using several messages of the same right segments.

Random variables $X_1, \ldots, X_m \in \mathbb{Z}_M \times \mathbb{Z}_N$ are t-*wise right-matching* if they satisfy the following constraints:

- If we partition $X_1, \ldots, X_m$ into groups $P_1, \ldots, P_d$ according to their right segments then $d \leq t$.
- Within each partition P_i, the left segments of the messages in P_i are uniformly distributed over $\mathbb{Z}_M$, subject to the constraint that those left segments are distinct.

Our attack LHD takes as input m messages $(X_1, \ldots, X_m)$ that are t-wise right-matching and their ciphertexts $(C_1, \ldots, C_m)$, where $m = \lceil \frac{p}{N} \cdot \lceil 32N^{1/6} \rceil \rceil$ and $t = \lceil \frac{mM}{p} \rceil$. The code of LHD is given in Fig. 5. Informally, LHD will compute *count*, the number of pairs X_i and X_j, with $i < j$, such that $\mathsf{RH}(X_i) = \mathsf{RH}(X_j)$ and $\mathsf{LH}(C_i) \boxminus \mathsf{LH}(C_j) = \mathsf{LH}(X_i) \boxminus \mathsf{LH}(X_j)$. If the ciphertexts are produced by a 4-round Feistel network then from Lemma 3, the expected value of *count* is $\frac{M+N-1}{MN} \cdot size$, where *size* is the number of pairs X_i, X_j such that $i < j$ and $\mathsf{RH}(X_i) = \mathsf{RH}(X_j)$. If the ciphertexts are produced by a truly random permutation on $\mathbb{Z}_M \times \mathbb{Z}_N$ then the expected value of *count* is $\frac{N}{MN-1} \cdot size$. The algorithm LHD will return 1 if *count* is greater than the weighted average $\left(\frac{1}{5} \cdot \frac{M+N-1}{MN} + \frac{4}{5} \cdot \frac{N}{MN-1} \right) size$, otherwise it will return 0.

Procedure $\mathsf{LHD}(X_1, \ldots, X_m, C_1, \ldots, C_m)$

Partition $X_1, \ldots, X_m$ by the right segments into groups $P_1, \ldots, P_d$

$count \leftarrow 0; \quad \Delta \leftarrow \frac{1}{5} \cdot \frac{M+N-1}{MN} + \frac{4}{5} \cdot \frac{N}{MN-1}; \quad size \leftarrow \sum_{\ell=1}^{d} \frac{|P_\ell|(|P_\ell|-1)}{2}$

For $\ell \leftarrow 1$ to d do

    For $X_i, X_j \in P_\ell$ with $i < j$ do

        If $\mathsf{LH}(C_i) \boxminus \mathsf{LH}(C_j) = \mathsf{LH}(X_i) \boxminus \mathsf{LH}(X_j)$ then $count \leftarrow count + 1$

If $count \geq \Delta \cdot size$ then return 1 else return 0

Fig. 5. Distinguishing attack LHD on four-round **Feistel**.

IMPLEMENTING LHD. The code in Fig. 5 describes just the conceptual view of LHD for ease of understanding. Implementing it efficiently requires some care. First, messages $X_1, \ldots, X_m$ will be grouped according to their right segments, by a one-time preprocessing that we will describe in Sect. 3.3. Thus the partitioning takes only linear time. Let $P_1, \ldots, P_d$ be the resulting partitions, and let $|m_\ell| = |P_\ell|$, for every $\ell \leq d$. In the for loops, if we naively follow the code, then the running time would be

$$\sum_{\ell=1}^{d} \Omega(m_\ell^2) = \Omega(m^2/d) = \Omega(M^{1/3}N^{7/6}),$$

which is expensive. Instead, we will execute as in Fig. 6. That is,

- For each fixed $\ell \leq d$, we want to find $count_\ell$, the number of pairs (i, j) such that $i < j$ and $X_i, X_j \in P_\ell$ and $\mathsf{LH}(C_i) \boxminus \mathsf{LH}(X_i) = \mathsf{LH}(C_j) \boxminus \mathsf{LH}(X_j)$. We then can compute $count$ via $count_1 + \cdots + count_d$.
- Thus for each $\ell \leq d$, we create an empty hash table H_ℓ of key-value pairs and initialize $count_\ell \leftarrow 0$. We process P_ℓ so that eventually, for each entry in H_ℓ, its key is a number $Z \in \mathbb{Z}_M$ and its value indicates how many $X_k \in P_\ell$ that $\mathsf{LH}(C_k) \boxminus \mathsf{LH}(X_k) = Z$.
- Finally, we iterate through all keys of H_ℓ. For each key Z, we find its value v and update $count_\ell \leftarrow count_\ell + \frac{v(v-1)}{2}$.

The total running time of this implementation is $O(m) = O(M^{2/3}N^{1/6})$.

ANALYSIS OF LHD. Lemma 4 below bounds the probability that LHD outputs 1 in the ideal world, for *generic* m and t, and also for a *generic* weighted average $\Delta = \lambda \cdot \frac{M+N-1}{MN} + (1-\lambda)\frac{N}{MN-1}$; see the full version for the proof. If we pick $m = \lceil \frac{p}{N} \cdot \lceil 32N^{1/6}\rceil \rceil$, $t = \lceil \frac{mM}{p} \rceil$, and $\lambda = \frac{1}{5}$ as suggested then this probability is about $\frac{N^{5/6}}{M^{4/3}}$.

Lemma 4. *Let $M \geq N \geq 8$ be integers, and let $0 < \lambda < 1$ be a real number. Let $m > t \geq 1$ be integers. Let $X_1, \ldots, X_m$ be t-wise right-matching messages, and let $C_1, \ldots, C_m$ be their ciphertexts, respectively, under an ideal random permutation on $\mathbb{Z}_M \times \mathbb{Z}_N$. Let V be the random variable of the number of pairs X_i and X_j, with*

Procedure $\mathsf{LHD}(X_1, \ldots, X_m, C_1, \ldots, C_m)$

// $X_1, \ldots, X_m$ are already grouped according to their right segments

Partition $X_1, \ldots, X_m$ by the right segments into groups $P_1, \ldots, P_d$

$count \leftarrow 0;\ \Delta \leftarrow \frac{1}{5} \cdot \frac{M+N-1}{MN} + \frac{4}{5} \cdot \frac{N}{MN-1};\ size \leftarrow \sum_{\ell=1}^{d} \frac{|P_\ell|(|P_\ell|-1)}{2}$

For $\ell \leftarrow 1$ to d do

    $count_\ell \leftarrow 0$; Initialize a hash table H_ℓ

    For $X_k \in P_\ell$ do

        $Z \leftarrow \mathsf{LH}(C_k) \boxminus \mathsf{LH}(X_k);\ v \leftarrow H_\ell[Z]$

        If $v = \bot$ then $H_\ell[Z] \leftarrow 1$ else $H_\ell[Z] \leftarrow v + 1$

        For each key Z in H_ℓ do $v \leftarrow H_\ell[Z];\ count_\ell \leftarrow count_\ell + \frac{v(v-1)}{2}$

    $count \leftarrow count_1 + \cdots + count_d$

If $count \geq \Delta \cdot size$ then return 1 else return 0

Fig. 6. Implementation of LHD.

$i < j$, such that $\mathsf{RH}(X_i) = \mathsf{RH}(X_j)$ and $\mathsf{LH}(C_i) \boxminus \mathsf{LH}(C_j) = \mathsf{LH}(X_i) \boxminus \mathsf{LH}(X_j)$. Let $size$ be the number of pairs X_i, X_j such that $i < j$ and $\mathsf{RH}(X_i) = \mathsf{RH}(X_j)$, and $\Delta = \lambda \cdot \frac{M+N-1}{MN} + (1-\lambda)\frac{N}{MN-1}$. Then

$$\Pr\left[V \geq \Delta \cdot size\right] \leq \frac{N^2}{\lambda^2(M-2)^2}\left(\frac{1}{MN-2} + \frac{2MN-2}{N(m^2/t - m)}\right).$$

Lemma 5 below bounds the probability that LHD fails to output 1 in the real world, again for generic m and t, and for a generic weighted average $\Delta = \lambda \cdot \frac{M+N-1}{MN} + (1-\lambda)\frac{N}{MN-1}$; see the full version for the proof. If we use $m = \lceil \frac{p}{N} \cdot \lceil 32N^{1/6} \rceil \rceil$, $t = \lceil \frac{mM}{p} \rceil$, and $\lambda = \frac{1}{5}$ as suggested then this probability is about $\frac{\sqrt{N}}{8M} + \frac{9.7}{M} + \frac{0.88N^{3/4}}{M^{3/2}}$.

Lemma 5. Let $M \geq N \geq 8$ be integers and let $0 < \lambda < 1$ be a real number. Let $m > t \geq 1$ be integers. Let $X_1, \ldots, X_m$ be t-wise right-matching messages and let $C_1, \ldots, C_m$ be their ciphertexts, respectively, under $\overline{\mathsf{F}} = \mathbf{Feistel}[4, M, N, \boxplus]$ with a uniformly random key. Let V be the random variable of the number of pairs X_i and X_j, with $i < j$, such that $\mathsf{RH}(X_i) = \mathsf{RH}(X_j)$ and $\mathsf{LH}(C_i) \boxminus \mathsf{LH}(C_j) = \mathsf{LH}(X_i) \boxminus \mathsf{LH}(X_j)$. Let $\Delta = \lambda \cdot \frac{M+N-1}{MN} + (1-\lambda)\frac{N}{MN-1}$, and let $size$ be the number of pairs X_i, X_j such that $i < j$ and $\mathsf{RH}(X_i) = \mathsf{RH}(X_j)$. Then

$$\Pr\left[V \leq \Delta \cdot size\right] \leq \frac{2(M+N-1)MN}{(1-\lambda)^2(m^2/t - m)(M-2)^2} + \frac{6.2(M-1)(N-1)}{(1-\lambda)^2 N(M-2)^2}$$
$$+ \frac{4MN}{(1-\lambda)^2(M-2)^2\sqrt{(m^2/t - m)}}.$$

USING LHD. The LHD attack requires m chosen plaintexts, but recall that for each slid-pair candidate, we only have p known plaintext/ciphertext pairs for f, and also p known pairs for g.[7] To find m messages that are $\lceil \frac{mM}{p} \rceil$-wise right-matching, the naive approach is to partition p given messages according to their right segments, and let $P_1, \ldots, P_M$ be the (possibly empty) partitions, with $|P_1| \geq \cdots \geq |P_M|$. We then output m messages from the first (and also biggest) $s = \lceil \frac{mM}{p} \rceil$ partitions. Since there are at most M partitions, our chosen partitions contain at least $\lceil s \cdot \frac{p}{M} \rceil \geq m$ messages. Moreover, as the given p messages are sampled uniformly without replacement from $\mathbb{Z}_M \times \mathbb{Z}_N$, within each partition P_i, the left segments of the messages in P_i are sampled uniformly without replacement from $\mathbb{Z}_M$.

The naive approach above is however very expensive. Totally, for $\Theta(MN)$ slid-pair candidates, it uses $\Omega(MNp) = \Omega(M^{5/3}N^2)$ time just to find their right-matching messages. In the next section we'll describe a one-time preprocessing of $O(MN)$ time such that later for each slid-pair candidate, we need only $O(m)$ time to find their right-matching messages to run LHD tests. Only for candidates that survive the LHD tests that we extract their p plaintext/ciphertext pairs from the corresponding chains to run TF.

ELIMINATING FALSE NEGATIVES. Since our distinguishing test of slid-pair candidates above might occasionally produce false negatives, we still have to use TF to eliminate the survived false candidates. The TF algorithm, after recovering the round functions (G_1, G_2, G_3, G_4), will compute the outputs of the first $3M \leq p$ plaintexts under a 4-round Feistel network with the round functions (G_1, G_2, G_3, G_4), and compare them with the corresponding ciphertexts. By a simple counting argument, one can show that it is extremely likely that TF will reject all these false candidates. Specifically, for a false candidate, view its $3M$ associated ciphertexts as the outputs of the $3M$ plaintexts under an ideal permutation on $\mathbb{Z}_M \times \mathbb{Z}_N$. On the one hand, there are at most $M^{2N}N^{2M} \leq M^{2M}N^{2N}$ choices of four round functions for a 4-round Feistel network on $\mathbb{Z}_M \times \mathbb{Z}_N$. On the other hand, if we sample $3M$ ciphertexts uniformly without replacement from $\mathbb{Z}_M \times \mathbb{Z}_N$, there are

$$MN \cdots (MN - 3M + 1) \geq (MN - 3M)^{3M} \geq M^{3M}(N - 3)^{3N}$$

equally likely outputs. Since we have to deal with at most MN false candidates, the chance that the TF algorithm fails to eliminate all false candidates is at most

$$\frac{MN \cdot M^{2M}N^{2N}}{M^{3M}(N-3)^{M+2N}} = \frac{N^{2N+1}}{M^{3M-1}(N-3)^{3N}} \leq \frac{1}{M^{3M-1}} .$$

RELATION TO PRIOR FEISTEL ATTACKS. Our LHD attack generalizes Patarin's distinguishing attack on 4-round balanced, boolean Feistel [24]; Patarin's result was later rediscovered by Aiello and Venkatesan [1]. To attack Feistel networks on $2n$-bit strings, those papers suggest using messages $X_1, \ldots, X_m$ of the same right half, with $m = \Theta(2^{n/2})$. However, both papers only compute the expected value

---

[7] Recall that in our attack, we require $M \geq N \geq 64$. This ensures that $m \leq p$, so that we can select m right-matching messages from p known messages.

of the number of pairs (i, j) such that $1 \leq i < j \leq m$ and $\mathsf{LH}(C_i) \oplus \mathsf{LH}(C_j) = \mathsf{LH}(X_i) \oplus \mathsf{LH}(X_j)$ in both the real and ideal worlds. Consequently, they cannot analyze the advantage of their attack, and can only suggest an asymptotic value of m. Our Lemmas 4 and 5 allow one to fill this gap. By using $N = M = 2^n$, $t = 1$, $m = c \cdot 2^{n/2}$, and $\lambda = \frac{1}{2}$, the attack in [1,24] achieves advantage around $1 - \frac{24}{c^2} - \frac{29}{2^n} - \frac{16}{c \cdot 2^{n/2}}$. Thus to achieve advantage $1/2$, for $n \geq 16$, we can use $c = 7$.

The attack in [1,24] however cannot be used in place of LHD. Recall that we want a distinguishing attack that outputs 1 with probability around $1/\sqrt{N}$ or smaller in the ideal world, so that it can be used to eliminate most false slid-pair candidates. Using $m = \Theta(\sqrt{N})$ messages as suggested in [1,24] does not meet this requirement, as the attack will output 1 with constant probability in the ideal world, according to Lemma 4.

3.3   Slide-then-Differential Attack

In this Section, we describe how to combine DV's slide attack with the LHD attack above, resulting in our Slide-then-Differential attack.

SPEEDING UP WITH PREPROCESSING. Recall that we have $\Theta(MN)$ slid-pair candidates, and for each such candidate we have to process $\Theta(p)$ pairs of plaintext/ciphertext. At the first glance it seems that we are doomed with $\Omega(pMN) = \Omega(M^{5/3}N^2)$ time. However, we will perform a one-time preprocessing using $O(MN)$ time. After this preprocessing, for every slid-pair candidate, we can extract m right-matching messages for f in $O(m) = O(M^{2/3}N^{1/6})$ time, and even better, those messages are already grouped according to their right segments. The same running time would be needed to extract messages for g. We then can run LHD to eliminate most false slid-pair candidates.

Proceeding into details, suppose that we have a U-chain $U_0 \rightarrow U_1 \rightarrow \cdots \rightarrow U_{2p}$ and a V-chain $V_0 \rightarrow V_1 \rightarrow \cdots \rightarrow V_{2p}$, and we want to check if (U_i, V_0) is a slid pair, for every $k \leq \{0, 1, \ldots, p-1\}$. For each slid-pair candidate (U_i, V_0), the known plaintext/ciphertext pairs for f are (U_{i+k}, V_k) for $k \leq 2p - i$, and the known plaintext/ciphertext pairs for g are (V_k, U_{i+k+1}), for $k \leq 2p - k - 1$.

- In order to preprocess these p slid-pair candidates for f, note that they all use plaintexts $U_{p+1}, \ldots, U_{2p}$. So we will partition these plaintexts by their right segments into (possibly empty) groups $P_1, \ldots, P_M$, with $|P_1| \geq \cdots \geq |P_M|$. We then store m messages U_j from the first $\lceil 32N^{1/6} \rceil$ partitions, together with their indices j, in a list L. Later, for a slid-pair candidate (U_i, V_0), we iterate through pairs (U_j, j) in L, and for each such pair, the corresponding ciphertext of U_j for f is V_{j-i}, which takes $O(1)$ time to find if we store $(U_0, \ldots, U_{2p})$ and $(V_0, \ldots, V_{2p})$ in arrays.
- Preprocessing for g is similar, but note that the p candidates all use plaintexts $V_0, V_1, \ldots, V_{p-1}$.

For a pair of U chain and V chain, partitioning takes $O(p + M)$ time, and by using a max-heap, we can find the $\lceil 32N^{1/6} \rceil$ biggest partitions in $O(M)$ time,

---

Procedure Slide(U, V)

$(U_0, \dots, U_{2p}) \leftarrow U$; $(V_0, \dots, V_{2p}) \leftarrow V$; $(Z_1, \dots, Z_{MN}) \leftarrow \mathbb{Z}_M \times \mathbb{Z}_N$

$L \leftarrow$ Process$((U_{p+1}, p+1), \dots, (U_{2p}, 2p))$

$L' \leftarrow$ Process$((V_0, 0), \dots, (V_{p-1}, p-1))$

For $i = 0$ to $p - 1$ do // Check if (U_i, V_0) is a slid pair

  If (Dist$(L, V, -i) \wedge$ Dist$(L', U, i+1)$) then

    $X \leftarrow (U_p, \dots, U_{2p-1})$; $Y \leftarrow (V_{p-i}, \dots, V_{2p-1-i})$

    $X^* \leftarrow (V_0, \dots, V_{p-1})$; $Y^* \leftarrow (U_{i+1}, \dots, U_{i+p})$

    $f \leftarrow_\$ $ Recover(X, Y); $g \leftarrow_\$ $ Recover(X^*, Y^*)

    If $f \neq \bot$ and $g \neq \bot$ then

      For $i = 1$ to MN do $C_i \leftarrow g(f(Z_i))$

      Return $(C_1, \dots, C_{MN})$ // Codebook is $(Z_1, C_1), \dots, (Z_{MN}, C_{MN})$

Procedure Process$((X_1, r_1), \dots, (X_p, r_p))$ // Preprocessing

$L \leftarrow \emptyset$; $m \leftarrow \lceil \frac{p}{N} \cdot \lceil 32N^{1/6} \rceil \rceil$

Partition $(X_1, r_1), \dots, (X_p, r_p)$ by the right segments

Let $P_1, \dots, P_M$ be the resulting partitions, with $|P_1| \geq \dots \geq |P_M|$

For $i = 1$ to M, $(X, r) \in P_i$ do: If $|L| < m$ then $L \leftarrow L \cup \{(X, r)\}$

Return L

Procedure Dist(L, V, k) // Running LHD with preprocessed list L

$i \leftarrow 1$, $(V_0, \dots, V_{2p}) \leftarrow V$; $m \leftarrow |L|$

For $(Z, r) \in L$ do $X_i \leftarrow Z$; $C_i \leftarrow V_{r+k}$; $i \leftarrow i + 1$

Return LHD$(X_1, \dots, X_m, C_1, \dots, C_m)$

Procedure Recover(X, Y)

$(X_0, \dots, X_{p-1}) \leftarrow X$; $(Y_0, \dots, Y_{p-1}) \leftarrow Y$

$(F_1, F_2, F_3, F_4) \leftarrow_\$ $ TF$(X_0, \dots, X_{p-1}, Y_0, \dots, Y_{p-1})$

Let f be the 4-found Feistel of round functions (F_1, F_2, F_3, F_4)

// Function f will be $\bot$ if TF does not fully recover (F_1, F_2, F_3, F_4)

Return f

Fig. 7. The implementation of procedure Slide in the SD attack. The numbers M and N are global parameters. We assume that there is a global total ordering on the domain $\mathbb{Z}_M \times \mathbb{Z}_N$, so that we can write $(Z_1, \dots, Z_{MN}) \leftarrow \mathbb{Z}_M \times \mathbb{Z}_N$.

and extracting m messages from those partitions takes $O(m)$ time. Summing up, for a pair of U chain and V chain, the running time of the preprocessing is $O(p)$. Hence, totally, for s^2 pairs of U chains and V chains, the overall running time of the preprocessing is $O(s^2 p) = O(MN)$.

PUTTING THINGS TOGETHER. By combining the LHD attack and the preprocessing, one can implement procedure Slide as in Fig. 7. Thus the SD attack uses $O(sp) = O(M^{5/6}N)$ queries and space, and its running time is $O(MN)$ for the preprocessing, $O(MN \cdot m) = O(M^{5/3}N^{7/6})$ for running LHD, and expectedly, $O(M^{5/3}N)$ for running TF. Hence the total running time of SD is $O(M^{5/3}N^{7/6})$.

IMPROVING THE RECOVERY RATE. To improve the recovery rate of SD, one can run the attack several times with different tweak pairs $(T_1, T_1 \oplus \mathsf{Mask}), \ldots, (T_r, T_r \oplus \mathsf{Mask})$, where $\mathsf{Mask} = [4]_\tau \parallel [4]_\tau$. If $(T_i \oplus T_j)[1 : \tau], (T_i \oplus T_j)[\tau + 1 : 2\tau] \notin \{[0]_\tau, \ldots, [7]_\tau\}$ for every $i \neq j$ then those r instances of SD will call AES on different τ-bit prefixes. If we model AES as a good PRF then the results of those SD instances are independent. Hence if the recovery rate of SD is ϵ then running it for r times will have recovery rate $1 - (1 - \epsilon)^r$.

4   Attacking 4-Round Feistel-Based Blockciphers

In this section, we generalize and improve DV's known-plaintext codebook-recovery attack on Feistel-based, 4-round blockciphers where the Feistel network might be unbalanced. In particular, on a four-round balanced Feistel of domain size N^2, DV's attack runs in $O(N^3)$ expected time, but our attack, which we name *Triangle-Finding* (TF), runs in only $O(N^{5/3})$ expected time. Both our attack and DV's rely on a conjecture of Feistel networks that DV empirically verified for balanced Feistel of domain $\{0, 1\}^{2n}$, for $n \in \{1, \ldots, 9\}$. We prove that this conjecture indeed holds, making both attacks unconditional.

Let

$$p = \max\left\{ \lfloor 2^{1/3}M^{2/3}N \rfloor, \lceil M(\ln(M) + 5) \rceil \right\} .$$

In our attack, we suppose that we are given p pairs $(X_1, C_1), \ldots, (X_p, C_p)$ of plaintext/ciphertext under a four-round Feistel network $\mathsf{F} = \mathbf{Feistel}[4, M, N, \boxplus]$ with a uniformly random key, where $M \geq N \geq 64$ and the plaintexts are chosen uniformly without replacement from $\mathbb{Z}_M \times \mathbb{Z}_N$. To simplify our asymptotic analysis, we assume that $N = \Omega(\sqrt{M})$, which applies to the setting of the FF3 scheme, since FF3 uses near-balanced Feistel. Thus $p \in O(M^{2/3}N)$.

In our attack, we need a graph representation of the first p plaintext/ciphertext pairs, and a few algorithms, which we will elaborate in Sect. 4.1. We then describe our TF attack in Sect. 4.2.

4.1   Differential Graph and Its Triangles

DIFFERENTIAL GRAPH. Let $\mathcal{G} = (V, E)$ be the following directed graph. First, for each $i \neq j$, we create a node $v_{i,j}$ with label $\mathsf{Label}(v_{i,j}) = \mathsf{RH}(C_i) \boxminus \mathsf{RH}(C_j)$ if $\mathsf{RH}(X_i) = \mathsf{RH}(X_j)$ and $\mathsf{LH}(C_i) \boxminus \mathsf{LH}(X_i) = \mathsf{LH}(C_j) \boxminus \mathsf{LH}(X_j)$. Next, for every two nodes $v_{i,j}$ and $v_{k,\ell}$ such that i, j, k, ℓ are distinct, we create a directed edge $(v_{i,j}, v_{k,\ell})$ if $\mathsf{LH}(C_j) = \mathsf{LH}(C_k)$ and the following non-degeneracy conditions hold:

(1) $\mathsf{LH}(C_i) \neq \mathsf{LH}(C_\ell)$,
(2) $\mathsf{RH}(X_i) \neq \mathsf{RH}(X_k)$,
(3) $\mathsf{RH}(X_j) \boxminus \mathsf{RH}(X_k) \neq \mathsf{RH}(C_j) \boxminus \mathsf{RH}(C_k)$, and
(4) $\mathsf{LH}(X_i) \boxminus \mathsf{LH}(X_j) \neq \mathsf{LH}(X_k) \boxminus \mathsf{LH}(X_\ell)$.

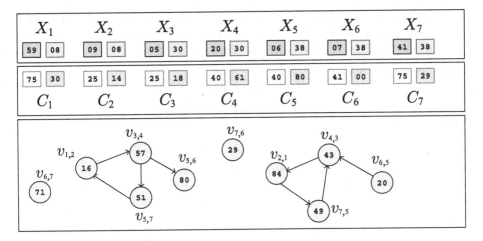

Fig. 8. Top: Seven pairs of plaintext/ciphertext $(X_1, C_1), \ldots, (X_7, C_7)$ on $\mathbb{Z}_M \times \mathbb{Z}_N$, with $M = N = 100$. Bottom: Differential graph $\mathcal{G}$ constructed from those seven pairs.

See Fig. 8 for an illustration of the graph $\mathcal{G}$. We call $\mathcal{G}$ the *differential graph* of the plaintext/ciphertexts pairs $(X_1, C_1), \ldots, (X_p, C_p)$.

<u>GOOD VERSUS BAD NODES.</u> For a node $v_{i,j}$ in the differential graph $\mathcal{G}$, we say that it is *good* if $\mathsf{RH}(X_i^2) = \mathsf{RH}(X_j^2)$; otherwise we say that it is *bad*. Lemma 6 below characterizes an important property of good nodes; it is a direct generalization of a result in DV's work; see the full version for the proof.

Lemma 6. *Let* $M, N \geq 10$ *be integers and let* $\mathsf{F} = \mathbf{Feistel}[4, M, N, \boxplus]$. *Let* $(X_1, C_1), \ldots, (X_p, C_p)$ *be plaintext/ciphertext pairs under* F *with a uniformly random key, where the plaintexts are chosen uniformly without replacement from* $\mathbb{Z}_M \times \mathbb{Z}_N$. *Let* $\mathcal{G} = (V, E)$ *be the differential graph of those pairs, and let* (G_1, G_2, G_3, G_4) *be the functions specified by the secret key of* F. *Then for any good node* $v_{i,j} \in V$, *we have* $\mathsf{Label}(v_{i,j}) = G_4(\mathsf{LH}(C_i)) \boxminus G_4(\mathsf{LH}(C_j))$.

The following Lemma 7 computes the average number of good and bad nodes, and estimates the average number of edges in the differential graph; see the full version for the proof.

Lemma 7. *Let* $M, N \geq 10$ *be integers and let* $\mathsf{F} = \mathbf{Feistel}[4, M, N, \boxplus]$. *Let* $(X_1, C_1), \ldots, (X_p, C_p)$ *be plaintext/ciphertext pairs under* F *with a uniformly random key, where the plaintexts are chosen uniformly without replacement from* $\mathbb{Z}_M \times \mathbb{Z}_N$. *Let* $\mathcal{G} = (V, E)$ *be the differential graph of those pairs, and let* Z *be the random variable for the number of good nodes in* $\mathcal{G}$. *Then*

(a) $\mathbf{E}[|V|] = \frac{p(p-1)(M-1)(M+N-1)}{MN(MN-1)}$.

(b) $\mathbf{E}[Z] = \frac{p(p-1)(M-1)}{(MN-1)N}$.

(c) $\mathbf{E}[|E|] \leq \frac{p!}{(p-4)!} \cdot \frac{(M+N)^2}{M^3 N^4}$.

Since $p = O(M^{2/3}N)$ and $M \geq N$, from Lemma 7, on average, the differential graph $\mathcal{G}$ contains about $O(M^{4/3})$ nodes, and the majority of them are good. In addition, there are on average $O(M^{5/3})$ edges in $\mathcal{G}$.

A FAST CONSTRUCTION OF DIFFERENTIAL GRAPHS. The naive approach to construct $\mathcal{G}$ would take $\Theta(p^2) = \Theta(M^{10/3})$ time just to construct the node set V. We now show how to build $\mathcal{G}$ in $O(M^{5/3})$ expected time; the code is given in Fig. 9.

- First, partition the pairs (X_i, C_i) based on $(\mathsf{LH}(C_i) \boxminus \mathsf{LH}(X_i), \mathsf{RH}(X_i))$. Using appropriate data structure, this takes $O(p)$ time.
- For each partition P, enumerate all distinct pairs $(X_i, C_i), (X_j, C_j) \in P$. Each such pair forms a node in V, and its label can be computed accordingly. This takes $O(|V|)$ time.
- Finally, partition V into M groups P_d with $d \in \mathbb{Z}_M$, such that each node $v_{i,j}$ goes to group $P_{\mathsf{LH}(C_j)}$. Also, partition V into M groups S_d with $d \in \mathbb{Z}_M$, such that each node $v_{k,\ell}$ goes to group $S_{\mathsf{LH}(C_k)}$. By enumerating elements in $P_d \times S_d$ (with some pruning) for every $d \in \mathbb{Z}_M$ via appropriate data structure, we can create the edge set E using

$$O\left(|V| + M + \sum_{i,j,k,\ell} D_{i,j,k,\ell}\right)$$

time, where $D_{i,j,k,\ell}$ is the indicator random variable for the event that (i) $v_{i,j} \in V$, (ii) $v_{k,\ell} \in V$, and (iii) $\mathsf{LH}(C_j) = \mathsf{LH}(C_k)$. The summation is taken over all distinct $i, j, k \in \{1, \ldots, p\}$ and $\ell \in \{1, \ldots, p\} \backslash \{k\}$. By pretending that (i), (ii), (iii) are independent, we can heuristically estimate that $\Pr[D_{i,j,k,\ell} = 1] \lesssim \frac{4}{MN^4}$.

Hence the total expected time is

$$O\left(p + M + \mathbf{E}[|V|] + \frac{4p^4}{MN^4}\right) = O(M^{5/3}) \ .$$

TRIANGLES. Recall that from Lemma 7, on average, the majority of nodes in the differential graph are good. We now describe a method to realize good nodes with high probability. A *triangle* of the graph $\mathcal{G}$ is a directed cycle of length 3. For a triangle $\mathcal{T} = (u_1, u_2, u_3, u_1)$, its *weight* $\mathsf{weight}(\mathcal{T})$ is defined as the sum of the labels, meaning that

$$\mathsf{weight}(\mathcal{T}) = \mathsf{Label}(u_1) \boxplus \mathsf{Label}(u_2) \boxplus \mathsf{Label}(u_3) \ .$$

DV observed that in the balanced setting, for a triangle $\mathcal{T}$, if all of its three nodes are good then its weight is 0. Lemma 8 below shows that their observation also holds in the unbalanced setting; see the full version for the proof.

Procedure BuildGraph($X_1, C_1, \ldots, X_p, C_p$)

$V, E \leftarrow \emptyset$; Initialize a hash table H

For $i = 1$ to p do $Z \leftarrow (\mathsf{LH}(C_i) \boxminus \mathsf{LH}(X_i), \mathsf{RH}(X_i))$; $P \leftarrow H[Z]$; $P \leftarrow P \cup \{i\}$

For each key Z in H do

   For $i, j \in H[Z]$, with $i \neq j$, do

      $v_{i,j} \leftarrow (i, j)$; $\mathsf{Label}(v_{i,j}) \leftarrow \mathsf{RH}(C_i) \boxminus \mathsf{RH}(C_j)$; $V \leftarrow V \cup \{v_{i,j}\}$

For $d \in \mathbb{Z}_M$ do $P_d, S_d \leftarrow \emptyset$

For $v \in V$ do $(i, j) \leftarrow v$; $P_{\mathsf{LH}(C_j)} \leftarrow P_{\mathsf{LH}(C_j)} \cup \{v\}$; $P_{\mathsf{LH}(C_i)} \leftarrow P_{\mathsf{LH}(C_i)} \cup \{v\}$

For $d \in \mathbb{Z}_M$ do

   Initialize a hash table H_d

   For $v = (k, \ell) \in S_d$ do $L \leftarrow H_d[k]$; $L \leftarrow L \cup \{v\}$

   For every $u = (i, j) \in P_d$ and every key $k \in H_d$ such that $k \notin \{i, j\}$ do

      For every $v = (k, \ell) \in H_d[k]$ such that $\ell \notin \{i, j\}$ do

      // Check non-degeneracy requirements

      If $\mathsf{LH}(C_i) \neq \mathsf{LH}(C_\ell)$ and $\mathsf{RH}(X_i) \neq \mathsf{RH}(X_k)$

        and $\mathsf{RH}(X_j) \boxminus \mathsf{RH}(X_k) \neq \mathsf{RH}(C_j) \boxminus \mathsf{RH}(C_k)$

        and $\mathsf{LH}(X_i) \boxminus \mathsf{LH}(X_j) \neq \mathsf{LH}(X_k) \boxminus \mathsf{LH}(X_\ell)$ then $E \leftarrow E \cup \{(u, v)\}$

Return $\mathcal{G} = (V, E)$

Fig. 9. Code for building the differential graph $\mathcal{G} = (V, E)$ of $X_1, C_1, \ldots, X_p, C_p$.

Lemma 8. *Let $M, N \geq 10$ be integers and let $\mathsf{F} = \mathbf{Feistel}[4, M, N, \boxplus]$. Let $(X_1, C_1), \ldots, (X_p, C_p)$ be plaintext/ciphertext pairs under F with a uniformly random key, where the plaintexts are chosen uniformly without replacement from $\mathbb{Z}_M \times \mathbb{Z}_N$. Let $\mathcal{G} = (V, E)$ be the differential graph of those pairs. For a triangle $\mathcal{T}$ in $\mathcal{G}$, if all the three nodes of $\mathcal{T}$ are good then $\mathsf{weight}(\mathcal{T}) = 0$.*

Above, we show that a triangle whose nodes are all good will have weight 0. The following Lemma 9 shows that the converse holds with very high probability; see the full version for the proof. This proves a conjecture in DV's work [15] that they empirically verified for the balanced, boolean case $M = N = 2^n$, with $n \in \{2, 3, \ldots, 9\}$. We also give a rigorous lower bound for the expected number of triangles whose all nodes are good, whereas DV could only give a heuristic estimation of this number.

Lemma 9. *Let $M, N \geq 19$ be integers and let $\mathsf{F} = \mathbf{Feistel}[4, M, N, \boxplus]$. Let $(X_1, C_1), \ldots, (X_p, C_p)$ be plaintext/ciphertext pairs under F with a uniformly random key, where the plaintexts are chosen uniformly without replacement from $\mathbb{Z}_M \times \mathbb{Z}_N$. Let $\mathcal{G} = (V, E)$ be the differential graph of those pairs.*

(a) For a triangle $\mathcal{T}$ in $\mathcal{G}$ of zero weight, the probability that all nodes in $\mathcal{T}$ are good is at least $\frac{1}{1+\epsilon}$, where $\epsilon = \frac{N}{M-9} \left(\frac{4}{M} + \frac{33N}{(N-2)M^2} + \frac{39}{M^2} \right)$.

```
Procedure GetTriangles(G, X_1, C_1, ..., X_p, C_p)
(V, E) ← G; L ← ∅; Initialize hash tables H_0, H_1, H_2
For every edge (u, v) ∈ E do
    d ← Label(u) ⊞ Label(v); P ← H_1[v, d]; P ← P ∪ {e}
    s ← 0 ⊟ Label(v); S ← H_2[u, s]; S ← S ∪ {e}; H_0[e] ← 1
For every key (v, d) of H_1, every u ∈ H_1[v, d], w ∈ H_2[v, d] do
    If H_0[(w, u)] = 1 then T ← (u, v, w); L ← L ∪ {T}
Return L
```

Fig. 10. Code to enumerate triangles of zero weight from the differential graph $G = (V, E)$ of $X_1, C_1, \ldots, X_p, C_p$.

(b) *The expected number of triangles in G whose all nodes are good is at least*
$$\frac{p!}{3(p-6)!} \cdot \frac{1}{M^3 N^6} \left(1 - \frac{7}{N} - \frac{14}{M}\right).$$

DISCUSSION. While the core idea of differential graphs is from DV's work, there are important differences between our definition and DV's:

- Because of the symmetry of Feistel, one can view $\mathsf{Rev}(X_1), \ldots, \mathsf{Rev}(X_p)$ as the "ciphertexts" of $\mathsf{Rev}(C_1), , \ldots, \mathsf{Rev}(C_p)$ under a four-round Feistel $\overline{\mathsf{F}} = \mathbf{Feistel}[4, N, M, \boxminus]$ where $\mathsf{Rev}(X) = (R, L)$ for any $X = (L, R) \in \mathbb{Z}_M \times \mathbb{Z}_N$. In this sense, DV's notion is the dual of ours. While the two definitions yield no difference in the balanced setting, if $M \gg N$, DV's notion would give a much poorer bound[8] in a dual version of Lemma 9, leading to an inferior recovery rate of the TF attack.
- Our notion also adds some non-degeneracy requirements, allowing us to prove DV's conjecture on differential graph in Lemma 9 further below.

ENUMERATING ZERO-WEIGHT TRIANGLES. From Lemma 9, a simple way to realize good nodes is to enumerate all triangles of zero weight. We now show how to do that in $O(M^{5/3})$ expected time; the code is given in Fig. 10.

- First, for each node $v \in V$, partition its set of incoming edges such that each edge (u, v) goes to group $P_{v,d}$, with $d = \mathsf{Label}(u) \boxplus \mathsf{Label}(v)$, and also partition the set of outgoing edges into groups such that each edge (v, w) goes to group $S_{v,s}$, where $s = 0 \boxminus \mathsf{Label}(w)$.
- Next for each (v, d), enumerate all pairs $(u, w) \in P_{v,d} \times S_{v,d}$ such that there is a directed edge $(w, u) \in E$. Each triple (v, u, w) is a triangle of zero weight.

---

[8] In fact, the dual version of Lemma 9 would yield the bound $\frac{1}{1+\epsilon^*}$, where $\epsilon^* = \frac{M}{N-9}\left(\frac{4}{N} + \frac{33M}{(M-2)N^2} + \frac{39}{N^2}\right)$. Concretely, for $M = 100$ and $N = 10$, the bound in our Lemma 9 is 0.9947, whereas its dual is much poorer, just 0.0089.

Using appropriate data structure, the first step takes $O(|V|+|E|)$ time, whereas the cost of the second step is in the order of

$$\sum_{v \in V} \sum_{d \in \mathbb{Z}_N} |P_{v,d}| \cdot (1 + |S_{v,d}|) \leq \left(\sum_{v \in V} \sum_{d \in \mathbb{Z}_N} |P_{v,d}|\right) + \left(\sum_{v \in V} \sum_{d \in \mathbb{Z}_N} |P_{v,d}| \cdot |S_{v,d}|\right)$$

$$= \left(\sum_{v \in V} \mathrm{indeg}(v)\right) + \left(\sum_{v \in V} \sum_{d \in \mathbb{Z}_N} |P_{v,d}| \cdot |S_{v,d}|\right)$$

$$= |E| + \left(\sum_{v \in V} \sum_{d \in \mathbb{Z}_N} |P_{v,d}| \cdot |S_{v,d}|\right) ,$$

where $\mathrm{indeg}(v)$ is the incoming degree of node v. Since $\mathbf{E}[|V|] \in O(M^{4/3})$ and $\mathbf{E}[|E|] \in O(M^{5/3})$, what remains is to show that

$$\mathbf{E}\left[\sum_{v \in V} \sum_{d \in \mathbb{Z}_N} |P_{v,d}| \cdot |S_{v,d}|\right] \in O(M^{5/3}) . \tag{2}$$

For each tuple $\mathcal{L} = (i,j,k,\ell,r,s,d) \in (\{1,\dots,p\})^6 \times \mathbb{Z}_N$ such that i,j,k,ℓ,r,s are distinct, let $B_{\mathcal{L}}$ denote the Bernoulli random variable such that $B_{\mathcal{L}} = 1$ if and only if $(v_{i,j}, v_{k,\ell})$ and $(v_{r,s}, v_{i,j})$ are edges of $\mathcal{G}$, and $v_{k,\ell} \in S_{v_{i,j},d}$ and $v_{r,s} \in P_{v_{i,j},d}$. Then

$$\mathbf{E}\left[\sum_{v \in V} \sum_{d \in \mathbb{Z}_N} |P_{v,d}| \cdot |S_{v,d}|\right] = \mathbf{E}\left[\sum_{\mathcal{L}} B_{\mathcal{L}}\right] = \sum_{\mathcal{L}} \mathbf{E}[B_{\mathcal{L}}] = \sum_{\mathcal{L}} \Pr[B_{\mathcal{L}} = 1] .$$

Note that for each $\mathcal{L} = (i,j,k,\ell,r,s,d)$, the event $B_{\mathcal{L}} = 1$ happens only if the following events happen: (1) $v_{i,j} \in V$, (2) $v_{j,k} \in V$, (3) $v_{r,s} \in V$, (4) $\mathsf{LH}(C_j) = \mathsf{LH}(C_k)$, (5) $\mathsf{LH}(C_s) = \mathsf{LH}(C_i)$, (6) $\mathsf{Label}(v_{i,j}) \boxplus \mathsf{Label}(v_{r,s}) = d$, and (7) $\mathsf{Label}(v_{k,\ell}) = 0 \boxminus d$. By pretending that these seven events are independent, from Lemma 3, we can heuristically estimate

$$\Pr[B_{\mathcal{L}} = 1] \lesssim \left(\frac{M+N-1}{MN} \cdot \frac{1}{N}\right)^3 \left(\frac{1}{M}\right)^2 \left(\frac{1}{N}\right)^2 \leq \frac{(M+N)^3}{M^5 N^8} \leq \frac{8}{M^2 N^8} .$$

Hence

$$\sum_{\mathcal{L}} \mathbf{E}[B_{\mathcal{L}}] \lesssim \frac{p! \cdot N}{(p-6)!} \cdot \frac{8}{M^2 N^8} \in O(M^2/N) .$$

Moreover, due to our assumption that $N = \Omega(\sqrt{M})$, it follows that $M^2/N \in O(M^{1.5})$. We then conclude that

$$\mathbf{E}\left[\sum_{v \in V} \sum_{d \in \mathbb{Z}_N} |P_{v,d}| \cdot |S_{v,d}|\right] \in O(M^{1.5})$$

and thus justify Eq. (2).

REMARKS. DV also considered the problem of finding zero-weight triangles for the balanced case $M = N$. They first enumerated all triangles via sparse matrix multiplications, and then computed the sum of labels for each triangle. In this balanced setting, DV's algorithm takes $O(N^3)$ time, whereas our algorithm takes $O(N^{5/3})$ time.

4.2   The TF Attack

We begin with a simple but useful observation of DV on four-round Feistel.

AN OBSERVATION. Let (F_1, F_2, F_3, F_4) be the round functions of F. For any $\Delta \in \mathbb{Z}_N$, let $\mathsf{Shift}(\mathsf{F}, \Delta)$ denote a 4-round Feistel network $\overline{\mathsf{F}} = \mathbf{Feistel}[4, M, N, \boxplus]$ of round functions $(\overline{F}_1, \overline{F}_2, \overline{F}_3, \overline{F}_4)$ such that $\overline{F}_1 = F_1$, $\overline{F}_2(K, x) = F_2(K, x) \boxminus \Delta$, $\overline{F}_3(K, y) = F_3(K, y \boxplus \Delta)$, and $\overline{F}_4(K, x) = F_4(K, x) \boxplus \Delta$, for any $x \in \mathbb{Z}_M, y \in \mathbb{Z}_N$, and any key K. Note that for any choice of Δ, scheme $\overline{\mathsf{F}} = \mathsf{Shift}(\mathsf{F}, \Delta)$ ensures that $\overline{\mathsf{F}}.\mathsf{E}(K, X) = \mathsf{F}.\mathsf{E}(K, X)$ for any key K and any $X \in \mathbb{Z}_M \times \mathbb{Z}_N$. Therefore, in a codebook recovery attack against $\mathsf{F}.\mathsf{E}(K, \cdot)$, without loss of generality, one can pick a designated point $x^* \in \mathbb{Z}_M$ and *assume* that $F_4(K, x^*) = 0$.

THE ATTACK. Let (G_1, G_2, G_3, G_4) be the functions specified by the secret key of F. We will recover even the tables of those functions, instead of just the codebook.

Our attack TF is based on a known-plaintext codebook-recovery attack RY on three-round Feistel that we describe in the full version. If we run RY on $\ell \geq \max\{\lceil N(\ln(N)+\ln(2)+\lambda)\rceil, \lceil M(\ln(M)+\delta)\rceil\}$ known plaintext/ciphertext pairs, for any $\lambda, \delta > 0$, the RY attack will take $O(\ell)$ time, and recovers all the round functions of the three-round Feistel with probability around $e^{-e^{-\lambda}} - e^{-\delta}$. If we just have $\ell \geq \lceil N(\ln(N)+\ln(2)+\lambda)\rceil$ then RY will recover the top round function with probability at least $e^{-e^{-\lambda}}$. We note that RY is used in a modular way; one does not need to know the technical details of RY to understand the TF attack.

While TF somewhat resembles DV's attack on 4-round Feistel, there are important changes to improve efficiency and recovery rate, which we will elaborate further below. The code of TF is given in Fig. 11; below we will describe the attack.

In the TF attack, we will first construct the differential graph $\mathcal{G} = (V, E)$ of the plaintext/ciphertext pairs $(X_1, C_1), \ldots, (X_p, C_p)$, and then enumerate all triangles of $\mathcal{G}$ of zero weight. Let S be the set of the nodes of those triangles; each node S is very likely to be good, due to Lemma 9. For each $v_{i,j} \in S$, from Lemma 6, if $v_{i,j}$ is indeed good then $\mathsf{Label}(v_{i,j}) = G_4(\mathsf{LH}(C_i)) \boxminus G_4(\mathsf{LH}(C_j))$.

Our first step is to recover several (but possibly not all) entries of G_4. In order to do that, construct the following undirected graph $\mathcal{G}^* = (V^*, E^*)$ of $|V^*| = M$ nodes. Nodes in V^* are distinctly labeled by elements of $\mathbb{Z}_M$. For each node $v_{i,j} \in S$, we create an edge between nodes $\mathsf{LH}(C_i)$ and $\mathsf{LH}(C_j)$ of V^*, indicating that we know the difference between $G_4(\mathsf{LH}(C_i))$ and $G_4(\mathsf{LH}(C_j))$. Once the graph $\mathcal{G}^*$ is constructed, we pick an arbitrary node $x^* \in V^*$ that belongs to the biggest connected component of $\mathcal{G}^*$, and set $G_4(x^*) \leftarrow 0$. We then recover $G_4(u)$ for every node $u \in V^*$ reachable from x^* using breadth-first search (BFS), but stop when $\left\lfloor \frac{3M}{\sqrt{N}} \right\rfloor$ entries of G_4 are recovered. Let $\mathcal{I} \subseteq \{1, 2, \ldots, p\}$ be the set of indices i such that $G_4(\mathsf{LH}(C_i))$ is recovered at this point.

Our next step is to recover the entire table of G_1 using RY. For each $i \in \mathcal{I}$, recover the round-3 intermediate output Y_i of X_i via $\mathsf{LH}(Y_i) = \mathsf{LH}(C_i)$ and

---

Procedure $\mathsf{TF}(X_1, C_1, \ldots, X_p, C_p)$

$\mathcal{G} \leftarrow \mathsf{BuildGraph}(X_1, C_1, \ldots, X_p, C_p)$ // Build the differential graph

$L \leftarrow \mathsf{GetTriangles}(\mathcal{G}, X_1, C_1, \ldots, X_p, C_p)$ // Enumerate zero-weight triangles

$V^*, E^*, S \leftarrow \emptyset$; Initialize a hash table H

For every $\mathcal{T} \in L$ do $(u, v, w) \leftarrow \mathcal{T}$; $S \leftarrow S \cup \{u, v, w\}$

For $i \in \mathbb{Z}_M$ do $V^* \leftarrow V^* \cup \{i\}$

For every $v \in S$ do

$\quad (i, j) \leftarrow v$; $e \leftarrow \{\mathsf{LH}(C_i), \mathsf{LH}(C_j)\}$; $E^* \leftarrow E^* \cup \{e\}$

$\quad H[e] \leftarrow (\mathsf{LH}(C_i), \mathsf{Label}(v))$

$\mathcal{G}^* \leftarrow (V^*, E^*)$; Let $\mathcal{C}$ be the biggest connected component of $\mathcal{G}^*$

For $i \leftarrow 1$ to μ do

$\quad (G_1, G_2, G_3, G_4) \leftarrow \mathsf{Restore}(X_1, C_1, \ldots, X_p, C_p, \mathcal{C})$

$\quad$ For $j \leftarrow 1$ to $3M$ do // Checking consistency of (G_1, G_2, G_3, G_4)

$\quad\quad (L, R) \leftarrow X_j$

$\quad\quad$ For $k \leftarrow 1$ to 4 do

$\quad\quad\quad$ If $(k \bmod 2 = 1)$ then $L \leftarrow L \boxplus G_k(R)$ else $R \leftarrow R \boxplus G_k(L)$

$\quad\quad C_j^* \leftarrow (L, R)$

$\quad$ If $(C_1, \ldots, C_{3M}) = (C_1^*, \ldots, C_{3M}^*)$ then return (G_1, G_2, G_3, G_4)

Fig. 11. The TF attack (parameterized by a small number μ) on 4-round Feistel, which is based on another attack RY on 3-round Feistel and a procedure Restore in Fig. 12.

$\mathsf{RH}(Y_i) = G_4(\mathsf{LH}(C_i)) \boxminus \mathsf{RH}(C_i)$. Then run RY on the pairs $\{(X_i, Y_i) \mid i \in \mathcal{I}\}$ to recover G_1, and then recover the round-1 intermediate outputs $Z_1, \ldots, Z_p$ of $X_1, \ldots, X_p$. In addition, observe that $\mathsf{Rev}(C_i)$ is the ciphertext of $\mathsf{Rev}(Z_i)$ under a 3-round Feistel $\overline{\mathsf{F}} = \mathbf{Feistel}[3, N, M, \boxplus]$ of round functions G_2, G_3, G_3 (note that the roles of M and N are now reversed), where for $Z = (A, B) \in \mathbb{Z}_M \times \mathbb{Z}_N$, we write $\mathsf{Rev}(Z)$ to denote the pair $(B, A) \in \mathbb{Z}_N \times \mathbb{Z}_M$. We then run RY on $\mathsf{Rev}(Z_1), \ldots, \mathsf{Rev}(Z_p), \mathsf{Rev}(C_1), \ldots, \mathsf{Rev}(C_p)$ to recover G_2, G_3, G_4.

To amplify the recovery rate, instead of using just one random node x^*, we try μ independent choices of x^*; in our implementation, we pick $\mu = 10$.[9] As analyzed in Sect. 3.3, we can decide which node yields a correct output by evaluating $3M$ plaintexts on a Feistel network with the recovered round functions, and comparing them with the corresponding ciphertexts.

ANALYSIS. We now analyze the advantage of the TF attack; the key ideas in our analysis are largely from DV's work. We however tighten some of their arguments to improve the bounds.

---

[9] We note that here $\mu = 10$ means that the attack will iterate up to 10 times, each time with an independent choice of the initial node x^*, until it succeeds in recovering the entire codebook. The expected number of the iterations is often smaller than 10. For example, with $M = 1000$ and $N = 100$, empirically the attack would succeed at the first iteration, and thus it only performs a single iteration.

Procedure Restore$(X_1, C_1, \ldots, X_p, C_p, \mathcal{C})$

Pick a node x^* uniformly at random from $\mathcal{C}$

Initialize tables G_1, G_2, G_3, G_4; $G_4(x^*) \leftarrow 0$; $count \leftarrow 0$

Run a breadth-first search on $\mathcal{C}$ from x^* and let T be the corresponding BFS tree

Let $(v_0, \ldots, v_t)$ be the visiting order of the nodes in the BFS above

For $(k = 1$ to $t)$ and $count \leq \lfloor 3M/\sqrt{N} \rfloor$ do

   $count \leftarrow count + 1$; Let u be the parent of v_k in T

   $e \leftarrow \{u, v_k\}$; $(w, R) \leftarrow H[e]$

   If $u = w$ then $G_4(v_k) \leftarrow G_4(u) \boxminus R$ else $G_4(v_k) \leftarrow G_4(u) \boxplus R$

$\mathcal{I} \leftarrow \{i \mid G_4(\mathsf{LH}(C_i)) \neq \perp\}$ // Consider the entire set $\{1, \ldots, p\}$ to find $\mathcal{I}$

// Recover the round-3 intermediate outputs for X_i with $i \in \mathcal{I}$

For $i \in \mathcal{I}$ do $Y_i \leftarrow \big(\mathsf{LH}(C_i), \mathsf{RH}(C_i) \boxminus G_4(\mathsf{LH}(C_i))\big)$

Run RY on (X_i, Y_i) for $i \in \mathcal{I}$ to recover G_1 // Here RY attacks domain $\mathbb{Z}_M \times \mathbb{Z}_N$

For $i = 1$ to p do // Recover round-1 intermediate outputs for every X_i

   $(L_0, R_0) \leftarrow X_i$; $L_1 \leftarrow L_0 \boxplus G_1(R_0)$; $R_1 \leftarrow R_0$; $Z_i \leftarrow (L_1, R_1)$

For $i = 1$ to p do $Z_i' \leftarrow \mathsf{Rev}(Z_i)$; $C_i' \leftarrow \mathsf{Rev}(C_i)$

Run RY on (Z_i', C_i') to recover (G_2, G_3, G_4) // Now RY attacks domain $\mathbb{Z}_N \times \mathbb{Z}_M$

Return (G_1, G_2, G_3, G_4)

Fig. 12. Procedure Restore in the TF attack. Here for $Z = (A, B) \in \mathbb{Z}_M \times \mathbb{Z}_N$, we write $\mathsf{Rev}(Z)$ to denote the pair $(B, A) \in \mathbb{Z}_N \times \mathbb{Z}_M$.

Table 3. Empirical estimation of $|E^*|$ over 100 trials. The first row indicates the values of M and N. The second row shows the 95% confidence interval of $|E^*|$.

(M, N)	$(2^7, 2^6)$	$(2^7, 2^7)$	$(2^8, 2^7)$	$(2^8, 2^8)$	$(2^9, 2^8)$	$(2^9, 2^9)$	$(10^2, 10^2)$	$(10^3, 10^2)$
$|E^*|$	217 ± 6	203 ± 6	270 ± 6	321 ± 6	802 ± 11	965 ± 11	156 ± 5	1576 ± 16

$\triangleright$ We begin by estimating $\mathbf{E}[|E^*|]$. A direct generalization of DV's analysis would yield $\mathbf{E}[|E^*|] \approx \frac{p^6}{M^3 N^6}$, which is rather loose. Consider the empirical estimation of $|E^*|$ in Table 3. For $M = N = 100$, the 95% confidence interval of $|E^*|$ is 156 ± 5, but the approximation above suggests that $\mathbf{E}[|E^*|] \approx 400$.

We now provide a tighter analysis. Let W be the number of triangles in $\mathcal{G}$ whose all three nodes are good. From part (a) of Lemma 9, when we enumerate zero-weight triangles in $\mathcal{G}$, most of them will have three good nodes. Thus those triangles will contribute approximately $3W$ distinct good nodes. Note that if $\mathcal{T} = (v_{i,j}, v_{k,\ell}, v_{r,s})$ is a triangle then $\mathcal{T}^* = (v_{j,i}, v_{s,r}, v_{\ell,k})$ is also a triangle, and $\mathsf{weight}(\mathcal{T}^*) = 0 \boxminus \mathsf{weight}(\mathcal{T})$. (See Fig. 9 for an illustration.) Hence if $\mathcal{T}$ is a zero-weight triangle then so is $\mathcal{T}^*$, but these two triangles will produce the same three edges for E^*. Taking into account this duplication, $\mathbf{E}[|E^*|] \approx \frac{3\mathbf{E}[W]}{2}$. From part (b) of Lemma 9,

$$\mathbf{E}[W] \gtrsim \frac{p^6}{3M^3N^6}\left(1 - \frac{7}{N} - \frac{14}{M}\right) \geq \frac{2p^6}{9M^3N^6}$$

for $M \geq N \geq 64$. Hence

$$\mathbf{E}[|E^*|] \gtrsim \frac{p^6}{3M^3N^6} \gtrsim \frac{4M}{3} \quad . \tag{3}$$

This lower bound is consistent with Table 3. For example, for $M = N = 100$, we estimate that $\mathbf{E}[|E^*|] \gtrsim 133$, and recall that empirically, the 95% confidence interval of $|E^*|$ is 156 ± 5.

▷ Next, following DV, we model the graph $\mathcal{G}^*$ as a random graph according to the Erdős-Rényi model, in which each of the $\binom{M}{2}$ possible edges will have probability ρ to appear in E^*, independent of other edges. To determine the parameter ρ, note that according to the model above, the expected number of edges in E^* is

$$\binom{M}{2}\rho = \frac{M(M-1)\rho}{2} \quad . \tag{4}$$

From Eqs. (3) and (4), we have $M\rho = \frac{2\mathbf{E}[|E^*|]}{M-1} \gtrsim \frac{8M}{3}$. From the theory of random graph (see, for example, Chapter 2 of Durrett's book [17]), the graph $\mathcal{G}^*$ will almost surely contain a giant component of size about $(1-c)M$ or bigger, where $c \approx 0.0878$ is the unique solution of the equation $e^{\frac{8}{3}(t-1)} = t$ in the interval $(0,1)$. In DV's attack, one recovers $G_4(u)$ for every node u in the giant component, but since S *may* contain a few bad nodes, some entries of G_4 that we recover might be incorrect. Instead, we only recover $G_4(u)$ for nodes u in a connected subgraph of the giant component of size $\left\lfloor \frac{3M}{\sqrt{N}} \right\rfloor$. Those nodes are produced by about $\left\lfloor \frac{M}{\sqrt{N}} \right\rfloor$ triangles of zero-weight, and thus from Lemma 9, the chance that the nodes of these triangles are good is at least $1 - \frac{\sqrt{N}}{(M-9)} \cdot \left(4 + \frac{33N}{(N-2)M} + \frac{39}{M}\right)$. On the other hand, expectedly, we obtain about

$$\frac{3p}{\sqrt{N}} \geq 3 \cdot 2^{1/3} M^{2/3}\sqrt{N} \geq N(\ln(N) + \ln(2) + 2.7)$$

pairs of plaintext/ciphertext for RY, where the second inequality is due to the fact that $M \geq N \geq 64$. Thus we can run RY to recover G_1 with probability at least $e^{-e^{-2.7}} > 0.935$. We then can run RY with

$$p \geq M(\ln(M) + 5) \geq \max\{N(\ln(N) + 5), M(\ln(M) + \ln(2) + 4.3)\}$$

inputs, and thus can recover (G_2, G_3, G_4) with probability at least $e^{-e^{-4.3}} - e^{-5} > 0.975$. Summing up, if we just try one node x^* then our recovery rate is at least $0.91 - \frac{\sqrt{N}}{(M-9)} \cdot \left(4 + \frac{33N}{(N-2)M} + \frac{39}{M}\right)$. Using μ independent choices of x^* can only

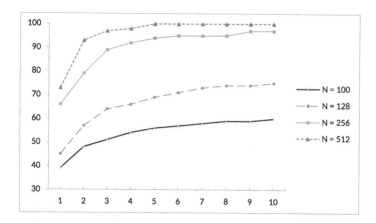

Fig. 13. The performance of TF with respect to μ, on balanced domains $\mathbb{Z}_N \times \mathbb{Z}_N$, over 100 trials. The x-axis indicates the values of μ, and the y-axis shows how many trials, out of 100 ones, that TF can recover the entire codebook.

improve the success probability. In fact, as illustrated in Fig. 13, empirically, the improvement when we increase μ from 1 to 10 is substantial.

▷ As mentioned in Sect. 4.1, constructing the differential graph $\mathcal{G}$ takes $O(M^{5/3})$ expected time, and so does enumerating zero-weight triangles. The graph $\mathcal{G}^*$ has M nodes, and expectedly, around $\frac{p^6}{3M^3N^6} \in O(M)$ edges. Thus identifying the connected components of $\mathcal{G}^*$ and doing BFS on its largest component takes $O(|V^*| + \mathbf{E}[|E^*|]) = O(M)$ expected time. Running RY takes $O(p)$ time. Hence the total running time is $O(M^{5/3})$.

COMPARISON WITH DV'S ATTACK. While our attack is inspired by DV's attack, there are important changes:

- First, as mentioned earlier, compared to DV's notion of differential graphs, we actually use a dual definition, for better recovery rate. Our notion also adds some non-degeneracy requirements, allowing us to find a proof for Lemma 9 and resolve DV's conjecture.
- Next, our attack has a much faster way to enumerate triangles of zero weight, reducing the running time from $O(N^3)$ to $O(N^{5/3})$ in the balanced setting.
- Recall that some zero-weight triangles may contain bad nodes, creating some noise in the attack. DV mentioned that their attack hardly succeeded for $2N^{5/3}$ or more plaintext/ciphertext pairs, and posed an open question to eliminate the noise. To resolve this issue, we introduce the trick of exploring the giant component from μ random places, and from each place, we stop after visiting $\lceil 3M/\sqrt{N} \rceil$ nodes.
- DV only run RY once to recover (G_1, G_2, G_3), and then derive the round-3 intermediate W_i values of X_i, and use (W_i, C_i) pairs to recover G_4. This works well for DV, as they consider just the balanced setting $M = N$. However, in unbalanced settings, for the first run of RY, we only have $3p/\sqrt{N} \ll M \ln(M)$

Table 4. Empirical performance of the LHD attack over 100 trials. The first row indicates the values of M and N. The second row shows how many times, over 100 trials, that LHD correctly outputs 1 in the real world. The last row shows how many times, again over 100 trials, that LHD incorrectly outputs 1 in the ideal world.

(M, N)	$(2^7, 2^6)$	$(2^7, 2^7)$	$(2^8, 2^7)$	$(2^8, 2^8)$	$(2^9, 2^8)$	$(2^9, 2^9)$	$(10^2, 10^2)$	$(10^3, 10^2)$
Real	100	100	100	100	100	100	100	100
Ideal	0	0	0	0	0	0	1	0

inputs. Thus the chance that one can recover (G_1, G_2, G_3) by using one RY call is poor. We therefore only use the first RY call to get G_1, and run RY another time to recover (G_2, G_3, G_4).

5   Experiments

In this section, we empirically evaluate the LHD, SD, and TF attacks.

BENCHMARKING ENVIRONMENT. We implemented our attack in C++, and ran experiments using 72 threads in a server of dual Intel(R) Xeon(R) CPU E5-2699 v3 2.30 GHz CPU and 256 GB RAM. We evaluate our attacks in both balanced and unbalanced settings, and for both binary and decimal domains. Specifically, we consider every $(M, N) \in \{(2^7, 2^6), (2^7, 2^7), (2^8, 2^7), (2^8, 2^8), (2^9, 2^8), (2^9, 2^9), (10^2, 10^2), (10^3, 10^2)\}$. For each choice of (M, N), we let

$$p = \max\left\{\lfloor 2^{1/3} M^{2/3} N \rfloor, \lceil M(\ln(M) + 5) \rceil\right\}$$

as specified in Eq. (1).

EVALUATING LHD. For each domain $\mathbb{Z}_M \times \mathbb{Z}_N$, we sample p messages uniformly without replacement from $\mathbb{Z}_M \times \mathbb{Z}_N$, and then extract $m = \lceil \frac{p}{N} \cdot \lceil 32 N^{1/6} \rceil \rceil$ t-wise right-matching plaintexts, with $t = \lceil \frac{mM}{p} \rceil$. In the real world, we encrypt the plaintexts using the 4-round version of FF3 with the all-zero tweak to produce m ciphertexts. In contrast, the ciphertexts are chosen uniformly without replacement from $\mathbb{Z}_M \times \mathbb{Z}_N$ in the ideal world. The results of our experiments, given in Table 4, show that LHD is nearly perfect, which is much better than our theoretical estimation in Sect. 3.2. This is not surprising, since our analysis is very conservative.

EVALUATING TF. For each domain $\mathbb{Z}_M \times \mathbb{Z}_N$, we sample p messages uniformly without replacement from $\mathbb{Z}_M \times \mathbb{Z}_N$, and generate p ciphertexts using 4-round FF3 with the all-zero tweak. We consider all choices of μ from 1 to 10. The results of our experiments, given in Table 5, are consistent with the theory. For example, with $M = N = 128$ and $\mu = 1$, the attack is supposed to recover the entire codebook with probability around

$$0.91 - \frac{\sqrt{N}}{M - 9} \cdot \left(4 + \frac{33N}{(N - 2)M} + \frac{39}{M}\right) \approx 47.6\%$$

Table 5. Empirical performance of the TF attack over 100 trials. The first row indicates the values of M and N. Each subsequent row shows how many trials, over 100 ones, that TF correctly recovers the entire codebook, for the given choice of μ.

(M, N)	$(2^7, 2^6)$	$(2^7, 2^7)$	$(2^8, 2^7)$	$(2^8, 2^8)$	$(2^9, 2^8)$	$(2^9, 2^9)$	$(10^2, 10^2)$	$(10^3, 10^2)$
Recover, $\mu = 1$	70	45	75	66	84	73	39	100
Recover, $\mu = 2$	77	57	88	79	93	93	48	100
Recover, $\mu = 3$	81	64	91	89	95	97	51	100
Recover, $\mu = 4$	84	66	94	92	99	98	54	100
Recover, $\mu = 5$	85	69	95	94	100	100	56	100
Recover, $\mu = 6$	86	71	96	95	100	100	57	100
Recover, $\mu = 7$	87	73	97	96	100	100	58	100
Recover, $\mu = 8$	87	74	97	97	100	100	59	100
Recover, $\mu = 9$	88	74	98	97	100	100	59	100
Recover, $\mu = 10$	88	75	98	97	100	100	60	100

Table 6. Empirical performance of the SD attack over 100 trials. The first row indicates the values of M and N. The second row shows how many trials, over 100 ones, have at least one slid pair, the third row shows how many of them survive the LHD tests, and the last row shows how many of them can successfully recover the entire codebook.

(M, N)	$(2^7, 2^6)$	$(2^7, 2^7)$	$(2^8, 2^7)$	$(2^8, 2^8)$	$(2^9, 2^8)$	$(2^9, 2^9)$	$(10^2, 10^2)$	$(10^3, 10^2)$
Have slid pairs	61	65	53	53	38	33	75	27
Survive LHD tests	61	65	53	53	38	33	75	27
Recover	50	39	51	50	38	33	22	27

Table 7. The total number of survived (possibly false) candidates after using LHD tests in SD, over 100 trials. The first row indicates the values of M and N. The second row shows the total number of survived (possibly false) candidates after using LHD tests in SD, over 100 trials.

(M, N)	$(2^7, 2^6)$	$(2^7, 2^7)$	$(2^8, 2^7)$	$(10^2, 10^2)$
No. of candidates	74	72	53	89

and in the experiments, 45 out of 100 trials yield the correct codebook. Increasing μ will improve the performance substantially. For example, with $\mu = 10$, the recovery rate goes up to 75%.

EVALUATING SD. To save time in evaluating SD, we use the FF3 key to find true candidates and run the LHD tests and the TF attack on them. Table 6 reports the empirical performance of SD over 100 trials, in which we use $\mu = 10$ for the underlying TF attack. The recovery rate is reasonable, ranging from 22% to 51%, and we never miss any true candidate using LHD tests. In addition, we also run the full SD attack on $(M, N) \in \{(2^7, 2^6), (2^7, 2^7), (2^8, 2^7), (10^2, 10^2)\}$ to evaluate the performance of LHD test on false slid-pair candidates. As shown in Table 7, our test is a nearly perfect filtering, leaving on average a single (possibly false) slid-pair candidate in each trial.

Acknowledgments. We thank anonymous reviewers of EUROCRYPT 2019 for insightful feedback. Viet Tung Hoang was supported by NSF grants CICI-1738912 and CRII-1755539. Ni Trieu was supported by NSF award #1617197.

References

1. Aiello, W., Venkatesan, R.: Foiling birthday attacks in length-doubling transformations. In: Maurer, U. (ed.) EUROCRYPT 1996. LNCS, vol. 1070, pp. 307–320. Springer, Heidelberg (1996). https://doi.org/10.1007/3-540-68339-9_27
2. Bar-On, A., Biham, E., Dunkelman, O., Keller, N.: Efficient slide attacks. J. Cryptol. **31**(3), 641–670 (2018)
3. Bellare, M., Hoang, V.T.: Identity-based format-preserving encryption. In: CCS 2017, pp. 1515–1532 (2017)
4. Bellare, M., Hoang, V.T., Tessaro, S.: Message-recovery attacks on Feistel-based format preserving encryption. In: Weippl, E.R., Katzenbeisser, S., Kruegel, C., Myers, A.C., Halevi, S. (eds.) ACM CCS 16, pp. 444–455. ACM Press, October 2016
5. Bellare, M., Hoang, V.T., Tessaro, S.: Message-recovery attacks on Feistel-based format preserving encryption. In: CCS 2016 (2016)
6. Bellare, M., Ristenpart, T., Rogaway, P., Stegers, T.: Format-preserving encryption. In: Jacobson, M.J., Rijmen, V., Safavi-Naini, R. (eds.) SAC 2009. LNCS, vol. 5867, pp. 295–312. Springer, Heidelberg (2009). https://doi.org/10.1007/978-3-642-05445-7_19
7. Bellare, M., Rogaway, P.: The security of triple encryption and a framework for code-based game-playing proofs. In: Vaudenay, S. (ed.) EUROCRYPT 2006. LNCS, vol. 4004, pp. 409–426. Springer, Heidelberg (2006). https://doi.org/10.1007/11761679_25
8. Biham, E., Biryukov, A., Dunkelman, O., Richardson, E., Shamir, A.: Initial observations on Skipjack: cryptanalysis of Skipjack-3XOR (invited talk). In: Tavares, S., Meijer, H. (eds.) SAC 1998. LNCS, vol. 1556, pp. 362–375. Springer, Heidelberg (1999). https://doi.org/10.1007/3-540-48892-8_27
9. Biryukov, A., Leurent, G., Perrin, L.: Cryptanalysis of Feistel networks with secret round functions. In: Dunkelman, O., Keliher, L. (eds.) SAC 2015. LNCS, vol. 9566, pp. 102–121. Springer, Cham (2016). https://doi.org/10.1007/978-3-319-31301-6_6
10. Biryukov, A., Wagner, D.: Slide attacks. In: Knudsen, L. (ed.) FSE 1999. LNCS, vol. 1636, pp. 245–259. Springer, Heidelberg (1999). https://doi.org/10.1007/3-540-48519-8_18
11. Biryukov, A., Wagner, D.: Advanced slide attacks. In: Preneel, B. (ed.) EUROCRYPT 2000. LNCS, vol. 1807, pp. 589–606. Springer, Heidelberg (2000). https://doi.org/10.1007/3-540-45539-6_41
12. Black, J., Rogaway, P.: Ciphers with arbitrary finite domains. In: Preneel, B. (ed.) CT-RSA 2002. LNCS, vol. 2271, pp. 114–130. Springer, Heidelberg (2002). https://doi.org/10.1007/3-540-45760-7_9

13. Brier, E., Peyrin, T., Stern, J.: BPS: a format-preserving encryption proposal. Submission to NIST (2010). http://csrc.nist.gov/groups/ST/toolkit/BCM/documents/proposedmodes/bps/bps-spec.pdf
14. Dara, S., Fluhrer, S.: FNR: arbitrary length small domain block cipher proposal. In: Chakraborty, R.S., Matyas, V., Schaumont, P. (eds.) SPACE 2014. LNCS, vol. 8804, pp. 146–154. Springer, Cham (2014). https://doi.org/10.1007/978-3-319-12060-7_10
15. Durak, F.B., Vaudenay, S.: Breaking the FF3 format-preserving encryption standard over small domains. In: Katz, J., Shacham, H. (eds.) CRYPTO 2017. LNCS, vol. 10402, pp. 679–707. Springer, Cham (2017). https://doi.org/10.1007/978-3-319-63715-0_23
16. Durak, F.B., Vaudenay, S.: Generic round-function-recovery attacks for Feistel networks over small domains. In: Preneel, B., Vercauteren, F. (eds.) ACNS 2018. LNCS, vol. 10892, pp. 440–458. Springer, Cham (2018). https://doi.org/10.1007/978-3-319-93387-0_23
17. Durrett, R.: Random Graph Dynamics. Cambridge University Press, Cambridge (2008)
18. Hoang, V.T., Morris, B., Rogaway, P.: An enciphering scheme based on a card shuffle. In: Safavi-Naini, R., Canetti, R. (eds.) CRYPTO 2012. LNCS, vol. 7417, pp. 1–13. Springer, Heidelberg (2012). https://doi.org/10.1007/978-3-642-32009-5_1
19. Hoang, V.T., Rogaway, P.: On generalized feistel networks. In: Rabin, T. (ed.) CRYPTO 2010. LNCS, vol. 6223, pp. 613–630. Springer, Heidelberg (2010). https://doi.org/10.1007/978-3-642-14623-7_33
20. Hoang, V.T., Tessaro, S., Trieu, N.: The curse of small domains: new attacks on format-preserving encryption. In: Shacham, H., Boldyreva, A. (eds.) CRYPTO 2018. LNCS, vol. 10991, pp. 221–251. Springer, Cham (2018). https://doi.org/10.1007/978-3-319-96884-1_8
21. Mattsson, U.: Format controlling encryption using datatype preserving encryption. Cryptology ePrint Archive, Report 2009/257 (2009). http://eprint.iacr.org/2009/257
22. Morris, B., Rogaway, P.: Sometimes-recurse shuffle. In: Nguyen, P.Q., Oswald, E. (eds.) EUROCRYPT 2014. LNCS, vol. 8441, pp. 311–326. Springer, Heidelberg (2014). https://doi.org/10.1007/978-3-642-55220-5_18
23. Motwani, R., Raghavan, P.: Randomized Algorithms. Cambridge University Press, Cambridge (1995)
24. Patarin, J.: New results on pseudorandom permutation generators based on the DES scheme. In: Feigenbaum, J. (ed.) CRYPTO 1991. LNCS, vol. 576, pp. 301–312. Springer, Heidelberg (1992). https://doi.org/10.1007/3-540-46766-1_25
25. Patarin, J.: Generic attacks on Feistel schemes. In: Boyd, C. (ed.) ASIACRYPT 2001. LNCS, vol. 2248, pp. 222–238. Springer, Heidelberg (2001). https://doi.org/10.1007/3-540-45682-1_14
26. Ristenpart, T., Yilek, S.: The mix-and-cut shuffle: small-domain encryption secure against N queries. In: Canetti, R., Garay, J.A. (eds.) CRYPTO 2013, Part I. LNCS, vol. 8042, pp. 392–409. Springer, Heidelberg (2013). https://doi.org/10.1007/978-3-642-40041-4_22
27. Saltykov, A.: The number of components in a random bipartite graph. Discrete Math. Appl. **5**(6), 515–524 (1995)
28. Vance, J., Bellare, M.: Delegatable Feistel-based format preserving encryption mode. Submission to NIST, November 2015

Session Resumption Protocols
and Efficient Forward Security
for TLS 1.3 0-RTT

Nimrod Aviram[1], Kai Gellert[2(✉)], and Tibor Jager[2]

[1] Tel Aviv University, Tel Aviv, Israel
nimrodav@mail.tau.ac.il
[2] Paderborn University, Paderborn, Germany
{kai.gellert,tibor.jager}@uni-paderborn.de

Abstract. The TLS 1.3 0-RTT mode enables a client reconnecting to a server to send encrypted application-layer data in "0-RTT" ("zero round-trip time"), without the need for a prior interactive handshake. This fundamentally requires the server to reconstruct the previous session's encryption secrets upon receipt of the client's first message. The standard techniques to achieve this are *Session Caches* or, alternatively, *Session Tickets*. The former provides forward security and resistance against replay attacks, but requires a large amount of server-side storage. The latter requires negligible storage, but provides no forward security and is known to be vulnerable to replay attacks.

In this paper, we first formally define *session resumption protocols* as an abstract perspective on mechanisms like Session Caches and Session Tickets. We give a new generic construction that provably provides forward security and replay resilience, based on puncturable pseudorandom functions (PPRFs). This construction can immediately be used in TLS 1.3 0-RTT and deployed unilaterally by servers, without requiring any changes to clients or the protocol.

We then describe two new constructions of PPRFs, which are particularly suitable for use for forward-secure and replay-resilient session resumption in TLS 1.3. The first construction is based on the strong RSA assumption. Compared to standard Session Caches, for "128-bit security" it reduces the required server storage by a factor of almost 20, when instantiated in a way such that key derivation and puncturing together are cheaper on average than one full exponentiation in an RSA group. Hence, a 1 GB Session Cache can be replaced with only about 51 MBs of storage, which significantly reduces the amount of secure memory required. For larger security parameters or in exchange for more expensive computations, even larger storage reductions are achieved. The second construction combines a standard binary tree PPRF with a new

Supported by the German Research Foundation (DFG), project JA 2445/2-1, scholarships from The Israeli Ministry of Science and Technology, The Check Point Institute for Information Security, and The Yitzhak and Chaya Weinstein Research Institute for Signal Processing. We thank Nick Sullivan, Sven N. Hebrok and all anonymous reviewers for their valuable comments.

Y. Ishai and V. Rijmen (Eds.): EUROCRYPT 2019, LNCS 11477, pp. 117–150, 2019.
https://doi.org/10.1007/978-3-030-17656-3_5

"domain extension" technique. For a reasonable choice of parameters, this reduces the required storage by a factor of up to 5 compared to a standard Session Cache. It employs only symmetric cryptography, is suitable for high-traffic scenarios, and can serve thousands of tickets per second.

1   Introduction

0-RTT Protocols. A major innovation of TLS 1.3 [39] is its *0-RTT* (zero round-trip time) mode, which enables the resumption of sessions with minimal latency and without the need for an interactive handshake. A 0-RTT protocol allows the establishment of a secure connection in "one-shot", that is, with a single message sent from a client to a server, such that cryptographically protected payload data can be sent immediately ("in 0-RTT") along with the key establishment message, without the need for a latency-incurring prior handshake protocol. This significant speedup of connection establishment yields a smoother Web browsing experience and, more generally, better performance for applications with low-latency requirements. This is particularly noticeable in networks with relatively high latency, such as mobile networks.

The huge practical demand for 0-RTT is exemplified by the fact that many large Internet companies have developed and experimented with such protocols in the recent past, for example Google's QUIC [13] and Facebook's Zero [27] protocols. The content distribution provider Cloudflare has deployed the 0-RTT mode of TLS 1.3 as early as March 2017 at large scale, long before the finalization of the standard [44]. Google and Facebook declared that the cryptography in QUIC and Zero will soon be replaced by TLS 1.3 0-RTT [4,27].

The TLS 1.3 0-RTT Handshake. A full TLS 1.3 handshake (not 0-RTT) is always used in the very first connection between a client and a server. If the server supports 0-RTT, then both the client and server can derive a *Resumption Secret* from their shared key and session parameters. The client will simply store this secret. Naturally, the server then needs to retrieve the Resumption Secret during a subsequent handshake. There are two standard approaches for this, *Session Caches* and *Session Tickets*, which have different advantages and drawbacks. During the first handshake, the server sends to the client either a lookup key pointing to an entry in the Session Cache of the server, or a Session Ticket - depending on the configuration of the server. These approaches essentially work as follows:

Session Caches: The server stores all Resumption Secrets of recent sessions in a local database and issues each client a unique lookup key. When a client reconnects, it includes that lookup key in its 0-RTT messages, enabling the server to retrieve and use the matching Resumption Secret.

Session Tickets: The server uses a long-term symmetric encryption key, called the *Session Ticket Encryption Key* (STEK). Instead of storing the Resumption Secret in a local database, the server encrypts it with the STEK to create a *Session Ticket*. The Session Ticket is stored by the client. When a client reconnects, it includes that Session Ticket in its 0-RTT messages, which enables the server to decrypt it and recover the Resumption Secret. Note that the same STEK is used for many sessions and clients.

On a subsequent 0-RTT handshake, the client will include in its first message either the lookup key or the encrypted Session Ticket, in addition to a Diffie-Hellman key exchange message. The client can also send, in the same message, encrypted application-layer data, termed 0-RTT data. This data will be encrypted with a key derived from the Resumption Secret and a public client random value, without any input from the server.[1]

In its reply, the server will typically include a Diffie-Hellman key exchange message, and further messages (in either direction) will be encrypted with a key derived also from the DH secret, not only the Resumption Secret. Hence, the only data protected by the Resumption Secret alone is the 0-RTT data. We note that the use of DH is not mandatory, and it is possible to rely only on the Resumption Secret for the security of the entire session; we expect most traffic will use DH as described above.

We stress that the use of Session Caches or Session Tickets is opaque to clients. That is, in either case the server sends a *New Session Ticket* message containing an opaque sequence of bytes, which may be either a lookup key for the Session Cache, or an encrypted Session Ticket, without specifying which is the case.[2] This property ensures that our proposed techniques are compatible with the final TLS 1.3 standard [39] and can be implemented on the server-side without requiring modifications to the protocol or to clients.

Forward Security and Replay Resilience of 0-RTT Protocols. Forward security essentially means that the protocol provides security of sessions, even if an attacker is able to corrupt one party after the session has terminated (e.g., by breaking into a Web server and learning the long-term secret key). Resilience to replay attacks is a fundamental, classical design goal of cryptographic protocols, which prevents an attacker from replaying the same payload data to a server repeatedly.

Both forward security and replay resilience are standard design goals of modern security protocols. However, achieving these properties is well-known to be difficult for 0-RTT protocols. This is because classical ("non-0-RTT") protocols

---

[1] The above describes typical modes of operation of TLS 1.3. The standard also allows for other modes, e.g. modes that include client authentication. We expect other modes will be used much less often, and therefore they are beyond the scope of this paper.

[2] Confusingly, the message containing this opaque sequence of bytes is always termed a "New Session Ticket Message", for both Session Caches and encrypted self-contained Session Tickets. To our knowledge there is no standard nomenclature, in [39] or elsewhere, for these two different approaches when used in TLS 1.3; see e.g. [39, § 8.1]. TLS 1.2 referred to "Session ID Resumption" and "Session Ticket Resumption", but these terms are not used in TLS 1.3.

include fresh input from the server (e.g., a Diffie-Hellman message) generated using ephemeral randomness, which provides a leverage to achieve forward security. However, there is no such interactivity in 0-RTT protocols. Furthermore, an attacker is able to replay the 0-RTT key establishment message along with the 0-RTT payload data over and over again to a server, which is not detectable without additional server-side countermeasures.

Forward Security and Replay Resilience of TLS 1.3 0-RTT. With Session Caches the server stores a "unique" Resumption Secret in a local database for each client. In most cases, it is able to delete the Resumption Secret immediately after retrieving it. This provides forward security, as an attacker obtaining the server state cannot decrypt past sessions. It also provides resilience against replay attacks, as the server is not able to decrypt replayed messages.

If Session Tickets are used, then an attacker that obtains access to the server can learn the STEK, and thus decrypt all tickets encrypted with this key to learn the Resumption Keys. Hence, servers using Session Tickets do not provide forward security. They are also generally vulnerable to replay attacks.[3] Since an attacker learning the STEK has catastrophic implications for security, large server operators usually rotate the STEK. Such deployments typically generate a new STEK roughly once per hour, and limit the STEK lifetime to roughly a day [34]. An attacker that learns one STEK can therefore decrypt approximately one hour's worth of traffic. However, most current TLS implementations do not provide out-of-the-box support for STEK rotation, and this (welcome) defensive measure is usually limited to large operators who can afford to modify TLS implementations [32,34]. Long-lived STEKs are unfortunately prevalent, and even among high-profile websites, some reuse the same STEK for many weeks, or even for many months [43].

To summarize, Session Caches are generally forward-secure and replay-resistant, while Session Tickets are not. Naïvely, it would therefore appear that Session Caches are the superior solution. However, Session Caches require the server to store the session state for each (recent) connection. This is often infeasible, in particular for high-traffic server operators. Such server operators often reluctantly use Session Tickets, knowingly forgoing forward secrecy. Additionally, even if forward security is not prioritized by a particular server operator and thus Session Tickets are used, the prevention of replay attacks may still require additional storage at the server, since the only way to prevent replay attacks in this case is to log used tickets.[4] In this context it is sometimes claimed that so-called *idempotent requests*, that is, requests that have the same effect on the server state whether they are served once or several times, are safe to use with TLS 1.3 0-RTT. However, it is well-known [36] and also discussed in the TLS

---

[3] Unless there is additional server-side logging of tickets that have already been used.

[4] When using resumption, the client must include in its first message the ticket's age, i.e. the time elapsed between receiving the ticket from the server in a previous session. The server expects this time interval to be precise up to a small window of error allowing for propagation delay, typically on the order of 10 s. An attacker can perform replay attacks within this time window.

1.3 specification [39] that even replays of idempotent requests may give rise to attacks that, e.g., reveal the target URL of HTTP requests.

All of these issues are well-known to apply to TLS 1.3 0-RTT and have raised significant concerns about its secure deployability in practice [36]. Eric Rescorla, the main author of the TLS 1.3 RFC draft, acknowledges that this poses a *"difficult application integration issue"* [38]. However, due to the huge practical demand, 0-RTT is also considered *"too big a win not to do"* [38]. Very recently, at EUROCRYPT 2017 [25] and 2018 [15,16], the first 0-RTT protocols that simultaneously achieve forward security and replay resilience were proposed, but these require relatively heavy cryptographic machinery, such as hierarchical or broadcast identity-based encryption, and thus are not yet suitable for large-scale deployment in TLS 1.3.

Our Contributions. We give the first formal definition for secure 0-RTT session resumption protocols, as an abstraction of the constructions currently used in practice in TLS 1.3. We propose new techniques to achieve forward security and replay resilience that are ready-to-use with TLS 1.3 as it is standardized, without any changes to the protocol. Our proposal is based on Session Tickets, and thus requires minimal storage at the server side, but we extend this approach with efficient puncturable pseudorandom functions (PPRFs) that enable us to achieve forward security and replay resilience for Session Tickets. We provide new constructions of PPRFs with short keys and formal security proofs based on standard hardness assumptions. We propose two variants:

1. The first variant is based on the strong RSA assumption. It reduces the server storage by a factor of at least 11 compared to a Session Cache, increases ticket size by a negligible length, and requires the server to perform two exponentiations (one per issuance and one per resumption).
2. The second variant reduces server storage by a factor of up to 5 compared to a Session Cache, while using tickets that are roughly 400 bytes longer than standard tickets. It extends a standard GGM-style [22] binary tree-based PPRF, as described in [10,11,30], with a new *domain extension* idea. It employs only symmetric cryptography, is suitable for very-high-traffic scenarios, and can serve thousands of tickets per second, at the cost of hundreds of megabytes in server storage.

Our Approach. At the base of our approach is the concept of *puncturing* a pseudorandom function (PRF) to obtain a puncturable symmetric-key encryption scheme. Puncturable PRFs are a special case of constrained PRFs [10,11,30], which make it possible to derive constrained keys that allow computation of PRF output only for certain inputs.

In our approach, a server initially maintains a STEK k that allows decryption of any Session Ticket; when receiving ticket t, the server uses k to decrypt t in order to recover the Resumption Secret. Using the puncturing feature of the

PPRF, it then derives from k a new key, k', that can decrypt any ticket *except* for t. The server then discards k and stores only k'. It repeats this process for every ticket received. This yields forward secrecy and replay-resistance: an attacker that compromises the server learns a key that is not capable of decrypting past tickets. Similarly, an attacker cannot successfully replay a message, since the server is only able to decrypt each ticket once.

The naïve way to employ this approach in TLS 1.3 0-RTT would be to use public-key puncturable encryption, as in [15,16,25]. However, this approach results in impractically long puncturing times or very long secret keys. Moreover, the most practical constructions require relatively expensive pairing-based cryptography by both the client and the server, thereby obviating a significant benefit of TLS 1.3 0-RTT. Rather than using public-key puncturable encryption, we observe that in TLS 1.3 0-RTT, the server itself generates the tickets it would later need to decrypt. It therefore suffices to use symmetric cryptography, and to maintain a key that allows decryption of only a limited set of ciphertexts, generated by the server itself. To achieve this, we use PPRFs to derive keys for standard TLS 1.3 tickets. Concretely, we describe two new PPRF constructions that are particularly suitable for our application:

- The first builds a new PPRF from the Strong RSA Assumption. The PPRF has a polynomially-bounded input size, but this is sufficient for our application (and probably for certain other PPRF applications as well). Its main distinguishing feature is that its secret key size is independent of the number of puncturings. It consists of an RSA modulus N, a number $g \in \mathbb{Z}_N$, and a bitfield, indicating positions where the PPRF was punctured. Due to the short secret key, our construction may find other applications in applied and theoretical cryptography. Since our primary objective is to provide an as-efficient-as-possible solution for practical protocols such as TLS 1.3 0-RTT, we describe a construction with security proof in the random oracle model [5]. It seems likely that our construction can be lifted to the standard model in a straightforward way, via standard techniques like hardcore predicates [6,8,23], but this would yield less efficient constructions and is therefore outside the scope of this paper.

- The second construction is based on a standard tree-based PPRF [10,11,30], instantiated with a cryptographic hash function, such as SHA-3.
  The size of punctured keys depends linearly on the depth of the tree, which in turn depends on the size of the domain of the PPRF. We describe a new *domain extension* technique that reduces the size of punctured keys by trading secret key size for ticket size, while preserving the puncturing functionality. Domain extension makes it possible to use a PPRF with a smaller domain (and thus smaller punctured keys). To save a factor of up to n in server-side storage, the ticket size rises roughly as $(n-1)!$. Thus, this is only useful for small values of n, but choosing e.g. $n = 5$ can yield significant savings with a modest increase in ticket size on the wire. Concretely, for $n = 5$

and "128-bit security", ticket size is increased by 384 bytes. As discussed in Sect. 5.5, experiments done by Google estimate that this will impose only a small impact on latency [33].

Large-Scale Server Clusters and Load Balancing. Large TLS server deployments typically consist of many servers that share the same public key. This complicates any logic that relies on the server storing some state, since these servers will typically not share a globally-consistent state. Such discussion is beyond the scope of this paper, and we will assume a single server with consistent storage throughout. When many servers share a Session Cache, the cache is likely to be distributed, and any logic relying on an atomic retrieve-and-delete operation becomes more complex. Therefore, distributed Session Caches are not necessarily replay-resistant nor forward-secure, as this requires synchronous deletion of Resumption Secrets at all servers, and thus synchronized state.[5] However, in such large-scale settings it is highly desirable to minimize the amount of memory that must be consistently synchronized across different servers. Our techniques are therefore useful to that end as well.

Further Applications to Devices with Restricted Resources. Our techniques may also be useful for devices with very restricted resources, such as battery-powered IoT devices with a wireless network connection. For such devices, it is usually extremely expensive to *send* data, because each transmitted bit costs energy, which limits the battery lifetime and thus the range of possible applications. In order to maximize the battery lifetime, it is useful to avoid expensive interactive handshakes and use a 0-RTT protocol whenever data is sent to such devices. Note that here the main gain from using 0-RTT is *not* minimal latency, but rather that no key exchange messages must be sent by the receiver. Ideally, transmitted data should be forward-secure, but such devices have low storage capacity and we cannot use large amounts of storage to achieve forward security.

For such devices, it is reasonable to relax the requirement for very efficient computation, since adding unnecessary transmissions to even a fraction of connections is likely more costly than using moderately more expensive computations. By instantiating our session resumption protocol in a way that puncturing is more expensive (say, five full RSA exponentiations, which may still be reasonable for most IoT devices), we achieve reductions in storage by factors close to 100. Thus, our techniques make it possible to use forward-secure 0-RTT protocols even on such devices. Instead of requiring, say, 1 GB of memory for a session cache, we need only about 10 MBs of memory.

Related Work. Puncturable encryption [24] was used to construct forward-secure instant messaging [24] and 0-RTT protocols [15,16,25], for instance. Green and

---

[5] When using Session Tickets, the same holds for mechanisms that store used tickets, which are likely to be distributed as well. See [39, §2.3, §8, §E.5], [36,37] for more in-depth discussion.

Miers [24] first proposed puncturable encryption as a practical building block for the case of asynchronous messaging. They used pairing-based puncturable encryption, and as a result observed impractically long processing times for their construction. Günther *et al.* [25] proposed using puncturable encryption for 0-RTT protocols, again proposing concrete constructions based on pairings that are also impractical for high-traffic scenarios. Derler *et al.* [15,16] proposed trading off space in exchange for processing time, with the use of their proposed Bloom Filter Encryption. Their construction essentially precomputes many already-punctured keys, and these keys are used only once, so puncturing becomes simply key deletion. Bloom Filter Encryption may be considered practical for low-traffic scenarios, but supporting a large number of puncturings per key requires precomputation and storage of keys on the order of many gigabytes.

Over the past years there have been several papers formally analyzing the security of TLS 1.2 [7,28,31] and TLS 1.3 [17,21]. Especially noteworthy are the analyses of the 0-RTT mode of TLS 1.3 [21] and QUIC [20] by Fischlin and Günther, who analyze both protocols in a multi-stage key exchange model [20]. Lychev *et al.* [35] further formally analyzed QUIC in a security model that additionally captures the secure composition of authenticated encryption and key exchange. A security definition and construction for QUIC-like 0-RTT protocols were given in [26]. However, all these publications do *not* consider forward secrecy for the very first message in their security models. Hence, we believe that our techniques may also influence the design of protocols providing a 0-RTT key exchange, such as TLS 1.3 and QUIC, in order to achieve forward secrecy for all messages.

Outline. The rest of this paper is organized as follows. In Sect. 2 we provide formal definitions for secure 0-RTT Session Resumption Protocols. In Sect. 3 we describe a generic construction, based on abstract PPRFs, and formally prove forward security and replay resilience. Section 4 describes the Strong-RSA-based PPRF and an analysis of the efficiency when used in the protocol construction in Sect. 3. Section 5 describes the tree-based PPRF and a novel "domain extension" technique for standard binary tree PPRFs, along with an efficiency analysis.

Notation. We denote the security parameter as λ. For any $n \in \mathbb{N}$ let 1^n be the unary representation of n and let $[n] = \{1, \ldots, n\}$ be the set of numbers between 1 and n. Moreover, $|x|$ denotes the length of a bitstring x, while $|\mathcal{S}|$ denotes the size of a set $\mathcal{S}$. We write $x \xleftarrow{\$} \mathcal{S}$ to indicate that we choose element x uniformly at random from set $\mathcal{S}$. For a probabilistic polynomial-time algorithm $\mathcal{A}$ we define $y \xleftarrow{\$} \mathcal{A}(a_1, \ldots, a_n)$ as the execution of $\mathcal{A}$ (with fresh random coins) on input $a_1, \ldots, a_n$ and assigning the output to y.

2   0-RTT Session Resumption Protocols and Their Security

In this section we provide formal definitions for *secure 0-RTT session resumption protocols*. These definitions capture well both our new techniques and the existing solutions already standardized in TLS 1.3. We also expect that the techniques used to formally analyze and verify TLS 1.3 0-RTT [14,18] can be extended to use our abstraction of a session resumption protocol within TLS 1.3.[6] This leads us to believe that our definitions capture a reasonable abstraction of the cryptographic core of the TLS 1.3 0-RTT mode (and likely also of similar protocols that may be devised in the future).

For simplicity, in the following we will refer to pre-shared values as *session keys*, as they are either previously established session keys, or a Resumption Secret derived from a session key, as e.g. in TLS 1.3. The details of how to establish a shared secret and potentially derive a session key from it are left to the individual protocol and are outside the scope of our abstraction. Session keys are elements of a keyspace $\mathcal{S}$.

Definition 1. *A 0-RTT session resumption protocol consists of three probabilistic polynomial-time algorithms* Resumption = (Setup, TicketGen, ServerRes) *with the following properties.*

- Setup(1^λ) *takes as input the security parameter λ and outputs the server's long-term key k.*
- TicketGen(k, s) *takes as input a long-term key k and a session key s, and outputs a ticket t and a potentially modified long-term key k'.*
- ServerRes(k, t) *takes as input the server's long-term key k and the ticket t, and outputs a session key s and a potentially modified key k', or a failure symbol $\perp$.*

Using a Session Resumption Protocol. A 0-RTT session resumption scheme is used by a set of clients C and a set of servers S. If a client and a server share a session key s, the session resumption is executed as follows (cf. Fig. 1).

1. The server uses its long-term key k and the session key s to generate a ticket t by running $(t, k') \xleftarrow{\$} $ TicketGen(k, s). The ticket is sent to the client. Additionally, the server replaces its long-term key k by k' and deletes the session key s and ticket t, i.e. it is not required to keep any session state.
2. For session resumption at a later point in time, the client sends the ticket t to the server.
3. Upon receiving the ticket t, the server runs $(s, k') := $ ServerRes(k, t) to retrieve the session key s. Additionally, k is deleted and replaced by the updated key k'.

---

[6] Obtaining a formal security proof for this would be an interesting direction for future research, but is beyond the scope of this work.

Client	Server
	// Generate long-term key
	$k \xleftarrow{\$} \mathsf{Setup}(1^\lambda)$
	// Issue ticket(s)
$\xleftarrow{\quad t \quad}$	$(t, k) \xleftarrow{\$} \mathsf{TicketGen}(k, s)$
store s, t	delete s, t
// Encrypt data with s	// Resume sessions
$c \xleftarrow{\$} \mathsf{Enc}(s, m)$ $\xrightarrow{\quad t, c \quad}$	$(s, k) \xleftarrow{\$} \mathsf{ServerRes}(k, t)$
	$m := \mathsf{Dec}(s, c)$

Fig. 1. Execution of a generic 0-RTT session resumption protocol with early data m, where client and server initially are in possession of a shared secret s. Note that procedures TicketGen and ServerRes both potentially modify the server's key k.

Compatibility with TLS 1.3. As explained in Sect. 1, using either Session Tickets or Session Caches in TLS 1.3 is transparent to clients, i.e. clients are generally unaware of which is used. In either case, the client stores a sequence of bytes which is opaque from the client's point of view. Since all algorithms of a session resumption protocol are executed on the server, while a client just has to store the ticket t (encoded as a sequence of bytes), this generic approach of TLS 1.3 is immediately compatible with our notion of session resumption protocols. Thus, a session resumption protocol can be used immediately in TLS 1.3, without requiring changes to clients or to the protocol. Furthermore, Session Tickets and Session Caches are specific examples of such protocols.

2.1 Security in the Single-Server Setting

We define the security of a 0-RTT session resumption protocol Resumption by a security game $\mathsf{G}^{\text{0-RTT-SR}}_{\mathcal{A},\text{Resumption}}(\lambda)$ between a challenger $\mathcal{C}$ and an adversary $\mathcal{A}$. For simplicity, we will start with a single-server setting and argue below that security in the single-server setting implies security in a multi-server setting. The security game is parametrized by the number of session keys μ (equal to the number of clients in the single-server setting).

1. $\mathcal{C}$ runs $k \xleftarrow{\$} \mathsf{Setup}(1^\lambda)$, samples a random bit $b \xleftarrow{\$} \{0, 1\}$ and generates session keys $s_i \xleftarrow{\$} \mathcal{S}$ for all clients $i \in [\mu]$. Furthermore, it generates tickets t_i and updates key k by running $(t_i, k) \xleftarrow{\$} \mathsf{TicketGen}(k, s_i)$ for all clients $i \in [\mu]$. The sequence of tickets $(t_i)_{i \in [\mu]}$ is sent to $\mathcal{A}$.
2. The adversary gets access to oracles it may query.
   (a) $\mathsf{Dec}(t)$ takes as input a ticket t. It computes $(s_i, k') := \mathsf{ServerRes}(k, t_i)$, returns the session key s_i and replaces $k := k'$. Note that ticket t can either be a ticket of the initial sequence of tickets $(t_i)_{i \in [\mu]}$ or an arbitrary ticket chosen by the adversary.

(b) Test(t) takes as input a ticket t. It computes $(s_i, k') :=$ ServerRes(k, t) and outputs $\perp$ if the output of ServerRes was $\perp$. Otherwise, it updates $k := k'$. If $b = 1$, then it returns the session key s_i. Otherwise, a random $r_i \xleftarrow{\$} S$ is returned. Note that ticket t can either be a ticket of the initial sequence of tickets $(t_i)_{i \in [\mu]}$ or an arbitrary ticket chosen by the adversary. The adversary is allowed to query Test only once.

(c) Corr returns the current long-term key k of the server. The adversary must not query Test after Corr, as this would lead to a trivial attack.

3. Eventually, adversary $\mathcal{A}$ outputs a guess b^*. Challenger $\mathcal{C}$ outputs 1 if $b = b^*$ and 0 otherwise.

Note that this security model reflects both forward secrecy and replay protection. Forward secrecy is ensured, as an adversary may corrupt the challenger after issuing the Test-query. If the protocol would not ensure forward secrecy, an attacker could corrupt its long-term key and trivially decrypt the challenge ticket. Replay protection is ensured, as an adversary is allowed to issue Dec(t_i) after already testing Test(t_i) (as both queries invoke the ServerRes algorithm). If the protocol would not ensure replay protection, an attacker could use the decryption oracle to distinguish a real or random session key of the Test-query.

Definition 2. *We define the advantage of an adversary $\mathcal{A}$ in the above security game* $\mathsf{G}^{\text{0-RTT-SR}}_{\mathcal{A},\text{Resumption}}(\lambda)$ *as*

$$\mathsf{Adv}^{\text{0-RTT-SR}}_{\mathcal{A},\text{Resumption}}(\lambda) = \left| \Pr\left[\mathsf{G}^{\text{0-RTT-SR}}_{\mathcal{A},\text{Resumption}}(\lambda) = 1\right] - \frac{1}{2} \right|.$$

We say a 0-RTT session resumption protocol is secure *in a single-server environment if the advantage* $\mathsf{Adv}^{\text{0-RTT-SR}}_{\mathcal{A},\text{Resumption}}(\lambda)$ *is a negligible function in λ for all probabilistic polynomial-time adversaries $\mathcal{A}$.*

3   Constructing Secure Session Resumption Protocols

In this section we will show how session resumption protocols providing full forward security and replay resilience can be constructed. We will start with a generic construction, based on authenticated encryption with associated data and any puncturable pseudorandom function that is invariant to puncturing. Later we describe new constructions of PPRFs, which are particularly suitable for use in session resumption protocols.

3.1   Building Blocks

We briefly recall the basic definition of puncturable pseudorandom functions and authenticated encryption with associated data.

Puncturable PRFs. A puncturable pseudorandom function is a special case of a pseudorandom function (PRF), where it is possible to compute punctured keys which do not allow evaluation on inputs that have been punctured. We recall the definition of puncturable pseudorandom functions and its security from [41].

Definition 3. *A puncturable pseudorandom function (PPRF) with keyspace $\mathcal{K}$, domain $\mathcal{X}$ and range $\mathcal{Y}$ consists of three probabilistic polynomial-time algorithms* $\mathsf{PPRF} = (\mathsf{Setup}, \mathsf{Eval}, \mathsf{Punct})$, *which are described as follows.*

- $\mathsf{Setup}(1^\lambda)$: *This algorithm takes as input the security parameter λ and outputs a description of a key $k \in \mathcal{K}$.*
- $\mathsf{Eval}(k, x)$: *This algorithm takes as input a key $k \in \mathcal{K}$ and a value $x \in \mathcal{X}$, and outputs a value $y \in \mathcal{Y}$.*
- $\mathsf{Punct}(k, x)$: *This algorithm takes as input a key $k \in \mathcal{K}$ and a value $x \in \mathcal{X}$, and returns a punctured key $k' \in \mathcal{K}$.*

Definition 4. *A PPRF is* correct *if for every subset $\{x_1, \ldots, x_n\} = \mathcal{S} \subseteq \mathcal{X}$ and all $x \in \mathcal{X} \setminus \mathcal{S}$, we have that*

$$\Pr\left[\mathsf{Eval}(k_0, x) = \mathsf{Eval}(k_n, x) : \begin{array}{l} k_0 \xleftarrow{\$} \mathsf{Setup}(1^\lambda); \\ k_i = \mathsf{Punct}(k_{i-1}, x_i) \text{ for } i \in [n]; \end{array}\right] = 1.$$

A new property of PPRFs that we will need is that puncturing is "commutative", i.e. the order of puncturing operations does not affect the resulting secret key. That is, for any $x_0, x_1 \in \mathcal{X}, x_0 \neq x_1$, if we first puncture on input x_0 and then on x_1, the resulting key is identical to the key obtained from first puncturing on x_1 and then on x_0. Formally:

Definition 5. *A PPRF is* invariant to puncturing *if for all keys $k \in \mathcal{K}$ and all elements $x_0, x_1 \in \mathcal{X}, x_0 \neq x_1$ it holds that*

$$\mathsf{Punct}(\mathsf{Punct}(k, x_0), x_1) = \mathsf{Punct}(\mathsf{Punct}(k, x_1), x_0).$$

We define two notions of PPRF security. The first notion represents the typical pseudorandomness security experiment with adaptive evaluation queries by an adversary. The second notion is a weaker, non-adaptive security experiment. We show that it suffices to prove security in the non-adaptive experiment if the PPRF is invariant to puncturing and has a polynomial-size domain.

Definition 6. *We define the advantage of an adversary $\mathcal{A}$ in the* rand *(resp.* na-rand*) security experiment $\mathsf{G}^{\mathsf{rand}}_{\mathcal{A},\mathsf{PPRF}}(\lambda)$ (resp. $\mathsf{G}^{\mathsf{na\text{-}rand}}_{\mathcal{A},\mathsf{PPRF}}(\lambda)$) defined in Fig. 2 as*

$$\mathsf{Adv}^{\mathsf{rand}}_{\mathcal{A},\mathsf{PPRF}}(\lambda) := \left| \Pr\left[\mathsf{G}^{\mathsf{rand}}_{\mathcal{A},\mathsf{PPRF}}(\lambda) = 1\right] - \frac{1}{2} \right|,$$

$$\mathsf{Adv}^{\mathsf{na\text{-}rand}}_{\mathcal{A},\mathsf{PPRF}}(\lambda) := \left| \Pr\left[\mathsf{G}^{\mathsf{na\text{-}rand}}_{\mathcal{A},\mathsf{PPRF}}(\lambda) = 1\right] - \frac{1}{2} \right|.$$

We say a puncturable pseudorandom function PPRF is rand-*secure (resp.* na-rand *-secure), if the advantage $\mathsf{Adv}^{\mathsf{rand}}_{\mathcal{A},\mathsf{PPRF}}(\lambda)$ (resp. $\mathsf{Adv}^{\mathsf{na\text{-}rand}}_{\mathcal{A},\mathsf{PPRF}}(\lambda)$) is a negligible function in λ for all probabilistic polynomial-time adversaries $\mathcal{A}$.*

$G^{\text{rand}}_{\mathcal{A},\text{PPRF}}(\lambda)$	$G^{\text{na-rand}}_{\mathcal{A},\text{PPRF}}(\lambda)$
$k \xleftarrow{\$} \text{Setup}(1^\lambda), b \xleftarrow{\$} \{0,1\}, \mathcal{Q} := \emptyset$	$k_0 \xleftarrow{\$} \text{Setup}(1^\lambda), b \xleftarrow{\$} \{0,1\}$
$x^* \xleftarrow{\$} \mathcal{A}^{\text{Eval}(k,\cdot)}(1^\lambda)$	$(x_1, \ldots, x_\ell) \xleftarrow{\$} \mathcal{A}(1^\lambda)$
$\quad$ where $\text{Eval}(k,x)$ behaves like Eval, but sets	$k_i := \text{Punct}(k_{i-1}, x_i)$ for all $i \in [\ell]$
$\quad \mathcal{Q} := \mathcal{Q} \cup \{x\}$, and runs $k := \text{Punct}(k,x)$	$y_{i,0} \xleftarrow{\$} \mathcal{Y}, y_{i,1} := \text{Eval}(k_0, x_i)$ for all $i \in [\ell]$
$y_0 \xleftarrow{\$} \mathcal{Y}, y_1 := \text{Eval}(k, x^*), k := \text{Punct}(k, x^*)$	$b^* \xleftarrow{\$} \mathcal{A}(k_\ell, (y_{i,b})_{i \in [\ell]})$
$b^* \xleftarrow{\$} \mathcal{A}(k, y_b)$	return 1 if $b = b^*$
return 1 if $b = b^* \wedge x^* \notin \mathcal{Q}$	return 0
return 0	

Fig. 2. Security experiments for PPRFs. The na-rand security experiment for PPRF is left and the rand security experiment is right.

It is relatively easy to prove that na-rand-security and rand-security are equivalent, up to a linear security loss in the size of the domain of the PPRF. In particular, if the PPRF has a polynomially-bounded domain size and is invariant to puncturing, then both are polynomially equivalent.

Theorem 1. *Let* PPRF *be a* na-rand-*secure PPRF with domain* $\mathcal{X}$. *If* PPRF *is invariant to puncturing, then it is also* rand-*secure with advantage*

$$\text{Adv}^{\text{rand}}_{\mathcal{A},\text{PPRF}}(\lambda) \leq \frac{\text{Adv}^{\text{na-rand}}_{\mathcal{A},\text{PPRF}}(\lambda)}{|\mathcal{X}|}.$$

Proof. The proof is based on a straightforward reduction. We give a sketch. Let $\mathcal{A}$ be an adversary against the rand security of PPRF. We guess $\mathcal{A}$'s challenge value in advance by sampling $\nu \xleftarrow{\$} \mathcal{X}$ uniformly at random. We initialize the na-rand challenger by sending it ν. In return we receive a challenge y (either computed via Eval or random) and a punctured key k that cannot be evaluated on input ν.

The punctured key k allows us to correctly answer all of $\mathcal{A}$'s Eval queries, except for ν. When the adversary outputs its challenge x^* we will abort if $x^* \neq \nu$. Otherwise, we forward y and a punctured key k' that has been punctured on all values of the Eval queries. Note that the key has a correct distribution, as we require that the PPRF is invariant to puncturing.

Eventually, $\mathcal{A}$ outputs a bit b^* which we forward to the na-rand challenger.

The simulation is perfect unless we abort it, which happens with polynomially-bounded probability $1/|\mathcal{X}|$, due to the fact that $|\mathcal{X}|$ is polynomially bounded. $\qquad\square$

Authenticated Encryption with Associated Data. We will furthermore need authenticated encryption with associated data (AEAD) [40], along with the standard notions of confidentiality and integrity.

Definition 7. *An* authenticated encryption scheme with associated data *is a tuple* AEAD $= (\text{KGen}, \text{Enc}, \text{Dec})$ *of three probabilistic polynomial-time algorithms:*

- KGen(1^λ) *takes as input a security parameter λ and outputs a secret key k.*
- Enc(k, m, ad) *takes as input a key k, a message m, associated data ad and outputs a ciphertext c.*
- Dec(k, c, ad) *takes as input a key k, a ciphertext c, associated data ad and outputs a message m or an failure symbol $\perp$.*

An AEAD scheme is called *correct* if for any key $k \xleftarrow{\$} \mathsf{KGen}(1^\lambda)$, any message $m \in \{0,1\}^*$, any associated data $ad \in \{0,1\}^*$ it holds that $\mathsf{Dec}(k, \mathsf{Enc}(k, m, ad), ad) = m$.

Definition 8. *We define the advantage of an adversary $\mathcal{A}$ in the* IND-CPA *experiment* $\mathsf{G}^{\mathsf{IND\text{-}CPA}}_{\mathcal{A},\mathsf{AEAD}}(\lambda)$ *defined in Fig. 3 as*

$$\mathsf{Adv}^{\mathsf{IND\text{-}CPA}}_{\mathcal{A},\mathsf{AEAD}}(\lambda) := \left| \Pr\left[\mathsf{G}^{\mathsf{IND\text{-}CPA}}_{\mathcal{A},\mathsf{AEAD}}(\lambda) = 1\right] - \frac{1}{2} \right|.$$

We say an AEAD scheme AEAD *is* indistinguishable under chosen-plaintext attacks *(*IND-CPA *-secure), if the advantage* $\mathsf{Adv}^{\mathsf{IND\text{-}CPA}}_{\mathcal{A},\mathsf{AEAD}}(\lambda)$ *is a negligible function in λ for all probabilistic polynomial-time adversaries $\mathcal{A}$.*

Definition 9. *We define the advantage of an adversary $\mathcal{A}$ in the* INT-CTXT *experiment* $\mathsf{G}^{\mathsf{INT\text{-}CTXT}}_{\mathcal{A},\mathsf{AEAD}}(\lambda)$ *defined in Fig. 3 as*

$$\mathsf{Adv}^{\mathsf{INT\text{-}CTXT}}_{\mathcal{A},\mathsf{AEAD}}(\lambda) := \left| \Pr\left[\mathsf{G}^{\mathsf{INT\text{-}CTXT}}_{\mathcal{A},\mathsf{AEAD}}(\lambda) = 1\right] \right|.$$

We say an AEAD scheme AEAD *provides* integrity of ciphertexts *(*INT-CTXT *-secure), if the advantage* $\mathsf{Adv}^{\mathsf{INT\text{-}CTXT}}_{\mathcal{A},\mathsf{AEAD}}(\lambda)$ *is a negligible function in λ for all probabilistic polynomial-time adversaries $\mathcal{A}$.*

Additionally, we will need the notion of ε-spreadness for AEAD. ε-spreadness captures the intuition that a ciphertext encrypted under a key k should not be valid under a random key $k' \neq k$.

Definition 10. *An AEAD scheme is ε-spread if for all messages m and all associated data ad it holds that*

$$\Pr_{\substack{k,k' \xleftarrow{\$} \mathsf{KGen}(1^\lambda) \\ k \neq k'}} [\mathsf{AEAD.Dec}(k', \mathsf{AEAD.Enc}(k, m, ad), ad) \neq \perp] \leq \varepsilon.$$

We note that one can easily prove that INT-CTXT-security implies ε-spreadness with negligible ε. However, the "statistical" formulation of Definition 10 will simplify parts of our proof significantly, and therefore we believe it reasonable to make it explicit.

3.2   Generic Construction

Now we are ready to describe our generic construction of a 0-RTT session resumption protocol, based on a PPRF and an AEAD scheme, and to prove its security.

$G_{\mathcal{A},\text{AEAD}}^{\text{IND-CPA}}(\lambda)$	$G_{\mathcal{A},\text{AEAD}}^{\text{INT-CTXT}}(\lambda)$
$k \xleftarrow{\$} \text{KGen}(1^\lambda), b \xleftarrow{\$} \{0,1\}$	$k \xleftarrow{\$} \text{KGen}(1^\lambda), \mathcal{Q} := \emptyset, \text{win} := 0$
$b^* \xleftarrow{\$} \mathcal{A}^{\text{LoR}(\cdot,\cdot,\cdot)}(1^\lambda)$	$\mathcal{A}^{\text{Enc}(\cdot,\cdot),\text{Dec}(\cdot,\cdot)}(1^\lambda)$
where $\text{LoR}(m_0, m_1, ad)$.	where $\text{Enc}(m, ad)$ returns $\text{Enc}(k, m, ad)$
returns $\text{Enc}(k, m_b, ad)$.	and sets $\mathcal{Q} := \mathcal{Q} \cup \{(c, ad)\}$,
return 1 if $b = b^*$	and where $\text{Dec}(c, ad)$ sets $\text{win} := 1$
return 0	if $\text{Dec}(k, c, ad) \neq \perp$ and $(c, ad) \notin \mathcal{Q}$.
	return win

Fig. 3. The IND-CPA and INT-CTXT security experiment for AEAD [40].

Construction 1. *Let* AEAD $=$ (KGen, Enc, Dec) *be an authenticated encryption scheme with associated data and let* PPRF $=$ (Setup, Eval, Punct) *be a PPRF with range* $\mathcal{Y}$*. Then we can construct a 0-RTT session resumption protocol* Resumption $=$ (Setup, TicketGen, ServerRes) *in the following way.*

- Setup(1^λ) *runs* $k_{\text{PPRF}} = \text{PPRF.Setup}(1^\lambda)$*, and outputs* $k := (k_{\text{PPRF}}, 0)$*, where "0" is a counter initialized to zero.*
- TicketGen(k, s) *takes a key* $k = (k_{\text{PPRF}}, n)$*. It computes* $\kappa = \text{PPRF.Eval}$ (k_{PPRF}, n)*. Then it encrypts the ticket as* $t' \xleftarrow{\$} \text{AEAD.Enc}(\kappa, s, n)$*. Finally, it defines* $t = (t', n)$ *and* $k := (k_{\text{PPRF}}, n + 1)$*, and outputs* (t, k)*.*
- ServerRes(k, t) *takes* $k = (k_{\text{PPRF}}, n)$ *and* $t = (t', n')$*. It computes a key* $\kappa :=$ $\text{PPRF.Eval}(k_{\text{PPRF}}, n')$*. If* $\kappa = \perp$*, then it returns* $\perp$*. Otherwise it computes a session key* $s := \text{AEAD.Dec}(\kappa, t', n')$*. If* $s = \perp$*, it returns* $\perp$*. Else it punctures* $k_{\text{PPRF}} := \text{PPRF.Punct}(k_{\text{PPRF}}, n')$*, and returns* $(s, (k_{\text{PPRF}}, n))$*.*

Note that the associated data n is sent in plaintext, posing a potential privacy leak. This can be circumvented by additionally encrypting n under a dedicated symmetric key. Compromise of this key would only allow an attacker to link sessions by the same returning client, not to decrypt past traffic, therefore this symmetric key needs not be punctured to achieve forward security.[7]

Theorem 2. *If* AEAD *is* ε*-spread and* PPRF *is invariant to puncturing, then from each probabilistic polynomial-time adversary* $\mathcal{A}$ *against the security of* Resumption *in a single-server environment with advantage* $\text{Adv}_{\mathcal{A},\text{Resumption}}^{\text{0-RTT-SR}}(\lambda)$*, we can construct four adversaries* $\mathcal{B}_{\text{PPRF1}}$*,* $\mathcal{B}_{\text{PPRF2}}$*,* $\mathcal{B}_{\text{AEAD1}}$*, and* $\mathcal{B}_{\text{AEAD2}}$ *such that*

---

[7] The natural solution would be to encrypt n using public-key puncturable encryption, but this would be costly, and obviate most of the efficiency benefits described in this work. We are unfortunately unaware of a good solution that achieves session unlinkability in the event of server compromise. We further note that TLS 1.3 0-RTT includes a mechanism named "obfuscated ticket age" that solves a similar session linkability concern; that mechanism as well is not applicable here.

$$\mathsf{Adv}^{\text{0-RTT-SR}}_{\mathcal{A},\text{Resumption}}(\lambda) \le \mathsf{Adv}^{\text{rand}}_{\mathcal{B}_{\text{PPRF1}},\text{PPRF}}(\lambda) + \varepsilon + \mu \cdot \Big(\mathsf{Adv}^{\text{na-rand}}_{\mathcal{B}_{\text{PPRF2}},\text{PPRF}}(\lambda)$$
$$+ \mathsf{Adv}^{\text{INT-CTXT}}_{\mathcal{B}_{\text{AEAD1}},\text{AEAD}}(\lambda) + \mathsf{Adv}^{\text{IND-CPA}}_{\mathcal{B}_{\text{AEAD2}},\text{AEAD}}(\lambda)\Big),$$

where μ is the number of clients.

Proof. We will conduct this proof in a sequence of games between a challenger $\mathcal{C}$ and an adversary $\mathcal{A}$. We start with an adversary playing the 0-RTT-SR security game. Over a sequence of hybrid arguments, we will stepwise transform the security game to a game where the Test-query is independent of the challenge bit b. The claim then follows from bounding the probability of distinguishing any two consecutive games. By Adv_i we denote $\mathcal{A}$'s advantage in the i-th game.

Game 0. We define Game 0 to be the original 0-RTT-SR security game. By definition we have
$$\mathsf{Adv}_0 = \mathsf{Adv}^{\text{0-RTT-SR}}_{\mathcal{A},\text{Resumption}}(\lambda).$$

Game 1. This game is identical to Game 0, except that we raise an event $\mathsf{abort}_{\text{PPRF}}$, abort the game, and output a random bit $b^* \xleftarrow{\$} \{0,1\}$, if the adversary $\mathcal{A}$ ever queries Test(t) for a ticket $t = (t', n')$ such that $n' \notin [\mu]$ and AEAD.Dec(κ, t', n') $\ne \bot$, where $\kappa := \mathsf{PPRF.Eval}(k_{\text{PPRF}}, n')$. Since both games proceed identical until abort, we have

$$|\mathsf{Adv}_1 - \mathsf{Adv}_0| \le \Pr[\mathsf{abort}_{\text{PPRF}}]$$

and we claim that we can construct an adversary $\mathcal{B}_{\text{PPRF1}}$ on the rand-security of the PPRF with advantage at least $\Pr[\mathsf{abort}_{\text{PPRF}}]$.

Construction of $\mathcal{B}_{\text{PPRF1}}$. $\mathcal{B}_{\text{PPRF1}}$ behaves like the challenger in Game 1, expect that it uses the Eval-oracle to generate the keys to encrypt the initial sequence of μ tickets and to answer all Dec-queries by $\mathcal{A}$. Eventually, $\mathcal{A}$ will query Test(t) for a ticket $t = (t', n')$. $\mathcal{B}_{\text{PPRF1}}$ outputs n' to its PPRF-challenger, which will respond with a punctured key $k := \mathsf{PPRF.Punct}(k, n')$ and a value γ, where either $\gamma := \rho \xleftarrow{\$} \mathcal{Y}$ or $\gamma := \mathsf{PPRF.Eval}(k, n')$.

$\mathcal{B}_{\text{PPRF1}}$ now tries to decrypt the challenge ticket by invoking AEAD.Dec (γ, t', n'). If $\gamma = \mathsf{PPRF.Eval}(k, n')$, the decryption will succeed by definition. If $\gamma = \rho$, the decryption will fail with probability $1 - \varepsilon$, since the ε-spreadness of AEAD ensures that AEAD.Dec(ρ, t', n') $\ne \bot$ for random ρ happens only with probability ε. Hence, $\mathcal{B}_{\text{PPRF1}}$ returns 1 if decryption succeeds and 0 otherwise. Thus, we have

$$\Pr[\mathsf{abort}_{\text{PPRF}}] \le \mathsf{Adv}^{\text{rand}}_{\mathcal{B}_{\text{PPRF1}},\text{PPRF}}(\lambda) + \varepsilon.$$

Game 2. This game is identical to Game 1, except for the following changes. At the beginning of the experiment the challenger picks an index $\nu \xleftarrow{\$} [\mu]$. It aborts the security experiment and outputs a random bit $b^* \xleftarrow{\$} \{0,1\}$, if the adversary queries Test(t) with $t = (t', i)$ such that $i \ne \nu$. Since the choice of $\nu \xleftarrow{\$} [\mu]$ is oblivious to $\mathcal{A}$ until an abort occurs, we have

$$\text{Adv}_2 \geq \frac{1}{\mu} \cdot \text{Adv}_1.$$

Game 3. This game is identical to Game 2, except that at the beginning of the game we compute $\kappa_\nu = \text{PPRF.Eval}(k, \nu)$ and then $k := \text{PPRF.Punct}(k, \nu)$. Furthermore, we replace algorithm PPRF.Eval with the following algorithm F_3:

$$F_3(k, i) := \begin{cases} \text{PPRF.Eval}(k, i) & \text{if } i \neq \nu \\ \kappa_\nu & \text{if } i = \nu \end{cases}$$

Everything else works exactly as before. Note that we have simply implemented algorithm PPRF.Eval in a slightly different way. Since PPRF is invariant to puncturing, the fact that κ_ν was computed early, immediately followed by $k := \text{PPRF.Punct}(k, \nu)$, is invisible to $\mathcal{A}$. Hence, Game 3 is perfectly indistinguishable from Game 2, and we have

$$\text{Adv}_3 = \text{Adv}_2.$$

Game 4. This game is identical to Game 3, except that the challenger now additionally picks a random key $\rho \xleftarrow{\$} \mathcal{Y}$ from the range of the PPRF. Furthermore, we replace algorithm F_3 with the following algorithm F_4:

$$F_4(k, i) := \begin{cases} \text{PPRF.Eval}(k, i) & \text{if } i \neq \nu \\ \rho & \text{if } i = \nu \end{cases}$$

Everything else works exactly as before. We will now show that any adversary that is able to distinguish Game 3 from Game 4 can be used to construct an adversary $\mathcal{B}_{\text{PPRF2}}$ against the na-rand-security of the PPRF. Concretely, we have

$$|\text{Adv}_4 - \text{Adv}_3| \leq \text{Adv}^{\text{na-rand}}_{\mathcal{B}_{\text{PPRF2}}, \text{PPRF}}(\lambda).$$

Construction of $\mathcal{B}_{\text{PPRF2}}$. $\mathcal{B}_{\text{PPRF2}}$ initially picks $\nu \xleftarrow{\$} [\mu]$ and outputs ν to its PPRF-challenger, which will respond with a punctured key $k := \text{PPRF.Punct}(k, \nu)$ and a value γ, where either $\gamma = \text{PPRF.Eval}(k, \nu)$ or $\gamma \xleftarrow{\$} \mathcal{Y}$. Now $\mathcal{B}_{\text{PPRF2}}$ simulates Game 4, except that it uses the following function F in place of F_4.

$$F(k, i) := \begin{cases} \text{PPRF.Eval}(k, i) & \text{if } i \neq \nu \\ \gamma & \text{if } i = \nu \end{cases}$$

Eventually, $\mathcal{A}$ will output a guess b^*. $\mathcal{B}_{\text{PPRF2}}$ forwards this bit to the PPRF-challenger. Note that if $\gamma = \text{Eval}(k, \nu)$, then function F is identical to F_3, while if $\gamma = \rho$ then it is identical to F_4. This proves the claim.

Game 5. This game is identical to Game 4, except that we raise an event $\text{abort}_{\text{AEAD}}$, abort the game, and output a random bit $b^* \xleftarrow{\$} \{0, 1\}$, if the adversary $\mathcal{A}$ ever queries $\text{Test}(t)$ for a ticket $t = (t', \nu) \neq t_\nu$, but $\text{AEAD.Dec}(\rho, t', \nu) \neq \perp$, where $\rho = F_4(k, \nu)$. We have

$$|\text{Adv}_5 - \text{Adv}_4| \leq \Pr[\text{abort}_{\text{AEAD}}]$$

and we claim that we can construct an adversary $\mathcal{B}_{\text{AEAD1}}$ on the INT-CTXT-security of the AEAD with advantage at least $\Pr[\text{abort}_{\text{AEAD}}]$.

Construction of $\mathcal{B}_{\mathsf{AEAD1}}$. $\mathcal{B}_{\mathsf{AEAD1}}$ proceeds exactly like the challenger in Game 5, except that it uses its challenger from the AEAD security experiment to create ticket t_ν. To this end, it outputs the tuple (s_ν, ν) for some $s_\nu \stackrel{\$}{\leftarrow} \mathcal{S}$. The AEAD challenger responds with $t'_\nu := \mathsf{AEAD.Enc}(\rho, s_\nu, \nu)$, computed with an independent AEAD key ρ. Finally, $\mathcal{B}_{\mathsf{AEAD1}}$ defines the ticket as $t_\nu = (t'_\nu, \nu)$. Apart from this, $\mathcal{B}_{\mathsf{AEAD1}}$ proceeds exactly like the challenger in Game 5.

Whenever the adversary $\mathcal{A}$ makes a query $\mathsf{Test}(t)$ with a ticket $t = (t', i)$ with $i \neq \nu$, then we abort, due to the changes introduced in Game 2. If it queries $\mathsf{Test}(t)$ with $t = (t', \nu)$ such that $t \neq t_\nu$, then $\mathcal{B}_{\mathsf{AEAD1}}$ responds with $\perp$ and outputs the tuple (t', ν) to its AEAD challenger. With probability $\Pr[\mathsf{abort}_{\mathsf{AEAD}}]$ this ticket is valid, which yields

$$\mathsf{Adv}^{\mathsf{INT\text{-}CTXT}}_{\mathcal{B}_{\mathsf{AEAD1}}, \mathsf{AEAD}}(\lambda) \geq \Pr[\mathsf{abort}_{\mathsf{AEAD}}].$$

Game 6. This game is identical to Game 5, except that when the adversary queries $\mathsf{Test}(t_\nu)$, then we will always answer with a random value, independent of the bit b. More precisely, recall that we abort if the adversary queries $\mathsf{Test}(t)$, $t = (t', \nu)$ such that $t \neq t_\nu$, due to the changes introduced in Game 5. If the adversary queries $\mathsf{Test}(t_\nu)$, then the challenger in Game 5 uses the bit $b \stackrel{\$}{\leftarrow} \{0,1\}$ sampled at the beginning of the experiment as follows. If $b = 1$, then it returns the session key s_ν. Otherwise, a random $r_\nu \stackrel{\$}{\leftarrow} \mathcal{S}$ is returned.

In Game 6, the challenger samples another random value $s'_\nu \stackrel{\$}{\leftarrow} \mathcal{S}$ at the beginning of the game. When the adversary queries $\mathsf{Test}(t_\nu)$, then if $b = 1$ the challenger returns s'_ν. Otherwise, it returns a random $r_\nu \stackrel{\$}{\leftarrow} \mathcal{S}$. Note that in either case the response of the $\mathsf{Test}(t_\nu)$-query is a random value, independent of b. Therefore the view of $\mathcal{A}$ in Game 6 is independent of b. Obviously, we have

$$\mathsf{Adv}_6 = 0.$$

We will now show that any adversary who is able to distinguish Game 5 from Game 6 can be used to construct an adversary $\mathcal{B}_{\mathsf{AEAD2}}$ against the IND-CPA-security of AEAD.

Construction of $\mathcal{B}_{\mathsf{AEAD2}}$. Recall that the key used to generate ticket t_ν is $\rho = F_4(k, \nu)$. By definition of F_4, ρ is an independent random string chosen at the beginning of the security experiment. This enables a straightforward reduction to the IND-CPA-security of the AEAD.

$\mathcal{B}_{\mathsf{AEAD2}}$ proceeds exactly like the challenger in Game 6, except for the way the ticket t_ν is created. $\mathcal{B}_{\mathsf{AEAD2}}$ computes $\rho_\nu = F_4(k, \nu)$. Then it outputs (s_ν, s'_ν, ν) to its challenger, which returns

$$t_\nu := \begin{cases} \mathsf{AEAD.Enc}(\rho, s_\nu, \nu) & \text{if } b' = 0 \\ \mathsf{AEAD.Enc}(\rho, s'_\nu, \nu) & \text{if } b' = 1 \end{cases}$$

where ρ is distributed identically to ρ_ν and b' is the hidden bit used by the challenger of the AEAD. Apart from this, $\mathcal{B}_{\mathsf{AEAD2}}$ proceeds exactly like the challenger in Game 6. Eventually, $\mathcal{A}$ will output a guess b^*. $\mathcal{B}_{\mathsf{AEAD2}}$ forwards this bit to its challenger.

Note that if $b' = 0$, then the view of $\mathcal{A}$ is perfectly indistinguishable from Game 5, while if $b' = 1$ then it is identical to Game 6. Thus, we have

$$|\mathsf{Adv}_6 - \mathsf{Adv}_5| \leq \mathsf{Adv}^{\mathsf{IND\text{-}CPA}}_{\mathcal{B}_{\mathsf{AEAD2}},\mathsf{AEAD}}(\lambda).$$

By summing up probabilities from Game 0 to Game 6, we obtain

$$\mathsf{Adv}^{\mathsf{0\text{-}RTT\text{-}SR}}_{\mathcal{A},\mathsf{Resumption}}(\lambda) \leq \mathsf{Adv}^{\mathsf{rand}}_{\mathcal{B}_{\mathsf{PPRF1}},\mathsf{PPRF}}(\lambda) + \varepsilon + \mu \cdot \left(\mathsf{Adv}^{\mathsf{na\text{-}rand}}_{\mathcal{B}_{\mathsf{PPRF2}},\mathsf{PPRF}}(\lambda) \right.$$
$$\left. + \mathsf{Adv}^{\mathsf{INT\text{-}CTXT}}_{\mathcal{B}_{\mathsf{AEAD1}},\mathsf{AEAD}}(\lambda) + \mathsf{Adv}^{\mathsf{IND\text{-}CPA}}_{\mathcal{B}_{\mathsf{AEAD2}},\mathsf{AEAD}}(\lambda) \right).$$

□

4   A PPRF with Short Secret Keys from Strong RSA

In order to instantiate our generic construction of forward-secure and replay-resilient session resumption protocol with minimal storage requirements, which is the main objective of this paper, it remains to construct suitable PPRFs with minimal storage requirements and good computational efficiency. Note that a computationally expensive PPRF may void all efficiency gains obtained from the 0-RTT protocol.

In this section we describe a PPRF based on the Strong RSA (sRSA) assumption with secret keys that only consist of three elements, even after an arbitrary number of puncturings. More precisely, a secret key consists of an RSA modulus N, an element $g \in \mathbb{Z}_N$ and a bitfield r, indicating positions where the PPRF was punctured. The secret key size is linear in the size of the PPRF's domain, since the bitfield needs to be of the same size as the domain (which is determined at initialization, and does not change over time). Hence, the PPRF's secret key size is independent of the number of puncturings. Moreover, for any reasonable choice of parameters, the bitfield is only several hundred bits long, yielding a short key in practice. Servers can use many instances in parallel with the instances sharing a single modulus, so it is only necessary to generate (and store) the modulus once, at initialization.

Since our primary objective is to provide an efficient practical solution for protocols such as TLS 1.3 0-RTT, the PPRF construction described below is analyzed in the random oracle model [5]. However, we note that we use the random oracle only to turn a "search problem" (sRSA) into a "decisional problem" (as required for a pseudorandom function). Therefore we believe that our construction can be lifted to the standard-model via standard techniques, such as hardcore predicates [6,8,23]. All of these approaches would yield less efficient constructions, and therefore are outside the scope of our work. Alternatively, one could formulate an appropriate "hashed sRSA" assumption, which would essentially boil down to assuming that our scheme is secure. Therefore we consider a random oracle analysis based on the standard sRSA problem as the cleanest and most insightful approach to describe our ideas.

Idea Behind the Construction. The construction is inspired by the *RSA accumulator* of Camenisch and Lysyanskaya [12]. The main idea is the following. Given a modulus $N = pq$, a value $g \in \mathbb{Z}_N$, and a prime number P, it is easy to compute

$g \mapsto g^P \mod N$, but hard to compute $g^P \mapsto g \mod N$ without knowing the factorization of N.

In the following let p_i be the i-th odd prime. That is, we have $(p_1, p_2, p_3, p_4, \ldots) = (3, 5, 7, 11, \ldots)$. Let n be the size of the domain of the PPRF. Our PPRF on input ℓ produces an output of the form $H(g^{p_1 \cdots p_n / p_\ell})$, where H is a hash function that will be modeled as a random oracle in the security proof. Note that g is raised to a sequence of prime numbers *except for* the ℓ-th prime number. As long as we have access to g, this is easy to compute. However, if we only have access to g^{p_ℓ} instead of g, we are unable to compute the PPRF output without knowledge of the factorization of N. This implies that by raising the generator to certain powers, we prevent the computation of specific outputs. We will use this property to puncture values of the PPRF's domain.

4.1   Formal Description of the Construction

Definition 11. *Let p, q be two random safe primes of bitlength $\lambda/2$ and let $N = pq$. Let $y \xleftarrow{\$} \mathbb{Z}_N^*$. We define the advantage of algorithm $\mathcal{B}$ against the Strong RSA Assumption [2] as*

$$\mathsf{Adv}_{\mathcal{B}}^{\mathsf{sRSA}}(\lambda) := \Pr\left[(x, e) \leftarrow \mathcal{A}(N, y) : x^e = y \mod N\right].$$

The following lemma, which is due to Shamir [42], is useful for the security proof of our construction.

Lemma 1. *There exists an efficient algorithm that, on input $y, z \in \mathbb{Z}_N$ and integers $e, f \in \mathbb{Z}$ such that $\gcd(e, f) = 1$ and $Z^e \equiv Y^f \mod N$, computes $X \in \mathbb{Z}_N$ satisfying $X^e = Y \mod N$.*

Construction 2. *Let $H : \mathbb{Z}_N \to \{0,1\}^\lambda$ be a hash function and let p_i be the i-th odd prime number. Then we construct a PPRF $\mathsf{PPRF} = (\mathsf{Setup}, \mathsf{Eval}, \mathsf{Punct})$ with polynomial-size $\mathcal{X} = [n]$ in the following way.*

- $\mathsf{Setup}(1^\lambda)$ *computes an RSA modulus $N = pq$, where p, q are safe primes. Next, it samples a value $g \xleftarrow{\$} \mathbb{Z}_N$ and defines $r := 0^n$ and $k = (N, g, r)$. The primes p, q are discarded.*
- $\mathsf{Eval}(k, x)$ *parses $k = (N, g, (r_1, \ldots, r_n))$. If $r_x = 1$, then it outputs $\bot$. Otherwise it computes and returns*

$$y := H\left(g^{P_x} \mod N\right).$$

where p_i is the i-th odd prime and

$$P_x := \prod_{i \in [n], i \neq x, r_i \neq 1} p_i$$

is the product of the first n odd primes, except for p_x.
- $\mathsf{Punct}(k, x)$ *parses $k = (N, g, (r_1, \ldots, r_n))$. If $r_x = 1$, then it returns k. If $r_x = 0$, it computes $g' := g^{p_x}$ and $r' = (r_1, \ldots, r_{x-1}, 1, r_{x+1}, \ldots, r_n)$ and returns $k' = (N, g', r')$.*

It is straightforward to verify the correctness of Construction 2 and that it is invariant to puncturing in the sense of Definition 5.

4.2 Security Analysis

We prove the following security theorem in the full version of this paper [1].

Theorem 3. *Let* PPRF = (Setup, Eval, Punct) *be as above with polynomial-size input space* $\mathcal{X} = [n]$. *From each probabilistic polynomial-time adversary* $\mathcal{A}$ *with advantage* $\mathsf{Adv}^{\mathsf{na\text{-}rand}}_{\mathcal{A},\mathsf{PPRF}}(\lambda)$ *against the* na-rand*-security (cf. Definition 6) we can construct an efficient adversary* $\mathcal{B}$ *with advantage* $\mathsf{Adv}^{\mathsf{sRSA}}_{\mathcal{B}}(\lambda)$ *against the Strong RSA problem, such that*

$$\mathsf{Adv}^{\mathsf{sRSA}}_{\mathcal{B}}(\lambda) \geq \mathsf{Adv}^{\mathsf{na\text{-}rand}}_{\mathcal{A},\mathsf{PPRF}}(\lambda).$$

4.3 Efficiency Analysis

Note that a server is able to create multiple instances of our construction to serve more tickets than one instance is able to. Using multiple instances allows using smaller exponents, but in return, the storage cost grows linearly in the number of instances.

Serving a ticket requires two exponentiations, one for computing the key and one for puncturing. Computing the key requires raising the state g to the power of $\prod_{p \in S} p$ for some subset of primes S. Puncturing requires exponentiating by a single prime. Therefore, all exponentiations feature exponents smaller than $\prod_{i=1}^{n} p_i$. We start by comparing to 2048-bit RSA, which according to the NIST key size recommendations [3] corresponds to "112-bit security", before comparing to larger RSA key sizes.

Worst-case Analysis. We compare to standard exponentiation in the group, i.e. raising to the power of $d \in \mathbb{N}$, where $\log d \approx 2048$. For puncturing to be comparable in the worst-case, we require $\log(\prod_{i=1}^{n} p_i) \leq 2048$. Choosing p_i to be the i-th odd prime yields $n \leq 232$. An economic server may store only one 2048-bit group element for the current state, and a bitfield indicating which of the 232 primes have been punctured, requiring 2280 bits in total. This allows serving 232 tickets, resulting in a storage cost of 1.22 bytes per ticket. Alternatively, a standard Session Cache would require $112 \cdot 232 = 25984$ bits to serve those 232 tickets, assuming symmetric keys of 112 bits. Therefore, our construction decreases storage size compared to a Session Cache by a factor of $25984/2280 = 11.4$.

Averaged Analysis. Note that in the above worst-case analysis we consider an *upper* bound on the exponentiation cost. That is, we guarantee that a puncturing and key derivation operation is *never* more expensive than a full exponentiation. Indeed, the first key computation raises to the power of $p_1 \cdot \ldots \cdot p_n/p_\ell$, i.e. to the product of $n - 1$ primes. However, subsequent key calculations raise to smaller powers, i.e. to the product of $n - 2$ primes, then $n - 3$, and so on. Therefore serving tickets arriving later is much cheaper than serving the first. In particular in settings where a server uses many PPRF instances in parallel, in order to deal with potentially thousands of simultaneously issued tickets, an alternative and more reasonable efficiency analysis considers the average cost of serving a ticket be comparable to exponentiation in the group. In the worst-case, primes are punctured in order, so p_n is included in the exponent in all key derivations, p_{n-1} in all derivations except the last, etc. Each prime is also used once for puncturing. Requiring $\sum_{i=1}^{n} i \cdot \log(p_i) \leq n \cdot 2048$ yields a maximum $n = 387$, and a savings factor of $112 \cdot 387/(2048 + 387) = 17.8$. The required storage is therefore 0.79 bytes per ticket.

Considering Other Security Parameters and Efficiency Requirements. Generalizing the above calculations, Table 1 gives concrete parameters for various security levels, following the NIST recommendations for key sizes [3]. Larger key sizes result in larger reductions in storage, especially when requiring average cost similar to exponentiation in the RSA group. We also show the improvement factor in storage when relaxing the above heuristic choice that serving a ticket must not cost more than one full RSA-exponentiation, by considering the case where serving a ticket is cheaper on average than 5 group exponentiations. This demonstrates that the proposed PPRF can yield very significant storage savings in general cryptographic settings, while keeping computation costs on the same order of magnitude as common public key operations. In the context of TLS, however, we expect most server operators would prefer parameters that keep processing time comparable to a single exponentiation. We emphasize that the improvement factor in storage is determined at initialization time, and is deterministic rather than probabilistic. The largest prime used in exponentiations determines how many tickets are served using a single group element. The worst-case and average-case refer to the processing time, not to the savings in storage.

Additional Storage for the Primes. The server will also need to store the first n primes, but this requires negligible additional storage. Storing the primes requires on the order of magnitude of ten kilobytes, where we expect typical caches to use many megabytes. For the minimal storage requirement, we consider 2048-bit RSA while requiring that the worst case puncturing time is cheaper than group exponentiation. In this case $n = 232$ and $p_n = 1471$, therefore all primes fit in 32-bit integers. Storing all the primes would require at most $4 \cdot 232 = 928$ bytes.

The largest value of n for the parameter choices presented in this work is $n = 9704$, for the "average cheaper than 5 exponentiations" case with 15360-bit RSA. $p_{9704} = 101341$. The required additional storage is therefore $4 \cdot 9704 = 38,816$ bytes. To reiterate, we expect typical caches to use many megabytes.

Concrete Benchmarks. We now give concrete performance estimates for this construction, using OpenSSL [45]. OpenSSL is a well-known production-grade library that implements the TLS and SSL protocols, as well as low-level cryptographic primitives. For each key size, we measure the computation time of exponentiating by all primes $\prod_{i=1}^{n} p_i$, by calling the OpenSSL "BigNum" exponentiating function. This is analogous to the computation required to serve the first ticket and then puncture the key: Serving requires exponentiating to the power of all primes except one, p_i, and puncturing requires exponentiating to the power of p_i. This is the worst-case, since serving later tickets is cheaper.

Table 1. Savings factors for various key sizes. Symmetric and asymmetric key sizes are matched according to the NIST recommendations [3]. Both savings factors denote the reduction in server-side storage required when using Construction 3. Column 3 denotes the reduction in storage achieved under the requirement that serving a single ticket is always cheaper than an exponentiation in the RSA group of respective key size. Column 4 denotes the reduction in storage achieved under the requirement that the average cost for serving a ticket is cheaper than a single exponentiation. Column 5 denotes the reduction in storage achieved under the requirement that the average cost for serving a ticket is cheaper than 5 group exponentiations.

		Storage Savings Factor		
Symmetric Key Size	Modulus Size	W.C. cheaper than exponentiation	Average cheaper than exponentiation	Average cheaper than 5 exponentiations
112	2048	11.40	17.80	48.92
128	3072	12.28	19.47	54.49
192	7680	16.37	26.52	77.36
256	15360	20.10	33.05	99.12

We measure the performance of this calculation for two of the above cases, which determine the value of n: (1) Worst-case is cheaper than exponentiation, and (2) The average case is cheaper than exponentiation. We note the latter case is slightly unintuitive: we measure *the worst-case performance, under the requirement that the average case is comparable to one exponentiation in the group.*

Table 2 gives our results. We observe that performance is comparable to, but slower than, RSA decryption. In typical cases, it requires only a few additional milliseconds compared to RSA decryption. We argue the additional latency and computation requirement are small enough to allow the construction to be deployed as-is, in current large scale TLS deployments. It is unsurprising that RSA decryption is faster than our construction, since OpenSSL performs RSA decryption using the Chinese Remainder Theorem.

Table 2. Worst-case running time for serving a single ticket using our construction, compared to RSA decryption. All times are measured in milliseconds. Measurements were performed on a standard workstation, with a 3.60 GHz Intel i7 CPU. All measurements used code from OpenSSL 1.0.2q, released in November 2018. To benchmark our construction we used a short piece of custom code, based on [9], to repeatedly call the OpenSSL exponentiating function. For each parameter choice, we generated 100 random moduli, and performed 100 exponentiations of random group elements to the power of $\prod_{i=1}^{n} p_i$. To benchmark RSA decryption, we used a built-in OpenSSL benchmarking command, "openssl speed" (after applying a small patch that adds support for 3072-bit RSA to the command [29]).

	Our construction: Decryption + Puncturing		
Modulus Size	W.C. cheaper than exponentiation	Average cheaper than exponentiation	RSA Decryption
2048	2.6	4.7	0.5
3072	8.3	15.2	2.5
4096	19.4	35.8	5.6

5   Tree-Based PPRFs

This section will consider a different approach to instantiating Construction 1 based on PPRFs using trees. At first we will recap the idea behind tree-based PPRFs and explain how we utilize tree-based PPRFs as an instantiation of our session resumption protocol and highlight implications. Finally, we will describe our new "domain extension" technique for PPRFs and analyze its efficiency.

5.1   Tree-Based PPRFs

We will briefly recap the main idea behind tree-based PPRFs. It is well known that the GGM tree-based construction of pseudorandom functions (PRFs) from one-way functions [22] can be modified to construct a puncturable PRF, as noted in [10,11,30]. It works as follows.

Let $G : \{0,1\}^\lambda \rightarrow \{0,1\}^{2\lambda}$ be a pseudorandom generator (PRG) and let $G_0(k)$, $G_1(k)$ be the first and second half of string $G(k)$, where k is a random seed. The GGM construction defines a binary tree on the PRF's domain, where each leaf represents an evaluation of the PRF. We label each edge with 0 if it connects to a left child, and 1 if it connects to a right child. We label each node with the binary string determined by the path from the root to the node. The PRF value of $x = x_1 \ldots x_n \in \{0,1\}^n$ is $(G_{x_n} \circ \ldots \circ G_{x_1})(k) \in \{0,1\}^\lambda$, i.e. we compose G according to the path from root to leaf x.

We will briefly describe how this construction can be transformed into a PPRF. In order to puncture the PPRF at input $x = x_1 \ldots x_n$ we compute a tuple of n intermediate node evaluations for prefixes $\overline{x_1}, x_1\overline{x_2}, \ldots, x_1 x_2 \ldots \overline{x_n}$ and discard the initial seed k. The intermediate evaluations enable us to still compute evaluations on all inputs but x. Successive puncturing is possible if we apply the above computations to an intermediate evaluation. Note that we have to compute at most $n \cdot m$ intermediate values if we puncture at random, where m is the number of puncturing operations performed.

The PPRF is secure if an adversary is not able to distinguish between a punctured point and a truly random value, even when given the values of all computed "neighbor nodes". This holds as long as the underlying PRG is indistinguishable from random [10,11,30].

5.2   Combining Tree-Based PPRFs with Tickets

In our session resumption scenario the tree-based PPRF will act as a puncturable STEK. That is, evaluating the PPRF returns a ticket encryption key. Upon resumption with a ticket we will retrieve the ticket encryption key from the PPRF by evaluating it and puncture the PPRF at that very value to ensure the ticket encryption key cannot be computed twice. Note that each ticket encryption key essentially corresponds to a leaf of the tree. Thus we will subsequently use the terms leaf and ticket (encryption key) interchangeably depending on the context.

For simplicity, we consider tickets which consist of a ticket number i and a ticket lifetime t. Following Construction 1 we will issue the tickets one after another while incrementing the ticket number for each. Note that the ticket number i corresponds to the i-th leftmost leaf of the tree. The ticket lifetime t determines how long an issued ticket is valid for resumption. That is, if $t' > t$ time has passed, the server will reject the ticket.

We assume that the rate at which tickets are issued is roughly the same as the rate tickets are used for session resumption. This holds as for each session resumption we will issue a new ticket to again resume the session at a later point in time. Similarly, we argue that tickets are roughly used in the same order for resumption as we issued them. Again, if we consider multiple users, repeatedly requesting tickets and resuming sessions, we are able to average the time a user takes until a session is resumed.[8] This yields an implicit window of tickets in usage. The window is bounded left by the ticket lifetime and bounded right by the last ticket the server issued. Within the lifetime of the tree-based PPRF this implicit window will shift from left to right over the tree's leaves. It immediately follows that tickets are also roughly used in that order.

5.3   Efficiency Analysis of the Tree-Based PPRF

We will now discuss how the performance of tree-based PPRFs depends on the ticket lifetime. We consider a scenario where the ticket lifetime t equals the number of leaves ℓ. It is also possible to consider a scenario where the ticket lifetime is smaller than the number of leaves. If both number of leaves ℓ and ticket lifetime t are powers of 2, we can divide the leaves in ℓ/t windows, which span a subtree each.[9] The subtrees are all linked with the "upper part" of the tree. A different approach would be to instantiate a new tree when a tree runs out of tickets. We stress that this does not affect our analysis. As soon as one subtree runs out of tickets, the next subtree is used. If the rate at which we issue tickets stays the same, we are able to delete parts of the former tree when issuing tickets of the next one. Hence, for analysis, it is sufficient to consider a single tree.

If we were to puncture leaves strictly from left to right, we would need to store at most $\log(\ell)$ leaves (one leaf per layer). Note that if we puncture leaves at random, we would need to store at most $p \cdot \log(\ell)$ nodes, where p is the number of punctures performed. We can also bound the number of nodes we need to store by $p \cdot \log(\ell) \leq \ell/2$. This is due to the tree being binary. Essentially each node (except for the lowest layer) represents at least two leaves. To be more precise, in a tree with L layers, storing a node on layer i allows evaluating its 2^{L-i} children. Thus it is preferable to store those nodes instead of storing leaves in order to save memory. In the worst-case only every second leaf is punctured.

---

[8] Cloudflare have suggested that these assumptions seem reasonable. Unfortunately, they cannot provide data on returning clients' behavior yet.

[9] When implementing tree-based PPRFs in session resumption scenarios, such windows should not be implemented as they only add management overhead to the algorithm instead of providing notable advantages. It is sufficient to use a tree-based PPRF as is and puncture leaves for which the ticket's lifetime has expired. This way we achieve an implicit implementation of a sliding window scenario that ensures all established bounds still hold.

This results in precomputation of all other leaves without being able to save memory by only storing an intermediary node. Note that this would actually resemble a Session Cache, where all issued tickets are stored. However, note that a session cache needs to store each ticket when it has been issued, whereas our construction only needs to increase its storage if a ticket is used for resumption. Thus, our tree-based construction performs (memory-wise) at least as well as a Session Cache. In practice, where user behavior is much more random, our approach is *always* better than Session Caches.

The tree-based PPRF performs more computations compared to a Session Cache. When issuing tickets we need to compute all nodes from the closest computed node to a leaf. For puncturing we need to compute the same, plus computation of some additional sibling nodes. However, when instantiating the construction with a cryptographic hash function, such as SHA-3, evaluation and puncturing of the PPRF consists only of several hash function evaluations. This makes our construction especially suitable for high-traffic scenarios.

Table 3 gives worst-case secret key sizes based on the above analysis. However, we expect the secret key size to be much smaller in practice. Unfortunately, we are not able to estimate the average key size as this would depend on the exact distribution of returning clients' arrival times.

Table 3. Worst-case size of secret key depending on the rate of tickets per second and the ticket lifetime assuming 128 bit ticket size. The worst-case secret key size is computed as $|k| = 128rt/2$.

Tickets per Second r	Ticket Lifetime t	Worst-case Secret Key Size $|k|$
16	1 hour	461 kB
16	1 day	11.06 MB
128	1 hour	3.69 MB
128	1 day	88.47 MB
1024	1 hour	29.49 MB
1024	1 day	707.79 MB

5.4   Generic Domain Extension for PPRFs

Most forward-secure and replay-resilient 0-RTT schemes come with large secret keys (possibly several hundred megabytes) when instantiated in a real-world environment [15,16,25]. This is especially problematic if the secret key needs to be synchronized across multiple server instances. Therefore it is often desirable to minimize the secret key size.

In this section we will describe a generic domain extension. In the context of our work, the domain extension reduces the size of punctured keys by trading secret key size for ticket size, while preserving the puncturing functionality.

Idea Behind the Construction. Our session resumption protocol uses the output of the PPRF as a ticket encryption key. Normally, a PPRF only allows one output per input as it is designed to be a function. Our protocol, however, does not rely on this property. Instead of only using one ticket encryption key we could generate multiple ticket encryption keys. Ticket issuing would work as follows. First, we generate an intermediary symmetric key to encrypt the Resumption Secret[10]. The intermediary symmetric key is then encrypted under each of the ticket encryption keys. The ticket will consist of one encryption of the Resumption Secret and several (redundant) encryptions of the intermediary symmetric key.

As long as the PPRF is able to recompute *at least one* of those ticket encryption keys, the server will still be able to resume the session. This allows us to construct a wrapper around the PPRF that extends the PPRF's domain by relaxing the requirement that every input has only a single output.

Before formally describing our construction, we will provide an example to illustrate the idea. Let $\mathcal{X}$ be the PPRF's domain. We will extend the domain to $\mathcal{X} \times [n]$ with a domain extension factor of n. That is, we will allow $(x, i), i \in [n]$ for any $x \in \mathcal{X}$ as input. Let $G : \{0, 1\}^\lambda \to \{0, 1\}^{n\lambda}$ be a pseudorandom generator and let $G_j(x)$ be the j-th bitstring of size λ of G on input x. We define the evaluation of (x, i) as all possible compositions of G_j which end with G_i. That is, for any input (x, i) there will be $(n - 1)!$ different outputs, as there are $(n - 1)!$ ways to compose G_j with $j \neq i$. The possible compositions of PRGs can be illustrated as a tree as shown in Fig. 4.

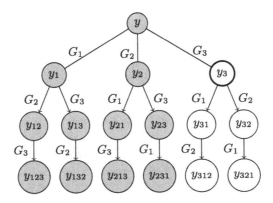

Fig. 4. Possible composition of PRGs for $n = 3$ illustrated as a tree. Each path from parent to child illustrates an evaluation of the PRG shown next to the path. Upon puncturing $(x, 3)$, the value y_3 is computed and stored and y is discarded. Thus, only the white nodes are computable, whereas the gray nodes cannot be computed without inverting G_3.

---

[10] Typically, a ticket contains not only the Resumption Secret but also the chosen cipher suite and other additional session parameters, and is thus larger than just the Resumption Secret. Therefore it is reasonable to encrypt this data only once, while encrypting the shorter intermediary symmetric key multiple times. This makes the ticket as short as possible.

After puncturing the PPRF's key for a value (x, i), it must not be possible to evaluate the value anymore. This requires a mechanism to ensure that composing the PRGs which end with G_i is no longer possible. We achieve this by forcing an evaluation of $y_i := G_i(y)$, where y is the evaluation of the underlying PPRF on input x. In order to render recomputation of y impossible, we additionally need to puncture the PPRF's key on value x and delete the computed y. Formally, the construction is defined as follows.

Construction 3. *Let $G : \{0,1\}^\lambda \to \{0,1\}^{n\lambda}$ be a PRG and let $G_i(k)$ be the i-th bitstring of size λ of G. Let $\mathsf{PPRF}' = (\mathsf{Setup}', \mathsf{Eval}', \mathsf{Punct}')$ be a PPRF with domain $\mathcal{X}$. We construct a domain extended PPRF $\mathsf{DE} = (\mathsf{Setup}, \mathsf{Eval}, \mathsf{Punct})$ with domain $\mathcal{X} \times [n]$ for $n \in \mathbb{N}$ as follows.*

- $\mathsf{Setup}(1^\lambda)$ *computes* $k_{\mathsf{PPRF}} := \mathsf{Setup}'(1^\lambda)$. *Next, it defines an empty list* $\mathcal{L} = \emptyset$. *Output is* $k = (k_{\mathsf{PPRF}}, \mathcal{L})$
- $\mathsf{Eval}(k, x)$ *parses* $x = (x_{\mathsf{PPRF}}, x_{\mathsf{ext}}) \in \mathcal{X} \times [n]$ *and* $k = (k_{\mathsf{PPRF}}, \mathcal{L})$. *It computes* $y := \mathsf{Eval}(k_{\mathsf{PPRF}}, x_{\mathsf{PPRF}})$.
  If $y = \bot$, *it checks whether* $\exists x_{\mathsf{PPRF}}$ *with* $(x_{\mathsf{PPRF}}, y', (r_1, \ldots, r_n)) \in \mathcal{L}$. *If it exists, assign* $y := y'$. *Otherwise it outputs* $\bot$.
  Furthermore, it defines a set $\mathcal{R} = \{i \in [n] \mid r_i = 1\}$. *If* r_i *are undefined, set* $\mathcal{R}$ *is empty. Next, it computes*

$$\mathcal{Y} = \{(G_{i_{n-|\mathcal{R}|-1}} \circ \ldots \circ G_{i_1})(y)\},$$

  where $(i_1, \ldots, i_{n-|\mathcal{R}|-1})$ *are all* $(n-|\mathcal{R}|-1)!$ *possible permutations of elements in* $[n] \setminus (\mathcal{R} \cup \{x_{\mathsf{ext}}\})$. *Output is* $\mathcal{Y}$.
- $\mathsf{Punct}(k, x)$ *parses* $k = (k_{\mathsf{PPRF}}, \mathcal{L})$ *and* $x = (x_{\mathsf{PPRF}}, x_{\mathsf{ext}}) \in \mathcal{X} \times [n]$. *It computes* $y := \mathsf{Eval}(k_{\mathsf{PPRF}}, x_{\mathsf{PPRF}})$. *If* $y \neq \bot$, *it appends* $\mathcal{L}' = \mathcal{L} \cup \{x_{\mathsf{PPRF}}, y, (r_1, \ldots, r_n)\}$, *where* $r_i = 0$, *but* $r_{x_{\mathsf{ext}}} = 1$. *Additionally, it punctures* $k'_{\mathsf{PPRF}} := \mathsf{Punct}'(k_{\mathsf{PPRF}}, x_{\mathsf{PPRF}})$.
  If $y = \bot$ *and* $\nexists x_{\mathsf{ext}}$ *with* $(x_{\mathsf{ext}}, y', r) \in \mathcal{L}$, *it outputs* k.
  Otherwise it retrieves $\ell = (x_{\mathsf{ext}}, y', (r_1, \ldots, r_n)) \in \mathcal{L}$. *If* $r_i = 1$ *for all* $i \in [n] \setminus \{x_{\mathsf{ext}}\}$, *remove* ℓ *from* $\mathcal{L}$. *Else it updates* $\ell \in \mathcal{L}$ *by computing* $\mathcal{L}' = (\mathcal{L} \setminus \{\ell\}) \cup \{\ell'\}$. *Output is* $k = (k'_{\mathsf{PPRF}}, \mathcal{L}')$.

5.5   Efficiency Analysis of the Generic Domain Extension

Increased Ticket Size. Note that a ticket is longer than a standard ticket by $(n - 1)!$ encrypted blocks. Assuming 128-bit AES, and choosing $n = 5$, this translates to $4! \cdot 16 = 384$ additional bytes. This is likely to be insignificant on the modern Internet. For example, Google has pushed for increasing the maximum initial flight from 4 TCP packets to 10 [19], as most server responses span several packets already (a typical full packet is about 1500 bytes). A basic experiment performed by Google and Cloudflare in 2018 measured a similar scenario: It added 400 bytes for both the client's and server's first flights [33]. They observed relatively small additional latencies: 2–4 ms in the median, and less than 20 ms for the 95th percentile.[11] However, choosing $n = 6$ or larger is likely to be not cost-effective. This would translate to $5! \cdot 16 = 1920$ additional bytes, larger than a standard TCP packet.

Storage Requirements. Comparing the storage requirements of the tree-based construction to standard Session Caches depends on the specific distribution of returning clients. In the best case, tickets arrive in large contiguous blocks. In this case, a tree-based construction uses negligible storage (logarithmic in the number of tickets), making the savings factor in storage huge. However, this is unrealistic in practice. In the worst-case, tickets arrive in blocks of $n-1$ tickets of the form (x_{PPRF}, i) for $i \in [n-1]$, adversarially rendering the domain extension technique useless as each subtree is reduced to a single node. As before, this is unrealistic in practice.

We have therefore resorted to simulations in order to assess the improvement in storage requirements. Our simulation constructs two trees: a standard binary tree with ℓ layers, and a domain-extended tree with $n = 4$. For the domain-extended tree, the first $\ell - 2$ layers are constructed as a standard binary tree, and the last $\log(4) = 2$ layers are represented by the domain extension.

We simulated the storage requirements for trees of 10,000 tickets.[12] We focused on the relationship between ticket puncturing rate and savings in storage. The ticket puncturing rate denotes the percentage of tickets that are punctured, out of the 10,000 outstanding tickets. This can also be thought of as the percentage of returning clients. After fixing the puncturing rate to r, we simulate the arrival of $r\%$ of clients according to two distributions: Gaussian and uniform. With the uniform distribution, the next ticket to be punctured is sampled uniformly out of the outstanding tickets. With the Gaussian distribution, the next ticket to be punctured is sampled using a discrete Gaussian distribution with mean $\mu = 5000$ and standard deviation σ (for varying values of σ). We then simulate the state of both trees after puncturing the sampled ticket. We repeatedly sample tickets and puncture them, until we reach the desired puncturing rate. We then report the ratio between the storage for the standard binary tree and the storage for the domain-extended tree, in their final states.

Intuitively, the Gaussian distribution aims to simulate the assumption where tickets arrive in some periodic manner. For example, assume the tickets most likely to arrive are the tickets issued roughly one hour ago. Then the distribution of arriving tickets will exhibit a noticeable mode ("peak"), where tickets close to the mode are much more likely to arrive than tickets far from it. The Gaussian distribution is a natural fit for this description. On the other hand, the uniform distribution makes no assumptions on which ticket is likely to arrive next. In

---

[11] The relevant experiment is denoted as "Phase Two"; "Phase One" only added bytes to the client's first flight.

[12] We note that results for trees of 10,000 tickets should closely follow results for larger tree sizes. Trees are quickly split into smaller sub-trees when puncturing, regardless of the initial tree size. In the first puncturing operation we delete the root and store smaller sub-trees with at most half the nodes in each, and so forth.

personal communication, Cloudflare have advised us that it is reasonable to assume tickets are redeemed roughly in order of issuance (they do not have readily-available data on returning clients' behavior). This motivated our use of Gaussian distributions. We hope to see additional research in this area. In particular, it would be helpful if large server operators could release anonymized datasets that allow simulating the behavior of returning clients in practice.

Using our domain extension technique with $n = 4$ results in a typical factor of 1.4 (or more) reduction in storage compared to a tree-based PPRF. Figure 5 plots the results when using the uniform distribution and a Gaussian distribution with $\sigma = 2000$. We encountered similar results when using other values for σ. We estimate ticket redeeming rates in large-scale deployments are roughly 50%. We therefore focus on cases where the puncturing rate is at least 40% and at most 60%. We note that in the worst-case, the domain extension performs as well as the binary tree.

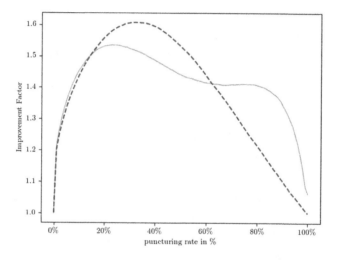

Fig. 5. Average storage improvement factor of the domain-extended binary tree (with $n = 4$) compared to a standard binary tree, depending on the ticket puncturing rate. All simulations used trees of 10,000 tickets. The dashed blue line (resp. continuous red line) shows the storage improvement when modeling client's arrivals with a uniform distribution (resp. discrete Gaussian distribution with mean $\mu = 5000$ and standard deviation $\sigma = 2000$). (Color figure online)

6   Comparison of Solutions and Conclusion

Comparison of Solutions. To summarize this work, Table 4 compares our two constructions with the standard solutions of Session Tickets and Session Caches.

Table 4. Comparison of security guarantees and dominant cost for Session Tickets, Session Caches, and our two constructions. For Session Tickets, we assume a deployment that rotates STEKs, as in [34]. For Session Caches, we assume each key is 128 bits (16 bytes) long. The unique ticket identifier, and other storage overhead, will typically require a few more bytes. We therefore estimate total storage per key as 20–30 bytes. For the Tree-based PPRF, actual storage per ticket highly depends on returning clients behavior. However, this solution always requires at most as much storage as a Session Cache.

Solution	Forward Security	Replay Protection	Storage per Ticket	Dominant Cost	See Section
Session Tickets	After $\approx$ 1 day	No	Negligible	Symmetric encryption	1
Session Caches	Yes	Yes	$\approx$20–30 bytes	Database access	1
sRSA-based PPRF	Yes	Yes	$\approx$0.8–1.2 bytes	Group exponentiation	4.3
Tree-based PPRF	Yes	Yes	$\leq$20–30 bytes	Database access	5.3

Conclusion. In most facets, TLS 1.3 offers significant improvements in security compared to earlier TLS versions. However, when 0-RTT mode is used, it surprisingly weakens standard security guarantees, namely forward security and replay resilience. This was noted as the protocol was standardized, but the latency reduction from 0-RTT was considered "too big a win not to do" [38].

This paper presented formal definitions for secure 0-RTT Session Resumption Protocols, and two new constructions that allow achieving the aforementioned security guarantees at a practical cost. We expect continued research in the coming years in this area, of achieving secure 0-RTT traffic as cheaply as possible. Currently, many large server operators serve 0-RTT traffic using STEK-encrypted Session Tickets. As more Internet traffic becomes 0-RTT traffic, this solution rolls back the security guarantees offered to everyday secure sessions.

References

1. Aviram, N., Gellert, K., Jager, T.: Session resumption protocols and efficient forward security for TLS 1.3 0-RTT. Cryptology ePrint Archive (2019). https://eprint.iacr.org
2. Barić, N., Pfitzmann, B.: Collision-free accumulators and fail-stop signature schemes without trees. In: Fumy, W. (ed.) EUROCRYPT 1997. LNCS, vol. 1233, pp. 480–494. Springer, Heidelberg (1997). https://doi.org/10.1007/3-540-69053-0_33
3. Barker, E.: Recommendation for key management part 1: general (revision 4). NIST special publication (2016)
4. Behr, M., Swett, I.: Introducing QUIC support for HTTPS load balancing (2018). https://cloudplatform.googleblog.com/2018/06/Introducing-QUIC-support-for-HTTPS-load-balancing.html
5. Bellare, M., Rogaway, P.: Random oracles are practical: a paradigm for designing efficient protocols. In: Ashby, V. (ed.) ACM CCS 1993, Fairfax, Virginia, USA, 3–5 November, pp. 62–73. ACM Press (1993)

6. Bellare, M., Stepanovs, I., Tessaro, S.: Poly-many hardcore bits for any one-way function and a framework for differing-inputs obfuscation. In: Sarkar, P., Iwata, T. (eds.) ASIACRYPT 2014, Part II. LNCS, vol. 8874, pp. 102–121. Springer, Heidelberg (2014). https://doi.org/10.1007/978-3-662-45608-8_6

7. Bhargavan, K., Fournet, C., Kohlweiss, M., Pironti, A., Strub, P.-Y., Zanella-Béguelin, S.: Proving the TLS handshake secure (as it is). In: Garay, J.A., Gennaro, R. (eds.) CRYPTO 2014, Part II. LNCS, vol. 8617, pp. 235–255. Springer, Heidelberg (2014). https://doi.org/10.1007/978-3-662-44381-1_14

8. Blum, L., Blum, M., Shub, M.: A simple unpredictable pseudo-random number generator. Siam J. Comput. **15**(2), 364–383 (1986). https://doi.org/10.1137/0215025

9. Böck, H.: Fuzz-compare the OpenSSL function BN_mod_exp() and the libgcrypt function gcry_mpi_powm(). https://github.com/hannob/bignum-fuzz/blob/master/openssl-vs-gcrypt-modexp.c

10. Boneh, D., Waters, B.: Constrained pseudorandom functions and their applications. In: Sako, K., Sarkar, P. (eds.) ASIACRYPT 2013, Part II. LNCS, vol. 8270, pp. 280–300. Springer, Heidelberg (2013). https://doi.org/10.1007/978-3-642-42045-0_15

11. Boyle, E., Goldwasser, S., Ivan, I.: Functional signatures and pseudorandom functions. In: Krawczyk, H. (ed.) PKC 2014. LNCS, vol. 8383, pp. 501–519. Springer, Heidelberg (2014). https://doi.org/10.1007/978-3-642-54631-0_29

12. Camenisch, J., Lysyanskaya, A.: Dynamic accumulators and application to efficient revocation of anonymous credentials. In: Yung, M. (ed.) CRYPTO 2002. LNCS, vol. 2442, pp. 61–76. Springer, Heidelberg (2002). https://doi.org/10.1007/3-540-45708-9_5

13. Chang, W.T., Langley, A.: QUIC crypto (2014). https://docs.google.com/document/d/1g5nIXAIkN_Y-7XJW5K45IblHd_L2f5LTaDUDwvZ5L6g

14. Cremers, C., Horvat, M., Scott, S., van der Merwe, T.: Automated analysis and verification of TLS 1.3: 0-RTT, resumption and delayed authentication. In: 2016 IEEE Symposium on Security and Privacy, San Jose, CA, USA, 22–26 May, pp. 470–485. IEEE Computer Society Press (2016)

15. Derler, D., Gellert, K., Jager, T., Slamanig, D., Striecks, C.: Bloom filter encryption and applications to efficient forward-secret 0-RTT key exchange. Cryptology ePrint Archive, Report 2018/199 (2018). https://eprint.iacr.org/2018/199

16. Derler, D., Jager, T., Slamanig, D., Striecks, C.: Bloom filter encryption and applications to efficient forward-secret 0-RTT key exchange. In: Nielsen, J.B., Rijmen, V. (eds.) EUROCRYPT 2018, Part III. LNCS, vol. 10822, pp. 425–455. Springer, Cham (2018). https://doi.org/10.1007/978-3-319-78372-7_14

17. Dowling, B., Fischlin, M., Günther, F., Stebila, D.: A cryptographic analysis of the TLS 1.3 handshake protocol candidates. In: Ray, I., Li, N., Kruegel, C. (eds.) ACM CCS 2015, Denver, CO, USA, 12–16 October, pp. 1197–1210. ACM Press (2015)

18. Dowling, B., Fischlin, M., Günther, F., Stebila, D.: A cryptographic analysis of the TLS 1.3 draft-10 full and pre-shared key handshake protocol. Cryptology ePrint Archive, Report 2016/081 (2016). http://eprint.iacr.org/2016/081

19. Dukkipati, N., et al.: An argument for increasing TCP's initial congestion window. Comput. Commun. Rev. **40**(3), 26–33 (2010)

20. Fischlin, M., Günther, F.: Multi-stage key exchange and the case of Google's QUIC protocol. In: Ahn, G.J., Yung, M., Li, N. (eds.) ACM CCS 2014, Scottsdale, AZ, USA, 3–7 November, pp. 1193–1204. ACM Press (2014)

21. Fischlin, M., Günther, F.: Replay attacks on zero round-trip time: the case of the TLS 1.3 handshake candidates. In: 2017 IEEE European Symposium on Security and Privacy, EuroS&P 2017, Paris, France, 26–28 April 2017, pp. 60–75. IEEE (2017). https://doi.org/10.1109/EuroSP.2017.18

22. Goldreich, O., Goldwasser, S., Micali, S.: How to construct random functions. J. ACM **33**(4), 792–807 (1986). https://doi.org/10.1145/6490.6503

23. Goldreich, O., Levin, L.A.: A hard-core predicate for all one-way functions. In: 21st ACM STOC, Seattle, WA, USA, 15–17 May, pp. 25–32. ACM Press (1989)

24. Green, M.D., Miers, I.: Forward secure asynchronous messaging from puncturable encryption. In: 2015 IEEE Symposium on Security and Privacy, San Jose, CA, USA, 17–21 May, pp. 305–320. IEEE Computer Society Press (2015)

25. Günther, F., Hale, B., Jager, T., Lauer, S.: 0-RTT key exchange with full forward secrecy. In: Coron, J.-S., Nielsen, J.B. (eds.) EUROCRYPT 2017, Part III. LNCS, vol. 10212, pp. 519–548. Springer, Cham (2017). https://doi.org/10.1007/978-3-319-56617-7_18

26. Hale, B., Jager, T., Lauer, S., Schwenk, J.: Simple security definitions for and constructions of 0-RTT key exchange. In: Gollmann, D., Miyaji, A., Kikuchi, H. (eds.) ACNS 2017. LNCS, vol. 10355, pp. 20–38. Springer, Cham (2017). https://doi.org/10.1007/978-3-319-61204-1_2

27. Iyengar, S., Nekritz, K.: Building zero protocol for fast, secure mobile connections (2017). https://code.fb.com/android/building-zero-protocol-for-fast-secure-mobile-connections/

28. Jager, T., Kohlar, F., Schäge, S., Schwenk, J.: On the security of TLS-DHE in the standard model. In: Safavi-Naini, R., Canetti, R. (eds.) CRYPTO 2012. LNCS, vol. 7417, pp. 273–293. Springer, Heidelberg (2012). https://doi.org/10.1007/978-3-642-32009-5_17

29. Kario, H.: Add 3072, 7680 and 15360 bit RSA tests to openssl speed. https://groups.google.com/forum/#!topic/mailing.openssl.dev/bv8t7QcXrqg

30. Kiayias, A., Papadopoulos, S., Triandopoulos, N., Zacharias, T.: Delegatable pseudorandom functions and applications. In: Sadeghi, A.R., Gligor, V.D., Yung, M. (eds.) ACM CCS 2013, Berlin, Germany, 4–8 November, pp. 669–684. ACM Press (2013)

31. Krawczyk, H., Paterson, K.G., Wee, H.: On the security of the TLS protocol: a systematic analysis. In: Canetti, R., Garay, J.A. (eds.) CRYPTO 2013, Part I. LNCS, vol. 8042, pp. 429–448. Springer, Heidelberg (2013). https://doi.org/10.1007/978-3-642-40041-4_24

32. Langley, A.: How to botch TLS forward secrecy (2013). https://www.imperialviolet.org/2013/06/27/botchingpfs.html

33. Langley, A.: Post-quantum confidentiality for TLS (2018). https://www.imperialviolet.org/2018/04/11/pqconftls.html

34. Lin, Z.: TLS session resumption: full-speed and secure (2015). https://blog.cloudflare.com/tls-session-resumption-full-speed-and-secure/

35. Lychev, R., Jero, S., Boldyreva, A., Nita-Rotaru, C.: How secure and quick is QUIC? Provable security and performance analyses. In: 2015 IEEE Symposium on Security and Privacy, pp. 214–231. IEEE Computer Society Press, San Jose, 17–21 May 2015

36. MacCarthaigh, C.: Security Review of TLS 1.3 0-RTT. https://github.com/tlswg/tls13-spec/issues/1001, Accessed 29 July 2018

37. Rescorla, E.: TLS 0-RTT and Anti-Replay (2015). https://www.ietf.org/mail-archive/web/tls/current/msg15594.html

38. Rescorla, E.: TLS 1.3 (2015). http://web.stanford.edu/class/ee380/Abstracts/151118-slides.pdf

39. Rescorla, E.: The Transport Layer Security (TLS) Protocol Version 1.3. RFC 8446 (2018). https://rfc-editor.org/rfc/rfc8446.txt

40. Rogaway, P.: Authenticated-encryption with associated-data. In: Atluri, V. (ed.) ACM CCS 2002, Washington D.C., USA, 18–22 November, pp. 98–107. ACM Press (2002)

41. Sahai, A., Waters, B.: How to use indistinguishability obfuscation: deniable encryption, and more. In: Shmoys, D.B. (ed.) 46th ACM STOC, New York, NY, USA, 31 May–3 June, pp. 475–484. ACM Press (2014)

42. Shamir, A.: On the generation of cryptographically strong pseudorandom sequences. ACM Trans. Comput. Syst. $\mathbf{1}$(1), 38–44 (1983). https://doi.org/10.1145/357353.357357

43. Springall, D., Durumeric, Z., Halderman, J.A.: Measuring the security harm of TLS crypto shortcuts. In: Proceedings of the 2016 Internet Measurement Conference, pp. 33–47. ACM (2016)

44. Sullivan, N.: Introducing Zero Round Trip Time Resumption (2017). https://blog.cloudflare.com/introducing-0-rtt/

45. The OpenSSL Project: OpenSSL: The open source toolkit for SSL/TLS. https://www.openssl.org/

An Analysis of NIST SP 800-90A

Joanne Woodage[1]([⊠]) and Dan Shumow[2]

[1] Royal Holloway, University of London, Egham, UK
joanne.woodage.2014@rhul.ac.uk
[2] Microsoft Research, Redmond, USA

Abstract. We investigate the security properties of the three deterministic random bit generator (DRBG) mechanisms in NIST SP 800-90A [2]. The standard received considerable negative attention due to the controversy surrounding the now retracted DualEC-DRBG, which appeared in earlier versions. Perhaps because of the attention paid to the DualEC, the other algorithms in the standard have received surprisingly patchy analysis to date, despite widespread deployment. This paper addresses a number of these gaps in analysis, with a particular focus on HASH-DRBG and HMAC-DRBG. We uncover a mix of positive and less positive results. On the positive side, we prove (with a caveat) the robustness [13] of HASH-DRBG and HMAC-DRBG in the random oracle model (ROM). Regarding the caveat, we show that if an optional input is omitted, then – contrary to claims in the standard—HMAC-DRBG does not even achieve the (weaker) property of forward security. We then conduct a more informal and practice-oriented exploration of flexibility in the standard. Specifically, we argue that these DRBGs have the property that partial state leakage may lead security to break down in unexpected ways. We highlight implementation choices allowed by the overly flexible standard that exacerbate both the likelihood, and impact, of such attacks. While our attacks are theoretical, an analysis of two open source implementations of CTR-DRBG shows that these potentially problematic implementation choices are made in the real world.

1 Introduction

Secure pseudorandom number generators (PRNGs) underpin the vast majority of cryptographic applications. From generating keys, nonces, and IVs, to producing random numbers for challenge responses, the discipline of cryptography—and hence system security—critically relies on these primitives. However, it has been well-established by a growing list of real-world failures [6,18,32,39], that when a PRNG is broken, the security of the reliant application often crumbles with it. Indeed, with much currently deployed cryptography being effectively 'unbreakable' when correctly implemented, exploiting a weakness in the underlying PRNG emerges as a highly attractive target for an attacker. As such, it is of paramount importance that standardized PRNGs are as secure as possible.

NIST Special Publication 800-90A "Recommendation for Random Number Generation Using Deterministic Random Bit Generators (NIST SP 800-90A) [2]

© International Association for Cryptologic Research 2019
Y. Ishai and V. Rijmen (Eds.): EUROCRYPT 2019, LNCS 11477, pp. 151–180, 2019.
https://doi.org/10.1007/978-3-030-17656-3_6

has had a troubled history. The first version of this standard included the now infamous DualEC-DRBG, which was long suspected to contain a backdoor inserted by the NSA [36]. This suspicion was confirmed by documents included in the Snowden leaks [29], leading to a revision of the document that removed the disgraced algorithm. The remaining DRBGs—which respectively use a hash function, HMAC, and a block cipher as their basic building blocks—are widely used. Indeed, any cryptographic software or hardware seeking FIPS certification *must* implement a PRNG from the standard [16,38]. While aspects of these constructions have been analyzed [9,19–21,33,35] and some implementation considerations discussed [5], these works make significant simplifying assumptions and/or treat certain algorithms rather than the constructions as a whole. To date, there has not been a deeper analysis of these standardized DRBGs, either investigating the stronger security properties claimed in the standard or taking into account the (considerable) flexibility in their specification.

The constructions provided in NIST SP 800-90A are nonstandard. Even the term DRBG is rare, if not absent from the literature, which favors the term PRNG. Similarly the NIST DRBGs—which return variable length (and sizable) outputs upon request, and support a variety of optional inputs and parameters—do not fit cleanly into the usual PRNG security models. With limited formal analysis to date, coupled with the fact that the standardization of these algorithms did not follow from a competition or widely publicly vetted process, this leaves large parts of software relying on relatively unanalyzed algorithms.

Security claims. The standard claims that each of the NIST DRBGs is 'backtracking resistant' and 'prediction resistant'. The former property guarantees that in the event of a state compromise, prior output remains secure. The latter property ensures that if a compromised state is reseeded with sufficient entropy then security will be recovered. To the best of our knowledge, neither of these properties have been formally investigated and proved. In fact, the NIST DRBG algorithms which are responsible for initial state generation and reseeding do not seem to have been analyzed at all in prior work.

A number of factors may have contributed to this lack of analysis. It seems likely that the attention given to the Dual EC resulted in the other PRNGs in NIST SP 800-90A being comparatively overlooked. Secondly, our understanding of what a PRNG should achieve has developed since the NIST DRBGs were standardized in 2006. Indeed, the concept of robustness for PRNGs [13] was not formalized until 2013. Finally, the NIST DRBGs are based on fairly run-of-the mill concepts such as running a hash function in counter mode, and yet simultaneously display design quirks which significantly complicate analysis and defy attempts at a modular treatment. As such, perhaps they have escaped deeper analysis by not being 'interesting' enough to tackle for an attention-grabbing result and yet too tricky for a straightforward proof.

The goal of this paper is to address some of these gaps in analysis.

1.1   Contributions

We conduct an investigation into the security of the DRBGs in NIST SP 800-90A, with a focus on HASH-DRBG and HMAC-DRBG. We pay particular attention to flexibilities in the specification of these algorithms, which are frequently abstracted away in previous analysis. We set out to analyze the algorithms as they are specified and used, and so sometimes make heuristic assumptions in our modeling (namely, working in the random oracle model (ROM) and assuming an oracle-independent entropy source). We felt this to be more constructive than modifying the constructions solely to derive a proof under weaker assumptions, and explain the rationale behind all such decisions.

Robustness proofs. Robustness, introduced by Dodis et al. [13], captures both backtracking and prediction resistance and is the 'gold-standard' for PRNG security. For our main technical results, we analyze HASH-DRBG and HMAC-DRBG within this framework. As a (somewhat surprising) negative result, we show that if optional strings of additional input are not always included in next calls (see Sect. 3), then HMAC-DRBG is not forward secure. This contradicts the claimed backtracking resistance of HMAC-DRBG. This highlights the importance of formally proving security claims which at first sight seem obviously correct, and of paying attention to implementation choices. As positive results, we prove that HASH-DRBG and HMAC-DRBG (called with additional input) are robust in the ROM. The first result is fully general, while the latter is for a class of entropy sources which includes those approved by the standard.

A key challenge is that the NIST DRBGs do not appear to have been designed with a security proof in mind. As such, seemingly innocuous design decisions turn out to significantly complicate matters. The first step is to reformulate robustness for the ROM. Our modeling is inspired by Gazi and Tessaro's treatment of robustness in the ideal permutation model [17]. We must make various adaptations to accommodate the somewhat unorthodox interface of the NIST DRBGs, and specifying the model requires some care. It is for this reason that we focus on HASH-DRBG and HMAC-DRBG in this work, since they map naturally into the same framework. Providing a similar treatment for CTR-DRBG would require different techniques, and is an important direction for future work.

At first glance, it may seem obvious that a PRNG built from a random oracle will produce random looking bits. However, formally proving that the constructions survive the strong forms of compromise required to be robust is far from trivial. While the proofs employ fairly standard techniques, certain design features of the algorithms introduce unexpected complexities and some surprisingly fiddly analysis. Throughout this process, we highlight points at which a minor design modification would have allowed for a simpler proof.

Implementation flexibilities. We counter these formal and (largely) positive results with a more informal discussion of flexibilities in the standard. We argue that when the NIST DRBGs are used to produce many blocks of output per request—a desirable implementation choice in terms of efficiency, and permitted by the standard—then the usual security models may overlook important attack vectors against these algorithms. Taking a closer look, we propose an informal security model in which an attacker compromises part of the state of the DRBG—for example through a side-channel attack—*during* an output generation request. Reconsidered within this framework, each of the DRBGs admits vulnerabilities which allow an attacker to recover unseen output. We find a further flaw in a certain variant of CTR-DRBG, which allows an attacker who compromises the state to potentially recover strings of additional input which are fed to the DRBG and which may contain secrets. While our attacks are theoretical in nature, we follow this up with an analysis of the open-source OpenSSL and mbed TLS CTR-DRBG implementations and find that these potentially problematic implementation decisions are taken by implementors in the real world. We conclude with reflections and recommendations for the safe use of these DRBGs.

Related work. The PRNGs in NIST SP 800-90A have received little formal analysis to date; we provide an overview here. A handful of prior works have analyzed the NIST DRBGs as deterministic PRGs with an idealized initialization procedure. i.e., they prove the next algorithm produces pseudorandom bits when applied to an *ideally random* initial state e.g., $S_0 = (K_0, V_0, cnt_0)$ for uniformly random K_0, V_0 in the case of CTR-DRBG and HMAC-DRBG. This is a substantial simplification; in reality, state components must be derived from the entropy source using the setup algorithm. Such proofs are provided for CTR-DRBG by Campagna [9] and Shrimpton and Terashima [35], and for HMAC-DRBG by Hirose [19] and Ye et al. [21]. A formal verification of the mbedTLS implementation of HMAC-DRBG is also provided in [21]. None of these works model setup or reseeding; as far as we know, ours is the first analysis of these algorithms for HASH-DRBG and HMAC-DRBG. With the exception of [19], they do not model the use of additional input. Moreover, pseudorandomness is a weaker property than robustness and does not model state compromise. Kan [20] considers assumptions underlying the security claims of the DRBGs; however, the analysis is informal and contains inaccuracies. To our knowledge, this is the only previous work to consider HASH-DRBG. Ruhault claims a potential attack against the robustness of CTR-DRBG [33]. However, the specification of CTR-DRBG's BCC function in [33] is different to that provided by the standard. In [33], BCC is defined to split the input $IV \| S$ into n 128-bit blocks ordered from right to left as $[B_n, \ldots, B_1]$. However, in the standard these blocks are ordered left to right $[B_1, \ldots, B_n]$. The attack from [33] does not work when the correct BCC function is used, and it does not seem possible to recover the attack.

2   Preliminaries

Notation. The set of binary strings of length n is denoted $\{0, 1\}^n$. We write $\{0, 1\}^*$ to denote the set of all binary strings, and $\{0, 1\}^{\leq n}$ to denote the set of binary strings of length at most n-bits; we include the empty string ε in both sets. We write $\{0, 1\}^{n \leq i \leq \overline{n}}$ to denote the set of binary strings of length between n and $\overline{n}$-bits inclusive. We convert binary strings to integers, and vice versa, in the standard way. We let $x \oplus y$ denote the exclusive-or (XOR) of two strings $x, y \in \{0, 1\}^n$, and write $x\|y$ to denote their concatenation. We write $\mathrm{left}(x, \beta)$ (resp. $\mathrm{right}(x, \beta)$) to denote the leftmost (resp. rightmost) β bits of string x, and $\mathrm{select}(x, \alpha, \beta)$ to denote the substring of x consisting of bits α to β inclusive. We let $[j_1, j_2]$ denote the set of integers between j_1 and j_2 inclusive. For an integer

$j \in \mathbb{N}$, we write $(j)_c$ to represent j encoded as a c-bit binary string. The notation $x \xleftarrow{\$} \mathcal{X}$ denotes sampling an element uniformly at random from the set $\mathcal{X}$. We let $\mathbb{N} = \{1, 2, \dots\}$ denote the set of natural numbers, and let $\mathbb{N}^{\leq n} = \{1, 2, \dots, n\}$.

Entropy and Cryptographic Components. In the full version, we recall the standard definitions of worst-case and average-case min-entropy, along with the usual definitions of pseudorandom functions (PRFs) and block ciphers.

PRNGs with input. A *pseudorandom number generator with input* (PRNG) [13] produces pseudorandom bits and offers strong security guarantees (see Sect. 4) when given continual access to an imperfect source of randomness. We define PRNGs, and discuss our choice of syntax, below.

Definition 1. *A PRNG with input is a tuple of algorithms* $\mathcal{G} =$ (setup, refresh, next) *with associated parameters* $(p, \overline{p}, \alpha, \beta_{max})$, *defined:*

- setup: Seed $\times \{0, 1\}^{p \leq i \leq \overline{p}} \times \mathcal{N} \to \mathcal{S}$ *takes as input a seed* $X \in$ Seed, *an entropy sample* $I \in \{0, 1\}^{p \leq i \leq \overline{p}}$, *and a nonce* $N \in \mathcal{N}$ *(where* Seed, $\mathcal{N}$, *and* $\mathcal{S}$ *denote the seed space, nonce space, and state space of the PRNG respectively), and returns an initial state* $S_0 \in \mathcal{S}$.
- refresh: Seed $\times \mathcal{S} \times \{0, 1\}^{p \leq i \leq \overline{p}} \to \mathcal{S}$ *takes as input a seed* $X \in$ Seed, *a state* $S \in \mathcal{S}$, *and an entropy sample* $I \in \{0, 1\}^{p \leq i \leq \overline{p}}$, *and returns a state* $S' \in \mathcal{S}$.
- next: Seed $\times \mathcal{S} \times \mathbb{N}^{\leq \beta_{max}} \times \{0, 1\}^{\leq \alpha} \to \{0, 1\}^{\leq \beta_{max}} \times \mathcal{S}$ *takes as input a seed* $X \in$ Seed, *a state* $S \in \mathcal{S}$, *a parameter* $\beta \in \mathbb{N}^{\leq \beta_{max}}$, *and a string of additional input* addin $\in \{0, 1\}^{\leq \alpha}$, *and returns an output* $R \in \{0, 1\}^{\beta}$ *and an updated state* $S' \in \mathcal{S}$.

If a PRNG always has $X = \varepsilon$ *or* addin $= \varepsilon$ *(indicating that, respectively, a seed or additional input is never used), then we omit these parameters.*

Discussion. Our definition follows that of Dodis et al. [13], with a number of modifications. The key differences are: **(1)** we extend the PRNG syntax to accommodate additional input, nonces and a parameter indicating the number of output bits requested, all of which are part of the NIST DRBG interface; **(2)** following Shrimpton et al. [34], we define setup to be the algorithm which constructs the initial state of the PRNG from a sample drawn from the entropy source, and assume that a random seed X is generated externally and supplied to the PRNG; and **(3)** we allow entropy samples and outputs to take any length in a range indicated by the parameters of the PRNG, rather than being of fixed length. We provide a full discussion of these modifications in the full version. NIST SP 800-90A uses the term *deterministic random bit generator* (DRBG) instead of the more familiar PRNG. We use these terms interchangeably.

3   The NIST SP 800-90A Standard

3.1   Overview of the Standard

NIST SP 800-90A defines three DRBG mechanisms, HASH-DRBG, HMAC-DRBG, and CTR-DRBG. The former two are based on an approved hash function (e.g., SHA-256), and the latter on an approved block cipher (e.g., AES-128); we include parameters for these variants in the full version.

Algorithms. The standard specifies (Instantiate, Reseed, Generate) algorithms for each of the DRBGs. These map directly into the (setup, refresh, next) algorithms in the PRNG model of Definition 1. Since the NIST DRBGs are not specified to take a seed (see Sect. 4), we take $X = \varepsilon$ and Seed $= \emptyset$ in this mapping, and omit these parameters from subsequent definitions. For consistency, we refer to the NIST DRBG algorithms as (setup, refresh, next) throughout. These (setup, refresh, next) algorithms underly (respectively) the (`Instantiate`, `Reseed`, `Generate`) functions of the DRBG. When called, these functions check the validity of the request (e.g., that the number of requested bits does not exceed β_{max}), and return an error if these checks fail. If not, the function fetches the internal state of the DRBG, plus any other required inputs (e.g., entropy samples, a nonce, etc.), and the underlying algorithm is applied to these inputs. The resulting outputs are returned and/or used to update the internal state, and the successful status of the call is indicated to the caller. To avoid cluttering our exposition we abstract away this process, assuming that required inputs are provided to algorithms (without modeling how these are fetched), and that all inputs and requests are valid, omitting success/error notifications.

The DRBG State. The standard defines the *working state* of a DRBG to be the set of stored variables which are used to produce pseudorandom output. The *internal state* is then defined to be the working state plus administrative information which indicates e.g., the security strength of the instantiation. We typically omit administrative information as this shall be clear from the context.

Entropy sources. A DRBG must have access to an *approved entropy source*[1] during initial state generation via setup, after which the DRBG is said to be *instantiated*. The function `Get_entropy_input`() is used to request an entropy sample I of length within range $[p, \overline{p}]$ containing a given amount of entropy (discussed further below). Nonces used by setup must either contain $\gamma^*/2$-bits of min-entropy or not be expected to repeat more than such a value would. Examples of suitable nonces given in the standard include strings drawn from the entropy source, time stamps, and sequence numbers.

Reseeding. Entropy samples drawn from the source are periodically incorporated into the DRBG state via refresh. A parameter *reseed_interval* indicates the maximum number of output generation requests allowed by an implementation before a reseed is forced; this is tracked by a state component called a reseed counter (cnt). The reseed counter is not a security critical state variable, and we assume that it is publicly known. For all approved DRBG variants (except CTR-DRBG based on 3-KeyTDEA) *reseed_interval* may be as large as 2^{48}. We assume here that reseeds are always explicitly requested by the caller; this is without loss of generality. A DRBG instantiation is parameterized by a *security*

---

[1] Either a live entropy source as approved in NIST SP 800-90B [37] or a (truly) random bit generator as per NIST SP 800-90C [3].

strength $\gamma^* \in \{112, 128, 192, 256\}$, where the highest supported security strength depends on the underlying primitive. Each entropy sample used by setup and refresh must contain at least γ^*-bits of entropy[2].

Output generation. The caller may request outputs of length up to β_{max} bits via the input β to the next function. For all approved DRBG variants (except CTR-DRBG based on 3-KeyTDEA), β_{max} may be as large as 2^{19}.

Additional input. Optional strings of additional input (denoted *addin*) may be provided to the DRBG by the caller during next calls. These inputs may be public or predictable (e.g., device serial numbers and time stamps), or may contain secrets. Optional additional input may also be provided in refresh calls and during setup; for brevity, we do not model this here and omit these inputs from the presentation of the algorithms.

3.2 HASH-DRBG

HASH-DRBG is built from an approved cryptographic hash function H : $\{0,1\}^{\leq \omega} \rightarrow \{0,1\}^{\ell}$. The working state is defined $S = (V, C, cnt)$, where the counter $V \in \{0,1\}^L$ and constant $C \in \{0,1\}^L$ are the security critical state variables. The standard does not explicitly state the role of C; however its purpose would appear to be preventing HASH-DRBG falling into a sequence of repeated states. We discuss this further in the full version.

Algorithms. The component algorithms for HASH-DRBG are shown in Fig. 1. Both setup and refresh derive a new state by applying the derivation function HASH-DRBG_df to the entropy input and (in the case of refresh) the previous counter. Output generation via next first incorporates any additional input into the counter V (lines 3–5). Output blocks are then produced by hashing V in CTR-mode (lines 7–10). At the conclusion of the call, V is hashed with a distinct prefix prepended, and the resulting string—along with the constant C and reseed counter cnt—are added into V (lines 12–13).

3.3 HMAC-DRBG

HMAC-DRBG is based on HMAC : $\{0,1\}^{\ell} \times \{0,1\}^{\leq \omega} \rightarrow \{0,1\}^{\ell}$, built from an approved hash function. The working state is of the form $S = (K, V, cnt)$, where the key $K \in \{0,1\}^{\ell}$ and counter $V \in \{0,1\}^{\ell}$ are security critical.

Algorithms. The component algorithms of HMAC-DRBG are shown in Fig. 1. Algorithms setup and refresh both use the update subroutine to incorporate an entropy sample I into K and V. For setup, these variables are initialized to $K = \text{0x00} \dots 00$ and $V = \text{0x01} \dots 01$ prior to this process. Output production via next first incorporates any additional input into K and V via the update

---

[2] In contrast, robustness [13] requires that a PRNG is secure when reseeded with a set of entropy samples which *collectively* has γ^*-bits of entropy. Looking ahead to Sect. 5, we analyze HASH-DRBG with respect to this stronger notion.

HASH-DRBG_df

Require: $inp\_str, (num\_bits)_{32}$
Ensure: req_bits
$temp \leftarrow \varepsilon$; $m \leftarrow \lceil num\_bits/\ell \rceil$
For $i = 1, \ldots, m$
  $temp \leftarrow temp \parallel \mathsf{H}((i)_8 \parallel (num\_bits)_{32} \parallel inp\_str)$
$req\_bits \leftarrow \mathrm{left}(temp, num\_bits)$
Return req_bits

HASH-DRBG setup

Require: I, N
Ensure: $S_0 = (V_0, C_0, cnt_0)$
$seed\_material \leftarrow I \parallel N$
$V_0 \leftarrow \mathsf{HASH\text{-}DRBG\_df}(seed\_material, L)$
$C_0 \leftarrow \mathsf{HASH\text{-}DRBG\_df}(0x00 \parallel V_0, L)$
$cnt_0 \leftarrow 1$
Return (V_0, C_0, cnt_0)

HASH-DRBG refresh

Require: $S = (V, C, cnt), I$
Ensure: $S' = (V', C', cnt')$
$seed\_material \leftarrow 0x01 \parallel V \parallel I$
$V' \leftarrow \mathsf{HASH\text{-}DRBG\_df}(seed\_material, L)$
$C' \leftarrow \mathsf{HASH\text{-}DRBG\_df}(0x00 \parallel V', L)$
$cnt' \leftarrow 1$
Return (V', C', cnt')

HASH-DRBG next

Require: $S = (V, C, cnt), \beta, addin$
Ensure: $R, S' = (V', C', cnt')$
1. If $cnt > reseed\_interval$
2.   Return reseed_required
3. If $addin \neq \varepsilon$
4. $w \leftarrow \mathsf{H}(0x02 \parallel V \parallel addin)$
5. $V \leftarrow (V + w) \bmod 2^L$
6. $data \leftarrow V$; $temp_R \leftarrow \varepsilon$; $n \leftarrow \lceil \beta/\ell \rceil$
7. For $j = 1, \ldots, n$
8. $r \leftarrow \mathsf{H}(data)$
9. $data \leftarrow (data + 1) \bmod 2^L$
10. $temp_R \leftarrow temp_R \parallel r$
11. $R \leftarrow \mathrm{left}(temp_R, \beta)$
12. $H \leftarrow \mathsf{H}(0x03 \parallel V)$
13. $V' \leftarrow (V + H + C + cnt) \bmod 2^L$
14. $C' \leftarrow C$; $cnt' \leftarrow cnt + 1$
15. Return $R, (V', C', cnt')$

CTR-DRBG update

Require: $provided\_data, K, V$
Ensure: K, V
$temp \leftarrow \varepsilon$; $m \leftarrow \lceil (\kappa + \ell)/\ell \rceil$
For $j = 1, \ldots, m$
  $V \leftarrow (V + 1) \bmod 2^\ell$; $Z \leftarrow \mathsf{E}(K, V)$
  $temp \leftarrow temp \parallel Z$
$temp \leftarrow \mathrm{left}(temp, (\kappa + \ell))$
$temp \leftarrow temp \oplus provided\_data$
$K \leftarrow \mathrm{left}(temp, \kappa)$
$V \leftarrow \mathrm{right}(temp, \ell)$
Return K, V

HMAC-DRBG update

Require: $provided\_data, K, V$
Ensure: K, V
$K \leftarrow \mathsf{HMAC}(K, V \parallel 0x00 \parallel provided\_data)$
$V \leftarrow \mathsf{HMAC}(K, V)$
If $provided\_data \neq \varepsilon$
  $K \leftarrow \mathsf{HMAC}(K, V \parallel 0x01 \parallel provided\_data)$
  $V \leftarrow \mathsf{HMAC}(K, V)$
Return (K, V)

HMAC-DRBG setup

Require: I, N
Ensure: $S_0 = (K_0, V_0, cnt_0)$
$seed\_material \leftarrow I \parallel N$
$K \leftarrow 0x00 \ldots 00$
$V \leftarrow 0x01 \ldots 01$
$(K_0, V_0) \leftarrow \mathsf{update}(seed\_material, K, V)$
$cnt_0 \leftarrow 1$
Return (K_0, V_0, cnt_0)

HMAC-DRBG refresh

Require: $S = (K, V, cnt), I$
Ensure: $S' = (K', V', cnt')$
$seed\_material \leftarrow I$
$(K_0, V_0) \leftarrow \mathsf{update}(seed\_material, K, V)$
$cnt_0 \leftarrow 1$
Return (K_0, V_0, cnt_0)

HMAC-DRBG next

Require: $S = (K, V, cnt), \beta, addin$
Ensure: $R, S' = (K', V', cnt')$
1. If $cnt > reseed\_interval$
2.   Return reseed_required
3. If $addin \neq \varepsilon$
4. $(K, V) \leftarrow \mathsf{update}(addin, K, V)$
5. $temp \leftarrow \varepsilon$; $n \leftarrow \lceil \beta/\ell \rceil$
6. For $j = 1, \ldots, n$
7. $V \leftarrow \mathsf{HMAC}(K, V)$
8. $temp \leftarrow temp \parallel V$
9. $R \leftarrow \mathrm{left}(temp, \beta)$
10. $(K', V') \leftarrow \mathsf{update}(addin, K, V)$
11. $cnt' \leftarrow cnt + 1$
12. Return $R, (K', V', cnt')$

CTR-DRBG next

Require: $S = (K, V, cnt), \beta, addin$
Ensure: $R, S' = (K', V', cnt')$
1. If $cnt > reseed\_interval$
2.   Return reseed_required
3. If $addin \neq \varepsilon$
4.   If derivation function used then
5. $addin \leftarrow \mathsf{CTR\text{-}DRBG\_df}(addin, (\kappa + \ell))$
6.   Else if $len(addin) < (\kappa + \ell)$ then
7. $addin \leftarrow addin \parallel 0^{(\kappa + \ell - len(addin))}$
8. $(K, V) \leftarrow \mathsf{update}(addin, K, V)$
9. Else $addin \leftarrow 0^{\kappa + \ell}$
10. $temp \leftarrow \varepsilon$; $n \leftarrow \lceil \beta/\ell \rceil$
11. For $j = 1, \ldots, n$
12. $V \leftarrow (V + 1) \bmod 2^\ell$; $r \leftarrow \mathsf{E}(K, V)$
13. $temp \leftarrow temp \parallel r$
14. $R \leftarrow \mathrm{left}(temp, \beta)$
15. $(K', V') \leftarrow \mathsf{update}(addin, K, V)$
16. $cnt' \leftarrow cnt + 1$
17. Return $R, (K', V', cnt')$

Fig. 1. Component algorithms for HASH-DRBG, HMAC-DRBG and CTR-DRBG.

function (lines 3–4). Output blocks r are generated by iteratively computing $V \leftarrow \mathsf{HMAC}(K, V)$ and setting $r = V$ (lines 6–8). At the conclusion of the call, both key and counter are updated via the update function (line 10).

3.4 CTR-DRBG

CTR-DRBG is built from an approved block cipher $\mathsf{E} : \{0,1\}^\kappa \times \{0,1\}^\ell \to \{0,1\}^\ell$. The working state is defined $S = (K, V, cnt)$, where key $K \in \{0,1\}^\kappa$ and counter $V \in \{0,1\}^\ell$ are the security critical state variables.

Algorithms. Component algorithms for CTR-DRBG are shown in Fig. 1[3]. Use of the derivation function CTR-DRBG_df is optional only if the implementation has access to a 'full entropy source' which returns statistically close to uniform strings. Output generation via next first incorporates any additional input *addin* into the state via the update function (line 8). If a derivation function is used, additional input *addin* is conditioned into a $(\kappa + \ell)$-bit string with CTR-DRBG_df prior to this process (line 5); otherwise *addin* is restricted to be at most $(\kappa + \ell)$-bits in length. Output blocks are then iteratively generated using the block cipher in CTR-mode (lines 11–13). At the conclusion of the call, both K and V are updated via an application of the update function (line 15).

4 Robustness in the Random Oracle Model

The security properties of backtracking and prediction resistance claimed in the standard have never been formally investigated. We address this be analyzing the robustness [13] of HASH-DRBG and HMAC-DRBG. This models a powerful attacker who is able to compromise the state and influence the entropy source of the PRNG, and is easily verifed to imply both backtracking and prediction resistance. In this section, we adapt the robustness model of [13] to accommodate the NIST DRBGs, and introduce the notion of robustness in the ROM.

Distribution sampler. We model the gathering of entropy inputs from the entropy source via a *distribution sampler* [13]. Formally, a distribution sampler $\mathcal{D} : \{0,1\}^* \to \{0,1\}^* \times \{0,1\}^* \times \mathbb{R}^{\geq 0} \times \{0,1\}^*$ is a stateful and probabilistic algorithm which takes as input its current state $\sigma \in \{0,1\}^*$ and outputs a tuple (σ', I, γ, z), where $\sigma' \in \{0,1\}^*$ denotes the updated state of the sampler, $I \in \{0,1\}^*$ denotes the entropy sample, $\gamma \in \mathbb{R}^{\geq 0}$ is an entropy estimate for the sample, and $z \in \{0,1\}^*$ denotes a string of side information about the sample. We say that a sampler $\mathcal{D}$ is $(q_\mathcal{D}^+, \gamma^*)$-*legitimate* if **(1)** for all $j \in [1, q_\mathcal{D} + 1]$:

$$\mathrm{H}_\infty(I_j | I_1, \ldots, I_{j-1}, I_{j+1}, \ldots, I_{q_\mathcal{D}+1}, \gamma_1, \ldots, \gamma_{q_\mathcal{D}+1}, z_1, \ldots, z_{q_\mathcal{D}+1}) \geq \gamma_j ,$$

where $\sigma_0 = \varepsilon$ and $(\sigma_j, I_j, \gamma_j, z_j) \leftarrow_\$ \mathcal{D}(\sigma_{j-1})$; and **(2)** it holds that $\gamma_1 \geq \gamma^*$. Here, condition **(2)** extends the definition of [13] to model the sample (which recall

---

[3] We do not directly analyze setup, refresh and CTR-DRBG_df in this work, and so defer their presentation to the full version.

must contain γ^* bits of entropy) with which the DRBG is initially seeded during setup. It is straightforward to see that to any sequence of $Get\_entropy\_input()$ calls made by the DRBG we can define an associated sampler[4].

4.1 Robustness and Forward Security in the Random Oracle Model

Our positive results about HASH-DRBG and HMAC-DRBG will be in the random oracle model (ROM). As such, the first step in our analysis is to adapt the security model of Dodis et al. [13] to the ROM.

Robustness. Consider game Rob shown in Fig. 2. The game is parameterized by an entropy threshold γ^*. We expect security when the entropy in the system exceeds this value. When analyzing the NIST DRBGs, we take γ^* to be the security strength of the instantiation. At the start of the game, we choose a random function $H \leftarrow\!\!s\ \mathcal{H}$ where $\mathcal{H}$ denotes the set of all functions of a given domain and range. All of the PRNG algorithms have access to H (indicated in superscript e.g., setup^H). For reasons discussed below, we do *not* give the sampler $\mathcal{D}$ access to H. To the best of our knowledge, this is the first treatment of robustness in the ROM, and our model may be useful beyond analyzing HASH-DRBG and HMAC-DRBG. We additionally modify game Rob from [13] to: **(1)** accommodate our PRNG syntax (including the use of additional input, discussed below); **(2)** remove the Next oracle, which was shown in [11] to be without loss of generality; and **(3)** generate the initial state with the deterministic setup^H algorithm using the first entropy sample output by the sampler, similarly to [34].

The game is implicitly parameterized by a nonce distribution $\mathcal{N}$, where we write $N \leftarrow \mathcal{N}$ to denote sampling a nonce. Since nonces may be predictable (e.g., if a sequence number is used) we assume that $\mathcal{N}$ is public and the nonce used during initalization is provided to $\mathcal{A}$. Similarly, we assume the attacker can choose the strings of additional input which may be included in next calls. These are conservative assumptions, since any entropy in these values can only make the attacker's job harder. With this in place, the Rob advantage of an adversary $\mathcal{A}$, and a $(q_{\mathcal{D}}^+, \gamma^*)$-legitimate sampler $\mathcal{D}$, is defined

$$\mathrm{Adv}_{\mathcal{G},\gamma^*}^{\mathrm{rob}}(\mathcal{A}, \mathcal{D}) = 2 \cdot |\mathrm{Pr}\left[\mathrm{Rob}_{\mathcal{G},\gamma^*}^{\mathcal{A},\mathcal{D}} \Rightarrow 1\right] - \frac{1}{2}|.$$

We say that $\mathcal{A}$ is a $(q_H, q_R, q_D, q_C, q_S)$-adversary if it makes q_H queries to the random oracle H, q_R queries to its RoR oracle, a total of q_S queries to its Get and Set oracles, and q_D queries to its Ref oracle of which at most q_C are consecutive.

Fixed length variant. While we define the general game Rob here, our robustness proofs will be in a slightly restricted variant Rob$_\beta$, in which the attacker may only request outputs of fixed length $\beta \leq \beta_{max}$ in RoR queries. This simplifying assumption is to avoid further complicating bounds with parameters

---

[4] NIST SP 800-90B [37] defines the entropy estimate of sample I as $H_\infty(I)$, rather than this conditioned on other samples and associated data. However, since the tests in NIST SP 800-90B estimate entropy using multiple samples drawn from the source, it seems reasonable to assume the conditional entropy condition is satisfied also.

indicating the length of each RoR query. Results for the general game Rob can be recovered as a straightforward extension of our proofs.

Standard model forward security. Our negative result about the forward security of HMAC-DRBG shall be in the standard model. We define game Fwd to be a restricted variant of Rob in which: **(1)** we remove oracle access to H from all algorithms; and **(2)** the attacker $\mathcal{A}$ is allowed no Set queries, and makes a single Get query after which they may make no further queries. The forward security advantage of $(\mathcal{A}, \mathcal{D})$ is defined

$$\mathrm{Adv}_{\mathcal{G},\gamma^*}^{\mathrm{fwd}}(\mathcal{A}, \mathcal{D}) = 2 \cdot |\mathrm{Pr}\left[\mathrm{Fwd}_{\mathcal{G},\gamma^*}^{\mathcal{A},\mathcal{D}} \Rightarrow 1\right] - \frac{1}{2}|.$$

The problem of seeding. Since deterministic extraction from imperfect sources is impossible in general, the PRNG in game Rob is initialized with a random public seed X which crucially is independent of the entropy source. Unfortunately (for our analysis), none of the NIST DRBGs are specified to take a seed, (i.e., $X = \varepsilon$ in our modeling). Moreover, all state components and inputs to HASH-DRBG and HMAC-DRBG may depend on the entropy source, and so cannot be reframed as a seed without adding substantial assumptions; we discuss this further in the full version. At this point we are faced with two choices. We either: **(1)** give sampler $\mathcal{D}$ access to H (as in the robustness in the IPM model of [17]), and either modify the NIST DRBGs to accommodate a random seed or restrict our analysis to implementations for which additional input is sufficiently independent of the source to suffice as a seed. Or: **(2)** do not give $\mathcal{D}$ access to H. In this case, the oracle H with respect to which security analysis is carried out is chosen randomly and independently of the entropy source, and so serves the same purpose as a seed. We take the latter approach for a number of reasons. Firstly, we wish to analyze the NIST DRBGs as they are specified and used, and so modifying the construction or restricting the implementations we can reason about (as per **(1)**) solely to facilitate the analysis seems counterproductive. Secondly, as pointed out in [35], generating a seed is challenging in practice due to the necessary independence from the entropy source. Moreover, given the litany of tests which approved entropy sources in NIST SP 800-90B are subjected to, it seems reasonable to assume that the pathological sources used to illustrate e.g., deterministic extraction impossibility results, are unlikely to pass these tests.

$\mathrm{Rob}_{\mathcal{G},\gamma^*}^{\mathcal{A},\mathcal{D}}$	Ref	RoR$(\beta, addin)$	Get
$H \leftarrow^\$ \mathcal{H}$; $b \leftarrow^\$ \{0,1\}$; $N \leftarrow \mathcal{N}$	$(\sigma, I, \gamma, z) \leftarrow^\$ \mathcal{D}(\sigma)$	$(R_0, S) \leftarrow \mathrm{next}^H(X, S, \beta, addin)$	Return S
$\sigma \leftarrow \varepsilon$; $X \leftarrow^\$ \mathrm{Seed}$	$S \leftarrow \mathrm{refresh}^H(X, S, I)$	If $c < \gamma^*$	$c \leftarrow 0$
$(\sigma, I, \gamma, z) \leftarrow^\$ \mathcal{D}(\sigma)$	$c \leftarrow c + \gamma$	$\quad$ Return R_0	$\mathrm{Set}(S^*)$
$S \leftarrow \mathrm{setup}^H(X, I, N)$	Return (γ, z)	$\quad c \leftarrow 0$	$S \leftarrow S^*$
$c \leftarrow \gamma^*$; $\overline{\gamma} \leftarrow (\gamma, z, N)$	proc. H(X)	Else $R_1 \leftarrow^\$ \{0,1\}^\beta$	$c \leftarrow 0$
$b^* \leftarrow^\$ \mathcal{A}^{\mathrm{Ref,RoR,Get,Set,H}}(X, \overline{\gamma})$	Return H(X)	Return R_b	
Return $(b = b^*)$			

Fig. 2. Security game Rob for a PRNG $\mathcal{G} = (\mathsf{setup}, \mathsf{refresh}, \mathsf{next})$.

Security games. A key insight of [13] is that the complex notion of robustness can be decomposed into two simpler notions called preserving and recovering security. The former models the PRNG's ability to maintain security if the state is secret but the attacker is able to influence the entropy source. The latter models the PRNG's ability to recover from state compromise after sufficient (honestly generated) entropy has entered the system. Here we will utilize the variants of these from [34], which extended the original definitions and added a new game Init modelling initial state generation.

Consider games Pres, Rec, and Init shown in Fig. 3, given here for Rob_β. Here we have adapted the notions of [34] in the natural way to accommodate: (1) a random oracle; and (2) our PRNG syntax. It is straightforward to extend our analysis to accommodate variable length outputs. All games are defined with respect to a *masking function*, which is a randomized function $\mathsf{M} : \mathcal{S} \cup \{\varepsilon\} \to \mathcal{S}$ where $\mathcal{S}$ denotes the state space of the PRNG. Here, we extend the definition of [34] to include ε in the domain (implicitly assuming that ε does not lie in the state space of the PRNG; if this is not the case then any distinguished symbol may be used instead). We discuss the reasons for this adaptation in Sect. 5. We give the masking function access to the random oracle, indicated by M^H. We make a number of further modifications. Firstly in Init, we require S_0^* to be indistinguishable from $\mathsf{M}^\mathsf{H}(\varepsilon)$ as opposed to $\mathsf{M}^\mathsf{H}(S_0^*)$ as in [34]. Secondly, during the computation of the challenge in Pres and Rec, we apply the masking function to the state S_d which was *input* to next^H as opposed to the state S^* which is *output* by next^H. Finally, in Pres, we allow $\mathcal{A}$ to output $S \in \mathcal{S} \cup \{\varepsilon\}$ at the start of his challenge rather than $S \in \mathcal{S}$. In all cases, this is to accommodate the somewhat complicated state distribution of HASH-DRBG (see Sect. 5). For all $\text{Gm}_y^x \in \{\text{Init}_{\mathcal{G},\mathsf{M},q_\mathcal{D},\gamma^*}^{\mathcal{A},\mathcal{D}}, \text{Pres}_{\mathcal{G},\mathsf{M},\beta}^{\mathcal{A}}, \text{Rec}_{\mathcal{G},\mathsf{M},\gamma^*,q_\mathcal{D},\beta}^{\mathcal{A},\mathcal{D}}\}$ we define

$$\text{Adv}_x^{\text{gm}}(y) = 2 \cdot |\Pr\left[\,\text{Gm}_y^x \Rightarrow 1\,\right] - \frac{1}{2}|.$$

An adversary in Init is said to be a q_H-adversary if it makes q_H queries to its H oracle. An adversary in game Pres or Rec is said to be a (q_H, q_C)-adversary if it makes q_H queries to its H oracle and always outputs $d \leq q_C$.

With this in place, the following theorem—which says that Init, Pres and Rec security collectively imply Rob security—is an adaptation of the analogous results from [17,34]. As a bonus, employing a slightly different line of argument with two series of hybrid arguments means our proof holds for arbitrary masking functions, lifting the restriction from [34] that masking functions possess a property called idempotence. The proof is given in the full version. We note that the original result [13] omits a factor of two from the right-hand side of the equation which we recover here.

Theorem 1. *Let $\mathcal{G} = (\mathsf{setup}^\mathsf{H}, \mathsf{refresh}^\mathsf{H}, \mathsf{next}^\mathsf{H})$ be a PRNG with input with associated parameter set $(p, \overline{p}, \alpha, \beta_{max})$, built from a hash function H which we model as a random oracle. Suppose that each invocation of $\mathsf{refresh}^\mathsf{H}$ and next^H makes at most q_{ref} and q_{nxt} queries to H respectively. Let $\mathsf{M}^\mathsf{H} : \mathcal{S} \cup \{\varepsilon\} \to \mathcal{S}$ be a masking function for which each invocation of M^H makes at most q_M H queries.*

Then for any $(q_H, q_R, q_D, q_C, q_S)$*-adversary* $\mathcal{A}$ *and* (q_D^+, γ^*)*-legitimate sampler* $\mathcal{D}$ *in game* Rob_β *against* $\mathcal{G}$*, there exists a* $(q_H + q_D \cdot q_{ref} + q_R \cdot q_{nxt})$*-adversary* $\mathcal{A}_1$ *and* $(q_H + q_D \cdot q_{ref} + q_R \cdot (q_{nxt} + q_M), q_C)$*-adversaries* $\mathcal{A}_2, \mathcal{A}_3$*, such that*

$$\mathrm{Adv}^{rob}_{\mathcal{G},\gamma^*,\beta}(\mathcal{A}, \mathcal{D}) \leq 2 \cdot \mathrm{Adv}^{init}_{\mathcal{G},\mathsf{M},\gamma^*,q_D}(\mathcal{A}_1, \mathcal{D})$$
$$+ 2q_R \cdot \mathrm{Adv}^{pres}_{\mathcal{G},\mathsf{M},\beta}(\mathcal{A}_2) + 2q_R \cdot \mathrm{Adv}^{rec}_{\mathcal{G},\mathsf{M},\gamma^*,q_D,\beta}(\mathcal{A}_3, \mathcal{D}).$$

Tightness. Unfortunately due to a hybrid argument taken over the q_R RoR queries made by $\mathcal{A}$, Theorem 1 is not tight. This is exacerbated in the ROM, since the attacker in each of q_R hybrid reductions must make enough H queries to simulate the whole of game Rob for $\mathcal{A}$. This hybrid argument accounts for the q_R coefficients in the bound and in the attacker query budgets. This seems inherent to the proof technique and is present in the analogous results of [13,17,34]. Developing a technique to obtain tighter bounds is an important open question.

$\mathrm{Init}^{\mathcal{A},\mathcal{D}}_{\mathcal{G},\mathsf{M},\gamma^*,q_D}$
$H \leftarrow\!\!\$\ \mathcal{H}$; $b \leftarrow\!\!\$\ \{0,1\}$; $N \leftarrow \mathcal{N}$
$\sigma_0 \leftarrow \varepsilon$; $X \leftarrow\!\!\$\ \mathrm{Seed}$
For $k = 1, \ldots, q_D + 1$
$\quad (\sigma_k, I_k, \gamma_k, z_k) \leftarrow \mathcal{D}(\sigma_{k-1})$
If $(b = 0)$ then $S_0^* \leftarrow \mathsf{setup}^H(X, I_1, N)$
Else $S_0^* \leftarrow\!\!\$\ \mathsf{M}^H(\varepsilon)$
$b^* \leftarrow\!\!\$\ \mathcal{A}^H(X, S_0^*, (I_i)_{i=2}^{q_D+1}, (\gamma_i, z_i)_{i=1}^{q_D+1}, N)$
Return $(b = b^*)$

$\mathrm{Pres}^{\mathcal{A}}_{\mathcal{G},\mathsf{M},\beta}$
$H \leftarrow\!\!\$\ \mathcal{H}$; $b \leftarrow\!\!\$\ \{0,1\}$
$X \leftarrow\!\!\$\ \mathrm{Seed}$
$(S_0', I_1, \ldots, I_d, addin) \leftarrow\!\!\$\ \mathcal{A}^H(X)$
If $S_0' \notin \mathcal{S} \cup \{\varepsilon\}$ return $\bot$
$S_0 \leftarrow\!\!\$\ \mathsf{M}^H(S_0')$
For $i = 1, \ldots, d$
$\quad S_i \leftarrow \mathsf{refresh}^H(X, I_i, S_{i-1})$
If $(b = 0)$ then $(R^*, S^*) \leftarrow \mathsf{next}^H(X, S_d, \beta, addin)$
Else $R^* \leftarrow \{0,1\}^\beta$; $S^* \leftarrow \mathsf{M}^H(S_d)$
$b^* \leftarrow\!\!\$\ \mathcal{A}^H(X, R^*, S^*)$
Return $(b = b^*)$

$\mathrm{Rec}^{\mathcal{A},\mathcal{D}}_{\mathcal{G},\mathsf{M},\gamma^*,q_D,\beta}$
$H \leftarrow\!\!\$\ \mathcal{H}$; $b \leftarrow\!\!\$\ \{0,1\}$
$\sigma \leftarrow \varepsilon$; $X \leftarrow\!\!\$\ \mathrm{Seed}$; $\mu \leftarrow 1$
For $k = 1, \ldots, q_D + 1$
$\quad (\sigma_k, I_k, \gamma_k, z_k) \leftarrow \mathcal{D}(\sigma_{k-1})$
$(S_0, d, addin) \leftarrow\!\!\$\ \mathcal{A}^{H,\mathrm{Sam}}(X, (\gamma_k, z_k))_{i=1}^{q_D+1})$
If $S_0 \notin \mathcal{S}$ return $\bot$
If $\mu + d > (q_D + 1)$ or $\sum_{i=\mu+1}^{\mu+d} \gamma_i < \gamma^*$
$\quad$ Return $\bot$
For $i = 1, \ldots, d$
$\quad S_i \leftarrow \mathsf{refresh}^H(X, I_{\mu+i}, S_{i-1})$
$\quad$ If $(b = 0)$ then $(R^*, S^*) \leftarrow \mathsf{next}^H(X, S_d, \beta, addin)$
$\quad$ Else $R^* \leftarrow \{0,1\}^\beta$; $S^* \leftarrow\!\!\$\ \mathsf{M}^H(S_d)$
$b^* \leftarrow\!\!\$\ \mathcal{A}(X, R^*, S^*, (I_k)_{k>\mu+d})$
Return $(b = b^*)$

$\mathrm{Sam}()$
$\mu = \mu + 1$
Return I_μ

proc. $H(X)$
Return $H(X)$

Fig. 3. Security games Init, Pres and Rec for a PRNG $\mathcal{G} = (\mathsf{setup}, \mathsf{refresh}, \mathsf{next})$ and $\mathsf{M} : \mathcal{S} \cup \{\varepsilon\} \to \mathcal{S}$.

5   Analysis of **HASH-DRBG**

We now present our analysis of the robustness of HASH-DRBG, in which the underlying hash function $H : \{0,1\}^{\leq \omega} \to \{0,1\}^\ell$ is modelled as a random oracle. Our proof is with respect to the masking function M^H shown in Fig. 4. To avoid further complicating security bounds we assume that HASH-DRBG is never called with additional input; we expect extending the proof to include additional input to be straightforward.

Challenges. Certain features of HASH-DRBG significantly complicate the proof, and necessitated adaptations in our security modeling (Sect. 4). Notice that the distributions of states returned by setup^H and $\mathsf{refresh}^H$ are quite different from the distribution of states S' where $(R, S') \leftarrow \mathsf{next}^H(S, \beta)$. To model this in games Init, Pres and Rec, we extended the domain of M to include the empty string ε to indicate that an idealized state of the first form should be returned (for example, when modeling initial state generation in Init), and extended Pres to allow $\mathcal{A}$ to output ε at the

```
M^H(S)
If S = ε
    V' ←$ {0,1}^L
    C' ← HASH-DRBG_df^H(0x00||V, L)
    cnt' ← 1
Else (V, C, cnt) ← S
    H ←$ {0,1}^ℓ
    V' ← (V + C + cnt + H) mod 2^L
    C' ← C ; cnt' ← cnt + 1
    S' ← (V', C', cnt')
Return S'
```

Fig. 4. Masking function for proof of Theorem 2

start of the challenge (which is required for the proof of Theorem 1, see the full version). Juggling these different state distributions complicates analysis, and introduces multiple cases into the proof of Pres security. Care is also required when analyzing the distribution of $S' = \mathsf{M}^H(S)$ for $S \in \mathcal{S}$, which idealizes the distribution of the state $S' = (V', C', cnt')$ as updated following an output generation request. It is straightforward to verify that V' is distributed uniformly over the range $[V + C + cnt, V + C + cnt + (2^\ell - 1)]$ where $S = (V, C, cnt)$ and $L > \ell + 1$. To accommodate this dependency between the updated state S' and the previous state S, we have modifed games Pres and Rec so that it is S which is masked instead of S'. More minor issues, such as: **(1)** not properly separating the domain of queries made by setup^H to produce the counter V from those made to produce the constant C; and **(2)** the way in which L is not a multiple of ℓ for the approved hash functions; make certain steps in the analysis more fiddly than they might have been. We discuss these issues further in the full version.

Parameter settings. We provide a general treatment into which any parameter setting may be slotted, subject to two restrictions which are utilized in the proof. Namely, we assume that $L > \ell + 1$ and $n < 2^L$, where $n = \lceil \beta/\ell \rceil$ is the number of output blocks produced by next^H to satisfy a request for β-bits (without this latter restriction HASH-DRBG is trivially insecure, as the same counter would be hashed twice during output production). We additionally require that $L < 2^{32}$ and $m < 2^8$ where $|V| = |C| = L$ and $m = \lceil L/\ell \rceil$ is the number of blocks hashed by $\mathsf{setup}^H/\mathsf{refresh}^H$ to produce a new counter. This is because these values have to be encoded as a 32-bit and an 8-bit string, respectively, by HASH-DRBG_df. All approved hash functions fall well within these parameters. (Indeed, for all of these, $L > 2\ell$, $n < 3277 \ll 2^L$, and $m \leq 3$.)

Robustness. With this in place, we present the following theorem bounding the Rob security of HASH-DRBG. The proof follows from a number of lemmas presented below, combined with Theorem 1. (When applying Theorem 1, it is readily verified that for HASH-DRBG $q_{nxt} = n + 1$, $q_{ref} = 2m$, and $q_{\mathsf{M}} = m$.) The proofs of all lemmas are given in the full version.

Theorem 2. *Let* $\mathcal{G}$ *be* HASH-DRBG *with parameters* $(p, \overline{p}, \alpha, \beta_{max})$, *built from a hash function* H $: \{0,1\}^{\leq \omega} \to \{0,1\}^{\ell}$ *which we model as a random oracle. Let* L *denote the state length of* HASH-DRBG *where* $L > \ell + 1$, $n = \lceil \beta/\ell \rceil$, *and* $m = \lceil L/\ell \rceil$. *Let* M^{H} *denote the masking function shown in Fig. 4, and suppose that* HASH-DRBG *is never called with additional input. Then for any* $(q_{\mathsf{H}}, q_R, q_{\mathcal{D}}, q_C, q_S)$-*attacker* $\mathcal{A}$ *in game* Rob_β *against* $\mathcal{G}$, *and any* $(q_{\mathcal{D}}^+, \gamma^*)$-*legitimate sampler* $\mathcal{D}$, *it holds that*

$$\mathrm{Adv}^{\mathrm{rob}}_{\mathcal{G}, \mathsf{M}, \gamma^*, \beta}(\mathcal{A}, \mathcal{D}) \leq \frac{q_R \cdot \overline{q}_{\mathsf{H}} + 2q_{\mathsf{H}}'}{2^{\gamma^* - 2}} + \frac{q_R \cdot \overline{q}_{\mathsf{H}} \cdot (2n + 1)}{2^{\ell - 2}}$$
$$+ \frac{q_R \cdot ((q_C - 1)(2\overline{q}_{\mathsf{H}} + q_C) + \overline{q}_{\mathsf{H}}^2) + 2}{2^{L - 2}}.$$

Moreover, $q_{\mathsf{H}}' = (q_{\mathsf{H}} + 2m \cdot q_{\mathcal{D}} + (n + 1) \cdot q_R)$ *and* $\overline{q}_{\mathsf{H}} = q_{\mathsf{H}}' + m \cdot q_R$.

Init security. We first argue that the states returned by $\mathsf{setup}^{\mathsf{H}}$ are indistinguishable from $\mathsf{M}^{\mathsf{H}}(\varepsilon)$. The $q_{\mathsf{H}} \cdot 2^{-\gamma^*}$ term follows since the initial state variable V_0 will be indistinguishable from random unless the attacker queries H on one of the points which was hashed to produce it. This in turn requires $\mathcal{A}$ to guess the value of the entropy sample I_1, which contains γ^*-bits of entropy. The additional 2^{-L} term arises since the queries made to compute the counter V_0 are not fully domain separated from those made to compute the constant C_0. Indeed, if it so happens that $I_1 || N = \mathsf{0x00} || V_0$ where I_1 and N denote the entropy input and nonce (an event which—while very unlikely—is not precluded by the parameter constraints in the standard), then the derived values of V_0 and C_0 will be equal, allowing the attacker to distinguish with high probability. A small tweak to the design of setup (e.g., prepending $\mathsf{0x01}$ to $I || N$ before hashing) would have avoided this. For implementations of HASH-DRBG for which such a collision is impossible (e.g., due to length restrictions on the input I) this additional term can be removed.

Lemma 1. *Let* $\mathcal{G} =$ HASH-DRBG *and masking function* M^{H} *be as specified in Theorem 2. Then for any* q_{H}-*adversary* $\mathcal{A}$ *in game* Init *against* $\mathcal{G}$, *and any* $(q_{\mathcal{D}}^+, \gamma^*)$-*legitimate sampler* $\mathcal{D}$, *it holds that*

$$\mathrm{Adv}^{\mathrm{init}}_{\mathcal{G}, \mathsf{M}, \gamma^*, q_{\mathcal{D}}}(\mathcal{A}, \mathcal{D}) \leq q_{\mathsf{H}} \cdot 2^{-\gamma^*} + 2^{-L}.$$

Pres security. At the start of game Pres, the (q_{H}, q_C)-attacker $\mathcal{A}$ outputs $(S_0', I_1, \ldots, I_d)$ where $d \leq q_C$. The game sets $(V_0, C_0, cnt_0) \leftarrow_{\$} \mathsf{M}^{\mathsf{H}}(S_0')$, and iteratively computes S_d via $S_i = \mathsf{refresh}^{\mathsf{H}}(S_{i-1}, I_i)$ for $i \in [1, d]$. The proof first argues that unless $\mathcal{A}$ queries H on the counter V_0, or any of the counters $V_1, \ldots, V_{d-1}$ passed through during reseeding, then (barring certain accidental collisions) $S_d = (V_d, C_d, cnt_d)$ is indistinguishable from a masked state. The proof then shows that, unless the attacker can guess V_d, the resulting output/state pair is indistinguishable from its idealized counterpart. We must consider a number of

cases depending on whether the tuple $(S'_0, I_1, \ldots, I_d)$ output by $\mathcal{A}$ is such that:
(1) $S'_0 \in \mathcal{S}$ or $S'_0 = \varepsilon$; and **(2)** $d \geq 1$ or $d = 0$; since these induce different
distributions on S_0 and S_d respectively.

Lemma 2. *Let* $\mathcal{G}$ = HASH-DRBG *and masking function* M^H *be as specified in
Theorem 2. Then for any* (q_H, q_C)-*adversary* $\mathcal{A}$ *in game* Pres *against* $\mathcal{G}$, *it holds
that*

$$\mathrm{Adv}^{\mathrm{pres}}_{\mathcal{G},\mathsf{M},\beta}(\mathcal{A}) \leq \frac{q_\mathsf{H} \cdot (n+1)}{2^{\ell-1}} + \frac{(q_C - 1)(2q_\mathsf{H} + q_C)}{2^L}.$$

Rec security. The first stage in the proof of Rec security argues that iteratively reseeding an adversarially chosen state S_0 with d entropy samples which
collectively have entropy γ^* yields a state $S_d = (V_d, C_d, cnt_d)$ which is indistinguishable from $\mathsf{M}^\mathsf{H}(\varepsilon)$. This represents the main technical challenge in the proof,
and uses Patarin's H-coefficient technique (see full version). Our proof is closely
based on the analogous result for sponge-based PRNGs in the ideal permutation
model (IPM) of Gazi and Tessaro [17], essentially making the same step-by-step
argument. However, making the necessary adaptations to analyze HASH-DRBG
is still non-trivial. As well as working in the ROM as opposed to the IPM, we
must adapt the proof to handle the state component C and the more involved
reseeding process, which concatenates the responses to multiple H queries to
derive updated counters. With this in place, an analogous argument to that
made for Pres security implies that an output/state pair produced by applying
next^H to this masked state is indistinguishable from its idealized counterpart.

Lemma 3. *Let* $\mathcal{G}$ = HASH-DRBG *and masking function* M^H *be as specified in
Theorem 2. Then for any* (q_H, q_C)-*adversary* $\mathcal{A}$ *in game* Rec *against* $\mathcal{G}$, *and any*
$(q_\mathcal{D}^+, \gamma^*)$-*legitimate sampler* $\mathcal{D}$, *it holds that*

$$\mathrm{Adv}^{\mathrm{rec}}_{\mathcal{G},\mathsf{M},\gamma^*,q_\mathcal{D},\beta}(\mathcal{A}, \mathcal{D}) \leq \frac{q_\mathsf{H}}{2^{\gamma^*-1}} + \frac{q_\mathsf{H} \cdot n}{2^{(\ell-1)}} + \frac{(q_C - 1) \cdot (2q_\mathsf{H} + q_C) + 2q_\mathsf{H}^2}{2^L}.$$

Using Theorem 1 to combine Lemmas 1, 2 and 3—which bound the Init, Pres and
Rec security of HASH-DRBG respectively—proves Theorem 2, and completes our
analysis of the robustness of HASH-DRBG in the ROM.

6   Analysis of HMAC-DRBG

In this section, we present our analysis of HMAC-DRBG. We give both positive
and negative results, showing that the security guarantees of HMAC-DRBG differ
depending on whether additional input is provided in next calls.

6.1   Negative Result: HMAC-DRBG Called Without Additional Input Is Not Forward Secure

We present an attack which breaks the forward security of HMAC-DRBG if
called without additional input. This contradicts the claim in the standard that

HMAC-DRBG is backtracking resistant. Since Rob security implies Fwd security, this rules out a proof of robustness in this case also.

The attack. Consider the application of update at the conclusion of a next call for HMAC-DRBG (Fig. 1). Notice that if $addin = \varepsilon$, then the final two lines of update are not executed. In this case, the updated state $S^* = (K^*, V^*, cnt^*)$ is of the form $V^* = \mathsf{HMAC}(K^*, r^*)$ where r^* is the final output block produced in the call. An attacker $\mathcal{A}$ in game Fwd who makes a RoR query with $addin = \varepsilon$ to request ℓ-bits of output, followed immediately by a Get query to learn S^*, can easily test this relation. If it does not hold, they know the challenge output is truly random. We note that the observation that V^* depends on r^* is also implicit in the proof of pseudorandomness by Hirose [19]; however, the connection to forward security is not made in that work. To concretely bound $\mathcal{A}$'s advantage we define game Fwd$^\$$, which is identical to game Fwd against HMAC-DRBG except the PRNG is initialized with an 'ideally distributed' state $S_0 = (K_0, V_0, cnt_0)$ where $K_0, V_0 \leftarrow_\$ \{0,1\}^\ell$ and $cnt_0 \leftarrow 1$. The attacker's job can only be harder in Fwd$^\$$, since they cannot exploit any flaws in the setup procedure.

Theorem 3. *Let $\mathcal{G}$ be HMAC-DRBG built from the function HMAC $: \{0,1\}^\ell \times \{0,1\}^{\leq \omega} \to \{0,1\}^\ell$, with parameters $(p, \overline{p}, \alpha, \beta_{max})$ such that $\beta_{max} \geq \ell$. Then there exist efficient adversaries $\mathcal{A}, \mathcal{B}$, such that for any sampler $\mathcal{D}$, it holds that*

$$\mathrm{Adv}_{\mathcal{G}, \gamma^*}^{\mathrm{fwd}\text{-}\$}(\mathcal{A}, \mathcal{D}) \geq 1 - 2 \cdot \mathrm{Adv}_{\mathsf{HMAC}}^{\mathrm{prf}}(\mathcal{B}, 2) - 2^{-(\ell-1)}.$$

$\mathcal{A}$ makes one RoR query in which additional input is not included, and one Get query. $\mathcal{B}$ runs in the same time as $\mathcal{A}$, and makes two queries to their real-or-random function oracle.

Discussion. The first negative term on the right-hand side of the above equation is the advantage of an attacker $\mathcal{B}$ who tries to break the PRF-security of HMAC given two queries to its real-or-random function oracle; since HMAC is widely understood to be a secure PRF, we expect this term to be small. Similarly, ℓ denotes the output length of HMAC and so the second negative term will be small for all commonly used hash functions. This implies that $\mathcal{A}$ succeeds with probability close to one, making this an effective attack. For simplicity, the theorem assumes that outputs of length ℓ bits may be requested (i.e., $\beta_{max} \geq \ell$); in the full version, we discuss how to relax this restriction.

6.2   Positive Result: Robustness of HMAC-DRBG Called with Additional Input in the ROM

In this section, we prove that HMAC-DRBG is robust in the ROM when additional input is always used, with respect to a restricted (but realistic) class of samplers. We model the function HMAC $: \{0,1\}^\ell \times \{0,1\}^{\leq \omega} \to \{0,1\}^\ell$ as a keyed random oracle, whereby each fresh query of the form $(K, X) \in \{0,1\}^\ell \times \{0,1\}^{\leq \omega}$ is answered with an independent random ℓ-bit string.

Rationale. While a standard model proof of Pres security is possible via a reduction to the PRF-security of HMAC, how to achieve the same for Init and Rec is unclear. These results require showing that HMAC is a good randomness extractor. In games Init and Rec, the HMAC key is chosen by or known to the attacker, and so we cannot appeal to PRF-security. Entropy samples are non-uniform, so a dual-PRF assumption does not suffice either. As such, some idealized assumption on HMAC or the underlying hash/compression func-

$$
\begin{array}{|l|}
\hline
\mathsf{M}^{\mathsf{HMAC}}(S) \\
\hline
\text{If } S = \varepsilon \\
\quad cnt \leftarrow 0 \\
\text{Else } (K, V, cnt) \leftarrow S \\
K', V' \leftarrow\!\!{\scriptstyle\$}\; \{0,1\}^{\ell} \\
cnt' \leftarrow cnt + 1 \\
S' \leftarrow (K', V', cnt') \\
\text{Return } S' \\
\hline
\end{array}
$$

Fig. 5. Masking function for proof of Theorem 4.

tion seems to be inherently required. The extraction properties of HMAC (under various idealized assumptions) were studied in [12]. However, these consider a single-use version of extraction which is weaker than what is required here, and typically require inputs containing much more entropy than is required by the standard, and so are not generally applicable to real-world implementations of HMAC-DRBG.

By modeling HMAC as a keyed RO, we can analyze HMAC-DRBG with respect to the entropy levels of inputs specified in the standard (and at levels which are practical for real-world applications). This is a fairly standard assumption, having been made in other works in which HMAC is used with a known key or to extract from lower entropy sources e.g., [23,24,32]. In [14], HMAC was proven to be indifferentiable from a random oracle for all commonly deployed parameter settings (although since robustness is a multi-stage game, the indifferentiability result cannot be applied generically here [31]).

Discussion. A standard model proof of Rec security for HMAC-DRBG would be a stronger and more satisfying result. While idealizing HMAC or the underlying hash/compression function seems inherent, a result under weaker idealized assumptions is an important open problem. Despite this, we feel our analysis is a significant forward step from existing works. Ours is the first analysis of the full specification of HMAC-DRBG; prior works [19,21] omit reseeding and initialization, assuming HMAC-DRBG is initialized with a state for which $K, V \leftarrow\!\!{\scriptstyle\$}\; \{0,1\}^{\ell}$, which is far removed from HMAC-DRBG in a real system. Our work is also the first to consider security properties stronger than pseudorandomness. We hope our result is a valuable first step to progress the understanding of this widely deployed (yet little analyzed) PRNG, and a useful starting point for further work to extend.

Sampler. Our proof is with respect to the class of samplers $\{\mathcal{D}\}_{\gamma^*}$, defined to be the set of $(q_{\mathcal{D}}^+, \gamma^*)$-legitimate samplers for which $\gamma_i \geq \gamma^*$ for $i \in [1, q_{\mathcal{D}} + 1]$ (i.e., each sample I contains γ^*-bits of entropy). This is a simplifying assumption, making the proof of Rec security less complex. However, we stress that this is the entropy level per sample required by the standard, and so this is precisely the restriction imposed on allowed entropy sources. An H-coefficient analysis, as in the proof of Lemma 3, seems likely to yield a fully general result.

Proof of robustness. We now present the following theorem bounding the robustness of HMAC-DRBG. The proof follows from the lemmas presented below, combined with Theorem 1. (It is straightforward to verify that $q_{ref} = 4$, $q_{nxt} = n + 8$, and $q_M = 0$). The proofs of all lemmas are given in the full version.

Theorem 4. *Let $\mathcal{G}$ be HMAC-DRBG with parameters $(p, \overline{p}, \alpha, \beta_{max})$, built from HMAC : $\{0,1\}^\ell \times \{0,1\}^{\leq \omega} \to \{0,1\}^\ell$ which we model as a keyed random oracle. Let M^{HMAC} be the masking function shown in Fig. 5 and $n = \lceil \beta/\ell \rceil$. Then for any $(q_H, q_R, q_D, q_C, q_S)$-attacker $\mathcal{A}$ in game Rob_β against HMAC-DRBG who always outputs addin $\neq \varepsilon$, and any (q_D^+, γ^*)-legitimate sampler $\mathcal{D} \in \{\mathcal{D}\}_{\gamma^*}$, it holds that*

$$
\mathrm{Adv}_{\mathcal{G},M,\gamma^*,\beta}^{rob}(\mathcal{A}) \leq q_R \cdot (\bar{q}_H \cdot \epsilon_1 + \epsilon_2) \cdot 2^{-(2\ell-1)}
$$
$$
+ \bar{q}_H \cdot 2^{-(\ell-2)} + q_R \cdot (\bar{q}_H \cdot (n+3) + \epsilon_3) \cdot 2^{-(\ell-2)}
$$
$$
+ (\bar{q}_H \cdot (2q_R + (1 + 2^{-2\ell})) \cdot 2^{-(\gamma^*-1)} + 2^{-(2\ell-1)}.
$$

Here $\epsilon_1 = 12q_C + 10 + (4q_C - 2) \cdot 2^{-\gamma^*}$, $\epsilon_2 = (q_C \cdot (10q_C + 4n + 18 + (q_C - 1) \cdot 2^{-(\gamma^*-1)}) + 6n + 16)$, $\epsilon_3 = n(n + 1)$, and $\bar{q}_H = (q_H + 4 \cdot q_D + (n + 8) \cdot q_R)$.

Concrete example. For HMAC-SHA-512, $\ell = 512$ and the bound is dominated by the $\mathcal{O}(\bar{q}_H \cdot q_R) \cdot 2^{-(\gamma^*-1)}$ term. Supposing $q_D \leq q_R$ (i.e., there are fewer Ref than RoR calls) and n is small, then $\bar{q}_H \cdot q_R \leq q_R \cdot (q_H + c \cdot q_R)$ for some small constant c. Now if HMAC-DRBG is instantiated at strength $\gamma^* = 256$, it achieves a good security margin up to fairly large q_H, q_R. At lower γ^* the margins are less good; however, this is likely an artefact of the proof technique.

Init security. The proof of Init security argues that unless attacker $\mathcal{A}$ queries HMAC on certain points which require guessing the value of either the input I_1 with which HMAC-DRBG was seeded, or an intermediate key/counter computed during setup, then—barring a collision in the inputs to the second and fourth HMAC queries made by setup, contributing $2^{-2\ell}$ to the bound—the resulting state is identically distributed to $M^{HMAC}(\varepsilon)$.

Lemma 4. *Let $\mathcal{G} = $ HMAC-DRBG and masking function M^{HMAC} be as specified in Theorem 4. Then for any q_H-adversary $\mathcal{A}$ in game Init against HMAC-DRBG, and any (q_D^+, γ^*)-legitimate sampler $\mathcal{D} \in \{\mathcal{D}\}_{\gamma^*}$, it holds that*

$$
\mathrm{Adv}_{\mathcal{G},M,\gamma^*,q_D}^{init}(\mathcal{A}, \mathcal{D}) \leq q_H \cdot ((1 + 2^{-2\ell}) \cdot 2^{-\gamma^*} + 2^{-(\ell-1)}) + 2^{-2\ell}.
$$

Pres and Rec security. The proofs of Pres and Rec security proceed by bounding: **(1)** the probability that two of the points queried to HMAC during the challenge computation collide; and **(2)** the probability that $\mathcal{A}$ queries HMAC on one of these points. We then argue that if neither of these events occur, then the challenge output/state are identically distributed to their idealized counterparts. However, this process is surprisingly delicate. Firstly, the domains of queries are not fully separated, so multiple collisions must be dealt with. Secondly, the guessing/collision probabilities of points from the same domain differ throughout the game. This rules out a modular treatment, and complicates the bound. A small modification to separate queries would simplify analysis.

Lemma 5. *Let* $\mathcal{G}$ = HMAC-DRBG *and masking function* $\mathsf{M}^{\mathsf{HMAC}}$ *be as specified in Theorem 4. Then for any* (q_{H}, q_C)-*adversary* $\mathcal{A}$ *in game* Pres *against* HMAC-DRBG *who always outputs addin* $\neq \varepsilon$, *it holds that*

$$\mathrm{Adv}^{\mathrm{pres}}_{\mathcal{G},\mathsf{M},\beta}(\mathcal{A}) \leq (q_{\mathsf{H}} \cdot (8q_C + 6) + \epsilon) \cdot 2^{-2\ell} + (q_{\mathsf{H}} \cdot (n+2) + n(n+1)) \cdot 2^{-\ell},$$

where $\epsilon = q_C \cdot (6q_C + 2n + 8) + 3n + 8$.

Lemma 6. *Let* $\mathcal{G}$ = HMAC-DRBG *and masking function* $\mathsf{M}^{\mathsf{HMAC}}$ *be as specified in Theorem 4. Then for any* (q_{H}, q_C)-*adversary* $\mathcal{A}$ *in game* Rec *against* HMAC-DRBG *who always outputs addin* $\neq \varepsilon$, *and any* $(q_{\mathcal{D}}^+, \gamma^*)$-*legitimate sampler* $\mathcal{D} \in \{\mathcal{D}\}_{\gamma^*}$, *it holds that*

$$\mathrm{Adv}^{\mathrm{rec}}_{\mathcal{G},\mathsf{M},\gamma^*,q_{\mathcal{D}},\beta}(\mathcal{A}, \mathcal{D}) \leq (q_{\mathsf{H}} \cdot (4q_C + 4 + (4q_C - 2) \cdot 2^{-\gamma^*}) + \epsilon') \cdot 2^{-2\ell}$$
$$+ (q_{\mathsf{H}} \cdot (n+4) + n(n+1)) \cdot 2^{-\ell} + q_{\mathsf{H}} \cdot 2^{-(\gamma^*-1)},$$

where $\epsilon' = (q_C \cdot (4q_C + 2n + 10 + (q_C - 1) \cdot 2^{-(\gamma^*-1)}) + 3n + 8)$.

This completes our analysis of the Init, Pres, and Rec security of HMAC-DRBG. Combining these results via Theorem 1 then proves Theorem 4.

7   Overlooked Attack Vectors

While the positive results of Sects. 5 and 6 are reassuring, the flexibility in the standard to produce variable length and *large* outputs (of up to 2^{19} bits) means that two implementations of the same DRBG may be very different depending on how limits on output production are set. While this is reflected in the security bounds of the previous sections (in terms of the parameter n denoting the number of output blocks computed per request), we argue that the standard security notion of robustness may overlook attack vectors against the (fairly non-standard) NIST DRBGs. The points made in this section do not contradict the results of the previous sections; rather we argue that in certain (realistic) scenarios—namely when the DRBG is used to produce many output blocks per next call—it is worth taking a closer look at which points during output generation a state may be compromised.

Iterative next algorithms. The next algorithm of each of the NIST DRBGs has the same high-level structure (modulo slight variations which again frustrate a modular treatment). First, any additional input provided in the call is incorporated into the state, and in the case of HASH-DRBG one of the state variables is copied into an additional variable in preparation for output generation (i.e., setting $data = V$, see Fig. 1). Output blocks are produced by iteratively applying a function to the state variables (or in the case of HASH-DRBG, the copy of the state variable). These blocks are concatenated and truncated to β-bits to form the returned output R, and a final state update is performed to produce S'.

Decomposition. We wish to track the evolution of the state variables during a next call relative to the production of different output blocks, in order to reason precisely about the effects of state component compromises at different points. We say that a DRBG has an *iterative* next *algorithm* if next may be decomposed into a tuple of subroutines $\mathcal{C} = (\mathsf{init}, \mathsf{gen}, \mathsf{final})$. Here $\mathsf{init} : \mathrm{Seed} \times \mathcal{S} \times \mathbb{N}^{\leq \beta_{max}} \times \{0,1\}^{\leq \alpha} \rightarrow \mathcal{S} \times \{0,1\}^*$ updates the state with additional input prior to output generation, and optionally sets a variable $data \in \{0,1\}^*$. Algorithm $\mathsf{gen} : \mathrm{Seed} \times \mathcal{S} \times \{0,1\}^* \rightarrow \{0,1\}^\ell \times \mathcal{S} \times \{0,1\}^*$ maps a state S and optional string $data$ to an output block $r \in \{0,1\}^\ell$, an updated state S', and string $data' \in \{0,1\}^*$. Finally $\mathsf{final} : \mathrm{Seed} \times \mathcal{S} \times \mathbb{N}^{\leq \beta_{max}} \times \{0,1\}^{\leq \alpha} \rightarrow \mathcal{S}$ updates the state post output generation. The next algorithm is constructed from these component parts as shown in the top left panel of Fig. 6. The decomposition algorithms $\mathcal{C} = (\mathsf{init}, \mathsf{gen}, \mathsf{final})$ for each of the NIST DRBGs are shown in remaining panels of Fig. 6. For CTR-DRBG and HMAC-DRBG, $data$ is not set during output generation (e.g., $data = \varepsilon$, and so we omit it from the discussion of these DRBGs. Similarly since none of the NIST DRBGs are specified to take a seed, we omit this parameter.) A diagrammatic depiction of output generation for each of the DRBGs is shown in the full version.

Variable length outputs. Within this iterative structure, the gen subroutine acts like the next algorithm of an internal PRNG, called multiple times within a single next call to produce output blocks. However, as we shall see, the state updates performed by gen do *not* provide forward security after each block [5]. This may not seem unreasonable if the DRBG produces only a handful of blocks per request; however since the standard allows for up to 2^{19} bits of output to be requested in each next call, there are situations in which the possibility of a partial state compromise occurring *during* output generation is worth considering.

Attack scenario: side channels. We consider an attacker who learns some information about the state variables being processed during output generation, but who is *not* able to perform a full memory compromise by which they would learn e.g., the output blocks $r^1, \ldots, r^n$ buffered in the internal memory, thereby compromising all output in the call. The natural scenario we consider here is a *side channel attack*. Generating multiple output blocks in a single next call results in a significant amount of computation going on 'under the hood' of next—e.g., up to $2^{12} = 4096$ AES-128 computations using a fixed key K^0 for CTR-DRBG with AES-128—which, given that AES invites leaky implementations [4,7,22,25,27,28], is concerning. Since robustness only allows the attacker to compromise the state *after* it has 'properly' updated (via the final process) at the conclusion of a next call, it does not model side channel *during* the call.

Use case: buffering output. As pointed out by Bernstein [5], the overhead incurred by the state update at the conclusion of a CTR-DRBG next call is undesirable. As such, an appealing usage choice is to generate a large output

---

[5] This is similar to an observation by Bernstein [5] criticizing the inefficiency of CTR-DRBG's update function which appeared concurrently to the production of the first draft of this work. We stress that our modelling of the attack scenario, and systematic treatment of how the issue affects each of the NIST DRBGs, is novel.

$\text{next}(X, S, \beta, addin)$	HMAC-DRBG init
If $cnt > reseed\_interval$	Require : $S = (K, V, cnt), \beta, addin$
Return reseed_required	Ensure: $S = (K, V, cnt)$
$(S^0, data^0) \leftarrow \text{init}(X, S, \beta, addin)$	If $addin \neq \varepsilon$
If $addin \leftarrow \varepsilon$ then $addin \leftarrow 0^n$	$(K, V) \leftarrow \text{update}(addin, K, V)$
$temp_R \leftarrow \varepsilon ; n \leftarrow \lceil \beta/\ell \rceil$	Return (K, V, cnt)
For $i = 1, \ldots, n$	HMAC-DRBG gen
$(r^i, S^i, data^i) \leftarrow \text{gen}(X, S^{i-1}, data^{i-1})$	Require (K, V, cnt)
$temp_R \leftarrow temp_R \parallel r^i$	Ensure $r, S = (K, V, cnt)$
$R \leftarrow \text{left}(temp_R, \beta)$	$V \leftarrow \text{HMAC}(K, V) ; r \leftarrow V$
$S' \leftarrow \text{final}(X, S^n, \beta, addin)$	Return $r, (K, V, cnt)$
Return (R, S')	HMAC-DRBG final
	Require : $S = (K, V, cnt), \beta, addin$
HASH-DRBG init	Ensure: $S = (K, V, cnt)$
Require: $S = (V, C, cnt), \beta, addin$	$(K, V) \leftarrow \text{update}(addin, K, V)$
Ensure: $S = (V, C, cnt), data$	$cnt \leftarrow cnt + 1$
If $addin \neq \varepsilon$	Return (K, V, cnt)
$w \leftarrow \text{H}(0x02 \parallel V \parallel addin)$	
$V \leftarrow (V + w) \bmod 2^L$	CTR-DRBG init
$data \leftarrow V$	Require: $S = (K, V, cnt), \beta, addin$
Return $(V, C, cnt), data$	Ensure: $S = (K, V, cnt)$
HASH-DRBG gen	If $addin \neq \varepsilon$
Require $S = (V, C, cnt), data$	If derivation function used then
Ensure: $r, S = (V, C, cnt), data$	$addin \leftarrow \text{CTR-DRBG\_df}(addin, (\kappa + \ell))$
$r \leftarrow \text{H}(data)$	Else if $len(addin) < (\kappa + \ell)$ then
$data \leftarrow (data + 1) \bmod 2^L$	$addin \leftarrow addin \parallel 0^{(\kappa + \ell - len(addin))}$
Return $r, (V, C, cnt), data$	$(K, V) \leftarrow \text{update}(addin, K, V)$
HASH-DRBG final	Return (K, V, cnt)
Require: $S = (V, C, cnt), \beta, addin$	CTR-DRBG gen
Ensure: $S = (V, C, cnt)$	Require: $S = (K, V, cnt)$
$H \leftarrow \text{H}(0x03 \parallel V)$	Ensure: $r, S = (K, V, cnt)$
$V \leftarrow (V + H + C + cnt) \bmod 2^L$	$V \leftarrow (V + 1) \bmod 2^\ell ; r \leftarrow \text{E}(K, V)$
$cnt \leftarrow cnt + 1$	Return $r, (K, V, cnt)$
Return (V, C, cnt)	CTR-DRBG final
	Require: $S = (K, V, cnt), \beta, addin$
	Ensure: $S = (K, V, cnt)$
	$(K, V) \leftarrow \text{update}(addin, K, V)$
	$cnt \leftarrow cnt + 1$
	Return (K, V, cnt)

Fig. 6. Top left: iterative next algorithm for a DRBG with associated decomposition $\mathcal{C} = (\text{init}, \text{gen}, \text{final})$. Boxed text included for CTR-DRBG only. Right and bottom left: $\mathcal{C} = (\text{init}, \text{gen}, \text{final})$ for HASH-DRBG, HMAC-DRBG and CTR-DRBG.

upfront in a single request, and buffer it to later be used for different purposes[6]. Our attack model investigates the soundness of this approach for scenarios in which partial state compromise during output generation via a side channel— which can only be exacerbated by such usage—is a realistic concern. Portions of the buffered output may be used for public values such as nonces, whereas other portions of the output from the same call may be used for e.g., secret keys. As such, our model assumes an attacker learns an output block sent in the clear as

---

[6] Indeed, NIST SP 800-90A says: "For large generate requests, CTR-DRBG produces outputs at the same speed as the underlying block cipher algorithm encrypts data", highlighting the efficiency of this approach.

e.g., a nonce, in conjunction with the partial state information gleaned via a side channel. The attacker's goal is to recover *unseen* output blocks used as security critical secrets, thereby breaking the security of the consuming application.

7.1   Attack Model

We now describe our attack model. We found that a more formal and/or code-based model using abstract leakage functions (in line with the literature on leakage-resilient cryptography e.g., [1,15,26]) introduced significant complexity, without clarifying the presentation of the attacks or providing further insight. We therefore opted for a more informal written definition of the attack model which is nonetheless sufficiently precise to capture e.g., exactly what the attacker may learn, what he is challenged to guess, and so on. We aim to demonstrate key attacks rather than providing an exhaustive treatment.

Attack setup and goals. Consider the next call shown in Fig. 6. Letting S denote the state input to next, then this defines a sequence of intermediate states/output blocks passed through during the course of the request:

$$(S, (S^0, data^0), r^1, (S^1, data^1), \ldots, r^n, (S^n, data^n), S'),$$

with the algorithm finally returning $(R, S') = (r^1 \parallel \ldots \parallel r^m, S')$. (For simplicity, we assume the requested number of bits is a multiple of the block length; it is straightforward to remove this assumption.) We consider an attacker $\mathcal{A}$ who is able to compromise a given component of an arbitrary intermediate state S^i (or in the case of HASH-DRBG, the additional state information $data^i$) for $i \in [0, n]$, in addition to an arbitrary output block r^j for $j \in [1, n]$ produced in the same call. We assume the indices (i, j)[7] are known to $\mathcal{A}$. We then assess the attacker's ability to achieve each of the following 'goals':

- **(1)** Recover unseen output blocks produced prior to the compromised block within the call $\{r^k\}_{k<j}$;
- **(2)** Recover unseen output blocks produced following the compromised block within the call $\{r^k\}_{k>j}$; and
- **(3)** Recover the state S' as updated at the conclusion of the call. This allows the attacker to run the generator forwards and recover future output.

Extensions. If $addin = \varepsilon$, then init returns the state unchanged, $S^0 = S$. As such, all attacks which succeed when S^0 is partially compromised in our model can also be executed if the relevant component of state S is compromised *prior* to the next call, creating a greater window of opportunity for the attacker.

---

[7] Here we assume the attacker learns a full block and knows its index. This seems reasonable; for example, a TLS client or server random will contain at least one whole block and 12 bytes of a second block (if 4 bytes of timestamp are used). These values would be generated early in a call to the DRBG, and so have a low index j. Both assumptions can be relaxed at the cost of the attacker performing more work to brute-force any missing bits and/or the index.

	(1) Past output within call	(2) Future output within call	(3) Updated state S'	Additional input
CTR-DRBG // compromised K	✓	✓	✓	✓*
HMAC-DRBG // compromised K	✗	✓	✓	✗
HASH-DRBG // compromised V	✓	✓	✗**	✗

(a) Table summarizing our analysis. The leftmost three columns correspond to Sections 7.2–7.4. The rightmost column corresponds to Section 7.5. A ✓ indicates that we demonstrate an attack. A ✗ indicates that we believe the DRBG is not vulnerable to such an attack, with justification given. * corresponds to an attack if CTR-DRBG is implemented without a derivation function. ** indicates an exception in the case that $cnt = 1$ at the point of compromise.

$$CTR\text{-}DRBG//\mathcal{A}(K^i, r^j, i, j)$$
$V^j \leftarrow \mathsf{E}^{-1}(K^i, r^j)$
$V^0 \leftarrow (V^j - j) \bmod 2^\ell$
For $k = 1, \ldots, n$
$\quad V^k \leftarrow (V^{k-1} + 1) \bmod 2^\ell$
$\quad r^k \leftarrow \mathsf{E}(K^i, V^k)$
$(K', V') \leftarrow \mathsf{update}(addin, K^i, V^n)$
$cnt' \leftarrow cnt + 1$
$S' \leftarrow (K', V', cnt')$
Return $(\{r^k\}_{k<j}, \{r^k\}_{k>j}, S')$

$$HMAC\text{-}DRBG//\mathcal{A}(K^i, r^j, i, j)$$
$V^j \leftarrow r^j$
For $k = j+1, \ldots, n$
$\quad V^k \leftarrow \mathsf{HMAC}(K^i, V^{k-1})$
$\quad r^k \leftarrow V^k$
$(K', V') \leftarrow \mathsf{update}(addin, K^i, V^n)$
$cnt' \leftarrow cnt + 1$
$S' \leftarrow (K', V', cnt')$
Return $(\bot, \{r^k\}_{k>j}, S')$

$$HASH\text{-}DRBG//\mathcal{A}(data^i, r^j, i, j)$$
$data^0 \leftarrow (data^i - i) \bmod 2^L$
For $k = 1, \ldots, n$
$\quad r^k \leftarrow \mathsf{H}(data^{k-1})$
$\quad data^k \leftarrow (data^{k-1} + 1)$
Return $(\{r^k\}_{k<j}, \{r^k\}_{k>j}, \bot)$

(b) Adversaries for Section 7.2-7.4.

Fig. 7. Summary of analysis (top) and adversaries (bottom) for Sect. 7.

Security analysis. We analyzed each of the NIST DRBGs with respect to our attack model, and found that each DRBG exhibited vulnerabilities, with CTR-DRBG faring especially badly. We summarize our findings in Fig. 7a.

7.2   CTR-DRBG with Compromised Key

Since each output block encrypts the secret counter V, leakage of the key component of the CTR-DRBG state is especially damaging. Consider attacker $\mathcal{A}$ shown in the left-hand panel of Fig. 7b. We claim that for all $i \in [0, n]$ and $j \in [1, n]$, if additional input is not used ($addin = \varepsilon$) then $\mathcal{A}$ achieves goals (1), (2) (recovery of all unseen output blocks produced in the next call) and (3) (recovery of the next state S') with probability one. If additional input is used ($addin \neq \varepsilon$) then the same statement holds for (1), (2), and the attacker's ability to satisfy (3) is equal to his ability to guess $addin$. To see this, notice that each block of output produced in the next call is computed as $r^k = \mathsf{E}(K^0, V^0 + k)$ for $k \in [1, n]$, where K^0, V^0 denote the key and counter as returned by init at the start of output generation. The key does not update through this process, and so whatever intermediate key K^i attacker $\mathcal{A}$ compromises, this is the key used for output generation. It is then trivial for $\mathcal{A}$ to decrypt the output block r^j received in his challenge to recover the secret counter, thereby possessing all security critical state variables. However if $addin \neq \varepsilon$, $\mathcal{A}$ must guess its value to compute S'.

Discussion. This attack is especially damaging, since target output blocks used as e.g., secret keys will be recovered *irrespective* of their position relative to the block learnt by the attacker, increasing the exploitability of the compromised CTR-DRBG. In comparison, the infamously backdoored DualEC-DRBG only allowed recovery of output produced *after* the compromised block, impacting its practical exploitability [10] (although, of course, the embedded backdoor in DualEC means the attack itself is far easier to execute).

7.3 HMAC-DRBG with Compromised Key

Consider the attacker $\mathcal{A}$ shown in the middle panel of Fig. 7b, who compromises the key component K^i of an intermediate state of HMAC-DRBG. We claim that for all $i \in [0, n]$ and $j \in [1, n]$, if $addin = \varepsilon$ then $\mathcal{A}$ achieves goals (2) and (3) with probability one. If $addin \neq \varepsilon$ then the same statement holds for (2), and the attacker's ability to satisfy (3) is equal to his ability to guess $addin$. To see this, let K^0, V^0 denote the state variables at the beginning of output generation. Output blocks are iteratively produced by computing $r^k = \mathsf{HMAC}(K^0, V^{k-1})$ for $k \in [1, n]$, and setting $r^k = V^k$. Since the key does not update during this process, the key K^i compromised by the attacker will be equal to the key K^0 used for output generation. Since the output block r^j which $\mathcal{A}$ receives in his challenge is equal to the secret counter V^j, $\mathcal{A}$ now knows all security critical state variables of intermediate state S^j. $\mathcal{A}$ can then run HMAC-DRBG forward to recover all output produced following the compromised block in the call, and the updated state S' (subject to guessing $addin$).

Past output in a compromised next call. It appears that even if an attacker learns the entirety of an intermediate state S^i for $i \in [0, n]$ in addition to an output block r^j for $j \in [1, n]$, then it is still infeasible to achieve goal (1) and recover the set of output blocks $\{r^k\}_{k<j}$ produced prior to the compromised block within the call. To see this, let V^0 denote the value of the counter at the start of output generation. For each $j \in [1, n]$, output block r^j takes the form:

$$r^j = V^j = \mathsf{HMAC}^j(K^0, V^0),$$

where $\mathsf{HMAC}^i(K, \cdot)$ denotes the ith iterate of $\mathsf{HMAC}(K, \cdot)$. As such, recovering prior blocks r^k for $k < j$ given K^0 and V^j corresponds to finding preimages of $\mathsf{HMAC}(K^0, \cdot)$. Since the key is known to the attacker, we clearly cannot argue that this is difficult based on the PRF-security of HMAC. However, modeling HMAC as a random oracle (Sect. 6), it follows that inverting HMAC for sufficiently high entropy V^0 is infeasible. Formalizing this intuition under a standard model assumption remains an interesting open question.

7.4 HASH-DRBG with Compromised Counter

For HASH-DRBG, it is straightforward to see that if $\mathcal{A}$ learns the counter V^i or its iterating copy $data^i$ for any $i \in [0, n]$, $j \in [1, n]$, then $\mathcal{A}$ achieves goals (1) and (2) with probability one. Knowledge of the counter is sufficient to execute the attack; no output block is needed. The case in which $data^i$ is compromised is shown in the rightmost panel of Fig. 7b. However, unlike CTR-DRBG and

HMAC-DRBG, without also learning the constant C, achieving goal **(3)** does not seem to be possible in general; we discuss this further in the full version.

7.5 Security of Additional Input

We present an additional attack against CTR-DRBG implemented without a derivation function; an appealing implementation choice in terms of efficiency due to the overhead incurred by CTR-DRBG_df. Under certain conditions, this allows an attacker who compromises the DRBG state to recover strings of additional input (which may contain secrets) fed to the DRBG during next calls.

The attack. Notice that if CTR-DRBG is implemented without a derivation function, then raw strings of additional input are XORed directly into the CTR-DRBG state during the application of the update function in next calls (Fig. 1, lines 8 and 15). Consider such an implementation of CTR-DRBG built from AES-128. We describe the attack with respect to the 'ideal' conditions. Suppose that attacker $\mathcal{A}$ has compromised the internal state $S = (K, V, cnt)$[8], and that the state compromise is followed by a next call in which additional input $addin$ is used. Moreover, suppose $addin$ has the form $addin = X_1 \| X_2$ where $X_1 \in \{0,1\}^{128}$ is known to the attacker and $X_2 \in \{0,1\}^{128}$ consists of 128 unknown bits. We assume X_2 includes a secret value such as a password which will be the target of the attack.

At the start of the next call, the state components K, V are updated with $addin$ via $(K^0, V^0) \leftarrow \mathsf{update}(addin, K, V)$. It is straightforward to verify that

$$K^0 \| V^0 = K^* \| V^* \oplus addin = (K^* \oplus X_1) \| (V^* \oplus X_2),$$

where $K^* \| V^* = \mathsf{E}(K, V+1) \| \mathsf{E}(K, V+2)$. Since $\mathcal{A}$ has compromised (K, V), they can compute (K^*, V^*). Moreover, since X_1 is known to $\mathcal{A}$, it follows that the updated key $K^0 = (K^* \oplus X_1)$ is known to $\mathcal{A}$ also. During output generation, output blocks are produced by encrypting the iterating counter under K^0. Therefore, the kth block of output is of the form:

$$r^k = \mathsf{E}(K^0, V^0 + k) = \mathsf{E}(\mathbf{K^0}, (\mathbf{V^*} \oplus X_2) + \mathbf{k}),$$

where the variables in bold are known to $\mathcal{A}$. As such, each block of output produced is effectively an encryption of the target secret X_2 under a known key. Given a single block of output r^k, $\mathcal{A}$ can *instantly* recover the target secret X_2—consisting of 128-bits of unknown and secret data—as $X_2 = (\mathsf{E}^{-1}(K^0, r^k) - k) \oplus V^*$. Moreover, it is straightforward to verify that $\mathcal{A}$ has sufficient information to compute the state as updated following the next call. As such, $\mathcal{A}$ can continue to execute the same attack against subsequent output generation requests for as long as the key component of the state evolves predictably.

Extensions. In the full version we describe how to generalize the attack, and discuss how use of the derivation function prevents it.

---

[8] Here we mean the working state of the PRNG, as opposed to the 'intermediate' states considered in the previous section.

8   Open Source Implementation Analysis

In Sect. 7, we showed that certain implementation decisions—permitted by the overly flexible standard—may influence the security guarantees of the NIST DRBGs. To determine if these decisions are taken by implementers in the real world, we investigated two open source implementations of CTR-DRBG, in OpenSSL [30] and mbed TLS [8]. We found that between the two libraries these problematic decisions have indeed been made.

Large output requests. Generating many blocks of output in a single request increases the likelihood and impact of our attacks. In OpenSSL, the next call of CTR-DRBG is implemented in the function drbg_ctr_generate in the file drbg_ctr.c. Interestingly—and contrary to the standard—this function does not impose *any limit* on the number of random bits which may be requested. As such, an arbitrarily large output may be generated using a single key, exacerbating the attacks of Sect. 7.2. More generally, exceeding the output generation limit increases the success probability of the well-known distinguishing attack against a block cipher in CTR-mode, which uses colliding blocks to determine if an output is truly random. The implementation of CTR-DRBG in mbed TLS limits the number of output blocks per next call to 64 blocks of 128-bits; much better for security in the context of our attacks than the 4,096 blocks allowed by the standard. Also, this implementation forces a reseed after 10,000 calls to next; much lower than the allowed maximum of 2^{48}.

Derivation function. In Sect. 7.5, we described a potential vulnerability in implementations of CTR-DRBG which do not use a derivation function. We found that the OpenSSL implementation of CTR-DRBG allows the generator to be called simultaneously without the derivation function and with additional input. Specifically, by setting the flags field of the RAND_DRBG_FLAG_CTR_NO_DF structure to RAND_DRBG_FLAG_CTR the caller may suppress calls to the derivation function, presumably for performance purposes. As such, the attack described in Sect. 7.5 may be possible in real world implementations.

Summary. Despite the high level and theoretical nature of our analysis, we found that the problematic implementation decisions which we highlight *are* made in the real world. While none of these decisions leads to an immediate vulnerability, both the implementation and usage of the functions may exacerbate other problems such as side channel or state compromise attacks. We hope that highlighting these issues will help implementers make informed decisions about how best to use these algorithms in the context of their implementation.

9   Conclusion

We conducted an in-depth analysis of NIST SP 800-90A, to investigate unproven security claims and explore flexibilities in the standard. On the positive side, we formally verify a number of the claimed—and yet, until now unproven—security properties in the standard. However, we argue that taking certain implementation choices permitted by the overly flexible standard may lead to vulnerabilities.

Design and prove. Certain design features of the NIST DRBGs complicate their analysis, and a small tweak in design would facilitate a far simpler proof. This emphasizes the importance of developing cryptographic algorithms alongside security proofs, and—more importantly—not standardizing algorithms with unproven security properties.

Flexibility. In Sect. 6, we saw how the option to call HMAC-DRBG without additional input changed the algorithm in a subtle way which lead to an attack. Similarly, the attacks of Sect. 7 are both facilitated, and exacerbated, by certain implementation choices allowed by the overly flexible standard. In Sect. 8, we confirmed that implementers do make these choices in the real world. These may be a warning to standard writers to avoid unnecessary flexibility which may lead to unintended vulnerabilities.

Recommendations. Because these vulnerabilities stem from implementation choices, we can offer recommendations to make the use of these algorithms more secure. First off, if the algorithms are being run in a setting where side channel attacks are a concern then CTR-DRBG should not be used. Additional input should be (safely) incorporated during output generation wherever possible and the DRBG should be reseeded with fresh entropy as often as is practical. While the standard allows outputs of sizeable length to be requested, users should not 'batch up' calls by making a single call for all randomness required for an application. Finally, the CTR-DRBG derivation function should always be used.

Future work. Analyzing the robustness of CTR-DRBG is an important direction for further work. More generally, the design flexibilities we critique above are related to efficiency savings. Designing PRNGs that achieve an optimal balance between security and efficiency is a key direction for future work. The gap between the specification of these DRBGs, which allows for various optional inputs and implementation choices, and the far simpler manner in which PRNGs are typically modeled in the literature could indicate that theoretical models are not adequately capturing real world PRNGs. Extending these models may help understand the limits and possibilities of what can be achieved.

Acknowledgements. The authors thank Kenny Paterson and the anonymous reviewers for their insightful comments which greatly improved the paper. The first author is supported by the EPSRC and the UK government as part of the Centre for Doctoral Training in Cyber Security at Royal Holloway, University of London (EP/K035584/1); much of this work was completed during an internship at Microsoft Research.

References

1. Abdalla, M., Belaïd, S., Pointcheval, D., Ruhault, S., Vergnaud, D.: Robust pseudo-random number generators with input secure against side-channel attacks. In: Malkin, T., Kolesnikov, V., Lewko, A.B., Polychronakis, M. (eds.) ACNS 2015. LNCS, vol. 9092, pp. 635–654. Springer, Cham (2015). https://doi.org/10.1007/978-3-319-28166-7_31

2. Barker, E., Kelsey, J.: NIST SP 800-90A Rev. 1 Recommendation for random number generation using deterministic random bit generators (2015)

3. Barker, E., Kelsey, J.: Draft NIST SP 800-90C. Recommendation for random bit generator (RBG) constructions (2012)

4. Bernstein, D.J.: Cache-timing attacks on AES (2005). https://cr.yp.to/antiforgery/cachetiming-20050414.pdf

5. Bernstein, D.J.: Fast-key-erasure random-number-generators (2017). https://blog.cr.yp.to/20170723-random.html

6. Bernstein, D.J., et al.: Factoring RSA keys from certified smart cards: Coppersmith in the wild. In: Sako, K., Sarkar, P. (eds.) ASIACRYPT 2013. LNCS, vol. 8270, pp. 341–360. Springer, Heidelberg (2013). https://doi.org/10.1007/978-3-642-42045-0_18

7. Bogdanov, A.: Improved side-channel collision attacks on AES. In: Adams, C., Miri, A., Wiener, M. (eds.) SAC 2007. LNCS, vol. 4876, pp. 84–95. Springer, Heidelberg (2007). https://doi.org/10.1007/978-3-540-77360-3_6

8. Butcher, S., Follath, J., García, A.A.: mbed TLS (2015–2018). https://tls.mbed.org/

9. Campagna, M.J.: Security bounds for the NIST codebook-based deterministic random bit generator. ePrint (2006)

10. Checkoway, S., et al.: On the practical exploitability of dual EC in TLS implementations. In: USENIX (2014)

11. Cornejo, M., Ruhault, S.: Characterization of real-life PRNGs under partial state corruption. In: ACM CCS (2014)

12. Dodis, Y., Gennaro, R., Håstad, J., Krawczyk, H., Rabin, T.: Randomness extraction and key derivation using the CBC, cascade and HMAC modes. In: Franklin, M. (ed.) CRYPTO 2004. LNCS, vol. 3152, pp. 494–510. Springer, Heidelberg (2004). https://doi.org/10.1007/978-3-540-28628-8_30

13. Dodis, Y., Pointcheval, D., Ruhault, S., Vergniaud, D., Wichs, D.: Security analysis of pseudo-random number generators with input:/dev/random is not robust. In: ACM CCS (2013)

14. Dodis, Y., Ristenpart, T., Steinberger, J.P., Tessaro, S.: To hash or not to hash again? (In)differentiability results for H^2 and HMAC. In: Safavi-Naini, R., Canetti, R. (eds.) CRYPTO 2012. LNCS, vol. 7417, pp. 348–366. Springer, Heidelberg (2012). https://doi.org/10.1007/978-3-642-32009-5_21

15. Dziembowski, S., Pietrzak, K.: Leakage-resilient cryptography. In: FOCS (2008)

16. FIPS PUB 140-2. Security Requirements for Cryptographic Modules (2001)

17. Gaži, P., Tessaro, S.: Provably robust sponge-based PRNGs and KDFs. In: Fischlin, M., Coron, J.-S. (eds.) EUROCRYPT 2016. LNCS, vol. 9665, pp. 87–116. Springer, Heidelberg (2016). https://doi.org/10.1007/978-3-662-49890-3_4

18. Heninger, N., Durumeric, Z., Wustrow, E., Halderman, J.A.: Mining your Ps and Qs: detection of widespread weak keys in network devices. In: USENIX (2012)

19. Hirose, S.: Security analysis of DRBG using HMAC in NIST SP 800-90. In: Chung, K.-I., Sohn, K., Yung, M. (eds.) WISA 2008. LNCS, vol. 5379, pp. 278–291. Springer, Heidelberg (2009). https://doi.org/10.1007/978-3-642-00306-6_21

20. Kan, W.: Analysis of underlying assumptions in NIST DRBGs (2007)
21. Katherine, Q.Y., Green, M., Sanguansin, N., Beringer, L., Petcher, A., Appel, A.W.: Verified correctness and security of mbedTLS HMAC-DRBG. In: ACM CCS (2017)
22. Kocher, P., Jaffe, J., Jun, B., Rohatgi, P.: Introduction to differential power analysis. JCEN **1**, 5–27 (2011)
23. Krawczyk, H.: Cryptographic extraction and key derivation: the HKDF scheme. In: Rabin, T. (ed.) CRYPTO 2010. LNCS, vol. 6223, pp. 631–648. Springer, Heidelberg (2010). https://doi.org/10.1007/978-3-642-14623-7_34
24. Krawczyk, H., Eronen, P.: HMAC-based extract-and-expand key derivation function (HKDF) (2010)
25. Mangard, S.: A simple power-analysis (SPA) attack on implementations of the AES key expansion. In: Lee, P.J., Lim, C.H. (eds.) ICISC 2002. LNCS, vol. 2587, pp. 343–358. Springer, Heidelberg (2003). https://doi.org/10.1007/3-540-36552-4_24
26. Micali, S., Reyzin, L.: Physically observable cryptography. In: Naor, M. (ed.) TCC 2004. LNCS, vol. 2951, pp. 278–296. Springer, Heidelberg (2004). https://doi.org/10.1007/978-3-540-24638-1_16
27. Osvik, D.A., Shamir, A., Tromer, E.: Cache attacks and countermeasures: the case of AES. In: Pointcheval, D. (ed.) CT-RSA 2006. LNCS, vol. 3860, pp. 1–20. Springer, Heidelberg (2006). https://doi.org/10.1007/11605805_1
28. Percival, C.: Cache missing for fun and profit (2005)
29. Perlroth, N.: Government announces steps to restore confidence on encryption standards (2013)
30. The OpenSSL Project: OpenSSL (1998–2018). https://www.openssl.org/
31. Ristenpart, T., Shacham, H., Shrimpton, T.: Careful with composition: limitations of the indifferentiability framework. In: Paterson, K.G. (ed.) EUROCRYPT 2011. LNCS, vol. 6632, pp. 487–506. Springer, Heidelberg (2011). https://doi.org/10.1007/978-3-642-20465-4_27
32. Ristenpart, T., Yilek, S.: When good randomness goes bad: virtual machine reset vulnerabilities and hedging deployed cryptography. In: NDSS (2010)
33. Ruhault, S.: SoK: security models for pseudo-random number generators. IACR Trans. Symmetric Cryptol. **2017**, 506–544 (2017)
34. Shrimpton, T., Terashima, R.S.: A provable-security analysis of Intel's secure key RNG. In: Oswald, E., Fischlin, M. (eds.) EUROCRYPT 2015. LNCS, vol. 9056, pp. 77–100. Springer, Heidelberg (2015). https://doi.org/10.1007/978-3-662-46800-5_4
35. Shrimpton, T., Terashima, R.S.: Salvaging weak security bounds for blockcipher-based constructions. In: Cheon, J.H., Takagi, T. (eds.) ASIACRYPT 2016. LNCS, vol. 10031, pp. 429–454. Springer, Heidelberg (2016). https://doi.org/10.1007/978-3-662-53887-6_16
36. Shumow, D., Ferguson, N.: On the possibility of a back door in the NIST SP800-90 Dual EC PRNG (2007)
37. Turan, M.S., Barker, E., Kelsey, J., McKay, K.A., Baish, M.L., Boyle, M.: SP 800-90B. Recommendation for the entropy sources used for random bit generation (2012)
38. Vassilev, A., May, W.: Annex C: approved random number generators for FIPS PUB 140-2, security requirements for cryptographic modules (2016)
39. Yilek, S., Rescorla, E., Shacham, H., Enright, B., Savage, S.: When private keys are public: results from the 2008 Debian OpenSSL vulnerability. In: ACM SIGCOMM (2009)

Searchable Encryption and ORAM

Computationally Volume-Hiding Structured Encryption

Seny Kamara$^{(\boxtimes)}$ and Tarik Moataz

Brown University, Providence, USA
{seny,tarik_moataz}@brown.edu

Abstract. We initiate the study of structured encryption schemes with computationally-secure leakage. Specifically, we focus on the design of volume-hiding encrypted multi-maps; that is, of encrypted multi-maps that hide the response length to computationally-bounded adversaries. We describe the first volume-hiding STE schemes that do not rely on naïve padding; that is, padding all tuples to the same length. Our first construction has efficient query complexity and storage but can be lossy. We show, however, that the information loss can be bounded with overwhelming probability for a large class of multi-maps (i.e., with lengths distributed according to a Zipf distribution). Our second construction is not lossy and can achieve storage overhead that is asymptotically better than naïve padding for Zipf-distributed multi-maps. We also show how to further improve the storage when the multi-map is highly *concentrated* in the sense that it has a large number of tuples with a large intersection. We achieve these results by leveraging computational assumptions; not just for encryption but, more interestingly, to hide the volumes themselves. Our first construction achieves this using a pseudo-random function whereas our second construction achieves this by relying on the conjectured hardness of the planted densest subgraph problem which is a planted variant of the well-studied densest subgraph problem. This assumption was previously used to design public-key encryptions schemes (Applebaum et al., *STOC '10*) and to study the computational complexity of financial products (Arora et al., *ICS '10*).

1 Introduction

A structured encryption (STE) scheme encrypts a data structure in such a way that it can be privately queried. An STE scheme is secure if it reveals nothing about the structure and query beyond a well-specified and "reasonable" leakage profile [12,15]. An important special case of STE is searchable symmetric encryption (SSE) which relies on encrypted multi-maps [4,5,7,8,10–12,15,16,18,28,29,36] to achieve optimal-time search. Another example is graph encryption which encrypts various kinds of graphs [12,33]. STE has received a lot of attention due to its potential applications to cloud storage and database security. In recent years, much of the work on STE has focused on supporting more complex queries like Boolean [11,21,25,37] and range queries [20,21,37,38],

© International Association for Cryptologic Research 2019
Y. Ishai and V. Rijmen (Eds.): EUROCRYPT 2019, LNCS 11477, pp. 183–213, 2019.
https://doi.org/10.1007/978-3-030-17656-3_7

more complex structures like relational databases [26] and on improving security, for example achieving forward-privacy [1,7,8,19,40].

Leakage. One aspect of STE that is still poorly understood is its leakage. There are currently two approaches to dealing with leakage. The first is cryptanalysis; that is, designing leakage attacks against various leakage profiles so that we can better understand their concrete security. This was initiated by Islam, Kuzu and Kantarcioglu in the context of SSE [24] and expanded to PPE by Naveed, Kamara and Wright [35] and to ORAM by Kellaris, Kollios, Nissim and O'Neill [30]. While there has been some progress on designing leakage attacks against STE [9,24,30,32], these attacks remain mostly of theoretical interest due to the strong assumptions they rely on. Assumptions like knowledge of at least 80%–90% of client data in addition to knowledge of 5% of client queries [9,24], or assuming clients make queries uniformly at random, often in addition to assumptions about how client data is distributed [30,32]. Nevertheless, these attacks do provide us with some guidance as to which leakage profiles to avoid when designing schemes. Another line of work related to leakage was initiated recently by Kamara, Moataz and Ohrimenko in [27] where they propose designing general-purpose techniques to suppress specific leakage patterns. In [27], they show how to do this for the query equality pattern (also known as the search pattern) without making use of ORAM simulation and, therefore, without incurring its poly-logarithmic multiplicative overhead.

Computationally-secure leakage. In this work, we consider a new approach to dealing with leakage. Our work starts from the observation that the presence of leakage does not necessarily imply that this leakage can be exploited. In fact, it could be that the leakage is not exploitable because it does not convey enough useful information to the adversary. Alternatively, it could be that the leakage does convey enough information but no computationally-bounded adversary can extract it. In other words, the leakage could be computationally-secure. The possibility of designing STE schemes with computationally-secure leakage patterns is interesting for several reasons. From a theoretical point of view, as far as we know, this question has never been considered before and it raises some intriguing foundational questions; like what kind of computational assumptions would lend themselves to the design of secure leakage patterns? The traditional assumptions used in cryptography are usually algebraic or number-theoretic in nature and it is not clear how such assumptions could be used. From a more practical perspective, the ability to leverage "computationally-secure leakage" in the design of STE schemes could lead to a whole new set of techniques and, ultimately, to highly-efficient zero- or low-leakage schemes—computationally speaking.

Volume-hiding EMMs. In this work, we initiate the study of computationally-secure leakage. In particular, we focus on the design of volume-hiding encrypted multi-maps or, more precisely, of encrypted multi-maps that hide the response length to computationally-bounded adversaries.[1] We focus on encrypted multi-

---

[1] Our constructions also reveal the query equality—even to a bounded adversary—but the latter can be suppressed using the cache-based transform from [27].

maps because they are by far the most important encrypted structure; this is illustrated by the fact that they are central to the design of optimal-time single-keyword SSE [7,8,10,12,15,19,29,34], of sub-linear Boolean SSE [11,25], of graph encryption [12,33], of encrypted range structures [17,20] and of encrypted relational databases [26]. We consider the response length leakage pattern for several reasons. The first is that it is a very difficult leakage pattern to suppress. In fact, though encrypted search has been investigated since 2000, the first non-trivial construction to even partially hide the response length is the recent PBS scheme of [27].[2] In fact, response lengths are leaked even by ORAM-based solutions. The second reason we focus on response lengths is because of the recent volume attacks of Kellaris et al. [30] or its extension by Grubbs et al. [23]. Again, while these attacks are mostly of theoretical interest, they do suggest that the design of volume-hiding encrypted structures is well-motivated.

1.1 Naïve Approaches

To better understand our techniques and the improvements they provide, we first describe two possible naïve approaches to designing volume-hiding EMMs. Recall that a multi-map is a data structure that stores a set of pairs $\{(\ell, \mathbf{v})\}$, where ℓ is a label from a label space $\mathbb{L}$ and $\mathbf{v}$ is a tuple of values from some value space $\mathbb{V}$. Multi-maps support get and put operations. Get takes as input a label ℓ and returns its associated tuple $\mathbf{v}$ whereas Put takes as input a label/value pair $(\ell, \mathbf{v})$ and stores it. We denote the get operation by $\mathbf{v} := \mathsf{MM}[\ell]$ and the put operation by $\mathsf{MM}[\ell] := \mathbf{v}$.

Naïve padding. The first approach to designing a volume-hiding multi-map encryption scheme is to pad the tuples of the plaintext multi-map MM to their maximum response length $t = \max_{\ell \in \mathbb{L}} \#\mathsf{MM}[\ell]$ and encrypt the padded multi-map with any standard multi-map encryption scheme [1,7,10,12]. It is easy to see that this hides the response lengths. Unfortunately, it also induces a non-trivial storage overhead.

Using ORAM. We now describe a volume-hiding construction based on ORAM. Note that, as far as we know, this construction has not appeared before and may be of independent interest.[3] We first represent the multi-map MM as a dictionary by generating $N \stackrel{def}{=} \sum_{\ell \in \mathbb{L}} \#\mathsf{MM}[\ell]$ pairs of the form $\{(\ell, v)_{\ell \in \mathbb{L}, v \in \mathsf{MM}[\ell]}\}$ and storing them in a dictionary DX. We then add $t - 1$ dummy label/value pairs to DX, where t is the maximum response length of a label in MM. DX is then stored and managed using ORAM. To get the tuple associated with a label ℓ, we first obliviously access DX. There are two

---

[2] The PBS construction has two variants. One can hide the response length on non-repeating sub-patterns but has a probability of failure in the sense that the client might not receive all its query responses. The second variant is always correct but reveals the sequence response length on non-repeating sub-patterns.

[3] Kellaris, Kollios, Nissim and O'Neil show in [31] how to use differential privacy to perturb the response length in ORAM. This is different from this naïve approach which completely hides the response length.

cases: if $\#\mathsf{MM}[\ell] = t$, then we retrieve all pairs associated with ℓ; otherwise if $\#\mathsf{MM}[\ell] < t$, we retrieve an additional $t - \#\mathsf{MM}[\ell]$ dummies.

It is clear that this hides the response length since the ORAM simulation hides the query equality and, therefore, an adversary can't distinguish between a dummy label and a real label. From an efficiency perspective, if we use a state-of-the-art ORAM [41] then the storage overhead is $O(N)$. The communication complexity, however, is $O(t \cdot \log^2 N)$ which includes a multiplicative poly-logarithmic factor in addition to logarithmic round complexity.

1.2   Our Techniques and Contributions

In this work, we describe two volume-hiding multi-map encryption schemes: VLH and AVLH. Both our constructions work by first transforming an input multi-map into a volume-hiding multi-map and encrypting the result with a custom multi-map encryption scheme that itself makes black-box use of a standard multi-map encryption scheme. These constructions avoid the limitations of the naïve approaches described above either by improving on the storage of naïve padding or avoiding the multiplicative poly-logarithmic overhead of the ORAM-based solution.

A time-efficient construction. Our first construction relies on a simple transformation we call the *pseudo-random transform* which is parameterized by a public parameter λ and makes use of a small-domain pseudo-random function as follows. Each tuple $\mathbf{v}$ in the multi-map is transformed into a new tuple $\mathbf{v}'$ of size $n' = \lambda + F_K(n)$, where $n = \#\mathbf{v}$. If $n' > n$, then the elements of $\mathbf{v}$ are stored in $\mathbf{v}'$ and the latter is padded to have length n'. If $n' \leq n$ then only the first n' items of $\mathbf{v}$ are stored in $\mathbf{v}'$ which effectively truncates $\mathbf{v}$ (we think of the case $n' = n$ as a padding). Note that the multi-map that results from this process is volume-hiding since each tuple has pseudo-random length. Perhaps surprisingly, we also show that if the lengths of the input multi-map are Zipf-distributed then the storage overhead and the number of truncations can be kept relatively small with overwhelming probability in the number of labels. More precisely, we show that the storage overhead is half that of naïve padding while the number of truncations is equal to $m/\log m$. Our scheme VLH essentially consists of transforming a multi-map using the pseudo-random transform and encrypting it with a standard multi-map encryption scheme. The query complexity of VLH is $O(\lambda + \nu)$, where ν is the largest value in the domain of F. While the pseudo-random transform leads to an efficient construction, it is lossy since tuples can be truncated. In many practical settings, however, truncations are not necessarily an issue. For example, in the case of SSE where EMMs are used to store document identifiers clients can rank the document ids (say, by relevance) at setup time so that truncations only affect the low-ranked documents. Nevertheless, we also consider the problem of designing non-lossy volume-hiding EMMs.

A non-lossy transform. Our second construction relies on a different transformation we call the *dense subgraph transform*. Unlike the pseudo-random transform which introduces truncations, this approach is non-lossy. On the other hand, it is less efficient in terms of query complexity. Note, however, that it is hard to imagine any non-lossy construction being able to hide the response length of a query and having query complexity $o(t)$, where t is the maximum response length. Our goal, therefore, is to design a non-lossy scheme that improves on the *storage overhead* of the naïve padding approach. At a high-level, our non-lossy transform works by re-arranging the data stored in the multi-map into bins according to a random bi-partite graph. Roughly speaking, we construct a random (regular) bi-partite graph with labels in one set and bins in the other. We then assign the values in a label's tuple to the bins that are incident to the label. The bins are then padded to hide their size. To ensure that this re-arrangement is still efficiently queryable, we show how to represent the structure encoded in the bi-partite graph and the data stored in the bins with a pair of standard data structures; specifically, a multi-map and a dictionary. We show that, with the right choice of parameters, this version of our transformation already yields a volume-hiding multi-map structure with better storage overhead than the naïve padding approach. More precisely, we show that the naïve approach produces a volume-hiding multi-map of size $S_{NV} = \Omega(N)$, where N is the size of the original multi-map, whereas our approach yields a volume-hiding multi-map of size $O(N)$ with overwhelming probability in N. Interestingly, we also show that if the tuple-lengths of the input multi-map are Zipf-distributed then our transformation yields a multi-map of size $o(S_{NV})$ with overwhelming probability. We note that this version of the transformation already makes use of computational assumptions. In particular, it uses a pseudo-random function to generate the edges of the random bi-partite graph which allows us to "compress" the size of our data structures by storing random seeds as opposed to all the graph's edges. To query our transformed multi-map on some label ℓ, it suffices to retrieve the bins incident to ℓ. Intuitively, this is volume-hiding because the bins are padded and the number of bins is fixed. Furthermore, it hides other leakage patterns because the tuple values are assigned to bins randomly.

Concentration and planted subgraphs. The version of the transformation described so far already improves over naïve padding (with overwhelming probability) but we show that for a certain class of multi-maps we can do even better—though at the cost of increased query complexity. Specifically, we consider multi-maps that have a large number of tuples with a large intersection. We refer to this property as *concentration* and describe a version of the dense subgraph transform that leverages the multi-map's concentration to improve storage efficiency even more. At a high-level, the idea is as follows. A concentrated multi-map has a number of redundant values which our transformation assigns to multiple bins. In our improved transform, we instead assign each of these redundant values to a single bin and add edges between these bins and a large subset of the labels whose tuples they appear in. The rest of the bi-partite graph is generated (pseudo-)randomly as above. This has the benefit of inducing smaller bins and, therefore, of requiring less padding. The bi-partite graph, however, is not random anymore (even ignoring our use of a pseudo-random function to generate edges). We observe, however, that by adding the edges to the bins of the redundant values, we are effectively *planting* a small dense subgraph inside of a larger random graph. And while the resulting graph is clearly not random anymore, it can be shown to be computationally indistinguishable from a random graph. In fact, this reduces to the *planted densest subgraph problem* which has been used in the past by Applebaum et al. in the context of cryptography [2] and by Arora et al. in the context of computational complexity [3]. Based on this assumption, we can show that for multi-maps with concentration parameters within a certain range (in turn determined by the densest subgraph assumption) the transformed multi-map is of size $O(N - m^{0.5+\delta} \cdot \text{polylog}(m))$ with overwhelming probability, where m is the number of labels in the original multi-map and $\delta \geq 0$. If the input multi-map is Zipf-distributed, then the output multi-map has size $o(S_{\text{NV}})$.

Our non-lossy construction. As mentioned above, the dense subgraph transform produces multi-maps that we represent using a combination of a dictionary and a standard multi-map. To encrypt this particular representation, we design a new scheme called AVLH. The resulting construction has query complexity $O\left(t \cdot \frac{N - m^{0.5+\delta} \cdot \text{polylog}(m)}{m \cdot \text{polylog}(m)} \right)$ for multi-maps with concentration parameters within a certain range.

Dynamism. Our VLH and AVLH constructions are for static multi-maps. While there are many important applications of static EMMs, we describe how to extend these constructions to handle updates. This results in two additional constructions, VLH^d and AVLH^d. The former handles three kinds of updates: tuple addition, tuple deletion and tuple edits; and the latter handles tuple edits.

1.3   Related Work

Structured encryption was introduced by Chase and Kamara [12] as a generalization of searchable symmetric encryption which was first considered by Song, Wagner and Perrig [39] and formalized by Curtmola, Garay, Kamara and Ostrovsky [15]. Multi-map encryption schemes are a special case of STE and have been used to achieve optimal-time single-keyword SSE [7,8,10,15,19,29], sublinear Boolean SSE [11,25], encrypted range search [17,20], encrypted relational databases [26] and graph encryption [12,33]. The first leakage attack against volume leakage was described by Kollios, Kellaris, Nissim and O'Neill [30] under the assumption of uniform query distributions. In [27], Kamara, Moataz and Ohrimenko describe an STE scheme called PBS which partially hides the volume pattern. More precisely, the first variant of PBS reveals only the sequence response length (i.e., the sum of the response lengths of a given query sequence) on non-repeating query sequences. The second variant reveals nothing (beyond a public parameter independent of the volume) on non-repeating query sequences. While there are schemes that hide the response length at setup time [42] or use differential privacy to perturb response lengths [31], our techniques hide the pattern entirely at query time. The planted densest graph problem was first used as a computational assumption by Applebaum, Barak and Wigderson in [2] for the purpose of designing public-key encryption schemes under new assumptions. It was later used by Arora, Barak, Brunnermeier and Ge to study the computational complexity of financial products [3].

2   Preliminaries

Notation. The set of all binary strings of length n is denoted as $\{0,1\}^n$, and the set of all finite binary strings as $\{0,1\}^*$. We write $x \leftarrow \chi$ to represent an element x being sampled from a distribution χ, and $x \overset{\$}{\leftarrow} X$ to represent an element x being sampled uniformly at random from a set X. The output x of an algorithm $\mathcal{A}$ is denoted by $x \leftarrow \mathcal{A}$. Given a sequence $\mathbf{v}$ of n elements, we refer to its ith element as v_i or $\mathbf{v}[i]$. If S is a set then $\#S$ refers to its cardinality and 2^S to its powerset.

Basic cryptographic primitives. A private-key encryption scheme is a set of three polynomial-time algorithms $\mathsf{SKE} = (\mathsf{Gen}, \mathsf{Enc}, \mathsf{Dec})$ such that Gen is a probabilistic algorithm that takes a security parameter k and returns a secret key K; Enc is a probabilistic algorithm takes a key K and a message m and returns a ciphertext c; Dec is a deterministic algorithm that takes a key K and a ciphertext c and returns m if K was the key under which c was produced. Informally, a private-key encryption scheme is secure against chosen-plaintext attacks (CPA) if the ciphertexts it outputs do not reveal any partial information about the plaintext even to an adversary that can adaptively query an encryption oracle. We say a scheme is random-ciphertext-secure against chosen-plaintext attacks (RCPA) if the ciphertexts it outputs are computationally indistinguishable from random even to an adversary that can adaptively query an encryption oracle. In addition to encryption schemes, we also make use of pseudo-random functions (PRF), which are polynomial-time computable functions that cannot be distinguished from random functions by any probabilistic polynomial-time adversary.

3   Definitions

Structured encryption schemes encrypt data structures in such a way that they can be privately queried. There are several natural forms of structured encryption. The original definition of [12] considered schemes that encrypt both a structure and a set of associated data items (e.g., documents, emails, user profiles etc.). In [13], the authors also describe *structure-only* schemes which only encrypt structures. Another distinction can be made between *interactive* and *non-interactive* schemes. Interactive schemes produce encrypted structures that are queried through an interactive two-party protocol, whereas non-interactive schemes produce structures that can be queried by sending a single message, i.e, the token. One can also distinguish between *response-hiding* and *response-revealing* schemes: the former reveal the response to queries whereas the latter do not. We recall here the syntax of an interactive response-hiding structured encryption scheme.

Definition 1 (Structured encryption). *An interactive response-hiding structured encryption scheme $\Sigma_{\mathsf{DS}} = (\mathsf{Setup}, \mathsf{Query})$ for data type DS consists of the following polynomial-time algorithms and protocols:*

– $(K, \mathsf{EDS}) \leftarrow \mathsf{Setup}_\mathbf{C}(1^k, \mathsf{DS})$: *is a probabilistic algorithm that takes as input a security parameter 1^k and a structure DS of type $\mathbb{DS}$ and outputs a secret key K and an encrypted structure EDS.*

– $(r, \perp) \leftarrow \mathsf{Query}_{\mathbf{C},\mathbf{S}}(\mathsf{tk}; \mathsf{EDS})$: *is an interactive protocol executed between a client $\mathbf{C}$ and a server $\mathbf{S}$. The client inputs a token tk and the server inputs an encrypted structure EDS. The client receives a response r and the server receives $\perp$.*

We refer the reader to, for example [1], for syntax definitions of dynamic STE.

Security. The standard notion of security for STE guarantees that: (1) an encrypted structure reveals no information about its underlying structure beyond the setup leakage $\mathcal{L}_\mathsf{S}$; (2) that the query protocol reveals no information about the structure and the queries beyond the query leakage $\mathcal{L}_\mathsf{Q}$. If this holds for non-adaptively chosen operations then the scheme is said to be non-adaptively secure. If, on the other hand, the operations can be chosen adaptively, the scheme is said to be adaptively-secure.

Definition 2 (Adaptive security of interactive STE). *Let $\Sigma = (\mathsf{Setup}, \mathsf{Query})$ be an interactive STE scheme and consider the following probabilistic experiments where $\mathcal{A}$ is a stateful semi-honest adversary, $\mathcal{S}$ is a stateful simulator, $\mathcal{L}_\mathsf{S}$ and $\mathcal{L}_\mathsf{Q}$ are leakage profiles and $z \in \{0,1\}^*$:*

Real$_{\Sigma,\mathcal{A}}(k)$*: given z the adversary $\mathcal{A}$ outputs a structure DS and receives EDS from the challenger, where $(K, \mathsf{EDS}) \leftarrow \mathsf{Setup}(1^k, \mathsf{DS})$. The adversary then* adaptively *chooses a polynomial number of queries and, for each, executes the* Query *protocol with the challenger, where the adversary plays the server and the challenger plays the client. Finally, $\mathcal{A}$ outputs a bit b that is output by the experiment.*

Ideal$_{\Sigma,\mathcal{A},\mathcal{S}}(k)$*: given z the adversary $\mathcal{A}$ generates a structure DS which it sends to the challenger. Given z and leakage $\mathcal{L}_\mathsf{S}(\mathsf{DS})$ from the challenger, the simulator $\mathcal{S}$ returns an encrypted structure EDS to $\mathcal{A}$. The adversary then* adaptively *chooses a polynomial number of queries and, for each one, executes the* Query *protocol with the simulator, where the adversary plays the server and the simulator plays the client (note that here, the simulator is allowed to deviate from* Query*). Finally, $\mathcal{A}$ outputs a bit b that is output by the experiment.*

We say that Σ is adaptively $(\mathcal{L}_\mathsf{S}, \mathcal{L}_\mathsf{Q})$-secure if there exists a PPT *simulator $\mathcal{S}$ such that for all* PPT *adversaries $\mathcal{A}$, for all $z \in \{0,1\}^*$,*

$$|\Pr[\mathbf{Real}_{\Sigma,\mathcal{A}}(k) = 1] - \Pr[\mathbf{Ideal}_{\Sigma,\mathcal{A},\mathcal{S}}(k) = 1]| \leq \mathsf{negl}(k).$$

Modeling leakage. Every STE scheme is associated with leakage which itself can be composed of multiple *leakage patterns*. The collection of all these leakage patterns forms the scheme's *leakage profile*. Leakage patterns are (families of) functions over the various spaces associated with the underlying data structure. For concreteness, we borrow the nomenclature introduced in [27] and recall some well-known leakage patterns that we make use of in this work:

Let $F : \{0,1\}^k \times \{0,1\}^* \to \{0,1\}^{\log \nu}$ be a pseudo-random function, $\mathsf{rank} : \mathbb{R}^n \to$ $\mathbb{R}^n$ be ranking function and $\lambda \in \mathbb{N}$ be a public parameter. Consider the transform PRT defined as follows:

- $\mathsf{PRT}(1^k, \lambda, \mathsf{MM})$:
    1. sample a key $K \xleftarrow{\$} \{0,1\}^k$;
    2. instantiate an empty multi-map MM';
    3. for all $\ell \in \mathbb{L}_{\mathsf{MM}}$,
        (a) let $\mathbf{r} := \mathsf{MM}[\ell]$ and $n_\ell = \#\mathbf{r}$;
        (b) compute $\mathbf{r}' := \mathsf{rank}(\mathbf{r})$;
        (c) let $n_\ell' = \lambda + F_K(\ell \| n_\ell)$;
        (d) if $n_\ell' > n_\ell$, set $\mathsf{MM}'[\ell] := (\mathbf{r}', \perp_1, \dots, \perp_{n_\ell' - n_\ell})$;
        (e) otherwise, set $\mathsf{MM}'[\ell] := (r_1', \cdots, r_{n_\ell'}')$;
    4. output MM'.
- $\mathsf{Get}(\ell, \mathsf{MM})$: output $\mathsf{MM}[\ell]$.

Fig. 1. The pseudo-random transform.

- the *query equality pattern* is the function family $\mathsf{qeq} = \{\mathsf{qeq}_{k,t}\}_{k,t \in \mathbb{N}}$ with $\mathsf{qeq}_{k,t} : \mathbb{D}_k \times \mathbb{Q}_k^t \to \{0,1\}^{t \times t}$ such that $\mathsf{qeq}_{k,t}(\mathsf{DS}, q_1, \dots, q_t) = M$, where M is a binary $t \times t$ matrix such that $M[i,j] = 1$ if $q_i = q_j$ and $M[i,j] = 0$ if $q_i \neq q_j$. The query equality pattern is referred to as the search pattern in the SSE literature;
- the *response identity pattern* is the function family $\mathsf{rid} = \{\mathsf{rid}_{k,t}\}_{k,t \in \mathbb{N}}$ with $\mathsf{rid}_{k,t} : \mathbb{D}_k \times \mathbb{Q}_k^t \to [2^{[n]}]^t$ such that $\mathsf{rid}_{k,t}(\mathsf{DS}, q_1, \dots, q_t) = (\mathsf{DS}[q_1], \dots, \mathsf{DS}[q_t])$. The response identity pattern is referred to as the access pattern in the SSE literature;
- the *response length pattern* is the function family $\mathsf{rlen} = \{\mathsf{rlen}_{k,t}\}_{k,t \in \mathbb{N}}$ with $\mathsf{rlen}_{k,t} : \mathbb{D}_k \times \mathbb{Q}_k^t \to \mathbb{N}^t$ such that $\mathsf{rlen}_{k,t}(\mathsf{DS}, q_1, \dots, q_t) = (|\mathsf{DS}[q_1]|, \dots, |\mathsf{DS}[q_t]|)$;
- the *domain size pattern* is the function family $\mathsf{dsize} = \{\mathsf{dsize}k, t\}_{k,t \in \mathbb{N}}$ with $\mathsf{dsize}k, t : \mathbb{D}_k \to \mathbb{N}$ such that $\mathsf{dsize}_{k,t}(\mathsf{DS}) = \#\mathbb{Q}$.
- the *total response length pattern* is the function family $\mathsf{trlen} = \{\mathsf{trlen}_k\}_{k \in \mathbb{N}}$ with $\mathsf{trlen}_k : \mathbb{D}_k \to \mathbb{N}$ such that $\mathsf{trlen}_k(\mathsf{DS}) = \sum_{q \in \mathbb{Q}_k} |\mathsf{DS}[q]|$;

4   The Pseudo-Random Transform

We describe the pseudo-random transform (PRT) in Fig. 1 and provide a high level description below.

Overview. PRT is a data structure transformation that takes as input a multi-map MM, a security parameter k and a public parameter λ. It first generates a random key K and initializes an empty multi-map MM'. For each label ℓ in the multi-map, it ranks the tuple $\mathbf{r} := \mathsf{MM}[\ell]$;[4] resulting in a ranked tuple $\mathbf{r}'$.

---

[4] The ranking function can be any ordering defined by the user; including standard ranking algorithms from information retrieval.

It then evaluates a PRF on the label ℓ concatenated to the length n_ℓ of $\mathbf{r}$. The output of this PRF evaluation is then added to λ in order to compute a new length n'_ℓ. There are two possible cases that can occur at this point: (1) if n'_ℓ is larger than n_ℓ, the ranked response is padded with dummies and inserted in $\mathsf{MM}'[\ell]$; (2) if n'_ℓ is at most n_ℓ, the ranked response is truncated to its first n'_ℓ elements and inserted in $\mathsf{MM}'[\ell]$. Note that for ease of exposition and without loss of generality, we consider the case where $n'_\ell = n_\ell$ a padding. Finally, the transform outputs the multi-map MM'. The get algorithm simply outputs the tuple corresponding to the label ℓ.

A note on probabilistic analysis. Throughout this work, we model pseudo-random functions as random functions for the purposes of probabilistic analysis. It should be understood that all our bounds will have an additional negligible value in the security parameter.

4.1 ꡃAnalyzing the Number of Truncations

For any label ℓ of the multi-map, the transform can pad or truncate its ranked response depending on the output of the PRF. In this Section, we will analyze the number of truncations induced by our transformation. The number of truncations is defined as

$$\#\{\ell \in \mathbb{L}_{\mathsf{MM}} : \#\mathsf{MM}'[\ell] < \#\mathsf{MM}[\ell]\}.$$

In the worst-case, the number of truncations can be $\#\mathbb{L}_{\mathsf{MM}}$ which occurs when every label in MM is truncated. We will show, however, that in practice this is very unlikely to occur. In particular, we will show that for real-world distributions of response lengths, the number of truncations is small with high probability. Note that if we set $\lambda \geq \max_{\ell \in \mathbb{L}} \#\mathsf{MM}[\ell]$, then truncations can never occur since $\#\mathsf{MM}'[\ell] \geq \max_{\ell \in \mathbb{L}} \#\mathsf{MM}[\ell] \geq \#\mathsf{MM}[\ell]$. We therefore only consider settings in which $\lambda < \max_{\ell \in \mathbb{L}} \#\mathsf{MM}[\ell]$.

Zipf-distributed multi-maps. To get a concrete bound on the number of truncations, we have to make an assumption on how the response lengths of the multi-map are distributed. Here, we will assume that they are distributed according to the Zipf distribution which is a standard assumption in information retrieval [14, 43]. We note that our analysis can be extended to any power-law distribution. More precisely, we say that a multi-map MM is $\mathcal{Z}_{a,b}$-distributed if its rth response has length

$$\frac{r^{-b}}{H_{a,b}} \cdot N$$

where $N \stackrel{def}{=} \sum_{\ell \in \mathbb{L}} \#\mathsf{MM}[\ell]$ is the volume of MM and $H_{a,b}$ is the harmonic number $\sum_{i=1}^{a} i^{-b}$. Throughout, we will consider multi-maps that are $\mathcal{Z}_{m,1}$-distributed where $m = \#\mathbb{L}_{\mathsf{MM}}$. From this assumption, it follows that the set of all response lengths is

$$L = (L_1, \ldots, L_m) = \left(\frac{N}{1 \cdot H_{m,1}}, \ldots, \frac{N}{m \cdot H_{m,1}} \right),$$

Note that we consider the case where $b = 1$ for ease of exposition but our analyses generalize to any b.

Theorem 1. *If MM is $\mathcal{Z}_{m,1}$-distributed, then with probability at least $1 - \varepsilon$ the number of truncations is at most*

$$m \cdot \left(\frac{N}{\nu \cdot H_{m,1}^2} \cdot \left(H_{\rho,2} - \frac{\lambda \cdot H_{m,1}}{N} \cdot H_{\rho,1} \right) + \sqrt{\frac{\ln(1/\varepsilon)}{2m}} \right),$$

where $\rho = \lfloor N/(\lambda \cdot H_{m,1}) \rfloor$.

Due to space limitations, the proof of Theorem 1 is in the full version of this work. Note that the worst case information loss (i.e., the total number of pairs lost due to truncations) can be computed as $\sum_{i \in [\sigma]}(N/(i \cdot H_{m,1})) - \lambda$, where σ is the number of truncations.

4.2 Analyzing the Storage Overhead

As detailed above, PRT can truncate or pad the responses in the multi-map. This has a direct impact on the storage overhead of the transformed multi-map since padding increases the storage overhead while truncations decrease it. In the following, we show that the size of the transformed multi-map MM$'$ can be upper bounded with high probability without any assumptions on the distribution of response lengths.

Theorem 2. *With probability at least $1 - \varepsilon$, the size of the transformed multi-map is at most*

$$m \cdot \left(\frac{\nu - 1}{2} + (\nu - 1) \cdot \sqrt{\frac{\ln(1/\varepsilon)}{2m}} + \lambda \right),$$

where $\lambda \geq 0$.

Due to space limitations, the proof of Theorem 2 is in the full version.

4.3 Concrete Parameters

In this Section, we will provide concrete parameters for PRT. Our goal is to find parameters that will provide a good balance between a small number of truncations and a small storage overhead. To study this, we first introduce two naïve transformations that achieve extreme tradeoffs between truncations and storage:

- the *naïve padding transform* is a transformation that pads the response of every label with dummies $\bot$ so that the length of the new responses are all set to the maximum response's length $\max_{\ell \in \mathbb{L}_{MM}} \#MM[\ell]$. Note that there are no truncations in this case and the size of the transformed multi-map is

$$S_{NV} \overset{def}{=} m \cdot \max_{\ell \in \mathbb{L}_{MM}} \#MM[\ell].$$

Let $\mathsf{STE_{EMM}} = (\mathsf{Setup}, \mathsf{Get})$ be a static multi-map encryption scheme and PRT be the pseudo-random transform. Consider the scheme $\mathsf{VLH} = (\mathsf{Setup}, \mathsf{Get})$ defined as follows:

- $\mathsf{Setup}(1^k, \lambda, \mathsf{MM})$:
  1. generate a PRT-transform of MM by computing $\mathsf{MM}' = \mathsf{PRT}(1^k, \lambda, \mathsf{MM})$;
  2. encrypt the transform by computing

$$(K, st, \mathsf{EMM}) \leftarrow \mathsf{STE_{EMM}}.\mathsf{Setup}(1^k, \mathsf{MM}');$$

  3. output (K, st, EMM).
- $\mathsf{Get_{C,S}}((K, st, \ell), \mathsf{EMM})$: $\mathbf{C}$ and $\mathbf{S}$ execute

$$(r, \bot) \leftarrow \mathsf{STE_{EMM}}.\mathsf{Get_{C,S}}((K, st, \ell), \mathsf{EMM}).$$

Fig. 2. VLH: A volume-hiding multi-map encryption scheme.

- the *naïve truncating transform* truncates the responses of every label to the minimum response length $\min_{\ell \in \mathbb{L}} \#\mathsf{MM}[\ell]$. Note that the number of truncations in this case is $T_{\mathsf{NV}} \overset{def}{=} \#\mathbb{L}_{\mathsf{MM}} = m$ and the storage overhead is $m \cdot \min_{\ell \in \mathbb{L}_{\mathsf{MM}}} \#\mathsf{MM}[\ell]$, which is optimal.

In the following Corollary, we set concrete values for λ and s so that we can achieve the best of both worlds. Specifically, we show that if the input multi-map is $\mathscr{Z}_{m,1}$-distributed, then by setting the output length of the PRF to $s = \log(L_1 + 1)$, where L_1 is the maximum response length, and setting $\lambda = O(\nu \cdot \alpha)$, where $1/2 < \alpha < 1$, then we can achieve storage overhead $\alpha \cdot S_{\mathsf{NV}}$ with $\beta \cdot T_{\mathsf{NV}}$ truncations with high probability, where β is a function of α and m.

Corollary 1. *Let $1/2 < \alpha < 1$. If* MM *is $\mathscr{Z}_{m,1}$-distributed and if*

$$\log \nu = \log(L_1 + 1) \quad \text{and} \quad \lambda = (\nu - 1) \cdot (2\alpha - 1)/4$$

then with probability at least:

- $1 - \exp(-m \cdot (2\alpha - 1)^2/8)$, *the total volume of the transformed multi-map is at most $\alpha \cdot S_{\mathsf{NV}}$.*
- $1 - \exp(-2m/\log^2 m)$, *the number of truncations is at most*

$$\frac{1}{\log m} \cdot H_{\lfloor \frac{4}{2\alpha - 1} \rfloor, 2} \cdot T_{\mathsf{NV}}.$$

Due to space limitations, the proof of Corollary 1 is in the full version.

5   A Volume-Hiding Multi-map Encryption Scheme

In this Section, we use the PRT to construct a volume-hiding multi-map encryption scheme. Our construction is described in detail in Fig. 2 and works as follows.

Overview. The construction, VLH = (Setup, Get), makes black-box use of an underlying multi-map encryption scheme $\mathsf{STE_{MM}}$ = (Setup, Get). VLH.Setup takes as input a security parameter k, a public parameter λ and a multi-map MM. It applies the PRT transform on MM which results in a new multi-map MM'. It then encrypts MM with $\mathsf{STE_{MM}}$, resulting in an encrypted multi-map EMM, a state st and a key K which it returns as its own output. To execute a Get query ℓ on EMM, the client and the server execute $\mathsf{STE_{MM}}$.Get on ℓ and EMM.

Efficiency. Assuming that $\mathsf{STE_{MM}}$ is an optimal-time multi-map encryption scheme [1,7,10,15], the get complexity of VLH is $O(\lambda + n'_\ell)$, where $n'_\ell \in \{0, \cdots, \nu - 1\}$. Therefore, the worst-case complexity is $O(\lambda + \nu)$ while the best-case complexity is $O(\lambda)$. The expected complexity is $O(\lambda + 2^{s-1})$.

The storage overhead of VLH is

$$O(N) = O\left(\sum_{\ell \in \mathbb{L}_{MM}} \#\mathsf{MM}'[\ell] \right) = O\left(\lambda \cdot m + \sum_{\ell \in \mathbb{L}_{MM}} n'_\ell \right),$$

where, again, $n'_\ell \in \{0, \cdots, \nu - 1\}$. So based on Corollary 1, when $\lambda = (\nu - 1) \cdot (2\alpha - 1)/4$ and $1/2 < \alpha < 1$, the storage overhead of VLH is

$$O\big(\alpha \cdot (\nu - 1) \cdot m\big)$$

with high probability.

Correctness. The correctness of VLH is affected by the number of truncations induced by PRT. Based on Corollary 1, we can show that the number of truncations performed by VLH is at most $O(m/\log m)$ under the same assumptions stated in the corollary.

Security. We now describe the leakage profile of VLH assuming $\mathsf{STE_{MM}}$ is instantiated with one of the standard optimal-time multi-map encryption schemes [1,7,10,15] all of which have leakage profile

$$\Lambda_{MM} = (\mathcal{L}_S, \mathcal{L}_Q) = \big(\mathsf{trlen}, (\mathsf{qeq}, \mathsf{rlen})\big).$$

Theorem 3. *If* $\mathsf{STE_{EMM}}$ *is a* $(\mathsf{trlen}, (\mathsf{qeq}, \mathsf{rlen}))$-*secure multi-map encryption scheme and* F *in* PRT *is a pseudo-random function, then* VLH *is a* $(\mathsf{dsize}, \mathsf{qeq})$-*secure multi-map encryption scheme.*

Due to space limitations, the proof of Theorem 3 is in the full version of this work. We observe that if we consider λ to be a public parameter, then dsize will leak an approximation of m rather than the exact value of m as stated in the theorem since $\lambda^{-1} \cdot \mathsf{trlen} = \lambda^{-1} \cdot \sum_{i \in [m]} \#\mathsf{MM}[\ell] = m + \lambda^{-1} \cdot \sum_{i \in [m]} r_i$, where r_i is generated uniformly at random. Note that this differs slightly from standard EMM schemes as their setup leakage is usually the sum of all pairs in the multi-map.

6   DST: The Densest Subgraph Transform

In this section, we introduce a new data structure transformation called the *dense subgraph transform* (DST). Unlike the PRT which achieves efficient storage by increasing truncations (and therefore losing information), this new transform improves on the storage complexity of PRT without losing any information. The transformation is randomized and, surprisingly, we can show that, with high probability, it incurs no asymptotic storage overhead. Furthermore, we can also show that, for the case of Zipf-distributed multi-maps, it produces multi-maps that are asymptotically-smaller than the naïve padding and truncating transforms described in Sect. 4.3.

The DST takes a multi-map MM as input and creates a new multi-map MM' that is volume-hiding. This new structure results from re-arranging the data in the input multi-map according to a random bi-partite graph. To ensure this re-arrangement is still efficiently queryable, we represent it using a pair of standard data structures which include a multi-map MM_G and a dictionary DX. As we will show, the storage complexity of the final representation depends on certain properties of the bi-partite graph which are, in turn, inherited from the original multi-map.

Below, we provide a high-level overview of our transformation. A more detailed description is given in Sect. 6.1. The overview is divided in two parts: (1) a variant for general multi-maps; and (2) a variant for what we refer to as *concentrated* multi-maps. Note that the transformation handles both cases but achieves better results for the later. We then provide a more detailed description in Fig. 6.

General multi-maps. Given a multi-map MM we begin creating a bi-partite graph with $\mathbb{L}_{MM}$ as the top vertices and a set of n empty bins as the bottom vertices. For each label/vertex ℓ in V_{top}, we randomly select t bins and insert in each bin a single value of the tuple $MM[\ell]$. Here, t is the maximum tuple size in the multi-map. If $\#MM[\ell] < t$, then some of the selected bins won't receive a value. At the end of this process, we pad all bins so that they all have the same size. Note that this process creates a bi-partite graph where the edges incident to some top vertex/label ℓ correspond to the bins selected for that label/vertex. We now create two data structures to represent and efficiently process this bi-partite graph. The first is a dictionary that maps bin identifiers to the bin's contents. The second is a multi-map that maps a label to the identifiers of the bins associated to it. To retrieve the values associated to a given label ℓ, we query the multi-map on ℓ to retrieve its t bin identifiers and then query the dictionary on each of the t bin identifier to retrieve the contents of the bins.

It is already clear from this high-level description that all labels will have exactly the same response length: $t \cdot \alpha$, where α is the maximum size of a bin. It can be shown that with the right choice of parameters, this transformation results in a small amount of padding compared to the naïve approach.

Concentrated multi-maps. The storage overhead of our approach can be greatly improved when the multi-map satisfies a certain property we refer to as *concentration.* At a high level, a multi-map is concentrated if there exists a large number of values that appear in the tuples of a large number of labels. More formally, we define this property as follows.

Definition 3. (Concentrated multi-maps). *Let $\mu, \nu > 0$. We say that a multi-map MM is (μ, ν)-concentrated if there exists a set of μ labels $\ell_1, \cdots, \ell_\mu \in \mathbb{L}_{MM}$ such that,*

$$\# \bigcap_{i=1}^{\mu} MM[\ell_i] = \nu.$$

We refer to this set of labels as MM*'s concentrated component and denote it* $\widehat{\mathbb{L}}_{MM}$. *Throughout, we will assume the existence of an efficient algorithm* FindComp *that takes as input a multi-map and outputs the multi-map's concentrated component* $\widehat{\mathbb{L}}_{MM}$. *If no such component exists, the algorithm outputs* $\perp$.

A (μ, ν)-concentrated multi-map has μ labels with an intersection of size ν which means that there is some redundancy in the structure. Unfortunately, the previous approach does not take advantage of this since it stores all the values in the multi-map independently. To exploit this redundancy, we proceed as follows. We dedicate a random subset of the bins to store the tuple values of the multi-map's concentrated component. Because the component's tuples have a large intersection, we will avoid storing the same values over and over again. At a high-level, we modify the process as follows. We first choose a random subset of ν bins and store, in each one, one value from the intersection $\cap_{\ell \in \widehat{\mathbb{L}}_{MM}} MM[\ell]$. We then add an edge between a random subset of size τ of these bins and the labels/vertices in the concentrated component. This results in μ labels/vertices sharing a large portion of the bins. In the special case of $\tau = \nu$, then this will result in μ labels/vertices that share the same bins. Notice the improved storage overhead as we don't store the values in the intersection in multiple bins. For the remaining labels, we follow a similar process to the one presented in the generic case. We sometimes refer to the value $(\mu - \frac{\nu}{\tau}) \cdot \tau$ as the multi-map's concentration, for $\tau > 0$.

Finding the concentration component. Our DST transform relies on an efficient algorithm FindComp to find the concentration component of a multi-map. We now describe such an algorithm. Informally, this algorithm will try different combinations of labels, compute the intersection of their tuples, and only retain the combination for which the intersection was the highest and that verify some specific conditions on its size. The algorithm first determines the set of labels $\widetilde{\mathbb{L}}_{MM}$ with tuples of size $\Omega(n^{0.5+\delta})$, for some positive n. For $i \in [\lambda]$, it selects μ labels uniformly at random with replacement from $\widetilde{\mathbb{L}}_{MM}$. We refer to this set as $\overline{\mathbb{L}}_{MM}^i$. The algorithm then computes $\nu_i = \# \left(\cap_{\ell \in \overline{\mathbb{L}}_{MM}^i} MM[\ell] \right)$ where λ is the number of times the random selection is computed. The algorithm finally determines $\rho = \text{argmax}_{i \in [\lambda]} \{\nu_i : \nu_i \in \Omega(n^{0.5+\delta})\}$ and outputs $\widehat{\mathbb{L}}_{MM} = \overline{\mathbb{L}}_{MM}^\rho$ if such ρ exists and $\widehat{\mathbb{L}}_{MM} = \perp$ otherwise. Notice that the algorithm runs in $O(\lambda \cdot \mu)$ time. So it is sufficient to choose λ and μ to be polynomial in m. In our setting, we need to set the parameters to align with the the densest subgraph assumption (described in Definition 5) so we need $\mu = \Omega(m^{0.5+\delta})$ and for some positive δ. We note that this algorithm is only an example and we believe that more efficient algorithms can be designed.

A storage optimization. Notice that the auxiliary multi-map MM_G associates to every label a randomly selected set of t bins. In particular this means that the size of MM_G is $O(m \cdot t)$ which could be rather large. Fortunately, storing the identifiers of each bin is not necessary. Instead, we can choose the bins to assign to a label using a pseudo-random function and store the key in MM_G. This will reduce the size of MM_G to $O(m)$.

6.1   Detailed Description

We now provide a detailed description of the DST. The pseudo-code is in Figs. 4 and 5. The transform makes black-box use of three pseudo-random functions F, H and G.

Setup. The Setup algorithm takes as input a security parameter 1^k, two integers n and τ and a multi-map MM. It instantiates a bi-partite graph $G = (V_{\mathsf{top}}, V_{\mathsf{bot}}, E)$ where the top vertices $V_{\mathsf{top}} = \mathbb{L}_{\mathsf{MM}}$ are the labels in MM, the bottom vertices V_{bot} are n empty bins denoted $\mathbf{B} = \{B_1, \ldots, B_n\}$ and the set of edges E is empty.

The set of edges are generated as follows. Setup first computes the concentrated component of the multi-map $\widehat{\mathbb{L}}_{\mathsf{MM}} := \mathsf{FindComp}(\mathsf{MM})$. If no concentrated component exists, $\mathsf{FindComp}$ outputs $\bot$. If $\#\widehat{\mathbb{L}}_{\mathsf{MM}} \neq \bot$, it then pseudo-randomly chooses ν bins $\mathbf{B}' = \{B'_1, \ldots, B'_\nu\}$, where $\nu = \#\left(\bigcap_{\ell \in \widehat{\mathbb{L}}_{\mathsf{MM}}} \mathsf{MM}[\ell]\right)$. More precisely, it samples a k-bit value $\mathsf{rand}^\star$ uniformly at random and chooses the bins indexed by the set

$$\left\{ F_{K_1}(\mathsf{rand}^\star \| 1), \ldots, F_{K_1}(\mathsf{rand}^\star \| \nu) \right\}.$$

Note that all these ν positions have to be *distinct*. If not, then it keeps resampling a new k-bit value $\mathsf{rand}^\star$ uniformly at random until no collisions are found. Note however that the probability p that no collision occurs, modeling F as a random function, is equal to

$$p = \prod_{i=0}^{\nu-1} \left(1 - \frac{i}{n}\right) \geq \left(1 - \frac{\nu}{n}\right)^\nu \approx e^{-\nu^2/n},$$

which tends to 1 when $\nu = o(n)$–which aligns with the concrete parameters that we will detail in Sect. 6.4.

For all $\ell \in \widehat{\mathbb{L}}_{\mathsf{MM}}$, it: (1) adds an edge between ℓ and $t-\tau$ bins outside of $\mathbf{B}'$; and (2) adds an edge between ℓ and τ bins in $\mathbf{B}'$. Note that this separation between the labels is necessarily for our reduction to the densest subgraph problem to hold.

To do the former, it indexes the bins in $\mathbf{B} \setminus \mathbf{B}'$ from 1 to $n - \tau$, samples a k-bit value $\mathsf{rand}_{\ell,1}$ uniformly at random, and chooses the bins indexed by the set

$$\left\{ H_{K_2}(\mathsf{rand}_{\ell,1}\|1) + \mathsf{slide}_1, \ldots, H_{K_2}(\mathsf{rand}_{\ell,1}\|t - \tau) + \mathsf{slide}_{t-\tau} \right\},$$

where slide_i, for $i \in \{1, \cdots, t - \tau\}$, is an integer used to deterministically map back the smaller output of H in $[n - \tau]$ to the corresponding bin identifier in $[n]$ and is computed as follows. First, it orders the set of bins in $\mathbf{B}'$ in a numerical order such that

$$\mathbf{B}' = \left(B_{\mathsf{pos}_1}, \cdots, B_{\mathsf{pos}_\nu} \right),$$

where $\mathsf{pos}_i < \mathsf{pos}_j$, for $i, j \in [\nu]$. Then it defines the following quantities based on which the slide value is determined– refer to Fig. 3 for an illustration of the computation,

$$\mathsf{gap}_i = \begin{cases} [1, \mathsf{pos}_i - 1] & \text{if } i = 1 \\ ]\mathsf{pos}_{i-1} - (i-1), \mathsf{pos}_i - (i-1)[& \text{if } i \in \{2, \cdots, \nu\} \\ ]\mathsf{pos}_{i-1} - (i-1), n - \nu] & \text{if } i = \nu + 1 \end{cases}$$

Then, for $i \in \{1, \cdots, n - \nu\}$, identify $j \in \{1, \cdots, \nu + 1\}$ such that $H_{K_2}(\mathsf{rand}_{\ell,1}\|i) \in \mathsf{gap}_j$, then set $\mathsf{slide}_i = j - 1$.

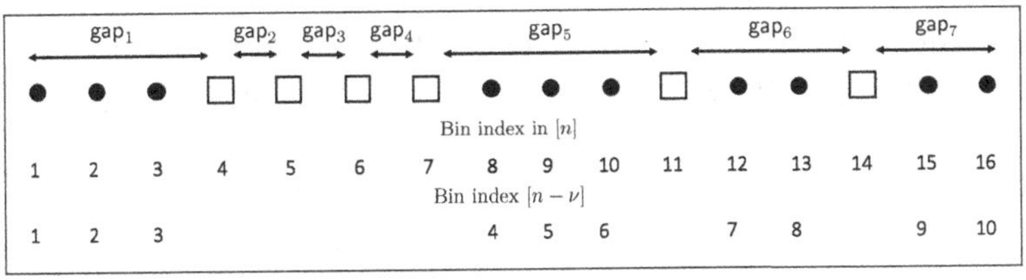

Fig. 3. Gaps computation for $n = 16$ and $\nu = 6$. $\square$ denotes bins being part of $\mathbf{B}' = \{B_4, B_5, B_6, B_7, B_{11}, B_{14}\}$ while $\bullet$ denotes bins in $\mathbf{B} \setminus \mathbf{B}'$.

Note that $\mathsf{rand}_{\ell,1}$ has also to be chosen in such a way that the selected $t - \tau$ positions are distinct. If not, similarly to above, it resamples a new k-bit value uniformly at random until no collision occurs. The probability that no collision occurs is approximately equal to $e^{-(t-\tau)^2/(n-\nu)}$ which tends to 1 when $t = o(n)$– which aligns with our concrete parameterization as we are going to detail in Sect. 6.4.

To do the latter, it samples $\mathsf{rand}_{\ell,2}$ uniformly at random and adds an edge between ℓ and all bins indexed by

$$\left\{ j_{G_{K_3}}(\mathsf{rand}_{\ell,2}\|1), \cdots, j_{G_{K_3}}(\mathsf{rand}_{\ell,2}\|\tau) \right\},$$

where $j_i = F_{K_1}(\text{rand}^\star \| i)$, for $i \in [\nu]$. If a collision is found, then it keeps resampling a new k-bit value uniformly at random until all τ positions are distinct. The probability that no collision occurs is approximately $e^{-\tau^2/\nu}$ which tends to 1 given our parametrization.

For each $\ell \notin \widehat{\mathbb{L}}_{\mathsf{MM}}$, it samples a k-bit value rand_ℓ uniformly at random and adds an edge between ℓ and all bins indexed by the set

$$\left\{ F_{K_1}(\text{rand}_\ell \| 1), \ldots, F_{K_1}(\text{rand}_\ell \| t) \right\}.$$

Again, if a collision is found, it keeps resampling a new k-bit value uniformly at random until all positions are distinct. The probability that no collision occurs is approximately equal to $e^{-t^2/n}$. Notice that at the end of this process, each vertex has degree exactly t.

Now Setup will use the graph to load the bins in V_{bot} as follows. For each $\ell \notin \widehat{\mathbb{L}}_{\mathsf{MM}}$, it stores one value from the tuple $\mathsf{MM}[\ell]$ in one of the bins that are incident to ℓ. When inserting into a bin, the algorithm concatenates each value with ℓ (this will be helpful at query time). If $\#\mathsf{MM}[\ell] < t$, then some of the incident bins will not receive any value. For all $\ell \in \widehat{\mathbb{L}}_{\mathsf{MM}}$, it stores one element from $\mathsf{MM}[\ell] \backslash \mathbf{r}_\ell$ in the bins from $\mathbf{B} \backslash \mathbf{B}'$ that are incident to ℓ—again concatenating each value with ℓ, where

$$\mathbf{r}_\ell = (r_1, \cdots, r_\tau) \subseteq \bigcap_{\ell \in \widehat{\mathbb{L}}_{\mathsf{MM}}} \mathsf{MM}[\ell]$$

Also if $\#\mathsf{MM}[\ell] < t$, then some of the incident bins will not receive any value. Finally, it stores each value from the set

$$\mathbf{r}' := \bigcap_{\ell \in \widehat{\mathbb{L}}_{\mathsf{MM}}} \mathsf{MM}[\ell]$$

in a distinct bin in $\mathbf{B}'$ in such a way that every bin in $\mathbf{B}'$ will contain one value in $\mathbf{r}'$. Here, the algorithm concatenates the values with $\star$. The algorithm then pads all the bins to have the same size.

Finally, it creates a dictionary DX and a multi-map MM_G. The dictionary maps bin identifiers to bin contents. The multi-map MM_G maps labels $\ell \notin \widehat{\mathbb{L}}_{\mathsf{MM}}$ to rand_ℓ and labels $\ell \in \widehat{\mathbb{L}}_{\mathsf{MM}}$ to $(\text{rand}_{\ell,1}, \text{rand}_{\ell,2}, \text{rand}^\star)$. It outputs $\mathsf{MM}' = (\mathsf{MM}_G, \mathsf{DX})$.

The storage complexity of MM' is $O(m + n \cdot \lambda)$, where λ is the maximum load of a bin.

Get. Get operations on $\mathsf{MM}' = (\mathsf{MM}_G, \mathsf{DX})$ work as follows. Given a label ℓ, we first query MM_G on ℓ. If $\ell \notin \widehat{\mathbb{L}}_{\mathsf{MM}}$, then MM_G returns rand_ℓ from which we compute the bin identifiers $\{F_{K_1}(\text{rand}_\ell \| i)\}_{i=1}^t$. We can then query DX on the bin identifiers to recover the bins and output the elements concatenated with ℓ. If $\ell \in \widehat{\mathbb{L}}_{\mathsf{MM}}$, MM_G returns a triple $(\text{rand}_{\ell,1}, \text{rand}_{\ell,2}, \text{rand}^\star)$ from which we can compute the sets

Let $n \in \mathbb{N}$ be a public parameter, $F : \{0,1\}^k \times \{0,1\}^* \to [n]$, $G : \{0,1\}^k \times \{0,1\}^* \to [n']$ and $H : \{0,1\}^k \times \{0,1\}^* \to [n'']$ be two pseudo-random functions with $n' < n'' < n$. Consider the transform DST defined as follows:

- DST(1^k, param, MM):
  1. parse param as (n, τ), instantiate an empty dictionary DX, an empty multi-map MM_G, and a bi-partite graph $G = ((\mathbb{L}_{MM}, \mathbf{B}), E)$ where $\mathbf{B} = (B_1, \cdots, B_n)$ and $E = \emptyset$;
  2. compute $\widehat{\mathbb{L}}_{MM} \leftarrow \mathsf{FindComp}(MM)$, set $\nu = \#(\bigcap_{\ell \in \widehat{\mathbb{L}}_{MM}} MM[\ell])$ and $t := \max_{\ell \in \mathbb{L}_{MM}} MM[\ell]$;
  3. sample three keys $K_1 \xleftarrow{\$} \{0,1\}^k$, $K_2 \xleftarrow{\$} \{0,1\}^k$ and $K_3 \xleftarrow{\$} \{0,1\}^k$;
  4. for all $\ell \in \mathbb{L}_{MM} \setminus \widehat{\mathbb{L}}_{MM}$,
     (a) sample $\mathsf{rand}_\ell \xleftarrow{\$} \{0,1\}^k$ and output
     $$(i_1, \cdots, i_t) := \left\{ F_{K_1}(\mathsf{rand}_\ell \| 1), \ldots, F_{K_1}(\mathsf{rand}_\ell \| t) \right\},$$
     if there exist distinct $i, j \in [t]$ for which $i_i = i_j$ redo the sampling. Add to E
     $$\left\{ (\ell, i_j) : j \in [t] \right\};$$
     (b) parse $MM[\ell]$ as $(r_1, \cdots, r_{n_\ell})$ and put $r_j \| \ell$ in B_{i_j} for all $j \in [n_\ell]$;
  5. if $\widehat{\mathbb{L}}_{MM} \neq \perp$, sample $\mathsf{rand}^\star \xleftarrow{\$} \{0,1\}^k$ and set $\mathbf{B}' = (B_{i_1}, \cdots, B_{i_\nu})$ where
     $$(i_1, \cdots, i_\nu) := \left\{ F_{K_1}(\mathsf{rand}^\star \| 1), \ldots, F_{K_1}(\mathsf{rand}^\star \| \nu) \right\},$$
     if there exist distinct $i, j \in [\tau]$ for which $i_i = i_j$ redo the sampling. Otherwise set $\mathbf{B}' = \perp$;
  6. compute
     $$\mathbf{r}' := \bigcap_{\ell \in \widehat{\mathbb{L}}_{MM}} MM[\ell] = (r'_1, \cdots, r'_\nu);$$
  7. put $r'_j \| \star$ in B_j for all $j \in [\nu]$ and $B_j \in \mathbf{B}'$;
  8. for all $\ell \in \widehat{\mathbb{L}}_{MM}$,
     (a) sample $\mathsf{rand}_{\ell,1} \xleftarrow{\$} \{0,1\}^k$ and output
     $$(i_1, \cdots, i_{t-\tau}) := \left\{ H_{K_2}(\mathsf{rand}_{\ell,1} \| 1), \ldots, H_{K_2}(\mathsf{rand}_{\ell,1} \| t - \tau) \right\},$$
     if there exist distinct $i, j \in [t-\tau]$ for which $i_i = i_j$, redo the sampling. Add to E
     $$\left\{ (\ell, i_j + \mathsf{slide}_j) : j \in [t - \tau] \right\};$$
     where slide_j is computed as follows
       i. order $\mathbf{B}'$ in a numerical order such that $\mathbf{B}' := (B_{\mathsf{pos}_1}, \cdots, B_{\mathsf{pos}_\nu})$;
       ii. if $i_j \in [1, \mathsf{pos}_1]$, set $\mathsf{slide}_j = 0$;
       iii. if $i_j \in]\mathsf{pos}_{i-1} - (i-1), \mathsf{pos}_i - (i-1)[$, set $\mathsf{slide}_j = i - 1$, for any $i \in \{2, \cdots, \nu\}$;
       iv. if $i_j \in]\mathsf{pos}_\nu - \nu, n - \nu[$, set $\mathsf{slide}_j = \nu$;

Fig. 4. DST: The Dense Subgraph Transform (Part 1).

- DST(1^k, param, MM):
    8. for all $\ell \in \widehat{\mathbb{L}}_{\mathsf{MM}}$,
       (b) sample $\mathsf{rand}_{\ell,2} \xleftarrow{\$} \{0,1\}^k$ and set $\mathbf{r}_\ell = (r_{i_1}, \cdots, r_{i_\tau}) \subseteq \mathbf{r'}$ where

$$(i_1, \cdots, i_\tau) := \left\{ G_{K_3}(\mathsf{rand}_{\ell,2}\|1), \ldots, G_{K_3}(\mathsf{rand}_{\ell,2}\|\tau) \right\},$$

       if there exist distinct $i, j \in [\tau]$ for which $i_i = i_j$, redo the sampling. Add

$$\left\{ (\ell, F_{K_1}(\mathsf{rand}^\star\|i_j) : j \in [\tau] \right\}$$

       to E;
       (c) parse MM[ℓ] as $(r_1, \cdots, r_{n_\ell})$;
       (d) for all $r_j \in \mathsf{MM}[\ell] \setminus \mathbf{r}_\ell$, then put $r_j\|\ell$ in $B_{i_j+\mathsf{slide}_j}$;
    9. set $\theta = \max_{i\in[n]} \#B_i$ and set for all $i \in [n]$

$$B_i = (B_i, \bot_1, \cdots, \bot_{\theta-\#B_i});$$

    10. for all $i \in [n]$, set DX[i] = B_i;
    11. for all $\ell \in \mathbb{L}_{\mathsf{MM}}$, if $\ell \in \widehat{\mathbb{L}}_{\mathsf{MM}}$ set MM[ℓ] := $(\mathsf{rand}_{\ell,1}, \mathsf{rand}_{\ell,2}, \mathsf{rand}^\star)$, otherwise set MM[$\ell$] := rand_ℓ;
    12. output the key $K = (K_1, K_2, K_3)$ and $\mathsf{MM'} = (\mathsf{DX}, \mathsf{MM}_G)$.
- Get(K, ℓ, MM):
    1. parse K as (K_1, K_2, K_3) and MM as $(\mathsf{DX}, \mathsf{MM}_G)$ and instantiate an empty set Result;
    2. if $\mathsf{MM}_G[\ell] = \mathsf{rand}$, then
       (a) add DX[ℓ_i] to Result, where for all $i \in [t]$,

$$\ell_i := F_{K_1}(\mathsf{rand}\|i);$$

       (b) keep all values of the form $\cdot\|\ell$;
    3. if $\mathsf{MM}_G[\ell] = (\mathsf{rand}_1, \mathsf{rand}_2, \mathsf{rand}^\star)$, then
       (a) add DX[ℓ_i] to Result, where for all $i \in [t - \tau]$,

$$\ell_i := H_{K_2}(\mathsf{rand}_1\|i) + \mathsf{slide}_i,$$

       and for all $i \in [\tau]$,

$$\ell_i := j_{G_{K_3}(\mathsf{rand}_2\|i)}$$

       where $(j_1, \cdots, j_\nu) = F_{K_1}(\mathsf{rand}^\star\|1), \ldots, F_{K_1}(\mathsf{rand}^\star\|\nu)$;
       (b) keep all values of the form $\cdot\|\ell$ or $\cdot\|\star$;
    4. output Result.

Fig. 5. DST: The Dense Subgraph Transform (Part 2).

$$\left\{ H_{K_2}(\mathsf{rand}_{\ell,1}\|i) + \mathsf{slide}_i \right\}_{i=1}^{t-\tau} \quad \text{and} \quad \left\{ j_{G_{K_3}(\mathsf{rand}_{\ell,2}\|i)} \right\}_{i=1}^{\tau}$$

where $j_i = F_{K_1}(\mathsf{rand}^\star\|i)$, for $i \in [\nu]$, which we, in turn, use to query DX and recover the bins. From these bins the algorithm recovers the elements concatenated with ℓ and $\star$. The complexity of gets is $O(t \cdot \lambda)$ where, again, λ is the maximum load of a bin.

6.2 Analyzing the Load of a Bin

As seen in the previous Section, an important quantity to evaluate the query and storage efficiency of our transformation is the maximum load of a bin. In this Section, we will show that, with high probability, the maximum load can be upper bounded by $(N - N_{ds})/n$ where N_{ds} is the size of the concentrated component and n is the number of bins. Before stating our result, we recall a generalization of Chernoff's inequality for the binomial distribution.

Lemma 1. *Let* $X_1, \ldots, X_m$ *be independent random variables over* $\{0, 1\}$ *such that* $\Pr[X_i = 1] = p_i$ *and* $\Pr[X_i = 0] = 1 - p_i$. *If* $X = X_1 + \cdots + X_m$, *then*

$$\Pr[X \geq E[X] + \theta] \leq \exp\left(-\frac{\theta^2}{2(E[X] + \theta/3)}\right).$$

Theorem 4. *With probability at least* $1 - \varepsilon$, *the maximum load of a bin is at most*

$$\frac{N - N_{ds}}{n} + \frac{\ln(1/\varepsilon)}{3}\left(1 + \sqrt{1 + \frac{18(N - N_{ds})}{n \cdot \ln(1/\varepsilon)}}\right),$$

where $N_{ds} = (\mu - \frac{\nu}{\tau}) \cdot \tau$, *for* $\tau > 0$.

Due to space limitations, the proof of Theorem 4 is in the full version.

6.3 Query and Storage Efficiency

We now give the storage and query efficiency of the DST transform.

Storage efficiency. The output of DST consists of a multi-map MM_G and a dictionary DX. The multi-map MM_G has tuples of size 1 or 3 depending on the label. That is, the size of the multi-map is upper bounded by $2m$. The dictionary DX stores the content of the padded bins. From Theorem 4 and the union bound, we have that the size of the dictionary is at most

$$N - N_{ds} + \frac{n \cdot \ln(1/\varepsilon)}{3} \cdot \left(1 + \sqrt{1 + \frac{18(N - N_{ds})}{n \cdot \ln(1/\varepsilon)}}\right)$$

with probability $1 - n \cdot \varepsilon$.

Get efficiency. The Get algorithm first retrieves either a random value or a pair of random values from MM_G. In the former case, t PRF evaluations are computed and t bins are retrieved. In the later case, $2t + \nu$ PRF evaluations are computed (using F, H and G) and t bins are retrieved.[5] Assuming that both MM_G and DX are data structures with optimal query complexity, the Get query complexity is at most

---

[5] Note that the computation of the $slide_i$'s is $O(\nu)$. These evaluations can be performed once and stored at the client which reduces the total PRF evaluations at query time to $2t$.

$$t \cdot \frac{N - N_{\mathsf{ds}}}{n} + \frac{t \cdot \ln(1/\varepsilon)}{3} \cdot \left(1 + \sqrt{1 + \frac{18(N - N_{\mathsf{ds}})}{n \cdot \ln(1/\varepsilon)}}\right)$$

with probability $1 - n \cdot \varepsilon$.

6.4   Concrete Parameters

In this Section, we propose concrete parameters for the DST. In particular, we will be interested in parameters that guarantee better storage overhead than the naïve padding transform. Note that we do not compare to the naïve truncating approach since the DST does not lose any information.

General multi-maps. Recall that the naïve padding transform has a storage overhead

$$S_{\mathsf{NV}} \stackrel{def}{=} m \cdot \max_{\ell \in \mathbb{L}_{\mathsf{MM}}} \#\mathsf{MM}[\ell] = \Omega(N),$$

where $N \stackrel{def}{=} \sum_{\ell \in \mathbb{L}} \#\mathsf{MM}[\ell]$. From Theorem 4, we have the following corollary.

Corollary 2. *Let $n \geq 1$ and $m \geq 0$. If $N > n \log n$, then with probability at least $1 - 1/e^{N/5n}$, the size of the resulting multi-map is at most $O(N)$.*

Notice that if the original multi-map is $\mathcal{Z}_{m,1}$-distributed (but not necessarily concentrated), then $S_{\mathsf{NV}} = N \cdot m/H_{m,1}$ where $H_{m,1} = \Theta(\log m)$ is the harmonic number (please refer to Sect. 4). It follows that, in this case, $N = o(S_{\mathsf{NV}})$ so the storage overhead of DST is *small-o* of the overhead of the naïve padding transform.

Concentrated multi-maps. We now consider a multi-map MM with a concentrated component of size $(\mu - \frac{\nu}{\tau}) \cdot \tau$. We show below that in this case, the storage overhead induced by DST can be considerably smaller than the storage overhead of the naïve padding transform. The following Corollary is a consequence of Theorem 4.

Corollary 3. *Let $n \geq 1$ and $m \geq 0$. If $N > n \log n$,*

$$\mu = O\left(m^{0.5+\delta} \cdot polylog(m)\right) \quad and \quad \tau = O\left(polylog(m)\right),$$

for some $\delta \geq 0$, then with probability at least $1 - 1/e^{N/5n}$, the size of the resulting multi-map is at most

$$O\left(N - m^{0.5+\delta} \cdot polylog(m)\right).$$

As above, if the original multi-map MM is $\mathcal{Z}_{m,1}$-distributed, then the storage overhead of DST is *small-o* of the overhead of the naïve padding transform.

A remark on security. As we will see in Sect. 7, the parameters μ and τ have to satisfy certain constraints for our multi-map encryption scheme to be secure. In particular, the parameters have to be chosen in such a way that they verify the densest subgraph assumption which we detail in Definition 5. We note here that to satisfy both this assumption and the constraints of Corollary 3, it is sufficient that for some positive δ,

$$\mu = O\left(m^{0.5+\delta} \cdot \mathrm{polylog}(m)\right), \quad \tau = O(t) = O\left(\mathrm{polylog}(m)\right)$$

and,

$$\nu = O\left(m^{0.5+\delta} \cdot \mathrm{polylog}(m)\right), \quad n = \Theta\left(m \cdot \mathrm{polylog}(m)\right).$$

Note that this is only an example and is not the only choice of parameters that can be used. The intuition is that the larger the multi-map's concentration is, the better storage overhead DST will achieve. More precisely, multi-maps with larger values of μ and τ will achieve better storage gain as long as the DSP problem is hard.

7   AVLH: Advanced Volume Hiding Multi-map Encryption Scheme

In this Section, we use the DST to construct a volume-hiding multi-map encryption scheme. Our construction is described in detail in Fig. 6 and works as follows.

Overview. The construction, AVLH = (Setup, Get), makes black-box use of an underlying response-hiding dictionary encryption scheme $\mathsf{STE}_{\mathsf{DX}}^{\mathsf{RH}}$ = (Setup, Get). AVLH.Setup takes as input a security parameter k, a public parameter param, and a multi-map MM. It first applies the DST transform on MM which results in a key $K_1 = (K_{1,1}, K_{1,2}, K_{1,3})$ and two structures: a multi-map $\mathsf{MM_G}$ and a dictionary DX. It then encrypts the dictionary DX, resulting in an encrypted dictionary EDX, a state st_{DX} and a key K_2. It finally outputs a key $K = (K_1, K_2)$, a state $st = (\mathsf{MM_G}, st_{\mathsf{DX}})$ and an encrypted multi-map EMM = EDX. To execute AVLH.Get, the client differentiates two cases: if $\mathsf{MM_G}[\ell]$ is a tuple composed of a single value rand, then the client and server execute the $\mathsf{STE}_{\mathsf{DX}}^{\mathsf{RH}}$.Get on ℓ_i where ℓ_i is a new label equal to $F_{K_{1,1}}(\mathrm{rand}\|i)$, for all $i \in [t]$, and $t = \max_{\ell \in \mathbb{L}_{\mathsf{MM}}} \#\mathsf{MM}[\ell]$. In this case the client $\mathbf{C}$ only outputs values of the form $\cdot\|\ell$. Otherwise, if $\mathsf{MM_G}[\ell]$ is a tuple composed of a triple $(\mathrm{rand}_1, \mathrm{rand}_2, \mathrm{rand}^\star)$, then the client and server execute $\mathsf{STE}_{\mathsf{DX}}^{\mathsf{RH}}$.Get on ℓ_i where now ℓ_i is equal to $j_{G_{K_{1,3}}(\mathrm{rand}_2)\|i}$ where $j_l = F_{K_{1,1}}(\mathrm{rand}^\star\|l)$, for $l \in [\nu]$ and $i \in [\tau]$, and $H_{K_{1,2}}(\mathrm{rand}\|i) + \mathrm{slide}_i$ for all $i \in \{1, \cdots, t - \tau\}$. Note that slide_i, for which the computation was detailed in Sect. 6.1, is used to deterministically map the smaller output of H in $[n - \tau]$ into a value in $[n]$. In this case, the client $\mathbf{C}$ only outputs values of the form $\cdot\|\ell$ or $\cdot\|\star$.

Efficiency. Assuming that $\mathsf{STE}_{\mathsf{DX}}^{\mathsf{RH}}$ is an optimal-time dictionary encryption scheme [1,7,10,15], the get complexity of AVLH is $O(t \cdot \lambda)$ where $t = \max_{\ell \in \mathbb{L}_{\mathsf{MM}}} \#\mathsf{MM}[\ell]$ and λ is the load of a bin. Given the parameters detailed in the previous section and if $N > n \cdot \log n$ then the get complexity is

$$O\left(t \cdot \frac{N - m^{0.5+\delta} \cdot \mathrm{polylog}(m)}{m \cdot \mathrm{polylog}(m)}\right).$$

for some $\delta > 0$, when

$$\mu = O\left(m^{0.5+\delta} \cdot \text{polylog}(m)\right), \quad \tau = O(t) = O\left(\text{polylog}(m)\right)$$

and,

$$\nu = O\left(m^{0.5+\delta} \cdot \text{polylog}(m)\right), \quad n = \Theta\left(m \cdot \text{polylog}(m)\right).$$

The storage overhead of AVLH is, with high probability,

$$O\left(\sum_{i=1}^{n} \#\mathsf{DX}[\ell_i]\right) = O(n \cdot \lambda) = O(N - m^{0.5+\delta} \cdot \text{polylog}(m)).$$

7.1   Security

We will now study the security of our construction. More precisely, we will show that it is volume-hiding in the sense that its query leakage does not include the response length. The proof relies on a computational assumption known as the densest subgraph assumption. We first recall this assumption and then proceed to stating our security theorem.

The densest subgraph problem. The hardness of the (decisional) *densest subgraph problem* problem was first used by Applebaum, Barak, and Wigderson in [2] to design public-key encryption schemes based on new assumptions. It was later used by Arora et al. [3] to study the hardness of financial products. Informally, the DSP asks whether it is possible to distinguish between a random regular bi-partite graph and a random regular bi-partite graph with a planted random subgraph.

Definition 4 (The (decision) densest subgraph problem). *Let* $m, n, t,$ $\mu, \nu, \tau > 0$. *The decisional unbalanced expansion problem is to distinguish between the two following distributions:*

- $\mathcal{R}$ *samples an* (m, n, t)-*bi-partite graph uniformly at random. In other words, for each vertex in* V_{top} *it samples* t *neighbors from* V_{bot} *uniformly at random.*
- $\mathcal{P}$ *is obtained as follows. First, two sets* $T \subset V_{\text{top}}$ *and* $B \subset V_{\text{bot}}$, *such that* $\#T = \mu$ *and* $\#B = 2\nu$, *are sampled uniformly at random. Then, for each vertex in* T, *we choose* $t-\tau$ *random neighbors in* V_{bot} *and* τ *random neighbors in* B. *For each vertex in* $V_{\text{top}} \setminus T$, *we choose* t *random neighbors in* V_{bot}.

The following hardness assumption, used in [2,3], is based on state-of-the-art algorithms of Bhaskara, Charikar, Chlamtac, Feige, and Vijayaraghavan in [6].

Definition 5 (The DSP assumption). *There is no* $\varepsilon > 0$ *and* PPT *adversary* $\mathcal{A}$ *that can distinguish between* $\mathcal{R}$ *and* $\mathcal{P}$ *with advantage* ε *when*

$$n = o(m \cdot t), \quad \left(\frac{\mu \cdot \tau^2}{\nu}\right)^2 = o\left(\frac{m \cdot t^2}{n}\right), \quad \nu = \Omega(n^{0.5+\delta}),$$

$$\mu = \Omega(m^{0.5+\delta}) \quad and \quad \tau = \tilde{O}(\sqrt{t})$$

for some positive δ.

Leakage profile. We now describe the leakage profile of AVLH assuming $\mathsf{STE}_{\mathsf{DX}}^{\mathsf{RH}}$ is instantiated with one of the standard optimal-time dictionary encryption schemes [1,7,10,15] all of which have leakage profile

$$\Lambda_{\mathsf{DX}} = (\mathcal{L}_{\mathsf{S}}, \mathcal{L}_{\mathsf{Q}}) = \big(\mathsf{trlen}, \mathsf{qeq}\big).$$

Theorem 5. *If* $\mathsf{STE}_{\mathsf{DX}}^{\mathsf{RH}}$ *is a* $(\mathsf{trlen}, \mathsf{qeq})$*-secure dictionary encryption scheme, F, G and H are pseudo-random functions, and the DSP assumptions holds, then* AVLH *is a* $\big((\mathsf{trlen}, \mathsf{conc}), \mathsf{qeq}\big)$*-secure multi-map encryption scheme; where* conc *is the leakage pattern that outputs a multi-map's concentration.*

Due to space limitations, the proof of Theorem 5 is in the full version of this work. The leakage pattern conc is due to the fact that we leak the size of the bins in trlen which is a function of the concentration.

Improving communication complexity. The communication (query) complexity of AVLH is equal to $O(t \cdot \lambda)$ where λ is the size of the bin and t the maximum response length. In the following we introduce a simple modification of AVLH such that the communication complexity becomes sub-linear in λ.

At a high level, the idea consists of replacing the retrieval of the entire bin's content by an oblivious retrieval that only fetches the value of interest (note that a bin will always contain at most one value associated to any label). Therefore this technique would reduce the overhead from λ to the overhead of a single oblivious access into an array of size λ. The (informal) modified AVLH works as follows. At setup time, we parse the content of each bin as an array (a RAM) and encrypt it using a computationally-secure state-of-the-art ORAM algorithm. Note that now, instead of using a response-hiding dictionary, we use a response-revealing one. The get algorithm works similarly to the one in AVLH except that the dictionary's get algorithm outputs an ORAM that we access separately. In terms of efficiency, the communication complexity becomes $O(t \cdot \sqrt{\lambda})$ assuming that we use square-root ORAM [22] as the underlying ORAM.[6] Note that we can achieve better communication complexity by leveraging techniques from [27]. The storage complexity however remains the same since square-root does not asymptotically increase the load of the bin.

8   Dynamic Volume Hiding Multi-map Encryption Schemes

In this section, we show how to extend both VLH and AVLH to be dynamic. In particular, we will be interested in the following class of updates:

- *tuple addition*: this update operation adds a new tuple $(\ell, \mathbf{v})$ to the multi-map where ℓ is a label that was not part of the original label space $\mathbb{L}_{\mathsf{MM}}$.

---

[6] Note that one cannot use tree-based ORAM schemes such as Path ORAM [41] as the security is function of the size of the RAM. In our case, under realistic parameters, the bin's load is very small to consider any of these schemes.

Let $\mathsf{STE}_{\mathsf{DX}}^{\mathsf{RH}} = (\mathsf{Setup}, \mathsf{Get})$ be a response-hiding dictionary encryption scheme and DST the densest subgraph transform. Consider the scheme $\mathsf{AVLH} = (\mathsf{Setup}, \mathsf{Get})$ defined as follows:

- $\mathsf{Setup}(1^k, \mathsf{param}, \mathsf{MM})$:
  1. generate a DST-transform of MM by computing

  $$(K_1, \mathsf{MM}_{\mathsf{G}}, \mathsf{DX}) \leftarrow \mathsf{DST}(1^k, \mathsf{param}, \mathsf{MM});$$

  2. encrypt DX by computing

  $$(K_2, st_{\mathsf{DX}}, \mathsf{EDX}) \leftarrow \mathsf{STE}_{\mathsf{DX}}^{\mathsf{RH}}.\mathsf{Setup}(1^k, \mathsf{DX});$$

  3. set $K = (K_1, K_2)$, $st = (\mathsf{MM}_{\mathsf{G}}, st_{\mathsf{DX}})$, and $\mathsf{EMM} = \mathsf{EDX}$ and output (K, st, EMM).
- $\mathsf{Get}_{\mathbf{C},\mathbf{S}}((K, st, \ell), \mathsf{EMM})$:
  1. $\mathbf{C}$ parses K as $((K_{1,1}, K_{1,2}, K_{1,3}), K_2)$, st as $(\mathsf{MM}_{\mathsf{G}}, st_{\mathsf{DX}})$ and $\mathbf{S}$ parses EMM as EDX;
  2. if $\mathsf{MM}_{\mathsf{G}}[\ell] = \mathsf{rand}$, then
     (a) $\mathbf{C}$ and $\mathbf{S}$ execute $\mathsf{STE}_{\mathsf{DX}}^{\mathsf{RH}}.\mathsf{Get}_{\mathbf{C},\mathbf{S}}((K_2, st_{\mathsf{DX}}, \ell_i), \mathsf{EDX})$, for all $i \in [t]$, where

     $$\ell_i := F_{K_{1,1}}(\mathsf{rand}\|i);$$

     (b) $\mathbf{C}$ outputs values of the form $\cdot\|\ell$;
  3. if $\mathsf{MM}_{\mathsf{G}}[\ell] = (\mathsf{rand}_1, \mathsf{rand}_2, \mathsf{rand}^\star)$, then
     (a) $\mathbf{C}$ and $\mathbf{S}$ execute $\mathsf{STE}_{\mathsf{DX}}^{\mathsf{RH}}.\mathsf{Get}_{\mathbf{C},\mathbf{S}}((K_2, st_{\mathsf{DX}}, \ell_i), \mathsf{EDX})$, where for all $i \in [\tau]$,

     $$\ell_i := j_{G_{K_{1,3}}(\mathsf{rand}_2\|i)},$$

     and $(j_1, \cdots, j_\nu) = (F_{K_{1,1}}(\mathsf{rand}^\star\|1), \cdots, F_{K_{1,1}}(\mathsf{rand}^\star\|\nu))$ and for all $i \in \{1, \cdots, t-\tau\}$,

     $$\ell_i := H_{K_{1,2}}(\mathsf{rand}_1\|i) + \mathsf{slide}_i,$$

     where slide_j is computed as follows
     i. order

     $$\left\{ F_{K_{1,1}}(\mathsf{rand}^\star\|i) \right\}_{i \in [\nu]}$$

     as $(\mathsf{pos}_1, \cdots, \mathsf{pos}_\nu)$;
     ii. if $H_{K_{1,2}}(\mathsf{rand}\|i) \in [1, \mathsf{pos}_1]$, set $\mathsf{slide}_i = 0$;
     iii. if $H_{K_{1,2}}(\mathsf{rand}\|i) \in]\mathsf{pos}_{j-1}-(j-1), \mathsf{pos}_j-(j-1)[$, set $\mathsf{slide}_i = j-1$, for any $j \in \{2, \cdots, \nu\}$;
     iv. if $H_{K_{1,2}}(\mathsf{rand}\|i) \in]\mathsf{pos}_\nu - \nu, n - \nu]$, set $\mathsf{slide}_i = \nu$;
     (b) $\mathbf{C}$ outputs all values of the form $\cdot\|\ell$ or $\cdot\|\star$.

Fig. 6. AVLH: An Advanced Volume Hiding Multi-Map Encryption Scheme.

- *tuple deletion*: this update operation removes an entire label/tuple pair $(\ell, \mathbf{v})$ from the multi-map.
- *editing*: this update operation modifies the content of a specific tuple $\mathbf{v}$ associated to ℓ by replacing an old value $v_{old} \in \mathbf{v}$ by a new one v_{new}.

In particular, we do not consider updates that add or remove a value to/from an existing tuple in the multi-map. In the following, we detail how to extend VLH to handle these three update operations and AVLH to handle the third update operation.

8.1 VLH^d: A Dynamic Variant of VLH

The pseudo-code of VLH^d is in the full version and it works as follows.

Overview. VLH^d = (Setup, Get, Put) makes black-box use of a dynamic response-hiding multi-map encryption scheme $\mathsf{STE}^{\mathsf{RH}}_{\mathsf{EMM}}$ = (Setup, Get, Put, Remove) and of the volume-hiding multi-map encryption scheme VLH = (Setup, Get).[7] Both the Setup algorithm and the Get protocol are exactly the same as of those of VLH. The Put algorithm takes as input an update u and processes it as follows. If $u = (\mathsf{add}, (\ell, \mathbf{v}))$, then the client first computes the PRT transform on a single-pair multi-map defined as $\{(\ell, \mathbf{v})\}$ and outputs a new single-pair multi-map $\{(\ell, \mathbf{v}')\}$. The client and server then execute $\mathsf{STE}^{\mathsf{RH}}_{\mathsf{EMM}}$.Put on the label/tuple pair $(\ell, \mathbf{v}')$. If $u = (\mathsf{rm}, \ell)$, then the client and server execute $\mathsf{STE}^{\mathsf{RH}}_{\mathsf{EMM}}$.Remove on the label ℓ. If $u = (\mathsf{edit}, (\ell, v_{old}, v_{new}))$, then the client and server first execute VLH.Get, the client receives the tuple $\mathbf{v}$ associated to the label ℓ. The client and server then execute $\mathsf{STE}^{\mathsf{RH}}_{\mathsf{EMM}}$.Remove on the label ℓ. The client locally replaces the value v_{old} by v_{new} in the tuple $\mathbf{v}$ and then executes $\mathsf{STE}^{\mathsf{RH}}_{\mathsf{EMM}}$.Put with the server on the modified label/tuple pair.

Efficiency analysis. In our analysis, assume $\mathsf{STE}^{\mathsf{RH}}_{\mathsf{EMM}}$ is an optimal-time dynamic multi-map encryption scheme [7,10,29]. It is clear that the get and storage complexity of VLH^d are exactly the same as VLH. The Put complexity varies depending on the type of the update operation. If u is a tuple addition or a tuple edit, then the Put complexity is $O(\lambda + n'_\ell)$ where $n'_\ell \in \{0, \cdots, 2^s - 1\}$. The worst-case is $O(\lambda + 2^s)$ while the best case is $O(\lambda)$. The expected complexity is $O(\lambda + 2^{s-1})$. If u is a *tuple deletion*, then the put complexity has constant time.

Security analysis. We now describe the leakage of VLH^d assuming that $\mathsf{STE}^{\mathsf{RH}}_{\mathsf{EMM}}$ is instantiated with one of the standard optimal-time forward-private multi-map encryption schemes [1,7,8] all of which have leakage profile

$$\Lambda_{\mathsf{MM}} = (\mathcal{L}_{\mathsf{S}}, \mathcal{L}_{\mathsf{Q}}, \mathcal{L}_{\mathsf{U}}) = (\mathsf{trlen}, (\mathsf{qeq}, \mathsf{rlen}), (\mathsf{op}, \mathsf{rlen}))$$

---

[7] Note that the same multi-map encryption scheme $\mathsf{STE}^{\mathsf{RH}}_{\mathsf{EMM}}$ = (Setup, Get, Put, Remove) has to be used as the underlying multi-map encryption scheme for VLH.

Theorem 6. *If* $\mathsf{STE}_{\mathsf{EMM}}^{\mathsf{RH}}$ *is a* $(\mathsf{trlen}, (\mathsf{qeq}, \mathsf{rlen}), (\mathsf{op}, \mathsf{rlen}))$*-secure multi-map encryption scheme,* F *in* PRT *is a pseudo-random function and* VLH *is a* (m, qeq)*-secure multi-map encryption scheme, then* $\mathsf{VLH}^{\mathsf{d}}$ *is a* $(m, \mathsf{qeq}, (\mathsf{op}, \mathsf{ueq}))$*-secure multi-map encryption scheme.*

The update equality pattern ueq leaks if and when a label edit has occurred. The proof of this theorem is similar to Theorem 3 and deferred to the full version of this work.

8.2 $\mathsf{AVLH}^{\mathsf{d}}$: Dynamic Variant of AVLH

The pseudo-code of $\mathsf{AVLH}^{\mathsf{d}}$ is in the full version and it works as follows.

Overview. The construction, $\mathsf{AVLH}^{\mathsf{d}}$, makes black box use of a dynamic response-hiding dictionary $\mathsf{STE}_{\mathsf{EDX}}^{\mathsf{RH}} = (\mathsf{Setup}, \mathsf{Get}, \mathsf{Put}, \mathsf{Remove})$ and of the volume hiding multi-map encryption scheme $\mathsf{AVLH} = (\mathsf{Setup}, \mathsf{Get})$.[8] The Setup algorithm and the Get protocol are exactly the same as of those of AVLH. The Put algorithm takes as input an update u and processes it as follows. Parse u as $(\mathsf{edit}, (\ell, \mathbf{v}))$, the client and server execute $(r, \perp) \leftarrow \mathsf{AVLH}.\mathsf{Get}_{\mathsf{C},\mathsf{S}}\big((K, st, \ell), \mathsf{EMM}\big)$ where $r = (B_{i_1}, \cdots, B_{i_t})$ and the client here does not dismiss any value from the retrieved bins. The client and server then execute $\mathsf{STE}_{\mathsf{EDX}}^{\mathsf{RH}}.\mathsf{Remove}$ on all retrieved bins. The client then identifies the bin that contains the value $\mathbf{v}_{\mathsf{old}}\|\ell$ (or $\mathbf{v}_{\mathsf{old}}\|\star$) that it replaces with $v_{\mathsf{new}}\|\ell$ (or by $\mathbf{v}_{\mathsf{new}}\|\star$ if concentrated). The client and server then execute $\mathsf{STE}_{\mathsf{EDX}}.\mathsf{Put}$ on the pairs (i_j, B_{i_j}), for all $j \in [t]$.

Efficiency analysis. We assume $\mathsf{STE}_{\mathsf{EDX}}^{\mathsf{RH}}$ is an optimal-time dynamic dictionary encryption scheme [7,10,29]. Clearly, the get and the storage complexity of $\mathsf{AVLH}^{\mathsf{d}}$ are exactly the same as AVLH. The Put complexity is equal to $O(t \cdot \lambda)$, where $t = \max_{\ell \in \mathbb{L}_{\mathsf{MM}}} \#\mathsf{MM}[\ell]$ is the maximum response length and λ is the size of the bin–which is the same as the get complexity. Refer to Sect. 7 for a more detailed and concrete analysis of the bin size λ.

Security analysis. We now describe the leakage of $\mathsf{AVLH}^{\mathsf{d}}$ assuming that $\mathsf{STE}_{\mathsf{EDX}}^{\mathsf{RH}}$ is instantiated with one of the standard optimal-time forward-private dictionary encryption scheme [1,7,8] all of which have a leakage profile at most

$$\Lambda_{\mathsf{DX}} = (\mathcal{L}_{\mathsf{S}}, \mathcal{L}_{\mathsf{Q}}, \mathcal{L}_{\mathsf{U}}) = (\mathsf{trlen}, \mathsf{qeq}, \mathsf{op})$$

Theorem 7. *If* $\mathsf{STE}_{\mathsf{EDX}}^{\mathsf{RH}}$ *is a* $(\mathsf{trlen}, \mathsf{qeq}, \mathsf{op})$*-secure dictionary encryption scheme and* AVLH *is a* $(\mathsf{trlen}, \mathsf{qeq})$*-secure multi-map encryption scheme, then* $\mathsf{VLH}^{\mathsf{d}}$ *is a* $(\mathsf{trlen}, \mathsf{qeq}, (\mathsf{op}, \mathsf{ueq}))$*-secure multi-map encryption scheme.*

The proof of this theorem is similar to Theorem 5 and deferred to the full version of this work.

---

[8] Note that the same dictionary encryption scheme $\mathsf{STE}_{\mathsf{EDX}}^{\mathsf{RH}} = (\mathsf{Setup}, \mathsf{Get}, \mathsf{Put}, \mathsf{Remove})$ has to be used as the underlying dictionary encryption scheme for AVLH.

References

1. Amjad, G., Kamara, S., Moataz, T.: Breach-resistant structured encryption. IACR Cryptology ePrint Archive, 2018:195 (2018)
2. Applebaum, B., Barak, B., Wigderson, A.: Public-key cryptography from different assumptions. In: Proceedings of the Forty-Second ACM Symposium on Theory of Computing, pp. 171–180. ACM (2010)
3. Arora, S., Barak, B., Brunnermeier, M., Ge, R.: Computational complexity and information asymmetry in financial products. Commun. ACM **54**(5), 101–107 (2011)
4. Asharov, G., Naor, M., Segev, G., Shahaf, I.: Searchable symmetric encryption: optimal locality in linear space via two-dimensional balanced allocations. In: STOC 2016, pp. 1101–1114. ACM, New York (2016)
5. Asharov, G., Segev, G., Shahaf, I.: Tight tradeoffs in searchable symmetric encryption. In: Shacham, H., Boldyreva, A. (eds.) CRYPTO 2018. LNCS, vol. 10991, pp. 407–436. Springer, Cham (2018). https://doi.org/10.1007/978-3-319-96884-1_14
6. Bhaskara, A., Charikar, M., Chlamtac, E., Feige, U., Vijayaraghavan, A.: Detecting high log-densities: an o (n 1/4) approximation for densest k-subgraph. In: Proceedings of the Forty-Second ACM Symposium on Theory of Computing, pp. 201–210. ACM (2010)
7. Bost, R.: Sophos - forward secure searchable encryption. In: ACM CCS 2016 (2016)
8. Bost, R., Minaud, B., Ohrimenko, O.: Forward and backward private searchable encryption from constrained cryptographic primitives. In: Proceedings of the 2017 ACM SIGSAC Conference on Computer and Communications Security, pp. 1465–1482. ACM (2017)
9. Cash, D., Grubbs, P., Perry, J., Ristenpart, T.: Leakage-abuse attacks against searchable encryption. In: ACM CCS 2015, pp. 668–679. ACM (2015)
10. Cash, D., et al.: Dynamic searchable encryption in very-large databases: data structures and implementation. In NDSS 2014 (2014)
11. Cash, D., Jarecki, S., Jutla, C., Krawczyk, H., Roşu, M.-C., Steiner, M.: Highly-scalable searchable symmetric encryption with support for boolean queries. In: Canetti, R., Garay, J.A. (eds.) CRYPTO 2013. LNCS, vol. 8042, pp. 353–373. Springer, Heidelberg (2013). https://doi.org/10.1007/978-3-642-40041-4_20
12. Chase, M., Kamara, S.: Structured encryption and controlled disclosure. In: Abe, M. (ed.) ASIACRYPT 2010. LNCS, vol. 6477, pp. 577–594. Springer, Heidelberg (2010). https://doi.org/10.1007/978-3-642-17373-8_33
13. Chase, M., Kamara, S.: Structured encryption and controlled disclosure. Technical Report 2011/010.pdf, IACR Cryptology ePrint Archive (2010)
14. Chaudhuri, S., Church, K.W., König, A.C., Sui, L.: Heavy-tailed distributions and multi-keyword queries. In: ACM SIGIR (2007)
15. Curtmola, R., Garay, J., Kamara, S., Ostrovsky, R.: Searchable symmetric encryption: improved definitions and efficient constructions. In: CCS 2006 (2006)
16. Demertzis, I., Papadopoulos, D., Papamanthou, C.: Searchable encryption with optimal locality: achieving sublogarithmic read efficiency. In: Shacham, H., Boldyreva, A. (eds.) CRYPTO 2018. LNCS, vol. 10991, pp. 371–406. Springer, Cham (2018). https://doi.org/10.1007/978-3-319-96884-1_13
17. Demertzis, I., Papadopoulos, S., Papapetrou, O., Deligiannakis, A., Garofalakis, M.: Practical private range search revisited. In: Proceedings of the 2016 International Conference on Management of Data, pp. 185–198. ACM (2016)

18. Demertzis, I., Papamanthou, C.: Fast searchable encryption with tunable locality. In: SIGMOD 2017 (2017)
19. Etemad, M., Küpçü, A., Papamanthou, C., Evans, D.: Efficient dynamic searchable encryption with forward privacy. Proc. Priv. Enhancing Technol. **2018**(1), 5–20 (2018)
20. Faber, S., Jarecki, S., Krawczyk, H., Nguyen, Q., Rosu, M., Steiner, M.: Rich queries on encrypted data: beyond exact matches. In: Pernul, G., Ryan, P.Y.A., Weippl, E. (eds.) ESORICS 2015. LNCS, vol. 9327, pp. 123–145. Springer, Cham (2015). https://doi.org/10.1007/978-3-319-24177-7_7
21. Fisch, B.A., et al.: Malicious-client security in blind seer: a scalable private DBMS. In: IEEE Symposium on Security and Privacy, pp. 395–410. IEEE (2015)
22. Goldreich, O., Ostrovsky, R.: Software protection and simulation on oblivious RAMs. J. ACM **43**(3), 431–473 (1996)
23. Grubbs, P., Lacharité, M., Minaud, B., Paterson, K.G.: Pump up the volume: practical database reconstruction from volume leakage on range queries. In: Lie, D., Mannan, M., Backes, M., Wang, X. (eds.) Proceedings of the 2018 ACM SIGSAC Conference on Computer and Communications Security, CCS 2018, Toronto, ON, Canada, 15–19 October 2018, pp. 315–331. ACM (2018)
24. Islam, M.S., Kuzu, M., Kantarcioglu, M.: Access pattern disclosure on searchable encryption: ramification, attack and mitigation. In: NDSS 2012 (2012)
25. Kamara, S., Moataz, T.: Boolean searchable symmetric encryption with worst-case sub-linear complexity. In: Coron, J.-S., Nielsen, J.B. (eds.) EUROCRYPT 2017. LNCS, vol. 10212, pp. 94–124. Springer, Cham (2017). https://doi.org/10.1007/978-3-319-56617-7_4
26. Kamara, S., Moataz, T.: SQL on structurally-encrypted databases. In: Peyrin, T., Galbraith, S. (eds.) ASIACRYPT 2018. LNCS, vol. 11272, pp. 149–180. Springer, Cham (2018). https://doi.org/10.1007/978-3-030-03326-2_6
27. Kamara, S., Moataz, T., Ohrimenko, O.: Structured encryption and leakage suppression. In: Shacham, H., Boldyreva, A. (eds.) CRYPTO 2018. LNCS, vol. 10991, pp. 339–370. Springer, Cham (2018). https://doi.org/10.1007/978-3-319-96884-1_12
28. Kamara, S., Papamanthou, C.: Parallel and dynamic searchable symmetric encryption. In: Sadeghi, A.-R. (ed.) FC 2013. LNCS, vol. 7859, pp. 258–274. Springer, Heidelberg (2013). https://doi.org/10.1007/978-3-642-39884-1_22
29. Kamara, S., Papamanthou, C., Roeder, T.: Dynamic searchable symmetric encryption. In: ACM CCS 2012 (2012)
30. Kellaris, G., Kollios, G., Nissim, K., O'Neill, A.; Generic attacks on secure outsourced databases. In: ACM Conference on Computer and Communications Security (CCS 2016) (2016)
31. Kellaris, G., Kollios, G., Nissim, K., O'Neill, A.: Accessing data while preserving privacy. CoRR, abs/1706.01552 (2017)
32. Lacharité, M.-S., Minaud, B., Paterson, K.G.: Improved reconstruction attacks on encrypted data using range query leakage. In: 2018 IEEE Symposium on Security and Privacy (SP), pp. 297–314. IEEE (2018)
33. Meng, X., Kamara, S., Nissim, K., Kollios, G.: GRECS: graph encryption for approximate shortest distance queries. In: CCS 15 (2015)
34. Miers, I., Mohassel, P.: IO-DSSE: scaling dynamic searchable encryption to millions of indexes by improving locality. Cryptology ePrint Archive, Report 2016/830 (2016). http://eprint.iacr.org/2016/830

35. Naveed, M., Kamara, S., Wright, C.V.: Inference attacks on property-preserving encrypted databases. In: ACM Conference on Computer and Communications Security (CCS), CCS 2015, pp. 644–655. ACM (2015)
36. Naveed, M., Prabhakaran, M., Gunter, C.: Dynamic searchable encryption via blind storage. In: IEEE Symposium on Security and Privacy (S&P 2014) (2014)
37. Pappas, V., et al.: Blind seer: a scalable private DBMS. In: 2014 IEEE Symposium on Security and Privacy (SP), pp. 359–374. IEEE (2014)
38. Poddar, R., Boelter, T., Popa, R.A.: Arx: a strongly encrypted database system. Technical Report 2016/591
39. Song, D., Wagner, D., Perrig, A.: Practical techniques for searching on encrypted data. In: IEEE S&P, pp. 44–55. IEEE Computer Society (2000)
40. Stefanov, E., Papamanthou, C., Shi, E.: Practical dynamic searchable encryption with small leakage. In: NDSS 2014 (2014)
41. Stefanov, E., et al.: Path ORAM: an extremely simple oblivious RAM protocol. In: CCS (2013)
42. Zhang, Y., O'Neill, A., Sherr, M., Zhou, W.: Privacy-preserving network provenance. PVLDB 10(11), 1550–1561 (2017)
43. Zipf, G.K.: The Psycho-Biology of Language (1935)

Locality-Preserving Oblivious RAM

Gilad Asharov[1]([⊠]), T.-H. Hubert Chan[2], Kartik Nayak[3], Rafael Pass[1], Ling Ren[4], and Elaine Shi[1]

[1] Cornell/Cornell Tech, New York, USA
asharov@cornell.edu
[2] The University of Hong Kong, Pok Fu Lam, Hong Kong
[3] University of Maryland, College Park, USA
[4] MIT, Cambridge, USA

Abstract. Oblivious RAMs, introduced by Goldreich and Ostrovsky [JACM'96], compile any RAM program into one that is "memory oblivious", i.e., the access pattern to the memory is independent of the input. All previous ORAM schemes, however, completely break the *locality* of data accesses (for instance, by shuffling the data to pseudorandom positions in memory).

In this work, we initiate the study of *locality-preserving ORAMs*— ORAMs that preserve locality of the accessed memory regions, while leaking only the lengths of contiguous memory regions accessed. Our main results demonstrate the existence of a locality-preserving ORAM with poly-logarithmic overhead both in terms of bandwidth and locality. We also study the tradeoff between locality, bandwidth and leakage, and show that any scheme that preserves locality and does not leak the lengths of the contiguous memory regions accessed, suffers from prohibitive bandwidth.

To the best of our knowledge, before our work, the only works combining locality and obliviousness were for symmetric searchable encryption [e.g., Cash and Tessaro (EUROCRYPT'14), Asharov et al. (STOC'16)]. Symmetric search encryption ensures obliviousness if each keyword is searched only once, whereas ORAM provides obliviousness to any input program. Thus, our work generalizes that line of work to the much more challenging task of preserving locality in ORAMs.

Keywords: Oblivious RAM · Locality · Randomized algorithms

1 Introduction

Oblivious RAM [23,25,36], originally proposed in the seminal work by Goldreich and Ostrovsky [23,25], allows a client to outsource encrypted data to an untrusted server, and access the data in a way such that the access patterns observed by the server are provably obfuscated.

G. Asharov—Currently a researcher at JP Morgan AI Research.
K. Nayak and L. Ren—Currently at VMware Research.

Y. Ishai and V. Rijmen (Eds.): EUROCRYPT 2019, LNCS 11477, pp. 214–243, 2019.
https://doi.org/10.1007/978-3-030-17656-3_8

Thus far, the primary metric used to analyze ORAM schemes has been bandwidth which is the number of memory blocks accessed for every logical access. After a long sequence of works (e.g., [25,34,36,37,41]) it is now understood that ORAM schemes can be constructed incurring only *logarithmic* bandwidth [6]; and moreover, this is asymptotically optimal [25,33].

An important performance metric that has been traditionally overlooked in the ORAM literature is *data locality*. The majority of real-world applications and programs exhibit a high-degree of data locality, i.e., if a program or application accesses some address it is very likely to access also a neighboring address. This observation has profoundly influenced the design of storage systems—for example, commodity hard-drive and SSD disks support sequential accesses faster than random accesses.

Unfortunately, existing ORAM schemes (e.g., [6,18,23,25,36,37,41]) are not locality-friendly. Randomization in ORAMs is inherent due to the requirement to hide the access pattern of the program, and ORAM schemes (pseudo-)randomly permute blocks and shuffle them in the memory. As a result, if a client wants to read a large file consisting of $\Theta(N)$ *contiguous* blocks, all known ORAM schemes would have to access more than $\Omega(N \log N)$ random (i.e., discontiguous) disk locations, introducing significant delays due to lack of locality.

In this paper, we ask the question: can we design ORAM schemes with data locality? At first sight, this seems impossible. Intuitively, an ORAM scheme must hide whether the client requests N random locations or a single contiguous region of size N. As a result, such a scheme cannot preserve locality, and indeed we formalize this intuition and formally show that any ORAM scheme that hides the differences between the above two extreme cases must necessarily suffer from either high bandwidth or bad locality.

However, this does not mean that providing oblivious data accesses and preserving locality simultaneously is a hopeless cause. In particular, in many practical applications, it may already be public knowledge that a user is accessing contiguous regions; e.g., consider the following two motivating scenarios:

- *Outsourced file server.* Imagine that a client outsources encrypted files to a server, and then repeatedly queries the server to retrieve various files. In this case, each file captures a contiguous region in logical memory. Note that unless we pad all files to the maximum size possible (which can be very expensive if files sizes vary greatly), we would already leak the file size (i.e., length of contiguous memory region visited) on each request.
- *Outsourced range query database.* Consider an outsourced (encrypted) database system where a client makes range queries on a primary search key, e.g., an IoT database that allows a client to retrieve all sensor readings during a specified time range. We would like to protect the client's access patterns from the server. As previous works argued [16,30], in this case one can leverage differential privacy to hide the number of matching records and it may be safe to reveal a noisy version of the length of the contiguous region accessed.

Note that in both of the above scenarios, some length leakage seems unavoidable unless we always pad to the maximum with every request—and this is true even if we employ ORAM to outsource the files/database! Further, disk IO may be more costly than network bandwidth depending on the deployment scenario: for example, if the server is serving many clients simultaneously (e.g., serving many users from the same organization sharing a secret key, or if the server has a trusted CPU such as Intel SGX and is serving multiple mutually distrustful clients), the system's bottleneck may well be the server's disk I/O rather than the server's aggregate bandwidth.

Motivated by these practical scenarios, we ask the following question.

Can we construct a bandwidth-efficient ORAM that preserves data locality while leaking only the lengths of contiguous regions accessed?

We answer the question in the affirmative and prove the following result:

Theorem 1.1 (Informal). *Let N be the size of the logical address space. There is an ORAM scheme that makes use of only 2 disks and $O(1)$ client storage, such that upon receiving a sufficiently long request sequence containing T logical addresses, the ORAM can correctly answer the requests paying only $T \cdot \mathrm{poly} \log N$ bandwidth; and moreover, if the T addresses requested contains ℓ discontiguous regions, the ORAM server visits only $\ell \cdot \mathrm{poly} \log N$ discontiguous regions on its 2 disks.*

To the best of our knowledge, we are the first to consider and formulate the problem of locality-friendly ORAM. Even formulating the problem turns out to be non-trivial, since it requires teasing out the boundaries between theoretical feasibility and impossibility, and capturing what kind of leakage is reasonable in practical applications and yet does not rule out constructions that are both bandwidth-efficient and locality-friendly. Besides the conceptual definitional contributions, we also describe novel algorithmic techniques that result in the first non-trivial locality-friendly ORAM construction.

To help the reader understand the technical nature of our work, we point out that our problem formulation in fact generalizes a line of work on optimizing locality in Searchable Symmetric Encryption (SSE) schemes. The issue of locality was encountered in recent implementations [12] of searchable symmetric encryption in real-world databases, showing that the practical performance of known schemes that overlook the issue of locality do not scale well to large data sizes. The problem of optimizing locality in searchable symmetric schemes has received considerable attention recently (see, e.g., [7, 8, 13, 20, 21]). Our problem generalizes this line of work, and achieving good locality in oblivious RAM is significantly more challenging due to the following reasons: (1) In SSE, obliviousness is guaranteed only if each "file" is accessed at most once (and the length of the file is also leaked in SSE)[1]; and (2) SSE assumes that rebuilding the "server-side oblivious data structure" happens on a powerful client with linear storage, and thus the rebuilding comes "for free". We show, for the first time, how to remove both of these above restrictions, and provide a generalized, full-fledged oblivious memory abstraction that supports unbounded polynomial accesses and yet preserves both bandwidth and locality.

---

[1] Intuitively, a file stores the identifiers of the documents matching a keyword search in SSE schemes.

2 Technical Roadmap

In the following we provide a summary of results and techniques. In Sect. 2.1 we discuss our modeling of locality. In Sect. 2.2 we discuss our lower bounds, providing tradeoffs between the locality of a program, leakage and bandwidth. Towards introducing our construction, we start in Sect. 2.3 with a warmup–oblivious sort with "good" locality. In Sect. 2.4 we introduce range ORAM, our core building block for achieving locality, in which in Sect. 2.5 we overview its construction. In Sect. 2.6 we overview a variant of Range ORAM, called "Online Range ORAM", which can also be viewed as a locality preserving ORAM.

2.1 A Generalized Model of Locality

How do we model locality of an algorithm (e.g., an ORAM or SSE algorithm)? A natural option is to use the well-accepted approach adopted by the SSE line of work [7,8,13,20]. Imagine that every time an algorithm (e.g., SSE or ORAM) needs to read an item from disk, it has two choices: (1) read the next contiguous address; and (2) jump to a new address (often called "seek" in the systems literature). While both types of operations contribute to the *bandwidth* measure; only the latter type contributes to the *locality* measure [7,8,13,20] since seeks are significantly more expensive than sequential reads on real-world disks. We point out that locality alone is not a meaningful measure since we can always achieve better locality and minimize jumps by scanning through the entire memory extracting the values we want along the way. Thus we always use locality in conjunction with a *bandwidth* metric too, i.e., how many blocks we must must fetch from the disk upon each request. This model was adopted by the SSE line of work, however, is very constraining in the sense that they assume that the server has access to only 1 disk. In practice, cloud-hosting services such as EC2 and Azure provide servers with multiple disks. Constraining to such a single-disk model might rule out interesting cryptographic algorithms of practical value. Therefore, we generalize the locality definition as follows.

Defining (D, ℓ)-locality. We consider the scenario where the ORAM server may have multiple (but ideally a small number) of disks, where eack disk still supports the aforementioned two types of instructions: "read the next contiguous address" and "jump to a new address". Henceforth, we say that an ORAM scheme satisfies (D, ℓ)-locality and β bandwidth cost iff for a sufficiently long input sequence containing B requests spanning L non-contiguous regions, the ORAM server, with access to D disks, may access at most $\beta \cdot B$ blocks and issue at most $\ell \cdot L$ jump instructions. Of course, the adversary can observe all disks, and all movements operations in these disks. We refer the readers to Sect. 3.1 for the formal definition.

Under these new definitions, our result can be stated technically as "an ORAM scheme with $(2, \text{poly} \log N)$-locality and $\text{poly} \log N$ bandwidth (amortized) cost" where N is the total number of logical blocks. Moreover, as mentioned, our ORAM scheme leaks only the length of each contiguous region in the request sequence and nothing else (and as mentioned, some leakage is inherent if we desire efficiency).

Open questions. Given our new modeling techniques and results, we also suggest several exciting open questions, e.g., is it possible to have an ORAM scheme that achieves $(1, \ell)$-locality and β bandwidth cost where ℓ and β are small? Can we compile source programs that exhibit (D, ℓ)-locality where $D > 1$ with meaningful leakage? For the former question, if there is a lower bound that shows a sharp separation between 1 and 2 disks, it would be technically really intriguing. For the latter question, the constructions in this paper directly imply that if one is willing to leak the disk each request wants to access, such schemes are possible. However, depending on the practical application such leakage vary from reasonable to extremely harmful. Thus the challenge is to understand the feasibility/infeasibility of achieving such compilation while hiding which disk each request wants to access. We refer the reader to Sect. 7 for other open problems.

2.2   Locality with No Leakage

As we already discussed, preserving both bandwidth and locality with no leakage is impossible. We formalize this claim, and study tradeoffs between leakage profiles and performance. We consider schemes that leak only the total number of accesses (just as in standard ORAM[2]) and show that a scheme with good locality must incur a high bandwidth, even when allowing large client-side space blowup. We prove the following:

Theorem 2.1. *For any $\ell, c \leq \frac{N}{10}$, any (D, ℓ)-local ORAM scheme with c blocks of client storage that leaks no information (besides the total number of requests) must incur $\Omega(\frac{N}{D})$ bandwidth.*

To intuitively understand the lower bound, consider a simplified case where the ORAM must satisfy $(1, 1)$-locality. Consider the following two scenarios: (1) requesting contiguous blocks at addresses $1, 2, \ldots N$; and (2) requesting blocks at random addresses. By the locality constraint, in the former scenario the ORAM scheme can access only 1 contiguous region on 1 disk. Now the oblivious requirement says that the address distributions under these two scenarios must be indistinguishable, and thus even for the second scenario the ORAM server can only access a single contiguous region too. Now, if each request's address is generated at random, in expectation the desired block is at least $N/2$ far from where the disk's head currently is—and this holds no matter how one arranges the contents stored on the disk, and even when the server's disk may be unbounded! Since the ORAM scheme must perform a single linear scan even in the second scenario, it must read in expectation $N/2$ locations to serve each randomized request. Note that one key idea in this lower bound proof is that we generate the request sequence at random in the second scenario, such that even if the ORAM scheme is allowed to perform arbitrary, possibly randomized setup, informally speaking it does not help. In Sect. 6, we make non-trivial generalizations to the above intuition and prove a lower bound for generalized choices of D and ℓ.

---

[2] We emphasize that many practical applications leak some more information even when using standard ORAM, e.g., in the form of communication volume. See discussion in below.

On leaking the lengths. Given our lower bound, our constructions presented next leak the lengths of the accessed regions to achieve good locality. Before proceeding with our construction, we remark the following points regarding this leakage: (1) The input program can always break locality (say, via fictitious non-contiguous accesses) and therefore our scheme can be viewed as a strict generalization of ordinary ORAM schemes. In other words, the user can choose to opt out of the locality feature. (2) As we mentioned above, in many applications it is already public knowledge that the client accesses contiguous regions. In those cases, the leakage is the same had we used an ordinary ORAM [29]. (3) Finally, we stress that just like the case of ordinary ORAM, our locality-friendly ORAM can be combined with differential privacy techniques as Kellaris et al. [30] suggested to offer strengthened privacy guarantees.

Despite these arguments, in some applications with good locality, such leakage might be harmful. For example, a program may access several regions of different lengths and which regions are accessed depend on some sensitive data. Whether the locality feature of our scheme should be used or not is application dependent, and we encourage using the locality feature only in places where the leakage pattern is clear and is public information to begin with.

2.3   Warmup: Locality-Friendly Oblivious Sort

Before describing our main construction, we first introduce a new building block called *locality-friendly oblivious sort* which we will repeatedly use. First, we observe that not all known oblivious sorting algorithms are "locality-friendly". For example, algorithms such as AKS sort [2] and Zig-zag sort [26] are described with a sorting circuit whose wiring has good randomness-like properties (e.g., in AKS the wiring involve expander graphs, which have proven random-walk properties), thus making these algorithms difficult to implement with small locality consuming a small number of disks (while preserving the algorithm's runtime).

Fortunately, we observe that there is a particular method to implement the Bitonic Sort [9] algorithm such that with only 2 disks, the algorithm can be accomplished using $O(\log^2 n)$ "jumps" (note also that "natural" implementations of the Bitonic Sort circuit do not seem to have such locality friendliness).

We defer the details of this specific locality-friendly implementation of Bitonic-Sort to Appendix A, stating only the theorem here:

Theorem 2.2 (Locality-friendly oblivious sort). *Bitonic sort (when implemented as in Appendix A) is a perfectly oblivious sorting algorithm that sorts n elements using $O(n \log^2 n)$ bandwidth and $(2, O(\log^2 n))$-locality.*

2.4   Range ORAM: An Intermediate, Relaxed Abstraction

We now start to give an informal exposition of our upper bound results. This is perhaps the most technically sophisticated part of our work.

To achieve the final result, we will do it in two steps. In our final ORAM scheme (henceforth called *Online Range ORAM*), the ORAM client receives the requests one by one in an online fashion, and it is not informed a-priori when a contiguous scan would occur in the request sequence. That is, it has

exactly the same syntax as an ordinary ORAM, but when the client accesses
contiguous addresses, the online range ORAM has to recognize this fact, and
fetch contiguous regions from the memory. To reach this final goal, however, we
need an intermediate stepping stone called *Range ORAM*, which is an "offline"
version of Online Range ORAM. In a Range ORAM, imagine that the ORAM
client receives a request sequence that can look ahead into the future, i.e., the
client is informed that the next len requests will scan contiguously through the
logical memory.

More formally, in a Range ORAM, the ORAM client receives requests of the
form Access(op, $[s, t]$, data), where op $\in$ {read, write}, $s, t \in [N]$, $s < t$, and
data $\in (\{0, 1\}^b)^{(t-s+1)}$ where b is the block size. Upon each request, the client
interacts with the server to update the server-side data structure and fetch the
data it needs:

- If op $=$ read, at the end of the request, all blocks whose logical addresses
  belong to the range $[s, t]$ are written down in server memory starting at a
  designated address; the server may then return the blocks to the client one-
  by-one in a single contiguous scan.
- If op $=$ write, then imagine that the client has already written down a data
  array consisting of $t - s + 1$ blocks on the server in a designated, contiguous
  region; the client and the server then perform interactions to update the
  server-side data structure to reflect that the logical address range $[s, t]$ should
  now store the contents of data.

Note that as described above, a Range ORAM is well-defined even for a client
that has only $O(1)$ blocks of storage—and indeed we give a more general formulation by assuming $O(1)$ client storage.

As for obliviousness, we require that the distribution of memory addresses
accessed by the Range ORAM can be simulated from the lengths of the accessed
ranges only, which implies that there is no other leakage other than these lengths.
We prove the following theorem:

Theorem 2.3. *There exists a perfectly secure Range ORAM construction consuming $O(N \log N)$ space with (amortized)* len $\cdot$ poly log N *bandwidth and* (2,
poly log N)*-locality, for accessing a range of length* len.

In comparison, for all existing ORAM schemes, accessing a single region of
len contiguous blocks involves accessing $\Omega(\text{len} \cdot \log N)$ blocks residing at discontiguous physical locations. We now overview the high level ideas behind our
range ORAM construction.

Strawman scheme: read-only Range ORAM. Assuming that the CPU
sends only read instructions, we can achieve locality and obliviousness as follows.
The idea is to make replications of a set of super-blocks that form contiguous
memory regions. Specifically, let N be a power of 2 that bounds the size of the
logical memory. A size-2^i super-block consists of 2^i consecutive blocks with the
starting address being a multiple of 2^i. We call size-1 blocks as "primitive blocks".
We store $\log N$ different ORAMs, where the i-th ORAM (for $i = 0, \ldots, \log N - 1$)
stores all size-2^i (super-)blocks (exactly $N/2^i$ blocks of size 2^i each). Since any
contiguous memory region of length 2^i is "covered" by two super-blocks of that

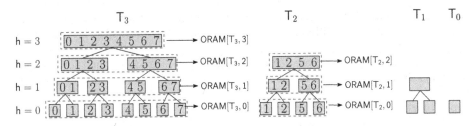

Fig. 1. Hierarchy of range trees. Logically, data is divided into trees of exponentially increasing sizes. In each tree block, a parent super-block stores the contents of both its children. If a block appears in more than one tree, the smallest tree contains the freshest copy. The above figure shows the state of the data structure after two accesses $(\text{read}, 5, 2, \bot)$ and $(\text{read}, 1, 2, \bot)$. h denotes height of a node in the Range Tree.

length, reading any contiguous memory of length 2^i region would boil down to making two accesses to the i-th ORAM.

However, this approach breaks down once we also need to support writes. The main challenge is to achieve data coherency in different ORAMs. Since there are multiple replicas of each data block, either a write must update all replicas, or a read must fetch all replicas to retrieve the latest copy. Both strategies break data locality.

2.5 Constructing Range ORAM

Range Trees. The aforementioned strawman scheme demonstrates the challenges we face if we want a Range ORAM supporting both reads and writes. To achieve this we need more sophisticated data structures.

We first describe a *logical* data structure called a *Range Tree* (without specifying at this point how to actually store this logical Range Tree on physical memory). A Range Tree of size 2^i is the following (logical) data-structure: the leaves store 2^i primitive blocks sorted by their (possibly non-contiguous) addresses, whereas each internal node replicates and stores all blocks contained in the leaves of its subtree. For example, in Fig. 1, each of $\mathsf{T}_0, \mathsf{T}_1, \mathsf{T}_2$ and T_3 is a logical Range Tree of sizes 1, 2, 4, 8 respectively. In such a Range Tree, each node at height j stores a super-block of size 2^j (leaves have height 0 and store primitive blocks).

Range ORAM's data structure. As shown in Fig. 1, our full Range ORAM (supporting both reads and writes) will *logically* contain a hierarchy of such Range Trees of sizes $1, 2, 4, 8, \ldots, N$, denoted $\mathsf{T}_0, \mathsf{T}_1, \ldots, \mathsf{T}_L$ respectively where $L = O(\log N)$. These trees form a hierarchy of stashes just like in hierarchical ORAM [23,25], i.e., each T_i is a stash for T_{i+1} which is twice as large. Thus, if a block at some logical address is replicated multiple times in multiple Range Trees, *the copy in a smaller Range Tree is always more fresh* (e.g., in Fig. 1, notice that the block at logical address 1 appears in both T_3 and T_2). Within each Range Tree, a logical block also appears multiple times within super-blocks (or primitive blocks) of different sizes, but all these copies within the same tree contain the same value.

We now specify how these logical Range Trees are stored in the physical memory. Basically, in each Range Tree, all super-blocks at the same height will be stored in a separate ORAM—thus an ORAM at height j of the tree stores super-blocks of size 2^j.

Besides the ORAMs storing each height of each Range Tree, we also need an auxiliary data structure that facilitates lookup. The client can access this data structure to figure out, for a requested range $[s, t]$, which super-blocks in a specific tree height intersect the request. This auxiliary data structure is stored on the server in an ORAM, and it can be viewed as a variant of "oblivious binary search tree".

Fetch phase of the Range ORAM. Let us now consider how to read and write contiguous ranges of blocks (i.e., implement the read and write operations of Range ORAM). Each request, no matter read or write requests, proceed in two phases, a fetch phase and a maintain phase. We first describe the fetch phase whose goal is to write down the requested range in a designated contiguous space on the server.

Suppose that the range $[s, t]$ is requested. Without loss of generality, assume that the length of the range $t - s + 1 = 2^i$ (otherwise round it up to the nearest power of 2). Roughly speaking, we would like to achieve the following effect:

- For every Range Tree at least 2^i in size, we would like to fetch all size-2^i super-blocks that intersect the range requested—it is not difficult to see that there are at most *two* such super-blocks.
- For every Range Tree smaller than 2^i in size, we simply fetch the root.
- Write down all these super-blocks fetched in a contiguous region on the server, and then obliviously reconstruct the freshest value of each logical address (using locality-friendly oblivious sort).

Henceforth we focus only on the Range Trees that are at least 2^i in size since for the smaller trees it is trivial to read the entire root. To achieve the above, roughly speaking, the client may proceed in the following steps. For each Range Tree that is not too small,

1. Look up the auxiliary data structure (stored on the server) to figure out which *two* super-blocks to request in the desired height that stores super-blocks of size 2^i;
2. Fetch these two desired super-blocks from the corresponding ORAM and write down the fetched super-block in a contiguous region (starting at a designated position) on the server's memory.

All these fetched super-blocks are written down on the server's memory contiguously (including the root nodes for the smaller Range Trees which we have ignored above). The client now relies on oblivious sorting to reconstruct the freshest copy of each logical address requested, and the result is stored in a designated contiguous region on the server.

Notice that the entire read procedure reads only polylogarithmically many contiguous memory regions:

- Queries to the oblivious auxiliary data structure accesses polylogarithmically many "small" metadata blocks using ordinary oblivious data structures;
- There are only logarithmically many requests to per-height ORAMs storing super-blocks of size 2^i. Using an ordinary ORAM scheme, this step requires reading polylogarithmically many regions of size 2^i. Here, since every super-block of size 2^i is bundled together, we do not need to read 2^i separate small blocks from an ORAM, and this is inherently why the algorithm's *locality is independent of the length of the range requested.*
- The oblivious sorting needed for reconstruction also consumes polylogarithmic locality as mentioned in Sect. 2.3.

Maintain phase of the Range ORAM. Inspired by the hierarchial ORAM [23,25], here a super-block fetched will be written to the smallest Range Tree that is large enough to fit this super-block. If this Range Tree is full, we will then perform a cascading merge to merge consecutive, full Range Trees into the next empty Range Tree.

During this rebuilding process, we must also maintain correctness, including but not restricted to the following:

- for duplicated copies of each block, figure out the freshest copy and suppress duplicates; and
- correctly rebuild the oblivious auxiliary data structure in the process.

Without going into algorithmic details at this point, most of this rebuilding process can be accomplished through a locality-friendly oblivious sorting procedure as mentioned earlier in Sect. 2.3. However, technically instantiating all the details and making everything work together is non-trivial. To enable this, we in fact introduce a new algorithmic abstraction, that is, *an ordinary ORAM scheme with a locality-friendly initialization procedure* (see Sect. 4.3). We will use this new building block to instantiate both the oblivious auxiliary data structure and each tree height's ORAM. In comparison with a traditional ORAM where rebuilding can be supported by writing the blocks one by one (which will consume super-linear locality), here we would like to rebuild the server-side ORAM data structure using a special locality-friendly algorithm upon receiving a possibly large input array of the blocks. In subsequent technical sections, we show how to have such a special ORAM scheme where initializing the server-side data structure can be accomplished using locality-friendly oblivious sorting as a building block. We refer the reader to Sect. 5 for the algorithmic details.

2.6   Online Range ORAM

Given our Range ORAM abstraction, we are now ready to construct Online Range ORAM. The difference is that now, when the client receives request, it is unaware whether the future requests will be contiguous. In fact, Online Range ORAM provides the same interface as an ordinary ORAM: each request the client receives is of the form $(\mathsf{op}, \mathsf{addr}, \mathsf{data})$ where $\mathsf{op} \in \{\mathtt{read}, \mathtt{write}\}$, and $\mathsf{addr} \in [N]$ specifies a single address to read or write (with data). Yet the Online Range ORAM must preserve the locality that is available in the request sequence up to polylogarithmic factors.

Roughly speaking, we can construct Online Range ORAM from Range ORAM as follows, by using a predictive prefetching idea: when a request (containing a single address) comes in, the client first requests that singe address. When a new request comes in, it checks whether the request is consecutive to the address of the previous request. If so, it requests 2 contiguous blocks – the specified address and also its next address. This can be done by requesting a range in Range ORAM. If the next 2 requests happen to be contiguous, then the client prefetches the next 4 blocks with Range ORAM; and if the requests are still contiguous, it will next prefetch 8 blocks with Range ORAM. At any time if the contiguous pattern stops, back off and start requesting a region of size 1 again. It is not hard to see that the Online Range ORAM still preserves polylogarithmic bandwidth blowup; moreover, if the request sequence contains a contiguous region of length len, it will be separated into at most $\log(\mathsf{len})$ Range ORAM requests. Thus the Online Range ORAM's locality is only a logarithmic factor worse than the Range ORAM. The reader is referred to Sect. 5.5 for further details.

2.7   Related Work

Related work on locality. Algorithmic performance with data stored on the disk has been studied in the external memory models (e.g., [4,35,39,40] and references within). Fundamental problems in this area include scanning, permuting, sorting, range searching, where there are known lower bounds and matching upper bounds.

Relationship to locality-preserving SSE. Searchable symmetric encryption (SSE) enables a client to encrypt an index of record/keyword pairs and later retrieve all records matching a keyword. The typical approach (e.g., [17,19,28, 31,38], and references within) is to store an inverted index. Our work is inspired by recent works that study locality in SSE schemes [7,8,13,20,21]. Our new locality ORAM formulation can be viewed as a generalization of the one-time ORAM (with free rebuild) construct adopted in recent SSE constructions.

In a concurrent work, Demertzis, Papadopoulos and Papamanthou [20] also consider such a one-time ORAM (with free rebuild) abstraction for an SSE application. In their construction, they leverage as a building block a perfectly secure (multi-use) ORAM with $O(1)$-locality, by blowing up the bandwidth to $O(\sqrt{N})$ and the client storage to $O(N^{2/3})$. This construction fails to preserve the locality of the input program, and when accessing a region of size len will result in $O(\mathsf{len})$-locality, and $O(\mathsf{len} \cdot \sqrt{N})$-bandwidth. In contrast, we achieve $\mathrm{poly} \log N$-locality and $\mathsf{len} \cdot \mathrm{poly} \log N$-bandwidth when accessing a region of size len, and with $O(1)$-client space.

Oblivious RAM (ORAM). Numerous works [6,27,32,34,36,37,41–45] construct ORAMs in different settings. Most of ORAM constructions follow one of two frameworks: the hierarchical framework, originally proposed by Goldreich and Ostrovsky [23,25], or the tree-based framework proposed by Shi et al. [36].

Up until recently, the asymptotically most efficient scheme was given by [32], providing $O(\log^2 N/\log \log N)$ bandwidth. A recent improvement was given by Patel et al. [34], reducing the bandwidth to $O(\log N \cdot \text{poly} \log \log N)$. The scheme of Asharov et al. [6] achieves $O(\log N)$ bandwidth, and matches the lower bounds given by Goldreich and Ostrovsky [23,25] and Larsen and Nielsen [33]. Further, the Goldreich-Ostrovsky lower bound is also known not to hold when the memory (i.e., ORAM server) is capable of performing computation [3,22], which is beyond the scope of this paper.

In a subsequent work, Chakraborti et al. [14] show an ORAM called rORAM with good locality and with $O(\log^2 N)$ bandwidth assuming $\Omega(\log^2 N)$ block size. Their scheme is based on tree-based ORAM. The construction works with large client storage (i.e., linear in the sequential data to be read/write), and reducing this client storage to $O(1)$ would incur multiplicative poly $\log N$ factors in locality and bandwidth in addition to using more disks to achieve locality.

3   Definitions

Notations and conventions. We let $[n]$ denote the set $\{1, \ldots, n\}$. We denote by p.p.t. probabilistic polynomial time Turing machines. A function $\text{negl}(\cdot)$ is called negligible if for any constant $c > 0$ and all sufficiently large λ's, it holds that $\text{negl}(\lambda) < \lambda^{-c}$. We let λ denote the security parameter. For an ensemble of distributions $\{D_\lambda\}$ (parametrized with λ), we denote by $x \leftarrow D_\lambda$ a sampling of an instance according to the distribution D_λ. Given two ensembles of distributions $\{X_\lambda\}$ and $\{Y_\lambda\}$, we use the notation $\{X_\lambda\} \overset{\epsilon(N)}{\equiv} \{Y_\lambda\}$ to say that the two ensembles are statistically (resp. computationally) indistinguishable if for any unbounded (resp. p.p.t.) adversary $\mathcal{A}$,

$$\left| \Pr_{x \leftarrow X_\lambda} \left[\mathcal{A}(1^\lambda, x) = 1 \right] - \Pr_{y \leftarrow Y_\lambda} \left[\mathcal{A}(1^\lambda, y) = 1 \right] \right| \leq \epsilon(\lambda)$$

Throughout this paper, for underlying building blocks, we will use n to denote the size of the instance and use λ to denote the security parameter. For our final ORAM constructions, we use N to denote the size of the total logical memory size as well as the security parameter—note that this follows the convention of most existing works on ORAMs [23,25,27,32,36,37,41].

3.1   Memory with Multiple Disks and Data Locality

To understand the notion of data locality, it may be convenient to view the memory as D rotational hard drives or other storage mediums where sequential accesses are faster than random accesses. The program interacting with the memory has to specify which disk to access. Each disk is equipped with one read/write head. In order to serve a read or write request with address addr in some disk $d \in [D]$, the memory has to move the read/write head of the disk d to the physical location addr to perform the operation. Any such movement of the head introduces cost and delays, and the machine that interacts with the memory would like to minimize the number of move head operations. Traditionally, the latter can be improved by ensuring that the program accesses contiguous regions of the memory. However, this poses a great challenge for oblivious computation in which data is often continuously shuffled across memory.

More formally, a memory is denoted as $\mathsf{mem}[N, b, \mathsf{D}]$, consisting of D disks, indexed by the address space $[N] = \{1, 2, \ldots, N\}$, where $\mathsf{D} \cdot N$ is the size of the logical memory. We refer to each memory word also as a *block* and we use b to denote the bit-length of each block. The memory supports the following two types of instructions.

- **Move head operation** $(\mathsf{move}, \mathsf{d}, \mathsf{addr})$ moves the head of the d-th disk $(\mathsf{d} \in [\mathsf{D}])$ to point to address addr within that disk.
- **A read/write operation** $(\mathsf{op}, \mathsf{d}, \mathsf{data})$, where $\mathsf{op} \in \{\mathsf{read}, \mathsf{write}\}$, $\mathsf{d} \in [\mathsf{D}]$ and $\mathsf{data} \in \{0, 1\}^b \cup \{\bot\}$. If $\mathsf{op} = \mathsf{read}$, then $\mathsf{data} = \bot$ and mem should return the content of the block pointed to by the d-th disk; If $\mathsf{op} = \mathsf{write}$, the block pointed to by the d-th disk is updated to data. The d-th head is then incremented to point to the next consecutive address, and wrapped around when the end of the disk is reached.

Locality. A sequence of memory operations has (D, ℓ) worst-case locality if it contains ℓ move operations to a memory that is equipped with D disks.

Examples. The above formalism enables us to distinguish between different degrees of locality, such that:

- An algorithm that just accesses an array sequentially can be described using a program that is $(1, O(1))$-local.
- An algorithm that computes the inner product of two vectors can be implemented with $(2, O(1))$-local (but cannot be implemented with $O(1)$ locality with 1 disk).
- An algorithm that merges two sorted arrays is $(3, O(1))$-local (and cannot be implemented with $O(1)$ locality with only 2 disks).
- An algorithm that makes N random accesses to an array is $(\mathsf{D}, \Theta(N))$-local for any constant number of D disks with overwhelming probability.

Relation to the standard memory definition. Instead of specifying which disk to read from/write to, we can define a memory of range $[\mathsf{D} \cdot N] = \{1, \ldots, \mathsf{D} \cdot N\}$. The address space determines the disk index, and therefore also whether or not to move the read/write head. Thus, one can consider the regular notion of a RAM program, and our definition provides a way to measure the locality of the program. Different implementations of the same functionality can have different locality, similarly to other metrics.

3.2   Oblivious Machines

In this section, we define oblivious simulation of functionalities, either stateless (non-reactive) or stateful (reactive). As most prior works, we consider oblivious simulation of deterministic functionalities only. We capture a stronger notion than what is usually considered, in which the adversary is adaptive and can issue request as a function of previously observed access pattern.

Warmup: Oblivious simulation of a stateless deterministic functionality. We consider machines that interact with the memory via move and read/write operations. In case of a stateless (non-reactive) functionality, the machine M receives one instruction I as input, interacts with the memory, computes the output and halts. Formally, we say that the stateless algorithm M obliviously simulates a stateless, deterministic functionality f w.r.t. to the leakage function leakage : $\{0,1\}^* \to \{0,1\}^*$, iff

- **Correctness:** there exists a negligible function $\mu(\cdot)$ such that for every λ and I, $M(1^\lambda, I) = f(I)$ except with $\mu(\lambda)$ probability.
- **Obliviousness:** there exists a stateless p.p.t. simulator Sim, such that for any λ and I, $\mathsf{Addr}(M(1^\lambda, I)) \overset{\epsilon(\lambda)}{\equiv} \mathsf{Sim}(1^\lambda, \mathsf{leakage}(I))$, where $\mathsf{Addr}(M(1^\lambda, I))$ is a random variable denoting the addresses incurred by an execution of M over the input I.

Depending on whether $\overset{\epsilon(\lambda)}{\equiv}$ refers to computational or statistical indistinguishability, we say M is computationally or statistically oblivious. If $\epsilon(\cdot) = 0$, we say M is perfectly oblivious. For example, an oblivious sorting algorithm is an oblivious simulation of the functionality that receives an array and sorts it (according to some specified preference function), where the leakage function contains only the length of the array being sorted.

Oblivious simulation of a stateful functionality. We often care about oblivious simulation of stateful functionalities. For example, the ordinary ORAM is an oblivious simulation of a logical memory abstraction. We define a composable notion of security for oblivious simulation of a stateful functionality below. This time, the machine M, the simulator Sim, the functionality f and the leakage function leakage are all interactive machines that might receive instructions as long as they are activated, and each might maintain a secret state. Moreover, we explicitly introduce the distinguisher $\mathcal{A}$, which is now also an interactive machine. In each step, the distinguisher $\mathcal{A}$ observes the access pattern and selects the next command to perform. We write $(\mathsf{out}_i, \mathsf{addr}_i) \leftarrow M(I_i)$, where out_i denotes the intermediate output of M for the instruction I_i, and addr_i denote the memory addresses accessed by M when answering the instruction I_i. We have:

Definition 3.1 (Adaptively secure oblivious simulation of stateful functionalities). *Let M, leakage, f be interactive machines. We say that M obliviously simulates a possibly randomized, stateful functionality f w.r.t. to the leakage function leakage iff there exists an (interactive) p.p.t. simulator Sim, such that for any non-uniform (interactive) p.p.t. adversary $\mathcal{A}$, $\mathcal{A}$'s view in the following two experiments, $\mathsf{Expt}_{\mathcal{A}}^{\mathrm{real}, M}$ and $\mathsf{Expt}_{\mathcal{A}, \mathsf{Sim}}^{\mathrm{ideal}, f}$ are computationally indistinguishable.*

$$\begin{array}{|l|}
\hline
\mathsf{Expt}_{\mathcal{A}}^{\mathrm{real},M}(1^\lambda): \\
\hline
\mathsf{out}_0 = \mathsf{addr}_0 = \bot \\
\text{For } i = 1, 2, \ldots \mathrm{poly}(\lambda): \\
\quad I_i \leftarrow \mathcal{A}(1^\lambda, \mathsf{out}_{i-1}, \mathsf{addr}_{i-1}) \\
\quad \mathsf{out}_i, \mathsf{addr}_i \leftarrow M(I_i) \\
\hline
\end{array}
\qquad
\begin{array}{|l|}
\hline
\mathsf{Expt}_{\mathcal{A},\mathsf{Sim}}^{\mathrm{ideal},f}(1^\lambda): \\
\hline
\mathsf{out}_0 = \mathsf{addr}_0 = \bot \\
\text{For } i = 1, 2, \ldots \mathrm{poly}(\lambda): \\
\quad I_i \leftarrow \mathcal{A}(1^\lambda, \mathsf{out}_{i-1}, \mathsf{addr}_{i-1}) \\
\quad \mathsf{out}_i \leftarrow f(I_i) \\
\quad \mathsf{addr}_i \leftarrow \mathsf{Sim}(\mathsf{leakage}(I_i)) \\
\hline
\end{array}$$

In the above definition, if we replace computational indistinguishability with statistical indistinguishability (or identically distributed resp.) and remove the requirement for the adversary to be polynomially bounded, then we say that the stateful machine M obliviously simulates the stateful functionality f with statistical (or perfect resp.) security. Besides the leakage of the individual instruction, the simulator might have some additional information in the form of the public parameters of the functionality. We also remark that Definition 3.1 captures correctness and obliviousness simultaneously, and capture both *deterministic* and *randomized* functionalities. We refer the reader to the relevant discussions in the literature of secure computation for the importance of capturing correctness and obliviousness simultaneously for the case of randomized functionalities [11,24].

Our definition of oblivious simulation is general and captures any stateless or stateful functionality, and thus later in the paper, whenever we define any oblivious algorithm, it suffices to state (1) what functionality it computes; (2) what is the leakage; and (3) what security (i.e., computational, statistical, or perfect) we achieve. We use ordinary ORAM as an example to show how to use our definitions.

Ordinary ORAM. As an example, a conventional ORAM, first proposed by Goldreich and Ostrovsky [23], is an oblivious simulation of a "logical memory functionality", parameterized by (N, b), where N is the size of the logical memory and b is the block size:

- **Functionality:** The internal state of the functionality consists of an array $\mathsf{mem} \in (\{0, 1\}^b)^N$. Upon each instruction of the form $(\mathsf{op}, \mathsf{addr}, \mathsf{data})$, with $\mathsf{op} \in \{\mathtt{read}, \mathtt{write}\}$, $\mathsf{addr} \in [N]$, and $\mathsf{data} \in \{0, 1\}^b \cup \{\bot\}$, the functionality proceeds as follows. If $\mathsf{op} = \mathtt{write}$, then $\mathsf{mem}[\mathsf{addr}] = \mathsf{data}$. In both cases, the functionality returns $\mathsf{mem}[\mathsf{addr}]$.
- **Leakage:** The simulator has the public parameters of the functionality – N and b. With each instruction $(\mathsf{op}, \mathsf{addr}, \mathsf{data})$, the leakage is just that an access has been performed.

We remark that previous constructions of ORAM [32,37,41] in fact satisfy Definition 3.1.

Bandwidth, and private storage of oblivious machines. Throughout the paper, we use the terminology *bandwidth* to denote the total number of memory read/write operations of size $\Omega(\log N)$ a machine needs to use. We assume the machine/algorithm has only $O(1)$ blocks of private storage.

Remark. In this paper, we focus on hiding the access patterns to the memory, but not the data contents. Therefore, we do not explicitly mention that data is (re-)encrypted when it is accessed, but encryption should be added since the adversary can observe memory contents. That is, while we assume that the adversary completely sees the instructions (move, d, addr) and (op, d, data) that are sent to the memory, data should be encrypted. Note, however, that the adversary sees in particular the contents and accesses of all disks.

4   Locality-Friendly Building Blocks

In this section, we describe several locality-friendly building blocks that are necessary for our constructions.

4.1   Oblivious Sorting Algorithms with Locality

An important building block for our construction is an oblivious sorting algorithm that is locality-friendly. In Appendix A, we describe an algorithm for Bitonic sort to achieve good locality, and provide a detailed analysis.

Theorem 4.1 ((Theorem 2.2, restated) Perfectly secure oblivious sort with locality). *Bitonic sort (when implemented as in Appendix A) is a perfectly oblivious sorting algorithm that sorts n elements using $O(n \log^2 n)$ bandwidth and $(2, O(\log^2 n))$ locality.*

4.2   Oblivious Deduplication with Locality

We define a handy subroutine that removes duplicates obliviously. $Y \leftarrow \mathsf{Dedup}(X, n_Y)$, where X contains some real elements and dummy elements, and n_Y is some target output length. It is assumed that each real element is of the form $((k, k'), v)$ where k is a primary key and k' is a secondary key. The subroutine outputs an array Y of length n_Y in which for each primary key k in X, only the element with the smallest secondary key k' remains (possibly with some dummies at the end). It is assumed that the number of primary keys k is bounded by n_Y.

Given a locality-friendly oblivious sort, we can easily realize oblivious Dedup with locality. We obliviously sort X by the (k, k') tuple, scan X to replace duplicates with dummies, and sort X again to move dummies towards the end. Finally, pad or truncate X to have length n_Y and output. The procedure is just few scans of the array and 2 invocations of oblivious sort, and therefore the bandwidth and locality is the same as the oblivious sort. Concretely, using Theorem 4.1 this can be implemented using $O(|X| \log^2 |X|)$-bandwidth and $(2, O(\log^2 |X|))$-locality.

4.3   Locally Initializable ORAM

In this section, we show that the oblivious sort can be utilized to define an (ordinary) ORAM scheme that is also locally initializable.

A *locally initializable ORAM* is an ORAM with the additional property that it can be initialized efficiently and in a locality-friendly manner given a batch of initial blocks. The syntax and definitions of a locally initializable ORAM is the same as a normal ORAM, except that the first operation in the sequence is a locality-friendly initialization procedure. More formally, a locally initializable ORAM is an oblivious implementation of the following functionality, parametrized by N and b:

- **Secret state:** an array mem of size N and block size b. Initially all are 0.
- T.Build(X) takes an input array X of $|X| < N$ blocks of the form $(\mathsf{addr}_i, \mathsf{data}_i)$ where each $\mathsf{addr}_i \in [N]$ and $\mathsf{data}_i \in \{0,1\}^b$. Blocks in X have distinct integer addresses that are not necessarily contiguous. The functionality has no output, but it updates its internal state: For every $i = 1, \ldots, |X|$ it writes $\mathsf{mem}[\mathsf{addr}_i] = \mathsf{data}_i$.
- $B \leftarrow$ T.Access$(\mathsf{op}, \mathsf{addr}, \mathsf{data})$ with $\mathsf{op} \in \{\mathtt{read}, \mathtt{write}\}$, $\mathsf{addr} \in [N]$, and $\mathsf{data} \in \{0,1\}^b$. If $\mathsf{op} = \mathtt{write}$ then $\mathsf{mem}[\mathsf{addr}] = \mathsf{data}$. In both cases of $\mathsf{op} = \mathtt{read}$ and $\mathsf{op} = \mathtt{write}$, return $\mathsf{mem}[\mathsf{addr}]$.

The leakage function of locally initializable ORAM reveals $|X|$ and the number of Access operations (as well as the public parameters N and b). Obliviousness is defined as in Definition 3.1 with the above leakage and functionality.

Locality-friendly initialization. We now show that the hierarchical ORAM by Goldreich and Ostrovsky [23] can be initialized in a locality-friendly manner, i.e., how to implement Build with $(2, O(\mathsf{poly} \log n))$ locality, where $n = |X|$. To initialize a hierarchical ORAM, it suffices to place all the n blocks in the largest level of capacity n. In the Goldreich and Ostrovsky ORAM, each block is placed into one of the n bins by applying a pseudorandom function $\mathsf{PRF}_K(\mathsf{addr})$ where K is a secret key known only to the CPU and addr is the block's address. By a simple application of the Chernoff bound, except with $\mathsf{negl}(\lambda)$ probability, each bin's utilization is upper bounded by $\alpha \log \lambda$ for any super-constant function α. Goldreich and Ostrovsky [23] show how to leverage oblivious sorting to obliviously initialize such a hash table. For us to achieve locality, it suffices to use a locality-friendly oblivious sort algorithm such as Bitonic sort. This gives rise to the following theorem:

Theorem 4.2 (Computationally secure, locally initializable ORAM). *Assuming one-way functions exist, there exists a computationally secure locally-initializable ORAM scheme that has* $\mathsf{negl}(\lambda)$ *failure probability, and can be initialized with n blocks using $(n + \lambda) \cdot \mathsf{poly} \log(n + \lambda)$ bandwidth and $(2, \mathsf{poly} \log(n + \lambda))$ locality, and can serve an access using $\mathsf{poly} \log(n + \lambda)$ bandwidth and $(2, \mathsf{poly} \log(n + \lambda))$ locality.*

Notice that for ordinary ORAMs, since the total work for accessing a singe block is only polylogarithmic, obtaining polylogarithmic locality per access is trivial. Our goal later is to achieve ORAMs where even if you access a large file or large region, the locality is still polylogarithmic, i.e., one does not need to split up the file into little blocks and access them one by one. Our constructions later will leverage a locally initializable, ordinary ORAM as a building block.

5   Range ORAM

In this section, we define range ORAM and present a construction with poly-logarithmic bandwidth and poly-logarithmic locality. The construction uses a building block which we call an oblivious range tree (Sect. 5.2). It supports read-only range lookup queries with low bandwidth and good locality. From an oblivious range tree, we show how to construct a range ORAM, which supports reads and updates (Sect. 5.3). Then, we discuss statistical and perfect security in Sect. 5.4. Finally, we extend Range ORAM to online Range ORAM (Sect. 5.5).

Our ORAM construction uses multiple disks only when it invokes an oblivious sort operation (and Dedup operation which invokes an oblivious sort). Thus, for the following algorithms, it can be assumed that the entire data is stored on a single disk. Multiple disks are used only transiently using during an oblivious sort or a Dedup operation.

5.1   Range ORAM Definition

A Range ORAM is an oblivious machine that supports read/write range instructions, and interacts with the memory while leaking only the size of the range. Formally, using Definition 3.1, Range ORAM is defined as follows, parameterized by N and b:

Functionality: The internal state is an array mem of size N and blocksize b. Range ORAM takes as input range requests in the form $\mathsf{Access}(\mathsf{op}, [s, t], \mathsf{data})$, where $\mathsf{op} \in \{\mathtt{read}, \mathtt{write}\}$, $s, t \in [N]$, $s < t$, and $\mathsf{data} \in (\{0, 1\}^b)^{(t-s+1)}$. If $\mathsf{op} = \mathtt{read}$, then it returns $\mathsf{mem}[s, \ldots, t]$. If $\mathsf{op} = \mathtt{write}$, then $\mathsf{mem}[s, \ldots, t] = \mathsf{data}$.

Leakage: With each instruction $\mathsf{Access}(\mathsf{op}_i, [s_i, t_i], \mathsf{data}_i)$, range ORAM leaks $t_i - s_i + 1$.

5.2   Oblivious Range Tree

A necessary building block for construction Range ORAM is a Range Tree. An oblivious Range Tree is a *read-only* Range ORAM with an initialization procedure from a list of blocks with possibly non-contiguous addresses. Formally, it is an oblivious simulation of the following reactive functionality with the following leakage (where obliviousness is defined using Definition 3.1):

Functionality: Formally, an oblivious Range Tree T supports the following operations:

- T.Build(X) takes in a list X of blocks of the form $(\mathsf{addr}, \mathsf{data})$. Blocks in X have distinct integer addresses that are not necessarily contiguous. Store X as the secret state. Build has no output.
- B $\leftarrow$ T.Access($\mathtt{read}, [s, t], \bot$) takes in a range $[s, t]$ and returns all (and only) blocks in X that has addr in the range $[s, t]$. We assume $\mathsf{len} = t - s + 1 = 2^i$ is a power of 2 for simplicity.

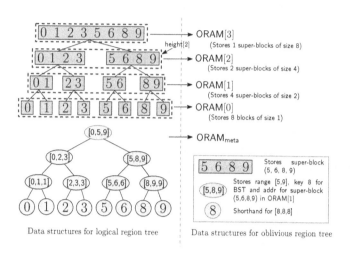

Fig. 2. An oblivious Range Tree with Locality.

Leakage: T.Build(X) leaks $|X|$. Each T.Access($\texttt{read}, [s,t], \perp$) leaks $t - s + 1$.

A logical Range Tree. For simplicity, assume $n := |X|$ is a power of 2; if not, we simply pad with dummy blocks that have $\texttt{addr} = \infty$. A logical Range Tree is a full binary tree with n leaves. Each leaf contains a block in X, sorted by $\texttt{addr}$ from left to right. Each internal node is a *super-block*, i.e., blocks from all leaves in its subtree concatenated and ordered by addresses. A height-i super-block thus has size 2^i. The leaves are at height 0, and the root is at height $\log_2 n$.

Metadata tree. Each super-block in the logical Range Tree defines a range: $[a_s, a_m, a_t]$ where a_s is the lowest address, a_t is the highest address, and a_m is the middle address (the address of the 2^{i-1}-th block for a height-i super-block). We use another full binary tree to store the range metadata of each super-block, henceforth referred to as the *metadata tree*. The metadata tree is a natural binary search tree that supports the following search operations:

– Given a request range $[s,t]$ with $\texttt{len} := t - s + 1 = 2^i$, find the leftmost and rightmost height-i (super-)blocks whose ranges intersect $[s,t]$, or return $\perp$ if none is found.

Since $t - s + 1 = 2^i$, the leftmost and rightmost height-i (super-)blocks that intersect $[s,t]$ (if they exist) are either contiguous or the same node.

Next, to achieve obliviousness, we will put the metadata tree and *each height* of the logical range tree into a separate ORAM, as shown in Fig. 2.

Algorithm 5.1: T.Build(X). The Build algorithm takes a list of blocks X, constructs the logical Range Tree and metadata tree, and then puts them into ORAMs through local initialization (Sect. 4.3).

1. **Create leaves.** Obliviously sort X by the addresses. Pad X to the nearest power of 2 with dummy blocks that have $\texttt{addr} = \infty$. Let $\texttt{height}[0]$ denote the sorted X, which will be the leaves of the logical Range Tree.

2. **Create super-blocks.** For each height $i = 1, 2, \ldots, L := \log_2 n$, create height-$i$ super-blocks by concatenating their two child nodes. Let height$[i]$ denote the set of height-i super-blocks. Tag each super-block with its offset in the height.

3. **Create metadata tree.** Let metadata be the resulting metadata tree represented as an array, i.e., metadata$[i]$ is the parent of metadata$[2i + 1]$ and metadata$[2i + 2]$. Tag each node in the metadata tree with its offset in metadata.

4. **Put each height and metadata tree in ORAMs.** For each height $i = 0, 1, \ldots, L$, let H_i be a locally initializable ORAM from Sect. 4.3, and call $\mathsf{H}_i.\mathsf{Build}(\mathsf{height}[i])$ in which each height-i super-block behaves as an atomic block. Let H_meta be a locally initializable ORAM, and call $\mathsf{H}_\mathrm{meta}.\mathsf{Build}(\mathsf{metadata})$.

Algorithm 5.2: $\mathsf{T}.\mathsf{Access}(\mathtt{read}, [s, t], \perp)$ (with $\mathsf{len} = t - s + 1 = 2^i$)

1. **Look up address.** Call $\mathsf{H}_\mathrm{meta}.\mathsf{Access}(\cdot)$ $2L$ times to obliviously search for the leftmost and rightmost height-i (super-)blocks in the logical Range Tree that intersects $[s, t]$. Suppose they have addresses addr_1 and addr_2 (which may be the same and may both be $\perp$).

2. **Retrieve super-blocks.** Call $\mathsf{B}_1 \leftarrow \mathsf{H}_i.\mathsf{Access}(\mathtt{read}, \mathsf{addr}_1, \perp)$ and $\mathsf{B}_2 \leftarrow \mathsf{H}_i.\mathsf{Access}(\mathtt{read}, \mathsf{addr}_2, \perp)$ to retrieve the two (super-)blocks.

3. **Output.** Remove blocks from B_1 and B_2 that are not in $[s, t]$. Output $\mathsf{B} = \mathsf{Dedup}(\mathsf{B}_1 \;\|\; \mathsf{B}_2, \mathsf{len})$.

We prove the following theorem in the full version.

Theorem 5.3 (Oblivious Range Tree). *Assuming one-way functions exist, there exists a computationally secure oblivious Range Tree scheme that has correctness except with $\mathsf{negl}(\lambda)$ probability, and*
- Build *requires* $n \cdot \mathsf{poly} \log(n + \lambda)$ *bandwidth and* $(2, \mathsf{poly} \log(n + \lambda))$ *locality,*
- Access *requires* $\mathsf{poly} \log(n + \lambda)$ *bandwidth and* $(2, \mathsf{poly} \log(n + \lambda))$ *locality.*

5.3   Range ORAM from Oblivious Range Tree

In this section, we show how to construct a Range ORAM from oblivious Range Tree scheme. Since the underlying oblivious Range Tree has good efficiency/locality, so will the resulting Range ORAM. The idea behind our construction is similar to that of the standard hierarchical ORAM [23,25]. Intuitively, where a standard hierarchical ORAM employs an oblivious hash table, we instead employ an oblivious Range Tree.

Data structure. We use N to denote both the total size of logical data blocks as well as the security parameter. There are $\log N + 1$ *levels* numbered $0, 1, \ldots, L$ respectively, where $L := \lceil \log_2 N \rceil$ is the maximum level. Each level is an oblivious Range Tree denoted $\mathsf{T}_0, \mathsf{T}_1, \ldots, \mathsf{T}_L$ where T_i has capacity 2^i. Data will be replicated across these levels. We maintain the invariant that data in lower levels are fresher. At any time, each T_i can be in two possible states, *non-empty* or *empty*. Initially, the largest level is marked non-empty, whereas all other levels are marked empty.

Algorithm 5.4: Range ORAM Access(op, $[s, t]$, data)   (with $t - s + 1 = 2^i$ for some i).

1. Retrieve all blocks in range trees of capacity no more than 2^i, i.e., fetched := $\cup_{j=0}^{i-1} T_j$. This can be easily done by fetching its root. Mark blocks in fetched that are not in the range $[s, t]$ as dummy.
   Each real block in fetched is tagged with its level number j as a secondary key so that later after calling Dedup(fetched, $t - s + 1$), where Dedup is defined in Sect. 4.2, only the most fresh version of each block remains. We assume each block also carries a copy of its address.
2. For each $j = i, i + 1, \ldots, L$, if T_j is non-empty, let fetched = fetched $\cup$ T_j.Access(read, $[s, t], \bot$).
3. Let data* := Dedup(fetched, 2^i). If op = read , then data* will be returned at the end of the procedure. Else, data* := data.
4. If all levels $\leq i$ are marked empty then perform T_i.Build(data*) and mark it as ready. Otherwise:
   (a) Let ℓ denote the smallest level greater than i that is empty. If no such level exists, let $\ell := L$.
   (b) Let $S := \cup_{j=0}^{\ell-1} T_j$. If $\ell = L$, additionally include $S := S \cup T_L$. Call T_ℓ.Build(Dedup($S, 2^\ell$)) and T_i.Build(data*). Mark levels ℓ and i as non-empty, and all other levels below ℓ as empty.

Example. We show a simple example for how levels are updated after some accessed. We assume initially that all blocks are stored in the largest Range Tree. Consider the following sequence of ranges $[1, 1], [2, 3], [4, 5], [6, 6]$.

- Access $[1, 1]$: A block of size 1. Added to T_0.
- Access $[2, 3]$: A block of size 2, and so $i = 1$. Levels $\leq i$ are not empty. The smallest empty level larger than $i = 1$ is 2. Thus, move $[1, 1]$ to T_2 (which has capacity 4), and then put $[2, 3]$ to T_1. At this point, T_0 is empty and T_1 and T_2 are occupied.
- Access $[4, 5]$: A block of size 2, and so $i = 1$. Levels $\leq i$ are not empty. The smallest empty level larger than $i = 1$ is 3. Thus, move $\{1, 2, 3\}$ to T_3 (which has capacity 8), and then put $[4, 5]$ to T_1. At this point, T_0 and T_2 are empty, and T_1 and T_3 are occupied.
- Access $[6, 6]$: A block of size 1, and so $i = 0$. Levels $\leq i$ are empty. $[6, 6]$ is added to T_0. At this point, T_2 is empty, and T_0, T_1 and T_3 are occupied.

The following theorem is proven in the full version of the paper.

Theorem 5.5 (Range ORAM). *Assuming one-way functions exist, there exists a computationally secure Range ORAM consuming $O(N \log N)$ space with* negl(N) *failure probability, and* len$\cdot$poly log N *bandwidth and* $(2, \text{poly} \log N)$ *locality for accessing a range of size* len.

We remark that the both bandwidth and locality are in an amortized sense: for sufficiently large amount of accesses of contiguous addresses len$_1, \ldots,$ len$_m$, the total bandwidth is $(\sum_{i=1}^m \text{len}_i) \cdot \text{poly} \log N$ and locality is $(2, m \cdot \text{poly} \log N)$.

5.4   Perfectly Security Range ORAM

The computational security in our construction is due to the use of a computationally secure locally initializable hierarchical ORAM (Theorem 4.2).

We can achieve perfect security by making the perfectly secure ORAM construction with polylogarithmic bandwidth in Chan et al. [15] locally initializable.

For a hierarchical ORAM, within each level, the position of a data block is determined by applying a PRF to the block's logical address. To achieve perfect security, Chan et al. [15] replace the PRF with a truly random permutation. To access a block within a level, the client must first figure out the block's correct location within the level. If the client had linear storage, it could simply store the locations (or position labels). To achieve small client storage, Chan et al. recursively store the position labels in a smaller ORAMs, similar to the idea of recursion in tree-based ORAMs [36]. Thus, there are logarithmically many ORAMs (each is a perfectly secure hierarchical ORAM), where the ORAM at depth d stores position labels for the ORAM at depth $d + 1$; and finally, the ORAM at the maximum depth $D = O(\log N)$ stores the real data blocks.

The Build procedure for one ORAM depth relies only on oblivious sorts and linear scans, and thus consumes $(2, \text{poly} \log N)$ locality using locality-preserving Bitonic sort. The Build procedure for one ORAM depth outputs its position map, which is subsequently used to initialize the next ORAM depth. Thus, all ORAM depths combined can be initialized with $(2, \text{poly} \log N)$ locality. Thus, we have the following theorem.

Theorem 5.6 (Perfectly secure Range ORAM). *There exists a perfectly secure Range ORAM consuming* $O(N \log N)$ *space,* len $\cdot$ poly $\log N$ *bandwidth and* $(2, \text{poly} \log N)$ *locality for accessing a range of size* len.

5.5   Online Range ORAM

So far, our range ORAM assumes an abstraction where we have foresight on how many contiguous locations of logical memory we wish to access. We now consider an *online* variant, where the memory requests arrive one by one just as in normal ORAM. Formally:

Functionality: A logical memory functionality that supports the following types of instructions:
- (op, addr, data): where op $\in$ {read, write}, addr $\in [N]$ and data $\in \{0, 1\}^b \cup \{\bot\}$. If op = write, then write mem[addr] = data. In both cases, return mem[addr].

Leakage: Consider a sequence of requests $\mathbf{I} = ((\text{op}_1, \text{addr}_1, \text{data}_1), \dots, (\text{op}_i, \text{addr}_i, \text{data}_i), \dots)$. Each instruction leaks one bit indicating whether the last instruction is contiguous, i.e., for every i, the leakage is 1 iff $\text{addr}_{i+1} = \text{addr}_i + 1$.

Blackbox construction of online range ORAM from range ORAM. Given a range ORAM construction, we can convert it to an online range ORAM scheme as follows, incurring only logarithmic further blowup. Intuitively, the idea is to prefetch a contiguous region of size 2^k every time a 2^k contiguous region has been accessed. That is, if a contiguous region of overall size 2^k is being read, then it is fetched as k distinct blocks of size $1, 2, 4, 8, \ldots, 2^k$. The detailed construction is given below:

Let prefetch be a dedicated location in memory storing prefetched contiguous memory regions. Initially, let rsize $:= 1$, $p = 1$, and let prefetch $:= \perp$. Upon receiving a memory request:

- If prefetch$[p]$ does not match the logical address requested, then do the following.
    1. First, write back the entire prefetch back into the range ORAM.
    2. Next, request a region of length 1 consisting of only the requested logical address, store the result in prefetch;
    3. Reset $p := 1$ and rsize $:= 1$;
- Read and write prefetch$[p]$, and let $p := p + 1$.
- If $p > $ rsize, then do the following.
    1. First, let rsize $:= 2 \cdot$ rsize.
    2. Next, write prefetch back into the range ORAM.
    3. Now, prefetch the next contiguous region containing rsize logical addresses, and store them in prefetch, and let $p := 1$.

It is not hard to see that given the above algorithm, accessing each range of size R will be broken up into at most $O(\log R)$ accesses, to regions of sizes $1, 2, 4, \ldots, R$ respectively, and each size has one read request and one write request. Security is straightforward as range ORAM is oblivious, and the transformation between the leakage profiles of online range ORAM and range ORAM is straightforward. Thus we have the following theorem.

Theorem 5.7 (Online Range ORAM). *There exists a perfectly secure online Range ORAM, which on receiving* len *consecutive memory locations online performs* len $\cdot$ poly log N *bandwidth and achieves* $(2, \mathsf{poly} \log N)$ *locality.*

6   Lower Bound for More Restricted Leakage

In Sect. 5.5, the online range ORAM leaks which instructions form a contiguous group of addresses. In this section, we show that if we restrict the leakage and do not allow the adversary to learn whether adjacent instructions access contiguous addresses, the lower bound for bandwidth to achieve locality will be significantly worse.

Model assumptions. We first clarify the model in which we prove the lower bound.

1. We restrict the leakage such that the adversary knows only the number N of logical blocks stored in memory, and the total number T of online operations, each of which has the form $(\mathsf{op}, \mathsf{addr}, \mathsf{data})$, where $\mathsf{op} \in \{\mathtt{read}, \mathtt{write}\}$, $\mathsf{addr} \in [N]$ and $\mathsf{data} \in \{0, 1\}^b \cup \{\perp\}$.

2. Just like earlier ORAM lower bounds [10,23,25]), we assume the so-called *balls-and-bins model*, i.e., the blocks are opaque objects and the algorithm, for instance, cannot use encoding techniques to combine blocks in the storage. Note that all known ORAM algorithms indeed fall within this model.
3. We assume that the algorithm has an offline phase in which it can preprocess memory before seeing any instructions. However, recall that the instructions are online, i.e., the algorithm must finish serving an instruction before seeing the next one.

Notation. Recall that we use D to denote the number of disks (each of which has a single head), ℓ to denote the locality (where we consider the very general case $\ell \leq \frac{N}{10}$), m to denote the memory size blowup[3], and β to denote the bandwidth. Moreover, suppose the CPU has only c block of local cache, where we just need a loose bound $c \leq \frac{N}{10}$. We shall prove the following theorem.

Theorem 6.1. *For any $\ell, c \leq \frac{N}{10}$, any Online Range ORAM satisfying the restricted leakage that has (D, ℓ)-locality with c blocks of cache storage will incur $\Omega(\frac{N}{D})$ bandwidth.*

Proof Intuition. By our leakage restriction assumption, the adversary cannot distinguish between the following two scenarios.

1. There are N operations that access contiguous addresses in the order from 0 to $N - 1$.
2. There are N operations, each of which access an address chosen independently uniformly at random from $[N]$.

Observe that to achieve (D, ℓ)-locality, in scenario 1, there can be at most ℓ jumping moves for the disk heads. Therefore, the same must hold for scenario 2. To serve an online request in scenario 2, we consider the following cases.

1. The block of the requested address is already in the cache. (However, the ORAM might still pretend to do some accesses.) Observe this happens with probability at most $\frac{c}{N} \leq \frac{1}{10}$, since the next requested address is chosen independently uniformly at random.
2. The online request is served by some disk head jump, which takes $O(1)$ physical accesses. Again, the ORAM might make other accesses to hide the access pattern. Observe at most $\ell \leq \frac{N}{10}$ requests can be served this way.
3. The online request is served by linear scan of the disk heads. By the Chernoff Bound, except with $e^{-\Theta(N)}$ probability, at least $\frac{N}{2}$ of the requests are served by linear scan. The following lemma gives a stochastic lower bound on the number of physical accesses in this case.

For ease of notation, we assume that $K := \frac{N-c}{D}$ is an integer.

---

[3] However, as we shall see, m does not play a role in the lower bound.

Lemma 6.2 (Stochastic Lower Bound on the Number of Physical Accesses). *Suppose in Scenario 2, the block of the next random address requested is not in the ORAM's cache. Moreover, suppose this request is served by only linear scan of disk heads, i.e., no jump move is made. Then, the random variable of the number of physical accesses for serving this request stochastically dominates the random variable with uniform distribution on $\{1, 2, \ldots, \frac{N-c}{D}\}$.*

Proof. Consider some configuration of the disk heads. Without loss of generality, assume that the cache currently stores the blocks for exactly c distinct addresses. For each of the remaining $N - c$ addresses, we can assign it to the disk head that takes a minimum number of accesses to reach a corresponding block by linear scan, where a tie can be resolved arbitrarily. For each $j \in [D]$, let a_j be the number of addresses assigned to disk head j; observe that we have $\sum_{j \in [D]} a_j = N - c$.

For each integer $1 \le i \le K = \frac{N-c}{D}$, observe that the number of addresses that take at least i physical accesses to reach is at least $\sum_{j \in [D]} \max\{0, a_j - i + 1\} \ge D \cdot (K - i + 1)$, where the last equality holds when all a_j's equal K.

Hence, the probability that at least i physical accesses is needed is at least $\frac{D \cdot (K-i+1)}{N-c} = \frac{K-i+1}{K}$, which implies the required result.   □

Lemma 6.3 (Lower Bound on Bandwidth). *Except with probability at most $e^{-\Theta(N)}$, the average number of physical accesses to serve each request in Scenario 2 is at least $\Omega(\frac{N}{D})$.*

Proof. As observed above, except with at most $e^{-\Theta(N)}$ probability, at least $\frac{N}{2}$ of the online requests must be served by linear scan of disk heads. By Lemma 6.2, the number of physical accesses for each such request stochastically dominates the uniform distribution on $\{1, 2, \ldots, \frac{N-c}{D}\}$, which has expectation $\Theta(\frac{N}{D})$, since we assume the cache size $c \le \frac{N}{10}$.

Since the addresses of the online requests are picked independently after the previous requests are served, by Chernoff bound, except with probability $e^{-\Theta(N)}$, the average number of physical accesses to serve each such online request is at least $\Omega(\frac{N}{D})$, as required.   □

7   Conclusions and Open Problems

We initiate a study of locality in oblivious RAM. For conclusion, we obtain the following results:

- There is an ORAM scheme that makes use of only 2 disks, that preserves the locality of the input program. Namely, if the input program accesses in total ℓ discontiguous regions, the ORAM scheme accesses at most $\ell \cdot \text{poly} \log N$ discontiguous regions. Moreover, if the program accesses in total T logical addresses, then the ORAM accesses in total $T \cdot \text{poly} \log N$ addresses. The ORAM leaks the sizes of the contiguous regions being accessed.
- Without leaking the sizes, we show a lower bound that the bandwidth of an oblivious program must be $\Omega(N)$, assuming $O(1)$-disks.

Open problems. We hope that our result will inspire future work on this topic. In the following, we provide several open questions on further understanding the trade-off between locality and bandwidth in oblivious compilation.

Preserving the number of disks. Our ORAM construction compiles $(1, \ell)$-local program into $(2, \text{poly} \log N)$-local program that is oblivious. Is it possible to achieve a compiler that preserve the number of disks? We emphasize that our construction uses the second disk only in the oblivious sorting, and it unclear whether sorting with $(1, \ell \cdot \text{poly} \log N)$-locality is possible to achieve.

Supporting more expressive input programs. Our motivated applications (e.g., outsourced file server, outsourced range query database), involve fetching some region from the memory and then accessing it in a streaming fashion. That is, we focused so far on supporting ORAM for $(1, \ell)$-local programs. A natural generalization is to construct an ORAM scheme that supports more expressive input programs, such as (D, ℓ)-local programs for $D \geq 2$. This allows, for instance, computing inner products of D-arrays, or merging D-arrays. The input program sends to the memory instructions that also specify which disks to access, i.e., instructions of the form $(\text{move}, d, \text{addr})$ and $(\text{op}, d, \text{data})$, as defined in Sect. 3. As we discuss further in the appendices of the online full version [5], depending on how we formulate the allowable leakage, the problem can be easy or an open challenge.

Locality preserving OPRAM. We have considered a single CPU in this work. A natural question is whether we can extend the construction to support multiple CPUs, namely, to construct an oblivious parallel RAM (OPRAM) that preserves locality.

Asymptotic efficiency. We have showed the theoretic feasibility of constructing a Range ORAM with poly-logarithmic work and locality. In this feasibility result, we favored conceptual simplicity over optimizing poly-logarithmic factors. Nevertheless, it is interesting to see to what extent the constructions can be optimized. Perhaps locality-preserving ORAM can be constructed with the same bandwidth efficiency as a regular ORAM?

Acknowledgments. This work was partially supported by a Junior Fellow award from the Simons Foundation to Gilad Asharov. This work was supported in part by NSF grants CNS-1314857, CNS-1514261, CNS-1544613, CNS-1561209, CNS-1601879, CNS-1617676, an Office of Naval Research Young Investigator Program Award, a Packard Fellowship, a Sloan Fellowship, Google Faculty Research Awards, a VMWare Research Award, and a Baidu Faculty Research Award to Elaine Shi. Kartik Nayak was partially supported by a Google Ph.D. Fellowship Award. T.-H. Hubert Chan was partially supported by the Hong Kong RGC under the grant 17200418.

A   Appendix: Locality of Bitonic Sort

In this section, we first analyze the locality of Bitonic sort, which runs in $O(n \log^2 n)$ time.

We call an array of numbers bitonic if it consists of two monotonic sequences, the first one ascending and the other descending, or vice versa. For an array S, we write it as $\widehat{S}$ if it is bitonic, as $\overrightarrow{S}$ (resp. $\overleftarrow{S}$) if it is sorted in an ascending (resp. descending) order.

The algorithm is based on a "bitonic split" procedure $\overrightarrow{\mathsf{Split}}$, which receives as input a bitonic sequence $\widehat{S}$ of length n and outputs a sorted sequence $\overrightarrow{S}$. $\overrightarrow{\mathsf{Split}}$ first separates $\widehat{S}$ into two bitonic sequences $\widehat{S}_1, \widehat{S}_2$, such that all the elements in S_1 are smaller than all the elements in S_2. It then calls $\overrightarrow{\mathsf{Split}}$ recursively on each sequence to get a sorted sequence.

Procedure A.1: $\overrightarrow{S} = \overrightarrow{\mathsf{Split}}(\widehat{S})$

- Let $\widehat{S}_1 = \langle \min(a_0, a_{n/2}), \min(a_1, a_{n/2+1}), \ldots, \min(a_{n/2-1}, a_{n-1}) \rangle$.
- Let $\widehat{S}_2 = \langle \max(a_0, a_{n/2}), \max(a_1, a_{n/2+1}), \ldots, \max(a_{n/2-1}, a_{n-1}) \rangle$.
- $\overrightarrow{S}_1 = \overrightarrow{\mathsf{Split}}(\widehat{S}_1)$, $\overrightarrow{S}_2 = \overrightarrow{\mathsf{Split}}(\widehat{S}_2)$ and $\overrightarrow{S} = (\overrightarrow{S}_1, \overrightarrow{S}_2)$.

Similarly, $\overleftarrow{S} = \overleftarrow{\mathsf{Split}}(\widehat{S})$ sorts the array in a descending order. We refer to [9] for details.

To sort an array S of n elements, the algorithm first converts S into a bitonic sequence using the Split procedures in a bottom up fashion, similar to the structure of merge-sort. Specifically, any size-2 sequence is a bitonic sequence. In each iteration $i = 1, \ldots, \log n - 1$, the algorithm merges each pair of size-2^i bitonic sequences into a size-2^{i+1} bitonic sequence. Towards this end, it uses the $\overrightarrow{\mathsf{Split}}$ and $\overleftarrow{\mathsf{Split}}$ alternately, as two sorted sequences $(\overrightarrow{S}_1, \overleftarrow{S}_2)$ form a bitonic sequence. The full bitonic sort algorithm is presented below:

Algorithm A.2: BitonicSort(S)

1. Convert S to a bitonic sequence: For $i = 1, \ldots, \log n - 1$:
   (a) Let $S = (\widehat{S}_0, \ldots, \widehat{S}_{n/2^i-1})$ be the size-2^i bitonic sequences from the previous iteration.
   (b) For $j = 0, \ldots, n/2^{i+1} - 1$, $\widehat{B}_j = (\overrightarrow{\mathsf{Split}}(\widehat{S}_{2j}), \overleftarrow{\mathsf{Split}}(\widehat{S}_{2j+1}))$.
   (c) Set $S = (\widehat{B}_0, \ldots, \widehat{B}_{n/2^{i+1}-1})$.
2. The array $\widehat{S}$ is now a bitonic sequence. Apply $\overrightarrow{S} = \overrightarrow{\mathsf{Split}}(\widehat{S})$ to obtain a sorted sequence.

Locality and obliviousness. It is easy to see that the sorting algorithm is oblivious, as all accesses to the memory are independent of the input data. For locality, first note that procedure $\overrightarrow{\mathsf{Split}}$ and $\overleftarrow{\mathsf{Split}}$ are $(2, O(\log n))$-local. No move operations are needed between instances of recursions, as these can be executed one after another as iterations (and using some vacuous reads). Thus, Algorithm A.2 is $(2, O(\log^2 n))$-local as it runs in $\log n$ iterations, each invoking $\overrightarrow{\mathsf{Split}}$ and $\overleftarrow{\mathsf{Split}}$. Figure 3 gives a graphic representation of the algorithm for input size 8 and Fig. 4 illustrates its locality. The $(2, O(\log^2 n))$ locality of Bitonic sort is also obvious from the figure.

Remark. Observe that in each pass of $\overrightarrow{\mathsf{Split}}$ (or $\overleftarrow{\mathsf{Split}}$), a min/max operation is a *read-compare-write* operation. Thus, strictly speaking, each memory location is accessed twice for this operation – once for reading and once for writing. When the write is performed, the read/write head has already moved forward and is thus not writing back to the same two locations that it read from.

Going back to the same two locations would incur an undesirable move head operation. However, we can easily convert this into a solution that still preserves $(2, O(1))$-locality for each pass of $\overrightarrow{\mathsf{Split}}$ by introducing a *slack* after every memory location (and thus using twice the amount of storage). In this solution, every memory location a_i is followed by a_i'; the entire array is stored as $((a_0, a_0'), \ldots, (a_{n-1}, a_{n-1}'))$ where a_i stores real blocks and a_i' is a slack location. When a_i and a_j are compared, the results can be written to a_i' and a_j' respectively without incurring a move operation. Before starting the next iteration, we can move the data from slack locations to the actual locations in a single pass, thus preserving $(2, O(1))$-locality for each pass of $\overrightarrow{\mathsf{Split}}$ (and $\overleftarrow{\mathsf{Split}}$).

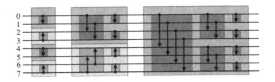

Fig. 3. Bitonic sorting network for 8 inputs. Input come in from the left end, and outputs are on the right end. When two numbers are joined by an arrow, they are compared, and if necessary are swapped such that the arrow points from the smaller number toward the larger number. This figure is modified from [1].

	Pass 1							Pass 2						Pass 3					
Disk 1	0	1	2	3	4	5	6	0	1	2	3	4	5	0	1	2	3	4	5	6
Disk 2	1	2	3	4	5	6	7	2	3	4	5	6	7	1	2	3	4	5	6	7
Operation	↓	⊥	↑	⊥	↓	⊥	↑	↓	↓	⊥	⊥	↑	↑	↓	⊥	↓	⊥	↑	⊥	↑

Fig. 4. Locality of Bitonic Sort for 8 elements. The figure shows the allocation of the data in the two disks for an 8 element array. For each input, either a compare-and-swap operation is performed in the specified direction or the input is ignored as denoted by ⊥. The figure shows the first 3 passes out of the required 6 passes for 8 elements (see Fig. 3).

References

1. Bitonic sorter. https://en.wikipedia.org/wiki/Bitonic_sorter. Accessed October 2018
2. Ajtai, M., Komlós, J., Szemerédi, E.: An $O(N \log N)$ sorting network. In: ACM Symposium on Theory of Computing (STOC 1983), pp. 1–9 (1983)
3. Apon, D., Katz, J., Shi, E., Thiruvengadam, A.: Verifiable oblivious storage. In: Krawczyk, H. (ed.) PKC 2014. LNCS, vol. 8383, pp. 131–148. Springer, Heidelberg (2014). https://doi.org/10.1007/978-3-642-54631-0_8
4. Arge, L., Ferragina, P., Grossi, R., Vitter, J.S.: On sorting strings in external memory (extended abstract). In: ACM Symposium on the Theory of Computing (STOC 1997), pp. 540–548 (1997)

5. Asharov, G., Chan, T.H.H., Nayak, K., Pass, R., Ren, L., Shi, E.: Locality-preserving oblivious ram. https://eprint.iacr.org/2017/772
6. Asharov, G., Komargodski, I., Lin, W.K., Nayak, K., Peserico, E., Shi, E.: OptORAMa: optimal oblivious RAM. Cryptology ePrint Archive, Report 2018/892
7. Asharov, G., Naor, M., Segev, G., Shahaf, I.: Searchable symmetric encryption: optimal locality in linear space via two-dimensional balanced allocations. In: ACM Symposium on Theory of Computing (STOC 2016), pp. 1101–1114 (2016)
8. Asharov, G., Segev, G., Shahaf, I.: Tight tradeoffs in searchable symmetric encryption. In: Shacham, H., Boldyreva, A. (eds.) CRYPTO 2018. LNCS, vol. 10991, pp. 407–436. Springer, Cham (2018). https://doi.org/10.1007/978-3-319-96884-1_14
9. Batcher, K.E.: Sorting networks and their applications. In: AFIPS 1968 (1968)
10. Boyle, E., Naor, M.: Is there an oblivious RAM lower bound? In: ACM Conference on Innovations in Theoretical Computer Science (ITCS 2016), pp. 357–368 (2016)
11. Canetti, R.: Security and composition of multiparty cryptographic protocols. J. Cryptol. **13**(1), 143–202 (2000)
12. Cash, D., Jarecki, S., Jutla, C.S., Krawczyk, H., Roşu, M.-C., Steiner, M.: Highly-scalable searchable symmetric encryption with support for boolean queries. In: Canetti, R., Garay, J.A. (eds.) CRYPTO 2013, Part 1. LNCS, vol. 8042, pp. 353–373. Springer, Heidelberg (2013). https://doi.org/10.1007/978-3-642-40041-4_20
13. Cash, D., Tessaro, S.: The locality of searchable symmetric encryption. In: Nguyen, P.Q., Oswald, E. (eds.) EUROCRYPT 2014. LNCS, vol. 8441, pp. 351–368. Springer, Heidelberg (2014). https://doi.org/10.1007/978-3-642-55220-5_20
14. Chakraborti, A., Aviv, A.J., Choi, S.G., Mayberry, T., Roche, D.S., Sion, R.: rORAM: efficient range ORAM with $O(\log^2 N)$ locality. In: Network and Distributed System Security (NDSS) (2019)
15. Chan, T.H., Nayak, K., Shi, E.: Perfectly secure oblivious parallel RAM. In: Theory of Cryptography Conference (TCC) (2018)
16. Chan, T.H., Chung, K.M., Maggs, B., Shi, E.: Foundations of differentially oblivious algorithms. In: Symposium on Discrete Algorithms (SODA) (2019)
17. Chase, M., Kamara, S.: Structured encryption and controlled disclosure. In: Abe, M. (ed.) ASIACRYPT 2010. LNCS, vol. 6477, pp. 577–594. Springer, Heidelberg (2010). https://doi.org/10.1007/978-3-642-17373-8_33
18. Chung, K.-M., Liu, Z., Pass, R.: Statistically-secure ORAM with $\tilde{O}(\log^2 n)$ overhead. In: Sarkar, P., Iwata, T. (eds.) ASIACRYPT 2014. LNCS, vol. 8874, pp. 62–81. Springer, Heidelberg (2014). https://doi.org/10.1007/978-3-662-45608-8_4
19. Curtmola, R., Garay, J.A., Kamara, S., Ostrovsky, R.: Searchable symmetric encryption: improved definitions and efficient constructions. In: ACM Conference on Computer and Communications Security (CCS 2006), pp. 79–88 (2006)
20. Demertzis, I., Papadopoulos, D., Papamanthou, C.: Searchable encryption with optimal locality: achieving sublogarithmic read efficiency. In: Shacham, H., Boldyreva, A. (eds.) CRYPTO 2018. LNCS, vol. 10991, pp. 371–406. Springer, Cham (2018). https://doi.org/10.1007/978-3-319-96884-1_13
21. Demertzis, I., Papamanthou, C.: Fast searchable encryption with tunable locality. In: SIGMOD Conference, pp. 1053–1067. ACM (2017)
22. Devadas, S., van Dijk, M., Fletcher, C.W., Ren, L., Shi, E., Wichs, D.: Onion ORAM: a constant bandwidth blowup oblivious RAM. In: Kushilevitz, E., Malkin, T. (eds.) TCC 2016. LNCS, vol. 9563, pp. 145–174. Springer, Heidelberg (2016). https://doi.org/10.1007/978-3-662-49099-0_6
23. Goldreich, O.: Towards a theory of software protection and simulation by oblivious RAMs. In: STOC (1987)
24. Goldreich, O.: The Foundations of Cryptography - Volume 2, Basic Applications. Cambridge University Press, Cambridge (2004)

25. Goldreich, O., Ostrovsky, R.: Software protection and simulation on oblivious RAMs. J. ACM **43**, 431–473 (1996)
26. Goodrich, M.T.: Zig-zag sort: a simple deterministic data-oblivious sorting algorithm running in $O(n \log n)$ time. In: STOC (2014)
27. Goodrich, M.T., Mitzenmacher, M.: Privacy-preserving access of outsourced data via oblivious RAM simulation. In: Aceto, L., Henzinger, M., Sgall, J. (eds.) ICALP 2011. LNCS, vol. 6756, pp. 576–587. Springer, Heidelberg (2011). https://doi.org/10.1007/978-3-642-22012-8_46
28. Kamara, S., Papamanthou, C.: Parallel and dynamic searchable symmetric encryption. In: Sadeghi, A.-R. (ed.) FC 2013. LNCS, vol. 7859, pp. 258–274. Springer, Heidelberg (2013). https://doi.org/10.1007/978-3-642-39884-1_22
29. Kellaris, G., Kollios, G., Nissim, K., O'Neill, A.: Generic attacks on secure outsourced databases. In: ACM CCS, pp. 1329–1340 (2016)
30. Kellaris, G., Kollios, G., Nissim, K., O'Neill, A.: Accessing data while preserving privacy. CoRR abs/1706.01552 (2017). http://arxiv.org/abs/1706.01552
31. Kurosawa, K., Ohtaki, Y.: How to update documents *verifiably* in searchable symmetric encryption. In: Abdalla, M., Nita-Rotaru, C., Dahab, R. (eds.) CANS 2013. LNCS, vol. 8257, pp. 309–328. Springer, Cham (2013). https://doi.org/10.1007/978-3-319-02937-5_17
32. Kushilevitz, E., Lu, S., Ostrovsky, R.: On the (in)security of hash-based oblivious RAM and a new balancing scheme. In: SODA (2012)
33. Larsen, K.G., Nielsen, J.B.: Yes, there is an oblivious RAM lower bound!. In: Shacham, H., Boldyreva, A. (eds.) CRYPTO 2018. LNCS, vol. 10992, pp. 523–542. Springer, Cham (2018). https://doi.org/10.1007/978-3-319-96881-0_18
34. Patel, S., Persiano, G., Raykova, M., Yeo, K.: Panorama: oblivious RAM with logarithmic overhead. In: FOCS (2018)
35. Ruemmler, C., Wilkes, J.: An introduction to disk drive modeling. IEEE Comput. **27**(3), 17–28 (1994)
36. Shi, E., Chan, T.-H.H., Stefanov, E., Li, M.: Oblivious RAM with $O((\log N)^3)$ worst-case cost. In: Lee, D.H., Wang, X. (eds.) ASIACRYPT 2011. LNCS, vol. 7073, pp. 197–214. Springer, Heidelberg (2011). https://doi.org/10.1007/978-3-642-25385-0_11
37. Stefanov, E., et al.: Path ORAM - an extremely simple oblivious RAM protocol. In: CCS (2013)
38. van Liesdonk, P., Sedghi, S., Doumen, J., Hartel, P., Jonker, W.: Computationally efficient searchable symmetric encryption. In: Jonker, W., Petković, M. (eds.) SDM 2010. LNCS, vol. 6358, pp. 87–100. Springer, Heidelberg (2010). https://doi.org/10.1007/978-3-642-15546-8_7
39. Vitter, J.S.: External memory algorithms and data structures. ACM Comput. Surv. **33**(2), 209–271 (2001)
40. Vitter, J.S.: Algorithms and data structures for external memory. Found. Trends Theor. Comput. Sci. **2**(4), 305–474 (2006)
41. Wang, X., Chan, T.H., Shi, E.: Circuit ORAM: on tightness of the Goldreich-Ostrovsky lower bound. In: ACM Conference on Computer and Communications Security, pp. 850–861. ACM (2015)
42. Wang, X.S., Huang, Y., Chan, T.H.H., Shelat, A., Shi, E.: SCORAM: oblivious RAM for secure computation. In: CCS (2014)
43. Williams, P., Sion, R.: Usable PIR. In: Network and Distributed System Security Symposium (NDSS) (2008)
44. Williams, P., Sion, R.: Round-optimal access privacy on outsourced storage. In: ACM Conference on Computer and Communication Security (CCS) (2012)
45. Williams, P., Sion, R., Carbunar, B.: Building castles out of mud: practical access pattern privacy and correctness on untrusted storage. In: CCS, pp. 139–148 (2008)

Private Anonymous Data Access

Ariel Hamlin[1], Rafail Ostrovsky[2], Mor Weiss[1,3(✉)], and Daniel Wichs[1]

[1] Department of Computer Science, Northeastern University,
Boston, MA, USA
{ahamlin,wichs}@ccs.neu.edu
[2] UCLA, Los Angeles, CA, USA
rafail@cs.ucla.edu
[3] Department of Computer Science, IDC Herzliya,
Herzliya, Israel
mor.weiss01@post.idc.ac.il

Abstract. We consider a scenario where a server holds a huge database that it wants to make accessible to a large group of clients. After an initial setup phase, clients should be able to read arbitrary locations in the database while maintaining *privacy* (the server does not learn which locations are being read) and *anonymity* (the server does not learn which client is performing each read). This should hold even if the server colludes with a subset of the clients. Moreover, the run-time of both the server and the client during each read operation should be low, ideally only poly-logarithmic in the size of the database and the number of clients. We call this notion *Private Anonymous Data Access* (PANDA). PANDA simultaneously combines aspects of *Private Information Retrieval* (PIR) and *Oblivious RAM* (ORAM). PIR has no initial setup, and allows anybody to privately and anonymously access a public database, but the server's run-time is linear in the data size. On the other hand, ORAM achieves poly-logarithmic server run-time, but requires an initial setup after which only a single client with a secret key can access the database. The goal of PANDA is to get the best of both worlds: allow many clients to privately and anonymously access the database as in PIR, while having an efficient server as in ORAM.

In this work, we construct *bounded-collusion* PANDA schemes, where the efficiency scales linearly with a bound on the number of corrupted clients that can collude with the server, but is otherwise poly-logarithmic in the data size and the total number of clients. Our solution relies on standard assumptions, namely the existence of fully homomorphic encryption, and combines techniques from both PIR and ORAM. We also extend PANDA to settings where clients can *write* to the database.

1 Introduction

As individuals and organizations increasingly rely on third party data stored remotely, there is often a need to access such data both privately and anonymously. For example, we can envision a service that has a large database of medical conditions, and allows clients to look up their symptoms; naturally clients do

Y. Ishai and V. Rijmen (Eds.): EUROCRYPT 2019, LNCS 11477, pp. 244–273, 2019.
https://doi.org/10.1007/978-3-030-17656-3_9

not want to reveal which symptoms they are searching for, or even the frequency with which they are performing such searches.

To address this, we consider a setting where a server holds a huge database that it wants to make accessible to a large group of clients. The clients should be able to read arbitrary locations in the database while hiding from the server which locations are being accessed (*privacy*), and which client is performing each access (*anonymity*). We call this *Private Anonymous Data Access* (PANDA).

In more detail, PANDA allows some initial setup phase, after which the server holds an encoded database, and each client holds a short key. The setup can be performed by a trusted third party, or via a multi-party computation protocol. After the setup phase, any client can execute a read protocol with the server, to retrieve an arbitrary location within the database. We want this protocol to be highly efficient, where both the server's and client's run-time during the protocol should be sub-linear (ideally, poly-logarithmic) in the database size and the total number of clients. For security, we consider an adversarial server that colludes with some subset of clients. We want to ensure that whenever an honest client performs a read access, the server learns nothing about the location being accessed, or the identity of the client performing the access beyond the fact that she belongs to the group of all honest clients. For example, the server should not learn whether two accesses correspond to *two different* clients reading *the same* location of the database, or *one client* reading *two different* database locations.[1]

We call the above a *read-only* PANDA, and also consider extensions that allow clients to write to the database, which we discuss below in more detail.

Connections to PIR and ORAM. PANDA combines aspects of both *Private Information Retrieval* (PIR) [CGKS95,KO97] and *Oblivious RAM* (ORAM) [GO96]. Therefore, we now give a high-level overview of these primitives, their goals, and main properties.

In a (single-database) PIR scheme [KO97], the server holds a public database in the clear. The scheme has no initial setup, and anybody can run a protocol with the server to retrieve an arbitrary location within the database. Notice that since there are no secret keys that distinguish one client from another, a PIR scheme also provides perfect anonymity. However, although the communication complexity of the PIR protocol is sub-linear in the data size, the server's run-time is *inherently* linear in the size of the data. (Indeed, if the server didn't read the entire database during the protocol, it would learn something about the location being queried, since it must be among the ones read.) Therefore, PIR does not provide a satisfactory answer to the PANDA problem, where we want sub-linear efficiency for the server.

In an ORAM scheme, there is an initial setup after which the server holds an encoded database, and a client holds a secret key. The client can execute a protocol with the server to privately read or write to arbitrary locations within the database, and the run-time of both the client and the server during each such protocol is sub-linear in the data size. However, only *a single client* in possession

---

[1] We assume clients have an anonymous communication channel with the server (e.g., using anonymous mix networks [Cha03] such as TOR [DMS04] or [BG12,LPDH17]).

of a secret key associated with the ORAM can access the database. Therefore, ORAM is also not directly applicable to the PANDA problem, where we want a large group of clients to access the database.

1.1 Prior Work Extending PIR and ORAM

Although neither PIR nor ORAM alone solve the PANDA problem, several prior approaches have considered extensions of PIR and ORAM, aimed to overcome their aforementioned limitations. We discuss these approached, and explain why they do not provide a satisfactory solution for PANDA.

ORAM with Multiple Clients. As mentioned above, in an ORAM scheme *only a single client* can access the database, whereas in PANDA we want multiple clients to access it. There are several natural ways that we can hope to extend ORAM to the setting of multiple clients.

The first idea is to store the data in a single ORAM scheme, and give all the clients the secret key for this ORAM. Although this solution provides anonymity (all clients are identical) it does not achieve privacy; if the server colludes with even a single client, the privacy of all other clients is lost.

A second idea is to store the data in a separate ORAM scheme for each client, and give the client the corresponding secret key. Each client then accesses the data using her own ORAM. This achieves privacy even if the server colludes with a subset of clients, but does not provide anonymity since the server sees which ORAM is being accessed.[2]

The third idea is similar to the previous one, where the data is stored in a separate ORAM scheme for each client, and the client accesses the data using her own ORAM. However, unlike the second idea, the client also performs a "dummy" access on the ORAM schemes of all other clients to hide her identity. This requires a special ORAM scheme where any client *without a secret key* can perform a "dummy" access which looks indistinguishable from a real access to someone that *does not* have the secret key. It turns out that existing ORAM schemes can be upgraded relatively easily to have this property (using re-randomizable encryption). Although this solution achieves privacy and anonymity, the efficiency of both the server and the client during each access is linear in the total number of clients.

Lastly, we can also store the data in a single ORAM scheme on the server, and distribute the ORAM secret key across several additional proxy servers. When a client wants to access a location of the data, she runs a multiparty computation protocol with the proxy servers to generate the ORAM access. Although this solution provides privacy, anonymity and efficiency, it requires having multiple non-colluding servers, whereas our focus is on the *single server* setting.

Variants of the above ideas have appeared in several prior works (e.g., [BMN17, MMRS15, KPK16, BHKP16, ZZQ16]) that explored multi-client

---

[2] Also, the server storage in this solution grows proportionally to the number of clients *times* the data size. Reducing the server storage, even without anonymity, is an interesting relaxation of PANDA which we explore in the full version.

ORAM. In particular, the work of Backes et al. [BHKP16] introduced the notion of Anonymous RAM (which is similar to our notion of secret-writes PANDA, discussed below), and proposed two solutions which can be seen as variants of the third and fourth ideas discussed above. Specifically, they are able to achieve security for up to all but one colluding clients in both schemes, one achieving linear storage in the number of users, the other relying on two non-colluding servers. Our solution, for the same collusion threshold, is able to achieve linear storage overhead in the number of users with only a single server, and for lower collusion thresholds we are more efficient (linear in the collusion threshold). We note, despite much research activity, no prior solution simultaneously provides privacy, anonymity and efficiency in the single-server setting.

Doubly Efficient PIR. As noted above, the server run-time in a PIR protocol is inherently linear in the data size, whereas in PANDA we want the run time of both the client *and the server* to be sub-linear. However, it may be possible to get a *doubly efficient PIR* (DEPIR) variant in which the server run-time is sub-linear, by relaxing the PIR problem to allow a *pre-processing* stage after which the server stores *an encoded version* of the database. This concept was first proposed by Beimel, Ishai and Malkin [BIM00], who showed how to construct information-theoretic DEPIR schemes in the *multi-server setting*, with several non-colluding servers. Two recent works, of Canetti et al. [CHR17] and Boyle et al. [BIPW17], give the first evidence that this notion may even be achievable in the *single-server* setting. Concretely, they consider DEPIR schemes with a pre-processing stage which generates an encoded database for the server, and a key that allows clients to query the database at arbitrary locations. They distinguish between *symmetric-key* and *public-key* variants of DEPIR, based on whether the key used to query the database needs to be kept secret or can be made public. Both works show how to construct symmetric-key DEPIR under new, previously unstudied, computational hardness assumptions relating to Reed-Muller codes. The work of [BIPW17] also shows how to extend this to get public-key DEPIR by also relying on a heuristic use of obfuscation. Unfortunately, both of the above assumptions are non-standard, poorly understood, and not commonly accepted.

In relation to PANDA, symmetric-key DEPIR suffers from the same drawbacks as ORAM, specifically, only a single client with a secret key can access the database.[3] If we were to give this key to several clients, then all privacy would be lost even if *only a single client* colludes with the server. On the other hand, public-key DEPIR immediately yields a solution to the PANDA problem, at least for the read-only variant. Moreover, it even has additional perks not required by PANDA, specifically: the set of clients does not need to be chosen ahead of time, anybody can use the system given only a public key, and the server is stateless. Unfortunately, we currently appear to be very far from being able to instantiate public-key DEPIR under any standard hardness assumptions.

---

[3] The main difference between symmetric-key DEPIR and ORAM is that in the former the server is stateless and only stores a static encoded database, while in the latter the server is stateful and its internal storage is continuously updated after each operation. In PANDA, we allow the server to be stateful.

1.2 Our Results

Read-Only PANDA. In this work, we construct a *bounded-collusion* PANDA scheme, where we assume some upper bound t on the number of clients that collude with the server. The client and server efficiency scales linearly with t, but is otherwise poly-logarithmic in the data size and the total number of clients. In particular, our PANDA scheme allows for up to a poly-logarithmic collusion size t while maintaining poly-logarithmic efficiency for the server and the client. Our construction relies on the generic use of (leveled) *Fully Homomorphic Encryption (FHE)* [RAD78, Gen09] which is in turn implied by the *Learning With Errors (LWE)* assumption [Reg09]. Our basic construction provides security against a semi-honest adversary, and we also discuss how to extend this to get security in the fully malicious setting. In summary, we get the following theorem.

Theorem 1 (Informal statement of Theorem 6**).** *Assuming the existence of FHE, there exists a (read-only) PANDA scheme with n clients, t collusion bound, database size L and security parameter λ such that, for any constant $\varepsilon > 0$, we get:*

– *The client/server run-time per* read *operation is* $t \cdot \mathsf{poly}(\lambda, \log L)$.
– *The server storage is* $t \cdot L^{1+\varepsilon} \cdot \mathsf{poly}(\lambda, \log L)$.

PANDA with Writes. We also consider extensions of PANDA to a setting that supports writes to the database. If the database is public and shared by all clients, then the location and content of write operations is inherently public as well. However, we still want to maintain privacy and anonymity for read operations, as well as anonymity for write operations. We call this a *public-writes* PANDA and it may, for example, be used to implement a public message board where clients can anonymously post and read messages, while hiding from the server which messages are being read. We also consider an alternate scenario where each client has her own individual private database which only she can access. In this case we want to maintain privacy and anonymity for *both the reads and writes* of each client, so that the server does not learn the content of the data, which clients are accessing their data, or what parts of their data they are accessing. We call this a *secret-writes* PANDA.[4] We show the following result.

Theorem 2 (Informal statement of Theorem 7**).** *Assuming the existence of FHE, there exists a* public-writes *PANDA with n clients, t collusion bound, database size L and security parameter λ such that, for any constant $\varepsilon > 0$, we get:*

– *The client/server run-time per* read *operation is* $t \cdot \mathsf{poly}(\lambda, \log L)$.

---

[4] Note that in the read-only setting, having a scheme for a shared public database is strictly more flexible than a scheme for individual private databases. We can always use the former to handle the latter by having clients encrypt their individual data and store it in a shared public database. However, once we introduce writes, these settings become incomparable.

- *The client run-time per* write *is* $O(\log L)$, *and the server run-time is* $t \cdot L^{\varepsilon} \cdot$ $\mathsf{poly}(\lambda, \log L)$.
- *The server storage is* $t \cdot L^{1+\varepsilon} \cdot \mathsf{poly}(\lambda, \log L)$.

The same results as above hold for *secret-writes* PANDA, except that the client run-time per write increases to $t \cdot \mathsf{poly}(\lambda, \log L)$, and L now denotes the sum of the initial database size and the total number of writes performed throughout the lifetime of the system.

Extensions. We also consider the PANDA problem in stronger security models in which the adversary can *adaptively* choose the access pattern, and maliciously corrupt parties. Our constructions are also secure in the adaptive setting. The read-only PANDA scheme is secure against maliciously-corrupted clients, and a variant of it (which employs Merkle hash trees and succinct interactive arguments of knowledge) is secure if the server is also maliciously corrupted. Finally, we discuss modifications of our PANDA with writes schemes that remain secure in the presence of malicious corruptions. See the full version [HOWW18] for further details.

1.3   Our Techniques

We now give a high-level overview of our PANDA constructions. We start with the read-only setting, and then discuss how to enable writes.

Read-Only PANDA. Our construction relies on *Locally Decodable Codes (LDCs)* [KT00], which have previously been used to construct multi-server PIR [CGKS95,WY05]. We first give an overview of what these are, and then proceed to use them to build our scheme in several steps.

Locally Decodable Codes (LDCs). An LDC consists of a procedure that *encodes* a message into a codeword, and a procedure that *locally decodes* any individual location in the message by reading only few locations in the codeword. We denote the locality by k. An LDC has *s-smoothness* if any s out of k of the codeword locations accessed by the local decoder are uniformly random and independent of the message location being decoded. Such LDCs (with good parameters) immediately give information-theoretic *multi-server doubly-efficient PIR* without any keys [BIM00]: each of the k servers holds a copy of the encoded database, and the client runs the local decoding procedure by reading each of the k queried codeword locations from a different server. Even if s out of k servers collude, they don't learn anything about the database location that the client is retrieving.[5] LDCs with sufficiently good parameters for our work can be constructed using Reed-Muller codes [Ree54,Mul54].

---

[5] In standard PIR schemes, the servers hold the original database, and each query is answered by computing the requested codeword symbol on the fly. However, if the codeword size is polynomial, then the servers can compute the codeword first in a preprocessing phase, and then use the pre-computed codeword to answer each query in sub-linear time.

Initial Idea: LDCs + ORAM. Although LDCs naturally only give a *multi-server* PIR, our initial idea is to think of these as "virtual servers" which will all be emulated by *a single real server* by placing each virtual server under a separate ORAM instance. Each client is assigned a random committee consisting of a small subset of these virtual servers, for which she gets the corresponding ORAM keys. When the adversary corrupts a subset of the clients, it gets all of their ORAM keys, and can therefore be seen as corrupting all the virtual servers that are on the committees of these clients. Nevertheless, we can ensure that the committee of any honest client has sufficiently few corrupted virtual servers for LDC smoothness to hide the client's queries.

In more detail, we think of having k' virtual servers, for some k' which is sufficiently larger than the locality k of the LDC. For each virtual server, we choose a fresh ORAM key, and store an LDC encoding of the database under this ORAM. Each client is assigned to a random committee consisting of k out of k' of the virtual servers, for which she gets the related ORAM keys. To read a database location, the client runs the LDC local decoding algorithm, which requests to see k codeword locations. The client then reads each of the k codeword locations from a different virtual server on her committee, by using the corresponding ORAM scheme. Notice that an adversary that corrupts some subset of t clients, thus obtaining all of their ORAM keys, can be seen as corrupting all the virtual servers on their committees. We can choose the parameters to ensure that the probability of the adversary corrupting more than s out of k of the virtual servers on the committee *of any honest client* is negligible (specifically, setting $k' = tk^2$ and s to be the security parameter). As long as this holds, our scheme guarantees *privacy*, since the server only learns at most s out of the k codeword locations being queried (by the security of ORAM), and these locations reveal nothing about the database location being read (by the LDC smoothness).

Although the above solution already gives a non-trivial multi-client ORAM with privacy and low server storage (see the full version [HOWW18]), it does not provide any *anonymity*. The problem is that each client only accesses the k out of k' ORAM schemes belonging to her committee, and doesn't have the keys needed to access the remaining ORAM schemes. Therefore, the server can distinguish between different clients based on which of the ORAMs they access.

One potential idea to fix this issue would be for the client to make some "dummy" accesses to the $k' - k$ remaining ORAM schemes (which are not on her committee) without knowing the corresponding keys. Most ORAM schemes can be easily modified to enable such "dummy" accesses without a key, that look indistinguishable from real accesses to a distinguisher *that doesn't have the key.* Unfortunately, in our case the adversarial colluding server *does have the keys* for many of these ORAM schemes. Therefore, to make this idea work in our setting, we would need an ORAM where clients can make a *"smart* dummy" access without a key that looks indistinguishable from a real access to a random location even to a "smart" distinguisher *that has the key.* The square-root ORAM scheme [Gol87, GO96] can be modified to have this property, but the overall client/server efficiency in the final solutions would be at least square-root of the data size. Unfortunately, more efficient ORAM schemes with poly-logarithmic overhead (such as hierarchical ORAM [Ost90, GO96] or tree-based ORAM [SvDS+13]) do not have this property, and it does not appear that they could be naturally modified to add it. Instead, we take a different approach and get rid of ORAM altogether.

Bounded-Access PANDA: LDCs + Permute. Our second idea is inspired by the recent works of Canetti et al. [CHR17] and Boyle et al. [BIPW17] on DEPIR, as well as earlier works of Hemenway et al. [HO08, HOSW11]. Instead of implementing the virtual servers by storing the LDC codeword under an ORAM scheme, we do something much simpler and use a *Pseudo-Random Permutation* (PRP) to permute the codeword locations. In particular, for each of the k' virtual servers we choose a different PRP key, and use it to derive a different permuted codeword. Each client still gets assigned a random committee consisting of k out of k' of the virtual servers, for which she gets the corresponding PRP keys. To retrieve a value from the database, the client runs the LDC local decoding algorithm, which requests to see k codeword locations, and reads these locations using the virtual servers on her committee by applying the corresponding PRPs. She also reads uniformly random locations from the $k' - k$ virtual servers that are not on her committee.

In relation to the first idea, we can think of the PRP as providing much weaker security than ORAM. Namely, it reveals when *the same* location is read multiple times, but hides *everything else* about the locations being read (whereas an ORAM scheme even hides the former). On the other hand, it is now extremely easy to perform a "smart dummy" access (as informally defined above) by reading a truly random location in the permuted codeword, which is something we don't know how to do with poly-logarithmic ORAM schemes.

It turns out that this scheme is already secure if we fix some a-priori bound B on the total number of read operations that honest clients will perform. We call this notion a *bounded-access PANDA*. Intuitively, even though permuting the codewords provides much weaker security than putting them in an ORAM, and leaks partial information about the access pattern to the codeword, the fact that this access pattern is sampled via a smooth LDC ensures that this leakage is harmless when the number of accesses is sufficiently small. More specifically, our proof follows the high-level approach of Canetti et al. [CHR17], who constructed a *bounded-access (symmetric-key) DEPIR* which is essentially equivalent to the above scheme in the setting with *a single honest* client and *exactly k* virtual servers, where the adversary *doesn't get any of the PRP keys*. In our case, we need to extend this proof to deal with the fact that the adversary colludes with some of the clients, and therefore learns some subset of the PRP keys.

Upgrading to Unbounded-Access PANDA. Our bounded-access PANDA scheme is only secure when the number of accesses is a-priori bounded by some bound B, and can actually be shown to be insecure for sufficiently many accesses beyond that bound (following the analysis of [CHR17]). In the work of Canetti et al. [CHR17] and Boyle et al. [BIPW17], going from bounded-access DEPIR to unbounded-access DEPIR required new non-standard computational hardness assumptions. In our case, we will convert bounded-access PANDA to unbounded-access PANDA using standard assumptions, namely *leveled* FHE (instantiatable under the LWE assumption). The main reason that we can use our approach for PANDA, but not DEPIR, is that it makes the server *stateful*. This is something we allow in PANDA, whereas the main goal of DEPIR was to avoid it.

Our idea is essentially to "refresh" the bounded-access PANDA after every B accesses. More specifically, we think of the execution as proceeding in *epochs*, each consisting of B accesses. We associate a different *Pseudo-Random Function* (PRF) key with each virtual server and, for epoch i, we derive an epoch-specific PRP key for each server by applying the corresponding PRF on i. We then use this PRP key to freshly permute the codeword in each epoch. The clients get the PRF keys for the virtual servers in their committee. This lets clients derive the corresponding epoch-specific PRP keys for any epoch, and they can then proceed as they would using the bounded-access PANDA. The only difficulty is making sure that the server can correctly permute the codeword belonging to each virtual server in each epoch without knowing the associated PRF/PRP keys. We do this by storing FHE encryptions of each of the PRF keys on the server and, at the beginning of each epoch, the server performs a homomorphic computation to derive an encryption of the correctly permuted codeword for each virtual server. The clients also get the FHE decryption keys for the virtual servers in their committee, and thus can decrypt the codeword symbols that they read from the virtual servers. Note that the server has to do a large amount of work, linear in the codeword size, at the beginning of each epoch. However, we can use *amortized accounting* to spread this cost over the duration of the epoch and get low amortized complexity. Alternately, the server can spread out the actual computation across the epoch by performing a few steps of it at a time during each access to get low *worst-case* complexity. (This is possible because the database is read-only, and so its contents at the onset of the next epoch are known in advance at the beginning of the current epoch.) The security of this scheme follows from that of the bounded-access PANDA since in each epoch, the read operations are essentially performed using a fresh copy of the bounded-access PANDA (with fresh PRP keys).

PANDA with Public Encoding. Our construction of (unbounded-access) PANDA scheme described above has some nice features beyond what is required by the definition. Specifically, although the server is stateful and its internal state is updated in each epoch, the state can be computed using public information (the FHE encryptions of the PRF keys), the database, and the epoch number.[6] We find it useful to abstract this property further as a *PANDA with public encoding*. Specifically, we think of the PANDA scheme as having a key generation algorithm

---

[6] For example, the state does not depend on the history of protocol executions with the clients, and is unaffected by client actions. This may be of independent interest even if we downgrade the scheme to the single client setting, and gives the first ORAM scheme we are aware of with this property.

which doesn't depend on the database, and generates a public-key for the server and secret-keys for each of the clients. The server can then use the public-key to create a fresh encoding of the database with respect to an arbitrary epoch identifier (which can be a number, or an arbitrary bit string). The clients are given the epoch identifier, and can perform read operations which consist of reading some subset of locations from the server. Security holds as long as the number of read operations performed by honest clients with respect to any epoch identifier is bounded by B. Such a scheme can immediately be used to get an unbounded-access PANDA by having the server re-encode the database at the beginning of each epoch with an incremented epoch counter.

Note that our basic security definition considers a semi-honest adversary who corrupts the server and some subset of the clients, but otherwise follows the protocol specification. However, with the above structure, it's also clear that fully malicious clients (who might not follow the protocol) have no affect on the server state, and therefore cannot violate security. A fully malicious server, on the other hand, can lie about the epoch number and cause honest clients to perform too many read operations in one epoch, which would break security. However, if we assume that the epoch number is independently known to honest clients (for example, epochs occur at regular intervals, and clients know the rate at which accesses occur and have synchronized clocks) then this attack is prevented. The only other potential attack for a fully malicious server is to give incorrect values for the locations accessed in the encoded database. We can also prevent this attack by using succinct (interactive) arguments to prove that the values were computed correctly.

PANDA with Writes. We also consider PANDA schemes where clients can write to the database, and discuss two PANDA variants in this setting which we call *public-writes* and *secret-writes*.

Public-Writes PANDA. In a public-writes PANDA, we consider a setting where the server holds a shared public database which should be accessible to all clients. Clients can write to arbitrary locations in the database but, since the database is public, the locations and the values being written are necessarily public as well. However, we still want to maintain anonymity for the write operations (i.e., the server does not learn which client is performing each write), and both privacy *and* anonymity for the read operations (namely, the server does not learn which client is performing each read, or the locations being read). Our write operation is extremely simple: the client just sends the location and value being written to the server. However, even if we use PANDA with public encoding, the server cannot simply update the value in the encoded database since this would require (at least) linear time to re-encode the entire database.

Instead, we use an idea loosely inspired by hierarchical ORAM [Ost90, GO96]. We will store the database on the server in a sequence of $\log L$ levels, where L is the database size. Each level i consists of a separate instance of a read-only PANDA with public encoding, and will contain at most $L_i = 2^i$ database values.

We think of the levels as growing from the top down, namely level-0 (the smallest) is the top-most level, and level-$\log L$ (the largest) is the bottom-most. Initially, all the data is stored in the bottom level $i = \log L$, and all the remaining levels are empty. When a client wants to read some location j of the database, she uses the read-only PANDA for each of the $\log L$ levels to search for location j, and takes the value found in the top-most level that contains it. When a client writes to some location j, the server will place that database value in the top level $i = 0$. The server knows (in the clear) which database values are stored at each level. After every 2^i write operations, the server takes all the values in levels $0, \ldots, i$ and moves them to level $i + 1$ by using the public encoding procedure of PANDA and incrementing the epoch counter; level $i + 1$ will contain all the values that were previously in levels $\leq i + 1$, and levels $0, \ldots, i$ will be emptied.[7] Although the cost of moving all the data to level $i + 1$ scales with the data size L_{i+1}, the *amortized* cost is low since this only happens once every 2^i writes.[8]

One subtlety that we need to deal with is that our read-only PANDA was designed as an array data structure which holds L items with addresses $1, \ldots, L$. However, the way we use it in this construction requires a map data structure where the intermediate levels store $L_i \ll L$ items with addresses corresponding to some subset of the values $1, \ldots, L$. We can resolve this using the standard data-structures trick of storing a map in an array by hashing the n addresses into n buckets where each bucket contains some small number of values (to handle collisions). Our final public-writes PANDA scheme can also be thought of as implementing a map data structure, where database entries can be associated with arbitrary bit-strings as addresses, and clients can read/write to the value at any address. We can also allow the total database size to grow dynamically by adding additional levels as needed.

Secret-Writes PANDA. In this setting, instead of having a shared public database, we think of each client as having an individual private database which only she can access. We want the clients to be able to read and write to locations in their own database, while maintaining privacy and anonymity so that the server doesn't learn the identity of the client performing each access, the location being accessed, or the content of the data.

Our starting point is the public-writes PANDA scheme, which already guarantees privacy and anonymity of read operations, and anonymity of write operations. The clients can also individually encrypt all their content to ensure that it remains private. Therefore, we only need to modify write operations to provide privacy for the underlying location being written. To achieve this, we rely on the fact that our public-writes PANDA scheme already supports a map data structure, where data can be associated with an arbitrary bit-string as an address. As a first idea, when a client wants to write to some location j in her database,

---

[7] Note that the epoch counters are also incremented, and the encodings are refreshed, when sufficiently many reads occur at that level, just like in the read-only case.

[8] The server complexity can actually be de-amortized using the pipelining trick of Ostrovsky and Shoup [OS97].

she can use a client-unique PRF, associating the data with the address $PRF(j)$, and then write it using the public-writes scheme. While this partially hides the location j, the server still learns when the same location is written repeatedly. To solve this problem, we also add a counter c, and set the address to be $PRF(j, c)$. Whenever a client wants to read some location j, she uses the read operation of the public-writes PANDA to perform a binary search, and find the largest count c such that there is a value at the address $PRF(j, c)$ in the database. Whenever a client wants to write to location j, she first finds the correct count c (as she would in a read access), and then writes the value to address $PRF(j, c + 1)$. This ensures that the address being written reveal no information about the underlying database location. The only downside to this approach is that the server storage grows with the total *number of writes*, rather than the total *data size*. Indeed, since the server cannot correlate different "versions" of the same database location, it cannot delete old copies. Although we view this as a negative, we note that many existing database systems only support "append only" operations, and keep (as a backup) all old versions of the data. Therefore, in such a setting the growth in server storage caused by our scheme does not in fact add any additional overhead.

2   Preliminaries

Throughout the paper λ denotes a security parameter. We use standard cryptographic definitions of Pseudo-Random Permutations (PRPs), Pseudo-Random Functions (PRFs), and Fully Homomorphic Encryption (FHE) (see, e.g., [Gol01, Gol04]). For a vector $\mathbf{a} = (a_1, \ldots, a_n)$, and a subset $S = \{i_1, \ldots, i_s\} \subseteq [n]$, we denote $\mathbf{a}_S = (a_{i_1}, \ldots, a_{i_s})$.

Parameter Names. For all variants of the PANDA problem, we will let n denote the number of clients, L denote the database size, and t denote a bound on the number of corrupted clients colluding with the server.

2.1   Locally Decodable Codes (LDCs)

Locally decodable codes were first formally introduced by [KT00]. We rely on the following definition of smooth LDCs.

Definition 1 (Smooth LDC). *An s-smooth, k-query locally decodable code with message length L, and codeword size M over alphabet Σ, denoted by $(s, k, L, M)_\Sigma$-smooth LDC, is a triplet (Enc, Query, Dec) of PPT algorithms with the following properties.*

Syntax. Enc is given a message $msg \in \Sigma^L$ and outputs a codeword $c \in \Sigma^M$, Query is given an index $\ell \in [L]$ and outputs a vector $\mathbf{r} = (r_1, \ldots, r_k) \in [M]^k$, and Dec is given $c_\mathbf{r} = (c_{r_1}, \ldots, c_{r_k}) \in \Sigma^k$ and outputs a symbol in Σ.

Local decodability. For every message $msg \in \Sigma^L$, and every index $\ell \in [L]$,

$$\Pr\left[\mathbf{r} \leftarrow Query(\ell) \; : \; \mathsf{Dec}\left(\mathsf{Enc}\left(msg\right)_\mathbf{r}\right) = msg_\ell\right] = 1.$$

Smoothness. *For every index $\ell \in [L]$, the distribution of $(r_1, \ldots, r_k) \leftarrow$ Query(ℓ) is s-wise uniform. In particular, for any subset $S \subseteq [k]$ of size $|S| = s$, the random variables $r_i : i \in S$ are uniformly random over $[M]$ and independent of each other.*

We will use the Reed-Muller (RM) family of LDCs [Ree54, Mul54] over a finite field $\mathbb{F}$ which, roughly, are defined by m-variate polynomials over $\mathbb{F}$. More specifically, to encode messages in $\mathbb{F}^L$, one chooses a subset $H \subseteq \mathbb{F}$ such that $|H|^m \geq L$. Encoding a message $\mathsf{msg} \in \mathbb{F}^L$ is performed by interpreting the message as a function $\mathsf{msg} : H^m \to \mathbb{F}$, and letting $\widetilde{\mathsf{msg}} : \mathbb{F}^m \to \mathbb{F}$ be the *low degree extension* of msg; i.e., the m-variate polynomial of individual degree $< |H|$ whose restriction to H^m equals msg. The codeword c consists of the evaluations of $\widetilde{\mathsf{msg}}$ at all points if $\mathbb{F}^m$. We can locally decode any coordinate $\ell \in [L]$ of the message by thinking of ℓ as a value in H^m. This is done by choosing a random degree-s curve $\varphi : \mathbb{F} \to \mathbb{F}^m$ such that $\varphi(0) = \ell$, and querying the codeword on $k \geq ms(|H| - 1)$ non-0 points on the curve. The decoder then uses the answers $a_1, \cdots, a_k$ to interpolate the (unique) univariate degree-$(k-1)$ polynomial $\widetilde{\varphi}$ such that $\widetilde{\varphi}(i) = a_i$ for every $1 \leq i \leq k$. It outputs $\widetilde{\varphi}(0)$ as the ℓ'th message symbol. To guarantee that the field contains sufficiently many evaluation points, the field is chosen such that $|\mathbb{F}| \geq k + 1$. The codeword length is $M = |\mathbb{F}|^m$. We will need the following theorem, whose proof appears in the full version [HOWW18].

Theorem 3. *For any constant $\varepsilon > 0$, there exist $(s, k, L, M)_\Sigma$-smooth LDCs with $|\Sigma| = \mathsf{poly}(s, \log L)$, $k = \mathsf{poly}(s, \log L)$ and $M = L^{1+\varepsilon} \cdot \mathsf{poly}(s, \log L)$. Furthermore, the encoding time is $\widetilde{O}(M)$ and the decoding time is $\widetilde{O}(k)$.*

3   Read-Only PANDA

In this section we describe our read-only PANDA scheme. We first formally define this notion. At a high level, a PANDA scheme is run between a server S and n clients $C_1, \cdots, C_n$, and allows clients to *securely* access a database DB, even in the presence of a (semi-honestly) corrupted coalition consisting of the server S and a subset of at most t of the clients. In this section, we focus on the setting of a *read-only, public* database, in which the security guarantee is that read operations of honest clients remain entirely private and anonymous, meaning the corrupted coalition learns nothing about the identity of the client performed the operation, or which location was accessed.

Definition 2 (RO-PANDA). *A Read-Only Private Anonymous Data Access (RO-PANDA) scheme consists of procedures (Setup, Read) with the following syntax:*

- *Setup$(1^\lambda, 1^n, 1^t, DB)$ is a function that takes as input a security parameter λ, the number of clients n, a collusion bound t, and a database $DB \in \{0, 1\}^L$, and outputs the initial server state st_S, and client keys $\mathsf{ck}_1, \cdots, \mathsf{ck}_n$. We require that the size of the client keys $|\mathsf{ck}_j|$ is bounded by some fixed polynomial in the security parameter λ, independent of $n, t, |DB|$.*

- **Read** *is a protocol between the server* S *and a client* C_j. *The client holds as input an address* addr $\in [L]$ *and the client key* ck_j, *and the server holds its current states* st_S. *The output of the protocol is a value* val *to the client, and an updated server state* st'_S.

We require the following correctness and security properties.

- **Correctness:** *In any execution of the* **Setup** *algorithm followed by a sequence of* **Read** *protocols between various clients and the server, each client always outputs the correct database value* val $= \mathsf{DB}_{addr}$ *at the end of each protocol.*
- **Security:** *Any PPT adversary* $\mathcal{A}$ *has only* $\mathsf{negl}(\lambda)$ *advantage in the following security game with a challenger* $\mathcal{C}$:
  - $\mathcal{A}$ *sends to* $\mathcal{C}$:
    * *The values* n, t *and the database* $DB \in \{0,1\}^L$.
    * *A subset* $T \subset [n]$ *of corrupted clients with* $|T| \leq t$.
    * *A pair of read sequences* $R^0 = \left(j^0_l, \mathsf{addr}^0_l\right)_{1 \leq l \leq q}, R^1 = \left(j^1_l, \mathsf{addr}^1_l\right)_{1 \leq l \leq q}$
      (for some $q \in \mathbb{N}$*), where* $\left(j^b_l, \mathsf{addr}^b_l\right)$ *denotes that client* $j^b_l \in [n]$ *reads address* $\mathsf{addr}^b_l \in [L]$.
    We require that $\left(j^0_l, \mathsf{addr}^0_l\right) = \left(j^1_l, \mathsf{addr}^1_l\right)$ *for every* $l \in [q]$ *such that* $j^0_l \in T \vee j^1_l \in T$.
  - $\mathcal{C}$ *performs the following:*
    * *Picks a random bit* $b \leftarrow \{0,1\}$.
    * *Initializes the scheme by computing* **Setup** $\left(1^\lambda, 1^n, 1^t, DB\right)$.
    * *Sequentially executes the sequence* R^b *of* **Read** *protocol executions between the honest server and clients. It sends to* $\mathcal{A}$ *the views of the server* S *and the corrupted clients* $\{C_j\}_{j \in T}$ *during these protocol executions, where the view of each party consists of its internal state, randomness, and all protocol messages received.*
  - $\mathcal{A}$ *outputs a bit* b'.
  The advantage $\mathsf{Adv}_{\mathcal{A}}(\lambda)$ *of* $\mathcal{A}$ *in the security game is defined as:* $\mathsf{Adv}_{\mathcal{A}}(\lambda) = |\Pr[b' = b] - \frac{1}{2}|$.

Efficiency Goals. Since a secure PANDA scheme can be trivially obtained by having the client store the entire database locally, or having the server send the entire database to the client in every read request, the *efficiency* of the scheme is our main concern. We focus on minimizing the client storage and the client/server run-time during each Read protocol. At the very least, we require these to be $t \cdot o(|\mathsf{DB}|)$.

Bounded-Access PANDA. We will also consider a weaker notion of a *bounded-access* RO-PANDA scheme, for which security is only guaranteed to hold as long as the total number of read operations q is a-priori bounded. Such schemes will be useful building blocks for designing RO-PANDA schemes with full-fledged security.

Definition 3 (B-access RO-PANDA). *Let B be an access bound. We say that (*Setup, Read*) is a B-access RO-PANDA scheme if the security property of Definition 2 is only guaranteed to hold for PPT adversaries that are restricted to choose read sequences R^0, R^1 of length $q \leq B$.*

Remark on Adaptive Security. Note that, for simplicity, our definition is selective, where the adversary chooses the entire read sequences R^0, R^1 ahead of time. We could also consider a stronger adaptive security definition where the adversary chooses the sequence of reads *adaptively* as the protocol progresses. Although our constructions are also secure in the stronger setting (with minimal modifications to the proofs), we chose to present our results in the selective setting to keep them as simple as possible.

3.1   A Bounded-Access Read-Only PANDA Scheme

As a first step, we now show how to construct a bounded-access RO-PANDA scheme, yielding the following theorem.

Theorem 4 (B-access RO-PANDA). *Assuming one-way functions exist, for any constant $\varepsilon > 0$ there is a B-bounded access RO-PANDA where, for n clients with t collusion bound and database size L:*

- *The client and server complexity during each **Read** protocol is $t \cdot \mathsf{poly}(\lambda, \log L)$.*
- *The client storage is $t \cdot \mathsf{poly}(\lambda, \log L)$.*
- *The server storage is $\alpha \leq t \cdot L^{1+\varepsilon} \cdot \mathsf{poly}(\lambda, \log L)$.*
- *The access bound is $B = \alpha/(t \cdot \mathsf{poly}(\lambda, \log L))$.*

Note that in the above theorem we can increase the access-bound B arbitrarily by artificially inflating the database size L to increase α. However, we will mainly be interested in having a small ratio α/B while keeping α as small as possible.

Construction Outline. As outlined in the introduction, our idea is inspired by the recent works of Canetti et al. [CHR17] and Boyle et al. [BIPW17] on DEPIR. We rely on an s-smooth, k-query LDC where $s = \lambda$ is set to be the security parameter. We think of the server S as consisting of $k' = k^2 t$ different "virtual servers", where t is the collusion bound. Each virtual server contains a permuted copy of the LDC codeword under a fresh PRP. Each client is assigned a random committee consisting of k out of k' of the virtual servers and gets the corresponding PRP keys. To retrieve an entry from the database, the client runs the LDC local decoding algorithm, which requests to see k codeword locations, and reads these locations using the virtual servers on its committee by applying the corresponding PRPs. It also reads uniformly random locations from the $k' - k$ virtual servers that are not on its committee.

Construction 1 (B-Access RO-PANDA). The scheme uses the following building blocks:

- An $(s, k, L, M)_\Sigma$-smooth LDC $(\mathsf{Enc}_{\mathsf{LDC}}, \mathsf{Query}_{\mathsf{LDC}}, \mathsf{Dec}_{\mathsf{LDC}})$ (see Definition 1, Theorem 3).
- A CPA-secure symmetric encryption scheme $(\mathsf{KeyGen}_{\mathsf{sym}}, \mathsf{Enc}_{\mathsf{sym}}, \mathsf{Dec}_{\mathsf{sym}})$.
- A pseudorandom permutation (PRP) family $P : \{0,1\}^\lambda \times [M] \to [M]$ where for every $K \in \{0,1\}^\lambda$ the function $P(K, \cdot)$ is a permutation.

The scheme consists of the following procedures:

- **Setup**$(1^\lambda, 1^n, 1^t, \mathsf{DB})$: Recall that n denotes the number of clients, t is the collusion bound, and $\mathsf{DB} \in \{0,1\}^L$. Instantiate the LDC with message size L and smoothness $s = \lambda$, and let k be the corresponding number of queries, M be the corresponding codeword size and Σ be the alphabet. Set $k' = k^2 t$ to be the number of virtual servers. Proceed as follows.
  - Database encoding. Generate the codeword $\widehat{\mathsf{DB}} = \mathsf{Enc}_{\mathsf{LDC}}(\mathsf{DB})$ with $\widehat{\mathsf{DB}} \in \Sigma^M$.
  - Virtual server generation. For every $1 \le i \le k'$:
    * Generate a PRP key $K_{\mathsf{PRP}}^i \leftarrow \{0,1\}^\lambda$, and an encryption key $K_{\mathsf{sym}}^i \leftarrow \mathsf{KeyGen}_{\mathsf{sym}}(1^\lambda)$.
    * Let $\widehat{\mathsf{DB}}^i \in \Sigma^M$ be a permuted database which satisfies $\widehat{\mathsf{DB}}^i_{P(K_{\mathsf{PRP}}^i, j)} = \widehat{\mathsf{DB}}_j$ for all $j \in [M]$.
    * Let $\widetilde{\mathsf{DB}}^i$ be the encrypted-permuted database with $\widetilde{\mathsf{DB}}^i_j = \mathsf{Enc}_{\mathsf{sym}}\left(K_{\mathsf{sym}}^i, \widehat{\mathsf{DB}}^i_j\right)$.
  - Committee generation. For every $j \in [n]$, pick a random size-k subset $\mathsf{S}_j \subseteq [k']$.
  - Output. For each client C_j, set the client key $\mathsf{ck}_j = (\mathsf{S}_j, \{K_{\mathsf{PRP}}^i, K_{\mathsf{sym}}^i : i \in \mathsf{S}_j\})$ to consist of the description of the committee and the PRP and encryption keys of the virtual servers on the committee. Set the server state $\mathsf{st}_S = \{\widetilde{\mathsf{DB}}^i : i \in [k']\}$ to consist of the encrypted-permuted databases of every virtual server.
- **The Read protocol.** To read database entry at location $\mathsf{addr} \in [L]$ from the server S, a client C_j with key $\mathsf{ck}_j = (\mathsf{S}_j, \{K_{\mathsf{PRP}}^i, K_{\mathsf{sym}}^i : i \in \mathsf{S}_j\})$ operates as follows.
  - (Query.) Denote $\mathsf{S}_j = \{v_1, \ldots, v_k\} \subseteq [k']$. Sample $(r_{v_1}, \cdots, r_{v_k}) \leftarrow \mathsf{Query}_{\mathsf{LDC}}(\mathsf{addr})$, and for each $v \in \mathsf{S}_j$ set $\hat{r}_v = P(K_{\mathsf{PRP}}^v, r_v)$ to be the query to virtual server v. For every $v \in [k'] \setminus \mathsf{S}_j$, pick $\hat{r}_v \in_R [M]$ uniformly.
  - (Recover.) Send $(\hat{r}_1, \cdots, \hat{r}_{k'})$ to the server S and obtain the answers $\left(\widetilde{\mathsf{DB}}^1_{\hat{r}_1}, \cdots, \widetilde{\mathsf{DB}}^{k'}_{\hat{r}_{k'}}\right)$. For every $v = v_h \in \mathsf{S}_j$, decrypt $a_h = \mathsf{Dec}_{\mathsf{sym}}\left(K_{\mathsf{sym}}^v, \widetilde{\mathsf{DB}}^v_{\hat{r}_v}\right)$, and output $\mathsf{Dec}_{\mathsf{LDC}}(a_1, \cdots, a_k)$.

Remark. Note that in the above construction the server is completely static and stateless. Indeed the **Read** protocol simply consists of the client retrieving some subset of the locations from the server.

Proof of Security. We prove the following claim about the above construction.

Claim 1. Assuming the security of all of the building blocks, Construction 1 is B-bounded-access RO-PANDA for $B = M/(2k^2)$.

Claim implies Theorem. It's easy to see that Claim 1 immediately implies Theorem 4 by plugging in the LDC parameters from Theorem 3. In particular, for n clients, t collusion bound and database size L:

- The client/server run-time is $k' \log |\Sigma| = tk^2 \log |\Sigma| = t \cdot \mathsf{poly}(\lambda, \log L)$.
- The client storage is $k' \cdot (\mathsf{poly}(\lambda) + \log L) = t \cdot \mathsf{poly}(\lambda, \log L)$.
- The server storage is $\alpha = k' \cdot M \cdot \log |\Sigma| = tk^2 M \cdot \log |\Sigma| = t \cdot L^{1+\varepsilon} \cdot \mathsf{poly}(\lambda, \log L)$.
- The bound B is $B = M/(2k^2) = \alpha/(t \cdot \mathsf{poly}(\lambda, \log L))$.

Background Lemmas. To show that the construction is secure, we rely on two lemmas. The first lemma comes from the work of Canetti et al. [CHR17].

Lemma 1 (Lemma 1 in [CHR17]). *Let* $X = (X_1, \cdots, X_m), Y = (Y_1, \cdots, Y_m)$ *be* l-*wise independent random variables such that for every* $1 \leq i \leq m$, X_i, Y_i *are identically distributed. Assume also that there is a value* $\star$ *such that* $\Pr[X_i = \star] \geq 1 - \delta$. *Then* $\mathsf{SD}(X, Y) \leq (m\delta)^{l/2} + m^{l-1}\delta^{l/2-1} \leq 2m^{l-1}\delta^{l/2-1} \leq 2m(m^2\delta)^{l/2-1}$.

The second lemma (whose proof appears in the full version [HOWW18]) deals with the intersection size of random sets.

Lemma 2. *Let* $T \subseteq [n]$ *be an arbitrary set of size* $|T| \leq t$. *Let* $S_1, \cdots, S_n$ *be chosen as random subsets* $S_j \subseteq [k']$ *of size* $|S_j| = k$, *where* $k' = k^2 t$. *Then, for all* $\rho > 2e$, *the probability that there exists some* $j \in [n] \setminus T$ *such that* $|(\cup_{i \in T} S_i) \cap S_j| \geq \rho$ *is at most* $n \cdot 2^{-\rho}$.

Proof of Claim. We are now ready to prove Claim 1.

Proof of Claim 1. The correctness of the scheme follows directly from the correctness of the LDC and the symmetric encryption scheme. We now argue security.

Let $\mathcal{A}$ be a PPT adversary corrupting the server and a subset T of at most t clients. Let R^0, R^1 be the two sequences of **read** operations of length $q \leq B$ which $\mathcal{A}$ chooses in the security game. Without loss of generality, we can assume that R^0, R^1 do not contain **read** operations by corrupted clients since $\mathcal{A}$ can generate the corresponding accesses itself (and it does not affect the server state in any way). Let $S_1, \ldots, S_n$ be the random committees chosen during Setup and let $E = \bigcup_{i \in T} S_i$. We proceed via a sequence of hybrids.

H$_1$ Hybrid H_1 is the security game as in Definition 2.

H$_2$ In hybrid H_2, for all $i \notin E$, we replace the encrypted database $\widetilde{\mathsf{DB}}^i$ by a dummy encryption (e.g.,) of the all 0 string.
    Hybrids H_1 and H_2 are computationally indistinguishable by CPA security of the encryption scheme.

H₃ In hybrid H_3, for all $i \notin E$ we replace all calls to the PRP $P(K_{\mathsf{PRP}}^i, \cdot)$ during the various executions of the Read protocol with a truly random permutation $\pi^i : [M] \to [M]$.
Hybrids H_2 and H_3 are computationally indistinguishable by PRP security.
Here we rely on the fact that in both hybrids the encrypted database $\widetilde{\mathsf{DB}}^i$ for $i \notin E$ is independent of the permutation.

H₄ In hybrid H_4, if during the committee selection in the Setup algorithm it occurs that there exists some $j \in [n] \setminus T$ such that $|E \cap S_j| \geq s/2$, then the game immediately halts.
Hybrids H_3 and H_4 are statistically indistinguishable by Lemma 2, where we set $\rho = s/2$. Recall that $s = \lambda$ and therefore $n \cdot 2^{-\rho} = \mathsf{negl}(\lambda)$.

H₅ In hybrid H_5, we replace the queries $(\hat{r}_1, \cdots, \hat{r}_{k'})$ created during the execution of each Read protocol with truly random values $(u_1, \cdots, u_{k'}) \leftarrow [M]^{k'}$.

The main technical difficulty is showing that hybrids H_4 and H_5 are (statistically) indistinguishable, which we do below. Once we do that, note that hybrid H_5 is independent of the challenge bit b and therefore in hybrid H_5 we have $\Pr[b = b'] = \frac{1}{2}$. Since hybrids H_1 and H_5 are indistinguishable, it means that in hybrid H_1 we must have $|\Pr[b = b'] - \frac{1}{2}| = \mathsf{negl}(\lambda)$ which proves the claim.

We are left to show that hybrids H_4 and H_5 are statistically indistinguishable. We do this by showing that for every Read protocol execution, even if we fix the entire view of the adversary prior to this protocol, the queries sent during the protocol in hybrid H_4 are statistically close to uniform. The protocol is executed by some honest client j with committee $S_j = \{v_1, \ldots, v_k\}$ and we know that $|S_j \cap E| \leq s/2$. Let $(\hat{r}_1, \ldots, \hat{r}_{k'})$ be the distribution on the client queries in the protocol.

(i) For all $v \notin S_j$ the values $\hat{r}_v$ are chosen uniformly at random and independently by the client.

(ii) For $v \in S_j \cap E$, the values $\hat{r}_v = P(K_{\mathsf{PRP}}^v, r_v)$ are uniformly random by the s-wise independence of $\{r_v\}_{v \in S_j}$ and the fact that $|S_j \cap E| \leq s/2$.

(iii) For $v \in S_j \setminus E$, we want to show that the values $\hat{r}_v = \pi^v(r_v)$ are statistically close to uniform, even if we condition on (i),(ii). Note that the values $\{r_v\}_{v \in S_j \setminus E}$ are $s/2$-wise independent even conditioned on the above, and therefore so are the values $\{\hat{r}_v\}_{v \in S_j \setminus E}$. For each v, let $\mathcal{Z}_v \subseteq [M]$ be the set of values $\pi^v(r)$ that were queried in some prior protocol execution by some client. Then $|\mathcal{Z}_v| \leq B$. Note that if $\hat{r}_v = \pi^v(r_v) \notin \mathcal{Z}_v$ then $\hat{r}_v$ is simply uniform over $[M] \setminus \mathcal{Z}_v$ by the randomness of the permutation π^v. We can define random variables X_v where $X_v = \hat{r}_v$ when $\hat{r}_v \in \mathcal{Z}_v$ and $X_v = \star$ otherwise. We can then think of sampling $\{\hat{r}_v\}_{v \in S_j \setminus E}$ by sampling $\{X_v\}_{v \in S_j \setminus E}$ and defining $\hat{r}_v = X_v$ when $X_v \neq \star$ and sampling $\hat{r}_v$ uniformly at random over $[M] \setminus \mathcal{Z}_v$ otherwise. Note that $\{X_v\}_{v \in S_j \setminus E}$ is a set of $|S_j \setminus E| \leq k$ variables which are $s/2$-wise independent and $\Pr[X_v = \star] \geq 1 - \delta$ where $\delta \leq |\mathcal{Z}_v|/M \leq B/M$. Therefore, by applying Lemma 1, the variables $\{X_v\}_{v \in S_j \setminus E}$ are statistically close to truly independent variables $\{Y_v\}_{v \in S_j \setminus E}$ such that each Y_v has the same marginal distribution as X_v, where the statistical distance is $2k(k^2 B/M)^{s/4-1} \leq 2k(1/2)^{s/4-1} = \mathsf{negl}(\lambda)$. Replacing the variables $\{X_v\}$ by $\{Y_v\}$ is equivalent to replacing the values $\{\hat{r}_v\}_{v \in S_j \setminus E}$ by truly uniform and independent values. □

3.2 Public-Encoding PANDA

In this section we describe a *public-encoding* variant of bounded-access RO-PANDA schemes, which will be used to construct an unbounded-access RO-PANDA as well as PANDA schemes that support writes. At a high level, a public-encoding bounded-access PANDA scheme contains a key-generation algorithm KeyGen that generates a public key pk and a set $\{ck_j\}$ of client secret keys. Any database owner can *locally* encode the database using only the public key. The scheme guarantees privacy and anonymity, even if the adversary obtains a subset of the secret keys, as long as the honest clients make at most B accesses to the database. Furthermore, we allow the server to create many encodings of the same, or different, databases with respect to some labels lab, and the clients can generate accesses using the corresponding label lab. As long as the clients make at most B accesses with respect to any *one* label, security is maintained.

Definition 4 (Public-Encoding PANDA). *A* public-encoding PANDA (PE-PANDA) *consists of a tuple of algorithms* (*KeyGen, Encode, Query, Recover*) *with the following syntax.*

- *KeyGen*$(1^\lambda, 1^n, 1^t, 1^L)$ *is a PPT algorithm that takes as input a security parameter λ, the number of clients n, and the collusion bound t, and a database size L. It outputs a public key* pk, *and a set of client secret keys* $\{ck_j\}_{j \in [n]}$.
- *Encode*(pk, DB, lab) *is a deterministic algorithm that takes as input a public-key* pk, *a database* DB, *and a label* lab, *and outputs an encoded database* $\widetilde{DB}$.
- *Query*$(ck_j, addr, lab)$ *is a PPT algorithm that takes as input a secret-key* ck_j, *an address* addr *in a database, and a label* lab, *and generates a list* $(q_1, \cdots, q_{k'})$ *of coordinates in the encoded database.*
- *Recover* $\left(ck_j, lab, \left(\widetilde{DB}_{q_1}, \cdots, \widetilde{DB}_{q_{k'}}\right)\right)$ *is a deterministic algorithm that takes as input a secret-key* ck, *a label* lab, *and a list* $\left(\widetilde{DB}_{q_1}, \cdots, \widetilde{DB}_{q_{k'}}\right)$ *of entries in an encoded database, and outputs a database value* val.

We require that it satisfies the following correctness and security properties.

- **Correctness:** *For every* $\lambda, n, t, L \in \mathbb{N}$, *every* DB $\in \{0,1\}^L$, *every label* lab $\in \{0,1\}^*$, *every address* addr $\in [L]$, *and every client* $j \in [n]$:

$$
\Pr \left[\begin{array}{c} \left(pk, \{ck_j\}_{j \in [n]}\right) \leftarrow KeyGen\left(1^\lambda, 1^n, 1^t, 1^L\right) \\ \widetilde{DB} = Encode\,(pk, DB, lab) \\ (q_1, \cdots, q_{k'}) \leftarrow Query\,(ck_j, addr, lab) \\ val = Recover\left(ck_j, lab, \left(\widetilde{DB}_{q_1}, \cdots, \widetilde{DB}_{q_{k'}}\right)\right) \end{array} : val = DB_{addr} \right] = 1.
$$

- B-**Bounded-Access Security:** *Every PPT adversary $\mathcal{A}$ has only $\mathsf{negl}\,(\lambda)$ advantage in the following security game with a challenger $\mathcal{C}$:*
  - *$\mathcal{A}$ sends to $\mathcal{C}$ values n, t, L, and a subset $T \subset [n]$ of size $|T| \leq t$.*
  - *$\mathcal{C}$ executes $\left(\mathsf{pk}, \{\mathsf{ck}_j\}_{j\in[n]}\right) \leftarrow \mathsf{KeyGen}\left(1^\lambda, 1^n, 1^t, 1^L\right)$ and sends pk and $\{\mathsf{ck}_j\}_{j\in T}$ to $\mathcal{A}$. Additionally, $\mathcal{C}$ picks a random bit b.*
  - *$\mathcal{A}$ is given access to the oracle $\mathsf{Query}^b_{\{\mathsf{ck}_j\}}$ that on input $(j_0, j_1, \mathsf{addr}_0, \mathsf{addr}_1, \mathsf{lab})$ such that $j_0, j_1 \notin T$, outputs $\mathsf{Query}(\mathsf{ck}_{j_b}, \mathsf{addr}_b, \mathsf{lab})$.*
    We restrict $\mathcal{A}$ to make at most B queries to the oracle with any given label lab, but allow it to make an unlimited number of queries in total.
  - *$\mathcal{A}$ outputs a bit b'.*
  The advantage $\mathsf{Adv}_\mathcal{A}\,(\lambda)$ of $\mathcal{A}$ in the security game is defined as: $\mathsf{Adv}_\mathcal{A}\,(\lambda) = \left|\Pr\left[b' = b\right] - \frac{1}{2}\right|$.

Next, we construct a public-encoding PANDA scheme, based on our bounded-access PANDA scheme (Construction 1 in Sect. 3.1). The high-level idea is to use fresh PRP keys for every label, by creating them via a PRF applied to the label. The public key of the server contains FHE encryptions of the PRF keys. This enables the server to create the encoded-permuted databases for each virtual server, as in Construction 1, by operating on the PRF keys under FHE.

Construction 2 (Public-Encoding PANDA). The scheme uses the same building blocks as Construction 1. In addition we rely on:

- A pseudo-random function $F : \{0,1\}^\lambda \times \{0,1\}^* \to \{0,1\}^\lambda$.
- The symmetric-key encryption scheme in Construction 1 will be replaced by a symmetric-key leveled FHE scheme ($\mathsf{KeyGen_{FHE}}, \mathsf{Enc_{FHE}}, \mathsf{Dec_{FHE}}, \mathsf{Eval_{FHE}}$).

The scheme consists of the following algorithms:

- **KeyGen** $\left(1^\lambda, 1^n, 1^t, 1^L\right)$ operates as follows:
  - Let the parameters s, k, M, k' be chosen the same way as in Construction 1.
  - For every virtual server $i \in [k']$:
    * Generates a random FHE key $K^i_{\mathsf{FHE}} \leftarrow \mathsf{KeyGen_{FHE}}\left(1^\lambda\right)$. We use a leveled FHE that can evaluate circuits up to some fixed polynomial depth $d = \mathsf{poly}(\lambda, \log M)$ specified later.
    * Generates a random PRF key $K^i_{\mathsf{PRF}} \leftarrow \{0,1\}^\lambda$.
    * Encrypts the PRF key: $\widetilde{K}^i_{\mathsf{PRF}} \leftarrow \mathsf{Enc_{FHE}}\left(K^i_{\mathsf{FHE}}, K^i_{\mathsf{PRF}}\right)$.
  - Generates the random size-k committee $\mathsf{S}_j \subseteq [k']$ for every $1 \leq j \leq n$.
  - Outputs the public key $\mathsf{pk} = \left(L, \left\{\widetilde{K}^i_{\mathsf{PRF}}\right\}_{i\in[k']}\right)$, and the secret keys $\left\{\mathsf{ck}_j = \left(\mathsf{S}_j, L, \left\{K^i_{\mathsf{PRF}}, K^i_{\mathsf{FHE}} : i \in \mathsf{S}_j\right\}\right)\right\}_{j\in[n]}$.

- **Encode** $\left(\mathbf{pk} = \left(L, \left\{\widetilde{K}^i_{\mathsf{PRF}}\right\}_{i\in[k']}\right), \mathbf{DB}, \mathsf{lab}\right)$ operates as follows:

  • Let $\widehat{\mathsf{DB}} = \mathsf{Enc_{LDC}}(\mathsf{DB})$ using and LDC with parameters s, k, L, M as in the Setup algorithm of Construction 1.
  • For every $i \in [k']$:
    * Generates an encrypted key $\widetilde{K}^i_{\mathsf{PRP}} = \mathsf{Eval_{FHE}}\left(C_{F,\mathsf{lab}}(\cdot), \widetilde{K}^i_{\mathsf{PRF}}\right)$, where $C_{F,x}(\cdot)$ is the circuit that on input K computes $F(K, x)$.
    * Generate an encrypted-permuted database $\widetilde{\mathsf{DB}}^i = \mathsf{Eval_{FHE}}\left(C_{P,\widehat{\mathsf{DB}}}(\cdot),\right.$
      $\left. \widetilde{K}^i_{\mathsf{PRP}}\right)$, where $C_{P,\widehat{\mathsf{DB}}}(\cdot)$ is the circuit that on input K computes the permuted database $\widehat{\mathsf{DB}}$ which satisfies $\widehat{\mathsf{DB}}^i_{P(K,j)} = \widehat{\mathsf{DB}}_j$ for all $j \in [M]$.
  • Outputs $\left(\widetilde{\mathsf{DB}}^1, \cdots, \widetilde{\mathsf{DB}}^{k'}\right)$.
- **Query, Recover.** These algorithms work the same way as the two stages of the Read protocol in Construction 1 where the client sets $K^i_{\mathsf{PRP}} := F\left(K^i_{\mathsf{PRF}}, \mathsf{lab}\right)$ and $K^i_{\mathsf{sym}} := K^i_{\mathsf{FHE}}$ for $i \in \mathsf{S}_j$.

Leveled FHE Remark. In the above construction we set parameter d representing the maximum circuit depth for the leveled FHE to be the combined depth of the circuits $C_{F,x}(\cdot)$ and $C_{P,\widehat{\mathsf{DB}}}(\cdot)$ defined above. Since we can use a permutation network which permutes data of size M in depth $\log M$, so we have $d = \mathsf{poly}(\lambda, \log M)$. We assume that the leveled FHE scheme allows us to compute circuits C of depth d in time $|C| \cdot \mathsf{poly}(\lambda, d)$.

In the full version [HOWW18] we prove the following theorem:

Theorem 5 (Public-Encoding PANDA). *Suppose leveled FHE exists. Then for any constant $\varepsilon > 0$ there is a PE-PANDA scheme with B-bounded access security, for n clients, t collusion bound and database size L where:*

- *The complexity of Query and Recover procedures is $t \cdot \mathsf{poly}(\lambda, \log L)$.*
- *The server public key and the client secret keys are each of size $t \cdot \mathsf{poly}(\lambda, \log L)$.*
- *The complexity of the encoding procedure and the size of the encoded database is $\alpha \le t \cdot L^{1+\varepsilon} \cdot \mathsf{poly}(\lambda, \log L)$.*
- *The access bound is $B = \alpha/(t \cdot \mathsf{poly}(\lambda, \log L))$.*

3.3   Read-Only PANDA with Unbounded Accesses

In this section we use the public-encoding PANDA scheme of Sect. 3.2, which has B-bounded-access security, to obtain a read-only PANDA scheme that is secure against any unbounded number of accesses.

The high-level idea of our construction is conceptually simple: after every B operations, the server re-encodes the database with a fresh label. We think of these sequences of B consecutive accesses as "epochs", and the label is simply a counter indicating the current epoch. The clients get the current epoch number by reading it from the server before performing an access.

Construction 3 (Read-only PANDA). The scheme uses a PE-PANDA scheme (KeyGen, Encode, Query, Recover) with B-bounded-access security as a building block. We define the following procedures.

- **Setup**$(1^\lambda, 1^n, 1^t, \mathsf{DB})$. Takes as input a security parameter λ, the number of clients n, a collusion bound t, and a database $\mathsf{DB} \in \{0, 1\}^L$. It does the following.
  - Counter initialization. Initializes an *epoch counter* count_e, and a *step counter* count_s, to 0.
  - Generating keys. Runs $\left(\mathsf{pk}, \{\mathsf{ck}_j\}_{j \in [n]}\right) \leftarrow \mathsf{KeyGen}\left(1^\lambda, 1^n, 1^t, 1^L\right)$.
  - Encoding the database. Runs $\widetilde{\mathsf{DB}} = \mathsf{Encode}\,(\mathsf{pk}, \mathsf{DB}, \mathsf{count}_e)$.
  - Output. For each client $C_j, 1 \le j \le n$ set the client key to $\mathsf{ck}_j := \mathsf{ck}_j$. For the server S set $\mathsf{st}_S := (\mathsf{pk}, \widetilde{\mathsf{DB}}, \mathsf{count}_e, \mathsf{count}_s)$.
- **The Read Protocol.** To read the data block at address addr from the server, a client C_j and the server S run the following protocol.
  - The client reads the epoch counter count_e from S.
  - The client runs $(q_1, \cdots, q_{k'}) \leftarrow \mathsf{Query}\,(\mathsf{ck}_j, \mathsf{addr}, \mathsf{count}_e)$, and sends $(q_1, \cdots, q_{k'})$ to S.
  - The server computes $a_i = \widetilde{\mathsf{DB}}_{q_i}$ and sends back the values $(a_1, \cdots, a_{k'})$ to the client.
  - The client recovers $\mathsf{DB}_{\mathsf{addr}} = \mathsf{Recover}\,(\mathsf{ck}_j, \mathsf{count}_e, (a_1, \cdots, a_{k'}))$.
  - The server S updates its state as follows: if $\mathsf{count}_s < B - 1$, S updates $\mathsf{count}_s := \mathsf{count}_s + 1$. Otherwise, S updates $\mathsf{count}_s := 0, \mathsf{count}_e := \mathsf{count}_e + 1$, and replaces $\widetilde{\mathsf{DB}} := \mathsf{Encode}\,(\mathsf{pk}, \mathsf{DB}, \mathsf{count}_e)$. If the complexity of the computation $\mathsf{Encode}\,(\mathsf{pk}, \mathsf{DB}, \mathsf{count}_e)$ is c_{Encode}, the server performs c_{Encode}/B steps of this computation during each protocol execution so that it is completed by the end of the epoch.

In the full version [HOWW18] we prove the following theorem:

Theorem 6 (Read-Only PANDA). *Suppose leveled FHE exists. Then for any constant $\varepsilon > 0$ there is a read-only PANDA, for n clients, t collusion bound and database size L where:*

- *The client/server complexity during each Read protocol is $t \cdot \mathsf{poly}(\lambda, \log L)$.*
- *The client keys are of size $t \cdot \mathsf{poly}(\lambda, \log L)$.*
- *The server state is of size $t \cdot L^{1+\varepsilon} \cdot \mathsf{poly}(\lambda, \log L)$.*

4   PANDA with Public-Writes

In this section we extend the read-only scheme of Sect. 3 to support writes in the *public* database setting. In the full version [HOWW18] we design a PANDA scheme that supports writes in the *private* database setting.

Our PANDA scheme for public databases supports write operations, but only guarantee privacy of read operations. We call this primitive a *Public-Writes PANDA (PW-PANDA)*. Notice that this is the "best possible" security guarantee when there is (even) a (single) corrupted client. (Indeed, as the database is public, a corrupted coalition can always learn what values were written to which locations by simply reading the entire database after every operation.) We note that it suffices to consider this weaker security guarantee *when all clients are honest*, since any public-writes PANDA scheme can be generically transformed into a PANDA scheme which guarantees the privacy of write operations *when all clients are honest*. Indeed, one can implement a (standard) single-client ORAM scheme on top of the public-writes PANDA scheme, for which all clients know the private client key. (We note that the transformation might require FHE-encrypting the PANDA, to allow the server to perform operations on the PANDA which are caused by client operations on the ORAM.)

We now formally define the notion of a public-writes PANDA scheme.

Definition 5 (Public-Writes PANDA (PW-PANDA)). *A public-writes PANDA (PW-PANDA) scheme consists of procedures (Setup, Read, Write), where Setup, Read have the syntax of Definition 2, and Write has the following syntax. It is a protocol between the server S and a client C_j. The client holds as input an address $addr \in [L]$, a value v, and the client key ck_j, and the server holds its current states st_S. The output of the protocol is an updated server state st'_S.*

We require the following correctness and security properties.

- **Correctness:** *In any execution of the Setup algorithm followed by a sequence of Read and Write protocols between various clients and the server, where the Write protocols were executed with a sequence Q of values, the output of each client in a read operation is the value it would have read from the database if (the prefix of) Q (performed before the corresponding Read protocol) was performed directly on the database.*

- **Security:** *Any PPT adversary $\mathcal{A}$ has only $\mathsf{negl}(\lambda)$ advantage in the following security game with a challenger $\mathcal{C}$:*
  - *$\mathcal{A}$ sends to $\mathcal{C}$:*
    * *The values n, t, and the database $DB \in \{0,1\}^L$.*
    * *A subset $T \subset [n]$ of corrupted clients with $|T| \leq t$.*
    * *A pair of access sequences $Q^0 = \left(op_l, \mathsf{val}_l^0, j_l^0, addr_l^0\right)_{1 \leq l \leq q}, Q^1 = \left(op_l, \mathsf{val}_l^1, j_l^1, addr_l^1\right)_{1 \leq l \leq q}$, where $\left(op_l, \mathsf{val}_l^b, j_l^b, addr_l^b\right)$ denotes that client j_l^b performs operation op_l at address $addr_l^b$ with value val_l^b (which, if $op_l = \text{read}$, is $\perp$).*

    We require that $\left(op_l, \mathsf{val}_l^0, j_l^0, addr_l^0\right) = \left(op_l, \mathsf{val}_l^1, j_l^1, addr_l^1\right)$ for every $l \in [q]$ such that $j_l^0 \in T \vee j_l^1 \in T$; and $\left(\mathsf{val}_l^0, addr_l^0\right) = \left(\mathsf{val}_l^1, addr_l^1\right)$ for every $l \in [q]$ such that $op_l = \text{write}$ (in particular, write operations differ only in the identity of the client performing the operation).
  - *$\mathcal{C}$ performs the following:*
    * *Picks a random bit $b \leftarrow \{0,1\}$.*
    * *Initializes the scheme by computing $\mathsf{Setup}\left(1^\lambda, 1^n, 1^t, DB\right)$.*

∗ *Sequentially executes the sequence Q^b of Read and Write protocol executions between the honest server and clients. It sends to $\mathcal{A}$ the views of the server S and the corrupted clients $\{C_j\}_{j \in T}$ during these protocol executions, where the view of each party consists of its internal state, randomness, and all protocol messages received.*

- $\mathcal{A}$ *outputs a bit b'.*

The advantage $Adv_{\mathcal{A}}(\lambda)$ of $\mathcal{A}$ in the security game is defined as: $Adv_{\mathcal{A}}(\lambda) = |\Pr[b' = b] - \frac{1}{2}|$.

Construction Outline. As outlined in the introduction, the public-writes PANDA scheme consists of $\log L$ levels of increasing size (growing from top to bottom), each containing size-λ "buckets" that hold several data blocks, and implemented with a B-bounded-access PE-PANDA scheme. To initialize our PANDA scheme, we generate PE-PANDA public- and secret-keys for every level. Initially, all levels are empty, except for the lowest level, which consists of a PE-PANDA for the database DB. read operations will look for the data block in all levels (returning the top-most copy),[9] whereas write operations will write to the top-most level, causing a reshuffle at predefined intervals to prevent levels from overflowing. We note that adding a *new copy* of the data block (instead of updating the existing data block wherever it is located) allows us to change *only* the content of the top level. This is crucial to obtaining a non-trivial scheme, since levels are implemented using a *read-only* PANDA, and so can only be updated by generating a *new* scheme for the *entire* content of the level, which might be expensive (and so must not be performed too often for lower levels).

Notice that since the levels are implemented using a PE-PANDA scheme (which, in particular, is only secure against a bounded number of accesses), security is guaranteed only as long as each level is accessed at most an a-priori bounded number of times. To guarantee security against *any* (polynomial) number of accesses, we "regenerate" each level when the number of times it has been accessed reaches the bound. This regeneration is performed by running the Encode algorithm of the PE-PANDA scheme with a new label, consisting of the epoch number of the current level and the number of regeneration operations performed during the current epoch (this guarantees that every label is used at most once in each level). In summary, each level can be updated in one of two forms: (1) through a reshuffle operation that merges an upper level into it; or (2) through a regenerate operation, in which the PE-PANDA of the level is updated (but the actual data blocks stored in it do not change). We note that (unlike standard hierarchical ORAM) the reshuffling and regeneration need not be done obliviously, since the server knows the contents of all levels.

As in the introduction, we associate a public hash function with each level, which is used to map data blocks into buckets, thus overcoming the issue that

---

[9] We note that in standard hierarchical ORAM, once the data block was found, the client should make "dummy" random accesses to lower levels. However, since in our construction each level is implemented as a PE-PANDA scheme which anyway hides the identity of the read operation, we can simply continue looking for the data block in the "right" locations at all levels.

the PE-PANDA scheme is designed for an array structure (in particular, reading a certain data block requires knowing its index in the array), whereas the hierarchical structure causes the structure implemented in each level to be a *map*, since levels contain a subset of (not necessarily consecutive) data blocks. (In particular, since this subset depends on previous write operations performed on the PANDA, a client does not know the map structure of the levels, and consequently will not know in which location to look for the desired data block.)

We now formally describe the construction. We assume for simplicity of the exposition that B is a multiple of λ.

Construction 4 (Public-writes PANDA). The scheme uses the following building blocks:

- A PE-PANDA scheme (KeyGen, Encode, Query, Recover).
- A hash function family h (used to map data blocks to buckets).

We define the following protocols.

- **Setup**$(1^\lambda, 1^n, 1^t, \mathbf{DB})$: Recall that n denotes the number of clients, t is the collusion bound, and $\mathbf{DB} \in \{0, 1\}^L$. It does the following.
  - Counter initialization. Initialize a counter count_W to 0. (count_W counts the total number of writes performed so far.)
  - Generating level counters and keys. For every $1 \le i \le \ell$, where $\ell = \log L$ is the number of levels:
    * Run $\left(\mathsf{pk}^i, \left\{ \mathsf{ck}^i_j \right\}_{j \in [n]} \right) \leftarrow \mathsf{KeyGen}\left(1^\lambda, 1^n, 1^t, 1^{2^i \cdot \lambda} \right)$.
    * Pick a random hash function h^i for level i.
    * Initialize a *write-epoch counter* count^i_W, a *read-epoch counter* count^i_R, and a *step counter* count^i_s, to 0.[10]
  - Initializing level ℓ. Generate an encoded database using the InitLevel procedure of Fig. 1:

$$\widetilde{\mathsf{DB}}^\ell \leftarrow \mathsf{InitLevel}\left(\ell, \mathsf{pk}^\ell, h^\ell, \text{count}^\ell_W, \text{count}^\ell_R, \mathsf{DB}' \right)$$

where $\mathsf{DB}' = ((1, b_1), \dots, (L, b_L))$,[11] and set level ℓ to be $\mathsf{L}^\ell = \left(\mathsf{DB}', \widetilde{\mathsf{DB}}^\ell \right)$.

---

[10] count^i_W represents the number of times the level was reshuffled into a lower level, i.e., the number of level-i epochs; count^i_R represents the number of times the underlying PE-PANDA scheme was refreshed, i.e., re-initialized, in the current level-i epoch; and count^i_s represents the number of read operations performed in level i since its underlying PE-PANDA was last refreshed. We note that though count^i_W can be computed from count_W, it is included for simplicity.

[11] This guarantees that each data block contains also the logical address of the block, which will be needed when blocks are mapped to buckets.

- **Output.** For each client C_j, set the client key $\mathsf{ck}_j = \left(\{ \mathsf{ck}_j^i \}_{i \in [\ell]}, \{ h^i \}_{i \in [\ell]} \right)$ to consist of its secret keys, and the hash functions, for all levels. Set the server state

$$\mathsf{st}_S = \left(\mathsf{count}_W, \{ \mathsf{count}_W^i, \mathsf{count}_R^i, \mathsf{count}_s^i \}_{i \in [\ell]}, \{ \mathsf{pk}^i \}_{i \in [\ell]}, \{ h^i \}_{i \in [\ell]}, \mathsf{L}^\ell \right)$$

to consist of all counters, its public keys and the hash functions of all levels, and the contents of level ℓ.

- **The Read protocol.** To read the database value at location $\mathsf{addr} \in [L]$ from the server S, a client C_j with key $\left(\{ \mathsf{ck}_j^i \}_{i \in [\ell]}, \{ h^i \}_{i \in [\ell]} \right)$ and the server S run the following protocol.
  - The client C_j initializes an output value val to $\bot$.
  - C_j performs the following for every non-empty level i from ℓ to 1:
    * Obtaining database label. Read $\mathsf{count}_W^i, \mathsf{count}_R^i$ from S.
    * Computing bucket index. Computes $l = h^i(\mathsf{addr})$. (If addr appears in level i, it will be in bucket Buc_l.)
    * Looking for data block addr in level i. Reads Buc_l from level i, namely for every $(l-1) \cdot \lambda + 1 \le m \le l \cdot \lambda$:
      · Runs $(q_1, \ldots, q_z) \leftarrow \mathsf{Query}\left(\mathsf{ck}_j^i, m, (\mathsf{count}_W^i, \mathsf{count}_R^i) \right)$, sends $(q_1, \ldots, q_z)$ to S, and obtains answers $(a_1, \ldots, a_z)$.
      · Runs $(\mathsf{addr}', \mathsf{val}') = \mathsf{Recover}\left(\mathsf{ck}_j^i, (\mathsf{count}_W^i, \mathsf{count}_R^i), (a_1, \ldots, a_z) \right)$.
      · If $\mathsf{addr}' = \mathsf{addr}$ then set $\mathsf{val} := \mathsf{val}'$.
  - The server S updates its state as follows: if $\mathsf{count}_s^i < B_i - \lambda$, S updates $\mathsf{count}_s^i \leftarrow \mathsf{count}_s^i + \lambda$.[12] Otherwise, S updates $\mathsf{count}_s^i = 0, \mathsf{count}_R^i \leftarrow \mathsf{count}_R^i + 1$, and sets $\widetilde{\mathsf{DB}}^i := \mathsf{Encode}\left(\mathsf{pk}^i, \mathsf{DB}^i, (\mathsf{count}_W^i, \mathsf{count}_R^i) \right)$ (where $\mathsf{L}^i = \left(\mathsf{DB}^i, \widetilde{\mathsf{DB}}^i \right)$).

- **The Write protocol.** To write value val at location $\mathsf{addr} \in [L]$ on the server S, a client C_j with key $\left(\{ \mathsf{ck}_j^i \}_{i \in [\ell]}, \{ h^i \}_{i \in [\ell]} \right)$ and the server S run the following protocol.
  - The client C_j generates a "dummy" level 0 which contains a single data block $(\mathsf{addr}, \mathsf{val})$, and sends it to the server.
  - The server S updates its state as follows:
    * $\mathsf{count}_W := \mathsf{count}_W + 1$.
    * For $i = 0, 1, \ldots, \ell$ such that 2^i divides count_W, S reshuffles level i into level $i + 1$ using the ReShuffle procedure of Fig. 1, namely executes

$$\mathsf{ReShuffle}(i, \mathsf{count}_W^i, \mathsf{count}_R^i, \mathsf{count}_s^i, \mathsf{count}_W^{i+1}, \mathsf{count}_R^{i+1}, \mathsf{count}_s^{i+1},$$
$$\mathsf{pk}^{i+1}, h^{i+1}, \mathsf{L}^i, \mathsf{L}^{i+1})$$

where $\mathsf{L}^i, \mathsf{L}^{i+1}$ are the contents of levels i and $i + 1$ (respectively).

---

[12] This is where we use the assumption that λ divides B, otherwise a regeneration of level i mights be needed *while* Buc_l is being read.

The InitLevel procedure

Inputs:

i: the index of a level to initialize.
$\mathsf{pk}^i, h^i, \mathsf{count}^i_W, \mathsf{count}^i_R$: the public key, hash function, and counters of level i.
DB: a database (of size at most 2^i), consisting of entries of the form $(\mathsf{addr}, \mathsf{val})$.

Operation:

- For every entry $(\mathsf{addr}, \mathsf{val}) \in \mathsf{DB}$, add $(\mathsf{addr}, \mathsf{val})$ to bucket $\mathsf{Buc}^i_{h^i(\mathsf{addr})}$.[a]
- Fill every bucket to size λ using "dummy" blocks of the form $(0, 0)$.[b]
- Run $\widetilde{\mathsf{DB}}^i \leftarrow \mathsf{Encode}\left(\mathsf{pk}^i, \left(\mathsf{Buc}^i_1, \ldots, \mathsf{Buc}^i_{2^i}\right), \left(\mathsf{count}^i_W, \mathsf{count}^i_R\right)\right)$, and output $\widetilde{\mathsf{DB}}^i$.

The ReShuffle procedure

Inputs:

i: the index of a level to reshuffle.
$\mathsf{count}^j_W, \mathsf{count}^j_R, \mathsf{count}^j_s, j \in \{i, i+1\}$: the counters of levels $i, i+1$.
$\mathsf{pk}^{i+1}, h^{i+1}$: the public-key and hash function of level $i+1$.
$\mathsf{L}^j = \left(\mathsf{DB}^j, \widetilde{\mathsf{DB}}^j\right), j \in \{i, i+1\}$: the contents of levels $i, i+1$.

Operation:

- <u>Removing duplicate entries.</u> Merge $\mathsf{DB}^i, \mathsf{DB}^{i+1}$ into a single database DB, where if $(\mathsf{addr}, \mathsf{val}) \in \mathsf{DB}^i, (\mathsf{addr}, \mathsf{val}') \in \mathsf{DB}^{i+1}$, then DB contains only $(\mathsf{addr}, \mathsf{val})$.[c]
- <u>Update level i.</u> Set level i to be empty, and update $\mathsf{count}^i_W := \mathsf{count}^i_W + 1$, and $\mathsf{count}^i_R = \mathsf{count}^i_s = 0$.
- <u>Update level $i+1$.</u> Run

$$\widetilde{\mathsf{DB}}^{i+1\prime} = \mathsf{InitLevel}\left(i+1, \mathsf{pk}^{i+1}, h^{i+1}, \mathsf{count}^{i+1}_W, \mathsf{count}^{i+1}_R, \mathsf{DB}\right).$$

Set the new level $i+1$ to be $\mathsf{L}^{i+1} := \left(\mathsf{DB}, \widetilde{\mathsf{DB}}\right)$, and update $\mathsf{count}^{i+1}_R = \mathsf{count}^{i+1}_s = 0$.

---

[a] We implicitly assume here that no bucket overflows. If a bucket overflows then the server simply aborts the computation. As we show below, this happens only with negligible probability.

[b] We note that "dummy" blocks are not required to be indistinguishable from real blocks, but rather all parties should be able to distinguish between the two. Here, we implicitly assume that "0" is not a valid address, but it could be replaced with any other non-valid address. Alternatively, one could concatenate a bit to every entry, indicating whether it is a real or "dummy" entry.

[c] This can be done in time $O\left(|\mathsf{DB}^i| + |\mathsf{DB}^{i+1}|\right)$ by putting both databases in a hash table, and then scanning the table for collisions, and checking, for every collision, whether it is due to multiple copies of the same data block.

Fig. 1. Procedures used in Construction 4

If before executing ReShuffle for level i, L^{i+1} is empty (following a previous reshuffle, or because it has not yet been initialized), then S first sets $L^{i+1} := \left(DB_\emptyset, \widetilde{DB}^{i+1}\right)$ where $DB_\emptyset$ is the empty database, and $\widetilde{DB}$ is generated using the InitLevel procedure of Fig. 1: $\widetilde{DB}^{i+1} :=$ InitLevel $\left(i+1, pk^{i+1}, h^{i+1}, count_W^{i+1}, count_R^{i+1}, DB_\emptyset\right)$.

In the full version [HOWW18] we prove the following theorem:

Theorem 7 (Public-writes PANDA). *Suppose leveled FHE exists. Then for any constant $\varepsilon > 0$ there is a PW-PANDA, for n clients, t collusion bound and database size L, where:*

- *The client/server complexity during each Read protocol is $t \cdot \text{poly}(\lambda, \log L)$.*
- *The client complexity during each Write protocol is $O(\log L)$, and the amortized server complexity is $t \cdot L^\varepsilon \cdot \text{poly}(\lambda, \log L)$.*
- *The client keys are of size $t \cdot \text{poly}(\lambda, \log L)$.*
- *The server state is $t \cdot L^{1+\varepsilon} \cdot \text{poly}(\lambda, \log L)$.*

Remark on De-amortization. We note that using a technique of Ostrovsky and Shoup [OS97], the server complexity in Theorem 7 can be de-amortized, by slightly modifying Construction 4 to allow the server to *spread-out* the reshuffling process. More specifically, we only need to modify the order in which reshuffles are performed in the Write algorithm, such that the operations needed for reshuffle can be executed over multiple accesses to the PANDA. (We note that the server complexity caused by Encode operations in the Read algorithm can be de-amortized as in Construction 3.) See the full version [HOWW18] for additional details.

Acknowledgements. Rafail Ostrovsky is supported in part by NSF-BSF grant 1619348, DARPA SafeWare subcontract to Galois Inc., DARPA SPAWAR contract N66001-15-1C-4065, US-Israel BSF grant 2012366, OKAWA Foundation Research Award, IBM Faculty Research Award, Xerox Faculty Research Award, B. John Garrick Foundation Award, Teradata Research Award, and Lockheed-Martin Corporation Research Award. The views expressed are those of the authors and do not reflect position of the Department of Defense or the U.S. Government. Mor Weiss is supported in part by ISF grants 1861/16 and 1399/17, and AFOSR Award FA9550-17-1-0069. Daniel Wichs and Ariel Hamlin are supported by NSF grants CNS-1314722, CNS-1413964, CNS-1750795 and the Alfred P. Sloan Research Fellowship.

References

[BG12] Bayer, S., Groth, J.: Efficient zero-knowledge argument for correctness of a shuffle. In: Pointcheval, D., Johansson, T. (eds.) EUROCRYPT 2012. LNCS, vol. 7237, pp. 263–280. Springer, Heidelberg (2012). https://doi.org/10.1007/978-3-642-29011-4_17

[BHKP16] Backes, M., Herzberg, A., Kate, A., Pryvalov, I.: Anonymous RAM. In: Askoxylakis, I., Ioannidis, S., Katsikas, S., Meadows, C. (eds.) ESORICS 2016, Part I. LNCS, vol. 9878, pp. 344–362. Springer, Cham (2016). https://doi.org/10.1007/978-3-319-45744-4_17

[BIM00] Beimel, A., Ishai, Y., Malkin, T.: Reducing the servers computation in private information retrieval: PIR with preprocessing. In: Bellare, M. (ed.) CRYPTO 2000. LNCS, vol. 1880, pp. 55–73. Springer, Heidelberg (2000). https://doi.org/10.1007/3-540-44598-6_4

[BIPW17] Boyle, E., Ishai, Y., Pass, R., Wootters, M.: Can we access a database both locally and privately? In: Kalai, Y., Reyzin, L. (eds.) TCC 2017, Part II. LNCS, vol. 10678, pp. 662–693. Springer, Cham (2017). https://doi.org/10.1007/978-3-319-70503-3_22

[BMN17] Blass, E.-O., Mayberry, T., Noubir, G.: Multi-client oblivious RAM secure against malicious servers. In: Gollmann, D., Miyaji, A., Kikuchi, H. (eds.) ACNS 2017. LNCS, vol. 10355, pp. 686–707. Springer, Cham (2017). https://doi.org/10.1007/978-3-319-61204-1_34

[CGKS95] Chor, B., Goldreich, O., Kushilevitz, E., Sudan, M.: Private information retrieval. In: 36th Annual Symposium on Foundations of Computer Science, Milwaukee, Wisconsin, 23–25 October 1995, pp. 41–50 (1995)

[Cha03] Chaum, D.: Untraceable electronic mail, return addresses and digital pseudonyms. In: Gritzalis, D.A. (ed.) Secure Electronic Voting. Advances in Information Security, vol. 7, pp. 211–219. Springer, Boston (2003). https://doi.org/10.1007/978-1-4615-0239-5_14

[CHR17] Canetti, R., Holmgren, J., Richelson, S.: Towards doubly efficient private information retrieval. In: Kalai, Y., Reyzin, L. (eds.) TCC 2017, Part II. LNCS, vol. 10678, pp. 694–726. Springer, Cham (2017). https://doi.org/10.1007/978-3-319-70503-3_23

[DMS04] Dingledine, R., Mathewson, N., Syverson, P.F.: Tor: the second-generation onion router. In: Proceedings of the 13th USENIX Security Symposium, San Diego, CA, USA, 9–13 August 2004, pp. 303–320 (2004)

[Gen09] Gentry, C.: Fully homomorphic encryption using ideal lattices. In: Proceedings of the 41st Annual ACM Symposium on Theory of Computing, STOC 2009, Bethesda, MD, USA, 31 May–2 June 2009, pp. 169–178 (2009)

[GO96] Goldreich, O., Ostrovsky, R.: Software protection and simulation on oblivious RAMs. J. ACM 43(3), 431–473 (1996)

[Gol87] Goldreich, O.: Towards a theory of software protection and simulation by oblivious RAMs. In: STOC 1987, pp. 182–194 (1987)

[Gol01] Goldreich, O.: The Foundations of Cryptography - Volume 1, Basic Techniques. Cambridge University Press, Cambridge (2001)

[Gol04] Goldreich, O.: The Foundations of Cryptography - Volume 2, Basic Applications, vol. 2. Cambridge University Press, Cambridge (2004)

[HO08] Hemenway, B., Ostrovsky, R.: Public-key locally-decodable codes. In: Wagner, D. (ed.) CRYPTO 2008. LNCS, vol. 5157, pp. 126–143. Springer, Heidelberg (2008). https://doi.org/10.1007/978-3-540-85174-5_8

[HOSW11] Hemenway, B., Ostrovsky, R., Strauss, M.J., Wootters, M.: Public key locally decodable codes with short keys. In: Goldberg, L.A., Jansen, K., Ravi, R., Rolim, J.D.P. (eds.) APPROX/RANDOM -2011. LNCS, vol. 6845, pp. 605–615. Springer, Heidelberg (2011). https://doi.org/10.1007/978-3-642-22935-0_51

[HOWW18] Hamlin, A., Ostrovsky, R., Weiss, M., Wichs, D.: Private anonymous data access. IACR Cryptology ePrint Archive 2018/363 (2018)

[KO97] Kushilevitz, E., Ostrovsky, R.: Replication is not needed: single database, computationally-private information retrieval. In: 38th Annual Symposium on Foundations of Computer Science, FOCS 1997, Miami Beach, Florida, USA, 19–22 October 1997, pp. 364–373 (1997)

[KPK16] Karvelas, N.P., Peter, A., Katzenbeisser, S.: Blurry-ORAM: a multi-client oblivious storage architecture. IACR Cryptology ePrint Archive 2016/1077 (2016)

[KT00] Katz, J., Trevisan, L.: On the efficiency of local decoding procedures for error-correcting codes. In: Proceedings of the Thirty-Second Annual ACM Symposium on Theory of Computing, Portland, OR, USA, 21–23 May 2000, pp. 80–86 (2000)

[LPDH17] Leibowitz, H., Piotrowska, A.M., Danezis, G., Herzberg, A.: No right to remain silent: isolating malicious mixes. IACR Cryptology ePrint Archive 2017/1000 (2017)

[MMRS15] Maffei, M., Malavolta, G., Reinert, M., Schröder, D.: Privacy and access control for outsourced personal records. In: 2015 IEEE Symposium on Security and Privacy, SP 2015, San Jose, CA, USA, 17–21 May 2015, pp. 341–358 (2015)

[Mul54] Muller, D.E.: Application of Boolean algebra to switching circuit design and to error detection. Trans. I.R.E. Prof. Group Electron. Comput. $3(3)$, 6–12 (1954)

[OS97] Ostrovsky, R., Shoup, V.: Private information storage (extended abstract). In: Proceedings of the Twenty-Ninth Annual ACM Symposium on the Theory of Computing, El Paso, Texas, USA, 4–6 May 1997, pp. 294–303 (1997)

[Ost90] Ostrovsky, R.: Efficient computation on oblivious RAMs. In: STOC 1990, pp. 514–523 (1990)

[RAD78] Rivest, R.L., Adleman, L., Dertouzos, M.L.: On data banks and privacy homomorphisms. Found. Secur. Comput. $4(11)$, 169–180 (1978)

[Ree54] Reed, I.S.: A class of multiple-error-correcting codes and the decoding scheme. Trans. IRE Prof. Group Inf. Theor. (TIT) 4, 38–49 (1954)

[Reg09] Regev, O.: On lattices, learning with errors, random linear codes, and cryptography. J. ACM $56(6)$, 34:1–34:40 (2009)

[SvDS+13] Stefanov, E., et al.: Path ORAM: an extremely simple oblivious RAM protocol. In: 2013 ACM SIGSAC Conference on Computer and Communications Security, CCS 2013, Berlin, Germany, 4–8 November 2013, pp. 299–310 (2013)

[WY05] Woodruff, D.P., Yekhanin, S.: A geometric approach to information-theoretic private information retrieval. In: 20th Annual IEEE Conference on Computational Complexity (CCC 2005), San Jose, CA, USA, 11–15 June 2005, pp. 275–284 (2005)

[ZZQ16] Zhang, J., Zhang, W., Qiao, D.: MU-ORAM: dealing with stealthy privacy attacks in multi-user data outsourcing services. IACR Cryptology ePrint Archive 2016/73 (2016)

Proofs of Work and Space

Reversible Proofs of Sequential Work

Hamza Abusalah[1](✉), Chethan Kamath[2], Karen Klein[2], Krzysztof Pietrzak[2], and Michael Walter[2]

[1] SBA Research, Vienna, Austria
habusalah@sba-research.org
[2] IST Austria, Am Campus 1, 3400 Klosterneuburg, Austria
{ckamath,kklein,pietrzak,mwalter}@ist.ac.at

Abstract. Proofs of sequential work (PoSW) are proof systems where a prover, upon receiving a statement χ and a time parameter T computes a proof $\phi(\chi, T)$ which is efficiently and publicly verifiable. The proof can be computed in T sequential steps, but not much less, even by a malicious party having large parallelism. A PoSW thus serves as a proof that T units of time have passed since χ was received.

PoSW were introduced by Mahmoody, Moran and Vadhan [MMV11], a simple and practical construction was only recently proposed by Cohen and Pietrzak [CP18].

In this work we construct a new simple PoSW in the random permutation model which is almost as simple and efficient as [CP18] but conceptually very different. Whereas the structure underlying [CP18] is a hash tree, our construction is based on skip lists and has the interesting property that computing the PoSW is a reversible computation.

The fact that the construction is reversible can potentially be used for new applications like constructing *proofs of replication*. We also show how to "embed" the sloth function of Lenstra and Weselowski [LW17] into our PoSW to get a PoSW where one additionally can verify correctness of the output much more efficiently than recomputing it (though recent constructions of "verifiable delay functions" subsume most of the applications this construction was aiming at).

1 Introduction

Timed-release cryptography was envisioned by May [May93] and realised by Rivest, Shamir and Wagner [RSW00] in the form of a "time-lock puzzle". For a time parameter T, such a puzzle can be efficiently sampled together with a solution. However, solving it requires T sequential computational steps, and this holds even for parties aided with massive parallelism. In other words, there are no "shortcuts" to the solution. The application envisioned in [RSW00] was "sending a message to the future": generate a puzzle, derive a symmetric key from the

---

H. Abusalah—Partially supported by FFG COMET under SBA-K1 grant.
C. Kamath, K. Klein, K. Pietrzak and M. Walter—Supported by the European Research Council, ERC consolidator grant (682815 - TOCNeT).

Y. Ishai and V. Rijmen (Eds.): EUROCRYPT 2019, LNCS 11477, pp. 277–291, 2019.
https://doi.org/10.1007/978-3-030-17656-3_10

solution, encrypt your message using that key, then release the ciphertext and the puzzle. Now everyone can decrypt by solving the puzzle which requires T sequential steps.

The construction put forward in [RSW00] is in the RSA setting: the puzzle is a tuple (N, x, T), where $N = p \cdot q$ is an RSA modulus and $x \in Z_N^*$ a group element, and the solution to the puzzle is $x^{2^T} \bmod N$. Although the solution can be computed efficiently *if* the factorisation of N is known, it is conjectured to require T sequential squarings given only N.

The assumption that underlies the soundness of the [RSW00] time-lock puzzle is rather non-standard (which is basically that the puzzle is sound, i.e., there's no shortcut in computing the solution) and it's an open problem to come up with constructions under more standard assumptions. In a negative result, Mahmoody, Moran and Vadhan [MMV11] show that there's no black-box construction of a time-lock puzzle in the random oracle model. In subsequent work the same authors [MMV13] propose and construct proofs of sequential work (PoSW). This is a proof system wherein a prover $\mathcal{P}$ can convince a verifier $\mathcal{V}$ that it spent T sequential time steps upon receiving some challenge χ. Even though PoSW seem related to time-lock puzzles, they are not directly comparable. In particular, a PoSW does not require that one can sample the solution together with an instance. On the other hand, a PoSW must be publicly verifiable[1] and sampling a challenge must be public-coin[2] so it can be made non-interactive by the Fiat-Shamir heuristic. [MMV13] construct a PoSW in the random oracle model (or under a standard model assumption on hash functions called "sequentiality").

As possible applications for PoSW [MMV13] suggest universally verifiable CPU benchmarks and non-interactive time-stamping. The construction given in [MMV13] is not practical as a prover needs not only T sequential time steps but also linear in T space to compute a proof. Cohen and Pietrzak [CP18] resolved this issue by constructing a PoSW where the prover requires just $\log(T)$ space.

More recently there has been renewed interest in time-delayed cryptography as it found applications in decentralized systems, including public randomness beacons (cf. discussion in [BBBF18]), blockchain designs like chia.net or proofs of replication [Fis19]. For the first two applications the PoSW need to be *unique*, which means it should not be possible (or at least computationally hard) to compute more than one accepting proof for the same challenge. This notion was introduced in [CP18] but constructing such a PoSW was left as an open problem.

Our Contribution. Constructions of hash-based PoSW start with some underlying graph structure, which in [MMV13] is a depth-robust graph and in [CP18] a binary tree with some extra edges. In this paper we construct a new PoSW which

---

[1] So everybody, not just the party who generated the challenge, can efficiently verify correctness. Note that in the RSW time-lock puzzle only the party who generated the challenge (which is called a puzzle in this context) and thus knows the factorization can verify the proof efficiently.

[2] This basically means that the challenge is just a uniformly random string. Note that the RSW time-lock puzzle is not public-coin as the coins used to sample the RSA modulus N must remain secret.

is as simple and almost as efficient as [CP18] with the underlying graph being a skip list. Our construction can be instantiated with permutations – instead of hash functions – and is "reversible".

Until recently the sloth hash function [LW17] was the closest we had to a *unique* PoSW. It's not a PoSW because the computation required for verification is linear in the time parameter T (albeit around a 1000 times faster). The fact that our PoSW is reversible allows us to "embed" sloth into our PoSW, this way we get a PoSW where verification (of the claim that T sequential time steps were spent) is very efficient (logarithmic in T), while verifying correctness can be done as efficiently as in sloth (in time $O(T)$ with a very small hidden constant). We outline this construction in more detail in Sect. 1.1 below.

For applications of unique PoSW our construction is by now mostly subsumed by very recent constructions of verifiable delay functions [BBBF18] (VDF). A VDF is defined almost like a PoSW, but the (non-interactive) proof does not only certify that T sequential time has been used to compute some value, but the stronger property that this value is actually the correct value. A VDF is thus basically a unique PoSW (the only reason it's not exactly a unique PoSW is that the proof itself could be malleable, but this doesn't matter for any of the applications). The notion of a VDF has been introduced by Boneh et al. in [BBBF18] who also construct a VDF using rather heavy machinery like incrementally verifiable computation. Subsequently two extremely simple and efficient VDFs have been proposed [Wes19, Pie19b], both papers basically show how to make the RSW time-lock puzzle [RSW00] publicly verifiable, that is, they give proof systems for showing that a given tuple (x, y, T) satisfies $x^{2^T} = y$ in a group of unknown order (e.g. Z_N^* as used in [RSW00]). These constructions are clearly favourable to ours as correctness (which here means that the output is correctly computed) can be verified much more efficiently, though as they are not post-quantum secure, ours is arguably still the best option in a post-quantum setting for some applications. This hopefully will change in the near future as research on post-quantum VDFs is ongoing [FMPS19].

The fact that our PoSW is reversible seems also useful in the context of proofs of replication [Fis19, Fis18, Pie19a] for similar reasons that "decodable" VDFs are useful in this context as discussed in [BBBF18], we are currently working towards constructing simple proofs of replication based on the skip list based PoSW presented in this paper.

1.1 Hash Chains and the Sloth Function

A simple construction which is not quite a PoSW is a hash chain, where on input $x = x_0$ one outputs as proof x_T which is recursively computed as $x_i = hash(x_{i-1})$. If *hash* is a bijection and can be efficiently evaluated in both directions (i.e., a permutation), then from a given state x_i, one can compute the previous state x_{i-1}, we call such a construction *reversible*.

Verifying that x_T has been correctly computed requires T hashes (so it's no a PoSW), but at least one can parallelize verification by additionally outputting some q intermediate values $x_0, x_{T/q}, x_{2T/q}, \ldots, x_T$ (then the proof can be verified in T/q time assuming one can evaluate q instantiations of *hash* in parallel: for every $i \in [q]$, verify that T/q times hashing $x_{(i-1)T/q}$ gives $x_{iT/q}$).

Lenstra and Wesolowski [LW17] suggest a construction called "sloth", which basically is a hash chain but with the additional property that it can be verified with a few hundred times less computation than what is required to compute it. The construction is based on the assumption that computing square roots in a field $\mathbb{F}p$ of size p is around $\log(p)$ times slower than the inverse operation, which is just squaring. A typical value would be $\log(p) \approx 1000$, going much higher is problematic as then fast multiplication methods (e.g., Karatsuba, Schönhage-Strassen) can be applied.

Their idea is to simply use a hash chain where the hash function is some permutation $\pi : \mathbb{F}p \rightarrow \mathbb{F}p$, where $\mathbb{F}p$ is a finite field of size p, followed by taking a square root: that is $x_i = \sqrt{\pi(x_{i-1})}$. Verification goes as for a standard hash chain, but one computes backwards, checking $x_{i-1} = \pi^{-1}(x_i^2)$, which – assuming computing π, π^{-1} is cheap compared to squaring, and squaring is $\log(p)$ times faster than taking square roots – gives the claimed speedup of $\approx \log(p)$ compared to a simple hash chain.

In Sect. 4 we show how sloth can be embedded into our skip list based PoSW to get a construction such that it remains a good PoSW, while correctness of the output can be verified as efficiently as in sloth, the constructions discussed are summarized in the table below.

2   Construction

2.1   Notation

Throughout we denote the time parameter of our construction by $N = 2^n$ with $n \in \mathbb{N}$ and assume it's a power of 2. We reserve $w, t \in \mathbb{N}$ to denote two statistical security parameters, w is the block size (say $w = 256$) and t denotes the number of challenges: a cheating prover who only makes $N(1-\epsilon)$ sequential steps (instead N) will pass verification with probability $(1 - \epsilon)^t$. For integers m, m' we denote with $[m, m'] = \{m, m+1, \ldots, m'\}$, $[m], [m]_0$ are short for $[1, m]$ and $[0, m]$.

We define $\tilde{0} = n + 1$ and for $i \geq 1$ we denote with $\tilde{i}$ the number of trailing zeros in the binary representation of i, plus 1

$$\tilde{0}, \tilde{1}, \tilde{2}, \tilde{3}, \tilde{4}, \tilde{5}, \tilde{6}, \tilde{7}, \tilde{8}, \tilde{9}, \ldots = n+1, 1, 2, 1, 3, 1, 2, 1, 4, 1, \ldots$$

For $\sigma \in \{0,1\}^{w \cdot i}$ we denote with $\sigma^{(j)}$ the jth w-bit block of σ, so that $\sigma = \sigma^{(1)} \| \ldots \| \sigma^{(i)}$. $\sigma^{(i \ldots j)}$ is short for $\sigma^{(i)} \| \ldots \| \sigma^{(j)}$.

For a permutation π over ℓ bit strings, we denote with $\dot{\pi}$ the function over bit stings of length $\geq \ell$ which simply applies π to the ℓ bit prefix of the input, and leaves the rest untouched.

construction (using time parameter T and statistical security parameter λ)	# of steps[d] to verify sequential computation $O(\cdot)$ of	# of steps[d] to verify if output is correct (uniqueness) $O(\cdot)$ of	assumption	step	post quantum	reversible
hash chain	T	T	random oracle[a]	RO call	yes	yes[b]
sloth [LW17]	$T/\log(p)$	$T/\log(p)$	$\log(p)$ gap computing $\sqrt{x}$ vs x^2 and random permutation[a,c]	$x \to \sqrt{x}$ & RP call	yes	yes
PoSW [CP18]	$\lambda \cdot \log(T)$	T	random oracle[a]	RO call	yes	no
PoSW Sect. 2	$\lambda \cdot \log^2(T)$[f]	T	random permutation[a,c]	RP call	yes	yes
Combined Sect. 4	$\lambda \cdot \log^2(T)$[f]	$T/\log(p)$	like sloth	$x \to \sqrt{x}$ & RP call	yes	yes
[Pie19b] VDF	$\lambda \cdot \log(T)$	$\lambda \cdot \log(T)$	$(x,T) \to x^{2^T}$ requires T sequential squarings[e]	$x \to x^2$	no	no
[Wes19] VDF	λ	λ	as above plus "root assumption"	$x \to x^2$	no	no

[a]Or a standard model assumption called "sequential hash function".
[b]If the function used is an efficiently invertible permutation.
[c]The random permutation model is equivalent to the random oracle model.
[d]What a step is depends on the construction, but evaluating the function is always assumed to require T sequential steps.
[e]This assumption can only hold in groups of unknown order.
[f]Strictly speaking, the number of oracle calls required to verify is just $\lambda \cdot \log T$ (as in [CP18]), but in our constructions the input consists of up to $\log T + 1$ blocks (unlike [CP18], where it is 2 blocks) and therefore to make a fairer comparison, we count the cost of an oracle call on an input of length k blocks as k calls.

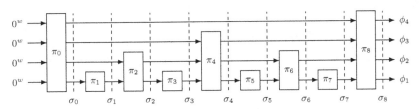

Fig. 1. Illustration of the computation of $\sigma_\Pi = (\sigma_0, \ldots, \sigma_N)$ with $n = 3, N = 2^n = 8$. The blocks represent the permutations, whereas the dashed vertical lines represent the states. Note that the structure of the graph is the same as a skip list with four layers, where a pointer in layer i, $i \in \{0, 1, 2, 3\}$, points to the 2^i-th element to its right on the list.

2.2   The Sequence σ_Π

At the core of our construction is a mapping based on the skip list data structure (see Fig. 1). It is built from a set of permutations $\Pi = \{\pi_i\}_{i \in [N]_0}$, where each π_i is over $\{0, 1\}^{w \cdot \tilde{i}}$, and defines a sequence of states $\sigma_\Pi = \sigma_0, \ldots, \sigma_N$, $\sigma_i \in \{0, 1\}^{(n+1) \cdot w}$, recursively as

$$\sigma_0 = \dot{\pi}_0(0^{w \cdot (n+1)}) \text{ and for } i > 0 \; : \; \sigma_i = \dot{\pi}_i(\sigma_{i-1}) \left(= \pi_i(\sigma_{i-1}^{(1 \ldots \tilde{i})}) \| \sigma_{i-1}^{((\tilde{i}+1) \ldots (n+1))} \right).$$

2.3 The DAG G_N

It will be convenient to consider the directed acyclic graph (DAG)

$$G_N = (V, E) , \quad V = [N]_0 , \quad E = \{(i, j) \in V^2 : \exists k \geq 0 : j - i = 2^k, 2^k | i\}$$

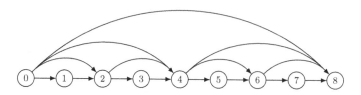

Fig. 2. The graph G_8 that corresponds to the computation of σ_Π with $n = 3$.

that is derived from the computation of σ_Π as follows: identify the permutation π_i with the node i and add a directed edge (i, j) if in the computation of σ_Π part of the output of π_i is piped through directly to π_j (see Fig. 2).

For $i \in [N - 1]$ we denote with $\mathsf{path}(i) \subseteq V$ the subgraph of V induced by the nodes on the shortest path in G_N which starts at 0, ends at N and passes through node i. For example, in Fig. 1,

$$\mathsf{path}(5) = (\{0, 4, 5, 6, 8\}, \{(0, 4), (0, 8), (4, 5), (4, 6), (4, 8), (5, 6), (6, 8)\}).$$

It's not hard to check that the number of vertices in $\mathsf{path}(i)$ is $n + 3 - \tilde{i}$, and in particular is never more than $n + 2$.

2.4 Consistent States/Paths

By construction, the $\sigma_i \in \sigma_\Pi$ satisfy $\sigma_i = \dot{\pi}_{i+1}^{-1}(\sigma_{i+1})$, and more generally, for every edge $(i, j) \in E$ and $d = \min(\tilde{i}, \tilde{j})$

$$\sigma_i^{(d...n+1)} = (\dot{\pi}_j^{-1}(\sigma_j))^{(d...n+1)}.$$

We say two strings are consistent for (i, j) if they satisfy this condition.

Definition 1 (Consistent States/Path). $\alpha_i, \alpha_j \in \{0, 1\}^{(n+1)\cdot w}$ *are consistent for edge* $(i, j) \in E$ *if with* $d = \min(\tilde{i}, \tilde{j})$

$$\alpha_i^{(d...n+1)} = (\dot{\pi}_j^{-1}(\alpha_j))^{(d...n+1)} .$$

We say $\alpha_i' \in \{0, 1\}^{\tilde{i}\cdot w}, \alpha_j' \in \{0, 1\}^{\tilde{j}\cdot w}$ *are consistent if they can be "padded" to consistent* α_i, α_j *as above, which is the case if*

$$\alpha_i'^{(d)} = \pi^{-1}(\alpha_j')^{(d)}.$$

We say $\{\alpha_i\}_{i \in \mathsf{path}(k)}$ *are consistent with* $\mathsf{path}(k)$ *if* α_i, α_j *are consistent for every edge* $(i, j) \in \mathsf{path}(k)$.

Note that if α_j is computed from α_i by applying $\dot{\pi}_{i+1}, \ldots, \dot{\pi}_j$ to α_i, then those α_i, α_j will be consistent with (i, j), but the converse is not true (except if $j = i + 1$).

2.5   PoSW Construction

The protocol between $\mathcal{P}, \mathcal{V}$ on common input $T = N = 2^n, w, t$ goes as follows

1. $\mathcal{V}$ samples $\chi \leftarrow \{0,1\}^{w \cdot n}$ and sends it to $\mathcal{P}$. This χ defines a fresh set of random permutations Π (cf. Remark 1 below).
2. $\mathcal{P}$ computes $\sigma_0, \ldots, \sigma_N$ and sends $\phi = \sigma_N$ to $\mathcal{V}$.
3. $\mathcal{V}$ samples t challenges $\gamma = (\gamma_1, \ldots, \gamma_t) \leftarrow [N-1]^t$ and sends them to $\mathcal{P}$.
4. $\mathcal{P}$ sends $\{\sigma_i\}_{i \in \mathsf{path}(j), j \in \gamma}$ to $\mathcal{V}$ (cf. Remark 2 below).
5. $\mathcal{V}$ verifies for every $j \in \gamma$ that $\{\sigma_i\}_{i \in \mathsf{path}(j)}$ is consistent as in Definition 1. If any check fails output reject, output accept otherwise.

Remark 1 (Seeding Random Oracles/Permutations). Ideal permutations can be constructed from random oracles [CPS08, HKT11, DSKT16] (formally, the ideal permutation model is indifferentiable from the random oracle model), so we can realize Π in the standard random oracle model.[3] Consider a fixed random oracle $\mathcal{H}(\cdot)$ about which a potential adversary has some auxiliary input (i.e., it has queried it on many inputs before, and stored some information aux). If one samples a random seed χ and uses it as a prefix to define the function $\mathcal{H}_\chi(x) = \mathcal{H}(\chi \| x)$, this $\mathcal{H}_\chi$ – from the adversaries' perspective – is a fresh random oracle as long as this seed is just a bit longer that $\log(|\mathsf{aux}|)$ [DGK17]. Thus, we can also sample a fresh Π by just sending a seed χ.

Remark 2 (P's Space Requirement). To avoid any extra computation in step 4., $\mathcal{P}$ would need to store the entire $\sigma_\Pi = \{\sigma_i\}_{i \in [N]_0}$. By using a bit of extra computation, one can reduce the space requirement (we remark that a similar trade off comes up in [CP18]). Concretely, for some $K = 2^k$, we let $\mathcal{P}$ only store σ_i where $2^k | i$, thus storing only N/K states. From this, every state σ_i can be computed making at most $K/2$ invocations to Π (and $K/2$ not K as we can also compute backwards).

3   Security Proof

Theorem 1. *Consider a malicious prover $\tilde{\mathcal{P}}$ which*

1. *makes at most $N - \Delta$ sequential queries to permutations in Π before sending $\phi = \sigma_N$ (in step 3 of the protocol); and*
2. *queries the permutations in Π on at most q inputs in total during execution of the protocol.*

Then $\tilde{\mathcal{P}}$ will win (i.e., make $\mathcal{V}$ output accept) with probability at most

$$\Pr\left(\tilde{\mathcal{P}} \ wins \ \right) \le \frac{2q^2(n+3)^2}{2^w} + \left(\frac{N-\Delta}{N}\right)^t. \tag{1}$$

---

[3] In practice, one could e.g. use χ to sample $N+1$ AES keys $k_0, \ldots, k_N$, and then use $AES(k_i, \cdot) : \{0,1\}^{256} \to \{0,1\}^{256}$ – i.e., AES with a fixed public key – to construct π_i, where for $\tilde{i} > 1$ one would use domain extension for random permutations to extend the domain to $256 \cdot \tilde{i}$ bits.

The proof of Theorem 1 mainly follows the intuition that sending ϕ "commits" the prover to a set of challenges it can respond to. We prove this fact formally in Lemma 7. If this set is a large fraction of the possible challenges, it implies the existence of a long sequence (as defined below) that necessarily requires many sequential steps.

To aid the proof we begin with a couple of definitions. The first one is merely for notational convenience.

Definition 2. *We use* $\sim$ *to denote that two strings (composed of w-bit blocks) contain an identical block*

$$\alpha \sim \alpha' \iff \exists i, j : \alpha^{(i)} = \alpha'^{(j)}.$$

We then say that α *and* α' *collide.*

The next definition characterizes the property that paths through our skiplist construction satisfy and that we rely on for the proof.

Definition 3 (Π-Sequence). *For a family of N permutations $\Pi = \{\pi_i\}_i$ a Π-sequence of length $N' < N$ is an N'-tuple of pairs of strings $((x_j, y_j))_j$ together with an N'-tuple of strictly increasing integers $(i_j)_j$ such that for all j*

$$\pi_{i_j}(x_j) = y_j \text{ and } y_j \sim x_{j+1}.$$

Below we show that Π-sequences are inherently sequential (cf. Lemma 6 and Corollary 1), but that requires a few technical lemmas, so we defer the details and proceed to the main proof. We now show how Lemma 7 and Corollary 1 imply Theorem 1.

Proof (of Theorem 1). Consider a malicious prover $\tilde{\mathcal{P}}$ that convinces the verifier on a random challenge with probability $\geq \frac{N-\Delta}{N}$. Since the correct response to any challenge is a distinct Π-sequence from 0 to ϕ, Lemma 7 implies that $\tilde{\mathcal{P}}$ must be able to respond to a fraction $\geq \frac{N-\Delta}{N}$ of the challenges. This is because the set of Π-sequences from 0 to ϕ that $\tilde{\mathcal{P}}$ can compute is essentially fixed after sending ϕ and thus independent of the choice of challenges. This means there must exist a set of $N - \Delta$ responses which can be pieced together to a Π-sequence of length $N - \Delta$ from 0 to ϕ (details below). Note that all responses can be obtained by sending all the challenges using rewinding (which does not increase the number of queries). Then the result follows from Corollary 1.

It remains to establish the fact that the responses can be merged to a long sequence. To see this, first assume that whenever two paths contain the same node, the corresponding responses have the same state at this node. If this is the case then merging the responses to k distinct challenges is easy: simply take the "union" of the responses, which will be a Π-sequence of length at least k.

Finally, we show that different verifying reponses must have the same state at intersecting nodes. The proof of this fact is recursive: consider the node $N/2$ and assume for contradiction that there are two paths that both verify and each contains a state $\sigma_{N/2}$ and $\sigma'_{N/2}$, respectively, with $\sigma_{N/2} \neq \sigma'_{N/2}$. First note that the states σ_0 and σ_0' must both be equal to $\pi_0(0)$, so they are equal to each other. Similarly, σ_N and σ_N' must both be equal to ϕ in order to both verify. Furthermore, verification ensures that $\sigma_{N/2} \sim \sigma_{N/2}'$. Specifically, they are equal in block $n - 1$, where they must both be equal to $\pi_N^{-1}(\sigma_N)^{(n-1)}$, since verification checks the edge $(N/2, N)$ for consistency (cf. Definition 1). Analogously, verification ensures that $\pi_{N/2}^{-1}(\sigma_{N/2}) \sim \pi_{N/2}^{-1}(\sigma_{N/2}')$, since this corresponds to the edge $(0, N/2)$. Note that the latter pair of values could be extracted from the prover by sending the appropriate challenges. By Lemma 5 (proved below) this can only happen with probability $\leq \frac{2q^2(n+3)^2}{2^w}$. We conclude that $\sigma_{N/2}$ is equal among all valid responses with overwhelming probability. This allows to recurse on the node $N/4$ and $3N/4$, etc. $\qquad\square$

We now establish the remaining lemmas used in the main proof. Throughout the rest of this section, w.l.o.g. we only consider algorithms that do not make redundant queries. In all results in this section pertaining to random permutations the probabilities are taken over the choice of the permutations.

First, we need a version of a PRP/PRF switching lemma that allows the adversary oracle access to the permutation and its inverse. We have not seen such a version in the literature so we prove it in the appendix.

Lemma 1. *Let $\pi : \{0,1\}^w \mapsto \{0,1\}^w$ be a random permutation and consider an algorithm $\mathsf{A}^{\pi,\pi^{-1}}$ with oracle access to π and π^{-1} that makes exactly q queries in total. Assume that A does not repeat any queries to π nor any queries to π^{-1}, and that if it queries π at x, it does not query π^{-1} at $\pi(x)$ and vice versa. Let $F_1, F_2 : \{0,1\}^w \mapsto \{0,1\}^w$ be independent random functions. Then for any event E over the output of A, we have $\Pr\left(\mathsf{A}^{\pi,\pi^{-1}} \in E\right) \leq \Pr\left(\mathsf{A}^{F_1,F_2} \in E\right) + \frac{q(q-1)}{2^w}$, where the first probability is over the choice of π and the second over the choice of F_1, F_2.*

The Lemma shows that in the analysis we can replace the random permutation and its inverse oracle with random functions. Note that by a simple hybrid argument, Lemma 1 also holds for families of permutations, where q is the sum over all queries and w is the minimal input/output length over all permutations.

We now show that Lemma 1 implies a few restrictions on what an algorithm can achieve when querying random permutations. Namely, we first show that input/output pairs are hard to guess (cf., Lemma 2), that preimages are hard to find without using the inverse oracle (cf., Lemma 3), and that it is hard to find queries that result in collisions with earlier queries (cf., Lemma 4).

Lemma 2. *Let $\Pi = \{\pi_i\}_i$ be a family of random permutations. For any oracle algorithm outputting a pair (x, y) and an integer i and making q queries to Π except x in forward or y in backward direction, the probability that $\pi_i(x) = y$ is $\leq \frac{q^2}{2^w}$.*

Proof. We are trying to bound the probability that the algorithm is able to guess the input/output pair of one of the permutations in Π (after making at most q

queries). If the π_i were random functions, this probability would be $\leq \frac{1}{2^w}$. By Lemma 1 the bound follows. □

Lemma 3. *Let $\Pi = \{\pi_i\}_i$ be a family of random permutations. For any algorithm taking y as input and making q queries to Π except querying π_i^{-1} for y and outputting some x and i, the probability that $\pi_i(x) = y$ is $\leq \frac{q}{2^w}$.*

Proof. If π_i and π_i^{-1} were random function, the probability of finding such an x would be $\frac{1}{2^w}$. Lemma 1 completes the proof. □

Lemma 4. *Let $\Pi = \{\pi_i\}_i$ be a family of random permutations. For any algorithm making q queries to Π the probability of a query to Π either in forward or backward direction resulting in a response z that collides (in the sense of $\sim$) with any of the previous queries (in either input or output) is $\leq \frac{2q^2(n+2)^2}{2^w}$.*

Proof. Assume we replace the permutations with random functions. The probability that the response to any query collides with a specific string is at most $(n+1)^2/2^w$, since there are at most $n+1$ blocks in each string. By union bound, the probability that a query collides with any of the previous queries is thus at most $2q(n+1)^2/2^w$, since there are two strings in each query (input and output). Applying a final union bound to all queries shows that the probability of this event is $2q^2(n+1)^2/2^w$. Lemma 1 now proves the result. □

Using the basic lemmas above, we can make statements about certain cyclic structures that are hard to find in random permutations and about the sequentiality of random permutations.

Lemma 5. *Let $\Pi = \{\pi_i\}_i$ be a family of random permutations. For any algorithm making q queries to Π and outputting two distinct values, x and x', and an integer i, the probability that $x \sim x'$ and $\pi_i(x) \sim \pi_i(x')$ is $\leq \frac{2q^2(n+3)^2}{2^w}$.*

Proof. Obtaining two such pairs requires to guess one of the two input/output pairs or find a colliding query. Union bound over the two events (which are bounded by Lemmas 2 and 4, respectively) yields the bound. □

Lemma 6 (Π-Sequentiality of Random Permutations). *Let $\Pi = \{\pi_i\}_i$ be a family of random permutations. For any algorithm A taking as input x and making a sequence Q of q queries to Π, and any Π-sequence s starting at x, the probability of A outputting s and Q not containing the pairs in s in order and in forward direction is $\leq \frac{2q^2(n+3)^2}{2^w}$.*

Proof. Producing a Π-sequence starting at a specific value without querying the pairs in order and in forward direction, requires to either guess some input/output pair (for some specific π_i) or find a colliding query, similarly to Lemma 5. □

Corollary 1. *Let $\Pi = \{\pi_i\}_i$ be a family of random permutations. For any algorithm taking as input x and making q sequential queries to Π, the probability of outputting a Π-sequence of length longer than q is $\leq \frac{2q^2(n+3)^2}{2^w}$.* □

We use the above observations to show that nothing the prover does after sending its commitment ϕ will help responding to challenges.

Lemma 7. *Let* $(\mathcal{P}_1^\Pi, \mathcal{P}_2^\Pi)$ *be a pair of algorithms such that*

- $\mathcal{P}_1$*, on input* x*, makes* q_1 *queries to* Π*, and outputs a state* s_1 *and some* y
- $\mathcal{P}_2$*, on input* s_1, x, y*, makes* q_2 *queries to* Π *and outputs a* Π*-sequence* s*.*

Let Q *be the set of queries (including responses) made by* $\mathcal{P}_1$*, and let* S *be the set of* Π*-Sequences between* x *and* y *computable[4] from* Q *without any further queries to* Π*. Then* $s \in S$ *except with probability* $\leq \frac{2q^2(n+3)^2}{2^w}$*, where* $q = q_1 + q_2$*.*

Proof. Assume $s \notin S$. Let (x', y) be the last pair in s. First consider the case that the query $(x', y) \in Q$. Since s is a new sequence not computable from S it must contain a pair $(x_i, y_i) \notin Q$, so the queries in s were not made in order and thus by Lemma 6 the probability of $\mathcal{P}_2$ outputting s is $\leq \frac{2q^2(n+3)^2}{2^w}$.

Now consider the case $(x', y) \notin Q$. If $\mathcal{P}_2$ did not query x' in forward direction, by Lemma 6 the probability of $\mathcal{P}_2$ outputting s is $\leq \frac{2q^2(n+3)^2}{2^w}$. Finally, if $\mathcal{P}_2$ queried x' in forward direction, it did not submit an equivalent query in the reverse direction by assumption. (Recall that we consider only algorithms that do not make redundant queries.) It follow from Lemma 3 that the probability of this event is $\leq \frac{q^2}{2^w}$. □

This completes the proof.

4   Embedding Sloth

As discussed in the introduction, we propose a reversible PoSW that is almost as efficient as the construction from [CP18] but achieves a larger time gap between the computation of the proof and the verification of correctness. To this aim, we embed the sloth hash function from [LW17] into construction 2.5.

The idea underlying sloth is to use the fact that the best known algorithms for computing modular square roots in a field $\mathbb{F}p$ takes $\approx \log(p)$ sequential squarings, whereas verification of the result only takes a single modular squaring. Thus, this gives a good candidate to build the *slow-timed hash* function *sloth*.

Let $p \equiv 3 \mod 4$ be a prime. We identify $x \in \mathbb{F}p^\times$ with its canonical representative in $[0, p-1]$. If $x \in \mathbb{F}p^\times$ is a quadratic residue, then there are two square roots $y, y' \in \mathbb{F}p^\times$, where $y' = p - y$, one of them being even, the other one odd. Let $\sqrt[+]{x}, \sqrt[-]{x}$ denote the (unique) even and odd square root of x, respectively. If $x \in \mathbb{F}p^\times$ is not a quadratic residue, then $-x$ is a quadratic residue, so it makes sense to define a permutation $\rho : \mathbb{F}p^\times \to \mathbb{F}p^\times$ as

$$\rho(x) = \begin{cases} \sqrt[+]{x}, & \text{if } x \text{ is a quadratic residue,} \\ \sqrt[-]{-x}, & \text{otherwise.} \end{cases}$$

---

[4] By "computable" we mean here that there exists an algorithm for which the output is correct with non-negligible probability.

Its inverse is defined by

$$\rho^{-1}(x) = \begin{cases} x^2, & \text{if } x \text{ is even,} \\ -x^2, & \text{otherwise.} \end{cases}$$

Unfortunately, one cannot directly build a hash chain by iterating ρ since reducing modulo $p-1$ in the exponent would yield a much faster computation than sequentially computing ρ. Lenstra and Wesolowski [LW17] solve this problem by prepending an easily computable (in both directions) permutation π on $\mathbb{F}p^\times$ to each iteration of the square rooting function ρ. Setting $\tau = \rho \circ \pi$, the sloth function is hence defined as τ^N for some appropriate chain length N. Verification can be done backwards by the computation $(\tau^N)^{-1} = (\sigma^{-1} \circ \rho^{-1})^N$, which is by a factor $\log p$ faster.

We now combine the ideas from [LW17] with our construction to achieve an efficient PoSW while preserving the fast verification of correctness obtained by the sloth construction. Let $\Pi = \{\pi_i\}_{i \in [N]_0}$ be a set of permutations where, for each $i \in [N]_0$, $\pi_i : \mathbb{F}p^\times \times \{0,1\}^{w \cdot (i-1)}$. We define the sequence $\bar{\sigma}_\Pi = \bar{\sigma}_0, \dots, \bar{\sigma}_N$ with $\bar{\sigma}_i \in \mathbb{F}p^\times \times \{0,1\}^{n \cdot w}$ recursively as

$$\bar{\sigma}_0 = \dot{\rho} \circ \pi_0((0, 0^{w \cdot n})) \quad \left(= \rho\big(\pi_0((0, 0^{w \cdot n}))^{(1)}\big) \| \pi_0((0, 0^{w \cdot n}))^{(2 \dots n+1)} \right) \text{ and}$$

for $i > 0$: $\bar{\sigma}_i = \dot{\rho} \circ \dot{\pi}_i(\bar{\sigma}_{i-1}) \left(= \rho\big(\pi_i(\bar{\sigma}_{i-1}^{(1 \dots \tilde{i})})^{(1)}\big) \| \pi_i(\bar{\sigma}_{i-1}^{(1 \dots \tilde{i})})^{(2 \dots \tilde{i})} \| \bar{\sigma}_{i-1}^{(\tilde{i}+1 \dots n+1)} \right).$

See Fig. 3 for an illustration of the computation of $\bar{\sigma}_\Pi$. Defining $\Pi' = \{\pi'_i\}_{i \in [N]_0}$ by $\pi'_i = \dot{\rho} \circ \pi_i$, it holds $\bar{\sigma}_\Pi = \sigma_{\Pi'}$. Thus, using $\bar{\sigma}_\Pi$ in our protocol results in a PoSW that is secure in the random permutation model, almost as efficient as the construction from [CP18], and at the same time achieves verification of correctness as efficient as in sloth. More formally, the efficiency of the combined scheme can be analysed as follows: First, consider the proof size:

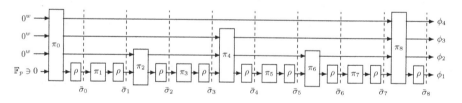

Fig. 3. Illustration of the computation of $\bar{\sigma}_\Pi = (\bar{\sigma}_0, \dots, \bar{\sigma}_N)$ with $n = 3, N = 2^n = 8$.

$$|\chi| = w \cdot n, \quad |\phi| = \log(p) + w \cdot n, \quad |\gamma| = t \cdot n, \quad |\{\sigma_i\}_{i,j}| \le t \cdot n(\log(p) + w \cdot n).$$

Hence, compared to [CP18], the proofs are by a factor $n = \log(N) = \log(T)$ larger. Next, consider the prover efficiency. To compute ϕ, the prover needs to sequentially evaluate $N+1$ permutations $\pi_i' = \dot\rho \circ \pi_i$, $i = 0, \ldots, N$. Storing only the N/K states σ_i with $K|i$ for some $K = 2^k$, the prover can answer the challenge γ after K parallel invocations to permutations $(\pi_i')^{-1}$. Note, by construction computing $(\pi_i')^{-1}$ is assumed to be by a factor $\log(p)$ faster than computing π_i'. The verifier, on the other hand, only needs $t \cdot n$ evaluations of $(\pi_i')^{-1}$ to verify the PoSW. Also verification of correctness can be done in backwards direction by sequentially invocating $(\pi_i')^{-1}$ for $i = N, \ldots, 0$, which is assumed to be by a factor $\log(p)$ faster than the computation of the prover.

When applied to a blockchain, our new PoSW allows extremely efficient rejection of wrong proofs while additionally providing sloth-like verification of correctness, which can be used whenever two or more distinct proofs pass the verification.

A   Proof of Lemma 1

Proof (of Lemma 1). Let $X = (X_1, \ldots, X_q)$ be the random variable corresponding to the responses to the queries of $\mathsf{A}^{\pi,\pi^{-1}}$ and $Y = (Y_1, \ldots, Y_q)$ the one corresponding to the responses to the queries of A^{F_1,F_2}. We will show that $\Delta_{SD}(X,Y) \leq \frac{q(q-1)}{2^w}$. The lemma then follows from standard properties of Δ_{SD}.

In the following, we will abbreviate the conditional distributions $(X_i|X_1 = x_1, \ldots, X_{i-1} = x_{i-1})$ as $(X_i|(x_1, \ldots, x_{i-1}))$ and similarly for Y. From subadditivity for joint distributions (a property of Δ_{SD}), we have

$$\Delta_{SD}(X,Y) \leq \sum_{i=1}^{q} \max_{x=(x_1, \ldots, x_{i-1})} \Delta_{SD}(X_i|x, Y_i|x).$$

For each particular i we have

$$\Delta_{SD}(X_i|x, Y_i|x) = \frac{1}{2} \sum_{y \in \{0,1\}^w} |\Pr(X_i = y|x) - \Pr(Y_i = y|x)|.$$

From the definition of F_1, F_2, it is clear that $\Pr(Y_i = y|x) = 2^{-w}$ for all $y \in \{0,1\}^w$ and $x \in (\{0,1\}^w)^{i-1}$. For the other case, notice that any query to π or π^{-1} fixes a particular input/output pair. Accordingly, X_i is uniform among the remaining $2^w - (i-1)$, no matter if π or π^{-1} was queried (recall that no input/output pair is repeated). It follows that

$$\Delta_{SD}(X_i|x, Y_i|x) = \frac{1}{2} \left[\frac{i-1}{2^w} + (2^w - (i-1)) \left(\frac{1}{2^w - (i-1)} - \frac{1}{2^w} \right) \right]$$

$$= \frac{i-1}{2^w}$$

for any x (in particular, the maximum). Summing over all i yields the final bound.   □

References

[BBBF18] Boneh, D., Bonneau, J., Bünz, B., Fisch, B.: Verifiable delay functions. In: Shacham, H., Boldyreva, A. (eds.) CRYPTO 2018, Part I. LNCS, vol. 10991, pp. 757–788. Springer, Cham (2018). https://doi.org/10.1007/978-3-319-96884-1_25

[CP18] Cohen, B., Pietrzak, K.: Simple proofs of sequential work. In: Nielsen, J.B., Rijmen, V. (eds.) EUROCRYPT 2018, Part II. LNCS, vol. 10821, pp. 451–467. Springer, Cham (2018). https://doi.org/10.1007/978-3-319-78375-8_15

[CPS08] Coron, J.-S., Patarin, J., Seurin, Y.: The random oracle model and the ideal cipher model are equivalent. In: Wagner, D. (ed.) CRYPTO 2008. LNCS, vol. 5157, pp. 1–20. Springer, Heidelberg (2008). https://doi.org/10.1007/978-3-540-85174-5_1

[DGK17] Dodis, Y., Guo, S., Katz, J.: Fixing cracks in the concrete: random oracles with auxiliary input, revisited. In: Coron, J.-S., Nielsen, J.B. (eds.) EUROCRYPT 2017. LNCS, vol. 10211, pp. 473–495. Springer, Cham (2017). https://doi.org/10.1007/978-3-319-56614-6_16

[DSKT16] Dachman-Soled, D., Katz, J., Thiruvengadam, A.: 10-round Feistel is indifferentiable from an ideal cipher. In: Fischlin, M., Coron, J.-S. (eds.) EUROCRYPT 2016. LNCS, vol. 9666, pp. 649–678. Springer, Heidelberg (2016). https://doi.org/10.1007/978-3-662-49896-5_23

[Fis18] Fisch, B.: PoReps: proofs of space on useful data. IACR Cryptology ePrint Archive 2018/678 (2018)

[Fis19] Fisch, B.: Tight proofs of space and replication. In: Advances in Cryptology - EUROCRYPT 2019 (2019)

[FMPS19] De Feo, L., Masson, S., Petit, C., Sanso, A.: Verifiable delay functions from supersingular isogenies and pairings. Cryptology ePrint Archive, Report 2019/166, 2019. https://eprint.iacr.org/2019/166

[HKT11] Holenstein, T., Künzler, R., Tessaro, S.: The equivalence of the random oracle model and the ideal cipher model, revisited. In: Proceedings of the Forty-third Annual ACM Symposium on Theory of Computing, STOC 2011, pp. 89–98, ACM, New York (2011)

[LW17] Lenstra, A.K., Wesolowski, B.: Trustworthy public randomness with sloth, unicorn, and trx. IJACT 3(4), 330–343 (2017)

[May93] May, T.C.: Timed-release crypto (1993). http://www.hks.net/cpunks/cpunks-0/1460.html

[MMV11] Mahmoody, M., Moran, T., Vadhan, S.: Time-lock puzzles in the random oracle model. In: Rogaway, P. (ed.) CRYPTO 2011. LNCS, vol. 6841, pp. 39–50. Springer, Heidelberg (2011). https://doi.org/10.1007/978-3-642-22792-9_3

[MMV13] Mahmoody, M., Moran, T., Vadhan, S.: Publicly verifiable proofs of sequential work. In: Proceedings of the 4th Conference on Innovations in Theoretical Computer Science, ITCS 2013, pp. 373–388, ACM, New York (2013)

[Pie19a] Pietrzak, K.: Proofs of catalytic space. In: 10th Innovations in Theoretical Computer Science Conference, ITCS 2019, 10–12 January 2019, San Diego, California, USA, pp. 59:1–59:25 (2019)

[Pie19b] Pietrzak, K.: Simple verifiable delay functions. In: 10th Innovations in The-
          oretical Computer Science Conference, ITCS 2019, 10–12 January 2019,
          San Diego, California, USA, pp. 60:1–60:15 (2019). https://eprint.iacr.org/
          2018/627
[RSW00]   Rivest, R.L., Shamir, A., Wagner, D.: Time-lock puzzles and timed-release
          crypto. Technical report MIT/LCS/TR-684, MIT, February 2000
[Wes19]   Wesolowski, B.: Efficient verifiable delay functions. In: Advances in Cryp-
          tology - EUROCRYPT 2019 (2019)

Incremental Proofs of Sequential Work

Nico Döttling[1], Russell W. F. Lai[2(✉)], and Giulio Malavolta[3]

[1] CISPA Helmholtz Center for Information Security, Saarbrücken, Germany
[2] Friedrich-Alexander-Universität Erlangen-Nürnberg, Erlangen, Germany
russell.lai@cs.fau.de
[3] Carnegie Mellon University, Pittsburgh, USA

Abstract. A proof of sequential work allows a prover to convince a verifier that a certain amount of sequential steps have been computed. In this work we introduce the notion of *incremental proofs of sequential work* where a prover can carry on the computation done by the previous prover incrementally, without affecting the resources of the individual provers or the size of the proofs.

To date, the most efficient instance of proofs of sequential work [Cohen and Pietrzak, Eurocrypt 2018] for N steps require the prover to have $\sqrt{N}$ memory and to run for $N + \sqrt{N}$ steps. Using incremental proofs of sequential work we can bring down the prover's storage complexity to $\log N$ and its running time to N.

We propose two different constructions of incremental proofs of sequential work: Our first scheme requires a single processor and introduces a poly-logarithmic factor in the proof size when compared with the proposals of Cohen and Pietrzak. Our second scheme assumes $\log N$ parallel processors but brings down the overhead of the proof size to a factor of 9. Both schemes are simple to implement and only rely on hash functions (modelled as random oracles).

1 Introduction

Imagine that you discover a candidate solution to a famous open problem (*e.g.*, the Riemann Hypothesis), and are *fairly* convinced that your solution is correct but not entirely. Before publishing your solution you want to scrutinize it further. However, fearing that someone else might make the same discovery, you need a way to timestamp yours. While there are many online timestamping services available[1], authenticity of such a timestamp depends on how much one trusts the service provider. Clearly, a solution independent of trust and resting only on a cryptographic assumption is more desirable.

Proofs of Sequential Work (PoSW) [10] is an emerging paradigm which offers a conceptually simple solution to the timestamping problem. Roughly speaking,

---

G. Malavolta—Work done while at Friedrich-Alexander-Universität Erlangen-Nürnberg.

---

[1] *e.g.*, https://www.freetsa.org.

Y. Ishai and V. Rijmen (Eds.): EUROCRYPT 2019, LNCS 11477, pp. 292–323, 2019.
https://doi.org/10.1007/978-3-030-17656-3_11

proofs of sequential work allow a prover $\mathcal{P}$ to convince a verifier $\mathcal{V}$ that *almost* time T elapsed since a certain event happened. A little more concretely, a PoSW system consists of a prover $\mathcal{P}$ and a verifier $\mathcal{V}$. The prover takes as input a statement χ and a time parameter N. The statement χ can be something like a hash of the file which one wants to timestamp. After terminating, the prover interacts with the verifier $\mathcal{V}$ to convince him that at least time N has elapsed since χ was sampled.

We require a PoSW to be complete, sound and efficient. Here completeness means that an honest prover will succeed in convincing the verifier that time N has elapsed since the sampling of χ. Soundness means that a cheating prover will not succeed in convincing the verifier that time N has elapsed if, in fact significantly less time has passed. Finally, efficiency means that time N is also sufficient for the prover to generate such a proof. Another practically important aspect is memory complexity of the prover, i.e. how much memory is required to compute a proof for time parameter N. Regardless of the requirements on prover efficiency, the verifier's runtime should be essentially independent of N. Finally, for practical reasons such a proof should be non-interactive. That is, after a proof π is computed by the prover $\mathcal{P}$ and published, no further interaction with $\mathcal{P}$ is necessary to verify the proof.

1.1 Incremental Proofs of Sequential Work

An aspect not considered in the original formulation of proofs of sequential work is whether a *still running* proof of sequential work can be migrated from one prover to another, or forked to two provers. This aspect becomes relevant when considering that real computers are not immune to hardware failure, so one may want to spawn clones of important proofs that have been running for a long time.

In this work, we introduce the notion of *incremental proofs of sequential work (iPoSW)*. Essentially, an iPoSW is a non-interactive PoSW with the additional feature that anyone who obtains a proof π for a statement χ and time parameter N can *resume* the computation of π, thereby generating a proof π' for χ with time parameter $N + N'$. More formally, we require that there exists an algorith Inc which takes as input a proof π for time N and a parameter N' and outputs a proof π'. We require that π' has the same distribution as a proof for χ for time $N + N'$.

One could imagine a direct construction of iPoSW from PoSW as follows. To increment a proof π for a statement χ and time N, first derive a new statement χ' from χ and π, e.g., by computing a hash $\chi' \leftarrow \mathsf{H}(\chi, \pi)$. Now compute a proof π' for statement χ' and time N' and then append π' to π, i.e., output (π, π'). To verify (π, π') that (π, π') is a proof for χ and time $N + N'$, compute $\chi' \leftarrow \mathsf{H}(\chi, \pi)$ and check whether π is a proof for χ and time N and π' is a proof for χ' and time N'.

This simple solution has, however, an obvious drawback: The size of the proof grows linearly in the number of increments, which very is undesirable if the proof is frequently passed on to new provers.

Moreover, if we look at existing constructions of PoSW [5,10], a prover $\mathcal{P}$ computing a proof π for a statement χ and time N needs to commit memory proportional to N. Cohen and Pietrzak [5] propose a tradeoff which reduces the memory requirement of p_i to a sublinear but still polynomial amount, however this comes at the expense of additional *sequential* computation time, *i.e.*, prover efficiency is affected by this tradeoff.

1.2   Our Results

In this work we provide constructions of incremental proofs of sequential work where the sequential runtime of an honest prover is N, while its memory complexity is $\mathsf{poly}(\log N)$.

We provide two instantiations, both based on the construction of Cohen and Pietrzak [5], which differ in terms of prover resources and the proof size.

- The first construction is single-threaded, *i.e.*, the prover needs a single processor. Compared to the construction of [5], the proof size grows by a factor of $(\log N)^2$.
- The second construction is multi-threaded, where the prover needs $\log N$ parallel processors. Compared to [5], the proof size grows by a factor of 9.

In particular, our results close the soundness gap between a prover with a large memory and a prover with a poly-logarithmic memory present in previous constructions.

We remark that from a technological point of view the assumption of prover parallelism is justified. For actual applications, the expression $\log N$ can be upper-bounded by 100, which corresponds to a processor capable of computing 100 hashes in parallel, a number well in the reach of modern GPUs.

1.3   Technical Overview

The starting point of our construction is the recent elegant PoSW construction of Cohen and Pietrzak [5]. We will henceforth refer to this scheme as the CP scheme, which is briefly reviewed below. The CP construction relies on properties of a special directed acyclic graph, which we will refer to as CP_n. This graph is constructed as follows: Let B_n be a complete binary tree of depth n, *i.e.*, the longest leaf-to-root path consists of n edges, with edges pointing from the leaves towards the root. Each node in B_n is indexed by a bit string of length at most n, while the root node is indexed by the empty string ϵ. The graph CP_n is constructed by adding edges from all nodes v to all leaves u such that v is a left-sibling of the path from u to the root.

The CP Approach. For a time parameter N, choose n such that CP_n contains (at least) N nodes. The prover is given a statement χ which is used to seed random oracles $\mathsf{H}_\chi(\cdot) := \mathsf{H}(\chi, \cdot)$ and $\mathsf{H}'_\chi(\cdot) := \mathsf{H}'(\chi, \cdot)$ given the random oracles

H and H' respectively. Using H_χ, the prover computes a label for each node v in CP_n by hashing the labels of all nodes with incoming edges to v. Starting from the leftmost leaf 0^n, which is assigned a label 0^λ, the prover iteratively computes the labels of all nodes in CP_n, completing each subtree before starting a new leaf. Eventually the prover obtains a label ℓ_ϵ for the root node.

Next, the prover computes $H'_\chi(\ell_\epsilon)$ which outputs a randomness for sampling t challenge leaves, where t is a statistical security parameter. The proof then consists of the labels of all t challenge leaves, as well as the labels of all siblings of the paths from the challenge leaves to the root. To verify a proof, the verifier recomputes $H'_\chi(\ell_\epsilon)$ to verify if the prover provided the correct paths, and if so checks that the t paths provided by the prover are consistent.

Note that in order to compute a proof, the prover has to either remember the N labels for the entire CP_n graph, or recompute the labels required in the proof once the challenge leaves are chosen, which requires N sequential hash computations. This introduces a soundness slack of $\frac{1}{2}$ between these two strategies, i.e., the memory efficient prover has to compute for time $2N$ to prove a statement for time N. This factor becomes particularly significant when large values of N are considered, e.g., a PoSW that 10 years of sequential operations have been performed may take between 10 and 20 years to be computed. To attenuate this problem, Cohen and Pietrzak propose a hybrid approach where the prover stores $\sqrt{N}$ nodes and can then recompute the challenge root-to-leaf paths in time $\sqrt{N}$.

At the Heart of the Problem. This soundness slack is clearly undesirable as the value of N grows: A prover with access to a large amount of memory can achieve a non-trivial speed up in the computation of the proof over a prover with polylogarithmic memory. As it turns out, this issue is tightly connected with the fact that the CP proofs cannot be extended incrementally: On a very high level, the crux of the problem is that the challenge leaves are determined solely by the root of the CP_n tree. Extending the tree causes the root to change and renders the previous challenge set obsolete.

The main idea in our first construction is to choose challenge leaves "on-the-fly" at *each node* of the tree and then gradually discard some of them as the tree grows. This will allow us to compute a proof π *in a single pass*.

More precisely, our selection mechanism works as follows: For any node v in CP_n which has at most t leaves, we assign all these leaves to be the challenge leaves for the node v. Let l and r be the children of a node v which has more than t leaves, and let S_l and S_r be the challenge leaves for l and r respectively. To determine the set S_v of challenge leaves for v, we first compute the label ℓ_v of v as in the CP scheme, and then hash the label ℓ_v with H'_χ to obtain random coins[2]. Using these random coins, we can sample S_v as a random subset of size t from the set $S_l \cup S_r$. This operation is visualized in Figs. 1 and 2.

---

[2] As we are working in the random oracle model, these coins can be taken directly from ℓ_v if we make the hashes sufficiently longer. However, for presentation purposes we use a separate hash function which hashes ℓ_v.

In a bit more detail, due to the way the graphs are traversed, we only need to store challenge-paths at what we call *unfinished nodes*. A node is unfinished if it has already been traversed/processed, but its right sibling has not yet been traversed. Consequently, only left siblings can be unfinished. Moreover, due to the structure of the graph CP_n and the way it is traversed, at each step the unfinished nodes are exactly the left siblings on the path from the root to the node which is currently processed. Consequently, at each step there are at most $\log N$ unfinished nodes. Essentially, when a node l becomes unfinished, it *waits* until its right sibling r is processed. By the way we traverse CP_n, the next node to be traversed is the parent v of l and r. Once the label of v has been computed, we can compute a set of challenge paths for v as described above and remove l from the list of unfinished nodes.

Observe that if a leaf previously chosen as a challenge leaf is dropped due to the above subset sampling, this leaf will not be chosen as a challenge leaf again in the rest of the computation. Therefore the prover can safely erase the labels of some of the nodes lying on the paths from these dropped challenge nodes to the root, which surely will not appear in the eventual proof. On the other hand the final challenge set is still unpredictable to the eyes of the prover since the decision which paths are discarded is uniquely determined by the complete labelling of the tree.

It is not immediately clear that the strategy we just described lead to a sound protocol. Infact, a malicious prover can already see a large fraction of the challenge path before the label of the root node is even computed and *adaptively* recompute parts of the proof. The main observation on which our analysis is based is that, once a node v becomes unfinished, its label commits to all the leafs under v, thus the challenge paths at v provide a good statistical sample of the overall fraction of invalid leafs in the subtree of v.

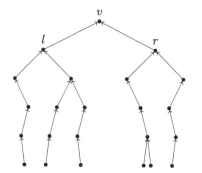

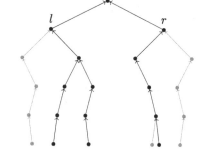

Fig. 1. Before choosing challenge subset.   **Fig. 2.** After choosing challenge subset.

Recomputation to the Rescue. The above strategy seems to solve all problems at once:

1. The prover algorithm can traverse the tree and remember the local challenge paths using poly-logarithmic memory in N and in sequential time N. Once the root is reached, the set of challenge paths is already in the memory of the prover! Therefore no recomputation is needed and the source of the slack is obliterated.

2. The proof is naturally incremental: Further iterations of the tree only shave off root-to-leaf paths in the challenge set, as opposed to determining a completely new set of challenges.

However there is still a challenge to be addressed: Due to the adaptivity of the adversary, our strategy introduces a factor of $\log N$ in the soundness loss. That is, if the CP scheme with a set of parameters achieves soundness α, *i.e.*, the prover cannot cheat by computing less than $(1 - \alpha)N$ steps, our scheme only achieves soundness $\log N \cdot \alpha$. This in turn means that in order to achieve the same soundness parameter as the CP scheme, we need to increase the number of challenge paths by a factor of $(\log N)^2$, which also results in an increase of the proof size by a factor of $(\log N)^2$. Although this does not affect the asymptotic performance of our scheme, it has an impact on the concrete proof sizes. For $N = 2^{40}$, our proofs are bigger than those obtained with the CP scheme by a factor of ~ 1600. To bring down the proof sizes to a practical regime, we reconcile CP scheme with our "on-the-fly" selection strategy. Our second construction assumes that the prover is a parallel machine, but we can show that the number of parallel processors required will never exceed $\log N$.

Our second scheme is based on the following observation. Let v be a node in CP_n, and assume that l is its left child and r is its right child. Further assume that the prover just finished traversing the tree under l, that is l becomes processed but unfinished. By the structure of CP_n, the prover next traverses the tree underneath r. In our first scheme the node l would just be on a *waiting list* of unfinished nodes and has to wait and remember its challenge paths until r is processed. However, due to symmetry it will take the prover the same amount of sequential steps to traverse the tree underneath r as it took to traverse the tree under l. This suggests a strategy (depicted in Fig. 3): While l is unfinished and waiting for the r to be processed, we can *recompute* the subtree underneath l in order to fetch fresh challenge paths using an additional parallel processor. By the time r is finished, this process will have terminated and we do not need to bear the above soundness loss for l.

Notice further that, to recompute the tree underneath l, all the prover needs is the labels of the currently unfinished nodes on the path from the root to v, which the prover needs to keep in memory regardless. This modification of the prover strategy must also be reflected by the verifier. When we verify a root-to-leaf path, the verification strategy will change once the path takes a left turn.

Note that the memory complexity of the main thread is unchanged and that at any point in time there are at most $\log N$ parallel processes. The parallel threads are identical to the recomputation step. Therefore, the complexity of each parallel thread is essentially the same as that of the CP scheme. This hybrid construction brings down the loss in soundness to a factor of 3, which

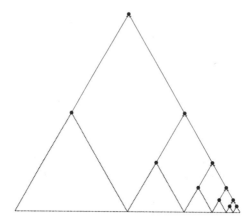

Fig. 3. Recomputation of Sub-Trees.

corresponds to an increase of the proof size by a factor of 9. We consider this to be a modest price to pay in exchange for getting the additional feature of incremental proofs and an essentially optimal prover complexity.

1.4 Perspectives

Merkle trees are ubiquitous in cryptographic protocol design, allowing to compress large amounts of data into a succinct digest. Membership proofs are particularly efficient as they usually consist of root-to-leaf paths and can be encoded with logarithmic-size strings. The de-facto methodology to non-interactively probe Merkle trees at random locations is to apply the Fiat-Shamir [7] transform, on input the root of the tree. This means that the challenge locations are determined only when the full tree is computed. Thus, the prover must either recompute the paths or store the full tree in its memory.

Using our techniques one can compress data and generate challenges in a single pass, without any memory blowup. This becomes particularly advantageous when computing over very large databases or data streams. Here we exemplify the applications of our methods to scenarios of interest.

Verifiable Probing. Consider a stream of data where some statistical measure is computed by an untrusted party. Using our approach we can increase the confidence in the validity of the statistics by probing the stream on random locations. The prover iteratively computes a Merkle commitment of the stream and selects random probes using our "on-the-fly" selection strategy. The verifier can then non-interactively check whether the distribution of the probes resembles the reported statistics.

Streaming Arguments. In Micali's CS proofs paradigm [8,11], witnesses for NP relations are encoded into probabilistically checkable proofs (PCP) [1] and then committed using a Merkle tree. The testing locations for the PCP are selected using Fiat-Shamir [7] and the corresponding root-to-leaf paths form the CS proof. Our techniques can be useful for memory-constrained provers that cannot store the complete PCP encoding in their memory. Our challenge-selection algorithm allows the provers to compute the CS proof using only one stream of the encoding.

1.5   Related Work

Proofs of work, a concept introduced by Dwork and Naor [6] and having become wildly popular in the context of cryptocurrencies, allow a prover to convince a verifier that a certain amount of computational effort has been invested in a certain task. However, the computation can be parallelized, thus it generates a mismatch among players which have different resource constraints. Mahmoody, Moran, and Vadhan [10] introduced the concept of PoSW and provided a construction based on depth-robust graphs. Very recently, Cohen and Pietrzak [5] provided an elegant construction based on a binary tree with some useful combinatorial properties. Their scheme improves over [10] in terms of conceptual simplicity, concrete efficiency, and can reduce the memory complexity of the prover up to $\log N$. A shortcoming of their approach is that, in order to achieve such a memory bound, one has to perform the same amount of computation twice.

Incrementally verifiable computation (IVC) was introduced by Valiant [14] and allows a machine to output short proofs that arbitrary parts of the computation have been done correctly without significantly affecting the resources of such a machine. As observed by Boneh et al. [4], IVC is a more general primitive than PoSW. The main construction paradigm for IVC consists of a recursive composition of succinct arguments of knowledge [11], which means that existing constructions of IVC either

 – make non-black-box use of random oracles [14], or
 – require a trusted setup [2].

In general, incremental PoSW appears to be an easier problem than IVC which justifies the existence of more efficient solutions based on weaker assumptions.

Verifiable delay functions (VDF) have been introduced by Boneh et al. [4] and can be seen as PoSW with unique proofs: The prover can only convince the verifier with a single value, which is uniquely determined by the time parameter N and by the statement. Thus VDF constitutes a stronger primitive than PoSW and as to our current understanding requires stronger cryptographic material: Known constructions [12,15] rely either on IVC or on specific number-theoretic assumptions related to factoring large integers.

Time-lock puzzles [13] encapsulate a secret information for a pre-determined amount of time. This primitive is tightly related to sequential computation as it needs to withstand attacks from highly parallel processors. Time lock-puzzles can be constructed assuming the hardness of a variant of the RSA assumption [13] or succinct randomized encodings and the existence of a worst case non-parallelizable language [3]. Unlike PoSW, no construction based on symmetric-key primitives is known and [9] gave a black-box separation for these two objects.

2 Preliminaries

2.1 Notations

Let $G = (V, E)$ be a graph where V is the set of nodes and E is the set of edges. If $v \in V$, we write also $v \in G$ for convenience. Let $\mathfrak{T}$ be a tree and $i \in \mathfrak{T}$ be a node. $\mathfrak{T}_i$ denotes the set of nodes in the subtree rooted at node i. leaf(i) denotes the set of all leaf nodes that are descendants of i. parent(i) and child(i) denote the parent of and the set of children of i, repectively. path(i) returns the set of nodes located at the (unique) path from the root (inclusive) to node i (inclusive). The notations are extended naturally to sets of nodes. Let $S \subseteq \mathfrak{T}$ be a set of nodes, then $\mathfrak{T}_S := \bigcup_{i \in S} \mathfrak{T}_i$, leaf$(S) := \{$leaf$(i) : i \in S\}$ and path$(S) := \{$path$(i) : i \in S\}$.

For a complete binary tree $B_n = (V, E')$ of $N = 2^{n+1} - 1$ nodes, we say that B_n is of depth n (counting the number of edges in the longest leaf-to-root path). The nodes $V = \{0, 1\}^{\leq n}$ are identified by binary strings of length at most n and the empty string ϵ represents the root. The edges $E' = \{(x||b, x) : b \in \{0, 1\}, x \in \{0, 1\}^i, i < n\}$ are directed from the leaves towards the root. Let $v \in \{0, 1\}^{n_v} \subseteq B_n$ be a node n_v edges away from the root. We say that v is of depth n_v or height $h_v := n - n_v$.

2.2 Statistical Distance

In the following we recall the definition of statistical distance.

Definition 1 (Statistical Distance). *Let X and Y be two random variables over a finite set $\mathcal{U}$. The statistical distance between X and Y is defined as*

$$\mathbb{SD}[X, Y] = \frac{1}{2} \sum_{u \in \mathcal{U}} |\Pr[X = u] - \Pr[Y = u]|$$

2.3 Tail Bound for the Hypergeometric Distributions

Here we introduce a useful inequality by Hoeffding.

Theorem 1 (Hoeffding Inequality). *Let X be distributed hypergeometrically with t draws. Then it holds that*

$$\Pr[X < \mathsf{E}[X] - \zeta] < e^{-2\zeta^2 t}.$$

3   Incremental Proofs of Sequential Work

Below we define incremental proof of sequential work in the same spirit as Cohen and Pietrzak [5], except that we state directly the non-interactive variant.

Definition 2. *A (non-interactive) incremental proof of sequential work (iPoSW) scheme consists of a tuple of* PPT *oracle-aided algorithms* (Prove, Inc, Vf), *executed by a prover* $\mathcal{P}$ *and a verifier* $\mathcal{V}$ *in the following fashion:*

Common Inputs. $\mathcal{P}$ *and* $\mathcal{V}$ *get as common input a computation security parameter* $\lambda \in \mathbb{N}$, *a statistical security parameter* $t \in \mathbb{N}$, *and a time parameter* $N \in \mathbb{N}$. *All parties have access to a random oracle* $H : \{0,1\}^* \to \{0,1\}^\lambda$.

Statement. $\mathcal{V}$ *samples a random statement* $\chi \leftarrow_\$ \{0,1\}^\lambda$ *and sends it to* $\mathcal{P}$.

Prove. $\mathcal{P}$ *computes* $\pi \leftarrow \mathsf{Prove}^H(\chi, N)$ *and sends* π *to* $\mathcal{V}$.

Increment. $\mathcal{P}$ *computes* $\pi' \leftarrow \mathsf{Inc}^H(\chi, N, N', \pi)$ *and sends* π' *to* $\mathcal{V}$.

Verify. $\mathcal{V}$ *computes and outputs* $\mathsf{Vf}^H(\chi, N, \pi)$.

We require a PoSW scheme to be complete in the following sense.

Definition 3. (Completeness). *For all* $\lambda \in \mathbb{N}$, *all* $N \in \mathbb{N}$, *all random oracles* H, *and all statements* $\chi \in \{0,1\}^\lambda$ *we say that a tuple* (χ, N, π) *is honest if*

$$\pi \in \mathsf{Prove}^H(\chi, N) \qquad or \qquad \pi \in \mathsf{Inc}^H(\chi, N', N'', \pi'),$$

where $N' + N'' = N$ *and the tuple* (χ, N', π') *is also honest. A (non-interactive) incremental proof of sequential work is complete if for all honest tuples* (χ, N, π) *it holds that*

$$\mathsf{Vf}^H(\chi, N, \pi) = 1.$$

In the following we define soundness for incremental proofs of sequential work.

Definition 4. (Soundness). *A (non-interactive) incremental proof of sequential work PoSW is sound if for all* $\lambda, N \in \mathbb{N}$, *for all* $\alpha > 0$, *for all adversaries* $\mathcal{A}$ *that make at most* $(1 - \alpha)N$ *sequential queries to* H, *it holds that*

$$\mu := \mathsf{Pr}\left[\chi \leftarrow \{0,1\}^\lambda; \pi \leftarrow \mathcal{A}^H(\chi, N) : \mathsf{Vf}^H(\chi, N, \pi) = 1\right] \in \mathsf{negl}(\lambda)$$

where μ *is called the soundness error.*

For our construction we recall the following directed acyclic graph constructed by Cohen and Pietrzak [5] which has some nice combinatorial properties.

Definition 5. (CP Graphs). *For* $n \in \mathbb{N}$, *let* $N = 2^{n+1} - 1$ *and* $B_n = (V, E')$ *be a complete binary tree of depth* n *with edges pointing from the leaves to the root. The graph* $CP_n = (V, E)$ *is a directed acyclic graph constructed from* $B_n = (V, E')$ *as follows. For any leaf* $u \in \{0,1\}^n$, *for any node* v *which is a left-sibling of a node on the path from* u *to the root* ϵ, *an edge* (v, u) *is appended to* E'. *Formally,* $E := E' \cup E''$ *where*

$$E'' := \{(v, u) : u \in \{0,1\}^n, u = a||1||a', v = a||0, \text{ for some } a, a' \in \{0,1\}^{\leq n}\}.$$

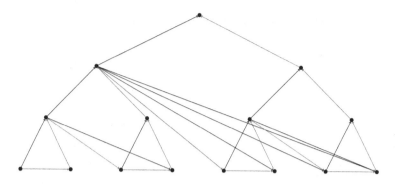

Fig. 4. CP_3 with traversal order highlighted in red. (Color figure online)

An illustration of CP_3 is in Fig. 4, with its traversal order (c.f. Lemma 2) highlighted in red. Here we recall some technical lemmas from [5].

Lemma 1. ([5]). *The labels of a CP_n graph can be computed in topological order using $\lambda(n+1)$ bits of memory.*

Let $\mathfrak{T}$ be a tree and $S \subseteq \mathfrak{T}$ be a subset of nodes. We denote by S^* the minimal set of nodes with exactly the same set of leaves as S, in other words, S^* is the smallest set such that $\mathsf{leaf}(S^*) = \mathsf{leaf}(S)$.

Lemma 2. ([5]). *For all $S \subseteq V$, the subgraph of $CP_n = (V, E)$ on vertex set $V \setminus \mathfrak{T}_{S^*}$ has a directed path going through all the $|V| - |\mathfrak{T}_{S^*}|$ nodes.*

Lemma 3. ([5]). *For all $S \subseteq V$, $\mathfrak{T}_{S^*}$ contains $\frac{|\mathfrak{T}_{S^*}| + |S|}{2}$ many leaves.*

4   Main Construction

For any $n \in \mathbb{N}$, we construct an incremental PoSW scheme based on the graph $CP_n = (V, E)$ as follows. We assume without loss of generality that, given a random oracle H, one can sample a fresh random oracle indexed by a string x, which we denote by H_x. This can *e.g.*, be done by prepending x and a special separator symbol to any query to H, *i.e.*, $\mathsf{H}_x(y) := \mathsf{H}(x\#y)$ for a separator symbol $\#$.

4.1   Parameters

Our incremental Proof-of-Sequential-Work system depends on the following parameters and objects.

- A time parameter N of the form $N = 2^{n+1} - 1$, for some integer $n \in \mathbb{N}$.
- A computational security parameter λ
- A statistical security parameter t

- A full-domain hash function $H : \{0,1\}^* \to \{0,1\}^\lambda$ modelled as a random oracle.
- A full-domain hash function $H' : \{0,1\}^* \to \{0,1\}^{3t}$ modelled as a random oracle.
- A sampler $\mathsf{RandomSubset}(M, m; r)$ which takes a universe size M, a sample size m and uniform random coins r and outputs a uniformly random subset $X \subseteq [M]$ such that $|X| = m$. In our application, we will always set $M = 2t$ and $m = t$. Since $\binom{2t}{t} < \left(\frac{2t \cdot e}{t}\right)^t = (2e)^t$, where $\log 2e \approx 2.44$, random coins of size $3t$ are sufficient to sample statistically close to a uniform subset.

Notation. Let ϵ be the root-node of CP_n and 0^n the left-most leaf in the tree or starting-node. We will call a left node $v \in V$ unfinished, if v has been traversed by the prover algorithm but $\mathsf{parent}(v)$ has not yet been. At any time, the prover will keep a list of the currently unfinished nodes U. At each unfinished node v, the prover will store $\mathcal{L}_v$, a set of extended labeled paths from v to $\mathsf{leaf}(v)$. An extended labeled path consists of a list of tuples of the form $(v_i, \ell_{l_i}, \ell_{r_i}, \mathsf{ind}_i)$, where v_i is the index/address of a node on the path, l_i is the left child of v_i, r_i is the right child of v_i and consequently ℓ_{l_i} is the label of l_i and ℓ_{r_i} is the label of r_i. Finally, ind_i is a local path index, the meaning of which will be explained later.

For simplicity of exposition, we assume that t is a power of 2. Our construction can be easily adapted to the more general case where t is arbitrarily chosen. For convenience, we denote by $n^* = n^*(n, t)$ the depth at which every node has exactly t leafs underneath, *i.e.*, it holds for every node v which is n^* edges from ϵ that $|\mathsf{leaf}(v)| = t$.

4.2   Scheme Description

$\underline{\mathsf{Prove}^{H,H'}(\chi, N)}$:

1. Initialize $U \leftarrow \emptyset$, the set of unfinished nodes.
2. Assign $\ell_{0^n} \leftarrow 0^\lambda$ as the label of the starting node.
3. Traverse the graph CP_n starting from 0^n. At every node $v \in V$ which is traversed, do the following:
   (a) Compute the label ℓ_v by

$$\ell_v \leftarrow H_{(\chi,v)}(\ell_{v_1}, \ldots, \ell_{v_d})$$

   where $v_1, \ldots, v_d \in V$ are all nodes with edges pointing to v, *i.e.*, $(v_i, v) \in E$.
   (b) Let l and r be the children of v.
   (c) If $|\mathsf{leafs}(v)| \leq t$, set $\mathcal{L}_v \leftarrow \{[(v, \ell_l, \ell_r, \bot)\|L]$ where $L \in \mathcal{L}_l \cup \mathcal{L}_r\}$.
   (d) Otherwise (*i.e.*, if $|\mathsf{leafs}(v)| \geq 2t$), do the following:
      i. Compute

$$r_v \leftarrow H'_{(\chi,v)}(\ell_v).$$

ii. Choose a random t-subset S_v of $[2t]$ via $S_v \leftarrow \mathsf{RandomSubset}(2t, t; r_v)$.

iii. For $j \in \{0, \ldots, t-1\}$, write $S_v[j] = at + b$ with $a \in \{0, 1\}$ and $0 \leq b < t$.

$$\mathcal{L}_v[j] \leftarrow \begin{cases} [(v, \ell_l, \ell_r, j) \| \mathcal{L}_l[b]], & \text{if } a = 0 \\ [(v, \ell_l, \ell_r, j) \| \mathcal{L}_r[b]], & \text{if } a = 1 \end{cases}$$

(e) Mark l as finished, *i.e.*, remove l from U and, if v is a left child, mark v as unfinished, *i.e.*, add v to U.

4. Once the set of unfinished nodes consists only of the root-node (*i.e.*, $U = \{\epsilon\}$), terminate and output $\pi \leftarrow (\ell_\epsilon, \mathcal{L}_\epsilon)$.

$\mathsf{Inc}^{\mathsf{H}, \mathsf{H}'}(\chi, N, N', \pi)$:

1. Initialize $U \leftarrow \emptyset$.
2. Parse π as $(\ell_\epsilon, \mathcal{L}_\epsilon)$
3. Assign $\ell_{0^{n'-n}} := \ell_\epsilon$ and $\mathcal{L}_{0^{n'-n}} := \mathcal{L}_\epsilon$.
4. Execute the algorithm $\mathsf{Prove}^{\mathsf{H}, \mathsf{H}'}(\chi, N')$ starting from step 3 with a slight change: Traverse the graph $CP_{n'}$ starting from $0^{n'-n-1} \| 1 \| 0^n$ (instead of from $0^{n'}$).

$\mathsf{Vf}^{\mathsf{H}, \mathsf{H}'}(\chi, N, \pi)$:

1. Parse π as $(\ell_\epsilon, \mathcal{L}_\epsilon)$.
2. For all paths $\mathsf{path} \in \mathcal{L}_\epsilon$ do the following:
   (a) Parse path as $[(v_0, \ell_{l_0}, \ell_{r_0}, \mathsf{ind}_0) \| \ldots \| (v_n, \ell_{l_n}, \ell_{r_n}, \mathsf{ind}_n)]$.
   (b) For every node $v \in \{v_0, \ldots, v_n\}$ on the path, check if the label ℓ_v was computed correctly. That is, for $v = 0^n$ check whether $\ell_v = 0^\lambda$, and for any other node $v \in V \setminus \{0^n\}$ check whether $\ell_v = \mathsf{H}_{(\chi, v)}(\ell_{v_1}, \ldots, \ell_{v_d})$, where $v_1, \ldots, v_d$ are all the nodes with edges pointing to v. The value ℓ_v can either be retrieved from the parent node of v, or is directly available for the case of the root-node ϵ. For the special case of leaf-nodes, the values $\ell_{v_1}, \ldots, \ell_{v_d}$ are not stored locally with the node v, but are stored at some other (a-priori known) nodes along the path path (refer to the structure of the graph CP_n).
   (c) For all $j \in \{0, \ldots, n^*\}$, compute $r_{v_j} \leftarrow \mathsf{H}'_{(\chi, v_j)}(\ell_{v_j})$ and $S_{v_j} \leftarrow \mathsf{RandomSubset}(2t, t; r_{v_j})$. If v_{j+1} is the left child of v_j, check if

$$S_v[\mathsf{ind}_j] = \mathsf{ind}_{j+1}.$$

Otherwise, if v_{j+1} is the right child of v_j, check if

$$S_v[\mathsf{ind}_j] = t + \mathsf{ind}_{j+1}.$$

3. If all checks pass output 1, otherwise 0.

Incomplete Trees. We briefly outline how to handle incomplete binary trees. If N does not define a complete tree, then at the end of the prover's iteration the list of unfinished nodes consists of several elements: $U = \{v_1, \ldots, v_n\}$. The new proof π consists of the tuples $(\ell_{v_1}, \mathcal{L}_{v_1}), \ldots, (\ell_{v_n}, \mathcal{L}_{v_n})$. The proof can be easily verified by running the standard verification algorithm on each pair $(\ell_{v_i}, \mathcal{L}_{v_i})$ separately and outputting 1 if all the verifications succeeds. In a similar way, one can increment the proof by recovering the trees computed so far, setting the labels of the unfinished nodes to $(\ell_{v_1}, \ldots, \ell_{v_n})$ and the corresponding sets to $(\mathcal{L}_{v_1}, \ldots, \mathcal{L}_{v_n})$. Given such a snapshot of the execution, one can continue the standard iteration and complete the proof for the new (larger) tree.

4.3   Efficiency Analysis

We now discuss the efficiency of our scheme in terms of proof size, computation and communication.

Proof Size. The proof consists of the root-label ℓ_ϵ and t challenge paths $\mathsf{path}_0, \ldots, \mathsf{path}_{t-1}$. Each $\mathsf{path} \in \{\mathsf{path}_0, \ldots, \mathsf{path}_{t-1}\}$ consists of n tuples of the form $(v, \ell_l, \ell_r, \mathsf{ind})$, where v is the index of a node, ℓ_l and ℓ_r are the labels of the left and right children of v, and $\mathsf{ind} \in [t]$ is the index of path in the challenge set S_v at v. The node index v can be stored using a single bit per node, indicating whether it is the left or right child of its parent. Each of ℓ_l and ℓ_r can be stored using λ bits, and ind can be represented using $\log t$ bits. Consequently, the entire proof has size at most $t \cdot n \cdot (1 + 2\lambda + \log t) = O(t \cdot \lambda \cdot n)$ (assuming $t \in \mathrm{poly}(\lambda)$). Later, in the soundness analysis, we will show that our construction is sound if $t \in O(\lambda \cdot n^2)$. With such choice of t, the proof size is bounded by $O(\lambda^2 \cdot n^3)$.

Prover Efficiency. The prover traverses the N nodes of the graph CP_n in the same manner as the prover algorithm of the CP scheme. Additionally, at each node the prover computes a challenge using the random oracle $\mathsf{H}'_{(\chi,v)}$.

The challenges $\mathsf{H}'_{(\chi,v)}$ can be computed in a way that does not increase the parallel time complexity of the prover. Specifically, instead of computing the randomness for the challenges via $r_v \leftarrow \mathsf{H}'_{(\chi,v)}(\ell_v)$, we can equivalently compute the r_v similar to ℓ_v via $r_v \leftarrow \mathsf{H}'_{(\chi,v)}(\ell_{v_1}, \ldots, \ell_{v_d})$. This is possible as both H and H' are random oracles. The proof changes only slightly, but we kept the naive version for presentation purposes. In the modified scheme H and H' can be evaluated in parallel. thus the parallel complexity is not increased by the evaluation of H'. To conclude, the parallel complexity of the prover is bounded by the time needed for $O(N)$ sequential calls to the random oracles.

For the memory complexity of the prover, Cohen and Pietrzak [5] show using a standard pebbling argument (c.f. Lemma 1) that the labels of CP_n can be computed in topological order storing at most $n + 1$ labels at any time, i.e., having at most $n + 1$ pebbles in the graphs at any time. This corresponds to the number of unfinished nodes, i.e. at every time-step there are at most $n + 1$ unfinished nodes. At each unfinished node $v \in U$, the prover keeps a list $\mathcal{L}_v$ consisting of t labeled paths. By the analysis above these t paths can be stored using $O(\lambda^2 \cdot n^3)$ bits. Consequently, the space complexity of the prover is bounded by $O(\lambda^2 \cdot n^4)$.

Verifier Efficiency. The verifier needs to check the consistency of t paths, each consisting n nodes. Checking a node incurs the computation of a hash using $\mathsf{H}_{(\chi,v)}$ and one using $\mathsf{H}'_{(\chi,v)}$. All nodes can be checked in parallel with by computing a constant number of hashes. After that, the verifier has to check whether all $t \cdot n$ checks are passed, which can be performed in parallel time $O(\log(t \cdot n)) = O(\log(\lambda \cdot n^3))$.

4.4 Soundness

We now establish soundness of our construction. Before proving the main theorem, we prove some useful lemmas. Throughout the following analysis, we always assume that N and t are powers of two, but the arguments naturally extend to the more generic case. We denote by $\mathsf{L} := (\mathfrak{T}_v, \{\ell_u\}_{u\in\mathfrak{T}_v})$ the labelling for a sub-tree $\mathfrak{T}_v$. We slightly abuse the notation and we say that $u \in \mathsf{L}$ if $u \in \mathfrak{T}_v$.

Lemma 4. *Let $\mathcal{A}$ be an algorithm with access to a random oracle $\mathsf{H} : \{0,1\}^* \to \{0,1\}^\lambda$ which outputs a root-hash of a Merkle tree of depth n and a (valid) root-to-leaf path path with siblings. Then there exists an efficient online extractor $\mathsf{Extract}$, which on input a node $v \in \mathfrak{T}$, a label ℓ^* and a list Q (of size q) of all H-queries of $\mathcal{A}$ so far, outputs a labelling L of the sub-tree $\mathfrak{T}_v$ rooted at v such that the following holds. Let path^* be the leaf-to-root path p truncated at v and let path_L be the same path in L, then $\mathsf{path}^* = \mathsf{path}_\mathsf{L}$, except with probability $\frac{1+q(q-1)}{2^\lambda}$, over the choice of H.*

Proof. We assume without loss of generality that the list Q is of the form $\{(\mathsf{in}, \mathsf{out})\}$, and that the depth n_v of a node is efficiently computable from its identifier. We define the algorithm $\mathsf{Extract}$ in the following.

$\mathsf{Extract}(v, \ell^*, Q)$: The root of the tree L set to be ℓ^* and the rest of the tree is recursively constructed applying $(n - n_v)$ times the following function $f(\mathsf{node})$: Parse Q for an entry of the form $(\mathsf{in}, \mathsf{node})$, if such an entry does not exist then return $\bot$. Else parse in as $\ell_0 \| \ell_1$, set ℓ_0 as the left child of node and ℓ_0 as the right child of node in L. Then run $f(\ell_0)$, $f(\ell_1)$ and return L.

The algorithm runs with a logarithmic factor in the size of $\mathfrak{T}_v$ (assuming an ordered list Q) and therefore it is efficient. Let BAD be the event such that there exists a node $v \in \mathsf{path}^*$ labelled ℓ'_v such that $\ell_v \neq \ell'_v$ and $\ell_{\mathsf{parent}(v)} = \ell'_{\mathsf{parent}(v)}$, where ℓ'_v and ℓ_v are the labelling output by $\mathcal{A}$ and by the extractor, respectively. By the law of total probability we have that

$$\Pr[\mathsf{BAD}] = \Pr[\mathsf{BAD} \mid \ell_v = \bot]\Pr[\ell_v = \bot] + \Pr[\mathsf{BAD} \mid \ell_v \neq \bot]\Pr[\ell_v \neq \bot]$$
$$\leq \Pr[\mathsf{BAD} \mid \ell_v = \bot] + \Pr[\mathsf{BAD} \mid \ell_v \neq \bot].$$

To bound the first summand observe that $\mathsf{H}(\ell'_v \| \ell'_{v'}) = \ell'_{\mathsf{parent}(v)}$, where v' is the sibling of v, since the path path needs to be valid. Further note that there exists no entry of the form $(\cdot, \ell'_{\mathsf{parent}(v)}) \in Q$, since ℓ_v is set to $\bot$ and $\ell'_{\mathsf{parent}(v)} = \ell_{\mathsf{parent}(v)}$. This implies that the adversary has correctly guessed a pre-image of $\ell'_{\mathsf{parent}(v)}$ without querying H, which happens with probability $2^{-\lambda}$. Thus we can bound from above

$$\Pr\left[\mathsf{BAD} \mid \ell_v = \bot\right] \leq 2^{-\lambda}.$$

For the second summand consider again that $\mathsf{H}(\ell'_v \| \ell'_{v'}) = \ell'_{\mathsf{parent}(v)}$ and that $\mathsf{H}(\ell_v \| \ell_{v'}) = \ell_{\mathsf{parent}(v)}$. Since $\ell'_{\mathsf{parent}(v)} = \ell_{\mathsf{parent}(v)}$ we have that $\mathsf{H}(\ell'_v \| \ell'_{v'}) = \mathsf{H}(\ell_v \| \ell_{v'})$, which is a valid collision for H since, by assumption, $\ell'_v \neq \ell_v$. Therefore we have that

$$\Pr\left[\mathsf{BAD} \mid \ell_v \neq \bot\right] \leq 1 - \prod_{k=0}^{q-1}\left(1 - \frac{k}{2^{\lambda}}\right) = 1 - \frac{2^{\lambda}}{2^{\lambda}} \cdot \frac{2^{\lambda}-1}{2^{\lambda}} \cdots \frac{2^{\lambda}-(q-1)}{2^{\lambda}}$$

$$\leq 1 - \left(\frac{2^{\lambda}-(q-1)}{2^{\lambda}}\right)^q = 1 - \left(1 - \frac{q-1}{2^{\lambda}}\right)^q \leq \frac{q(q-1)}{2^{\lambda}}$$

where the last inequality is due to Bernoulli. Thus by triangle inequality we have that

$$\Pr\left[\mathsf{BAD}\right] \leq \frac{1}{2^{\lambda}} + \frac{q(q-1)}{2^{\lambda}} = \frac{1 + q(q-1)}{2^{\lambda}},$$

which implies that the complementary event happens with all but negligible probability. That is, for all nodes in $v \in \mathsf{path}^*$ labelled ℓ_v such that and $\ell'_{\mathsf{parent}(v)} = \ell_{\mathsf{parent}(v)}$ it holds that $\ell'_v = \ell_v$. Since L is rooted at ℓ^* and path^* and L have the same depth, it follows by induction that path^* must be identical to path_L, with the same probability.   □

Given a labeled tree L, we say that a node $v \in \mathsf{L}$ is inconsistent if it holds that $\ell_v \neq \mathsf{H}(\ell_{v_1}, \ldots, \ell_{v_d})$, where $(v_1, \ldots, v_d)$ are the nodes with an incoming edge to v. Let $n(\mathsf{L})$ be the depth of L, then L has $2^{n(\mathsf{L})}$-many paths (or, equivalently, leaves) and we define C as the set of paths which contain at least one inconsistent node. Note that L uniquely defines a set of t challenge paths (as specified in the description of the prover algorithm) which we denote by Z. For convenience we define the functions $\gamma(\mathsf{L}) := \frac{|C|}{2^{n(\mathsf{L})}}$ and $\delta(\mathsf{L}) := \frac{|Z \cap C|}{|Z|}$.

Lemma 5. *Let v be a node and let l and r be the left and right child of v, respectively. If*

$$\delta(\mathsf{L}_l) \geq \gamma(\mathsf{L}_l) - \eta_l \text{ and } \delta(\mathsf{L}_r) \geq \gamma(\mathsf{L}_r) - \eta_r$$

then it holds that

$$\Pr\left[\gamma(\mathsf{L}_v) \leq \delta(\mathsf{L}_v) + \eta_v\right] \geq \left(1 - e^{-2\left(\eta_v - \frac{(\eta_l + \eta_r)}{2}\right)^2 t}\right).$$

Proof. Recall that $\gamma(L_v)$ counts the fraction of inconsistent paths of v. Since l and r are the children of v it holds that

$$\gamma(L_v) = \frac{(\gamma(L_l) + \gamma(L_r))}{2}. \tag{1}$$

Rearranging the terms we have that

$$\gamma(L_l) \le \delta(L_l) + \eta_l \tag{2}$$
$$\gamma(L_r) \le \delta(L_r) + \eta_r, \tag{3}$$

thus combining (1), (2), and (3) we obtain

$$\gamma(L_v) \le \frac{(\delta(L_l) + \eta_l + \delta(L_r) + \eta_r)}{2} = \frac{(\delta(L_l) + \delta(L_r))}{2} + \frac{(\eta_l + \eta_r)}{2}. \tag{4}$$

Let Z'_v be the set of all paths in $Z_l \cup Z_r$ extended to v, i.e. $Z'_v = \{(v,p) \mid p \in Z_l \cup Z_r\}$. By construction, the set Z_v is a random t-subset of Z'_v (where the randomness for this choice is taken from $H'_{(\chi,v)}(\ell_v)$). Assume now that there are s_l rejecting paths in Z_l and s_r rejecting paths in Z_r, i.e. it holds that $\delta(L_l) = \frac{s_l}{t}$ and $\delta(L_r) = \frac{s_r}{t}$. That is, there are $s_l + s_r$ rejecting paths in Z'_v. Consequently, the expected number of rejecting paths in Z_v is $\frac{s_l + s_r}{2t} \cdot t = \frac{1}{2}(\delta(L_l) + \delta(L_r)) \cdot t$, that is

$$\mathsf{E}[\delta(L_v)] = \frac{(\delta(L_l) + \delta(L_r))}{2}, \tag{5}$$

where the expectation is taken over the random choice $H'_{(\chi,v)}(\ell_v)$. Thus we can rewrite

$$\begin{aligned}
\Pr\left[\gamma(L_v) > \delta(L_v) + \eta_v\right] &= \Pr\left[\delta(L_v) < \gamma(L_v) - \eta_v\right] \\
&< \Pr\left[\delta(L_v) < \frac{(\delta(L_l) + \delta(L_r))}{2} + \frac{(\eta_l + \eta_r)}{2} - \eta_v\right] \\
&= \Pr\left[\delta(L_v) < \mathsf{E}[\delta(L_v)] + \frac{(\eta_l + \eta_r)}{2} - \eta_v\right] \\
&< e^{-2\left(\eta_v - \frac{(\eta_l + \eta_r)}{2}\right)^2 t}
\end{aligned}$$

where the first inequality holds by (4), the second equality holds by (5), and the last inequality is a direct application of the Hoeffding inequality for hypergeometric distributions (Theorem 1). □

We are now ready to state and prove the main theorem.

Theorem 2. *The construction given in Sect. 4.2 is sound for any $t \in O(\lambda \cdot n^2)$, and the soundness error is given by $\frac{1+q(q-1)}{2^\lambda} + q \cdot e^{-2(\frac{\alpha}{n})^2 t}$.*

Proof. Let χ be the challenge statement and let q_v be the number of calls of $\mathcal{A}$ to the random oracle $H'_{(\chi,v)}$, i.e., the adversary makes at most $q = \sum_{v \in \mathfrak{T}} q_v$ calls to H' in total.

By Lemma 4, there exists an (efficient) algorithm Extract which on input a node $v \in \mathfrak{T}$, a label ℓ_v and a list Q of all query to H by $\mathcal{A}$ with their responses, outputs a a labelling L_v of the sub-tree $\mathfrak{T}_v$ rooted at v. For $i = \{0, \dots, n\}$ and for $j = \{1, \dots, 2^i\}$, let $v_{i,j}$ be the j-th node at layer i of the tree (counting from the root towards the leaves).

Consider the following sequence of hybrids.

– Hybrid $\mathcal{H}_0$: This is identical to the real experiment.
– Hybrid $\mathcal{H}_1$: The same as $\mathcal{H}_0$, except for the following modifications.
  • The experiment records a list Q of all H queries made by $\mathcal{A}$ with their responses.
  • Every time $\mathcal{A}$ queries $H'_{(\chi,v)}$ for a $v \in V$ with a label ℓ_v, a labelling L_v for the sub-tree under v is computed via $L_v \leftarrow \mathsf{Extract}(v, \ell_v, Q)$.
  • If it holds for any path opened by $\mathcal{A}$ that the labels on the path are different from the labels in L_ϵ (where ϵ is the root), then $\mathcal{H}_1$ aborts and outputs 0.

Let BAD_v be the following event: $\mathcal{A}$ queries $H'_{(\chi,v)}$ with a query $\hat{\ell}_v$ corresponding to a labeled sub-tree $L_v \leftarrow \mathsf{Extract}(v, \hat{\ell}_v, Q)$ for which it holds that $\delta(L_v) < \gamma(L_v) - \frac{n^* - n_v}{n^*} \cdot \alpha$, where n_v is the depth of v (i.e. the distance between the root-node ϵ and v) and n^* is the depth at which every node has exactly t leafs underneath.

For $i = n^*, \dots, 0$ and $j = 1, \dots, 2^i$ define the following hybrids.

– Hybrid $\mathcal{H}_{i,j}$: The same as the previous hybrid, except that the experiment outputs 0 if the event $\mathsf{BAD}_{v_{i,j}}$ happens (Recall that $v_{i,j}$ is the j-th node at layer i of the tree, counting from the root towards the leaves).

We will now show indistinguishability between the hybrids. By Lemma 4 it holds that $\mathcal{H}_0$ and $\mathcal{H}_1$ are indistinguishable. We now turn to the indistinguishability of hybrids $\mathcal{H}_{i,j}$. For notational convenience, let $\mathcal{H}_{i,j}^\downarrow$ be the hybrid before $\mathcal{H}_{i,j}$.

First consider $i = n^*$. It holds for each node v at level i that the set Z_v of challenge paths consists of all paths from v to the leaves under v. Consequently, it holds for all v at level i that $\delta(L_v) = \gamma(L_v)$ and therefore BAD_v happens with probability 0.

Now consider the case of $i < n^*$ and let $v = v_{i,j}$. Moreover, let l and r be the the left and right children of v.

First notice that, conditioned on that the event BAD_v does not happen, hybrid $\mathcal{H}_{i,j}$ is distributed identically to the previous hybrid, i.e. $\Pr[\mathcal{H}_{i,j}(\mathcal{A}) = 1|\neg\mathsf{BAD}_v] = \Pr\left[\mathcal{H}_{i,j}^\downarrow(\mathcal{A}) = 1|\neg\mathsf{BAD}_v\right]$. Therefore

$$\mathbb{SD}[\mathcal{H}_{i,j}, \mathcal{H}_{i,j}^\downarrow] = \Pr[\mathsf{BAD}_v] \cdot \underbrace{\left|\Pr[\mathcal{H}_{i,j}(\mathcal{A}) = 1|\mathsf{BAD}_v] - \Pr\left[\mathcal{H}_{i,j}^\downarrow(\mathcal{A}) = 1|\mathsf{BAD}_v\right]\right|}_{\leq 1}$$

$$\leq \Pr[\mathsf{BAD}_v]$$

It is thus sufficient to bound the probability for the event BAD_v. $\mathcal{A}$ queries the random oracle $\mathsf{H}'_{(\chi,v)}$ with at most q_v distinct queries. Fix a query $\hat{\ell}_v$, and let $\hat{\ell}_l$ and $\hat{\ell}_r$ be the corresponding labels of the children l and r of v. It holds that

$$\delta(\mathsf{L}_l) \geq \gamma(\mathsf{L}_l) - \frac{n^* - (i+1)}{n^*} \cdot \alpha$$

$$\delta(\mathsf{L}_r) \geq \gamma(\mathsf{L}_r) - \frac{n^* - (i+1)}{n^*} \cdot \alpha,$$

as otherwise one of the events BAD_l or BAD_r would have happened and the experiment would have aborted. We can now rewrite

$$\Pr[\mathsf{BAD}_v] = \Pr\left[\delta(\mathsf{L}_v) < \gamma(\mathsf{L}_v) - \frac{n^* - i}{n^*} \cdot \alpha\right]$$

$$= 1 - \Pr\left[\gamma(\mathsf{L}_v) \leq \delta(\mathsf{L}_v) + \frac{n^* - i}{n^*} \cdot \alpha\right]$$

$$< e^{-2\left(\frac{n^*-i}{n^*} \cdot \alpha - \frac{n^*-(i+1)}{n^*} \cdot \alpha\right)^2 t}$$

$$= e^{-2\left(\frac{\alpha}{n^*}\right)^2 t} \qquad\qquad \backslash$$

by Lemma 5. A union-bound over all queries to $\mathsf{H}_{(\chi,v)}$ yields

$$\Pr[\mathsf{BAD}_v] < q_v \cdot e^{-2\left(\frac{\alpha}{n^*}\right)^2 t}.$$

Thus we conclude that the statistical distance between $\mathcal{H}_{i,j}$ and $\mathcal{H}^{\downarrow}_{i,j}$ is at most $q_v \cdot e^{-\left(\frac{\alpha}{n^*}\right)^2 t}$. Consequently, we can bound the statistical distance between the first hybrid $\mathcal{H}_0$ and the last hybrid $\mathcal{H}_{0,1}$ by

$$\mathbb{SD}[\mathcal{H}_0, \mathcal{H}_{0,1}] = \frac{1 + q(q-1)}{2^\lambda} + \sum_{v \in \mathfrak{T}} q_v \cdot e^{-2\left(\frac{\alpha}{n^*}\right)^2 t} = \frac{1 + q(q-1)}{2^\lambda} + q \cdot e^{-2\left(\frac{\alpha}{n^*}\right)^2 t}.$$

We will finally bound the success probability of $\mathcal{A}$ in the last hybrid $\mathcal{H}_{0,1}$. This is in fact identical to the analysis of [5]. Let S denote the set of all inconsistent nodes in the tree output by $\mathcal{A}$ in $\mathcal{H}_{0,1}$. Then by Lemma 2 there exists a path going though all the nodes in $V \setminus \mathfrak{T}_{S^*}$. We distinguish two cases

1. $|\mathfrak{T}_{S^*}| \leq \alpha N$
2. $|\mathfrak{T}_{S^*}| > \alpha N$

For the first case $\mathcal{A}$ must have done at least $(1-\alpha)N$ sequential queries, so we are left with a bound on the second case. By Lemma 3 $\mathfrak{T}_{S^*}$ (and therefore S^*) contains at least $\frac{|S^*| + |\mathfrak{T}_{S^*}|}{2} > \alpha 2^n$ leaves. However, note that in the experiment $\mathcal{H}_{0,1}$ the challenger aborts whenever the adversary satisfies the winning conditions, since

$$\gamma(\mathsf{L}_\epsilon) > \frac{\alpha 2^n}{2^n} = \alpha$$

and therefore

$$\delta(\mathsf{L}_\epsilon) \geq \gamma(\mathsf{L}_\epsilon) - \alpha > 0.$$

Consequently, as $\delta(\mathsf{L}_\epsilon) = \frac{|Z \cap C|}{|Z|}$, this implies that $|Z \cap C| > 0$ and therefore at least one of the paths in Z is also in C and therefore we detect an inconsistent node. This however implies that the proof is always rejected by the verifier. So in the final experiment $\mathcal{H}_{0,1}$ the success-probability of the adversary is exactly 0. This concludes our proof. $\qquad\square$

5 Multi-thread Construction

In this section we show how to improve the concrete efficiency of incremental proofs of sequential work by assuming some parallel capability of the prover. More specifically, we assume that the prover can spawn n parallel threads, where n denotes the depth of the graph CP_n. Note that we can upper bound n by $\lambda = 100$, since we require the prover to be polynomial time.

5.1 Parameters

Throughout the following section we use the same parameters and notation of Sect. 4.1 and we define the following additional subroutines.

- A full-domain hash function $\mathsf{H}'' : \{0,1\}^* \to \{0,1\}^{t(n+2)}$ modelled as a random oracle.
- A sampler $\mathsf{RandomPath}(v; \mathsf{r})$ which takes as input a node v and uniform random coins r, and outputs a set of t uniformly random paths with common prefix v. Since $\log \binom{2^{h_v}}{t} < \log \left(\frac{2^{h_v} \cdot e}{t} \right)^t < t(h_v + 2) \leq t(n + 2)$, random tapes of size $t(n + 2)$ always suffice to sample a uniform set.
- A function $\mathsf{FetchPath}(S_v, U, \{\ell_v : v \in U\})$ which takes as input a set S_v of t paths with common prefix v, a set of $U = \{u : \exists v' \in \mathfrak{T}_v \text{ s.t. } (u, v') \in E\}$ where with edges pointing to $\mathfrak{T}_v$, and the set $\{\ell_v : v \in U\}$ of labels of all nodes in U. The function recomputes the labelling of $\mathfrak{T}_v$ using the labels of nodes in U. The output of the function is the labelling of all paths in S_v. Note that such a function can be computed in time $O(2^{h_v})$ and with memory $O(t \cdot h_v)$.

5.2 Scheme Description

$\underline{\mathsf{Prove}^{\mathsf{H},\mathsf{H}',\mathsf{H}''}(\chi, N)}$:

1. Initialize $U \leftarrow \emptyset$ to be the set of unfinished nodes.
2. Assign $\ell_{0^n} \leftarrow 0^\lambda$.
3. Traverse the graph CP_n starting from 0^n. At every node $v \in V$ which is traversed, do the following:

(a) Compute the label ℓ_v by

$$\ell_v \leftarrow \mathsf{H}_{(\chi,v)}(\ell_{v_1},\ldots,\ell_{v_d})$$

where $v_1,\ldots,v_d \in V$ are all nodes v is adjacent with, *i.e.*, $(v_i,v) \in E$.
(b) Let l and r be the children of v.
(c) If $|\mathsf{leafs}(v)| \leq t$, set $\mathcal{L}_v \leftarrow \{[(v,\ell_l,\ell_r,\bot)\|L]$ where $L \in \mathcal{L}_l \cup \mathcal{L}_r\}$.
(d) Otherwise (*i.e.*, if $|\mathsf{leafs}(v)| \geq 2t$), do the following:
   i. Compute

$$r_u \leftarrow \mathsf{H}'_{(\chi,u)}(\ell_u).$$

   ii. Choose a random t-subset S_v of $[2t]$ via $S_v \leftarrow \mathsf{RandomSubset}(2t,t;r_v)$.
   iii. For $j \in \{0,\ldots,t-1\}$, write $S_v[j] = at + b$ where $a \in \{0,1\}$ and $0 \leq b < t$. Set

$$\mathcal{L}_u[j] := \begin{cases} [(u,\ell_l,\ell_r,j)\|\mathcal{L}_l[b]], & \text{if } a = 0 \\ [(u,\ell_l,\ell_r,j)\|\mathcal{L}_r[b]], & \text{if } a = 1 \end{cases}$$

(e) If v is a left node (*i.e.*, it is the left child of its parent):
   i. Compute

$$r_u \leftarrow \mathsf{H}''_{(\chi,u)}(\ell_u).$$

   ii. Choose a random t-set of paths with prefix v via $S_v \leftarrow \mathsf{RandomPath}(v;r_v)$.
   iii. Execute in a parallel thread $\mathcal{L} \leftarrow \mathsf{FetchPath}(S_v,U,\{\ell_v : v \in U\})$ and set $\mathcal{L}_v := \{[(v,\ell_l,\ell_r,\bot)\|L]$ where $L \in \mathcal{L}\}$.
   iv. Mark l as finished, *i.e.*, remove l from U and mark v as unfinished, *i.e.*, add v to U.
4. Once the set of unfinished nodes consists only of the root-node (*i.e.*, $U = \{\epsilon\}$), terminate and output $\pi \leftarrow (\ell_\epsilon, \mathcal{L}_\epsilon)$.

$\underline{\mathsf{Inc}^{\mathsf{H},\mathsf{H}',\mathsf{H}''}(\chi, N, N', \pi)}$: Defined as in Sect. 4.2.
$\underline{\mathsf{Vf}^{\mathsf{H},\mathsf{H}',\mathsf{H}''}(\chi, N, \pi)}$:

1. Parse π as $(\ell_\epsilon, \mathcal{L}_\epsilon)$.
2. For all paths $\mathsf{path} \in \mathcal{L}_\epsilon$ do the following:
  (a) Parse path as $[(v_0, \ell_{l_0}, \ell_{r_0}, \mathsf{ind}_0)\| \ldots \|(v_n, \ell_{l_n}, \ell_{r_n}, \mathsf{ind}_n)]$.
  (b) For every node $v \in \{v_0,\ldots,v_n\}$ on the path, check if the label ℓ_v was computed correctly. That is, for $v = 0^n$ check whether $\ell_v = 0^\lambda$, and for any other node $v \in V\backslash\{0^n\}$ check whether $\ell_v = \mathsf{H}_{(\chi,v)}(\ell_{v_1},\ldots,\ell_{v_d})$, where $v_1,\ldots,v_d$ are the nodes with edges pointing to v. The value ℓ_v can either be retrieved from the parent node of v, or is directly available for the case of the root-node ϵ. For the special case of leaf-nodes, the values $\ell_{v_1},\ldots,\ell_{v_d}$ are not stored locally with the node v, but are stored at some other (a-priori known) nodes along path (refer to the structure of the graph CP_n).

(c) For all $j \in \{0, \ldots, n^*\}$:

    i. If v_j is a right node or $j = 0$: Compute $r_{v_j} \leftarrow H'_{(\chi, v_j)}(\ell_{v_j})$ and $S_{v_j} \leftarrow$ RandomSubset$(2t, t; r_{v_j})$. If v_{j+1} is the left child of v_j, check if

$$S_u[\text{ind}_j] = \text{ind}_{j+1}.$$

Otherwise, if v_{j+1} is the right child of v_j, check if

$$S_u[\text{ind}_j] = t + \text{ind}_{j+1}.$$

    ii. If v_j is a left node: Compute $r_{v_j} \leftarrow H''_{(\chi, v_j)}(\ell_{v_j})$ and $S_{v_j} \leftarrow$ RandomPath$(v_j; r_{v_j})$. Check if all paths in S_{v_j} are present in $\mathcal{L}_\epsilon$.

3. If all checks pass output 1, otherwise 0.

5.3 Efficiency Analysis

The verifier efficiency is essentially unchanged from the construction in Sect. 4.2.

Prover Efficiency. For the main thread the prover complexity is identical to our construction in Sect. 4.2. For the parallel threads the prover has to recompute a CP_n graph of size at most n, so we can again upper bound their memory complexity to $\lambda(n + 1)$ by Lemma 1.

In the following we argue that the number of parallel threads of our protocol is upper-bounded by n. Recall that a new thread is spawned each time the main thread traverses a left node v (*i.e.*, a node which is the left child of its parent). The complexity of each parallel thread is dominated by the factor $O(2^{h_v})$ of the function FetchPath, where h_v is the height at which the thread was spawned. However, note that the main thread must perform at least $O(2^{h_v})$ steps before spawning a new sub-thread at height h_v. This implies that for each $h_v = 1, \ldots, n$ there can be at most one parallel thread running. It follows that n parallel processors are sufficient to run the prover algorithm.

Proof Size. As for our construction in Sect. 4.2, the proof size is $O(t \cdot \lambda \cdot n)$. Theorem 3 shows that our construction is sound if $t = O(\lambda)$, which gives proofs of size $O(\lambda^2 \cdot n)$. Concretely, our proofs are larger than those of the CP scheme by a factor of roughly 9.

5.4 Soundness

Theorem 3. *The construction given in Sect. 5.2 is sound for any $t \in O(\lambda)$, and the soundness error is given by $\frac{1 + q(q-1)}{2^\lambda} + q \cdot e^{-\frac{2\alpha^2 t}{9}}$.*

Proof. Let χ be the challenge statement and let q_v be the number of calls of $\mathcal{A}$ to the random oracle $H'_{(\chi, v)}$, *i.e.*, the adversary makes at most $q = \sum_{v \in \mathfrak{T}} q_v$ calls to H' in total. Let η be a free (positive) variable to be fixed later.

Consider the following sequence of hybrids.

– Hybrid $\mathcal{H}_0$: This is identical to the real experiment.

- Hybrid $\mathcal{H}_1$: The same as $\mathcal{H}_0$, except for the following modifications.
  - The experiment records a list Q of all H queries made by $\mathcal{A}$ with their responses.
  - Every time $\mathcal{A}$ queries $H'_{(\chi,v)}$ for a $v \in V$ with a label ℓ_v, a labelling L_v for the sub-tree under v is computed via $L_v \leftarrow \text{Extract}(v, \ell_v, Q)$.
  - If it holds for any path opened by $\mathcal{A}$ that the labels on the path are different from the labels in L_ϵ (where ϵ is the root), then $\mathcal{H}_1$ aborts and outputs 0.

Let BAD_v be the following event: $\mathcal{A}$ queries $H'_{(\chi,v)}$ with a query $\hat{\ell}_v$ corresponding to a labeled sub-tree $L_v \leftarrow \text{Extract}(v, \hat{\ell}_v, Q)$ for which it holds that $\delta(L_v) < \gamma(L_v) - \eta$.
For $v \in \{1^{n^*-1}\|0, \ldots, 10, 0\}$ define the following hybrids.

- Hybrid $\mathcal{H}_1^v$: The same as the previous hybrid, except that the experiment outputs 0 if the event BAD_v happens.

Let $\hat{\text{BAD}}_v$ be the following event: $\mathcal{A}$ queries $H'_{(\chi,v)}$ with a query $\hat{\ell}_v$ corresponding to a labeled sub-tree $L_v \leftarrow \text{Extract}(v, \hat{\ell}_v, Q)$ for which it holds that $\delta(L_v) < \gamma(L_v) - \left(3\eta - 2^{n^*-n_v}\eta\right)$, where n_v is the depth of v (i.e., the distance between the root-node ϵ and v).
For $v \in \{1^{n^*}, \ldots, 1, \epsilon\}$ define the following hybrids.

- Hybrid $\mathcal{H}_2^v$: The same as the previous hybrid, except that the experiment outputs 0 if the event $\hat{\text{BAD}}_v$ happens.

We will now show indistinguishability between the hybrids. By Lemma 4 it holds that $\mathcal{H}_0$ and $\mathcal{H}_1$ are indistinguishable. We now turn to the indistinguishability of hybrids $\mathcal{H}_1^v$. For notational convenience, let $\mathcal{H}_1^{v\downarrow}$ be the hybrid before $\mathcal{H}_1^v$.

First consider $v = 1^{n^*-1}\|0$. For each node v at level n it holds that the set Z_v of challenge paths consists of all paths from v to the leaves under v. Consequently, it holds that $\delta(L_v) = \gamma(L_v)$ and therefore BAD_v happens with probability 0.

First notice that, conditioned on that the event BAD_v does not happen, hybrid $\mathcal{H}_1^v$ is distributed identically to the previous hybrid, i.e., $\Pr\left[\mathcal{H}_1^v(\mathcal{A}) = 1 | \neg\text{BAD}_v\right] = \Pr\left[\mathcal{H}_1^{v\downarrow}(\mathcal{A}) = 1 | \neg\text{BAD}_v\right]$. Therefore

$$\mathbb{SD}[\mathcal{H}_1^v, \mathcal{H}_1^{v\downarrow}] = \Pr\left[\text{BAD}_v\right] \cdot \underbrace{\left|\Pr\left[\mathcal{H}_1^v(\mathcal{A}) = 1 | \text{BAD}_v\right] - \Pr\left[\mathcal{H}_1^{v\downarrow}(\mathcal{A}) = 1 | \text{BAD}_v\right]\right|}_{\leq 1}$$

$$\leq \Pr\left[\text{BAD}_v\right]$$

It is thus sufficient to bound the probability for the event BAD_v. $\mathcal{A}$ queries the random oracle $H'_{(\chi,v)}$ with at most q_v distinct queries. Note that v is always a left node and therefore the challenge set Z is chosen uniformly at random for each label. Hence we have that $E[\delta(L_v)] = \gamma(L_v)$, i.e., the fraction of inconsistent paths is preserved in expectation, over the random coins of $H'_{(\chi,v)}$. We can then rewrite

$$
\begin{aligned}
\Pr\left[\mathsf{BAD}_v\right] &= \Pr\left[\delta(\mathsf{L}_v) < \gamma(\mathsf{L}_v) - \eta\right] \\
&= \Pr\left[\delta(\mathsf{L}_v) < \mathsf{E}[\delta(\mathsf{L}_v)] - \eta\right] \\
&< e^{-2\eta^2 t}
\end{aligned}
$$

by Theorem 1. A union-bound over all queries to $H_{(\chi,v)}$ yields

$$
\Pr\left[\mathsf{BAD}_v\right] < q_v \cdot e^{-2\eta^2 t}.
$$

Thus we conclude that the statistical distance between $\mathcal{H}_1^v$ and $\mathcal{H}_1^{v\downarrow}$ is at most $q_v \cdot e^{-\eta^2 t}$. We now turn to the indistinguishability of hybrids $\mathcal{H}_2^v$. Again we use the convention that $\mathcal{H}_2^{v\downarrow}$ denotes the hybrid before $\mathcal{H}_2^v$.

First consider $v = 1^{n^*}$. As argued above, for each node at depth n it holds that $\delta(\mathsf{L}_v) = \gamma(\mathsf{L}_v)$ and therefore $\hat{\mathsf{BAD}}_v$ happens with probability 0. For the rest of the cases, bounding the probability that $\hat{\mathsf{BAD}}_v$ happens suffice, since, if $\hat{\mathsf{BAD}}_v$ does not happen, the hybrids are identical. We bound the probability that $\hat{\mathsf{BAD}}_v$ happens with an inductive argument over $v \in \{1^{n^*}, \ldots, 1, \epsilon\}$. The base case $v = 1^{n^*}$ is settled above.

For any node $v \in \{1^{n^*-1}, \ldots, 1, \epsilon\}$, fix a query $\hat{\ell}_v$ and let l and r be the left and right child of v. Since l is a left node, we have that

$$
\delta(\mathsf{L}_l) \geq \gamma(\mathsf{L}_l) - \eta \tag{6}
$$

as otherwise BAD_l would be triggered. For the right node r we have that

$$
\delta(\mathsf{L}_r) \geq \gamma(\mathsf{L}_r) - \left(3\eta - 2^{n^*-(n_v+1)}\eta\right) \tag{7}
$$

by induction hypothesis, as otherwise $\hat{\mathsf{BAD}}_r$ would be triggered. We can now rewrite

$$
\begin{aligned}
\Pr\left[\hat{\mathsf{BAD}}_v\right] &= \Pr\left[\delta(\mathsf{L}_v) < \gamma(\mathsf{L}_v) - \left(3\eta - 2^{n^*-n_v}\eta\right)\right] \\
&= 1 - \Pr\left[\gamma(\mathsf{L}_v) \leq \delta(\mathsf{L}_v) + \left(3\eta - 2^{n^*-n_v}\eta\right)\right] \\
&< e^{-2\left(\left(3\eta - 2^{n^*-n_v}\eta\right) - \frac{\eta + \left(3\eta - 2^{n^*-(n_v+1)}\eta\right)}{2}\right)^2 t} \\
&= e^{-2\left(3\eta - \frac{(\eta + 3\eta)}{2}\right)^2 t} \\
&= e^{-2\eta^2 t}
\end{aligned}
$$

by (6), (7), and Lemma 5. A union-bound over all queries to $H_{(\chi,v)}$ yields

$$
\Pr\left[\hat{\mathsf{BAD}}_v\right] \leq q_v \cdot e^{-2\eta^2 t}.
$$

This bounds the statistical distance between $\mathcal{H}_2^v$ and $\mathcal{H}_2^{v\downarrow}$ by $q_v \cdot e^{-2\eta^2 t}$.

We are now in the position to bound the statistical distance between the first hybrid $\mathcal{H}_0$ and the last hybrid $\mathcal{H}_2^\epsilon$. Let $\mathfrak{T}_l$ be the set $\{1^{n^*-1}\|0, \ldots, 10, 0\}$ and let $\mathfrak{T}_r$ be the set $\{1^{n^*}, \ldots, 1, \epsilon\}$

$$\mathbb{SD}[\mathcal{H}_0, \mathcal{H}_2^\epsilon] = \frac{1 + q(q-1)}{2^\lambda} + \sum_{v \in \{\mathfrak{T}_l \cup \mathfrak{T}_r\}} q_v \cdot e^{-2\eta^2 t}$$

$$\leq \frac{1 + q(q-1)}{2^\lambda} + q \cdot e^{-2\eta^2 t}.$$

Setting $\eta := \frac{\alpha}{3}$ we obtain

$$\mathbb{SD}[\mathcal{H}_0, \mathcal{H}_2^\epsilon] \leq \frac{1 + q(q-1)}{2^\lambda} + q \cdot e^{-2\frac{\alpha^2 t}{9}}.$$

What is left to be shown is that $\mathcal{A}$ cannot win in $\mathcal{H}_2^\epsilon$. Note that in the latter experiment we have that for all L_ϵ computed via Extract we have that

$$\delta(\mathsf{L}_\epsilon) \geq \gamma(\mathsf{L}_\epsilon) - \left(3\eta - 2^{n^*}\eta\right) \geq \gamma(\mathsf{L}_\epsilon) - 3\eta = \gamma(\mathsf{L}_\epsilon) - \alpha.$$

The same argument as in the proof of Theorem 2 can be used to show that the success probability of $\mathcal{A}$ is exactly 0. □

General Arity Trees. Both schemes presented in this work can be generalized to work over p-ary trees, for any $p \geq 2$. By adjusting the value p, we can achieve slightly better concrete proof sizes and prover efficiency. We refer the reader to Sect. A for an extensive treatment on the matter.

A   General Arity Constructions

The schemes described in Sects. 4.2 and 5.2 can be generalized rather easily to work with p-ary trees for any $p \geq 2$.

A.1   Generalized CP Graphs

We begin by describing the generalized CP graph CP_n^p, and generalizing Lemmas 1, 2, and 3.

Definition 6 (Generalized CP Graphs). *For $n \in \mathbb{N}$, let $N = p^{n+1} - 1$ and $T_{p,n} = (V, E')$ be a complete p-ary tree of depth n. Let $\Sigma := \{0, \ldots, p-1\}$ be an alphabet set of size p. The nodes $V = \Sigma^{\leq n}$ are identified by p-ary strings of length at most n and the empty string ϵ represents the root. The edges $E' = \{(x\|s, x) : s \in \Sigma, x \in \Sigma^i, i < n\}$ are directed from the leaves towards the root.*

The graph $CP_n^p = (V, E)$ is a DAG constructed from $T_{p,n} = (V, E')$ as follows. For any leaf $u \in \Sigma^n$, for any node v which is a left-sibling of a node on the path from u to the root ϵ, an edge (v, u) is appended to E'. Formally, $E := E' \cup E''$ where

$$E'' := \{(v, u) : u \in \Sigma^n, u = a\|r\|a', v = a\|s, r > s \text{ for some } a, a' \in \Sigma^{\leq n}\}.$$

We state and prove the generalizations of Lemmas 1, 2, and 3.

Lemma 6. *The labels of a CP_n^p graph can be computed in topological order using $\lambda((p-1)n+1)$ bits of memory.*

Proof. We prove by induction on n. Let $0, \ldots, p-1$ be the children of ϵ. For $i \in \Sigma = \{0, \ldots, p-1\}$, let $\mathfrak{T}_i$ be the subtree rooted at the i. Note that $\mathfrak{T}_i$ is isomorphic to CP_{n-1}^p. To compute the labels of CP_n^p, we first compute the labels of $\mathfrak{T}_0$. Upon completion, we store only the label of 0, denoted ℓ_0. Next, we compute the labels of $\mathfrak{T}_1$ using ℓ_0. This is possible since all edges start from the node 0. Upon completion, we store the label ℓ_1. Now suppose that for some $i \in \{1, \ldots, p\}$ the labels of $\mathfrak{T}_0, \ldots, \mathfrak{T}_{i-1}$ are computed, and we have stored $\ell_0, \ldots, \ell_{i-1}$. The labels of $\mathfrak{T}_i$ can be computed since all edges start from the nodes $0, \ldots, i-1$. Eventually, we obtain the last label ℓ_{p-1}. Using this with $\ell_0, \ldots, \ell_{p-2}$ stored in the memory, we can compute the label of ϵ.

Since for each $i \in \Sigma$, storing ℓ_i requires λ bits of memory, the memory required for computing the label of CP_n^p equals to that of CP_{n-1}^p plus $\lambda(p-1)$ extra bits. Furthermore, CP_0^p has exactly 1 node and its label can be computed using λ bits of memory. Solving the recursion gives the claimed bound.

Lemma 7. *For all $S \subseteq V$, the subgraph of $CP_n^p = (V, E)$ on vertex set $V \setminus \mathfrak{T}_{S^*}$, has a directed path going through all the $|V| - |\mathfrak{T}_{S^*}|$ nodes.*

Proof. We prove by induction on n. The lemma is trivial for CP_0^p as it contains only 1 node. Now, suppose the lemma is true for CP_{n-1}^p. Consider CP_n^p, and let $0, \ldots, p-1$ be the children of ϵ. For $i \in \Sigma = \{0, \ldots, p-1\}$, let $\mathfrak{T}_i$ be the subtree rooted at the i. Note that $\mathfrak{T}_i$ is isomorphic to CP_{n-1}^p. CP_n^p consists of the root ϵ, the subtrees $\mathfrak{T}_0, \ldots, \mathfrak{T}_{p-1}$, and edges going from i to the leaves of $\mathfrak{T}_j$ for all $i < j$ and $i, j \in \Sigma$.

The lemma is true if $\epsilon \in S^*$, as $|V| - |\mathfrak{T}_{S^*}| = 0$. Otherwise, let $I := S^* \cap \Sigma$ be the subset of children of ϵ which are in S^*. For concreteness, we write $I = \{i_1, \ldots, i_k\}$ for some $k \in \{1, \ldots, p\}$. We apply the lemma to $\mathfrak{T}_i$ for all $i \in \Sigma \setminus I$, so that for each $\mathfrak{T}_i$ there exists a directed path going from the left-most leaf of $\mathfrak{T}_i$, i.e., $i0 \ldots 0$, to i. Since for all $i, j \in \Sigma$ where $i < j$, there exists an edge from i to $j0 \ldots 0$, it means that for each $i' \in I$, there exists a edge $(i'-1, (i'+1)0 \ldots 0)$ which "skips" $\mathfrak{T}_{i'}$. Formally, the following edges exist:

$$(0, 10 \ldots 0), \ldots, (i_1 - 2, (i_1 - 1)0 \ldots 0),$$
$$(i_1 - 1, (i_1 + 1)0 \ldots 0), \ldots, (i_k - 1, (i_k + 1)0 \ldots 0),$$
$$(i_k + 1, (i_k + 2)0 \ldots 0), \ldots, (p-1, p0 \ldots 0).$$

Finally, we note that there also exists an edge (i^*, ϵ) where $i^* := \max_{i \notin I}(i \in \Sigma)$, which completes the path from $0 \ldots 0$ to ϵ, passing through all $|V| - |\mathfrak{T}_{S^*}|$ nodes.

Lemma 8. *For all $S \subset V$, $\mathfrak{T}_{S^*}$ contains $\frac{|\mathfrak{T}_{S^*}| + |S|}{p}$ many leaves.*

Proof. Let $S^* = \{v_1, \ldots, v_k\}$. Since S^* is minimal, it holds that $\mathfrak{T}_{v_i} \cap \mathfrak{T}_{v_j} = \emptyset$ for all $i, j \in \{1, \ldots, k\}$ with $i \neq j$. Therefore we can write

$$|\Sigma^n \cap \mathfrak{T}_{S^*}| = \sum_{i=1}^{k} |\Sigma^n \cap \mathfrak{T}_{v_i}|.$$

As for all $i \in \{1, \ldots, k\}$, $\mathfrak{T}_{v_i}$ is a complete p-ary tree, it has $(|\mathfrak{T}_{v_i}| + 1)/p$ many leaves. Thus,

$$\sum_{i=1}^{k} |\Sigma^n \cap \mathfrak{T}_{v_i}| = \sum_{i=1}^{k} \frac{|\mathfrak{T}_{v_i}| + 1}{p} = \frac{|\mathfrak{T}_{S^*}| + |S|}{p}.$$

A.2   Generalized Single-Thread Construction

The generalized construction is almost identical to the basic one presented in Sect. 4.2, except the graph CP_n is replaced with CP_n^p, and the computation of the labels is changed accordingly.

$\mathsf{Prove}^{\mathsf{H},\mathsf{H}'}(\chi, N):$

1. Initialize $U \leftarrow \emptyset$.
2. Assign $\ell_{0^n} \leftarrow 0^\lambda$.
3. Traverse the graph $CP_n^p = (V, E)$ starting from 0^n. At every node $v \in V$ which is traversed, do the following:
   (a) Compute the label ℓ_v by $\ell_v \leftarrow \mathsf{H}_{(\chi,v)}(\ell_{v_1}, \ldots, \ell_{v_d})$, where $v_1, \ldots, v_d \in V$ are all nodes with edges pointing to v, i.e., $(v_i, v) \in E$.
   (b) Let $c_0, \ldots, c_{p-1}$ be the children of v.
   (c) If $|\mathsf{leafs}(v)| \leq t$, set

   $$\mathcal{L}_v \leftarrow \{[(v, \ell_{c_0}, \ldots, \ell_{c_{p-1}}, \perp) \| L] \text{ where } L \in \mathcal{L}_{c_0} \cup \ldots \cup \mathcal{L}_{c_{p-1}}\}.$$

   (d) Otherwise (i.e., if $|\mathsf{leafs}(v)| \geq pt$), do the following:
      i   Compute $r_v \leftarrow \mathsf{H}'_{(\chi,v)}(\ell_v)$.
      ii Choose a random t-subset S_v of $[pt]$ via $S_v \leftarrow \mathsf{RandomSubset}(pt, t; r_v)$.
      iii For $j \in \{0, \ldots, t - 1\}$, write $S_v[j] = at + b$ where $0 \leq a < p$ and $0 \leq b < t$ and set $\mathcal{L}_v[j] \leftarrow (v, \ell_{c_0}, \ldots, \ell_{c_{p-1}}, j) \| \mathcal{L}_{c_a}[b]$.
   (e) Mark $c_0, \ldots, c_{p-2}$ as finished, i.e., remove $c_0, \ldots, c_{p-2}$ from U and, if v is not the right-most child of its parent, mark v as unfinished, i.e., add v to U.
4. Once the set of unfinished nodes consists only of the root-node (i.e., $U = \{\epsilon\}$), terminate and output $\pi \leftarrow (\ell_\epsilon, \mathcal{L}_\epsilon)$.

$\mathsf{Inc}^{\mathsf{H},\mathsf{H}'}(\chi, N, N', \pi):$

1. Initialize $U := \emptyset$.
2. Parse π as $(\ell_\epsilon, \mathcal{L}_\epsilon)$
3. Assign $\ell_{0^{n'-n}} := \ell_\epsilon$ and $\mathcal{L}_{0^{n'-n}} := \mathcal{L}_\epsilon$.

4. Execute the algorithm $\mathsf{Prove}^{\mathsf{H},\mathsf{H}'}(\chi, N')$ starting from step 3 with a slight change: Traverse the graph $CP^p_{n'}$ starting from $0^{n'-n-1}\|1\|0^n$ (instead of from $0^{n'}$).

$\underline{\mathsf{Vf}^{\mathsf{H},\mathsf{H}'}(\chi, N, \pi)}$:

1. Parse $\pi = (\ell_\epsilon, \mathcal{L}_\epsilon)$.
2. For all paths $\mathsf{path} \in \mathcal{L}_\epsilon$ do the following:
   (a) Parse path as $[(v_0, \ell_{c_{0,0}}, \ldots, \ell_{c_{0,p-1}}, \mathsf{ind}_0), \ldots, (v_n, \ell_{c_{n,0}}, \ldots, \ell_{c_{n,p-1}}, \mathsf{ind}_n)]$.
   (b) For every node $v \in \{v_0, \ldots, v_n\}$ on the path, check if the label ℓ_v was computed correctly. That is, for $v = 0^n$ check whether $\ell_v = 0^\lambda$, and for any other node $v \in V\backslash\{0^n\}$ check whether $\ell_v = \mathsf{H}_{(\chi,v)}(\ell_{v_1}, \ldots, \ell_{v_d})$, where $\ell_{v_1}, \ldots, \ell_{v_d}$ are the nodes with edges pointing to v. The value ℓ_v can either be retrieved from the parent node of v, or is directly available for the case of the root-node ϵ. For the special case of leaf-nodes, the values $\ell_{v_1}, \ldots, \ell_{v_d}$ are not stored locally with the node v, but are stored at some other (a-priori known) nodes along the path path (refer to the structure of the graph CP^p_n).
   (c) For all $j \in \{0, \ldots; n^*\}$, compute $\mathsf{r}_{v_j} \leftarrow \mathsf{H}'_{(\chi,v_j)}(\ell_{v_j})$ and $S_{v_j} \leftarrow \mathsf{RandomSubset}(pt, t; \mathsf{r}_{v_j})$. Let $i \in \{0, \ldots, p-1\}$ so that v_{j+1} is the i-th child of v_j. Check if $S_v[\mathsf{ind}_j] = i \cdot t + \mathsf{ind}_{j+1}$.
3. If all checks pass then output 1. Otherwise output 0.

We state the soundness error and the efficiency of the generalized construction. The analysis is essentially identical to that in Sect. 4.3 and is therefore omitted.

Soundness. Here we state a generalized version of Lemma 5 for p-ary trees.

Lemma 9. *Let v be a node and let $(v_1, \ldots, v_p)$ the set of children of v. If for all $i \in \{1, \ldots, p\}$ we have*

$$\delta(\mathsf{L}_{v_i}) \geq \gamma(\mathsf{L}_{v_i}) - \eta_{v_i}$$

then it holds that

$$\mathsf{Pr}\left[\gamma(\mathsf{L}_v) \leq \delta(\mathsf{L}_v) + \eta_v\right] \geq 1 - e^{-2\left(\eta_v - \frac{\sum_{i\in p} \eta_{v_i}}{p}\right)^2 t}.$$

The bound for the soundness error has the same form as that in the basic construction, except that $n = \log_p(N+1) - 1$. Previously, $n = \log(N+1) - 1$. The proof is identical to that of Theorem 2, except that we apply Lemma 9 instead of Lemma 5.

Theorem 4. *The construction given in Sect. A.2 is sound for any $t \in O(\lambda \cdot n^2)$, and the soundness error is given by $\frac{1+q(q-1)}{2^\lambda} + q \cdot e^{-2(\frac{\alpha}{n})^2 t}$.*

Efficiency. In the following, we set $t = O(\lambda \cdot n^2)$ and $n = \log_p N$. The parallel time complexity of the prover remains unchanged at $O(N)$. The parallel time complexity of the verifier is $O(\log(\frac{1}{\log^3 p} \cdot \lambda \cdot \log^3 N))$, which decreases at p increases.

The proof size and the space complexity of the prover are $O(\frac{p}{\log^3 p} \cdot \lambda^2 \cdot \log^3 N)$ and $O(\frac{p^2}{\log^4 p} \cdot \lambda^2 \cdot \log^4 N)$ respectively. The fractions $\phi_p := \frac{p}{\log^3 p}$ and $\theta_p := \frac{p^2}{\log^4 p}$ are minimized at $p = 20$ and $p = 7$ respectively. Compared to $p = 2$, we have $\phi_{20}/\phi_2 \approx 0.124$ and $\theta_7/\theta_2 \approx 0.197$.

A.3   Generalized Multi-thread Construction

Similar to the above, we present a generalization of the construction in Sect. 5.2.
$\underline{\mathsf{Prove}^{\mathsf{H},\mathsf{H}',\mathsf{H}''}(\chi, N)}$:

1. Initialize $U \leftarrow \emptyset$ to be the set of unfinished nodes.
2. Assign $\ell_{0^n} \leftarrow 0^\lambda$.
3. Traverse the graph CP_n^p starting from 0^n. At every node $v \in V$ which is traversed, do the following:
   (a) Compute the label ℓ_v by $\ell_v \leftarrow \mathsf{H}_{(\chi,v)}(\ell_{v_1}, \ldots, \ell_{v_d})$, where $v_1, \ldots, v_d \in V$ are all nodes nodes v is adjacent with, $i.e.$, $(v_i, v) \in E$.
   (b) Let $c_0, \ldots, c_{p-1}$ be the children of v.
   (c) If $\|\mathsf{leafs}(v)\| \le t$, set

   $$\mathcal{L}_v \leftarrow \{[(v, \ell_{c_0}, \ldots, \ell_{c_{p-1}}, \bot)\|L] \text{ where } L \in \mathcal{L}_{c_0} \cup \ldots \cup \mathcal{L}_{c_{p-1}}\}.$$

   (d) Otherwise ($i.e.$, if $\|\mathsf{leafs}(v)\| \ge pt$), do the following:
      i. Compute $r_v \leftarrow \mathsf{H}'_{(\chi,v)}(\ell_v)$.
      ii. Choose a random t-subset S_v of $[pt]$ via $S_v \leftarrow \mathsf{RandomSubset}(pt, t; r_v)$.
      iii. For $j \in \{0, \ldots, t-1\}$, write $S_v[j] = at + b$ where $0 \le a < p$ and $0 \le b < t$. Set $\mathcal{L}_v[j] := [(v, \ell_l, \ell_r, j)\|\mathcal{L}_{c_a}[b]]$.
   (e) If v is not a right node ($i.e.$, it is not the right-most child of its parent):
      i. Compute $r_v \leftarrow \mathsf{H}''_{(\chi,v)}(\ell_v)$.
      ii. Choose a random t-set of paths with prefix v via $S_v \leftarrow \mathsf{RandomPath}(v; r_v)$.
      iii. Execute in a parallel thread $\mathcal{L} \leftarrow \mathsf{FetchPath}(S_v, U, \{\ell_v : v \in U\})$ and set $\mathcal{L}_v := \{[(v, \ell_l, \ell_r, \bot)\|L] \text{ where } L \in \mathcal{L}\}$.
      iv. Mark $c_0, \ldots, c_{p-2}$ as finished, $i.e.$, remove $c_0, \ldots, c_{p-2}$ from U and mark v as unfinished, $i.e.$, add v to U.
4. Once the set of unfinished nodes consists only of the root-node ($i.e.$, $U = \{\epsilon\}$), terminate and output $\pi \leftarrow (\ell_\epsilon, \mathcal{L}_\epsilon)$.

$\underline{\mathsf{Inc}^{\mathsf{H},\mathsf{H}',\mathsf{H}''}(\chi, N, N', \pi)}$: Defined as in Sect. A.2.
$\underline{\mathsf{Vf}^{\mathsf{H},\mathsf{H}',\mathsf{H}''}(\chi, N, \pi)}$:

1. Parse π as $(\ell_\epsilon, \mathcal{L}_\epsilon)$.
2. For all paths $\mathsf{path} \in \mathcal{L}_\epsilon$ do the following:
   (a) Parse path as $[(v_0, \ell_{c_{0,0}}, \ldots, \ell_{c_{0,p-1}}, \mathsf{ind}_0)\| \ldots \|(v_n, \ell_{c_{n,0}}, \ldots, \ell_{c_{n,p-1}}, \mathsf{ind}_n)]$.

(b) For every node $v \in \{v_0, \ldots, v_n\}$ on the path, check if the label ℓ_v was computed correctly. That is, for $v = 0^n$ check whether $\ell_v = 0^\lambda$, and for any other node $v \in V \setminus \{0^n\}$ check whether $\ell_v = H_{(\chi,v)}(\ell_{v_1}, \ldots, \ell_{v_d})$, where $v_1, \ldots, v_d$ are the nodes with edges pointing to v. The value ℓ_v can either be retrieved from the parent node of v, or is directly available for the case of the root-node ϵ. For the special case of leaf-nodes, the values $\ell_{v_1}, \ldots, \ell_{v_d}$ are not stored locally with the node v, but are stored at some other (a-priori known) nodes along path (refer to the structure of the graph CP_n^p).

(c) For all $j \in \{0, \ldots, n^*\}$:

    i. If v_j is the right-most child of its parent or $j = 0$: Compute $r_{v_j} \leftarrow H'_{(\chi,v_j)}(\ell_{v_j})$ and $S_{v_j} \leftarrow \mathsf{RandomSubset}(pt, t; r_{v_j})$. Let v_{j+1} be the i-th child of v_j, check if $S_v[\mathsf{ind}_j] = i \cdot t + \mathsf{ind}_{j+1}$.

    ii. If v_j is not the right-most child of its parent: Compute $r_{v_j} \leftarrow H''_{(\chi,v_j)}(\ell_{v_j})$ and $S_{v_j} \leftarrow \mathsf{RandomPath}(v_j; r_{v_j})$. Check if all paths in S_{v_j} are present in $\mathcal{L}_\epsilon$.

3. If all checks pass output 1, otherwise 0.

Next we state the soundness error and the efficiency.

Soundness. The soundness analysis requires some tweaking of the argument.

Theorem 5. *The construction given in Sect. A.3 is sound for any $t \in O((1 + \frac{p}{p-1})^2 \cdot \lambda)$, and the soundness error is given by $\frac{1+q(q-1)}{2^\lambda} + q \cdot e^{-\left(\frac{\alpha}{1 + \frac{p}{p-1}}\right)^2 t}$.*

Proof. The proof follows the blueprint of the proof of Theorem 3, except for the following changes. First we add a hybrid $\mathcal{H}_1^v$ for each sibling of the nodes $\{1^{n^*}, \ldots, 1, \epsilon\}$. The indistinguishability arguments are identical.

Then we define the event $\hat{\mathsf{BAD}}_v$ as follows: $\mathcal{A}$ queries $H'_{(\chi,v)}$ with a query $\hat{\ell}_v$ corresponding to a labeled sub-tree $L_v \leftarrow \mathsf{Extract}(v, \hat{\ell}_v, Q)$ for which it holds that $\delta(L_v) < \gamma(L_v) - \left(2\eta + \eta \sum_{i=1}^{n^*-n_v} \frac{1}{p^i}\right)$, where n_v is the depth of v.

We bound the probability that $\hat{\mathsf{BAD}}_v$ happens with an inductive argument over $v \in \{1^{n^*}, \ldots, 1, \epsilon\}$. For the base case $v = 1^{n^*}$ is enough to observe that $\delta(L_v) = \gamma(L_v)$ and therefore $\hat{\mathsf{BAD}}_v$ happens with probability 0.

For any node $v \in \{1^{n^*-1}, \ldots, 1, \epsilon\}$, fix a query $\hat{\ell}_v$ and let $(v_1, \ldots, v_p)$ be the children of v. For all $i \in \{1, \ldots, p-1\}$ we have that

$$\delta(L_{v_i}) \geq \gamma(L_{v_i}) - \eta \tag{8}$$

as otherwise BAD_{v_i} would be triggered. For the node v_p we have that

$$\delta(L_{v_p}) \geq \gamma(L_{v_p}) - \left(2\eta + \eta \sum_{i=1}^{n^*-n_v-1} \frac{1}{p^i}\right) \tag{9}$$

by induction hypothesis, as otherwise $\hat{\mathsf{BAD}}_{v_p}$ would be triggered. We can now rewrite

$$\Pr\left[\hat{\mathsf{BAD}}_v\right] = \Pr\left[\delta(\mathsf{L}_v) < \gamma(\mathsf{L}_v) - \left(2\eta + \eta \sum_{i=1}^{n^*-n_v} \frac{1}{p^i}\right)\right]$$

$$= 1 - \Pr\left[\gamma(\mathsf{L}_v) \le \delta(\mathsf{L}_v) + \left(2\eta + \eta \sum_{i=1}^{n^*-n_v} \frac{1}{p^i}\right)\right]$$

$$< e^{-2\left(\left(2\eta+\eta\sum_{i=1}^{n^*-n_v}\frac{1}{p^i}\right)-\frac{\eta(p-1)+\left(2\eta+\eta\sum_{i=1}^{n^*-n_v-1}\frac{1}{p^i}\right)}{p}\right)^2 t}$$

$$= e^{-2\eta^2 t}$$

by (8), (9), and Lemma 9. For $p > 1$ we can bound

$$2\eta + \eta\sum_{i=1}^{n^*}\frac{1}{p^i} = \eta + \eta\sum_{i=0}^{n^*}\frac{1}{p^i} \le \left(1 + \frac{p}{p-1}\right)\eta.$$

since it is a geometric series. Thus we can set $\eta := \frac{\alpha}{\left(1+\frac{p}{p-1}\right)}$ and derive

$$\mathbb{SD}[\mathcal{H}_0, \mathcal{H}_2^\epsilon] \le \frac{1 + q(q-1)}{2^\lambda} + q \cdot e^{-\frac{2\alpha^2 t}{\left(1+\frac{p}{p-1}\right)^2}}.$$

The remainder of the analysis is unchanged. $\square$

Efficiency. In the following, we set $t = O\left(\left(1 + \frac{p}{p-1}\right)^2 \cdot \lambda\right)$ and $n = \log_p N$. The parallel time complexity of the prover remains unchanged at $O(N)$. The number of parallel threads is bounded by $O(p \log_p N)$, which is minimized at $p = 3$. The parallel time complexity of the verifier is $O(\log(\frac{(1+\frac{p}{p-1})^2}{\log p} \cdot \lambda \cdot \log N))$, which decreases at p increases. The proof size and the space complexity of the prover are $O(\frac{p(1+\frac{p}{p-1})^2}{\log p} \cdot \lambda^2 \cdot \log N)$ and $O(\frac{p^2(1+\frac{p}{p-1})^2}{\log^2 p} \cdot \lambda^2 \cdot \log^2 N)$ respectively. The fractions $\phi'_p := \frac{p(1+\frac{p}{p-1})^2}{\log p}$ and $\theta'_p := \frac{p^2(1+\frac{p}{p-1})^2}{\log^2 p}$ are both minimized at $p = 4$. Compared to $p = 2$, we have $\phi'_4/\phi'_2 \approx 0.605$ and $\theta'_7/\theta'_2 \approx 0.605$.

References

1. Arora, S., Safra, S.: Probabilistic checking of proofs: a new characterization of NP. J. ACM (JACM) **45**(1), 70–122 (1998)
2. Bitansky, N., Canetti, R., Chiesa, A., Tromer, E.: Recursive composition and bootstrapping for SNARKS and proof-carrying data. In: Boneh, D., Roughgarden, T., Feigenbaum, J. (eds.) 45th ACM STOC, Palo Alto, CA, USA, 1–4 June, pp. 111–120. ACM Press (2013)

3. Bitansky, N., Goldwasser, S., Jain, A., Paneth, O., Vaikuntanathan, V., Waters, B.: Time-lock puzzles from randomized encodings. In: Sudan, M. (ed.) ITCS 2016, Cambridge, MA, USA, 14–16 January, pp. 345–356. ACM (2016)

4. Boneh, D., Bonneau, J., Bünz, B., Fisch, B.: Verifiable delay functions. In: Shacham, H., Boldyreva, A. (eds.) CRYPTO 2018, Part I. LNCS, vol. 10991, pp. 757–788. Springer, Cham (2018). https://doi.org/10.1007/978-3-319-96884-1_25

5. Cohen, B., Pietrzak, K.: Simple proofs of sequential work. In: Nielsen, J.B., Rijmen, V. (eds.) EUROCRYPT 2018, Part II. LNCS, vol. 10821, pp. 451–467. Springer, Cham (2018). https://doi.org/10.1007/978-3-319-78375-8_15

6. Dwork, C., Naor, M.: Pricing via processing or combatting junk mail. In: Brickell, E.F. (ed.) CRYPTO 1992. LNCS, vol. 740, pp. 139–147. Springer, Heidelberg (1993). https://doi.org/10.1007/3-540-48071-4_10

7. Fiat, A., Shamir, A.: How to prove yourself: practical solutions to identification and signature problems. In: Odlyzko, A.M. (ed.) CRYPTO 1986. LNCS, vol. 263, pp. 186–194. Springer, Heidelberg (1987). https://doi.org/10.1007/3-540-47721-7_12

8. Kilian, J.: A note on efficient zero-knowledge proofs and arguments (extended abstract). In: 24th ACM STOC, Victoria, British Columbia, Canada, 4–6 May, pp. 723–732. ACM Press (1992)

9. Mahmoody, M., Moran, T., Vadhan, S.P.: Time-lock puzzles in the random oracle model. In: Rogaway, P. (ed.) CRYPTO 2011. LNCS, vol. 6841, pp. 39–50. Springer, Heidelberg (2011). https://doi.org/10.1007/978-3-642-22792-9_3

10. Mahmoody, M., Moran, T., Vadhan, S.P.: Publicly verifiable proofs of sequential work. In: Kleinberg, R.D. (ed.) ITCS 2013, Berkeley, CA, USA, 9–12 January, pp. 373–388. ACM (2013)

11. Micali, S.: CS proofs (extended abstracts). In: 35th FOCS, Santa Fe, New Mexico, 20–22 November, pp. 436–453. IEEE Computer Society Press (1994)

12. Pietrzak, K.: Simple verifiable delay functions. Cryptology ePrint Archive, Report 2018/627 (2018). https://eprint.iacr.org/2018/627

13. Rivest, R.L., Shamir, A., Wagner, D.A.: Time-lock puzzles and timed-release crypto (1996)

14. Valiant, P.: Incrementally verifiable computation or proofs of knowledge imply time/space efficiency. In: Canetti, R. (ed.) TCC 2008. LNCS, vol. 4948, pp. 1–18. Springer, Heidelberg (2008). https://doi.org/10.1007/978-3-540-78524-8_1

15. Wesolowski, B.: Efficient verifiable delay functions. Cryptology ePrint Archive, Report 2018/623 (2018). https://eprint.iacr.org/2018/623

Tight Proofs of Space and Replication

Ben Fisch$^{(\boxtimes)}$

Stanford University, Stanford, USA
benafisch@gmail.com

Abstract. We construct a concretely practical *proof-of-space* (PoS) with arbitrarily *tight* security based on stacked depth robust graphs and constant-degree expander graphs. A *proof-of-space* (PoS) is an interactive proof system where a prover demonstrates that it is persistently using space to store information. A PoS is arbitrarily tight if the honest prover uses exactly N space and for any $\epsilon > 0$ the construction can be tuned such that no adversary can pass verification using less than $(1 - \epsilon)N$ space. Most notably, the degree of the graphs in our construction are independent of ϵ, and the number of layers is only $O(\log(1/\epsilon))$. The proof size is $O(d/\epsilon)$. The degree d depends on the depth robust graphs, which are only required to maintain $\Omega(N)$ depth in subgraphs on 80% of the nodes. Our tight PoS is also secure against parallel attacks.

Tight proofs of space are necessary for *proof-of-replication* (PoRep), which is a publicly verifiable proof that the prover is dedicating unique resources to storing one or more retrievable replicas of a specified file. Our main PoS construction can be used as a PoRep, but data extraction is as inefficient as replica generation. We present a second variant of our construction called ZigZag PoRep that has fast/parallelizable data extraction compared to replica generation and maintains the same space tightness while only increasing the number of levels by roughly a factor two.

1 Introduction

Proof-of-space (PoS) has been proposed as an alternative to proof-of-work (PoW) for applications such as SPAM prevention, DOS attacks, and Sybil resistance in blockchain-based consensus mechanisms [8, 11, 16]. Several industry projects[1] are underway to deploy cryptocurrencies similar to Bitcoin that use proof-of-space instead of proof-of-work. Proof-of-space is promoted as more egalitarian and eco-friendly that proof-of-work because it is ASIC-resistant and does not consume its resource (space instead of energy), but rather reuses it.

A PoS is an interactive protocol between a prover and verifier in which the prover uses a minimum specified amount of space in order to pass verification. The protocol must have compact communication relative to the prover's space requirements and efficient verification. A PoS is *persistent* if repeated audits force the prover to utilize this space over a period of time. More precisely, there is an

---

[1] https://chia.net/, https://spacemesh.io/, https://filecoin.io/.

© International Association for Cryptologic Research 2019
Y. Ishai and V. Rijmen (Eds.): EUROCRYPT 2019, LNCS 11477, pp. 324–348, 2019.
https://doi.org/10.1007/978-3-030-17656-3_12

"offline" phase in which the prover obtains challenges from a verifier, generates a (long) string σ that it stores, and outputs a compact verification tag τ to the verifier. (The offline phase can be made non-interactive using the Fiat-Shamir transform). This is followed by an "online" challenge-response protocol in which the verifier uses τ to generate challenges and the prover uses σ to efficiently compute responses to the verifier's challenges.

The soundness of the PoS relies on a time bound on the online prover that is enforced by frequent verifier audits. A time bound is necessary as otherwise the prover could store its compact transcript and simulate the setup to re-derive the advice whenever it needs to pass an online proof. If the PoS guarantees that an adversary must use the minimum amount of space to pass challenges within the wall-clock time allotted no matter how much (polynomially bounded) computation it expends then the PoS is said to resist parallelization attacks. A PoS must resist parallelization attacks in order to be considered unconditionally secure. Otherwise, the security may still be reasoned through a cost benefit analysis for a *rational* prover, who will not expend significant computation to save a relatively small fraction of space.

More formally, correctness and (S, T, μ)-soundness for a PoS protocol is defined as follows. First, if the prover commits to persistently utilize N blocks of space then the honest prover algorithm defined by the protocol must use $O(N)$ persistent space and must succeed in passing the verifier's challenges without error. Next, a pair of offline/online adversaries is considered. The "offline" adversary generates an adversarial string σ' and offline proof π'. An (S, T, μ)-sound protocol guarantees that if the string length of σ' output by the online adversary is less than S then either the verifier accepts π' with negligible probability or otherwise any online adversary who runs in time less than T on the input σ' and the verifier's challenge will fail verification with probability at least $1 - \mu$.

There is generally a gap between the honest space utilization and the lower bound S on the adversary's space. If the honest prover uses S' space and some adversary is able to use $(1 - \epsilon)S'$ space then this PoS protocol has at least an ϵ *space gap*. Loosely speaking, a *tight PoS* construction makes ϵ arbitrarily small. The construction is allowed to involve ϵ as a parameter, and the value of ϵ may impact efficiency. All else equal, a tighter PoS is obviously more desirable as it has tighter provable security. Nearly all existing PoS constructions have enormous space gaps, including those that are currently being used in practice [2,8]. The one exception is a recent PoS protocol by Pietrzak [18]. Although this PoS construction is provably tight, the concrete parameters required in the analysis result in an impractically large offline proof.

Proof-of-replication (PoRep) [1,9,10,18] is a recently proposed variant of PoS. A PoRep demonstrates that the prover is dedicating unique resources to storing a retrievable copy of a committed data file, and is therefore a *useful proof of space*. It has been proposed as an alternative Sybil resistance mechanism (e.g. for a blockchain) that is not only ASIC resistant and eco-friendly, but also has a useful side-effect: file storage. Furthermore, since the prover may run several independent PoReps for the same file that each require unique resources, PoReps may be used as a publicly verifiable proof of data replication/duplication.

Unfortunately, it is not possible to cryptographically guarantee that a prover is persistently storing data in a replicated format. A prover can always sabotage the format (e.g. by encrypting it and storing the key separately) and can then recover the original format quickly when challenged. A recently proposed security model for PoReps is ϵ-*rational replication* [9], which says that an adversary can save at most an ϵ fraction of its space by deviating from storing the data in a replicated format. A PoRep that satisfies ϵ-rational replication is also a PoS with an ϵ space gap. Intuitively, if a PoRep is not a tight proof of space then there may exist some adversary that would be rationally incentivized to deviate from honest behavior in a way that also destroys the replication format. In fact, if the input file is incompressible then any adversary who manages to saves an ϵ fraction of the claimed space cannot be storing the data in the replicated format. Thus, tight proofs of space are necessary for PoReps because a PoRep construction is only meaningfully secure when ϵ is very small.

The goal of this work is to construct a practical and provably tight PoS that can also be used as a PoRep that satisfies ϵ-rational replication for arbitrarily small ϵ.

1.1   Related Work

The original PoS of Dziembowski et al. [8] was based on hard to pebble directed acyclic graphs (DAGs), using a blend of techniques from superconcentrators, random bipartite expander graphs and depth robust graphs [17]. During the offline initialization the prover computes a *labeling* of the graph using a collision-resistant hash function where the label e_v on each node $v \in \mathcal{G}$ of the graph is the output of the hash function on the labels of all parent nodes of v. It outputs a commitment to this labeling along with a proof that the committed labeling was "mostly" correct. This offline proof consists of randomly sampled labels and their parent labels, which the verifier checks for consistency. During the online challenge-response phase the verifier simply asks for random labels that the prover must produce along with a standard proof that these labels are consistent with the commitment. The construction leaves a space gap of at least $1 - \frac{1}{512}$.

Ren and Devadas [19] construct a PoS from stacked bipartite expander graphs that dramatically improved on the space gap, although it is not secure against parallel attacks. Their construction involves λ levels $V_1, ..., V_\lambda$ consisting of n nodes each, with edges between the layers defined by the edges of a constant-degree bipartite expander. The prover computes a labeling of the graph just as in the Dziembowski et al. PoS, however it only stores the labels on the final level. Their construction still leaves a space gap of at least[2] $1/2$.

---

[2] For practical parameters, the Ren and Devadas construction has a space gap larger than $1/2$. For example, it requires a graph of degree at least 40 in order to achieve a space gap of less than $2/3$.

Recently, Abusalah et al. [2] revived the simple PoS approach based on storing tables of random functions. The basic idea is for the prover to compute and store the function table of a random function $f : [N] \rightarrow [N]$ where f is chosen by the verifier or a random public challenge. During the online challenge-response the verifier asks the prover to invert f on a randomly sampled point $x \in [n]$. Intuitively, a prover who has not stored most of the function table will likely have to brute force $f^{-1}(x)$, performing $\Omega(N)$ work. This simple approach fails to be a PoS due to Hellman's time/space tradeoffs, which enable a prover to succeed with S space and T computation for any $ST = O(N)$. However, Abusalah et al. build on this approach to achieve a provable time/space tradeoff of $S^k T = \Omega(\epsilon^k N^k)$. This PoS is not secure against parallel attacks, and also has a very large (even asymptotic) space gap.

Pietrzak [18] and Fisch et al. [9,10] independently proposed simpler variants of the graph labeling PoS by Dziembowski et al. based solely on pebbling a depth robust graph (DRG). A degree d DAG on n nodes is (α, β)-depth robust if any subgraph on αn nodes contains a path of at least length βn. It is trivial to construct DRGs of large degree (a complete DAG is depth robust), but much harder to construct DRGs with small degree. Achieving constant α, β is only possible asymptotically with degree $\Omega(\log N)$. The graph labeling PoS on a DRG results in a PoS with an α space gap that is also secure against parallel attacks. Fisch et al. also suggested combining this labeling PoS with a *verifiable delay function* (VDF) [5] to increase the expense of labeling the graph without increasing the size of the proof verification complexity. The delay on the VDF can be tuned depending on the value of n. Both of these constructions were proposed for PoReps. In this variant of the PoS protocol, the prover uses the labeling of the graph to encode a data file on n blocks $D = d_1, ..., d_n$. The ith label e_i is computed by first deriving a key k_i by hashing the labels on the parents of the ith node, and then setting $e_i = k_i \oplus d_i$. If all the labels are stored then any data block can be quickly extracted from e_i by recomputing k_i. More generally, this *DAG encoding* of the data input could use any encoding scheme (enc, dec), where enc is sequentially slow and dec is fast, in order to derive $e_i = \mathsf{enc}(k_i, d_i)$. The data is decoded by computing $d_i = \mathsf{dec}(k_i, e_i)$.

The labeling PoS on a DRG is not technically a tight PoS because decreasing α also decreases the time bound βn on the prover's required computation to defeat the PoS. Moreover, while there exist constructions of (α, β)-DRGs for arbitrarily small α, these constructions have concretely very high degrees and are thus not useful for building a practical PoS. Pietrzak [18] improved on the basic construction by relying on a stronger property of special DRGs [4,17] that have degree $\Omega((\log n)/\epsilon)$ and are (α, β)-depth robust for all (α, β) such that $1 - \alpha + \beta \geq 1 - \epsilon$. This DRG can be constructed for any value of $\epsilon < 1$. In Pietrzak's PoS, the prover builds a DRG on $4n$ nodes and only stores the labels on the topologically last n nodes. This can similarly be used as a PoRep where the data is encoded only on the last level and the labels on previous levels are just used as keys. This PoRep has a slow data extraction time because extracting the data requires recomputing most of the keys from scratch, which is as expensive as the PoS initialization.

Pietrzak shows that a prover who deletes an ϵ' fraction of the labels on the last n nodes will not be able to re-derive them in fewer than n sequential steps. The value ϵ' can be made arbitrarily small, but at the expense of increasing the degree of the graph proportionally to $1/\epsilon'$. The resulting proof has asymptotic size $\Omega((\log N)/\epsilon^2)$. Moreover, although these special DRGs achieve asymptotic efficiency, their current analysis requires the graphs to have impractically large degrees. According to the analysis in [4], achieving just a $1/2$ space gap would require instantiating these graphs with degree at least $2,760 \log N$. The proof size is proportional to the graph degree, so to achieve the space gap $\epsilon = 1/2$ with soundness $\mu = 2^{-10}$ and $N = 2^{30}$ the proof size would be at least 26 MB.

Boneh et al. [5] describe a simple PoRep (also a PoS) just based on storing the output of a *verifiable delay function* (VDF) on N randomly sampled points, which generalizes an earlier proposal by Sergio Demian Lerner [13]. This is in fact an arbitrarily tight PoS with very practical proof sizes (essentially optimal). However, the time complexity of initializing the prover's $O(N)$ storage is $O(N^2)$, and therefore is not practically feasible for large N. This construction is similar to the PoS based on storing function tables [2], but uses the VDF as a moderately hard (non-parallelizable) function on a much larger domain (exponential in the security parameter) and stores a random subset of its function table. The reason for the large initialization complexity is that the prover cannot amortize its cost of evaluating the VDF on the entire subset of points.

1.2 Summary of Contributions

We construct a new tight PoS based on graph labeling with asymptotic proof size $O(\log N/\epsilon)$ where ϵ is the achieved space gap. We can instantiate this construction with relatively weak[3] depth robust graphs that do not require any special properties other than retaining $\Omega(N)$ depth in subgraphs on some constant fraction of the nodes bounded away from 1 (e.g. our concrete analysis assumes 80%).

PoS from Stacked DRGs. Our basic approach is a combination of the stacked bipartite expanders of Ren and Devadas [19] with depth robust graphs. Instead of stacking λ path graphs we stack $O(\log(1/\epsilon))$ levels of fixed-degree DRGs where ϵ is a construction parameter. We refer to this graph construction as **Stacked DRGs**. We are able to show that this results in a PoS that has only an ϵ space gap. Intuitively, the expander edges between layers amplify the dependence of nodes on the last layer and nodes on earlier layers so that deletion of a small ϵ fraction of node labels on the last level will require re-derivation of nearly all the node labels on the first several layers. Thus, since every layer is a DRG, recomputing the missing ϵ fraction of labels requires $\Omega(N)$ sequential computation. It is easy

---

[3] There is experimental evidence that a simple DRG construction with concretely small constant degree (even degree 2) has this property on a graph of size $N = 2^{20}$ [3].

to see that this would be the case if the prover were only storing $(1-\epsilon)n$ labels on the last level and none of the labels on earlier levels, however the analysis becomes much more difficult when the prover is allowed to store any arbitrary $(1-\epsilon)n$ labels. This analysis is the main technical contribution of this work. Concretely, we analyze the construction with an $(n, 0.80n, \Omega(n))$ DRG, i.e. deletion of 20% of nodes leaves a high depth graph on the 80% remaining nodes, regardless of the value of ϵ.

Our construction is efficient compared to prior constructions of tight PoS primarily because we can keep the degree of the graphs fixed for arbitrary ϵ while keeping the number of levels proportional to $\log(1/\epsilon)$. In a graph labeling PoS, the offline PoS proofs sample $O(1/\epsilon)$ labels along with their parent labels, which the verifier checks for consistency. Thus, any construction based on this approach that requires scaling the degree of graphs by $1/\epsilon$ also scales the proof size by $1/\epsilon$, resulting in a proof complexity of at least $O(1/\epsilon^2)$. In our stacked DRG PoS construction the offline proof must include queries from each level to prove that each level of computed labels are "mostly" correct. If done naively, $O(1/\epsilon)$ challenge labels are sampled from each level, resulting in a proof complexity $O(d/\epsilon \cdot \log(1/\epsilon))$ where d is the degree of the level graphs. This is already an improvement, however with a more delicate analysis we are able to go even further and show that the total number of queries over all layers can be kept at $O(1/\epsilon)$, achieving an overall proof complexity $O(d/\epsilon)$.[4]

The PoS on **Stacked DRGs** can also be used as the basis for a PoRep that satisfies ϵ-rational replication for arbitrarily small ϵ. The PoRep simply uses the labels on the $\ell - 1$st level as keys to encode the n-block data input $D = d_1, ..., d_n$ on the ℓth (last) level, using the same method described earlier for encoding data into the labels of a PoS (see Related Work, [9,10,18]). However, extracting data from this PoRep is as expensive as initializing the PoRep space because it requires recomputing the keys on the $\ell - 1$st level (Fig. 1).

PoRep from ZigZag Expander DRGs. Our second contribution is a variant of the PoS on **Stacked DRGs** that compromises slightly on initialization efficiency and proof size (requires doubling the number of levels for the same security guarantee) but improves the efficiency of extracting data when this is used as a PoRep. Instead of adding bipartite expander edge dependencies between the layers, these edges are mapped into each layer itself. Specifically, an edge from the ith node of one layer to the jth node of the next is replaced with edges between the ith and jth nodes in each layer. The directionality of these mapped edges alternates between layers, forming a "zig-zag". The only edges retained between layers are between nodes at the same indices. As an undirected graph, every layer is the union of a DRG and a constant degree non-bipartite expander graph. As a directed graph, each layer forms a DAG where the union of any subset with its dependencies and targets is a constant fraction larger than the subset itself.

---

[4] Asymptotically, this is close to the optimal proof complexity achievable for any PoS based on graph labeling that has an ϵ space gap. If the prover claims to be storing n labels and the proof queries less than $1/\epsilon$ then a random deletion of an ϵ fraction of these labels evades detection with probability at least $(1-\epsilon)^{1/\epsilon} \approx 1/e$.

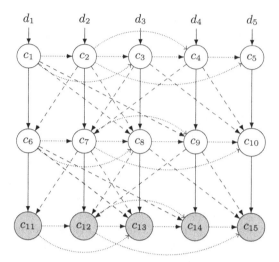

Fig. 1. Stacked DRGs. Dotted edges are the DRG edges and dashed edges are expander edges. In the PoS on Stacked DRGs the prover computes a labeling of the graph and stores the labels on the nodes in green.

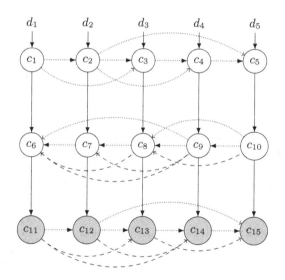

Fig. 2. ZigZag DRGs. The dashed edges in ZigZag DRGs are the same as in Stacked DRGs but projected into the layers. Dashed edges in ZigZag DRGs are reversed every other layer while dotted edges are redefined by reversing the order of the nodes. Dashed edges correspond to encoding instead of key-derivation dependencies. In the PoRep on ZigZag DRGs each labeling on a layer encodes the previous layer and the prover stores only the encoding labels of the green nodes.

By alternating the direction of the edges between layers, the dependencies of a subset in one layer become targets of the same subset in the adjacent layer, and the dependencies between layers expands. We refer to this graph construction as ZigZag DRGs (Fig. 2).

The PoRep on ZigZag DRGs encodes in the labels of each layer the labels of the previous levels. The edges within a layer enforce dependencies between labels by deriving a key for each encoding using a cryptographic hash function. A special key is derived for the encoding on each ith node from the labels on the parents of the ith node within the same layer. Essentially, this construction on ℓ DAG layers iterates the basic DAG encoding of the data inputs ℓ times rather than performing a long key derivation. The labels in any given layer can be decoded (in parallel) from the labels in the preceding layer.

2   Preliminaries

2.1   Proofs of Space

A PoS interactive protocol has three procedures:

1. Setup The setup runs on security parameters λ and outputs public parameters pp for the scheme. The public parameters are implicit inputs to the next two protocols.
2. Initialization is an interactive protocol between a prover P and verifier V that run on shared input (id, N). P outputs Φ and S, where S is its storage advice of length N and Φ is a compact $O(\text{polylog}(N))$ length string given to the verifier.
3. Execution is an interactive protocol between P and V where P runs on input S and V runs on input Φ. V sends challenges to P, obtains back a proof π, and outputs accept or reject.

Efficiency. The commitment Φ is $O(\text{polylog}(N))$ size, the storage S is size N, and the verifier runs in time $O(\text{polylog}(N))$.

Completeness. The prover succeeds with probability 1 (causes verifier to accept) if it follows the protocol honestly.

Soundness. The PoS is (s, t, μ)-sound if for all adversaries P^* running in time t and storing advice of size s during Execution, P^* passes verification with probability at most μ. The PoS is parallel (s, t, μ)-sound if P^* may run in parallel time t. We say that a PoS construction is *tight* if for any constant $\epsilon < 1$ the PoS can be parametrized so that the resulting PoS is $(\epsilon N, t, \mu)$-sound for $t \in \Omega(N)$ and $\mu = \text{negl}(\lambda)$.

Amortization-free soundness. An (s, t, μ) PoS is *amortization-free* if for any k distinct ids $id_1, ..., id_k$, the modified PoS protocol that runs Initialization on k independent inputs (id_i, N) for each i to get outputs (S_i, Φ_i) and then runs Execution independently on each (S_i, Φ_i) is $(ks, t, k\mu)$-sound.

2.2   Graph Pebbling Games

Pebbling games are the main analytical tool used in graph-based proofs of space and memory hard functions.

Black pebbling game. The black pebbling game is a single-player game on a DAG $\mathcal{G} = (V, E)$. At the start of the game the player chooses a starting configuration of $P_0 \subseteq V$ of vertices that contain black pebbles. The game then proceeds in rounds where in each round the player may place a black pebble on a vertex only if all of its parent vertices currently contain pebbles placed in some prior round. In this case we say that the vertex is *available*. Placing a pebble constitutes a *move*, whereas placing pebbles on all simultaneously available vertices consumes a *round*. The adversary may also remove any black pebble at any point. The game stops once the adversary has placed pebbles on all vertices in some target/challenge set $V_C \subseteq V$.

Pebbling complexity. The pebbling game on graph G with vertex set V and target set $V_C \subseteq V$ is (s, t)-*hard* if no player can pebble the set V_C in t moves (or fewer) starting from s initial pebbles, and is (s, t)-*parallel-hard* if no player can complete the pebbling in t rounds (or fewer) starting from an initial configuration of at most s pebbles. If a $1 - \alpha$ fraction of the nodes in V_C each require t rounds to pebble then the pebbling game on (G, V_C) is (s, t, α)-parallel-hard, i.e. every subset containing more than an α fraction of the nodes in VC requires t rounds to pebble.

In a random pebbling game a challenge node is sampled randomly from V_C after the player commits to the initial configuration P_0 of s vertices, and the hardness measure includes the adversary's probability of success. The random pebbling game is (s, t, ϵ)-(parallel)-hard if from any s fixed initial pebbles the probability that a uniformly sampled challenge node can be pebbled in t or fewer moves (resp. t or fewer rounds) is less than ϵ.

The following facts are easy to prove:

Fact 1. *The random pebbling game on a DAG G on n nodes with target set V_C is (s, t, α)-parallel-hard if and only if the deterministic pebbling game on G with target set V_C is (s, t, α)-parallel-hard.*

Fact 2. *A random pebbling game with a single challenge is (s, t, α)-parallel-hard if and only if the the random pebbling with κ challenges is (s, t, α^κ)-parallel-hard.*

DAG labeling game. A labeling game on a degree d DAG $\mathcal{G}$ is analogous to the pebbling game, but involves a cryptographic hash function $H : \{0, 1\}^{dm} \rightarrow \{0, 1\}^m$, often modeled as a random oracle. The vertices of the graph are indexed in $[n]$ and each ith vertex associated with the label c_i where $c_i = H(i)$ if i is a source vertex, or otherwise $c_i = H(i || c_{\mathsf{parents}}(i))$ where $c_{\mathsf{parents}}(i) = \{c_{v_1}, ..., c_{v_d}\}$ if $v_1, ..., v_d$ are the parents of the ith vertex, i.e. the vertices with a directed edge to vertex i. The game ends when the player has computed all the labels on a target/challenge set of vertices V_C. A "fresh" labeling of $\mathcal{G}$ could be derived by choosing a salt id for the hash function so that $H_{id}(x) = H(id||x)$, and the labeling may be associated with the identifier id.

The complexity of the labeling game (on a fresh identifier id) is measured in queries to the hash function instead of pebbles. This includes the number of labels initially stored, the total number of queries, and the total rounds of sequential queries, etc. The labeling game is $(s, r, q, \epsilon, \delta)$-labeling-hard if no algorithm that stores initial advice of size s and after receiving a uniform random challenge node $v \in [n]$ makes a total of q queries to H in r sequential rounds can output the correct label on v with probability greater than ϵ over the challenge v and δ over the random oracle H.

Random oracle query complexity. A general correspondence between the complexity of the black pebbling game on the underlying graph $\mathcal{G}$ and the random oracle labeling game is not yet known. However, Pietrzak [18] recently proved an equivalence between the parallel hardness of the randomized pebbling game and the parallel hardness of the random oracle labeling game for arbitrary initial configurations S_0 adapting the "ex post facto" technique from [7].

Theorem 1 (Pietrzak [18]). *If the random pebbling game on a DAG G with n nodes and in-degree d is (s, r, ϵ)-parallel-hard then the labeling game on G with a random oracle $H : \{0,1\}^{md} \rightarrow \{0,1\}^m$ is $(s', r, \epsilon, \delta, q)$-labeling-hard with $s' = s(m - 2(\log n + \log q)) - \log(1/\delta)$.*

Generic PoS from graph labeling game. Many PoS constructions are based on the graph labeling game [8,18,19]. Let $\mathcal{G}(\cdot)$ be a family of d-in-regular DAGs such that $G_n \leftarrow \mathcal{G}(n)$ is a d-in-regular DAG on $N > n$ nodes and $V_C(n)$ is a subset of n nodes from G_n. Let $H : \{0,1\}^{dm} \rightarrow \{0,1\}^m$ be a collision-resistant hash function (or random oracle). Let $\mathsf{Chal}(n, \Lambda)$ denote a distribution over challenge vectors in $[N]^\lambda$. For each $n \in \mathbb{N}$, the generic PoS based on the labeling game with G_n and target set $V_C(n)$ is as follows:

Initialization: The prover plays the labeling game on G_n using a hash function $H_{id} = H(id||\cdot)$. The prover does the following:

1. Computes the labels $c_1, ..., c_N$ on all nodes of $\mathcal{G}$ and commits to them in com using any vector commitment scheme.
2. Obtains vector of λ challenges $\boldsymbol{r} \xleftarrow{\text{R}} \mathsf{Chal}(n)$ from the verifier (or non-interactively derives them using as a seed $H_{id}(com)$).
3. For challenges $r_1, ..., r_\lambda$, the prover opens the label on the r_ith node of G_n, which was committed in com, as well as the labels $c_{\mathsf{parents}}(r_i)$ of all its parent nodes. The labels are added to a list L with corresponding opening proofs in a list Λ and the prover outputs the proof $\Phi = (com, L, \Lambda)$.

The verifier checks the openings Λ with respect to com. It also checks for each challenge specifying an index $v \in [N]$, the label c_v in L label and its parent labels $c_{\mathsf{parents}}(c_v)$, that $c_v = H_{id}(v||c_{\mathsf{parents}}(c_v))$. Finally, the prover stores as S only the n labels in V_C.

Execution: The verifier selects κ challenge nodes $v_1, ..., v_\kappa$ uniformly at random from V_C. The online prover uses its input S to respond with the label on v and an opening of *com* at the appropriate index. The verifier can repeat this sequentially, or ask for a randomly sampled vector of challenge vertices to amplify soundness.

Red-black pebbling game. An adversary places both black and red pebbles on the graph initially. The red pebbles correspond to incorrect labels that the adversary computes during Initialization and the black pebbles correspond to labels the adversary stores in its advice S. Without loss of generality, an adversary that cheats generates some label that does not require any space to store, which is why red pebbles will be "free" pebbles and counted separately from black pebbles. The adversary's choice of red pebble placements (specifically how many to place in different regions of the graph) is constrained by the λ non-interactive challenges, which may catch these red pebbles and reveal them to the verifier. The formal description of the red-black pebbling security game for a graph labeling PoS construction with $\mathcal{G}(n)$, $V_C(n)$, and $\mathsf{Chal}(n)$ is as follows.

Red-Black-Pebbles$^\mathcal{A}(\mathcal{G}, V_C, \mathsf{Chal}, t)$:

1. $\mathcal{A}$ outputs a set $R \subseteq [N]$ (of red pebble indices) and $S \subseteq [N]$ (of black pebble indices).
2. The challenger samples $c_1, ..., c_\lambda \xleftarrow{\text{R}} \mathsf{Chal}(n)$. If $c_i \in R$ for some i then $\mathcal{A}$ immediately loses. The challenger additionally samples $v_1, ..., v_\kappa$ uniformly at random from indices in $V_C(n)$ and sends these to $\mathcal{A}$.
3. $\mathcal{A}$ plays the random (black) pebbling game on $\mathcal{G}(n)$ with the challenges $v_1, ..., v_\kappa$ and initial pebble configuration $P_0 = R \cup S$. It runs for t parallel rounds and outputs its final pebble configuration P_t. $\mathcal{A}$ wins if P_t contains pebbles on all of $v_1, ..., v_\kappa$.

Graph labeling PoS soundness. Given the correspondence between the hardness of the random oracle labeling game and parallel black pebbling game, we can entirely capture the soundness of the graph labeling PoS in terms of the complexity of Red-Black-Pebbles$^\mathcal{A}(\mathcal{G}, V_C, t)$. Let $c : \mathbb{N} \to N$ denote a cost function $c : \mathbb{N} \to \mathbb{N}$ representing the parallel time cost (e.g. in sequential steps on a PRAM machine) of computing a label on a node of $\mathcal{G}(n)$ for each $n \in \mathbb{N}$.

Definition 1. *A graph labeling PoS with $\mathcal{G}(n), V_C(n), \mathsf{Chal}(n)$ and cost function $c(n)$ is parallel $(s, c(n) \cdot t, \mu)$-sound if and only if the probability that any $\mathcal{A}$ wins* Red-Black-Pebbles$^\mathcal{A}(\mathcal{G}, V_C, \mathsf{Chal}, t)$ *is bounded by μ where $|S| = s$.*

2.3 Depth Robust Graphs

A directed acyclic graph (DAG) on n nodes with d-indegree is (n, α, β, d) *depth robust graph* (DRG) if every subgraph of αn nodes contains a path of length at least βn.

DRGs have been constructed for constant α, β and $d = O(\log n)$ using extreme constant-degree bipartite expander graphs, or local expanders [4,14,17].

Explicit constructions of local expanders exist [15], however they are complicated to implement and their concrete practicality is hindered by very large hidden constants. The most efficient way to instantiate these extreme expander graphs is probabilistically. A probabilistic DRG construction outputs a graph that is a DRG with overwhelming probability. The most efficient probabilistic construction to date is due to Alwen et al. [3]. The analysis still leaves large gaps between security and efficiency although was shown to resist depth-reducing attacks empirically. Their construction is also *locally navigatable*, meaning that it comes with an efficient parent function to derive the parents of any node in the graph using polylogarithmic time and space.

2.4   Expander Graphs

The vertex expansion of a graph $\mathcal{G}$ on vertex set V characterizes the size of the boundary of vertex subsets $S \subseteq V$ (i.e. the number of vertices in $V \setminus S$ that are neighbors with vertices in S). In the case of directed bipartite graphs, vertex expansion is defined by the minimum number of sources connected to any given number of sinks.

Definition 2. *For any constants α, β where $0 < \alpha < \beta < 1$ and integer $n \in \mathbb{N}$, an (n, α, β) bipartite expander is a directed bipartite graph with n sources and n sinks such that any subset of αn sinks are connected to at least βn sources. For any $\delta > 0$, a subset S of sinks is called $(1 + \delta)$-expanding if it is connected to at least $(1 + \delta)|S|$ sources.*

Chung's bipartite expander. The randomized construction of Chung [6] defines the edges of a d-regular bipartite expander on $2n$ vertices by connecting the dn outgoing edges of the sources to the dn incoming edges of the sinks via a random permutation $\Pi : [d] \times [n] \rightarrow [d] \times [n]$. The ith source is connected to the jth sink if there is some $k_1, k_2 \in [d]$ such that $\Pi(k_1, i) = (k_2, j)$.

Lemma 1 (RD [19]). *The Chung random bipartite graph construction is a d-regular (n, α, β) expander with probability $1 - \mathsf{negl}(nH_b(\alpha))$ for all d, α, β satisfying:*

$$H_b(\alpha) + H_b(\beta) + d(\beta H_b(\alpha/\beta) - H_b(\alpha)) < 0 \qquad (2.1)$$

where $H_b(x) = -x \log_2 x - (1 - x) \log_2(1 - x)$ is the binary entropy function.

For example, the above formula shows that for $\alpha = 1/2$ and $\beta = 0.80$ Chung's construction gives an $(n, 0.5, 0.80)$ expander for $d \geq 8$, meaning any subset of 50% of the sinks are connected to at least 80% of the sources when the degree is at least 8.

The following lemmas establish further properties of Chung's bipartite expander construction that will be used in the analysis of our PoS. The proofs are included in the full version of this paper. Let $\beta_G(\alpha)$ denote the smallest expansion of a subset of αn sources in a bipartite graph G, i.e. every subset of αn sources is connected to at least $\beta_G(\alpha)$ sinks.

Lemma 2. *For any $k > 1$ and $d > 2$, if the output of Chung's construction is a d-regular $(n, \alpha, k\alpha)$ bipartite expander for some $\alpha < \frac{d-k-1}{k(d-2)}$ with probability $1 - negl(nH_b(\alpha))$ then $\beta_G(\alpha') \geq k\alpha'$ for every $\alpha' < \alpha$ with probability $1 - negl(nH_b(\alpha'))$.*

Corollary 1. *For $d = 8$ Chung's construction is an 8-regular bipartite graph such that every subset of at most $1/3$ of the nodes is 2-expanding, i.e. it is an $(n, \alpha, 2\alpha)$-bipartite expander for every $\alpha \leq 1/3$ with overwhelming probability.*

Proof. Plugging $\alpha = 1/3$ and $\beta = 2/3$ into the formula for degree (Eq. 2.1) gives $d = 7.21 < 8$. With $d = 8$ and $k = 2$ the condition in Lemma 2 is satisfied: $\alpha = 1/3 < (d - k - 1)/k(d - 2) = 5/12$.

For fixed d the expansion improves further as α decreases. Figure 3 provides a table of expansion factors over a range of α with fixed degree $d = 8$. Figure 4 plots the expansion as a function of subset size.

Size (α)	0.01	0.10	0.20	0.30	0.40	0.45	0.50	0.55	0.60	0.65	0.70	0.75	0.80
Expansion (β)	0.04	0.33	0.53	0.65	0.75	0.78	0.81	0.84	0.88	0.89	0.91	0.93	0.94
Factor (β/α)	4	3.3	2.65	2.1	1.8	1.73	1.62	1.53	1.47	1.37	1.3	1.24	1.17

Fig. 3. A table of the maximum expansion (β) satisfying the condition from Lemma 1 for Chung's construction with fixed degree $d = 8$ over a range of subset sizes (α).

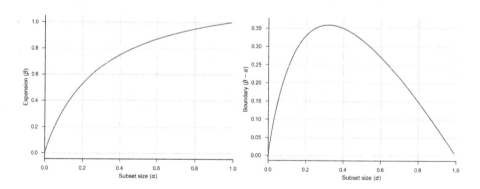

Fig. 4. The graph on the left plots the lower bound from Lemma 1 on the expansion β as a function of the subset size α (in fractions of the sources/sinks) for Chung's construction with fixed $d = 8$. The graph on the right plots the corresponding lower bound on $\beta - \alpha$, which is the analog of the subgraph boundary in non-bipartite expanders. Specifically, this is a lower bound on the fraction of sinks connected to an α fraction of sources that have distinct index labels from the sources.

In a bipartite expanders, the "boundary" of a set of sources is the set of sinks connected to these sources that have distinct index labels from the sources, which is at least $\beta_G(\alpha) - \alpha$. Lemma 3 gives a smooth lower bound on $\beta_G(\alpha) - \alpha$ for Chung's bipartite expander graphs that we can show has a unique local maximum in $(0,1)$. To simplify the analysis, we look at the function defined by the zeros of $\phi(\alpha, \beta) = d(\beta H_b(\alpha/\beta) - H_b(\alpha)) + 2 = 0$. Any α, β satisfying this relation also satisfies the relation in Lemma 1 because $H_b(\alpha) + H_b(\beta) < 2$ when $\beta > \alpha$. This implicitly defines β as a function of α, as well as the function $\hat{\beta} = \beta - \alpha$, by pairs of points $(\alpha, \hat{\beta}(\alpha))$ such that $\phi(\alpha, \alpha + \hat{\beta}(\alpha)) = 0$ is a lower bound to the boundary of subsets of size α, which holds at any point α with probability at least $1 - \mathsf{negl}(n H_b(\alpha))$.

Lemma 3. *Define $\phi(x, y) = d(y H_b(x/y) - H_b(x)) + c$ where c is any constant and let $\hat{\beta}$ be the function on $(0,1)$ defined by pairs of points $(\alpha, \beta - \alpha)$ such that $\phi(\alpha, \beta) = 0$ and $0 < \alpha < \beta < 1$. The function $\hat{\beta}$ is continuously differentiable on $(0,1)$ and has a unique local maximum.*

Corollary 2. *With overwhelming probability in n, Chung's construction (with $d = 8$) is an 8-regular bipartite graph on n sinks and n sources each indexed in $[n]$ such that for all $\alpha \in (0.10, 0.80)$ every αn sinks are connected to at least $0.12n$ sources with distinct indices.*

Lemma 4. *For any $d \geq 4$, Chung's construction yields a d-regular bipartite graph that is an $(n, \alpha, (d/3)\alpha)$ bipartite expander for every $\alpha \leq \frac{3}{2d}$ with probability $1 - \mathsf{negl}(n H_b(\alpha))$.*

3   Stacked DRG Proof of Space

In this section we show that stacking DRGs with bipartite expander edges between layers yields an arbitrarily tight proof of space with the number of layers increasing as $O(\log_2(1/\epsilon))$ where ϵ is the desired space gap. Moreover, the proof size is also $O(\log_2(1/\epsilon))$, which is asymptotically optimal. Our proofs attempt a tight analysis as well, e.g. showing that just 10 layers achieve a PoS with a 1% space gap, degree $8 + d$ graphs where d is the degree of the DRG, and relies only on a DRG that retains depth in 80% subgraphs.

3.1   Review of the Stacked-Expander PoS

The PoS construction by Ren and Devadas [19] based on stacked bipartite expander graphs is a building block towards our tight PoS construction. Their construction uses a layered graph where each layer is a directed line on n nodes and the directed edges of a bipartite expander graph are placed between layers. This was shown to be an $(\epsilon\gamma n, (1 - 2\epsilon)\gamma n)$-sound PoS for parameters $\epsilon < 1/2$ and $\gamma < 1$ [19].

The graph $\mathcal{G}_{SE}$. The stacked-expander PoS uses the same underlying graph as the Balloon Hash memory hard function [12]. The graph $\mathcal{G}_{SE}$ consists of $\ell = O(\lambda)$ layers $V_1, ..., V_\ell$ consisting each of n vertices indexed in each level by the integers $[n]$, and where λ is a security parameter. First directed edges are placed from each kth vertex to the $k + 1$st vertex in each level, i.e. forming a directed line. Next directed edges are placed from V_{i-1} to V_i according to the edges of an (n, α, β) bipartite expander on (V_{i-1}, V_i). Finally a "localization" operation is applied so that each kth vertex u_k in V_{i-1} is connected to the kth vertex v_k in V_i and any directed edge from the kth vertex of V_{i-1} to some jth vertex of V_i where $j > k$ is replaced with a directed edge from the kth vertex of V_i to the jth vertex of V_i. $\mathcal{G}_{SE}$ can be pebbled in $n\ell$ steps using a total of n pebbles.

Stacked-expander PoS. The PoS follows the generic PoS based on graph labeling. We remark only on several nuances. Due to the topology of $\mathcal{G}_{SE}$ after localization, the prover only needs to use a buffer of size n and deletes the labels of V_{i-1} as it derives the labels of V_i. After completing the labels C_i in the ith level it computes a vector commitment (e.g. Merkle commitment) to the labels in C_i denoted com_i. Once it has derived the labels C_ℓ of the final level V_ℓ it computes $com = H_{id}(com_1 || \cdots || com_\ell)$ and uses $H_{id}(com||j)$ to derive λ non-interactive challenges for each jth level.

3.2 A New Tight PoS from Stacked DRGs

By simply replacing each of the path graphs V_i in the stacked-expander PoS construction with a depth robust graph results in an arbitrarily tight PoS. Specifically, only $O(\log 1/\epsilon)$ layers are needed to achieve a $((1 - \epsilon)n, \Omega(n))$-parallel-sound PoS. We demand only very basic properties from the DRG, e.g. that any subgraph on 80% of the nodes contains a long path of $\Omega(n)$ length.

Construction of $\mathcal{G}_{SDR}[\ell]$. The graph $\mathcal{G}_{SDR}[\ell]$ will be exactly like $\mathcal{G}_{SE}$ only each of the ℓ layers $V_1, ..., V_\ell$ contains a copy of an $(n, 0.80n, \beta n)$-depth-robust graph for some constant β. For concreteness, we define the directed edges between the layers using the degree 8 Chung random bipartite graph construction. For simplicity we will analyze the construction without applying localization to the expander edges between layers. Even without localization this is already a valid PoS, only the initialization requires a buffer of size $2n$ rather than n. The PoS is still "tight" with respect to the persistent space storage.

Vector commitment storage. If the vector commitment storage overhead required for the PoS is significant then this somewhat defeats the point of a tight PoS. Luckily this is not the case. Most vector commitment protocols, including the standard Merkle tree, offer smooth time/space tradeoffs. With a Merkle tree the honest prover can delete the hashes on nodes on the first k levels of the tree to save a factor 2^k space and re-derive all hashes along a Merkle path by reading at most 2^k nodes and computing at most 2^k hashes. If $k = 7$ this is less than a 1% overhead in space, and requires at most 128 additional hashes and reads. Furthermore, as remarked in [18] these 2^k reads are sequential memory reads, which in practice are inexpensive compared to the random reads for challenge labels.

Proof size. We show that $\ell = O(\log(\frac{1}{3(\epsilon-2\delta)}))$ suffices to achieve $\mathsf{negl}(\lambda)$ soundness against any prover running in parallel time less than βn rounds of queries where Chal samples λ/δ nodes in each layer. This would result in a proof size of $O((1/\epsilon)\log(1/\epsilon))$, which is already a major improvement on any PoS involving a graph of degree $O(1/\epsilon)$ (recall that the only previously known tight PoS construction relied on very special DRGs whose degree must scale with $1/\epsilon$, which results in a total proof size of $O(1/\epsilon^2)$). However, we are able to improve the result even further and show that only $O(1/\delta)$ challenge queries are required overall, achieving proof complexity $O(1/\epsilon)$. This is the optimal proof complexity for the generic pebbling-based PoS with at most an ϵ space gap. If the prover claims to be storing n pebbles and the proof queries less than $1/\epsilon$ then a random deletion of an ϵ fraction of these pebbles evades detection with probability at least $(1-\epsilon)^{1/\epsilon} \approx 1/e$. The same applies if a random ϵ fraction of the pebbles the prover claims to be storing are red (i.e. errors).

Analysis outline. We prove the hardness of the red-black pebbling game Red-Black-Pebbles$^{\mathcal{A}}(\mathcal{G}_{SDR}[\ell], V_\ell, \mathsf{Chal})$ where Chal samples λ_i uniform challenges over V_i. We first show that it suffices to consider the parallel complexity of pebbling the set $U_\ell \subseteq V_\ell$ of all unpebbled nodes on V_ℓ from an initial configuration of γn black pebbles overall and $\delta_i n$ red pebbles in each layer where $\delta_\ell < \epsilon/2$.

As a shorthand notation, we will say that $\mathcal{G}_{SDR}[\ell]$ is $(\gamma, \boldsymbol{\delta}, t, \mu)$-hard if every subset containing a μ fraction of the nodes in V_ℓ require t rounds to pebble (i.e. greater than a $1 - \mu$ fraction of the nodes each individually require t rounds to pebble) from an initial configuration of γn black pebbles overall and $\delta_i n$ red pebbles in each layer where $\delta_\ell < \epsilon/2$. In Claim 1 we show that if $\mathcal{G}_{SDR}[\ell]$ is $(\gamma, \boldsymbol{\delta}, t, \mu)$-hard then the labeling PoS on $\mathcal{G}_{SDR}[\ell]$ is $(\gamma n, t, max\{p^*, \mu^\kappa\})$-sound where $p^* = max_i(1 - \delta_i)^{\lambda_i}$. (Recall that κ and $\boldsymbol{\lambda}$ are parameters defined in the game).

For $\mu = 1$, $(\gamma, \boldsymbol{\delta}, t, 1)$-hardness is nearly equivalent to the standard parallel pebbling complexity of U_ℓ. The one distinction[5] is its dependency on the restriction to δ_i red pebbles in each layer, counted separately from black pebbles. In Claim 3 we show that if $\mathcal{G}_{SDR}[\ell]$ is $(1 - \epsilon + 2\delta_\ell, \boldsymbol{\delta}, t, 1)$-hard then $\mathcal{G}_{SDR}[\ell + 1]$ is $(1 - \epsilon, \boldsymbol{\delta}^*, t, 1 - \epsilon/2)$-hard where $\boldsymbol{\delta}^*$ is equal to $\boldsymbol{\delta}$ on all common indices and $\delta_{\ell+1} = \delta_\ell$.

Finally, we analyze the complexity of pebbling all of U_ℓ, i.e. the $(\gamma, \boldsymbol{\delta}, t, 1)$-hardness of $\mathcal{G}_{SDR}[\ell]$. We show in Claim 5 that when the adversary uses at most $\gamma < 1 - \epsilon$ black pebbles and δ red pebbles in each layer then pebbling all the unpebbled nodes in layer V_ℓ (for ℓ dependent on ϵ and δ) requires pebbling $0.80n$ unpebbled nodes (including both red and black pebbles) in some layer V_i. Since the layer V_i contains a $(n, 0.80, \beta n)$-depth-robust graph, this takes at least βn

---

[5] In prior uses of the red-black pebbling game to analyze proofs of space, it sufficed to consider parallel black pebbling complexity because replacing red pebbles with "free" black pebbles only increases the adversary's power. Our more refined analysis requires analyzing the weaker adversary who is restricted to a maximum number of red pebbles on each level of the graph, enforced by the construction.

rounds. We then generalize this analysis (Claim 6) to apply when δ_i is allowed to increase from level ℓ to 1 by a multiplicative factor such that $\sum_i \delta_i = O(\delta_\ell)$.

Theorem 2 ties everything together, taking into account the constraints of each claim to derive the PoS soundness of the labeling PoS on $\mathcal{G}_{SDR}[\ell]$.

Theorem 2. *The labeling PoS on $\mathcal{G}_{SDR}[\ell]$ with Chal sampling λ_i challenges in each level V_i and κ online challenges in V_ℓ is $((1 - \epsilon - \delta)n, \beta n, e^{-\lambda})$-sound with $\kappa = 2\lambda/\epsilon$ if either of the following conditions are met for $\epsilon \leq 0.24$:*

(a) $\ell = max(8, \log_2(\frac{1}{3(\epsilon - 2\delta)}) + 4)$ and each $\lambda_i = \lambda/\delta$ and $\delta < min(0.01, \epsilon/3)$
(b) $\ell = max(14, \log_2(\frac{1}{3(\epsilon - 3\delta)}) + 5)$ and each $\lambda_i = \lambda/\delta_i$ where $\delta_\ell = \delta < min(0.01, \epsilon/2)$ and $\delta_i = min(0.05, \frac{2}{3}\delta_{i-1})$

Proof. For any $\mathcal{G}_{SDR}[\ell]$, if the set of unpebbled nodes in V_ℓ are connected via unpebbled paths to at least $0.80n$ unpebbled nodes (including red and black) in some prior level V_i, then pebbling all of V_ℓ requires pebbling all these $0.80n$ unpebbled nodes, which requires βn rounds due to the fact that V_i is $(n, 0.80n, \beta n)$ depth robust. Claim 5 implies that $\mathcal{G}_{SDR}[\ell]$ is $(1 - \epsilon, \delta, \beta n, 1)$-hard for δ and ℓ such that $\delta_i = \delta < \epsilon/2$ for all i and $\ell = max(7, \log_2(\frac{1}{3(\epsilon - 2\delta)} + 3))$.

Claim 6 gives a different tradeoff between ℓ and δ, showing that the same hardness holds for δ and ℓ such that $\delta_\ell = \delta < \epsilon/3$ and $\delta_i = min(0.05, \frac{2}{3}\delta_{i-1})$ and $\ell = max(13, \log_2(\frac{1}{3(\epsilon - 3\delta)}) + 4)$.

Assuming $\epsilon \leq 0.24$, Claim 3 implies that $\mathcal{G}_{SDR}[\ell+1]$ is $(1 - \epsilon - \delta, \delta, \beta n, 1 - \epsilon/2)$-hard extending δ so that $\delta_{\ell+1} = \delta_\ell = \delta$. Finally, by Claim 1, the labeling PoS on $\mathcal{G}_{SDR}[\ell + 1]$ with challenge set V_ℓ and Chal sampling λ_i in each level V_i is $((1 - \epsilon - \delta)n, \beta n, max\{p^*, (1 - \epsilon/2)^\kappa\})$-sound where $p* = max_i(1 - \delta_i)^{\lambda_i}$. Setting $\lambda_i = \lambda/\delta_i$ and $\kappa = 2\lambda/\epsilon$, the PoS is $((1 - \epsilon - \delta)n, \beta n, e^{-\lambda})$-sound.

Notation 1 (Common analysis notations). *Let U_i denote the entire index set of nodes that are unpebbled in V_i and P_i the set that are pebbled. The total number of pebbles placed in the initial configuration is γn. Each level initially has $\rho_i n$ black pebbles and $\delta_i n$ red pebbles. Finally, $\gamma_i n = \sum_{j<i} \rho_i n$ is the number of black pebbles placed before level i.*

Claim 1. *If $\mathcal{G}_{SDR}[\ell]$ is (γ, δ, t, μ)-hard then the labeling PoS on $\mathcal{G}_{SDR}[\ell]$ is $(\gamma n, t, max\{p^*, \mu^\kappa\})$-sound where $p^* = max_i(1 - \delta_i)^{\lambda_i}$.*

Proof. Fix $\gamma = 1 - \epsilon$ and δ_i for each i. The λ_i challenges during Initialization in each level ensure that $\mathcal{A}$ wins with at most probability $(1 - \delta_i)^{\lambda_i}$ if it places more than δ_i red pebbles on V_i. If $\mathcal{A}$ has exceeded the δ_i bound in more than one level this only increases its probability of failure. Thus, in case 1 ($\mathcal{A}$ places more than δ_i red pebbles on some level i), $\mathcal{A}$'s success probability is bounded by maximum value of $(1 - \delta_i)^{\lambda_i}$ over all i. In case 2 ($\mathcal{A}$ places fewer than δ_i on each ith level), the fact that $\mathcal{G}_{SDR}[\ell]$ is (γ, δ, t, μ)-hard implies that at most a μ fraction of the nodes on V_ℓ can individually be pebbled from the starting configuration in t rounds, hence $\mathcal{A}$'s success probability of answering κ independent challenges is bounded by μ^κ. The success probability is bounded by the maximum of these two cases.

Claim 2 (trivial). $\mathcal{G}_{SDR}[\ell]$ *is* $(\gamma, \delta, t, 1)$*-hard if and only if given any initial configuration* P_0 *of* $\gamma_\ell n$ *black pebbles placed on layers* $V_1, ..., V_{\ell-1}$ *at most* $\delta_i n$ *red pebbles in each layer, and any set* $U \subseteq V_\ell$ *of an unpebbled nodes in* V_ℓ *such that* $\alpha - \gamma_\ell \geq 1 - \gamma - \delta$, *no adversary can pebble* U *in fewer than* t *rounds.*

The proof of this claim is in the full version of the paper.

Claim 3. *For any* $\epsilon \leq 0.24$, *if* $\mathcal{G}_{SDR}[\ell - 1]$ *is* $(1 - \epsilon + \delta_{\ell-1}, \delta, t - 1, 1)$*-hard then* $\mathcal{G}_{SDR}[\ell]$ *is* $(1 - \epsilon, \delta^*, min(\beta n, t), 1 - \epsilon/2)$*-hard where* δ^* *is identical to* v *on all common indices and* $\delta_\ell = \delta_{\ell-1} \leq \epsilon/2$.

Proof. Refer to Notation 1. Consider the graph $\mathcal{G}_{SDR}[\ell]$ with $\gamma n = (1-\epsilon)n$ black pebbles initially placed. Let $\delta = \delta_\ell = \delta_{\ell-1} \leq \epsilon/2$. Let $\alpha_\ell = |U_\ell|/|V_\ell|$ denote the fraction of nodes in V_ℓ that are unpebbled. Every $1 - \epsilon/2$ fraction of V_ℓ contains at least $\alpha^* n = (\alpha_\ell - \epsilon/2)n$ unpebbled nodes. If $\alpha^* \geq 0.80$, then every $1 - \epsilon/2$ fraction of V_ℓ contains a path of length βn because V_ℓ is a $(n, 0.80n, \beta n)$-depth robust graph. We consider the two other cases next:

Case $\alpha^* < 1/3$: The $\alpha^* n$ unpebbled nodes have dependencies on at least a $2\alpha^* = 2\alpha_\ell - \epsilon$ fraction of nodes in $V_{\ell-1}$ (Corollary 1, bipartite expansion). These contain at least $\alpha' n = (2\alpha_\ell - \epsilon - \rho_{\ell-1} - \delta)n$ unpebbled nodes because in the worst case they include $\rho_{\ell-1}n$ black pebbles and δn red pebbles. There are $\gamma_{\ell-1}n = (\gamma - \rho_{\ell-1} - \rho_\ell)n$ pebbles placed on all prior levels. By definition $\rho_\ell = 1 - \alpha_\ell - \delta$ and $\gamma = 1 - \epsilon$. Substituting $\alpha_\ell \geq 1 - \gamma - \delta$ shows that $\alpha' - \gamma_{\ell-1} = 2\alpha_\ell - \gamma + \rho_\ell - \epsilon - \delta = \alpha_\ell + 1 - \gamma - \epsilon - 2\delta \geq 1 - 2\gamma - 3\delta + 1 - \epsilon = 1 - \gamma - 3\delta$. Setting $\gamma' = 1 - \epsilon + \delta = \gamma + \delta$ gives the relation $\alpha' - \gamma_{\ell-1} \geq 1 - \gamma' - \delta$. It then follows from Claim 2 that if $\mathcal{G}_{SDR}[\ell - 1]$ is $(\gamma', \delta, t - 1, 1)$-hard then the $\alpha' n$ unpebbled nodes in $V_{\ell-1}$ require $t - 1$ rounds to pebbled. Thus, the $\alpha^* n$ unpebbled nodes in V_ℓ require t rounds to pebble.

Case $\alpha^* \geq 1/3$: In this case $\alpha^* \in (0.33, 0.80)$. It is connected to $\beta^* n$ nodes in $V_{\ell-1}$. Among these at least $\alpha' n$ for $\alpha' \geq \beta^* - \rho_{\ell-1} - \delta$ are unpebbled. Since $\gamma_{\ell-1} = \gamma - \rho_\ell - \rho_{\ell-1}$ we get $\alpha' - \gamma_{\ell-1} \geq \beta^* - \gamma + \rho_\ell - \delta$. According to Corollary 2 on the bipartite expander boundary $\beta^* - \alpha^* \geq 0.12$. Therefore, $\rho_\ell = 1 - \alpha_\ell - \delta \geq 1 - \alpha^* - \epsilon/2 - \delta$ so $\alpha' - \gamma_{\ell-1} \geq \beta^* - \gamma + 1 - \alpha^* - \epsilon/2 - 2\delta = \beta^* - \alpha^* + \epsilon/2 - 2\delta \geq 0.12 + \epsilon/2 - 2\delta$. If $\mathcal{G}_{SDR}[\ell - 1]$ is $(1 - \epsilon + \delta, \delta, t - 1, 1)$-hard, then by Claim 2 the $\alpha' n$ unpebbled nodes require $t - 1$ rounds as long as $0.12 + \epsilon/2 - 2\delta \geq \epsilon - 2\delta$, which is true for $\epsilon \leq 0.24$.

Claim 4. *If* $\mathcal{G}_{SDR}$ *initially has at most* γn *black pebbles for* $\gamma \leq 1 - \epsilon$ *and at most* $\delta n < \epsilon n/2$ *red pebbles in each layer then for* $\ell = \log_2(\frac{1}{3(\epsilon - 2\delta)})$ *the unpebbled nodes in* V_ℓ *have unpebbled paths from at least* $n/3$ *unpebbled nodes in some layer* V_i.

Proof. Refer to Notations 1. Let $\alpha_i n$ denote the number of unpebbled dependencies of U_ℓ in V_i, i.e. the number of nodes in U_i that have unpebbled paths to U_ℓ. Suppose that α_i is bounded by $1/3$ for all levels up to $\ell - k$, i.e. $\alpha_\ell < 1/3, ..., \alpha_{\ell-k} < 1/3$. We will prove the following bound:

$$\alpha_{\ell-k} \geq 2^k(\alpha_\ell - \gamma_\ell/2 - \delta) \geq 2^{k-1}(\alpha_\ell + \epsilon - 3\delta) \geq 2^k(\epsilon - 2\delta) \qquad (3.1)$$

Before proving this bound let us note its implication. For $k = \log_2(\frac{1}{3(\epsilon - 2\delta)})$ this implies $\alpha_{\ell-k} \geq 1/3$, which contradicts $\alpha_{\ell-k} < 1/3$. Therefore, it follows that $\alpha_i \geq 1/3$ at some index $i > \ell - \log_2(\frac{1}{3(\epsilon - 2\delta)})$, which leads to the conclusion that for $\ell \geq \log_2(\frac{1}{3(\epsilon - 2\delta)})$ there exists some level V_i with at least $n/3$ unpebbled nodes that have unpebbled dependency paths to the set X_ℓ of unpebbled nodes in V_ℓ.

Let $j = \ell - i$. From Corollary 1 (bipartite expansion), if $\alpha_j \leq 1/3$ then X_j is connected to at least $2\alpha_j$ nodes in V_{j-1}. At most $(\rho_{j-1} + \delta)n$ of these are pebbled. Therefore, $\alpha_{j-1} \geq 2\alpha_j - \rho_{j-1} - \delta$. Now we show by induction that $\alpha_{\ell-k} \geq 2^k \alpha_\ell - 2^{k-1}\rho_{\ell-1} - (2^k - 1)\delta$. The base case $k = 0$ is trivial. Assuming this holds for k:

$$\alpha_{\ell-k-1} \geq 2\alpha_{\ell-k} - \rho_{\ell-k-1} - \delta \geq 2(2^k \alpha_\ell - 2^{k-1}\rho_{\ell-1} - (2^k - 1)\delta) - \rho_{\ell-k-1} - \delta$$
$$\geq 2^{k+1}\alpha_\ell - 2^k \rho_{\ell-1} - (2^{k+1} - 1)\delta$$

The last inequality used the fact that $\sum_{i=1}^k \rho_{\ell-i} \leq \gamma_\ell$ and therefore $\sum_{i=1}^k 2^{k-i}\rho_{\ell-i}$ is maximized by setting $\rho_{\ell-1} = \gamma_\ell$ and $\rho_{\ell-i} = 0$ for all $i > 1$.

From the identities $\gamma_\ell = \gamma - \rho_\ell$ and $\alpha_\ell = 1 - \rho_\ell - \delta$ we derive $\gamma_\ell = \gamma + \alpha_\ell - 1 + \delta \leq \alpha_\ell + \delta - \epsilon$. Finally, inserting this into the bound above and using the fact that $\alpha_\ell \geq \epsilon - \delta$ gives:

$$\alpha_{\ell-k} \geq 2^{k-1}(2\alpha_\ell - \gamma_\ell - 2\delta) \geq 2^{k-1}(\alpha_\ell + \epsilon - 3\delta) \geq 2^k(\epsilon - 2\delta)$$

We could stop here as we have already shown unpebbled dependency paths from the unpebbled sinks in V_ℓ to a $1/3$ fraction of nodes in some level for $\ell = O(\log(1/(\epsilon - 2\delta))$ and the remainder of our PoS analysis could rely on a graph that is $(n, 0.33n, \Omega(n))$-depth-robust. However, we can tighten the analysis further so that we only need to assume the graph is $(n, 0.80, \Omega(n))$-depth robust.

Claim 5. *If $\mathcal{G}_{SDR}$ initially has at most γn black pebbles for $\gamma \leq 1 - \epsilon$ and at most $\delta < \epsilon/2$ red pebbles in each layer then the unpebbled nodes in V_ℓ have unpebbled paths to at least $0.80n$ unpebbled nodes in some layer V_i for $\ell = max(\frac{0.68 - \epsilon + \delta}{0.12 - \delta}, \log_2(\frac{1}{3(\epsilon - 2\delta)})) + 3)$. In particular, $\ell = max(7, \log_2(\frac{1}{3(\epsilon - 2\delta)})) + 3)$ when $\delta \leq 0.01$.*

Proof. In Claim 4 we showed that for $\ell \geq \log_2(\frac{2}{3(\alpha_\ell + \epsilon - 3\delta)})$ there exists an index i where $\alpha_i \geq 1/3$ and $\alpha_\ell + \epsilon - 3\delta \geq 2\epsilon - 4\delta$ (Eq. 3.1). Picking up from here, we consider what happens once $\alpha_i \geq 1/3$. We break the analysis into two cases: in the first case $\alpha_\ell < 1/3$ and in the second case $\alpha_\ell \geq 1/3$.

In both cases we will use a different bound on α_{i-k} because once $\alpha_i > 1/3$ the unpebbled sets may not be 2-expanding. Define the function $\beta(\alpha)$ to be the minimum bipartite expansion of a set of fractional size α, i.e. every set of αn nodes is connected to at least $\beta(\alpha)n$ nodes in the previous level. Let $\hat{\beta}(\alpha) = \beta(\alpha) - \alpha$. Using the relation $\alpha_{i-1} \geq \beta(\alpha_i) - \rho_{i-1} - \delta$ we derive that $\alpha_{i-2} \geq \hat{\beta}(\alpha_{i-1}) + \beta(\alpha_i) - \rho_{i-1} - \rho_{i-2} - 2\delta$ and more generally, since $\sum_{j=1}^k \rho_{i-j} \leq \gamma_i$:

$$\alpha_{i-k} \geq \sum_{j=1}^{k-1} \hat{\beta}(\alpha_{i-j}) + \beta(\alpha_i) - k\delta - \sum_{j=1}^{k} \rho_{i-j}$$

$$\geq (k-1)(min_{j<k}\hat{\beta}(\alpha_{i-j}) - \delta) + \beta(\alpha_i) - \gamma_i - \delta$$

The final ingredient is the bound $\gamma_i \leq \alpha_i - \epsilon + \delta$ for all i. To see this, first observe that $\alpha_\ell - \gamma_\ell = 1 - \rho_\ell - \delta - (\gamma - \rho_\ell) = \epsilon - \delta$. If $\alpha_i - \gamma_i \geq \epsilon - \delta$ and $\epsilon > 2\delta$, then $0.80 > \alpha_i > \delta$, and so the $\alpha_i n$ dependencies are connected to $\beta(\alpha_i)n > (\alpha_i + \delta)n$ nodes in level V_{i-1}. Therefore $\alpha_{i-1} \geq \alpha_i - \rho_{i-1} = \alpha_i - (\gamma_i - \gamma_{i-1})$. In words, decreasing the number of dependencies requires using black pebbles 1-to-1, so $\alpha_{i-1} - \gamma_{i-1} \geq \alpha_i - \gamma_i$.

$$\alpha_{i-k} \geq (k-1)(min_{j<k}\hat{\beta}(\alpha_{i-j}) - \delta) + \hat{\beta}(\alpha_i) + \epsilon - 2\delta \qquad (3.2)$$

By Corollary 2 to Lemma 3, $\hat{\beta}(\alpha) \geq 0.12$ for $\alpha \in (0.10, 0.80)$.

Case $\alpha_\ell \geq 1/3$: First we claim that $\alpha_{\ell-i} \geq 0.12$ for all i. If $\alpha_i \geq 0.12$ then as shown above $\alpha_{i-1} \geq \hat{\beta}(\alpha_i) + \epsilon - 2\delta \geq 0.12$ because $\hat{\beta}(\alpha) \geq 0.12$ for all $\alpha \in (0.10, 0.80)$ and $\epsilon > 2\delta$. Our claim thus follows by induction. Therefore, for all $j \leq k$ we derive that $min_{j \leq k}(\hat{\beta}(\alpha_{\ell-j})) \geq 0.12$. Equation 3.2 then shows that $\alpha_{\ell-k-1} \geq k(0.12 - \delta) + 0.12 + \epsilon - \delta$, or $\alpha_{\ell-k-1} \geq 0.80$ at $k \geq (0.68 - \epsilon + \delta)/(0.12 - \delta)$ (e.g. $k = 7$ when $\delta \leq 0.01$).

Case $\alpha_\ell < 1/3$: From Eq. 3.1, $\alpha_i \geq 1/3$ at some index $i \geq \ell - k$ for $k = log_2(\frac{2}{3(\alpha_\ell + \epsilon - 3\delta)})$. At this point $\alpha_i \geq 1/3$ and $\gamma_i < \gamma_\ell \leq \alpha_\ell - \epsilon + \delta$. Combining this with Eq. 3.2, we can apply the same analysis as in the previous case to first show by induction that $\alpha_{i-k'} \geq 0.12$ for all k' and then more generally: $\alpha_{i-k'} \geq (k'-1)(0.12 - \delta) + \beta(\alpha_i) - \alpha_\ell + \epsilon - 2\delta \geq k'(0.12 - \delta) + 0.68 - \alpha_\ell + \epsilon - 2\delta$. We used the fact that $\beta(\alpha_i) \geq \beta(0.33) \geq 0.68$. Therefore, $\alpha_{i-k'-1} \geq 0.80$ when $k' \geq (0.80 - 0.68 + \alpha_\ell)/(0.12 - \delta)$. This shows that the total number of levels where $\alpha_i < 0.80$ is at most:

$$\ell = k + k' + 1 \leq 1 + log_2\left(\frac{2}{3(\alpha_\ell + \epsilon - 3\delta)}\right) + \frac{0.12 + \alpha_\ell}{0.12 - \delta}$$

The derivative of this expression with respect to α_ℓ is $\frac{1}{0.12-\delta} - \frac{1}{\ln(2)(\alpha_\ell + \epsilon - 3\delta)}$, which is initially decreasing when $\ln(2)(\alpha_\ell + \epsilon - 3\delta) < 0.12 - \delta$ and then increasing for larger α_ℓ. Therefore, the maxima are on the endpoints of the interval $\alpha_\ell \in (\epsilon - \delta, 0.33)$. We already considered the case $\alpha_\ell = 0.33$. When $\alpha_\ell = \epsilon - \delta$ then the number of levels is at most $1 - log_2(3(\epsilon - 2\delta)) + \frac{0.12}{0.12-\delta}$.

In conclusion, the total number of levels before $\alpha_i \geq 0.80$ is at most:

$$\ell \leq max\left((0.68 - \epsilon + \delta)/(0.12 - \delta), 1 - log_2\left(3(\epsilon - 2\delta)\right) + 1/(1 - \delta/12)\right)$$

In particular, when $\delta \leq 0.01$ this becomes $max(7, -log_2(3(\epsilon - 2\delta)) + 3)$.

Relaxing δ. Claim 5 improved on Claim 4 to show unpebbled dependency paths to 80% of the subgraph in some layer. The final improvement is to redistribute the δ_i such that $\sum_i \delta_i = O(\delta)$ but security is still maintained. Intuitively, ensuring $\delta < \epsilon$ is necessary on level V_ℓ as otherwise $\gamma + \delta \geq 1$ and there are no unpebbled nodes on level V_ℓ (all the missing black pebbles can be covered with red pebbles). However, as the dependencies expand between levels a larger δ can be tolerated as well. Although the number of black pebbles the prover will place on each level isn't fixed a priori, we show that if $\delta < \epsilon/2$ in level V_ℓ then we can tolerate a factor $3/2$ increase between levels as long as $\delta \leq 0.05$ in any layer.

That is, if δ_i denotes the bound on the number of red pebbles in the ith layer then our new analysis requires $\delta_\ell < \epsilon/2$ and $\delta_i = min(0.05, (3/2)\delta_{i+1})$. This means that the total number of queries in the PoS over all levels is $O(1/\epsilon)$ because $\sum_{i=1}^{\ell} 1/\delta_i \leq max(0.10\ell, \frac{3}{2\delta_\ell})$.

Claim 6. *For any $\gamma \leq 1 - \epsilon$ and $\delta < \epsilon/3$, if $\mathcal{G}_{SDR}$ initially has at most γn black pebbles, $\delta_\ell = \delta$ red pebbles in layer V_ℓ, and $\delta_i = min(0.05, (2/3)\delta_{i-1})$ red pebbles in layer V_i, then for $\ell = max(13, \log_2(\frac{1}{3(\epsilon - 3\delta)})) + 4)$ the unpebbled nodes in V_ℓ have unpebbled paths to at least $0.80n$ unpebbled nodes in some layer V_i.*

Proof. Modifying Eq. 3.1 to account for the different values of δ_i gives:

$$\alpha_{\ell-k} \geq 2^k \alpha_\ell - 2^{k-1}\gamma_\ell - (2^{k-1}\delta_{\ell-1} + 2^{k-2}\delta_{\ell-2} + \cdots + \delta_{\ell-k})$$

$$\geq 2^k \alpha_\ell - 2^{k-1}\gamma_\ell - \sum_{i=1}^{k} 2^{k-i}(3/2)^{i-1}\delta_\ell$$

Let $\sigma_k = \sum_{i=1}^{k} 2^{k-i}(3/2)^{i-1}$. Then $(4/3)\sigma_k = \sigma_k + 2^{k+1}/3 - (3/2)^{k-1}$. Therefore $\sigma_k = 2^{k+1} - 3^k/2^{k-1} < 2^{k+1}$. Using $\gamma_\ell \leq \alpha_\ell + \delta_\ell - \epsilon$ and $\alpha_\ell \geq \epsilon - \delta_\ell$ we derive the new bound:

$$\alpha_{\ell-k} \geq 2^k \alpha_\ell - 2^{k-1}\gamma_\ell - 2^{k+1}\delta_\ell \geq 2^{k-1}(\alpha_\ell + \epsilon - 5\delta_\ell) \geq 2^k(\epsilon - 3\delta) \qquad (3.3)$$

This shows that if $\ell \geq \log_2(\frac{1}{3(\epsilon - 3\delta)})$ then there is some level V_i where $\alpha_i \geq 1/3$.

We must also modify Eq. 3.2 using $\sum_{j=1}^{k} \rho_{i-j} \leq \gamma_i \leq \alpha_i - \epsilon + \delta_i$:

$$\alpha_{i-k} \geq (k-1)min_{j<k}\hat{\beta}(\alpha_{i-j}) - \sum_{j=0}^{k} \delta_{i-j} + \hat{\beta}(\alpha_i) + \epsilon \qquad (3.4)$$

When $i = \ell$ and k is small $\delta_{\ell-k} = (3/2)^k \delta_\ell$ and $\delta_\ell \leq \epsilon/3$ implies:

$$\alpha_{\ell-k} \geq (k-1)min_{j<k}\hat{\beta}(\alpha_{i-j}) + \hat{\beta}(\alpha_\ell) + (\frac{2}{3} - \frac{3^k}{2^{k+1}})\epsilon$$

Otherwise, we can use $\delta_i \leq 0.05$.

$$\alpha_{i-k} \geq (k-1)(min_{j<k}\hat{\beta}(\alpha_{i-j}) - 0.05) + \hat{\beta}(\alpha_i) + \epsilon - 0.10$$

Now we turn back to the two cases for $\alpha_i \geq 1/3$.

Case $\alpha_\ell \geq 1/3$: We claim that $\alpha_{\ell-k} \geq 0.11$ for all k. This is true for α_ℓ by hypothesis. From the equation above and the bound $\hat{\beta}(\alpha) \geq 0.12$ for all $\alpha \in (0.10, 0.80)$ (Corollary 2), $\alpha_{\ell-1} \geq \hat{\beta}(\alpha_\ell) - \epsilon/12 > 0.11$. Therefore, $\alpha_{\ell-2} \geq (0.12 - 0.05) + 0.12 - (11/24)\epsilon \geq 0.18$. Now assume that $\alpha_{\ell-2-j} \geq 0.12$ for all $j < k$, then $\alpha_{\ell-2-j} \geq (k-1)0.07 + 0.12 - (11/24)\epsilon > 0.11$. The claim follows by induction. This also shows that $\alpha_{\ell-k} \geq (k-3)0.07 + 0.11 > 0.80$ when $k = 13$.

Case $\alpha_\ell < 1/3$: From Eq. 3.3, $\alpha_i \geq 1/3$ at some index $i \geq \ell - k$ for $k = \log_2(\frac{2}{3(\alpha_\ell + \epsilon - 5\delta_\ell)})$. At this point $\alpha_i \geq 1/3$ and $\gamma_i < \gamma_\ell \leq \alpha_\ell - \epsilon + \delta_\ell$. Combining this with Eq. 3.4 gives:

$$\alpha_{i-k'} \geq (k'-1)(min_{j<k'}\hat{\beta}(\alpha_{i-j}) - 0.05) + \beta(\alpha_i) - \alpha_\ell + \epsilon - 0.05 - \delta_\ell$$

We claim that $\alpha_{i-k'} \geq 0.30$ for all k'. Observe that $\alpha i - 1 \geq \beta(\alpha_i) - \alpha_\ell + \epsilon - 0.05 - \delta_\ell \geq 0.68 - 0.38 + \epsilon - \delta_\ell ll \geq 0.30$ for any value $\alpha_\ell < 0.33$ because $\beta(\alpha_i) \geq \beta(0.33) \geq 0.68$. Assuming this is true for all α_{i-j} where $1 < j \leq k'$ implies $\alpha_{i-k'} \geq (k'-1)0.07 + 0.30 \geq 30$. Therefore, we can state more generally that $\alpha_{i-k'} \geq (k'-1)0.07 + 0.68 - \alpha_\ell$ and $\alpha_{i-k'-1} \geq 0.80$ when $k' = (0.12 + \alpha_\ell)/0.07$. The total number of levels where $\alpha_i < 0.80$ is thus at most:

$$k + k' + 1 \leq 1 - \log_2((3/2)(\alpha_\ell + \epsilon - 5\delta_\ell)) + 2 + \alpha_\ell/0.07$$

Differentiating this expression with respect to α_ℓ shows that the maxima over $\alpha_\ell \in (\epsilon - \delta_\ell, 0.33)$ are on the endpoints. The endpoint $\alpha_\ell = 0.33$ coincides with the case above. At the endpoint $\epsilon - \delta_\ell$ the number of levels is bounded by $3 - \log_2(3(\alpha_\ell + \epsilon - 5\delta)) + \epsilon/0.07$.

In conclusion, considering both cases, the total number of levels before $\alpha_i \geq 0.80$ is at most:

$$\ell \geq max(13, 3 - \log_2(3(\epsilon - 3\delta_\ell)) + \epsilon/0.07)$$

In particular, when $\epsilon \leq 0.07$ and $\delta_\ell = \delta$ then $\ell \leq max(12, 4 - \log_2(3(\epsilon - 3\delta)))$.

4   "ZigZag" DRG PoS/PoRep

The Stacked-DRG PoS can be adapted into a PoRep which encodes input data D using the labels on the last level V_ℓ as encoding keys. However, decoding the data requires first re-computing the PoS labels, which is by design expensive.

The basic idea of the ZigZag PoS/PoRep is to layer DRGs so that each layer "encodes" the previous layer. The critical desired property to achieve is: if *all* the labels on a given level are available in memory then the labels can decoded in parallel. To achieve this, instead of adding edge dependencies between the layers, we add the edges of a constant degree expander graph in each layer so that every layer is both depth-robust and high "expansion". Technically, the graph we construct in each layer is an expander as an undirected graph. As a

DAG this means that the union of the dependencies and targets of any subset is large. By alternating the direction of the edges between layers, forming a "zig-zag", we are able to show that the dependencies between layers expand. Now the only edges between layers are between nodes at the same index, and the label on each node encodes the label on the node at the same index in the previous level. The dependencies used for keys are all contained in the same layer. Thus, the labels in any layer are sufficient to recover the labels in the preceding layer. Moreover, the decoding step can be done in parallel.

Without alternating the direction of the edges between layers this construction would fail to be a tight proof of space because the topologically last ϵn nodes in a layer would only depend on the topologically last ϵn nodes in the previous layer. Moreover, if the prover stores the labels on the topologically first $(1-\epsilon)n$ nodes it can quickly recover the labels on the topologically first $(1-\epsilon)n$ nodes in the preceding level, allowing it to recover the missing ϵn labels as well in parallel-time $O(\epsilon n)$.

Construction of $\mathcal{G}_{ZZ}[\ell]$. Similar to $\mathcal{G}_{SDE}$, the graph $\mathcal{G}_{ZZ}[\ell]$ contains a copy of an $(n, 0.80n, \beta n)$-depth-robust graph for some constant β in each of the ℓ layers $V_1, ..., V_\ell$. The nodes in each layer are indexed in $[n]$. Every odd layer overlays the edges of the DRG in the forward direction (edges go from lower to higher indices) while every even layer the edges of the DRG in the reverse direction (edges go from higher indices to lower indices).

Edges are added between same index nodes in adjacent layers (i.e. the ith node in layer V_k is connected to the ith node in layer V_{k+1} for all i, k). Next, the edges that were between layers in $\mathcal{G}_{SDE}[\ell]$ are projected into each layer of $G_{ZZ}[\ell]$ with the direction of each edge determined by the parity of the layer. We call these *expander edges* to distinguish[6] them from other edges. More precisely, if $G_{SDE}[\ell]$ has an edge from the ith node of a layer V_k to the jth node of layer V_{k+1} then $G_{ZZ}[\ell]$ has an edge between the ith node of V_{k+1} and the jth node of V_{k+1}. The direction of the edges added to V_{k+1} is from lower indices to higher indices when $k+1$ is odd, and from higher indices to lower indices when $k+1$ is even. (For concreteness in the analysis, the edges between layers in the reference graph $\mathcal{G}_{SDE}$ are assumed to be constructed using the degree 8 Chung random bipartite graph construction).

DAG encoding. Instead of the standard DAG labeling, the ZigZag PoS/PoRep uses a DAG encoding scheme. It takes in a data file X on n blocks $x_1, ..., x_n$, a salt σ for a collision-resistant hash function $H : \{0,1\}^{md} \to \{0,1\}^m$, and a d-irregular DAG on n nodes together with its parent function $\mathsf{Parents}(i)$ which outputs the parent nodes of the ith node. It uses a randomized encoding scheme $\mathsf{Enc}, \mathsf{Dec}$ to derive the label c_i on each node as $\mathsf{Enc}(k_i, x_i)$ where $k_i \leftarrow H(\sigma||c_{v_1}||\cdots||c_{v_d})$ for $(v_1, ..., v_d) \leftarrow \mathsf{Parents}(i)$. This encoding scheme may be as simple as the identity function, or could use sequentially slow encoding for added delay.

---

[6] The distinction between expander edges and all other edges is important in the analysis. In particular, the expander edges are between the same index nodes in every layer and differ only in their directionality.

4.1   PoS Analysis of ZigZag PoRep

Invertible pebbling game. The red-black pebbling game no longer entirely captures the PoS security of the ZigZag PoRep due to the involvement of the encoding scheme (Enc, Dec) in the labeling rather than purely a collision-resistant hash function. Most significantly, the labels are now invertible. In terms of the dependency graph of the labeling computation, the keys in each layer V_i still need to be computed in topological order, however the labels may either be derived by decoding labels in layer V_{i+1} or encoding labels in layer V_{i-1}. We modify the black pebbling game to capture invertibility of labels by coloring edges.

White & green colored edges. White edges are "one-way streets" corresponding to edge dependencies involved in deriving keys via calls to the random oracle and are treated like normal pebbling game edges. Green edges are "two-way street", but still have a direction and different rules in either direction. If there is a directed green edge from u to v then a pebble can be placed on v if and only if u and all nodes with white edges to v have pebbles. A pebble can be placed on u if and only if v and all nodes *with white edges to* v have pebbles.

We still analyze the soundness of a PoS with invertible labels through the game Red-Black-Pebbles as in Definition 1, however with the modification that the adversary plays the black pebbling game with white/green edges instead of the plain black pebbling game. Specifically we analyze the hardness of a modification of Red-Black-Pebbles$^{\mathcal{A}}(\mathcal{G}_{ZZ}[\ell], V_\ell, \mathsf{Chal})$ using green/white edges where the directed edges within every layer V_i are white and the directed edges between the same index nodes in adjacent layers are green. Our analysis, included in the full version of this paper, will demonstrate in this model that the ZigZag PoRep is an arbitrarily tight PoS with only $\ell = O(\log(1/\epsilon))$ layers.

Acknowledgments. This research was generously supported by an NSF Graduate Fellowship. Joseph Bonneau, Nicola Greco, and Juan Benet provided critical input throughout the development of this work and are coauthors on a related systems project prototyping practical implementations of PoReps, including the constructions presented at BPASE 2018 and discussed in further detail in this work. Many others have contributed through helpful comments and conversations, including Dan Boneh, Rafael Pass, Ethan Cecchetti, Benedikt Bünz, and Florian Tramer.

References

1. Proof of replication. Protocol Labs (2017). https://filecoin.io/proof-of-replication.pdf
2. Abusalah, H., Alwen, J., Cohen, B., Khilko, D., Pietrzak, K., Reyzin, L.: Beyond Hellman's time-memory trade-offs with applications to proofs of space. In: Takagi, T., Peyrin, T. (eds.) ASIACRYPT 2017. LNCS, vol. 10625, pp. 357–379. Springer, Cham (2017). https://doi.org/10.1007/978-3-319-70697-9_13
3. Alwen, J., Blocki, J., Harsha, B.: Practical graphs for optimal side-channel resistant memory-hard functions. In: CCS (2017)

4. Alwen, J., Blocki, J., Pietrzak, K.: Sustained space complexity. In: Nielsen, J.B., Rijmen, V. (eds.) EUROCRYPT 2018. LNCS, vol. 10821, pp. 99–130. Springer, Cham (2018). https://doi.org/10.1007/978-3-319-78375-8_4

5. Boneh, D., Bonneau, J., Bünz, B., Fisch, B.: Verifiable delay functions. In: Shacham, H., Boldyreva, A. (eds.) CRYPTO 2018. LNCS, vol. 10991, pp. 757–788. Springer, Cham (2018). https://doi.org/10.1007/978-3-319-96884-1_25

6. Chung, F.R.K.: On concentrators, superconcentrators, generalizers, and nonblocking networks. Bell Syst. Tech. J. **58**, 1765–1777 (1979)

7. Dwork, C., Naor, M., Wee, H.: Pebbling and proofs of work. In: Shoup, V. (ed.) CRYPTO 2005. LNCS, vol. 3621, pp. 37–54. Springer, Heidelberg (2005). https://doi.org/10.1007/11535218_3

8. Dziembowski, S., Faust, S., Kolmogorov, V., Pietrzak, K.: Proofs of space. In: Gennaro, R., Robshaw, M. (eds.) CRYPTO 2015. LNCS, vol. 9216, pp. 585–605. Springer, Heidelberg (2015). https://doi.org/10.1007/978-3-662-48000-7_29

9. Fisch, B.: Poreps: proofs of space on useful data. Cryptology ePrint Archive, Report 2018/678 (2018). https://eprint.iacr.org/2018/678

10. Fisch, B., Bonneau, J., Benet, J., Greco, N.: Proofs of replication using depth robust graphs. In: Presentation at Blockchain Protocol Analysis and Security Engineering (2018). https://cyber.stanford.edu/bpase2018

11. Ateniese, G., Bonacina, I., Faonio, A., Galesi, N.: Proofs of space: when space is of the essence. In: Abdalla, M., De Prisco, R. (eds.) SCN 2014. LNCS, vol. 8642, pp. 538–557. Springer, Cham (2014). https://doi.org/10.1007/978-3-319-10879-7_31

12. Boneh, D., Corrigan-Gibbs, H., Schechter, S.: Balloon hashing: a provably memory-hard function with a data-independent access pattern. In: Asiacrypt (2016)

13. Lerner, S.D.: Proof of unique blockchain storage (2014). https://bitslog.wordpress.com/2014/11/03/proof-of-local-blockchain-storage/

14. Mahmoody, M., Moran, T., Vadhan, S.P.: Time-lock puzzles in the random oracle model. In: Rogaway, P. (ed.) CRYPTO 2011. LNCS, vol. 6841, pp. 39–50. Springer, Heidelberg (2011). https://doi.org/10.1007/978-3-642-22792-9_3

15. Vadhan, S., Reingold, O., Wigderson, A.: Entropy waves, the zig-zag graph product, and new constant-degree expanders and extractors. In: FOCS (2000)

16. Park, S., Pietrzak, K., Kwon, A., Alwen, J., Fuchsbauer, G., Gǎzi, P.: Spacemint: a cryptocurrency based on proofs of space. Cryptology ePrint Archive, Report 2015/528 (2015). http://eprint.iacr.org/2015/528

17. Graham, R.L., Erdös, P., Szemeredi, E.: On sparse graphs with dense long paths. In: Computers & Mathematics with Applications (1975)

18. Pietrzak, K.: Proofs of catalytic space. Cryptology ePrint Archive # 2018/194 (2018)

19. Ren, L., Devadas, S.: Proof of space from stacked expanders. In: Hirt, M., Smith, A. (eds.) TCC 2016. LNCS, vol. 9985, pp. 262–285. Springer, Heidelberg (2016). https://doi.org/10.1007/978-3-662-53641-4_11

Secure Computation

Founding Secure Computation on Blockchains

Arka Rai Choudhuri[1]([envelope])[ORCID], Vipul Goyal[2], and Abhishek Jain[1]

[1] Johns Hopkins University, Baltimore, USA
{achoud,abhishek}@cs.jhu.edu
[2] Carnegie Mellon University, Pittsburgh, USA
goyal@cs.cmu.edu

Abstract. We study the foundations of secure computation in the *blockchain-hybrid model*, where a blockchain – modeled as a *global* functionality – is available as an Oracle to all the participants of a cryptographic protocol. We demonstrate both destructive and constructive applications of blockchains:

- We show that classical rewinding-based simulation techniques used in many security proofs fail against *blockchain-active* adversaries that have read and post access to a global blockchain. In particular, we show that zero-knowledge (ZK) proofs with black-box simulation are impossible against blockchain-active adversaries.
- Nevertheless, we show that achieving security against blockchain-active adversaries is possible if the honest parties are also blockchain active. We construct an $\omega(1)$-round ZK protocol with black-box simulation. We show that this result is tight by proving the impossibility of constant-round ZK with black-box simulation.
- Finally, we demonstrate a novel application of blockchains to overcome the known impossibility results for concurrent secure computation in the plain model. We construct a concurrent self-composable secure computation protocol for general functionalities in the blockchain-hybrid model based on standard cryptographic assumptions.

We develop a suite of techniques for constructing secure protocols in the blockchain-hybrid model that we hope will find applications to future research in this area.

1 Introduction

Blockchain is an exciting new technology which is having a profound impact on the world of cryptography. Blockchains provide both: new applications of existing cryptographic primitives (such as hash function, or zero-knowledge proofs), as well as, novel foundations on which new cryptographic primitives can be realized (such as fair-secure computation [2,10,24], or, one-time programs [36]). In this work, we seek to examine the foundations of secure computation protocols in the context of blockchains. More concretely, we study what we call the

© International Association for Cryptologic Research 2019
Y. Ishai and V. Rijmen (Eds.): EUROCRYPT 2019, LNCS 11477, pp. 351–380, 2019.
https://doi.org/10.1007/978-3-030-17656-3_13

blockchain-hybrid model and examine constructions of zero-knowledge and secure computation in this model.

The Blockchain-Hybrid Model. In order to facilitate the use of blockchains in secure computation, we study the blockchain-hybrid model, where the blockchain – modeled as a *global* ledger functionality – is available to all the participants of a cryptographic protocol. The parties can access the blockchain by posting and reading content, but no single party has any control over the blockchain. Our modeling follows previous elegant works on formalizing the blockchain functionality [3,4,47]. In particular, our model is based on the global blockchain ledger model from Badertscher et al. [4].

We study simulation-based security in the blockchain-hybrid model. In our model, *the simulator does not have any control over the blockchain*, and simply treats it as an oracle just like protocol participants. Thus, unlike traditional trusted setup models such as common reference string, the blockchain-hybrid model does not provide any new "power" to the simulator. In particular, the simulator is restricted to its plain model capabilities such as resetting the adversary or using knowledge of its code. Thus, in our model, the blockchain can be *global*, in that it can be used by multiple different protocols at the same time. This is reminiscent of simulation in the global UC framework [15,18,43]. A related model is the global Random Oracle model [18] where the simulator can only observe the queries made by the adversary to the random oracle, but cannot program the random oracle (since it is global and therefore shared across many protocols).

Secure Computation based on Blockchains. We study the foundations of secure computation in the presence of the global blockchain functionality. Interestingly, we demonstrate both destructive and constructive applications of blockchains to cryptography. Primitives which were earlier possible to realize now become impossible. At the same, working in this model allows us to overcome previously established deep impossibility results in cryptography. Interestingly, we also utilize mining delays – typically viewed as a negative feature of blockchains – for constructive purposes in this work. Our main results as discussed next.

1.1   Our Results

Simulation Failure in the Presence of Blockchains. We consider a new class of adversaries that we refer to as *blockchain-active adversaries*. These adversaries are similar to usual cryptographic adversaries, except that they have user access to a blockchain, i.e., they can post on the blockchain and read its state at any point.

We observe that such adversaries can foil many existing simulation techniques that are used for proving security of standard cryptographic schemes. To illustrate the main idea, let us consider rewinding-based black-box simulation techniques that are used, e.g., in zero-knowledge (ZK) proofs [35], secure multiparty computation [34,59], and signature schemes in the random oracle model constructed via the Fiat-Shamir heuristic [29]. A crucial requirement for the success of rewinding-based simulation is that the adversary should be *oblivious* to the rewinding. Usually, this requirement can be easily met since the simulator can simply "reset" the code of the adversary, which prevents it from keeping state across the rewindings.

A blockchain-active adversary, however, can periodically post on the blockchain and use it to maintain state across rewindings, and therefore detect that it is being rewound. In this case, the adversary can simply abort and therefore fail the simulation process.[1] It is not too difficult to turn the above idea into a formal impossibility result for ZK proofs against blockchain-active adversaries, when the simulation is required to be *black-box*.

Theorem 1 (Informal). *There does not exist an interactive argument in the plain model which is zero-knowledge w.r.t. black-box simulation against blockchain-active adversaries.*

The above impossibility result extends to secure multiparty computation and other natural cryptographic primitives whose security is proven via a rewinding simulator.

Constructing Zero-Knowledge Protocols. To overcome the above problems posed by blockchains, we look towards blockchains for a solution as well. Our idea is to make the *protocol* blockchain active as well. That is, in addition to the adversary, the honest parties would have access to the blockchain as well.

Our first positive result is an $\omega(1)$-round ZK proof system in the blockchain-hybrid model whose security is proven w.r.t. black-box simulation.

Theorem 2 (Informal). *Assuming collision-resistant hash functions, there exists an $\omega(1)$-round ZK proof system in the blockchain-hybrid model w.r.t. black-box simulation.*

Interestingly, in our construction, *the honest parties do not post any message on the blockchains*. Instead, they only keep a "tab" on the current state of the blockchain in order to decide whether or not to continue the protocol.

We also show that the above result is tight. Namely, we show that using black-box simulation, constant-round ZK is impossible in the blockchain-hybrid model.

Theorem 3 (Informal). *Assuming one-way functions, there does not exist an $O(1)$-round ZK argument system in the blockchain-hybrid model w.r.t. a (expected probabilistic polynomial time) black-box simulator.*

---

[1] This is reminiscent to the problems that arise in the context of UC security, where the adversary cannot be rewound since it can communicate with an external environment, leading to broad impossibility results for zero-knowledge and secure computation [14,16,19].

This is in sharp contrast to the plain model where there are a number of classical constant round zero-knowledge protocols that are proven secure w.r.t. a black-box simulator [9, 28, 33].

Concurrent Secure Computation using Blockchains. Classical secure computation protocols such as [34, 59] only achieve "stand-alone" security, and fail in the setting of *concurrent self-composition*, where multiple copies of a protocol may be executed concurrently, under the control of an adversary. In fact, achieving concurrent secure computation in the plain model has been shown to be impossible [1, 8, 13, 32, 37, 50–52]. The above impossibility results are far reaching and rule out secure computation for a large class of functionalities in a variety of settings.

Interestingly, we show that concurrent self-composition is possible in the blockchain-hybrid model w.r.t. standard real/ideal model notion of security with a PPT simulator. Thus, our results (put together) show that designing cryptographic primitives in the blockchain-hybrid model is, in some sense, harder and easier at the same time.

Theorem 4 (Informal). *Assuming collision-resistant hash functions and oblivious transfer, there exists a concurrent self-composable secure computation protocol for all polynomial-time functionalities in the blockchain-hybrid model.*

In our protocol, each party is required to post an initial message (which corresponds to a commitment to its input and randomness) on the blockchain. However, an honest party can simply perform this posting in an "offline" phase prior to the start of the protocol. In particular, once the protocol starts, an honest party is not required to post any additional message on the blockchain.

A number of prior beautiful works have constructed concurrent (and universally composable) secure computation in various setup models such as the trusted common reference string model [21], the registered public-key model [6], the tamper-proof hardware model [22, 39, 45], and the physically uncloneable functions model [5, 12, 25]. We believe that the blockchain model provides an appealing *decentralized* alternative to these models since there are no physical assumptions or centralized trusted parties involved. Moreover, it allows for basing concurrent security on an already existing and widely used infrastructure. Further, it is possible to obtain strong guarantees of the following form: an adversary who can break our construction can also break the security of the underlying blockchain (potentially allowing it to gain large amounts of cryptocurrency), or the underlying cryptographic assumptions (oblivious transfer and collision-resistant hash functions in our case).

Impossibility of UC Security. While Theorem 4 establishes the feasibility of concurrent self-composition, we show that universal composition security [14] is impossible in the blockchain-hybrid model:

Theorem 5 (Informal). *Universally composable commitments are impossible in the blockchain-hybrid model.*

We prove the above result via a simple adaptation of the impossibility result of [16] to the blockchain-hybrid model. The main intuition behind this result is that a simulator in the blockchain-hybrid model has the same capabilities as in the plain model, namely, the ability to rewind the adversary or using knowledge of its code. Crucially, (unlike the non-programmable random oracle model [18]) the ability to see the queries made to the blockchain do not constitute a new capability for the simulator since *everyone* can see those queries.

1.2 Technical Overview

We start with the observation that if an adversary is blockchain-active, it can "detect" that it is being rewound by posting the transcript of the interaction so far on the blockchain. In more detail, upon getting an incoming message, the adversary concatenates the entire transcript with a session ID and submits it to the blockchain Oracle. Before giving a response, the adversary waits for the next block to be mined and checks the following: the transcript it posted on the blockchain has indeed appeared, and, no such transcript (for the same session and the same round) appeared on any of the prior blocks. If the check passes (which is guaranteed in the real execution), the adversary proceeds honestly with computing and sending the next protocol message. We show that it would be impossible for any polynomial-time simulator to rewind this adversary which forms the basis of our black-box impossibility result for zero-knowledge.

Constructing Black-Box Zero-Knowledge Protocols. To overcome the above problems posed by blockchains, we look towards blockchains for a solution as well. Our idea is to make the *protocol* blockchain active as well. Specifically, we let the honest prover keep track of the blockchain state, and, if the number of new blocks mined since the beginning of the protocol exceed a fixed number k, abort. Thus, the honest parties use the blockchain to implement a time-out mechanism. We emphasize, however, that we do not require the honest parties to have synchronized clocks. The only requirement placed is that the protocol must be finished in an a priori bounded amount of time, *as measured by the progress of the blockchain*. For example, while using Bitcoin, if k is set to 20, this gives the parties nearly 3.5 h to finish the zero-knowledge protocol before a time-out occurs (since a block is mined roughly every 10 min in Bitcoin). For simplicity, we will treat the parameter k as a constant (even though our constructions can handle an arbitrary value of k by scaling the round complexity of the protocol appropriately).

We devise a construction for black-box zero-knowledge proofs where the number of "slots" (or rewinding opportunities) in the protocol is higher than k. While the adversary can send any information to the blockchain Oracle at any point of time, there can be at most k points in the protocol execution where the adversary actually *receives* from the Oracle a new (unforgeable) mined block. However by our construction, this would still leave several slots in the protocol where the simulator is free to rewind (without having to forge the blockchain state).

A potentially complication in the design of the simulator arises from the fact that, apart from the newly mined blocks, the adversary can also "listen in" on the network communication in real time. This could consist of various (honest party) transactions currently outstanding on the network and waiting to be included in the next block. This is formalized by *buffer reads* in the model of Badertscher et al. [4]. We handle this problem by having the simulator simply replay the honest-party outstanding transactions since they could not have changed from the main thread to the look-ahead thread. The adversarial outstanding transactions (which might change from thread to thread) in the current thread are already known to the simulator since the simulator can read all outgoing messages from the adversary. The above ideas form the basis of our first positive result modulo the issue of simulation time which is discussed next.

The Issue of Simulation Time. Interestingly, the fact that blockchains can be used to implement a global unforgeable clock presents a novel challenge in proving security against blockchain-active adversaries, that to the best of our knowledge, does not arise elsewhere in cryptography. Typically in cryptography, the running time of the simulator is larger than the running time of the adversary. This means that the number of blocks mined during a simulated execution may be higher than the number of blocks mined during a real execution. Then, the number of mined blocks can be used as "side-channel" information to distinguish real and simulated executions, if the adversary and the distinguisher are blockchain-active! Such a difficulty does not arise in the plain model since the simulator is assumed to have complete control over the clock of the adversary (including the ability to freeze it).

To address this issue, we seek to construct a simulator whose running time is the same as the real protocol execution. Towards that end, we build upon techniques from the notion of precise zero-knowledge [53]. To start with, it would seem that we need to construct a simulator with precision exactly 1, something that is currently not known to be possible. To resolve this problem, our key observation is that there is a crucial difference between the *time that the simulator takes to finish* and *the number of computation steps it executes*. In particular, if the simulator can execute a number of computations *in parallel*, it could potentially perform more computations than the prover in the real execution, and yet, finish in the same amount of time. Our rewinding strategy would run several threads of execution in parallel (e.g., by making several copies of the adversary code) and ensure that by the time the *main*[2] thread finishes, all the rewound execution threads have finished as well. To ensure that the simulation succeeds, our simulator is necessarily required to have a super-constant number of rewinding opportunities (which can be pursued in parallel). Such a simulator would give a guarantee of the following form: any information learnt by an adversarial verifier in the protocol could also be produced from scratch by an algorithm which is capable of running sufficient (polynomial) number of computations in parallel. For example, a quad core processor is capable of running 4 parallel computations.

We believe that the issue of simulation time is one of independent interest. In particular, developing an understanding of the time required by the simulator

---

[2] The thread output by the simulator is referred to as the main thread.

(as opposed to the number of computation steps) could shed additional light on the knowledge complexity of cryptographic constructions as well as motivate the study of strong notions of security.

Lower Bound on Round Complexity of Black-Box Zero-Knowledge. We prove that constant round ZK arguments are impossible w.r.t black-box simulation in the blockchain-hybrid model. Our impossibility result holds even for expected polynomial-time simulators.

Consider an adversarial verifier that waits for a fixed constant time c before responding to any message from the prover. Our proof works in two steps:

1. Recall that black-box simulators can only query the adversarial verifier as an Oracle. However, the simulator may choose to make these queries *in parallel* rather than sequentially by making several copies of the adversary state (and hence, increasing the number of available Oracles).

   In the first step, we assume that the simulator is *memory bounded*. This means that at any given time, the simulator may only have a bounded (strict polynomial) number of copies (say) $q(\cdot)$ of the adversary. Furthermore, since the verifier takes time c to answer each query, the total number of queries the simulator may make to the adversary in a given time t can be bounded by $\frac{q \cdot t}{c}$ (an a priori bounded strict polynomial). Now we observe the following:
   - The simulator must terminate within roughly t steps where t is the time an honest prover takes to complete the proof. To see this, let r be an upper bound on the number of blocks that can created in the time taken by the honest prover to complete the proof. We consider a blockchain active adversary that observes the state of the blockchain when the protocol starts, and posts a transcript on the completion of the proof. If it notices that more than r blocks have been created since the protocol started, it concludes that it is interacting with the simulator.
   - Thus, the overall number of queries (and hence) the running time of the simulator is a strict polynomial. Now, we can directly invoke the result of Barak and Lindell [7] that rules out constant-round ZK arguments with strict polynomial-time black-box simulation.

2. The above only rules out a simulator with "a priori bounded parallelism". However what if, e.g., the number of parallel queries the simulator may make to the verifier cannot be a priori bounded (and instead we only require that the simulator finish in a priori bounded number of computational steps)? In particular, the simulator may see the responses to the queries made so far, and, *adaptively* decide to increase the number of parallel queries (i.e., the number of copies of the adversary)? This case is more tricky and as such, the ideas from the work of [7] don't apply.

   To resolve this issue, we crucially rely upon the fact that by carefully choosing the delay parameter c and an aborting probability for the adversary, the number of such "adaptive steps" can be bounded by a constant. Thereafter, we argue that in each adaptive step, if the simulator increases the number of parallel copies by more than an a priori bounded polynomial factor, it runs the risks of blowing the number of computation steps to beyond expected polynomial. On the other hand if the number of parallel copies blow up by at most a fixed polynomial factor, since the number of adaptive steps is a constant, the simulator is still using "bounded parallelism" (a case already covered by our previous step). The full proof is delicate and can be found in the full version.

Concurrent Secure computation. We now proceed to describe the main ideas behind our positive result for concurrent self-composable secure computation. We start by recalling the intuition behind the impossibility of concurrent secure computation w.r.t. black-box simulation in the plain model.

A primary task of a simulator for a secure computation protocol is to extract the adversary's input. A black-box simulator extracts the input of the adversary by rewinding. However, in the concurrent setting, extracting the input of the adversary in each session is a non-trivial task. In particular, given an adversarial scheduling of the messages of concurrent sessions, it may happen that in order to extract the input of the adversary in a given session s, the simulator rewinds past the beginning of another session s' that is interleaved inside the protocol messages of session s. When this happens, the adversary may change its input in session s'. Thus, the simulator would be forced to query the ideal functionality more than once for the session s'.

Indeed, as shown in [51], this intuition can be formalized to obtain a black-box impossibility result for concurrent self-composition w.r.t. the standard definition of secure computation, where only one query per session is allowed. While Lindell's impossibility result is only w.r.t. black-box simulation, subsequent works have shown impossibility of concurrent secure computation even w.r.t. non-black-box simulation [1,8,32,37].

In order to overcome the impossibility results, our starting idea is the following: prior to the start of a protocol, each party must commit to its input and randomness on the blockchain. It must then wait for its commitment string to be posted on the blockchain before sending any further message in the protocol. Similar to our ZK protocols (with stand-alone security), we use a time-out mechanism to place an upper bound on the number of blocks that can be mined during a session. Then, by using sufficiently many rewinding "slots," we can ensure that there exist some slots in each session where the adversary is guaranteed to not see new block (and hence no new interleaved sessions), making them "safe" for rewinding. Note, however, that this mechanism does not bound the overall number of concurrent sessions since an adversary can start any polynomial number of sessions *in parallel*.

Once we have the above protocol template, the key technical challenge is to perform concurrent extraction of the adversary's inputs in all of the sessions. Since there are multiple "unsafe" rewinding slots in every session (wherever a new block is mined), we need to extract adversary's inputs in all of the sessions under the constraint that only the safe slots are rewound. Unfortunately, commonly known rewinding strategies in the concurrent setting [49,57,58] rewind all parts of the protocol transcripts (potentially multiple times). Therefore, they immediately fail in our setting.

In order to solve this problem, we develop a new concurrent rewinding strategy. The starting idea towards developing this rewinding strategy is the observation that our particular setting has some similarities to the work of Goyal et al. [41] who were interested in a seemingly unrelated problem: designing commitment schemes that are secure w.r.t. chosen commitment attacks [20]. Goyal et al. introduced what they call the "robust extraction lemma" that guarantees concurrent extraction even if a *constant* number of "breakpoints" – that cannot be rewound – are interspersed throughout the overall transcript of the concurrent sessions. These breakpoints are analogous to the unsafe points in our setting.

While this serves as a useful starting point, robust extraction is not directly applicable to our setting since overall, the number of external blocks seen by the adversary (the equivalent of breakpoints in [41]) cannot be bounded. Indeed, if the number of sessions is T, the number of blocks can only be upper bounded by $T \cdot k$ (if e.g., all the sessions are sequential).

Our main observation is that the concurrent adversary can only choose one of the following: either too much concurrency, or too many newly mined blocks, but not both. This allows us to come up with a new variant and analysis of the robust extraction lemma which we believe could be of independent interest. In particular, our new variant uses twice as many slots as the one used by the robust extraction lemma. We refer the reader to the technical sections for more details.

1.3 Related Work

Blockchains and Cryptography. In a recent work, [36] used blockchains to construct non-interactive zero-knowledge (NIZK) arguments and selectively-secure one-time programs. Their model, however, is fundamentally different from ours in that they rely on a much stronger notion of simulation where the simulator controls all the honest miners in the blockchain. Intuitively, this is somewhat similar to the honest majority model used to design (universally composable) secure multiparty computation protocols. Due to the power given to the simulator, their model necessitates the blockchain to be "local" (i.e., private) to the protocol. In contrast, our model allows for the blockchain to be a "global" setup since the simulator has no extra power over the blockchain compared to the adversary. This is similar to the difference between universal composability framework [14] and global universal composability framework [15], where in the former model, a setup (such as a common reference string) cannot be reused by different protocols, whereas in the latter model, a common setup can be used across multiple protocols. Indeed, since the simulator has no additional power except the ability to reset the adversary or use knowledge of its code, NIZKs are impossible in our model, similar to the plain model. Unlike our work, [36] do not consider interactive ZK proofs or any notion of secure multiparty computation.

In another recent work, [24] study the problem of fair multiparty computation in a "bulletin-board" model that can be implemented with blockchains. Similar to [36], however, their model provides the simulator the ability to control the blockchain. Prior to their work, multiple works [2,10] studied the problem of fairness with penalties using cryptocurrencies.

Several elegant works have conducted a formal study of various properties of blockchains [4,30,31,46,56]. Most relevant to our work is that of Badertscher et al. [4] whose modeling of the blockchain ideal functionality we closely follow.

Concurrent Security. The study of concurrent security for cryptographic protocols was initiated by Dwork et al. [27] who also introduced a timing model for constructing concurrent ZK. In this model, the parties have synchronized clocks and are required to insert "delays" at appropriate points in the protocol. A refined version of their model was later considered in [44] for the problem of concurrent secure computation. We note that while our approach to concurrent secure computation in the blockchain-hybrid model appears to bear some similarity to the timing model, there are fundamental differences that separate these models. For example, the simulator can fully control the clock of the adversary in the timing model, while this is not possible in our setting since the blockchains provide an unforgeable clock to the adversary. More importantly, in the timing model, there are no "unsafe" points, and the simulator can rewind anywhere. For this reason, the timing model does not require developing new concurrent extraction techniques, and instead standard rewinding techniques for the stand-alone setting are applicable there. Finally, in the timing model, honest parties insert artificial delays in the protocol based on their clocks, while in our constructions, an honest party responds immediately to messages received from the other (possibly adversarial) party.

1.4 Organization

We start with our model of the blockchain in Sect. 2, and all subsequent results are in this model. In Sect. 4 we describe a $\omega(1)$ round black-box zero-knowledge protocol. We describe our concurrently extractable commitment scheme in Sect. 5 and use our constructed commitment scheme to achieve a concurrently secure two-party computation protocol described in Sect. 5.2.

1.5 Full Version

Due to space constraints, our impossibility results and the security proofs are omitted from this manuscript, and appear in the full version of this paper [23].

2 Blockchain Model

Blockchains. In a blockchain protocol, the goal of all parties is to maintain a global ordered set of records that are referred to as *blocks*. New blocks can only be added using a special mining procedure that simulates a puzzle-solving race between participants and can be run by any party (called miner) executing the blockchain protocol. Presently, two broad categories of puzzles are used: Proof-of-Work (PoW) and Proof-of-Stake (PoS).

Following the works of [3,4,47], we model the blockchain as a global ledger $\mathcal{G}_{\mathsf{ledger}}$ that internally keeps a state state which is the sequence of all the blocks in the ledger. Parties interact with the ledger by making one of many queries described by the functionality.

We reproduce here the ledger functionality described in [4] with a few minor modifications to be described subsequent to the description.

The ledger maintains a central and unique permanent state denoted by state. When data/transactions are sent to $\mathcal{G}_{\mathsf{ledger}}$, they are validated using a Validate predicate and added to a buffer buffer. The buffer is meant to indicate those transactions that are not sufficiently deep to become permanent. The Blockify function creates a block including some transactions from buffer and extends state. The decision of when the state is extended is left to the adversary. The adversary proposes a next block candidate NxtBC containing the transactions from the buffer it wants included in the state. An empty NxtBC is used to indicate that the adversary does not want the state to be updated at the current clock tick. To restrict the behavior of the adversary, there is a ledger algorithm ExtendPolicy that enforces a state-update policy restriction. Further discussion on the ExtendPolicy can be found in the full version.

Each registered party can see the state, but is guaranteed only a sufficiently long prefix of it. This is implemented by monotonically increasing pointers pt_i, defining the prefix $\mathsf{state}|_{\mathsf{pt}_i}$, for each party that the adversary can manipulate with some restrictions. This can be viewed as a sliding window over the state, wherein the adversary can only set pointers to be within this window starting from the head of state. The size of the sliding window is denoted by windowSize. It should be noted that the prefix view guarantees that the value at position k will appear in position i in every party's state.

A party is said to be desynchronized if the party recently registered or recently got de-registered from the clock. At this point, due to network delays, the adversary can make the parties believe in any value of the state up until the party gets messages from the network. This time period is denoted by the parameter Delay, wherein the desynchronized parties are practically under the control of the adversary. A timed honest input sequence $\overrightarrow{\mathcal{I}}_H^T$, is a vector of the form $((x_1, P_1, \tau_1), \cdots, (x_m, P_m, \tau_m))$, used to denote the inputs received by the parties from the environment, where P_i is the player that received the input and τ_i was the time of the clock when the environment handed the input to P_i. The ledger uses the function predict-time to ensure that the ideal world execution advances with the same pace (relative to the clock) as the protocol does. $\overrightarrow{\tau}_{\mathsf{state}}$ denotes the block-insertion times vector, which lists the times each block was inserted into state.

Functionality $\mathcal{G}_{\mathsf{ledger}}$

$\mathcal{G}_{\mathsf{ledger}}$ is parameterized by found algorithms, Validate, ExtendPolicy, Blockify, and predict-time: windowSize, Delay$\in \mathbb{N}$. The functionality manages variables state, NxtBCbuffer, τ_L, and $\overrightarrow{\mathcal{T}}_{\mathsf{state}}$ as described above. The variables are initialized as follows: state $:= \overrightarrow{\mathcal{T}}_{\mathsf{state}} := \mathsf{NxtBC} := \varepsilon$, buffer $:= \emptyset$, $\tau_L = 0$.

The functionality maintains the set of registered parties $\mathcal{P}$, the subset of honest parties $\mathcal{H} \subseteq \mathcal{P}$ and the subset of de-synchronized honest parties $\mathcal{P}_{DS} \subset \mathcal{H}$. The sets $\mathcal{P}, \mathcal{H}, \mathcal{P}_{DS}$ are all initially set to $\emptyset$. When a new honest party is registered at the ledger, if it is registered with the clock already then it added to the party sets $\mathcal{H}$ and $\mathcal{P}$ and the current time of registration is also recorded if the current time $\tau_L > 0$, it is also added to $\mathcal{P}_{DS}$. Similarly, when a party is deregistered, it is removed from both $\mathcal{P}$ (and therefore also from $\mathcal{P}_{DS}$ or $\mathcal{H}$). The ledger maintains the invariant that it is registered (as a functionality) to the clock whenever $\mathcal{H} \neq \emptyset$.

For each party $P_i \in \mathcal{P}$ the functionality maintains a pointer pt_i (initially set to 1) and a current-state view state$_i := \varepsilon$ (initially set to empty). The functionality also keeps track of the timed honest-input sequence in a vector $\overrightarrow{\mathcal{I}}_H^T$ (initially $\overrightarrow{\mathcal{I}}_H^T := \varepsilon$)

Upon receiving any input I from any party or from the adversary, send $(\mathsf{CLOCK\text{-}READ}, \mathsf{sid}_C)$ to $\mathcal{G}_{\mathsf{clock}}$ and upon receiving the response $(\mathsf{CLOCK\text{-}READ}, \mathsf{sid}_C, \tau)$, set $\tau_L := \tau$ and do the following:

1. Let $\widehat{\mathcal{P}} \subseteq \mathcal{P}_{DS}$ denote the set of desynchronized honest parties that have been registered continuously since time $\tau' < \tau_L - \mathsf{Delay}$ (to both ledger and clock). Set $\mathcal{P}_{DS} := \mathcal{P}_{DS} \setminus \widehat{\mathcal{P}}$.

2. If I was received from an honest party $P_i \in \mathcal{P}$:
   (a) Set $\overrightarrow{\mathcal{I}}_H^T := \overrightarrow{\mathcal{I}}_H^T \| (I, P_i, \tau_L)$;
   (b) Compute $\overrightarrow{N} = (\overrightarrow{N}_1, \cdots, \overrightarrow{N}_\ell) := \mathsf{ExtendPolicy}\left(\overrightarrow{\mathcal{I}}_H^T, \mathsf{state}, \mathsf{NxtBC}, \mathsf{buffer}, \overrightarrow{\mathcal{T}}_{\mathsf{state}}\right)$ and if $\overrightarrow{N} \neq \varepsilon$ set $\mathsf{state} := \mathsf{state} \| \mathsf{Blockify}(\overrightarrow{N}_1) \| \cdots \| \mathsf{Blockify}(\overrightarrow{N}_\ell)$ and $\overrightarrow{\mathcal{T}}_{\mathsf{state}} := \overrightarrow{\mathcal{T}}_{\mathsf{state}} \| \tau_L^\ell$ where $\tau_L^\ell = \tau_L \| \cdots \| \tau_L$.
   (c) If there exists $P_j \in \mathcal{H} \setminus \mathcal{P}_{DS}$ such that $|\mathsf{state}| - \mathsf{pt}_j > \mathsf{windowSize}$ or $\mathsf{pt}_j < |\mathsf{state}_j|$, then set $\mathsf{pt}_k := |\mathsf{state}|$ for all $P_k \in \mathcal{H} \setminus \mathcal{P}_{DS}$.
   (d) If $\overrightarrow{N} \neq \varepsilon$, send (state) to $\mathcal{A}$; else send (I, P_i, τ_L) to $\mathcal{A}$

3. Depending on the above input I and its sender's ID, $\mathcal{G}_{\mathsf{ledger}}$ executes the corresponding code from the following list:
   – *Submitting data:*
     If $I = (\mathsf{SUBMIT}, \mathsf{sid}, x)$ and is received from a party $P_i \in \mathcal{P}$ or from $\mathcal{A}$ (on behalf of corrupted party P_i) do the following
     (a) Choose a unique identifier uid and set $y := (x, \mathsf{uid}, \tau_L, P_i)$
     (b) $\mathsf{buffer} := \mathsf{buffer} \cup \{y\}$.

(c) Send (SUBMIT, y) to $\mathcal{A}$ if not received from $\mathcal{A}$.

- *Reading the state:*
  If $I = (\mathsf{READ}, \mathsf{sid})$ is received from a party $P_i \in \mathcal{P}$ then set $\mathsf{state}_i :=$ $\mathsf{state}|_{\min\{\mathsf{pt}_i, |\mathsf{state}|\}}$ and return $(\mathsf{READ}, \mathsf{sid}, \mathsf{state}_i)$ to the requestor. If the the requestor is $\mathcal{A}$ then send $(\mathsf{state}, \mathsf{buffer})$.

- *Maintain the ledger state:*
  If $I = (\mathsf{MAINTAIN\text{-}LEDGER}, \mathsf{sid})$ is received by an honest $P_i \in \mathcal{P}$ and $\mathsf{predict\text{-}time}(\overrightarrow{\mathcal{I}}_H^T) = \tilde{\tau} > \tau_L$ then send $(\mathsf{CLOCK\text{-}UPDATE}, \mathsf{sid}_C)$ to $\mathcal{G}_{\mathsf{clock}}$. Else send I to $\mathcal{A}$.

- *The adversary proposing the next block:*
  If $I = (\mathsf{NEXT\text{-}BLOCK}, \mathsf{hflag}, (\mathsf{uid}_1, \cdots, \mathsf{uid}_\ell))$ is sent from the adversary, update NxtBC as follows:
  (a) Set $\mathsf{listOfUid} \leftarrow \varepsilon$

  (b) For $i \in [\ell]$, if there exists $y := (x, \mathsf{uid}, \tau_L, P_i) \in \mathsf{buffer}$ with ID $\mathsf{uid} = \mathsf{uid}_i$ then set $\mathsf{listOfUid} := \mathsf{listOfUid}\|\mathsf{uid}_i$.

  (c) Finally, set $\mathsf{NxtBC} := \mathsf{NxtBC}\|(\mathsf{hflag}, \mathsf{listOfUid})$.

- *The adversary setting state-slackness:*
  If $I = (\mathsf{SET\text{-}SLACK}, (P_{i_1}, \widehat{\mathsf{pt}}_{i_1}), \cdots, (P_{i_\ell}, \widehat{\mathsf{pt}}_{i_\ell}))$ with $\{P_{i_1}, \cdots, P_{i_\ell}\} \subseteq \mathcal{H} \setminus \mathcal{P}_{DS}$ is received from the adversary, do the following:
  (a) If $\forall j \in [\ell] : |\mathsf{state}| - \widehat{\mathsf{pt}}_{i_j} \leq \mathsf{windowSize}$ and $\widehat{\mathsf{pt}}_{i_1} \geq |\mathsf{state}_{i_j}|$, set $\mathsf{pt}_{i_j} := \widehat{\mathsf{pt}}_{i_j}$ for every $j \in [\ell]$.

  (b) Otherwise set $\mathsf{pt}_{i_j} := |\mathsf{state}|$ for all $j \in [\ell]$

- *The adversary setting the state for desynchronized parties:*
  If $I = (\mathsf{DESYNC\text{-}STATE}, (P_{i_1}, \mathsf{state}'_{i_1}), \cdots, (P_{i_1}, \mathsf{state}'_{i_\ell}))$ with $\{P_{i_1}, \cdots, P_{i_\ell}\} \subseteq \mathcal{P}_{DS}$ is received from the adversary, set $\mathsf{state}_{i_j} := \mathsf{state}'_{i_j}$ for every $j \in [\ell]$.

The work of Badertscher et al. [4] show that under appropriate assumptions, Bitcoin realizes the ledger functionality described enforcing the ExtendPolicy described in the full version of our paper. For convenience we've made a few syntactic changes to the $\mathcal{G}_{\mathsf{ledger}}$ functionality as described in [4]:

- Firstly, the Validate predicate is not relevant in our setting since parties will use ledger to post data, and these should be trivially validated. Hence, we've abstracted out the Validate predicate from the description of the model.
- We require that the adversary cannot invalidate data sent by other parties, thereby denying data from ever making it on to the ledger. For transactions, the adversary can invalidate honest transactions. This can be remedied using a strong variant of $\mathcal{G}_{\mathsf{ledger}}$ described in [4].
- Every time that the size of the state increases, the adversary is notified of the new state by $\mathcal{G}_{\mathsf{ledger}}$.

The changed functionality the same properties of the ideal $\mathcal{G}_{\text{ledger}}$ functionality as described in [4].

Remarks. We point out a few properties of the $\mathcal{G}_{\text{ledger}}$ functionality and its use case in our setting.

- As described in [4], we can achieve a strong liveness guarantee by slightly modifying the above ledger functionality which guarantees that posted information will make it on to the view of other parties within $\Delta := 4 \cdot \text{windowSize}$ number of blocks (relative to the view of the submitting party).
- There are occasions wherein we will run parallel executions of the adversary, and one thread will be assigned to be the main execution thread while the others will be denoted as "look-ahead" threads. In an effort to make the adversary oblivious to rewinding, we cannot allow messages from these "look-ahead" threads to make its way to $\mathcal{G}_{\text{ledger}}$. Drop messages sent by the adversary to $\mathcal{G}_{\text{ledger}}$ and will have to abort the thread if $\mathcal{G}_{\text{ledger}}$ sends a state with an increased size.
- We require that for a READ query, buffer is efficiently simulatable, while state is not. This is a reasonable assumption to make given that the state indicates the permanent component of the blockchain, and simulating this would requiring forging the state. On the other hand, the buffer consists of outstanding queries from both honest and adversarial parties. From the description of $\mathcal{G}_{\text{ledger}}$, each time a SUBMIT query is made to $\mathcal{G}_{\text{ledger}}$, the information is passed along to the adversary, and the adversary's own outstanding queries are known. Looking ahead, a READ query can be answered without making a query to $\mathcal{G}_{\text{ledger}}$. The honest outstanding queries are replayed on each thread since they could not have changed across threads, while the adversarial queries local to that thread are known to the simulator.
- We wait for Delay time before the start of any protocol to ensure all parties are synchronized. Moving ahead, for simplicity of exposition, the notion of de-synchronised parties is ignored.
- While the works of [3, 4, 47] use $\mathcal{G}_{\text{clock}}$ functionality, we do not require parties to have access to a clock and can consider this to be local to $\mathcal{G}_{\text{ledger}}$. In fact our positive results do not rely on parties having access to a clock.
- Additionally, we require that a locally initialized $\mathcal{G}_{\text{ledger}}$ is efficiently simulatable to any adversary that does not have additional access to the global $\mathcal{G}_{\text{ledger}}$. These local $\mathcal{G}_{\text{ledger}}$ will be useful in establishing certain properties of our protocol.

Blockchain active (BCA) adversaries. Consider an adversary that has access to $\mathcal{G}_{\text{ledger}}$, and thus can post to and access the state (the entire blockchain) at any time. In fact its strategy in any protocol may be a function of the state. We refer to any such adversary that actively uses the $\mathcal{G}_{\text{ledger}}$ as a *blockchain active adversary (BCA)*.

Simulation in the Blockchain-hybrid model. Moving ahead, we interchangeably use blockchain-hybrid and $\mathcal{G}_{\text{ledger}}$-hybrid, while preferring the later for our formal descriptions. A simulator has the same power as other parties while accessing the global functionality $\mathcal{G}_{\text{ledger}}$. In addition, it acts as an interface between the party and $\mathcal{G}_{\text{ledger}}$, and thus can choose what messages between the party and the functionality it wants delivered. This is unlike the setting considered in [24,36] where the simulator has control of the blockchain, and thus can "rewind" the blockchain by discarding and re-creating blocks. This is reminiscent of the difference between simulation in Universal Composability (UC) framework [14] and simulation in the global UC framework [15,18,43].

Our simulator can use arbitrary polynomial amount of parallelism. Although arbitrary, the polynomial is fixed in advance. We will use this modeling to run parallel invocations of the adversary by making copies.

At this point we would like to emphasize the need for considering this model for the simulator. We start off by mentioning that any party can use the state obtained from $\mathcal{G}_{\text{ledger}}$ as the basis for its execution. Importantly, the adversary's view is now no longer determined solely by the message it receives from the simulator since the $\mathcal{G}_{\text{ledger}}$ state gives it an additional auxiliary input. In the plain model, if we wanted to rewind the adversary back to a specific point in the execution, we could restart the adversary and send the same messages up to the specific point. And we were guaranteed that the adversary's responses would be identical. But now since the adversary has access to $\mathcal{G}_{\text{ledger}}$, its responses could depend on the state of $\mathcal{G}_{\text{ledger}}$.

Let us consider such an adversary. Now when the simulator tries to restart the adversary, suppose the state has expanded since. Even if the simulator provides the same messages as a previous execution, the adversary's behavior now may be drastically different and of potentially no use to the simulator. The simulator could ensure identical behavior by providing it the earlier truncated view of the state, but moving forward with this execution would be problematic since any message that the adversary wants to post will no longer appear on the state within the promised time period, and thus the adversary will notice that the $\mathcal{G}_{\text{ledger}}$ no longer follows the model specified. Thus it is imperative that executions are run in parallel to ensure that views across multiple threads are identical if the same inputs are provided.

The above modeling is crucial for rewinding when we prove security of our protocols. We will work with this modeling unless otherwise specified. Looking ahead, our construction of the zero-knowledge proof in the non-black-box setting will use a modified variant of this model.

Security. Since the distinguisher attempting to distinguish between views of the adversary in the real and simulated setting has access to $\mathcal{G}_{\text{ledger}}$, the simulator cannot create an isolated view of $\mathcal{G}_{\text{ledger}}$ for the adversary. But as it turns out, the ability to initialize a local $\mathcal{G}_{\text{ledger}}$ is a useful property useful in certain situations that we will leverage in our work.

Protocols in the plain model are a reference to any protocol that does not require its participants to interact with $\mathcal{G}_{\text{ledger}}$ in any form. These protocols are proven secure without considering the presence of $\mathcal{G}_{\text{ledger}}$. Given such a protocol, a blockchain active adversary may try to leverage access to this global functionality

$\mathcal{G}_{\text{ledger}}$ to gain undue advantage over the setting where it did not have such access. We are interested in such adversaries since we want to see how the security of known protocols or primitives fare when the adversary has access to the $\mathcal{G}_{\text{ledger}}$.

3   Definitions and Preliminaries

Unless otherwise specified, we consider the adversaries that have access to the global functionality $\mathcal{G}_{\text{ledger}}$, and thus the view includes messages received from and sent to $\mathcal{G}_{\text{ledger}}$. Thus, when we denote that two distributions representing the views of parties with access to $\mathcal{G}_{\text{ledger}}$ are computationally indistinguishable in the $\mathcal{G}_{\text{ledger}}$-hybrid model, we give distinguisher access to the global $\mathcal{G}_{\text{ledger}}$ functionality. An immediate consequence of this is that, any view generated by a simulator using a privately initialized $\mathcal{G}_{\text{ledger}}$ functionality will be trivially distinguished from the real execution by the distinguisher that views the state of the global $\mathcal{G}_{\text{ledger}}$.

3.1   Zero Knowledge in the $\mathcal{G}_{\text{ledger}}$-hybrid Model

Definition 1. *An interactive protocol* (P, V) *for a language* L *is zero knowledge in the* $\mathcal{G}_{\text{ledger}}$-*hybrid model if the following properties hold:*

– **Completeness.** *For every* $x \in L$,

$$\Pr\Big[\mathsf{out}_{\mathsf{V}}\left[\mathsf{P}(x, w) \leftrightarrow \mathsf{V}(x)\right] = 1\Big] = 1$$

– **Soundness.** *There exists a negligible function* $\mathsf{negl}(\cdot)$ *s.t.* $\forall x \notin L$ *and for all adversarial prover* P^*.

$$\Pr\Big[\mathsf{out}_{\mathsf{V}}\left[\mathsf{P}^*(x) \leftrightarrow \mathsf{V}(x)\right] = 1\Big] \leq \mathsf{negl}(n)$$

– **Zero Knowledge.** *For every PPT adversary* V^*, *there exists a PPT simulator* Sim *such that the probability ensembles*
  – $\Big\{\mathsf{view}_{\mathsf{V}}\left[\mathsf{P}(x, w) \leftrightarrow \mathsf{V}(x, z)\right]\Big\}_{x \in L, w \in R_L(x), z \in \{0,1\}^*}$
  – $\Big\{\mathsf{Sim}(x, z)\Big\}_{x \in L, w \in R_L(x), z \in \{0,1\}^*}$
  are computationally indistinguishable *in the* $\mathcal{G}_{\text{ledger}}$-*model.*

3.2   Concurrently Secure Computation in the $\mathcal{G}_{\text{ledger}}$-hybrid Model

In this work, we consider a malicious, static adversary that chooses whom to corrupt before the execution of the protocol. The adversary controls the scheduling of the concurrent executions. We only consider *computational* security and therefore restrict our attention to adversaries running in probabilistic polynomial time. We denote computational indistinguishability by $\approx_c$, and the security parameter by n. We do not require fairness and hence in the ideal model, we allow a corrupt party to receive its output in a session and then optionally block the output from being delivered to the honest party, in that session. Further, we only consider "security with abort". To formalize the above requirement and define security, we follow the standard paradigm for defining secure computation (see also [52]). We define an ideal model of computation and a real model of computation, and require that any adversary in the real model can be *emulated* by an adversary in the ideal model. More details follow.

IDEAL MODEL. We first define the ideal world experiment, where there is a trusted party for computing the desired two-party functionality $\mathcal{F} : \{0,1\}^{r_1} \times \{0,1\}^{r_2} \to \{0,1\}^{s_1} \times \{0,1\}^{s_2}$. Let P_1 and P_2 denote the two parties in a single execution. In total. let there be k parties $Q_1, Q_2, \cdots, Q_k$, where each party may be involved in multiple sessions with possibly interchangeable roles, i.e. Q_i may play the role of P_1 in one session and P_2 in some other session. Let the total number of executions be $m = m(n)$. For each $\ell \in [m]$, we will denote by P_1^ℓ, the party playing the role of P_1 in session ℓ. P_2^ℓ is defined analogously. The adversary may corrupt any subset of the parties in $Q_1, \ldots, Q_k$. The ideal world execution proceeds as follows:

I **Inputs:** There is a PPT *usage scenario* which gives inputs to all the parties. For each session $\ell \in [m]$, it gives inputs $x_\ell \in X \subseteq \{0,1\}^{r_1}$ to P_1^ℓ and $y_\ell \in Y \subseteq \{0,1\}^{r_2}$ to P_2^ℓ. The adversary is given auxiliary input $z \in \{0,1\}^*$, and chooses the subset of the parties to corrupt, say M. The adversary receives the inputs of the corrupted parties.

II **Session initiation:** When the adversary wishes to initiate the session number ℓ, it sends a (start-session, ℓ) message to the trusted party. On receiving a message of the form (start-session, ℓ), the trusted party sends (start-session, ℓ) to both P_1^ℓ and P_2^ℓ.

III **Honest parties send inputs to the trusted party:** Upon receiving (start-session, ℓ) from the trusted party, an honest party P_i^ℓ sends its real input along with the session identifier. More specifically, if P_1^ℓ is honest, it sends (ℓ, x_ℓ) to the trusted party. Similarly, an honest P_2^ℓ sends (ℓ, y_ℓ) to the trusted party.

IV **Corrupted parties send inputs to the trusted party:** At any point during execution, a corrupted part P_1^ℓ may send a message (ℓ, x_ℓ') to the trusted party, for any string x_ℓ' (of appropriate length) of its choice. Similarly, a corrupted party P_2^ℓ sends (ℓ, y_ℓ') to the trusted party, for any string y_ℓ' (of appropriate length) of its choice.

V **Trusted party sends results to the adversary:** For a session ℓ, when the trusted party has received messages from both P_1^ℓ and P_2^ℓ, it computes the output for that session. Let x_ℓ' and y_ℓ' be the inputs received from P_1^ℓ and P_2^ℓ, respectively. It computes the output $\mathcal{F}(x_\ell', y_\ell')$. If either P_1^ℓ or P_2^ℓ is corrupted, it sends $(\ell, \mathcal{F}(x_\ell', y_\ell'))$ to the adversary. If neither of the parties is corrupted, then the trusted party sends the output message $(\ell, \mathcal{F}(x_\ell', y_\ell'))$ to both P_1^ℓ and P_2^ℓ.

VI **Adversary instructs the trusted party to answer honest players:** For a session ℓ, where exactly one of the party is corrupted, the adversary, depending on its view up to this point, may send the message (output, ℓ) to the trusted party. Then, the trusted party sends the output $(\ell, \mathcal{F}(x'_\ell, y'_\ell))$, computed in the previous step, to the honest party in session ℓ.

VII **Outputs:** An honest party always outputs the value that it received from the trusted party. The adversary outputs an arbitrary (PPT computable) function of its entire view (including the view of all corrupted parties) throughout the execution of the protocol including messages exchanged with the $\mathcal{G}_{\mathsf{ledger}}$ functionality.

The ideal execution of a function $\mathcal{F}$ with security parameter n, input vectors $\overrightarrow{x}, \overrightarrow{y}$, auxiliary input z to Sim and the set of corrupted parties M, denoted by $\mathsf{IDEAL}^{\mathcal{F}}_{M,\mathsf{Sim}}(n, \overrightarrow{x}, \overrightarrow{y}, z)$, is defined as the output pair of the honest parties and the ideal world adversary Sim from the above ideal execution.

REAL MODEL. We now consider the real model in which a real two-party protocol is executed (and there exists no trusted third party). Let $\mathcal{F}, \overrightarrow{x}, \overrightarrow{y}, z$ be as above and let Π be a two-party protocol for computing $\mathcal{F}$. Let $\mathcal{A}$ denote a non-uniform probabilistic polynomial time adversary that controls any subset M of parties $Q_1, \ldots, Q_k$. The parties run concurrent executions of the protocol Π, where the honest parties follow the instructions of Π in all executions. The honest party initiates a new session ℓ, using the input provided whenever it receives a start-session message from $\mathcal{A}$. The scheduling of all messages throughout the execution is controlled by the adversary. That is, the execution proceeds as follows: the adversary sends a message of the form (ℓ, msg) to the honest party. The honest party then adds msg to its view of session ℓ and replies according to the instructions of Π and this view in that session. At the conclusion of the protocol, an honest party computes its output as prescribed by the protocol. Without loss of generality, we assume the adversary outputs exactly its entire view in the execution of the protocol, which includes messages exchanged with the $\mathcal{G}_{\mathsf{ledger}}$ functionality.

The real concurrent execution of Π with security parameter n, input vectors $\overrightarrow{x}, \overrightarrow{y}$, auxiliary input z to $\mathcal{A}$ and the set of corrupted parties M, denoted by $\mathsf{REAL}^{\mathcal{F}}_{M,\mathcal{A}}(n, \overrightarrow{x}, \overrightarrow{y}, z)$, is defined as the output pair of the honest parties and the real world adversary $\mathcal{A}$ from the above real world process.

Definition 2. *Let $\mathcal{F}$ and Π be as above. Then protocol Π for computing $\mathcal{F}$ is a concurrently secure computation protocol in the $\mathcal{G}_{\mathsf{ledger}}$-hybrid model if for every probabilistic polynomial time adversary $\mathcal{A}$ in the real model, there exists a probabilistic polynomial time adversary Sim in the ideal model such that for every polynomial $m = m(n)$, every input vectors $\overrightarrow{x} \in X^m, \overrightarrow{y} \in Y^m$, every $z \in \{0,1\}^*$, and every subset of corrupt parties M, the following*

$$\left\{ \mathsf{IDEAL}^{\mathcal{F}}_{M,\mathsf{Sim}}(n, \overrightarrow{x}, \overrightarrow{y}, z) \right\}_{n \in \mathbb{N}} \approx_c \left\{ \mathsf{REAL}^{\mathcal{F}}_{M,\mathcal{A}}(n, \overrightarrow{x}, \overrightarrow{y}, z) \right\}_{n \in \mathbb{N}}$$

holds in the $\mathcal{G}_{\mathsf{ledger}}$-hybrid model.

4   Black-Box Zero Knowledge

In this section we will describe a $\omega(1)$ round zero-knowledge protocol that can be proven secure using a black-box simulator. We build our protocol atop the protocol for graph hamiltonicity proof.

4.1   Our Protocol

The high level idea for our protocol is that the verifier commits to its challenge via a multi-round extractable commitment, and reveals the challenge in place of the second round of the Hamiltonicity proof system. Since we are constructing a proof system where the prover has unbounded computational power, we require the commitment by the verifier to be statistically hiding so that an unbounded adversarial prover is not able to guess the challenge. We refer to the multi-round extractable commitment as the preamble.

In the preamble, the challenge committed to by the verifier is retrieved by rewinding the verifier in each of the slots. As long as the rewinding is successful in one of the slots, the committed challenge can be extracted. But in the presence of the blockchain (abstracted by the $\mathcal{G}_{\text{ledger}}$ functionality) this becomes difficult. Consider a verifier that sends the challenge received by the prover in a given slot to $\mathcal{G}_{\text{ledger}}$, and waits for the state to expand to include the challenge before responding to the challenge. It then checks in the state if there is another challenge from the prover for the same slot. If this is the case, it knows that it has been rewound, and will abort the protocol. Thus, in the simulated setting, the verifier will abort with a disproportionate probability in comparison to the real execution.

The trivial solution of not relaying messages from the verifier to $\mathcal{G}_{\text{ledger}}$ on the look ahead threads does not work because the verifier can refuse to respond unless the state expands.

Thus, to overcome this issue, we design a protocol in the blockchain-hybrid model, where the protocol requires all parties to access $\mathcal{G}_{\text{ledger}}$ in order to participate in the protocol. In our protocol, we just require that during the preamble, the local state of each party increases by at most k. But since parties may have different views of thus state, we must be careful when we claim the state size increase for other parties. But since $\mathcal{G}_{\text{ledger}}$ guarantees that $|\text{state}_P - \text{state}_V| \leq \text{windowSize}$, we are guaranteed that if the size of the state of one party increases by k, the size of the state of any other party can increase by at most $\text{windowSize} + k$ (with maximum when both parties point to the head of the state initially).

If we set the number of rounds of the preamble to be $m > k + \text{windowSize}$, we are guaranteed to have at least $m - (k + \text{windowSize})$ slots where the state does not expand during the slot. For simplicity we assume k to be a constant, but our protocol can handle arbitrary k by scaling the number of rounds accordingly. The high level idea then is to just rewind in the slots where the state has not expanded, and thus the verifier does not expect the state to expand before it responds, and thus messages to or from $\mathcal{G}_{\text{ledger}}$ can be kept from the verifier on the look ahead threads. Of course the exact number of rounds would depend on the exact simulator strategy. In our protocol, the number of rounds in the preamble is set to be $m = \omega(1)$. We should point out that $k >$ windowSize to avoid trivial aborts in an honest execution of the protocol since otherwise the parties may start off with states that may then be k behind the head of the state, and in one computation step catches up to the head, thereby increasing local state size by k, and thus causing an abort. The complete protocol is presented in Fig. 1.

Protocol BCA-ZK

Common Input: An instance x of a language L with witness relation R_L, the security parameter n, the time out parameter k and the round parameter $m := m(n)$.

Auxiliary Input for Prover: a witness w, such that $(x, w) \in R_L$, size of local state from the ledger $i_P := |\text{state}_P|$.

Auxiliary Input for Verifier: size of local state from the ledger $i_V := |\text{state}_V|$.

Phase I: Prior to each message sent in this phase, the respective party checks if the size of the state is such that $|\text{state}_P| < i_P + k$ (correspondingly $|\text{state}_V| < i_V + k$ for the verifier). If not, the party aborts.

1. Prover uniformly select a first message for a two round *statistically hiding commitment scheme* and send it to the verifier.
2. Verifier uniformly selects $\sigma \in \{0, 1\}^n$, and mn pairs of n-bit strings $(\sigma_{\ell,p}^0, \sigma_{\ell,p}^1)$ for $\ell \in [n], p \in [m]$ such that $\forall \ell, p : \sigma_{\ell,p}^0 \oplus \sigma_{\ell,p}^1 = \sigma$. It commits to all $2mn + 1$ selected strings using the statistically hiding commitment scheme. The commitments are denoted by $\alpha, \{\alpha_{\ell,p}^b\}_{b \in \{0,1\}, \ell \in [n], p \in [m]}$.
3. For $p = 1$ to m:
   (a) Prover sends an n-bit challenge string $r_p = r_{1,p}, \ldots, r_{n,p}$ to the verifier.
   (b) Verifier decommits $\alpha_{1,p}^{r_{1,p}}, \ldots, \alpha_{n,p}^{r_{n,p}}$ to $\sigma_{1,p}^{r_{1,p}}, \ldots, \sigma_{n,p}^{r_{n,p}}$.
4. The prover proceeds with the execution if and only if all the decommitments send by the verifier are valid.

Phase II: The prover and verifier engage in n parallel executions of the Hamiltonicity protocol as described below:

1. The prover sends the first message of the Hamiltonicity proof system.
2. The verifier decommits α to σ. And also reveals all mn commitments not decommitted to in the earlier phase.
3. The prover checks if decommitted values $\sigma, \{\sigma_{\ell,p}^b\}_{b \in \{0,1\}, \ell \in [n], p \in [m]}$ are valid decommitments. Additionally, check if $\forall \ell, p : \sigma_{\ell,p}^0 \oplus \sigma_{\ell,p}^1 = \sigma$. If any of the checks fail, abort. Else, send the third message of the Hamiltonicity proof system.
4. Verifier checks if all conditions of the Hamiltonicity proof system are met. It accepts if and only if this is the case.

Fig. 1. Protocol for zero-knowledge proof in the blockchain aware setting.

Theorem 6. *The protocol BCA-ZK is a Zero-Knowledge Proof with black-box simulation in the $\mathcal{G}_{ledger}$-hybrid model.*

5   Concurrent Self Composable Secure Computation

In this section, we will construct a two-party protocol that is secure under concurrent self composition. We follow the line of works [17,38,40] that rely on realizing an extractable commitment scheme that remains extractable even when there are multiple concurrent copies of this scheme in execution. Thus we construct our protocol in a two-step process. First, we describe a modified version of the multi-round extractable commitment preamble in the blockchain-hybrid model and show that we can extract from each session when multiple sessions are executed concurrently. Next, we plug our constructed concurrently extractable commitments into the compilers constructed in [17,38,40] to achieve a concurrently secure two-party computation protocol.

5.1   Concurrently Extractable Commitment

In this section we present our construction of the concurrently extractable commitment scheme in the blockchain-hybrid model. We will refer to this as the modified PRS preamble. The idea for the modified PRS preamble is quite simple. Prior to starting the preamble, the party needs to post the first message to $\mathcal{G}_{\text{ledger}}$. It is guaranteed that it will appear in the view of every party within the next $\Delta := 4 \cdot \text{windowSize}$ blocks. Once the local state increase by Δ blocks, it sends the same message to the receiver. Posting to $\mathcal{G}_{\text{ledger}}$ gives the party an "expiry period" of k-blocks after the Δ wait i.e., all slots of the preamble must be completed before the size of the state increases by a total of $\Delta + k$. As in the case of zero-knowledge, if the size of the state of a party increases by $\Delta + k$, for any other party the size of the state can have increased by at most $\Delta + k + \text{windowSize}$, which is a constant when k is a constant. This needs to be taken into account when choosing the parameters ℓ and k. The formal description of the protocol is given below.

Protocol $\langle C, R \rangle_{\text{BCA}}$

Common Input: The security parameter n, the time-out parameter k, and the round parameter $2 \cdot \ell := \ell(n)$.

Input to the Committer: the value σ to be committed, size of local state from the ledger $i_C := |\text{state}_C|$.

Input to the Receiver: size of local state from the ledger $i_R := |\text{state}_R|$.

Commitment:
1. Committer uniformly selects $\sigma \in \{0, 1\}^n$, and $2 \cdot \ell \cdot n$ pairs of n-bit strings $(\sigma_{\ell,p}^0, \sigma_{\ell,p}^1)$ for $\ell \in [n], p \in [2 \cdot \ell]$ such that $\forall \ell, p: \sigma_{\ell,p}^0 \oplus \sigma_{\ell,p}^1 = \sigma$. It generate commitments to all $2(2 \cdot \ell) \cdot n + 1$ selected strings using the statistically binding commitment scheme. The commitments are denoted

by α, $\{\alpha_{\ell,p}^b\}_{b\in\{0,1\},\ell\in[n],p\in[2\cdot\ell]}$. Send a SUBMIT query of these commitments to $\mathcal{G}_{\mathsf{ledger}}$. By our assumption, these will be guaranteed to appear in every party's state (at the same position) when $|\mathsf{state}_C| = i_C + \Delta$. Let it appear in index i of the state.

2. The committer sends to receiver the commitments along with the index i of the state that it appears in. The receiver verifies if the commitments were indeed in the designated index of the state.

3. Prior to each message subsequently sent, the respective party checks if the size of the state is such that $|\mathsf{state}_R| < i_R + k + \Delta$ (correspondingly $|\mathsf{state}_C| < i_C + k + \Delta$ for the committer). If not, the party aborts. For $p = 1$ to m:

   (a) Receiver sends an n-bit challenge string $r_p = r_{1,p}, \ldots, r_{1^n,p}$ to the committer.

   (b) Committer decommits $\alpha_{1,p}^{r_{1,p}}, \ldots, \alpha_{n,p}^{r_{n,p}}$ to $\sigma_{1,p}^{r_{1,p}}, \ldots, \sigma_{n,p}^{r_{n,p}}$.

In the full version of our paper, we describe the simulation-extraction strategy to extract values committed by an adversary in every session of multiple concurrent executions of the modified preamble described above.

5.2 Protocol for Concurrent Self Composable Secure Computation

We now describe our concurrent secure computation protocol Π in the $\mathcal{G}_{\mathsf{ledger}}$-hybrid model for a general functionality $\mathcal{F}$. Our protocol is, in fact, the same as the one presented in [17,38,40], except that we use the concurrently extractable commitment from Sect. 5.1. Indeed, the core ingredient of the compiler in [40] (which is also used in [17,38]) is a concurrently extractable commitment, and in particular, it follows from these works that if there exists a concurrent simulator for the extractable commitment, then the resultant compiled protocol securely evaluates the function $\mathcal{F}$.

For completeness, we recall the protocol here. The proof of security for our case follows in essentially an identical fashion to [40] with the main difference being that our simulator only performs a single ideal world query per session (while the simulator performs multiple ideal world queries per session in their work). We discuss other minor differences below.

Building Blocks

Statistical Binding String Commitments. We will use a (2-round) statistically binding string commitment scheme, e.g., a parallel version of Naor's bit commitment scheme [54] based on one-way functions. For simplicity of exposition, however, in the presentation of our results, we will use a non-interactive perfectly binding string commitment. Let $\mathsf{com}(\cdot)$ denote the commitment function of the string commitment scheme.

Statistical Witness Indistinguishable Arguments. We shall use a statistically witness indistinguishable (SWI) argument $\langle P_{\mathsf{swi}}, V_{\mathsf{swi}} \rangle$ for proving membership in any **NP** language with perfect completeness and negligible soundness error. Such a scheme can be constructed by using $\omega(\log k)$ copies of Blum's Hamiltonicity protocol [11] in parallel, with the modification that the prover's commitments in the Hamiltonicity protocol are made using a statistically hiding commitment scheme [42,55].

Semi-Honest Two Party Computation. We will also use a semi-honest two party computation protocol $\langle P_1^{\mathsf{sh}}, P_2^{\mathsf{sh}} \rangle$ that emulates the ideal functionality $\mathcal{F}$ in the stand-alone setting. The existence of such a protocol $\langle P_1^{\mathsf{sh}}, P_2^{\mathsf{sh}} \rangle$ follows from [34,48,59].

Concurrent Non-Malleable Zero Knowledge Argument. Concurrent non-malleable zero knowledge (CNMZK) considers the setting where a man-in-the-middle adversary is interacting with several honest provers and honest verifiers in a concurrent fashion: in the "left" interactions, the adversary acts as verifier while interacting with honest provers; in the "right" interactions, the adversary tries to prove some statements to honest verifiers. The goal is to ensure that such an adversary cannot take "help" from the left interactions in order to succeed in the right interactions. This intuition can be formalized by requiring the existence of a machine called the simulator-extractor that generates the view of the man-in-the-middle adversary and additionally also outputs a witness from the adversary for each "valid" proof given to the verifiers in the right sessions.

Barak, Prabhakaran and Sahai [8] gave the first construction of a concurrent non-malleable zero knowledge (CNMZK) argument for every language in **NP** with perfect completeness and negligible soundness error. In our construction, we will use a specific CNMZK protocol, denoted $\langle P, V \rangle$, based on the CNMZK protocol of Barak et al. [8] to guarantee non-malleability. Specifically, we will make the following two changes to Barak et al's protocol: (a) Instead of using an $\omega(\log n)$-round extractable commitment scheme [57], we will use the N-round extractable commitment scheme $\langle C, R \rangle$ (described in the full version). (b) Further, we require that the non-malleable commitment scheme being used in the protocol be public-coin w.r.t. receiver[3]. We now describe the protocol $\langle P, V \rangle$.

Let P and V denote the prover and the verifier respectively. Let L be an NP language with a witness relation R. The common input to P and V is a statement $x \in L$. P additionally has a private input w (witness for x). Protocol $\langle P, V \rangle$ consists of two main phases: (a) the *preamble phase*, where the verifier commits to a random secret (say) σ via an execution of $\langle C, R \rangle$ with the prover, and (b) the *post-preamble phase*, where the prover proves an NP statement. In more detail, protocol $\langle P, V \rangle$ proceeds as follows.

---

[3] The original NMZK construction only required a public-coin extraction phase inside the non-malleable commitment scheme. We, however, require that the entire commitment protocol be public-coin. We note that the non-malleable commitment protocol of [26] only consists of standard perfectly binding commitments and zero knowledge proof of knowledge. Therefore, we can easily instantiate the DDN construction with public-coin versions of these primitives such that the resultant protocol is public-coin.

PREAMBLE PHASE.

1. P and V engage in execution of $\langle C, R \rangle$ where V commits to a random string σ.

POST-PREAMBLE PHASE.

2. P commits to 0 using a statistically-hiding commitment scheme. Let c be the commitment string. Additionally, P proves the knowledge of a valid decommitment to c using a statistical zero-knowledge argument of knowledge (SZKAOK).

3. V now reveals σ and sends the decommitment information relevant to $\langle C, R \rangle$ that was executed in step 1.

4. P commits to the witness w using a public-coin non-malleable commitment scheme.

5. P now proves the following statement to V using SZKAOK:
   (a) *either* the value committed to in step 4 is a valid witness to x (i.e., $R(x, w) = 1$, where w is the committed value), *or*
   (b) the value committed to in step 2 is the trapdoor secret σ.

   P uses the witness corresponding to the first part of the statement.

Modified Extractable Commitment Scheme. $\langle C', R' \rangle$ Due to technical reasons, in our secure computation protocol, we will also use a minor variant, denoted $\langle C', R' \rangle_{\mathsf{BCA}}$, of the extractable commitment scheme presented in 5.1. Protocol $\langle C', R' \rangle_{\mathsf{BCA}}$ is the same as $\langle C, R \rangle_{\mathsf{BCA}}$, except that for a given receiver challenge string, the committer does not "open" the commitments, but instead simply reveals the appropriate committed values (without revealing the randomness used to create the corresponding commitments). More specifically, in protocol $\langle C', R' \rangle_{\mathsf{BCA}}$, on receiving a challenge string $v_j = v_{1,j}, \ldots, v_{\ell,j}$ from the receiver, the committer uses the following strategy: for every $i \in [\ell]$, if $v_{i,j} = 0$, C' sends $\alpha_{i,j}^0$, otherwise it sends $\alpha_{i,j}^1$ to R'. Note that C' does not reveal the decommitment values associated with the revealed shares.

When we use $\langle C', R' \rangle_{\mathsf{BCA}}$ in our main construction, we will require the committer C' to prove the "correctness" of the values (i.e., the secret shares) it reveals in the last step of the commitment protocol. In fact, due to technical reasons, we will also require the committer to prove that the commitments that it sent in the first step are "well-formed".

We remark that the extraction proof for the simulation-extraction procedure also holds for the $\langle C', R' \rangle_{\mathsf{BCA}}$ commitment scheme.

Protocol Description Notation. Let $\mathsf{com}(\cdot)$ denote the commitment function of a non-interactive perfectly binding commitment scheme. Let $\langle C, R \rangle_{\mathsf{BCA}}$ denote the N-round extractable commitment scheme and $\langle C', R' \rangle_{\mathsf{BCA}}$ be its modified version as described above. For the description, we drop the subscript and refer to them as $\langle C, R \rangle$ and $\langle C', R' \rangle$ respectively. Let $\langle P, V \rangle$ denote the modified version of the CNMZK argument of Barak et al. [8]. Further, let $\langle P_{\mathsf{swi}}, V_{\mathsf{swi}} \rangle$ denote a SWI argument and let $\langle P_1^{\mathsf{sh}}, P_2^{\mathsf{sh}} \rangle$ denote a semi-honest two party computation protocol $\langle P_1^{\mathsf{sh}}, P_2^{\mathsf{sh}} \rangle$ that securely computes $\mathcal{F}$ in the stand-alone setting as per the standard definition of secure computation.

Let P_1 and P_2 be two parties with inputs x_1 and x_2. Let n be the security parameter. The protocol proceeds as follows.

Protocol BCA-CONC

I. Trapdoor Creation Phase.
1. $P_1 \Rightarrow P_2$: P_1 creates a commitment $\mathsf{Com}_1 = \mathsf{com}(0)$ to bit 0 and sends Com_1 to P_2. P_1 and P_2 now engage in the execution of $\langle P, V \rangle$ where P_1 proves that Com_1 is a commitment to 0.
2. $P_2 \Rightarrow P_1$: P_2 now acts symmetrically. That is, it creates a commitment $\mathsf{Com}_2 = \mathsf{com}(0)$ to bit 0 and sends Com_2 to P_1. P_2 and P_1 now engage in the execution of $\langle P, V \rangle$ where P_2 proves that Com_2 is a commitment to 0.

Informally speaking, the purpose of this phase is to aid the simulator in obtaining a "trapdoor" to be used during the simulation of the protocol.

II. Input Commitment Phase. In this phase, the parties commit to their inputs and random coins (to be used in the next phase) via the commitment protocol $\langle C', R' \rangle$.

1. $P_1 \Rightarrow P_2$: P_1 first samples a random string r_1 (of appropriate length, to be used as P_1's randomness in the execution of $\langle P_1^{\mathsf{sh}}, P_2^{\mathsf{sh}} \rangle$ in Phase III) and engages in an execution of $\langle C', R' \rangle$ (denoted as $\langle C', R' \rangle_{1 \to 2}$) with P_2, where P_1 commits to $x_1 \| r_1$. Next, P_1 and P_2 engage in an execution of $\langle P_{\mathsf{swi}}, V_{\mathsf{swi}} \rangle$ where P_1 proves the following statement to P_2: (a) *either* there exist values $\hat{x}_1, \hat{r}_1$ such that the commitment protocol $\langle C', R' \rangle_{1 \to 2}$ is *valid* with respect to the value $\hat{x}_1 \| \hat{r}_1$, *or* (b) Com_1 is a commitment to bit 1.
2. $P_2 \Rightarrow P_1$: P_2 now acts symmetrically. Let r_2 (analogous to r_1 chosen by P_1) be the random string chosen by P_2 (to be used in the next phase).

Informally speaking, the purpose of this phase is aid the simulator in extracting the adversary's input and randomness.

III. Secure Computation Phase. In this phase, P_1 and P_2 engage in an execution of $\langle P_1^{\mathsf{sh}}, P_2^{\mathsf{sh}} \rangle$ where P_1 plays the role of P_1^{sh}, while P_2 plays the role of P_2^{sh}. Since $\langle P_1^{\mathsf{sh}}, P_2^{\mathsf{sh}} \rangle$ is secure only against semi-honest adversaries, we first enforce that the coins of each party are truly random, and then execute $\langle P_1^{\mathsf{sh}}, P_2^{\mathsf{sh}} \rangle$, where with every protocol message, a party gives a proof using $\langle P_{\mathsf{swi}}, V_{\mathsf{swi}} \rangle$ of its honest behavior "so far" in the protocol. We now describe the steps in this phase.

1. $P_1 \leftrightarrow P_2$: P_1 samples a random string r_2' (of appropriate length) and sends it to P_2. Similarly, P_2 samples a random string r_1' and sends it to P_1. Let $r_1'' = r_1 \oplus r_1'$ and $r_2'' = r_2 \oplus r_2'$. Now, r_1'' and r_2'' are the random coins that P_1 and P_2 will use during the execution of $\langle P_1^{\mathsf{sh}}, P_2^{\mathsf{sh}} \rangle$.

2. Let t be the number of rounds in $\langle P_1^{\mathsf{sh}}, P_2^{\mathsf{sh}} \rangle$, where one round consists of a message from P_1^{sh} followed by a reply from P_2^{sh}. Let transcript $T_{1,j}$ (resp., $T_{2,j}$) be defined to contain all the messages exchanged between P_1^{sh} and P_2^{sh} before the point P_1^{sh} (resp., P_2^{sh}) is supposed to send a message in round j. For $j = 1, \ldots, t$:

   (a) $P_1 \Rightarrow P_2$: Compute $\beta_{1,j} = P_1^{\mathsf{sh}}(T_{1,j}, x_1, r_1'')$ and send it to P_2. P_1 and P_2 now engage in an execution of $\langle P_{\mathsf{swi}}, V_{\mathsf{swi}} \rangle$, where P_1 proves the following statement:

   i. *either* there exist values $\hat{x}_1$, $\hat{r}_1$ such that (a) the commitment protocol $\langle C', R' \rangle_{1 \to 2}$ is *valid* with respect to the value $\hat{x}_1 \| \hat{r}_1$, and (b) $\beta_{1,j} = P_1^{\mathsf{sh}}(T_{1,j}, \hat{x}_1, \hat{r}_1 \oplus r_1')$

   ii. *or*, Com_1 is a commitment to bit 1.

   (b) $P_2 \Rightarrow P_1$: P_2 now acts symmetrically.

Proof of Security. Our proof of security follows in almost an identical fashion to [17,38,40]. The main difference is that due to the property of our concurrent extractor (discussed in the full version), our simulator only needs to make one ideal world query per session (as opposed to multiple ideal world queries). Indeed, this is why we achieve standard concurrent security, while [17,38,40] achieve security in the so-called multiple-ideal-query model.

Our indistinguishability hybrids also follow in the same manner as in [17,38, 40]. There is one minor difference that we highlight. The hybrids of [17,38,40] maintain a "soundness invariant", where roughly speaking, it is guaranteed that whenever an honest party changes its input in any sub-protocol used within the secure computation protocol, the value committed by the adversary in the non-malleable commitment (inside the CNMZK) does not change, except with negligible probability. In some hybrids, this property is argued via extraction from the non-malleable commitment.

In our setting, we have to be careful with such an extraction since a blockchain-active adversary may try to keep state using $\mathcal{G}_{\mathsf{ledger}}$. However, the key point is that for such a soundness argument, the reduction can use a locally initialized $\mathcal{G}_{\mathsf{ledger}}$ that it controls (and can therefore modify arbitrarily). This follows from the fact that we do not care about the view of an adversary in such a reduction to be indistinguishable to a distinguisher that has access to $\mathcal{G}_{\mathsf{ledger}}$. In fact, it will trivially be distinguishable. But since a locally initialized $\mathcal{G}_{\mathsf{ledger}}$ is indistinguishable to the adversary that is simply allowed to interact using the given interface (i.e. efficiently simulatable), the adversary's behavior does not change. Using this idea, we can perform extraction as in the plain model.

Acknowledgments. The second author's research was supported in part by a grant from Northrop Grumman, a gift from DOS Networks, and, a Cylab seed funding award. The first and third authors' research was supported in part by a DARPA/ARL Safeware Grant W911NF-15-C-0213, and a subaward from NSF CNS-1414023.

References

1. Agrawal, S., Goyal, V., Jain, A., Prabhakaran, M., Sahai, A.: New impossibility results for concurrent composition and a non-interactive completeness theorem for secure computation. In: Safavi-Naini, R., Canetti, R. (eds.) CRYPTO 2012. LNCS, vol. 7417, pp. 443–460. Springer, Heidelberg (2012). https://doi.org/10.1007/978-3-642-32009-5_26

2. Andrychowicz, M., Dziembowski, S., Malinowski, D., Mazurek, L.: Secure multi-party computations on bitcoin. In: 2014 IEEE Symposium on Security and Privacy, SP 2014, Berkeley, CA, USA, 18–21 May, pp. 443–458 (2014)

3. Badertscher, C., Gaži, P., Kiayias, A., Russell, A., Zikas, V.: Ouroboros genesis: composable proof-of-stake blockchains with dynamic availability. Cryptology ePrint Archive, Report 2018/378 (2018). https://eprint.iacr.org/2018/378

4. Badertscher, C., Maurer, U., Tschudi, D., Zikas, V.: Bitcoin as a transaction ledger: a composable treatment. In: Katz, J., Shacham, H. (eds.) CRYPTO 2017, Part I. LNCS, vol. 10401, pp. 324–356. Springer, Cham (2017). https://doi.org/10.1007/978-3-319-63688-7_11

5. Badrinarayanan, S., Khurana, D., Ostrovsky, R., Visconti, I.: Unconditional UC-secure computation with (stronger-malicious) PUFs. In: Coron, J.-S., Nielsen, J.B. (eds.) EUROCRYPT 2017, Part I. LNCS, vol. 10210, pp. 382–411. Springer, Cham (2017). https://doi.org/10.1007/978-3-319-56620-7_14

6. Barak, B., Canetti, R., Nielsen, J.B., Pass, R.: Universally composable protocols with relaxed set-up assumptions. In: 45th FOCS, 17–19 October, pp. 186–195. IEEE Computer Society Press, Rome (2004)

7. Barak, B., Lindell, Y.: Strict polynomial-time in simulation and extraction. In: 34th ACM STOC, 19–21 May, pp. 484–493. ACM Press, Montréal (2002)

8. Barak, B., Prabhakaran, M., Sahai, A.: Concurrent non-malleable zero knowledge. In: 47th FOCS, 21–24 October, pp. 345–354. IEEE Computer Society Press, Berkeley (2006)

9. Bellare, M., Jakobsson, M., Yung, M.: Round-optimal zero-knowledge arguments based on any one-way function. In: Fumy, W. (ed.) EUROCRYPT 1997. LNCS, vol. 1233, pp. 280–305. Springer, Heidelberg (1997). https://doi.org/10.1007/3-540-69053-0_20

10. Bentov, I., Kumaresan, R.: How to use bitcoin to design fair protocols. In: Garay, J.A., Gennaro, R. (eds.) CRYPTO 2014, Part II. LNCS, vol. 8617, pp. 421–439. Springer, Heidelberg (2014). https://doi.org/10.1007/978-3-662-44381-1_24

11. Blum, M.: How to prove a theorem so no one else can claim it. In: International Congress of Mathematicians, pp. 1444–1451 (1987)

12. Brzuska, C., Fischlin, M., Schröder, H., Katzenbeisser, S.: Physically uncloneable functions in the universal composition framework. In: Rogaway, P. (ed.) CRYPTO 2011. LNCS, vol. 6841, pp. 51–70. Springer, Heidelberg (2011). https://doi.org/10.1007/978-3-642-22792-9_4

13. Canetti, R., Kushilevitz, E., Lindell, Y.: On the limitations of universally composable two-party computation without set-up assumptions. J. Cryptol. **19**(2), 135–167 (2006)

14. Canetti, R.: Universally composable security: a new paradigm for cryptographic protocols. In: 42nd FOCS, 14–17 October, pp. 136–145. IEEE Computer Society Press, Las Vegas (2001)

15. Canetti, R., Dodis, Y., Pass, R., Walfish, S.: Universally composable security with global setup. In: Vadhan, S.P. (ed.) TCC 2007. LNCS, vol. 4392, pp. 61–85. Springer, Heidelberg (2007). https://doi.org/10.1007/978-3-540-70936-7_4

16. Canetti, R., Fischlin, M.: Universally composable commitments. In: Kilian, J. (ed.) CRYPTO 2001. LNCS, vol. 2139, pp. 19–40. Springer, Heidelberg (2001). https://doi.org/10.1007/3-540-44647-8_2

17. Canetti, R., Goyal, V., Jain, A.: Concurrent secure computation with optimal query complexity. In: Gennaro, R., Robshaw, M.J.B. (eds.) CRYPTO 2015, Part II. LNCS, vol. 9216, pp. 43–62. Springer, Heidelberg (2015). https://doi.org/10.1007/978-3-662-48000-7_3

18. Canetti, R., Jain, A., Scafuro, A.: Practical UC security with a global random oracle. In: Ahn, G.J., Yung, M., Li, N. (eds.) ACM CCS 14, 3–7 November, pp. 597–608. ACM Press, Scottsdale (2014)

19. Canetti, R., Kushilevitz, E., Lindell, Y.: On the limitations of universally composable two-party computation without set-up assumptions. In: Biham, E. (ed.) EUROCRYPT 2003. LNCS, vol. 2656, pp. 68–86. Springer, Heidelberg (2003). https://doi.org/10.1007/3-540-39200-9_5

20. Canetti, R., Lin, H., Pass, R.: Adaptive hardness and composable security in the plain model from standard assumptions. In: 51st FOCS, 23–26 October, pp. 541–550. IEEE Computer Society Press, Las Vegas (2010)

21. Canetti, R., Lindell, Y., Ostrovsky, R., Sahai, A.: Universally composable two-party and multi-party secure computation. In: 34th ACM STOC, 19–21 May, pp. 494–503. ACM Press, Montréal (2002)

22. Chandran, N., Goyal, V., Sahai, A.: New constructions for UC secure computation using tamper-proof hardware. In: Smart, N.P. (ed.) EUROCRYPT 2008. LNCS, vol. 4965, pp. 545–562. Springer, Heidelberg (2008). https://doi.org/10.1007/978-3-540-78967-3_31

23. Choudhuri, A.R., Goyal, V., Jain, A.: Founding secure computation on blockchains. Cryptology ePrint Archive, Report 2019/253 (2019). https://eprint.iacr.org/2019/253

24. Choudhuri, A.R., Green, M., Jain, A., Kaptchuk, G., Miers, I.: Fairness in an unfair world: fair multiparty computation from public bulletin boards. In: Thuraisingham, B.M., Evans, D., Malkin, T., Xu, D. (eds.) ACM CCS 17, October 31–2 November, pp. 719–728. ACM Press, Dallas (2017)

25. Dachman-Soled, D., Fleischhacker, N., Katz, J., Lysyanskaya, A., Schröder, D.: Feasibility and infeasibility of secure computation with malicious PUFs. In: Garay, J.A., Gennaro, R. (eds.) CRYPTO 2014, Part II. LNCS, vol. 8617, pp. 405–420. Springer, Heidelberg (2014). https://doi.org/10.1007/978-3-662-44381-1_23

26. Dolev, D., Dwork, C., Naor, M.: Non-malleable cryptography (extended abstract). In: 23rd ACM STOC, 6–8 May, pp. 542–552. ACM Press, New Orleans (1991)

27. Dwork, C., Naor, M., Sahai, A.: Concurrent zero-knowledge. In: 30th ACM STOC, 23–26 May, pp. 409–418. ACM Press, Dallas (1998)

28. Feige, U., Shamir, A.: Witness indistinguishable and witness hiding protocols. In: 22nd ACM STOC, 14–16 May, pp. 416–426. ACM Press, Baltimore (1990)

29. Fiat, A., Shamir, A.: How to prove yourself: practical solutions to identification and signature problems. In: Odlyzko, A.M. (ed.) CRYPTO 1986. LNCS, vol. 263, pp. 186–194. Springer, Heidelberg (1987). https://doi.org/10.1007/3-540-47721-7_12

30. Garay, J., Kiayias, A., Leonardos, N.: The bitcoin backbone protocol: analysis and applications. In: Oswald, E., Fischlin, M. (eds.) EUROCRYPT 2015, Part II. LNCS, vol. 9057, pp. 281–310. Springer, Heidelberg (2015). https://doi.org/10.1007/978-3-662-46803-6_10

31. Garay, J., Kiayias, A., Leonardos, N.: The bitcoin backbone protocol with chains of variable difficulty. In: Katz, J., Shacham, H. (eds.) CRYPTO 2017, Part I. LNCS, vol. 10401, pp. 291–323. Springer, Cham (2017). https://doi.org/10.1007/978-3-319-63688-7_10

32. Garg, S., Kumarasubramanian, A., Ostrovsky, R., Visconti, I.: Impossibility results for static input secure computation. In: Safavi-Naini, R., Canetti, R. (eds.) CRYPTO 2012. LNCS, vol. 7417, pp. 424–442. Springer, Heidelberg (2012). https://doi.org/10.1007/978-3-642-32009-5_25

33. Goldreich, O., Krawczyk, H.: On the composition of zero-knowledge proof systems. SIAM J. Comput. 25(1), 169–192 (1996)

34. Goldreich, O., Micali, S., Wigderson, A.: How to play any mental game or a completeness theorem for protocols with honest majority. In: Aho, A. (ed.) 19th ACM STOC, 25–27 May, pp. 218–229. ACM Press, New York City (1987)

35. Goldwasser, S., Micali, S., Rackoff, C.: The knowledge complexity of interactive proof-systems (extended abstract). In: Proceedings of the 17th Annual ACM Symposium on Theory of Computing, 6–8 May 1985, Providence, Rhode Island, USA, pp. 291–304 (1985)

36. Goyal, R., Goyal, V.: Overcoming cryptographic impossibility results using blockchains. In: Kalai, Y., Reyzin, L. (eds.) TCC 2017, Part I. LNCS, vol. 10677, pp. 529–561. Springer, Cham (2017). https://doi.org/10.1007/978-3-319-70500-2_18

37. Goyal, V.: Positive results for concurrently secure computation in the plain model. In: 53rd FOCS, 20–23 October, pp. 41–50. IEEE Computer Society Press, New Brunswick (2012)

38. Goyal, V., Gupta, D., Jain, A.: What information is leaked under concurrent composition? In: Canetti, R., Garay, J.A. (eds.) CRYPTO 2013, Part II. LNCS, vol. 8043, pp. 220–238. Springer, Heidelberg (2013). https://doi.org/10.1007/978-3-642-40084-1_13

39. Goyal, V., Ishai, Y., Sahai, A., Venkatesan, R., Wadia, A.: Founding cryptography on tamper-proof hardware tokens. In: Micciancio, D. (ed.) TCC 2010. LNCS, vol. 5978, pp. 308–326. Springer, Heidelberg (2010). https://doi.org/10.1007/978-3-642-11799-2_19

40. Goyal, V., Jain, A., Ostrovsky, R.: Password-authenticated session-key generation on the internet in the plain model. In: Rabin, T. (ed.) CRYPTO 2010. LNCS, vol. 6223, pp. 277–294. Springer, Heidelberg (2010). https://doi.org/10.1007/978-3-642-14623-7_15

41. Goyal, V., Lin, H., Pandey, O., Pass, R., Sahai, A.: Round-efficient concurrently composable secure computation via a robust extraction lemma. In: Dodis, Y., Nielsen, J.B. (eds.) TCC 2015, Part I. LNCS, vol. 9014, pp. 260–289. Springer, Heidelberg (2015). https://doi.org/10.1007/978-3-662-46494-6_12

42. Haitner, I., Horvitz, O., Katz, J., Koo, C.-Y., Morselli, R., Shaltiel, R.: Reducing complexity assumptions for statistically-hiding commitment. In: Cramer, R. (ed.) EUROCRYPT 2005. LNCS, vol. 3494, pp. 58–77. Springer, Heidelberg (2005). https://doi.org/10.1007/11426639_4

43. Hazay, C., Polychroniadou, A., Venkitasubramaniam, M.: Composable security in the tamper-proof hardware model under minimal complexity. In: Hirt, M., Smith, A. (eds.) TCC 2016, Part I. LNCS, vol. 9985, pp. 367–399. Springer, Heidelberg (2016). https://doi.org/10.1007/978-3-662-53641-4_15

44. Kalai, Y.T., Lindell, Y., Prabhakaran, M.: Concurrent general composition of secure protocols in the timing model. In: Gabow, H.N., Fagin, R. (eds.) 37th ACM STOC, 22–24 May, pp. 644–653. ACM Press, Baltimore (2005)

45. Katz, J.: Universally composable multi-party computation using tamper-proof hardware. In: Naor, M. (ed.) EUROCRYPT 2007. LNCS, vol. 4515, pp. 115–128. Springer, Heidelberg (2007). https://doi.org/10.1007/978-3-540-72540-4_7

46. Kiayias, A., Russell, A., David, B., Oliynykov, R.: Ouroboros: a provably secure proof-of-stake blockchain protocol. In: Katz, J., Shacham, H. (eds.) CRYPTO 2017, Part I. LNCS, vol. 10401, pp. 357–388. Springer, Cham (2017). https://doi.org/10.1007/978-3-319-63688-7_12

47. Kiayias, A., Zhou, H.-S., Zikas, V.: Fair and robust multi-party computation using a global transaction ledger. In: Fischlin, M., Coron, J.-S. (eds.) EUROCRYPT 2016, Part II. LNCS, vol. 9666, pp. 705–734. Springer, Heidelberg (2016). https://doi.org/10.1007/978-3-662-49896-5_25

48. Kilian, J.: Founding cryptography on oblivious transfer. In: 20th ACM STOC, 2–4 May, pp. 20–31. ACM Press, Chicago (1988)

49. Kilian, J., Petrank, E.: Concurrent and resettable zero-knowledge in polyloalgorithm rounds. In: STOC, pp. 560–569 (2001)

50. Lindell, Y.: General composition and universal composability in secure multi-party computation. In: FOCS, pp. 394–403 (2003)

51. Lindell, Y.: Lower bounds for concurrent self composition. In: Naor, M. (ed.) TCC 2004. LNCS, vol. 2951, pp. 203–222. Springer, Heidelberg (2004). https://doi.org/10.1007/978-3-540-24638-1_12

52. Lindell, Y.: Lower bounds and impossibility results for concurrent self composition. J. Cryptol. **21**(2), 200–249 (2008)

53. Micali, S., Pass, R.: Local zero knowledge. In: Proceedings of the 38th Annual ACM Symposium on Theory of Computing, Seattle, WA, USA, 21–23 May 2006, pp. 306–315 (2006). https://doi.org/10.1145/1132516.1132561

54. Naor, M.: Bit commitment using pseudorandomness. J. Cryptol. **4**(2), 151–158 (1991)

55. Naor, M., Ostrovsky, R., Venkatesan, R., Yung, M.: Perfect zero-knowledge arguments for NP using any one-way permutation. J. Cryptol. **11**(2), 87–108 (1998)

56. Pass, R., Seeman, L., Shelat, A.: Analysis of the blockchain protocol in asynchronous networks. In: Coron, J.-S., Nielsen, J.B. (eds.) EUROCRYPT 2017, Part II. LNCS, vol. 10211, pp. 643–673. Springer, Cham (2017). https://doi.org/10.1007/978-3-319-56614-6_22

57. Prabhakaran, M., Rosen, A., Sahai, A.: Concurrent zero knowledge with logarithmic round-complexity. In: 43rd FOCS, 16–19 November, pp. 366–375. IEEE Computer Society Press, Vancouver (2002)

58. Richardson, R., Kilian, J.: On the concurrent composition of zero-knowledge proofs. In: Stern, J. (ed.) EUROCRYPT 1999. LNCS, vol. 1592, pp. 415–431. Springer, Heidelberg (1999). https://doi.org/10.1007/3-540-48910-X_29

59. Yao, A.C.C.: How to generate and exchange secrets (extended abstract). In: 27th FOCS, 27–29 October, pp. 162–167, IEEE Computer Society Press, Toronto (1986)

Uncovering Algebraic Structures
in the MPC Landscape

Navneet Agarwal[(✉)], Sanat Anand, and Manoj Prabhakaran

Indian Institute of Technology Bombay, Mumbai, India
{navneet,sanat,mp}@cse.iitb.ac.in

Abstract. A fundamental problem in the theory of secure multi-party computation (MPC) is to characterize functions with *more than 2 parties* which admit MPC protocols with information-theoretic security against passive corruption. This question has seen little progress since the work of Chor and Ishai (1996), which demonstrated difficulties in resolving it. In this work, we make significant progress towards resolving this question in the important case of aggregating functionalities, in which m parties $P_1, \ldots, P_m$ hold inputs $x_1, \ldots, x_m$ and an aggregating party P_0 must learn $f(x_1, \ldots, x_m)$.

We uncover a rich class of algebraic structures that are closely related to secure computability, namely, "Commuting Permutations Systems" (CPS) and its variants. We present an extensive set of results relating these algebraic structures among themselves and to MPC, including new protocols, impossibility results and separations. Our results include a necessary algebraic condition and slightly stronger sufficient algebraic condition for a function to admit information-theoretically secure MPC protocols.

We also introduce and study new models of minimally interactive MPC (called UNIMPC and UNIMPC*), which not only help in understanding our positive and negative results better, but also open up new avenues for studying the cryptographic complexity landscape of multi-party functionalities. Our positive results include novel protocols in these models, which may be of independent practical interest.

Finally, we extend our results to a definition that requires UC security as well as semi-honest security (which we term *strong security*). In this model we are able to carry out the characterization of *all* computable functions, except for a gap in the case of aggregating functionalities.

1 Introduction

Secure Multi-Party Computation (MPC) is a central and unifying concept in modern cryptography. The foundations, as well as the applications, of MPC have been built up over a period of almost four decades of active research since the initial ideas emerged [SRA79, Blu81, Yao82]. Yet, some of the basic questions

Supported by the Dept. of Science and Technology, India via the Ramanujan Fellowship and an Indo-Israel Joint Research Project grant, 2018.

© International Association for Cryptologic Research 2019
Y. Ishai and V. Rijmen (Eds.): EUROCRYPT 2019, LNCS 11477, pp. 381–406, 2019.
https://doi.org/10.1007/978-3-030-17656-3_14

in MPC remain open. Specifically, the following basic problem remains open to this day for various standard notions of security (when there are no restrictions like honest majority):

Which multi-party functions admit information-theoretically secure MPC?

Indeed, one of the most basic forms of this problem remains wide open: for the case of security against passive corruption, a characterization of securely realizable functions is known only for 2-party functions [Kus89]. Chor and Ishai pointed out the difficulty of this problem, by disproving a natural conjecture for characterizing securely realizable k-party functionalities in terms of functionalities involving fewer parties [CI96]. Since then, very little progress has been made on this problem.

In this work, we make significant progress towards resolving this question in the important case of *aggregating functionalities*: In an aggregating functionality, there are m parties $P_1, \ldots, P_m$ with inputs $x_1, \ldots, x_m$ and an aggregating party P_0 must learn $f(x_1, \ldots, x_m)$. Aggregating functionalities form a practically and theoretically important class. In particular, it has been the subject of an influential line of study that started with the *minimal model for secure computation* of Feige, Kilian and Naor [FKN94]. This model – also referred to as the Private Simultaneous Messages (PSM) model [IK97] – served as a precursor of important concepts like randomized encodings [IK00] that have proven useful in a variety of cryptographic applications. Recently, a strengthening of this model, called Non-Interactive MPC (NIMPC) was introduced by Beimel et al. [BGI+14], which is closer to standard MPC in terms of the security requirements.[1] However, these models do not address the question of secure realizability in the standard model, because due to weakened security requirements, all aggregating functions are securely realizable in these models.

Towards characterizing secure realizability under (the standard model of) MPC, we uncover and examine a rich class of algebraic structures of aggregating functionalities. We exploit these structures to give new positive and negative results for MPC. Further, we also put forth new minimalistic, yet natural models of secure computation that arise from these results. These new models and algebraic structures, in tandem, open up new avenues for investigating the landscape of secure multiparty computation involving many parties.

Commuting Permutations Systems. We identify an algebraic-combinatorial structure called Commuting Permutations System (CPS) and interesting subclasses thereof. CPS generalizes the function of abelian group summation to a

---

[1] Both PSM and NIMPC consider protocols of the following form: a coordinator sends a private message to each of $P_1, \ldots, P_m$; each P_i uses this message and its input to compute a single message which it sends to P_0; P_0 computes an output. PSM has a corruption model in which only P_0 could be corrupted, whereas NIMPC allows any subset of the parties (other than the coordinator) to be corrupted. But when such corruption takes place, NIMPC allows the adversary to learn the *residual function* determined by the honest parties' inputs – i.e., the output for each possible setting of the inputs for the corrupt parties (unlike in MPC, where the output for only a given input of the corrupt parties is learned).

less structured class of functions. Indeed, as a function of two inputs (denoted as $m = 2$), a CPS can be identified with a *quasigroup* operation, or equivalently the function specified by a minor of a Latin square. (For $m > 2$ inputs, CPS imputes more structure than m-dimensional Latin hypercubes.)

We define **CPS** as the class of all aggregating functions which *embed* into a CPS functionality (Definition 2). We also identify two interesting sub-classes of CPS that (as we shall see) are closely related to secure computability, corresponding to Commuting Permutation *Subgroup* Systems (CPSS) and *Complete* CPS (CCPS).

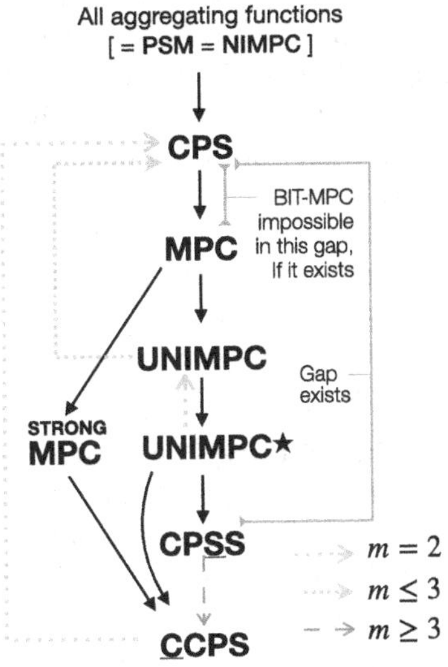

Minimal Models of MPC. In a parallel thread, we develop new minimalistic models of MPC, that help us study feasibility of information-theoretic MPC. These models (called UNIMPC* and UNIMPC) admit secure protocols only for functions which have secure protocols in the standard MPC model. We remark that ours is perhaps the first significant minimalistic model with this property, as previous minimalistic models – PSM [FKN94] and NIMPC [BGI+14] – admit secure protocols for all functions.

UNIMPC stands for *Unassisted NIMPC* and, as the name suggests, removes the assistance from the trusted party in NIMPC: Instead the parties should securely compute

Fig. 1. The m-PC landscape of aggregating functions. The classes in blue typeface are defined in terms of algebraic/combinatorial properties, and the others in terms of secure computability. Arrow **A** → **B** indicates **A** ⊇ **B**. (Color figure online)

the correlated randomness by themselves, in an offline phase. Unlike PSM and NIMPC, which have an incorruptible party, *UNIMPC retains the standard security model of MPC*, allowing corruption of any set of parties, and requiring the adversary to learn nothing more than the output of the function.

A UNIMPC protocol is an MPC protocol and can also be immediately interpreted as an NIMPC protocol.[2]

Note that MPC and NIMPC are incomparable in the sense that an MPC protocol does not yield an NIMPC protocol (because of the general communication pattern) and an NIMPC protocol does not yield an MPC protocol (because of the use of a trusted party, and because the adversary is allowed to learn potentially more than the output of the function). Thus UNIMPC could be seen as a common denominator of these two secure computation models.

UNIMPC* corresponds to a minimalistic version of UNIMPC, with protocols which have a single round of (simultaneous) communication among the parties

---

[2] Replacing the views from the pre-processing phase of a UNIMPC protocol with correlated randomness from a trusted party turns it into an NIMPC protocol.

before they get their inputs, followed by a single message from each party to the aggregator after they receive their input. (UNIMPC allows arbitrarily many rounds of communication prior to receiving inputs.)

Strongly Secure MPC. We also study feasibility under a *stronger* model of MPC, which requires both UC security and passive security to hold simultaneously (information theoretically). Traditionally, UC security refers to the setting of active corruption, in which the security guarantees are relative to an ideal model where too the corrupt parties are actively corrupt. While stronger in general, this gives a weaker guarantee than security against passive corruption, when the corrupt parties are indeed only passively corrupt.[3] From a practical point of view, strong security (possibly weakened to hold only against PPT adversaries) is important, and arguably the "right" notion in many cases. Here we initiate the study of characterizing multi-party functionalities that are strongly securely realizable.

Relating Secure Computation to the Algebraic Classes. Our results show the rich connections between the cryptographic complexity landscape of MPC and the combinatorial/algebraic structures of the functions, as summarized in Fig. 1. We briefly point out the several results that go into making this map. All results relate to the information-theoretic setting with finite functions.

☐ **MPC ⊆ CPS**: This result hinges on characterizing the following cryptographic property algebraically: given any subset of the inputs and the output of the function, the *residual function* of the remaining inputs can be determined. (Theorem 2).

☐ **CPSS ⊆ UNIMPC⋆**: We establish this by developing a novel MPC protocol that generalizes the simple abelian group summation protocol to a certain class of (non-abelian) group actions (Theorem 3).

☐ **CPSS ⊊ CPS**: We give a concrete family of functions that fall into the gap between these two classes (Theorem 1). Combined with the above results, this separation leaves an intriguing gap between the necessary and sufficient conditions for MPC. (But we show in Theorem 4, that this gap disappears/reduces for a small number of input parties.)

☐ **CCPS ⊆ UNIMPC⋆**: The class **CCPS** (for Complete CPS) consists of the "Latin Hypercube" functionalities that fall within **CPS**. We show that all such functions, in more than two dimensions, are highly structured and in particular fall within **CPSS** (Theorem 5). For two dimensions, i.e., Latin squares, this is not true; but in this case a UNIMPC⋆ protocol can be directly given for all Latin squares. Further, in this case, due to a classical result of Ryser [Rys51], **CPS = CCPS** (see Sect. 1.2).

☐ **UC security results**: The characterization of UC securely realizable functions has been resolved for *2 and 3-party functionalities* [CKL06,PR08], but

---

[3] E.g., a 2-party functionality in which Bob receives $a \vee b$, where $a, b \in \{0, 1\}$ are inputs to Alice and Bob respectively, has no protocol secure against passive corruption; but a protocol in which Alice simply sends a to Bob is UC secure. Also see $\mathcal{F}_{\text{AND}}$ discussed in Sect. 8.1..

remains open for more than 3 parties. Prabhakaran and Rosulek [PR08] showed that there are only two classes of secure function evaluation functionalities – aggregating and disseminating – that can possibly have UC secure protocols. They also gave a UC secure protocol for the "disseminated OR" functionality for 3 parties. We build on this further to show that:

- Disseminated OR functionality with any number of players is UC securely realizable. Further, every disseminating functionality is UC securely realizable by a reduction to the disseminated OR functionality (Sect. 8.2).
- Every aggregating functionality in **CCPS** has a UC secure protocol; this relies on a compiler from a strongly secure protocol for $\mathcal{F}$ (which exists only if $\mathcal{F}$ is a CPS functionality) to one for $\mathcal{F}$ restricted to a domain D (Sect. 8.1).
- In both these positive results, we obtain strong security (Theorem 7). Combined with the negative results (Theorem 6), this shows that

$$\textbf{CCPS} \cup \textbf{DISS} \subseteq \textsc{strongMPC} \subseteq \textbf{CPS} \cup \textbf{DISS}$$

where **strongMPC** denotes the class of *all* functionalities (not just aggregating functionalities) that have strongly secure protocols, and **DISS** and **CCPS** are interpreted as all functionalities "isomorphic" to functionalities that are disseminating or functionalities that embed into a CCPS functionality. In Fig. 1, this relationship is indicated restricted to aggregating functionalities (in which, case the extension to isomorphism – which allows all parties to have inputs and outputs – can be ignored).

☐ **Additional Results and Implications:**

- Recently, Halevi et al. introduced the notion of "Best Possible Information-Theoretic MPC" (BIT-MPC) [HIKR18], by removing the trusted party and the non-interactive structure in the NIMPC model, but retaining the provision that (in the ideal-world) the adversary is allowed to learn the residual function of the honest parties' inputs. While the set of functions for which BIT-MPC is possible is a strict superset of **MPC**, the main open problem posed in [HIKR18] is whether all functions have BIT-MPC protocols. We note that for all functions in **CPS**, BIT-MPC protocols are automatically MPC protocols (because for them the residual function can be deduced from the output and the corrupt parties' own inputs). Thus if **CPS** \ **MPC** $\neq \emptyset$, then there exist functions which do not have a BIT-MPC protocol.
- Our necessity result – that **MPC** $\subseteq$ **CPS** – can be extended in a couple of ways (Sect. 5.1): Firstly, the necessity condition continues to hold even if the corruption model allowed the corruption of at most one party other than the aggregating party, if we require a UNIMPC protocol (this model could be called 1-Robust UNIMPC). Secondly, the necessity condition holds even for NIMPC (even 1-Robust NIMPC), if we required an additional security property that the adversary learns only what the output reveals (like in MPC) rather than the residual function of the honest parties (as NIMPC does).

- While our focus is on aggregating functionalities, our positive results for passive-secure MPC do yield new protocols for *symmetric functionalities* wherein all parties get the same output – as considered in [CI96]. This is because a passive-secure MPC protocol for an aggregating functionality can be readily converted into one for a symmetric functionality computing the same function.
- Since one of our results (Theorem 4) depends on the existence of NIMPC protocols, we present a simple NIMPC protocol for general functionalities in the full version. This protocol is a generalization of an NIMPC protocol in [HIJ+16] to arbitrary input domains, presented more directly in terms of the function matrix. This NIMPC protocol is more efficient and much simpler than the earlier ones in the literature [BGI+14, OY16].

We present more details of our results and techniques in Sect. 1.2. In the full version, we also discuss several problems that are left open by this work.

1.1   Related Work

There has been a large body of work aimed at characterizing functionalities with MPC protocols in various models (see, e.g., a survey [MPR13]). For some important classes, exact characterizations are known: this includes passive and active (stand-alone) security for 2-party deterministic functions [Kus89, KMR09, MPR09], multi-party functions with restricted adversary structures [BGW88, CCD88, HM97], multi-party functions with binary alphabet [CK91], multi-party protocols which only have public communication [KMR09], and UC security for 2-party functions [CKL06, PR08].

The characterization question for the multi-party setting (with point-to-point channels and no honest majority, for passive security) was explicitly considered in [CI96]. It was shown there that there exist m-party functions which do not have any passive-secure protocol such that the $m - 1$-party function obtained by merging any two parties results in a securely realizable functionality. This problem in the context of UC security was studied in [PR08], where the terms aggregating functionality and disseminating functionality were coined.

The NIMPC model was introduced by Beimel et al. [BGI+14], inspired by the earlier work of Feige et al. [FKN94]. This was generalized to other patterns of interaction in [HIJ+16]. A computational version of UNIMPC (but with a public-key infrastructure) was recently explored in [HIJ+17].

A recent independent and concurrent work by Halevi et al. [HIKR18] overlaps with some of our results. Specifically, they also observe the fact that an MPC protocol must reveal the residual function of the honest parties to an adversary corrupting the output party, which is the staring point of our proof of Theorem 2 (they do not derive the combinatorial characterization of CPS). The transformation from NIMPC to UNIMPC we use to prove Theorem 4 is a special case of the NIMPC to MPC compiler of [HIKR18], which forms the main tool for their positive results. Finally, as pointed out above, the main open problem left in [HIKR18] is whether there are functions with no BIT-MPC protocol, and this relates to an open problem we leave, namely whether **CPS** = **MPC**: A negative answer to our question answers that of [HIKR18] in the negative.

1.2 Technical Overview

We give a brief overview of CPS functions, and a couple of our protocols that exploit this structure.

An $m + 1$ aggregating functionality involves parties $P_1, \cdots, P_m$ with inputs and an aggregator P_0 who learns the output. A classical example of an aggregating functionality that admits secure computation is the summation operation in an abelian group. As a starting point to understanding all securely computable functions, one could try to generalize this function. Consider the 3-party version of this problem, involving two input parties P_1, P_2 and an output party P_0. W.l.o.g. we can consider computing a function $f : [n_1] \times [n_2] \to [n]$, given an as a matrix M with $M_{ij} = f(i, j)$. Suppose there is a passive secure protocol Π for computing f. From the results on 2-party MPC we know that an adversary which passively corrupts $\{P_0, P_1\}$ must learn P_2's input fully (up to equivalent inputs). Then, for this protocol to be secure, even given an ideal functionality, an adversary who passively corrupts $\{P_0, P_1\}$ should be able to learn P_2's input. A passive adversary is not allowed to change the parties' inputs. Hence, for any inputs $x_1 \in [n_1], x_2 \in [n_2]$, it must be the case that $(x_1, f(x_1, x_2))$ uniquely determines x_2. Symmetrically, $(x_2, f(x_1, x_2))$ uniquely determines x_1. We refer to this as the *Latin property* of M, named after Latin squares. (Latin squares are $n \times n$ square matrices in which each row and each column is a permutation of $[n]$. Note that a square matrix with the Latin property is the same as a Latin square.)

It is easy to see that any 3-party aggregating functionality $f : [n] \times [n] \to [n]$ which is a Latin square has a passive secure protocol: P_1 and P_2 privately agree on a random permutation σ over $[n]$, and then P_1 sends P_0 the row indexed by its input x_1, but with positions permuted according to σ: i.e., a vector $(z_1, \cdots, z_n)$ where $z_{\sigma(j)} = M_{x_1,j}$. P_2 sends $k = \sigma(x_2)$ to P_0, and P_0 outputs $z_k = M_{x_1,x_2}$. Note that the security of this protocol relies on not only the Latin property, but also on the fact that each row has all n elements. However, since any rectangle with the Latin property can be embedded into an (at most quadratically larger) Latin square [Rys51], any function f which has the Latin property does indeed have a passive secure protocol.

This might suggest that for arbitrary number of parties, an analogous Latin hypercube property would be a tight characterization of secure computability. Interestingly, this is not the case. With m input clients, the 2-party results imply that an adversary corrupting a subset of the m input parties and the aggregator P_0 can learn the *residual function* of the honest parties' inputs. Since the passive adversary cannot change the input of the corrupt parties even in the ideal world, this means that any choice of the corrupt parties' inputs should reveal the residual function of the honest parties. We identify an algebraic formulation in terms of a "Commuting Permutation System" (CPS) that captures this condition tightly.

A CPS over the output alphabet $[n]$ has input sets $X_i \subseteq S_n$, for $i = 1$ to m, where S_n is the group of all permutations of $[n]$. On input $(\pi_1, \cdots, \pi_m) \in X_1 \times \cdots \times X_m$, the output is defined as $\pi_1 \circ \cdots \circ \pi_m(1)$. The "commuting" property is the requirement that this output is invariant to the order in which the m permutations are applied to 1. Note that the commutativity needs to hold only when applied to 1. Also, it holds only *across* the sets $X_1, \cdots, X_m$. That is if $\pi, \pi' \in X_i$, it is not necessary that $\pi \circ \pi'(1)$ equals $\pi' \circ \pi(1)$. The function table of a CPS functionality is indeed a Latin hypercube, but the converse does not hold.

Being a CPS functionality is necessary to have an MPC protocol (let alone a UNIMPC protocol). Unfortunately, we do not know if this is also a sufficient condition. But given some additional structure in a CPS, we are able to give a new protocol. The additional structure that we can exploit is that each X_i is a *subgroup* of S_n, in which case we call the system a Commuting Permutation *Subgroups* System or CPSS. Exploiting this property, we design a protocol for computing CPSS functions, as discussed below.

UNIMPC Protocol for CPSS Functionalities. We present a novel protocol with perfect, information-theoretic security against passive corruption for all CPSS functionalities (and, further, is in fact, UC secure for a sub-class). Recall that the goal is to let P_0 learn $\pi_1 \circ \cdots \circ \pi_m(1)$, where π_i is a permutation that P_i receives as input. At first glance, our protocol may appear similar in structure to a protocol for an abelian group sum: each party P_i shares its input π_i as $\pi_i = \sigma_{i,0} \circ \sigma_{i,1} \circ \cdots \circ \sigma_{i,m}$, where each of the shares itself belongs to X_i. It will be helpful to visualize these shares as forming the ith row in a matrix of shares. The shares in each column $(\sigma_{1,j}, \cdots, \sigma_{m,j})$ for $j \in [m]$ will be correlated with each other in some manner, so that the output can be reconstructed by aggregating only the shares $(\sigma_{1,0}, \cdots, \sigma_{m,0})$. (An analogy for the case of the abelian group would be to choose the shares in each column to sum up to the identity element.) These shares will be sent to P_0.

But there are a couple of major differences. Firstly, permutations do not commute in general, and it is not clear how the shares can be meaningfully combined. Secondly, we *must not reveal* the composition of the inputs – i.e., the permutation $\pi_1 \circ \cdots \circ \pi_m$ – to the aggregator; only the result of applying this composition to 1 should be revealed. So, choosing the column shares to "add up to" the identity permutation would be problematic, not to mention that there may not be any such choice other than choosing all the shares to be the identity element.

In our protocol, we choose the column shares such that their composition has 1 as a fixed point (there is at least one such choice, since the each entry can be chosen as the identity permutation). Then, using the CPSS property, it can be shown that $(\prod_{i \in [m]} \sigma_{i,0})(1) = (\prod_{i \in [m]} \pi_i)(1)$ (see Fig. 2). It turns out that we can use the subgroup structure in CPSS to argue that if the shares are chosen uniformly at random subject to the above constraint, then $(\sigma_{1,0}, \cdots, \sigma_{m,0})$ reveals nothing more than $\pi_1 \circ \cdots \circ \pi_m(1)$.

Further, even if we consider all the shares $\sigma_{i,j}$ except for $(i,j) \in S \times S$ for some $S \subseteq [m]$, we show that they reveal nothing more than the residual function $(\prod_{i \in S} \pi_i)(1)$. The need to consider revealing this set of shares comes from the fact that our protocol

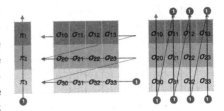

Fig. 2. Elements in the ith row belong to a subgroup X_i in a CPSS. The subgroup structure enables secret-sharing as $\pi_i = \prod_{j=m}^{0} \sigma_{i,j}$. Then the illustrated quantities are equal: $(\prod_{i \in [m]} \pi_i)(1) = (\prod_{i \in [m]} \prod_{j=m}^{0} \sigma_{i,j})(1) = (\prod_{j=m}^{0} \prod_{i \in [m]} \sigma_{i,j})(1)$. The last equality relies on the closure property in the subgroup, as well as the commutativity guarantee (when applied to 1). In our protocol, for each $j > 0$, $(\prod_{i \in [m]} \sigma_{i,j})(1) = 1$, and hence this also equals $(\prod_{i \in [m]} \sigma_{i,0})(1)$.

is not an NIMPC protocol (where a trusted dealer could compute $\sigma_{i,j}$ for all $(i,j) \in [m]^2$ and send only $(\sigma_{i,1}, \cdots, \sigma_{i,m})$ to each party P_i); instead we require the parties to compute all the shares themselves, which is achieved by each party P_j computing the jth column of shares, and distributing it among all the parties P_i. Thus when we consider a set S of honest parties, only the shares $\sigma_{i,j}$ where $(i,j) \in S^2$ remain hidden from the adversary.

UC-secure Protocols. It turns out that the above protocol for aggregating functions is UC secure if the function is a Complete CPSS (CCPSS) function. For $m \geq 3$, a Complete CPS is always a Complete CPSS, and hence this gives a UC secure (in fact, strongly secure) protocol for all CCPS functionalities. (The case of $m = 2$ is handled separately.)

However, for a function that is only *embedded in* a CCPS functionality, this protocol is not necessarily UC secure (because nothing prevents an adversary from using an input from the full domain of the CCPS functionality). We give a compiler that can take a UC secure protocol for a CCPS functionality, and transform it into a UC secure protocol for the functionality restricted to a smaller domain. The main idea of the compiler is to run several instances of the original protocol with the parties using random inputs from the restricted domain. That they used inputs from the restricted domain is then verified using a cut-and-choose phase. Then, an aggregated AND functionality is used to identify instances among the unopened executions to obtain the output. Plugging in a simple UC secure protocol for aggregated AND, this compiler yields a UC secure protocol. Interestingly, though aggregated AND itself has no strongly secure protocol (or passive-secure protocol, for that matter) as it is not a CPS functionality, the resulting protocol above is a strongly secure protocol.

We remark that this is a feasibility result that relies on the domains being finite (small) as the compiler's overhead is polynomial in the domain size.

We also present a reduction from any disseminating function to the disseminated-OR functionality. This is also a feasibility result that relies on the number of parties being finite (small) as the protocol is exponential in the number of parties. To complete establishing the realizability of all disseminating functions, we give a UC secure protocol for the disseminated-OR functionality (extending a 3-party protocol for the same functionality in [PR08]).

2   Preliminaries

We write $[n]$ to denote the set $\{1, \cdots, n\}$. S_n denotes the symmetric group over $[n]$, namely, the group of all permutations of $[n]$. In our proofs, we shall use the product notation $\prod$ to denote the composition operation of permutations. Note that composition of permutations is a non-commutative operation in general, and hence the order of the indices is important (as in $\prod_{i=1}^{t} \rho_i$). When the order is not important, we denote the indices by a set (as in $\prod_{i \in [t]} \rho_i$).

Below we define notions referred to through out the paper. Additional notions relevant to strong security are deferred to Sect. 8.

We adapt the definition of an *aggregating functionality* from [PR08].[4]

Definition 1 (Aggregating Functionality). *An* $(m + 1)$ *party Aggregating functionality accepts inputs* $x_i \in X_i$ *from* P_i *for* $i = 1$ *to* m, *and sends* $f(x_1, \cdots, x_m)$ *to party* P_0, *where* $f : X_1 \times \cdots \times X_m \to \Omega$ *is a fixed function.*

Consistent with the literature on feasibility questions, we consider the functions to have constant-sized domains (rather than infinite domains or domains expanding with the security parameter). Also, in all our positive results, the security obtained is perfect and hence the protocols themselves do not depend on the security parameter. Our negative results do allow protocols to have a negligible statistical error in security.

Definition 2 (Embedding). *An aggregating functionality* $f : X_1 \times \cdots \times X_m \to [n]$ *is said to* embed *into a functionality* $g : X'_1 \times \cdots \times X'_m \to [n']$ *if there exist functions* $\phi_i : X_i \to X'_i$ *for* $i \in [m]$, *and an injective function* $\phi_0 : [n] \to [n']$ *such that for all* $(x_1, \cdots, x_m) \in X_1 \times \cdots \times X_m$,

$$\phi_0(f(x_1, \cdots, x_n)) = g(\phi_1(x_1), \cdots, \phi_m(x_m)). \tag{1}$$

Below, $\mathcal{A} \cong \mathcal{B}$ denotes that the statistical difference between the two distributions $\mathcal{A}$ and $\mathcal{B}$ is negligible as a function of a (statistical) security parameter.

---

[4] We allow only the aggregating party P_0 to have an output. The original definition in [PR08] allows all the parties to have outputs, but requires that for each party other than P_0, its output is a function only of its own input. Such a function is "isomorphic" to an aggregated functionality as we define here.

Definition 3 (Passive Secure MPC). *An $(m+1)$-party protocol Π with parties $P_1, \cdots, P_m, P_0$ is said to be an information-theoretically secure MPC protocol for an $(m+1)$-party aggregating functionality f against passive corruption, if for any subset $T \subseteq [m] \cup \{0\}$, there exists a simulator S s.t. for any input $x \in X$:*

$$\mathrm{VIEW}_{\Pi(x)}(\{P_i | i \in T\}) \cong \begin{cases} S(x_T, f(x)) & \text{if } 0 \in T \\ S(x_T, \bot) & \text{otherwise} \end{cases}$$

where $\mathrm{VIEW}_{\Pi(x)}(\{P_i | i \in T\})$ represents the view of the parties $\{P_i | i \in T\}$ in an execution of Π with input x and $\bot$ represents an empty input.

We shall use the following result for 2-party MPC, obtained from the general characterization in [KMR09].

Lemma 1 (2-Party MPC with one-sided output [KMR09]). *If a finite 2-party functionality which takes inputs $x \in X$ and $y \in Y$ from Alice and Bob respectively and outputs $f(x, y)$ to Bob for some function $f : X \times Y \to Z$ has a statistically secure protocol against passive adversaries, then $\forall x, x' \in X$ it holds that $\exists y \in Y, f(x, y) = f(x', y) \Rightarrow \forall y \in Y, f(x, y) = f(x', y)$.*

We refer the reader to [BGI+14] for a definition of NIMPC and PSM.

3   New Models

In this section we define UNIMPC and UNIMPC*, which are models of secure computation, as well as combinatorial objects CPS and CPSS. For simplicity, we define UNIMPC and UNIMPC* for fixed functions rather than function families (though the definitions can be easily extended to function families, where all the input players receive the function as an input).

Definition 4 (UNIMPC). *We define an* Unassisted Non-Interactive Secure Multi-party Computation (UNIMPC) *protocol Π for an $(m+1)$-party aggregating functionality $f : X \to \Omega$ as $\Pi = (\mathcal{R}, \mathrm{Enc}, \mathrm{Dec})$ where:*

- *$\mathcal{R}$ is an m-party randomized protocol (without inputs), generating correlated views $(r_1, \cdots, r_m) \in R_1 \times \cdots \times R_m$.*
- *Enc is an m-tuple of deterministic functions $(\mathrm{Enc}_1, \cdots, \mathrm{Enc}_m)$ where $\mathrm{Enc}_i : X_i \times R_i \to M_i$.*
- *Dec $: M_1 \times \cdots \times M_m \to \Omega$ is a deterministic function satisfying the following correctness requirement: for any $(x_1, \cdots, x_m) \in X$ and any view $(r_1, \cdots, r_m)$ which $\mathcal{R}$ generates with positive probability,*

$$\mathrm{Dec}((\mathrm{Enc}_1(x_1, r_1), \cdots, \mathrm{Enc}_m(x_m, r_m)) = f(x_1, \cdots, x_m).$$

It is identified with a two-phase MPC protocol where:

1. **Offline Phase:** *The parties* $P_i : i \in [m]$ *run* $\mathcal{R}$ *(without any input) so that each* P_i *obtains the view* r_i.
2. **Online Phase:** *Every* P_i *encodes its input* x_i *as* $z_i = \mathrm{Enc}_i(x_i, r_i)$ *and sends it to the aggregator* P_0. P_0 *outputs* $\mathrm{Dec}(z_1, \cdots, z_m)$.

Security: A UNIMPC protocol Π *for* $f : X \to \Omega$ *is said to be* T-*secure (for* $T \subseteq [m]$*) if there exists a simulator* S *s.t. for any* $x \in X$:

$$\mathrm{VIEW}_{\Pi(x)}(\{P_i | i \in T\} \cup \{P_0\}) \cong S(x_T, f(x))$$

where $\mathrm{VIEW}_{\Pi(x)}(\cdot)$ *represents the view of a given set of parties in the two-phase protocol above, with input* x.
 For any $t \in [m]$, Π *is said to be* t-*robust if it is* T-*secure* $\forall T \subseteq [m]$ *s.t.* $|T| \leq t$. *A UNIMPC protocol* Π *is said to be secure if it is* m-*robust.*

We point out that a secure UNIMPC protocol as defined above is a passive secure MPC protocol for f (as in Definition 3). Note that in defining T-security we considered only the case when the set of corrupt parties includes the aggregator. But when the aggregator is honest, security is automatically guaranteed by the structure of the UNIMPC protocol (the view of the adversary being derived completely from the offline phase).

Definition 5 (UNIMPC*). *We define an Unassisted Non-Interactive Secure Multi-party Computation protocol with Non-Interactive Pre-Processing (UNIMPC* protocol)* Π *for a functionality* $f : X \to \Omega$ *as a UNIMPC protocol* $\Pi = (\mathcal{R}, \mathrm{Enc}, \mathrm{Dec})$ *for* f *where* $\mathcal{R}$ *consists of a single round (i.e., each party simply sends messages to the others, and then receives all the messages sent to it).*

We define classes **MPC, UNIMPC, UNIMPC*** as the class of aggregating functionalities that have (information-theoretically) passive secure MPC, UNIMPC and UNIMPC* protocols, respectively.

4 Commuting Permutations System

In this section, we define the new algebraic-combinatorial classes.

Definition 6 (CPS and CPSS). *An* (n, m)-*Commuting Permutations System (CPS) is a collection* $(X_1, \cdots, X_m)$ *where for all* $i \in [m]$, $X_i \subseteq S_n$ *contains the identity permutation, and for any collection* $(\pi_1, \cdots . \pi_m)$ *with* $\pi_i \in X_i$, *and* $\rho \in S_m$, $\pi_1 \circ \cdots \circ \pi_m(1) = \pi_{\rho(1)} \circ \cdots \circ \pi_{\rho(m)}(1)$.[5]
 It is called an (n, m)-*Commuting Permutation Subgroups System (CPSS) if each* X_i *is a subgroup of* S_n.

---

[5] Choice of 1 is arbitrary. Requiring identity permutation to always be part of each X_i is w.l.o.g., as a CPS without it will remain a CPS on adding it.

Note that given a CPS $(X_1, \cdots, X_m)$, for any $(\pi_1, \cdots, \pi_m) \in X_1 \times \cdots \times X_m$, the expression $(\prod_{i \in [m]} \pi_i)(1)$ is well-defined as the order of composition is not important.

Definition 7 (CCPS). *An (n, m)-CPS $(X_1, \cdots, X_m)$ is said to be complete in dimension i if $\{\pi(1) \mid \pi \in X_i\} = [n]$. If it is complete in all m dimensions, it is called a* Complete CPS *(CCPS).*

Definition 8. *An $(m + 1)$-party aggregating functionality $f : X_1 \times \cdots \times X_m \to [n]$ is said to be a* CPS functionality *(resp.,* CPSS *and* CCPS *functionality) if $(X_1, \cdots, X_m)$ is an (n, m)-CPS (resp., (n, m)-CPSS and (n, m)-CCPS), and for all $(\pi_1, \cdots, \pi_m) \in X_1 \times \cdots \times X_m$, $f(\pi_1, \cdots, \pi_m) = (\prod_{i \in [m]} \pi_i)(1)$.*

CPS *(resp.,* **CPSS** *and* **CCPS***) is defined as the class of all aggregating functionalities that embed into a CPS functionality (resp., CPSS functionality and CCPS functionality).*

A CPSS enjoys a certain (non-abelian) group structure. More specifically, the CPSS $(G_1, \cdots, G_m)$ can be identified with a group, with the set of elements $G_1 \times \cdots \times G_m$ and group operation $*$ defined as $(\sigma_1, \ldots, \sigma_m) * (\sigma'_1, \ldots, \sigma'_m) = (\sigma_1 \circ \sigma'_1, \ldots, \sigma_m \circ \sigma'_m)$. This is captured in the following lemma.

Lemma 2. *Suppose $(G_1, \cdots, G_m)$ is a CPSS. Then, for any set of mt permutations $\{\sigma_{i,j} \mid i \in [m], j \in [t]\}$ such that $\sigma_{i,j} \in G_i$, it holds that*

$$\left(\prod_{j=1}^{t} \prod_{i=1}^{m} \sigma_{i,j} \right)(1) = \left(\prod_{i \in [m]} \prod_{j=1}^{t} \sigma_{i,j} \right)(1).$$

Proof: Consider $\rho \circ \prod_{i=1}^{m} \rho_i(1)$, where $\rho_i \in G_i$ for each i, and $\rho \in G_{i_0}$ for some $i_0 \in [m]$. Note that the order of composition is not important in $\prod_{i=1}^{m} \rho_i(1)$, since $(G_1, \cdots, G_m)$ is a CPS(S), and we may write it as $\prod_{i \in [m]} \rho_i(1)$. Also, define ρ'_i as

$$\rho'_i = \begin{cases} \rho \circ \rho_{i_0} & \text{if } i = i_0 \\ \rho_i & \text{otherwise.} \end{cases}$$

Since G_{i_0} is a group, we have $\rho'_i \in G_i$ for all $i \in [m]$ (including i_0). Then,

$$\left(\rho \circ \prod_{i=1}^{m} \rho_i \right)(1) = \left(\rho \circ \rho_{i_0} \circ \prod_{i \in [m] \backslash \{i_0\}} \rho_i \right)(1) = \left(\rho'_{i_0} \circ \prod_{i \in [m] \backslash \{i_0\}} \rho'_i \right)(1) = \left(\prod_{i \in [m]} \rho'_i \right)(1)$$

where in the last step, we again used the CPS property. The claim follows by repeatedly using the above equality. □

Our first result is a separation:

Theorem 1. CPSS $\subsetneq$ CPS.

Proof: We prove this by giving an explicit $(5,3)$-CPS (X_1, X_2, X_3), and showing that the corresponding CPS functionality does not embed into any $(n, 3)$-CPSS functionality. (In the full version we give instances of (n, m)-CPS that cannot be embedded into a CPSS, for every value of $m \geq 2$.) Let

$$X_1 = \{\pi_0, \pi_1\}, X_2 = \{\pi_0, \pi_2\}, X_3 = \{\pi_0, \pi_3\},$$

where (using the standard cycle notation for permutations), $\pi_0 = (1)(2)(3)(4)(5)$, $\pi_1 = (1\ 2\ 5)(3\ 4)$, $\pi_2 = (1\ 3\ 5)(2\ 4)$ and $\pi_3 = (1\ 4\ 5)(2\ 3)$. It can be verified that this is a CPS by computing all non-trivial applications of these permutations on 1: $\prod_{i \in \{1,2,3\}} \pi_i(1) = 5$, $\prod_{i \in \{1,2\}} \pi_i(1) = 4$, $\prod_{i \in \{1,3\}} \pi_i(1) = 3$, and $\prod_{i \in \{2,3\}} \pi_i(1) = 2$.

We argue that this cannot be embedded into a CPSS. Suppose, for some n, there is an $(n, 3)$-CPSS, (G_1, G_2, G_3), and functions $\phi_i : X_i \to G_i$ and an injective function $\phi_0 : [5] \to [n]$, as specified in Definition 2. Let $\phi_i(\pi_0) = \sigma_i$ and $\phi_i(\pi_i) = \rho_i$. First, we argue that w.l.o.g., we can require all σ_i to be the identity function. This is because, otherwise, $\hat{\phi}_i(\pi) = \sigma_i^{-1} \circ \phi_i(\pi)$ and $\hat{\phi}_0 = (\sigma_1 \circ \sigma_2 \circ \sigma_3)^{-1} \circ \phi_0$ is a valid embedding, with $\hat{\phi}_i(\pi_0)$ being the identity function. This follows from the fact (see Lemma 2) that in a CPSS with $\{\alpha_i, \beta_i\} \subseteq G_i$,

$$(\alpha_1 \circ \beta_1) \circ \cdots \circ (\alpha_m \circ \beta_m)(1) = (\alpha_1 \circ \cdots \circ \alpha_m) \circ (\beta_1 \circ \cdots \circ \beta_m)(1).$$

Next we argue that (with σ_i being identity), w.l.o.g., ϕ_0 is the identity function as well. This is because $\hat{\phi}_i(\pi) = \phi_0 \circ \phi_i(\pi)\phi_0^{-1}$, along with $\hat{\phi}_0$ being the identity function yields an embedding. This relies on the fact that $\phi_0(1) = 1$ (as implied by Eq. 1 of Definition 2, by considering $x_1 = x_2 = x_3 = \pi_0$).

Now, we derive a contradiction from the following two requirements:

- From Eq. 1, we get that $\pi_i(a) = \rho_i(a)$ for all i and $a \in \{1, 2, 3, 4\}$ (but not necessarily for $a = 5$).
- Since (G_1, G_2, G_3) is a CPSS, we require that $\rho_2^2 \in G_2$. Then, we require that $\rho_2^2 \circ \rho_3(1) = \rho_3 \circ \rho_2^2(1)$.

Using the first condition, we derive three equalities: $\rho_3(1) = 4$, $\rho_2^2 \circ \rho_3(1) = 4$ and $\rho_3 \circ \rho_2^2(1) = \rho_3(5)$. From the last two equalities, and the second condition, we find that $\rho_3(5) = 4$, yielding a contradiction with the first equality. $\square$

5   Only CPS Functionalities Have (UNI)MPC Protocols

We show that if an aggregating functionality has a statistically secure MPC protocol against semi-honest adversaries (without honest majority or setups), then it must be a CPS functionality. Since UNIMPC protocols are MPC protocols, this applies to UNIMPC as well.

Theorem 2. *If an aggregating functionality has an information-theoretically secure MPC protocol against semi-honest adversaries, then it embeds into a CPS functionality.*

Proof: Suppose an $(m + 1)$-party aggregating functionality $f : X_1 \times \cdots \times X_m \to [n]$ is semi-honest securely realizable. Denote the aggregating party as P_0 and for each $i \in [m]$, the party with input domain X_i as P_i.

Firstly, w.l.o.g., we may assume that no party has two *equivalent inputs*, by considering an embedding if necessary. Further, we may let $X_i = [n_i]$ for each i, and $f(1, \cdots, 1) = 1$, by relabeling the inputs and the outputs.

Now, for each $i \in [m]$, consider the 2-party SFE functionality obtained by grouping parties $\{P_j | j \in [m] \backslash \{i\}\}$ as a single party Alice, and the parties $\{P_i, P_0\}$ as a single party Bob. This functionality has the form in Lemma 1, namely, only Bob has any output. Then applying the lemma, we get the following (where the notation $\mathbf{x}[i : \ell]$ denotes the vector obtained from $\mathbf{x}$ by setting x_i to ℓ): $\forall \mathbf{x}, \mathbf{x}' \in X_1 \times \cdots \times X_m$,

$$f(\mathbf{x}) = f(\mathbf{x}') \text{ and } x_i = x_i' \Rightarrow \forall \ell \in X_i, f(\mathbf{x}[i : \ell]) = f(\mathbf{x}'[i : \ell]). \qquad (2)$$

We use this to prove the following claim.

Claim. For each $i \in [m]$ and $\ell \in X_i$, there exists a permutation $\pi_\ell^{(i)}$ such that, for all $\mathbf{x} \in X_1 \times \cdots \times X_m$ with $x_i = 1$,

$$\pi_\ell^{(i)}(f(\mathbf{x})) = f(\mathbf{x}[i : \ell]). \qquad (3)$$

Proof: Fix $i \in [m]$, $\ell \in X_i$. Now, consider defining a (partial) function $\pi_\ell^{(i)}$ using Eq. 3. This is well-defined thanks to Eq. 2: Even though there could be multiple $\mathbf{x}$ with $x_i = 1$ and the same value for $f(\mathbf{x})$, Eq. 2 ensures that they all lead to the same value for $f(\mathbf{x}[i : \ell])$.

Further, with this definition, if $\pi_\ell^{(i)}(a) = \pi_\ell^{(i)}(b)$, this means that there exist $\mathbf{x}, \mathbf{x}'$ with $x_i = x_i' = 1$, $f(\mathbf{x}) = a$, $f(\mathbf{x}') = b$ and $f(\mathbf{x}[i : \ell]) = f(\mathbf{x}'[i : \ell])$. But by considering $\mathbf{z} = \mathbf{x}[i : \ell]$, $\mathbf{z}' = \mathbf{x}'[i : \ell]$, we have $z_i = z_i'$ and $f(\mathbf{z}) = f(\mathbf{z}')$. Hence, by Eq. 2, we have $f(\mathbf{z}[i : 1]) = f(\mathbf{z}'[i : 1]$. But since $\mathbf{x} = \mathbf{z}[i : 1]$ and $\mathbf{x}' = \mathbf{z}'[i : 1]$, this means that $a = f(\mathbf{x}) = f(\mathbf{x}') = b$. Hence, $\pi_\ell^{(i)}$ is a one-to-one function, from $\{a | \exists \mathbf{x}, x_i = 1, f(\mathbf{x}) = a\} \subseteq [n]$ to $[n]$. We can arbitrarily extend this to be a permutation over $[n]$ to meet the condition in the claim. $\qquad \square$

Finally, for any $\mathbf{x}$ such that $x_{i_1} = \cdots = x_{i_t} = 1$, and distinct $i_1, \cdots, i_t$, by iteratively applying Eq. 3, $\pi_{\ell_t}^{(i_t)} \circ \cdots \circ \pi_{\ell_1}^{(i_1)}(f(\mathbf{x})) = f(\mathbf{x}[i_1 : \ell_1] \cdots [i_t : \ell_t])$. Taking $(i_k, \ell_k) = (\rho(k), z_{\rho(k)})$ for any permutation $\rho \in S_m$ and any $\mathbf{z} \in X_1 \times \cdots \times X_m$, we have $\mathbf{x}[i_1 : \ell_1] \cdots [i_m : \ell_m] = \mathbf{z}$, for any $\mathbf{x}$. Then, with $\mathbf{x} = (1, \cdots, 1)$ we get that

$$f(\mathbf{z}) = \pi_{z_{\rho(1)}}^{(\rho(1))} \circ \cdots \circ \pi_{z_{\rho(m)}}^{(\rho(m))}(1),$$

where we substituted $f(\mathbf{x}) = 1$. This concludes the proof that f embeds into the CPS functionality with input domains $\hat{X}_i = \{\pi_\ell^{(i)} | \ell \in [n_i]\}$. $\qquad \square$

5.1   Extensions to 1-Robust UNIMPC and NIMPC

Since every secure UNIMPC protocol is a secure MPC protocol, Theorem 2 applies to UNIMPC as well. But it extends to UNIMPC in a stronger manner than it holds for MPC. Note that if we restrict the number of corrupt parties to be at most $m/2$, then every $m + 1$ party functionality has a passive secure MPC protocol, even if the functionality is a non-CPS aggregating functionality. But we show that as long as the adversary can corrupt just two parties (the aggregator and one of the input parties), the only aggregating functionalities that have secure UNIMPC protocols are CPS functionalities.

To see this, we consider how Eq. 2 was derived in the proof of Theorem 2 (the rest of the argument did not rely on the protocol). We used the given $(m + 1)$-party protocol to derive a secure 2-party protocol to which Lemma 1 was applied. In arguing that this 2-party protocol is secure we considered two corruption patterns in the original protocol: the adversary could corrupt $\{P_0, P_i\}$ (Bob) or $\{P_j \mid j \in [m] \setminus \{i\}\}$ (Alice). Now, if we allow only corruption of up to two parties, we cannot in general argue that the resulting two party protocol is secure when Alice is corrupted. However, if the starting protocol was a UNIMPC protocol, then in the resulting 2-phase protocol, there is an offline phase when Alice and Bob interact without using their inputs, and after that Alice sends a single message to Bob in the second phase. *Any such protocol* is secure against the corruption of Alice, as Alice's view can be perfectly simulated without Bob's input. Thus, when the starting protocol is a UNIMPC protocol that is T-secure for every T of the form $\{0, i\}$ ($i \in [m]$), then Lemma 1 applies to the 2-party protocol constructed, and the rest of the proof goes through unchanged. Thus, an aggregating functionality f has a 1-robust UNIMPC protocol only if it is a CPS functionality.

The above argument extends in a way to 1-robust NIMPC as well. Of course, every function has a secure NIMPC protocol [BGI+14], and we cannot require all such functions to be CPS. But we note that NIMPC turned out to be possible for all functions not only because a trusted party is allowed (to generate correlated randomness), but also because NIMPC allows the adversary (corrupting the aggregator and some set of parties) to learn the residual function of the honest parties' inputs. So, one may ask for which functionalities does the adversary *learn nothing more than the output of the function* on any input (just as in the security requirement for MPC), even as we allow a trusted party to generate correlated randomness. Here, we note that the above argument in fact extends to the NIMPC setting with the trusted party: We simply include the trusted party as part of Alice in the above 2-party protocol. Since the security of the 2-party protocol relied only on security against Bob (and the 2-phase nature of the protocol), including the trusted party as part of Alice does not affect our proof. Thus we conclude that only CPS functionalities have 1-robust NIMPC where the simulator takes only the input of the corrupt parties and the output of the function (rather than the residual function of the honest parties' inputs).

6 UNIMPC Protocols

In this section we present our positive results for UNIMPC* and UNIMPC (Theorems 3 and 4).

Theorem 3. *Any function embeddable in a CPSS function has a UNIMPC* protocol with perfect security.*

To prove Theorem 3 it is enough to present a perfectly secure protocol for a CPSS function: the protocol retains security against passive corruption when the input domains are restricted to subsets.

UNIMPC* Protocol for CPSS Function.

For $i \in [m]$, party P_i has input $\pi_i \in G_i$, where $(G_1, \cdots, G_m)$ is an (n, m)-CPSS. Party P_0 will output $\pi_1 \circ \cdots \circ \pi_m(1)$.

1. **Randomness Computation:** For each $j \in [m]$, P_j samples $(\sigma_{1j}, \cdots, \sigma_{mj})$ uniformly at random from $G_1 \times \cdots \times G_m$, conditioned on

$$\sigma_{1j} \circ \sigma_{2j} \circ \cdots \circ \sigma_{mj}(1) = 1. \tag{4}$$

   For each $i, j \in [m]$, P_j sends σ_{ij} to P_i.
2. **Input Encoding:** P_i computes $\sigma_{i0} := \pi_i \circ (\sigma_{i1} \circ \cdots \circ \sigma_{im})^{-1}$, and sends it to P_0. Note that $(\sigma_{i0}, \cdots, \sigma_{im})$ is an additive secret-sharing of π_i in the group G_i.
3. **Output Decoding:** P_0 outputs $\sigma_{1,0} \circ \sigma_{2,0} \circ \cdots \circ \sigma_{m,0}(1)$.

By construction, the protocol has the structure of a UNIMPC* protocol. Indeed, it is particularly simple for a UNIMPC* protocol in that the randomness computation protocol in offline phase is a single round protocol. Below we argue that this protocol is indeed a perfectly secure protocol for computing $\left(\prod_{i \in [m]} \pi_i \right)(1)$ against passive corruption of any subset of parties.

Perfect Correctness: The output of P_0 is $\prod_{i=0}^{m} \sigma_{i,0}(1)$. By Eq. 4 (applied to $j = 1$) we may write $1 = \prod_{i=1}^{m} \sigma_{i1}(1)$. We further expand 1 in this expression again by applying Eq. 4 successively for $j = 2, \cdots, m$ to obtain $1 = \prod_{j=1}^{m} \prod_{i=1}^{m} \sigma_{ij}(1)$. Hence, the output of P_0 may be written as $\prod_{j=0}^{m} \prod_{i=1}^{m} \sigma_{i,j}(1)$. By Lemma 2, this equals $\prod_{i \in [m]} \prod_{j=0}^{m} \sigma_{ij}(1)$. By the definition of $\sigma_{i,0}$ this in turn equals $\prod_{i \in [m]} \pi_i(1)$, as desired.

Perfect Semi-Honest Security: A protocol with the UNIMPC structure is always perfectly semi-honest secure as long as the aggregator is honest, or if all the input parties are corrupt. Hence we focus on the case when the aggregator P_0 is corrupt and there is at least one honest party. Suppose the adversary corrupts P_0 and $\{P_i \mid i \in S\}$ for some set $S \subsetneq [m]$. Below, we write $\overline{S} := [m] \setminus S$ to denote the set of indices of the honest parties. Recall that an execution of the protocol .(including the inputs) is fully determined by the $m \times (m+1)$ matrix $\boldsymbol{\sigma}$, with $(i, j)^{\text{th}}$ entry $\sigma_{ij} \in G_i$, for $(i, j) \in [m] \times ([m] \cup \{0\})$. The input determined by

σ is defined by $\text{input}(\sigma) = (\pi_1, \cdots, \pi_m)$, where $\pi_i = \prod_{j=0}^{m} \sigma_{ij}$. We say that σ is valid if for every $j \in [m]$, $\prod_{i \in [m]} \sigma_{ij}(1) = 1$.

When the functionality is invoked with inputs $\boldsymbol{\pi} = (\pi_1, \cdots, \pi_m)$, in the ideal world, the adversary learns only the corrupt parties' inputs $\boldsymbol{\pi}|_S$ and the residual function of the honest parties' inputs $\pi_{\overline{S}}(1)$, where $\pi_{\overline{S}} := (\prod_{i \in \overline{S}} \pi_i)$. But in the real world its view consists also $\langle \sigma \rangle_S := \{\sigma_{ij} \mid i \in S \vee j \in S \cup \{0\}\}$. We need to show that for any two input vectors $\boldsymbol{\pi}, \boldsymbol{\pi}'$ with identical ideal views for the adversary – i.e., $\boldsymbol{\pi}|_S = \boldsymbol{\pi}'|_S$, and $\pi_{\overline{S}}(1) = \pi'_{\overline{S}}(1)$ – the distribution of $\langle \sigma \rangle_S$ is also identical. For this we shall show a bijective map $\phi_S^{\boldsymbol{\pi}'}$ between valid matrices σ consistent with $\boldsymbol{\pi}$ and those consistent with $\boldsymbol{\pi}'$, which preserves $\langle \sigma \rangle_S$. Since σ is distributed uniformly over all valid matrices consistent with the input in the protocol, this will establish that the distribution of $\langle \sigma \rangle_S$ is identical for $\boldsymbol{\pi}$ and $\boldsymbol{\pi}'$. More precisely, the following claim completes the proof.

Claim. For any $S \subsetneq [m]$, and any $\boldsymbol{\pi}, \boldsymbol{\pi}' \in G_1 \times \cdots \times G_m$ such that $\boldsymbol{\pi}|_S = \boldsymbol{\pi}'|_S$ and $\pi_S(1) = \pi'_S(1)$, there is is a bijection $\phi_S^{\boldsymbol{\pi}'}$ from $\{\sigma \mid \text{input}(\sigma) = \boldsymbol{\pi} \wedge \sigma \text{ valid}\}$ to $\{\sigma \mid \text{input}(\sigma) = \boldsymbol{\pi}' \wedge \sigma \text{ valid}\}$, such that $\langle \sigma \rangle_S = \langle \phi_S^{\boldsymbol{\pi}'}(\sigma) \rangle_S$.

Proof: Let $S, \boldsymbol{\pi}, \boldsymbol{\pi}'$ be as in the lemma. We shall first define $\phi_S^{\boldsymbol{\pi}'}$ for all $m \times (m+1)$ matrices σ, with $\sigma_{ij} \in G_i$, and then prove the claimed properties when restricted to the domain in the claim. Fix $h \in \overline{S}$ as (say) the smallest index in $\overline{S}$. Given σ, $\phi_S^{\boldsymbol{\pi}'}$ maps it to σ' as follows.

$$\sigma'_{ij} = \begin{cases} \sigma_{ij} & \text{if } j \neq h \\ \alpha_i^{-1} \circ \pi'_i \circ \beta_i^{-1} & \text{if } j = h \end{cases}$$

where $\alpha_i := \prod_{j=0}^{h-1} \sigma_{ij}$ and $\beta_i := \prod_{j=h+1}^{m} \sigma_{ij}$. Note that like σ, σ' also satisfies the condition that $\sigma'_{ij} \in G_i$ for all $j = 0 \cup [m]$, because $\alpha_i, \beta_i, \pi'_i \in G_i$.

By construction, $\prod_{j=0}^{m} \sigma'_{ij} = \pi'_i$, and hence the image of $\phi_S^{\boldsymbol{\pi}'}$ is contained in $\{\sigma' \mid \text{input}(\sigma') = \boldsymbol{\pi}'\}$. Also, when the domain is $\{\sigma \mid \text{input}(\sigma) = \boldsymbol{\pi}\}$, the mapping is invertible since $\phi_S^{\boldsymbol{\pi}}(\phi_S^{\boldsymbol{\pi}'}(\sigma)) = \sigma$, when $\text{input}(\sigma) = \boldsymbol{\pi}$. Hence, by symmetry, this is a bijection from $\{\sigma \mid \text{input}(\sigma) = \boldsymbol{\pi}\}$ to $\{\sigma \mid \text{input}(\sigma) = \boldsymbol{\pi}'\}$. Further, for $i \in S$, $\pi_i = \pi'_i$ and hence $\sigma'_{ih} = \sigma_{ih}$, so that $\langle \sigma' \rangle_S = \langle \sigma \rangle_S$.

It remains to prove that the map is a bijection when the domain and range are restricted to *valid* matrices. So, suppose σ is a valid matrix. Then we have

$$\left(\prod_{i \in [m]} \sigma_{ij} \right)(1) = 1 \qquad\qquad \forall j \in [m] \qquad (5)$$

$$\left(\prod_{i \in [m]} \beta_i \right)(1) = \left(\prod_{j=h+1}^{m} \prod_{i \in [m]} \sigma_{ij} \right)(1) = 1. \qquad (6)$$

where the first equality in (6) is obtained by applying Lemma 2, and the second by applying the validity condition (5) successively for $j = m, \cdots, h+1$.

To verify that $\sigma' = \phi_S^{\boldsymbol{\pi}'}(\sigma)$ is valid, we only need to verify that $(\prod_{i \in [m]} \sigma'_{ih})(1) = 1$ (as the other columns of σ' are the same as in σ). This we show as follows (where for brevity, we write $\alpha := \prod_{i \in [m]} \alpha_i$ and $\beta := \prod_{i \in [m]} \beta_i$):

$$\prod_{i\in[m]} \pi_i'(1) = \prod_{i\in[m]} \pi_i(1)$$

$$\Rightarrow (\prod_{i\in[m]} \alpha_i \circ \sigma_{ih}' \circ \beta_i)(1) = (\prod_{i\in[m]} \alpha_i \circ \sigma_{ih} \circ \beta_i)(1)$$

$$\Rightarrow \alpha \circ (\prod_{i\in[m]} \sigma_{ih}') \circ \beta(1) = \alpha \circ (\prod_{i\in[m]} \sigma_{ih}) \circ \beta(1) \qquad \text{by Lemma 2}$$

$$\Rightarrow (\prod_{i\in[m]} \sigma_{ih}') \circ \beta(1) = (\prod_{i\in[m]} \sigma_{ih}) \circ \beta(1) \qquad \alpha \text{ a permutation}$$

$$\Rightarrow (\prod_{i\in[m]} \sigma_{ih}')(1) = (\prod_{i\in[m]} \sigma_{ih})(1) = 1 \qquad \text{by (6) and (5).}$$

$\square$

Theorem 4. *Any CPS functionality with 4 or fewer parties has a UNIMPC protocol with perfect security. Further, any CPS functionality with 3 or fewer parties has a UNIMPC* protocol with perfect security.*

We present the full proof in the full version. In particular, for the case of 4 parties, we describe a UNIMPC protocol, which uses an NIMPC scheme (Gen, Enc, Dec), but implements Gen using a 3-party perfectly secure protocol for general functions that is secure against passive corruption of 1 party (e.g., the passive-secure protocol in [BGW88]). This transformation has appeared in a recent, independent work [HIKR18].

7   Latin Hypercubes

CPS functions are closely related to Latin Squares, and more generally, *Latin Hypercubes*. An n-ary Latin Square is an $n \times n$ matrix with entries from $[n]$ such that each row and column has all elements of $[n]$ appearing in it. The m-dimensional version is similarly a tensor indexed by m-dimensional vectors, so that every "row" (obtained by going through all values for one coordinate of the index, keeping the others fixed) is a permutation of $[n]$. We can associate an m-input functionality with a Latin hypercube, which maps the index vector to the corresponding entry in the hypercube.

In the case of $m = 2$, an n-ary Latin square functionality f always is (or, technically, embeds into) an $(n, 2)$-CPS (X_1, X_2).[6] However, this is not true in higher dimensions (see the full version for an explicit counter example). So not all Latin hypercube functions can have MPC protocols. We obtain an exact characterization of all Latin hypercube functionalities that have UNIMPC*

---

[6] We let $X_1 = \{\pi_i \mid \pi_i(f(1,j)) = f(i,j) \ \forall j \in [n]\}$, and $X_2 = \{\rho_j \mid \rho_j(f(i,1)) = f(i,j) \ \forall i \in [n]\}$. These functions are well-defined permutations because of f being a Latin square functionality, and it is a CPS because, $\pi_i \circ \rho_j(f(1,1)) = \rho_j \circ \pi_i(f(1,1)) = f(i,j)$. With a bijective embedding that relabels the outputs of f so that $f(1,1) = 1$, this meets the definition of a CPS.

(or MPC) protocols. Recall that by Theorem 2 only CPS functionalities can have UNIMPC* (or even MPC) protocols. We show that *all Latin hypercube functionalities that are CPS functionalities indeed have UNIMPC* protocols*. To prove this, we relate this class—Latin hypercube functionalities that are CPS functionalities—to CPSS functionalities (which have UNIMPC* protocols). Firstly, a Latin hypercube functionality that is a CPS functionality forms a Complete CPS (CCPS) functionality, as defined in Definition 7. Then we use the following theorem:

Theorem 5. *For $m > 2$, an (n, m)-CCPS is an (n, m)-CPSS.*

The proof of this theorem, given in the full version, has two parts: Firstly, we show that for $m > 2$, the permutations in an (n, m)-CCPS enjoy *"full-commutativity,"* rather than commutativity when applied to 1. Then we show that any (n, m)-CPS functionality with such full-commutativity embeds into an (n, m)-CPSS. Further, since a CCPS has the maximal number of possible inputs for every party in a CPS (namely, n), this embedding must use a surjective mapping for the inputs, making the original CCPS itself a CPSS.

The following can be stated as a corollary of the above theorems (see the full version).

Corollary 1. *A Latin hypercube functionality has a UNIMPC* protocol if and only if it is a CPS functionality.*

8   Towards a Characterization of Strong Security

While security against active corruption is often stronger than security against passive corruption, this is not always the case. This is because, in the ideal world model for active corruption, the adversary (i.e., simulator) is allowed to send any inputs of its choice to the functionality, the adversary in the passive corruption setting is required to send the same input as the corrupt parties received. To reconcile this discrepancy, one could weaken the notion of passive security by allowing the simulator to change the input sent to the functionality. However, the resulting security guarantee is quite pessimistic, as it assumes that even passively corrupt parties will alter their inputs, and may not be appropriate in scenarios where the passively corrupt parties will not do so (see Footnote 3). Instead, we propose using a stronger definition – which we simply call *strong security* – which requires the simulator to not alter the inputs if the parties are corrupted passively, but allows it to use arbitrary inputs if they are corrupted actively. Formally, we use the following information-theoretic security definition:

Definition 9 (Strong security). *A protocol Π is said to be a strongly secure protocol for a functionality $\mathcal{F}$ if it is both passive secure and UC secure (with selective abort) for $\mathcal{F}$ against computationally unbounded adversaries.*

Note that strong security admits composition as both semi-honest security and UC security are composable. From a practical point of view, strong security (possibly weakened to hold only against PPT adversaries) is important, and arguably the "right" notion in many cases. Here we initiate the study of characterizing multi-party functionalities that are strongly securely realizable. Clearly, the impossibility results for both UC security and passive security apply to strong security.

To state our results for *all* multi-party functions, we need to go beyond aggregating functionalities. Firstly, we shall need the notion of disseminating functionalities: An $(m + 1)$-party disseminating functionality $f = (f_1, \cdots, f_m)$ has a single party P_0 with an input x, so that every other party P_i receives the output $f_i(x)$. The class of disseminating functions is denoted by **DISS**. Secondly, we need to consider functions which are "essentially" aggregating or disseminating, but not strictly so because of the presence of additional information in each party's local output which is derived solely from its own inputs. The idea that a function can be *essentially the same* as another function is captured using the notion of isomorphism among functionalities, as defined in [MPR13]. We reproduce this below, adapted to strong security. Here, a protocol $\pi_{\mathcal{F}}^{\mathcal{G}}$ for $\mathcal{F}$, using $\mathcal{G}$ as a setup, is said to be *local* if each party (deterministically) maps its input to an input for the functionality $\mathcal{G}$, then calls $\mathcal{G}$ once with that input and, based on their private input and the output obtained from $\mathcal{G}$, locally computes the final output (deterministically), without any other communication.

Definition 10 (Isomorphism [MPR13]). *We say $\mathcal{F}$ and $\mathcal{G}$ are isomorphic to each other if there exist two local protocols $\pi_{\mathcal{F}}^{\mathcal{G}}$ and $\pi_{\mathcal{G}}^{\mathcal{F}}$ that strongly securely realize $\mathcal{F}$ and $\mathcal{G}$ respectively.*

Now we are ready to state and prove our main results regarding strongly secure MPC.

Theorem 6. *If a functionality has a strongly secure protocol, then it is isomorphic to a functionality in* **DISS** $\cup$ **CPS**.

Proof: It follows from [PR08] that all strongly securely realizable functionalities are isomorphic to a *disseminating* functionality (i.e., a functionality in **DISS**), or an *aggregating* functionality (as defined in here). Further, if a functionality $\mathcal{F}$ that has a strongly secure protocol is isomorphic to an aggregating functionality $\mathcal{F}'$, then from the definition of isomorphism, $\mathcal{F}'$ too has a strongly secure (and in particular, a passive secure) protocol. Then, by Theorem 2, $\mathcal{F}' \in$ **CPS**. $\square$

We contrast this with our positive result below, which refers to **CCPS** (Definition 7), instead of **CPS**. We point out that our protocols below are efficient in the sense of having polynomial complexity in the statistical security parameter, but can be polynomial (rather than logarithmic) in the domain sizes or exponential in the number of parties.

Theorem 7. *If a functionality is isomorphic to one in* **DISS** $\cup$ **CCPS**, *then it has a strongly secure protocol.*

Proof: We show in Sect. 8.2 that every disseminating functionality has a UC secure protocol. A UC secure protocol for a disseminating functionality is always passive secure as well: only the disseminator has any input, and if the disseminator is passively corrupt, the correctness guarantee under UC security (when no party is corrupt) ensures that the simulator can send the disseminator's actual input to the functionality.

In the full version, we prove that the UNIMPC* protocol in Sect. 6 is UC secure for every Complete CPSS functionality. By Theorem 5, this covers all Complete CPS functionalities of more than 2 dimensions. For 2-dimensional Complete CPS functionalities (which are precisely Latin Squares), we give a UC secure protocol in the full version. In Sect. 8.1, we show a compiler that extends these results to functionalities embedded in a CCPS functionality.

Finally, we note that for aggregating CPS functionalities too, UC security implies strong security: If the aggregator is honest, the correctness guarantee under UC security allows the simulator to send the corrupt parties' actual input to the functionality; if the aggregator is corrupt, a simulator which sends the correct inputs of the passively corrupt players obtains the honest parties' residual function, and can internally execute the UC simulator (which may send arbitrary inputs to the functionality and expect the output). □

8.1   Restricting Input Domains While Retaining UC Security

In this section we give a compiler to transform a UC secure protocol for a CPS functionality $\mathcal{F}$ to a UC secure protocol for the same functionality, but with restricted input domains for each party. To illustrate the need for this compiler, suppose m input parties wish to total their votes (0 or 1) and provide it to an aggregator, securely. We do have a UC secure protocol for addition modulo $m+1$, and this functionality can correctly compute the total of m bits. However, this is not a UC secure protocol for our functionality, as the corrupt parties can provide inputs other than 0 or 1. Nevertheless, we show that the original protocol can be transformed into one which restricts the domain as desired.

Definition 11 (Domain Restriction). *Given a functionality $\mathcal{F}$ with input domain $X = X_1 \times \cdots \times X_m$, we define a domain restriction of $\mathcal{F}$ to $D = D_1 \times \cdots \times D_m \subseteq X$ as a functionality $\mathcal{F}_D$ which is defined only on inputs in D, where it behaves identically as $\mathcal{F}$.*

We give a compiler that transforms a UC secure protocol for a CPS functionality $\mathcal{F}$ to a UC secure protocol for $\mathcal{F}_D$ for any $D = D_1 \times \cdots \times D_m$. Our compiler can be presented as a protocol $\mathsf{RDom}_D^{\mathcal{F}, \mathcal{F}_{\mathrm{AND}}}$ – a protocol in a hybrid model with access to the ideal functionalities $\mathcal{F}$ and (m-input) aggregating functionality $\mathcal{F}_{\mathrm{AND}}$. We note that while $\mathcal{F}_{\mathrm{AND}}$ is not a CPS functionality (and hence cannot have a passive secure protocol), it does have a UC secure protocol. Specifically, one can reduce $\mathcal{F}_{\mathrm{AND}}$ to summation over an exponentially large abelian group, where each party P_i maps its input x_i to a group element g_i as follows: if $x_i = 0$, let g_i be random, and if $x_i = 1$, let $g_i = 0$. The aggregator receives $\sum_i g_i$ and outputs 1 if the sum is 0, and 0 otherwise.

Protocol RDom$_D^{\mathcal{F},\mathcal{F}_{\text{AND}}}$. The high-level idea of this protocol is to first invoke $\mathcal{F}$ on random inputs from the domain D, and use a cut-and-choose phase to verify that indeed most of the invocations used inputs in the domain D. Then, using access to $\mathcal{F}_{\text{AND}}$, the executions involving the correct input from all the parties are isolated, and the aggregator P_0 outputs what it received from $\mathcal{F}$ in those executions (if there is a consistent output). The formal description follows.

Let $\mathcal{F}$ represent the functionality to be realized and k be the security parameter. Let $\mathcal{E}$ be the input domain of $\mathcal{F}$ and D be the desired domain. Let $P_i, i \in [m] \cup \{0\}$ be the set of parties with inputs $\{x_i\}_{i \in [m]}$. Let P_0 be the aggregator with output space $[n]$.

1. **Random Execution:** Invoke k sessions of the functionality $\mathcal{F}$ with domain $\mathcal{E}$. Each honest party $P_i, i \in [m]$ chooses input uniformly at random from domain D. Let u_{ij} be the input used by party P_i in the jth execution and let v_j be its output.
2. **Opening:** P_0 chooses $S \subseteq [k]$, where every element has a probability of 0.5 of being picked up (thus $\mathbb{E}(|S|) = k/2$), and announces it. Every party $P_i, i \in [m]$ sends $u_{ij}, \forall j \in S$ to P_0. Then, P_0 checks the consistency of all the inputs and outputs it received: i.e., if $\forall j \in S$, $\mathcal{F}(\{u_{ij}\}_{i \in [m]}) = v_j$. It also confirms that each input is chosen from the domain D. Otherwise P_0 aborts.
3. **Tallying with actual inputs:** Invoke $k - |S|$ sessions of the $\mathcal{F}_{\text{AND}}$ functionality, indexed by $\bar{S} = [m] \setminus S$. Each honest party P_i sets its input to session j of $\mathcal{F}_{\text{AND}}$ a_{ij} as

$$a_{ij} = \begin{cases} 1 & \text{if } v_{ij} = x_i \\ 0 & \text{otherwise} \end{cases}$$

   and let the output for j^{th} $\mathcal{F}_{\text{AND}}$ be b_j. Also let $T = \{j : b_j = 1\}$.
4. **Computing the result:** If $|T| \geq t/2$ where $t = k/(2 \cdot \prod_{i \in [m]} |X_i|)$ is the expected size of T, and if $\exists v \forall j \in T$, $v_j = v$, then P_0 outputs v. Otherwise P_0 Aborts.

In the full version, we prove the following.

Theorem 8. *If $\mathcal{F}$ is an m-input CPS functionality, and $D = D_1 \times \cdots \times D_m$ is a subset of its domain, then RDom$_D^{\mathcal{F},\mathcal{F}_{\text{AND}}}$ is a UC secure protocol for $\mathcal{F}_D$.*

8.2 Disseminating Functionalities

We rely on the disseminated-OR functionality $\mathcal{D}_{\text{OR}}$ to show that all disseminated functionalities are UC secure. The functionality $\mathcal{D}_{\text{OR}}$ takes $(x_1, \cdots, x_m)$ from the disseminator P_0 and outputs (b, x_i) to P_i where $b = x_1 \vee \cdots \vee x_m$. We start by giving a UC secure protocol for $\mathcal{D}_{\text{OR}}$.

Protocol for $\mathcal{D}_{OR}$. In [PR08] a UC secure protocol for 3-party $\mathcal{D}_{OR}$ was given. We present a variant that works for all values of m (please see the full version for the proof).

1. P_0 broadcasts (UC-securely [GL02]) $b := \bigvee_{i>0} x_i$ to all P_i.
2. If $b = 0$, for each $i > 0$, P_i outputs $(0,0)$ and halts. Else, they continue.
3. P_0 sends x_i to each P_i.
4. For $i \in [m]$, $j \in [k]$, P_0 samples r_{ij} from a large group (e.g., k-bit strings) s.t. $\forall j, \sum_i r_{ij} = 0$.
5. For each i, if $x_i = 0$, P_0 sends r_{ij} for all j to P_i (and otherwise sends nothing to P_i).
6. Cut-and-choose:
   (a) P_1 picks a random subset $S \subset [k]$ of size $k/2$ and sends it to P_0.
   (b) For all $j \in S$, P_0 broadcasts r_{ij} for all i, and all parties verify that $\sum_i r_{ij} = 0$. P_1 verifies that the set S used is what it picked.
   (c) Any P_i with $x_i = 0$ aborts if it sees that for some j, r_{ij} broadcast by P_0 is not equal to r_{ij} it received.
7. For each $j \notin S$, $P_1, \cdots, P_m$ do the following:
   (a) For each i, if $x_i = 0$, P_i sets $s_{ij} = r_{ij}$, and otherwise samples s_{ij} randomly.
   (b) They use the standard semi-honest secure protocol to compute $\sum_i s_{ij}$.
   (c) Each P_i aborts if it gets the sum as 0.
8. If no abort has been observed, each P_i outputs $(1, x_i)$, where x_i is as received from P_0 in the beginning. Otherwise it aborts.

In the full version we prove that this protocol is secure. The interesting cases are when (1) a corrupt P_0 attempts to make all (honest) P_i's output $(1,0)$ (thwarted by the summation evaluating to 0, or the cut-and-choose failing), and (2) when P_0 is honest and a set of corrupt P_i's may learn *all* s_{ij} (thwarted by s_{ij} being distributed uniformly, either because a corrupt P_i does not know r_{ij} as $x_i = 1$, or because an honest P_i used a random s_{ij}).

Protocol for any disseminating functionality. A disseminating functionality $\mathcal{F}$ with m output parties is specified by a function $F : X \to Y_1 \times \cdots \times Y_m$, for some finite domains X and Y_i. We consider a boolean function $\mathrm{Inv}_{[m]}^F :$ $Y_1 \times \cdots \times Y_m \to \{0,1\}$ (for "invalid") as follows: $\mathrm{Inv}_{[m]}^F(y_1, \cdots, y_n) = 1$ iff $\nexists x \in X$ s.t. $F(x) = (y_1, \ldots, y_n)$.

More generally, for any $S \subseteq [m]$, define $\mathrm{Inv}_S^F : Y_S \to \{0,1\}$ as follows (denoting by Y_S the input combinations of parties indexed by S): for $y_S \in Y_S$, $\mathrm{Inv}_S^F(y) = 1$ iff $\nexists x \in X, y_{\overline{S}} \in Y_{\overline{S}}$ s.t. $F(x) = (y_S, y_{\overline{S}})$ (with the output tuple understood as being sorted appropriately by the indices).

Protocol $\mathsf{Diss}_{\mathcal{F}}^{\mathcal{D}_{OR}}$ (for disseminating functionality $\mathcal{F}$ computing F):

1. On input x, P_0 sends y_i to each P_i, where $F(x) = (y_1, \cdots, y_m)$.
2. For each subset $S \subseteq [m]$
   – For each $\tilde{y}_S \in Y_S$ such that $\mathrm{Inv}_S^F(\tilde{y}_S) = 1$:
     (a) Invoke $\mathcal{D}_{OR}$, with P_0's input being $(a_1, \cdots, a_m)$, where $a_i = 0$ iff $\tilde{y}_i = y_i$ and 1 otherwise.

(b) Each P_i receives (b, a_i). If $b = 0$, or if $a_i = 1$ but $\tilde{y}_i = y_i$, then abort.
3. If no abort has been observed, each P_i outputs y_i, and else aborts.

We point out that it is important to have the protocol consider all subsets $S \subseteq [m]$ (which makes it take time exponential in m), and not just the whole set $[m]$, as otherwise P_0 can collude with a corrupt P_{i^*} (who never aborts), and ensure that $b = 1$ always, by setting $a_{i^*} = 1$. Then P_0 can make the honest parties accept any combination of outputs, valid or not. In the full version we prove that the above protocol UC securely realizes $\mathcal{F}$.

References

[BGI+14] Beimel, A., Gabizon, A., Ishai, Y., Kushilevitz, E., Meldgaard, S., Paskin-Cherniavsky, A.: Non-interactive secure multiparty computation. In: Garay, J.A., Gennaro, R. (eds.) CRYPTO 2014, Part II. LNCS, vol. 8617, pp. 387–404. Springer, Heidelberg (2014). https://doi.org/10.1007/978-3-662-44381-1_22

[BGW88] Ben-Or, M., Goldwasser, S., Wigderson, A.: Completeness theorems for non-cryptographic fault-tolerant distributed computation. In: Proceedings of the 20th STOC, pp. 1–10 (1988)

[Blu81] Blum, M.: Three applications of the oblivious transfer: part I: coin flipping by telephone; part II: how to exchange secrets; part III: how to send certified electronic mail. Technical report, University of California, Berkeley (1981)

[CCD88] Chaum, D., Crépeau, C., Damgård, I.,: Multiparty unconditionally secure protocols. In: Proceedings of the 20th STOC, pp. 11–19 (1988)

[CI96] Chor, B., Ishai, Y., On privacy and partition arguments. In: Proceedings of the Fourth Israel Symposium on Theory of Computing and Systems, ISTCS 1996, Jerusalem, Israel, 10–12 June 1996, pp. 191–194 (1996). Journal version appears in Inf. Comput. **167**(1)

[CK91] Chor, B., Kushilevitz, E.: A zero-one law for Boolean privacy. SIAM J. Discrete Math. **4**(1), 36–47 (1991)

[CKL06] Canetti, R., Kushilevitz, E., Lindell, Y.: On the limitations of universally composable two-party computation without set-up assumptions. J. Cryptol. **19**(2), 135–167 (2006)

[FKN94] Feige, U., Kilian, J., Naor, M.: A minimal model for secure computation (extended abstract). In: STOC, pp. 554–563 (1994)

[GL02] Goldwasser, S., Lindell, Y.: Secure computation without agreement. In: Malkhi, D. (ed.) DISC 2002. LNCS, vol. 2508, pp. 17–32. Springer, Heidelberg (2002). https://doi.org/10.1007/3-540-36108-1_2

[HIJ+16] Halevi, S., Ishai, Y., Jain, A., Kushilevitz, E., Rabin, T.: Secure multiparty computation with general interaction patterns. In: Proceedings of the 2016 ACM Conference on Innovations in Theoretical Computer Science, Cambridge, MA, USA, 14–16 January 2016, pp. 157–168 (2016)

[HIJ+17] Halevi, S., Ishai, Y., Jain, A., Komargodski, I., Sahai, A., Yogev, E.: Non-interactive multiparty computation without correlated randomness. In: Takagi, T., Peyrin, T. (eds.) ASIACRYPT 2017, Part III. LNCS, vol. 10626, pp. 181–211. Springer, Cham (2017). https://doi.org/10.1007/978-3-319-70700-6_7

[HIKR18] Halevi, S., Ishai, Y., Kushilevitz, E., Rabin, T.: Best possible information-theoretic MPC. In: Proceedings of Theory of Cryptography - 16th Theory of Cryptography Conference, TCC (2018, to appear)

[HM97] Hirt, M., Maurer, U.M.: Complete characterization of adversaries tolerable in secure multi-party computation (extended abstract). In: PODC, pp. 25–34 (1997)

[IK97] Ishai, Y., Kushilevitz, E.: Private simultaneous messages protocols with applications. In: Israel Symposium on the Theory of Computing and Systems, ISTCS, pp. 174–184 (1997)

[IK00] Ishai, Y., Kushilevitz, E.: Randomizing polynomials: a new representation with applications to round-efficient secure computation. In: FOCS, pp. 294–304 (2000)

[KMR09] Künzler, R., Müller-Quade, J., Raub, D.: Secure computability of functions in the IT setting with dishonest majority and applications to long-term security. In: Reingold, O. (ed.) TCC 2009. LNCS, vol. 5444, pp. 238–255. Springer, Heidelberg (2009). https://doi.org/10.1007/978-3-642-00457-5_15

[Kus89] Kushilevitz, E.: Privacy and communication complexity. In: FOCS, pp. 416–421 (1989)

[MPR09] Maji, H.K., Prabhakaran, M., Rosulek, M.: Complexity of multi-party computation problems: the case of 2-party symmetric secure function evaluation. In: Reingold, O. (ed.) TCC 2009. LNCS, vol. 5444, pp. 256–273. Springer, Heidelberg (2009). https://doi.org/10.1007/978-3-642-00457-5_16

[MPR13] Maji, H., Prabhakaran, M., Rosulek, M.: Complexity of multi-party computation functionalities. In: Secure Multi-Party Computation. Cryptology and Information Security Series, vol. 10, pp. 249–283. IOS Press, Amsterdam (2013)

[OY16] Obana, S., Yoshida, M.: An efficient construction of non-interactive secure multiparty computation. In: Foresti, S., Persiano, G. (eds.) CANS 2016. LNCS, vol. 10052, pp. 604–614. Springer, Cham (2016). https://doi.org/10.1007/978-3-319-48965-0_39

[PR08] Prabhakaran, M., Rosulek, M.: Cryptographic complexity of multi-party computation problems: classifications and separations. In: Wagner, D. (ed.) CRYPTO 2008. LNCS, vol. 5157, pp. 262–279. Springer, Heidelberg (2008). https://doi.org/10.1007/978-3-540-85174-5_15. Full version available as ECCC Report TR08-050 from https://eccc.weizmann.ac.il

[Rys51] Ryser, H.J.: A combinatorial theorem with an application to Latin rectangles. Proc. Am. Math. Soc. 2(4), 550–552 (1951)

[SRA79] Shamir, A., Rivest, R.L., Adleman, L.M.: Mental poker. Technical report LCS/TR-125, Massachusetts Institute of Technology, April 1979

[Yao82] Yao, A.C.-C.: Protocols for secure computation. In: Proceedings of the 23rd FOCS, pp. 160–164 (1982)

Quantum I

Quantum Circuits for the CSIDH: Optimizing Quantum Evaluation of Isogenies

Daniel J. Bernstein[1]($\boxtimes$), Tanja Lange[2]($\boxtimes$), Chloe Martindale[2]($\boxtimes$), and Lorenz Panny[2]($\boxtimes$)

[1] Department of Computer Science, University of Illinois at Chicago, Chicago, IL 60607-7045, USA
djb@cr.yp.to
[2] Department of Mathematics and Computer Science, Technische Universiteit Eindhoven, P.O. Box 513, 5600 MB Eindhoven, The Netherlands
tanja@hyperelliptic.org, chloemartindale@gmail.com, lorenz@yx7.cc

Abstract. Choosing safe post-quantum parameters for the new CSIDH isogeny-based key-exchange system requires concrete analysis of the cost of quantum attacks. The two main contributions to attack cost are the number of queries in hidden-shift algorithms and the cost of each query. This paper analyzes algorithms for each query, introducing several new speedups while showing that some previous claims were too optimistic for the attacker. This paper includes a full computer-verified simulation of its main algorithm down to the bit-operation level.

Keywords: Elliptic curves · Isogenies · Circuits · Constant-time computation · Reversible computation · Quantum computation · Cryptanalysis

1 Introduction

Castryck, Lange, Martindale, Panny, and Renes recently introduced CSIDH [15], an isogeny-based key exchange that runs efficiently and permits non-interactive key exchange. Like the original CRS [20,64,68] isogeny-based cryptosystem, CSIDH has public keys and ciphertexts only about twice as large as traditional

---

Author list in alphabetical order; see https://www.ams.org/profession/leaders/culture/CultureStatement04.pdf. This work was supported in part by the Commission of the European Communities through the Horizon 2020 program under project number 643161 (ECRYPT-NET), 645622 (PQCRYPTO), 645421 (ECRYPT-CSA), and CHIST-ERA USEIT (NWO project 651.002.004); the Netherlands Organisation for Scientific Research (NWO) under grants 628.001.028 (FASOR) and 639.073.005; and the U.S. National Science Foundation under grant 1314919. "Any opinions, findings, and conclusions or recommendations expressed in this material are those of the author(s) and do not necessarily reflect the views of the National Science Foundation" (or other funding agencies). Permanent ID of this document: 9b88023d7d9ef3f55b11b6f009131c9f. Date of this document: 2019.02.28.

Y. Ishai and V. Rijmen (Eds.): EUROCRYPT 2019, LNCS 11477, pp. 409–441, 2019.
https://doi.org/10.1007/978-3-030-17656-3_15

elliptic-curve keys and ciphertexts for a similar security level against all pre-quantum attacks known. CRS was accelerated recently by De Feo, Kieffer, and Smith [23]; CSIDH builds upon this and chooses curves in a different way, obtaining much better speed.

For comparison, the SIDH (and SIKE) isogeny-based cryptosystems [22,36, 37] are somewhat faster than CSIDH, but they do not support non-interactive key exchange, and their public keys and ciphertexts are 6 times larger[1] than in CSIDH. Furthermore, there are concerns that the extra information in SIDH keys might allow attacks; see [58].

These SIDH disadvantages come from avoiding the commutative structure used in CRS and now in CSIDH. SIDH deliberately avoids this structure because the structure allows *quantum* attacks that asymptotically take subexponential time; see below. The CRS/CSIDH key size thus grows superlinearly in the post-quantum security level. For comparison, if the known attacks are optimal, then the SIDH key size grows linearly in the post-quantum security level.

However, even in a post-quantum world, it is not at all clear how much weight to put on these asymptotics. It is not clear, for example, how large the keys will have to be before the subexponential attacks begin to outperform the exponential-time non-quantum attacks or an exponential-time Grover search. It is not clear when the superlinear growth in CSIDH key sizes will outweigh the factor 6 mentioned above. For applications that need non-interactive key exchange in a post-quantum world, the SIDH/SIKE family is not an option, and it is important to understand what influence these attacks have upon CSIDH key sizes. The asymptotic performance of these attacks is stated in [15], but it is challenging to understand the concrete performance of these attacks for specific CSIDH parameters.

1.1 Contributions of This Paper.

The most important bottleneck in the quantum attacks mentioned above is the cost of evaluating a group action, a series of isogenies, in superposition. Each quantum attack incurs this cost many times; see below. The goals of this paper are to analyze and optimize this cost. We focus on CSIDH because CSIDH is much faster than CRS.

Our main result has the following shape: the CSIDH group action can be carried out in B nonlinear bit operations (counting ANDs and ORs, allowing free XORs and NOTs) with failure probability at most ϵ. (All of our algorithms know when they have failed.) This implies a reversible computation of the CSIDH group action with failure probability at most ϵ using at most $2B$ Toffoli gates (allowing free NOTs and CNOTs). This in turn implies a quantum computation of the CSIDH group action with failure probability at most ϵ using at most $14B$

---

[1] When the goal is for pre-quantum attacks to take 2^λ operations (without regard to memory consumption), CRS, CSIDH, SIDH, and SIKE all choose primes $p \approx 2^{4\lambda}$. The CRS and CSIDH keys and ciphertexts use (approximately) $\log_2 p \approx 4\lambda$ bits, whereas the SIDH and SIKE keys and ciphertexts use $6\log_2 p \approx 24\lambda$ bits for 3 elements of $\mathbb{F}_{p^2}$. There are compressed variants of SIDH that reduce $6\log_2 p$ to $4\log_2 p \approx 16\lambda$ (see [1]) and to $3.5\log_2 p \approx 14\lambda$ (see [19] and [75]), at some cost in run time.

T-gates (allowing free Clifford gates). Appendix A reviews these cost metrics and their relationships.

We explain how to compute pairs (B, ϵ) for any given CSIDH parameters. For example, we show how to compute CSIDH-512 for uniform random exponent vectors in $\{-5, \dots, 5\}^{74}$ using

- 1118827416420 $\approx 2^{40}$ nonlinear bit operations using the algorithm of Sect. 7, or
- 765325228976 $\approx 0.7 \cdot 2^{40}$ nonlinear bit operations using the algorithm of Sect. 8,

in both cases with failure probability below 2^{-32}. CSIDH-512 is the smallest parameter set considered in [15]. For comparison, computing the same action with failure probability 2^{-32} using the Jao–LeGrow–Leonardi–Ruiz-Lopez algorithm [38], with the underlying modular multiplications computed by the same algorithm as in Roetteler–Naehrig–Svore–Lauter [63], would use approximately 2^{51} nonlinear bit operations.

We exploit a variety of algorithmic ideas, including several new ideas pushing beyond the previous state of the art in isogeny computation, with the goal of obtaining the best pairs (B, ϵ). We introduce a new constant-time variable-degree isogeny algorithm, a new application of the Elligator map, new ways to handle failures in isogeny computations, new combinations of the components of these computations, new speeds for integer multiplication, and more.

1.2 Impact upon Quantum Attacks. Kuperberg [46] introduced an algorithm using $\exp\big((\log N)^{1/2+o(1)}\big)$ queries and $\exp\big((\log N)^{1/2+o(1)}\big)$ operations on $\exp((\log N)^{1/2+o(1)})$ qubits to solve the order-N dihedral hidden-subgroup problem. Regev [61] introduced an algorithm using only a polynomial number of qubits, although with a worse $o(1)$ for the number of queries and operations. A followup paper by Kuperberg [47] introduced further algorithmic options.

Childs, Jao, and Soukharev [17] pointed out that these algorithms could be used to attack CRS. They analyzed the asymptotic cost of a variant of Regev's algorithm in this context. This cost is dominated by queries, in part because the number of queries is large but also because the cost of each query is large. Each query evaluates the CRS group action using a superposition of group elements.

We emphasize that computing the exact attack costs for any particular set of CRS or CSIDH parameters is complicated and requires a lot of new work. The main questions are (1) the exact number of queries for various dihedral-hidden-subgroup algorithms, not just asymptotics; and (2) the exact cost of each query, again not just asymptotics.

The first question is outside the scope of our paper. Some of the simpler algorithms were simulated for small sizes in [46], [10], and [11], but Kuperberg commented in [46, p. 5] that his "experiments with this simulator led to a false conjecture for [the] algorithm's precise query complexity".

Our paper addresses the second question for CSIDH: the concrete cost of quantum algorithms for evaluating the action of the class group, which means computing isogenies of elliptic curves in superposition.

1.3 Comparison to Previous Claims Regarding Query Cost. Bonnetain and Schrottenloher claim in [11, online versions 4, 5, and 6] that CSIDH-512 can be broken in "only" 2^{71} quantum gates, where each query uses 2^{37} quantum gates ("Clifford+T" gates; see Appendix A.4).

We work in the same simplified model of counting operations, allowing any number of qubits to be stored for free. We further simplify by counting only T-gates. We gain considerable performance from optimizations not considered in [11]. We take the best possible distribution of input vectors, disregarding the 2^2 overhead estimated in [11]. Our final gate counts for each query are nevertheless much higher than the 2^{37} claimed in [11]. Even assuming that [11] is correct regarding the number of queries, the cost of each query pushes the total attack cost above 2^{80}.

The query-cost calculation in [11] is not given in enough detail for full reproducibility. However, some details are provided, and given these details we conclude that costly parts of the computation are overlooked in [11] in at least three ways. First, to estimate the number of quantum gates for multiplication in $\mathbb{F}_p$, [11] uses a count of nonlinear bit operations for multiplication in $\mathbb{F}_2[x]$, not noticing that all known methods for multiplication in $\mathbb{Z}$ (never mind reduction modulo p) involve many more nonlinear bit operations than multiplication in $\mathbb{F}_2[x]$. Second, at a higher level, the strategy for computing an ℓ-isogeny requires first finding a point of order ℓ, an important cost not noticed in [11]. Third, [11] counts the number of operations in a *branching* algorithm, not noticing the challenge of building a *non-branching* (constant-time) algorithm for the same task, as required for computations in superposition. Our analysis addresses all of these issues and more.

1.4 Memory Consumption. We emphasize that our primary goal is to minimize the number of bit operations. This cost metric pays no attention to the fact that the resulting quantum algorithm for, e.g., CSIDH-512 uses a quantum computer with 2^{40} qubits.

Most of the quantum-algorithms literature pays much more attention to the number of qubits. This is why [17], for example, uses a Regev-type algorithm instead of Kuperberg's algorithm. Similarly, [15] takes Regev's algorithm "as a baseline" given "the larger memory requirement" for Kuperberg's algorithm.

An obvious reason to keep the number of qubits under control is the difficulty of scaling quantum computers up to a huge number of qubits. Post-quantum cryptography starts from the assumption that there will be enough scalability to build a quantum computer using thousands of logical qubits to run Shor's algorithm, but this does not imply that a quantum computer with millions of logical qubits will be only 1000 times as expensive, given limits on physical chip size and costs of splitting quantum computation across multiple chips.

On the other hand, [11] chooses Kuperberg's algorithm, and claims that the number of qubits used in Kuperberg's algorithm is not a problem:

> The algorithm we consider has a subexponential memory cost. More precisely, it needs exactly one qubit per query, plus the fixed overhead of the oracle, which can be neglected.

Concretely, for CSIDH-512, [11, online versions 1, 2, 3] claim $2^{29.5}$ qubits, and [11, online versions 4, 5, 6] claim 2^{31} qubits. However, no justification is provided for the claim that the number of qubits for the oracle "can be neglected". There is no analysis in [11] of the number of qubits used for the oracle.

We are not saying that our techniques *need* 2^{40} qubits. On the contrary: later we mention various ways in which the number of qubits can be reduced with only moderate costs in the number of operations. However, one cannot trivially extrapolate from the memory consumption of CSIDH software (a few kilobytes) to the number of qubits used in a quantum computation. The requirement of reversibility makes it more challenging and more expensive to reduce space, since intermediate results cannot simply be erased. See Appendix A.3.

Furthermore, even if enough qubits are available, simply counting qubit operations ignores critical bottlenecks in quantum computation. Fault-tolerant quantum computation corrects errors in every qubit at every time step, even if the qubit is merely being stored; see Appendix A.5. Communicating across many qubits imposes further costs; see Appendix A.6. It is thus safe to predict that the actual cost of a quantum CSIDH query will be much larger than indicated by our operation counts. Presumably the gap will be larger than the gap for, e.g., the AES attack in [28], which has far fewer idle qubits and much less communication overhead.

1.5   Acknowledgments. Thanks to Bo-Yin Yang for suggesting factoring the average over vectors of the generating function in Sect. 7.3. Thanks to Joost Renes for his comments.

2   Overview of the Computation

We recall the definition of the CSIDH group action, focusing on the computational aspects of the concrete construction rather than discussing the general case of the underlying algebraic theory.

Parameters. The only parameter in CSIDH is a prime number p of the form $p = 4 \cdot \ell_1 \cdots \ell_n - 1$, where $\ell_1 < \cdots < \ell_n$ are (small) odd primes and $n \geq 1$. Note that $p \equiv 3 \pmod 8$ and $p > 3$.

Notation. For each $A \in \mathbb{F}_p$ with $A^2 \neq 4$, define E_A as the Montgomery curve $y^2 = x^3 + Ax^2 + x$ over $\mathbb{F}_p$. This curve E_A is supersingular, meaning that $\#E_A(\mathbb{F}_p) \equiv 1 \pmod p$, if and only if it has trace zero, meaning that $\#E_A(\mathbb{F}_p) = p + 1$. Here $E_A(\mathbb{F}_p)$ means the group of points of E_A with coordinates in $\mathbb{F}_p$, including the neutral element at ∞; and $\#E_A(\mathbb{F}_p)$ means the number of points.

Define S_p as the set of A such that E_A is supersingular. For each $A \in S_p$ and each $i \in \{1, \ldots, n\}$, there is a unique $B \in S_p$ such that there is an ℓ_i-isogeny from E_A to E_B whose kernel is $E_A(\mathbb{F}_p)[\ell_i]$, the set of points $Q \in E_A(\mathbb{F}_p)$ with $\ell_i Q = 0$. Define $\mathcal{L}_i(A) = B$. One can show that $\mathcal{L}_i$ is invertible: specifically, $\mathcal{L}_i^{-1}(A) = -\mathcal{L}_i(-A)$. Hence $\mathcal{L}_i^e$ is defined for each integer e.

Inputs and Output. Given an element $A \in S_p$ and a list $(e_1, \ldots, e_n)$ of integers, the CSIDH group action computes $\mathcal{L}_1^{e_1}(\mathcal{L}_2^{e_2}(\cdots (\mathcal{L}_n^{e_n}(A)) \cdots)) \in S_p$.

2.1   Distribution of Exponents. The performance of our algorithms depends on the distribution of the exponent vectors $(e_1, \ldots, e_n)$, which in turn depends on the context.

Constructively, [15] proposes to sample each e_i independently and uniformly from a small range $\{-C, \ldots, C\}$. For example, CSIDH-512 in [15] has $n = 74$ and uses the range $\{-5, \ldots, 5\}$, so there are $11^{74} \approx 2^{256}$ equally likely exponent vectors. We emphasize, however, that all known attacks actually use considerably larger exponent vectors. This means that the distribution of exponents $(e_1, \ldots, e_n)$ our quantum oracle has to process is *not* the same as the distribution used constructively.

The first step in the algorithms of Kuperberg and Regev, applied to a finite abelian group G, is to generate a uniform superposition over all elements of G. CRS and CSIDH define a map from vectors $(e_1, \ldots, e_n)$ to elements $\mathfrak{l}_1^{e_1} \cdots \mathfrak{l}_n^{e_n}$ of the ideal-class group G. This map has a high chance of being surjective but it is far from injective: its kernel is a lattice of rank n. Presumably taking, e.g., 17^{74} length-74 vectors with entries in the range $\{-8, \ldots, 8\}$ produces a close-to-uniform distribution of elements of the CSIDH-512 class group, but the literature does not indicate how Kuperberg's algorithm behaves when each group element is represented as many different strings.

In his original paper on CRS, Couveignes [20] suggested instead generating a unique vector representing each group element as follows. Compute a basis for the lattice mentioned above; on a quantum computer this can be done using Shor's algorithm [67] which runs in polynomial time, and on a conventional computer this can be done using Hafner and McCurley's algorithm [29] which runs in subexponential time. This basis reveals the group size $\#G$ and an easy-to-sample set R of representatives for G, such as $\{(e_1, 0, \ldots, 0) : 0 \leq e_1 < \#G\}$ in the special case that $\mathfrak{l}_1$ generates G; for the general case see, e.g., [50, Sect. 4.1]. Reduce each representative to a short representative, using an algorithm that finds a close lattice vector. If this algorithm is deterministic (for example, if all randomness used in the algorithm is replaced by pseudorandomness generated from the input) then applying it to a uniform superposition over R produces a uniform superposition over a set of short vectors uniquely representing G.

The same idea was mentioned in the Childs–Jao–Soukharev paper [17] on quantum attacks against CRS, and in the description of quantum attacks in the CSIDH paper. However, close-vector problems are not easy, even in dimensions as small as 74. Bonnetain and Schrottenloher [11] estimate that CSIDH-512 exponent vectors can be found whose 1-norm is 4 times larger than vectors used constructively. They rely on a very large precomputation, and they do not justify their assumption that the 1-norm, rather than the ∞-norm, measures the cost of a class-group action in superposition. Jao, LeGrow, Leonardi, and Ruiz-Lopez [38] present an algorithm that guarantees $(\log p)^{O(1)}$ bits in each exponent, i.e., in the ∞-norm, but this also requires a subexponential-time precomputation, and the exponents appear to be rather large.

Perhaps future research will improve the picture of how much precomputation time and per-vector computation time is required for algorithms that find vectors of a specified size; or, alternatively, will show that Kuperberg-type algorithms can handle non-unique representatives of group elements. The best conceivable case for the attacker is the distribution used in CSIDH itself, and we choose this distribution as an illustration in analyzing the concrete cost of our algorithms.

2.2 Verification of Costs. To ensure that we are correctly computing the number of bit operations in our group-action algorithms, we have built a bit-operation simulator, and implemented our algorithms inside the simulator. The simulator is available from https://quantum.isogeny.org/software.html.

The simulator has a very small core that implements—and counts the number of—NOT, XOR, AND, and OR operations. Higher-level algorithms, from basic integer arithmetic up through isogeny computation, are built on top of this core.

The core also encapsulates the values of bits so that higher-level algorithms do not accidentally inspect those values. There is an explicit mechanism to break the encapsulation so that output values can be checked against separate computations in the Sage computer-algebra system.

2.3 Verification of Failure Probabilities. Internally, each of our group-action algorithms moves the exponent vector $(e_1, \ldots, e_n)$ step by step towards 0. The algorithm fails if the vector does not reach 0 within the specified number of iterations. Analyzing the failure probability requires analyzing how the distribution of exponent vectors interacts with the distribution of curve points produced inside the algorithm; each e_i step relies on finding a point of order ℓ_i.

We mathematically calculate the failure probability in a model where each generated curve point has probability $1 - 1/\ell_i$ of having order divisible by ℓ_i, and where these probabilities are all independent. The model would be exactly correct if each point were generated independently and uniformly at random. We actually generate points differently, so there is a risk of our failure-probability calculations being corrupted by inaccuracies in the model. To address this risk, we have carried out various point-generation experiments, suggesting that the model is reasonably accurate. Even if the model is inaccurate, one can compensate with a minor increase in costs. See Sects. 4.3 and 5.2.

There is a more serious risk of errors in the failure-probability calculations that we carry out within the model. To reduce this risk, we have carried out 10^7 simple trials of the following type for each algorithm: generate a random exponent vector, move it step by step towards 0 the same way the algorithm does (in the model), and see how many iterations are required. The observed distribution of the number of iterations is consistent with the distribution that we calculate mathematically. Of course, if there is a calculation error that somehow affects only very small probabilities, then this error will not be caught by only 10^7 experiments.

2.4 Structure of the Computation. We present our algorithms from bottom up, starting with scalar multiplication in Sect. 3, generation of curve points in Sect. 4, computation of $\mathcal{L}_i$ in Sect. 5, and computation of the entire CSIDH group action in Sects. 6, 7, and 8. Lower-level subroutines for basic integer and modular arithmetic appear in Appendices B and C respectively.

Various sections and subsections mention ideas for saving time beyond what we have implemented in our bit-operation simulator. These ideas include low-level speedups such as avoiding exponentiations in inversions and Legendre-symbol computations (see Appendix C.4), and higher-level speedups such as using division polynomials (Sect. 9) and/or modular polynomials (Sect. 10) to eliminate failures for small primes. All of the specific bit-operation counts that we state, such as the $1118827416420 \approx 2^{40}$ nonlinear bit operations mentioned above, are fully implemented.

3   Scalar Multiplication on an Elliptic Curve

This section analyzes the costs of scalar multiplication on the curves used in CSIDH, supersingular Montgomery curves $E_A : y^2 = x^3 + Ax^2 + x$ over $\mathbb{F}_p$.

For CSIDH-512, our simulator shows (after our detailed optimizations; see Appendices B and C) that a squaring $\mathbf{S}$ in $\mathbb{F}_p$ can be computed in 349596 nonlinear bit operations, and that a general multiplication $\mathbf{M}$ in $\mathbb{F}_p$ can be computed in 447902 nonlinear bit operations, while addition in $\mathbb{F}_p$ takes only 2044 nonlinear bit operations. We thus emphasize the number of $\mathbf{S}$ and $\mathbf{M}$ in scalar multiplication (and in higher-level operations), although in our simulator we have also taken various opportunities to eliminate unnecessary additions and subtractions.

3.1   How Curves Are Represented. We consider two options for representing E_A. The **affine** option uses $A \in \mathbb{F}_p$ to represent E_A. The **projective** option uses $A_0, A_1 \in \mathbb{F}_p$, with $A_0 \neq 0$, to represent E_A where $A = A_1/A_0$.

The formulas to produce a curve in Sect. 5 naturally produce (A_0, A_1) in projective form. Dividing A_1 by A_0 to produce A in affine form costs an inversion and a multiplication. Staying in projective form is an example of what Appendix C.5 calls "eliminating inversions", but this requires some extra computation when A is used, as we explain below.

The definition of the class-group action requires producing the output A in affine form at the end of the computation. It could also be beneficial to convert each intermediate A to affine form, depending on the relative costs of the inversion and the extra computation.

3.2   How Points Are Represented. As in [51, p. 425, last paragraph] and [53, p. 261], we avoid computing the y-coordinate of a point (x, y) on E_A. This creates some ambiguity, since the points (x, y) and $(x, -y)$ are both represented as $x \in \mathbb{F}_p$, but the ambiguity does not interfere with scalar multiplication.

We again distinguish between affine and projective representations. As in [5], we represent both $(0, 0)$ and the neutral element on E_A as $x = 0$, and (except where otherwise noted) we allow $X/0$, including $0/0$, as a projective representation of $x = 0$. The projective representation thus uses $X, Z \in \mathbb{F}_p$ to represent $x = X/Z$ if $Z \neq 0$, or $x = 0$ if $Z = 0$. These definitions eliminate branches from the scalar-multiplication techniques that we use.

3.3   Computing nP. We use the Montgomery ladder to compute nP, given a b-bit exponent n and a curve point P. The Montgomery ladder consists of b

"ladder steps" operating on variables (X_2, Z_2, X_3, Z_3) initialized to $(1, 0, x_1, 1)$, where x_1 is the x-coordinate of P. Each ladder step works as follows:

- Conditionally swap (X_2, Z_2) with (X_3, Z_3), where the condition bit in iteration i is bit n_{b-1-i} of n. This means computing $X_2 \oplus X_3$, ANDing each bit with the condition bit, and XORing the result into both X_2 and X_3; and similarly for Z_2 and Z_3.
- Compute $Y = X_2 - Z_2$, Y^2, $T = X_2 + Z_2$, T^2, $X_4 = T^2 Y^2$, $E = T^2 - Y^2$, and $Z_4 = E(Y^2 + ((A+2)/4)E)$. This is a **point doubling**: it uses $2S + 3M$ and a few additions (counting subtractions as additions). We divide $A + 2$ by 4 modulo p before the scalar multiplication, using two conditional additions of p and two shifts.
- Compute $C = X_3 + Z_3$, $D = X_3 - Z_3$, DT, CY, $X_5 = (DT + CY)^2$, and $Z_5 = x_1(DT - CY)^2$. This is a **differential addition**: it also uses $2S + 3M$ and a few additions.
- Set $(X_2, Z_2, X_3, Z_3) \leftarrow (X_4, Z_4, X_5, Z_5)$.
- Conditionally swap (X_2, Z_2) with (X_3, Z_3), where the condition bit is again n_{b-1-i}. We merge this conditional swap with the conditional swap at the beginning of the next iteration by using $n_{b-i-i} \oplus n_{b-i-2}$ as condition bit.

Then nP has projective representation (X_2, Z_2) by [9, Theorem 4.5]. The overall cost is $4bS + 6bM$ plus a small overhead for additions and conditional swaps.

Representing the input point projectively as X_1/Z_1 means computing $X_5 = Z_1(DT + CY)^2$ and $Z_5 = X_1(DT - CY)^2$, and starting from $(1, 0, X_1, Z_1)$. This costs bM extra. Beware that [9, Theorem 4.5] requires $Z_1 \neq 0$.

Similarly, representing A projectively as A_1/A_0 means computing $X_4 = T^2(4A_0Y^2)$ and $Z_4 = E(4A_0Y^2 + (A_1 + 2A_0)E)$, after multiplying Y^2 by $4A_0$. This also costs bM extra.

Other Techniques. The initial $Z_2 = 0$ and $Z_3 = 1$ (for an affine input point) are small, and remain small after the first conditional swap, saving time in the next additions and subtractions. Our framework for tracking sizes of integers recognizes this automatically. The framework does not, however, recognize that half of the output of the last conditional swap is unused. We could use dead-value elimination and other standard peephole optimizations to save bit operations.

Montgomery [53, p. 260] considered carrying out many scalar multiplications at once, using affine coordinates for intermediate points inside each scalar multiplication (e.g., $x_2 = X_2/Z_2$), and batching inversions across the scalar multiplications. This could be slightly less expensive than the Montgomery ladder for large b, depending on the S/M ratio. Our computation of a CSIDH group action involves many scalar multiplications, but not in large enough batches to justify considering affine coordinates for intermediate points. Computing the group action for a batch of inputs might change the picture, but for simplicity we focus on the problem of computing the group action for one input.

A more recent possibility is scalar multiplication on a birationally equivalent Edwards curve. Sliding-window Edwards scalar multiplication is somewhat less expensive than the Montgomery ladder for large b; see generally [8] and [34].

On the other hand, for constant-time computations it is important to use fixed windows rather than sliding windows. Despite this difficulty, we estimate that small speedups are possible for $b = 512$.

3.4   Computing $P, 2P, 3P, \ldots, kP$.
An important subroutine in isogeny computation (see Sect. 5) is to compute the sequence $P, 2P, 3P, \ldots, kP$ for a constant $k \geq 1$.

We compute $2P$ by a doubling, $3P$ by a differential addition, $4P$ by a doubling, $5P$ by a differential addition, $6P$ by a doubling, etc. In other words, each multiple of P is computed by the Montgomery ladder as above, but these computations are merged across the multiples (and conditional swaps are eliminated). This takes $2(k-1)\mathbf{S} + 3(k-1)\mathbf{M}$ for affine P and affine A. Projective P adds $\lfloor (k-1)/2 \rfloor \mathbf{M}$, and projective A adds $\lfloor k/2 \rfloor \mathbf{M}$.

We could instead compute $2P$ by a doubling, $3P$ by a differential addition, $4P$ by a differential addition, $5P$ by a differential addition, $6P$ by a differential addition, etc. This again takes $2(k-1)\mathbf{S} + 3(k-1)\mathbf{M}$ for affine P and affine A, but projective P and projective A now have different effects: projective P adds $(k-2)\mathbf{M}$ if $k \geq 2$, and projective A adds $\mathbf{M}$ if $k \geq 2$. The choice here also has an impact on metrics beyond bit operations: doublings increase space requirements but allow more parallelism.

4   Generating Points on an Elliptic Curve

This section analyzes the cost of several methods to generate a random point on a supersingular Montgomery curve $E_A : y^2 = x^3 + Ax^2 + x$, given $A \in \mathbb{F}_p$. As in Sect. 2, p is a standard prime congruent to 3 modulo 8.

Sometimes one instead wants to generate a point on the twist ·of the curve. The **twist** is the curve $-y^2 = x^3 + Ax^2 + x$ over $\mathbb{F}_p$; note that -1 is a non-square in $\mathbb{F}_p$. This curve is isomorphic to E_{-A} by the map $(x, y) \mapsto (-x, y)$. Beware that there are several slightly different concepts of "twist" in the literature; the definition here is the most useful definition for CSIDH, as explained in [15].

4.1   Random Point on Curve or Twist.
The conventional approach is as follows: generate a uniform random $x \in \mathbb{F}_p$; compute $x^3 + Ax^2 + x$; compute $y = (x^3 + Ax^2 + x)^{(p+1)/4}$; and check that $y^2 = x^3 + Ax^2 + x$.

One always has $y^4 = (x^3 + Ax^2 + x)^{p+1} = (x^3 + Ax^2 + x)^2$ so $\pm y^2 = x^3 + Ax^2 + x$. About half the time, y^2 will match $x^3 + Ax^2 + x$; i.e., (x, y) will be a point on the curve. Otherwise (x, y) will be a point on the twist.

Since we work purely with x-coordinates (see Sect. 3.2), we skip the computation of y. However, we still need to know whether we have a curve point or a twist point, so we compute the Legendre symbol of $x^3 + Ax^2 + x$ as explained in Appendix C.4.

The easiest distribution of outputs to mathematically analyze is the uniform distribution over the following $p + 1$ pairs:

- $(x, +1)$ where x represents a curve point;
- $(x, -1)$ where x represents a twist point.

One can generate outputs from this distribution as follows: generate a uniform random $u \in \mathbb{F}_p \cup \{\infty\}$; set x to u if $u \in \mathbb{F}_p$ or to 0 if $u = \infty$; compute the Legendre symbol of $x^3 + Ax^2 + x$; and replace symbol 0 with $+1$ if $u = 0$ or -1 if $u = \infty$.

For computations, it is slightly simpler to drop the two pairs with $x = 0$: generate a uniform random $x \in \mathbb{F}_p^*$ and compute the Legendre symbol of the value $x^3 + Ax^2 + x$. This generates a uniform distribution over the remaining $p - 1$ pairs.

4.2 Random Point on Curve. What if twist points are useless and the goal is to produce a point specifically on the curve (or vice versa)? One approach is to generate, e.g., 100 random curve-or-twist points as in Sect. 4.1, and select the first point on the curve. This fails with probability $1/2^{100}$. If a computation involves generating 2^{10} points in this way then the overall failure probability is $1 - (1 - 1/2^{100})^{2^{10}} \approx 1/2^{90}$. One can tune the number of generated points according to the required failure probability.

We save time by applying "Elligator" [7], specifically the Elligator 2 map. Elligator 2 is defined for all the curves E_A that we use, *except* the curve E_0, which we discuss below. For each of these curves E_A, Elligator 2 is a fast injective map from $\{2, 3, \ldots, (p - 1)/2\}$ to the set $E_A(\mathbb{F}_p)$ of curve points. This produces only about half of the curve points; see Sect. 5.2 for analysis of the impact of this nonuniformity upon our higher-level algorithms.

Here are the details of Elligator 2, specialized to these curves, further simplified to avoid computing y, and adapted to allow twists as an option:

- Input $A \in \mathbb{F}_p$ with $A^2 \neq 4$ and $A \neq 0$.
- Input $s \in \{1, -1\}$. This procedure generates a point on E_A if $s = 1$, or on the twist of E_A if $s = -1$.
- Input $u \in \{2, 3, \ldots, (p - 1)/2\}$.
- Compute $v = A/(u^2 - 1)$.
- Compute e, the Legendre symbol of $v^3 + Av^2 + v$.
- Compute x as v if $e = s$, otherwise $-v - A$.

To see that this works, note first that v is defined since $u^2 \neq 1$, and is nonzero since $A \neq 0$. One can also show that $A^2 - 4$ is nonsquare for all of the CSIDH curves, so $v^3 + Av^2 + v \neq 0$, so e is 1 or -1. If $e = s$ then $x = v$ so $x^3 + Ax^2 + x$ is a square for $s = 1$ and a nonsquare for $s = -1$. Otherwise $e = -s$ and $x = -v - A$ so $x^3 + Ax^2 + x = -u^2(v^3 + Av^2 + v)$, which is a square for $s = 1$ and a nonsquare for $s = -1$. This uses that v and $-v - A$ satisfy $(-v - A)^2 + A(-v - A) + 1 = v^2 + Av + 1$ and $-v - A = -u^2 v$.

The $(p - 3)/2$ different choices of u produce $(p - 3)/2$ different curve points, but we could produce any particular x output twice since we suppress y.

The Case $A = 0$. One way to extend Elligator 2 to E_0 is to set $v = u$ when $A = 0$ instead of $v = A/(u^2 - 1)$. The point of the construction of v is that $x^3 + Ax^2 + x$ for $x = -v - A$ is a non-square times $v^3 + Av^2 + v$, i.e., that $(-v - A)/v$ is a non-square; this is automatic for $A = 0$, since -1 is a non-square.

We actually handle E_0 in a different way: we precompute a particular base point on E_0 whose order is divisible by $(p+1)/4$, and we always return this point if $A = 0$. This makes our higher-level algorithms slightly more effective (but we disregard this improvement in analyzing the success probability of our algorithms), since this point guarantees a successful isogeny computation starting from E_0; see Sect. 5. The same guarantee removes any need to generate other points on E_0, and is also useful to start walks in Sect. 10.

4.3 Derandomization. Rather than generating random points, we generate a deterministic sequence of points by taking $u = 2$ for the first point, $u = 3$ for the next point, etc. We precompute the inverses of $1 - 2^2$, $1 - 3^2$, etc., saving bit operations.

An alternative, saving the same number of bit operations, is to precompute inverses of $1 - u^2$ for various random choices of u, embedding the inverses into the algorithm. This guarantees that the failure probability of the outer algorithm for any particular input A, as the choices of u vary, is the same as the failure probability of an algorithm that randomly chooses u upon demand for each A.

We are heuristically assuming that failures are not noticeably correlated across choices of A. To replace this heuristic with a proof, one can generate the u sequence randomly for each input. This randomness, in turn, is indistinguishable from the output of a cipher, under the assumption that the cipher is secure. In this setting one cannot precompute the reciprocals of $1 - u^2$, but one can still batch the inversions.

5   Computing an ℓ-isogenous Curve

This section analyzes the cost of computing a single isogeny in CSIDH. There are two inputs: A, specifying a supersingular Montgomery curve E_A over $\mathbb{F}_p$; and i, specifying one of the odd prime factors ℓ_i of $(p+1)/4 = \ell_1 \cdots \ell_n$. The output is $B = \mathcal{L}_i(A)$. We abbreviate ℓ_i as ℓ and $\mathcal{L}_i$ as $\mathcal{L}$.

Recall that B is characterized by the following property: there is an ℓ-isogeny from E_A to E_B whose kernel is $E_A(\mathbb{F}_p)[\ell]$. Beyond analyzing the costs of computing $B = \mathcal{L}(A)$, we analyze the costs of applying the ℓ-isogeny to a point on E_A, obtaining a point on E_B. See Sect. 5.4.

The basic reason that CSIDH is much faster than CRS is that the CSIDH construction allows (variants of) Vélu's formulas [18,62,72] to use points in $E_A(\mathbb{F}_p)$, rather than points defined over larger extension fields. This section focuses on computing B via these formulas. The cost of these formulas is approximately linear in ℓ, assuming that a point of order ℓ is known. There are two important caveats here:

- Finding a point of order ℓ is not easy to do in constant time. See Sect. 5.1. We follow the obvious approach, namely taking an appropriate multiple of a random point; but this is expensive—recall from Sect. 3 that a 500-bit Montgomery ladder costs $2000\mathbf{S} + 3000\mathbf{M}$ when A and the input point are affine— and has failure probability approximately $1/\ell$.

- In some of our higher-level algorithms, i is a *variable*. Then $\ell = \ell_i$ is also a variable, and Vélu's formulas are variable-time formulas, while we need constant-time computations. Generic branch elimination produces a constant-time computation taking time approximately linear in $\ell_1 + \ell_2 + \cdots + \ell_n$, which is quite slow. However, we show how to do much better, reducing $\ell_1 + \ell_2 + \cdots + \ell_n$ to $\max\{\ell_1, \ell_2, \ldots, \ell_n\}$, by exploiting the internal structure of Vélu's formulas. See Sect. 5.3.

There are other ways to compute isogenies, as explored in [23, 42]:

- The "Kohel" strategy: Compute a univariate polynomial whose roots are the x-coordinates of the points in $E_A(\mathbb{F}_p)[\ell]$. Use Kohel's algorithm [45, Sect. 2.4], which computes an isogeny given this polynomial. This strategy is (for CSIDH) asymptotically slower than Vélu's formulas, but could nevertheless be faster when ℓ is very small. Furthermore, this strategy is deterministic and always works.
- The "modular" strategy: Compute the possible j-invariants of E_B by factoring modular polynomials. Determine the correct choice of B by computing the corresponding isogeny kernels or, on subsequent steps, simply by not walking back.

We analyze the Kohel strategy in Sect. 9, and the modular strategy in Sect. 10.

5.1   Finding a Point of Order ℓ. We now focus on the problem of finding a point of order ℓ in $E_A(\mathbb{F}_p)$. By assumption $(p+1)/4$ is a product of distinct odd primes $\ell_1, \ldots, \ell_n$; $\ell = \ell_i$ is one of those primes; and $\#E_A(\mathbb{F}_p) = p + 1$. One can show that $E_A(\mathbb{F}_p)$ has a point of order 4 and is thus cyclic:

$$E_A(\mathbb{F}_p) \cong \mathbb{Z}/(p+1) \cong \mathbb{Z}/4 \times \mathbb{Z}/\ell_1 \times \cdots \times \mathbb{Z}/\ell_n.$$

We *try* to find a point Q of order ℓ in $E_A(\mathbb{F}_p)$ as follows:

- Pick a random point $P \in E_A(\mathbb{F}_p)$, as explained in Sect. 4.
- Compute a "cofactor" $(p+1)/\ell$. To handle the case $\ell = \ell_i$ for variable i, we first use bit operations to compute the list $\ell'_1, \ldots, \ell'_n$, where $\ell'_j = \ell_j$ for $j \neq i$ and $\ell'_i = 1$; we then use a product tree to compute $\ell'_1 \cdots \ell'_n$. (Computing $(p+1)/\ell$ by a general division algorithm could be faster, but the product tree is simpler and has negligible cost in context.)
- Compute $Q = ((p+1)/\ell)P$ as explained in Sect. 3.

If P is a uniform random element of $E_A(\mathbb{F}_p)$ then Q is a uniform random element of $E_A(\mathbb{F}_p)[\ell] \cong \mathbb{Z}/\ell$. The order of Q is thus the desired ℓ with probability $1 - 1/\ell$. Otherwise Q is ∞, the neutral element on the curve, which is represented by $x = 0$. Checking for $x = 0$ is a reliable way to detect this case: the only other point represented by $x = 0$ is $(0, 0)$, which is outside $E_A(\mathbb{F}_p)[\ell]$.

Different Concepts of Constant Time. Beware that there are two different notions of "constant time" for cryptographic algorithms. One notion is that the time for each operation is independent of *secrets*. This notion allows the CSIDH user to generate a uniform random element of $E_A(\mathbb{F}_p)[\ell]$ and try again if the point is ∞, guaranteeing success with an average of $\ell/(\ell-1)$ tries. The time varies, but the variation is independent of the secret A.

A stricter notion is that the time for each operation is independent of *all* inputs. The time depends on parameters, such as p in CSIDH, but does not depend on random choices. We emphasize that a quantum circuit operating on many inputs in superposition is, by definition, using this stricter notion. We thus choose the sequence of operations carried out by the circuit, and analyze the probability that this sequence fails.

Amplifying the Success Probability. Having each 3-isogeny fail with probability $1/3$, each 5-isogeny fail with probability $1/5$, etc. creates a correctness challenge for higher-level algorithms that compute many isogenies.

A simple workaround is to generate many points $Q_1, Q_2, \ldots, Q_N$, and use bit operations on the points to select the first point with $x \neq 0$. This fails if all of the points have $x = 0$. Independent uniform random points have overall failure probability $1/\ell^N$. One can make $1/\ell^N$ arbitrarily small by choosing N large enough: for example, $1/3^N$ is below $1/2^{32}$ for $N \geq 21$, and is below $1/2^{256}$ for $N \geq 162$.

We return to the costs of generating so many points, and the costs of more sophisticated alternatives, when we analyze algorithms to compute the CSIDH group action.

5.2 Nonuniform Distribution of Points. We actually generate random points using Elligator (see Sect. 4.2), which generates only $(p-3)/2$ different curve points P. At most $(p+1)/\ell$ of these points produce $Q = \infty$, so the failure chance is at most $(2/\ell)(p+1)/(p-3) \approx 2/\ell$.

This bound cannot be simultaneously tight for $\ell = 3$, $\ell = 5$, and $\ell = 7$ (assuming $3 \cdot 5 \cdot 7$ divides $p+1$): if it were then the Elligator outputs would include all points having orders dividing $(p+1)/3$ or $(p+1)/5$ or $(p+1)/7$, but this accounts for more than 54% of all curve points, contradiction.

Points generated by Elligator actually appear to be much better distributed modulo each ℓ, with failure chance almost exactly $1/\ell$. Experiments support this conjecture. Readers concerned with the gap between the provable $2/\ell$ and the heuristic $1/\ell$ may prefer to add or subtract a few Elligator 2 outputs, obtaining a distribution provably close to uniform (see [70]) at a moderate cost in performance. A more efficient approach is to accept a doubling of failure probability and use a small number of extra iterations to compensate.

We shall later see other methods of obtaining rational ℓ-torsion points, e.g., by pushing points through ℓ'-isogenies. This does not make a difference in the analysis of failure probabilities.

For comparison, generating a random point on the curve or twist (see Sect. 4.1) has failure probability above $1/2$ at finding a curve point of order ℓ. See Sect. 6.2 for the impact of this difference upon higher-level algorithms.

5.3 Computing an ℓ-isogenous Curve from a Point of Order ℓ.

Once we have the x-coordinate of a point Q of order ℓ in $E_A(\mathbb{F}_p)$, we compute the x-coordinates of the points $Q, 2Q, 3Q, \ldots, ((\ell-1)/2)Q$. We use this information to compute $B = \mathcal{L}(A)$, the coefficient determining the ℓ-isogenous curve E_B.

Recall from Sect. 3.4 that computing $Q, 2Q, 3Q, \ldots, ((\ell-1)/2)Q$ costs $(\ell - 3)\mathbf{S} + 1.5(\ell - 3)\mathbf{M}$ for affine Q and affine A, and just $1\mathbf{M}$ extra for affine Q and projective A. The original CSIDH paper [15] took more time here, namely $(\ell - 3)\mathbf{S} + 2(\ell - 3)\mathbf{M}$, to handle projective Q and projective A. We decide, based on comparing ℓ to the cost of an inversion, whether to spend an inversion converting Q to affine coordinates.

Given the x-coordinates of $Q, 2Q, 3Q, \ldots, ((\ell-1)/2)Q$, the original CSIDH paper [15] took approximately $3\ell\mathbf{M}$ to compute B. Meyer and Reith [49] pointed out that CSIDH benefits from Edwards-coordinate isogeny formulas from Moody and Shumow [54]; we reuse this speedup. These formulas work as follows:

- Compute $a = A + 2$ and $d = A - 2$.
- Compute the Edwards y-coordinates of $Q, 2Q, 3Q, \ldots, ((\ell-1)/2)Q$. The Edwards y-coordinate is related to the Montgomery x-coordinate by $y = (x - 1)/(x + 1)$. We are given each x projectively as X/Z, and compute y projectively as Y/T where $Y = X - Z$ and $T = X + Z$. Note that Y and T naturally occur as intermediate values in the Montgomery ladder.
- Compute the product of these y-coordinates: i.e., compute $\prod Y$ and $\prod T$. This uses a total of $(\ell - 3)\mathbf{M}$.
- Compute $a' = a^\ell (\prod T)^8$ and $d' = d^\ell (\prod Y)^8$. Each ℓth power takes a logarithmic number of squarings and multiplications; see Appendix C.4.
- Compute, projectively, $B = 2(a' + d')/(a' - d')$. Subsequent computations decide whether to convert B to affine form.

These formulas are almost three times faster than the formulas used in [15]. The total cost of computing B from Q is almost two times faster than in [15].

Handling Variable ℓ. We point out that the isogeny computations for $\ell = 3$, $\ell = 5$, $\ell = 7$, etc. have a Matryoshka-doll structure, allowing a constant-time computation to handle many different values of ℓ with essentially the same cost as a single computation for the largest value of ℓ.

Concretely, the following procedure takes approximately $\ell_n \mathbf{S} + 2.5\ell_n \mathbf{M}$, and allows any $\ell \le \ell_n$. If the context places a smaller upper bound upon ℓ then one can replace ℓ_n with that upper bound, saving time; we return to this idea later.

Compute the Montgomery x-coordinates and the Edwards y-coordinates of $Q, 2Q, 3Q, \ldots, ((\ell_n - 1)/2)Q$. Use bit operations to replace each Edwards y-coordinate with 1 after the first $(\ell - 1)/2$ points. Compute the product of these modified y-coordinates; this is the desired product of the Edwards y-coordinates of the first $(\ell - 1)/2$ points. Finish computing B as above. Note that the exponentiation algorithm in Appendix C.4 allows variable ℓ.

5.4 Applying an ℓ-isogeny to a Point.

The following formulas define an ℓ-isogeny from E_A to E_B with kernel $E_A(\mathbb{F}_p)[\ell]$. The x-coordinate of the image of a point $P_1 \in E_A(\mathbb{F}_p)$ under this isogeny is

$$x(P_1) \prod_{j \in \{1,2,\dots,(\ell-1)/2\}} \left(\frac{x(P_1)x(jQ) - 1}{x(P_1) - x(jQ)} \right)^2.$$

Each $x(jQ)$ appearing here was computed above in projective form X/Z. The ratio $(x(P_1)x(jQ) - 1)/(x(P_1) - x(jQ))$ is $(x(P_1)X - Z)/(x(P_1)Z - X)$. This takes 2**M** to compute projectively if $x(P_1)$ is affine, and thus $(\ell - 1)$**M** across all j. Multiplying the numerators takes $((\ell-3)/2)$**M**, multiplying the denominators takes $((\ell-3)/2)$**M**, squaring both takes 2**S**, and multiplying by $x(P_1)$ takes 1**M**, for a total of $(2\ell - 3)$**M** $+ 2$**S**.

If $x(P_1)$ is instead given in projective form as X_1/Z_1 then computing $X_1X - Z_1Z$ and $X_1Z - Z_1X$ might seem to take 4**M**, but one can instead compute the sum and difference of $(X_1 - Z_1)(X + Z)$ and $(X_1 + Z_1)(X - Z)$, using just 2**M**. The only extra cost compared to the affine case is four extra additions. This speedup was pointed out by Montgomery [53] in the context of the Montgomery ladder. The initial CSIDH software accompanying [15] did not use this speedup but [49] mentioned the applicability to CSIDH.

In the opposite direction, if inversion is cheap enough to justify making $x(P_1)$ *and* every $x(jQ)$ affine, then 2**M** drops to 1**M**, and the total cost drops to approximately 1.5ℓ**M**.

As in Sect. 5.3, we allow ℓ to be a variable. The cost of variable ℓ is the cost of a single computation for the maximum allowed ℓ, plus a minor cost for bit operations to select relevant inputs to the product.

6   Computing the Action: Basic Algorithms

Jao, LeGrow, Leonardi, and Ruiz-Lopez [38] suggested a three-level quantum algorithm to compute $\mathcal{L}_1^{e_1} \cdots \mathcal{L}_n^{e_n}$. This section shows how to make the algorithm an order of magnitude faster for any particular failure probability.

6.1   Baseline: Reliably Computing Each $\mathcal{L}_i$. The lowest level in [38] *reliably* computes $\mathcal{L}_i$ as follows. Generate r uniform random points on the curve or twist, as in Sect. 4.1. Multiply each point by $(p + 1)/\ell_i$, as in Sect. 5.1, hoping to obtain a point of order ℓ_i on the curve. Use Vélu's formulas to finish the computation, as in Sect. 5.3.

Each point has success probability $(1/2)(1 - 1/\ell_i)$, where $1/2$ is the probability of obtaining a curve point (rather than a twist point) and $1 - 1/\ell_i$ is the probability of obtaining a point of order ℓ_i (rather than order 1). The chance of all r points failing is thus $(\ell_i + 1)^r/(2\ell_i)^r$, decreasing from $(2/3)^r$ for $\ell_i = 3$ down towards $(1/2)^r$ as ℓ_i grows. One chooses r to obtain a failure probability as small as desired for the isogeny computation, and for the higher levels of the algorithm.

The lowest level optionally computes $\mathcal{L}_i^{-1}$ instead of $\mathcal{L}_i$. The approach in [38], following [15], is to use points on the twist instead of points on the curve; an alternative is to compute $\mathcal{L}_i^{-1}(A)$ as $-\mathcal{L}_i(-A)$.

The middle level of the algorithm computes $\mathcal{L}_i^e$, where e is a variable whose absolute value is bounded by a constant C. This level calls the lowest level exactly C times, performing a series of C steps of $\mathcal{L}_i^{\pm 1}$, using bit operations on e to decide whether to retain the results of each step. The ± 1 is chosen as the sign of e, or as an irrelevant 1 if $e = 0$.

The highest level of the algorithm computes $\mathcal{L}_1^{e_1} \cdots \mathcal{L}_n^{e_n}$, where each e_i is between $-C$ and C, by calling the middle level n times, starting with $\mathcal{L}_1^{e_1}$ and ending with $\mathcal{L}_n^{e_n}$. Our definition of the action applied $\mathcal{L}_n^{e_n}$ first, but the $\mathcal{L}_i$ operators commute with each other, so the order does not matter.

Importance of Bounding Each Exponent. We emphasize that this algorithm requires each exponent e_i to be between $-C$ and C, i.e., requires the vector $(e_1, \ldots, e_n)$ to have ∞-norm at most C.

We use $C = 5$ for CSIDH-512 as an illustrative example, but all known attacks use larger vectors (see Sect. 2.1). C is chosen in [38] so that every input, every vector in superposition, has ∞-norm at most C; smaller values of C create a failure probability that needs to be analyzed.

We are not saying that the ∞-norm is the only important feature of the input vectors. On the contrary: our constant-time subroutine to handle variable-ℓ isogenies creates opportunities to share work between separate exponents. See Sects. 5.3 and 7.

Concrete Example. For concreteness we consider uniform random input vectors $e \in \{-5, \ldots, 5\}^{74}$. The highest level calls the middle level $n = 74$ times, and the middle level calls the lowest level $C = 5$ times. Taking $r = 70$ guarantees failure probability at most $(2/3)^{70}$ at the lowest level, and thus failure probability at most $1 - (1 - (2/3)^{70})^{74 \cdot 5} \approx 0.750 \cdot 2^{-32}$ for the entire algorithm.

This type of analysis is used in [38] to select r. We point out that the failure probability of the algorithm is actually lower, and a more accurate analysis allows a smaller value of r. One can, for example, replace $(1 - (2/3)^r)^{74}$ with $\prod_i (1 - (\ell_i + 1)^r/(2\ell_i)^r)$, showing that $r = 59$ suffices for failure probability below 2^{-32}. With more work one can account for the distribution of input vectors e, rather than taking the worst-case e as in [38]. However, one cannot hope to do better than $r = 55$ here: there is a $10/11$ chance that at least one 3-isogeny is required, and taking $r \leq 54$ means that this 3-isogeny fails with probability at least $(2/3)^{54}$, for an overall failure chance at least $(10/11)(2/3)^{54} > 2^{-32}$.

With the choice $r = 70$ as in [38], there are $74 \cdot 5 \cdot 70 = 25900$ iterations, in total using more than 100 million multiplications in $\mathbb{F}_p$. In the rest of this section we will reduce the number of iterations by a factor 30, and in Sect. 7 we will reduce the number of iterations by another factor 3, with only moderate increases in the cost of each iteration.

6.2 Fewer Failures, and Sharing Failures.

We now introduce Algorithm 6.1, which improves upon the algorithm from [38] in three important ways. First, we use Elligator to target the curve (or the twist if desired); see Sect. 4.2. This reduces the failure probability of r points from $(2/3)^r$ to, heuristically, $(1/3)^r$ for $\ell_i = 3$; from $(3/5)^r$ to $(1/5)^r$ for $\ell_i = 5$; from $(4/7)^r$ to $(1/7)^r$ for $\ell_i = 7$; etc.

Algorithm 6.1: Basic class-group action evaluation.

Parameters: Odd primes $\ell_1 < \cdots < \ell_n$, a prime $p = 4\ell_1 \cdots \ell_n - 1$, and positive
integers $(r_1, \ldots, r_n)$.
Input: $A \in S_p$, integers $(e_1, \ldots, e_n)$.
Output: $\mathcal{L}_1^{e_1} \cdots \mathcal{L}_n^{e_n}(A)$ or "fail".

for $i \leftarrow 1$ **to** n **do**
    for $j \leftarrow 1$ **to** r_i **do**
        Let $s = \text{sign}(e_i) \in \{-1, 0, +1\}$.
        Find a random point P on E_{sA} using Elligator.
        Compute $Q \leftarrow ((p+1)/\ell_i)P$.
        Compute B with $E_B \cong E_{sA}/\langle Q \rangle$ if $Q \neq \infty$.
        Set $A \leftarrow sB$ if $Q \neq \infty$ and $s \neq 0$.
        Set $e_i \leftarrow e_i - s$ if $Q \neq \infty$.

Set $A \leftarrow$ "fail" if $(e_1, \ldots, e_n) \neq (0, \ldots, 0)$.
Return A.

Second, we allow a separate r_i for each ℓ_i. This lets us exploit the differences in failure probabilities as ℓ_i varies.

Third, we handle failures at the middle level instead of the lowest level. The strategy in [38] to compute $\mathcal{L}_i^e$ with $-C \leq e \leq C$ is to perform C iterations, where each iteration builds up many points on one curve and *reliably* moves to the next curve. We instead perform r_i iterations, where each iteration *tries* to move from one curve to the next by generating just one point. For $C = 1$ this is the same, but for larger C we obtain better tradeoffs between the number of points and the failure probability.

As a concrete example, generating 20 points on one curve with Elligator has failure probability $(1/3)^{20}$ for $\ell_i = 3$. A series of 5 such computations, overall generating 100 points, has failure probability $1 - (1 - (1/3)^{20})^5 \approx 2^{-29.37}$. If we instead perform just 50 iterations, where each iteration generates one point to move 1 step with probability $2/3$, then the probability that we move fewer than 5 steps is just $3846601/3^{50} \approx 2^{-57.37}$; see Sect. 6.3. Our iterations are more expensive than in [38]—next to each Elligator computation, we always perform the steps for computing an ℓ_i-isogeny, even if $Q = \infty$—but (for CSIDH-512 etc.) this is not a large effect: the cost of each iteration is dominated by scalar multiplication.

We emphasize that all of our algorithms take constant time. When we write "Compute $X \leftarrow Y$ if c" we mean that we always compute Y and the bit c, and we then replace the jth bit X_j of X with the jth bit Y_j of Y for each j if c is set, by replacing X_j with $X_j \oplus c(X_j \oplus Y_j)$. This is why Algorithm 6.1 always carries out the bit operations for computing an ℓ_i-isogenous curve, as noted above, even if $Q = \infty$.

Table 6.1. Examples of choices of r_i for Algorithm 6.1 for three levels of failure probability for uniform random CSIDH-512 vectors with entries in $\{-5, \ldots, 5\}$. Failure probabilities ϵ are rounded to three digits after the decimal point. The "total" column is $\sum r_i$, the total number of iterations. The "[38]" column is $74 \cdot 5 \cdot r$, the number of iterations in the algorithm of [38], with r chosen as in [38] to have $1 - (1 - (2/3)^r)^{74.5}$ at most 2^{-1} or 2^{-32} or 2^{-256}. Compare Table 6.2 for $\{-10, \ldots, 10\}$.

ϵ \ ℓ_i	3	5	7	11	13	17	...	359	367	373	587	total	[38]
$0.499 \cdot 2^{-1}$	11	9	8	7	7	6	...	5	5	5	5	406	5920
$0.178 \cdot 2^{-32}$	36	25	21	18	17	16	...	10	10	10	9	869	25900
$0.249 \cdot 2^{-256}$	183	126	105	85	80	73	...	37	37	37	34	3640	167610

6.3 Analysis. We consider the inner loop body of Algorithm 6.1 for a fixed i, hence write $\ell = \ell_i$, $e = e_i$, and $r = r_i$ for brevity.

Heuristically (see Sect. 5.2), we model each point Q as independent and uniform random in a cyclic group of order ℓ, so Q has order 1 with probability $1/\ell$ and order ℓ with probability $1 - 1/\ell$. The number of points of order ℓ through r iterations of the inner loop is binomially distributed with parameters r and $1 - 1/\ell$. The probability that this number is $|e|$ or larger is $\text{prob}_{\ell,e,r} = \sum_{t=|e|}^{r} \binom{r}{t} (1 - 1/\ell)^t / \ell^{r-t}$. This is exactly the probability that Algorithm 6.1 successfully performs the $|e|$ desired iterations of $\mathcal{L}^{\text{sign}(e)}$.

Let C be a nonnegative integer. The overall success probability of the algorithm for a particular input vector $(e_1, \ldots, e_n) \in \{-C, \ldots, C\}^n$ is

$$\prod_{i=1}^{n} \text{prob}_{\ell_i, e_i, r_i} \geq \prod_{i=1}^{n} \text{prob}_{\ell_i, C, r_i}.$$

Average over vectors to see that the success probability of the algorithm for a uniform random vector in $\{-C, \ldots, C\}^n$ is $\prod_{i=1}^{n} \left(\sum_{-C \leq e \leq C} \text{prob}_{\ell_i, e, r_i} / (2C+1) \right)$.

6.4 Examples of Target Failure Probabilities. The acceptable level of failure probability for our algorithm depends on the attack using the algorithm. For concreteness we consider three possibilities for CSIDH-512 failure probabilities, namely having the algorithm fail for a uniform random vector with probabilities at most 2^{-1}, 2^{-32}, and 2^{-256}.

Our rationale for considering these probabilities is as follows. Probabilities around 2^{-1} are easy to test, and may be of interest beyond this paper for constructive scenarios where failing computations can simply be retried. If each computation needs to work correctly, and there are many computations, then failure probabilities need to be much smaller, say 2^{-32}. Asking for every input in superposition to work correctly in one computation (for example, [38] asks for this) requires a much smaller failure probability, say 2^{-256}. Performance results for these three cases also provide an adequate basis for estimating performance in other cases.

Table 6.1 presents three reasonable choices of $(r_1, \ldots, r_n)$, one for each of the failure probabilities listed above, for the case of CSIDH-512 with uniform random

Table 6.2. Examples of choices of r_i for Algorithm 6.1 for three levels of failure probability for uniform random CSIDH-512 vectors with entries in $\{-10, \ldots, 10\}$. Failure probabilities ϵ are rounded to three digits after the decimal point. The "total" column is $\sum r_i$, the total number of iterations. The "[38]" column is $74 \cdot 10 \cdot r$, the number of iterations in the algorithm of [38], with r chosen as in [38] to have $1 - (1 - (2/3)^r)^{74 \cdot 10}$ at most 2^{-1} or 2^{-32} or 2^{-256}. Compare Table 6.1 for $\{-5, \ldots, 5\}$.

ϵ \ ℓ_i	3	5	7	11	13	17	...	359	367	373	587	total	[38]
$0.521 \cdot 2^{-1}$	20	15	14	13	12	12	...	10	10	10	10	786	13320
$0.257 \cdot 2^{-32}$	48	34	30	25	24	22	...	15	15	15	14	1296	52540
$0.215 \cdot 2^{-256}$	201	139	116	96	90	82	...	43	43	43	41	4185	335960

vectors with entries in $\{-5, \ldots, 5\}$. For each target failure probability δ and each i, the table chooses the minimum r_i such that $\sum_{-C \le e \le C} \mathrm{prob}_{\ell_i, e, r_i} / (2C + 1)$ is at least $(1 - \delta)^{1/n}$. The overall success probability is then at least $1 - \delta$ as desired. The discontinuity of choices of $(r_1, \ldots, r_n)$ means that the actual failure probability ϵ is somewhat below δ, as shown by the coefficients $0.499, 0.178, 0.249$ in Table 6.1. We could move closer to the target failure probability by choosing successively $r_n, r_{n-1}, \ldots$, adjusting the probability $(1 - \delta)^{1/n}$ at each step in light of the overshoot from previous steps. The values r_i for $\epsilon \approx 0.499 \cdot 2^{-1}$ have been experimentally verified using a modified version of the CSIDH software. To illustrate the impact of larger vector entries, we also present similar data in Table 6.2 for uniform random vectors with entries in $\{-10, \ldots, 10\}$.

The "total" column in Table 6.1 shows that this algorithm uses, e.g., 869 iterations for failure probability $0.178 \cdot 2^{-32}$ with vector entries in $\{-5, \ldots, 5\}$. Each iteration consists mostly of a scalar multiplication, plus some extra cost for Elligator, Vélu's formulas, etc. Overall there are roughly 5 million field multiplications, accounting for roughly 2^{41} nonlinear bit operations, implying a quantum computation using roughly 2^{45} T-gates.

As noted in Sect. 1, using the algorithm of [38] on top of the modular-multiplication algorithm from [63] would use approximately 2^{51} nonlinear bit operations for the same distribution of input vectors. We save a factor 30 in the number of iterations compared to [38], and we save a similar factor in the number of bit operations for each modular multiplication compared to [63].

We do not analyze this algorithm in more detail: the algorithms we present below are faster.

7   Reducing the Top Nonzero Exponent

Most of the iterations in Algorithm 6.1 are spent on exponents that are already 0. For example, consider the 869 iterations mentioned above for failure probability $0.178 \cdot 2^{-32}$ for uniform random CSIDH-512 vectors with entries in $\{-5, \ldots, 5\}$.

Algorithm 7.1: Evaluating the class-group action by reducing the top nonzero exponent.

Parameters: Odd primes $\ell_1 < \cdots < \ell_n$ with $n \geq 1$, a prime $p = 4\ell_1 \cdots \ell_n - 1$, and a positive integer r.

Input: $A \in S_p$, integers $(e_1, \ldots, e_n)$.

Output: $\mathcal{L}_1^{e_1} \cdots \mathcal{L}_n^{e_n}(A)$ or "fail".

for $j \leftarrow 1$ **to** r **do**

> Let $i = \max\{k : e_k \neq 0\}$, or $i = 1$ if each $e_k = 0$.
>
> Let $s = \text{sign}(e_i) \in \{-1, 0, +1\}$.
>
> Find a random point P on E_{sA} using Elligator.
>
> Compute $Q \leftarrow ((p+1)/\ell_i)P$.
>
> Compute B with $E_B \cong E_{sA}/\langle Q \rangle$ if $Q \neq \infty$, using the ℓ_i-isogeny formulas from Section 5.3 with maximum degree ℓ_n.
>
> Set $A \leftarrow sB$ if $Q \neq \infty$ and $s \neq 0$.
>
> Set $e_i \leftarrow e_i - s$ if $Q \neq \infty$.

Set $A \leftarrow$ "fail" if $(e_1, \ldots, e_n) \neq (0, \ldots, 0)$.

Return A.

Entry e_i has absolute value $30/11$ on average, and needs $(30/11)\ell_i/(\ell_i - 1)$ iterations on average, for a total of $\sum_i (30/11)\ell_i/(\ell_i - 1) \approx 206.79$ useful iterations on average. This means that there are 662.21 useless iterations on average, many more than one would expect to be needed to guarantee this failure probability.

This section introduces a constant-time algorithm that achieves the same failure probability with far fewer iterations. For example, in the above scenario, just 294 iterations suffice to reduce the failure probability below 2^{-32}. Each iteration becomes (for CSIDH-512) about 25% more expensive, but overall the algorithm uses far fewer bit operations.

7.1 Iterations Targeting Variable ℓ. It is obvious how to avoid useless iterations for variable-time algorithms: when an exponent reaches 0, move on to the next exponent. In other words, always focus on reducing a nonzero exponent, if one exists.

What is new is doing this in constant time. This is where we exploit the Matryoshka-doll structure from Sect. 5.3, computing an isogeny for variable ℓ in constant time. We now pay for an ℓ_n-isogeny in each iteration rather than an ℓ-isogeny, but the iteration cost is still dominated by scalar multiplication. Concretely, for CSIDH-512, an average ℓ-isogeny costs about 600 multiplications, and an ℓ_n-isogeny costs about 2000 multiplications, but a scalar multiplication costs about 5000 multiplications.

We choose to reduce the top exponent that is not 0. "Top" here refers to position, not value: we reduce the nonzero e_i where i is maximized. See Algorithm 7.1.

7.2   Upper Bounds on the Failure Probability. One can crudely estimate the failure probability of Algorithm 7.1 in terms of the 1-norm $E = |e_1| + \cdots + |e_n|$ as follows. Model each iteration as having failure probability $1/3$ instead of $1/\ell_i$; this produces a loose upper bound for the overall failure probability of the algorithm.

In this model, the chance of needing exactly r iterations to find a point of order ℓ_i is the coefficient of x^r in the power series

$$(2/3)x + (2/9)x^2 + (2/27)x^3 + \cdots = 2x/(3-x).$$

The chance of needing exactly r iterations to find all E points is the coefficient of x^r in the Eth power of that power series, namely $c_r = \binom{r-1}{E-1} 2^E / 3^r$ for $r \geq E$. See generally [74] for an introduction to the power-series view of combinatorics; there are many other ways to derive the formula $\binom{r-1}{E-1} 2^E / 3^r$, but we make critical use of power series for fast computations in Sects. 7.3 and 8.3.

The failure probability of r iterations of Algorithm 7.1 is at most the failure probability of r iterations in this model, namely $f(r, E) = 1 - c_E - c_{E+1} - \cdots - c_r$. The failure probability of r iterations for a uniform random vector with entries in $\{-C, \ldots, C\}$ is at most $\sum_{0 \leq E \leq nC} f(r, E) g[E]$. Here $g[E]$ is the probability that a vector has 1-norm E, which we compute as the coefficient of x^E in the nth power of the polynomial $(1 + 2x + 2x^2 + \cdots + 2x^C)/(2C + 1)$. For example, with $n = 74$ and $C = 5$, the failure probability in this model (rounded to 3 digits after the decimal point) is $0.999 \cdot 2^{-1}$ for $r = 302$; $0.965 \cdot 2^{-2}$ for $r = 319$; $0.844 \cdot 2^{-32}$ for $r = 461$; and $0.570 \cdot 2^{-256}$ for $r = 823$. As a double-check, we observe that a simple simulation of the model for $r = 319$ produces 241071 failures in 1000000 experiments, close to the predicted $0.965 \cdot 2^{-2} \cdot 1000000 \approx 241250$.

7.3   Exact Values of the Failure Probability. The upper bounds from the model above are too pessimistic, except for $\ell_i = 3$. We instead compute the exact failure probabilities as follows.

The chance that $\mathcal{L}_1^{e_1} \cdots \mathcal{L}_n^{e_n}$ requires exactly r iterations is the coefficient of x^r in the power series

$$\left(\frac{(\ell_1 - 1)x}{\ell_1 - x} \right)^{|e_1|} \cdots \left(\frac{(\ell_n - 1)x}{\ell_n - x} \right)^{|e_n|}.$$

What we want is the average of this coefficient over all vectors $(e_1, \ldots, e_n) \in \{-C, \ldots, C\}^n$. This is the same as the coefficient of the average, and the average factors nicely as

$$\left(\sum_{-C \leq e_1 \leq C} \frac{1}{2C+1} \left(\frac{(\ell_1 - 1)x}{\ell_1 - x} \right)^{|e_1|} \right) \cdots \left(\sum_{-C \leq e_n \leq C} \frac{1}{2C+1} \left(\frac{(\ell_n - 1)x}{\ell_n - x} \right)^{|e_n|} \right).$$

We compute this product as a power series with rational coefficients: for example, we compute the coefficients of $x^0, \ldots, x^{499}$ if we are not interested in 500 or more iterations. We then add together the coefficients of $x^0, \ldots, x^r$ to find the exact success probability of r iterations of Algorithm 7.1.

As an example we again take CSIDH-512 with $C = 5$. The failure probability (again rounded to 3 digits after the decimal point) is $0.960 \cdot 2^{-1}$ for $r = 207$; $0.998 \cdot 2^{-2}$ for $r = 216$; $0.984 \cdot 2^{-32}$ for $r = 294$; $0.521 \cdot 2^{-51}$ for $r = 319$; and $0.773 \cdot 2^{-256}$ for $r = 468$. We double-checked these averages against the results of Monte Carlo calculations for these values of r. Each Monte Carlo iteration sampled a uniform random 1-norm (weighted appropriately for the initial probability of each 1-norm), sampled a uniform random vector within that 1-norm, and computed the failure probability for that vector using the single-vector generating function.

7.4 Analysis of the Cost. We have fully implemented Algorithm 7.1 in our bit-operation simulator. One iteration for CSIDH-512 uses $9208697761 \approx 2^{33}$ bit operations, including $3805535430 \approx 2^{32}$ nonlinear bit operations. More than 95% of the cost is explained as follows:

- Each iteration uses a Montgomery ladder with a 511-bit scalar. (We could save a bit here: the largest useful scalar is $(p+1)/3$, which is below 2^{510}.) We use an affine input point and an affine A, so this costs $2044\mathbf{S} + 3066\mathbf{M}$.
- Each iteration uses the formulas from Sect. 5.3 with $\ell = 587$. This takes $602\mathbf{S} + 1472\mathbf{M}$: specifically, $584\mathbf{S} + 876\mathbf{M}$ for multiples of the point of order ℓ (again affine); $584\mathbf{M}$ for the product of Edwards y-coordinates; $18\mathbf{S} + 10\mathbf{M}$ for two ℓth powers; and $2\mathbf{M}$ to multiply by two 8th powers. (We merge the $6\mathbf{S}$ for the 8th powers into the squarings used for the ℓth powers.)
- Each iteration uses two inversions to obtain affine Q and A, each $507\mathbf{S}+97\mathbf{M}$, and one Legendre-symbol computation, $506\mathbf{S} + 96\mathbf{M}$.

This accounts for $4166\mathbf{S}+4828\mathbf{M}$ per iteration, i.e., $4166 \cdot 349596 + 4828 \cdot 447902 = 3618887792 \approx 2^{32}$ nonlinear bit operations.

The cost of 294 iterations is simply $294 \cdot 3805535430 = 1118827416420 \approx 2^{40}$ nonlinear bit operations. This justifies the first (B, ϵ) claim in Sect. 1.

7.5 Decreasing the Maximum Degrees. Always performing isogeny computations capable of handling degrees up to ℓ_n is wasteful: With overwhelming probability, almost all of the 294 iterations required for a failure probability of less than 2^{-32} with the approach discussed so far actually compute isogenies of degree (much) less than ℓ_n. For example, with e uniformly random in $\{-5, \ldots, 5\}$, the probability that 10 iterations are not sufficient to eliminate all 587-isogenies is approximately 2^{-50}. Therefore, using smaller upper bounds on the isogeny degrees for later iterations of the algorithm will not do much harm to the success probability while significantly improving the performance. We modify Algorithm 7.1 as follows:

- Instead of a single parameter r, we use a list $(r_1, \ldots, r_n)$ of non-negative integers, each r_i denoting the number of times an isogeny computation capable of handling degrees up to ℓ_i is performed.
- The loop iterating from 1 through r is replaced by an outer loop on u from n down to 1, and inside that an inner loop on j from 1 up to r_u. The loop body is unchanged, except that the maximum degree for the isogeny formulas is now ℓ_u instead of ℓ_n.

Table 7.1. Examples of choices of $r_i, \ldots, r_i$ for Algorithm 7.1 with reducing the maximal degree in Vélu's formulas for uniform random CSIDH-512 vectors with entries in $\{-5, \ldots, 5\}$. Failure probabilities ϵ are rounded to three digits after the decimal point.

ϵ	$r_n \ldots r_1$	$\sum r_i$	avg. ℓ
$0.594 \cdot 2^{-1}$	5 3 4 5 3 5 5 4 3 5 4 3 4 4 3 4 3 4 3 3 3 4 3 3 3 4 3 3 3 3 3 3 3 3 3 3 3 3 3 3 3 3 3 3 3 2 4 2 3 3 3 3 2 3 3 3 2 3 3 2 3 2 3 2 3 2 2 3 2 2 2 1 1 1 0 0	218	205.0
$0.970 \cdot 2^{-32}$	9 5 5 5 5 5 4 5 5 5 4 5 4 5 5 4 5 4 4 5 5 4 4 4 5 4 4 4 4 4 3 5 3 4 4 4 3 4 4 4 4 3 4 4 3 4 3 4 3 4 3 4 4 4 3 4 3 3 4 4 3 3 4 3 3 4 3 4 3 3 3 4 3 3 3 4	295	196.0
$0.705 \cdot 2^{-256}$	34 8 6 6 5 6 6 6 5 5 6 5 6 5 5 5 5 6 5 5 5 5 6 5 5 5 5 5 5 5 5 5 5 5 5 5 5 5 5 5 5 5 5 5 5 5 5 4 6 5 5 5 5 5 4 6 5 5 5 5 5 5 6 5 5 6 6 6 6 7 7 11 16 38	469	182.7

For a given sequence $(r_1, \ldots, r_n)$, the probability of success can be computed as follows:

- For each $i \in \{1, \ldots, n\}$, compute the generating function

$$\phi_i(x) = \sum_{-C \le e_i \le C} \frac{1}{2C+1} \left(\frac{(\ell_i - 1)x}{\ell_i - x} \right)^{|e_i|}$$

  of the number of ℓ_i-isogeny steps that have to be performed.
- Since we are no longer only interested in the total number of isogeny steps to be computed, but also in their degrees, we cannot simply take the product of all ϕ_i as before. Instead, to account for the fact that failing to compute a ℓ_i-isogeny before the maximal degree drops below ℓ_i implies a total failure, we iteratively compute the product of the ϕ_i from $k = n$ down to 1, but truncate the product after each step. Truncation after some power x^t means eliminating all branches of the probabilistic process in which more than t isogeny steps are needed for the computations so far. In our case we use $t = \sum_{j=i}^{n} r_j$ after multiplying by ϕ_i, which removes all outcomes in which more isogeny steps of degree $\ge \ell_i$ would have needed to be computed.
- After all ϕ_i have been processed (including the final truncation), the probability of success is the sum of all coefficients of the remaining power series.

Note that we have only described a procedure to compute the success probability once $r_1, \ldots, r_n$ are known. It is unclear how to find the optimal values r_i which minimize the cost of the resulting algorithm, while at the same time respecting a certain failure probability. We tried various reasonable-looking choices of strategies to choose the r_i according to certain prescribed failure probabilities after each individual step. Experimentally, a good rule seems to be that the failure probability after processing ϕ_i should be bounded by $\epsilon \cdot 2^{2/i-2}$, where ϵ is the overall target failure probability. The results are shown in Table 7.1.

The average degree of the isogenies used constructively in CSIDH-512 is about 174.6, which is not much smaller than the average degree we achieve.

Since we still need to control the error probability, it does not appear that one can expect to get much closer to the constructive case.

Also note that the total number of isogeny steps for $\epsilon \approx 2^{-32}$ and $\epsilon \approx 2^{-256}$ is each only one more than the previous number r of isogeny computations, hence one can expect significant savings using this strategy. Assuming that about $1/4$ of the total time is spent on Vélu's formulas (which is close to the real proportion), we get a speedup of about 16% for $\epsilon \approx 2^{-32}$ and about 17% for $\epsilon \approx 2^{-256}$.

8 Pushing Points Through Isogenies

Algorithms 6.1 and 7.1 spend most of their time on scalar multiplication. This section pushes points through isogenies to reduce the time spent on scalar multiplication, saving time overall.

The general idea of balancing isogeny computation with scalar multiplication was introduced in [22] in the SIDH context, and was reused in the variable-time CSIDH algorithms in [15]. This section adapts the idea to the context of constant-time CSIDH computation.

8.1 Why Pushing Points Through Isogenies Saves Time. To illustrate the main idea, we begin by considering a sequence of just two isogenies with the same sign. Specifically, assume that, given distinct ℓ_1 and ℓ_2 dividing $p + 1$, we want to compute $\mathcal{L}_1\mathcal{L}_2(A) = B$. Here are two different methods:

- **Method 1.** The method of Algorithm 6.1 uses Elligator to find $P_1 \in E_A(\mathbb{F}_p)$, computes $Q_1 \leftarrow [(p + 1)/\ell_1]P_1$, computes $E_{A'} = E_A/\langle Q_1 \rangle$, uses Elligator to find $P_2 \in E_{A'}(\mathbb{F}_p)$, computes $Q_2 \leftarrow [(p + 1)/\ell_2]P_2$, and computes $E_B = E_{A'}/\langle Q_2 \rangle$. Failure cases: if $Q_1 = \infty$ then this method computes $A' = A$, failing to compute $\mathcal{L}_1$; similarly, if $Q_2 = \infty$ then this method computes $B = A'$, failing to compute $\mathcal{L}_2$.
- **Method 2.** The method described in this section instead uses Elligator to find $P \in E_A(\mathbb{F}_p)$, computes $R \leftarrow [(p+1)/\ell_1\ell_2]P$, computes $Q \leftarrow [\ell_2]R$, computes $\varphi : E_A \to E_{A'} = E_A/\langle Q \rangle$ and $Q' = \varphi(R)$, and computes $E_B = E_{A'}/\langle Q' \rangle$. Failure cases: if $Q = \infty$ then this method computes $Q' = R$ (which has order dividing ℓ_2) and $A' = A$, failing to compute $\mathcal{L}_1$; if $Q' = \infty$ then this method computes $B = A'$, failing to compute $\mathcal{L}_2$.

For concreteness, we compare the costs of these methods for CSIDH-512. The rest of this subsection uses approximations to the costs of lower-level operations to simplify the analysis. The main costs are as follows:

- For p a 512-bit prime, Elligator costs approximately 600**M**.
- Given $P \in E(\mathbb{F}_p)$ and a positive integer k, the computation of $[k]P$ via the Montgomery ladder, as described in Sect. 3.3, costs approximately $10(\log_2 k)$**M**, i.e., approximately $(5120 - 10\log_2 \ell)$**M** if $k = (p + 1)/\ell$.
- The computation of a degree-ℓ isogeny via the method described in Sect. 5.3 costs approximately $(3.5\ell + 2\log_2 \ell)$**M**.

- Given an ℓ-isogeny $\varphi_\ell : E \to E'$ and $P \in E(\mathbb{F}_p)$, the computation of $\varphi_\ell(P)$ via the method described in Sect. 5.4 costs approximately $2\ell\mathbf{M}$.

Method 1 costs approximately

$$(1200 + 10240 + 3.5\ell_1 + 3.5\ell_2 - 8\log_2 \ell_1 - 8\log_2 \ell_2)\mathbf{M},$$

while Method 2 costs approximately

$$(600 + 5120 + 5.5\ell_1 + 3.5\ell_2 - 8\log_2 \ell_1 + 2\log_2 \ell_2)\mathbf{M}.$$

The savings of $(600 + 5120)\mathbf{M}$ clearly outweighs the loss of $(2\ell_1 + 10\log_2 \ell_2)\mathbf{M}$, since the largest value of ℓ_i is 587.

There are limits to the applicability of Method 2: it cannot combine two isogenies of opposite signs, it cannot combine two isogenies using the same prime, and it cannot save time in applying just one isogeny. We will analyze the overall magnitude of these effects in Sect. 8.3.

8.2   Handling the General Case, Two Isogenies at a Time. Algorithm 8.1 computes $\mathcal{L}_1^{e_1} \cdots \mathcal{L}_n^{e_n}(A)$ for any exponent vector $(e_1, \ldots, e_n)$. Each iteration of the algorithm tries to perform two isogenies: one for the top nonzero exponent (if the vector is nonzero), and one for the next exponent having the same sign (if the vector has another exponent of this sign). As in Sect. 7, "top" refers to position, not value.

The algorithm pushes the first point through the first isogeny, as in Sect. 8.1, to save the cost of generating a second point. Scalar multiplication, isogeny computation, and isogeny application use the constant-time subroutines described in Sects. 3.3, 5.3, and 5.4 respectively. The cost of these algorithms depends on the bound ℓ_n for the prime for the top nonzero exponent and the bound ℓ_{n-1} for the prime for the next exponent. The two prime bounds have asymmetric effects upon costs; we exploit this by applying the isogeny for the top nonzero exponent *after* the isogeny for the next exponent.

Analyzing the correctness of Algorithm 8.1—assuming that there are enough iterations; see Sect. 8.3—requires considering three cases. The first case is that the exponent vector is 0. Then i, i', s are initialized to $0, 0, 1$ respectively, and i, i' stay 0 throughout the iteration, so A does not change and the exponent vector does not change.

The second case is that the exponent vector is nonzero and the top nonzero exponent e_i is the only exponent having sign s. Then i' is 0 throughout the iteration, so the "first isogeny" portion of Algorithm 8.1 has no effect. The point $Q = R$ in the "second isogeny" portion is cP where $c = (p+1)/\ell_i$, so $\ell_i Q = \infty$. If $Q = \infty$ then i is set to 0 and the entire iteration has no effect, except for setting A to sA and then back to $s(sA) = A$. If $Q \neq \infty$ then i stays nonzero and A is replaced by $\mathcal{L}_i(A)$, so A at the end of the iteration is $\mathcal{L}_i^s$ applied to A at the beginning of the iteration, while s is subtracted from e_i.

The third case is that the exponent vector is nonzero and that $e_{i'}$ is the next exponent having the same sign s as the top nonzero exponent e_i. By construction $i' < i \leq n$ so $\ell_{i'} \leq \ell_{n-1}$. Now $R = cP$ where $c = (p+1)/(\ell_i \ell_{i'})$. The first isogeny

---

Algorithm 8.1: Evaluating the class-group action by reducing the top nonzero exponent and the next exponent with the same sign.

Parameters: Odd primes $\ell_1 < \cdots < \ell_n$ with $n \geq 1$, a prime $p = 4\ell_1 \cdots \ell_n - 1$, and a positive integer r.

Input: $A \in S_p$, integers $(e_1, \ldots, e_n)$.

Output: $\mathcal{L}_1^{e_1} \cdots \mathcal{L}_n^{e_n}(A)$ or "fail".

for $j \leftarrow 1$ **to** r **do**

> Set $I \leftarrow \{k : 1 \leq k \leq n \text{ and } e_k \neq 0\}$.
> Set $i \leftarrow \max I$ and $s \leftarrow \text{sign}(e_i) \in \{-1, 1\}$, or $i \leftarrow 0$ and $s \leftarrow 1$ if $I = \{\}$.
> Set $I' \leftarrow \{k : 1 \leq k < i \text{ and } \text{sign}(e_k) = s\}$.
> Set $i' \leftarrow \max I'$, or $i' \leftarrow 0$ if $I' = \{\}$.
> **Twist.** Set $A \leftarrow sA$.
> **Isogeny preparation.** Find a random point P on E_A using Elligator.
> Compute $R \leftarrow cP$ where $c = 4 \prod_{1 \leq j \leq n, j \neq i, j \neq i'} \ell_j$.
> **First isogeny.** Compute $Q \leftarrow \ell_i R$, where ℓ_0 means 1.
> [Now $\ell_{i'} Q = \infty$ if $i' \neq 0$.] Set $i' \leftarrow 0$ if $Q = \infty$.
> Compute B with $E_B \cong E_A / \langle Q \rangle$ if $i' \neq 0$, using the $\ell_{i'}$-isogeny formulas from
> > Section 5.3 with maximum degree ℓ_{n-1}.
> Set R to the image of R in E_B if $i' \neq 0$, using the $\ell_{i'}$-isogeny formulas from
> > Section 5.4 with maximum degree ℓ_{n-1}.
> Set $A \leftarrow B$ and $e_{i'} \leftarrow e_{i'} - s$ if $i' \neq 0$.
> **Second isogeny.** Set $Q \leftarrow R$.
> [Now $\ell_i Q = \infty$ if $i \neq 0$.] Set $i \leftarrow 0$ if $Q = \infty$.
> Compute B with $E_B \cong E_A / \langle Q \rangle$ if $i \neq 0$, using the ℓ_i-isogeny formulas from
> > Section 5.3 with maximum degree ℓ_n.
> Set $A \leftarrow B$ and $e_i \leftarrow e_i - s$ if $i \neq 0$.
> **Untwist.** Set $A \leftarrow sA$.

Set $A \leftarrow$ "fail" if $(e_1, \ldots, e_n) \neq (0, \ldots, 0)$.

Return A.

---

uses the point $Q = \ell_i R$, which is either ∞ or a point of order $\ell_{i'}$. If Q is ∞ then i' is set to 0; both A and the vector are unchanged; the point R must have order dividing ℓ_i; and the second isogeny proceeds as above using this point. If Q has order $\ell_{i'}$ then the first isogeny replaces A with $\mathcal{L}_{i'}(A)$, while subtracting s from $e_{i'}$ and replacing R with a point of order dividing ℓ_i on the new curve (note that the $\ell_{i'}$-isogeny removes any $\ell_{i'}$ from orders of points); again the second isogeny proceeds as above.

8.3   Analysis of the Failure Probability. Consider a modified dual-isogeny algorithm in which the isogeny with a smaller prime is saved to handle later:

- Initialize an iteration counter to 0.
- Initialize an empty bank of positive isogenies.
- Initialize an empty bank of negative isogenies.
- For each ℓ in decreasing order:
  - While an ℓ-isogeny needs to be done and the bank has an isogeny of the correct sign: Withdraw an isogeny from the bank, apply the isogeny, and adjust the exponent.
  - While an ℓ-isogeny still needs to be done: Apply an isogeny, adjust the exponent, deposit an isogeny with the bank, and increase the iteration counter.

This uses more bit operations than Algorithm 8.1 (since the work here is not shared across two isogenies), but it has the same failure probability for the same number of iterations. We now focus on analyzing the distribution of the number of iterations used by this modified algorithm.

We use three variables to characterize the state of the modified algorithm before each ℓ:

- $i \geq 0$ is the iteration counter;
- $j \geq 0$ is the number of positive isogenies in the bank;
- $k \geq 0$ is the number of negative isogenies in the bank.

The number of isogenies actually applied so far is $2i - (j+k) \geq i$. The distribution of states is captured by the three-variable formal power series $\sum_{i,j,k} s_{i,j,k} x^i y^j z^k$ where $s_{i,j,k}$ is the probability of state (i, j, k). Note that there is no need to track which primes are paired with which; this is what makes the modified algorithm relatively easy to analyze.

If there are exactly h positive ℓ-isogenies to perform then the new state after those isogenies is $(i, j - h, k)$ if $h \leq j$, or $(i + h - j, h - j, k)$ if $h > j$. This can be viewed as a composition of two operations on the power series. First, multiply by y^{-h}. Second, replace any positive power of y^{-1} with the same power of xy; i.e., replace $x^i y^j z^k$ for each $j < 0$ with $x^{i-j} y^{-j} z^k$.

We actually have a distribution of the number of ℓ-isogenies to perform. Say there are h isogenies with probability q_h. We multiply the original series by $\sum_{h \geq 0} q_h y^{-h}$, and then eliminate negative powers of y as above. We similarly handle $h < 0$, exchanging the role of (j, y) with the role of (k, z).

As in the analyses earlier in the paper, we model each point Q for an ℓ-isogeny as having order 1 with probability $1/\ell$ and order ℓ with probability $1 - 1/\ell$, and we assume that the number of ℓ-isogenies to perform is a uniform random integer $e \in \{-C, \ldots, C\}$. Then q_h for $h \geq 0$ is the coefficient of x^h in $\sum_{0 \leq e \leq C} (((\ell - 1)x)/(\ell - x))^e / (2C + 1)$; also, $q_{-h} = q_h$.

We reduce the time spent on these computations in three ways. First, we discard all states with $i > r$ if we are not interested in more than r iterations. This leaves a cubic number of states for each ℓ: every i between 0 and r inclusive, every j between 0 and i inclusive, and every k between 0 and $i - j$ inclusive.

Second, we use fixed-precision arithmetic, rounding each probability to an integer multiple of (e.g.) 2^{-512}. We round down to obtain lower bounds on success probabilities; we round up to obtain upper bounds on success probabilities; we choose the scale 2^{-512} so that these bounds are as tight as desired. We could save more time by reducing the precision slightly at each step of the computation, and by using standard interval-arithmetic techniques to merge computations of lower and upper bounds.

Third, to multiply the series $\sum_{i,j,k} s_{i,j,k} x^i y^j z^k$ by $\sum_{h \geq 0} q_h y^{-h}$, we actually multiply $\sum_j s_{i,j,k} y^j$ by $\sum_{h \geq 0} q_h y^{-h}$ for each (i, k) separately. We use Sage for these multiplications of univariate polynomials with integer coefficients. Sage, in turn, uses fast multiplication algorithms whose cost is essentially bd for d b-bit coefficients, so our total cost for n primes is essentially bnr^3.

Concretely, we use under two hours on one core of a 3.5GHz Intel Xeon E3-1275 v3 to compute lower bounds on all the success probabilities for CSIDH-512 with $b = 512$ and $r = 349$, and under three hours[2] to compute upper bounds. Our convention of rounding failure probabilities to 3 digits makes the lower bounds and upper bounds identical, so presumably we could have used less precision.

We find, e.g., failure probability $0.943 \cdot 2^{-1}$ after 106 iterations, failure probability $0.855 \cdot 2^{-32}$ after 154 iterations, and failure probability $0.975 \cdot 2^{-257}$ after 307 iterations. Compared to the $207{,}294{,}468$ single-isogeny iterations required in Sect. 7.3, the number of iterations has decreased to 51.2%, 52.3%, 65.6% respectively.

8.4   Analysis of the Cost. We have fully implemented Algorithm 8.1 in our bit-operation simulator. An iteration of Algorithm 8.1 uses $4969644344 \approx 2^{32}$ nonlinear bit operations, about 1.306 times more expensive than an iteration of Algorithm 7.1.

If the number of iterations were multiplied by exactly 0.5 then the total cost would be multiplied by 0.653. Given the actual number of iterations (see Sect. 8.3), the cost is actually multiplied by 0.669, 0.684, 0.857 respectively. In particular, we reach failure probability $0.855 \cdot 2^{-32}$ with $154 \cdot 4969644344 = 765325228976 \approx 0.7 \cdot 2^{40}$ nonlinear bit operations. This justifies the second (B, ϵ) claim in Sect. 1.

8.5   Variants. The idea of pushing points through isogenies can be combined with the idea of gradually reducing the maximum prime allowed in the Matryoshka-doll isogeny formulas. This is compatible with our techniques for analyzing failure probabilities.

A dual-isogeny iteration very late in the computation is likely to have a useless second isogeny. It should be slightly better to replace some of the last dual-isogeny iterations with single-isogeny iterations. This is also compatible with our techniques for analyzing failure probabilities.

There are many different possible pairings of primes: one can take any two distinct positions where the exponents have the same sign. Possibilities include reducing exponents from the bottom rather than the top; reducing the top nonzero exponent and the bottom exponent with the same sign; always pairing

---

[2] It is unsurprising that lower bounds are faster: many coefficients q_h round down to 0. We could save time in the upper bounds by checking for stretches of coefficients that round up to, e.g., $1/2^{512}$, and using additions to multiply by those stretches.

"high" positions with "low" positions; always reducing the largest exponents in absolute value; always reducing e_i where $|e_i|\ell_i/(\ell_i - 1)$ is largest. For some of these ideas it is not clear how to efficiently analyze failure probabilities.

This section has focused on reusing an Elligator computation and large scalar multiplication for (in most cases) two isogeny computations, dividing the scalar-multiplication cost by (nearly) 2, in exchange for some overhead. We could push a point through more isogenies, although each extra isogeny has further overhead with less and less benefit, and computing the failure probability becomes more expensive. For comparison, [15] reuses one point for every ℓ_i where e_i has the same sign; the number of such ℓ_i is variable, and decreases as the computation continues. For small primes it might also save time to push multiple points through one isogeny, as in [22].

9   Computing ℓ-isogenies Using Division Polynomials

As the target failure probability decreases, the algorithms earlier in this paper spend more and more iterations handling the possibility of repeated failures for small primes ℓ. This section presents and analyzes an alternative: a deterministic constant-time subroutine that uses division polynomials to *always* compute ℓ-isogenies. See full version of paper online at https://ia.cr/2018/1059.

10   Computing ℓ-isogenies Using Modular Polynomials

Modular polynomials, like division polynomials, give a deterministic subroutine to compute ℓ-isogenies. The advantage of modular polynomials over division polynomials is that modular polynomials are smaller for all $\ell \geq 5$. However, using modular polynomials requires solving two additional problems. See full version of paper online at https://ia.cr/2018/1059.

References

[1] Azarderakhsh, R., Jao, D., Kalach, K., Koziel, B., Leonardi, C.: Key compression for isogeny-based cryptosystems. In: AsiaPKC@AsiaCCS, pp. 1–10. ACM (2016)

[5] Bernstein, D.J.: Curve25519: new Diffie-Hellman speed records. In: Yung, M., Dodis, Y., Kiayias, A., Malkin, T. (eds.) PKC 2006. LNCS, vol. 3958, pp. 207–228. Springer, Heidelberg (2006). https://doi.org/10.1007/11745853_14

[7] Bernstein, D.J., Hamburg, M., Krasnova, A., Lange, T.: Elligator: elliptic-curve points indistinguishable from uniform random strings. In: ACM Conference on Computer and Communications Security, pp. 967–980. ACM (2013)

[8] Bernstein, D.J., Lange, T.: Analysis and optimization of elliptic-curve single-scalar multiplication. In: Finite Fields and Applications 2007, pp. 1–19. AMS (2008)

[9] Bernstein, D.J., Lange, T.: Montgomery curves and the Montgomery ladder. In: Bos, J.W., Lenstra, A.K. (eds.) Topics in Computational Number Theory Inspired by Peter L. Montgomery, pp. 82–115. Cambridge University Press, Cambridge (2017)

[10] Bonnetain, X., Naya-Plasencia, M.: Hidden shift quantum cryptanalysis and implications. In: Peyrin, T., Galbraith, S. (eds.) ASIACRYPT 2018. LNCS, vol. 11272, pp. 560–592. Springer, Cham (2018). https://doi.org/10.1007/978-3-030-03326-2_19

[11] Bonnetain, X., Schrottenloher, A.: Quantum security analysis of CSIDH and ordinary isogeny-based schemes (2018). IACR Cryptology ePrint Archive 2018/537

[15] Castryck, W., Lange, T., Martindale, C., Panny, L., Renes, J.: CSIDH: an efficient post-quantum commutative group action. In: Peyrin, T., Galbraith, S. (eds.) ASIACRYPT 2018. LNCS, vol. 11274, pp. 395–427. Springer, Cham (2018). https://doi.org/10.1007/978-3-030-03332-3_15

[17] Childs, A.M., Jao, D., Soukharev, V.: Constructing elliptic curve isogenies in quantum subexponential time. J. Math. Cryptol. **8**(1), 1–29 (2014)

[18] Costello, C., Hisil, H.: A simple and compact algorithm for SIDH with arbitrary degree isogenies. In: Takagi, T., Peyrin, T. (eds.) ASIACRYPT 2017, Part II. LNCS, vol. 10625, pp. 303–329. Springer, Cham (2017). https://doi.org/10.1007/978-3-319-70697-9_11

[19] Costello, C., Jao, D., Longa, P., Naehrig, M., Renes, J., Urbanik, D.: Efficient compression of SIDH public keys. In: Coron, J.-S., Nielsen, J.B. (eds.) EUROCRYPT 2017, Part I. LNCS, vol. 10210, pp. 679–706. Springer, Cham (2017). https://doi.org/10.1007/978-3-319-56620-7_24

[20] Couveignes, J.-M.: Hard Homogeneous Spaces (1997). IACR Cryptology ePrint Archive 2006/291

[22] De Feo, L., Jao, D., Plût, J.: Towards quantum-resistant cryptosystems from supersingular elliptic curve isogenies. J. Math. Cryptol. **8**(3), 209–247 (2014). IACR Cryptology ePrint Archive 2011/506

[23] De Feo, L., Kieffer, J., Smith, B.: Towards practical key exchange from ordinary isogeny graphs. In: Peyrin, T., Galbraith, S. (eds.) ASIACRYPT 2018. LNCS, vol. 11274, pp. 365–394. Springer, Cham (2018). https://doi.org/10.1007/978-3-030-03332-3_14

[28] Grassl, M., Langenberg, B., Roetteler, M., Steinwandt, R.: Applying Grover's algorithm to AES: quantum resource estimates. In: Takagi, T. (ed.) PQCrypto 2016. LNCS, vol. 9606, pp. 29–43. Springer, Cham (2016). https://doi.org/10.1007/978-3-319-29360-8_3

[29] Hafner, J.L., McCurley, K.S.: A rigorous subexponential algorithm for computation of class groups. J. Am. Math. Soc. **2**(4), 837–850 (1989)

[34] Hişil, H.: Elliptic curves, group law, and efficient computation. Ph.D. thesis, Queensland University of Technology (2010). https://eprints.qut.edu.au/33233/

[36] Jao, D., Azarderakhsh, R., Campagna, M., Costello, C., De Feo, L., Hess, B., Jalali, A., Koziel, B., LaMacchia, B., Longa, P., Naehrig, M., Renes, J., Soukharev, V., Urbanik, D.: SIKE. Submission to [55]. http://sike.org

[37] Jao, D., De Feo, L.: Towards quantum-resistant cryptosystems from supersingular elliptic curve isogenies. In: Yang, B.-Y. (ed.) PQCrypto 2011. LNCS, vol. 7071, pp. 19–34. Springer, Heidelberg (2011). https://doi.org/10.1007/978-3-642-25405-5_2

[38] Jao, D., LeGrow, J., Leonardi, C., Ruiz-Lopez, L.: A subexponential-time, polynomial quantum space algorithm for inverting the CM group action. J. Math. Cryptol. (2018, to appear)

[42] Kieffer, J.: Étude et accélération du protocole d'échange de clés de Couveignes-Rostovtsev-Stolbunov. Mémoire du Master 2, Université Paris VI (2017). https://arxiv.org/abs/1804.10128

[45] Kohel, D.: Endomorphism rings of elliptic curves over finite fields. Ph.D. thesis, University of California at Berkeley (1996). http://iml.univ-mrs.fr/~kohel/pub/thesis.pdf

[46] Kuperberg, G.: A subexponential-time quantum algorithm for the dihedral hidden subgroup problem. SIAM J. Comput. **35**(1), 170–188 (2005)

[47] Kuperberg, G.: Another subexponential-time quantum algorithm for the dihedral hidden subgroup problem. In: TQC. LIPIcs, vol. 22, pp. 20–34. Schloss Dagstuhl – Leibniz-Zentrum für Informatik (2013)

[49] Meyer, M., Reith, S.: A faster way to the CSIDH (2018). IACR Cryptology ePrint Archive 2018/782

[50] Micciancio, D.: Improving lattice based cryptosystems using the Hermite normal form. In: Silverman, J.H. (ed.) CaLC 2001. LNCS, vol. 2146, pp. 126–145. Springer, Heidelberg (2001). https://doi.org/10.1007/3-540-44670-2_11

[51] Miller, V.S.: Use of elliptic curves in cryptography. In: Williams, H.C. (ed.) CRYPTO 1985. LNCS, vol. 218, pp. 417–426. Springer, Heidelberg (1986). https://doi.org/10.1007/3-540-39799-X_31

[53] Montgomery, P.L.: Speeding the Pollard and elliptic curve methods of factorization. Math. Comput. **48**(177), 243–264 (1987)

[54] Moody, D., Shumow, D.: Analogues of Vélu's formulas for isogenies on alternate models of elliptic curves. Math. Comput. **85**(300), 1929–1951 (2016)

[55] NIST. Post-quantum cryptography. https://csrc.nist.gov/Projects/Post-Quantum-Cryptography/Post-Quantum-Cryptography-Standardization

[58] Petit, C.: Faster algorithms for isogeny problems using torsion point images. In: Takagi, T., Peyrin, T. (eds.) ASIACRYPT 2017, Part II. LNCS, vol. 10625, pp. 330–353. Springer, Cham (2017). https://doi.org/10.1007/978-3-319-70697-9_12

[61] Regev, O.: A subexponential time algorithm for the dihedral hidden subgroup problem with polynomial space (2004). https://arxiv.org/abs/quant-ph/0406151

[62] Renes, J.: Computing isogenies between Montgomery curves using the action of (0, 0). In: Lange, T., Steinwandt, R. (eds.) PQCrypto 2018. LNCS, vol. 10786, pp. 229–247. Springer, Cham (2018). https://doi.org/10.1007/978-3-319-79063-3_11

[63] Roetteler, M., Naehrig, M., Svore, K.M., Lauter, K.: Quantum resource estimates for computing elliptic curve discrete logarithms. In: Takagi, T., Peyrin, T. (eds.) ASIACRYPT 2017, Part II. LNCS, vol. 10625, pp. 241–270. Springer, Cham (2017). https://doi.org/10.1007/978-3-319-70697-9_9

[64] Rostovtsev, A., Stolbunov, A.: Public-key cryptosystem based on isogenies (2006). IACR Cryptology ePrint Archive 2006/145

[67] Shor, P.W.: Polynomial-time algorithms for prime factorization and discrete logarithms on a quantum computer. SIAM J. Comput. **26**(5), 1484–1509 (1997)

[68] Stolbunov, A.: Constructing public-key cryptographic schemes based on class group action on a set of isogenous elliptic curves. Adv. Math. Commun. **4**(2), 215–235 (2010)

[70] Tibouchi, M.: Elligator squared: uniform points on elliptic curves of prime order as uniform random strings. In: Christin, N., Safavi-Naini, R. (eds.) FC 2014. LNCS, vol. 8437, pp. 139–156. Springer, Heidelberg (2014). https://doi.org/10.1007/978-3-662-45472-5_10

[72] Vélu, J.: Isogénies entre courbes elliptiques. Comptes Rendus de l'Académie des Sciences de Paris **273**, 238–241 (1971)

[74] Wilf, H.S.: generatingfunctionology. Academic Press (1994). https://www.math.upenn.edu/~wilf/DownldGF.html

[75] Zanon, G.H.M., Simplicio Jr., M.A., Pereira, G.C.C.F., Doliskani, J., Barreto, P.S.L.M.: Faster isogeny-based compressed key agreement. In: Lange, T., Steinwandt, R. (eds.) PQCrypto 2018. LNCS, vol. 10786, pp. 248–268. Springer, Cham (2018). https://doi.org/10.1007/978-3-319-79063-3_12

A   Cost Metrics for Quantum Computation

See full version of paper online at https://ia.cr/2018/1059.

B   Basic Integer Arithmetic

See full version of paper online at https://ia.cr/2018/1059.

C   Modular Arithmetic

See full version of paper online at https://ia.cr/2018/1059.

A Quantum-Proof Non-malleable Extractor
With Application to Privacy Amplification Against Active Quantum Adversaries

Divesh Aggarwal[1]($\boxtimes$), Kai-Min Chung[2], Han-Hsuan Lin[3]($\boxtimes$),
and Thomas Vidick[4]

[1] Center of Quantum Technologies, and Department of Computer Science, NUS,
Singapore, Singapore
dcsdiva@nus.edu.sg
[2] Institute of Information Science, Academia Sinica, Taipei 11529, Taiwan
kmchung@iis.sinica.edu.tw
[3] Department of Computer Science, The University of Texas at Austin, Austin, USA
linhh@cs.utexas.edu
[4] Department of Computing and Mathematical Sciences,
California Institute of Technology, Pasadena, USA
vidick@cms.caltech.edu

Abstract. In privacy amplification, two mutually trusted parties aim
to amplify the secrecy of an initial shared secret X in order to establish
a shared private key K by exchanging messages over an insecure communication channel. If the channel is authenticated the task can be solved
in a single round of communication using a strong randomness extractor; choosing a quantum-proof extractor allows one to establish security
against quantum adversaries.

In the case that the channel is not authenticated, this simple solution
is no longer secure. Nevertheless, Dodis and Wichs (STOC'09) showed
that the problem can be solved in two rounds of communication using a
non-malleable extractor, a stronger pseudo-random construction than a
strong extractor.

We give the first construction of a non-malleable extractor that is
secure against quantum adversaries. The extractor is based on a construction by Li (FOCS'12), and is able to extract from source of min-entropy

D. Aggarwal—This research was further partially funded by the Singapore Ministry of
Education and the National Research Foundation under grant R-710-000-012-135.

K.-M. Chung—This research is partially supported by the 2016 Academia Sinica Career
Development Award under Grant no. 23-17, and MOST QC project under Grant no.
MOST 107-2627-E-002-002.

H.-H. Lin—This material is based on work supported by the Singapore National
Research Foundation under NRF RF Award No. NRF-NRFF2013-13.

T. Vidick—Supported by NSF CAREER Grant CCF-1553477, AFOSR YIP award
number FA9550-16-1-0495, and the IQIM, an NSF Physics Frontiers Center (NSF Grant
PHY-1125565) with support of the Gordon and Betty Moore Foundation (GBMF-
12500028).

Y. Ishai and V. Rijmen (Eds.): EUROCRYPT 2019, LNCS 11477, pp. 442–469, 2019.
https://doi.org/10.1007/978-3-030-17656-3_16

rates larger than $1/2$. Combining this construction with a quantum-proof variant of the reduction of Dodis and Wichs, due to Cohen and Vidick (unpublished) we obtain the first privacy amplification protocol secure against active quantum adversaries.

1   Introduction

Privacy amplification. We study the problem of *privacy amplification* [4,5,30,31] (PA). In this problem, two parties, Alice and Bob, share a weak secret X (a random variable with min-entropy at least k). Using X and an insecure communication channel, Alice and Bob would like to securely agree on a secret key R that is ϵ-close to uniformly random even to an adversary Eve who may have full control over their communication channel. This elegant problem has multiple applications including biometric authentication, leakage-resilient cryptography, and quantum cryptography.

If the adversary Eve is passive, i.e., she is only able to observe the communication but may not alter the messages exchanged, then there is a direct solution based on the use of a strong seeded randomness extractor Ext [33]. This can be done by Alice selecting a uniform seed Y for the extractor, and sending the seed to Bob; Alice and Bob both compute the key $R = \text{Ext}(X, Y)$, which is close to being uniformly random and independent of Y by the strong extractor property. The use of a quantum-proof extractor suffices to protect against adversaries holding quantum side information about the secret X.

Privacy amplification is substantially more challenging when the adversary is active, i.e. Eve can not only read but also modify messages exchanged across the communication channel. This problem has been studied extensively in several works including [2,3,8,9,12,13,16,17,19,20,25–29,31,35], yielding constructions that are optimal or near-optimal in any of the parameters involved in the problem, including the min-entropy k, the error ϵ, and the communication complexity of the protocol.

Active adversaries with quantum side information. We consider the problem of active attacks by quantum adversaries. This question arises naturally when privacy amplification is used as a sub-protocol, e.g., as a post-processing step in quantum key distribution (QKD), when it may not be safe to assume that the classical communication channel is authenticated.[1] To the best of our knowledge the question was first raised in [7], whose primary focus is privacy amplification with an additional property of source privacy. Although the authors of [7] initially claimed that their construction is secure against quantum side information, they later realized that there was an issue with their argument, and withdrew their claim of quantum security. The only other work we are aware of approaching the question of privacy amplification in the presence of active quantum adversaries

---

[1] QKD relies on an authenticated channel at other stages of the protocol, and here we only address the privacy amplification part: indeed, PA plays an important role in multiple other cryptographic protocols, and it is a fundamental task that it is useful to address first.

is [14]. In this paper it is shown that a classical protocol for PA introduced by Dodis and Wichs [19] remains secure against active quantum attacks when the main tool used in the protocol, a non-malleable extractor, is secure against quantum side information (a notion that is also formally introduced in that paper, and to which we return shortly). Unfortunately, the final contribution of [14], a construction of a quantum-proof non-malleable extractor, also had a flaw in the proof, invalidating the construction. Thus, the problem of quantum-secure active privacy amplification remained open.

It may be useful to discuss the difficulty faced by both these previous works, as it informed our own construction. The issue is related to the modeling of the side information held by the adversary Eve, and how that side information evolves as messages are being exchanged, and possibly modified, throughout the privacy amplification protocol. To explain this, consider the setting for a non-malleable extractor, whose security property can be defined without referring to the way the extractor is used for privacy amplification. Here, Alice initially has a secret X (the source), while Eve holds side information E, a quantum state, correlated with X. Alice selects a uniformly random seed Y and computes $\text{Ext}(X, Y)$. However, in addition to receiving Y (as would already be the case for a strong randomness extractor), Eve is also given the possibility to select an arbitrary $Y' \neq Y$ and receive $\text{Ext}(X, Y')$ as "advice" to help her break the extractor—i.e., distinguish $\text{Ext}(X, Y)$ from uniform. Now, clearly in any practical scenario the adversary may use her side information E in order to guide her choice of Y'; thus Y' should be considered as the outcome of a measurement $\{M_y^{y'}\}$, depending on $Y = y$ and performed on E, which returns an outcome $Y' = y'$ and a post-measurement state E'. This means that the security of the extractor should be considered with respect to the side information E'. But due to the measurement, E' may be correlated with both X and Y in a way that cannot be addressed by standard techniques for the analysis of strong extractors. Indeed, even if E' is classical, so that we can condition on its value, X and Y may not be independent after conditioning on $E' = e'$; due to the lack of independence it is unclear whether extraction works. (Classical proofs condition on $E = e$ at the outset, which does preserve independence.)

The issue seems particularly difficult to accommodate when analyzing extractors based on the technique of "alternate extraction", as was attempted in [7,14]. In fact, in the original version of [7] the issue is overlooked, resulting in a flawed security proof. In [14] the authors attempted to deal with the difficulty by using the formalism of quantum Markov chains; unfortunately, there is a gap in the argument and it does not seem like the scenario can be modeled using the Markov chain formalism. Note that in the classical setting the issue does not arise: having fixed $E = e$ we can consider Y' to be a fixed, deterministic function of Y—there is no E' to consider, and X is independent of both Y and Y' conditioned on $E = e$. In this paper we do not address the issue, but instead focus on a specific construction of non-malleable extractor whose security can be shown by algebraic techniques sidestepping the difficulty; we explain our approach in more detail below.

Our results. We show that a non-malleable extractor introduced by Li [27] in the classical setting is secure against quantum side information. Combining this construction with the protocol of Dodis and Wichs and its proof of security from [14], we obtain the first protocol for privacy amplification that is secure against active quantum adversaries.

Before describing our results in more detail we summarize Li's construction and its analysis for the case of classical side information. The construction is based on the inner product function. Let p be a prime, $\mathbb{F}_p$ the finite field with p elements, and $\langle \cdot, \cdot \rangle$ the inner product over $\mathbb{F}_p$. Consider the function Ext : $\mathbb{F}_p^n \times \mathbb{F}_p^n \to \mathbb{F}_p$ given by $\mathrm{Ext}(X, Y) := \langle X, Y \rangle$, where $X \in \mathbb{F}_p^n$ is a weak secret with min-entropy (conditioned on the adversary's side information) assumed to be greater than $(n \log p)/2$, and Y is a uniformly random and independent seed. For this function to be a non-malleable extractor, it is required that $\mathrm{Ext}(X, Y)$ is close to uniform and independent of $\mathrm{Ext}(X, f(Y))$, where f is any adversarially chosen function such that $f(Y) \neq Y$ for all Y. This is clearly not true, since if $f(Y) = cY$ for some $c \in \mathbb{F}_p \setminus \{1\}$, then $\mathrm{Ext}(X, f(Y)) = c\mathrm{Ext}(X, Y)$, and hence we don't get the desired independence. Thus, for such a construction to work, it is necessary to encode the source Y as $\mathrm{Enc}(Y)$, for a well-chosen function Enc, in such a way that $\langle X, \mathrm{Enc}(Y) \rangle - c \cdot \langle X, \mathrm{Enc}(f(Y)) \rangle$ is hard to guess. The non-uniform XOR lemma [3,13,17] shows that it is sufficient to show that $\langle X, \mathrm{Enc}(Y) \rangle - c \cdot \langle X, \mathrm{Enc}(f(Y)) \rangle = \langle X, \mathrm{Enc}(Y) - c \cdot \mathrm{Enc}(f(Y)) \rangle$ is close to uniform conditioned on Y and E. The encoding that we use in this paper (which is almost the same as the encoding chosen by Li) is to take $Y \in \mathbb{F}_p^{n/2}$, and encode it as $Y \| Y^2$, which we view as an n-character string over $\mathbb{F}_p$, with the symbol $\|$ denoting concatenation of strings and the square taken by first interpreting Y as an element of $\mathbb{F}_{p^{n/2}}$. Then it is not difficult to show that for any function f such that $f(Y) \neq Y$ and any c, we have that $(Y \| Y^2) - (c \cdot f(Y) \| c \cdot f(Y)^2)$ (taking the addition coordinatewise) has min-entropy almost $(n \log p)/2$. Thus, provided X has sufficiently high min-entropy and using the fact that X and $(Y \| Y^2) - (c \cdot f(Y) \| c \cdot f(Y)^2)$ are independent conditioned on E, the strong extractor property of the inner product function gives the desired result.[2]

Our main technical result is a proof of security of Li's extractor, against quantum side information. We show the following (we refer to Definition 5 for the formal definition of a quantum-proof non-malleable extractor):

Theorem 1. *Let $p \neq 2$ be a prime. Let n be an even integer. Then for any $\epsilon > 0$ the function $\mathrm{nmExt}(X, Y) : \mathbb{F}_p^n \times \mathbb{F}_p^{n/2} \to \mathbb{F}_p$ given by $\langle X, Y \| Y^2 \rangle$ is an $\left(\left(\frac{n}{2} + 6 \right) \log p - 1 + 4 \log \frac{1}{\epsilon}, \epsilon \right)$ quantum-proof non-malleable extractor.*

We give the main ideas behind our proof of security for this construction, highlighting the points of departure from the classical analysis. Subsequently, we explain the application to privacy amplification.

Proof ideas. We begin by generalizing the first step of Li's argument, the reduction provided by the non-uniform XOR lemma, to the quantum case. An XOR

---

[2] This description is a little different from Li's description since he was working with a field of size 2^n, but we find it more convenient to work with a prime field.

lemma with quantum side information is already shown in [22], where the lemma is used to show security of the inner product function as a two-source extractor against quantum side information. This version is not sufficient for our purposes, and we establish the following generalization, which may be of independent interest (we refer to Sect. 3 for relevant definitions):

Lemma 1. *Let p be a prime power and t an integer. Let $\rho_{X_0 X E}$ be a ccq state with $X_0 \in \mathbb{F}_p$ and $X = (X_1, \ldots, X_t) \in \mathbb{F}_p^t$. For all $a = (a_1, \ldots, a_t) \in \mathbb{F}_p^t$, define a random variable $Z = X_0 + \langle a, X \rangle = X_0 + \sum_{i=1}^t a_i X_i$. Let $\epsilon \geq 0$ be such that for all a, $\frac{1}{2} \| \rho_{ZE}^a - U_Z \otimes \rho_E \|_1 \leq \epsilon$. Then*

$$\frac{1}{2} \| \rho_{X_0 X E} - U_{X_0} \otimes \rho_{XE} \|_1 \leq p^{\frac{t+1}{2}} \sqrt{\frac{\epsilon}{2}}. \tag{1.1}$$

XOR lemmas are typically proved via Fourier-based techniques (including the one in [22]). Here we instead rely on a collision probability-based argument inspired from [3]. We prove Lemma 1 by observing that such arguments generalize to the quantum setting, as in the proof of the quantum leftover hash lemma in [36].

Based on the XOR lemma (used with $t = 1$), following Li's arguments it remains to show that the random variable $\langle X, g(Y, Y') \rangle \in \mathbb{F}_p$, where $g(Y, Y') = Y \| Y^2 - c(Y' \| Y'^2) \in \mathbb{F}_p^n$, is close to uniformly distributed from the adversary's point of view, specified by side information E', for every $c \neq 0 \in \mathbb{F}_p$. As already mentioned earlier, this cannot be shown by a reduction to the security proof of the inner product function as a two-source extractor against side information, as X and $g(Y, Y')$ are *not* independent (not even conditioned on the value of E' when E' is classical).

Instead, we are led to a more direct analysis which proceeds by formulating the problem as a communication task.[3] We relate the task of breaking our construction—distinguishing $\langle X, g(Y, Y') \rangle$ from uniform—to success in the following task. Alice is given access to a random variable X, and Bob is given a uniformly random Y. Alice is allowed to send a quantum message E, correlated with X, to Bob. Bob then selects a $Y' \neq Y$ and returns a value $b \in \mathbb{F}_p$. The players win if $b = \langle X, g(Y, Y') \rangle$. Based on our previous reductions it suffices to show that no strategy can succeed with probability substantially higher than random in this game, unless Alice's initial message to Bob contains a large amount of information about X; more precisely, unless the min-entropy of X, conditioned on E, is less than half the length of X.

Note that the problem as we formulated it does not fall in standard frameworks for communication complexity. In particular, it is a relation problem, as Bob is allowed to choose the value Y' to which his prediction b applies. This seems to prevent us from using any prior results on the communication complexity of the inner product function, and we develop an ad-hoc proof which may be of independent interest. We approach the problem using the "reconstruction paradigm"

---

[3] The correspondence between security of quantum-proof strong extractors and communication problems has been used repeatedly before, see e.g. [21,22].

(used in e.g. [15]), which amounts to showing that from any successful strategy of the players one may construct a measurement for Bob which completely "reconstructs" X, given E; if this can be achieved with high enough probability it will contradict the min-entropy assumption on X, via its dual formulation as a guessing probability [23]. We show this by running Bob's strategy "in superposition", and applying a Fourier transform to recover a guess for X. This argument is similar to one introduced in [11,32]. We refer to Sect. 4.1 for more detail.

Application to privacy amplification. Finally we discuss the application of our quantum-proof non-malleable extractor to the problem of privacy amplification against active quantum attacks, which is our original motivation. The application is based on a breakthrough result by Dodis and Wichs [19], who were first to show the existence of a two-round PA protocol with optimal (up to constant factors) entropy loss $L = \Theta(\log(1/\epsilon))$, for any initial min-entropy k. This was achieved by defining and showing the existence of non-malleable extractors with very good parameters.

The protocol from [19] is recalled in Sect. 5. The protocol proceeds as follows. Alice sends a uniformly random seed Y to Bob over the communication channel, which is controlled by Eve. Bob receives a possibly modified seed Y'. Then Alice computes a key $K = \mathrm{nmExt}(X, Y)$, and Bob computes $K' = \mathrm{nmExt}(X, Y')$. In the second round, Bob generates another uniformly random seed W', and sends W' together with $T' = \mathrm{MAC}_{K'}(W')$ to Alice, where MAC is a one-time message authentication code. Alice receives a possibly modified T, W and checks whether $T = \mathrm{MAC}_K(W)$. If yes, then the shared secret between Alice and Bob is $\mathrm{Ext}(X, W) = \mathrm{Ext}(X, W')$ with overwhelming probability, where Ext is any strong seeded extractor.

The security of this protocol intuitively follows from the following simple observation. If the adversary does not modify Y, then $K' = K$, and so W' must be equal to W by the security of the MAC. If $Y' \neq Y$, then by the non-malleability property of nmExt, K is uniform and independent of K', and so it is impossible for the adversary to predict $\mathrm{MAC}_K(W)$ for any W even given K' and W'.

Since [19] could not construct an explicit non-malleable extractor, they instead defined and constructed a so called a look-ahead extractor, which can be seen as a weakening of the non-malleability requirement of a non-malleable extractor. This was done by using the alternating extraction protocol by Dziembowski and Pietrzak [18].

In [14], Dodis and Wichs' reduction is extended to the case of quantum side information, provided that the non-malleable Extractor nmExt used in the protocol satisfies the appropriate definition of quantum non-malleability, and Ext is a strong quantum-proof extractor. Based on our construction of a quantum-proof non-malleable extractor (Theorem 1) we immediately obtain a PA protocol that is secure as long as the initial secret X has a min-entropy rate of (slightly more than) half. The result is formalized as Corollary 1 in Sect. 5.

In Sect. 5.2 we additionally prove security of a one-round protocol due to Dodis et al. [16] against active quantum attacks. The protocol has the advantage of being single-round, but it induces a significantly higher entropy loss, $(n/2) + \log(1/\epsilon)$, than the Dodis-Wichs protocol, for which the loss is independent of n.

Future work. There have been a series of works in the classical setting [3,9,12,13, 17,20,25,27–29] that have given privacy amplification protocols (via constructing non-malleable extractors or otherwise) that achieve near-optimal parameters. In particular, Li [29] constructed a non-malleable extractor that works for min-entropy $k = \Omega(\log n + \log(1/\epsilon) \log \log(1/\epsilon))$, where ϵ is the error probability.

Our quantum-proof non-malleable extractor requires the min-entropy rate of the initial weak secret to be larger than $1/2$. We leave it as an open question whether one of the above-mentioned protocols that work for min-entropy rate smaller than $1/2$ in the classical setting can be shown secure against quantum side information.

2 Preliminaries

2.1 Notation

For p a prime power we let $\mathbb{F}_p$ denote the finite field with p elements. For any positive integer n, there is a natural bijection $\phi : \mathbb{F}_p^n \mapsto \mathbb{F}_{p^n}$ that preserves group addition and scalar multiplication, i.e., the following hold:

- For all $c \in \mathbb{F}_p$, and for all $x \in \mathbb{F}_p^n$, $\phi(c \cdot x) = c \cdot \phi(x)$.
- For all $x_1, x_2 \in \mathbb{F}_p^n$, $\phi(x_1) + \phi(x_2) = \phi(x_1 + x_2)$.

We use this bijection to define the square of an element in $\mathbb{F}_p^n$, e.g. for $y \in \mathbb{F}_p^n$

$$y^2 = \phi^{-1}\left((\phi(y))^2\right) . \tag{2.1}$$

We write $\langle \cdot, \cdot \rangle$ for the inner product over $\mathbb{F}_p^n$. log denotes the logarithm with base 2.

We write $\mathcal{H}$ for an arbitrary finite-dimensional Hilbert space, $L(\mathcal{H})$ for the linear operators on $\mathcal{H}$, $Pos(\mathcal{H})$ for positive semidefinite operators, and $D(\mathcal{H}) \subset Pos(\mathcal{H})$ for positive semidefinite operators of trace 1 (*density matrices*). A linear map $T : L(\mathcal{H}) \to L(\mathcal{H}')$ is CPTP if it is completely positive, i.e. $T \otimes \mathrm{Id}(A) \geq 0$ for any $d \geq 0$ and $A \in Pos(\mathcal{H} \otimes \mathbb{C}^d)$, and trace-preserving.

We use capital letters $A, B, E, X, Y, Z, \ldots$ to denote quantum or classical random variables. Generally, the letters near the beginning of the alphabet, such as A, B, E, represent quantum variables (density matrices on a finite-dimensional Hilbert space), while the letters near the end, such as X, Y, Z represent classical variables (ranging over a finite alphabet). We sometimes represent classical random variables as density matrices diagonal in the computational basis, and write e.g. $(A, B, \ldots, E)_\rho$ for the density matrix $\rho_{A,B,\ldots,E}$. For a quantum random variable A, we denote $\mathcal{H}_A$ the Hilbert space on which the associated density matrix ρ_A is supported, and d_A its dimension. If X is classical we loosely identify its range $\{0, \ldots, d_X - 1\}$ with the space $\mathcal{H}_X$ spanned by $\{|0\rangle_X, \ldots, |d_X - 1\rangle_X\}$. We denote I_A the identity operator on $\mathcal{H}_A$. When an identity operator is tensor producted with another matrix, we sometimes omit the identity operator for brevity, e.g. writing $I_A \otimes B$ as B. When a density matrix specifies the states of two random variables, one of which is classical and the other is quantum, we call it a classical-quantum(cq)-state. A cq state $(X, E)_\rho$ takes the form

$$\rho_{XE} = \sum_x |x\rangle\langle x|_X \otimes \rho_E^x \, ,$$

where the summation is over all x in the range of X and $\{\rho_E^x\}$ are positive semidefinite matrices with $\text{Tr}\,\rho_E^x = p_x$, where p_x is the probability of getting the outcome x when measuring the X register. Similarly, a ccq state $(X, Y, E)_\sigma$ is a density matrix over two classical variables and one quantum variable, e.g. $\sigma_{XYE} = \sum_{x,y} |x\rangle\langle x|_X \otimes |y\rangle\langle y|_Y \otimes \sigma_E^{xy}$. We will sometimes add or remove random variables from an already-specified density matrix. When we omit a random variable, we mean the reduced density matrix, e.g. $(Y, E)_\sigma = \text{Tr}_X(\sigma_{XYE})$. When we introduce a classical variable, we mean that the classical variable is computed into another classical register. For example, for a function $F(\cdot, \cdot)$ on variables X, Y,

$$(F(X, Y), X, Y, E)_\sigma = \sum_{f,x,y} \delta(f, F(x,y))|f\rangle\langle f| \otimes |x\rangle\langle x| \otimes |y\rangle\langle y| \otimes \sigma_E^{xy} \, ,$$

where $\delta(\cdot, \cdot)$ is the Kronecker delta function, and the summation over f is taken over the range of F. When F is a random function, the density matrix is averaged over the appropriate probability distribution.

We use U_Σ to denote the uniform distribution over a set Σ. For m-bit string $\{0,1\}^m$, we abbreviate $U_{\{0,1\}^m}$ as U_m. For a classical random variable X, U_X denote the uniform distribution over the range of X.

For $p \geq 1$ we write $\|\cdot\|_p$ for the Schatten p-norm (this is the p-norm of the vector of singular values). We write $\|\cdot\|$ for the operator norm.

We write $\approx_\epsilon$ to denote that two density matrices are ϵ-close to each other in trace distance. For example, $(X, E)_\rho \approx_\epsilon (U_X, E)_\rho$ means $\frac{1}{2}\|\rho_{XE} - U_X \otimes \rho_E\|_1 \leq \epsilon$. Note that in case both X and E are classical random variables, this reduces to the statistical distance.

2.2   Quantum Information

The min-entropy of a classical random variable X conditioned on quantum side information E is defined as follows.

Definition 1 (Min-entropy). *Let $\rho_{XE} \in D(\mathcal{H}_X \otimes \mathcal{H}_E)$ be a cq state. The min-entropy of X conditioned on E is defined as*

$$H_{\min}(X|E)_\rho = \max\{\lambda \geq 0 : \exists \sigma_E \in \text{Pos}(\mathcal{H}_E), \text{Tr}(\sigma_E) \leq 1, \text{ s.t. } 2^{-\lambda}I_X \otimes \sigma_E \geq \rho_{XE}\}.$$

When the state ρ with respect to which the entropy is measured is clear from context we simply write $H_{\min}(X|E)$ for $H_{\min}(X|E)_\rho$.

Definition 2 $((n, k)$ -source). *A cq state ρ_{XE} is an (n, k)-source if $n = \log d_X$ and $H_{\min}(X|E))_\rho \geq k$.*

Rather than using Definition 1, we will most often rely on an operational expression for the min-entropy stated in the following lemma from [23].

Lemma 2 (Min-entropy and guessing probability). *For a cq state $\rho_{XE} \in D(\mathcal{H}_X \otimes \mathcal{H}_E)$, the guessing probability is defined as the probability to correctly guess X with the optimal strategy to measure E, i.e.*

$$p_{guess}(X|E)_\rho = \sup_{\{M_x\}} \sum_x p_x \operatorname{Tr}(M_x \rho_E^x) \,, \tag{2.2}$$

where $\{M_x\}$ is a positive operator-valued measure (POVM) on $\mathcal{H}_E$. Then the guessing probability is related to the min-entropy by

$$p_{guess}(X|E)_\rho = 2^{-H_{\min}(X|E)_\rho} \,. \tag{2.3}$$

2.3 Extractors

We first give the definition of a strong quantum-proof extractor. Recall the notation $(X, E)_\rho \approx_\epsilon (X', E')_\rho$ for $\frac{1}{2}\|\rho_{XE} - \rho_{X'E'}\|_1 \leq \epsilon$, and U_m for a random variable uniformly distributed over m-bit strings.

Definition 3. *Let k be an integer and $\epsilon \geq 0$. A function $\operatorname{Ext} : \mathcal{H}_X \times \mathcal{H}_Y \to \mathcal{H}_Z$ is a strong (k, ϵ) quantum-proof extractor if for all cq states $\rho_{XE} \in D(\mathcal{H}_X \otimes \mathcal{H}_E)$ with $H_{\min}(X|E) \geq k$, and for a classical uniform $Y \in \mathcal{H}_Y$ independent of ρ_{XE},*

$$(\operatorname{Ext}(X, Y), Y, E)_\rho \approx_\epsilon (U_Z, Y, E)_\rho \,.$$

There are known explicit constructions of strong quantum-proof extractors.

Theorem 2 ([36]). *For any integers d_X, k and for any $\epsilon > 0$ there exists an explicit strong (k, ϵ) quantum-proof extractor $\operatorname{Ext} : \{0, \ldots, d_X - 1\} \times \{0, \ldots, d_Y - 1\} \to \{0, \ldots, d_Z - 1\}$ with $\log d_Y = O(\log d_X)$ and $\log d_z = k - O(\log(1/\epsilon)) - O(1)$.*

We use the same definition of non-malleable extractor against quantum side information that was introduced in the work [14]. The definition is a direct generalization of the classical notion of non-malleable extractor introduced in [19]. The first step is to extend the notion that the adversary may query the extractor on any *different* seed Y' than the seed Y actually used to the case where Y' may be generated from Y as well as quantum side information held by the adversary.

Definition 4 (Map with no fixed points). *Let $\mathcal{H}_Y$, $\mathcal{H}_E$ and $\mathcal{H}_{E'}$ be finite-dimensional Hilbert spaces. We say that a CPTP map $T : L(\mathcal{H}_Y \otimes \mathcal{H}_E) \to L(\mathcal{H}_Y \otimes \mathcal{H}_{E'})$ has no fixed points if for all $\rho_E \in D(\mathcal{H}_E)$ and all computational basis states $|y\rangle \in \mathcal{H}_Y$ it holds that*

$$\langle y|_Y \operatorname{Tr}_{\mathcal{H}_{E'}}\left(T\left(|y\rangle\langle y|_Y \otimes \rho_E\right)\right)|y\rangle_Y = 0 \,.$$

The following definition is given in [14]:

Definition 5 (Non-malleable extractor). *Let $\mathcal{H}_X$, $\mathcal{H}_Y$, $\mathcal{H}_Z$ be finite-dimensional Hilbert spaces, of respective dimension d_X, d_Y, and d_Z. Let $k \leq \log d_X$ and $\epsilon > 0$. A function*

$$\operatorname{nmExt} : \{0, \ldots, d_X - 1\} \times \{0, \ldots, d_Y - 1\} \to \{0, \ldots, d_Z - 1\}$$

is a (k, ϵ) quantum-proof non-malleable extractor if for every cq-state $(X, E)_\rho$ on $\mathcal{H}_X \otimes \mathcal{H}_E$ such that $H_{\min}(X|E)_\rho \geq k$ and any CPTP map $\mathrm{Adv} : \mathrm{L}(\mathcal{H}_Y \otimes \mathcal{H}_E) \rightarrow \mathrm{L}(\mathcal{H}_Y \otimes \mathcal{H}_{E'})$ with no fixed points,

$$\left\| \sigma_{\mathrm{nmExt}(X,Y)\mathrm{nmExt}(X,Y')YY'E'} - U_Z \otimes \sigma_{\mathrm{nmExt}(X,Y')YY'E'} \right\|_1 \leq \epsilon \,,$$

where

$$\sigma_{YY'XE'} = \frac{1}{d_Y} \sum_y |y\rangle\langle y|_Y \otimes (I_X \otimes \mathrm{Adv})(|y\rangle\langle y|_Y \otimes \rho_{XE}) \qquad (2.4)$$

and $\sigma_{\mathrm{nmExt}(X,Y)\mathrm{nmExt}(X,Y')YY'E'}$ is obtained from $\sigma_{YY'XE'}$ by (classically) computing $\mathrm{nmExt}(X, Y)$ and $\mathrm{nmExt}(X, Y')$ in ancilla registers and tracing out X.

2.4   Hölder's Inequality

We use the following Hölder's inequality for matrices. For a proof, see e.g. [6].

Lemma 3 (Hölder's inequality). *For any $n \times n$ matrices A, B, C with complex entries, and real numbers $r, s, t > 0$ satisfying $\frac{1}{r} + \frac{1}{s} + \frac{1}{t} = 1$,*

$$\|ABC\|_1 \leq \||A|^r\|_1^{1/r} \||B|^s\|_1^{1/s} \||C|^t\|_1^{1/t} \,. \qquad (2.5)$$

3   Quantum XOR Lemma

In this section we prove two XOR lemmas with quantum side information. We prove a non-uniform version, Lemma 1, in Sect. 3.1. In the full version of the paper [1], we also prove a more standard XOR lemma with quantum side information.[4] Since XOR lemmas often play a fundamental role, they might be of independent interest. Our proofs are based on quantum collision probability techniques[5] from [36] to transform a classical collision probability-based proof into one that also allows for quantum side information. The idea of non-uniform XOR lemma is natural in the context of non-malleable extractors, and has been explored in [3,13,27]. Our non-uniform XOR lemma generalizes a restricted version of Lemma 3.15 of [27] to $\mathbb{F}_p$ with quantum side information.[6]

The quantum collision probability is defined as follows.

Definition 6 (Quantum collision probability). *Let $\rho_{AB} \in \mathrm{D}(\mathcal{H}_A \otimes \mathcal{H}_B)$ and $\sigma_B \in \mathrm{D}(\mathcal{H}_B)$. The collision probability of ρ_{AB}, conditioned on σ_B, is defined as*

$$\Gamma_c(\rho_{AB}|\sigma_B) \equiv \mathrm{Tr}\left(\rho_{AB}(I_A \otimes \sigma_B^{-1/2}) \right)^2 \,, \qquad (3.1)$$

where $\sigma_B \in \mathrm{D}(\mathcal{H}_B)$.

---

[4] When restricted to $\mathbb{F}_2$, our standard XOR lemma is very similar to Lemma 10 of [22], although the result from [22] provides a tighter bound in this case. [22] provides a bound of $p^{2t}\epsilon^2$, while ours scales as $p^t\epsilon$, a quadratic loss. However our result applies to $\mathbb{F}_p$, while it is unclear whether the proof of [22] generalizes to $p > 2$. [22] obtains the result by Fourier analysis.

[5] The term "quantum collision probability" is ours.

[6] Compared to [27, Lemma 3.15], we have $m = 1$ and $n = t$.

A careful reader might notice that $\Gamma_c \leq 1$ is not generally true, so calling Γ_c collision *probability* seems misleading. We give a general definition which allows arbitrary states ρ_{AB} and σ_B to match the existing literature, but here we always consider cq states ρ_{AB} and take $\sigma_B = \rho_B$. We prove in the full version [1] that $\Gamma_c \leq 1$ in such cases. $\Gamma_c(\rho_{AB}|\sigma_B)$ also reduces to the classical collision probability when both of A, B are classical and $\sigma_B = \rho_B$.

We will often use the following relation, also taken from [36], valid for any $\rho_{AB} \in D(\mathcal{H}_A \otimes \mathcal{H}_B)$:

$$\mathrm{Tr}\left((\rho_{AB} - U_A \otimes \rho_B)(I_A \otimes \rho_B^{-1/2})\right)^2 = \Gamma_c(\rho_{AB}|\rho_B) - \frac{1}{d_A}, \qquad (3.2)$$

which can be verified by expanding the square:

$$\mathrm{Tr}\left((\rho_{AB} - U_A \otimes \rho_B)(I_A \otimes \rho_B^{-1/2})\right)^2$$

$$= \mathrm{Tr}\left(\rho_{AB}\,\rho_B^{-1/2}\right)^2 - 2\,\mathrm{Tr}\left(\rho_{AB}\,\rho_B^{-1/2}(U_A\rho_B)\rho_B^{-1/2}\right) + \mathrm{Tr}\left((U_A\rho_B)\rho_B^{-1/2}\right)^2$$

$$= \Gamma_c(\rho_{AB}|\rho_B) - \frac{1}{d_A}.$$

3.1 Non-uniform XOR Lemma

Our non-uniform XOR lemma bounds the distance to uniform of a ccq state, a state with two classical registers and one quantum register. Roughly speaking, the lemma states that given two random variables $X_0 \in \mathbb{F}_p$ and $X \in \mathbb{F}_p^t$, if $X_0 + \langle a, X \rangle$ is close to uniform, then X_0 is close to uniform given X.

Lemma 1 (restated). *Let p be a prime power, t an integer and $\epsilon \geq 0$. Let ρ_{X_0XE} be a ccq state with $X_0 \in \mathbb{F}_p$ and $X = (X_1, \ldots, X_t) \in \mathbb{F}_p^t$. For all $a = (a_1, \ldots, a_t) \in \mathbb{F}_p^t$, define a random variable $Z = X_0 + \langle a, X \rangle = X_0 + \sum_{i=1}^t a_iX_i$. If for all a, $\frac{1}{2}\|\rho_{ZE}^a - U_Z \otimes \rho_E\|_1 \leq \epsilon$, then*

$$\frac{1}{2}\|\rho_{X_0XE} - U_{X_0} \otimes \rho_{XE}\|_1 \leq \frac{p^{(t+1)/2}}{\sqrt{2}}\sqrt{\epsilon}. \qquad (3.3)$$

The proof of the non-uniform XOR lemma has the following structure: we bound the collision probability by the trace distance in Lemma 5, then prove the non-uniform XOR lemma based on that. First we establish that for any ccq state ρ_{XZE}:

$$\mathrm{Tr}\left((\rho_{XZE} - U_X \otimes \rho_{ZE})(I_{XZ} \otimes \rho_E^{-1/2})\right)^2$$

$$= \mathrm{Tr}\left(\rho_{XZE}\,\rho_E^{-1/2}\right)^2 - 2\,\mathrm{Tr}\left(\rho_{XZE}\,\rho_E^{-1/2}(U_X\rho_{ZE})\rho_E^{-1/2}\right) + \mathrm{Tr}\left((U_X\rho_{ZE})\rho_E^{-1/2}\right)^2$$

$$= \Gamma_c(\rho_{XZE}|\rho_E) - \frac{1}{d_X}\Gamma_c(\rho_{ZE}|\rho_E). \qquad (3.4)$$

We need the following lemma to bound the collision probability by the trace distance in Lemma 5.

Lemma 4. *Let ρ_{XZE} be a ccq state. Then*

$$-\frac{1}{d_X}I_{XZE} \leq \left(I_{XZ} \otimes \rho_E^{-\frac{1}{2}}\right)(\rho_{XZE} - U_X \otimes \rho_{ZE})\left(I_{XZ} \otimes \rho_E^{-\frac{1}{2}}\right) \leq \left(1 - \frac{1}{d_X}\right)I_{XZE}.$$

(3.5)

Proof. We bound the eigenvalues of the middle expression. Since ρ_{XZE} is a ccq state, we know that the middle expression

$$\left(I_{XZ} \otimes \rho_E^{-1/2}\right)(\rho_{XZE} - U_X \otimes \rho_{ZE})\left(I_{XZ} \otimes \rho_E^{-1/2}\right)$$
$$= \sum_{x,z} |x\rangle\langle x| \otimes |z\rangle\langle z| \otimes \rho_E^{-1/2}\left(\rho_E^{xz} - \frac{1}{d_X}\rho_E^z\right)\rho_E^{-1/2}$$

(3.6)

is block diagonal, where $\rho_E^z = \sum_x \rho_E^{xz}$ and $\rho_E = \sum_{x,z}\rho_E^{xz}$. For any state $|\phi\rangle \in \mathcal{H}_E$ and x, z in the range of X, Z,

$$\langle\phi|\rho_E^{-1/2}\left(\rho_E^{xz} - \frac{1}{d_X}\rho_E^z\right)\rho_E^{-1/2}|\phi\rangle \geq \langle\phi|\rho_E^{-1/2}\left(-\frac{1}{d_X}\rho_E^z\right)\rho_E^{-1/2}|\phi\rangle \geq -\frac{1}{d_X}.$$

(3.7)

This proves the first inequality. We also have

$$\langle\phi|\rho_E^{-1/2}\left(\rho_E^{xz} - \frac{1}{d_X}\rho_E^z\right)\rho_E^{-1/2}|\phi\rangle$$
$$= \langle\phi|\rho_E^{-1/2}\left(\rho_E^{xz} - \frac{1}{d_X}\sum_{x'}\rho_E^{x'z}\right)\rho_E^{-1/2}|\phi\rangle$$
$$= \left(1 - \frac{1}{d_X}\right)\langle\phi|\rho_E^{-1/2}\rho_E^{xz}\rho_E^{-1/2}|\phi\rangle - \frac{1}{d_X}\sum_{x'\neq x}\langle\phi|\rho_E^{-1/2}\rho_E^{xz}\rho_E^{-1/2}|\phi\rangle$$
$$\leq \left(1 - \frac{1}{d_X}\right).$$

(3.8)

This proves the second inequality.

We then bound the collision probability by the trace distance.

Lemma 5 (Bounding collision probability with trace distance, non-uniform). *Let ρ_{XZE} be a ccq state. If*

$$\frac{1}{2}\|\rho_{XZE} - U_X\rho_{ZE}\|_1 = \epsilon,$$

(3.9)

then

$$\frac{4\epsilon^2}{d_Xd_Z} \leq \Gamma_c(\rho_{XZE}|\rho_E) - \frac{1}{d_X}\Gamma_c(\rho_{ZE}|\rho_E) \leq 2\epsilon\left(1 - \frac{1}{d_X}\right).$$

(3.10)

Proof. For the first inequality, we use Hölder's inequality (Lemma 3) with $r = t = 4$, $s = 2$, $A = C = I_{XZ} \otimes \rho_E^{1/4}$, and $B = \left(I_{XZ} \otimes \rho_E^{-1/4}\right)(\rho_{XZE} - U_X \rho_{ZE}) \left(I_{XZ} \otimes \rho_E^{-1/4}\right)$. This leads to

$$
\begin{aligned}
2\epsilon &= \|\rho_{XZE} - U_X \rho_{ZE}\|_1 \\
&= \|ABC\|_1 \\
&\leq \|A^4\|_1^{1/4} \|B^2\|_1^{1/2} \|C^4\|_1^{1/4} \\
&= \sqrt{d_X d_Z \operatorname{Tr}\left((\rho_{XZE} - U_X \otimes \rho_{ZE})\left(I_{XZ} \otimes \rho_E^{-1/2}\right)\right)^2} \\
&= \sqrt{d_X d_Z \left(\Gamma_c(\rho_{XZE}|\rho_E) - \frac{1}{d_X}\Gamma_c(\rho_{ZE}|\rho_E)\right)},
\end{aligned}
\tag{3.11}
$$

where we used Eq. (3.4) in the last line. Squaring both sides and dividing by $d_X d_Z$, we get the desired inequality. For the second inequality, we use Lemma 4 to show that

$$
-\frac{1}{d_X} I_{XZE} \leq \left(I_{XZ} \otimes \rho_E^{-\frac{1}{2}}\right)(\rho_{XZE} - U_X \otimes \rho_{ZE})\left(I_{XZ} \otimes \rho_E^{-\frac{1}{2}}\right) \leq \left(1 - \frac{1}{d_X}\right) I_{XZE}
$$

$$
\Rightarrow \left|\left(I_{XZ} \otimes \rho_E^{-1/2}\right)(\rho_{XZE} - U_X \otimes \rho_{ZE})\left(I_{XZ} \otimes \rho_E^{-1/2}\right)\right| \leq \left(1 - \frac{1}{d_X}\right) I_{XZE}.
\tag{3.12}
$$

Starting with Eq. (3.4), we have

$$
\begin{aligned}
&\Gamma_c(\rho_{XZE}|\rho_E) - \frac{1}{d_X}\Gamma_c(\rho_{ZE}|\rho_E) \\
&= \operatorname{Tr}\left((\rho_{XZE} - U_X \otimes \rho_{ZE})\left(I_{XZ} \otimes \rho_E^{-1/2}\right)\right)^2 \\
&\leq \operatorname{Tr}\left(|\rho_{XZE} - U_X \rho_{ZE}|\left|\left(I_{XZ} \otimes \rho_E^{-1/2}\right)(\rho_{XZE} - U_X \otimes \rho_{ZE})\left(I_{XZ} \otimes \rho_E^{-1/2}\right)\right|\right) \\
&\leq \operatorname{Tr}\left(|\rho_{XZE} - U_X \rho_{ZE}|\left(1 - \frac{1}{d_X}\right) I_{XZE}\right) \\
&= 2\epsilon\left(1 - \frac{1}{d_X}\right),
\end{aligned}
\tag{3.13}
$$

where we used Eq. (3.12) on the fourth line. □

Now we restate and prove the non-uniform XOR lemma. The proof idea is to start from the trace distance of X_0 given X to uniform, apply Lemma 5 to get an upper bound in terms of the collision probability of X_0 given X, apply Eq. (3.4) and expand the square to express the collision probability of X_0 given X in terms of the collision probability of $X_0 + \langle a, X \rangle$, and finally apply Lemma 5 again to get an upper bound in terms of the trace distance of $X_0 + \langle a, X \rangle$ to uniform.

Lemma 1 (restated). *Let p be a prime power, t an integer and $\epsilon \geq 0$. Let $\rho_{X_0 XE}$ be a ccq state with $X_0 \in \mathbb{F}_p$ and $X = (X_1, \ldots, X_t) \in \mathbb{F}_p^t$. For all $a = (a_1, \ldots, a_t) \in \mathbb{F}_p^t$,*

define a random variable $Z = X_0 + \langle a, X \rangle = X_0 + \sum_{i=1}^{t} a_i X_i$. *If for all* a,
$\frac{1}{2} \left\| \rho_{ZE}^a - U_Z \otimes \rho_E \right\|_1 \leq \epsilon$, *then*

$$\frac{1}{2} \left\| \rho_{X_0 X E} - U_{X_0} \otimes \rho_{XE} \right\|_1 \leq \frac{p^{(t+1)/2}}{\sqrt{2}} \sqrt{\epsilon} . \tag{3.14}$$

Proof. We start by relating the collision probability of Z and $X_0 + \langle a, X \rangle$:

$$\Gamma_c(\rho_{ZE}^a | \rho_E) - \frac{1}{p}$$

$$= \mathrm{Tr} \left[(\rho_{ZE}^a - U_Z \rho_E) I_Z \otimes \rho_E^{-1/2} \right]^2$$

$$= \mathrm{Tr} \left[\sum_z |z\rangle\langle z| \sum_{x,x_0} \left(\delta\left(z - x_0 - \langle a, x \rangle, 0\right) - \frac{1}{p} \right) \rho_E^{x_0 x} I_Z \rho_E^{-1/2} \right]^2$$

$$= \sum_z \mathrm{Tr} \left[\sum_{x_0 x} \left(\delta\left(z - x_0 - \langle a, x \rangle, 0\right) - \frac{1}{p} \right) \rho_E^{x_0 x} \rho_E^{-1/2} \right]^2$$

$$= \sum_{z,x_0,x_0',x,x'} \left[\delta\left(z - x_0 - \langle a, x \rangle, 0\right) \delta\left(z - x_0' - \langle a, x' \rangle, 0\right) \right.$$

$$\left. - \frac{2}{p} \delta\left(z - x_0 - \langle a, x \rangle, 0\right) + \frac{1}{p^2} \right] \mathrm{Tr} \left(\rho_E^{x_0 x} \rho_E^{-1/2} \rho_E^{x_0' x'} \rho_E^{-1/2} \right)$$

$$= \sum_{x_0,x_0',x,x'} \left[\delta\left(x_0 - x_0' + \langle a, x - x' \rangle, 0\right) - \frac{1}{p} \right] \mathrm{Tr} \left(\rho_E^{x_0 x} \rho_E^{-1/2} \rho_E^{x_0' x'} \rho_E^{-1/2} \right)$$

$$= \sum_{x_0,x_0',x} \left(\delta\left(x_0 - x_0', 0\right) - \frac{1}{p} \right) \mathrm{Tr} \left(\rho_E^{x_0 x} \rho_E^{-1/2} \rho_E^{x_0' x} \rho_E^{-1/2} \right)$$

$$+ \sum_{x_0,x_0',x \neq x'} \left[\delta\left(x_0 - x_0' + \langle a, x - x' \rangle, 0\right) - \frac{1}{p} \right] \mathrm{Tr} \left(\rho_E^{x_0 x} \rho_E^{-1/2} \rho_E^{x_0' x'} \rho_E^{-1/2} \right)$$

$$= \sum_{x_0,x} \mathrm{Tr} \left(\rho_E^{x_0 x} \rho_E^{-1/2} \rho_E^{x_0 x} \rho_E^{-1/2} \right) - \frac{1}{p} \sum_{x_0,x_0',x} \mathrm{Tr} \left(\rho_E^{x_0 x} \rho_E^{-1/2} \rho_E^{x_0' x} \rho_E^{-1/2} \right)$$

$$+ \sum_{x_0,x_0',x \neq x'} \left[\delta\left(x_0 - x_0' + \langle a, x - x' \rangle, 0\right) - \frac{1}{p} \right] \mathrm{Tr} \left(\rho_E^{x_0 x} \rho_E^{-1/2} \rho_E^{x_0' x'} \rho_E^{-1/2} \right)$$

$$= \Gamma_c(\rho_{X_0 X E} | \rho_E) - \frac{1}{p} \Gamma_c(\rho_{XE} | \rho_E)$$

$$+ \sum_{x_0,x_0',x \neq x'} \left[\delta\left(x_0 - x_0' + \langle a, x - x' \rangle, 0\right) - \frac{1}{p} \right] \mathrm{Tr} \left(\rho_E^{x_0 x} \rho_E^{-1/2} \rho_E^{x_0' x'} \rho_E^{-1/2} \right) .$$

$$\tag{3.15}$$

When we average over a, the last term vanishes,

$$\mathrm{E}_a \left(\Gamma_c(\rho_{ZE}^a | \rho_E) - \frac{1}{p} \right) = \Gamma_c(\rho_{X_0 X E} | \rho_E) - \frac{1}{p} \Gamma_c(\rho_{XE} | \rho_E) . \tag{3.16}$$

With the heavy work done, we put everything together and prove the lemma

$$\frac{\|\rho_{X_0 XE} - U_{X_0}\rho_{XE}\|_1^2}{p^{t+1}} \leq \Gamma_c(\rho_{X_0 XE}|\rho_E) - \frac{1}{p}\Gamma_c(\rho_{XE}|\rho_E)$$

$$= E_a\left(\Gamma_c(\rho_{ZE}^a|\rho_E) - \frac{1}{p}\right)$$

$$\leq 2\epsilon, \tag{3.17}$$

where we used Lemma 5 one the first line, Eq. (3.16) on the second line, Lemma 5 and the assumption of the lemma on the third line. Multiplying both sides by $\frac{p^{t+1}}{2}$ and take a square root, we get the desired result:

$$\frac{1}{2}\|\rho_{X_0 XE} - U_{X_0}\rho_{XE}\|_1 \leq \frac{p^{(t+1)/2}}{\sqrt{2}}\sqrt{\epsilon}. \tag{3.18}$$

$\square$

4   Quantum-Proof Non-malleable Extractor

In this section we introduce our non-malleable extractor and prove its security. The extractor was first considered by Li [27]. We use the symbol $\|$ for concatenation of strings, and for $a, b \in \mathbb{F}_p^n$ write $\langle a, b \rangle$ for the standard inner product over $\mathbb{F}_p^n$.

Definition 7 (Inner product-based non-malleable extractor). *Let $p \neq 2$ be a prime. For any even integer n, define a function $\text{nmExt} : \mathbb{F}_p^n \times \mathbb{F}_p^{n/2} \to \mathbb{F}_p$ by $\text{nmExt}(X, Y) = \langle X, Y\|Y^2 \rangle$, where Y^2 is defined as in Sect. 2.1.*

Theorem 1. *Let $p \neq 2$ be a prime. Let n be an even integer. Then for any $\epsilon > 0$ the function $\text{nmExt}(X, Y) = \langle X, Y\|Y^2 \rangle$ is an $\left(\left(\frac{n}{2} + 6\right)\log p - 1 + 4\log\frac{1}{\epsilon}, \epsilon\right)$ quantum-proof non-malleable extractor.*

The proof of Theorem 1 is based on a reduction showing that any successful attack for an adversary to nmExt leads to a good strategy for the players in a certain communication game, that we introduce next.

4.1   A Communication Game

Let $p \neq 2$ be a prime. Let n be an even integer, and $g : \mathbb{F}_p^{n/2} \times \mathbb{F}_p^{n/2} \to \mathbb{F}_p^n$ an arbitrary function such that for any $z \in \mathbb{F}_p^n$ there are at most two possible pairs (y, y') such that $y \neq y'$ and $g(y, y') = z$. Consider the following communication game, called $\text{GUESS}(n, p, g)$, between two players Alice and Bob.

1. Bob receives $y \in \mathbb{F}_p^{n/2}$ from the referee.
2. Alice creates a cq state ρ_{XE}, where $X \in \mathbb{F}_p^n$, and sends the quantum register E to Bob.
3. Bob returns $y' \in \mathbb{F}_p^{n/2}$ and $b \in \mathbb{F}_p$.

The players win if and only if $b = \langle x, g(y, y') \rangle$ and $y' \neq y$.

Note that Alice does not receive anything from the referee and is completely free in what state she wants to create, so it is easy for the players to win with probability 1 by creating a trivial state, e.g. $\rho_{XE} = |0\rangle\langle 0| \otimes |0\rangle\langle 0|$. Therefore we benchmark the success probability of a strategy by the min-entropy of Alice's "input" X, conditioned on her message E to Bob. The following lemma bounds the players' maximum success probability in this game over uniformly random input y and quantum measurements as a function of the min-entropy of Alice's input X, conditioned on her message E to Bob.

Lemma 6 (Success probability of the communication game). *Suppose there exists a communication protocol for Alice and Bob in* GUESS(n, p, g) *that succeeds with probability at least $\frac{1}{p} + \epsilon$, on average over a uniformly random choice of input y to Bob. Then $H_{\min}(X|E)_\rho \leq \frac{n}{2}\log p + 1 + 2\log\frac{1}{\epsilon}$.*

Proof. Let $\rho_{XE} = \sum_x |x\rangle\langle x|_X \otimes \rho_E^x$ be the cq state prepared by Alice. A strategy for Bob is a family of POVM $\{M_y^{y',b}\}_{y',b}$, indexed by $y \in \mathbb{F}_p^{n/2}$ and with outcomes $(y', b) \in \mathbb{F}_p^{n/2} \times \mathbb{F}_p$. We can assume that $\{M_y^{y',b}\}_{y',b}$ is projective, since Alice can send ancilla qubits along with ρ and allow Bob to apply Naimark's theorem to his POVM in order to obtain a projective measurement; this will change neither his success probability nor the min-entropy of Alice's state. By definition, the players' success probability in GUESS(n, p, g) is

$$\frac{1}{p} + \epsilon = \sum_x p^{-\frac{n}{2}} \sum_y \sum_{y'} \sum_b \delta(b, \langle x, g(y, y')\rangle) \operatorname{Tr}\left(M_y^{y',b} \rho_E^x\right). \tag{4.1}$$

For each $u \in \mathbb{F}_p$ let $A_{y,u}^{y'} = \sum_b \omega^{ub} M_y^{y',b}$, where $\omega = e^{\frac{2i\pi}{p}}$. By inversion, $M_y^{y',b} = \frac{1}{p}\sum_u \omega^{-ub} A_{y,u}^{y'}$. Replacing this into (4.1) we obtain

$$\frac{1}{p} + \epsilon = \frac{1}{p} \sum_u p^{-\frac{n}{2}} \sum_y \sum_{y'} \sum_b \delta(b, \langle x, g(y, y')\rangle) \omega^{-ub} \operatorname{Tr}\left(A_{y,u}^{y'} \rho_E^x\right)$$

$$\leq \frac{1}{p} + \left(1 - \frac{1}{p}\right) \max_{u \neq 0} \left| p^{-\frac{n}{2}} \sum_y \sum_{y'} \sum_b \delta(b, \langle x, g(y, y')\rangle) \omega^{-ub} \operatorname{Tr}\left(A_{y,u}^{y'} \rho_E^x\right)\right|, \tag{4.2}$$

where for the second line we used that $\sum_{y'} A_{y,0}^{y'} = \sum_{y',b} M_y^{y',b} = I_E$.

Fix $u \neq 0$ that achieves the maximum in (4.2). For fixed y, define the map $T_{y,u}$ on $\mathcal{H}_E$ by

$$T_{y,u} : |\psi\rangle \mapsto \sum_{y'} |y'\rangle A_{y,u}^{y'}|\psi\rangle . \tag{4.3}$$

$T_{y,u}$ has norm at most 1, since

$$T_{y,u}^\dagger T_{y,u} = \sum_{y'} (A_{y,u}^{y'})^\dagger A_{y,u}^{y'} = \sum_{y'} \sum_b \left(M_y^{y',b}\right)^2 = I_E .$$

For the second equality we used that $\{M_y^{y',b}\}_{y',b}$ is projective. Therefore $T_{y,u}$ is a physical operation.

Consider the following guessing strategy for an adversary holding side information ρ_E^x about x. The adversary first prepares a uniform superposition over y. Conditioned on y, it applies the map $T_{y,u}$. It computes $g(y,y')$ in an ancilla register, and erases (y,y'), except for one bit of information $r(y,y') \in \{0,1\}$, which specifies which pre-image (y,y') is, given $g(y,y')$ (this is possible by the 2-to-1 assumption on g). The adversary applies a Fourier transform on the register containing $g(y,y')$, using $\omega_u = \omega^{-u}$ as primitive p-th root of unity (this is possible since $u \neq 0$ and p is prime). It measures the result and outputs it as a guess for x. Formally, the transformation this implements is

$$|\psi\rangle \mapsto p^{-\frac{n}{4}} \sum_y |y\rangle \sum_{y'} |y'\rangle A_{y,u}^{y'} |\psi\rangle$$

$$\mapsto p^{-\frac{n}{4}} \sum_{y,y'} |g(y,y')\rangle |r(y,y')\rangle A_{y,u}^{y'} |\psi\rangle$$

$$\mapsto \sum_v |v\rangle \left(p^{-\frac{3n}{4}} \sum_{y,y'} \omega_u^{\langle v,g(y,y')\rangle} |r(y,y')\rangle A_{y,u}^{y'} \right) |\psi\rangle .$$

The adversary's success probability in guessing $v = x$ on input ρ_E^x is therefore

$$p_s = \sum_x \mathrm{Tr}\left(\left(p^{-\frac{3n}{4}} \sum_{y,y'} \omega_u^{\langle x,g(y,y')\rangle} |r(y,y')\rangle \otimes A_{y,u}^{y'} \right) \rho_E^x \right.$$

$$\left. \cdot \left(p^{-\frac{3n}{4}} \sum_{y,y'} \omega_u^{-\langle x,g(y,y')\rangle} \langle r(y,y')| \otimes (A_{y,u}^{y'})^\dagger \right) \right)$$

$$= \frac{1}{p^{\frac{3n}{2}}} \sum_x \sum_{r \in \{0,1\}} \mathrm{Tr}\left(\left(\sum_{y,y':r(y,y')=r} \omega_u^{\langle x,g(y,y')\rangle} A_{y,u}^{y'} \right)^\dagger \right.$$

$$\left. \cdot \left(\sum_{y,y':r(y,y')=r} \omega_u^{\langle x,g(y,y')\rangle} A_{y,u}^{y'} \right) \rho_E^x \right)$$

$$\geq \frac{1}{p^{\frac{3n}{2}}} \sum_x \frac{1}{2} \mathrm{Tr}\left(\left(\sum_{y,y'} \omega_u^{\langle x,g(y,y')\rangle} A_{y,u}^{y'} \right)^\dagger \left(\sum_{y,y'} \omega_u^{\langle x,g(y,y')\rangle} A_{y,u}^{y'} \right) \rho_E^x \right) , \quad (4.4)$$

where for the last line we used $\mathrm{Tr}(A^\dagger A \rho) + \mathrm{Tr}(B^\dagger B \rho) \geq \frac{1}{2}\mathrm{Tr}((A+B)^\dagger (A+B)\rho)$ if ρ is positive semidefinite. Now, recall from (4.2) and our choice of u that

$$\epsilon \leq p^{-\frac{n}{2}} \left| \sum_{x,y,y'} \omega^{-u(\langle x,g(y,y')\rangle)} \mathrm{Tr}\left(A_{y,u}^{y'} \rho_E^x \right) \right|$$

$$\leq p^{-\frac{n}{2}} \left(\sum_x \mathrm{Tr}(\rho_E^x) \right)^{1/2}$$

$$\cdot \left(\sum_x \mathrm{Tr}\left(\left(\sum_{y,y'} \omega^{-u(\langle x,g(y,y')\rangle)} A_{y,u}^{y'} \right) \rho_E^x \left(\sum_{y,y'} \omega^{-u(\langle x,g(y,y')\rangle)} A_{y,u}^{y'} \right)^\dagger \right) \right)^{1/2} ,$$

$$(4.5)$$

where the inequality is Cauchy-Schwarz. Comparing (4.4) and (4.5) gives

$$p_s \geq \frac{1}{2} p^{-\frac{n}{2}} \epsilon^2 .$$

We conclude using that by Lemma 2, $H_{\min}(X|E) \leq -\log p_s$.   □

4.2   Proof of Theorem 1

In this section we give the proof of Theorem 1. Towards this we first prove a preliminary lemma showing that a certain function, based on the definition of nmExt, has few collisions.

Lemma 7. *Let $p \neq 2$ be a prime and n an even integer. For $a \in \mathbb{F}_p$ define a function $g_a : \mathbb{F}_p^{n/2} \times \mathbb{F}_p^{n/2} \to \mathbb{F}_p^n$ by*

$$g_a(y, y') = y + ay' \| y^2 + ay'^2 , \tag{4.6}$$

where y^2 is defined in Sect. 2.1. Then for any $a \in \mathbb{F}_p, a \neq 0$ and $z \in \mathbb{F}_p^n$ there are at most 2 distinct pairs (y, y') such that $y' \neq y$ and $g_a(y, y') = z$.

Proof. We use the bijection defined in Sect. 2.1 to interpret y and y' in $\mathbb{F}_{p^{n/2}}$. For $a \neq 0$, we fix an image $g_a = (c, d)$, where c, d are interpreted as elements of $\mathbb{F}_{p^{n/2}}$, and solve for (y, y') in $\mathbb{F}_{p^{n/2}} \times \mathbb{F}_{p^{n/2}}$ satisfying

$$y + ay' = c , \tag{4.7}$$

$$y^2 + ay'^2 = d . \tag{4.8}$$

Using (4.7) to eliminate y we get

$$(c - ay')^2 + ay'^2 = d$$
$$\Rightarrow (a + a^2)y'^2 + (-2ca)y' + (c^2 - d) = 0 . \tag{4.9}$$

Since (4.9) is a quadratic equation, there are at most two solutions unless all coefficients are zero. Since $p \neq 2$, $-2 \neq 0$. If all coefficients are zero, $-2 \neq 0$, and $a \neq 0$, then $c = d = 0, a = -1$, which implies $y' = y$ by (4.7) and contradicts our assumption. So there are at most two different y' that can be mapped to (c, d). By (4.7) each y' corresponds to a unique y, so there are at most two pre-images.   □

We are ready to give the proof of Theorem 1. The proof depends on a simple lemma relating trace distance and guessing measurements, Lemma 8, which is stated and proved after the proof of the theorem.

Proof of Theorem 1. Let $k = \left(\frac{n}{2} + 6\right) \log p - 1 + 4 \log \frac{1}{\epsilon}$ and $\rho_{XE} \in D(\mathbb{C}^{p^n} \otimes \mathcal{H}_E)$ an $(n \log p, k)$-source. Fix a CPTP map $\mathrm{Adv} : L(\mathbb{C}^{p^{n/2}} \otimes \mathcal{H}_E) \to L(\mathbb{C}^{p^{n/2}} \otimes \mathcal{H}_{E'})$

with no fixed points, and define $\sigma_{\text{nmExt}(X,Y)\text{nmExt}(X,Y')YY'E'}$ as in Definition 5. Given the definition of nmExt, to prove the theorem we need to show that

$$(\langle X, Y\|Y^2\rangle, \langle X, Y'\|Y'^2\rangle, Y', Y, E')_\sigma \approx_\epsilon (U_{\mathbb{F}_p}, \langle X, Y'\|Y'^2\rangle, Y', Y, E')_\sigma . \quad (4.10)$$

Applying the XOR lemma, Lemma 1, with $X_0 = \langle X, Y\|Y^2\rangle$, $X = \langle X, Y'\|Y'^2\rangle$, $E = (Y', Y, E')$ and $t = 1$, (4.10) will follow once it is shown that

$$(\langle X, Y\|Y^2\rangle + a\langle X, Y'\|Y'^2\rangle, Y', Y, E')_\sigma \approx_{\frac{2\epsilon^2}{p^2}} (U_{\mathbb{F}_p}, Y', Y, E')_\sigma , \quad (4.11)$$

for all $a \in \mathbb{F}_p$. For $a = 0$, (4.11) follows from the fact that inner product is a quantum-proof two source extractor, which can be shown by the combination of Theorem 5.3 of [10] and Lemma 1 in [24]. For non-zero $a \in \mathbb{F}_p$, recall the function $g_a : \mathbb{F}_p^{n/2} \times \mathbb{F}_p^{n/2} \to \mathbb{F}_p^n$ defined in (4.6). Lemma 7 shows that for any $a \neq 0$, the restriction of g_a to $\{(y, y') : y \neq y'\}$ is at most 2-to-1, and $y \neq y'$ is ensured by the fact that Adv has no fixed points. We establish (4.11) by contradiction. Assume thus that

$$(\langle X, g_a(Y, Y')\rangle, Y', Y, E')_\sigma \approx_{\frac{2\epsilon^2}{p^2}} (U_{\mathbb{F}_p}, Y', Y, E')_\sigma \quad (4.12)$$

does not hold, for some non-zero $a \in \mathbb{F}_p$. Fix such an a and write g_a for g. From Lemma 8 it follows that there exists a POVM measurement $\{M^z\}_{z \in \mathbb{F}_p}$ on $\sigma_{Y'YE'}$ such that

$$\sum_{z \in \mathbb{F}_p} \text{Tr}\left(M^z \sigma_{YY'E}^z\right) \geq \frac{1}{p} + \frac{2\epsilon^2}{p^3} , \quad (4.13)$$

where $\sigma_{YY'E}^z$ is the reduced density of σ on $YY'E$ conditioned on $\langle X, g(Y, Y')\rangle = z$. To conclude the proof of the theorem we show that the adversary's map Adv and the POVM $\{M^z\}$ can be combined to give a "successful" strategy for the players in the communication game introduced in Sect. 4.1. To see this, consider the state ρ_{XE} that is instantiated as the source for the extractor; by definition $H_{\min}(X|E)_\rho = k = \left(\frac{n}{2} + 6\right)\log p - 1 + 4\log\frac{1}{\epsilon}$. In the third step of the game, Bob applies the map Adv to the registers Y and E containing his input Y and the state sent by Alice, and measures to obtain an outcome Y'. He then applies the measurement $\{M^z\}$ on his registers (Y, Y', E) to obtain a value $b = z \in \mathbb{F}_p$ that he provides as his output in the game. By (4.13) it follows that this strategy succeeds in the game with probability at least $\frac{1}{p} + \frac{2\epsilon^2}{p^3}$, which by Lemma 6 implies $H_{\min}(X|E) \leq \frac{n}{2}\log p + 1 + 2\log\frac{p^3}{2\epsilon^2}$, contradicting our choice of k. This proves (4.11) and thus the theorem. $\square$

The following lemma is used in the proof of the theorem.

Lemma 8. Let $\rho_{XE} = \sum_x |x\rangle\langle x| \otimes \rho_E^x$ be such that

$$\frac{1}{2}\|(X, E) - (U, E)\|_1 = \frac{1}{2}\|\rho_{XE} - U_X \otimes \rho_E\|_1 = \epsilon ,$$

where U_X is the totally mixed state on X and $\rho_E = \sum_x \rho_E^x$. Then there exists a POVM $\{M_x\}$ on ρ_E such that

$$\sum_x \mathrm{Tr}(M_x \rho_E^x) = \frac{1}{d_X} + \frac{\epsilon}{d_X} .$$

Proof. Since ρ_{XE} is a cq state, $\|\rho_{XE} - U_X \otimes \rho_E\|_1 = \sum_x \|\rho_E^x - \frac{1}{d_X}\rho_E\|_1$. For each x, let M_x' be the projector onto the positive eigenvalues of $\rho_E^x - \frac{1}{d_X}\rho_E$, so

$$\sum_x \mathrm{Tr}(M_x'(\rho_E^x - \frac{1}{d_X}\rho_E)) = \frac{1}{2}\sum_x \|\rho_E^x - \frac{1}{d_X}\rho_E\|_1 . \qquad (4.14)$$

Let $M' = \sum_x M_x'$ and $M_x = \frac{1}{d_X}(M_x' + (I_E - \frac{1}{d_X}M'))$. Then $M_x \geq 0$ and $\sum_x M_x = \frac{1}{d_X}(M' + d_X I_E - M') = I_E$. Moreover,

$$\begin{aligned}
\sum_x \mathrm{Tr}(M_x \rho_E^x) &= \sum_x \mathrm{Tr}\left[\frac{1}{d_X}(M_x' + (I_E - \frac{1}{d_X}M'))\rho_E^x\right] \\
&= \frac{1}{d_X}\left[\sum_x (\mathrm{Tr}(M_x'\rho_E^x)) + \mathrm{Tr}\left((I_E - \frac{1}{d_X}M')\rho_E\right)\right] \\
&= \frac{1}{d_X} + \frac{1}{d_X}\sum_x \left(\mathrm{Tr}(M_x'\rho_E^x) - \frac{1}{d_X}\mathrm{Tr}(M_x'\rho_E)\right) \\
&= \frac{1}{d_X} + \frac{1}{d_X}\left(\sum_x \mathrm{Tr}\left(M_x'(\rho_E^x - \frac{1}{d_X}\rho_E)\right)\right) \\
&= \frac{1}{d_X} + \frac{1}{2d_X}\sum_x \left\|\rho_E^x - \frac{1}{d_X}\rho_E\right\|_1
\end{aligned}$$

by (4.14). □

5   Privacy Amplification

Dodis and Wichs [19] introduced a framework for constructing a two-message privacy amplification protocol from any non-malleable extractor. In [14] it is shown that the same framework, when instantiated with a quantum-proof non-malleable extractor nmExt as defined in Definition 5, leads to a protocol that is secure against active quantum adversaries. In Sect. 5.1 we recall the Dodis-Wichs protocol, and state the security guarantees that follow by plugging in our non-malleable extractor construction. The guarantees follows from the quantum extension of the Dodis-Wichs results in [14]; since that work has not been published we include their results regarding the Dodis-Wichs protocol in Appendix A.

In Sect. 5.2 we show that a different protocol for privacy amplification due to Dodis et al. [16], whose main advantage is of being a one-round protocol, is also quantum-proof. The construction and analysis of the protocol of [16] is simple, with the drawback of a large entropy loss.

We start with the definition of a quantum-secure privacy amplification protocol against active adversaries. A privacy amplification protocol (P_A, P_B) is defined as follows. The protocol is executed by two parties Alice and Bob sharing a secret $X \in \{0, 1\}^n$, whose actions are described by P_A, P_B respectively.[7] In addition there is an active, computationally unbounded adversary Eve, who might have some quantum side information E correlated with X but satisfying $H_{\min}(X|E)_\rho \geq k$, where ρ_{XE} denotes the initial state at beginning of the protocol.

Informally, the goal for the protocol is that whenever a party (Alice or Bob) does not reject, the key R output by this party is random and statistically independent of Eve's view. Moreover, if both parties do not reject, they must output the same keys $R_A = R_B$ with overwhelming probability.

More formally, we assume that Eve is in full control of the communication channel between Alice and Bob, and can arbitrarily insert, delete, reorder or modify messages sent by Alice and Bob to each other. At the end of the protocol, Alice outputs a key $R_A \in \{0, 1\}^m \cup \{\bot\}$, where $\bot$ is a special symbol indicating rejection. Similarly, Bob outputs a key $R_B \in \{0, 1\}^m \cup \{\bot\}$. The following definition generalizes the classical definition in [17].

Definition 8. *Let k, m be integer and $\epsilon \geq 0$. A privacy amplification protocol (P_A, P_B) is a (k, m, ϵ)-privacy amplification protocol secure against active quantum adversaries if it satisfies the following properties for any initial state ρ_{XE} such that $H_{\min}(X|E)_\rho \geq k$, and where σ be the joint state of Alice, Bob, and Eve at the end of the protocol:*

1. Correctness. *If the adversary does not interfere with the protocol, then $\Pr[R_A = R_B \wedge R_A \neq \bot \wedge R_B \neq \bot] = 1$.*
2. Robustness. *This property comes in two flavors. The first is* pre-application *robustness, which states that even in the presence of an active adversary, $\Pr[R_A \neq R_B \wedge R_A \neq \bot \wedge R_B \neq \bot] \leq \epsilon$. The second is* post-application *robustness, which is defined similarly, except the adversary is additionally given the key R_A that is the result of the interaction (P_A, P_E), and the key R_B that results from the interaction (P_E, P_B), where P_E denotes the adversary's actions in its interaction with Alice and Bob.*
3. Extraction. *Given a string $r \in \{0, 1\}^m \cup \{\bot\}$, let $\mathsf{purify}(r)$ be a random variable on m-bit strings that is deterministically equal to $\bot$ if $r = \bot$, and is otherwise uniformly distributed. Let V denotes the transcript of an execution of the protocol execution, and $\rho_{E'}$ the final quantum state possessed by Eve. Then the following should hold:*

$$(R_A, V, E')_\sigma \approx_\epsilon (\mathsf{purify}(R_A), V, E')_\sigma \quad and \quad (R_B, V, E')_\sigma \approx_\epsilon (\mathsf{purify}(R_B), V, E')_\sigma .$$

---

[7] It is not necessary for the definition to specify exactly how the protocols are formulated; informally, each player's actions is described by a sequence of efficient algorithms that compute the player's next message, given the past interaction.

In other words, whenever a party does not reject, the party's key is indistinguishable from a fresh random string to the adversary.

The quantity $k - m$ is called the entropy loss.

5.1   Dodis-Wichs Protocol with Non-malleable Extractor

Here we first recall the Dodis-Wichs protocol for privacy amplification (hereafter called *Protocol DW*), which is summarized in Fig. 1, and the required security definitions, taken from [14]. We then state the result obtained by instantiating the protocol with the quantum-proof non-malleable extractor from Theorem 1.

Protocol DW

Let $d_X, d_Y, d_2, \ell, d_Z, t, k$ be integers and $\epsilon_{\text{MAC}}, \epsilon_{\text{Ext}}, \epsilon_{\text{nmExt}} > 0$.
Let MAC $: \{0, \ldots, d_Z - 1\} \times \{0,1\}^{d_2} \to \{0,1\}^t$ be a one-time ϵ_{MAC}-information-theoretically secure message authentication code.
Let Ext $: \{0, \ldots, d_X - 1\} \times \{0,1\}^{d_2} \to \{0,1\}^m$ be a strong $(k - \ell - \log(1/\epsilon_{\text{Ext}}), \epsilon_{\text{Ext}})$ quantum-proof extractor.
Let nmExt $: \{0, \ldots, d_X - 1\} \times \{0,, \ldots, d_Y - 1\} \to \{0, \ldots, d_Z - 1\}$ be a $(k, \epsilon_{\text{nmExt}})$ quantum-proof non-malleable extractor.
It is assumed that both parties, Alice and Bob, have access to a shared random variable $X \in \{0, \ldots, d_X - 1\}$.

1. Alice samples a Y_A uniformly from $\{0,, \ldots, d_Y - 1\}$. She sends Y_A to Bob. She computes $Z = \text{nmExt}(X, Y_A)$.
2. Bob receives Y'_A from Alice. He samples a uniform $Y_B \sim U_{d_2}$, and computes $Z' = \text{nmExt}(X, Y'_A)$ and $W = \text{MAC}(Z', Y_B)$. He sends (Y_B, W) to Alice. Bob then reaches the KEYDERIVED state and outputs $R_B = \text{Ext}(X, Y_B)$.
3. Alice receives (Y'_B, W') from Bob. If $W' = \text{MAC}(Z, Y'_B)$ she reaches the KEYCONFIRMED state and outputs $R_A = \text{Ext}(X, Y'_B)$. Otherwise she outputs $R_A = \perp$.

Fig. 1. The Dodis-Wichs privacy amplification protocol.

Aside from the use of a strong quantum-proof extractor (Definition 3) and a quantum-proof non-malleable extractor (Definition 5), the protocol relies on an information-theoretically secure one-time message authentication codes, or MAC. This security notion is defined as follows.

Definition 9. *A function* MAC $: \{0, \ldots, d_Z - 1\} \times \{0,1\}^d \to \{0,1\}^t$ *is an ϵ_{MAC}-information-theoretically secure one-time message authentication code if for any function* $\mathcal{A} : \{0,1\}^d \times \{0,1\}^t \to \{0,1\}^d \times \{0,1\}^t$ *it holds that for all $m \in \{0,1\}^d$*

$$\Pr_{k \leftarrow U_Z} \left[(\text{MAC}(k, m') = \sigma') \wedge (m' \neq m) : (m', \sigma') \leftarrow \mathcal{A}(m, \text{MAC}(k, m)) \right] \leq \epsilon_{\text{MAC}}.$$

Efficient constructions of MAC satisfying the conditions of Definition 9 are known. The following proposition summarizes some parameters that are achievable using a construction based on polynomial evaluation.

Proposition 1 (Proposition 1 in [34]). *For any $\epsilon_{\mathrm{MAC}} > 0$, integer $d > 0$, $d_Z \geq \frac{d^2}{\epsilon_{\mathrm{MAC}}^2}$, there exists an efficient family of ϵ_{MAC}-information-theoretically secure one-time message authentication codes*

$$\{\mathrm{MAC} : \{0, \ldots, d_Z - 1\} \times \{0, 1\}^d \to \{0, 1\}^t\}_{d \in \mathbb{N}}$$

with $t \leq \log d + \log(1/\epsilon_{\mathrm{MAC}})$.

The correctness and security requirements for the protocol are natural extensions of the classical case (see Definition 18 in [19]). Informally, the adversary has the following control over the outcome of the protocol. First, it possess initial quantum side information E about the weak secret X shared by Alice and Bob. That is, it has a choice of a cq source ρ_{XE}, under the condition that $H_{\min}(X|E)$ is sufficiently large. Second, the adversary may intercept and modify any of the messages exchanged. In Protocol DW there are only two messages exchanged, Y_A from Alice to Bob and (Y_B, σ) from Bob to Alice. To each of these messages the adversary may apply an arbitrary transformation, that may depend on its side information E. We model the two possible attacks, one for each message, as CPTP maps $T_1 : \mathrm{L}(\mathcal{H}_Y \otimes \mathcal{H}_E) \to \mathrm{L}(\mathcal{H}_Y \otimes \mathcal{H}_{E'})$ and $T_2 : \mathrm{L}(\mathbb{C}^{2^{d_2}} \otimes \mathcal{H}_{2^t} \otimes \mathcal{H}_{E'}) \to \mathrm{L}(\mathbb{C}^{2^{d_2}} \otimes \mathbb{C}^{2^t} \otimes \mathcal{H}_{E''})$, where $\mathcal{H}$ denotes the Hilbert space associated with system E. Note that we may always assume that $\mathcal{H}$ is large enough for the adversary to keep a local copy of the messages it sees, if it so desires.

The following result on the security of protocol DW is shown in [14]. We include the proof in Appendix A.

Theorem 3. *Let k, t, d_Z and $\epsilon_{\mathrm{MAC}}, \epsilon_{\mathrm{Ext}}, \epsilon_{\mathrm{nmExt}}$ be parameters of Protocol DW, as specified in Fig. 1. Let nmExt be a $(k, \epsilon_{\mathrm{nmExt}})$ quantum-proof non-malleable extractor, Ext a strong $(k - \log d_Z - \log(1/\epsilon_{\mathrm{Ext}}), \epsilon_{\mathrm{Ext}})$ quantum-proof extractor, and MAC an ϵ_{MAC}-information-theoretically secure one-time message authentication code. Then for any active attack (ρ_{XE}, T_1, T_2) such that $H_{\min}(X|E)_\rho \geq k$, the DW privacy amplification protocol described in Fig. 1 is (k, m, ϵ)-secure as defined in Definition 8 with $\epsilon = O(\epsilon_{\mathrm{Ext}} + \epsilon_{\mathrm{nmExt}} + \epsilon_{\mathrm{MAC}})$.*

Combined with Theorem 1 stating the security of our construction of a quantum-proof non-malleable extractor, Theorem 3 provides a means to obtain privacy amplification protocol secure against active attacks for a range of parameters. Due to the limitations of our non-malleable extractor we are only able to extract from sources whose entropy rate is at least $\frac{1}{2}$. This is a typical setting in the case of quantum key distribution, where the initial min-entropy satisfies $H_{\min}(X|E) \geq \alpha \log d_X$ for some constant α which depends on the protocol and the noise tolerance, but is generally larger than $3/4$. Specifically, we obtain the following:

Corollary 1. *For any $\epsilon > 0$, there exists a constant $c > 0$, such that the following holds. For any active attack (ρ_{XE}, T_1, T_2) such that $H_{\min}(X|E)_\rho = k \geq \frac{1}{2} \log d_X + c \cdot \log(1/\epsilon)$, there is an $O(\epsilon)$-secure DW protocol that outputs a key of length $m = k - O(\log(1/\epsilon))$.*

Proof. Let p be a prime and n a positive integer such that $\log p = \Theta(\log(1/\epsilon))$ and $d_X = p^n$. Let $d_Y = p^{n/2}$, and $d_Z = p$. Also, let $d_2 = O(\log d_X)$, $m = k - O(\log(1/\epsilon))$, and $t = O(\log(1/\epsilon))$. We instantiate Theorem 3 with the following.

- Let $\text{Ext} : \{0, \ldots, d_X - 1\} \times \{0, 1\}^{d_2} \to \{0, 1\}^m$ be the $(k - O(\log(1/\epsilon)), \epsilon)$ strong quantum-proof extractor from Theorem 2.
- Let $\text{nmExt} : \{0, \ldots, d_X - 1\} \times \{0, , \ldots, d_Y - 1\} \to \{0, \ldots, d_Z - 1\}$ be the $(\frac{1}{2} \cdot \log d_X + O(\log(1/\epsilon)), \epsilon)$ non-malleable extractor from Theorem 1.
- Let $\text{MAC} : \{0, \ldots, d_Z - 1\} \times \{0, 1\}^{d_2} \to \{0, 1\}^t$ be the one-time ϵ-information-theoretically secure message authentication code from Proposition 1.

The result follows. $\qquad\qquad\qquad\qquad\qquad\qquad\qquad\qquad\qquad\qquad\qquad$ $\square$

5.2 One-Round Privacy Amplification Protocol

In this section we show that the one-round protocol of Dodis et al. [16] is also quantum-proof. This protocol has significantly higher entropy loss, $(n/2) + \log(1/\epsilon)$, than the DW protocol we presented in the previous section.

One-round Privacy Amplification Protocol

Let n, k be integers and $\epsilon > 0$. Let $v = n - k + \log(1/\epsilon)$ and $m = (n/2) - v$.

It is assumed that both parties, Alice and Bob, have access to a shared random variable $X \in \{0, 1\}^n$. They interpret X as a pair $X = (X_1, X_2)$ where X_1, X_2 are identified as elements in $\mathbb{F}_{2^{n/2}}$.

1. Alice samples a Y uniformly from $\mathbb{F}_{2^{n/2}}$ and computes $Z = YX_1 + X_2$. Let $W = [Z]_1^v$ be the first v bits of Z. She sends (Y, W) to Bob and outputs $R_A = [Z]_{v+1}^{n/2}$, the remaining part of Z.
2. Bob receives (Y', W') from Alice and computes $Z' = Y'X_1 + X_2$. If $W' = [Z']_1^v$, then Bob outputs $R_B = [Z']_{v+1}^{n/2}$. Otherwise he outputs $\bot$.

Fig. 2. The one-round privacy amplification protocol from [16].

Theorem 4. *For any integer n and $k > n/2$, and any $\epsilon > 0$, the protocol in Fig. 2 is a one-round (k, m, ϵ)-quantum secure privacy amplification protocol with post-application robustness and entropy loss $k - m = (n/2) + \log(1/\epsilon)$.*

Proof. Correctness and extraction follow as in the classical proof by observing that $\text{Ext}(X, Y) = YX_1 + X_2$ is a quantum-proof extractor since $h_Y(X_1, X_2) = YX_1 + X_2$ is a family of universal hash function, which is shown to be a quantum-proof strong extractor in [36]. For robustness, the classical proof does not generalize directly. We prove post-application robustness as follows.

We proceed by contradiction. Suppose post-application robustness is violated, i.e. $\Pr[R_A \neq R_B \wedge R_A \neq \bot \wedge R_B \neq \bot] > \epsilon$. Then there is an initial state ρ_{XE} with $H_{\min}(X|E)_\rho \geq k$ and a CPTP map $T : L(\mathcal{H}_Y \otimes \mathcal{H}_W \otimes \mathcal{H}_{R_A} \otimes \mathcal{H}_E) \to$

$L(\mathcal{H}_Y \otimes \mathcal{H}_W \otimes \mathcal{H}_{E'})$ that can be applied by an adversary Eve to produce a modified message that is accepted by Bob with probability greater than ϵ. Note that T has R_A as input since we consider post-application robustness. Let $(Y', W', E') = T(Y, W, R_A, E)$. If post-application robustness is violated, then $\Pr[W' = [Y'X_1 + X_2]_1^v] > \epsilon$.

Consider the following communication game: Alice has access to a cq-state ρ_{XE}. Alice samples a uniformly random Y, computes $W = [YX_1 + X_2]_1^v$, $R_A = [YX_1 + X_2]_{v+1}^{n/2}$, and sends E, Y, W, and R_A to Bob. They win if Bob guesses X correctly from E, Y, W, and R_A. Using the map T introduced above, Bob can execute the following strategy. First, apply T on Alice's message to generate a guess (Y', W'). Second, guess a uniformly random R'_B. Third, use $Y, Y', (W, R_A) = YX_1 + X_2$, and $(W', R'_B) = Y'X_1 + X_2$ to solve for a unique $X = (X_1, X_2)$. Note that Bob succeeds if the guesses (Y', W') and R'_B in the first two steps are both correct (i.e., $(W', R'_B) = Y'X_1 + X_2$), which has probability greater than $\epsilon \cdot 2^{-((n/2)-v)}$. On the other hand, we can upper bound the winning probability of the communication game using the min entropy assumption $H(X|E)_\rho \geq k$. Since Y is independent of X and the length of (W, R_A) is $n/2$, $H_{\min}(X|E, Y, W)_\rho \geq k - (n/2)$. Thus the winning probability is less than $2^{-(k-(n/2))}$. Putting the two calculations together we have

$$\epsilon \cdot 2^{-((n/2)-v)} \leq \Pr[\text{ Bob wins }] \leq 2^{-(k-(n/2))},$$

which implies $v < n - k - \log(1/\epsilon)$, a contradiction.   □

A   The Dodis-Wichs Protocol

In this appendix we reproduce the proof of Theorem 3, taken from [14].

Proof of Theorem 3. Let an *active attack* on Protocol DW be specified by

- A cq state $\rho_{XE} \in D(\mathcal{H}_X \otimes \mathcal{H}_E)$ such that $H_{\min}(X|E)_\rho \geq k$;
- A CPTP map $T_1 : L(\mathcal{H}_Y \otimes \mathcal{H}_E) \to L(\mathcal{H}_Y \otimes \mathcal{H}_{E'})$ whose output on the first registered is systematically decohered in the computational basis; formally, for any ρ_{YE}, $T_1(\rho_{YE}) = \sum_y (|y\rangle\langle y|_Y \otimes \mathrm{Id}_E) T_1(\rho_{YE})(|y\rangle\langle y|_Y \otimes \mathrm{Id}_E)$;
- A CPTP map $T_2 : L(\mathbb{C}^{2^{d_2}} \otimes \mathbb{C}^{2^t} \otimes \mathcal{H}_{E'}) \to L(\mathbb{C}^{2^{d_2}} \otimes \mathbb{C}^{2^t} \otimes \mathcal{H}_{E''})$.

Given an active attack (ρ_{XE}, T_1, T_2) we instantiate random variables $Y_A, Z, Y'_A, Y_B, Z', \sigma, Y'_B, \sigma'$ and R_A, R_B in the obvious way, as defined in the protocol and taking into account the maps T_1 and T_2, applied successively to determine Y'_A and (Y'_B, σ').

The correctness of the protocol is clear.

To show robustness, let $\sigma_{Y'_A Y_A X E'}$ denote the joint state of Y'_A, Y_A (which represents a local copy of Y_A kept by Alice), X, and Eve's registers after her first map T_1 has been applied. Further decompose ρ as a sum of sub-normalized densities $\sigma^=_{Y'_A Y_A X E'}$, corresponding to conditioning on $Y'_A = Y_A$, and $\sigma^\perp_{Y'_A Y_A X E'}$, corresponding to conditioning on $Y'_A \neq Y_A$.

Conditioned on $Y_A' = Y_A$, by definition of a MAC the probability that $(Y_B', W') \neq (Y_B, W)$ and Alice reaches the KEYCONFIRMED state is at most ϵ_{MAC}. If $(Y_B', W') = (Y_B, W)$ then $R_A = R_B$, so that in this case robustness holds with error at most ϵ_{MAC}.

Now suppose $Y_A' \neq Y_A$. Consider a modified adversary Adv$'$ that keeps a copy of Y_A, applies the map T_1, and if $Y_A' = Y_A$ replaces Y_A' with a uniformly random string that is distinct from Y_A. This adversary implements a CPTP map T_1' that has no fixed point. By the assumption that nmExt is a quantum-proof non-malleable extractor,

$$\sigma'_{\mathrm{nmExt}(X,Y_A)\mathrm{nmExt}(X,Y_A')Y_A Y_A' E'} \approx_{\epsilon_{\mathrm{nmExt}}} U_m \otimes \sigma'_{\mathrm{nmExt}(X,Y_A')Y_A Y_A' E'}, \qquad (A.1)$$

where here $Y_A' E'$ is defined as the output system of the map T_1' implemented by Adv$'$. Conditioned on $Y_A \neq Y_A'$ the maps T_1 and T_1' are identical, thus it follows from (A.1) and the definition of $\rho^\perp$ that

$$\sigma^\perp_{\mathrm{nmExt}(X,Y_A)\mathrm{nmExt}(X,Y_A')Y_A Y_A' E'} \approx_{\epsilon_{\mathrm{nmExt}}} U_m \otimes \sigma^\perp_{\mathrm{nmExt}(X,Y_A')Y_A Y_A' E'},$$

where now the states are sub-normalized. Since $Z' = \mathrm{nmExt}(X, Y_A')$ this means that the key used by Alice to verify the signature in Step 3. of Protocol DW is (up to statistical distance $\epsilon_{\mathrm{nmExt}}$) uniform and independent of the key used by Bob to make the MAC. By the security of MAC, the probability for Alice to reach the KEYCONFIRMED state in this case is at most $\epsilon_{\mathrm{nmExt}} + \epsilon_{\mathrm{MAC}}$. Adding both parts together, $\Pr(R_A \notin \{R_B, \perp\}) \leq \epsilon_{\mathrm{nmExt}} + 2\epsilon_{\mathrm{MAC}}$. Since R_B is never $\perp$, this implies the robustness property.

For the extraction property, it is sufficient to show that $(R_B, V, E) \approx_\epsilon (U_m, V, E)$ since then key extraction property follows from the robustness and the fact that R_B is never $\perp$. We have that $R_B = \mathrm{Ext}(X, Y_B)$ is close to uniform given $V = Y_A Y_B W$ and E', and we need to establish two properties: first, independence between X and Y_B given $Y_A Z' E'$ and second, that the source has enough entropy conditioned on $Y_A Z' E'$. Regarding the first property, observe that conditioned on $Y_A Z'$, X and Y_B are independent given E'. Regarding the source entropy, by the chain rule for the (smooth) min-entropy [37], it follows that $H_{\min}^{\epsilon_{\mathrm{Ext}}}(X | Y_A Z' E') \geq k - \log d_Z - c \log(1/\epsilon_{\mathrm{Ext}})$ for some constant $c > 0$. Note that

$$\left\| (R_B, V, E')_\sigma - (U_m, V, E')_\sigma \right\|_1 \leq \left\| (R_B, Y_A, Y_B, Z', E')_\sigma - (U_m, Y_A, Y_B, Z', E')_\sigma \right\|_1,$$

which follows since W is a deterministic function Y_B and Z'. Using that Ext is a strong quantum-proof extractor, we conclude that $(R_B, V, E) \approx_\epsilon (U_m, V, E)$, as long as ϵ is such that $\epsilon > \epsilon_{\mathrm{Ext}}$. $\qquad \square$

References

1. Aggarwal, D., Chung, K.-M., Lin, H.-H., Vidick, T.: A quantum-proof non-malleable extractor, with application to privacy amplification against active quantum adversaries. arXiv preprint arXiv:1710.00557 (2017)
2. Aggarwal, D., Dodis, Y., Jafargholi, Z., Miles, E., Reyzin, L.: Amplifying privacy in privacy amplification. In: Garay, J.A., Gennaro, R. (eds.) CRYPTO 2014. Part II. LNCS, vol. 8617, pp. 183–198. Springer, Heidelberg (2014). https://doi.org/10.1007/978-3-662-44381-1_11
3. Aggarwal, D., Hosseini, K., Lovett, S.: Affine-malleable extractors, spectrum doubling, and application to privacy amplification. In: 2016 IEEE International Symposium on Information Theory (ISIT), pp. 2913–2917. IEEE (2016)
4. Bennett, C.H., Brassard, G., Crépeau, C., Maurer, U.M.: Generalized privacy amplification. IEEE Trans. Inf. Theory 41(6), 1915–1923 (1995)
5. Bennett, C.H., Brassard, G., Robert, J.-M.: Privacy amplification by public discussion. SIAM J. Comput. 17(2), 210–229 (1988)
6. Bhatia, R.: Matrix Analysis. Graduate Texts in Mathematics. Springer, Heidelberg (1997). https://doi.org/10.1007/978-1-4612-0653-8
7. Bouman, N.J., Fehr, S.: Secure authentication from a weak key, without leaking information. In: Paterson, K.G. (ed.) EUROCRYPT 2011. LNCS, vol. 6632, pp. 246–265. Springer, Heidelberg (2011). https://doi.org/10.1007/978-3-642-20465-4_15
8. Chandran, N., Kanukurthi, B., Ostrovsky, R., Reyzin, L.: Privacy amplification with asymptotically optimal entropy loss. In: Proceedings of the 42nd ACM Symposium on Theory of Computing, STOC 2010, Cambridge, Massachusetts, USA, 5–8 June 2010, pp. 785–794 (2010)
9. Chattopadhyay, E., Goyal, V., Li, X.: Non-malleable extractors and codes, with their many tampered extensions. arXiv preprint arXiv:1505.00107 (2015)
10. Chung, K.-M., Li, X., Wu, X.: Multi-source randomness extractors against quantum side information, and their applications (2014)
11. Cleve, R., van Dam, W., Nielsen, M., Tapp, A.: Quantum entanglement and the communication complexity of the inner product function. In: Williams, C.P. (ed.) QCQC 1998. LNCS, vol. 1509, pp. 61–74. Springer, Heidelberg (1999). https://doi.org/10.1007/3-540-49208-9_4
12. Cohen, G.: Non-malleable extractors - new tools and improved constructions. Electron. Colloq. Comput. Complex. (ECCC) 22, 183 (2015)
13. Cohen, G., Raz, R., Segev, G.: Non-malleable extractors with short seeds and applications to privacy amplification. In: 2012 IEEE 27th Annual Conference on Computational Complexity (CCC), pp. 298–308. IEEE (2012)
14. Cohen, G., Vidick, T.: Privacy amplification against active quantum adversaries (2016)
15. De, A., Portmann, C., Vidick, T., Renner, R.: Trevisan's extractor in the presence of quantum side information. SIAM J. Comput. 41(4), 915–940 (2012)
16. Dodis, Y., Kanukurthi, B., Katz, J., Reyzin, L., Smith, A.: Robust fuzzy extractors and authenticated key agreement from close secrets. IEEE Trans. Inf. Theory 58(9), 6207–6222 (2012)
17. Dodis, Y., Li, X., Wooley, T.D., Zuckerman, D.: Privacy amplification and non-malleable extractors via character sums. SIAM J. Comput. 43(2), 800–830 (2014)
18. Dziembowski, S., Pietrzak, K.: Leakage-resilient cryptography. In: 2008 49th Annual IEEE Symposium on Foundations of Computer Science, pp. 293–302. IEEE (2008)

19. Dodis, Y., Wichs, D.: Non-malleable extractors and symmetric key cryptography from weak secrets. In: Mitzenmacher, M (ed.) Proceedings of the 41st Annual ACM Symposium on Theory of Computing, Bethesda, MD, USA, pp. 601–610. ACM (2009)
20. Dodis, Y., Yu, Y.: Overcoming weak expectations. In: Sahai, A. (ed.) TCC 2013. LNCS, vol. 7785, pp. 1–22. Springer, Heidelberg (2013). https://doi.org/10.1007/978-3-642-36594-2_1
21. Gavinsky, D., Kempe, J., Kerenidis, I., Raz, R., De Wolf, R.: Exponential separations for one-way quantum communication complexity, with applications to cryptography. In: Proceedings of the Thirty-Ninth Annual ACM Symposium on Theory of Computing, pp. 516–525. ACM (2007)
22. Kasher, R., Kempe, J.: Two-source extractors secure against quantum adversaries. Theory Comput. 8(1), 461–486 (2012)
23. Koenig, R., Renner, R., Schaffner, C.: The operational meaning of min-and max-entropy. IEEE Trans. Inf. Theory 55(9), 4337–4347 (2009)
24. Lee, C.-J., Lu, C.-J., Tsai, S.-C., Tzeng, W.-G.: Extracting randomness from multiple independent sources. IEEE Trans. Inf. Theory 51(6), 2224–2227 (2005)
25. Li, X.: Design extractors, non-malleable condensers and privacy amplification. In: Proceedings of the 44th Symposium on Theory of Computing Conference, STOC 2012, New York, NY, USA, 19–22 May 2012, pp. 837–854 (2012)
26. Li, X.: Non-malleable condensers for arbitrary min-entropy, and almost optimal protocols for privacy amplification. CoRR, abs/1211.0651 (2012)
27. Li, X.: Non-malleable extractors, two-source extractors and privacy amplification. In: FOCS, pp. 688–697 (2012)
28. Li, X.: Non-malleable condensers for arbitrary min-entropy, and almost optimal protocols for privacy amplification. In: Dodis, Y., Nielsen, J.B. (eds.) TCC 2015. Part I. LNCS, vol. 9014, pp. 502–531. Springer, Heidelberg (2015). https://doi.org/10.1007/978-3-662-46494-6_21
29. Li, X.: Improved non-malleable extractors, non-malleable codes and independent source extractors. In: Proceedings of the 49th Annual ACM SIGACT Symposium on Theory of Computing, STOC 2017, Montreal, QC, Canada, 19–23 June 2017, pp. 1144–1156 (2017)
30. Maurer, U.: Conditionally-perfect secrecy and a provably-secure randomized cipher. J. Cryptol. 5(1), 53–66 (1992)
31. Maurer, U., Wolf, S.: Privacy amplification secure against active adversaries. In: Kaliski Jr., B.S. (ed.) CRYPTO 1997. LNCS, vol. 1294, pp. 307–321. Springer, Heidelberg (1997). https://doi.org/10.1007/BFb0052244
32. Nayak, A., Salzman, J.: Limits on the ability of quantum states to convey classical messages. J. ACM (JACM) 53(1), 184–206 (2006)
33. Nisan, N., Zuckerman, D.: Randomness is linear in space. J. Comput. Syst. Sci. 52(1), 43–53 (1996)
34. Renner, R., König, R.: Universally composable privacy amplification against quantum adversaries. In: Kilian, J. (ed.) TCC 2005. LNCS, vol. 3378, pp. 407–425. Springer, Heidelberg (2005). https://doi.org/10.1007/978-3-540-30576-7_22
35. Renner, R., Wolf, S.: Unconditional authenticity and privacy from an arbitrarily weak secret. In: Boneh, D. (ed.) CRYPTO 2003. LNCS, vol. 2729, pp. 78–95. Springer, Heidelberg (2003). https://doi.org/10.1007/978-3-540-45146-4_5
36. Tomamichel, M., Schaffner, C., Smith, A.D., Renner, R.: Leftover hashing against quantum side information. IEEE Trans. Inf. Theory 57(8), 5524–5535 (2011)
37. Vitanov, A., Dupuis, F., Tomamichel, M., Renner, R.: Chain rules for smooth min-and max-entropies. IEEE Trans. Inf. Theory 59(5), 2603–2612 (2013)

Secure Computation and NIZK

A Note on the Communication Complexity of Multiparty Computation in the Correlated Randomness Model

Geoffroy Couteau[✉]

KIT, Karlsruhe, Germany
geoffroy.couteau@kit.edu

Abstract. Secure multiparty computation (MPC) addresses the challenge of evaluating functions on secret inputs without compromising their privacy. A central question in multiparty computation is to understand the amount of communication needed to securely evaluate a circuit of size s. In this work, we revisit this fundamental question in the setting of information-theoretically secure MPC in the correlated randomness model, where a trusted dealer distributes correlated random coins, independent of the inputs, to all parties before the start of the protocol. This setting is of strong theoretical interest, and has led to the most practically efficient MPC protocols known to date.

While it is known that protocols with optimal communication (proportional to input plus output size) can be obtained from the LWE assumption, and that protocols with sublinear communication $o(s)$ can be obtained from the DDH assumption, the question of constructing protocols with $o(s)$ communication remains wide open for the important case of information-theoretic MPC in the correlated randomness model; all known protocols in this model require $O(s)$ communication in the online phase.

In this work, we exhibit the first generic multiparty computation protocol in the correlated randomness model with communication sublinear in the circuit size, for a large class of circuits. More precisely, we show the following: any size-s *layered* circuit (whose nodes can be partitioned into layers so that any edge connects adjacent layers) can be evaluated with $O(s/\log\log s)$ communication. Our results holds for both boolean and arithmetic circuits, in the honest-but-curious setting, and do not assume honest majority. For boolean circuits, we extend our results to handle malicious corruption.

Keywords: Multiparty computation · Correlated randomness model · Information-theoretic security · Sublinear communication

1 Introduction

Secure multiparty computation (MPC) allows n players with inputs $(x_1, \cdots, x_n)$ to jointly evaluate a function f, while leaking no information on their

© International Association for Cryptologic Research 2019
Y. Ishai and V. Rijmen (Eds.): EUROCRYPT 2019, LNCS 11477, pp. 473–503, 2019.
https://doi.org/10.1007/978-3-030-17656-3_17

own input beyond the output of the function. It is a fundamental problem in cryptography, which has received a considerable attention since its introduction in the seminal works of Yao [Yao86], and Goldreich, Micali, and Wigderson [GMW87b, GMW87a] (GMW). One of the core questions in secure multiparty computation is to understand the amount of communication needed to securely compute a function. For almost three decades after the protocols of Yao and GMW, all known constructions of secure computation protocols required a communication proportional to the circuit size of the function, and understanding whether this was inherent was a major open problem.

Secure Computation with Sublinear Communication. In 2009, this situation changed with the introduction by Gentry of the first fully-homomorphic encryption scheme [Gen09] (FHE), which led to secure computation protocols with communication independent of the size of the function (proportional only to its input size and its output size), under (a circular-security variant of) the LWE assumption. This resolved the long-standing open problem of designing MPC protocols with optimal (asymptotic) communication, although only under a specific assumption. More recently, the circuit-size barrier was broken again under the DDH assumption in [BGI16], for a large class of structured circuits[1] and in the two-party case. However, while these results are of strong theoretical interest, they require expensive computations.

Secure Computation in the Correlated Randomness Model. While secure computation (with no honest majority) is known to require computational assumptions, it was observed in several works (e.g. [IPS08, DPSZ12]) that executing a pre-computation phase independent of the inputs to the protocol, during which correlated random bits are distributed to the parties, allows to make the online phase both information-theoretically secure and significantly more efficient, by removing any expensive cryptographic operation from the online computation phase. These observations led to the development of increasingly efficient secure computation protocols in the correlated randomness model, e.g. [KOS16, DNNR17], which are currently considered the most practical secure computation protocols. Yet, unlike computationally secure protocols, all known unconditionally secure protocols in the correlated randomness model (with computation and storage polynomial in the circuit size) require communication proportional to the circuit size of the function. Therefore, the major question of understanding the communication required for multiparty computation remains wide open for the important case of MPC in the correlated randomness model, which captures the best candidates for practical secure computation. This is the question we address in this work: must MPC protocols in the correlated randomness model inherently use a communication linear in the size of the circuit? Or, in other words, can we get the best of both worlds: unconditional security with high practical efficiency, and sublinear communication?

---

[1] The work of [BGI16] considered, as we will do in this work, boolean circuits which can be divided into layers such as any edge connects adjacent layers. Such circuits are called *layered boolean circuits*.

On the Communication of Secure Computation in the Correlated Randomness Model. A partial answer to this question was given in [IKM+13], where the authors designed a *one-time truth-table* protocol, which allows to evaluate any function $f : \{0,1\}^n \mapsto \{0,1\}^m$ with unconditional security in the correlated randomness model, with optimal communication $O(n+m)$. However, this protocol requires storing an exponential number (in n) of correlated random bits (polynomial in the size of the entire truth-table of f), which makes it practical only for boolean functions with very small inputs. Furthermore, it was argued in [IKM+13] that reducing the amount of correlated random coins from exponential to polynomial (in the input size) for any function f is unlikely to be feasible, as it would imply an unexpected breakthrough for long-standing open problems related to private information retrieval.

While this negative result does not rule out a sublinear-communication protocol with small storage for circuits, this observation and the fact that all known protocols (with polynomial storage) have communication proportional to the circuit size s of the function have been seen as indications that breaking the circuit-size barrier for multiparty computation in the correlated randomness model might be non-trivial. For instance, it was mentioned in [DZ13] that "the results and evidence we know suggest that getting constant overhead [over the circuit size of the function] is the goal we can realistically hope to achieve". More recently in [DNPR16], the authors mentioned that "whether we can have constant round protocols and/or communication complexity much smaller than the size of the circuit and still be efficient (polynomial-time) in the circuit size of the function is a long-standing open problem".

In [DNPR16], the authors made progresses toward understanding why existing protocols have been stuck at the circuit-size barrier, by identifying a property shared by all known efficient protocols in the correlated randomness model, which states (informally) that they evaluate the function in a "gate-by-gate" fashion, and require communication for every multiplication gate. They demonstrated that all protocols following this approach (with passive security and dishonest majority) must inherently have communication proportional to the circuit size of the function. They concluded that improving the communication complexity of secure computation in the correlated randomness model requires a fundamentally new approach, and mentioned that the main question left open in their work is to find out whether their bound does hold for any protocol which is efficient in the circuit size of the function. This is the problem we address in this work.

1.1   Our Contribution

In this paper, we construct for the first time protocols with polynomial storage and communication sublinear in the circuit size, for a large class of circuit. Perhaps surprisingly, our results turn out to be relatively simple to obtain; it appears however that this simple solution was missed in previous works.

Sublinear Protocol for Structured Circuits. We exhibit a generic secure computation protocol in the correlated randomness model, with communication sublinear in the circuit size. More specifically, we consider *layered boolean circuits* (LBC), whose nodes can be arranged into layers so that any edge connects adjacent layers. We prove the following: for any N, there is an unconditionally secure N-party protocol that evaluates an arbitrary LBC of size s with n inputs and m outputs, with total communication

$$O\left(n + N \cdot \left(m + \frac{s}{\log\log s}\right)\right),$$

sublinear in the size of the circuit, and polynomial storage $O(s^2/\log\log s)$, in the correlated randomness model against semi-honest adversaries, with dishonest majority. While this requires an arguably large storage, it can be reduced to being only slightly superlinear in s, namely

$$O\left(s \cdot \frac{2^{(\log s)^{1/c}}}{\log\log s}\right),$$

at the cost of increasing the communication to $O(n + N \cdot (m + c \cdot s/\log\log s))$ (for an arbitrary $c = o(\log\log n)$). Our protocol enjoys perfect security, computational complexity $O(s\log s/\log\log s + n + m)$, and round complexity $d/\log\log s$, where d is the depth of the circuit. All the constants involved are very small (in fact equal to one, up to low order terms), and the computation involves solely searching lookup tables.

Extensions. We generalize our result to secure evaluation of arbitrary *layered arithmetic circuits* (LAC) over any (possibly exponentially large) field $\mathbb{F}$, by relying on a connection between MPC with correlated randomness and the classical notion of private simultaneous message protocols [FKN94]. The resulting protocol for arithmetic circuits has costs comparable to the boolean version. Furthermore, we show that all our results can be extended to the stronger function-independent preprocessing model, where only a bound on the size of the circuit is known in the preprocessing phase, and that the communication can be improved for "tall and narrow" circuits. Eventually, using the techniques of [DNNR17,KOR+17], our protocols directly extend to the malicious setting for boolean circuits, at an additive cost of $N \cdot \kappa$ bits of communication (for some statistical security parameter κ), and a $O(\kappa)$ overhead in computation and correlated randomness (more advanced techniques from [DNNR17] can be used to make this overhead constant).

Static vs Adaptive Setting. While we focus for simplicity on the static setting in this work, where the adversary decides before the protocol which parties to corrupt, our protocols can be proven to also satisfy adaptive security in a relatively straightforward way. Indeed, when it must reveal the input of a party which is being corrupted by the adversary, the simulator of our main protocol (and its variants) can easily explain the view of the adversary as being consistent with any input of its choice, by choosing the preprocessing material in an appropriate way. As the view of the adversary will always consist of values perfectly masked by random coins generated in the preprocessing phase, there will always be a choice of preprocessing material which "unmask" the values known to the adversary to any value chosen by the simulator.

1.2   Our Method

Perhaps surprisingly, our method does not depart significantly from existing techniques in secure computation. Our starting point is the one-time truth-table (OTTT) protocol of [IKM+13], which has optimal communication but requires an exponential amount of data. It has been observed in several works that using OTTT as an internal component in secure protocols can be used to reduce their communication. For example, it was suggested to use OTTT to securely compute S-boxes in AES in [DNNR17, KOR+17], as they can be efficiently represented as small lookup-tables. More recently, the work of [DKS+17] developped methods to automatically create tradeoffs between communication and computation in secure protocols, by relying on a compiler that transforms high-level descriptions of a function into a lookup-table-based representation of the function. All these works rely on the fact that, for functions that can be broken into small interconnected lookup-tables, the protocol of [IKM+13] can be used to save some communication.

Dividing Layered Boolean Circuits into Local Functions. In this work, we show that this intuition can in fact be extended to *arbitrary* layered boolean circuit of size s, and that the savings obtained this way lead to a protocol with $o(s)$ communication. Our protocol builds upon a variant of the result of [IKM+13], which states that every function can be securely evaluated in the correlated randomness model with perfect security, optimal communication, and exponential storage. Our variant relies on the observation that when evaluating *local functions*, where each output bit depends on a number c of input bits, we can reduce the storage cost of the protocol of [IKM+13] from being exponential in the input size to being only exponential in the locality parameter c. Indeed, consider the task of securely evaluating a function with n input bits, and m output bits. The protocol of [IKM+13] (called OTTT, for one-time truth table) requires the parties to store shares of (a shifted version of) the truth table of the function, which has size $m \cdot 2^n$, exponential in the input size. When the function is c-local, however, there is a better solution: the parties can store shares of (shifted variants of) truth tables corresponding to each function mapping c input bits to a given output bit, for a total storage cost of $m \cdot 2^c$. Some care must be taken, as doing straightfoward parallel repetitions of the OTTT protocol for each subfunction would increase the communication from $O(n)$ to $O(c \cdot m)$; we show that carefully avoiding redundancies in the secret-shared representation of the input allows to bring this cost back to $O(n)$. We formally state this result in a lemma, which we call *core lemma*.

Given the core lemma, our result is obtained by breaking an arbitrary layered circuit into chunks, each chunk containing some number k of consecutive layers.

We observe that, as the underlying directed graph of the circuit has indegree 2, each value associated to the last layer of a chunk can be computed as a function of at most 2^k values on the last layer of the previous chunk. Therefore, computing all the values on the last layer of a chunk can be reduced to evaluating a 2^k-local function of the values on the last layer of the previous chunk. Using the core lemma, this can be done using $O(w \cdot 2^{2^k})$ bits of preprocessing material, where w is the width of the input layer, with a communication proportional to w only. If the circuit has size s, width w, and depth d, this means that the circuit can be securely evaluated in a chunk-by-chunk fashion, with total communication $O((d/k) \cdot w) = O(s/k)$, using $O((d/k) \cdot 2^{2^k})$ bits of correlated randomness; setting $k \leftarrow \log \log s$ gives the claimed result.[2]

Extending the Result to Arithmetic Circuit. The above method breaks down in the case of arithmetic circuits over large order fields. While we can decompose an arbitrary LBC into polynomial-size truth tables (by breaking it into interconnected functions operating on logarithmically many inputs), this is not true anymore for arithmetic circuit over fields of exponential size, where even a function with a single input will have an exponential-size truth table. We nevertheless obtain a comparable result for arithmetic circuit, building upon a relation with the notion of private simultaneous message (PSM) protocols [FKN94], which establishes that PSM protocols with some additional decomposability property can be used to build two-party secure computation protocols in the correlated randomness model. This link was indirectly established in [BIKK14], where a connection was drawn both between PSM and PIR, and between MPC in the correlated randomness and PIR. Building upon a recent PSM protocol of [LVW17] for multivariate polynomial evaluation, we get an arithmetic analogue of the protocol of [IKM+13], which relies on the representation of arithmetic functions as multivariate polynomials. From this protocol, we derive a new version of our core lemma, tailored to the arithmetic setting, which directly leads to a secure computation protocol with communication $O(s/\log \log s)$ for layered arithmetic circuits over arbitrary fields.

We note that, while lookup-table-based secure computation protocols for boolean circuits have been investigated, the extension of this approach to the arithmetic setting was (to our knowledge) never observed before. As a minor side contribution of independent interest, we further observe that our generalization to the arithmetic setting does in fact also provide some improvement over the original TinyTable protocol [DNNR17] in the boolean setting: by replacing the lookup-table-based representation of boolean gates by a multivariate-polynomial-based representation, we show that the storage requirement of their protocol can be reduced by 25%.

On the Lower Bounds of [IKM+13, DNPR16]. It should be noted that our protocols do not follow the standard gate-by-gate design of unconditionally

---

[2] We assume $w \cdot d = O(s)$ in this high level explanation for simplicity only, this is not a necessary condition in the actual construction.

secure protocols in the correlated randomness model, hence our result does not contradict the lower bound of [DNPR16]. Moreover, our results apply only to circuits, while the implausibility result of [IKM+13] assumes the existence of a low-storage protocols for evaluating *any function*, which our results do not provide. Therefore, they do not lead to unexpected breakthroughs for information-theoretic private information retrieval.

1.3 Related Work

The possibility of securely computing functions given access to a source of correlated random coins was first studied in the work of Beaver for the (MPC-complete) oblivious-transfer functionality in [Bea95], and later generalized to the *commodity-based* model, where multiple servers generate correlated random coins in a honest majority setting in [Bea97]. The study of multiparty computation in the *preprocessing model*, where the correlated-randomness coin-generation phase is implemented with a computationally secure MPC protocol, was initiated in [Kil88, Bea92, IPS08]. These works started a rich line of work on increasingly efficient MPC protocols in the preprocessing model [IPS09, BDOZ11, NNOB12, DPSZ12, DZ13, DLT14, LOS14, FKOS15, BLN+15, DZ16, KOS16, DNNR17].

The quest for secure multiparty computation protocols with low-communication was initiated in [BFKR91], which gave a protocol with optimal communication, albeit with exponential computation and only for a number of party linear in the input size. An optimal communication protocol with exponential complexity was also given in [NN01]. The work of [BI05] gives a low-communication protocol for constant-depth circuit, for a number of parties polylogarithmic in the circuit size. The breakthrough result of Gentry [Gen09] led to optimal communication protocols in the computational setting [DFH12, AJL+12] under the LWE assumption.[3] More recently, computationally secure MPC protocols with sublinear communication were achieved from the DDH assumption in [BGI16].

The study of low-communication protocols in the correlated randomness model was initiated in [IKM+13], where a protocol with optimal communication and exponential storage complexity was presented. The same paper showed that improving the storage requirement for all functions would imply a breakthrough in information-theoretic PIR. The work of [BIKK14] reduces the storage requirement for functions with n inputs to $2^{O(\sqrt{n})}$, at the cost of increasing the communication complexity to $2^{O(\sqrt{n})}$. The work of [BIKO12] leads to low-communication protocols in the correlated randomness model for the special case of depth-2 circuits with a layer of OR gates and a layer of gates computing a sum modulo m, for composite m. All known protocols for evaluating arbitrary circuits in the correlated randomness model (with polynomial computation and storage) use communication linear in the circuit size. This limitation was formally studied

---

[3] More precisely, the protocol needs to assume the circular security of an LWE-based encryption scheme; alternatively, it can be based on the LWE assumption only, but the communication will grow with the depth of the circuit.

recently in [DNPR16], where it was shown that it is inherent in the setting of gate-by-gate protocols.

The idea of using truth-table representation to reduce the communication of secure computation protocols first arose in [CDv88], and was developped in [IKM+13]. It was later used implicitly in [KK13], to construct one-out-of-two oblivious transfer for short string from one-out-of-N oblivious transfer, and in the works of [DNNR17, KOR+17, DKS+17] to evaluate circuits with an appropriate structure.

On the Relation to [DNNR17]. At a late stage of our work on this paper, it was brought to our attention that the main techniques underlying the proof of our core lemma – informally, breaking a function into interconnected truth-tables, representing each outgoing wires from a table with secret-shared values, and carefully avoiding all redundancies for wires which are used by several tables – are already implicitly present in [DNNR17]. Indeed, [DNNR17] already explored the possibility of breaking a circuit into small interconnected truth table, avoiding redundancies in the secret-shared representation of the values associated to each wire, and envisionned the possibility of generalizing this to larger tables. However, it appears that the authors of [DNNR17] have overlooked the surprising potential consequences of these techniques, which we explore in this paper. Therefore, our work can be seen as indentifying and abstracting out the technical ideas underlying our main result (as well as providing additional contributions, such as the extension to the arithmetic setting), but while the core lemma is new to our work, we cannot (and do not) claim the novelty of the techniques used in its proof, which should be credited to [DNNR17]. Still, we believe that our result remains interesting and surprising, and that it deserves to be explicitly presented.

1.4 On the Practical Efficiency of Our Protocols

In spite of its theoretical nature, our result can in fact lead to concrete efficiency improvements for secure multiparty computation. We focus for simplicity on the case of two-party computation, and argue that our protocols can lead to improved efficiency, for useful types of computation. The state-of-the-art protocol for secure two-party computation in the correlated randomness model is, to our knowledge, the protocol of [DNNR17] (in both the passive setting and the active setting), which also relies on an OTTT-based evaluation of a boolean circuit. In the online phase, the protocol of [DNNR17] communicates 2 bits per AND gate (one from each player), and no bit at all for XOR and NOT gates (we note that our protocols can be readily adapted to allow for free XOR and NOT gates as well).

Concrete Efficiency. Using our protocol with $k = 2$, we get a two-party protocol which communicates on average a *single bit* per AND gate, improving over the protocol of [DNNR17] by 50% in both the passive and the active setting, for arbitrary layered circuits. This comes at the cost of storing 8 times more preprocessed data (a factor $2^{2^k}/k = 8$ for $k = 2$), and a factor $2^k/k = 2$ in computation (which comes from the need to search four-times larger lookup-tables). As noted in [DNNR17], the limiting factor in a concrete implementation of TinyTable is the bandwidth, hence we expect that an implementation of our protocol would result in concrete improvements over [DNNR17] in the speed of the online phase.

On the Generality of Layered Boolean Circuits. Unlike [DNNR17], however, our construction is restricted to layered boolean circuits. While this is a large class of circuits, and getting improved secure computation protocols for this class was already seen as an interesting goal in previous papers [BGI16], one might wonder whether this class captures *useful* circuits, ones that arise naturally in some applications. We argue that it is the case, by providing a (non-exhaustive) list of types of circuits that are well-suited for our protocols. We stress that this list is only for illustration purpose; many more examples can be found.

- *FFT circuit.* The circuit for the fast Fourier transform, which is used in signal processing and integer multiplication, and the circuit for permutation networks [Wak68], which allow to compute arbitrary permutations of the input, have the exact same structure and are layered. For these circuits, which occur naturally in many applications, our protocol leads to an online communication of $O(n \log n / \log \log n)$ instead of $O(n \log n)$.
- *Symmetric crypto primitives.* It was already observed previously that any computation involving large truth tables, such as block ciphers (e.g. AES), have the appropriate structure to be evaluated efficiently with our approach. More generally, algorithms that proceed in sequences of low-complexity rounds, where each round requires only the state of the previous round (and the input), are naturally "layered by blocks", which suffices for our result to apply. This structure is common to many primitives in symmetric cryptography.
- *Circuits for problems with a dynamic-programming algorithm.* Dynamic programming algorithm naturally proceed in stages, such that the computation at each stage depends on a (usually small) state of values stored after the previous stage. Such dynamic programming algorithms arise for example in various useful types of distance measures used in genetic computation, such as the Smith-Waterman distance [SW81], or the Levenshtein distance [Lev66] and its variants (LCS, weighted Levenshtein distance, etc). Privacy-preserving genomic computations are an important application of secure computation, hence the secure computation of the aforementioned measures (which are among the fundamental building blocks of computational biology) has been considered at length (see e.g. [AKD03, SPO+06, JKS08, HEKM11, ALSZ13, CKL15]). The natural circuit for computing Levenshtein and Smith-Waterman distances have size $O(n^2 \log n)$, but can be computed with online communication $O(n^2)$ with our protocol (the $\log n$ shaving comes from the high locality of dynamic programming algorithms; our result leads to better sublinearity guarantee for very local computations).

1.5   On Implementing the Correlated Randomness Model

It is well known that the distribution of correlated random coins in the preprocessing phase can be implemented by any generic MPC protocol. However, in our setting, generic approaches would require a communication superlinear in the circuit size. We note that, for the specific case of generating random shares of correlated strings, there are better (theoretical) solutions: under the learning with error assumption, or under (variants of) the decisional Diffie-Hellman assumption in the two-party case, the preprocessing phase of our protocols can be implemented with *constant* communication $\mathsf{poly}(\lambda)$ (where λ is a security parameter), independent of the size of the circuit, resulting in protocols with sublinear total communication $O(s/\log\log s + \mathsf{poly}(\lambda))$, and information-theoretically secure online phase.

We briefly sketch how the preprocessing phase can be implemented with constant communication. The main technical tool is a primitive known as *homomorphic secret sharing* [BGI16] (HSS); the idea of using HSS to the implement preprocessing phase of MPC protocols was suggested in [BGI17,BCG+17]. Informally, an HSS scheme for a class of functions F allows to secretly share an input x between several parties, such that given its share, each party can *locally* compute an additive share of $f(x)$, for any $f \in F$. Given an HSS scheme for all circuits, the preprocessing phase can be implemented as follows: we assume without loss of generality that the trusted dealer first samples a long random string x, computes $f(x)$ for some specified function f, and distributes random additive shares of $f(x)$ to the parties (e.g. in our protocol, f would output $\approx s/\log\log s$ shifted truth-tables). To implement this preprocessing phase, the parties jointly and securely construct, using a general purpose MPC protocol, an homomorphic secret sharing of a random PRF key K. Then, all parties locally evaluate the function f' that takes some counter c, generates pseudorandom coins x from this counter using the PRF with key K (e.g. by computing $\mathrm{PRF}(K,c)$, $\mathrm{PRF}(K,c+1)$, and so on), and returns $f(x)$. This way, with no further communication except for a one-time generation of the sharing of K (which takes communication $\mathsf{poly}(\lambda)$, independently of s), the parties obtain correlated (pseudo) random coins. An HSS scheme for all functions (and a PRF) can be constructed under the LWE assumption [BGI15,JRS17]. With a more involved construction, a protocol can also be obtained from DDH: under the DDH assumption, there exists an approximately-correct HSS scheme for NC_1 [BGI16], in the two-party setting. Noting that the preprocessing function is parallelizable (in NC_0) and that there exists PRFs in NC_1 under the DDH assumption, we can implement the previous strategy from DDH. The correlated random coins obtained this way are not all correct, but the approximately-correct HSS scheme of [BGI16] allows the parties to make the error probability arbitrarily small, and to detect when an output is erroneous. By setting the error parameter so that, with overwhelming probability, a small (constant) number of correlated random coins will be erroneous, and by introducing some redundancy in the coins generated this way, the parties can simply reveal to each other which correlated coins are susceptible to be erroneous (indicating the position of erroneous bits only requires $O(\log s)$ communication), and locally delete them. To prove security in spite of this small leakage, we need to rely on slightly leakage-resilient PRF and HSS, which can both be constructed from DDH-based primitives using standard approaches. We refer the reader to the full version [BCG+18] of [BCG+17] for a detailed overview of this approach.

1.6 Organization

Section 2 introduces our notations, and recalls standard preliminaries on circuits. In Sect. 3, we summarize the contributions of this paper in the form of a list of theorems, formally state the core lemma on which these theorems are based, and prove it. Section 4 builds upon the core lemma; it introduces our main protocol and several variants, and proves its security. In Sect. 5, we discuss the extension of our protocols to the malicious setting. Eventually, Sect. 6 lists some questions left open by our work, that we believe to be of interest for future works.

2 Preliminaries

Notations. Let k be an integer. We let $\{0,1\}^k$ denote the set of bitstrings of length k. For two strings (x, y) in $\{0,1\}^k$, we denote by $x \oplus y$ their bitwise xor. Given a subset S of $[k]$, $x[S]$ denotes the subsequence of the bits of x with indices from S. We use bold letters to denote vector; for a vector $\boldsymbol{x} = (x_1, \cdots, x_N)$, $\boldsymbol{x}[S]$ denotes the vector $(x_1[S], \cdots, x_N[S])$. For a matrix M, we denote $M|_{i,j}$ its entry (i, j).

2.1 Circuits

Boolean Circuits. A boolean circuit C with n inputs and m outputs is a directed acyclic graph with two types of nodes:

- The *input nodes* are labelled according to variables $\{x_1, \cdots, x_n\}$;
- The *gates* are labelled according to a base B of boolean functions.

In this work, we will focus on boolean circuits with indegree two (hence, B contains boolean functions with domain $\{0,1\}$ or $\{0,1\}^2$). C contains m gates with no children, which are called *output gates*. If there is a path between two nodes (v, v'), we say that v is an *ancestor* of v'. The *size* $\mathsf{size}(C)$ of C is the number of its nodes; its *depth* $\mathsf{depth}(C)$ is the length of the longest path from an input node to an output gate. The *width* of a circuit $C = (V, E)$ is defined as $\mathsf{width}(C) = \max_{1 \le i \le \mathsf{depth}(C)} \#\{v \in V \mid (0 \le \mathsf{depth}(v) \le i) \land (\exists w, (v, w) \in E \land \mathsf{depth}(w) > i)\}$.

Layered Boolean Circuits. In this work, we will consider a special type of boolean circuits, called *layered boolean circuits* (LBC). An LBC is a boolean circuit C whose nodes can be partitioned into $d = \mathsf{depth}(C)$ layers $(L_1, \cdots, L_d)$, such that any edge (u, v) of C satisfies $u \in L_i$ and $v \in L_{i+1}$ for some $i \le d - 1$.

Note that the width of a layered boolean circuit is also the maximal number of non-output gates contained in any single layer. Evaluating a circuit C on input $x \in \{0,1\}^n$ is done by assigning the bits of x to the variables $\{x_1, \cdots, x_n\}$, and then associating to each gate g of C (seen as a boolean function) the bit obtained by evaluating g on the values associated to its parent nodes. The output of C on input x, denoted $C(x)$, is the bit-string associated to the output gates.

Arithmetic Circuits. We define arithmetic circuits over a field $\mathbb{F}$ comparably to boolean circuits, as directed acyclic graphs with input nodes and arithmetic gates. Input nodes are labeled with variables $\{x_1, \cdots, x_n\}$ over $\mathbb{F}$, and the gates compute negation, addition, or multiplication over $\mathbb{F}$. Note that boolean circuits correspond to the special case of arithmetic circuits over the field $\mathbb{F}_2$; we extend layered boolean circuits to layered arithmetic circuits (LAC) in a similar way.

2.2 One-Time Truth Tables

We recall the one-time truth-table protocol of [IKM+13], which is at the heart of our protocols. It allows multiple parties to jointly evaluate a function $f : X_1 \times X_2 \times \cdots \times X_N \mapsto Z$, by sharing between all parties a scrambled version of the truth table of f. We focus for simplicity on a scenario where all parties receive the same output, but the protocol can be trivially generalized to a setting where the parties receive different outputs. The protocol is represented on Fig. 1; it has optimal communication $\sum_i \log |X_i| + N \cdot \log |Z|$, and exponential storage complexity $|Z| \cdot \prod_i |X_i|$ per party.

3   Theorems and Core Lemma

In this section, we formally introduce the theorems which we will prove in this work, state the core lemma from which we will derive them, and prove it.

Network Model. We consider protocols involving N parties communicating over synchronous and authenticated broadcast channel. Note that broadcasts channels can be unconditionally implemented from (insecure) point-to-point channels in the correlated randomness model.

Functionalities. An N-party functionality $F : X_1 \times X_2 \times \cdots \times X_n \mapsto Z_1 \times Z_2 \times \cdots \times Z_N$ specifies a mapping from the N input of each party to N outputs (one for each party). Such functionalities capture arbitrary non-reactive computation tasks. A useful special case of (randomized) N-party functionalities are *secret sharing functionalities* for functions over an abelian group $(\mathbb{G}, +)$: a protocol computes secret shares of a function $g : \mathbb{G} \mapsto \mathbb{G}$ if it computes the (randomized) N-party functionality which, on input $(x_1, \cdots, x_N) \in \mathbb{G}^N$, outputs N uniformly random group elements $(z_1, \cdots, z_N) \in \mathbb{G}^N$ subject to $\sum_{i=1}^N z_i = g(\sum_{i=1}^N x_i)$. This captures the situation where the parties hold secret shares of an input to a (deterministic) function, and want to receive secret shares of the output of the function.

Protocol OTTT

Functionality:
- Public parameters: an N-party functionality $f : X_1 \times X_2 \times \cdots \times X_N \mapsto Z$, where the $(X_i, +)$ and $(Z, +)$ are groups.
- The parties $(P_1, \cdots, P_N)$ hold respective inputs $\boldsymbol{x} = (x_1, \cdots, x_N)$;
- Output: each party P_i learns $z = f(\boldsymbol{x})$.

Preprocessing :
1. Sample $\boldsymbol{r} = (r_1, \cdots, r_N) \xleftarrow{\$} X_1 \times X_2 \times \cdots \times X_N$. Let M denote the truth-table of f permuted with the shifts $\boldsymbol{r}$, i.e., for any $\boldsymbol{x} \in X_1 \times X_2 \times \cdots \times X_N$, $M|_{\boldsymbol{x}+\boldsymbol{r}} = f(\boldsymbol{x})$.
2. Let $(M_i)_{i \leq N}$ be a random (N-out-of-N) secret sharing of M. Output (r_i, M_i) to each party P_i.

Protocol($\boldsymbol{x}$) :
1. Each party P_i with input x_i broadcasts $u_i \leftarrow x_i + r_i$.
2. Each party P_i broadcasts $z_i \leftarrow M_i|_{\boldsymbol{u}}$. All parties reconstruct $z \leftarrow \sum_{i=1}^{N} z_i$.

Fig. 1. Protocol OTTT for evaluating an arbitrary N-party functionality f in the correlated randomness model, against a passively corrupted majority

3.1 Theorems

Following is a summary of the results that we obtain in the subsequent sections.

Theorem 1. *For any N-party functionality f represented by a layered (boolean or arithmetic) circuit C of size s with n inputs and m outputs, and for any integer k, there is a perfectly secure protocol which realizes f in the preprocessing model against semi-honest parties, without honest majority, with communication $n + N \cdot (m + \lceil s/k \rceil)$ and storage $n/N + (m + \lceil s/k \rceil) \cdot (2^{2^k} + 1)$.*

In the above theorem, "storage" refers to the number of correlated random coins stored by each party at the end of the preprocessing phase (counted as a number of bits in the boolean case, and as a number of field elements in the arithmetic case). This gives, setting $k = \log \log s$,

Corollary 2. *There is a protocol that perfectly realizes any N-party functionality f (in the function-dependent preprocessing model and against semi-honest parties, without honest majority) represented by a layered (boolean or arithmetic) circuit C of size s with n inputs and m outputs, with communication $O(n + N \cdot (m + s/\log \log s))$ and polynomial storage.*

Building on the same techniques, we can also obtain a comparable result in the stronger *function-independent* correlated randomness model, where the correlated randomness is not allowed to depend on the target functionality (but is only given a bound on its size):

Theorem 3. *For any N-party functionality f represented by a layered (boolean or arithmetic) circuit C of size s with n inputs and m outputs, and for any integer k, there is a perfectly secure protocol which realizes f in the function-independent preprocessing model against semi-honest parties, without honest majority, with communication $n + N \cdot (m + \lceil s/k \rceil)$ and storage $n/N + (m + \lceil s/k \rceil) \cdot (2^{k+2^{2^k}} + 1)$.*

Setting $k = \log \log \log s$ gives us

Corollary 4. *There is a protocol that perfectly realizes any N-party functionality f (in the function-independent preprocessing model and against semi-honest parties, without honest majority) represented by a layered (boolean or arithmetic) circuit C of size s with n inputs and m outputs, with communication $O(n + N \cdot (m + s/\log \log \log s))$ and polynomial storage.*

Finally, we can obtain a stronger sublinearity guarantee for "tall and narrow" layered circuits:

Theorem 5. *For any N-party functionality f represented by a layered (boolean or arithmetic) circuit C of size s and width w with n inputs and m outputs, and for any integer k, there is a perfectly secure protocol which realizes f in the preprocessing model against semi-honest parties, without honest majority, with communication $n + N \cdot (m + \lceil s/k \rceil)$ and storage $n/N + (m + \lceil s/k \rceil) \cdot (2^{w \cdot k} + 1)$.*

For example, setting $k = \sqrt{\log s}$ gives us

Corollary 6. *There is a protocol that perfectly realizes any N-party functionality f (in the preprocessing model and against semi-honest parties, without honest majority) represented by a "tall and narrow" layered (boolean or arithmetic) circuit C of size s and width $w = O(\sqrt{\log s})$ with n inputs and m outputs, with communication $O(n + N \cdot (m + s/\sqrt{\log s}))$ and polynomial storage.*

Alternatively, we get a protocol with communication $O(s/\log s)$ for constant-width circuit (which corresponds to the complexity class SC_0). This can again be generalized to the stronger function-independent correlated randomness model. In the next section, we proceed with the description of our protocol. We first focus on the case of layered boolean circuits, and then discuss our extension to the case of arithmetic circuits.

3.2 Core Lemma

In this section, we state and prove the core lemma which underlies our results.

Definition 7 (Local Function). *A Function $g : \mathbb{F}_2^n \mapsto \mathbb{F}_2^m$ is c-local (for some integer $c \leq n$) if on any input $x \in \mathbb{F}_2^n$, any output bit of $g(x)$ depends on at most c bits from x.*

Lemma 8 (Core Lemma). *For any c-local function* $g : \mathbb{F}_2^n \mapsto \mathbb{F}_2^m$, *there is an information-theoretic semi-honest N-party secure computation protocol (with dishonest majority) in the correlated randomness model for computing secret shares of g with total online communication* $N \cdot n$ *bits, and correlated randomness* $m \cdot 2^c + n$ *bits per party.*

Before proving Lemma 8, it is instructive to compare its guarantees to the protocol obtained by applying directly the one-time truth-table protocol of [IKM+13] to the N-party functionality computing secret shares of g. Applying the OTTT protocol to the N-party functionality which sums its entries (over $\mathbb{F}_2$) before evaluating g, we get a protocol with total communication $N \cdot n$ and correlated randomness $m \cdot 2^{N \cdot n}$. However, it is straightforward to improve this protocol, by applying the OTTT protocol to the 1-party functionality g, and letting the trusted dealer distribute random shares of the shift r to all parties in the preprocessing phase: in the online phase, each party broadcasts his share of the input x, masked with his share of the shift r; this allows all parties to reconstruct $x + r$. With this modification, the parties need only to store a share of the one-dimensional truth-table of g, of size $m \cdot 2^n$.

Therefore, Lemma 8 can be seen as an generalization of the result of [IKM+13], which shifts the exponential cost of the correlated randomness from the input size to the locality parameter of the function. In the most general case, when $c = n$, we recover the result of [IKM+13] (for the special case of the secret sharing functionalities, and up to an additive factor n); when $c < n$, however, this leads to a protocol which uses a smaller amount of correlated randomness.

Proof. Let $g : \mathbb{F}_2^n \mapsto \mathbb{F}_2^m$ be a c-local function. Without loss of generality, we assume that each output bit of g depends on exactly c input bits. For $j = 1$ to m, we denote by $S_j \subset [n]$ the size-c subset of the bits of the input on which the j'th output bit depends. We denote by $g_j \leftarrow \mathsf{restrict}(g, j)$ the following function: $g_j : \mathbb{F}_2^c \mapsto \mathbb{F}_2$ is the function which, for any $x \in \mathbb{F}_2^n$, computes the j'th output bit of $g(x)$ when given the appropriate subset $x[S_j]$ of the bits of x as input.

We describe on Fig. 2 the protocol Π_{local}, which allows N parties holding shares of an input x to securely compute (in the semi-honest model, with correlated randomness) shares of $g(x)$, for some c-local function g. Below, we prove that Π_{local} satisfies all the properties of Lemma 8. It follows immediately by inspection that the total communication of Π_{local} is $N \cdot n$ bits, and that the amount of preprocessing material stored by each party is $m \cdot 2^c + n$. We now turn our attention to correctness and security.

Claim. The protocol Π_{local} is correct.

Protocol Π_{local}

Functionality:
- *Public parameters:* a c-local function $g : \mathbb{F}_2^n \mapsto \mathbb{F}_2^m$, and the m size-c subsets $S_j \subset [n]$ of the bits of the input on which the j'th output bit of g depends.
- *Input:* the parties $(P_1, \cdots, P_N)$ hold random shares $(x_1, \cdots, x_N)$ of an input x over $\mathbb{F}_2^n$;
- *Output:* the parties output uniformly random shares of $g(x)$.

$\Pi_{\text{local}}.\text{Preprocessing}(g)$:
1. Sample $(r_1, \cdots, r_N) \xleftarrow{\$} \mathbb{F}_2^n \times \cdots \times \mathbb{F}_2^n$. Set $r \leftarrow \sum_{i=1}^N r_i$.
2. For $j = 1$ to m, let $g_j \leftarrow \text{restrict}(g, j)$.
3. Let M_j denote the truth-table of g_j permuted with the shift $r[S_j]$, i.e., for any $y \in \mathbb{F}_2^c$, $M_j|_{y+r[S_j]} = g_j(y)$. Note that M_j is of size 2^c.
4. Let $(M_j^i)_{i \leq N, j \leq m}$ be random (N-out-of-N) secret sharings of the M_j. Output $(r_i, (M_j^i)_{j \leq m})$ to each party P_i for $i = 1$ to N.

$\Pi_{\text{local}}.\text{Protocol}(g, x)$:
1. Each party P_i with share x_i broadcasts $u_i \leftarrow x_i + r_i$. Let $u \leftarrow \sum_{i=1}^N u_i$.
2. *Output:* each party P_i outputs, for $j = 1$ to m, $z_{i,j} \leftarrow M_j^i|_{u[S_j]}$.

Fig. 2. Protocol Π_{local} for securely computing secret shares of a function g between N-party, with semi-honest and information-theoretic security in the correlated randomness model.

Proof: for any $j \in [m]$,

$$
\begin{aligned}
\sum_{i=1}^N z_{i,j} &= \sum_{i=1}^N M_j^i|_{u[S_j]} \\
&= M_j|_{u[S_j]} \text{ by definition of the } M_j^i \\
&= M_j|_{\sum_i x_i[S_j] + r_i[S_j]} \text{ by definition of } u \\
&= M_j|_{x[S_j] + r[S_j]} \\
&= g_j(x[S_j]) \text{ by definition of } M_j \\
&= g(x)[j] \text{ by definition of } g_j.
\end{aligned}
$$

We now turn our attention to security. We represent on Fig. 3 the ideal secret-sharing functionality for g. Note that the functionality explicitly allows the adversary to choose the output of the corrupted parties; this is a standard (and minor) technicality of protocols whose output is secret shared between the parties. An alternative is to let the functionality pick the output of all parties at random; however, to realize this functionality, we would need to add a (simple) resharing step at the end of the protocol Π_{local}, which would add unnecessary communication to the protocol.

Claim. The protocol Π_{local} implements the ideal functionality $\mathcal{F}_{\text{local}}$ with perfect security against a semi-honest corruption of a majority of the parties.

Ideal Functionality $\mathcal{F}_{\text{local}}$

The functionality is parametrized with the description of a function $g : \mathbb{F}_2^n \mapsto \mathbb{F}_2^m$, and the identities of and adversary $\mathscr{A}$ and N parties $P_1, \cdots, P_N$. The functionality aborts if it receives any incorrectly formatted message.

1. On input a message (corrupt, C) with $C \subsetneq [N]$ from $\mathscr{A}$, set $H \leftarrow [N] \setminus C$ and store (H, C).
2. On input a message (input, x_i) from each party P_i for $i \in [N]$, store $z \leftarrow g(\sum_{i \in N} x_i)$ and send ready to $\mathscr{A}$.
3. On input a message (set-output, $(z_i)_{i \in C}$) from $\mathscr{A}$, pick $|H|$ uniformly random values $(z_i)_{i \in H} \in (\mathbb{F}_2^n)^{|H|}$ under the constraint $\sum_{i \in H} z_i = z - \sum_{i \in C} z_i$.
4. On input a message (send, R) from $\mathscr{A}$ with $R \in [N]$, send z_i to each party P_i with $i \in R$, and $\perp$ to all other parties, then terminate.

Fig. 3. Ideal Functionality $\mathcal{F}_{\text{local}}$ for the secure computation of secret shares of $g(x)$ on an input $x \in \mathbb{F}_2^n$ shared between N parties.

Proof: let $H \subset [N]$ denote the subset of honest parties, and let $C \leftarrow [N] \setminus H$ denote the subset of (passively) corrupted parties; the simulator Sim first sends (corrupt, C) to $\mathcal{F}_{\text{local}}$ on behalf of the ideal aversary $\mathscr{A}$. Sim simulates the preprocessing phase by distributing uniformly random coins $(r_i, (M_j^i)_{j \leq m})_{i \in C}$ to all corrupted parties. In the online phase, the simulator picks random u_i in $\mathbb{F}_2^n$ for every $i \in H$, and broascasts them on behalf of the honest parties. When he receives $(u_i)_{i \in C}$, he computes for each $i \in C$ $x_i \leftarrow u_i - r_i$, and $z_i \leftarrow (M_j^i|_{u[S_j]})_{j \leq m} \in \mathbb{F}_2^m$. He sends (input, x_i) on behalf of each corrupted party P_i to the ideal functionality $\mathcal{F}_{\text{local}}$, and wait until he receives ready from $\mathcal{F}_{\text{local}}$. Then, he sends (set-output, $(z_i)_{i \in C}$) and (send, R) on behalf of $\mathscr{A}$ to $\mathcal{F}_{\text{local}}$, where R is the set of parties that can obtain the output (which Sim can obtain by observing which corrupted parties aborted early). It is immediate to see that the view of the environment (which consists of the preprocessing material, the u_i, and the outputs of the parties) in the ideal world with Sim is perfecty distributed as its view in the real world. This concludes the proof of the core lemma.

4   A Sublinear Protocol for Layered Circuits

In this section, we prove Theorem 1, by exhibiting a generic secure multiparty computation protocol in the correlated randomness model against passive corruption of a majority of the parties, for any layered boolean circuit, with sublinear communication in the circuit size s. Informally, the construction proceeds by breaking the layered circuit into chunks, each chunk containing $k = k(s)$ consecutive layers, for some function k. The parties will evaluate the circuit by computing shares of the values carried by the wires leaving a chunk, given as input shares of the values carried by the wires entering the chunk. As a chunk contains k layers and the directed graph of the circuit has indegree 2, this task corresponds to the secure evaluation of (shares of) a 2^k-local function, with (approximately) w inputs and w outputs (where w is the width of the circuit). By the core lemma (Lemma 8), this can be done with communication $O(w)$ and using $O(w \cdot 2^{2^k})$ bits of correlated randomness per party. After d/k chunk evaluations (d is the depth of the circuit), the parties end up with shares of the values the output wires, which they can broadcast to reconstruct the output. The total communication involved is $O(dw/k) = O(s/k)$, with $O(2^{2^k} \cdot s/k)$ bits of correlated randomness per party.

4.1   Construction

Let C be a layered boolean circuit with n inputs and m outputs, of size s and depth $d = d(n)$, with layers $(L_1, \cdots, L_d)$. For $i = 1$ to d, we let w_i denote the width of the layer L_i. We fix an arbitrary ordering of the nodes.

Let k be an integer. We divide C into $d' = \lceil d/k \rceil$ chunks $(\mathsf{ch}_i)_{i \leq d'}$, each chunk containing k consecutive layers (the last chunk contains less layers is k does not divide d). Let $t \in [k]$ be chosen so that the sum of the widths of the t'th layer of each chunk is bounded by $\lceil s/k \rceil$ (such a t necessarily exists, otherwise, we would get a contradiction: $s = \sum_{i=1}^{d} |L_i| = \sum_{i=1}^{k}(\sum_{j=1}^{d/k} |L_{jk+i}|) > \sum_{i=1}^{k} \lceil s/k \rceil \geq s$). For $i = 1$ to d', we denote t_i the index of the t'th layer in ch_i; it holds that $\sum_{i=1}^{d'} w_{t_i} \leq \lceil s/k \rceil$.

For $i = 1$ to d', we let m_i denote the number of output nodes between the layers $L_{t_{i-1}}$ and L_{t_i} ($\sum_i m_i = m$). For any $i \leq d'$, and $j \leq w_{t_i} + m_i$, we denote $\mathsf{n}_{i,j}$ the j'th node of the layer $L_{t_i} \in \mathsf{ch}_i$ if $j \leq w$, and the $(j - w)$'th output node between the layers $L_{t_{i-1}}$ and L_{t_i} otherwise. We associate two sets to each $\mathsf{n}_{i,j}$: we let $A_{i,j}$ denote the set of ancestors of $\mathsf{n}_{i,j}$ which belong to $L_{t_{i-1}}$ ($A_{1,j}$ is empty for all $j \leq w_{t_1} + m_1$), and we let $I_{i,j}$ denote the set of input nodes between the layers $L_{t_{i-1}}$ and L_{t_i} which are ancestors of $\mathsf{n}_{i,j}$. We let $\alpha_{i,j}$ (resp. $\iota_{i,j}$) denote the size of the set $A_{i,j}$ (resp. $I_{i,j}$). We illustrate this construction on Fig. 4. Observe that C has indegree 2, which implies that any node $\mathsf{n}_{i,j}$ of the t'th layer of a chunk can have at most 2^k ancestors in the t'th layer of the previous chunk, hence $\alpha_{i,j} + \iota_{i,j} \leq 2^k$.

Our protocol proceeds by evaluating the circuit C on an input $\boldsymbol{x}$ (seen as a size-N vector $(x_1, \cdots, x_N)$ over $\{0,1\}^{n/N}$, where x_j is the input of the party P_j) in a chunk-by-chunk fashion. We say that the parties *evaluate a chunk* i when they compute (shares of) all the values associated to the nodes of the layer L_{t_i}, as well as (shares of) all the values associated to the output nodes between the layers $L_{t_{i-1}}$ and L_{t_i}. Each chunk will be evaluated during a round. We will denote by $y_{i,\ell}$ the bitstring of the shares of the values on L_{t_i} computed by the party P_ℓ in the i'th round, and $y_i = \bigoplus_{\ell=1}^{N} y_{i,\ell}$ the reconstructed value. Similarly, we denote by $z_{i,\ell}$ the bitstring of the shares of the values on the output wires between $L_{t_{i-1}}$ and L_{t_i} computed by the party P_ℓ in the i'th round, and $z_i = \bigoplus_{\ell=1}^{N} z_{i,\ell}$ the reconstructed output string. For simplicity, for any $\ell \leq N$, we denote by $y_{0,\ell}$ an arbitrary dummy string (this is just to simplify the description of the protocol; as the $A_{1,j}$ are empty, these strings will not have any effect on the protocol anyway).

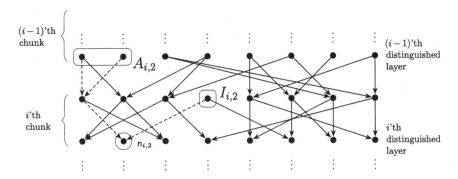

Fig. 4. Illustration of the construction of the sets $(A_{i,j}, I_{i,j})$ for a node $\mathsf{n}_{i,j}$ on a layered directed acyclic graph. The index j is taken equal to 2 on this figure. The dashed edges denote the paths of the graph that end at $\mathsf{n}_{i,2}$.

For any $i \leq d'$ and $j \leq w_{t_i} + m_i$, we let $f_{i,j}$ denote the following function: on input the substring $\boldsymbol{x}[I_{i,j}]$ of the input string $\boldsymbol{x}$, and the bitstring $y_{i-1}[A_{i,j}]$ (whose bits form a substring of the values in $L_{t_{i-1}}$), $f_{i,j}$ outputs the value associated to the node $\mathsf{n}_{i,j}$. We let $\delta_i \leftarrow w_{t_i} + m_i$ denote the number of functions $f_{i,j}$ for a fixed i. Finally, we denote by $f_i : \mathbb{F}_2^{w_{t_i}+n} \mapsto \mathbb{F}_2^{\delta_i}$ the following function: on input the string y_{i-1} associated to the distinguished layer of the $(i-1)$'th chunk and the input string $\boldsymbol{x}$, f_i outputs $(f_{i,j}(\boldsymbol{x}[I_{i,j}], y_{i-1}[A_{i,j}]))_{j \leq \delta_i} = (y_i, z_i)$. Observe that, by construction, f_i is a 2^k-local function (the j'th output bit of f_i depends on $\alpha_{i,j} + \iota_{i,j} \leq 2^k$ input bits). The full protocol is represented on Fig. 5.

4.2   Proof of Theorem 1

We now argue that the protocol Π_{sub} satisfies all the properties outlined in Theorem 1.

Correctness. It follows immediately by inspection: by the correctness of Π_{local}, the values $y_{i,\ell}$ computed by the parties form shares of the outputs of the functions $f_{i,j}$ evaluated on the ancestors (in $L_{t_{i-1}}$) of the nodes of layer L_{t_i} (and the ancestors in $L_{t_{i-1}}$ of the output nodes between the layers $L_{t_{i_1}}$ and L_{t_i}), as well as on the input nodes between the layers $L_{t_{i_1}}$ and L_{t_i}. By definitions, those values are exactly the values associated to the output nodes between the layers $L_{t_{i_1}}$ and L_{t_i} and the nodes in the layer L_{t_i}. From there, it immediately follows that the reconstructed outputs $(z_1, \cdots, z_m)$ are correct.

Security. We prove that the protocol Π_{sub} is perfectly secure against an adversary passively corrupting a majority of the parties. The ideal functionality $\mathcal{F}_{\mathsf{sub}}$ that Π_{sub} must realize is straightforward; it is represented on Fig. 6. The simulator Sim simply simulates Π_{sub} in the $\mathcal{F}_{\mathsf{local}}$ hybrid model, relying on the simulator for Π_{local} to interface with the real protocol. As Π_{sub} is a simple sequential composition of executions of Π_{local}, security follows immediately.

Protocol Π_{sub}

Functionality:
  – Public parameters: a layered boolean circuit C of size s and depth d, with n input gates and m output gates, and an integer k.
  – The parties $(P_1, \cdots, P_N)$ hold respective inputs $\boldsymbol{x} = (x_1, \cdots, x_N)$ of length n/N (we assume inputs of equal length for simplicity, but the protocol can be adapted to inputs of different lengths in a straightfoward way);
  – Output: all the parties learn $C(\boldsymbol{x})$.
Π_{sub}.Preprocessing(C) : for $i = 1$ to $d' = \lceil d/k \rceil$, execute Π_{local}.Preprocessing(f_i).
Π_{sub}.Protocol$(C, \boldsymbol{x})$:
  – For $i = 1$ to d', all parties execute Π_{local}.Protocol$(f_i, (y_{i-1}, \boldsymbol{x}))$ (using their shares of y_{i-1} and $\boldsymbol{x}$; note that $y_0 = $ by definition). Each party P_ℓ gets as output $(y_{i,\ell}, z_{i,\ell})$.
  – Output: all the parties broadcast the $z_{i,\ell}$ for $i = 1$ to d'. All the parties reconstruct the output $z = (\sum_\ell z_{i,\ell})_{i \leq d'}$.

Fig. 5. Protocol Π_{sub} for evaluating a layered boolean circuit C of size s and depth d, with n input gates and m output gates, in the correlated randomness model against passive corruption of up to $N - 1$ parties.

Complexity. We now analyze the communication, storage, computation, and interaction of the protocol Π_{sub}. We first outline a straightforward optimization: observe that each execution of Π_{local} to evaluate (shares of) the output of one of the f_i operates, in particular, on the input $\boldsymbol{x}$ (whose length is n). Instead of using independent executions of Π_{local}, where the input vector $\boldsymbol{x}$ ends up being re-shared between the parties for each execution, the parties can share it only once in an "input sharing step", before the execution of the first instance of Π_{local}, and reuse these shares in each execution. With this optimization, the parties exchange n bits in the input sharing step, and $N \cdot (\delta_i)$ bits during the i'th round of the circuit evaluation step, for $i = 1$ to $d' = \lceil d/k \rceil$. Therefore, the total number of bits exchanged is

$$n + N \cdot \sum_{i=1}^{d'} w_{t_{i-1}} + m_{i-1} \leq n + N \cdot (m + \lceil s/k \rceil)$$

(note that the additive factor n would be $n \cdot d'$ without the simple optimization outlined above). The amount of correlated randomness stored by each party can be upper bounded by

$$n/N + \sum_{i=1}^{d'} \sum_{j=1}^{\delta_i} 2^{\alpha_{i,j} + \iota_{i,j}} \leq n/N + \sum_{i=1}^{d'} \sum_{j=1}^{\delta_i} 2^{2^k} \leq n/N + (m + \lceil s/k \rceil) \cdot 2^{2^k},$$

where the first inequality comes from the fact that any node $\mathsf{n}_{i,j}$ of the t'th layer of a chunk can have at most 2^k ancestors in the t'th layer of the previous chunk, which leads to the claimed total storage. Eventually, the round complexity of

---

Ideal Functionality $\mathcal{F}_{\mathsf{sub}}$

The functionality is parametrized with the description of a circuit $C : \mathbb{F}_2^n \mapsto \mathbb{F}_2^m$, and the identities of and adversary $\mathscr{A}$ and N parties $P_1, \cdots, P_N$. The functionality aborts if it receives any incorrectly formatted message.

1. On input a message (corrupt, C) with $C \subsetneq [N]$ from $\mathscr{A}$, set $H \leftarrow [N] \setminus C$ and store (H, C).
2. On input a message (input, x_ℓ) from each party P_ℓ for $\ell \in [N]$, store $z \leftarrow C(x_1, \cdots, x_\ell)$ and send ready to $\mathscr{A}$.
3. On input a message (send, R) from $\mathscr{A}$ with $R \in [N]$, send z to all parties P_ℓ with $\ell \in R$, and $\perp$ to all other parties, then terminate.

Fig. 6. Ideal Functionality $\mathcal{F}_{\mathsf{sub}}$ for the secure computation of secret shares of $g(x)$ on an input $x \in \mathbb{F}_2^n$ shared between N parties.

the protocol is $d' + 1 = O(s/k)$, and the computation performed by each party essentially boils down to performing $m + \lceil s/k \rceil$ searches in lookup tables of size bounded by 2^{2^k}, which takes time $(m + \lceil s/k \rceil) \cdot 2^k$.

4.3 Extension to Layered Arithmetic Circuits

So far, our protocol does not readily extend to arithmetic circuits over (exponentially large) finite fields. The main obstacle toward getting an arithmetic analogue of the protocol Π_{sub} lies in the generalization of the core lemma to the arithmetic setting: our proof of Lemma 8 relies on the fact that we can use the OTTT protocol of [IKM+13] to evaluate functions with a "small enough" truth-table. While in the boolean case, any functionality with c input bits has a truth table of size 2^c, this is not true anymore for arithmetic functionalities over large fields, where even single-input functions have truth table of exponential size. In addition, the standard conversion of arithmetic circuits into boolean circuits would blow up the size too much: any size-s arithmetic circuit can be securely evaluated (in the correlated randomness model) with communication $O(s)$ (counting the number of field elements), but the conversion to a boolean circuit will in general blow up the circuit size by a $\log s$ factor, while our protocol only saves a factor $\log \log s$, and does therefore not lead to a sublinear communication protocol for arithmetic circuits.

Nevertheless, we show that our protocol can be extended to the arithmetic setting, by exhibiting a natural analogue of the OTTT protocol, tailored to arithmetic functions. Our starting point is the recent work of [LVW17], on conditional disclosure of secret and private simultaneous message (PSM) protocols. The authors of [LVW17] build an elegant PSM protocol for multivariate polynomial evaluation. The protocol has the following features: Alice holds an n-variate polynomial P of degree $\deg$, Bob holds a vector of input $\boldsymbol{x} \in \mathbb{F}^n$, and both parties share a common random string. They send a single simultaneous message to a third player, Charlie, with optimal communication (Alice's message has size

$O(\binom{n+\deg}{\deg})$, Bob's message has size $O(n)$. This allows Charlie to learn $P(\boldsymbol{x})$, and nothing more. The protocol works as follows:

- The shared randomness is $\boldsymbol{r} \in \mathbb{F}^n$ and a random n-variate degree-deg polynomial R.
- Alice sends $(\boldsymbol{x}', u) \leftarrow (\boldsymbol{x} + \boldsymbol{r}, R(\boldsymbol{x} + \boldsymbol{r}))$.
- Bob sends the polynomial $Q(\boldsymbol{X}) = P(\boldsymbol{X} - \boldsymbol{r}) + R(\boldsymbol{X})$.
- Charlie outputs $Q(\boldsymbol{x}') - u$.

The correctness follows immediately by inspection, and security follows by the argument of [LVW17, Sect. 5.1]. The above PSM can be readily converted into a 2-player arithmetic analogue of the OTTT protocol, which relies on a multivariate polynomial representation of an arithmetic circuit (instead of a truth table representation). We represent on Fig. 7 a variant of the protocol Π_{local}, tailored to the arithmetic setting (over an arbitrary field $\mathbb{F}$). Each party sends n field elements (which is essentially optimal), and stores $O(\binom{n+\deg}{\deg})$ field elements.

Using the above protocol, we immediately get a generalization of Lemma 8:

Lemma 9. *For any depth-k arithmetic circuit $f : \mathbb{F}^n \mapsto \mathbb{F}^m$, there is an information-theoretic semi-honest N-party secure computation protocol (with dishonest majority) in the correlated randomness model for computing secret shares of f with total online communication $N \cdot n$ elements of $\mathbb{F}$, and correlated randomness $m \cdot \binom{2^{k+1}}{2^k} + n \approx m \cdot 2^{2^{k+1}}/\sqrt{\pi 2^k} + n$ elements of $\mathbb{F}$ per party.*

Therefore, we get polynomial storage (in s) by setting $k \leftarrow \log \log s$, as before. This leads to a protocol for arithmetic circuits of size s, with n inputs and m outputs, with polynomial storage and total communication $O(n + N \cdot (m + s/\log\log s))$.

Reducing Storage in TinyTable. While the idea of using a multivariate-polynomial representation instead of a truth-table representation seems relatively natural and is the key to extend the construction to the arithmetic setting, it was not explicitly observed before. Somewhat surprisingly, we observe that even in the original (boolean) setting of the TinyTable paper [DNNR17] (which uses truth-table representation at the gate level, for two-party evaluation of AND gates in boolean circuits), replacing truth-tables by multivariate polynomials in normal form improves the construction: it reduces the storage of the parties by 25%. We sketch this observation below. The TinyTable protocol maintains the following invariant: the parties know masked representation of all inputs to some gate of the circuit, and will compute a masked representation of the output. Typically, for a two-input AND gate, both parties will know $u = x+r$ and $v = y + s$, where x, y are the inputs to the gate, and r, s are random masks. In addition, the parties hold random shares of the truth-table of the function

$$F_{r,s,t} : (u,v) \to (u-r) \cdot (v-s) + t,$$

Protocol POLY

Functionality:
- Public parameters: an arithmetic function $f : \mathbb{F}^n \mapsto \mathbb{F}^m$ of depth k over a finite field $\mathbb{F}$, and the m size-2^k subsets $S_j \in [n]$ of the coordinates of the input on which the j'th coordinate of the output of f depends.
- The parties $(P_1, \cdots, P_N)$ hold additive shares $(\boldsymbol{x}_1, \cdots, \boldsymbol{x}_N)$ of an input $\boldsymbol{x} \in \mathbb{F}^n$;
- Output: the parties output uniformly random shares of $f(\boldsymbol{x})$.

Preprocessing :
1. Sample $(\boldsymbol{r}_1, \cdots, \boldsymbol{r}_N) \xleftarrow{\$} \mathbb{F}^n \times \cdots \times \mathbb{F}^n$. Set $\boldsymbol{r} \leftarrow \sum_{i=1}^N \boldsymbol{r}_i$.
2. For $j = 1$ to m, let $f_j \leftarrow \mathsf{restrict}(f, j)$.
3. For $j = 1$ to m, Let $P_j(\boldsymbol{X})$ denote the normal-form of f_j, seen as a 2^k-variate polynomial of degree 2^k over $\mathbb{F}$.
4. For $j = 1$ to m, let $Q_j(\boldsymbol{X}) \leftarrow P(\boldsymbol{X} - \boldsymbol{r}[S_j])$ denote the polynomial P shifted with $\boldsymbol{r}[S_j]$.
5. For $j = 1$ to m, sample $N - 1$ uniformly random degree-2^k n-variate polynomials $(R_j^i(\boldsymbol{X}))_{i \leq N-1}$, and set $R_j^N(\boldsymbol{X}) \leftarrow Q_j(\boldsymbol{X}) + \sum_{i=1}^{N-1} R_j^i(\boldsymbol{X})$. Output $(\boldsymbol{r}_i, (R_j^i(\boldsymbol{X}))_{j \leq m})$ to each party P_i.

Protocol($\boldsymbol{x}$) :
1. Each party P_i with share $\boldsymbol{x}_i$ broadcasts $\boldsymbol{u}_i \leftarrow \boldsymbol{x}_i + \boldsymbol{r}_i$. Let $\boldsymbol{u} \leftarrow \sum_i \boldsymbol{u}_i$.
2. Each party P_i outputs, for $j = 1$ to m, $z_{i,j} \leftarrow R_j^i(\boldsymbol{u}[S_j])$.

Fig. 7. Protocol POLY for evaluating an arithmetic function f over a finite field $\mathbb{F}$ in the correlated randomness model, against a passively corrupted majority

where t is another fresh random coin. Observe that $F_{r,s,t}(x+r, y+s) = x \cdot y + t$, maintaining the appropriate invariant. In the TinyTable paper, each party knows a share F_0, F_1 of the truth-table of a function of this form, for each AND gate of the circuit, and the output is computed by broadcasting $F_0(u, v), F_1(u, v)$ and reconstructing $w = F_0(u, v) \oplus F_1(u, v)$. This represent a total storage of $4s$ bits per party (and $2s$ bits of communication), where s is the number of AND gates of the circuit.

Now, if we view instead $F_{r,s,t}$ as a degree-2 polynomial in two variables, we have $F_{r,s,t} = uv + \alpha u + \beta v + \gamma$ for some appropriate $(\alpha, \beta, \gamma) = (-s, -r, t+rs)$. Observe that to randomly share $F_{r,s,t}$ *viewed as a multivariate polynomial*, it suffices to share additively each of its coefficients randomly; furthermore, the leading coefficient of $F_{r,s,t}$ is always one. Hence, we can improve the TinyTable AND gate evaluation protocol as follows: the parties receive shares $(\alpha_0, \beta_0, \gamma_0)$ and $(\alpha_1, \beta_1, \gamma_1)$ of (α, β, γ) (this is identical to giving a random degree-one bivariate polynomial R to one party, and $F_{r,s,t} + R$ to the other party; note that R needs only having degree one since it needs not hide the leading coefficient of $F_{r,s,t}$, which is 1). Given public values $u = x + r$ and $v = y + s$, the parties exchange $w_0 = \alpha_0 u + \beta_0 v + \gamma_0$ and $w_1 = \alpha_1 u + \beta_1 v + \gamma_1$, and publicly reconstruct $w = uv + w_0 + w_1$. The communication and computation are essentially the same as in [DNNR17], but the parties must now only store three bits per AND gate, hence $3s$ bits in total, reducing the amount of storage required by the protocol by 25%.

4.4   Further Extensions

We sketch in this section how to extend our protocol to the case of function-independent correlated randomness, and to the case of tall-and-narrow circuits.

Function-Independent Preprocessing. We introduce below a variant of the core lemma, tailored to function-independent preprocessing. Theorem 3 follows immediately from this variant.

Lemma 10. *For any c-local function $g : \mathbb{F}_2^n \mapsto \mathbb{F}_2^m$, there is an information-theoretic semi-honest N-party secure computation protocol (with dishonest majority) in the* function-independent *correlated randomness model for computing secret shares of g with total online communication $N \cdot n$ bits, and correlated randomness $m \cdot 2^{c+2^c} + n$ bits per party.*

Proof. To prove Lemma 10, we modify Π_{local} as follows: instead of computing shares of the truth table M_j of g_j (which is of size 2^c) permuted with the shift $r[S_j]$, we consider the list $(M_{j,q})_{q \leq 2^{2^c}}$ of all possible truth tables, corresponding to a lexicographic ordering of all possible functions g_j, each table being shifted with the same $r[S_j]$. Each party P_i receives $(r_i, (M_{j,q}^i)_q)$, which amounts to $n + 2^c \cdot 2^{2^c}$ bits of correlated randomness. In the online protocol Π_{local}.Protocol, when the functions g_j are revealed, the parties locally drop all unnecessary shares of shifted truth tables, keeping only the one corresponding to g_j. The security analysis immediately follows from the analysis of Π_{local}.

Tall-and-Narrow Circuits. For tall-and-narrow circuits, whose width w is small, the proof follows by observing that in this situation the bound on the size of the sets $A_{i,j}$ and $I_{i,j}$ can be refined to $|A_{i,j}| + |I_{i,j}| \leq w \cdot k$, hence the $f_{i,j}$ have truth tables of size bounded by $2^{w \cdot k}$. Theorem 5 follows immediately.

5   Malicious Setting

In the two-party case, combining our passively secure protocol Π_{sub} with the techniques of [DNNR17] directly implies the existence of a (statistical) unconditionally secure two-party protocol secure against malicious adversaries, with communication $O(n + m + \frac{s}{\log \log s} + \kappa)$ for a layered boolean circuit of size s, where κ is a statistical security parameter. Indeed, the protocol of [DNNR17] has a structure similar to our protocol: it decomposes the circuit into tables, and distributes scrambled version of these tables to the parties in the preprocessing phase. Each gate of the circuit is evaluated using the OTTT protocol to obliviously select the output of the gate from its corresponding scrambled truth-table.

To enhance this protocol to security against malicious adversaries, [DNNR17] uses a simple and natural information-theoretic authentication procedure. Namely, for each entry $b \in \{0,1\}$ of a table given to the first party (let us call it A), the trusted dealer additionally generates two κ-bit string (x_0, x_1), hands x_b to A, and (x_0, x_1) to the other player B. This way, when A must send the entry b of the table to B, she can authenticate b by sending it along with x_b. B then retrieves the corresponding value x_b and checks that A honestly opened b; if A is dishonest, she will be caught with probability $1 - 2^{-\kappa}$. Note that the only local computation performed by the parties are searches through lookup table, hence authenticating each entry this way suffices to guarantee security of the entire protocol.

However, directly applying this approach would require transmitting κ bits per output of a table, which would increase the total communication by a factor of κ. To avoid this overhead, the authors of [DNNR17] observe that it is not necessary to explicitly authenticate each entry sent by a party. Instead, each time A reveals an entry b corresponding to some authentication string x_b, she locally updates a "global MAC key" $\Delta_A \leftarrow \Delta_A \oplus x_b$, where Δ_A is set to 0 at the start of the protocol. Simultaneously, when he receives an entry b from A, B retrieves the pair (x_0, x_1) corresponding to this entry, and locally updates $\Gamma_B \leftarrow \Gamma_B \oplus x_b$, where Γ_B is set to 0 at the start of the protocol. The parties proceed symmetrically, with Δ_B, Γ_A, when B sends an entry to A. At the end of the protocol, A reveals Δ_A and B reveals Δ_B. If $\Delta_B \neq \Gamma_A$, A aborts the protocol; B does the same if $\Delta_A \neq \Gamma_B$. If both checks passed, the parties reconstruct the output. The analysis of [DNNR17] shows that this guarantees that no party can cause its opponent to accept an incorrect output, except with probability $2^{-\kappa}$. It increases the amount of preprocessed material by a factor κ,[4] but only adds 2κ bits to the total communication.

For completeness, we provide a full self-contained description of the maliciously-secure two-party version of our protocol on Fig. 8. We refer the reader to Theorem 1 of [DNNR17] for a detailed proof of security against malicious adversaries; it is straightforward to adapt the proof to our protocol (we note that, while [DNNR17] focuses on small tables implementing standard two-input boolean gates, [DNNR17, Sect. 2.3] already observes that this mechanisms can be directly generalized to protocols evaluating larger tables).

Extension to N Parties. While the work of [DNNR17] focused only on (maliciously secure) two-party computation, it was subsequently observed in [KOR+17] that the techniques used in [DNNR17] can be easily generalized to the multiparty setting, for an arbitrary number N of parties. We refer the reader to [KOR+17] for more details; this directly gives:

---

[4] A technique to amortize this overhead, using a linear MAC scheme, is described in [DNNR17]; it applies to our setting as well, and allows to remove this factor κ overhead in the storage complexity, but we focus on the more naive approach in this work for simplicity.

Theorem 11. *For any N-party functionality f represented by a layered boolean circuit C of size s with n inputs and m outputs, and for any integer k and statistical security parameter κ, there is a κ-secure protocol which realizes f in the preprocessing model against malicious adversaries with adaptive corruption (of up to $N-1$ parties), with communication $n + N \cdot (m + \lceil s/k \rceil + \kappa)$ and correlated randomness $n/N + (3\kappa + 1) \cdot (m + \lceil s/k \rceil) \cdot (2^{2^k} + 1)$ per party.*

6   Open Questions

While our work shows that a large class of circuits of size s can be securely evaluated in the correlated randomness model using $o(s)$ communication, many questions related to the communication of MPC in the correlated randomness model remain open.

Question 1. Can our protocols be extended to arbitrary non-layered circuits?

It is immediate to extend our protocol to any circuit that is layered "by blocks" of depth c, in the sense that no edge crosses more than c consecutive layers, for any $c = o(\log \log s)$. However, generalizing our result to all circuits remains an interesting open question.

Question 2. Can we achieve better sublinearity for unconditional MPC in the correlated randomness model, in general or for specific circuits?

It is known that some specific functions can be evaluated in the correlated randomness model, with stronger sublinearity guarantees than those obtained in this work. In particular, *matrix multiplication* can be computed with communication linear in the size n^2 of the matrices, while the best known algorithm for multiplying matrices of size n requires $O(n^t)$ communication, with $t \approx 2.3$. The work of [BIKO12] also implies the existence of low-communication protocols in the correlated randomness model, for $N \geq 3$ parties, for specific types of constant-depth circuits. It would be interesting to improve the sublinearity of our work, and to characterize the functions for which better sublinearity can be achieved.

Question 3. Can we achieve sublinear communication and linear storage at the same time?

By a lower bound of [WW10], linear storage is the best we can hope for. Our protocols only achieve slightly superlinear storage; in the regime where the $1/\log \log s$ factor would give non-trivial communication savings, this implies that a rather large storage is required. Protocols for specific functions, such as matrix multiplication, achieve both sublinearity and linear storage, but the question remains open for more general functions.

Protocol $\Pi_{\text{sub}}^{\text{mal}}$

Functionality:
- Public parameters: a layered boolean circuit C of size s and depth d, with n input gates and m output gates, and an integer k. We let κ denote a statistical security parameter.
- The parties (P_1, P_2) hold respective inputs $\boldsymbol{x} = (x_1, x_2)$ of length $n/2$.

$\Pi_{\text{sub}}^{\text{mal}}$.Preprocessing :
- Sample $\boldsymbol{\rho} = (\rho_1, \rho_2) \xleftarrow{\$} (\{0,1\}^{n/2})^2$ and for $i = 1$ to d', sample δ_i bits $\boldsymbol{r}_i = (r_{i,1}, \cdots, r_{i,\delta_i})$ such that $r_{i,j}$ is 0 if (i, j) corresponds to an output gate, and random otherwise (looking ahead, the bits $r_{i,j}$ will be used to mask the output value of the function $f_{i,j}$).
- For $i = 1$ to d', for $j = 1$ to δ_i, let $M_{i,j}$ denote the permuted truth table of $f_{i,j}$ with shifts $(\boldsymbol{\rho}[I_{i,j}], \boldsymbol{r}_{i-1}[A_{i,j}])$ and output masked with $r_{i,j}$, i.e.:

$$M_{i,j}|_{(\boldsymbol{x} \oplus \boldsymbol{\rho})[I_{i,j}], (y \oplus \boldsymbol{r}_{i-1})[A_{i,j}]} = f_{i,j}(\boldsymbol{x}[I_{i,j}], y[A_{i,j}]) \oplus r_{i,j}.$$

- For $i = 1$ to d', for $j = 1$ to δ_i, sample a random truth-table $R_{i,j}^1$ over $\{0,1\}^{2^{\alpha_{i,j} + \iota_{i,j}}}$, and let $R_{i,j}^2 \leftarrow R_{i,j}^1 \oplus M_{i,j}$.
- For $\ell = 1, 2$, for $i = 1$ to d', for $j = 1$ to δ_i, for $q = 1$ to $2^{\alpha_{i,j} + \iota_{i,j}}$, we denote $R_{i,j,q}^\ell$ the q'th entries of $R_{i,j}^\ell$. Sample 2 random κ-bit strings $(s_{\ell,i,j,q}^0, s_{\ell,i,j,q}^1)$ (looking ahead, these values will allow to authenticate the value $R_{i,j,q}^\ell$). To simplify notations, we denote by $s'_{\ell,i,j,q}$ the value

$$s'_{\ell,i,j,q} \leftarrow s_{3-\ell,i,j,q}^b, \text{ with } b = R_{i,j,q}^{3-\ell}.$$

- For $\ell = 1, 2$, output to P_ℓ

$$\left(\rho_\ell, (R_{i,j}^\ell)_{i \le d', j \le \delta_i}, \left(s_{\ell,i,j,q}^0, s_{\ell,i,j,q}^1, s'_{\ell,i,j,q} \right)_{i,j,q} \right).$$

$\Pi_{\text{sub}}^{\text{mal}}$.Protocol$(\boldsymbol{x})$:
- Initialization: for $\ell = 1, 2$, P_ℓ sets $\Delta_\ell = \Gamma_\ell = 0^\kappa$.
- Input Sharing: for $\ell = 1, 2$, P_ℓ broadcasts $u_\ell \leftarrow x_\ell \oplus \rho_\ell$. Let $\boldsymbol{u} \leftarrow (u_1, u_2)$. Set v_0 to be an arbitrary dummy string.
- Circuit Evaluation: for $i = 1$ to d',
  - For $\ell = 1, 2$, P_ℓ sets

  $$\boldsymbol{v}_{i,\ell} \leftarrow \left(R_{i,1}^\ell |_{\boldsymbol{u}[I_{i,1}], v_{i-1}[A_{i,1}]}, \cdots, R_{i,\delta_i}^\ell |_{\boldsymbol{u}[I_{i,\delta_i}], v_{i-1}[A_{i,\delta_i}]} \right).$$

  - For $\ell = 1, 2$, P_ℓ broadcasts $\boldsymbol{v}_{i,\ell}$; let $\boldsymbol{v}_i \leftarrow \bigoplus_{\ell=1}^N \boldsymbol{v}_{i,\ell}$.
  - For $\ell = 1, 2$, P_ℓ sets $q_{i,j}$ to be the string $(\boldsymbol{u}[I_{i,j}], \boldsymbol{v}_{i-1}[A_{i,j}])$, and sets

  $$\Delta_\ell \leftarrow \Delta_\ell \oplus s'_{3-\ell,i,j,q_{i,j}}, \Gamma_\ell \leftarrow \Gamma_\ell \oplus s_{\ell,i,j,q_{i,j}}^{v_{i,3-\ell,j}}.$$

- Verification of all opened bits: for $\ell = 1, 2$, P_ℓ sends Δ_ℓ to $P_{3-\ell}$, and $P_{3-\ell}$ checks that $\Gamma_{3-\ell} = \Delta_\ell$.
- Output: For $\ell = 1, 2$, P_ℓ broadcasts the δ_i-bit string $v_{i,\ell,j}$ for every $i \le d'$ and $j > w_{t_i}$. All the parties reconstruct $\boldsymbol{z} = (z_1, \cdots, z_m) = (\bigoplus_\ell v_{i,\ell,j})_{i \le d', j > w_{t_i}}$.

Fig. 8. Two-party protocol $\Pi_{\text{sub}}^{\text{mal}}$ for evaluating a layered boolean circuit C of size s and depth d, with n input gates and m output gates, in the correlated randomness model against active corruption of one of the parties.

Acknowledgements. We thank Yuval Ishai for helpful comments and pointers. Work supported by ERC grant 724307 (project PREP-CRYPTO).

References

[AJL+12] Asharov, G., Jain, A., López-Alt, A., Tromer, E., Vaikuntanathan, V., Wichs, D.: Multiparty computation with low communication, computation and interaction via threshold FHE. In: Pointcheval, D., Johansson, T. (eds.) EUROCRYPT 2012. LNCS, vol. 7237, pp. 483–501. Springer, Heidelberg (2012). https://doi.org/10.1007/978-3-642-29011-4_29

[AKD03] Atallah, M.J., Kerschbaum, F., Du, W.: Secure and private sequence comparisons. In: Proceedings of the 2003 ACM Workshop on Privacy in the Electronic Society, pp. 39–44. ACM (2003)

[ALSZ13] Asharov, G., Lindell, Y., Schneider, T., Zohner, M.: More efficient oblivious transfer and extensions for faster secure computation. In: ACM CCS 2013, pp. 535–548. ACM Press, November 2013

[BCG+17] Boyle, E., Couteau, G., Gilboa, N., Ishai, Y., Orrù, M.: Homomorphic secret sharing: optimizations and applications. In: ACM CCS 2017, pp. 2105–2122. ACM Press (2017)

[BCG+18] Boyle, E., Couteau, G., Gilboa, N., Ishai, Y., Orrù, M.: Homomorphic secret sharing: optimizations and applications. Cryptology ePrint Archive, Report 2018/419 (2018). https://eprint.iacr.org/2018/419

[BDOZ11] Bendlin, R., Damgård, I., Orlandi, C., Zakarias, S.: Semi-homomorphic encryption and multiparty computation. In: Paterson, K.G. (ed.) EURO-CRYPT 2011. LNCS, vol. 6632, pp. 169–188. Springer, Heidelberg (2011). https://doi.org/10.1007/978-3-642-20465-4_11

[Bea92] Beaver, D.: Efficient multiparty protocols using circuit randomization. In: Feigenbaum, J. (ed.) CRYPTO 1991. LNCS, vol. 576, pp. 420–432. Springer, Heidelberg (1992). https://doi.org/10.1007/3-540-46766-1_34

[Bea95] Beaver, D.: Precomputing oblivious transfer. In: Coppersmith, D. (ed.) CRYPTO 1995. LNCS, vol. 963, pp. 97–109. Springer, Heidelberg (1995). https://doi.org/10.1007/3-540-44750-4_8

[Bea97] Beaver, D.: Commodity-based cryptography (extended abstract). In: 29th ACM STOC, pp. 446–455. ACM Press, May 1997

[BFKR91] Beaver, D., Feigenbaum, J., Kilian, J., Rogaway, P.: Security with low communication overhead. In: Menezes, A.J., Vanstone, S.A. (eds.) CRYPTO 1990. LNCS, vol. 537, pp. 62–76. Springer, Heidelberg (1991). https://doi.org/10.1007/3-540-38424-3_5

[BGI15] Boyle, E., Gilboa, N., Ishai, Y.: Function secret sharing. In: Oswald, E., Fischlin, M. (eds.) EUROCRYPT 2015. LNCS, vol. 9057, pp. 337–367. Springer, Heidelberg (2015). https://doi.org/10.1007/978-3-662-46803-6_12

[BGI16] Boyle, E., Gilboa, N., Ishai, Y.: Breaking the circuit size barrier for secure computation under DDH. In: Robshaw, M., Katz, J. (eds.) CRYPTO 2016, Part I. LNCS, vol. 9814, pp. 509–539. Springer, Heidelberg (2016). https://doi.org/10.1007/978-3-662-53018-4_19

[BGI17] Boyle, E., Gilboa, N., Ishai, Y.: Group-based secure computation: optimizing rounds, communication, and computation. In: Coron, J.-S., Nielsen, J.B. (eds.) EUROCRYPT 2017. LNCS, vol. 10211, pp. 163–193. Springer, Cham (2017). https://doi.org/10.1007/978-3-319-56614-6_6

[BI05] Barkol, O., Ishai, Y.: Secure computation of constant-depth circuits with applications to database search problems. In: Shoup, V. (ed.) CRYPTO 2005. LNCS, vol. 3621, pp. 395–411. Springer, Heidelberg (2005). https://doi.org/10.1007/11535218_24

[BIKK14] Beimel, A., Ishai, Y., Kumaresan, R., Kushilevitz, E.: On the cryptographic complexity of the worst functions. In: Lindell, Y. (ed.) TCC 2014. LNCS, vol. 8349, pp. 317–342. Springer, Heidelberg (2014). https://doi.org/10.1007/978-3-642-54242-8_14

[BIKO12] Beimel, A., Ishai, Y., Kushilevitz, E., Orlov, I.: Share conversion and private information retrieval. In: 2012 IEEE 27th Annual Conference on Computational Complexity (CCC), pp. 258–268. IEEE (2012)

[BLN+15] Burra, S.S., et al.: High performance multi-party computation for binary circuits based on oblivious transfer. Cryptology ePrint Archive, Report 2015/472 (2015). http://eprint.iacr.org/2015/472

[CDv88] Chaum, D., Damgård, I.B., van de Graaf, J.: Multiparty computations ensuring privacy of each party's input and correctness of the result. In: Pomerance, C. (ed.) CRYPTO 1987. LNCS, vol. 293, pp. 87–119. Springer, Heidelberg (1988). https://doi.org/10.1007/3-540-48184-2_7

[CKL15] Cheon, J.H., Kim, M., Lauter, K.: Homomorphic computation of edit distance. In: Brenner, M., Christin, N., Johnson, B., Rohloff, K. (eds.) FC 2015. LNCS, vol. 8976, pp. 194–212. Springer, Heidelberg (2015). https://doi.org/10.1007/978-3-662-48051-9_15

[DFH12] Damgård, I., Faust, S., Hazay, C.: Secure two-party computation with low communication. In: Cramer, R. (ed.) TCC 2012. LNCS, vol. 7194, pp. 54–74. Springer, Heidelberg (2012). https://doi.org/10.1007/978-3-642-28914-9_4

[DKS+17] Dessouky, G., Koushanfar, F., Sadeghi, A.-R., Schneider, T., Zeitouni, S., Zohner, M.: Pushing the communication barrier in secure computation using lookup tables. In: Network and Distributed System Security Symposium (NDSS 2017). The Internet Society (2017)

[DLT14] Damgård, I., Lauritsen, R., Toft, T.: An empirical study and some improvements of the MiniMac protocol for secure computation. In: Abdalla, M., De Prisco, R. (eds.) SCN 2014. LNCS, vol. 8642, pp. 398–415. Springer, Cham (2014). https://doi.org/10.1007/978-3-319-10879-7_23

[DNNR17] Damgård, I., Nielsen, J.B., Nielsen, M., Ranellucci, S.: The TinyTable protocol for 2-party secure computation, or: gate-scrambling revisited. In: Katz, J., Shacham, H. (eds.) CRYPTO 2017. LNCS, vol. 10401, pp. 167–187. Springer, Cham (2017). https://doi.org/10.1007/978-3-319-63688-7_6

[DNPR16] Damgård, I., Nielsen, J.B., Polychroniadou, A., Raskin, M.: On the communication required for unconditionally secure multiplication. In: Robshaw, M., Katz, J. (eds.) CRYPTO 2016, Part II. LNCS, vol. 9815, pp. 459–488. Springer, Heidelberg (2016). https://doi.org/10.1007/978-3-662-53008-5_16

[DPSZ12] Damgård, I., Pastro, V., Smart, N.P., Zakarias, S.: Multiparty computation from somewhat homomorphic encryption. In: Safavi-Naini, R., Canetti, R. (eds.) CRYPTO 2012. LNCS, vol. 7417, pp. 643–662. Springer, Heidelberg (2012). https://doi.org/10.1007/978-3-642-32009-5_38

[DZ13] Damgård, I., Zakarias, S.: Constant-overhead secure computation of boolean circuits using preprocessing. In: Sahai, A. (ed.) TCC 2013. LNCS, vol. 7785, pp. 621–641. Springer, Heidelberg (2013). https://doi.org/10.1007/978-3-642-36594-2_35

[DZ16]   Damgård, I., Zakarias, R.W.: Fast oblivious AES a dedicated application of the MiniMac protocol. In: Pointcheval, D., Nitaj, A., Rachidi, T. (eds.) AFRICACRYPT 2016. LNCS, vol. 9646, pp. 245–264. Springer, Cham (2016). https://doi.org/10.1007/978-3-319-31517-1_13

[FKN94]   Feige, U., Kilian, J., Naor, M.: A minimal model for secure computation (extended abstract). In: 26th ACM STOC, pp. 554–563. ACM Press, May 1994

[FKOS15] Frederiksen, T.K., Keller, M., Orsini, E., Scholl, P.: A unified approach to MPC with preprocessing using OT. In: Iwata, T., Cheon, J.H. (eds.) ASIACRYPT 2015. LNCS, vol. 9452, pp. 711–735. Springer, Heidelberg (2015). https://doi.org/10.1007/978-3-662-48797-6_29

[Gen09]   Gentry, C.: Fully homomorphic encryption using ideal lattices. In: 41st ACM STOC, pp. 169–178. ACM Press, May/June 2009

[GMW87a] Goldreich, O., Micali, S., Wigderson, A.: How to play any mental game or A completeness theorem for protocols with honest majority. In: 19th ACM STOC, pp. 218–229. ACM Press, May 1987

[GMW87b] Goldreich, O., Micali, S., Wigderson, A.: How to prove all NP statements in zero-knowledge and a methodology of cryptographic protocol design (extended abstract). In: Odlyzko, A.M. (ed.) CRYPTO 1986. LNCS, vol. 263, pp. 171–185. Springer, Heidelberg (1987). https://doi.org/10.1007/3-540-47721-7_11

[HEKM11] Huang, Y., Evans, D., Katz, J., Malka, L.: Faster secure two-party computation using garbled circuits. In: USENIX Security Symposium, pp. 331–335 (2011)

[IKM+13] Ishai, Y., Kushilevitz, E., Meldgaard, S., Orlandi, C., Paskin-Cherniavsky, A.: On the power of correlated randomness in secure computation. In: Sahai, A. (ed.) TCC 2013. LNCS, vol. 7785, pp. 600–620. Springer, Heidelberg (2013). https://doi.org/10.1007/978-3-642-36594-2_34

[IPS08]   Ishai, Y., Prabhakaran, M., Sahai, A.: Founding cryptography on oblivious transfer – efficiently. In: Wagner, D. (ed.) CRYPTO 2008. LNCS, vol. 5157, pp. 572–591. Springer, Heidelberg (2008). https://doi.org/10.1007/978-3-540-85174-5_32

[IPS09]   Ishai, Y., Prabhakaran, M., Sahai, A.: Secure arithmetic computation with no honest majority. In: Reingold, O. (ed.) TCC 2009. LNCS, vol. 5444, pp. 294–314. Springer, Heidelberg (2009). https://doi.org/10.1007/978-3-642-00457-5_18

[JKS08]   Jha, S., Kruger, L., Shmatikov, V.: Towards practical privacy for genomic computation. In: IEEE Symposium on Security and Privacy, SP 2008, pp. 216–230. IEEE (2008)

[JRS17]   Jain, A., Rasmussen, P.M.R., Sahai, A.: Threshold fully homomorphic encryption. Cryptology ePrint Archive, Report 2017/257 (2017). http://eprint.iacr.org/2017/257

[Kil88]   Kilian, J.: Founding cryptography on oblivious transfer. In: 20th ACM STOC, pp. 20–31. ACM Press, May 1988

[KK13]   Kolesnikov, V., Kumaresan, R.: Improved OT extension for transferring short secrets. In: Canetti, R., Garay, J.A. (eds.) CRYPTO 2013, Part II. LNCS, vol. 8043, pp. 54–70. Springer, Heidelberg (2013). https://doi.org/10.1007/978-3-642-40084-1_4

[KOR+17] Keller, M., Orsini, E., Rotaru, D., Scholl, P., Soria-Vazquez, E., Vivek, S.: Faster secure multi-party computation of AES and DES using lookup tables. In: Gollmann, D., Miyaji, A., Kikuchi, H. (eds.) ACNS 2017. LNCS, vol. 10355, pp. 229–249. Springer, Cham (2017). https://doi.org/10.1007/978-3-319-61204-1_12

[KOS16] Keller, M., Orsini, E., Scholl, P.: MASCOT: faster malicious arithmetic secure computation with oblivious transfer. In: ACM CCS 2016, pp. 830–842. ACM Press (2016)

[Lev66] Levenshtein, V.I.: Binary codes capable of correcting deletions, insertions, and reversals. In: Soviet Physics Doklady, pp. 707–710 (1966)

[LOS14] Larraia, E., Orsini, E., Smart, N.P.: Dishonest majority multi-party computation for binary circuits. In: Garay, J.A., Gennaro, R. (eds.) CRYPTO 2014, Part II. LNCS, vol. 8617, pp. 495–512. Springer, Heidelberg (2014). https://doi.org/10.1007/978-3-662-44381-1_28

[LVW17] Liu, T., Vaikuntanathan, V., Wee, H.: Conditional disclosure of secrets via non-linear reconstruction. In: Katz, J., Shacham, H. (eds.) CRYPTO 2017. LNCS, vol. 10401, pp. 758–790. Springer, Cham (2017). https://doi.org/10.1007/978-3-319-63688-7_25

[NN01] Naor, M., Nissim, K.: Communication preserving protocols for secure function evaluation. In: 33rd ACM STOC, pp. 590–599. ACM Press, July 2001

[NNOB12] Nielsen, J.B., Nordholt, P.S., Orlandi, C., Burra, S.S.: A new approach to practical active-secure two-party computation. In: Safavi-Naini, R., Canetti, R. (eds.) CRYPTO 2012. LNCS, vol. 7417, pp. 681–700. Springer, Heidelberg (2012). https://doi.org/10.1007/978-3-642-32009-5_40

[SPO+06] Szajda, D., Pohl, M., Owen, J., Lawson, B.G., Richmond, V.: Toward a practical data privacy scheme for a distributed implementation of the smith-waterman genome sequence comparison algorithm. In: NDSS (2006)

[SW81] Smith, T., Waterman, M.: Identification of common molecular subsequences. J. Mol. Biol. **147**(1), 195–197 (1981)

[Wak68] Waksman, A.: A permutation network. J. ACM (JACM) **15**(1), 159–163 (1968)

[WW10] Winkler, S., Wullschleger, J.: On the efficiency of classical and quantum oblivious transfer reductions. In: Rabin, T. (ed.) CRYPTO 2010. LNCS, vol. 6223, pp. 707–723. Springer, Heidelberg (2010). https://doi.org/10.1007/978-3-642-14623-7_38

[Yao86] Yao, A.C.-C.: How to generate and exchange secrets (extended abstract). In: 27th FOCS, pp. 162–167. IEEE Computer Society Press, October 1986

Degree 2 is Complete for the Round-Complexity of Malicious MPC

Benny Applebaum[1], Zvika Brakerski[2]([⊠]), and Rotem Tsabary[2]

[1] Tel-Aviv University, Tel Aviv, Israel
bennyap@post.tau.ac.il
[2] Weizmann Institute of Science, Rehovot, Israel
{zvika.brakerski,rotem.tsabary}@weizmann.ac.il

Abstract. We show, via a non-interactive reduction, that the existence of a secure multi-party computation (MPC) protocol for degree-2 functions implies the existence of a protocol with the *same round complexity* for general functions. Thus showing that when considering the round complexity of MPC, it is sufficient to consider very simple functions.

Our completeness theorem applies in various settings: information theoretic and computational, fully malicious and malicious with various types of aborts. In fact, we give a master theorem from which all individual settings follow as direct corollaries. Our basic transformation does not require any additional assumptions and incurs communication and computation blow-up which is polynomial in the number of players and in $S, 2^D$, where S, D are the circuit size and depth of the function to be computed. Using one-way functions as an additional assumption, the exponential dependence on the depth can be removed.

As a consequence, we are able to push the envelope on the state of the art in various settings of MPC, including the following cases.

– 3-round perfectly-secure protocol (with guaranteed output delivery) against an active adversary that corrupts less than 1/4 of the parties.
– 2-round statistically-secure protocol that achieves security with "selective abort" against an active adversary that corrupts less than half of the parties.
– Assuming one-way functions, 2-round computationally-secure protocol that achieves security with (standard) abort against an active adversary that corrupts less than half of the parties. This gives a new and conceptually simpler proof to the recent result of Ananth et al. (Crypto 2018).

Technically, our non-interactive reduction draws from the encoding method of Applebaum, Brakerski and Tsabary (TCC 2018). We extend

Full version available at https://eprint.iacr.org/2019/200.

B. Applebaum—Supported by the European Union's Horizon 2020 Programme (ERC-StG-2014-2020) under grant agreement no. 639813 ERC-CLC, and the Check Point Institute for Information Security.

Z. Brakerski and R. Tsabary—Supported by the Israel Science Foundation (Grant No. 468/14), Binational Science Foundation (Grants No. 2016726, 2014276), and by the European Union Horizon 2020 Research and Innovation Program via ERC Project REACT (Grant 756482) and via Project PROMETHEUS (Grant 780701).

Y. Ishai and V. Rijmen (Eds.): EUROCRYPT 2019, LNCS 11477, pp. 504–531, 2019.
https://doi.org/10.1007/978-3-030-17656-3_18

these methods to ones that can be meaningfully analyzed even in the presence of malicious adversaries.

1   Introduction

A secure multi-party computation (MPC) allows a collection of n parties to jointly compute a function f of their joint inputs without leaking additional information other than the output. The focus of this work is on the so-called "malicious" setting, where security should be guaranteed even if an adversary, that controls up to t parties, *actively* deviates from the protocol instructions. Security is usually formalized using the ideal vs. real paradigm, essentially translating an adversarial behavior in the protocol (the real model) into an indistinguishable behavior in a model where f is computed by a trusted party (the ideal model). Various flavors of this notion have been considered in the literature, but since our results apply in multiple settings we wish to keep the discussion general at this point and not commit to a specific ideal model (however, we do focus on the case where there are private channels between the parties).

A significant resource to optimize when designing an MPC protocol is the *round complexity*, the number of communication rounds that are required in order to complete the protocol (as usual, we assume simultaneous message transmission in each round). Efforts to minimize round complexity started as soon as MPC was introduced [24] and are receiving a lot of attention recently as well (e.g. [1,6,9,10] and many others). A prominent approach to reducing round complexity is tied to reducing the *algebraic degree* of the function to be computed.[1] This can be traced back to the work of Beaver, Micali and Rogaway [4,22] and to the *randomizing polynomials* approach of Ishai and Kushilevitz [15,16]. The latter work is based on the following paradigm: Reduce the task of securely computing the function f to the task of securely computing a different function h, such that h has low algebraic degree, and such that the output of h can be decoded to produce the appropriate output for f. Once such reduction exists, with adequate security guarantees (as we elaborate below), one can focus on providing a secure MPC protocol for h, a task that usually gets easier as the degree of h drops.

In this context, it is most desirable to present a *non-interactive* reduction. Such a reduction yields a function h together with a set of (possibly randomized) local preprocessing function ℓ_i, and a method to decode the value $h(\ell_1, \ldots, \ell_n)$ to recover the output of f. In terms of security, one has to show that the protocol where each party computes ℓ_i locally, sends the value to a trusted realization of h ("an h oracle"), and then performs the decoding of the output locally, is a secure MPC protocol for computing f, respective to a security model specified in the proof.

The resemblance of the h-oracle-aided protocol to the "ideal model" described above allows to compose the reduction with a secure implementation of h, resulting in a secure realization of f. Since the reduction is non-interactive, the round

---

[1] In this work we consider the algebraic degree over the binary field. This is the common setting, but one could consider working over other fields as well.

complexity of the resulting protocol is the same as the round-complexity of (the low degree function) h.

Given this paradigm, it is natural to ask how low can the degree of h be while still allowing a reduction from any arbitrary f. It is not hard to verify that a linear (i.e. degree 1) h cannot be used to compute general functions. However, the randomizing polynomials approach seems to only imply h of degree 3.[2] Recently, Applebaum, Brakerski and Tsabary [3], showed how to reduce any function to degree-2, but their reduction was only secure in a semi-honest setting, where the adversary is required to follow the protocol (i.e. to compute the functions ℓ_i correctly using a properly sampled random tape). Nevertheless, the [3] reduction allowed them to improve the round complexity of semi-honest MPC with perfect security and with honest majority to the optimum of 2 rounds. The question whether there is non-interactive reduction to a quadratic function in the malicious setting, and the implications of such reduction on the round complexity of malicious MPC, remained open and is addressed in this work.

1.1   Our Results

We show that in various settings, a non-interactive reduction to a degree-2 function is possible. This means that it is sufficient to design protocols for degree-2 functions in order to optimize round complexity. We then design round-optimal protocols by constructing round-efficient protocols for degree-2 functions in some of these settings. Our results are all derived using a single "master theorem", which we believe can serve as basis for deriving additional results in other settings as well. We elaborate on these contributions below.

A Master Non-Interactive Reduction (Sect. 4). The technical heart of our result is a generic non-interactive "master reduction" from any function f to a degree-2 function h. Methodologically, we show how to convert any protocol Π for computing f (irrespective of round complexity) into a non-interactive h-oracle-aided protocol $\hat{\Pi}$ (we denote this by $\hat{\Pi}^h$), while preserving the security properties of Π. Specifically, we show that any adversarial strategy in the h-oracle-aided protocol can be (perfectly) simulated by an adversary against the protocol Π. In terms of the ideal/real paradigm, we show that for any $\hat{\Pi}^h$ adversary there exists a Π adversary with an identical real model view. We believe that this could be an instrumental tool in constructing and analyzing MPC protocols, since it allows to translate arbitrary protocols to ones that make a single oracle call to a fairly simple function.

We note that the conversion between Π and $\hat{\Pi}^h$ incurs a communication and computation overhead that is polynomial in (roughly) the total computational complexity of Π (i.e. the sum of computational complexities of all parties participating in Π throughout the execution of the protocol) and exponential in the depth of Π (roughly the longest computational path between an input

---

[2] It is known that general functions cannot be represented by degree-2 *perfectly-private* randomizing polynomials [15]. The existence of statistically-private degree-2 randomizing polynomials has been open for nearly two decades.

and an output in the protocol). Therefore, if communication and computational complexity are of importance, we must be careful to only apply our theorem on fairly "shallow" protocols Π.

Our master theorem generalizes the "multi-party randomized encoding" (MPRE) approach of [3], both in terms of theorem statement and in terms of techniques. The notion of h-oracle-aided protocol converges in the semi-honest setting to MPRE, but allows to handle malicious adversarial behavior whereas MPRE is by default a "passive" notion. Our master theorem, while building on the techniques of [3], also implies their MPRE result as a special case since a semi-honest $\hat{\Pi}^h$ adversary translates to a semi-honest Π adversary.

A Completeness Theorem (Sect. 5). To illustrate the power of our master theorem, we show a non-interactive reduction from the task of computing an arbitrary function f to the task of computing degree-2 functions, in the context of full security (guaranteed output delivery):

- A perfectly secure reduction, assuming more than 2/3 of the parties are honest.
- A statistically secure reduction, assuming more than 1/2 of the parties are honest.
- A computationally secure reduction, assuming more than 1/2 of the parties are honest, and assuming the existence of one-way functions (that are used in a black-box manner).

All of those reductions incur a communication and computation overhead. In all reductions this overhead is polynomial in the number of parties and in the size of the circuit computing f, and in the first two reductions (i.e. without making computational assumptions) it is also exponential in the depth of f. We note that these results are optimal in terms of the size of the adversarial coalition achievable in each of these settings.

Optimal Round-Complexity Results (Sects. 6, 7). Finally, we obtain new protocols with low round complexity for general functions in various MPC settings. We believe that numerous results can be derived using our techniques. For concreteness, we focus on achieving perfect malicious security with optimal round complexity (i.e. 3 round). We then consider the setting of 2 round protocols, where malicious security is not achievable, and instead show statistical security with selective aborts, and (assuming one way functions) computational security with aborts. To this end, we devise round-efficient protocols for degree-2 functions with different malicious security guarantees and derive the following corollaries.

- **Fully Malicious Security in Three Rounds (Sect. 6).** For all f, there exists a 3 round protocol which is secure against fully malicious adversarial coalitions containing less than 1/4 fraction of the parties. For NC^1 the protocol is perfectly secure and for arbitrary polynomial-time functions the protocol has computational security with black-box access to one-way functions.

In both cases the protocol provides *full security* (i.e. no abort). This is the optimal round complexity, as Gennaro et al. [12] showed that full security cannot be achieved in less than three rounds even when the adversary is allowed to corrupt at most 2 players. Prior to our work, it was known that general 3-round MPC with perfect security can be achieved with security threshold of $t = \alpha n$ for some small (unspecified) constant $\alpha \ll 1/4$ [11]. Our protocol shows that the threshold can be improved to $n/4$.

- **Security with Aborts in Two Rounds (Sect. 7).** For all f, there exists a 2 round protocol (not requiring broadcast) which is secure with *selective aborts*[3] against any adversarial coalition containing a minority of the parties. For NC^1 the protocol is statistically secure and for arbitrary polynomial-time functions the protocol has computational security with black-box access to one-way functions. This improves over the result of Ishai et al. [17], that achieve, in the same setting, security against an adversary that corrupts less than 1/3-fraction of the parties.

We further show that in the computational setting (for polynomial-size functions) the protocol can be modified to be secure with (unanimous) *aborts*[4] at the expense of using a broadcast channel. A result with similar parameters was already shown by Ananth et al. [1]. Recently [20] showed that selective abort is the best possible security for two-round protocols that only use secure channels. Concurrently and independently from our work, Ananth et al. [2] presented a two-round protocol achieving statistical security with (unanimous) abort. Contrary to our work, they do not propose a general framework and do not achieve our general degree-2 completeness theorem or our results in the fully malicious setting.

1.2   Technical Overview

We now provide a high level overview of our techniques.

Our Master Theorem. Recall that we want to show how to encode an arbitrary protocol Π by an oracle aided protocol $\hat{\Pi}$ that uses a quadratic oracle h, while preserving the security properties of Π. As mentioned above, our techniques extend those of [3] to the malicious setting.

In [3], the authors consider the boolean circuit induced by an execution of the protocol Π, with wires corresponding to the internal values computed by all parties throughout the protocols, and gates that represent either local computation performed by a certain party, or a transmission of a value from one party to another. Their encoding constitutes of an information-theoretic point-and-permute garbled circuit [4,16,22] for this induced circuit. The encoding of

---

[3] Security with selective aborts is a notion where in the ideal model the adversary can prevent some of the honest parties of his choice from learning the output.

[4] Security with aborts is a notion where in the ideal model the adversary can prevent either all or none of the honest parties from receiving the output (but cannot allow only some of them to receive it). We specify "unanimous aborts" in places where there is a risk of confusion with the aforementioned notion of selective aborts.

Π is a protocol where the parties jointly compute this garbled circuit using their inputs and local randomness.

The randomness required for computing the garbled circuit is distributed between the parties in a clever way that ensures that the garbled circuit can be written as $h(\ell_1, \ldots, \ell_n)$ for a quadratic h and for values ℓ_i that only depend on the local input and randomness of each party. This allows to derive a protocol encoding theorem for the semi-honest setting, where parties in $\hat{\Pi}$ simply compute their local ℓ_i and send these values to the h oracle. Garbled circuit security ensures that *any* adversarial coalition can only learn from the garbled circuit their respective views in an honest execution of Π (assuming that the ℓ_i values were computed correctly).

However, the aforementioned approach relies on the values ℓ_i being computed honestly. In contrast, a malicious adversary in this $\hat{\Pi}$ can compute the ℓ_i values belonging to the parties under its control arbitrarily, and thus a-priori we are not guaranteed that $h(\ell_1, \ldots, \ell_n)$ is even a garbled circuit at all, not to mention that it does not reveal "forbidden" values to the adversary.

Our main insight is that if the garbled circuit and the manner of distributing randomness between parties are properly defined, such a malicious modification must lead to $h(\ell_1, \ldots, \ell_n)$ being a secure garbled circuit, but one that does not encode an honest execution of Π. Instead, $h(\ell_1, \ldots, \ell_n)$ can encode an execution of Π where the parties under the adversary's control may deviate from the protocol. In other words, any cheating strategy in the compiled protocol $\hat{\Pi}$ (i.e. some adversarial modification of the values ℓ_i controlled by the adversary) translates into some cheating strategy against Π with the same adversarial coalition. Hence, $\hat{\Pi}$ inherits the security properties of Π. More details follow.

Let us first be more specific about the "partition of work" between the local functions ℓ_i and the quadratic function h. The local function ℓ_i takes the input of the ith party x_i, and two types of random tapes which we denote s_i, α_i. The function performs some (deterministic) preprocessing on α_i, producing $\mathsf{pre}_i(\alpha_i)$, and outputs $(x_i, s_i, \alpha_i, \mathsf{pre}_i(\alpha_i))$. Our adversary is allowed to arbitrarily modify all of these values, let us examine the effects of such modification on each of these components.

- Changing the value x_i is equivalent to selecting a different input for the ith party, which cannot be avoided in any model of secure computation.
- The random string s_i is used by h as shares for *wire keys* of the garbled circuit. The exact functionality does not matter for this outline, but the important thing is that h XORs these values among all parties. Thus, choosing s_i maliciously does not buy the adversary any leverage, since h only uses the aggregate value $(\oplus_i s_i)$, which is uniform from the adversary's viewpoint (so long as there is at least one honest party).
- The random string α_i is used to produce *mask bits* for the values of the wires in the garbled circuit. Essentially, the evaluation of a point and permute garbled circuit results in producing, for each wire of the circuit that was

garbled, the value of this wire XORed with a random mask bit. Crucially, the string α_i contains the mask bits for the wires of the circuit whose values party i is *allowed to see*. (The definition of the induced circuit for a protocol guarantees that there is a disjoint partitioning of wires between the parties). Hence, an adversarial choice α_i gives the adversary no leverage. Such behavior can only hurt the privacy of the adversary's own wires, and has no effect on the privacy of honest parties.

- The crux of the matter is $\text{pre}_i(\alpha_i)$. The preprocessing pre_i is in fact what allows to reduce the degree of h to 2. A malicious adversary can certainly damage these computed values and indeed effect the resulting garbled circuit. What we show next is that the damage of such malleability is controllable. To explain, we go into a little more detail about the functionality of pre_i.

Recall that each gate in the circuit to be garbled represents either a local computation by a party or communication from one party to another. The function pre_i only computes on bits in α_i that are associated with inputs of local computation gates. For each such gate $\text{pre}_i(\alpha_i)$ contains four evaluations of the gate function (say NAND, w.l.o.g), on the four possible inputs in a specific permuted order, where the permutation is determined by respective α_i bits. Specifically, the permutation is obtained by taking the canonical sequence $00, 01, 10, 11$, and XOR-ing it with the respective mask bits of the input wires of that gate.

The adversary might plug in 4 arbitrary bits instead of the correct values to be computed by pre_i, regardless of the actual values it chose for the mask bits α_i, and possibly depending on any other value that the adversary might have. The crucial part of our argument is to notice that any change in the preprocessing can equivalently be described as a change in the *gate function*, e.g. computing OR instead of NAND, but executing this gate on the *correct* mask bits. Once this is established, we can take a step back and notice that in fact, all that the adversary can do by corrupting its $\text{pre}_i(\alpha_i)$ values is equivalent to changing the garbled circuit from one that corresponds to an honest execution of the protocol Π, to an execution of Π where the parties that are controlled by the adversary change the functionality of their local computation gates!

We conclude that even if the ℓ_i values controlled by the adversary are maliciously corrupted, h will still output a garbled circuit which corresponds to an execution of Π, possibly with some parties behaving dishonestly (the parties corresponding to the corrupted ℓ_i values). The security of this garbled circuit (which follows from the fact that the wire keys and the mask keys for honest parties remain random, as we described above) guarantees that the parties in $\hat{\Pi}^h$ learn the exact same information as they do in an execution of Π with the respective adversary (we use a perfectly secure garbled circuit which implies that the adversary's views in the two cases are identical).[5]

Lastly, we note that an additional modification to the [3] approach is required in order to handle broadcast channel, i.e. the possibility of a party to send a message to all other parties, such that parties are guaranteed that the same message was sent to all. This is a useful component that aids in the design of maliciously secure protocols, and is not needed in the semi-honest setting (since parties there follow the protocol specifications, so if a party is instructed to send the same value to all others, that is what it will do). If the underlying protocol Π is one that uses broadcast, this needs to be enforced by the "induced circuit" for which a garbled circuit is computed. Fortunately this is easy to handle by generalizing the point-to-point transmission gates into fan-out gates with a single input and multiple outputs. The way such gates are garbled guarantees that it is impossible to produce an execution where the outputs are inconsistent (i.e. where different parties receive different values).

The Completeness Theorem. Applying the master theorem is, on the face of it, straightforward. Instantiating Π with a protocol that is secure in the malicious setting, should immediately imply the theorem statement, and indeed the fraction of honest parties required exactly matches that of best known malicious MPC protocols with many rounds. However, there is one caveat that requires careful consideration. The encoding theorem induces a blowup in the communication and computational complexity of the protocol $\hat{\Pi}$, which is related to the size of the (information theoretic) garbled circuit of the circuit induced by Π. In particular, the size of the garbled circuit scales exponentially with the depth. We want to argue that our reduction scales with the properties of the *target function* f to be computed. Thus, for example, using an underlying Π whose depth is (say) n times the depth of f will incur an exponential cost in the parameters of the reduction. We thus carefully analyze existing protocols so as to guarantee that there exists Π where the depth of the induced circuit only relates linearly to the depth of the function f being evaluated.

One observation that proves very helpful is that there is no need to encode local postprocessing that takes place after all the messages has been sent. That is, given a protocol Π it is sufficient to apply our master theorem on a truncated protocol Π' in which the parties send all messages as in Π, but instead of performing the final postprocessing computation they just output their view in the execution. This modification leads to a much shallower circuit for our encoding theorem and at the same time allows to achieve the required functionality and security. Functionality is maintained since the postprocessing in the final step

---

[5] In fact, the adversary in Π is somewhat weaker than a full malicious adversary. First, the adversarial parties are required to have the same circuit topology as honest parties, since only gate functionality changes and not the interconnection of gates. Second, the adversary cannot adjust the behavior of party i under its control based on a message received by a different party j under its control during the execution of the protocol. We find this property quite interesting and potentially useful, although we do not need to exploit it to derive the consequences in the cases analyzed in this paper.

can be done on the output of the garbled circuit evaluation, rather than being included in the garbled circuit itself. See Sect. 5 for more details.

Optimal Round-Complexity Results. As explained above, these are achieved by plugging in secure protocols for evaluating degree-2 functions in various models. Such protocols are usually not made explicit in the literature (as computing degree-2 functions was not a major goal until this work). However, known techniques do seem to become monotonously more round-efficient as the degree drops. We apply modifications on top of existing methods in order to reduce the round complexity to the very optimum.

Three-Round Protocols with Full Security. In Sect. 6 we implement MPC with full security for degree-2 functions f via the following template:

1. Each party shares each of its inputs between all the parties using a sub-protocol for Verifiable Secret Sharing (VSS).
2. Each party locally computes the degree-2 functionality f over its shares and gets a share of the outputs. To enable this computation, the underlying secret sharing scheme has to be 2-multiplicative over the binary field.
3. The parties broadcast the result (after some randomization) and apply a correction procedure for handling malformed shares.

The template can be instantiated with different ingredients (e.g., for the VSS and for the recovery step). The security and round complexity of the resulting protocol depend on the corresponding properties of the underlying building blocks.

We instantiate the above template with the standard polynomial-based Shamir secret sharing scheme [23]. Gennaro et al. [11] showed that the sharing phase of this secret sharing scheme can be perfectly realized (with full security) in 2 rounds for our security threshold. This VSS natively supports secrets that are taken from a medium-size field of size at least $n + 1$, and we show how to modify it into a binary VSS.[6] Eventually, we get a 2-round binary VSS with the guarantee that at the end of the sharing phase, the honest parties hold shares that are consistent with some binary secret, even if the dealer was malicious. We observe that for our security threshold, after the third round (in which the parties broadcast their shares of the output), the honest parties can recover the output via the standard Reed-Solomon decoding algorithm.

Two-Round Protocols with Selective Abort. Here we rely on two results from [17]. In their work, they consider a weaker notion of security, *Privacy with Knowledge of Outputs* (PKO)[7], and show that:

---

[6] In particular, we use an extension field of $GF(2)$, and add a mechanism that forces the adversary to use binary inputs. Implementing this mechanism without increasing the round complexity is somewhat challenging, and for this, we rely on some specific properties of the [11] scheme. See Sect. 6 and full version for details.

[7] Intuitively, this means that the correctness of honest parties may be violated, but the adversary is required to "know" the (possibly incorrect) outputs of the honest parties. Formally, in the ideal model, the ideal functionality first delivers the outputs of the corrupted parties to the simulator, and then receives from the simulator an output to deliver to each of the uncorrupted parties.

1. Any r-rounds protocol with PKO security for functions in NC^1 (resp. polynomial-size functions) induces a r-rounds protocol with selective abort security for functions in NC^1 (resp. polynomial-size functions).
2. Any degree-2 function can be efficiently computed in two rounds with statistical PKO security for threshold $n/2$, without a broadcast channel.

To complete the proof, we show that our completeness theorem maintains PKO security. This turns to be somewhat subtle since, as we observe, PKO security is not always preserved under composition. (See full version for details.)

Two-Round Protocols with Abort. Lastly we use a modification from [19] which shows how a 2-round protocol with SSA security for polynomail-size functions can be converted to a 2-round protocol with SA security of similar complexity and security guarantees, at the expense of using a broadcast channel and one way functions. The general reduction, however, involves a reduction to a functionality f' that invokes the signing algorithm of a digital signature scheme. When instantiated with an arbitrary signature scheme, computing f' results in a non black-box use of a one-way function. We observe that the transformation of [19] requires only one-time secure signatures, and therefore can be instantiated with Lamport's one-time signatures (cf. [13, Chap. 6.4.1]), in which the one-way function is used only in the key-generation and verification algorithms, but not in the signing algorithm. See full version for details.

2 Preliminaries

In this section, we define Boolean circuits, multi-party protocols, oracle-aided protocols and security for multi-party computation. Briefly, we consider an active non-adaptive rushing adversary that may be computationally unbounded or computationally bounded (depending in the context) and, unless stated otherwise, assume a fully connected network with point-to-point private channels and a broadcast.

2.1 Boolean Circuits

In this work, we consider Boolean circuits containing two types of gates:

- A *(p-ary) fan-out gate*, sometimes denoted as a *transmission gate*, that has a single input and p outputs, its functionality is to copy its input to all outputs.
- A *local gate* has two input wires and one output wire. It computes some arbitrary function $G : \{0,1\}^2 \to \{0,1\}$ (that can vary from one gate to the next).

For purposes of analysis, we define the *depth* of a p-ary transmission gate to be $\lceil \log p \rceil$, and the depth of a local gate to be 1. The depth of a circuit C is the computed by considering the cumulative depth of gates along each path from an input wire to an output wire in C, and taking the maximum among all paths.

The *size* of a circuit, m, is the number of wires in the circuit (including input and output wires). We assume a topological ordering of the wires in $[m]$.

We say that two circuit C_1, C_2 are *topologically equivalent* (or have the same topology) if they are identical, except perhaps in the functions G associated with local gates.

2.2 Functionalities and Protocols

It will be convenient to treat functionalities and protocols as finite (fixed length) objects. The infinite versions of these objects will be defined and discussed later in Sect. 2.3. We continue with a formal definition.

Notation. For any set $T \subseteq [n]$, we denote $\overline{T} = [n]/T$ where n, which denotes the number of players, will be clear from the context. For any sequence $\boldsymbol{x} = (x_1, \ldots, x_n)$ and any $S \subseteq [n]$ let $\boldsymbol{x}[S]$ denote the ordered set $\{x_i\}_{i \in S}$.

Definition 2.1 (multi-party functionality). *An n-party functionality $f : (\{0,1\}^*)^n \to (\{0,1\}^*)^n$ is a (possibly randomized) function that maps a sequence of n inputs $\boldsymbol{x} = (x_1, \ldots, x_n)$ to a sequence of n outputs $\boldsymbol{y} = (y_1, \ldots, y_n)$. If f sends the same output to all parties then we denote its output as a scalar, i.e. we use the shorthand $f : (\{0,1\}^*)^n \to \{0,1\}^*$ and $y = f(x_1, \ldots, x_n)$.*

Next we define a multi-party protocol in a non-asymptotic setting.

Definition 2.2 (multi-party protocol, oracles). *An n-party, r-round protocol Π is a tuple of $n \cdot r$ boolean circuits $\{C_{j,i}\}_{j \in [r+1], i \in [n]}$ that correspond to the computation that party i in the protocol performs before the j-th communication round (or after the last round if $j = r + 1$). Each $C_{i,j}$ (except for $j = 1$ and $j = r + 1$, see below) takes n input strings, and outputs n output strings. The i'-th output of $C_{j,i}$ is the message sent from party i to party i' at round j of the protocol. If $i = i'$ then the respective output is the state of party i after the j-th round of communication. We therefore require that for all i, i', j the i'-th output of $C_{j,i}$ has the same length as the i-th input of $C_{j+1,i'}$. In the first round of communication $C_{1,i}$ only takes one input x_i to be interpreted as party i-th input for the protocol, and possibly an additional random tape. In the last round of communication $C_{r+1,i}$ only has one output which should be interpreted as the output of party i in the protocol, sometimes denoted y_i. We let M_i denote the collection of circuits associated with party i, i.e. $M_i = (C_{1,i}, \ldots, C_{r+1,i})$ and thus denote $\Pi = (M_1, \ldots, M_n)$. The view of the party in the protocol contains its input, randomness and all messages it received during the execution.*

Let h be an n-party functionality. A protocol Π with oracle h, which we denote by Π^h, is one that allows to replace some of the communication rounds with calls to the functionality h (i.e. the circuits respective to this round each produce one output that is sent to the oracle h as input, the outputs of h is then fed as a single input to the next round circuit).

A protocol with broadcast is one with access to the broadcast functionality that on input $\boldsymbol{x} = (x_1, \ldots, x_n)$ outputs $\boldsymbol{x}$ to all parties. More generally, the framework in this paper can handle any oracle functionality that delivers the same output (originating from a designated party) to a subset of parties. We note that in circuit terminology this can be described as $C_{j,i}$ producing an output associated with sets of parties rather than a single party.

A non-interactive h-oracle-aided protocol is one that consists only of a single round of oracle call, and no other communication between the parties.

Consistently with the above formal description, we often refer to M_i as an *interactive circuit* that sends and receives messages (and maintains a state throughout the execution), until finally producing an output after the r-th round of communication.

2.3   Correctness and Security of Protocols

Security of multi-party computations is analyzed via the real vs. ideal paradigm. The real model captures the information that can be made accessible to the adversary in an actual execution of the protocol, which includes an arbitrary function of the view of the corrupted parties, as well as honest parties' input and output (but not their internal state during the execution). The ideal model considers a case where the target functionality is computed using oracle access. The protocol is secure if the view of every real adversary can be simulated by an ideal adversary.

We first define the notion of an adversary, note that we slightly deviate from the standard notation and explicitly include the description of the set of corrupted parties as a part of the definition of the adversary. This will be useful for stating our results. We also note that the current definition is syntactic and non-asymptotic and does not address the adversary's having an efficient implementation.

Definition 2.3 (adversaries, the real model, ideal model). *An adversary (A, T) for an n-party protocol $\Pi = (M_1, \ldots, M_n)$ consists of an interactive circuit A (sometimes called the adversarial strategy), and a set $T \subseteq [n]$. The parties in T (resp. $\overline{T}$) are the* dishonest *(resp. honest) parties.*

The execution of Π with input $\boldsymbol{x}$ under (T, A) is as follows. The input to A is the set of inputs $\boldsymbol{x}[T]$ (the inputs for the parties in T). In each round, A first receives all messages sent to parties in T from parties in $\overline{T}$, and then outputs messages to be sent to the parties in $\overline{T}$ from the parties in T (i.e. A is rushing*). At the end of the protocol A produces outputs on behalf of all parties in T. The parties in $\overline{T}$ execute according to their respective prescribed M_i algorithms.*

A semi-honest *adversary is one in which A executes according to the parties $\{M_i\}_{i \in T}$, and outputs some function of the views of $\{M_i\}_{i \in T}$ in the protocol as the outputs of the parties in T.*

The ordered sequence of outputs of all parties in the execution above is called the output of the real-model execution and denoted $\mathsf{REAL}_{\Pi,T,A}(\boldsymbol{x})$. The ideal-model is defined by considering the trivial non-interactive f-oracle-aided protocol

Υ^f *in which each party simply sends its input x_i to the f-oracle, gets the output y_i from the oracle, and terminates with this output. For an adversary (A, T) and vector of inputs $\boldsymbol{x}$ we denote the output of the ideal-model execution by* IDEAL$_{f,T,A}(\boldsymbol{x})$.

In the ideal-model with selective abort *(SSA), the f-oracle first delivers the outputs of the corrupted parties to the adversary, which then can decide for each uncorrupted party whether this party will receive its output or a special abort symbol. The ideal-model with* abort *(SA) is similar to SSA except that when the adversary decides to abort, all of the honest parties receive a special abort symbol.*[8]

Asymptotic Versions. A sequence of functionalities $F = \{f_\kappa\}_{\kappa \in \mathbb{N}}$ is efficiently generated if there exists a polynomial time algorithm that on input 1^κ outputs a circuit that computes the $n(\kappa)$-party functionality f_κ. A sequence of protocols $\Pi = \{\Pi_\kappa\}$ is efficiently generated if there exists a polynomial time algorithm that takes 1^κ as input and outputs all circuits $C_{j,i}$ associated with Π_κ. A sequence of adversaries $A = \{A_\kappa\}$ is (non-uniformly) efficient if there exists a polynomial $p(\cdot)$ such that for every κ the size of the circuit A_κ is at most $p(\kappa)$. We often abbreviate "efficient functionality/protocol/algorithm" and not refer to the sequence explicitly. Throughout this work, we will be concerned with constructing efficiently generated protocols for efficiently generated function ensembles. In fact, our results (implicitly) give rise to a compiler that efficiently converts a finite functionality into a finite protocol.

Definition 2.4 (correctness and security of protocols). *Let $f = \{f_\kappa\}$ be an $n(\kappa)$-party functionality and $\Pi = \{\Pi_\kappa\}$ a (possibly oracle-aided) $n(\kappa)$-party protocol. We say that Π $t(\kappa)$-securely computes f if for every probabilistic non-uniform algorithm $A = \{A_\kappa\}$ and every infinite sequence of sets $\{T_\kappa\}$ where $T_\kappa \subseteq [n(\kappa)]$ is of cardinality at most $t(\kappa)$, there exists a probabilistic non-uniform algorithm $B = \{B_\kappa\}$ and a polynomial $p(\cdot)$ so that the complexity of B_κ is at most $p(|A_\kappa|)$, such that for every infinite sequence of inputs $\{\boldsymbol{x}_\kappa\}$, the distribution ensembles (indexed by κ)*

$$\text{IDEAL}_{f_\kappa, T_\kappa, B_\kappa}(\boldsymbol{x}_\kappa) \quad and \quad \text{REAL}_{\Pi_\kappa, T_\kappa, A_\kappa}(\boldsymbol{x}_\kappa)$$

are either identical (this is called perfect security), statistically close (this is called statistical security), or computationally indistinguishable (this is called computational security). In the latter case, A is assumed to be asymptotically efficient.

Note that for an efficiently generated protocol it follows from the definition that the number of parties n, and the input lengths are polynomial in the security parameter κ.

---

[8] The terminology of "security with abort" and "security with selective abort" is borrowed from [17] and [19] and it corresponds to the notions of "security with unanimous abort and no fairness" and "security with abort and no fairness" from [14].

Definition 2.5 (secure reductions, non-interactive reductions). *If there exists a secure h-oracle-aided protocol for computing f, we say that f is reducible to h. If the aforementioned oracle-aided protocol is non-interactive (i.e. only consists of non-adaptive calls to h) we say that the reduction is non-interactive.*

Appropriate composition theorems, e.g., [13, Theorems 7.3.3, 7.4.3], guarantee that the call to h can be replaced by any secure protocol realizing g, without violating the security of the high-level protocol for f. (In the case of computational security the reduction is required, of course, to be efficient.)

3 Building Blocks

In this section we define building blocks that rely on previous works and will be used for our master theorem in Sect. 4. Circuit representation of protocols is defined in Sect. 3.1, and our presentation of point and permute garbled circuits follows in Sect. 3.2.

3.1 Circuit Representation of a Protocol

Recall that a protocol $\Pi = (M_1, \ldots, M_n)$, is a sequence of interactive circuits. It will be convenient to collapse all these circuits to a single "circuit representation" of a protocol. (A similar abstraction appears in [3], but some of the details differ, e.g., the treatment of fan-out gates which are needed for handling protocol that employ broadcast.)

Informally, we consider the computation of all parties throughout the protocol as parts of one large computation. Each wire of the new circuit is associated with an index corresponding to the party in the protocol that computes this value. This includes the local computations performed by parties throughout the protocol, which are represented as gates whose inputs and outputs are associated with the party who is performing the local computation, and also message transmissions between parties, that are modeled as gates that simply copy their input to the output, where the inputs are associated with the sender and outputs are associated with the receiver.

Our definition only considers circuits corresponding to *deterministic* protocols. This is both for the sake of simplicity (since we can always consider parties' randomness as a part of their input) and since we will only apply this definition to deterministic protocols in our results.

Definition 3.1 (Circuit Representation of a Protocol). *The circuit representation of a deterministic n-party protocol Π is a pair (C, P), where C is a Boolean circuit of size m as defined in Sect. 2.1, and $P : [m] \to [n]$ is a mapping from the wires in C to the n parties.*

Given a protocol $\Pi = (M_1, \ldots, M_n)$, the circuit C and the mapping P are defined as follows.

1. *Recalling Definition 2.2, Π consists of a sequence of circuits $C_{j,i}$ which represent the local computation of party i before the j-th round of communication (and a final circuit $C_{r+1,i}$ for the local computation after the last round of communication), we call this the j-th computational step of the protocol.*
2. *All input wires of sub-circuits that correspond to the first step in the protocol are defined as input wires of C. All output wires of sub-circuits that correspond to the last step in the protocol are defined as outputs wires of C.*
3. *The input wires representing the input state of $C_{j,i}$ are connected to the wires representing the output state of $C_{j-1,i}$ via a unary transmission gate. That is, the state of party i in the beginning of computation step j is identical to its state in the end of computation step $j-1$.*
4. *If party i expects a message in step j from party i', then the respective output wire of $C_{j-1,i'}$ is connected to the respective input wire of $C_{j,i}$ via a unary transmission gate. If party i_0 was supposed to send some value via broadcast to multiple parties $i_1, \ldots, i_p$ then a p-ary transmission gates connects the respective output wire in C_{j-1,i_0} to the input wires in $C_{j,i_1}, \ldots, C_{j,i_p}$.*
5. *Note that by the description above, the set of wires in C is exactly the union of wires of all circuits $C_{j,i}$. The mapping P associates with party i the wires of circuits $C_{j,i}$, for all j.*

We note that this description implies that for any local gate, all inputs and outputs have the same association. We say that a local gate g belongs to player i if all g-adjacent wires are associated with i.

The following is an observation that will be useful for our construction. Essentially it says that if we switch some of the gates that belong to some party with different gates, then the resulting circuit still represents a protocol.

Lemma 3.1. *Let $\Pi = (M_1, \ldots, M_n)$ be a protocol, and let (C, P) be its circuit representation. Let $T \subseteq [n]$ be some set and let H be a subset of the local gates of T such that every gate in H belongs to a party $i \in T$. Consider a circuit C' topologically equivalent to C, which is identical to C except on the gates in H. Then (C', P) is a circuit representation of a protocol $\Pi' = (M_1', \ldots, M_n')$ with the same round complexity and message pattern as Π, and where $M_i' = M_i$ for all $i \notin T$.*

Proof. This follows almost by definition. Define the sub-circuits $C_{j,i}'$ of C' according to their isomorphic counterparts in C. Since only local gates belonging to parties in T are changed, it follows that $C_{j,i}' = C_{j,i}$ for all j and for all $i \notin T$. Now define the party M_i' for $i \in T$ to have the functionality that in computation step j it runs the cub-circuit $C_{j,i}'$ on its state from the previous step and incoming messages, to produce the next state and outgoing messages. By definition (C', P) is the circuit representation of Π'.

3.2 Point and Permute Garbled Circuits

We present an information theoretic variant of the point-and-permute construction of [4, 22]. Our variant extends the information theoretic garble circuits of [16] to handle (p-ary) transmission gates as in Definition 3.1. In addition, we slightly modify the encoding and decoding procedures, Encode and Decode, as follows. The encoding procedure Encode is decomposed into two parts, Permute and Encrypt, where Permute shuffles the truth tables of each gate based only on the mask bits α (and is independent from the randomness s that is being used to generate the gate keys) and the second part Encrypt (for "Table encryption") generates the encrypted gate tables (based on all the randomness and on the outcome of the first part). We also modify the decoding procedure so that it outputs the masked bits of all wires of the circuits (instead of outputting only the un-masked bits of the outputs). We begin with a detailed description of the encoding and decoding procedures, and continue by analyzing their properties.

The Construction Randomness of the Encoder. The encoder Encode$(C, \boldsymbol{x}; \boldsymbol{s}, \boldsymbol{\alpha})$ takes as input a circuit C with local and transmission gates, with m wires in total, as well as an input $\boldsymbol{x}$ for C and random tape consisting of two strings: a vector $\boldsymbol{\alpha} = (\alpha_j)_{j \in [m]} \in \{0, 1\}^m$ of masks (one for each wire), and a vector of "wire keys" $\boldsymbol{s} = (s_j^0, s_j^1)_{j \in [m]}$. The keys of the j-th wire $s_j^0, s_j^1 \in \{0, 1\}^{\omega_j}$ are of length ω_j which is defined recursively. If j is an output wire then $\omega_j = 0$. If j is an input wire of local gate whose output wire is k, then $\omega_j = 2(\omega_k + 1)$. If j is an input wire of a p-ary transmission gate whose output wires are $k_1, \ldots, k_p$ then $\omega_j = \sum_{i \in [p]} (\omega_{k_i} + 1)$. By our definition of depth, the total length of $\boldsymbol{s}$, denoted by $\omega_C = 2 \cdot \sum_{j \in [m]} \omega_j$, is polynomial in m and 2^d where d is the depth of C. Lastly, if j is an input wire for a local gate and s is one of its wire keys, we let $s[0], s[1]$ denote the first and second half of s respectively.

The Encoding. We now turn to the encoding procedure, which is divided into two parts, first we compute a sequence $\boldsymbol{\Gamma}$ by running a subroutine $\boldsymbol{\Gamma} =$ Permute$(C, \boldsymbol{\alpha})$ (note that this subroutine depends only on $\boldsymbol{\alpha}$ and not on any of the other input values). Then we apply Encrypt$(C, \boldsymbol{x}, \boldsymbol{s}, \boldsymbol{\alpha}, \boldsymbol{\Gamma})$, which outputs the final encoding. The procedures are described below.

– The procedure Permute$(C, \boldsymbol{\alpha})$ operates as follows. For every local gate g in C, with input wires $c, d \in [m]$, compute the (ordered) set

$$\Gamma_g := \left\{ \gamma_g^{\beta_c, \beta_d} := G(\alpha_c \oplus \beta_c, \alpha_d \oplus \beta_d) \right\}_{\beta_c, \beta_d \in \{0,1\}} \tag{1}$$

where $G : \{0, 1\}^2 \to \{0, 1\}$ is the function that the gate g computes. Let $\boldsymbol{\Gamma}$ denote the (ordered) set $\{\Gamma_g\}_g$ for all local gates in C, output $\boldsymbol{\Gamma}$.
– The procedure Encrypt$(C, \boldsymbol{x}, \boldsymbol{s}, \boldsymbol{\alpha}, \boldsymbol{\Gamma})$ operates as follows. For every gate g in C, construct its *gate table* Q_g:
  • If g is a local gate, with incoming wires c, d and outgoing wire k, the gate table of g consist of four values. For every $\beta_c, \beta_d \in \{0, 1\}$, compute $Q_g^{\beta_c, \beta_d}$ by setting $\gamma := \gamma_g^{\beta_c, \beta_d}$ and computing:

$$\underbrace{Q_g^{\beta_c, \beta_d}}_{\text{"ciphertext"}} := \underbrace{(s_k^\gamma \| \gamma \oplus \alpha_k)}_{\text{"message"}} \oplus s_c^{\alpha_c \oplus \beta_c}[\beta_d] \oplus s_d^{\alpha_d \oplus \beta_d}[\beta_c].$$

One can view $Q_g^{\beta_c, \beta_d}$ as a *one-time pad ciphertext*, encrypted using the wire keys of the input wires.

- Transmission gates are treated analogously. That is, if g is a transmission gate g with incoming wire c and outgoing wires $k_1, \ldots, k_p$, the table of g consists of two values. For every $\beta_c \in \{0, 1\}$, set $\gamma = \beta_c \oplus \alpha_c$ and define

$$Q_g^{\beta_c} := \left((s_{k_1}^{\gamma} \| \gamma \oplus \alpha_{k_1}) \| \ldots \| (s_{k_p}^{\gamma} \| \gamma \oplus \alpha_{k_p}) \right) \oplus s_c^{\alpha_c \oplus \beta_c}.$$

Finally, the encoding includes the sequence Q containing all gate table values $Q_g^{\beta_c, \beta_d}, Q_g^{\beta_c}$, as well as a sequence σ containing the wire keys and masked values $(s_j^{x_j}, x_j \oplus \alpha_j)$ for every input wire j.

Decoding. The decoding procedure $\mathsf{Decode}(Q, \sigma)$ takes as input a sequence of gate tables, and pairs $(s_j, \hat{v}_j)$ for the input wires. It outputs a sequence $\hat{v}_j$ for all $j \in [m]$ by traversing the gate tables in topological order. (Here we slightly deviate from the standard convention in randomized encoding literature that the decoder outputs the unmasked values of the output wires.) This is done by traversing the circuit from the inputs to the outputs as follows. For input wires j the pair $\hat{v}_j, s_j$ is given explicitly the input. For an internal wire that is an output wire of a local gate g with incoming wires c, d, this is done by using the masked bits $\hat{v}_c, \hat{v}_d$ to select the ciphertext $Q_g^{\hat{v}_c, \hat{v}_d}$ and then decrypting (i.e., XOR-ing) it with $s_c[\hat{v}_d] \oplus s_d[\hat{v}_c]$. The recovered value is then denoted $(s_k, \hat{v}_k)$. Transmission gates are treated similarly: use the masked bit $\hat{v}_c$ of the input wire to select the ciphertext $Q_g^{\hat{v}_c}$ and then XOR it with s_c to obtain $(s_{k_i}, \hat{v}_{k_i})$ for $i = 1, \ldots, p$.

Useful Properties. We first state properties of $\mathsf{Encrypt}$ that will be important for our purposes.

Proposition 3.1. *The function $\mathsf{Encrypt}$ has algebraic degree 2 when written as a polynomial over the binary field in its inputs.*

Proof. This property is straightforward from the definition, since the only non-linear components in $\mathsf{Encrypt}$ are ones that require making a selection of the form s^z, where z is some variable from α or Γ (or a linear shift thereof), such a value can be expressed as $s^0 \oplus z \cdot (s^0 \oplus s^1)$, i.e. a quadratic function. (Note that all β values are fixed and known whenever they are used.)

The next proposition follows by definition.

Proposition 3.2. *The function $\mathsf{Encrypt}$ is only dependent on the topology of C and not on the functionality G of local gates.*

The following Propositions (3.3, 3.4) have been proven multiple times in the garbled circuit literature (cf. [16]).

Proposition 3.3 (Efficiency). *For all C, x, s, α, where C if of depth d and size m, the computational complexity of $\mathsf{Encode}(C, x; s, \alpha)$ is a fixed polynomial in $m, 2^d$.*

For every circuit C and input $\boldsymbol{x}$, we define for all $j \in [m]$ the value v_j as the value that the wire takes when evaluating C on $\boldsymbol{x}$ (in particular for input wires, $v_j = x_j$).

Proposition 3.4 (Correctness). *For all $C, \boldsymbol{x}, \boldsymbol{s}, \boldsymbol{\alpha}$, setting $z = \mathsf{Encode}(C, \boldsymbol{x}; \boldsymbol{s}, \boldsymbol{\alpha})$, and then $\{\hat{v}_j\}_{j \in [m]} = \mathsf{Decode}(z)$, it holds that $\hat{v}_j = v_j \oplus \alpha_j$.*

In Sect. 3.3 we will prove the following, somewhat non-standard, simulation property of garbled circuit. (The case where W is taken to be set of output wires corresponds to the standard simulation property of information-theoretic garbled circuits.)

Proposition 3.5. *There exists a* PPT *simulator* Sim *that takes as an input a circuit C, a subset of wires W, and for every wire $j \in W$ a mask-bit/intermediate-value pair $(\alpha_j, v_j) \in \{0, 1\} \times \{0, 1\}$ such that the following holds. For every C, W and $\{\alpha_j\}_{j \in W}$ and every input $\boldsymbol{x}$ the random variable*

$$\mathsf{Sim}(C, W, \{\alpha_j, v_j\}_{j \in W}),$$

where the value v_j is the value induced on the j-th wire of C by the input $\boldsymbol{x}$, is distributed identically to the random variable

$$\mathsf{Encode}(C, \boldsymbol{x}; \boldsymbol{s}, \boldsymbol{\alpha}),$$

where $\boldsymbol{s}$ is uniformly random and $\boldsymbol{\alpha} = \{\alpha_j\}_{j \in W} \cup \{\alpha_j\}_{j \notin W}$ for a uniformly random $\{\alpha_j\}_{j \notin W}$.

Recall that the outcome $\boldsymbol{\Gamma}$ of $\mathsf{Permute}(C, \boldsymbol{\alpha})$ is a vector that is indexed by the gates of C where for each gate g the entry Γ_g is a four-bit string as defined in Eq. (1). The following key lemma shows that a corruption of some entries of $\boldsymbol{\Gamma}$ corresponds to applying $\mathsf{Permute}$ to a corrupted version of the circuit C with the same randomness $\boldsymbol{\alpha}$.

Lemma 3.2 (Corruption Lemma). *For all $C, \boldsymbol{\alpha}$, let $\boldsymbol{\Gamma} = \mathsf{Permute}(C, \boldsymbol{\alpha})$, let H be a subset of the gates of C, and let $\boldsymbol{\Gamma}'$ be a vector for which $\Gamma_g' = \Gamma_g$ for all gates $g \notin H$. Then there exists a circuit C' which is obtained from C by (possibly) modifying only gates in H, such that $\boldsymbol{\Gamma}' = \mathsf{Permute}(C', \boldsymbol{\alpha})$. Moreover, C' can be efficiently computed based on $C, H, \{\Gamma_g'\}_{g \in H}$ and based on the values of the masked bits α_i for all wires i that enter the gates in H.*

Proof. We define C' by modifying the H gates of the circuit C as follows. For all $g \in H$, consider $\Gamma_g' = \{\gamma'^{\beta_1, \beta_2}\}_{\beta_1, \beta_2 \in \{0,1\}}$. Let α_1, α_2 denote the $\boldsymbol{\alpha}$ values corresponding to the input wires of g. Define a new gate functionality G_g' : $\{0, 1\}^2 \rightarrow \{0, 1\}$ as

$$G'(\beta_1, \beta_2) = \gamma'^{(\beta_1, \beta_2) \oplus (\alpha_1, \alpha_2)}.$$

The property $\boldsymbol{\Gamma}' = \mathsf{Permute}(C', \boldsymbol{\alpha})$ follows by definition.

3.3   Proof of Proposition 3.5

We begin with the following standard proposition that captures the privacy property of garbled circuits. Note that our Encode procedure as defined above does not release any of the $\boldsymbol{\alpha}$ values in the clear.

Proposition 3.6 (Simulation). *Let* $(C^{(1)}, \boldsymbol{x}^{(1)})$, $(C^{(2)}, \boldsymbol{x}^{(2)})$ *be such that* $C^{(1)}, C^{(2)}$ *are topologically equivalent. Then for uniformly random* $\boldsymbol{s}, \boldsymbol{\alpha}$, *the random variables* $\mathsf{Encode}(C^{(1)}, \boldsymbol{x}^{(1)}; \boldsymbol{s}, \boldsymbol{\alpha})$ *and* $\mathsf{Encode}(C^{(2)}, \boldsymbol{x}^{(2)}; \boldsymbol{s}, \boldsymbol{\alpha})$ *are identically distributed.*

The following claim will be useful for deriving Proposition 3.5 below.

Claim 1. *Let* $(C^{(1)}, \boldsymbol{x}^{(1)}, \boldsymbol{\alpha}^{(1)})$, $(C^{(2)}, \boldsymbol{x}^{(2)}, \boldsymbol{\alpha}^{(2)})$ *be such that* $C^{(1)}, C^{(2)}$ *are topologically equivalent, and such that* $\boldsymbol{v}^{(2)} \oplus \boldsymbol{y}^{(2)} = \boldsymbol{v}^{(2)} \oplus \boldsymbol{y}^{(2)}$. *Then, considering a uniformly random* $\boldsymbol{s}$, *the distributions* $\mathsf{Encode}(C^{(1)}, \boldsymbol{x}^{(1)}; \boldsymbol{s}, \boldsymbol{\alpha}^{(1)})$ *and* $\mathsf{Encode}(C^{(2)}, \boldsymbol{x}^{(2)}; \boldsymbol{s}, \boldsymbol{\alpha}^{(2)})$ *are identical distributed.*

Proof. Let $\boldsymbol{y}^{(1)}, \boldsymbol{y}^{(2)}, \boldsymbol{s}^{(1)}, \boldsymbol{s}^{(2)}$ be uniform and independent, and define, for $i \in \{1, 2\}$, random variables $\zeta^{(i)} = \mathsf{Encode}(C^{(i)}, \boldsymbol{x}^{(i)}; \boldsymbol{s}^{(i)}, \boldsymbol{\alpha}^{(i)})$. Then by Proposition 3.6, the two random variables are identically distributed: $\zeta^{(1)} \equiv \zeta^{(2)}$. We recall that by the definition of Decode, there exists a deterministic function d s.t. $d(\zeta^{(i)}) = \boldsymbol{v}^{(i)} \oplus \boldsymbol{y}^{(i)}$. As for any deterministic function, we have $(\zeta^{(1)}, d(\zeta^{(1)})) \equiv (\zeta^{(2)}, d(\zeta^{(2)}))$, i.e. $(\zeta^{(1)}, \boldsymbol{v}^{(1)} \oplus \boldsymbol{y}^{(1)}) \equiv (\zeta^{(2)}, \boldsymbol{v}^{(2)} \oplus \boldsymbol{y}^{(2)})$. This implies that for any value of $\boldsymbol{y}^*$ it holds that $(\zeta^{(1)} | \boldsymbol{v}^{(1)} \oplus \boldsymbol{y}^{(1)} = \boldsymbol{y}^*) \equiv (\zeta^{(2)} | \boldsymbol{v}^{(2)} \oplus \boldsymbol{y}^{(2)} = \boldsymbol{y}^*)$. That is, the conditional distributions are identical.

Now set $\boldsymbol{y}^* = \boldsymbol{v}^{(1)} \oplus \boldsymbol{y}^{(1)} = \boldsymbol{v}^{(2)} \oplus \boldsymbol{y}^{(2)}$, and notice that the random variable $(\zeta^{(i)} | \boldsymbol{v}^{(i)} \oplus \boldsymbol{y}^{(i)} = \boldsymbol{y}^*)$ is distributed identically to $\mathsf{Encode}(C^{(i)}, \boldsymbol{x}^{(i)}; \boldsymbol{s}, \boldsymbol{\alpha}^{(i)})$. The claim follows.

We can now prove Proposition 3.5.

Proof (Proof of Proposition 3.5). The simulator $\mathsf{Sim}(C, W, \{\alpha_j, v_j\}_{j \in W})$ considers a circuit C' topologically equivalent to C, but such that the values on the wires W are always fixed to the respective v_j regardless of the input. This can be done by fixing some of the local gates to always output the desired values. This is possible since in $\boldsymbol{v}[\overline{W}]$, for any fan-out gate, the input and all outputs take the same value. The simulator then samples random values for $\boldsymbol{\alpha}'[\overline{W}]$, and creates $\boldsymbol{\alpha}'$ by merging them with the values $\boldsymbol{\alpha}[W]$. Finally it samples a random $\boldsymbol{s}'$ and outputs $z' \leftarrow \mathsf{Encode}(C', \boldsymbol{0}; \boldsymbol{s}', \boldsymbol{\alpha}')$.

Consider now $z \leftarrow \mathsf{Encode}(C, \boldsymbol{x}; \boldsymbol{s}, \boldsymbol{\alpha})$, where $\boldsymbol{s}$ is uniformly random and $\boldsymbol{\alpha} = \{\alpha_j\}_{j \in W} \cup \{\alpha_j\}_{j \notin W}$ for a uniformly random $\{\alpha_j\}_{j \notin W}$. Recall that C, C' are topologically equivalent, so have the same set of wires, and let v_j, v_j' denote the values of wire j in the executions $C(\boldsymbol{x})$ and $C'(\boldsymbol{0})$ respectively. We note that $\boldsymbol{\alpha}[W] = \boldsymbol{\alpha}'[W]$, and that by the definition of C' it holds that $\boldsymbol{v}[W] = \boldsymbol{v}'[W]$. Since $\boldsymbol{\alpha}[\overline{W}], \boldsymbol{\alpha}'[\overline{W}]$ are uniformly random, it must be the case that $\boldsymbol{v} \oplus \boldsymbol{\alpha}$ and $\boldsymbol{v}' \oplus \boldsymbol{\alpha}'$ are identically distributed. Invoking Claim 1 concludes the proof.

4 Our Master Theorem

We show how to convert any protocol Π for computing a functionality f (irrespective of round complexity) into a non-interactive h-oracle-aided protocol $\hat{\Pi}^h$, where h is a quadratic function and while preserving the security properties of Π. A formal statement follows.

Theorem 4.1 (Master Theorem). *For every n-party protocol Π there exists an n-party non interactive oracle-aided protocol $\hat{\Pi}^h$ and an $L = \mathrm{poly}(2^d, n, m)$, where d (resp. m) is the depth (resp. size) of the circuit representation of Π (see Definition 3.1), with the following properties.*

1. **Efficiency.** *The communication and computational complexity of $\hat{\Pi}^h$ is at most L times larger than that of Π.*
2. **Quadratic Oracle.** *The oracle h is a quadratic function.*
3. **Simulation.** *For every strategy $\hat{A}$ acting on $\hat{\Pi}^h$, there exists a strategy A of complexity at most L times larger acting on Π, such that for all $T \subseteq [n]$ and for all $\boldsymbol{x} = (x_1, \ldots, x_n)$, the distributions $\mathsf{REAL}_{\Pi,T,A}(\boldsymbol{x})$ and $\mathsf{REAL}_{\hat{\Pi}^h,T,\hat{A}}(\boldsymbol{x})$ are identical. Furthermore, if $\hat{A}$ is semi-honest (i.e. follows the protocol) then so is A.*

Note that the simulation property also guarantees that the functionality of $\hat{\Pi}$ is the same as that of Π since the outputs of honest parties is included in the real model distribution.

Remark 4.1. It suffices to prove Theorem 4.1 only for *deterministic* protocols Π, since for a randomized protocol we can always consider Π to be the induced deterministic protocol where the parties' coins are treated as part of their input. Since our theorem quantifies over all inputs $\boldsymbol{x}$, this will also capture the case where part of the input (corresponding to the random tapes of the randomized protocol) is uniformly sampled.

Remark 4.2. Interestingly, we are able to prove the theorem using strategies A that are somewhat weaker than the most general conceivable malicious strategy, in the following sense. The colluding parties can only communicate before the execution of Π starts. That is, they cannot change their strategy in intermediate rounds of Π according to messages that were received by other parties in the collusion, however they share their initial views after seeing their inputs and before the first round begins.

In the remainder of the section we prove Theorem 4.1. We note that Lemmas 3.1 and 3.2 are new observations made in this work and they constitute a fundamental part of this proof.

Proof. As explained in Remark 4.1, we may assume that Π is deterministic. Our protocol essentially computes the point-and-permute encoding (Sect. 3.2) of the circuit representation of the protocol Π (Definition 3.1). The oracle h will correspond to the procedure Encrypt, and the Γ values are to be precomputed by the parties. Details follow.

Let (C, P) be the circuit representation of Π, and consider the computation $\mathsf{Encode}(C, \boldsymbol{x}; \boldsymbol{s}, \boldsymbol{\alpha})$. The protocol $\hat{\Pi}^h$ is a non-interactive oracle-aided protocol, i.e. it contains a pre-processing step where each party locally computes a message to send to the oracle, followed by an oracle response and local post-processing.

- **Preprocessing.** Each party i, on input x_i, samples a uniform vector $\boldsymbol{s}_i \in \{0,1\}^{\omega_C}$ (see the description of Encode in Sect. 3.2 for the definition of ω_C), and uniform values α_j for all j for which $P(j) = i$ (i.e. for all wires "belonging" to player i). Then for each local gate g that belongs to player i, the party computes the values Γ_g as in Eq. (1) (note that since g belongs to i, then i possesses all α values required for this computation). Finally, player i sends $\ell_i = (x_i, \boldsymbol{s}_i, \boldsymbol{\alpha}_i, \boldsymbol{\Gamma}_i)$ to the oracle h.
- **Oracle.** The oracle h takes all messages $\ell_i = (x_i, \boldsymbol{s}_i, \boldsymbol{\alpha}_i, \boldsymbol{\Gamma}_i)$. It concatenates all x_i into a joint input $\boldsymbol{x}$ for C, unites all $\boldsymbol{\alpha}_i$ into a vector $\boldsymbol{\alpha}$ containing a value for every wire, and unites all $\boldsymbol{\Gamma}_i$ into a single $\boldsymbol{\Gamma}$ containing a set Γ_g for every local gate g. Finally, it XORs the $\boldsymbol{s}_i$ values into a single string $\boldsymbol{s} = \oplus_i \boldsymbol{s}_i$. Note that all of these are linear operations.
  Finally it computes $z = \mathsf{Encrypt}(C, \boldsymbol{x}, \boldsymbol{s}, \boldsymbol{\alpha}, \boldsymbol{\Gamma})$ and sends (the same) z to all parties as response to their query.
- **Postprocessing.** Upon receiving z, each party i applies $\mathsf{Decode}(z)$ to obtain the sequence $\hat{v}_j$ for all $j \in [m]$. Then, for any output wire j belonging to party i, it computes $\hat{v}_j \oplus \alpha_j$ to obtain the output value (recall that for a wire j belonging to party i, the value α_j was locally generated by party i and is therefore available for postprocessing). Its output contains the collection of values on these output wires.

Properties 1, 2 in the theorem follow immediately from the properties of the point-and-permute encoding (Propositions 3.3 and 3.1). It remains to prove Property 3.

Let $(\hat{A}, T)$ be an adversary for $\hat{\Pi}$. Since $\hat{\Pi}$ is non-interactive, then $\hat{A}$ only gets to choose the values $\boldsymbol{\ell}[T] = \{\ell_i\}_{i \in T}$ based on the inputs $\boldsymbol{x}[T]$, and then postprocess the oracle response z. We can further simplify and consider w.l.o.g only adversaries $\hat{A}$ that are deterministic (since our simulation is perfect and therefore holds even conditioned on any random string) and do not perform any postprocessing but instead just output z (since any postprocessing results in a deterministic function of z, thus simulating z allows to simulate any such value).

Our Simulator. Our task is to produce an adversary (A, T) for the original protocol Π with the same real-model distribution as our (deterministic, no-postprocessing) $\hat{A}$. We assume throughout that $T \neq [n]$ (i.e. there exist honest parties) otherwise the result is trivial. The adversary A first runs $\hat{A}$ on $\boldsymbol{x}[T]$ to obtain the values $\boldsymbol{\ell}[T]$. Let us denote by W all wires j s.t. $P(j) \in T$, i.e. all wires that belong to parties controlled by the adversary (and $\overline{W}$ the complement set of wires), and by H all local gates that are controlled by parties in T (and $\overline{H}$ the complement set of local gates). By parsing $\boldsymbol{\ell}[T]$ appropriately, we derive the values $\boldsymbol{x}'[T]$, $\boldsymbol{\alpha}[W]$ and $\boldsymbol{\Gamma}[H]$, namely all α and Γ values associated with adversarially controlled parts of C. (Note that $\boldsymbol{x}'[T]$ is not necessarily identical to $\boldsymbol{x}[T]$ since the adversary is allowed to "change its input".)

By the Corruption Lemma (3.2) we can efficiently generate a circuit C' that is topologically equivalent to C and only differs from it in local gates controlled by the adversary. By Lemma 3.1, (C', P) is a protocol representation of a protocol Π' where all M_i' for $i \notin T$ are the same as in Π, but M_i' for $i \in T$ might differ.

The adversary A now sets each party $i \in T$ to (honestly) execute the protocol Π' (i.e. the machine M_i') using its respective input x_i'. Since for honest parties $M_i = M_i'$, we have that the parties jointly execute Π' on input $\boldsymbol{x}' = \boldsymbol{x}[\overline{T}] \cup \boldsymbol{x}'[T]$. Notice that if $\hat{A}$ is semi-honest then $\boldsymbol{\Gamma}' = \boldsymbol{\Gamma}$ and thus $C' = C$, which, in turn, implies that $\Pi' = \Pi$ and therefore A is semi-honest as well.

After the end of the execution of Π', the parties under the adversary's control do not return their prescribed output in Π'. Instead, the adversary A collects the views of all parties under its control, which correspond to the set of values $\boldsymbol{v}[W]$, i.e. the values on the W wires of C' when computed on $\boldsymbol{x}'$ (however A does not know $\boldsymbol{x}[\overline{T}]$ or any of the values $\boldsymbol{v}[\overline{W}]$). Lastly we apply the simulator from Proposition 3.5, i.e. the adversary A executes $\mathsf{Sim}(C', W, \boldsymbol{\alpha}[W], \boldsymbol{v}[W]) \to z'$ and sets the outputs of all parties in T to be z'.

Proof of Simulation. It remains to show that indeed $\mathsf{REAL}_{\hat{\Pi}^h, T, \hat{A}}(\boldsymbol{x}) \equiv \mathsf{REAL}_{\Pi, T, A}(\boldsymbol{x})$. Let us fix a value for $\boldsymbol{x}$ throughout the proof. Since we assume w.l.o.g that $\hat{A}$ is deterministic, this also fixes values for $\boldsymbol{x}'[T]$, $\boldsymbol{\alpha}[W]$, $\boldsymbol{\Gamma}[H]$, and $\{\boldsymbol{s}_i\}_{i \in T}$. Recall that $\boldsymbol{x}' = \boldsymbol{x}[\overline{T}] \cup \boldsymbol{x}'[T]$ (again a fixed value).

We start by noting that in $\mathsf{REAL}_{\hat{\Pi}^h, T, \hat{A}}$ the parties in T all output the same value z, and in $\mathsf{REAL}_{\Pi, T, A}$ they all output the same z'. Letting $\boldsymbol{y}[\overline{T}]$ denote the output of $\overline{T}$ parties in $\mathsf{REAL}_{\hat{\Pi}^h, T, \hat{A}}$, and $\boldsymbol{y}'[\overline{T}]$ denote the outputs of these parties in $\mathsf{REAL}_{\Pi, T, A}$, we conclude that our goal is to prove that $(\boldsymbol{y}[\overline{T}], z)$ is distributed identically to $(\boldsymbol{y}'[\overline{T}], z')$.

Consider the distribution $(\boldsymbol{y}[\overline{T}], z)$, and note that $z = \mathsf{Encrypt}(C, \boldsymbol{x}', \boldsymbol{s}, \boldsymbol{\alpha}, \boldsymbol{\Gamma})$. The vector $\boldsymbol{s}$ is random since it is XOR of all parties' $\boldsymbol{s}_i$ and there exists at least one honest party that samples its $\boldsymbol{s}_i$ uniformly. The vector $\boldsymbol{\alpha}$ is the union of $\boldsymbol{\alpha}[W]$ and a uniformly sampled $\boldsymbol{\alpha}[\overline{W}]$. The vector $\boldsymbol{\Gamma}$, by Lemma 3.2, is equal to $\mathsf{Permute}(C', \boldsymbol{\alpha})$. Since $\mathsf{Encrypt}$ only cares about the topology of its input circuit (Proposition 3.2), then in fact

$$
\begin{aligned}
z &= \mathsf{Encrypt}(C, \boldsymbol{x}', \boldsymbol{s}, \boldsymbol{\alpha}, \boldsymbol{\Gamma}) \\
&= \mathsf{Encrypt}(C', \boldsymbol{x}', \boldsymbol{s}, \boldsymbol{\alpha}, \boldsymbol{\Gamma}) \\
&= \mathsf{Encode}(C', \boldsymbol{x}'; \boldsymbol{s}, \boldsymbol{\alpha}) ,
\end{aligned}
$$

where the last inequality is because Encode by definition first generates $\boldsymbol{\Gamma} = \mathsf{Permute}(C', \boldsymbol{\alpha})$, and then applies $\mathsf{Encrypt}$.

Defining z in this way will allow us to show that the marginal distributions of $\boldsymbol{y}[\overline{T}]$ and $\boldsymbol{y}'[\overline{T}]$ are both identical and in fact *fixed* (having fixed $\boldsymbol{x}$, deterministic $\hat{A}$). To see this, first note that by Proposition 3.4 (correctness of garbled circuit), $\boldsymbol{y}[\overline{T}]$ is determined by the values of the output wires belonging to $\overline{T}$ parties in the evaluation of C' on x'. These values are determined by C', x' regardless of randomness. Likewise, $y'[\overline{T}]$ by definition is the output of the honest parties during the execution of Π' on x', and since Π' is represented by C', the output values are exactly the output values of C'. Lastly, $z \equiv z'$ since by Proposition 3.5

$$\mathsf{Encode}(C', x'; s, \alpha) \equiv \mathsf{Sim}(C', W, \alpha[W], v[W])$$

where the randomness is taken over s, $\alpha[\overline{W}]$ and the coins of Sim. This finalizes the proof of the theorem.

5   Completeness Theorems

In this section we prove that degree-2 functionalities are complete under non-interactive reductions. We say that a protocol has a security loss of L if any viable real-world adversary A can be simulated by an ideal-world adversary B whose complexity is at most L times larger than the complexity of A. We prove the following theorem.

Theorem 5.1 (Completeness of quadratic functions). *Let f be an n-party functionality computable by a circuit of size S and depth D. Then there exists a non-interactive reduction from the task of securely computing f to the task of computing a degree-2 functionality over $\mathbb{F}_2$. The reduction can take any of the following forms:*

1. *Perfectly secure reduction with threshold of $t = \lceil \frac{n}{3} - 1 \rceil$ and computational complexity and security loss of $\mathrm{poly}(n, S, 2^D)$.*
2. *Statistically secure reduction with threshold of $t = \lceil \frac{n}{2} - 1 \rceil$ and computational complexity and security loss of $\mathrm{poly}(n, S, 2^D)$.*
3. *Assuming one-way functions, computationally secure reduction with threshold of $t = \lceil \frac{n}{2} - 1 \rceil$ and computational complexity and security loss of $\mathrm{poly}(n, S)$. Furthermore, the reduction makes a black-box use of the one-way function (as part of the preprocessing and postprocessing phases).[9]*

The protocols are employed over synchronous network with pairwise private channels and a broadcast channel (which is our default setting). In all three settings, we require full security (in particular, the adversary cannot abort the honest players). It is well known that in this case the best achievable threshold is $\lceil (n/3) - 1 \rceil$ for perfect MPC (cf. [5]) and $\lceil (n/2) - 1 \rceil$ for statistical, or even computational, MPC [21]. Hence, the theorem achieves optimal security thresholds in all three cases.

As usual in the context of constant-round information-theoretic MPC, our information-theoretic protocols are efficient only for NC^1 functionalities.[10] Nevertheless, even for general functions, for which our perfect and statistical reductions are inefficient, the result remains meaningful since the protocols resist computationally unbounded adversaries.

See full version for proof of Theorem 5.1.

---

[9] In the computational setting, we let the circuit size S play the role of the security parameter, and assume that n is at most polynomial in S.

[10] This can be slightly pushed to log-space computation via standard techniques.

6   Perfect Three-Round MPC

In this section we obtain a 3-round protocol with full security (i.e. no abort) for general functions. Namely, we prove the following theorem.

Theorem 6.1 (Perfect 3-round MPC with threshold of quarter).

- *Every NC^1 functionality can be securely computed in 3 rounds with perfect security and threshold of $t = \lceil \frac{n}{4} - 1 \rceil$.*
- *Given a black-box access to a one-way function, the above extends to arbitrary polynomial-time functionalities at the expense of downgrading security to computational.*

By item 1 and item 3 of Theorem 5.1, the design of such protocols reduces to the design of a protocol with similar security properties for degree-2 functionalities. It therefore suffices to prove the following proposition.

Proposition 6.1. *Let f be an n-party functionality with complexity S and degree 2 (over the binary field). Then f can be perfectly computed in 3 rounds with security threshold of $t = \lceil \frac{n}{4} - 1 \rceil$ and complexity of $\mathrm{poly}(n, S)$.*

The proof of Proposition 6.1 appears in Sect. 6.1.

6.1   Proof of Proposition 6.1

VSS and Friends. A key component in the proof is *verifiable secret sharing* (VSS) [7]. In such secret sharing schemes, even if the dealer acts maliciously while sharing the secret s, all of the honest parties end up with shares that are consistent with some secret s'. We consider the Shamir-based VSS for threshold t where $n = 4t + 1$. The VSS will be implemented over an extension field $\mathbb{F}$ of $GF(2)$ of size at least $n + 1$, e.g., $\mathbb{F} = GF(2^{\lfloor \log n + 1 \rfloor})$. In particular, we will need 2-round protocols that realize the following functionalities with perfect security and threshold of t.

- The functionality Share_d in which a single designated party (denoted as the dealer) holds as an input a degree d univariate polynomial P over $\mathbb{F}$ (whose zero coefficient s plays the role of the secret) and all other parties have no input. The functionality delivers to the i-th party the value $s[i] = P(i)$.[11] We refer to $(s[1], \ldots, s[n])$ as a degree-d sharing of s. Note that security guarantees that for any adversarial set T of cardinality at most t, after the execution of Share_d the outputs of honest parties, i.e. $s[\overline{T}]$, lie on a single polynomial P' of degree d, and, if the dealer is honest (i.e. not in T) then $P' = P$. For every degree-bound $d \le t$, Gennaro et al. [11] describe a 2-round n-party protocol that perfectly realizes Share_d.

---

[11] As usual we assume that every $i \in [n]$ is associated with some public distinct field element $\alpha_i \ne 0$ and, by abuse of notation, we denote this element by i.

- The functionality $\mathsf{Share}_{d,0}$ which is defined similarly to Share_d except that the free coefficient of the dealer's input polynomial P must be zero. This functionality will be employed with degree $d = 2t$, and we can realize it in 2 rounds with perfect security and threshold t via the following standard reduction to Share_t. The dealer decomposes her polynomial $P(Z)$ into

$$\sum_{j=1}^{t} Z^j R_j(Z) \qquad (2)$$

where $R_1, \ldots, R_t$ are degree t polynomial that are chosen uniformly at random subject to the above constraint. Then the dealer shares each R_j via Share_t, and the i-th party gets $R_1(i), \ldots, R_t(i)$ and locally set his output to $i^1 R_1(i) + \cdots + i^t R_t(i)$.
- The functionality $\mathsf{BinShare}_t$ which is defined similarly to Share_t except that the free coefficient of the dealer's input polynomial P must be zero or one. We realize this functionality in 2 rounds with perfect security and threshold t via a reduction to the Share_t protocol of Gennaro et al. [11]. This reduction, described in the full version, is non-black-box and it relies on some concrete properties of the protocol. (To the best of our knowledge this reduction has not appeared in the literature.)

Given the above ingredients the protocol is quite straightforward. In particular, we rely on the following two standard properties of polynomial-based secret sharing: (1) *2-multiplicative*: If the parties share the secrets $(s_1, \ldots, s_k)$ via a degree-d sharing then, for every degree-2 mapping f over $\mathbb{F}$, we can get a degree $2d$-sharing of the secret $f(s_1, \ldots, s_k)$ by locally applying the degree-2 mapping f to the shares of each party. (2) *Noisy interpolation*: Given N points $(y_1, \ldots, y_N) \in \mathbb{F}^N$ with the promise that there exists a degree-D polynomial P for which $P(i) = y_i$ for all but $\lfloor (N - D)/2 \rfloor$ of $i \in [N]$, we can efficiently recover the polynomial P (and this polynomial is unique) via the standard Reed-Solomon decoder.

The Protocol. Let f be a degree-2 n-party functionality. We view f as a formal degree-2 polynomial over $\mathbb{F}$ with 0-1 coefficients. For ease of notation, assume that each party holds a single input x_i, and that the functionality has a single output that is delivered to all parties. (The protocol can be easily modified to handle the more general case.)

1. In parallel, every party $i \in [n]$ that holds an input $x_i \in \{0,1\}$ samples a random degree-t polynomial P_i over $\mathbb{F}$ whose free coefficient is x_i. The party invokes the 2-round protocol that implements $\mathsf{BinShare}_t$ as a dealer whose input is P_i. All parties receive the shares $(x_i[1], \ldots, x_i[n])$. In addition, every party $i \in [n]$ chooses a random degree-$2t$ polynomial R_i whose free coefficient is zero and distribute it to all the parties using via $\mathsf{Share}_{2t,0}$.
2. Each party j computes f over its shares, i.e. $f(x_1[j], \ldots, x_n[j]) \to y[j]$. It then randomizes the result by adding the value $R_1(j) + \cdots + R_n(j)$ and broadcasts the randomized share $\tilde{y}[i]$.

3. Each party interpolates a degree $2t$ polynomial Y which is consistent with at least $n - t$ of the points $(\tilde{y}[1], \ldots, \tilde{y}[n])$, the party outputs the value $Y(0)$.

Standard analysis (cf. [8, Sect. 2.2]) shows that the above protocol perfectly computes f with threshold t.   □

7 Two-Round MPC with Abort

We move on to the case of two-round protocols. As already mentioned, even if a broadcast channel is given we cannot hope for full security (or even fairness) when more than a single party is corrupted [12]. We therefore consider two standard relaxations of security with abort. As explained in Sect. 3, both notions are formalized by modifying the ideal-model in a way that grants the adversary additional power. We repeat the definition for the convenience of the reader.

- *Security with Selective Abort* (SSA) allows the adversary to selectively abort some of the honest parties (after the adversary learns his output). Formally, the ideal functionality first delivers the outputs of the corrupted parties to the simulator, which then can decide for each uncorrupted party whether this party will receive its output or a special abort symbol.
- *Security with Abort* (SA) allows the adversary to abort the honest parties even after the adversary learns his output. This is formalized similarly to SSA except that when the adversary decides to abort, *all* the honest parties receive a special abort symbol.

In the remainder of this section we prove the following theorems.

Theorem 7.1 (2-Round MPC with selective abort).

- *Every* NC^1 *functionality can be computed in 2 rounds with statistical security, selective abort and security threshold of* $t = \left\lceil \frac{n}{2} - 1 \right\rceil$. *The protocol does not use a broadcast channel.*
- *Given a black-box access to a one-way function, the above extends to arbitrary polynomial-time functionalities at the expense of downgrading security to computational.*

Theorem 7.2 (Computational 2-Round MPC with abort). *Given a black-box access to a one-way function, every polynomial-time functionality can be computed in 2 rounds with computational security, standard abort and security threshold of* $t = \left\lceil \frac{n}{2} - 1 \right\rceil$.

Theorem 7.2 and the computational part in Theorem 7.1 both introduce 2-round protocols with black-box access to one-way functions for polynomial-time functions and the same security threshold. They differ, however, since the protocols in Theorem 7.2 guarantee the stronger security notion (SA) at the expense of using a broadcast channel. (Indeed, the proof of Theorem 7.2 relies on Item 2 in Theorem 7.1.). By [20], selective abort is the best possible security for 2-round protocols that only use secure channels.

For proofs of Theorems 7.1, 7.2, see full version.

Acknowledgements. We thank Yuval Ishai for helpful discussions, for providing us several useful pointers, and for sharing with us the full version of [11].

References

1. Ananth, P., Choudhuri, A.R., Goel, A., Jain, A.: Round-optimal secure multiparty computation with honest majority. In: Shacham, H., Boldyreva, A. (eds.) CRYPTO 2018, Part II. LNCS, vol. 10992, pp. 395–424. Springer, Cham (2018). https://doi.org/10.1007/978-3-319-96881-0_14
2. Ananth, P., Choudhuri, A.R., Goel, A., Jain, A.: Two round information-theoretic MPC with malicious security. Cryptology ePrint Archive, Report 2018/1078 (2018). https://eprint.iacr.org/2018/1078
3. Applebaum, B., Brakerski, Z., Tsabary, R.: Perfect secure computation in two rounds. In: Beimel, A., Dziembowski, S. (eds.) TCC 2018, Part I. LNCS, vol. 11239, pp. 152–174. Springer, Cham (2018). https://doi.org/10.1007/978-3-030-03807-6_6. https://eprint.iacr.org/2018/894
4. Beaver, D., Micali, S., Rogaway, P.: The round complexity of secure protocols (extended abstract). In: Ortiz, H. (ed.) Proceedings of the 22nd Annual ACM Symposium on Theory of Computing, Baltimore, Maryland, USA, 13–17 May 1990, pp. 503–513. ACM (1990)
5. Ben-Or, M., Goldwasser, S., Wigderson, A.: Completeness theorems for non-cryptographic fault-tolerant distributed computation (extended abstract). In: Simon, J. (ed.) Proceedings of the 20th Annual ACM Symposium on Theory of Computing, Chicago, Illinois, USA, 2–4 May 1988, pp. 1–10. ACM (1988)
6. Benhamouda, F., Lin, H.: k-round multiparty computation from k-round oblivious transfer via garbled interactive circuits. In: Nielsen and Rijmen [18], pp. 500–532 (2018)
7. Chor, B., Goldwasser, S., Micali, S., Awerbuch, B.: Verifiable secret sharing and achieving simultaneity in the presence of faults (extended abstract). In: 26th Annual Symposium on Foundations of Computer Science, Portland, Oregon, USA, 21–23 October 1985, pp. 383–395. IEEE Computer Society (1985)
8. Damgård, I., Ishai, Y.: Constant-round multiparty computation using a black-box pseudorandom generator. In: Shoup, V. (ed.) CRYPTO 2005. LNCS, vol. 3621, pp. 378–394. Springer, Heidelberg (2005). https://doi.org/10.1007/11535218_23
9. Garg, S., Srinivasan, A.: Garbled protocols and two-round MPC from bilinear maps. In: Umans, C. (ed.) 58th IEEE Annual Symposium on Foundations of Computer Science, FOCS 2017, Berkeley, CA, USA, 15–17 October 2017, pp. 588–599. IEEE Computer Society (2017)
10. Garg, S., Srinivasan, A.: Two-round multiparty secure computation from minimal assumptions. In: Nielsen and Rijmen [18], pp. 468–499 (2018)
11. Gennaro, R., Ishai, Y., Kushilevitz, E., Rabin, T.: The round complexity of verifiable secret sharing and secure multicast. In: Vitter, J.S., Spirakis, P.G., Yannakakis, M. (eds.) Proceedings on 33rd Annual ACM Symposium on Theory of Computing, Heraklion, Crete, Greece, 6–8 July 2001, pp. 580–589. ACM (2001)
12. Gennaro, R., Ishai, Y., Kushilevitz, E., Rabin, T.: On 2-round secure multiparty computation. In: Yung, M. (ed.) CRYPTO 2002. LNCS, vol. 2442, pp. 178–193. Springer, Heidelberg (2002). https://doi.org/10.1007/3-540-45708-9_12
13. Goldreich, O.: The Foundations of Cryptography - Volume 2, Basic Applications. Cambridge University Press, Cambridge (2004)

14. Goldwasser, S., Lindell, Y.: Secure computation without agreement. In: Malkhi, D. (ed.) DISC 2002. LNCS, vol. 2508, pp. 17–32. Springer, Heidelberg (2002). https://doi.org/10.1007/3-540-36108-1_2

15. Ishai, Y., Kushilevitz, E.: Randomizing polynomials: a new representation with applications to round-efficient secure computation. In: 41st Annual Symposium on Foundations of Computer Science, FOCS 2000, Redondo Beach, California, USA, 12–14 November 2000, pp. 294–304. IEEE Computer Society (2000)

16. Ishai, Y., Kushilevitz, E.: Perfect constant-round secure computation via perfect randomizing polynomials. In: Widmayer, P., Eidenbenz, S., Triguero, F., Morales, R., Conejo, R., Hennessy, M. (eds.) ICALP 2002. LNCS, vol. 2380, pp. 244–256. Springer, Heidelberg (2002). https://doi.org/10.1007/3-540-45465-9_22

17. Ishai, Y., Kushilevitz, E., Paskin, A.: Secure multiparty computation with minimal interaction. In: Rabin, T. (ed.) CRYPTO 2010. LNCS, vol. 6223, pp. 577–594. Springer, Heidelberg (2010). https://doi.org/10.1007/978-3-642-14623-7_31

18. Nielsen, J.B., Rijmen, V. (eds.): EUROCRYPT 2018, Part II. LNCS, vol. 10821. Springer, Cham (2018). https://doi.org/10.1007/978-3-319-78375-8

19. Paskin-Cherniavsky, A.: Secure computation with minimal interaction. Ph.D. thesis, Technion – Israel Institute of Technology (2012)

20. Patra, A., Ravi, D.: On the exact round complexity of secure three-party computation. In: Shacham, H., Boldyreva, A. (eds.) CRYPTO 2018, Part II. LNCS, vol. 10992, pp. 425–458. Springer, Cham (2018). https://doi.org/10.1007/978-3-319-96881-0_15

21. Rabin, T., Ben-Or, M.: Verifiable secret sharing and multiparty protocols with honest majority (extended abstract). In: Johnson, D.S. (ed.) Proceedings of the 21st Annual ACM Symposium on Theory of Computing, Seattle, Washigton, USA, 14–17 May 1989, pp. 73–85. ACM (1989)

22. Rogaway, P.: The round-complexity of secure protocols. Ph.D. thesis, MIT (1991)

23. Shamir, A.: How to share a secret. Commun. ACM **22**(11), 612–613 (1979)

24. Yao, A.C.-C.: How to generate and exchange secrets (extended abstract). In: FOCS, pp. 162–167 (1986)

Two Round Information-Theoretic MPC with Malicious Security

Prabhanjan Ananth[1(✉)], Arka Rai Choudhuri[2], Aarushi Goel[2], and Abhishek Jain[2]

[1] Massachusetts Institute of Technology, Cambridge, USA
prabhanjan@csail.mit.edu
[2] Johns Hopkins University, Baltimore, USA
{achoud,aarushig,abhishek}@cs.jhu.edu

Abstract. We provide the first constructions of *two round* information-theoretic (IT) secure multiparty computation (MPC) protocols in the plain model that tolerate any $t < n/2$ malicious corruptions. Our protocols satisfy the strongest achievable standard notions of security in two rounds in different communication models.

Previously, IT-MPC protocols in the plain model either required a larger number of rounds, or a smaller minority of corruptions.

1 Introduction

The ability to securely compute on private datasets of individuals has wide applications of tremendous benefits to society. The notion of secure multiparty computation (MPC) [9,14,26,37] provides a solution to the problem of computing on private data by allowing a group of mutually distrusting parties to jointly evaluate any function over their private inputs in a manner that reveals nothing beyond the output of the function.

Information-Theoretic MPC. Over the years, a large body of works have investigated the design of MPC protocols against computationally bounded as well as computationally unbounded adversaries. In this work, we focus on the latter, namely, MPC with *information-theoretic* (IT) security.

The seminal works of [9,14] established the first feasibility results for IT-MPC for general functionalities. These works also established that IT security for non-trivial functions is only possible when at most $t < n/2$ of the n parties are corrupted. In scenarios where honest majority is a viable assumption, IT-MPC protocols are extremely appealing over their computational counterparts. In particular, they are typically more efficient since they do not use any computational primitives. Furthermore, IT-MPC protocols achieve security in models such as concurrent composition [11] without relying on external trust [12].

Round Complexity. In this work, we investigate the minimal conditions necessary for IT-MPC in the plain model. We focus on *round complexity* – a well studied complexity measure in distributed protocol design. We consider the standard

© International Association for Cryptologic Research 2019
Y. Ishai and V. Rijmen (Eds.): EUROCRYPT 2019, LNCS 11477, pp. 532–561, 2019.
https://doi.org/10.1007/978-3-030-17656-3_19

simultaneous-message model of communication for MPC where in any round, each party can send messages to other parties, depending upon the communication from the previous rounds. We consider security against *malicious* adversaries who may corrupt any subset of $t < n/2$ parties and use arbitrary strategy to decide their protocol messages.

It is well known that two rounds of communication are necessary for MPC [28]. We ask whether two rounds are *sufficient* for achieving IT security:

Does there exist two round IT-MPC for any $t < n/2$ corruptions?

The above question has remained open for the last three decades. In particular, while constant round IT-MPC protocols are known for any $t < n/2$ corruptions (e.g., [6,31]), the only known two round IT-MPC protocols are due to [29,31,34] who require two-thirds honest majority (as opposed to standard honest majority). We refer the reader to Sect. 1.3 for a comprehensive survey of prior work, and Sect. 1.1 for comparison with the recent works of [3,4,19].

1.1 Our Results

In this work, we resolve the above question in the affirmative.

I. Main Result. Our first result is a two-round IT-MPC protocol for NC^1 functions that tolerates any $t < n/2$ corruptions. In the case of *malicious* adversaries, our protocol achieves statistical security with abort – the standard notion of security (c.f. [25]) where an adversary may prevent the honest parties from learning the output by aborting the computation. In the setting of two rounds, this is known to be the strongest achievable standard notion of security [24].

In the case of *semi-honest* adversaries, our protocol achieves perfect security.

Theorem 1. *There exists a two round MPC protocol for NC^1 functions that achieves:*

- *Statistical security with abort against $t < n/2$ malicious corruptions.*
- *Perfect security against $t < n/2$ semi-honest corruptions.*

II. Protocols over P2P Channels. Our protocol in Theorem 1 necessarily uses both broadcast and private point-to-point (P2P) channels for achieving security against malicious adversaries.[1] We next investigate whether it is possible to construct two round IT-MPC against malicious adversaries by using *only* P2P channels.[2]

Our second result is a two round IT-MPC protocol over P2P channels that achieves statistical *security with selective abort* against any $t < n/2$ malicious corruptions. This notion [27] is a weakening of the standard notion of security of

---

[1] In the case of semi-honest adversaries, broadcasts can be trivially emulated over P2P channels without any increase in round complexity.

[2] Note that the complementary goal of IT-MPC over only broadcast channels is known to be impossible.

(unanimous) abort in that it allows the adversary to separately decide for each honest party whether it will receive the correct output or ⊥. Achieving security with abort in two rounds over P2P channels is known to be impossible in general [18,35]. This establishes security with selective abort as the strongest achievable standard notion of security in two rounds.

Theorem 2. *There exists a two round MPC protocol over P2P channels for NC^1 functions that achieves statistical security with selective abort against $t < n/2$ malicious corruptions.*

Put together, Theorems 1 and 2 fully resolve the round complexity of maliciously secure IT-MPC (for NC^1 functions).

Comparison with [3,4,19]. Recently, Applebaum et al. [3] constructed two round perfectly secure MPC for NC^1 against any $t < n/2$ semi-honest corruptions. Garg et al. [19] achieve a similar result; however, the communication complexity of their protocols grows super-polynomially with the number of parties. Neither of these works consider security against malicious adversaries, which is the main focus of our work. A recent independent and concurrent work of Applebaum et al. [4] also considers the case of malicious adversaries. Similar to our work, they also construct a two-round statistically secure protocol for NC^1 functionalities that achieves security with selective abort. However, they do not achieve our main result, namely a two-round information-theoretic protocol for security with (unanimous) abort.

1.2   Technical Overview

We first focus on achieving two-round IT-secure MPC in the presence of both broadcast and point to point communication channels.

Recent works on two-round secure MPC [10,21,22] follow a common blueprint of squishing an arbitrary round secure protocol, referred to as *inner* protocol, into a two round secure protocol, referred to as *outer* protocol using garbled circuits. Roughly speaking, every party in the outer protocol computes t garbled circuits, one for every round of the inner protocol. The job of the jth garbled circuit computed by the ith party is to emulate the computation of the next message function of the ith party in the jth round. Every party sends the generated t garbled circuits to the other parties.

The main challenge here is to ensure that the garbled circuits can talk to each other the same way the parties in the inner protocol talk to each other. The tools used to address this challenge differs from one work to another: [21] use bilinear maps, [22] use two-round oblivious transfer, [10,20] use two-round oblivious oblivious transfer and additionally garbled circuits and finally, [1,3,19] use information-theoretic MPC protocols. Of particular interest to us is the work of Ananth et al. [1] who show how to achieve maliciously secure two-round secure MPC in the honest majority setting for polynomial-sized circuits assuming only one-way functions.

Background on [1]. They propose the following template: The first step is to construct *helper* protocols that enable communication between garbled circuits in the outer protocol. The helper protocols they consider are delayed-function two-round MPC protocols, handling malicious adversaries, for two functionalities defined below. In a delayed-function two-round MPC protocol, the functionality is only available to the parties after the first round.

- The first functionality, parameterized by a bit v, is defined as follows: it takes as input r_1 from the first party, r_2 from the second party and outputs $r_1 \oplus r_2 \oplus v$.
- The second functionality, parameterized by two bits (v_1, v_2), is defined as follows: it takes as input a string K from the first party (interpreted as an input wire label of a garbled circuit), three bits (r_1, r_2, r_3) from the second party and outputs $K_{r_3 \oplus NAND(v_1 \oplus r_1, v_2 \oplus r_3)}$.

Observe that both these functionalities can be represented by quadratic polynomials over $\mathbb{F}_2$ and there exist two-round protocols for quadratic polynomials in the literature (see [34]). While these protocols do not achieve full-fledged malicious security, they achieve a weaker property termed as *privacy with knowledge of outputs* and [1] show how this weaker property is sufficient for their goal.

The next step is to transform the inner interactive protocol into an outer two-round protocol using the helper protocols. Since the helper protocols can only compute restricted functionalities, they impose a restriction on the "structure" of the inner protocol. In particular, every round of the inner interactive protocol is forced to only perform a single NAND computation. The term conforming protocols (originally coined by [22]) was used to described such interactive protocols.

Informally, a conforming protocol proceeds in a sequence of rounds. In every round, a party, termed as "receiver", obtains a global state from another party, termed as "sender", that encodes information about the current states of all the parties. Every party possesses a decryption key that lets it decode only a certain section of the global state. Once the party decodes the appropriate information, it then performs some local computation and then re-encodes the result and the resulting updated global state will be broadcasted to the rest of the parties, termed "listeners". Thus in every round, there is a sender, receiver and the rest of the parties are listeners.

At first, it might seem unclear as to why conforming protocols should exist at all. Luckily, an arbitrary round information-theoretically secure protocol can be transformed into a conforming protocol. However, the transformation demonstrated by [1] blows up the round complexity of the conforming protocol. In particular, *even if the original protocol had a constant number of rounds, the corresponding conforming protocol will now have round complexity proportional to the size of the circuit being securely computed.* Nevertheless, their transformation from a conforming protocol into the two-round outer protocol for polynomial-sized circuits is unaffected by the round complexity of the underlying conforming protocol.

Limitations on Extending [1] to IT Setting. To construct maliciously secure information-theoretically secure MPC protocols for NC^1 circuits, a natural direction to explore is to adapt the construction of [1] to the information-theoretic setting. The only part in the construction where one-way functions are used is in the generation of garbled circuits. If we restrict to NC^1 circuits, we could hope to use garbling schemes with perfect security [32]. These garbling schemes have the property that the size of the wire labels for the input wires grows exponentially in the depth of the circuit being garbled and linearly in the size of the garbled circuit.

This results in a fundamental issue in using information-theoretic garbling schemes to replace the garbled circuits based on one-way functions in [1]: as part of the outer protocol, every party sends a sequence of garbled circuits, where every garbled circuit encodes wire labels for the next garbled circuit. Recall that every garbled circuit emulates the next message function in a round and it needs to encode the wire labels for the next garbled circuit to enable transferring information from one round to the next. Once we use information-theoretically secure garbling schemes, the communication complexity now blows up exponentially in the length of the chain of garbled circuits. Since the length of the chain is the round complexity of the underlying conforming protocol, this results in *exponential communication complexity* even for NC^1 functionalities.

Our Approach. As a first step towards achieving our goal, we consider conforming protocols that do not restrict every round in the outer protocol to be just a single NAND computation. More generally, we allow the next message in every round of the conforming protocol to be a polynomial-size NC^1 circuit. We term this class of protocols to be *generalized* conforming protocols. On the one hand, the advantage of considering generalized conforming protocols is that we can construct this in constant number of rounds for NC^1 which makes it suitable to use it towards constructing a two-round protocol in the information-theoretic setting. On the other hand, the helper protocols designed in [1] are no longer compatible with our notion of generalized conforming protocols; recall that since the helper protocols in [1] were associated with quadratic polynomials, they imposed the requirement that every round in the conforming protocol is a single NAND computation.

To address this issue, we design *new* helper protocols that are "compatible" with generalized conforming protocols. Specifically, we require that the helper protocols are associated with functionalities computable in NC^1. By carefully examining the interiors of [1], it can be observed that it suffices to construct helper protocols for *three*-input functionalities computable in NC^1; these are the functionalities where only three parties have inputs. Informally, the three parties correspond to a sender party that sends a message in a round, a receiver party that receives a message in a round and finally, a listener party that listens to the communication from the sender to the receiver. Even though there are multiple listeners in every round in the conforming protocol, it suffices to design helper protocols for every listener separately. In the helper protocol, the inputs of the sender and the receiver are their private states[3] and the listener's input would be the wire labels for its garbled circuits. Note, however, that *the functionality associated with the helper protocol is as complex as the next message function of the conforming protocol.*

As such, it is unclear how to construct helper protocols even for three-input functionalities; in fact, if we had a two-round secure protocol for the three-input functionality that outputs the product of its inputs, then it could be bootstrapped to achieve two-round secure protocols for arbitrary functionalities via randomized encodings [31]. In light of this, the problem of constructing two-round secure protocols for three-input functionalities seems as hard as constructing two-round secure protocols for all functionalities computable in NC^1.

We resolve this dilemma in two main steps:

- We first focus on a weaker goal: constructing two-round information theoretically secure protocols for *two-input* (as opposed to three-input) functionalities.
- We then go back to our definition of generalized conforming protocols and impose additional structure on generalized conforming protocols – without blowing up their round complexity – to make them compatible with helper protocols for two-input functionalities.

We start by defining and constructing helper protocols for two-input functionalities.

Helper Protocols for Two-Input Functionalities. A two-input multiparty functionality, as the name suggests, is a functionality where only the first two parties get inputs while the rest of the parties are input-less. We consider two-input functionalities of the following form: these functionalities $\mathcal{U}$ are parameterized by two NC^1 functions f, G such that $\mathcal{U}(x_1, x_2, \bot, \cdots, \bot) = G(x_1, f(x_2))$. At first sight, this representation may seem unnecessary since one can rewrite $\mathcal{U}$ as another NC^1 function G' such that $\mathcal{U}(x_1, x_2, \bot, \cdots, \bot) = G'(x_1, x_2)$. However, the functions G and f we use to express $\mathcal{U}$ makes a difference when we state the security guarantees. Moreover, we require that the resulting helper protocol satisfies *delayed-function* property, meaning that the functionalities is only available to the parties after the first round.

Informally, we require the following asymmetric security guarantees:

- If the first party is honest then no information about its input x_1 should be leaked beyond $G(x_1, y^*)$. Ideally, we would require y^* to be the output of f on some input x_2^*. Here, we relax the security requirement to allow y^* to not even belong in the range of f.
- If the second party is honest then no information about its input x_2 should be leaked beyond $f(x_2)$. In particular, we allow the adversary to learn the

---

[3] Since the listener listens to the conversation, the receiver and the sender would share a secret string in order to emulate communication over *private channels (which are necessary for information-theoretic security)*. This is the reason why the receiver should also input its private state.

value $f(x_2)$ during the execution of the protocol. In addition, we only require that the simulator extracts the implicit input (interpreted as $f(x_2)$) and not x_2 itself.

Both the security requirements are non-standard and indeed, its should not be clear in what context these two security properties would be useful. To answer this, lets recall the structure of the conforming protocol: in every round, every party receives a global state, decodes a portion of the global state, computes on it and re-encodes the result. Looking ahead, when the conforming protocol is used alongside the helper protocols, the function f would have the global state hardwired inside its code; it takes as input private state of the party, represented by x_2, performs computation and then re-encodes the result. So the output $f(x_2)$ denotes the resulting global state.

Let us revisit the security requirements stated above. Allowing for y^* to not be in the range of f reduces to allowing for the second party to be malicious in the conforming protocol. We handle this by designing conforming protocols already secure against malicious parties. Regarding the second security requirement, revealing the value $f(x_2)$ reduces to the party revealing the updated global state. Since a party anyways has to broadcast the entire global state in the conforming protocol, its perfectly safe to reveal $f(x_2)$.

We now give a glimpse of our construction of two-round protocol for two-input functionalities. Our construction is heavily inspired by the techniques introduced in the work of Benhamouda and Lin [10].

– In the first round, the second party holding the input x_2, sends a garbling GC_2 of a universal circuit with x_2 hardwired inside it. The first party, holding the input x_1, receives GC_2 and computes another garbling GC_1 of a circuit, with x_1 hardwired inside it, that is defined as follows: it takes as input, wire labels of GC_2 with respect to input f, evaluates GC_2 using these input wire labels to obtain $f(x_2)$ and finally outputs $G(x_1, f(x_2))$.
– Simultaneously, all the parties execute a secure MPC protocol for quadratic polynomials, that takes as input wire labels of GC_2 from the second party, input wire labels of GC_1 from the first party and finally, computes GC_1 input wire labels associated with the input which is in turn defined to be the GC_2 input wire labels associated with f.

At the end of the second round, every party evaluates GC_1 to obtain $G(x_1, f(x_2))$.

We briefly describe the simulation strategy for arguing security of the above construction. If the second party is corrupted then the simulator extracts all the wire labels of GC_2 and then evaluates GC_2 using the wire labels of f to obtain the value y^*. The simulator then sends y^* to the ideal functionality, which responds back with $G(x_1, y^*)$. The simulator cannot verify that the second party indeed sent a valid garbling of the universal circuit. However, this still satisfies our security definition since the simulator is not required to extract x_2 but only the value y^*.

The case when the first party is corrupted can similarly be argued by designing a simulator that first extracts all the wire labels of GC_1 and then simulates GC_2 using the value $f(x_2)$.

CLC Property of Generalized Conforming Protocols. As explained earlier, helper protocols for two-input functionalities is as such incompatible with our current definition of generalized conforming protocols. Recall that the reason for incompatibility was that in every round of the generalized conforming protocol there were three parties participating. To remedy this situation, we introduce a new structural property for generalized conforming protocols, that we refer to as *copy-local-copy* (CLC) property. Specifically, we require that a party in every round, behaves as follows:

- Copy operation: first, every party copies the information transferred on the communication channels onto its own private state.
- Local computation: then it performs computation on its own local state.
- Copy operation: finally, it copies the result obtained onto the communication channel.

The CLC property effectively "breaks down" each three-input computation required in the earlier notion of generalized conforming protocol into three different operations. Now, given a generalized conforming protocol that satisfies the CLC property, it suffices to devise helper protocols for the above three operations.

The helper protocols for the first copy operation, and also the third copy operation, are associated with three parties: speaker, receiver and the listener. However, since the copy operation is a simple function, we observe that it suffices to use helper protocols for quadratic polynomials to implement this. The helper protocol for the local computation, however, is only associated with two parties: the party performing the local computation and the listener. Now, we use the delayed-function secure protocol for two-input functionalities constructed earlier to realize helper protocols associated with the local computation operation.

Since we divide every round of the protocol into three parts, a party sends three garbled circuits for every round of the conforming protocol, instead of just one.

Summary. We now summarize the main steps in the construction of maliciously secure information-theoretically secure multiparty protocols for NC^1 functionalities.

- First, we consider delayed-function two round secure MPC protocols for quadratic polynomials in Sect. 3.1.
- Then we define the notion of delayed-function two round secure MPC protocols for two-input NC^1 functionalities in Sect. 3.2. We define the security requirements in Sect. 3.2. This is followed by a construction of this notion in Sect. 3.2.
- In Sect. 4, we define the notion of generalized conforming protocols. We state the CLC property in Definition 7.
- Finally, we present the main construction in Sect. 5.

Protocol over P2P Channels. Next, we focus on designing a two-round protocol over P2P channels that achieves security with selective abort against malicious adversaries. Recall that in security with selective abort, the adversary can selectively decide which of the honest parties can receive the output while the rest of them abort. However, the adversary cannot force an "invalid" output on any of the honest parties.

To achieve our goal, we start with a two-round protocol Π_{in} over broadcast and P2P channels satisfying security with (unanimous) abort. A naive attempt would be as follows: start with Π_{in} and whenever a party has to send a broadcast message, he instead sends this message over P2P channels to all the other parties. Note that the resulting protocol is over P2P channels. However, this doesn't work: there is no mechanism in place to ensure that a malicious party indeed sends the *same* message, originally a broadcast message in Π_{in}, to all the other parties over P2P channels. The protocol Π_{in} might not be resilient to such attacks which would result in our resulting protocol to be insecure.

We introduce mechanisms to prevent this attack. Towards this, our idea is to require each party to send a garbled circuit of (a slightly modified version of) their second round next message function in Π_{in} in the second round of the P2P channel protocol. This (modified) next message function has the party's input and randomness, and the private channel messages that the party received in the first round of Π_{in} hard-wired inside its description. It additionally takes the first round broadcast channel messages of Π_{in} as input. To enable other parties to evaluate this garbled circuit, we require each party to send additive secret shares of all the labels for its garbled circuit over private channels (in particular, each party only receives one of the shares for each label) in the first round itself. In the second round, each party simply reveals the appropriate shares for each garbled circuit based on the messages received in the first round. If the adversary does not send the same set of broadcast messages to all parties, each party will end up revealing shares corresponding to a different label. In this case, we rely on the security of garbled circuits to ensure that nobody (including the adversary) is able to evaluate any of the honest party garbled circuits.

However, there are some subtle issues that crop when implementing this approach:

- Since we want the resulting protocol to satisfy information-theoretic security, we require the next-message function of Π_{in} to be computable in NC^1.
- The transformation sketched above does not handle the case when Π_{in} sends messages over private channels in the second round.

Fortunately, the information-theoretically secure MPC protocol over broadcast and P2P channels that we constructed earlier satisfies both the above properties and thus can be used to instantiate Π_{in} in the above approach. This gives us a P2P channel two-round MPC protocol that achieves security with selective abort against malicious adversaries. We present the construction of this protocol in Sect. 6.

1.3   Related Work

Since the initial feasibility results [9,14,26,37], a long sequence of works have investigated the round complexity of MPC. Here, we focus on protocols in the honest majority setting, and refer the reader to [5] for a survey of related works in the dishonest majority setting.

Information-Theoretic MPC. The seminal works of [9,14] provided the first constructions of polynomial-round IT-MPC protocols for general functionalities. These results were further improved upon in [7,13,36] w.r.t. malicious corruption threshold.

Bar-Ilan and Beaver [6] initiated the study of constant-round IT-MPC protocols. Subsequently, further improvements were obtained by [15,17,30]. The work of [31] provided the first constructions of two and three round IT-MPC protocols against $t < n/3$ and $t < n/2$, respectively, semi-honest corruptions. In the three round setting, their work was extended to handle a constant fraction of malicious adversaries by [23]. [32] constructed constant round perfectly secure protocols, improving upon the work of [6]. More recently, two round IT-MPC protocols that achieve security with selective abort against $t < n/3$ malicious corruptions were constructed by [34] and [29]. In fact, [34] and [29], put together, also achieve the stronger notion of security with guaranteed output delivery for the specific case of $n \geqslant 4$ parties and $t = 1$ corruptions which is not covered by the impossibility results of [18,35]. All of these positive results are for NC^1 functions; [33] established the difficulty of constructing constant-round IT-MPC protocols for general functionalities.

We also highlight the work of [27] who provided a general compiler to transform protocols over broadcast channels that achieve security with abort into protocols over P2P channels that achieve security with selective abort. Their transformation is unconditional, and increases the round-complexity by a multiplicative factor of three.

Computationally Secure MPC. The study of constant-round computationally secure MPC protocols in the honest majority setting was initiated by Beaver et al. [8] who constructed such protocols for general functionalities based on one-way functions. Damgård and Ishai [16] provided improved constructions based on only black-box use of one-way functions.

Two round protocols for general functionalities against $t < n/3$ malicious corruptions were constructed by [34] and [29] based on one-way functions. Very recently, Ananth et al. [1] constructed two round protocols for general functionalities that achieve security with abort against any $t < n/2$ malicious corruptions based on black-box use of one-way functions. Applebaum et al. [3] and Garg et al. [19] also achieve similar results, albeit only against semi-honest adversaries.

2 Preliminaries

We denote the statistical security parameter by **k**. We use the standard notion of security with abort for multi-party computation against malicious adversaries. For our second result over P2P channels, we consider a weaker notion of security, call security with selective abort. In security with selective abort, the adversary can selective cause some honest parties to output ⊥. Note that this is slightly different from the standard notion of security with abort, where the adversary can only either allow all honest parties to learn the output or cause all the honest parties to output $\bot$.

We also consider an even weaker notion of security called privacy with knowledge of outputs where the privacy of honest parties' inputs is ensured but the correctness of output for the honest parties is not guaranteed. We also use statistically secure garbled circuits [37] in our protocols.

3   Helper Two-Round Secure Protocols

We consider two types of helper protocols towards achieving our main goal:

- First, we consider a two-round secure multiparty computation protocol for NC^1 two-input functionalities; that is, only two of the parties have inputs. We consider this notion in the delayed-function setting.
- Next, we consider a two-round secure multiparty computation protocol for quadratic polynomials, also in the delayed function setting.

3.1   Delayed-Function Two-Round Secure MPC for Quadratic Polynomials

A delayed-function two-round secure MPC protocol is a special case of maliciously secure two-round secure MPC where the functionality is available to the parties only after the first round. One of the helper tools we use is a two-round secure MPC protocol for quadratic polynomials in the delayed function setting. Such a result was already shown by Ishai et al. [34]. Formally, they prove the following lemma.

Lemma 1 ([34]). *Let $n > 0$ and $\ell_{out} > 0$. Consider a n-party functionality $G : \{0,1\} \times \cdots \times \{0,1\} \to \mathcal{Y}^{\ell_{out}}$, where $\mathcal{Y} = \{(0,\ldots,0),(1,\ldots,1)\}$, and every output bit of G is computable by an n-variate quadratic polynomial over $\mathbb{F}_2$. There is a delayed-function two-round MPC protocol for G satisfying perfect privacy with knowledge of outputs property in the honest majority setting. Moreover, the next message of this protocol can be represented by a $O(\log(n))$-depth $(\ell_{out} \cdot n)^c$-sized circuit, for some constant c.*

Remark. The protocol of [34] only guarantees a weaker variant of privacy with knowledge of outputs where the adversary can force different honest parties to output different values. However if we use a broadcast channel in the second round, their protocol achieves a stronger variant of privacy with knowledge of outputs, where all honest parties learn the same output.

3.2   Delayed-Function Two-Round Secure MPC

The other helper tool we require is a delayed-function secure MPC protocol for arbitrary functionalities, but where only two parties have inputs. In particular, we are interested in the class of functionalities $\{F_{G,f}\}$: each functionality $F_{G,f}$ is parameterized by two functions G, f; it takes as input $(x_1, x_2, \bot, \ldots, \bot)$ and outputs $F_{G,f}(x_1, x_2, \bot, \ldots, \bot) = G(x_1, f(x_2))$. That is, party P_1 gets as input x_1 and party P_2 gets as input x_2. If the functionality $F_{G,f}$ were to be available to the parties before the protocol begins then securely computing $F_{G,f}$ would reduce to securely computing G since P_2 can pre-compute $f(x_2)$ and then run the secure protocol for G. However, we consider delayed-function setting and so this would not work.

In terms of security, we require the following informal guarantees.

- Security against P_2: unlike the standard simulation-based paradigm, in the ideal world, the honest parties and the simulator only have oracle access to G. In particular, the simulator only has to extract the value y (termed as *true* input of P_2), interpreted as the output of f on some input x_2 (also called *implicit* input of P_2), from the adversary.
- Security against P_1: we require that the implicit input x_2 of P_2 is hidden from P_1. However, we don't enforce that the output $f(x_2)$ is hidden from P_1. Moreover, we require the input privacy of P_2 to hold even if P_1's behaviour deviates from the protocol.

In particular, we require different security guarantees depending on which party the adversary corrupts.

Two-Input Multiparty Functionalities. We consider delayed-function two-round secure MPC protocols, where the parties determine the functionality (to be computed on their private inputs) only after the first round. This notion is referred as delayed-function secure MPC protocols in the literature. We describe the class of functionalities that we are interested in. Later, we define the security properties associated with delayed-function secure MPC protocols for this class of functionalities.

Two-Input n-Party Functionalities. A two-input n-party functionality is an n-party functionality where only two parties receive inputs from the environment.

Definition 1 (Two-Input n-Party Functionality). *Let $n, \ell_1, \ell_2, \ell' > 0$. We define an n-party functionality G to be a two-input functionality if its of the following form: it takes as input from the domain $\{0,1\}^{\ell_1} \times \{0,1\}^{\ell_2} \times \bot \times \cdots \times \bot$ and outputs a value in $\{(y, \ldots, y)\}_{y \in \{0,1\}^{\ell'}}$.*

We are interested in a sub-class of two-input functionalities that we refer to as specialized two-input n-party functionalities. Every functionality in this class, on input $(x_1, x_2, \bot, \ldots, \bot)$, first performs pre-processing on one of the inputs, say x_2, and then performs computation on the preprocessed result and x_1. The reason why we differentiate between pre-processing and post-processing becomes clear later on, when we define security against adversarial P_2.

Definition 2 (Specialized Two-Input n-Party Functionality). *Let $n, \ell_1, \ell_2, \ell' > 0$. We define an n-party functionality mapping $\{0,1\}^{\ell_1} \times \{0,1\}^{\ell_2} \times \bot \cdots \times \bot$ to $\{(y,\ldots,y)\}_{y \in \{0,1\}^{\ell'}}$ (parameterized by a functions G and f) to be a specialized two-input functionality if its of the following form: it takes as input $(x_1, x_2, \bot, \ldots, \bot)$ and outputs $G(x_1, f(x_2))$.*

Security. Let $P_1, \ldots, P_n$ be the parties participating in the delayed-function secure MPC protocol. We consider three cases and define separate security properties for each of these three cases: (i) P_1 is in the corrupted set while P_2 is not, (ii) P_2 is in the corrupted set while P_1 is not and, (iii) neither P_1 nor P_2 is in the corrupted set. Note that we don't consider the case when P_1 and P_2 are both in the corrupted set because P_1 and P_2 are the only parties receiving inputs in the protocol. We note that in all the three cases we are required to handle adversaries that deviate from the behavior of the protocol.

We define the following set systems.

- $\mathcal{S}_1 = \left\{ T \subseteq \{P_1, \ldots, P_n\} : |T| < \lfloor \frac{n}{2} \rfloor, P_1 \in T, P_2 \notin T \right\}$
- $\mathcal{S}_2 = \left\{ T \subseteq \{P_1, \ldots, P_n\} : |T| < \lfloor \frac{n}{2} \rfloor, P_1 \notin T, P_2 \in T \right\}$
- $\mathcal{S}_3 = \left\{ T \subseteq \{P_1, \ldots, P_n\} : |T| < \lfloor \frac{n}{2} \rfloor, P_1 \notin T, P_2 \notin T \right\}$

We now handle the three cases below. Denote S to be the corrupted set of parties. Let x_1 and x_2 be the inputs of P_1 and P_2 respectively.

Case 1. $S \in \mathcal{S}_1$. To define the security property for this case, we consider two experiments Expt_0 and Expt_1. In Expt_0, the honest parties and the adversary execute the protocol (real world). The output of Expt_0 is the view of the adversary and the outputs of the honest parties.

In Expt_1, the corrupted set of parties execute the protocol with the rest of the parties, simulated by a PPT algorithm Sim. In the first round, the simulator does not get any input and after the first round, the simulator gets as input $f(x_2)$, where $F_{G,f}$ is the n-party functionality associated with the protocol. The output of Expt_1 is the view of the adversary and the output of the simulator.

We require that the output distributions of the experiments Expt_0 and Expt_1 are identically distributed.

Definition 3 (Security Against $\mathcal{S}_1$). *Consider a delayed-function n-party protocol Π for a class of specialized two-input n-party functionalities $\{F_{G,f}\}$ mapping $\{0,1\}^{\ell_1} \times \{0,1\}^{\ell_2} \times \bot \cdots \times \bot$ to $\{(y,\ldots,y)\}_{y \in \{0,1\}^{\ell'}}$. We say that Π is secure against $\mathcal{S}_1$ if for every adversary corrupting a set of parties $S \in \mathcal{S}_1$, there exists a PPT simulator Sim such that the output distributions of Expt_0 and Expt_1 are identically distributed.*

Case 2. $S \in \mathcal{S}_2$. We handle this case using the real world-ideal world paradigm. In the real world, the corrupted parties and the honest parties execute the protocol. The output of the real world is the view of the adversary and the outputs of the honest parties. In the ideal world, the honest parties and the simulator have oracle access to the n-party functionality G^4. The output of the ideal world are the outputs of the honest parties and the output of the simulator.

More formally, we can define the real world process $\mathsf{Real}^{\mathcal{A},F}$ and the ideal world process $\mathsf{Ideal}^{\mathsf{Sim},G}$ – in particular, as in the definition of privacy with knowledge of outputs property, the simulator directs the trusted party to deliver outputs, of its choice, to the honest parties.

We define security of delayed-function secure MPC protocols against $\mathcal{S}_2$.

Definition 4 (Security Against $\mathcal{S}_2$). *Consider a delayed-function n-party protocol Π for a class of specialized two-input n-party functionalities $\{F_{G,f}\}$ mapping $\{0,1\}^{\ell_1} \times \{0,1\}^{\ell_2} \times \perp \cdots \times \perp$ to $\{(y,\ldots,y)\}_{y \in \{0,1\}^{\ell'}}$. We say that Π is secure against $\mathcal{S}_2$ if for every adversary $\mathcal{A}$ corrupting a set of parties $S \in \mathcal{S}_2$, there exists a PPT simulator Sim such that the output distributions of $\mathsf{Real}^{\mathcal{A},F}(x_1,\ldots,x_n)$ and $\mathsf{Ideal}^{\mathsf{Sim},G}(x_1,\ldots,x_n)$ are identically distributed.*

Remark 1. Since the simulator only has access to the ideal functionality of G (and not F) in the ideal world, this means that the simulator is required to only extract the *implicit* input (and not the *true* input) of the adversary. In particular, if f is the identity function, then this security notion implies the standard simulation-based security.

Case 3. $S \in \mathcal{S}_3$. In this case, we require the protocol to satisfy privacy with knowledge of outputs property. Formally, we can analogously define the real world process $\mathsf{Real}^{\mathcal{A},F}$ and ideal world process $\mathsf{Ideal}^{\mathsf{Sim},F}$. We define the security property below.

Definition 5 (Security Against $\mathcal{S}_3$). *Consider a delayed-function n-party protocol Π for a class of specialized two-input n-party functionalities $\{F_{G,f}\}$ mapping $\{0,1\}^{\ell_1} \times \{0,1\}^{\ell_2} \times \perp \cdots \times \perp$ to $\{(y,\ldots,y)\}_{y \in \{0,1\}^{\ell'}}$. We say that Π is secure against $\mathcal{S}_3$ if for every adversary $\mathcal{A}$ corrupting a set of parties $S \in \mathcal{S}_3$, there exists a PPT simulator Sim such that the output distributions of $\mathsf{Real}^{\mathcal{A},F}(x_1,\ldots,x_n)$ and $\mathsf{Ideal}^{\mathsf{Sim},F}(x_1,\ldots,x_n)$ are identically distributed.*

We are now ready to formally define a delayed-function secure MPC protocol for specialized two-input functionalities.

Definition 6. *Consider a delayed-function n-party protocol Π for a specialized two-input n-party functionality. We say that Π is secure if Π is secure against $\mathcal{S}_1$ (Definition 3), secure against $\mathcal{S}_2$ (Definition 4) and secure against $\mathcal{S}_3$ (Definition 5).*

Construction. We prove the following lemma.

Lemma 2. *Let $n, \ell_1, \ell_2, \ell' > 0$. Consider a two-input n-party functionality $G : \{0,1\}^{\ell_1} \times \{0,1\}^{\ell_2} \times \perp \times \cdots \times \perp \rightarrow \{(y,\ldots,y)\}_{y \in \{0,1\}^{\ell'}}$ computable by a depth-d circuit of size s. There is a delayed-input two-round MPC protocol for*

---

[4] We emphasize that the parties have oracle access to G and not F.

a specialized two-input functionality G (Definition 2) satisfying perfect privacy with knowledge of outputs property in the honest majority setting. Moreover, the next message function of the every party in the protocol can be represented by a circuit of depth $O(d + \log(s))$ and size $s^c 2^{c \cdot (d + \log(s))}$, for some constant c.

Proof. The main tools used in the construction are a perfectly secure garbling scheme and a secure MPC protocol for quadratic polynomials in the honest majority setting satisfying privacy with knowledge of outputs property (Lemma 1). We denote the garbling scheme by (Gen, Garb, Eval). We denote the secure MPC protocol for quadratic polynomials by Π_{Quad}.

We construct a delayed-function secure MPC protocol for a class of specialized two-input functionalities $\{F_{G,f}\}$, each functionality implementable by a circuit of size s and depth d. Our construction is heavily inspired by the techniques introduced in the work of Benhamouda and Lin [10]. Suppose P_1 has input x_1, P_2 has input x_2 and the rest of the parties don't receive any input. The protocol proceeds as follows: set the statistical security parameter, $\mathbf{k} = 1$.

Round 1.

- P_1 generates $\mathsf{Gen}(1^{\mathbf{k}}, 1^{L'}, 1^{d'})$ to obtain $(\mathsf{gk}_1, \mathbf{K}_I^1)$, where L' and d' are defined below. It also generates the first round messages of Π_{Quad}. In Π_{Quad}, its input is $\mathbf{K}_I^1$. It sends the first round messages of Π_{Quad} to other parties.
- P_2 generates $\mathsf{Gen}(1^{\mathbf{k}}, 1^{L''}, 1^{d''})$ to obtain $(\mathsf{gk}_2, \mathbf{K}_I^2)$, where L'' and d'' (defined in first of Round 2). It also generates the first round messages of Π_{Quad}. It also generates a random string R (we define its length below). In Π_{Quad}, its input is $(\mathbf{K}_I^2 \circ R)$. It generates $\mathsf{Garb}(\mathsf{gk}_2, U_{x_2})$ to obtain GC_2, where U_{x_2} is a universal circuit with x_2 hardwired in it, it takes as input a circuit of size s, depth d and outputs a single bit. Set $|R| = |\mathsf{GC}_2|$. Note that U_{x_2} can be implemented by a circuit of size $L'' = O(s)$ and depth $d'' = O(d)$. It sends $\mathsf{GC}_2 \oplus R$ along with the first round messages of Π_{Quad} to other parties.
- P_i, for $i \neq 1, i \neq 2$, generates the first round messages of Π_{Quad}. It sends the first round messages to other parties.

Round 2. At the end of round 1, the parties receive the function f as input.

- P_1 generates the second round messages of Π_{Quad}. The protocol Π_{Quad} is associated with a function that takes as input $(\mathbf{K}_I^1, \mathbf{K}_I^2, \bot, \ldots, \bot)$ and outputs $\mathbf{K}_I^1 \left[\mathbf{K}_I^2[f] \circ R \right]$[5]. We note that this function can be implemented by a system of quadratic polynomials over $\mathbb{F}_2$. It generates $\mathsf{Garb}(\mathsf{gk}_1, \widehat{G})$ to obtain GC_1, where $\widehat{G}$ (with $\mathsf{GC}_2 \oplus R$ hardwired) is defined as follows: it takes as input $(\mathbf{K}_I^2[f], R)$, computes $y \leftarrow \mathsf{Eval}(\mathsf{GC}_2, \mathbf{K}_I^2[f])$ and finally it outputs $G(x_1, y)$. $\widehat{G}$ can be implemented by a circuit of size $L' = O(s)$ and depth $d' = O(d)$. P_1 sends the second round messages of Π_{Quad} along with GC_1.

---

[5] Recall that the notation $\mathbf{K}_I^1 \left[\mathbf{K}_I^2[f] \circ R \right]$ refers to the input wire labels for GC_1 corresponding to the input $(\mathbf{K}_I^2[f] \circ R)$. Moreover, $\mathbf{K}_I^2[f]$ refers to the input wire labels for GC_2 corresponding to the input f.

- P_2 generates the second round messages of Π_{Quad} and sends them to other parties.
- P_i, for $i \neq 1$ and $i \neq 2$, computes the second round messages of Π_{Quad} and sends them to other parties.

Reconstruction. All the parties compute the output of Π_{Quad} to learn the output $\mathbf{K}^1[\mathbf{K}^2[f]]$. They then evaluate GC_1 to obtain $G(x_1, f(x_2))$.

If any point in time, if one of the parties abort, the rest of the parties abort as well. This completes the description of the protocol.

We now argue security. However, we refer the reader to the full version of our paper [2] for more details on the security, correctness and efficiency of this construction.

Security. We consider the following cases. Let S be the set of parties corrupted by the adversary.

$\underline{P_1 \in S \text{ and } P_2 \notin S.}$ The simulator is defined as follows:

- **Round 1.**
  - Simulating on behalf of P_2: Execute the simulator of Π_{Quad} to obtain the first round messages of Π_{Quad}. Generate $R \xleftarrow{\$} \{0,1\}^{|\mathsf{GC}_2|}$. Send the first round messages of Π_{Quad} along with R, intended for the parties in S, to the adversary.
  - Simulating on behalf of parties in $\overline{S}\backslash\{P_2\}$: Execute the simulator of Π_{Quad} to obtain the first round messages of Π_{Quad}. Send the first round messages, intended for the parties in S, to the adversary.

  Also, extract the input $\mathbf{K}_I^1$ of P_1 in Π_{Quad} from the first round messages of Π_{Quad}.
- **Round 2.** At the end of Round 1, the simulator receives $(f, \widehat{y})$ from the environment.
  - Simulating on behalf of P_2: Execute the simulator Sim_{GC} of the garbling scheme $(\mathsf{Gen}, \mathsf{Garb}, \mathsf{Eval})$; generate $\left(\widehat{\mathsf{GC}_2}, \widehat{\mathbf{K}^2}\right) \leftarrow \mathsf{Sim}_{GC}(1^{\mathbf{k}}, \varphi(U_{x_2}), \widehat{y})$, where $\varphi(U_{x_2})$ is the topology of U_x. Execute the simulator of Π_{Quad}[6], with the output of Π_{Quad} set to be $\mathbf{K}_I^1\left[\widehat{\mathbf{K}_I^2} \circ R \oplus \widehat{\mathsf{GC}_2}\right]$, to generate the second round messages of Π_{Quad}. Send the second round messages of Π_{Quad}, intended for the parties in S, to the adversary.
  - Simulating on behalf of parties in $\overline{S}\backslash\{P_2\}$: Similar to simulation on behalf of P_2, Execute the simulator of Π_{Quad}, with the output of Π_{Quad} set to be $\mathbf{K}_I^1\left[\widehat{\mathbf{K}_I^2} \circ R \oplus \widehat{\mathsf{GC}_2}\right]$, to generate the second round messages of Π_{Quad}. Send the second round messages of Π_{Quad}, intended for the parties in S, to the adversary.

---

[6] By the privacy with knowledge of outputs property, the simulator of Π_{Quad} directs the ideal functionality to deliver outputs (of its choice) to honest parties. However, the outer simulator (i.e., the simulator of Π_{DFunc}), which is running the simulator of Π_{Quad} as a subroutine, discards these outputs.

- **Reconstruction.** Receive the second round messages of Π_{Quad} from the corrupted parties in S. Also receive $\widehat{\text{GC}_1}$ from P_1. Reconstruct the output $\mathbf{K}_I^1$ from the second round messages of Π_{Quad}. Evaluate $\text{Eval}\left(\widehat{\text{GC}_1}, \mathbf{K}_I^1\right)$ to obtain $\widehat{b}$. Output $\widehat{b}$.

  If at any point in time, the adversary aborts, the simulator aborts as well.

$\underline{P_2 \in S \text{ and } P_1 \notin S.}$ The simulator is defined as follows:

- **Round 1.**
  - Simulating on behalf of P_1: Execute the simulator of Π_{Quad} to obtain the first round messages. Send the messages intended for the parties in S to the adversary.
  - Simulating on behalf of parties in $\overline{S} \backslash P_1$: This is identical to the simulation on behalf of P_1.

  Receive the first round messages of Π_{Quad} from the adversary. Additionally receive $\widehat{R}$ (masked garbled circuit) from P_2. Extract the input $(\mathbf{K}_I^2 \circ R)$ of P_2 from the first round messages of Π_{Quad} generated by P_2.

- **Round 2.** At the end of first round, the simulator receives f from the environment. Compute $\text{Eval}\left(\widehat{\text{GC}_2}, \mathbf{K}_I^2[f]\right)$ to obtain $\widehat{y}$, where $\widehat{\text{GC}_2} = \widehat{R} \oplus R$. Send $\widehat{y}$ to the ideal functionality to receive $\widehat{b}$.
  - Simulating on behalf of P_1: Execute the simulator Sim_{GC} of the garbling scheme $(\text{Gen}, \text{Garb}, \text{Eval})$; generate $\left(\widehat{\text{GC}_1}, \mathbf{K}_I^1\right) \leftarrow \text{Sim}_{GC}(1^k, \varphi(\widehat{G}), \widehat{b})$.

    Execute the simulator of Π_{Quad}, with the output of Π_{Quad} set to be $\mathbf{K}_I^1$, to generate the second round messages of Π_{Quad}. Send the second round messages of Π_{Quad} along with the simulated garbled circuit $\widehat{\text{GC}_1}$, intended for the parties in S, to the adversary.
  - Simulating on behalf of parties in $\overline{S} \backslash P_1$: Execute the simulator of Π_{Quad}, with the output of Π_{Quad} set to be $\mathbf{K}_I^1$, to generate the second round messages of Π_{Quad}. Send the second round messages of Π_{Quad} intended for the parties in S to the adversary.

- **Reconstruction.** Receive the second round messages of Π_{Quad} from the corrupted parties in S. Reconstruct the output $\mathbf{K}_I^1$ from the second round messages of Π_{Quad}. Evaluate $\text{Eval}\left(\widehat{\text{GC}_1}, \mathbf{K}_I^1\right)$ to obtain $\widehat{b}'$. Direct the ideal functionality to deliver the output $\widehat{b}'$ to the honest parties. Output of the simulator is the view of the adversary.

  If any point in time, the adversary aborts, the simulator aborts as well.

$\underline{P_1 \notin S \text{ and } P_2 \notin S.}$ The simulator is defined as follows:

- **Round 1.**
  - Simulating on behalf of P_1: Execute the simulator of Π_{Quad} to generate the first round messages; send the messages intended for the parties in S to the adversary.

- Simulating on behalf of P_2: Execute the simulator of Π_{Quad} to generate the first round messages; send the messages intended for the parties in S to the adversary. Also, send a string $R \xleftarrow{\$} \{0,1\}^{|\mathsf{GC}_2|}$.
- Simulating on behalf of parties in $\overline{S}\backslash\{P_1, P_2\}$: This is identical to the simulation on behalf of P_1.

- **Round 2.** The simulator receives the value $\widehat{b}$ from the ideal functionality.
  - Simulating on behalf of P_1: Execute the simulator Sim_{GC} of $(\mathsf{Gen}, \mathsf{Garb}, \mathsf{Eval})$; compute $\left(\widehat{\mathsf{GC}_1}, \widehat{\mathbf{K}^1}\right) \leftarrow \mathsf{Sim}_{GC}\left(1^{\mathsf{k}}, \varphi(\widehat{G}), \widehat{b}\right)$. Execute the simulator of Π_{Quad}, with the output of Π_{Quad} set to be $\widehat{\mathbf{K}^1}$, to generate the second round messages; send the messages intended for the parties in S to the adversary.
  - Simulating on behalf of P_2: Execute the simulator of Π_{Quad}, with the output of Π_{Quad} set to be $\widehat{\mathbf{K}^1}$, to generate the second round messages; send the messages intended for the parties in S to the adversary.
  - Simulating on behalf of parties in $\overline{S}\backslash\{P_1, P_2\}$: This is identical to the simulation on behalf of P_2.

- **Reconstruction.** Receive the second round messages of Π_{Quad} from the corrupted parties in S. Reconstruct the output $\widehat{\mathbf{K}_I^1}$ from the second round messages of Π_{Quad}. Evaluate $\mathsf{Eval}\left(\widehat{\mathsf{GC}_1}, \widehat{\mathbf{K}_I^1}\right)$ to obtain $\widehat{b'}$. Direct the ideal functionality to deliver the output $\widehat{b'}$ to the honest parties. Output of the simulator is the view of the adversary.

If any point in time, the adversary aborts, the simulator aborts as well.

$\square$

4 Generalized Conforming Protocols

The notion of conforming protocols was first defined in [22] as an intermediate tool to construct two-round secure MPC from two-round oblivious transfer. Their notion as-is is insufficient to achieve our goal of constructing an information-theoretic multiparty computation protocol secure against malicious adversaries. To get around this, we define the notion of *generalized conforming protocols*.

Syntax. An n-party generalized conforming protocol Φ for an n-party functionality F is specified by the parameters $\left(n, N, \{\Phi_{i,j}\}_{i \in [n], j \in [t+1]}, \mathcal{P}\right)$, where n is the number of parties in the system, N denotes the size of the global state $\mathbf{Z}$, $\Phi_{i,j}$ is a set of actions and $\mathcal{P}$ is a set of $(2 \cdot \binom{n}{2} + n)$ partitions of $[N]$. We denote $\mathcal{P} = (S_1, \ldots, S_n, \{T_{i_1, i_2}\}_{i_1, i_2 \in [n], i_1 \neq i_2}, U)$. One can think of S_i as the set of locations reserved for private computation for party P_i, T_{i_1, i_2} as the space allocated to party P_{i_1} for communicating private messages to party P_{i_2} and U as the space allocated for storing broadcast messages of each party. A generalized conforming protocol proceeds as follows. Let $x_1, \ldots, x_n$ be the respective inputs of all parties.

- **Pre-processing Phase.** For each $i \in [n]$, party P_i defines $\mathbf{st}_i$ to be the list: $\mathbf{st}_i := \left(R_k : \forall k \in S_i \bigcup_{i \neq i'} T_{i,i'} \bigcup_{i \neq i'} T_{i',i} \right)$ where $\forall k \in S_i \bigcup_{i \neq i'} T_{i,i'}$ it samples each bit R_k uniformly at random. Compute an N-sized list $\mathbf{Z}^{i,1}$ as follows:

  - For each $k \in [N]$, initialize $\mathbf{Z}_k^{i,1} = 0$. Here $\mathbf{Z}_k^{i,1}$ denotes the kth bit of $\mathbf{Z}^{i,1}$
  - Compute $\{(z_k, k) : k \in L_i\} \leftarrow \mathsf{Pre}(1^k, i, x_i, \mathbf{st}_i)$ where L_i is a subset of $S_i \bigcup_{i \neq i'} T_{i,i'}$.
  - For every $k \in L_i$, set the kth location $\mathbf{Z}_k^{i,1}$ in $\mathbf{Z}^{i,1}$ to have the value z_k.
  - For each $i' \in [n] \setminus \{i\}$, it sends $(R_k : \forall k \in T_{i,i'})$ to party $P_{i'}$ over private channels.
  - It broadcasts $\mathbf{Z}^{i,1}$ to all other parties.

  We require that there does not exist $k \in [N]$ such that for any $i_1 \neq i_2$, the set output by $\mathsf{Pre}(1^k, i_1, x_{i_1}, \mathbf{st}_{i_1})$ contains $(\cdot, k)$ and the set output by $\mathsf{Pre}(1^k, i_2, x_{i_2}, \mathbf{st}_{i_2})$ also contains $(\cdot, k)$. This means that there is no location in the global state $\mathbf{Z}$ that gets overwritten twice.

  At the end of the pre-processing phase, P_i receives $(R_k : \forall k \in T_{i',i})$ from all other parties $P_{i'}$ $(i' \in [n] \setminus i)$. It includes this as a part of $\mathbf{st}_i$. It retains $\mathbf{st}_i$ as private information.

- **Computation Phase.** For each $i \in [n]$, party P_i sets $\mathbf{Z}^1 = \bigoplus_{i-1}^n \mathbf{Z}^{i,1}$ For each $j \in [t+1]$, it proceeds as follows:
  - Parse the action $\Phi_{i,j}$ as $(\mathbf{L}_{i,j}^I, \mathbf{C}_{i,j}, \mathbf{L}_{i,j}^O)$.
  - If $j \neq 1$, for $\{(k, z_k)\}_{\forall i' \neq i, \, k \in \mathbf{L}_{i',j-1}^O}$, update k^{th} location in $\mathbf{Z}^{i,j}$ with value z_k. Call the resulting state $\mathbf{Z}^j$.
  - Take as input values in the locations of $\mathbf{Z}^j$ specified by the set $\mathbf{L}_{i,j}^I$ along with $\mathbf{st}_i$, compute $\mathbf{C}_{i,j}$ and update the locations in $\mathbf{Z}^j$ specified by the set $\mathbf{L}_{i,j}^O$. Call the resulting state $\mathbf{Z}^{i,j+1}$.
  - Send all the updated values and locations $\{(k, z_k)\}_{k \in \mathbf{L}_{i,j}^O}$ to all other parties.

  As before, we require that there is no location in $\mathbf{Z}$, where two parties simultaneously write to this location in any given round. At the end of all the rounds, the output of the computation for party P_i is in the last ℓ_i' locations of S_i.

- **Reconstruction.** For every $i \in [n]$, party P_i unmasks the last ℓ_i' locations of S_i to learn the output.

In terms of correctness, we require that at the end of the above protocol, the last ℓ_i' locations of S_i contains masked (y_i), where $F(x_1, \ldots, x_n) = (y_1, \ldots, y_n)$. Since a generalized conforming protocol is a special instance of a secure multiparty computation protocol, the security notions for generalized conforming protocols can be defined analogously.

Definition 7 (CLC Property). *An n-party generalized conforming protocol, specified by the parameters $\left(n, N, \{\Phi_{i,j}\}_{i \in [n], j \in [t+1]}, \mathcal{P} \right)$, for an n-party functionality F satisfies CLC property if the following holds: every $\Phi_{i,j}$ can be parsed* as $(\mathbf{L}_{i,j}^I = \mathbf{L}_{i,j}^{I\rightarrow} \cup \mathbf{L}_{i,j}^{I\leftarrow}, \mathbf{C}_{i,j}, \mathbf{L}_{i,j}^O = \mathbf{L}_{i,j}^{O\rightarrow} \cup \mathbf{L}_{i,j}^{O\leftarrow})$. We require that $\mathbf{C}_{i,j}$, for every $i \in [n], j \in [t+1]\backslash\{1\}$ (that is, all rounds except the first), is defined as follows: it takes as input values in the locations of $\mathbf{Z}$ specified by the locations $\mathbf{L}_{i,j}^I = \mathbf{L}_{i,j}^{I\rightarrow} \cup \mathbf{L}_{i,j}^{I\leftarrow}$ and state $\mathbf{st}_i$,

- **Copy Operation:** For every $k \in \mathbf{L}_{i,j}^{I\rightarrow}$, there exists a unique $k' \in \mathbf{L}_{i,j}^{I\leftarrow} \subset S_i$, copy $z_k \oplus R_k \oplus R_{k'}$ to $(k')^{th}$ location in $\mathbf{Z}$, where z_k is the value in the kth location of $\mathbf{Z}$. Note that $R_k, R_{k'}$ are values in the list $\mathbf{st}_i$ and hence, $\mathbf{L}_{i,j}^{I\rightarrow} \subset U \bigcup_{i' \in [n]\backslash\{i\}} T_{i,i'} \bigcup_{i' \in [n]\backslash\{i\}} T_{i',i}$.

- **Local Computation:** Take as input a set of values in $\mathbf{Z}$, indexed by a subset of S_i, $\mathbf{st}_i$, and compute a polynomial-sized circuit on these values. The output of this computation is written to a subset of locations, indexed by S_i, in $\mathbf{Z}$.

- **Copy Operation:** For every $k' \in \mathbf{L}_{i,j}^{O\rightarrow} \subset S_i$, there exists a unique $k \in \mathbf{L}_{i,j}^{O\leftarrow}$, copy $z_{k'} \oplus R_k \oplus R_{k'}$ to k^{th} location in $\mathbf{Z}$, where z_k is the value in the kth location of $\mathbf{Z}$. As before, $R_k, R_{k'}$ are values in the list $\mathbf{st}_i$ and hence, $\mathbf{L}_{i,j}^{O\leftarrow} \subset U \bigcup_{i' \in [n]\backslash\{i\}} T_{i,i'}$.

For the first round, we require $\mathbf{C}_{i,1}$ to be defined as follows: it takes as input $\mathbf{Z}^1$, computes a circuit $\widehat{\mathbf{C}}_{i,1}$ on $\mathbf{Z}^1$ to obtain $\{v_k\}_{k \in \mathbf{L}_{i,1}^O}$ and finally, it updates the kth location in $\mathbf{Z}^{i,2}$ with the value $z_k = v_k \oplus R_k$ for every $k \in \mathbf{L}_{i,1}^O$.

Lemma 3. Let $n, \ell_1, \ell'_1, \ldots, \ell_n, \ell'_n > 0$. Consider an n-party functionality $F : \{0,1\}^{\ell_1} \times \cdots \times \{0,1\}^{\ell_n} \to \{0,1\}^{\ell'_1} \times \cdots \times \{0,1\}^{\ell'_n}$ computable by a depth-d circuit of size s. There is a maliciously secure t-round generalized conforming protcol for F, for some constant t, satisfying CLC property with perfect security in the honest majority setting. Moreover, the next message function of every party can be implemented by a circuit of depth $O(d + \log(s))$ and size $s^c 2^{c\cdot(d+\log(s))}$, for some constant c.

We defer the proof of this lemma to the full-version of our paper [2].

5 Two-Round MPC over Broadcast and P2P: Security with Abort

In this section, we show how to construct a two-round MPC in the honest majority setting and satisfying statistical malicious security.

Lemma 4. Let $n, \ell_1, \ell'_1, \ldots, \ell_n, \ell'_n > 0$. Consider an n-party functionality $F : \{0,1\}^{\ell_1} \times \cdots \times \{0,1\}^{\ell_n} \to \{0,1\}^{\ell'_1} \times \cdots \times \{0,1\}^{\ell'_n}$ computable by a depth-d circuit of size s.

Fix a statistical security parameter $k > 0$. There is a malicious two-round MPC protocol for F with $\mathsf{negl}(k)$-statistical security in the honest majority setting, for some negligible function negl. Moreover, the computational complexity of this protocol is polynomial in s and exponential in d.

Construction. Let $F' = (F, \tau)$ be an augmented single functionality that computes F along with a multi-key MAC τ on the output of F. At a high level, a multi-key MAC corresponding to n keys allows each party to locally verify the MAC using their own key. We give a full definition and construction of this primitive in the full version of our paper. We list the ingredients for our two round MPC construction:

- A t-round Generalized Conforming protocol for the augmented functionality F', guaranteed by Lemma 3. Denote this by Π_{GConf}. Let Π_{GConf} be parameterized by $\left(n, N, \{\Phi_{i,j}\}_{i \in [n], j \in [t+1]}, \mathcal{P} \right)$.
- Delayed-function two-round secure n-party MPC for quadratic polynomials, as guaranteed by Lemma 1.
- Delayed-function two-round secure n-party MPC for specialized two-input functionalities, as guaranteed by Lemma 2.
- Information-theoretic garbling scheme (Gen, Garb, Eval).
- A multi-key MAC scheme (KeyGen, Sign, Verify).

We now describe a two-round secure MPC protocol for F.

Round 1.

- **Generation of Initial Global State:** For every $i \in [n]$, the ith party samples a key K_i for the mult-key MAC scheme. It sets it's input to $x_i' = (x_i, K_i)$. It then computes the pre-processing phase of Π_{GConf}. In particular it does the following: it defines $\mathbf{st}_i := (R_k : \forall k \in S_i \bigcup_{i \neq i'} T_{i,i'} \bigcup_{i \neq i'} T_{i',i})$, where $\forall k \in S_i \bigcup_{i \neq i'}$, it samples the bit R_k uniformly at random. It computes $\mathsf{Pre}(1^\mathbf{k}, i, x_i', \mathbf{st}_i)$ to obtain the set $\{(z_k, k) : k \in L_i\}$. It computes a N-sized list $\mathbf{Z}^{i,1}$ as follows: initialize $\mathbf{Z}^{i,1}$ to consist of only zeroes. It sets the kth location in $\mathbf{Z}^{i,1}$ to have the value z_k. Broadcast $\mathbf{Z}^{i,1}$ and sends $(R_k : \forall k \in T_{i,i'})$ to party $P_{i'}$ for each $i' \in [n] \setminus i$ over a private channel.
- **Generation of Garbling Wire Labels:** $(\mathsf{gk}_{i,1}, \mathbf{K}_{i,1}) \leftarrow \mathsf{Gen}(1^\mathbf{k}, 1^L, 1^d)$, where L is the number of leaves and d is the depth of the formula in Fig. 1a.
- For every $j \in [t+1] \setminus \{1\}$, the ith party computes the following:

    - $(\mathsf{gk}_{i,j}[\mathsf{copy1}], \mathbf{K}_{i,j}[\mathsf{copy1}]) \leftarrow \mathsf{Gen}(1^\mathbf{k}, 1^L, 1^d)$, where L is the number of leaves and d is the depth of the formula in Fig. 1b.
    - $(\mathsf{gk}_{i,j}[\mathsf{copy2}], \mathbf{K}_{i,j}[\mathsf{copy2}]) \leftarrow \mathsf{Gen}(1^\mathbf{k}, 1^L, 1^d)$, where L is the number of leaves and d is the depth of the formula in Fig. 2b.

- **First Round Messages of Delayed-Function MPC for Quadratic Polynomials:** All the parties participate in $O(n^3 t)$ executions of delayed-function two-round secure n-party MPC for quadratic polynomials, as guaranteed by Lemma 1. Each of these instantiations are denoted as follows:
    - For every $i_1, i_2 \in [n]$ and $i_1 \neq i_2$, the input of the ith party in $\Pi_{\mathsf{Quad}}[i_1, i_2, 1, 1]$ is the following:

* If $i = i_1$ then the ith party inputs $\{v_k\}_{k \in \mathbf{L}^O_{i_1,1}}$, $\{R_k\}_{k \mathbf{L}^O_{i_1,1}}$[7], where $\{v_k\}_{k \in \mathbf{L}^O_{i_1,1}}$ is the output of circuit $\widehat{C}_{i_1,1}$[8] on $\mathbf{Z}^1$, as defined in the Definition 7.

* If $i = i_2$ then the ith party inputs $\mathbf{K}_{i_2,2}[\mathsf{copy}1]$.

* If $i \neq i_1, i \neq i_2$ then the ith party doesn't have any input.

Denote $\Pi_{\mathsf{Quad}}[i_1, i_2, 1, 1].\mathsf{msg}_{1,i_4 \to i_5}$ to be the first round message of $\Pi_{\mathsf{Quad}}[i_1, i_2, 1j, 1]$ sent by the $(i_4)^{th}$ party to the $(i_5)^{th}$ party. We don't require that i_4 or i_5 be distinct from i_1, i_2. We define similar notation for the other Π_{Quad} instantiations. Let the total randomness used by the ith party in all these executions be R^i_Q.

- $\{\Pi_{\mathsf{Quad}}[i_1, i_2, i_3, j, 1]\}_{i_1,i_3 \in [n], i_1 \neq i_3, i_2 \in [n+1]}$ $j \in [t+1] \backslash \{1\}$

  For $i_1, i_3 \in [n], i_1 \neq i_3, i_2 \in [n+1], j \in [t+1] \backslash \{1\}$, the input of the ith party in $\Pi_{\mathsf{Quad}}[i_1, i_2, i_3, j, 1]$ is the following:

  * If $i = i_1$, then the ith party inputs $\{R_k\}_{k \in S_{i_1}}$

  * If $i = i_1 = i_2$, then the ith party additionally inputs $\{\{R_{k'}\}_{k \in T_{i_1,i'}}\}_{i' \in [n] \backslash \{1\}}$.

  * If $i = i_2 \neq i_1$, then the ith party inputs $\{R_{k'}\}_{k' \in T_{i_2,i_1}}$.
    If $i_2 = n + 1$, then party P_{i_2} has no input. This corresponds to the copy operations from locations in U to locations in S_{i_1}.

  * If $i = i_3$ then the ith party inputs $\mathbf{K}_{i_3,j}[\mathsf{local}]$

  * If $i \neq i_1, i \neq i_2, i \neq i_3$ then the ith party doesn't have any input.

- $\{\Pi_{\mathsf{Quad}}[i_1, i_2, j, 2]\}_{i_1,i_2 \in [n], i_1 \neq i_2, \ j \in [t] \backslash \{1\}}$

  For $i_1, i_2 \in [n], i_1 \neq i_2, j \in [t] \backslash \{1\}$, the input of the ith party in $\Pi_{\mathsf{Quad}}[i_1, i_2, j, 1]$ is the following:

  * If $i = i_1$, then the ith party inputs $\{R_{k'}\}_{k' \in S_{i_1}}$,
    $\{R_k\}_{k \in U \bigcup_{i' \in [n] \backslash \{i_1\}} T_{i_1,i'}}$

  * If $i = i_2$ then the ith party inputs $\mathbf{K}_{i_2,j+1}[\mathsf{copy}1]$

  * If $i \neq i_1, i \neq i_2$ then the ith party doesn't have any input.

The functionalities associated with each of these protocols are determined in the second round.

- **First Round Messages of Delayed-Function MPC for Two-Input Functionalities:** All the parties participate in $O(n^2 t)$ executions of delayed-function two-round secure n-party MPC, as guaranteed by Lemma 2. Denote these instantiations to be $\{\Pi_{\mathsf{DFunc}}[i_1, i_2, j]\}_{i_1,i_2 \in [n], j \in [t+1] \backslash \{1\}}$. For every

---

[7] Recall that $\mathbf{L}^O_{i_1,1}$ consists of a subset of locations in $S_{i_1}, \bigcup_{i' \in [n] \backslash \{i_1\}} T_{i_1,i'}$ and U and the locations in U are not a part of $\mathbf{st}_{i_1}$. But since R_k is not a part of $\mathbf{st}_{i'}$ for any $k \in U$ and $i' \in [n]$. Hence this is equivalent to every party setting $R_k = 0$ for all $k \in U$.

[8] Note that the only values from $\mathbf{Z}^1$ that $\widehat{C}_{i_1,1}$ computes on are known to P_{i_1} in the first round itself. Hence even if it does not know the entire value of $\mathbf{Z}^1$ in the first round, values $\{v_k\}_{k \in \mathbf{L}^O_{i_1,1}}$ can still be computed.

$i_1, i_2 \in [n]$ and $i_1 \neq i_2, j \in [t+1] \backslash \{1\}$, the input of ith party in $\Pi_{\mathsf{DFunc}}[i_1, i_2, j]$ is the following:

- If $i = i_1$ then the ith party inputs $\mathbf{K}_{i_1, j}[\mathsf{copy2}]$.
- If $i = i_2$ then the ith party inputs $\{R_k\}_{k \in S_{i_2}}$.
- If $i \neq i_1, i \neq i_2$ then the ith party doesn't have any input.

Denote $\Pi_{\mathsf{DFunc}}[i_1, i_2, j].\mathsf{msg}_{1, i \to i''}$ to be the first message of $\Pi_{\mathsf{DFunc}}[i', j]$ sent by the ith party to $(i'')^{th}$ party. Let the total randomness used by the ith party in all the executions be R_{DF}^i.

Round 2.

- **Compute Joint Global State:** All the parties compute $\mathbf{Z}^1 = \bigoplus_{i=1}^{n} \mathbf{Z}^{i,1}$.
- **Updates Private State:** It updates st_i to include $(R_k : \forall k \in T_{i',i})$ received from party $P_{i'} (\forall i' \in [n] \backslash i)$ in the first round.
- **Generate Input Wire Labels for First Garbled Circuit:** The ith party computes $(\mathsf{GC}_{i,1}, \mathbf{K}_{i,1}) \leftarrow \mathsf{Garb}(\mathsf{gk}_{i,1}, C_{i,1})$, where $C_{i,1}$ is defined in Fig. 1a. Let $\mathbf{K}_{i,1}[\mathbf{Z}^1]$ be the set of wire keys corresponding to the input $\mathbf{Z}^1$.
- **Generate Garbled Circuits for every round of Generalized Conforming Protocol:** For every $j \in [t+1] \backslash \{1\}$, the ith party computes:
  - $(\mathsf{GC}_{i,j}[\mathsf{copy1}], \mathbf{K}_{i,j}[\mathsf{copy1}]) \leftarrow \mathsf{Garb}(\mathsf{gk}_{i,j}[\mathsf{copy1}], C_{i,j}[\mathsf{copy1}])$, where $C_{i,j}$ [copy1] is defined in Fig. 1b.
  - $(\mathsf{GC}_{i,j}[\mathsf{local}], \mathbf{K}_{i,j}[\mathsf{local}]) \leftarrow \mathsf{Garb}(\mathsf{gk}_{i,j}[\mathsf{local}], C_{i,j}[\mathsf{local}])$, where $C_{i,j}[\mathsf{local}]$ is defined in Fig. 2a.
  - $(\mathsf{GC}_{i,j}[\mathsf{copy2}], \mathbf{K}_{i,j}[\mathsf{copy2}]) \leftarrow \mathsf{Garb}(\mathsf{gk}_{i,j}[\mathsf{copy2}], C_{i,j}[\mathsf{copy2}])$, where $C_{i,j}$ [copy2] is defined in Fig. 2b.
- The ith party broadcasts the following message:
$$\left(\mathsf{GC}_{i,1}, \ \mathbf{K}_{i,1}[\mathbf{Z}^1], \ \{\mathsf{GC}_{i,j}[\mathsf{copy1}], \mathsf{GC}_{i,j}[\mathsf{local}], \mathsf{GC}_{i,j}[\mathsf{copy2}]\}_{j \in [t+1]} \right)$$

Evaluation. To compute the output of the protocol, each party P_i does the following:

- For each $i' \in [n]$, let $\mathbf{K}_{i',1}[\mathbf{Z}^1]$ be the labels received from party $P_{i'}$ at the end of round 2.
- Obtain For each $i' \in [n]$, compute $\mathsf{Eval}(\mathsf{GC}_{i',1}, \mathbf{K}_{i',1}[\mathbf{Z}^1])$ to obtain labels in $\mathbf{K}_{i',2}[\mathsf{copy1}]$ corresponding to $\mathbf{Z}^{i',2}[\mathsf{copy1}]$ and second round messages $\{\Pi_{\mathsf{Quad}}[i_1, i_2, 1, 1].\mathsf{msg}_{2, i' \to i''}\}_{i_1, i_2, i', i'' \in [n], i_1 \neq i_2}$. Use these second round messages to reconstruct the remaining labels in $\mathbf{K}_{i',2}[\mathsf{copy1}]$ corresponding to $\{(k, z_k)\}_{\forall i'' \neq i', k \in \mathbf{L}_{i,1}^O}$.
- For each j from 2 to $(t+1)$ do the following:
  * For each $i' \in [n]$, compute $\mathsf{Eval}(\mathsf{GC}_{i',j}[\mathsf{copy1}], \mathbf{K}_{i',j}[\mathsf{copy1}][\mathbf{Z}^{i',j}[\mathsf{copy1}]$ $\| \ \{(k, z_k)\}_{\forall i'' \neq i', k \in \mathbf{L}_{i'',j-1}^{O \leftarrow}}])$ (if $j = 2$, $\mathbf{L}_{i'',j-1}^{O \leftarrow} = \mathbf{L}_{i'',j-1}^{O}$) to obtain labels in $\mathbf{K}_{i',j}[\mathsf{local}]$ corresponding to $\mathbf{Z}^{i',j}[\mathsf{local}]$ and second round messages $\{\Pi_{\mathsf{Quad}}[i_1, i_2, i_3, j, 1].\mathsf{msg}_{2, i' \to i''}\}_{i_1, i_3, i', i'' \in [n], i_1 \neq i_3, i_2 \in [n+1]}$.

* Use these second round messages to reconstruct the remaining labels in $\mathbf{K}_{i',j}[\mathsf{local}]$ corresponding to $\{(k, z_k)\}_{\forall i'' \neq i', \ k \in \mathbf{L}^{I-}_{i'',j}}$.

* For each $i' \in [n]$, compute $\mathsf{Eval}(\mathsf{GC}_{i',j}[\mathsf{local}], \mathbf{K}_{i',j}[\mathsf{local}][\mathbf{Z}^{i',j}[\mathsf{local}] \ || \ \{(k, z_k)\}_{\forall i'' \neq i', \ k \in \mathbf{L}^{I-}_{i'',j}}])$ to obtain labels in $\mathbf{K}_{i',j}[\mathsf{copy2}]$ corresponding to $\mathbf{Z}^{i',j}[\mathsf{copy2}]$ and second round messages $\{\Pi_{\mathsf{DFunc}}[i_1, i_2, j, 1].\mathsf{msg}_{2,i' \rightarrow i''}\}_{i_1, i_2, i', i'' \in [n], i_1 \neq i_2}$.

* Use these second round messages to reconstruct the remaining labels in $\mathbf{K}_{i,j}[\mathsf{copy2}]$ corresponding to $\{(k, z_k)\}_{\forall i'' \neq i', \ k \in S_{i''}}$.

* For each $i' \in [n]$, if ($j \neq t+1$), compute $\mathsf{Eval}(\mathsf{GC}_{i',j}[\mathsf{copy2}], \mathbf{K}_{i',j}[\mathsf{copy2}][\mathbf{Z}^{i',j}[\mathsf{copy2}] || \ \{(k, z_k)\}_{\forall i'' \neq i', \ k \in S_{i''}}])$ to obtain labels in $\mathbf{K}_{i',j+1}[\mathsf{copy1}]$ corresponding to $\mathbf{Z}^{i',j+1}[\mathsf{copy1}]$ and second round messages $\{\Pi_{\mathsf{Quad}}[i_1, i_2 j, 1].\mathsf{msg}_{2,i' \rightarrow i''}\}_{i_1, i_2, i', i'' \in [n], i_1 \neq i_2}$.

* Use these second round messages to reconstruct the remaining labels in $\mathbf{K}_{i',j}[\mathsf{local}]$ corresponding to $\{(k, z_k)\}_{\forall i'' \neq i', \ k \in \mathbf{L}^{O-}_{i'',j-1}}$.

* If $j = t + 1$ compute $\mathsf{Eval}(\mathsf{GC}_{i',j}[\mathsf{copy2}], \mathbf{K}_{i',j}[\mathsf{copy2}][\mathbf{Z}^{i',j}[\mathsf{copy2}] || \ \{(k, z_k)\}_{\forall i'' \neq i', \ k \in S_{i''}}])$ to obtain $\mathbf{Z}_{fin}$.

- Use st_i to unmask the last ℓ_i locations of S_i in $\mathbf{Z}_{fin}$ to compute the output (y, τ). Use key K_i and the verification algorithm of the multi-key MAC scheme to verify if τ is a valid multi-key MAC on y. If it verifies, output y, else output $\perp$.

We defer the security of this protocol to the full-version of our paper [2].

6 Two-Round MPC over P2P: Security with Selective Abort

In this section we describe a general compiler to obtain a two-round information-theoretic MPC protocol satisfying malicious security with selective abort over point-point channels from any two-round information-theoretic MPC satisfying security with abort against malicious adversaries.

Theorem 3. *There exists a general information theoretic compiler that transforms any two round maliciously secure MPC protocol (whose second round messages are computable in NC^1) over broadcast and private channels that achieves security with abort into a two-round protocol over private channels that achieves selective security with abort against malicious adversaries.*

Building Blocks. We use the following ingredients in our construction. A two-round MPC Π, in the honest majority setting satisfying perfect malicious security. We additionally want the second round next-message function of each party in this protocol to be implementable using an NC^1 circuit from Sect. 5. Information-theoretic garbling scheme ($\mathsf{Gen}, \mathsf{Garb}, \mathsf{Eval}$).

Input: $\mathbf{Z}^1$
Hardwired Values: action $\Phi_{i,1} = (\mathbf{L}^I_{i,1}, \mathbf{C}_{i,1}, \mathbf{L}^O_{i,1})$, state $\mathbf{st}_i$, wire labels $\mathbf{K}_{i,2}[\text{copy1}]$

- Parse $\Phi_{i,1} = (\mathbf{L}^I_{i,1}, \mathbf{C}_{i,1}, \mathbf{L}^O_{i,1})$
- Compute $\widehat{\mathbf{C}}_{i,1}$, where $\widehat{\mathbf{C}}_{i,1}$ is the circuit associated with $\mathbf{C}_{i,1}$ (see Definition 7), on the values in $\mathbf{Z}^1$ indexed by $\mathbf{L}^I_{i,1}$ along with $\mathbf{st}_i$. The output of the computation is written to locations in $\mathbf{Z}^1$ indexed by $\mathbf{L}^O_{i,1}$. Call the resulting state $\mathbf{Z}^{i,2}[\text{copy1}]$.
- The input to $C_{i,2}[\text{copy1}]$ is of the form $\left(\mathbf{Z}^{i,2}[\text{copy1}], \{(k, z_k)\}_{\forall i' \neq i,\ k \in \mathbf{L}^O_{i',1}} \right)$. Thus, the wire labels $\mathbf{K}_{i,2}[\text{copy1}]$ can be divided into two parts: the first part corresponds to $\mathbf{Z}^{i,2}[\text{copy1}]$ and the second part corresponds to $\{(k, z_k)\}_{\forall i' \neq i,\ k \in \mathbf{L}^O_{i',1}}$.
- For every i_1, i_2 with $i_1 \neq i_2$, compute the second round messages of $\Pi_{\text{Quad}}[i_1, i_2, 1, 1]$. The n-party functionality associated with $\Pi_{\text{Quad}}[i_1, i_2, 1, 1]$ is $Q_{i_1, i_2, 1, 1}$, defined below. The $(i_1)^{th}$ party has input $\{v_k\}_{k \in \mathbf{L}^O_{i,1}}$, $\{R_k\}_{k \in S_i \cup_{i' \in [n] \setminus \{i_1\}} T_{i_1, i'}}$, the $(i_2)^{th}$ party has input $\mathbf{K}_{i_2, 2}[\text{copy1}]$, the rest of the parties don't have any input and the output of this function are the labels in $\mathbf{K}_{i_2, 2}[\text{copy1}]$ corresponding to $\{(k, R_k \oplus v_k)\}$, for every location $k \in \mathbf{L}^O_{i_1, 1}$; recall that $\mathbf{L}^O_{i_1, 1}$ is the set of the locations written to at the end of first round in the conforming protocol.

  Output the second round messages of all these protocols. Also, output the labels in $\mathbf{K}_{i,2}[\text{copy1}]$ with respect to updated state $\mathbf{Z}^{i,2}[\text{copy1}]$.

(a) Description of $C_{i,1}$

Input: $\left(\mathbf{Z}^{i,j}[\text{copy1}], \{(k, z_k)\}_{\forall i' \neq i,\ k \in \mathbf{L}^{O \leftarrow}_{i',j-1}} \right)$ (If $j = 2$, then $\mathbf{L}^{O \leftarrow}_{i',1} = \mathbf{L}^O_{i',1}$)
Hardwired Values: action $\Phi_{i,j} = (\mathbf{L}^I_{i,j}, \mathbf{C}_{i,j}, \mathbf{L}^O_{i,j})$, state $\mathbf{st}_i$, wire labels $\mathbf{K}_{i,j}[\text{local}]$.

- For every $i' \neq i$, every $k \in \mathbf{L}^O_{i',j-1}$, update the kth location in $\mathbf{Z}^{i,j}[\text{copy1}]$ with the value z_k. Call the resulting state $\mathbf{Z}^j[\text{copy1}]$.
- Compute the first copy operation of $\Phi_{i,j}$ on the global state $\mathbf{Z}^j[\text{copy1}]$. Call the resulting state $\mathbf{Z}^{i,j}[\text{local}]$.
- The input to $C_{i,j}[\text{local}]$ is of the form $\left(\mathbf{Z}^{i,j}[\text{local}], \{(k, z_k)\}_{\forall i' \neq i,\ k \in \mathbf{L}^{I \leftarrow}_{i',j}} \right)$ Thus, the wire labels $\mathbf{K}_{i,j}[\text{local}]$ can be divided into two parts: the first part corresponds to $\mathbf{Z}^{i,j}[\text{local}]$ and the second part corresponds to $\{(k, z_k)\}_{\forall i' \neq i,\ k \in \mathbf{L}^{I \leftarrow}_{i',j}}$.
- For every $i_1, i_3 \in [n]$ with $i_1 \neq i_3$ and $i_2 \in [n+1]$, compute the second round messages of $\Pi_{\text{Quad}}[i_1, i_2, i_3, j, 1]$. The n-party functionality associated with $\Pi_{\text{Quad}}[i_1, i_2, i_3, j, 1]$ is $Q_{i_1, i_2, i_3, j, 1}$, defined below. The $(i_1)^{th}$ party has input $\{R_{k'}\}_{k' \in S_{i_1}}$, the $(i_2)^{th}$ party has input $\{R_k\}_{k \in T_{i_2, i_1}}$ (if $i_1 = i_2$, the i_1^{th} party additionally has input $\{\{R_k\}_{k \in T_{i_1, i'}}\}_{i' \in [n] \setminus \{i_1\}}$ and if $i_2 = n+1$, the $(i_2)^{th}$ party has no input), the $(i_3)^{th}$ party has input $\mathbf{K}_{i_3, j}[\text{local}]$, the rest of the parties don't have any input and the output of this function are the labels in $\mathbf{K}_{i_3, j}[\text{local}]$ corresponding to $\{(k', R_{k'} \oplus R_k \oplus z_k)\}$, for every location $k \in \mathbf{L}^{I \leftarrow}_{i_1, j}$ and $k' \in \mathbf{L}^{I \leftarrow}_{i_1, j}$ such that k' is the unique location associated with k as guaranteed by Definition 7.

  Output the second round messages of all these protocols. Also, output the labels in $\mathbf{K}_{i,j}[\text{local}]$ with respect to updated state $\mathbf{Z}^{i,j}[\text{local}]$.

(b) Description of $C_{i,j}[\text{copy1}]$, for $j > 1$.

Fig. 1. Descriptions of $C_{i,1}$ and $C_{i,j}[\text{copy1}]$ for $j > 1$

Input: $\left(\mathbf{Z}^{i,j}[\text{local}], \{(k, z_k)\}_{\forall i' \neq i, \ k \in \mathsf{L}_{i',j}^{I \leftarrow}} \right)$

Hardwired Values: action $\Phi_{i,j} = (\mathbf{L}_{i,j}^{I}, \mathbf{C}_{i,j}, \mathbf{L}_{i,j}^{O})$, state st_i, wire labels $\mathbf{K}_{i,j}[\text{copy2}]$.

- For every $i' \neq i$, for every $k \in \mathsf{L}_{i',j}^{I \leftarrow}$, update the kth location in $\mathbf{Z}^{i',j}[\text{local}]$ with the value z_k. Call the resulting state $\mathbf{Z}^{j}[\text{local}]$.
- Compute the local operation of $\Phi_{i,j}$ on the global state $\mathbf{Z}^{j}[\text{local}]$. Call the resulting state $\mathbf{Z}^{i,j}[\text{copy2}]$.
- The input to $C_{i,j}[\text{copy2}]$ is of the form $\left(\mathbf{Z}^{i,j}[\text{copy2}], \{(k, z_k)\}_{\forall i' \neq i, \ k \in S_{i'}} \right)$. Thus, the wire labels $\mathbf{K}_{i,j}[\text{copy2}]$ can be divided into two parts: the first part corresponding to $\mathbf{Z}^{i,j}[\text{copy2}]$ and the second part corresponding to $\{(k, z_k)\}_{\forall i' \neq i, \ k \in S_{i'}}$.
- For every i_1, i_2 with $i_1 \neq i_2$, compute the second round messages of $\Pi_{\mathsf{DFunc}}[i_1, i_2, j]$. The n-party functionality associated with $\Pi_{\mathsf{DFunc}}[i_1, i_2, j]$ is $DF_{i_1, i_2, j}$, defined below. The i_1^{th} party has the input $\mathbf{K}_{i_1, j}[\text{copy2}]$, the i_2^{th} party has the input $\{R_k\}_{k \in S_{i_2}}$, the rest of the parties don't have any input and the output of the function is computed as follows: compute the local operation of $\Phi_{i_2, j}$ on the global state $\mathbf{Z}^{j}[\text{local}]$ and the output of the function are the labels in $\mathbf{K}_{i_1, j}[\text{copy2}]$ corresponding to $\{(k, z_k)\}$, for every location $k \in S_{i_2}$.

Output the second round messages of all these protocols. Also, output the labels in $\mathbf{K}_{i,j}[\text{copy2}]$ with respect to updated state $\mathbf{Z}^{i,j}[\text{copy2}]$.

(a) Description of $C_{i,j}[\text{local}]$

Input: $\left(\mathbf{Z}^{i,j}[\text{copy2}], \{(k, z_k)\}_{\forall i' \neq i, \ k \in S_{i'}} \right)$

Hardwired Values: action $\Phi_{i,j} = (\mathbf{L}_{i,j}^{I}, \mathbf{C}_{i,j}, \mathbf{L}_{i,j}^{O})$, state st_i, wire labels $\mathbf{K}_{i+1,j}[\text{copy1}]$.

- For every $i' \neq i$, for every $k \in S_{i'}$, update the kth location in $\mathbf{Z}^{i',j}[\text{copy2}]$ with the value v_k. If $j \neq t+1$, call the resulting state $\mathbf{Z}^{j}[\text{copy2}]$, otherwise call the resulting state $\mathbf{Z}_{fin}$.
- If $j = t+1$, output $\mathbf{Z}_{fin}$, else continue to next step.
- Compute the second copy operation of $\Phi_{i,j}$ on the global state $\mathbf{Z}^{j}[\text{copy2}]$. Call the resulting state $\mathbf{Z}^{i,j+1}[\text{copy1}]$.
- The input to $C_{i,j+1}[\text{copy1}]$ is of the form $\left(\mathbf{Z}^{i,j+1}[\text{copy1}], \{(k, z_k)\}_{\forall i' \neq i, \ k \in \mathsf{L}_{i',j}^{O \leftarrow}} \right)$. Thus, the wire labels $\mathbf{K}_{i,j+1}[\text{copy1}]$ can be divided into two parts: the first part corresponding to $\mathbf{Z}^{i,j+1}[\text{copy1}]$ and the second part corresponding to $\{(k, z_k)\}_{\forall i' \neq i, \mathsf{L}_{i',j}^{O \leftarrow}}$.
- For every i_1, i_2 with $i_1 \neq i_2$, compute the second round messages of $\Pi_{\mathsf{Quad}}[i_1, i_2, j, 2]$. The n-party functionality associated with $\Pi_{\mathsf{Quad}}[i_1, i_2, j, 2]$ is $Q_{i_1, i_2, j, 2}$, defined below. The $(i_1)^{th}$ party has input st_i, the $(i_2)^{th}$ party has input $\mathbf{K}_{i_2, j+1}[\text{copy1}]$, the rest of the parties don't have any input and the output of this function are the labels in $\mathbf{K}_{i_2, j+1}[\text{copy1}]$ corresponding to $\{(k, R_k \oplus R_{k'} \oplus z_k)\}$, for every location $k \in \mathsf{L}_{i_1, j}^{O \leftarrow}$ and $k' \in \mathsf{L}_{i_1, j}^{O \rightarrow}$ such that k' is the unique location associated with k as guaranteed by Definition 7.

Output the second round messages of all these protocols. Also, output the labels in $\mathbf{K}_{i,j+1}[\text{copy1}]$ with respect to updated state $\mathbf{Z}^{i,j+1}[\text{copy1}]$.

(b) Description of $C_{i,j}[\text{copy2}]$.

Fig. 2. Descriptions of $C_{i,j}[\text{local}]$ and $C_{i,j}[\text{copy2}]$

Protocol. Let $\mathcal{P} = \{P_1, \ldots, P_n\}$ be the set of parties in the protocol. Let $\{x_1, \ldots, x_n\}$ be their respective inputs and $r_1, \ldots, r_n$ be their respective randomness used in the underlying protocol Π. Let $\mathbf{k}$ be the statistical security parameter.

Round 1. For each $i \in [n]$, party P_i does the following in the first round.

- Compute the first round messages of Π.
  $(\Pi.\mathsf{msg}_{1,i\to B}, \{\Pi.\mathsf{msg}_{1,i\to j}\}_{j\in[n]}) := \Pi_1(1^{\mathbf{k}}, i, x_i; r_i)$ where Π_1 is the first round next message function of the protocol Π.
  $\Pi.\mathsf{msg}_{1,i\to B}$ is the message that is broadcast by P_i in the first round of Π and $\Pi.\mathsf{msg}_{1,i\to j}$ is the message that is sent to party P_j over a private channel.
- Computes $(\mathsf{gk}_i, \mathbf{K}_i) \leftarrow \mathsf{Gen}(1^{\mathbf{k}}, 1^L, 1^d)$ where L is the number of leaves and d is the depth of the second round next message function of Π as defined in the next round. We parse $\mathbf{K}_i$ as $(K_{i,1}^0, K_{i,1}^1, \ldots, K_{i,L}^0, K_{i,L}^1)$
- Compute shares $\{\{K_{i,\ell}^{b,j}\}_{j\in[n]}\}_{\ell\in[L],b\in\{0,1\}}$ such that for each $\ell \in [L]$ and $b \in \{0,1\}$, $K_{i,\ell}^b := \bigoplus_{j\in[n]} K_{i,\ell}^{b,j}$
- For each $j \in [n] \setminus \{i\}$, it sends $(\Pi.\mathsf{msg}_{1,i\to B}, \Pi.\mathsf{msg}_{1,i\to j}, \{K_{i,\ell}^{b,j}\}_{\ell\in[L],b\in\{0,1\}})$ to party P_j over a private channel.

Round 2. For each $i \in [n]$, party P_i does the following.

- Computes a garbled circuit as follows:
  $\mathsf{GC}_i \leftarrow \mathsf{Garb}(\mathsf{gk}_i, \Pi_2(1^{\mathbf{k}}, i, x_i, \{\Pi.\mathsf{msg}_{1,j\to i}, \Pi.\mathsf{msg}_{1,i\to j}\}_{j\in[n]}, .; r_i))$
  where $\Pi_2(1^{\mathbf{k}}, i, x_i, \{\Pi.\mathsf{msg}_{1,j\to i}\}_{j\in[n]}, .; r_i)$ is the second round next message function of party P_i in Π that takes the messages $\{\Pi.\mathsf{msg}_{1,j\to B}\}_{j\in[n]}$ that were broadcast in the first round as input.
- For each $j \in [n] \setminus \{i\}$, it sends $(\mathsf{GC}_i, \{\{K_{j,\ell}^{X_\ell,i}\}_{j\in[n]}\}_{\ell\in[L]})$ to party P_j over a private channel. Here $X = \Pi.\mathsf{msg}_{1,1\to B}||\ldots||\Pi.\mathsf{msg}_{1,n\to B}$ and X_ℓ denotes the ℓth bit of X.

Reconstruction. Each party does the following.

- It reconstructs the input wire keys received in the previous round. For each $i \in [n]$ and $\ell \in [L]$, it computes the following. $K_{j,\ell}^{X_\ell} := \bigoplus_{i\in[n]} K_{j,\ell}^{X_\ell,i}$ and for each $j \in [n]$, it sets $\mathbf{K}_j[\Pi.\mathsf{msg}_{1,1\to B}||\ldots||\Pi.\mathsf{msg}_{1,n\to B}] := (K_{j,1}^{X_1}, \ldots, K_{j,1}^{X_L})$
- For each $j \in [n]$ it evaluates the garbled circuit received in the previous round.
  $\Pi.\mathsf{msg}_{2,j\to B} := \mathsf{Eval}(\mathsf{GC}_j, \mathbf{K}_j[\Pi.\mathsf{msg}_{1,1\to B}||\ldots||\Pi.\mathsf{msg}_{1,n\to B}])$
- It runs the reconstruction algorithm of Π on $\{\Pi.\mathsf{msg}_{2,j\to B}\}_{j\in[n]}$ to compute the output.

We defer the security of this protocol to the full-version of our paper [2].

Acknowledgments. The last three authors were supported in part by a DARPA/ARL Safeware Grant W911NF-15-C-0213, and a subaward from NSF CNS-1414023.

References

1. Ananth, P., Choudhuri, A.R., Goel, A., Jain, A.: Round-optimal secure multiparty computation with honest majority. In: Shacham, H., Boldyreva, A. (eds.) CRYPTO 2018. Part II. LNCS, vol. 10992, pp. 395–424. Springer, Cham (2018). https://doi.org/10.1007/978-3-319-96881-0_14
2. Ananth, P., Choudhuri, A.R., Goel, A., Jain, A.: Two round information-theoretic MPC with malicious security. IACR Cryptology ePrint Archive 2018, 1078 (2018). https://eprint.iacr.org/2018/1078
3. Applebaum, B., Brakerski, Z., Tsabary, R.: Perfect secure computation in two rounds. In: 16th International Conference on Theory of Cryptography, TCC 2018 (2018). https://eprint.iacr.org/2018/894
4. Applebaum, B., Brakerski, Z., Tsabary, R.: Degree 2 is complete for the round-complexity of malicious MPC (2019). https://eprint.iacr.org/2019/200
5. Badrinarayanan, S., Goyal, V., Jain, A., Kalai, Y.T., Khurana, D., Sahai, A.: Promise zero knowledge and its applications to round optimal MPC. In: Shacham, H., Boldyreva, A. (eds.) CRYPTO 2018. Part II. LNCS, vol. 10992, pp. 459–487. Springer, Cham (2018). https://doi.org/10.1007/978-3-319-96881-0_16
6. Bar-Ilan, J., Beaver, D.: Non-cryptographic fault-tolerant computing in constant number of rounds of interaction. In: Rudnicki, P. (ed.) 8th ACM Symposium Annual on Principles of Distributed Computing, Edmonton, Alberta, Canada, 14–16 August 1989, pp. 201–209. Association for Computing Machinery (1989)
7. Beaver, D.: Multiparty protocols tolerating half faulty processors. In: Brassard, G. (ed.) CRYPTO 1989. LNCS, vol. 435, pp. 560–572. Springer, New York (1990). https://doi.org/10.1007/0-387-34805-0_49
8. Beaver, D., Micali, S., Rogaway, P.: The round complexity of secure protocols (extended abstract). In: 22nd Annual ACM Symposium on Theory of Computing, Baltimore, MD, USA, 14–16 May 1990, pp. 503–513. ACM Press (1990)
9. Ben-Or, M., Goldwasser, S., Wigderson, A.: Completeness theorems for non-cryptographic fault-tolerant distributed computation (extended abstract). In: 20th Annual ACM Symposium on Theory of Computing, Chicago, IL, USA, 2–4 May 1988, pp. 1–10. ACM Press (1988)
10. Benhamouda, F., Lin, H.: k-round MPC from k-round OT via garbled interactive circuits. Technical report (2018)
11. Canetti, R.: Universally composable security: a new paradigm for cryptographic protocols. In: 42nd Annual Symposium on Foundations of Computer Science, Las Vegas, NV, USA, 14–17 October 2001, pp. 136–145. IEEE Computer Society Press (2001)
12. Canetti, R., Kushilevitz, E., Lindell, Y.: On the limitations of universally composable two-party computation without set-up assumptions. In: Biham, E. (ed.) EUROCRYPT 2003. LNCS, vol. 2656, pp. 68–86. Springer, Heidelberg (2003). https://doi.org/10.1007/3-540-39200-9_5
13. Chaum, D.: The spymasters double-agent problem. In: Brassard, G. (ed.) CRYPTO 1989. LNCS, vol. 435, pp. 591–602. Springer, New York (1990). https://doi.org/10.1007/0-387-34805-0_52
14. Chaum, D., Crépeau, C., Damgård, I.: Multiparty unconditionally secure protocols (extended abstract). In: 20th Annual ACM Symposium on Theory of Computing, Chicago, IL, USA, 2–4 May 1988, pp. 11–19. ACM Press (1988)
15. Cramer, R., Damgård, I.: Secure distributed linear algebra in a constant number of rounds. In: Kilian, J. (ed.) CRYPTO 2001. LNCS, vol. 2139, pp. 119–136. Springer, Heidelberg (2001). https://doi.org/10.1007/3-540-44647-8_7

16. Damgård, I., Ishai, Y.: Constant-round multiparty computation using a black-box pseudorandom generator. In: Shoup, V. (ed.) CRYPTO 2005. LNCS, vol. 3621, pp. 378–394. Springer, Heidelberg (2005). https://doi.org/10.1007/11535218_23

17. Feige, U., Kilian, J., Naor, M.: A minimal model for secure computation (extended abstract). In: 26th Annual ACM Symposium on Theory of Computing, Montréal, Québec, Canada, 23–25 May 1994, pp. 554–563. ACM Press (1994)

18. Fischer, M.J., Lynch, N.A.: A lower bound for the time to assure interactive consistency. Inf. Process. Lett. **14**(4), 183–186 (1982). https://doi.org/10.1016/0020-0190(82)90033-3

19. Garg, S., Ishai, Y., Srinivasan, A.: Two-round MPC: information-theoretic and black-box. In: 16th International Conference on Theory of Cryptography, TCC 2018 (2018). https://eprint.iacr.org/2018/909

20. Garg, S., Miao, P., Srinivasan, A.: Two-round multiparty secure computation minimizing public key operations. In: Shacham, H., Boldyreva, A. (eds.) CRYPTO 2018. LNCS, vol. 10993, pp. 273–301. Springer, Cham (2018). https://doi.org/10.1007/978-3-319-96878-0_10

21. Garg, S., Srinivasan, A.: Garbled protocols and two-round MPC from bilinear maps. In: 2017 IEEE 58th Annual Symposium on Foundations of Computer Science (FOCS), pp. 588–599. IEEE (2017)

22. Garg, S., Srinivasan, A.: Two-round multiparty secure computation from minimal assumptions. In: Nielsen, J.B., Rijmen, V. (eds.) EUROCRYPT 2018. Part II. LNCS, vol. 10821, pp. 468–499. Springer, Cham (2018). https://doi.org/10.1007/978-3-319-78375-8_16

23. Gennaro, R., Ishai, Y., Kushilevitz, E., Rabin, T.: The round complexity of verifiable secret sharing and secure multicast. In: 33rd Annual ACM Symposium on Theory of Computing, Crete, Greece, 6–8 July 2001, pp. 580–589. ACM Press (2001)

24. Gennaro, R., Ishai, Y., Kushilevitz, E., Rabin, T.: On 2-round secure multiparty computation. In: Yung, M. (ed.) CRYPTO 2002. LNCS, vol. 2442, pp. 178–193. Springer, Heidelberg (2002). https://doi.org/10.1007/3-540-45708-9_12

25. Goldreich, O.: The Foundations of Cryptography - Volume 2, Basic Applications. Cambridge University Press, Cambridge (2004)

26. Goldreich, O., Micali, S., Wigderson, A.: How to play ANY mental game or A completeness theorem for protocols with honest majority. In: Aho, A. (ed.) 19th Annual ACM Symposium on Theory of Computing, New York City, NY, USA, 25–27 May 1987, pp. 218–229. ACM Press (1987)

27. Goldwasser, S., Lindell, Y.: Secure multi-party computation without agreement. J. Cryptol. **18**(3), 247–287 (2005)

28. Halevi, S., Lindell, Y., Pinkas, B.: Secure computation on the web: computing without simultaneous interaction. In: Rogaway, P. (ed.) CRYPTO 2011. LNCS, vol. 6841, pp. 132–150. Springer, Heidelberg (2011). https://doi.org/10.1007/978-3-642-22792-9_8

29. Ishai, Y., Kumaresan, R., Kushilevitz, E., Paskin-Cherniavsky, A.: Secure computation with minimal interaction, revisited. In: Gennaro, R., Robshaw, M.J.B. (eds.) CRYPTO 2015. Part II. LNCS, vol. 9216, pp. 359–378. Springer, Heidelberg (2015). https://doi.org/10.1007/978-3-662-48000-7_18

30. Ishai, Y., Kushilevitz, E.: Private simultaneous messages protocols with applications. In: Proceedings of Fifth Israel Symposium on Theory of Computing and Systems, ISTCS 1997, Ramat-Gan, Israel, 17–19 June 1997, pp. 174–184 (1997). https://doi.org/10.1109/ISTCS.1997.595170

31. Ishai, Y., Kushilevitz, E.: Randomizing polynomials: a new representation with applications to round-efficient secure computation. In: 41st Annual Symposium on Foundations of Computer Science, Redondo Beach, CA, USA, 12–14 November 2000, pp. 294–304. IEEE Computer Society Press (2000)

32. Ishai, Y., Kushilevitz, E.: Perfect constant-round secure computation via perfect randomizing polynomials. In: Widmayer, P., Eidenbenz, S., Triguero, F., Morales, R., Conejo, R., Hennessy, M. (eds.) ICALP 2002. LNCS, vol. 2380, pp. 244–256. Springer, Heidelberg (2002). https://doi.org/10.1007/3-540-45465-9_22

33. Ishai, Y., Kushilevitz, E.: On the hardness of information-theoretic multiparty computation. In: Cachin, C., Camenisch, J.L. (eds.) EUROCRYPT 2004. LNCS, vol. 3027, pp. 439–455. Springer, Heidelberg (2004). https://doi.org/10.1007/978-3-540-24676-3_26

34. Ishai, Y., Kushilevitz, E., Paskin, A.: Secure multiparty computation with minimal interaction. In: Rabin, T. (ed.) CRYPTO 2010. LNCS, vol. 6223, pp. 577–594. Springer, Heidelberg (2010). https://doi.org/10.1007/978-3-642-14623-7_31

35. Patra, A., Ravi, D.: On the exact round complexity of secure three-party computation. In: Shacham, H., Boldyreva, A. (eds.) CRYPTO 2018. Part II. LNCS, vol. 10992, pp. 425–458. Springer, Cham (2018). https://doi.org/10.1007/978-3-319-96881-0_15

36. Rabin, T., Ben-Or, M.: Verifiable secret sharing and multiparty protocols with honest majority (extended abstract). In: 21st Annual ACM Symposium on Theory of Computing, Seattle, WA, USA, 15–17 May 1989, pp. 73–85. ACM Press (1989)

37. Yao, A.C.C.: How to generate and exchange secrets. In: 1986 27th Annual Symposium on Foundations of Computer Science, pp. 162–167. IEEE (1986)

Designated-Verifier Pseudorandom Generators, and Their Applications

Geoffroy Couteau[✉] and Dennis Hofheinz

KIT, Karlsruhe, Germany
geoffroy.couteau@kit.edu

Abstract. We provide a generic construction of non-interactive zero-knowledge (NIZK) schemes. Our construction is a refinement of Dwork and Naor's (FOCS 2000) implementation of the hidden bits model using verifiable pseudorandom generators (VPRGs). Our refinement simplifies their construction and relaxes the necessary assumptions considerably.

As a result of this conceptual improvement, we obtain interesting new instantiations:

- A designated-verifier NIZK (with unbounded soundness) based on the computational Diffie-Hellman (CDH) problem. If a pairing is available, this NIZK becomes publicly verifiable. This constitutes the first fully secure CDH-based designated-verifier NIZKs (and more generally, the first fully secure designated-verifier NIZK from a non-generic assumption which does not already imply publicly-verifiable NIZKs), and it answers an open problem recently raised by Kim and Wu (CRYPTO 2018).
- A NIZK based on the learning with errors (LWE) assumption, and assuming a non-interactive witness-indistinguishable (NIWI) proof system for bounded distance decoding (BDD). This simplifies and improves upon a recent NIZK from LWE that assumes a NIZK for BDD (Rothblum et al., PKC 2019).

Keywords: Non-interactive zero-knowledge ·
Computational Diffie-Hellman · Learning with errors ·
Verifiable pseudorandom generators

1 Introduction

Zero-knowledge proof systems allow a prover to convince someone of the truth of a statement, without revealing anything beyond the fact that the statement is true. After their introduction in the seminal work of Goldwasser, Micali, and Rackoff [20], they have proven to be a fundamental primitive in cryptography. Among them, *non-interactive zero-knowledge proofs* [5] (NIZK proofs), where the proof consists of a single flow from the prover to the verifier, are of particular

G. Couteau—Supported by ERC grant "PREP-CRYPTO" (724307).
D. Hofheinz—Supported by ERC grant "PREP-CRYPTO" (724307) and DFG project GZ HO 4304/4-2.

Y. Ishai and V. Rijmen (Eds.): EUROCRYPT 2019, LNCS 11477, pp. 562–592, 2019.
https://doi.org/10.1007/978-3-030-17656-3_20

interest, in part due to their tremendous number of applications in cryptographic primitives and protocols, and in part due to the theoretical and technical challenges that they represent.

On Building Non-Interactive Zero-Knowledge Proofs. It is known that zero-knowledge proofs for arbitrary NP languages can be constructed from any one-way function [19], and that this is a minimal assumption [30,32,39]. In contrast, non-interactive zero-knowledge proofs have proven to be considerably harder to construct. NIZKs in the plain model can only exist for trivial languages [31]; therefore, NIZKs for non-trivial languages are typically constructed in the *common reference string model*, where the prover and the verifier are given access to a common string honestly generated ahead of time in a setup phase. Generic constructions of NIZK proof systems for NP in the CRS model have been described from primitives such as doubly-enhanced trapdoor permutations [17], invariant signatures [21], and verifiable pseudorandom generators [16], where the last two are known to be also necessary for NIZKs. However, concrete instantiations of these primitives are currently known only from factorization-related assumption [5], pairing-based assumptions [8], and indistinguishability obfuscation [4,9] (together with injective one-way functions). More recently, direct constructions of NIZKs in the CRS model have been given from pairings [24–26], or from strong and less-understood assumptions such as indistinguishability obfuscation [4,38] and exponentially-strong KDM-secure encryption [7].

A fundamental and intriguing open question remains: is it possible to build NIZKs from other classical and well-established assumptions, such as discrete-logarithm-type assumptions, or lattice-based assumptions? Faced with the difficulty of tackling this hard problem upfront, the researchers have investigated indirect approaches, which can be divided into two main categories: the *bottom-up* approach, and the *top-down* approach.

The Bottom-Up Approach. This line of research fundamentally asks the following: starting from classical assumptions, either generic (OWF, public-key encryption) or concrete (CDH, LWE), how close to full-fledged NIZKs in the CRS model can we get, in terms of functionality? Early results in this direction have established the existence of NIZKs for NP in the *preprocessing model* (where the prover and the verifier execute ahead of time a preprocessing phase to generate respectively a secret proving key and a secret verification) assuming any one-way function [15], and *designated-verifier* NIZKs for NP (where anyone can compute a proof, but a secret verification key is required to verify a proof) from any semantically-secure public-key encryption scheme [33]. In addition to requiring the prover and/or the verifier to hold a secret key, these early results all suffered from a severe limitation: they only achieve a bounded form of soundness, where forging a proof for an incorrect statement is hard only if the prover is not given access to a verification oracle. This strongly limits their usability as a replacement for full-fledged NIZKs in most applications. More recently, various NIZK proof systems with unbounded soundness have been proposed, from the LPN assumption in the preprocessing model [6], from strong form of partially homomorphic encryption in the designated-verifier setting [11], and from LWE in the designated-prover setting [29] (where a secret key is required to compute a proof but anyone can verify a proof; the latter work also implies a NIZK with unbounded soundness in the preprocessing model from a strong variant of the Diffie-Hellman assumption). A slightly different approach was taken in [2], where the authors introduce (and construct from the DDH assumption) implicit zero-knowledge proofs, which are not NIZKs, but can replace them in applications related to secure computation.

The Top-Down Approach. This line of research tackles the problem from another angle: sticking with the goal of building full-fledged NIZKs in the CRS model, it attempts to identify the minimal "missing piece" which would allow to build NIZKs from classical assumptions. The work of [34] conjectured that a NIZK proof system for a specific language (GapSVP) would allow to build a NIZK proof system for all of NP from lattice assumptions, and the work of [37] almost confirmed this conjecture, by establishing that a non-interactive zero-knowledge proof for a specific language (bounded distance decoding, BDD) would imply the existence of a full-fledged NIZK proof system for NP in the CRS model, from the LWE assumption.

1.1 Our Contribution

In this paper, we revisit the problem of building non-interactive zero-knowledge proofs for NP from classical assumptions, investigating both the bottom-up approach and the top-down approach.

Our starting point is a fresh view on the work [16] of Dwork and Naor. In a nutshell, they construct a NIZK proof for NP by implementing the hidden bits model (HBM, [17]) using a tool they call verifiable pseudorandom generator (VPRG).[1] Intuitively, a VPRG is a pseudorandom generator (PRG) that allows to selectively prove that certain parts of the PRG output are consistent (relative to a commitment to the PRG input).

In the first part of our work, we relax the definition of VPRGs, and show that the relaxed definition is still sufficient to implement the HBM (and thus to obtain NIZK proofs for NP). Unlike the definition of [16], our definition also generalizes to the designated-verifier setting. In the second part of our work, we show that our new definition allows for considerably simple and new instantiations, both in the designated-verifier and standard (publicly verifiable) setting. We obtain instantiations from computational assumptions which were so far not known to imply NIZKs for NP. Specifically, we provide:

- A designated-verifier NIZK (DVNIZK) system for NP from the CDH assumption (with adaptive unbounded soundness and adaptive multi-theorem zero-knowledge). If the underlying group allows for a (symmetric) pairing, our construction can be made publicly verifiable. This is the first DVNIZK for NP from a concrete (i.e., non-generic) assumption that is not already known to

---

[1] In the HBM, there exist unconditionally secure NIZK proofs [17].

imply publicly verifiable NIZKs for NP. Our result resolves an open problem recently raised by Kim and Wu in [29], regarding the possibility of building multi-theorem NIZKs from DDH in the preprocessing model. Note that our result achieves a strictly stronger form of NIZK and under a weaker assumption.

– A NIZK system for NP (satisfying adaptive soundness and adaptive multi-theorem zero-knowledge) that assumes LWE and a non-interactive witness-indistinguishable proof system Π' for BDD. If Π' is designated-verifier, resp. publicly verifiable, then so is our NIZK system for NP. Our scheme improves the mentioned work of [37] that requires non-interactive *zero-knowledge* proof system for BDD. (We comment below on what allows us to avoid the need for simulation inherent in the approach of [37].)

1.2 Our Approach

The Proof System of Dwork and Naor. To outline our conceptual contribution, we provide more background on the definitions and model of Dwork and Naor [16]. First, the hidden bits model (HBM) is an abstract model of computation for a prover and a verifier that allows to formulate the NIZK protocol for graph Hamiltonicity from [17] in a convenient way. In the HBM, the prover receives an ideally random string $hb = (hb_i)_{i=1}^t \in \{0,1\}^t$ of bits, as well as an NP-statement x with witness w. In order to prove x, the prover then selects a subset $S \subset [t]$ of bit indices and auxiliary information M. The verifier is then invoked with $hb[S] = (hb_i)_{i \in S}$ and M, and outputs 1 if it is convinced of the truth of x. [17] provide a NIZK proof in the HBM that is statistically sound and statistically zero-knowledge. (Of course, at least one of those properties will have to become computational when implementing the HBM.)

Now Dwork and Naor [16] implement the HBM using VPRGs. Formally, a VPRG is a pseudorandom generator $G : \{0,1\}^\lambda \to \{0,1\}^m$ which allows to construct commitments pvk to seeds (i.e., G-inputs) s and publicly verifiable openings of individual bits of $G(s)$ (relative to pvk). [16] require the following:

1. pvk information-theoretically determines a unique value y *in the image of G*,
2. valid openings to bits not consistent with the y determined by pvk *do not exist*, and
3. an opening computationally leaks nothing about unopened bits of y.

Given a VPRG, [16] implement the HBM as follows. The prover initially selects a seed $s \xleftarrow{\$} \{0,1\}^\lambda$ and then generates a commitment pvk to s. This implicitly sets $hb = G(s)$. After selecting S, the prover then sends to the verifier pvk and an opening of $hb[S]$.

Observe that this protocol may still allow the prover to cheat by choosing a "bad" seed s that might allow breaking soundness. However, since the HBM protocol of [17] is statistically sound, there can be only comparatively few "bad" HBM strings hb that allow cheating. Hence, the probability that there *exists* a seed s such that $hb = G(s)$ is bad will be negligible.[2]

---

[2] A formal argument requires a little care in choosing parameters, and in randomizing hb with an additional component in the NIZK common reference string.

Our Conceptual Improvement. We show that points 1. and 2. from the VPRG definition can be simplified. Specifically:

- We require that pvk uniquely determines some y, but we do not require that y is in the image of G. Instead, we require that the bitlength $|\text{pvk}|$ of pvk is short (i.e., independent of m). Observe that now up to $2^{|\text{pvk}|}$ "bad" y (and thus "bad" hb) might exist. However, since $|\text{pvk}|$ is still short (compared to m), essentially the original proof strategy of [17] applies.
- We only require that it is computationally infeasible to come up with an opening not consistent with y. This relaxation requires a careful tracking of "bad events" during the security proof, but is essentially compatible with the proof strategy from [16].

Our first change allows us to omit an explicit proof of consistency of pvk that was necessary in [16]. This simplification will be highly useful in our concrete instantiations. Furthermore, our second change allows to consider *designated-verifier* NIZKs. Indeed, observe that the original requirement 2. above states that no valid openings inconsistent with y exist. This excludes designated-verifier realizations of VPRGs in which the verifier secret key can be used to forge proofs. However, since most existing DVNIZK proofs have this property (otherwise, they could be made publicly verifiable by making the secret verification key public), they are not helpful to construct VPRGs in the sense of [16]. In contrast, our relaxation is compatible with existing DVNIZKs (and indeed our first VPRG instantiation crucially relies on DVNIZKs).

Concrete Constructions. We offer two VPRG constructions from concrete assumptions. The first construction assumes a CDH group $\mathbb{G} = \langle g \rangle$ of (not necessarily prime) order n. A seed is an exponent $s \in \mathbb{Z}_n$, and a commitment to s is g^s. Given public $u_i, v_i \in \mathbb{G}$ (for $i \in [t]$), the i-th bit $G(s)_i$ of the PRG image is $B(u_i^s, v_i^s)$, where B is a hard-core predicate of the CDH function. A proof π_i that certifies a given $G(s)_i$ consists of u_i^s, v_i^s, as well as proofs that both (g, g^s, u_i, u_i^s) and (g, g^s, v_i, v_i^s) are Diffie-Hellman tuples. In a designated-verifier setting, such proofs are known from hash proof systems [10,13]. Alternatively, a symmetric pairing $\mathbb{G} \times \mathbb{G} \to \mathbb{G}_T$ can be used to check the Diffie-Hellman property of these tuples even without explicit proof.

Our second construction assumes LWE and uses the notion of homomorphic commitments from [22]. These commitments have a "dual-mode" flavor, much like the commitments from [14,26]. Specifically, under LWE, the public parameters of these commitments can be switched between a "binding" mode (in which commitments are perfectly binding) and a "hiding" mode (in which commitments are statistically hiding). Furthermore, given a commitment com_s to s, it is possible to publicly compute a commitment $\text{com}_{C(s)}$ to $C(s)$ for any (a-priori bounded) circuit C.

In our construction, we will assume any PRG G, and set com_s to be a commitment to a PRG seed $s \in \{0,1\}^\lambda$. Let G_i be a circuit that computes the i-th bit of G. An opening of the i-th bit is then an opening of the commitment $\text{com}_{G_i(s)}$

to $G_i(s)$. (Note that $\text{com}_{G_i(s)}$ can be publicly computed from com_s.) Unfortunately, in the construction of [22], the opening of commitments may reveal sensitive information about intermediate computation results (or even about s in our case).

Hence, we will have to assume an additional proof system to open commitments without revealing additional information. For the commitments of [22], the corresponding language is the language of a BDD problem. Fortunately, the strong secrecy properties of these commitments allow us to restrict ourselves to a witness-indistinguishable (and not necessarily zero-knowledge) proof system for BDD.[3]

Relation to [37]. We note that recently, [37] established a reduction from NIZKs for NP from LWE to the existence of an *NIZK* for BDD, also through implementing the HBM. Informally, and casting the construction of their work in the language of VPRGs,[4] the core reason why a NIZK was required in [37] is the need for a consistency proof for pvk (as in [16]). Since the consistency of pvk is a unique-witness relation, and since the proof must hide predicates of the seed, it does not seem feasible to replace this NIZK, e.g., by a NIWI or a witness-hiding proof. We note that although NIWIs for NP imply the existence of NIWIs for NP, it is not clear whether a NIWI for a simple language such as BDD can be used to build a NIZK for BDD.

Relation to [1]. We also note that Abusalah [1] also implements the HBM using Diffie-Hellman-related assumptions (such as CDH in a pairing-friendly group). However, he does not follow the PRG-based paradigm of [16] that we refine. Instead, he directly generalizes the original HBM implementation of [17] to generalizations of trapdoor permutations.

1.3 Concurrent Works

Concurrently and independently to our work, two other works [27,35] have achieved a result comparable to the first of our two main contributions, namely, designated-verifier non-interactive zero-knowledge proofs for NP from CDH. In all three works, the construction proceeds in a comparable way, by designing a CDH-based primitive which allows to compile the hidden-bit model into a designated-verifier NIZK. We summarize below the main differences between our works.

---

[3] One might wonder why we do not follow another route to obtain VPRGs from NIWI proofs for BDD. Specifically, [3,23] construct even verifiable random *functions* from a NIWI for a (complex) LWE-related language. However, these constructions inherently use disjunctions, and it seems unlikely that the corresponding NIWIs can be reduced to the BDD language.

[4] The actual construction of [37] relies on a new notion of public-key encryption with prover-assisted oblivious ciphertext sampling, but the high level idea is comparable to the VPRG-based approach. A side contribution of our construction is that it is conceptually much simpler and straightforward than the construction of [37].

- The work of [35] provides in addition a construction of *malicious* designated-verifier NIZK for NP, where the setup consists of an (honestly generated) common random string and the verifier then gets to choose his own (potentially malicious) public/secret key pair to generate and verify proofs. The assumption underlying their construction is a stronger "one-more type" variant of CDH (i.e., the hardness of solving $n + 1$ CDH challenges given n calls to an oracle solving CDH).
- The work of [27] provides two relatively different additional constructions of NIZKs: a *designated-prover* NIZK for NP with proofs of size $|C| + \mathsf{poly}(\lambda)$ (where C is the circuit checking the NP relation), under a strong Diffie-Hellman-type assumption over pairing groups, and a *preprocessing NIZK* for NP with proofs of size $|C| + \mathsf{poly}(\lambda)$ from the DDH assumption over pairing-free groups.
- The construction of NIZKs for NP assuming LWE and a NIWI for BDD is new to our work.

1.4   Organization

Section 2 introduces necessary preliminaries about non-interactive proof systems. Section 3 formally introduces designated-verifier pseudorandom generators, and defines their security properties. Section 4 provides a generic construction of a (designated-verifier) non-interactive zero-knowledge proof system for NP from our relaxed and generalized notion of DVPRGs, by instantiating the hidden bit model. Section 5 provides two instantiations of DVPRGs, from the CDH assumption in arbitrary group, and from the LWE assumption assuming in addition a NIWI proof system for BDD (where the resulting scheme is publicly verifiable iff the NIWI scheme is publicly verifiable).

2   Preliminaries

Notation. Throughout this paper, λ denotes the security parameter. A probabilistic polynomial time algorithm (PPT, also denoted *efficient* algorithm) runs in time polynomial in the (implicit) security parameter λ. A function f is *negligible* if for any positive polynomial p there exists a bound $B > 0$ such that, for any integer $k \geq B$, $|f(k)| \leq 1/|p(k)|$. An event occurs with *overwhelming probability* when its probability is at least $1 - \mathsf{negl}(\lambda)$ for a negligible function negl. Given a finite set S, the notation $x \xleftarrow{\$} S$ means a uniformly random assignment of an element of S to the variable x. We represent adversaries as interactive probabilistic Turing machines; the notation $\mathsf{Adv}^{\mathcal{O}}$ indicates that the machine Adv is given oracle access to $\mathcal{O}$. Adversaries will sometimes output an arbitrary state st to capture stateful interactions. For an integer n, $[n]$ denotes the set of integers from 1 to n. Given a string x of length n, we denote by x_i its ith bit (for any $i \leq n$), and by $x[S]$ the subsequence of the bits of x indexed by a subset S of $[n]$.

2.1   Non-Interactive Zero-Knowledge

We recall the definition of non-interactive zero-knowledge (NIZK) proofs and argument.

Definition 1 (Non-Interactive Zero-Knowledge Argument System).
A non-interactive zero-knowledge argument system for an NP-*language $\mathscr{L}$ with relation $R_{\mathscr{L}}$ is a triple of probabilistic polynomial-time algorithms* (Setup, Prove, Verify) *such that*

- Setup(1^λ), *outputs a common reference string* crs *and a trapdoor $\mathcal{T}$,*
- Prove(crs, x, w), *on input the crs* crs, *a word x, and a witness w, outputs a proof π,*
- Verify(crs, x, π, $\mathcal{T}$), *on input the crs* crs, *a word x, a proof π, and the trapdoor $\mathcal{T}$, outputs $b \in \{0,1\}$,*

which satisfies the completeness, soundness, and zero-knowledge properties defined below.

If the trapdoor $\mathcal{T}$ of the non-interactive proof system is set to $\perp$ (or, alternatively, if it is included in the crs), we call the argument system *publicly verifiable*. Otherwise, we call it a *designated-verifier non-interactive argument system*. If the soundness guarantee holds with respect to computationally unbounded adversary, we have a NIZK *proof* system.

Definition 2 (Perfect Completeness). *A non-interactive argument system* (Setup, Prove, Verify) *for an* NP-*language $\mathscr{L}$ with witness relation $R_{\mathscr{L}}$ satisfies perfect completeness if for every $x \in \mathscr{L}$ and every witness w such that $R_{\mathscr{L}}(x, w) = 1$,*

$$\Pr[(\mathsf{crs}, \mathcal{T}) \xleftarrow{\$} \mathsf{Setup}(1^\lambda), \pi \leftarrow \mathsf{Prove}(\mathsf{crs}, x, w) : \mathsf{Verify}(\mathsf{crs}, x, \pi, \mathcal{T}) = 1] = 1.$$

The soundness notion can come in several flavors: it is *non-adaptive* if the adversary must decide on a word on which to forge a proof before the common reference string is drawn, and it is *adaptive* if the adversary can dynamically choose the word given the common reference string. We will consider a strong variant of adaptive soundness, denoted *unbounded adaptive soundness*, where the adversary is given oracle access to a verification oracle. Note that in the publicly-verifiable setting, this is equivalent to the standard soundness notion, where the adversary must forge a valid proof on an incorrect statement without the help of any oracle. However, in the designated-verifier setting, this is a strictly stronger notion: the standard soundness notion only guarantees, in this setting, that the argument system remains sound as long as the prover receives at most logarithmically many feedback on previous proofs. On the other hand, if the argument system satisfies unbounded soundness, its soundness is maintained even if the adversary receives an arbitrary (polynomial) number of feedback on previous proofs.

Definition 3 (Unbounded Adaptive Soundness). *A non-interactive argument system* (Setup, Prove, Verify) *for an* NP*-language* $\mathscr{L}$ *with relation* $R_{\mathscr{L}}$ *satisfies unbounded adaptive soundness if for any PPT* $\mathcal{A}$,

$$\Pr\left[\begin{array}{l}(\mathsf{crs}, \mathcal{T}) \xleftarrow{\$} \mathsf{Setup}(1^\lambda), \\ (x, \boldsymbol{\pi}) \xleftarrow{\$} \mathcal{A}^{\mathcal{O}(\mathsf{crs}, \cdot, \cdot, \mathcal{T})}(\mathsf{crs}) \; : \mathsf{Verify}(\mathsf{crs}, x, \boldsymbol{\pi}, \mathcal{T}) = 1 \land x \notin \mathscr{L}\end{array}\right] \approx 0,$$

where $\mathcal{A}$ *can make polynomially many queries to an oracle* $\mathcal{O}(\mathsf{crs}, \cdot, \cdot, \mathcal{T})$ *which, on input* $(x, \boldsymbol{\pi})$, *outputs* $\mathsf{Verify}(\mathsf{crs}, x, \boldsymbol{\pi}, \mathcal{T})$.

We now define zero-knowledge, which can again come in several flavors. We will consider *adaptive* zero-knowledge argument systems, where the adversary is allowed to pick a word on which to forge a proof after seeing the common reference string. We will also distinguish *single-theorem* zero-knowledge, in which the prover generates a single proof (and the length of the common reference string can be larger than the length of the statement to prove) and *multi-theorem* zero-knowledge (where the adversary can adaptively ask for polynomially many proofs on arbitrary pairs (x, w) for the same common reference string).

Definition 4 (Adaptive Single-Theorem Zero-Knowledge). *A non-interactive argument system* (Setup, Prove, Verify) *for an* NP*-language* $\mathscr{L}$ *with relation* $R_{\mathscr{L}}$ *satisfies (adaptive) single-theorem zero-knowledge if for any stateful PPT algorithm* $\mathcal{A}$, *there exists a simulator* $(\mathsf{Sim}_0, \mathsf{Sim}_1)$ *such that*

$$\left| \Pr\left[\begin{array}{l}(\mathsf{crs}, \mathcal{T}) \xleftarrow{\$} \mathsf{Setup}(1^\lambda), \\ (x, w) \xleftarrow{\$} \mathcal{A}(\mathsf{crs}, \mathcal{T}), \quad : (R_{\mathscr{L}}(x, w) = 1) \land (\mathcal{A}(\pi) = 1) \\ \pi \xleftarrow{\$} \mathsf{Prove}(\mathsf{crs}, x, w)\end{array}\right] - \right.$$
$$\left. \Pr\left[\begin{array}{l}(\mathsf{crs}, \mathcal{T}) \xleftarrow{\$} \mathsf{Sim}_0(1^\lambda), \\ (x, w) \xleftarrow{\$} \mathcal{A}(\mathsf{crs}, \mathcal{T}), \quad : (R_{\mathscr{L}}(x, w) = 1) \land (\mathcal{A}(\pi) = 1) \\ \pi \xleftarrow{\$} \mathsf{Sim}_1(\mathsf{crs}, \mathcal{T}, x)\end{array}\right] \right| \approx 0.$$

Definition 5 (Adaptive Multi-Theorem Zero-Knowledge). *A non-interactive argument system* (Setup, Prove, Verify) *for an* NP*-language* $\mathscr{L}$ *with relation* $R_{\mathscr{L}}$ *satisfies (adaptive) multi-theorem zero-knowledge if for any stateful PPT algorithm* $\mathcal{A}$, *there exists a simulator* $(\mathsf{Sim}_0, \mathsf{Sim}_1)$ *such that* $\mathcal{A}$ *has negligible advantage in distinguishing the experiments* $\mathsf{Exp}_{\mathcal{A}}^{zk,0}(1^\lambda)$ *and* $\mathsf{Exp}_{\mathcal{A}}^{zk,1}(1^\lambda)$ *given on Fig. 1.*

Note that $\mathcal{O}_{\mathsf{sim}}$ is only given the witness w to artificially enforce that $\mathcal{A}$ queries only words x in the language $\mathscr{L}$.

Zero-knowledge is a strong, simulation-style security notion. A common relaxation of zero-knowledge to an indistinguishability-based security notion is known as *witness-indistinguishability*.

Definition 6 (Computational Witness-Indistinguishability). *A non-interactive proof system* (Setup, Prove, Verify) *for an* NP*-language* $\mathscr{L}$ *with relation* $R_{\mathscr{L}}$ *is (computationally) witness-indistinguishable if for any PPT algorithm* $\mathcal{A}$,

$$
\begin{array}{|l|}
\hline
\mathsf{Exp}_{\mathcal{A}}^{\mathsf{zk},0}(1^\lambda): \\
\quad (\mathsf{crs}, \mathcal{T}) \xleftarrow{\$} \mathsf{Setup}(1^\lambda) \\
\quad \mathbf{return}\ b \xleftarrow{\$} \mathcal{A}^{\mathcal{O}_{\mathsf{prove}}(\mathsf{crs},\cdot,\cdot)}(\mathsf{crs}) \\
\hline
\end{array}
\qquad
\begin{array}{|l|}
\hline
\mathsf{Exp}_{\mathcal{A}}^{\mathsf{zk},1}(1^\lambda): \\
\quad (\mathsf{crs}, \mathcal{T}) \xleftarrow{\$} \mathsf{Sim}_0(1^\lambda) \\
\quad \mathbf{return}\ b \xleftarrow{\$} \mathcal{A}^{\mathcal{O}_{\mathsf{sim}}(\mathsf{crs},\mathcal{T},\cdot,\cdot)}(\mathsf{crs}) \\
\hline
\end{array}
$$

$$
\begin{array}{|l|}
\hline
\underline{\mathcal{O}_{\mathsf{prove}}(\mathsf{crs}, x, w):} \\
\quad \mathbf{if}\ R_{\mathscr{L}}(x, w) = 1\ \mathbf{then} \\
\qquad \mathbf{return}\ \mathsf{Prove}(\mathsf{crs}, x, w) \\
\quad \mathbf{else} \\
\qquad \mathbf{return}\ \bot \\
\quad \mathbf{end\ if} \\
\hline
\end{array}
\qquad
\begin{array}{|l|}
\hline
\underline{\mathcal{O}_{\mathsf{sim}}(\mathsf{crs}, \mathcal{T}, x, w):} \\
\quad \mathbf{if}\ R_{\mathscr{L}}(x, w) = 1\ \mathbf{then} \\
\qquad \mathbf{return}\ \mathsf{Sim}_1(\mathsf{crs}, \mathcal{T}, x) \\
\quad \mathbf{else} \\
\qquad \mathbf{return}\ \bot \\
\quad \mathbf{end\ if} \\
\hline
\end{array}
$$

Fig. 1. Experiments $\mathsf{Exp}_{\mathcal{A}}^{\mathsf{zk},0}(1^\lambda)$ and $\mathsf{Exp}_{\mathcal{A}}^{\mathsf{zk},1}(1^\lambda)$, and oracles $\mathcal{O}_{\mathsf{prove}}(\mathsf{crs}, x, w)$ and $\mathcal{O}_{\mathsf{sim}}(\mathsf{crs}, \mathcal{T}, x, w)$, for the (adaptive) multi-theorem zero-knowledge property of a non-interactive argument system. $\mathcal{A}$ outputs $b \in \{0, 1\}$.

$$
\left| \Pr \left[\begin{array}{l}
(\mathsf{crs}, \mathcal{T}) \xleftarrow{\$} \mathsf{Setup}(1^\lambda), \quad \mathcal{A}(\mathsf{crs}, \boldsymbol{\pi}) = 1 \\
(x, w_0, w_1) \xleftarrow{\$} \mathcal{A}(\mathsf{crs}), : \wedge R_{\mathscr{L}}(x, w_0) = 1 \\
\boldsymbol{\pi} \xleftarrow{\$} \mathsf{Prove}(\mathsf{crs}, x, w_0) \quad \wedge R_{\mathscr{L}}(x, w_1) = 1
\end{array} \right] \right.
$$
$$
\left. - \Pr \left[\begin{array}{l}
(\mathsf{crs}, \mathcal{T}) \xleftarrow{\$} \mathsf{Setup}(1^\lambda), \quad \mathcal{A}(\mathsf{crs}, \boldsymbol{\pi}) = 1 \\
(x, w_0, w_1) \xleftarrow{\$} \mathcal{A}(\mathsf{crs}), : \wedge R_{\mathscr{L}}(x, w_0) = 1 \\
\boldsymbol{\pi} \xleftarrow{\$} \mathsf{Prove}(\mathsf{crs}, x, w_1) \quad \wedge R_{\mathscr{L}}(x, w_1) = 1
\end{array} \right] \right| \approx 0
$$

We call such a proof system a non-interactive witness-indistinguishable (NIWI) proof system.

It is known that the existence of a NIWI proof system for NP implies the existence of a NIZK proof system for NP in the CRS model [17]. However, this does not extend to proof systems for specific languages: the existence of a NIWI proof system for a language $\mathscr{L}$ does not generally imply the existence of a NIZK proof system in the CRS model for the same language.

3   Designated-Verifier Pseudorandom Generators

Verifiable pseudorandom generators (VPRG) have been introduced in the seminal paper of Dwork and Naor [16], as a tool to construct non-interactive witness-indistinguishable proofs and NIZKs in the CRS model. Informally, a VPRG enhances a PRG with verifiability properties: the prover can compute a kind commitment to the seed (called the *verification key*), and issue proofs that a given position i of the pseudorandom string stretched from the committed seed is equal to a given bit. Furthermore, this proof does not leak anything about the output values at positions $j \neq i$.

In this section, we revisit the notion of verifiable pseudorandom generators. Toward our goal of building VPRGs from new assumptions, we significantly weaken the binding property of VPRGs (which states, informally, that the verification key binds the prover to the seed) to a security notion that is simpler to achieve and still allows to build NIZKs in the CRS model, and we extend the definition to the more general setting of *designated-verifier* VPRGs (DVPRGs) (this strictly encompasses public VPRGs since we recover the standard notion by restricting the secret verification key to be $\perp$).

3.1   On Defining DVPRGs

A natural attempt to define the binding property of a DVPRG would be as follows: it should be infeasible, for any polytime adversary, to output two accepting proofs π_0 and π_1 that a given output of the PRG is equal to 0 and 1 respectively (relative to the same committed parameters). However, this security notion turns out to be too weak for the construction of non-interactive witness-indistinguishable proofs from VPRG of [16]. Intuitively, this stems from the fact that a cheating prover will never send more than a single proof for a given output, hence we cannot extract two contradictory proofs from this adversary. Instead, the argument of [16] crucially rely on the following stronger definition: a VPRG is binding if for every (possibly malicious) public verification key pvk, there exists a single associated string x in the range of the stretching algorithm of the DVPRG, and for any accepting proof π of correct opening to a subset $y[I]$ of the bits of a string y, it must hold that $y[I] = x[I]$.

Unfortunately, this binding property turns out to be too strong for our purpose. The reason is that we seek to build candidate DVPRGs from assumptions such as LWE, where natural approaches lead to schemes where there exists malicious public verification keys associated to strings which are *not* in the range of the DVPRG, and which cannot be distinguished from honest verification keys (typically, in our LWE-based construction, an honest verification key will be a list of LWE samples, which are indistinguishable from random samples). A comparable issue arose in the work of [37], which tackled this issue by appending to the verification key (or, in their language, the public key of an obliviously-sampleable encryption scheme) a NIZK proof of validity.

Instead, we opt for a different approach and introduce a weaker binding property for DVPRGs, which does not require assuming any specific structure of the public verification key *beyond its length*. Namely, we consider the following notion: a DVPRG is binding if there exists a (possibly inefficient) extractor Ext such that no PPT adversary can output a triple (pvk, i, π) where π is a proof of correct opening of position i to $1 - x_i$, and $x = \mathsf{Ext}(\mathsf{pvk})$. Note that our definition does only consider verification keys generated by a computationally bounded adversary (instead of arbitrary pvk), and does not require pvk to be in the range of the DVPRG. This binding notion would in fact be trivial to achieve without further constraints (e.g. one could define pvk to be a sequence of extractable commitments to each bits of the pseudorandom string stretched from the seed), hence we further require that pvk must be short (of size $s(\lambda)$, for a polynomial s independent of the stretch of the DVPRG). Afterward, we prove that this weaker notion still suffices to build NIZKs for NP in the CRS model.

Generalizing to the designated-verifier setting, where verification can involve a secret-verification key, we strengthen the above property to the *unbounded binding* property, which states that no PPT adversary can produce a triple (pvk, i, π) as above, even given oracle access to a verification oracle (which has the secret verification key hardcoded). We note that the above weakening of the binding notion is also necessary for our generalization to the designated-verifier setting: in this setting, the stronger binding notion of [16] does typically *not* hold, since there always exists accepting proofs of opening to an incorrect bit (if this was not the case, we could safely make the secret verification key public, since it would not allow to find proofs of opening to incorrect values); however, it is infeasible to *find* such proof (without knowing the secret verification key). Below, we formally introduce designated-verifier pseudorandom generators and the corresponding security notions.

3.2 Definition

Definition 7 (Designated-Verifier Pseudorandom Generator). *A designated-verifier pseudorandom generator (*DVPRG*) is a four-tuple of efficient algorithms* (Setup, Stretch, Prove, Verify) *such that*

- Setup$(1^\lambda, m)$, *on input the security parameter (in unary) and a bound $m(\lambda) = \mathsf{poly}(\lambda)$, outputs a pair* $(\mathsf{pp}, \mathcal{T})$ *where* pp *is a set of public parameters (which contains 1^λ), and $\mathcal{T}$ is a trapdoor;*
- Stretch(pp), *on input the public parameters, outputs a triple* $(\mathsf{pvk}, x, \mathsf{aux})$, *where* pvk *is a public verification key of polynomial length $s(\lambda)$ independent of m, x is an m-bit pseudorandom string, and* aux *is an auxiliary information;*
- Prove$(\mathsf{pp}, \mathsf{aux}, i)$, *on input the public parameters, auxiliary informations* aux, *an index $i \in [m]$, outputs a proof π;*
- Verify$(\mathsf{pp}, \mathsf{pvk}, \mathcal{T}, i, b, \pi)$, *on input the public parameters, a public verification key* pvk, *a trapdoor $\mathcal{T}$, a position $i \in [m]$, a bit b, and a proof π, outputs a bit β;*

which is in addition complete, hiding, *and* binding, *as defined below.*

Note that the above definition also captures publicly verifiable pseudorandom generators, which are DVPRGs where we restrict Setup$(1^\lambda, m)$ to always output pairs of the form $(\mathsf{pp}, \bot)$.

Definition 8 (Completeness of a DVPRG). *For any $i \in [m]$, a perfectly complete DVPRG scheme* (Setup, Stretch, Prove, Verify) *satisfies:*

$$\Pr \left[\begin{array}{l} (\mathsf{pp}, \mathcal{T}) \xleftarrow{\$} \mathsf{Setup}(1^\lambda, m), \\ (\mathsf{pvk}, x, \mathsf{aux}) \xleftarrow{\$} \mathsf{Stretch}(\mathsf{pp}), : \mathsf{Verify}(\mathsf{pp}, \mathsf{pvk}, \mathcal{T}, i, x_i, \pi) = 1 \\ \pi \xleftarrow{\$} \mathsf{Prove}(\mathsf{pp}, \mathsf{aux}, i), \end{array} \right] = 1.$$

We now define the binding property of a DVPRG. We consider a flavor of the binding property which is significantly weaker than the one considered in [16], yet still suffices for the application to NIZKs (see the discussion in Sect. 3.1).

Definition 9 (Binding Property of a DVPRG). *Let* (Setup, Stretch, Prove, Verify) *be a* DVPRG. *A* DVPRG *is* binding *if there exists a (possibly inefficient) extractor* Ext *such that for any PPT* $\mathcal{A}$, *it holds that*

$$\Pr\left[\begin{array}{l}(\mathsf{pp}, \mathcal{T}) \xleftarrow{\$} \mathsf{Setup}(1^\lambda, m), \\ (\mathsf{pvk}, i, \pi) \xleftarrow{\$} \mathcal{A}(\mathsf{pp}), \\ x \leftarrow \mathsf{Ext}(\mathsf{pvk})\end{array} : \mathsf{Verify}(\mathsf{pp}, \mathsf{pvk}, \mathcal{T}, i, 1 - x_i, \pi) = 1\right] \approx 0.$$

As for non-interactive zero-knowledge proofs, the designated-verifier setting requires to explicitly consider whether the adversary is given access to a verification oracle. We therefore extend the above definition and consider the *unbounded* binding property:

Definition 10 (Unbounded Binding Property of a DVPRG). *Let* (Setup, Stretch, Prove, Verify) *be a* DVPRG. *A* DVPRG *satisfies* unbounded binding *if there exists a (possibly inefficient) extractor* Ext *such that for any PPT* $\mathcal{A}$, *it holds that*

$$\Pr\left[\begin{array}{l}(\mathsf{pp}, \mathcal{T}) \xleftarrow{\$} \mathsf{Setup}(1^\lambda, m), \\ (\mathsf{pvk}, i, \pi) \xleftarrow{\$} \mathcal{A}^{\mathsf{Verify}(\mathsf{pp}, \cdot, \mathcal{T}, \cdot, \cdot, \cdot)}(\mathsf{pp}), : \mathsf{Verify}(\mathsf{pp}, \mathsf{pvk}, \mathcal{T}, i, 1 - x_i, \pi) = 1 \\ x \leftarrow \mathsf{Ext}(\mathsf{pvk})\end{array}\right] \approx 0.$$

Note that in the case of publicly verifiable pseudorandom generators, where $\mathcal{T}$ is set to $\bot$, this security notion is equivalent to the binding property.

We now define equivocability. Intuitively, it states that no computationally bounded adversary can distinguish honestly generated proofs of correctness for bits of the pseudorandom sequence from simulated proofs (using $\mathcal{T}$) of opening to true random bits.

Definition 11 (Equivocability of a DVPRG). *A designated-verifier pseudorandom generator* (Setup, Stretch, Prove, Equivocate, Verify) *is* equivocable *if there are two additional algorithms* (SimSetup, Equivocate) *such that*

- SimSetup$(1^\lambda, m)$, *on input the security parameter in unary, outputs a triple* $(\mathsf{pp}, \mathcal{T}, \mathcal{T}_s)$,
- Equivocate$(\mathsf{pp}, \mathsf{pvk}, i, b, \mathcal{T}_s)$, *on input the public parameters, a public verification key* pvk, *an index* $i \in [m]$, *a bit* b, *and a simulation trapdoor* $\mathcal{T}_s$, *outputs a simulated proof* π';

such that the following distributions are computationally indistinguishable:

$$\left\{\begin{array}{l}(\mathsf{pp}, \mathcal{T}) \xleftarrow{\$} \mathsf{Setup}(1^\lambda, m), \\ (\mathsf{pvk}, x, \mathsf{aux}) \xleftarrow{\$} \mathsf{Stretch}(\mathsf{pp}) \quad : (\mathsf{pp}, \mathsf{pvk}, \mathcal{T}, x, \boldsymbol{\pi}) \\ \boldsymbol{\pi} \xleftarrow{\$} (\mathsf{Prove}(\mathsf{pp}, \mathsf{aux}, i))_i\end{array}\right\} = D_0$$

$$\approx \left\{\begin{array}{l}(\mathsf{pp}, \mathcal{T}, \mathcal{T}_s) \xleftarrow{\$} \mathsf{SimSetup}(1^\lambda, m), \\ (\mathsf{pvk}, x', \mathsf{aux}) \xleftarrow{\$} \mathsf{Stretch}(\mathsf{pp}), x \xleftarrow{\$} \{0, 1\}^m, \quad : (\mathsf{pp}, \mathsf{pvk}, \mathcal{T}, x, \boldsymbol{\pi}) \\ \boldsymbol{\pi} \xleftarrow{\$} (\mathsf{Equivocate}(\mathsf{pp}, \mathsf{pvk}, i, x_i, \mathcal{T}_s))_i\end{array}\right\} = D_1.$$

A weaker variant of equivocability is the following *hiding* property, which states that an adversary cannot guess the value of a particular output (with non-negligible advantage over the random guess), even if he is given the values of all other outputs together with proofs of correct opening. This notion is implied by the equivocability property, and it suffices for the Dwork and Naor construction of a NIZK proof system for NP; however, equivocable DVPRGs allow for a simpler and more direct construction of NIZKs, without having to rely on the FLS transform which constructs NIZKs from NIWI [17].

Definition 12 (Hiding Property of a DVPRG). *A* DVPRG *scheme* (Setup, Stretch, Prove, Verify) *is hiding if for any $i \in [m]$ and any PPT adversary $\mathcal{A}$ that outputs bits, it holds that:*

$$\Pr \left[\begin{array}{l} (\mathsf{pp}, \mathcal{T}) \xleftarrow{\$} \mathsf{Setup}(1^\lambda, m), \\ (\mathsf{pvk}, x, \mathsf{aux}) \xleftarrow{\$} \mathsf{Stretch}(\mathsf{pp}), : \mathcal{A}(\mathsf{pp}, \mathsf{pvk}, i, (x_j, \pi_j)_{j \neq i}) = x_i \\ (\pi_j \xleftarrow{\$} \mathsf{Prove}(\mathsf{pp}, \mathsf{aux}, j))_j \end{array} \right] \approx 1/2.$$

Eventually, we define an additional security notion, the *consistency*, which will prove useful to analyze the unbounded binding property of one of our candidates:

Definition 13 (Consistency of a DVPRG). *Given a* DVPRG (Setup, Stretch, Prove, Verify) *and a pair* $(\mathsf{pp}, \mathcal{T}) = \mathsf{Setup}(1^\lambda, m; r)$ *for some random coin r, we define for any ε the set ε-Good(r) to be the set of 4-tuples $(\mathsf{pvk}, i, \pi, x_i)$ satisfying*

$$\Pr \left[\mathcal{T}' \xleftarrow{\$} Dist(r) : \mathsf{Verify}(\mathsf{pp}, \mathsf{pvk}, \mathcal{T}', i, x_i, \pi) = 1 \right] \geq \varepsilon,$$

where $Dist(r)$ samples random pairs $(\mathsf{pp}', \mathcal{T}')$ with $\mathsf{Setup}(1^\lambda, m)$ subject to the constraint $\mathsf{pp}' = \mathsf{pp}$, and outputs $\mathcal{T}'$. Note that for any $\varepsilon' \geq \varepsilon$, it holds that ε'-Good$(r) \subset \varepsilon$-Good(r). Then, we say that a DVPRG is consistent if there exists a negligible function ε such that for any PPT adversary $\mathcal{A}$,

$$\Pr \left[\begin{array}{l} r \xleftarrow{\$} R, (\mathsf{pp}, \mathcal{T}) \leftarrow \mathsf{Setup}(1^\lambda, m; r), (\mathsf{pvk}, i, \pi, b) \xleftarrow{\$} \mathcal{A}(\mathsf{pp}) : \\ (\mathsf{pvk}, i, \pi, b) \in \varepsilon\text{-Good}(r) \setminus 1\text{-Good}(r) \end{array} \right] \approx 0.$$

In the full version of this paper [12], we prove the following:

Theorem 14. *Let $\mathcal{G} = $ (Setup, Stretch, Prove, Verify) be a binding and consistent* DVPRG, *such that for any r, the distribution $Dist(r)$ is efficiently sampleable. Then $\mathcal{G}$ is unbounded binding.*

4   DVNIZK Proof for NP from DVPRG

4.1   The Hidden Bit Model, and HB Proofs

The hidden bit model is an ideal formalization of a scenario in which both the prover and the verifier have access to a long string of hidden random bits (let us denote with hb the random bits and $t = t(\lambda)$ the length of the hidden string). In this idealized model, the prover can send to the verifier a subset $S \subset [t]$ of the positions of the hidden bits (together with additional informations). The verifier is restricted to inspecting only the bits of hb residing in the locations specified by the prover, while the prover can see hb entirely.

Definition 15. *A non-interactive proof system* HB *in the hidden bit model is a pair of PPT algorithms* (HB.Prove, HB.Verify) *such that*

- HB.Prove(hb, x, w), *on input a random bit string* hb $\in \{0,1\}^t$, *and a word $x \in \mathscr{L}$ with witness w, outputs a subset $S \subset [t]$ together with a string M of auxiliary informations,*
- HB.Verify(x, hb[S], M), *on input a word x, the subsequence of* hb *indexed by S, and an auxiliary information M, outputs $b \in \{0,1\}$,*

which satisfies the following perfect completeness, ε-soundness, and (adaptive, single-theorem) zero-knowledge properties:

- **Perfect Completeness.** *For any $x \in \mathscr{L}$ with witness w, any* hb $\in \{0,1\}^t$, *and for $(S, M) \xleftarrow{\$}$ HB.Prove(hb, x, w), it holds that* HB.Verify(x, hb[S], M) $= 1$.
- ε-**Soundness.** *For any (possibly unbounded) adversary $\mathcal{A}$,*

$$\Pr\begin{bmatrix} \text{hb} \xleftarrow{\$} \{0,1\}^t, \\ (x, S, M) \xleftarrow{\$} \mathcal{A}(\text{hb}) \end{bmatrix} : \text{HB.Verify}(x, \text{hb}[S], M) = 1 \wedge x \notin \mathscr{L} \end{bmatrix} \leq \varepsilon.$$

- **Single-Theorem Zero-Knowledge.** *For any (possibly unbounded) stateful adversary $\mathcal{A}$, there exists a simulator* (Sim$_{zk}$, Sim$'_{zk}$) *such that for every $x \in \mathscr{L}$ and any w satisfying $R_\mathscr{L}(x, w) = 1$,*
  - *the distributions*

$$\{(\text{hb}[S], S, M) : \text{hb} \xleftarrow{\$} \{0,1\}^t, (S, M) \xleftarrow{\$} \text{HB.Prove}(\text{hb}, x, w)\}$$

  and $\{\text{Sim}_{zk}(x)\}$ *are perfectly indistinguishable;*
  - *the distributions*

$$\{(\text{hb}, S, M) : \text{hb} \xleftarrow{\$} \{0,1\}^t, (S, M) \xleftarrow{\$} \text{HB.Prove}(\text{hb}, x, w)\}$$

  and

$$\{(\text{hb}, S, M) : (\text{hb}[S], S, M) \xleftarrow{\$} \text{Sim}_{zk}(x), \text{hb} \xleftarrow{\$} \text{Sim}'_{zk}(\text{hb}[S], S, M, x, w)\}$$

  are perfectly indistinguishable. That is, the simulator can generate (hb[S], S, M) *without a witness, and find a completion of the hidden string* hb *given a witness w, which is identically distributed to an honestly generated hidden string and proof with w.*

Note that in the hidden bit model, the parties do not have access to a common random string, but to a string of bits which are perfectly hidden to the verifier until the prover opens a subsequence of them. Therefore, the adaptive and non-adaptive formulations of zero-knowledge are equivalent since the verifier does not get to see anything about hb before producing a word x with a witness w (put differently, it is equivalent to define zero-knowledge for all $x \in \mathscr{L}$ or with respect to adversarially chosen x). Examples of non-interactive proof systems in the hidden-bit model can be found in [17,28]. We stress that the security of these proof systems is unconditional (although a specific implementation of the HB model can involve cryptography).

4.2   A DVNIZK for NP from Any DVPRG

We describe on Fig. 2 a general transformation that converts any (unconditional) proof system in the HB model into a DVNIZK for the same language, given any DVPRG. The DVNIZK inherits the specificities of the DVPRG: it satisfies unbounded soundness and/or statistical soundness whenever the DVPRG is unbounded binding and/or statistically binding. At the exception of using a DVPRG instead of a VPRG, the proof system is identical to the one of [16, Section 5.1] (actually, [16] provides a ZAP in the plain model where the first flow can be fixed non-uniformly, which immediately implies a NIZK in the CRS model. Our construction does not imply a ZAP in the plain model, as we need to setup a CRS containing, in particular, the public parameters of the DVPRG. These public parameters must be honestly sampled to maintain the hiding property, hence they cannot be picked by the verifier in the first round.)

While the scheme is almost identical to the scheme of [16], the proof of soundness is more involved, as it must cope with the weaker binding property of our PRGs. To prove soundness, we proceed as follows: we identify a "bad event", which occurs whenever the adversary outputs pvk and a proof π for some position i of correct opening to $1 - x_i$, where $x = \text{Ext}(\text{pvk})$ (Ext being the possibly inefficient extractor guaranteed by the unbounded binding security notion of the DVPRG). We show that when this bad even does *not* happen, then there is a string (essentially $x \oplus \rho$, where ρ is a long random string which is part of the CRS) which is a *bad string*, in the sense that if this string is used as the hidden bit string of the HB proof system, there exists accepting proofs of incorrect statement with respect to this hidden string. Then, we rely on the statistical soundness of the HB proof system to argue that only a tiny fraction of all possible strings (of a given length) are bad strings. Since ρ is random and x is uniquely defined given pvk, we can rely on the fact that pvk is short to argue, with a counting argument, that there is a negligible probability (over the random choice of ρ) that there exists a short pvk such that $\rho \oplus \text{Ext}(\text{pvk})$ is a bad string. Hence, this situation is statistically unlikely, and we must be in the case where the bad event happens; then, we conclude the proof by observing that an occurrence of this bad event directly contradicts the unbounded binding property of the DVPRG. In contrast, the argument of [16] uses a counting argument over all possible *seeds* of the VPRG, which crucially relies on their stronger binding property which states that any possible pvk is in the stretch of the PRG, and is bound to a seed (while this seed need not be unique, all seeds associated to a given pvk must lead to the same pseudorandom string).

DVNIZK Proof System Π

Let $\mathscr{L}$ be a language and let $y \in \mathscr{L}$ be a word with witness w. Let $\lambda \leftarrow |y|$. The DVNIZK relies on an HB proof system HB for the statement $y \in \mathscr{L}$ which uses $\ell = \ell(\lambda)$ hidden bits, and achieves $2^{-\lambda}$-statistical soundness. Let $\mathcal{G} = (\mathcal{G}.\mathsf{Setup}, \mathcal{G}.\mathsf{Stretch}, \mathcal{G}.\mathsf{Prove}, \mathcal{G}.\mathsf{Verify})$ be a DVPRG, with public verification key size $s(\lambda)$ and output size $m(\lambda)$, satisfying $m > (1 + s/\lambda)\ell + \ell^2/\lambda$. In the following, we consider the HB proof system HB' obtain by executing HB m/ℓ times in parallel (with independent hidden bits) and accepting only if all executions are accepted. Note that HB' uses m hidden bits and achieves $2^{-\lambda m/\ell}$-statistical soundness.

- $\Pi.\mathsf{Setup}(1^\lambda)$: on input the security parameter in unary, compute $(\mathsf{pp}, \mathcal{T}) \xleftarrow{\$} \mathcal{G}.\mathsf{Setup}(1^\lambda, m(\lambda))$, and $\rho \xleftarrow{\$} \{0,1\}^m$. Output $\mathsf{crs} \leftarrow (\mathsf{pp}, \rho)$ and $\mathcal{T}$.
- $\Pi.\mathsf{Prove}(\mathsf{crs}, y, w)$: parse crs as (pp, ρ). Compute $(\mathsf{pvk}, x, \mathsf{aux}) \xleftarrow{\$} \mathcal{G}.\mathsf{Stretch}(\mathsf{pp})$. Pick $\theta \xleftarrow{\$} \{0,1\}^\ell$. For $i = 1$ to m, set $\mathsf{hb}_i \leftarrow x_i \oplus \rho_i \oplus \theta_{(i-1 \bmod \ell)+1}$. Define $\mathsf{hb} = (\mathsf{hb}_i)_i$ to be the hidden string of HB'. Compute an HB proof $(S, M) \xleftarrow{\$} \mathsf{HB}'.\mathsf{Prove}(\mathsf{hb}, y, w)$. For every $i \in S$, compute $\pi_i \xleftarrow{\$} \mathcal{G}.\mathsf{Prove}(\mathsf{pp}, \mathsf{aux}, i)$. Output $(\mathsf{pvk}, \theta, S, \mathsf{hb}[S], M, (\pi_i)_{i \in S})$.
- $\Pi.\mathsf{Verify}(\mathsf{crs}, G, \boldsymbol{\pi}, \mathcal{T})$: parse crs as (pp, ρ) and $\boldsymbol{\pi}$ as $(\mathsf{pvk}, \theta, S, \mathsf{hb}[S], M, (\pi_i)_{i \in S})$. For every $i \in S$, set $x_i \leftarrow \mathsf{hb}_i \oplus \rho_i \oplus \theta_{(i-1 \bmod \ell)+1}$ and check that $\mathcal{G}.\mathsf{Verify}(\mathsf{pp}, \mathsf{pvk}, \mathcal{T}, i, x_i, \pi_i)$. Check that $\mathsf{HB}'.\mathsf{Verify}(y, \mathsf{hb}[S], M)$ returns 1. Output 1 if all checks succeeded, and 0 otherwise.

Fig. 2. Designated-verifier non-interactive zero-knowledge proof system Π for a language $\mathscr{L}$ using a DVPRG $\mathcal{G}$ and an HB proof system HB

Theorem 16. *Let $\mathcal{G}$ be a hiding unbounded binding DVPRG, and let $(\Pi.\mathsf{Setup}, \Pi.\mathsf{Prove}, \Pi.\mathsf{Verify})$ be the DVNIZK proof system given on Fig. 2. Then Π satisfies computational witness-indistinguishability and unbounded adaptive soundness. Furthermore, if $\mathcal{G}$ is equivocable, Π satisfies (adaptive, single-theorem) zero-knowledge.*

The completeness of Π follows immediately from the completeness of HB and $\mathcal{G}$. In the remainder of this section, we prove Theorem 16. The proof of witness-indistinguishability is similar to the one given in [16], but the proof of soundness is more involved (see the previous discussion).

4.3   Witness Indistinguishability of Π

We prove the witness-indistinguishability of Π through a sequence of hybrids. Let $\mathcal{A}$ be a PPT adversary; assume toward contradiction that

$$\left| \Pr \left[\begin{array}{ll} (\mathsf{crs}, \mathcal{T}) \xleftarrow{\$} \Pi.\mathsf{Setup}(1^\lambda), & \mathcal{A}(\mathsf{crs}, \boldsymbol{\pi}) = 1 \\ (y, w_0, w_1) \xleftarrow{\$} \mathcal{A}(\mathsf{crs}), & : \wedge R_{\mathscr{L}}(y, w_0) = 1 \\ \boldsymbol{\pi} \xleftarrow{\$} \Pi.\mathsf{Prove}(\mathsf{crs}, y, w_0) & \wedge R_{\mathscr{L}}(y, w_1) = 1 \end{array} \right] \right.$$
$$\left. - \Pr \left[\begin{array}{ll} (\mathsf{crs}, \mathcal{T}) \xleftarrow{\$} \Pi.\mathsf{Setup}(1^\lambda), & \mathcal{A}(\mathsf{crs}, \boldsymbol{\pi}) = 1 \\ (y, w_0, w_1) \xleftarrow{\$} \mathcal{A}(\mathsf{crs}), & : \wedge R_{\mathscr{L}}(y, w_0) = 1 \\ \boldsymbol{\pi} \xleftarrow{\$} \Pi.\mathsf{Prove}(\mathsf{crs}, y, w_1) & \wedge R_{\mathscr{L}}(y, w_1) = 1 \end{array} \right] \right| \geq \varepsilon$$

for some non-negligible quantity ε. Let us denote H_b for $b \in \{0,1\}$ the experiment in which we set $(\mathsf{crs}, \mathcal{T}) \xleftarrow{\$} \Pi.\mathsf{Setup}(1^\lambda)$, $(y, w_0, w_1) \xleftarrow{\$} \mathcal{A}(\mathsf{crs})$, $\boldsymbol{\pi} \xleftarrow{\$} \Pi.\mathsf{Prove}(\mathsf{crs}, y, w_b)$, and output $b' \xleftarrow{\$} \mathcal{A}(\mathsf{crs}, \boldsymbol{\pi})$.

Recall that HB' consists of m/ℓ parallel repetitions of HB (with independent hidden bits hb^j). We consider a sequence of intermediate hybrids $H_{0.j}$ for $j = 0$ to m/ℓ, in which we use the witness w_1 for the j first repetitions (computing (S_j, M_j) as $\mathsf{HB}.\mathsf{Prove}(\mathsf{hb}^j, y, w_0)$) and the witness w_0 for the repetitions $j + 1$ to m/ℓ. By a standard pigeonhole argument, there exists a j such that the advantage of $\mathcal{A}$ in distinguishing $H_{0.j}$ from $H_{0.j+1}$ is at least $\varepsilon\ell/m$. We further divide $H_{0.j}$ in the following sub-hybrids:

- $H_{0.j.0}$. In this hybrid, we modify the generation of $(\mathsf{hb}^{j+1}, S_{j+1}, M_{j+1})$. Namely, we compute $(\mathsf{pvk}, x, \mathsf{aux}) \xleftarrow{\$} \mathcal{G}.\mathsf{Stretch}(\mathsf{pp})$ (let x^{j+1}, ρ^{j+1} denote the $(j + 1)$-th block of ℓ bits of x, ρ), generate $(\mathsf{hb}^{j+1}[S_{j+1}], S_{j+1}, M_{j+1}) \xleftarrow{\$}$ $\mathsf{Sim}_{\mathsf{zk}}(y)$, $\mathsf{hb}^{j+1} \xleftarrow{\$} \mathsf{Sim}'_{\mathsf{zk}}(\mathsf{hb}^{j+1}[S_{j+1}], S_{j+1}, M_{j+1}, y, w_1)$, and set $\theta \leftarrow x^{j+1} \oplus \rho_{j+1} \oplus \mathsf{hb}^{j+1}$. The other repetitions of HB are executed as before; note that it holds that $\mathsf{hb}_i = x_i \oplus \rho_i \oplus \theta_{(i-1) \bmod \ell}$ for every $i \leq m$. By the (perfect) single-theorem zero-knowledge property of HB, the distribution of $(\mathsf{crs}, \boldsymbol{\pi})$ in $H_{0.j.1}$ is identical to its distribution in $H_{0.j}$, hence the advantage of $\mathcal{A}$ in distinguishing $H_{0.j.1}$ from $H_{0.j+1}$ is at least $\varepsilon\ell/m$.
- $H_{0.j.k}$. We denote by $\ell - r$ the size of S_{j+1} (r is the size of the "unopened" subsequence of hb^{j+1}). For $k = 0$ to r, we modify the generation of hb^{j+1} as follows: we generate as before $(\mathsf{hb}^{j+1}[S_{j+1}], S_{j+1}, M_{j+1}) \xleftarrow{\$} \mathsf{Sim}_{\mathsf{zk}}(y)$, denote R_{j+1} the set $[\ell] \setminus S_{j+1}$ of unopened positions of hb^{j+1}, and compute
  - $\mathsf{hb}^{j+1.0}[R_{j+1}] \xleftarrow{\$} \mathsf{Sim}'_{\mathsf{zk}}(\mathsf{hb}^{j+1}[S_{j+1}], S_{j+1}, M_{j+1}, y, w_0)$,
  - $\mathsf{hb}^{j+1.1}[R_{j+1}] \xleftarrow{\$} \mathsf{Sim}'_{\mathsf{zk}}(\mathsf{hb}^{j+1}[S_{j+1}], S_{j+1}, M_{j+1}, y, w_1)$.
  Then, we define $\mathsf{hb}^{j+1}[R_{j+1}]$ to be the string that agrees with $\mathsf{hb}^{j+1.1}[R_{j+1}]$ for positions 1 to k, and with $\mathsf{hb}^{j+1.0}[R_{j+1}]$ for positions $k + 1$ to r. By a standard pigeonhole argument, there exists a k such that $\mathcal{A}$ distinguishes $H_{0.j.k}$ from $H_{0.j.k+1}$ with probability at least $\varepsilon\ell/(mr) \geq \varepsilon/m$.

Note that the string hb^{j+1} differs by a single bit between $H_{0.j.k}$ and $H_{0.j.k+1}$. From there, we immediately reach a contradiction to the hiding property of

$\mathcal{G}$: denoting $i \leftarrow (j + 1)\ell + k + 1$, we receive $(\mathsf{pp}, \mathsf{pvk}, i, (x_t, \pi_t)_{t \neq i}$, compute $(\mathsf{hb}^{j+1}[S_{j+1}], S_{j+1}, M_{j+1}) \xleftarrow{\$} \mathsf{Sim}_{\mathsf{zk}}(y)$, guess the value x_i at random (completing the string x), and set $\theta \leftarrow x^{j+1} \oplus \rho_{j+1} \oplus \mathsf{hb}^{j+1}$. Depending on our guess of x_i, the distribution of (crs, π) is either identical to its distribution in $H_{0.j.k}$ or in $H_{0.j.k+1}$, hence we distinguish between $x_i = 0$ and $x_i = 1$ with probability at least ε/m. This concludes the proof.

4.4   Adaptive Single-Theorem Zero-Knowledge of Π

A witness-indistinguishable NIZK proof system for NP implies an adaptive zero-knowledge proof system for NP, by the transformation of [17]. However, if $\mathcal{G}$ is equivocable, there is a more direct construction: we prove that in this case, the DVNIZK Π is adaptive single-theorem zero knowledge (and can be made adaptive multi-theorem zero-knowledge using [17]); the argument is simpler than for witness indistinguishability, does only use $\mathsf{Sim}_{\mathsf{zk}}$ (the simulator $\mathsf{Sim}'_{\mathsf{zk}}$ is not needed), and does not require θ (which can be removed from the construction – we keep it in the proof below for simplicity). Let $\mathcal{A}$ be a PPT adversary against the (adaptive) single-theorem zero-knowledge of Π. Let $\mathsf{Sim} = (\mathsf{Sim}_0, \mathsf{Sim}_1)$ be the following simulator:

- On input 1^λ, Sim_0 computes $(\mathsf{pp}, \mathcal{T}) \xleftarrow{\$} \mathcal{G}.\mathsf{SimSetup}(1^\lambda, m)$, and $\rho \xleftarrow{\$} \{0, 1\}^m$. He outputs $\mathsf{crs} \leftarrow (\mathsf{pp}, \rho)$ and $\mathcal{T}$.
- On input $(\mathsf{crs}, \mathcal{T}, y)$, Sim parses crs as (pp, ρ) and computes $(\mathsf{pvk}, x', \mathsf{aux}) \xleftarrow{\$} \mathcal{G}.\mathsf{Stretch}(\mathsf{pp})$. Then, Sim_1 runs $(\mathsf{hb}[S], S, M) \xleftarrow{\$} \mathsf{Sim}_{\mathsf{zk}}(y)$, where $\mathsf{Sim}_{\mathsf{zk}}$ is the simulator of the zero-knowledge property of HB'. Sim_1 picks $\theta \xleftarrow{\$} \{0, 1\}^\ell$. For every $i \in S$, he sets $x_i \leftarrow \mathsf{hb}_i \oplus \rho_i \oplus \theta_{(i-1 \bmod \ell)+1}$ and computes $\pi_i \xleftarrow{\$} \mathcal{G}.\mathsf{Equivocate}(\mathsf{pp}, \mathsf{pvk}, i, x_i, \mathcal{T})$. Sim_1 outputs $(\mathsf{pvk}, \theta, S, \mathsf{hb}[S], M, (\pi_i)_{i \in S})$.

We prove that

$$\left| \Pr \begin{bmatrix} (\mathsf{crs}, \mathcal{T}) \xleftarrow{\$} \mathsf{Setup}(1^\lambda), \\ (y, w) \xleftarrow{\$} \mathcal{A}(\mathsf{crs}, \mathcal{T}), & : (R_{\mathscr{L}}(y, w) = 1) \wedge (\mathcal{A}(\pi) = 1) \\ \pi \xleftarrow{\$} \Pi.\mathsf{Prove}(\mathsf{crs}, y, w) \end{bmatrix} - \right.$$
$$\left. \Pr \begin{bmatrix} (\mathsf{crs}, \mathcal{T}) \xleftarrow{\$} \mathsf{Sim}_0(1^\lambda), \\ (y, w) \xleftarrow{\$} \mathcal{A}(\mathsf{crs}, \mathcal{T}), & : (R_{\mathscr{L}}(y, w) = 1) \wedge (\mathcal{A}(\pi) = 1) \\ \pi \xleftarrow{\$} \mathsf{Sim}_1(\mathsf{crs}, \mathcal{T}, y) \end{bmatrix} \right| \approx 0,$$

through a sequence of hybrids.

- **Game H_0.** This is the real game, where we generate $(\mathsf{crs}, \mathcal{T}) \xleftarrow{\$} \mathsf{Setup}(1^\lambda)$, run $(y, w) \xleftarrow{\$} \mathcal{A}(\mathsf{crs}, \mathcal{T})$, $\pi \xleftarrow{\$} \Pi.\mathsf{Prove}(\mathsf{crs}, y, w)$, and $b \xleftarrow{\$} \mathcal{A}(\pi)$.
- **Game H_1.** In this game, we generate instead $(\mathsf{crs}, \mathcal{T})$ as $\mathsf{Sim}_0(1^\lambda)$ (that is, we compute $(\mathsf{pp}, \mathcal{T}) \xleftarrow{\$} \mathcal{G}.\mathsf{SimSetup}(1^\lambda, m)$ and $\rho \xleftarrow{\$} \{0, 1\}^m$). Furthermore, we modify $\Pi.\mathsf{Prove}(\mathsf{crs}, y, w)$ as follow: after computing $(\mathsf{pvk}, x', \mathsf{aux}) \xleftarrow{\$} \mathcal{G}.\mathsf{Stretch}(\mathsf{pp})$, we pick $x \xleftarrow{\$} \{0, 1\}^m$ and set $\mathsf{hb}_i \leftarrow x_i \oplus \rho_i \oplus \theta_{(i-1 \bmod \ell)+1}$. We

compute the HB proof (S, M) honestly using (y, w) and the hidden string hb. Finally, we compute the π_i as $\mathcal{G}.\mathsf{Equivocate}(\mathsf{pp}, \mathsf{pvk}, \mathcal{T}, i, x_i)$.

By the equivocability of $\mathcal{G}$, the distribution of $(\mathsf{pp}, \mathsf{pvk}, \mathcal{T}, x, (\pi_i)_{i \in S})$ in H_1 is computationally indistinguishable from its distribution in H_0, and the rest of the proof is computed from $(\mathsf{pp}, \mathsf{pvk}, \mathcal{T}, x)$ identically in both games, hence there is a direct reduction from breaking the equivocability of $\mathcal{G}$ to distinguishing H_0 and H_1.

- **Game H_2.** In this game, instead of picking $x \xleftarrow{\$} \{0,1\}^m$ and setting $\mathsf{hb}_i \leftarrow x_i \oplus \rho_i \oplus \theta_{(i-1 \bmod \ell)+1}$, we first pick $\mathsf{hb} \xleftarrow{\$} \{0,1\}$ and set $x_i \leftarrow \mathsf{hb}_i \oplus \rho_i \oplus \theta_{(i-1 \bmod \ell)+1}$ for every $i \le m$. Note that this is a purely syntactic change, since hb and x are just a uniformly random sharing of $(\rho_i \oplus \theta_{(i-1 \bmod \ell)+1})_{i \le m}$, hence this game is perfectly indistinguishable from the previous one.

- **Game H_3.** In this game, we play as in Game H_2 except that we compute $(\mathsf{hb}[S], S, M) \xleftarrow{\$} \mathsf{Sim}_{\mathsf{zk}}(y)$ instead (note that the remaining hidden bits of hb are never used). By the single-theorem zero-knowledge property of HB, this game is perfectly indistinguishable from the previous one. Note that Game H_3 does exactly correspond to the simulation with $(\mathsf{Sim}_0, \mathsf{Sim}_1)$. This concludes the proof.

4.5   Unbounded Adaptive Soundness of Π

Let $\mathcal{A}$ be a PPT adversary against the soundness of Π, which is given oracle access to a verification oracle $\mathcal{O}(\mathsf{crs}, \cdot, \cdot, \mathcal{T})$. Let $(\mathsf{crs}, \mathcal{T}) \xleftarrow{\$} \Pi.\mathsf{Setup}(1^\lambda)$, and parse crs as (pp, ρ). Run $(y, \pi) \xleftarrow{\$} \mathcal{A}(\mathsf{crs})$. Let ε denote the probability (over the coins of $\Pi.\mathsf{Setup}$) that $\Pi.\mathsf{Verify}(\mathsf{crs}, y, \pi, \mathcal{T}) = 1$ and $y \notin \mathscr{L}$:

$$\Pr\left[\begin{array}{l}(\mathsf{crs}, \mathcal{T}) \xleftarrow{\$} \mathsf{Setup}(1^\lambda), \\ (y, \pi) \xleftarrow{\$} \mathcal{A}^{\mathcal{O}(\mathsf{crs}, \cdot, \cdot, \mathcal{T})}(\mathsf{crs})\end{array} : \Pi.\mathsf{Verify}(\mathsf{crs}, y, \pi, \mathcal{T}) = 1 \wedge y \notin \mathscr{L}\right] = \varepsilon.$$

In the following, we assume for the sake of contradiction that ε is non-negligible. We will construct from $\mathcal{A}$ an adversary $\mathcal{B}$ which contradicts the unbounded binding of $\mathcal{G}$. $\mathcal{B}$ interacts with $\mathcal{A}$ in the unbounded soundness security experiment of Π. The challenger of the unbounded binding property of $\mathcal{G}$ samples $(\mathsf{pp}, \mathcal{T}) \xleftarrow{\$} \mathsf{Setup}(1^\lambda, m)$. $\mathcal{B}$ receives pp and is given oracle access to $\mathcal{G}.\mathsf{Verify}(\mathsf{pp}, \cdot, \mathcal{T}, \cdot, \cdot, \cdot)$. It picks $\rho \xleftarrow{\$} \{0,1\}^m$, sets $\mathsf{crs} \leftarrow (\mathsf{pp}, \rho)$, and runs $\mathcal{A}(\mathsf{crs})$. Let q be the number of queries that $\mathcal{A}$ asks to $\mathcal{O}(\mathsf{crs}, \cdot, \cdot, \mathcal{T})$ in the unbounded soundness security experiment of Π. $\mathcal{B}$ simulates the answers of $\mathcal{O}(\mathsf{crs}, \cdot, \cdot, \mathcal{T})$ as follows: on input $\pi = (\mathsf{pvk}, \theta, S, \mathsf{hb}[S], M, (\pi_i)_{i \in S})$ it sets for every $i \in S$ $x_i \leftarrow \mathsf{hb}_i \oplus \rho_i \oplus \theta_{(i-1 \bmod \ell)+1}$ and calls $\mathcal{G}.\mathsf{Verify}(\mathsf{pp}, \cdot, \mathcal{T}, \cdot, \cdot, \cdot)$ on input $(\mathsf{pvk}, i, x_i, \pi_i)$. Then, it verifies the HB proof $(S, \mathsf{hb}[S], M)$ for the statement $y \in \mathscr{L}$ and outputs 1 iff all checks succeeded. Then, $\mathcal{A}$ outputs a pair (y, π). Since $\mathcal{B}$ perfectly simulates crs and the answers of $\mathcal{O}(\mathsf{crs}, \cdot, \cdot, \mathcal{T})$, it holds that $\Pi.\mathsf{Verify}(\mathsf{crs}, y, \pi, \mathcal{T}) = 1 \wedge y \notin \mathscr{L}$ with probability ε over the coins of the challenger and $\mathcal{A}, \mathcal{B}$. Finally, $\mathcal{B}$ parses π as $(\mathsf{pvk}^*, \theta, S, \mathsf{hb}[S], M, (\pi_i)_{i \in S})$, picks $i^* \xleftarrow{\$} S$, and outputs $(\mathsf{pvk}^*, i^*, \pi_{i^*})$. To simplify the analysis in the following, we assume that $\mathcal{B}$ also outputs (crs, y, π) in addition to $(\mathsf{pvk}^*, i^*, \pi_{i^*})$ (it is only a syntactic modification that will make it more convenient to describe the probability experiments).

We analyze the probability that $\mathcal{G}.\mathsf{Verify}(\mathsf{pp}, \mathsf{pvk}^*, \mathcal{T}, i^*, 1 - x_{i^*}, \pi_{i^*}) = 1$. Let us call 'bad' a string hb for which there exists $y \notin \mathscr{L}$ and an accepting proof of $y \in \mathscr{L}$ under the HB proof system HB'. Under the $2^{-\lambda m/\ell}$-statistical soundness of HB', the ratio of bad strings must be at most $2^{-\lambda m/\ell} < 2^{\lambda + s + \ell}$. Let us say that a string hb' is "close to a bad string w.r.t. pp" hb if there exists $\theta \in \{0,1\}^\ell$ and a public verification key $\mathsf{pvk} \in \{0,1\}^{s(\lambda)}$ such that $\mathsf{hb}' = (x_i \oplus \mathsf{hb}_i \oplus \theta_{(i-1 \bmod \ell)+1})_i$ is a bad string, where $x = \mathsf{Ext}(\mathsf{pp}, \mathsf{pvk})$. As there are at most $2^{\ell + s}$ possible choices of (θ, pvk), for any choice of public parameters pp, the ratio of strings which are close to a bad string w.r.t. pp must be at most $2^{-\lambda - s - \ell} \cdot 2^{\ell + s} = 2^{-\lambda}$. Therefore, with overwhelming probability $1 - 2^{-\lambda}$ over the distribution of ρ, ρ is not close to a bad string w.r.t. pp, hence there does not exist a string $(y, \mathsf{pvk}, \theta)$ with $y \notin \mathscr{L}$ such that $(\mathsf{hb}_i)_i = (x_i \oplus \rho_i \oplus \theta_{(i-1 \bmod \ell)+1})_i$ is a bad string.

We consider two complementary cases, one of which must necessarily occur:

Case 1. With probability at least $\varepsilon/2$, the output (y, π) of $\mathcal{A}$ satisfies $\Pi.\mathsf{Verify}(\mathsf{crs}, y, \pi, \mathcal{T}) = 1 \wedge y \notin \mathscr{L}$, and for every $i \in S$, it holds that $\mathcal{G}.\mathsf{Verify}(\mathsf{pp}, \mathsf{pvk}^*, \mathcal{T}, i^*, 1 - x_i, \pi_i) = 1$. That is,

$$\Pr\left[\begin{array}{l}(\mathsf{pp}, \mathcal{T}) \xleftarrow{\$} \mathsf{Setup}(1^\lambda, m), \\ (\mathsf{pvk}^*, i^*, \pi_{i^*}, \mathsf{crs}, y, \pi) \xleftarrow{\$} \mathcal{B}(\mathsf{pp}), \\ x \leftarrow \mathsf{Ext}(\mathsf{pp}, \mathsf{pvk}^*)\end{array} \middle| \begin{array}{l}\Pi.\mathsf{Verify}(\mathsf{crs}, y, \pi, \mathcal{T}) = 1 \\ : \wedge\, y \notin \mathscr{L} \wedge \forall i \in S, \\ \mathcal{G}.\mathsf{Verify}(\mathsf{pp}, \mathsf{pvk}^*, \mathcal{T}, i, 1 - x_i, \pi_i) = 1\end{array}\right] \geq \frac{\varepsilon}{2},$$

where $\mathcal{B}$ is given oracle access to $\mathcal{G}.\mathsf{Verify}(\mathsf{pp}, \cdot, \mathcal{T}, \cdot, \cdot, \cdot)$. Now, parse π as $(\mathsf{pvk}^*, \theta, S, \mathsf{hb}[S], M, (\pi_i)_{i \in S})$. Let x' denote $\mathsf{Ext}(\mathsf{pp}, \mathsf{pvk}^*)$, and let $(\mathsf{hb}'_i)_i = (x'_i \oplus \rho_i \oplus \theta_{(i-1 \bmod \ell)+1})_i$. Since a random ρ has probability at most $1/2^\lambda$ to be close to a bad string w.r.t. pp, $(\mathsf{hb}'_i)_i$ has probability at most $1/2^\lambda$ of being a bad string. Therefore, if case 1 happens, we necessarily have (denoting $\mu = \varepsilon/2 - 1/2^\lambda$):

$$\Pr\left[\begin{array}{l}(\mathsf{pp}, \mathcal{T}) \xleftarrow{\$} \mathsf{Setup}(1^\lambda, m), \\ (\mathsf{pvk}^*, i^*, \pi_{i^*}, \mathsf{crs}, y, \pi) \xleftarrow{\$} \mathcal{B}(\mathsf{pp}), \\ x \leftarrow \mathsf{Ext}(\mathsf{pp}, \mathsf{pvk}^*)\end{array} \middle| \begin{array}{l}\Pi.\mathsf{Verify}(\mathsf{crs}, y, \pi, \mathcal{T}) = 1 \\ \wedge\, y \notin \mathscr{L} \wedge \forall i \in S, \\ : \mathcal{G}.\mathsf{Verify}(\mathsf{pp}, \mathsf{pvk}^*, \mathcal{T}, i, 1 - x_i, \pi_i) = 1 \\ \wedge\, \mathsf{hb}' \text{ is not a bad string}\end{array}\right] \geq \mu.$$

Note that the condition $\Pi.\mathsf{Verify}(\mathsf{crs}, y, \pi, \mathcal{T}) = 1$ implies that for all $i \in S$, denoting $x_i \leftarrow \mathsf{hb}_i \oplus \rho_i \oplus \theta_{(i-1 \bmod \ell)+1}$, $\mathcal{G}.\mathsf{Verify}(\mathsf{pp}, \mathsf{pvk}^*, \mathcal{T}, i, x_i, \pi_i) = 1$. Denoting $x' = \mathsf{Ext}(\mathsf{pp}, \mathsf{pvk}^*)$, it holds by assumption that $\mathcal{G}.\mathsf{Verify}(\mathsf{pp}, \mathsf{pvk}^*, \mathcal{T}, i, 1 - x_i, \pi_i) = 1$ for every $i \in S$, hence $\mathcal{G}.\mathsf{Verify}(\mathsf{pp}, \mathsf{pvk}^*, \mathcal{T}, i, 1 - x'_i, \pi_i) = 0$. This implies that for any $i \in S$, $x_i \neq 1 - x'_i$, hence that $(x_i)_{i \in S} = (x'_i)_{i \in S}$, which in turns implies that $\mathsf{hb}[S] = \mathsf{hb}'[S]$. Therefore, if case 1 happens, with probability at least $\varepsilon/2 - 1/2^\lambda$ we have the following:

- $y \notin \mathcal{L}$,
- $\Pi.\mathsf{Verify}(\mathsf{crs}, y, \boldsymbol{\pi}, \mathcal{T}) = 1$,
- $\mathsf{hb}[S] = \mathsf{hb}'[S]$ is not a bad string.

However, by the soundness of the HB proof system HB', there cannot exist any accepting proof $(S, \mathsf{hb}[S], M)$ for a statement $y \notin \mathcal{L}$ unless $\mathsf{hb}[S]$ is a bad string. Since $\Pi.\mathsf{Verify}$ does also check the HB proof, this event can never happen and we get:

$$\frac{\varepsilon}{2} - \frac{1}{2^\lambda} = 0,$$

contradicting our assumption that ε is non-negligible. Hence, case 1 never happens and the following case necessarily happens:

Case 2. There exists $i \in S$ such that $\mathcal{G}.\mathsf{Verify}(\mathsf{pp}, \mathsf{pvk}^*, \mathcal{T}, i, 1 - x_i, \pi_i) = 1$ with probability at least $\varepsilon/2$. That is,

$$\Pr\left[\begin{array}{l}(\mathsf{pp}, \mathcal{T}) \xleftarrow{\$} \mathsf{Setup}(1^\lambda, m), \\ (\mathsf{pvk}^*, i^*, \pi_{i^*}, \mathsf{crs}, y, \boldsymbol{\pi}) \xleftarrow{\$} \mathcal{B}(\mathsf{pp})\end{array} \begin{array}{l} \Pi.\mathsf{Verify}(\mathsf{crs}, y, \boldsymbol{\pi}, \mathcal{T}) = 1 \\ : \wedge\ y \notin \mathcal{L} \wedge \exists i \in S, \\ \mathcal{G}.\mathsf{Verify}(\mathsf{pp}, \mathsf{pvk}^*, \mathcal{T}, i, 1 - x_i, \pi_i) = 1\end{array}\right] \geq \frac{\varepsilon}{2}.$$

Since $\mathcal{B}$ picks i^* at random in a set S of size at most m, this gives us in particular

$$\Pr\left[\begin{array}{l}(\mathsf{pp}, \mathcal{T}) \xleftarrow{\$} \mathsf{Setup}(1^\lambda, m), \\ (\mathsf{pvk}^*, i^*, \pi_{i^*}, \mathsf{crs}, y, \boldsymbol{\pi}) \xleftarrow{\$} \mathcal{B}(\mathsf{pp})\end{array} : \mathcal{G}.\mathsf{Verify}(\mathsf{pp}, \mathsf{pvk}^*, \mathcal{T}, i^*, 1 - x_i, \pi^*) = 1\right] \geq \frac{\varepsilon}{2m},$$

which immediately gives a contradiction to the unbounded binding of $\mathcal{G}$, concluding the proof.

Impact on Our LWE-Based Instantiation. Note that our alternative proof strategy, which does not use any assumed structure for pvk except a bound on its length, is the key to our LWE-based instantiation. Indeed, if we had to assume some structure of pvk (such as "pvk was honestly generated"), we would have to include a NIZK proof of validity of pvk in our instantiation (which is similar to the NIZK proof of validity for the public key used in [37]). Since there might not exist more than a single witness for the validity of pvk, it seems unlikely that we could use a NIWI instead of a NIZK here. By removing entirely the need for proving validity of pvk in our LWE-based instantiation, we enable the construction of a NIZK for NP from LWE using only a NIWI for a simple language (bounded distance decoding), improving over the result of [37].

5 Constructions of Designated-Verifier Pseudorandom Generators

5.1 A DVPRG from the CDH Assumption

Assumptions. Let DHGen denote a PPT algorithm which, on input 1^λ, outputs an integer n, the description of a group $\mathbb{G}$ of order n, and a generator g of $\mathbb{G}$.

The *computational Diffie-Hellman assumption* (CDH), with respect to g over $\mathbb{G}$, states that it is computationally infeasible, for any PPT algorithm which is given $(n, g, \mathbb{G})$ and a random pair (g^a, g^b) from $\mathbb{G}^2$, to compute g^{ab}. The *decisional Diffie-Hellman assumption* (DDH), with respect to g over $\mathbb{G}$, states that it is computationally infeasible for any PPT algorithm to distinguish the distribution $\{(g^a, g^b, g^{ab}) \mid (a, b) \xleftarrow{\$} \mathbb{Z}_n^3\}$ of random DDH tuples from the uniform distribution over $\mathbb{G}^3$.

The *twin (computational or decisional) Diffie-Hellman assumption* (twin-CDH and twin-DDH), defined in [10], are variants of the CDH and DDH assumptions. The twin-CDH problem with respect to g over $\mathbb{G}$ states that it is computationally infeasible, for any PPT algorithm which is given $(n, g, \mathbb{G})$ and a random triple (g^a, g^b, g^c) from $\mathbb{G}^3$, to compute (g^{ab}, g^{ac}); twin-DDH is its natural decisional variant. Twin-CDH (resp. twin-DDH) is equivalent to the standard CDH (resp. DDH) assumption. However, there is a natural trapdoor test that allows to check the correctness of twin-DDH tuples: let (α, β) be a random pair of exponents satisfying $g^c = g^\alpha (g^b)^{-\beta}$ (note that many such pairs exist). Then given an input (g^a, g^b, g^c), the probability for an arbitrary (possibly unbounded) adversary $\mathcal{A}$ to output a pair (h_1, h_2) such that the truth value of $h_1^\beta h_2 = (g^a)^\alpha$ does not agree with the truth value of $(h_1 = g^{ab}) \wedge (h_2 = g^{ac})$ is at most $1/n$ (see [10]). Therefore, a verifier which is given the trapdoor (α, β) can check the correctness of a twin Diffie-Hellman tuple, with negligible error probability. This trapdoor test implies that the *gap* twin Diffie-Hellman problem, which states that solving the twin-CDH problem is hard even given an oracle that solves the twin-DDH problem, is at least as hard as the standard CDH problem.

Our Construction. Our construction will rely on the conjectured hardness of the computational Diffie-Hellman (CDH) assumption. Let $B : \mathbb{G}^3 \mapsto \{0, 1\}$ be a predicate satisfying the following property: given (g^a, g^b, g^c), computing $B(g^a, g^{ab}, g^{ac})$ should be as hard (up to polynomial factors) as computing (g^a, g^{ab}, g^{ac}). Note that this implies that distinguishing $B(g^a, g^{ab}, g^{ac})$ from a random bit given a random triple (g^a, g^b, g^c) is as hard as solving CDH. There are standard method to build this predicate using e.g. the Goldreich-Levin construction [18], see e.g. [10] for an illustration in the specific case of CDH. Our construction proceeds as follows:

- Setup$(1^\lambda, m)$: sample $(n, \mathbb{G}, g) \xleftarrow{\$} \mathsf{DHGen}(1^\lambda)$. For $i = 1$ to m, pick $(a_i, b_i) \xleftarrow{\$} \mathbb{Z}_n^2$ and set $(u_i, v_i) \leftarrow (g^{a_i}, g^{b_i})$. Set $\mathsf{pp} = (u_i, v_i)_{i \leq m}$. For $i = 1$ to m, pick $\beta_i \xleftarrow{\$} \mathbb{Z}_n$ and set $\alpha_i \leftarrow b_i + a_i \beta_i$ (observe that (α_i, β_i) are uniformly distributed exponents subject to $v_i = g^{\alpha_i} u_i^{-\beta_i}$). Output pp and $\mathcal{T} \leftarrow (\alpha_i, \beta_i)_{i \leq m}$. We also define $\mathsf{SimSetup}(1^\lambda, m)$ to be identical to $\mathsf{Setup}(1^\lambda, m)$ and define $\mathcal{T}_s = \mathcal{T}$.
- Stretch(pp) : pick $r \xleftarrow{\$} \mathbb{Z}_n$, set $\mathsf{pvk} \leftarrow g^r$, and for $i = 1$ to m, set $x_i \xleftarrow{\$} B(\mathsf{pvk}, u_i^r, v_i^r)$. Output $(\mathsf{pvk}, x, \mathsf{aux} = r)$.
- Prove$(\mathsf{pp}, \mathsf{aux}, i)$: output $\pi \leftarrow (u_i^r, v_i^r)$.
- Equivocate$(\mathsf{pp}, \mathsf{pvk}, \mathcal{T}_s, i, \sigma)$: pick $u' \xleftarrow{\$} \mathbb{G}$, set $v' = \mathsf{pvk}^{\alpha_i}(u')^{-\beta_i}$, and check whether $B(\mathsf{pvk}, u', v') = \sigma$; if it does not hold, start again. Output $\pi \leftarrow (u', v')$.

– Verify$(\mathsf{pp}, \mathsf{pvk}, \mathcal{T}, i, \sigma, \pi)$: parse π as (u', v'), check whether $B(\mathsf{pvk}, u', v') = \sigma$ and check whether $(u')^{\beta_i} v' = \mathsf{pvk}^{\alpha_i}$. If both checks pass, output 1; otherwise, output 0.

This construction follows the twin Diffie-Hellman paradigm of Cash, Kiltz, and Shoup [10], which relies on the fact that computing the *twin Diffie-Hellman function*, which on input (g^x, g^{y_1}, g^{y_2}) outputs (g^{xy_1}, g^{xy_2}), is at least as hard as solving the CDH problem, even given an oracle for twin-DDH.

Theorem 17. *If the CDH assumption holds over* $\mathbb{G}$, *then the above construction is a computationally hiding unbounded statistically binding* DVPRG. *Furthermore, if the DDH assumption holds over* $\mathbb{G}$, *the above construction is also equivocable.*

Proof. Completeness follows easily by inspection. We now look at the unbounded binding property; by Theorem 14, it suffices to show that the scheme is binding, consistent, and that we can efficiently sample trapdoors consistent with pp (a proof of Theorem 14 is given in the full version of this paper [12]). From the analysis of [10, Section 2], as (g^x, g^{y_1}, g^{y_2}) uniquely define the pair (h_1, h_2) such that $(h_1 = g^{xy_1}) \wedge (h_2 = g^{xy_2})$ and any adversary has negligible probability $1/n$ of outputting a non-twin-DH pair (h_1, h_2) that fools the test, it follows that the scheme is statistically binding (the inefficient extractor Ext simply extracts r from $\mathsf{pvk} = g^r$ and computes the string x as $x_i \overset{\$}{\leftarrow} B(\mathsf{pvk}, u_i^r, v_i^r)$ for $i = 1$ to m). Second, observe that we can efficiently sample trapdoors consistent with pp, by storing the random values $(a_i, b_i)_i$ and sampling each trapdoor $\mathcal{T} = (\alpha_i, \beta_i)$ as $\beta_i \overset{\$}{\leftarrow} \mathbb{Z}_n$ and $\alpha_i \leftarrow b_i + a_i \beta_i$. Therefore, this defines an efficiently sampleable distribution $\mathtt{Dist}((a_i, b_i)_i)$.

We now show that our construction satisfies consistency. Let $\varepsilon \leftarrow 2/n^5$ and let $\mathcal{A}$ be an adversary that, on input $\mathsf{pp} = (u_i, v_i)_i = (g^{a_i}, g^{b_i})_i$, outputs a 4-tuple $(\mathsf{pvk}, i, \pi = (u', v'), \sigma)$ such that

$$\Pr[\beta_i \overset{\$}{\leftarrow} \mathbb{Z}_n : \mathsf{Verify}(\mathsf{pp}, \mathsf{pvk}, (\alpha_i, b_i + a_i\beta_i), i, \sigma, (u', v')) = 1] \geq \varepsilon.$$

The above implies that $\mathcal{A}$ outputs (pvk, u', v') such that $(u')^{\beta_i} v' = \mathsf{pvk}^{\alpha_i}$ holds with probability at least $2/n$. Suppose now that $(\mathsf{pvk}, u_i, v_i, u', v')$ is not a twin-DH tuple; let us denote $\mathsf{pvk} = g^r$ and $(u', v') = (g^s, g^t)$ with $s \neq a_i r$ or $t \neq b_i r$. Then the previous equation becomes $g^{s\beta_i + t} = g^{r\alpha_i}$, which gives

$$\beta_i(s - a_i r) = rb_i - t.$$

However, if $s - a_i r \neq 0$ or $rb_i - t \neq 0$, then this equation holds with probability at most $1/n$ over the random choice of β_i, hence since we assumed that this equation is satisfied with probability at least $2/n$, it must be that $s - a_i r = rb_i - t = 0$, hence $(\mathsf{pvk}, u_i, v_i, u', v')$ is a twin-DH tuple. But then, it immediately follows that the above equation is *always* satisfied, independently of the choice of β_i:

$$\Pr[\beta_i \overset{\$}{\leftarrow} \mathbb{Z}_n : \mathsf{Verify}(\mathsf{pp}, \mathsf{pvk}, (\alpha_i, b_i + a_i\beta_i), i, \sigma, (u', v')) = 1] = 1,$$

---

[5] Since n is the order of $\mathbb{G}$ and $\mathbb{G}$ is a group in which CDH is assumed to hold, $2/n$ is negligible in the security parameter.

which concludes the proof of consistency. Since the DVPRG is also statistically binding (in a bounded sense), we use Theorem 14 to conclude that the above construction satisfies (statistical) unbounded binding.

We now discuss the hiding property. We show that a PPT adversary against the hiding property of the above scheme implies the existence of a PPT adversary that solves the computational twin Diffie-Hellman problem. The result follows from the proof of [10] that the computational twin Diffie-Hellman problem is at least as hard as the CDH problem. The reduction is relatively straightforward: given a position $i \leq m$, we pick $(a_j, b_j)_{j \neq i}$, receive a computational twin-DH challenge (c_0, c_1, c_2), and set $\mathsf{pvk} \leftarrow c_0$, $(u_j, v_j) \leftarrow (g^{a_j}, g^{b_j})$ for every $j \neq i$, and $(u_i, v_i) \leftarrow (c_1, c_2)$. We output $\mathsf{pp} \leftarrow (u_j, v_j)_{j \leq m}$, pvk, and $(x_j, \pi_j) \xleftarrow{\$} (B(\mathsf{pvk}, c_0^{a_i}, c_0^{b_j}), (c_0^{a_i}, c_0^{b_j}))$ for every $j \neq i$. Note that $\mathsf{pp}, \mathsf{pvk}$ and the x_j, π_j are distributed exactly as in an honest execution of the experiment. Then, we run $\mathcal{A}(\mathsf{pp}, \mathsf{pvk}, (x_j, \pi_j)_{j \neq i})$ and get a bit b. If $\mathcal{A}$ guesses the value of $x_i = B(\mathsf{pvk}, c_1', c_2')$, where $(c_1', c_2') = (c_1^r, c_2^r)$ for the value r such that $c_0 = \mathsf{pvk} = g^r$, then we efficiently find a hardcore bit for the twin-DH problem with non-negligible probability. As guessing a hardcore bit for twin-DH is at least as hard as solving the computational twin-DH problem, the proof follows.

Regarding equivocability, the reduction gets a DDH challenge (c_0, c_1, c_2). It sets $\mathsf{pvk} \leftarrow g^r$, samples $(\alpha_i, \beta_i)_i \xleftarrow{\$} \mathbb{Z}_n^{2m}$ and $\mathsf{pp} = (u_i, v_i)_i$ as $(c_1^{\alpha_i}, g^{\alpha_i} u_i^{-\beta_i})_i$ with random a_i's. It computes each proof π_i as $(u', v') \leftarrow (c_2^{\alpha_i}, \mathsf{pvk}^{\alpha_i}(u')^{-\beta_i})$. Observe that the distribution of $(\mathsf{pp}, \mathcal{T}, \mathsf{pvk}, (\pi_i)_i)$ is identical to the distribution obtained with an honest run of the DVPRG when (c_0, c_1, c_2) is a DDH tuple, and identical to a run of the DVPRG with the algorithm Equivocate when (c_0, c_1, c_2). Hence, distinguishing honest proofs from equivocated proofs is equivalent to breaking the DDH assumption.

Corollary 18. *Assuming the computational Diffie-Hellman assumption, there exists an unbounded designated-verifier non-interactive (adaptive, multi-theorem) zero-knowledge proof system for NP.*

Note that the above construction also implies that the existence of a (publicly verifiable) NIZK proof system for the DDH language (together with the CDH assumption) would imply a NIZK proof system for NP.

5.2   A DVPRG from the LWE Assumption

We also give a construction of a DVPRG in the LWE setting. Our construction already assumes a designated-verifier NIWI proof system Π for the LWE language. We stress, however, that Π does not have to enjoy zero-knowledge; witness-indistinguishability is sufficient. We also note that Π can be publicly verifiable, in which case the DVPRG becomes publicly verifiable.

Algebraic setting. We largely follow the presentation of [22] and abstract the setting as far as possible. In the following, let $n, m = \mathsf{poly}(\lambda)$ and $q, \beta = 2^{\mathsf{poly}(\lambda)}$

with $m > n$ and $q > \beta$ be suitable integers. We also assume an error distribution χ that outputs integers e with $|e| < \beta$.

The *Learning With Errors (LWE)* problem (relative to n, m, q, β) is to distinguish access to an oracle $\mathcal{O}^{\mathsf{lwe}}_{\mathsf{real},\mathbf{s}}$ (with hardwired uniform $\mathbf{s} \in \mathbb{Z}_q^n$) from access to another oracle $\mathcal{O}^{\mathsf{lwe}}_{\mathsf{rand}}$. Here, $\mathcal{O}^{\mathsf{lwe}}_{\mathsf{real},\mathbf{s}}$ (parameterized over $\mathbf{s} \in \mathbb{Z}_q^n$) outputs samples $(\mathbf{a}, \mathbf{s}^\top \mathbf{a} + e)$ for fresh $\mathbf{a} \xleftarrow{\$} \mathbb{Z}_q^n$ and $e \leftarrow \chi$, and $\mathcal{O}^{\mathsf{lwe}}_{\mathsf{real},\mathbf{s}}$ outputs $(\mathbf{a}, r)$ with fresh $\mathbf{a} \xleftarrow{\$} \mathbb{Z}_q^n$ and $r \xleftarrow{\$} \mathbb{Z}_q$. The LWE assumption is that for every PPT adversary $\mathcal{A}$,

$$\left| \Pr\left[\mathcal{A}^{\mathcal{O}^{\mathsf{lwe}}_{\mathsf{real},\mathbf{s}}}(1^\lambda) = 1 \right] - \Pr\left[\mathcal{A}^{\mathcal{O}^{\mathsf{lwe}}_{\mathsf{rand}}}(1^\lambda) = 1 \right] \right| \approx 0,$$

where the probability is over $\mathbf{s} \xleftarrow{\$} \mathbb{Z}_q^n$ and the random coins of $\mathcal{A}$ and the oracles.

In the following, let $\mathbf{A} \in \mathbb{Z}_q^{n \times m}$, and consider the language

$$\mathcal{L}_{\mathbf{A}} := \left\{ \mathbf{Au} \mid \mathbf{u} \in \mathbb{Z}_q^m \text{ with } \|\mathbf{u}\|_\infty < \beta \right\}.$$

Depending on $\mathbf{A}$, $\mathcal{L}_{\mathbf{A}}$ may be trivial. However, if $\mathbf{A}$ can be written as $\mathbf{A} = \begin{pmatrix} \mathbf{A}' \\ \mathbf{s}^\top \mathbf{A}' + \mathbf{e} \end{pmatrix}$ with $\|\mathbf{e}\|_\infty < \beta$, then $\mathcal{L}_{\mathbf{A}}$ consists of all zero-encryptions under Regev's encryption scheme [36]. In that case, $\mathcal{L}_{\mathbf{A}}$ is hard to decide under the LWE assumption.

Homomorphic commitments. In the setting above, Gorbunov, Vaikuntanathan, and Wichs [22] construct homomorphic trapdoor functions (HTDFs). As they point out, HTDFs can also be viewed as homomorphic commitments. Formally, an HTDF HF consists of the following PPT algorithms:

Key generation. $\mathsf{HF.Setup}(1^\lambda)$ outputs a keypair $(\mathsf{pk}, \mathsf{sk})$. We require that pk defines input, output, and index sets $\mathcal{U}$, $\mathcal{V}$, and $\mathcal{X}$. These sets must be efficiently decidable, and we assume are efficiently samplable distributions $D_{\mathcal{U}}$ and $D_{\mathcal{V}}$ over $\mathcal{U}$ and $\mathcal{V}$.

Function evaluation. $f_{\mathsf{pk},x}$ evaluates a deterministic function from $\mathcal{U}$ to $\mathcal{V}$. We can view $f_{\mathsf{pk},x}(u)$ as a commitment under pk to x with random coins u.

Function inversion. $\mathsf{Inv}_{\mathsf{sk},x}$ probabilistically samples a preimage of $f_{\mathsf{pk},x}$. We require that for every $(\mathsf{pk}, \mathsf{sk})$ in the range of $\mathsf{HF.Setup}$, every $x \in \mathcal{X}$, and every v in the range of $f_{\mathsf{pk},x}$, the value $\mathsf{Inv}_{\mathsf{sk},x}(v)$ is distributed statistically close to a random preimage of v under $f_{\mathsf{pk},x}$ sampled from $D_{\mathcal{U}}$.

Homomorphic evaluation. Eval^{in} and Eval^{out} allow homomorphic computations on inputs and outputs, in the following sense. For all pk in the range of $\mathsf{HF.Setup}$, all $\ell \in \mathbb{N}$, all functions g (represented as circuits), all $(x_i, u_i, v_i) \in \mathcal{X} \times \mathcal{U} \times \mathcal{V}$ ($1 \le i \le \ell$) with $v_i = f_{\mathsf{pk},x_i}(u_i)$, and for $u^* := \mathsf{Eval}^{in}_{\mathsf{pk}}(g, (x_i, u_i)_{i=1}^\ell)$ and $v^* := \mathsf{Eval}^{out}_{\mathsf{pk}}(g, (v_i)_{i=1}^\ell)$, we have

$$f_{\mathsf{pk},g(x_1,\ldots,x_\ell)}(u^*) = v^*.$$

Dual-mode homomorphic commitments. For security, [22] require that it is computationally hard to find (x, u, x', u') with $x \neq x'$ and $f_{\mathsf{pk}, x}(u) = f_{\mathsf{pk}, x'}(u')$. When viewing HTDFs as commitment schemes, this corresponds to a computational binding property. For our purposes, however, we require a stronger property that [22] mention but do not formally define or use. Namely, in analogy to dual-mode commitment schemes [26], we require that there are two computationally indistinguishable ways to sample public keys: one way leads to a statistically hiding commitment scheme, and the other to a statistically binding scheme. In the HTDF setting, this translates to the following requirements:

Statistically hiding. For any fixed pk in the range of HF.Setup, and any $x, x' \in \mathcal{X}$, the random variables $f_{\mathsf{pk}, x}(u)$ and $f_{\mathsf{pk}, x'}(u)$ (for random $u \leftarrow D_{\mathcal{U}}$) are statistically close.

Perfectly binding under alternate key generation. There exists a PPT algorithm $\mathsf{HF.Setup_{bind}}$ that outputs public keys $\mathsf{pk_{bind}}$ with the following properties:

- $\mathsf{pk_{bind}} \overset{c}{\approx} \mathsf{pk}$ for public keys pk output by HF.Setup,
- the "function evaluation" and "homomorphic evaluation" properties above also hold (perfectly) for public keys $\mathsf{pk_{bind}}$,
- for all $\mathsf{pk_{bind}}$ in the range of $\mathsf{HF.Setup_{bind}}$, and all $x, x' \in \mathcal{X}$ with $x \neq x'$, the sets $\{f_{\mathsf{pk_{bind}}, x}(u) \mid u \in \mathcal{U}\}$ and $\{f_{\mathsf{pk_{bind}}, x'}(u) \mid u \in \mathcal{U}\}$ are disjoint. In other words, there are no (x, u, x', u') with $x \neq x'$ and $f_{\mathsf{pk_{bind}}, x}(u) = f_{\mathsf{pk_{bind}}, x'}(u')$.

The instantiation of Gorbunov, Vaikuntanathan, and Wichs. [22] offer a *leveled* instantiation of dual-mode homomorphic commitments. That is, their construction only allows for an arbitrary, but a-priori bounded number of homomorphic base operations on commitments. If this number of operations is exceeded, correctness will cease to hold. For our purposes, this leveled construction is sufficient, since the number and type of homomorphic operations is known in advance.

We further note that their HTDF application does not require any dual-mode features. However, in [22, App. B], they explicitly describe and analyze what we call $\mathsf{HF.Setup_{bind}}$ above. They show that their construction is secure (in the sense above) under the LWE assumption.

We will not need to consider any specifics of their construction, except for one. Namely, in their scheme, $\{0, 1\} \subset \mathcal{X} \subset \mathbb{Z}$, and commitments to x are of the form $f_{\mathsf{pk}, x}(u) = \mathbf{A}\mathbf{U} + x\mathbf{G}$ for fixed $\mathbf{A}, \mathbf{G} \in \mathbb{Z}_q^{n \times m}$ defined in pk, and a short $\mathbf{U} \in \mathbb{Z}_q^{m \times m}$ with $\|\mathbf{U}\|_\infty < \beta$. In other words, commitments to $x = 0$ are composed of m elements of the language $\mathcal{L}_{\mathbf{A}}$ defined above. Furthermore, a preimage u is the corresponding witness $\mathbf{U}$. Hence, given an argument system for $\mathcal{L}_{\mathbf{A}}$, we can prove that a given commitment v commits to a given x by proving that $v - x\mathbf{G} \in \mathcal{L}_{\mathbf{A}}^m$.

Our construction. We can now give our construction of a DVPRG. We assume dual-mode homomorphic commitments HF as described above, and any family of PRGs $G_m : \{0, 1\}^\lambda \to \{0, 1\}^m$. In the following, let $G_{m,i}$ denote the circuit that computes the i-th output bit of G_m. Furthermore, we assume a NIWI proof system $\Pi = (\Pi.\text{Gen}, \Pi.\text{Prove}, \Pi.\text{Verify})$ for the language $\mathcal{L}_\mathbf{A}$. Slightly abusing notation, we will use NIWI as an argument system for the language $(\mathcal{L}_\mathbf{A})^m$.

- Setup$(1^\lambda, m)$ runs $\text{pk}_{\text{bind}} \xleftarrow{\$} \text{HF.Setup}_{\text{bind}}(1^\lambda)$ and $(\text{crs}, \mathcal{T}) \xleftarrow{\$} \Pi.\text{Gen}(1^\lambda)$ (for the language $\mathcal{L}_\mathbf{A}$ given by the matrix $\mathbf{A}$ defined in pk_{bind}), and outputs public parameters $\text{pp} = (\text{pk}_{\text{bind}}, \text{crs})$ and a trapdoor $\mathcal{T}$.
- Stretch(pp) samples $s = (s_1, \ldots, s_\lambda) \xleftarrow{\$} \{0,1\}^\lambda$ and $u_1, \ldots, u_\lambda \xleftarrow{\$} D_\mathcal{U}$, then computes $v_i = f_{\text{pk}_{\text{bind}}, s_i}(u_i)$, and finally outputs $\text{pvk} = (v_i)_{i=1}^\lambda$, $x = G_m(s)$, and $\text{aux} = (s, (u_i)_{i=1}^\lambda)$. Observe that the size of pvk does not depend on m.[6]
- Prove$(\text{pp}, \text{aux}, i)$ (for $\text{aux} = (s, (u_j)_{j=1}^\lambda)$) computes $v_i = f_{\text{pk}_{\text{bind}}, s_i}(u_i)$ exactly as Stretch, and derives a witness $u^* = \text{Eval}_{\text{pk}_{\text{bind}}}^{in}(G_{m,i}, (s_j, u_j)_{j=1}^\lambda)$ that explains $v^* = \text{Eval}_{\text{pk}_{\text{bind}}}^{out}(G_{m,i}, (v_j)_{j=1}^\lambda)$ as $v^* = f_{\text{pk}_{\text{bind}}, b}(u^*)$. By our discussion above, we have that hence $v^* - b_i \mathbf{G} \in \mathcal{L}_\mathbf{A}^m$ with witness u^*. Hence, Prove next computes and outputs a proof $\pi \xleftarrow{\$} \Pi.\text{Prove}(\text{crs}, v^* - b\mathbf{G}, u^*)$.
- Verify$(\text{pp}, \text{pvk}, \mathcal{T}, i, b, \pi)$ parses $\text{pvk} = (v_i)_{i=1}^\lambda$, then computes

$$v^* = \text{Eval}_{\text{pk}_{\text{bind}}}^{out}\left(G_{m,i}, (v_j)_{j=1}^\lambda\right),$$

and finally returns $\Pi.\text{Verify}(\text{crs}, v^* - b_i \mathbf{G}, \pi, \mathcal{T})$.

Theorem 19. *Assume that LWE holds for the parameters from [22], that G_m is pseudorandom, and that Π is perfectly complete, computationally witness-indistinguishable and satisfies unbounded adaptive soundness. Then the above DVPRG is perfectly complete, equivocable, and has the unbounded binding property.*

We provide a proof of Theorem 19 in the full version [12].

References

1. Abusalah, H.: Generic instantiations of the hidden bits model for non-interactive zero-knowledge proofs for NP. Master's thesis, RWTH Aachen (2013)
2. Benhamouda, F., Couteau, G., Pointcheval, D., Wee, H.: Implicit zero-knowledge arguments and applications to the malicious setting. In: Gennaro, R., Robshaw, M. (eds.) CRYPTO 2015. Part II. LNCS, vol. 9216, pp. 107–129. Springer, Heidelberg (2015). https://doi.org/10.1007/978-3-662-48000-7_6
3. Bitansky, N.: Verifiable random functions from non-interactive witness-indistinguishable proofs. In: Kalai, Y., Reyzin, L. (eds.) TCC 2017. Part II. LNCS, vol. 10678, pp. 567–594. Springer, Cham (2017). https://doi.org/10.1007/978-3-319-70503-3_19

---

[6] Strictly speaking, this may not be true, depending on G_m: the LWE parameters may depend on the size of the $G_{m,i}$, which in turn may depend on m. However, here we implicitly assume that $m \leq 2^\lambda$ and a suitable G_m (e.g., one in which $G_{m,i}(s)$ is of the form $F_s(i)$ for a PRF $F : \{0,1\}^\lambda \to \{0,1\}$).

4. Bitansky, N., Paneth, O.: ZAPs and non-interactive witness indistinguishability from indistinguishability obfuscation. In: Dodis, Y., Nielsen, J.B. (eds.) TCC 2015. Part II. LNCS, vol. 9015, pp. 401–427. Springer, Heidelberg (2015). https://doi.org/10.1007/978-3-662-46497-7_16

5. Blum, M., Feldman, P., Micali, S.: Non-interactive zero-knowledge and its applications (extended abstract). In: 20th ACM STOC, pp. 103–112. ACM Press, May 1988

6. Boyle, E., Couteau, G., Gilboa, N., Ishai, Y.: Compressing vector OLE. In: Lie, D., Mannan, M., Backes, M., Wang, X. (eds.) ACM CCS 2018, pp. 896–912. ACM Press, October 2018

7. Canetti, R., Chen, Y., Reyzin, L., Rothblum, R.D.: Fiat-Shamir and correlation intractability from strong KDM-secure encryption. In: Nielsen, J.B., Rijmen, V. (eds.) EUROCRYPT 2018. Part I. LNCS, vol. 10820, pp. 91–122. Springer, Cham (2018). https://doi.org/10.1007/978-3-319-78381-9_4

8. Canetti, R., Halevi, S., Katz, J.: A forward-secure public-key encryption scheme. In: Biham, E. (ed.) EUROCRYPT 2003. LNCS, vol. 2656, pp. 255–271. Springer, Heidelberg (2003). https://doi.org/10.1007/3-540-39200-9_16

9. Canetti, R., Lichtenberg, A.: Certifying trapdoor permutations, revisited. In: Beimel, A., Dziembowski, S. (eds.) TCC 2018. Part I. LNCS, vol. 11239, pp. 476–506. Springer, Cham (2018). https://doi.org/10.1007/978-3-030-03807-6_18

10. Cash, D., Kiltz, E., Shoup, V.: The twin Diffie-Hellman problem and applications. In: Smart, N. (ed.) EUROCRYPT 2008. LNCS, vol. 4965, pp. 127–145. Springer, Heidelberg (2008). https://doi.org/10.1007/978-3-540-78967-3_8

11. Chaidos, P., Couteau, G.: Efficient designated-verifier non-interactive zero-knowledge proofs of knowledge. In: Nielsen, J.B., Rijmen, V. (eds.) EUROCRYPT 2018. Part III. LNCS, vol. 10822, pp. 193–221. Springer, Cham (2018). https://doi.org/10.1007/978-3-319-78372-7_7

12. Couteau, G., Hofheinz, D.: Designated-verifier pseudorandom generators, and their applications. Cryptology ePrint Archive (2019)

13. Cramer, R., Shoup, V.: Universal hash proofs and a paradigm for adaptive chosen ciphertext secure public-key encryption. In: Knudsen, L.R. (ed.) EUROCRYPT 2002. LNCS, vol. 2332, pp. 45–64. Springer, Heidelberg (2002). https://doi.org/10.1007/3-540-46035-7_4

14. Damgård, I., Nielsen, J.B.: Perfect hiding and perfect binding universally composable commitment schemes with constant expansion factor. In: Yung, M. (ed.) CRYPTO 2002. LNCS, vol. 2442, pp. 581–596. Springer, Heidelberg (2002). https://doi.org/10.1007/3-540-45708-9_37

15. De Santis, A., Micali, S., Persiano, G.: Non-interactive zero-knowledge with pre-processing. In: Goldwasser, S. (ed.) CRYPTO 1988. LNCS, vol. 403, pp. 269–282. Springer, New York (1990). https://doi.org/10.1007/0-387-34799-2_21

16. Dwork, C., Naor, M.: Zaps and their applications. In: 41st FOCS, pp. 283–293. IEEE Computer Society Press, November 2000

17. Feige, U., Lapidot, D., Shamir, A.: Multiple non-interactive zero knowledge proofs based on a single random string (extended abstract). In: 31st FOCS, pp. 308–317. IEEE Computer Society Press, October 1990

18. Goldreich, O., Levin, L.A.: A hard-core predicate for all one-way functions. In: 21st ACM STOC, pp. 25–32. ACM Press, May 1989

19. Goldreich, O., Micali, S., Wigderson, A.: Proofs that yield nothing but their validity and a methodology of cryptographic protocol design (extended abstract). In: 27th FOCS, pp. 174–187. IEEE Computer Society Press, October 1986

20. Goldwasser, S., Micali, S., Rackoff, C.: The knowledge complexity of interactive proof systems. SIAM J. Comput. **18**(1), 186–208 (1989)
21. Goldwasser, S., Ostrovsky, R.: *Invariant* signatures and non-interactive zero-knowledge proofs are equivalent. In: Brickell, E.F. (ed.) CRYPTO 1992. LNCS, vol. 740, pp. 228–245. Springer, Heidelberg (1993). https://doi.org/10.1007/3-540-48071-4_16
22. Gorbunov, S., Vaikuntanathan, V., Wichs, D.: Leveled fully homomorphic signatures from standard lattices. In: Servedio, R.A., Rubinfeld, R. (eds.) 47th ACM STOC, pp. 469–477. ACM Press, June 2015
23. Goyal, R., Hohenberger, S., Koppula, V., Waters, B.: A generic approach to constructing and proving verifiable random functions. In: Kalai, Y., Reyzin, L. (eds.) TCC 2017. Part II. LNCS, vol. 10678, pp. 537–566. Springer, Cham (2017). https://doi.org/10.1007/978-3-319-70503-3_18
24. Groth, J., Ostrovsky, R., Sahai, A.: Non-interactive zaps and new techniques for NIZK. In: Dwork, C. (ed.) CRYPTO 2006. LNCS, vol. 4117, pp. 97–111. Springer, Heidelberg (2006). https://doi.org/10.1007/11818175_6
25. Groth, J., Ostrovsky, R., Sahai, A.: Perfect non-interactive zero knowledge for NP. In: Vaudenay, S. (ed.) EUROCRYPT 2006. LNCS, vol. 4004, pp. 339–358. Springer, Heidelberg (2006). https://doi.org/10.1007/11761679_21
26. Groth, J., Sahai, A.: Efficient non-interactive proof systems for bilinear groups. In: Smart, N. (ed.) EUROCRYPT 2008. LNCS, vol. 4965, pp. 415–432. Springer, Heidelberg (2008). https://doi.org/10.1007/978-3-540-78967-3_24
27. Katsumata, S., Nishimaki, R., Yamada, S., Yamakawa, T.: Designated verifier/prover and preprocessing NIZKs from Diffie-Hellman assumptions. In: Eurocrypt 2019 (2019)
28. Kilian, J., Petrank, E.: An efficient noninteractive zero-knowledge proof system for NP with general assumptions. J. Cryptol. **11**(1), 1–27 (1998)
29. Kim, S., Wu, D.J.: Multi-theorem preprocessing NIZKs from lattices. In: Shacham, H., Boldyreva, A. (eds.) CRYPTO 2018. Part II. LNCS, vol. 10992, pp. 733–765. Springer, Cham (2018). https://doi.org/10.1007/978-3-319-96881-0_25
30. Ong, S.J., Vadhan, S.P.: Zero knowledge and soundness are symmetric. In: Naor, M. (ed.) EUROCRYPT 2007. LNCS, vol. 4515, pp. 187–209. Springer, Heidelberg (2007). https://doi.org/10.1007/978-3-540-72540-4_11
31. Oren, Y.: On the cunning power of cheating verifiers: some observations about zero knowledge proofs (extended abstract). In: 28th FOCS, pp. 462–471. IEEE Computer Society Press, October 1987
32. Ostrovsky, R., Wigderson, A.: One-way functions are essential for non-trivial zero-knowledge. In: 1993 Proceedings of the 2nd Israel Symposium on the Theory and Computing Systems, pp. 3–17. IEEE (1993)
33. Pass, R., Shelat, A., Vaikuntanathan, V.: Construction of a non-malleable encryption scheme from any semantically secure one. In: Dwork, C. (ed.) CRYPTO 2006. LNCS, vol. 4117, pp. 271–289. Springer, Heidelberg (2006). https://doi.org/10.1007/11818175_16
34. Peikert, C., Vaikuntanathan, V.: Noninteractive statistical zero-knowledge proofs for lattice problems. In: Wagner, D. (ed.) CRYPTO 2008. LNCS, vol. 5157, pp. 536–553. Springer, Heidelberg (2008). https://doi.org/10.1007/978-3-540-85174-5_30
35. Quach, W., Rothblum, R.D., Wichs, D.: Reusable designated-verifier NIZKs forall NP from CDH. In: Eurocrypt 2019 (2019)

36. Regev, O.: On lattices, learning with errors, random linear codes, and cryptography. In: Gabow, H.N., Fagin, R. (eds.) 37th ACM STOC, pp. 84–93. ACM Press, May 2005
37. Rothblum, R.D., Sealfon, A., Sotiraki, K.: Towards non-interactive zero-knowledge for NP from LWE. In: PKC 2019 (2019)
38. Sahai, A., Waters, B.: How to use indistinguishability obfuscation: deniable encryption, and more. In: Shmoys, D.B. (ed.) 46th ACM STOC, pp. 475–484. ACM Press, May/June 2014
39. Vadhan, S.P.: An unconditional study of computational zero knowledge. In: 45th FOCS, pp. 176–185. IEEE Computer Society Press, October 2004

Reusable Designated-Verifier NIZKs for all NP from CDH

Willy Quach[1(✉)], Ron D. Rothblum[2], and Daniel Wichs[1]

[1] Northeastern University, Boston, USA
quach.w@husky.neu.edu, wichs@ccs.neu.edu
[2] Technion, Haifa, Israel
rothblum@cs.technion.ac.il

Abstract. Non-interactive zero-knowledge proofs (NIZKs) are a fundamental cryptographic primitive. Despite a long history of research, we only know how to construct NIZKs under a few select assumptions, such as the hardness of factoring or using bilinear maps. Notably, there are no known constructions based on either the computational or decisional Diffie-Hellman (CDH/DDH) assumption without relying on a bilinear map.

In this paper, we study a relaxation of NIZKs in the *designated verifier* setting (DV-NIZK), in which the public common-reference string is generated together with a secret key that is given to the verifier in order to verify proofs. In this setting, we distinguish between *one-time* and *reusable* schemes, depending on whether they can be used to prove only a single statement or arbitrarily many statements. For reusable schemes, the main difficulty is to ensure that soundness continues to hold even when the malicious prover learns whether various proofs are accepted or rejected by the verifier. One-time DV-NIZKs are known to exist for general NP statements assuming only public-key encryption. However, prior to this work, we did not have any construction of reusable DV-NIZKs for general NP statements from any assumption under which we didn't already also have standard NIZKs.

In this work, we construct reusable DV-NIZKs for general NP statements under the CDH assumption, without requiring a bilinear map. Our construction is based on the *hidden-bits paradigm*, which was previously used to construct standard NIZKs. We define a cryptographic primitive called a *hidden-bits generator (HBG)*, along with a designated-verifier variant (DV-HBG), which modularly abstract out how to use this paradigm to get both standard NIZKs and reusable DV-NIZKs. We construct a DV-HBG scheme under the CDH assumption by relying on techniques from the Cramer-Shoup hash-proof system, and this yields our reusable DV-NIZK for general NP statements under CDH.

We also consider a strengthening of DV-NIZKs to the *malicious designated-verifier* setting (MDV-NIZK) where the setup consists of an honestly generated common random string and the verifier then gets to choose his own (potentially malicious) public/secret key pair to generate/verify proofs. We construct MDV-NIZKs under the "one-more CDH" assumption without relying on bilinear maps.

© International Association for Cryptologic Research 2019
Y. Ishai and V. Rijmen (Eds.): EUROCRYPT 2019, LNCS 11477, pp. 593–621, 2019.
https://doi.org/10.1007/978-3-030-17656-3_21

1   Introduction

(Non-Interactive) Zero-Knowledge. Zero-knowledge proofs, introduced in the seminal work of Goldwasser, Micali, and Rackoff [GMR85, GMR89], allow a prover to convince a verifier that a statement is valid without revealing anything beyond its validity. Standard zero-knowledge proof systems are interactive. Blum, Feldman, and Micali [BFM88] introduced the concept of non-interactive zero-knowledge (NIZK) proofs, which consist of a single message from the prover to the verifier. Such NIZKs cannot exist in the plain model, and are therefore considered in the *common reference string (CRS)* model, where a trusted third party chooses some common string (either uniformly at random or from some designated distribution) which is given to both the prover and the verifier. Such NIZKs for general **NP** statements have been constructed from a few select assumptions such as: (doubly-enhanced) trapdoor permutations which can be instantiated from factoring [BFM88, DMP88, FLS99, Gol11], the Diffie-Hellman assumption over bilinear groups [CHK03, GOS06] indistinguishability obfuscation [SW14] or fully exponential KDM hardness [CCRR18]. We also have such NIZKs in the random-oracle model [FS87].[1] However, despite a long history of research, we don't have any constructions based on several common standard assumptions: most notably the computational or decisional Diffie-Hellman assumptions (CDH, DDH) without requiring a bilinear map, or the learning-with-errors (LWE) assumption.

Designated-Verifier NIZK. In this work, we focus on a relaxed notion of NIZKs in the *designated-verifier* setting (DV-NIZK). In this model a trusted-third party generates a CRS together with secret key which is given to the verifier and is used to verify whether proofs are accepting or rejecting. We distinguish between schemes having *one-time* (a.k.a. single-theorem) security versus *reusable* (a.k.a. multi-theorem) security. One-time secure schemes only guarantee soundness for a single proof of a single statement. However, since the verifier's decision whether to accept or reject a proof depends on the secret key, a malicious prover may be able to learn something about the secret key over time by producing many proofs and seeing whether they are accepted or rejected by the verifier. Reusable DV-NIZKs ensure that soundness continues to hold even in such settings, where a prover can test whether the verifier accepts or rejects various proofs. In terms of constructions, there appears to be a huge gap between these notions. One-time secure DV-NIZKs were constructed for general **NP** statements assuming only the existence of public-key encryption [PsV06]. On the other hand, prior to this work, we did not have any constructions of reusable DV-NIZKs for general **NP** statements based on any assumptions under which we don't already also have standard NIZKs.

---

[1] Additionally, we have constructions of NIZKs with an inefficient prover based on one-way permutations [FLS99]. In this work, we restrict ourselves to NIZKs where the prover can generate proofs efficiently given an **NP** witness.

Malicious-Designated-Verifier NIZK. We also consider a strengthening of (reusable) DV-NIZKs to the malicious-designated-verifier setting (MDV-NIZKs). In this setting, the trusted party only generates a common uniformly random string. The verifier then gets to choose a public/secret key pair where the public key is used by the prover to generate proofs and the secret key is used by the verifier to verify proofs. The main difference between DV-NIZKs and MDV-NIZKs is that, in the latter, we require zero-knowledge to hold even if the public key is chosen maliciously by the verifier. Therefore, an MDV-NIZK is similar to standard NIZKs in that the only *trusted setup* consists of a common random string, but an MDV-NIZK also requires additional *potentially untrusted setup* where the verifier publishes a public-key for which it keeps the corresponding secret key.

The notion of (reusable) MDV-NIZKs is equivalent to 2-round malicious-verifier ZK protocols in the common random string model (where the verifier's first-round message is reusable) by thinking of the verifier's public key as the first-round message. It is easy to see that the construction of *non-reusable* DV-NIZKs of [PsV06] extends naturally to yield *non-reusable* MDV-NIZKs assuming 2-round maliciously secure oblivious transfer in the common random string model. However, prior to this work, we did not have any constructions of *reusable* MDV-NIZKs for general **NP** statements based on any assumptions under which we don't already also have standard NIZKs.

Prior Work on DV-NIZKs and NIZKs with Pre-processing. In prior work, the notion of DV-NIZKs was mainly studied in the context of non-malleable and CCA secure encryption. It is known that one-time DV-NIZKs allow us to compile any CPA secure (public-key) encryption scheme into a non-malleable one [PsV06] and reusable DV-NIZKs can compile it into a CCA secure one (by adapting the [NY90, DDN91] paradigm to the designated-verifier case). In this context, the work of Cramer and Shoup [CS98, CS02] constructed "hash-proof systems" which are unconditionally secure reusable DV-NIZKs for specific "algebraic" languages (e.g., the equality of two discrete logarithms) and used them to get practical CCA secure encryption. However, reusable DV-NIZKs have received surprisingly little attention as a general primitive. We believe that this notion is naturally interesting beyond its applications to non-malleable and CCA encryption. For example, it can take the place of standard NIZKs in the context of multiparty computation in scenarios where there is some (reusable) trusted setup.

DV-NIZKs can be thought of as a special case of a more general notion of "NIZKs with preprocessing" in which a trusted-third party creates a CRS together with two secret key: td_V given to the verifier and td_P given to the prover. We can consider two special cases of such NIZKs with preprocessing: if td_P is empty then this corresponds to the "designated-verifier" (DV-NIZK) setting that we study in this work, and if td_V is empty then we can think of this as a "designated-prover" (DP-NIZK). Several prior works study NIZKs with preprocessing [DMP90, KMO90, LS91, Dam93, DFN06, CC18] but all either (1) only consider specific "algebraic" languages rather than general **NP**, (2) are not reusable or (3) require assumptions such as factoring from which we already have standard NIZKs. The one exception is a very recent work of Kim and Wu [KW18] (CRYPTO 2018), which gave a novel construction of reusable DP-NIZKs for general **NP** languages under the LWE assumption. In that work, they explicitly asked the question whether one can construct reusable NIZKs in the preprocessing model under the CDH/DDH assumption. We answer their open question positively in this paper by constructing reusable DV-NIZKs under CDH. It remains a fascinating open question whether one can construct reusable DV-NIZKs under LWE, and conversely, whether one can construct reusable DP-NIZKs under CDH/DDH.

1.1   Our Results

In this work, we construct reusable DV-NIZKs for general **NP** languages under the computational Diffie-Hellman (CDH) assumption without requiring a bilinear map.

Theorem 1.1. *Under the CDH assumption, there exists an (adaptively secure, statistically sound) reusable DV-NIZK proof system for all NP.*

We also construct reusable MDV-NIZKs for general **NP** languages under the one-more CDH (OM-CDH) assumption without requiring a bilinear map.

Theorem 1.2. *Under the One-More CDH assumption (Definition 6.3), there exists an (adaptively secure, statistically sound) reusable MDV-NIZK proof system for all NP.*

Our construction goes through the *hidden-bits* paradigm introduced by Feige, Lapidot and Shamir [FLS99] (see also [Gol01, Gol11]) to construct standard NIZKs. This paradigm consists of two steps. First, construct a NIZK for general **NP** statements in an idealized model called the "hidden-bits model" where the prover is given a long string of uniformly random bits and can choose to reveal some subset of them to the verifier. Such NIZKs in the hidden-bits model were constructed unconditionally with statistical soundness and zero knowledge. Second, use a cryptographic tool to compile NIZKs in the hidden-bits model to NIZKs in the CRS model. Such a compiler was constructed concretely using (doubly enhanced) trapdoor permutations, which can be instantiated based on factoring.

We generalize the second step of the hidden bits paradigm by defining a cryptographic primitive called a "hidden-bits generator" (HBG) which can be used to compile NIZKs in the hidden-bits model into ones in the CRS model.[2] This primitive modularizes the "hidden-bits paradigm" and simplifies the task of constructing NIZKs by reducing it to the task of constructing a HBG. We also clarify how to use HBG to get adaptive ZK security via the "hidden bits

---

[2] A similar primitive called a "verifiable pseudorandom generator" was defined by [DN00] for the purpose of constructing ZAPs, which also lead to a construction of NIZKs.

paradigm", which turns out to be surprisingly subtle and was not very clear from prior presentations of this paradigm. To get our main result, we generalize the hidden bits paradigm even further by extending the notion of HBG to the designated-verifier setting (DV-HBG) and the malicious-designated-verifier setting (MDV-HBG) and showing that the same compiler allows us to go from DV-HBG (resp. MDV-HBG) to reusable DV-NIZKs (resp. MDV-NIZKs). We then show how to construct DV-HBG from the computational Diffie-Hellman (CDH) assumption without bilinear maps. The last step uses the Cramer-Shoup hash-proof system, which can be thought of as a reusable DV-NIZK for equality of two discrete logarithms. Therefore we are in some sense bootstrapping a reusable DV-NIZK for this specific language to get a reusable DV-NIZK for all of **NP**. Finally, we show how to construct MDV-HBG from the one-more CDH (OM-CDH) assumption. This essentially starts with our construction of DV-HBG, which is clearly insecure in the malicious-designated-verifier setting, and shows how to immunize it against malicious attacks. While the high level idea is simple, the proof of security is quite involved and uses techniques which may be of independent interest.

1.2 Technical Overview

NIZKs via the Hidden-Bits Paradigm. We first review the "hidden-bits paradigm" proposed by [FLS99]; see [Gol01,Gol11] for a modern presentation which we follow here.

The starting point of this paradigm is a construction of NIZKs in an idealized model called the "hidden-bits model". In this model, there is a trusted third party that generates uniformly random bits $r_1, \ldots, r_k$ and gives them to the prover. The prover outputs a proof π along with a subset $I \subseteq [k]$ of the bits to open. The verifier gets (I, π) from the prover together with the bits $\{r_i\}_{i \in I}$ from the trusted third party. Note that the verifier does not learn anything about the unopened bits $\{r_i\}_{i \notin I}$ and the prover cannot modify the values of the opened bits $\{r_i\}_{i \in I}$. Such NIZKs in the hidden-bits model can be constructed unconditionally with security against an unbounded prover/verifier where the soundness error can be made exponentially small.

The second step compiles NIZKs in the hidden-bits model into NIZKs in the CRS model. Such a compiler was presented by [FLS99,Gol01,Gol11] using doubly-enhanced trapdoor permutations (TDPs) (see also [BY93,GR13,CL17]). On a high level, the CRS consists of random values $y_1, \ldots, y_k$ in the range of the TDP. The prover chooses a random permutation f_{com} along with an inversion trapdoor sk and inverts all of the values in the CRS to get preimages $x_1, \ldots, x_k$. Define $r_1, \ldots, r_k$ to be hardcore bits of $x_1, \ldots, x_k$. The prover then runs the hidden-bits prover with $r_1, \ldots, r_k$ to generate some proof (π, I) to which it appends the values $\mathsf{com}, \{x_i\}_{i \in I}$. The verifier checks $y_i = f_{\mathsf{com}}(x_i)$, computes $\{r_i\}_{i \in I}$ to be the hardcore bit of x_i and then runs the hidden bits verifier on (π, I). Intuitively, a malicious prover has a extremely limited ability to control the randomness $r_1, \ldots, r_k$ by choosing com; by relying on an exponentially small soundness error of the hidden-bits proofs which survives a union-bound over all such com's, this flexibility is insufficient to break soundness. On the other hand, the verifier does not learn anything about the values $\{r_i\}_{i \notin I}$ by the security of the TDP.[3] While this is the high level approach, there are some subtleties involved; see [BY93, Gol01, Gol04, Gol11, GR13, CL17].

Hidden-Bits Generator (HBG). We begin by defining an abstract cryptographic primitive, which we call a hidden-bits generator (HBG), that can be used to compile NIZKs in the hidden-bits model to NIZKs in the CRS model. An HBG that generates k bits consists of three algorithms:

- Setup creates a crs.
- GenBits(crs) outputs a *short* commitment com whose size is much smaller than k, along with hidden-bits $\{r_i\}_{i \in [k]}$, and certificates $\{\pi_i\}_{i \in [k]}$.
- Verify(crs, com, i, r_i, π_i) checks the certificate π_i to verify that r_i is indeed the i'th hidden bit.

An HBG should satisfy two simple properties. Firstly, we require the scheme to be *statistically binding*, meaning that (crs, com) together completely determine some sequence of bits $r_1, \ldots, r_k$ and no (even inefficient) prover can come up with a valid certificate π_i' for the wrong bit $r_i' \neq r_i$. Intuitively, by combining the above property together with the requirement that com is short, we ensure that the prover does not have much control over the bits r_i that he can open and the limited control that he does have is insufficient to break the soundness of the hidden-bits NIZK (by amplifying its soundness sufficiently so that it survives a union bound over all the com's that the prover can choose). Secondly, we require the scheme to be *computationally hiding*, meaning that for any set $I \subseteq [k]$, if we are given honestly generated crs, com, $\{r_i, \pi_i\}_{i \in I}$ then the "unopened" hidden bits $\{r_j\}_{j \notin I}$ are computationally indistinguishable from uniform.

Compiling from Hidden-Bits Model to CRS Model. Intuitively, we would like to use HBG to compile NIZKs from the hidden-bits model to the CRS model by letting the prover generate the hidden-bits via the HBG GenBits algorithm. There are two issues with this basic approach:

- For soundness, if the malicious prover chooses a "bad" (not uniformly random) com then the HBG abstraction does not provide any guarantees that the bits r_i to which he is committed are random and hence we cannot rely on the soundness of the hidden-bits NIZK.
- For zero-knowledge, we notice that the honest hidden-bits prover may choose the set I adaptively depending on all of the bits $\{r_i\}_{i \in [k]}$ (and indeed this is the case for the hidden-bits NIZK constructed in [FLS99]) and we still need to argue that the unopened bits $\{r_j\}_{j \notin I}$ are hidden. The hiding property of HBG only guarantees that the unopened bits are hidden when I is chosen ahead of time.

---

[3] The basic compiler only achieves zero-knowledge for a single theorem and [FLS99] then relies on another generic compiler via the "or trick" to go from single-theorem to multi-theorem zero-knowledge.

To fix both of the above issues we add additional uniformly random bits $s_1, \ldots, s_k$ to the CRS of the NIZK and define the hidden-bits to be $r_i \oplus s_i$ where r_i comes from the HBG. This ensures that for any fixed com chosen by a malicious prover the hidden-bits that he can open are uniform over the choice of s_i.[4] It also ensures that the choice of the set I chosen by the honest hidden-bits prover is independent of the outputs r_i of the HBG and therefore allows us to rely on HBG security.

We uncover an additional complication when proving *adaptive* ZK, where the malicious verifier can choose the statement to be proven adaptively depending on the CRS. The work of [FLS99] showed adaptive ZK for their particular protocol (using particular hidden-bits NIZK) but it did not give a modular proof. Indeed, our attempts to prove that the compiler can generically start with any hidden-bits NIZK and achieve adaptive ZK failed for subtle reasons involving "selective opening" failures. Instead, we were able to abstract out a special property of the hidden-bits NIZK of [FLS99] which we call "special ZK", which we show to be sufficient to get adaptive ZK in the CRS model via the above compiler.

Using the compiler, we reduce the task of constructing NIZKs to that of constructing an HBG, which is a conceptually much simpler primitive.

Designated Verifier Setting: (M)DV-HBG to (M)DV-NIZK. We generalize the notion of HBG to the designated-verifier setting (DV-HBG). The only differences are that: (1) the Setup algorithm generates a crs together with a trapdoor td which is given to the verifier and the Verify algorithm takes the trapdoor td as an input, (2) we modify the statistically binding security property to hold even if a computationally unbounded prover can make polynomially many queries to the Verify(crs, td, $\cdots$) oracle which allows it to check whether various certificates are valid or invalid, and (3) we modify the computationally hiding property to hold even given td. To get our main result, we naturally extend our compiler to show that DV-HBG allows us to compile NIZKs in the hidden-bits model into reusable DV-NIZKs. Therefore, we reduce the task of constructing reusable DV-NIZKs to that of constructing DV-HBG.

We further generalize the notion of HBG to the malicious-designated-verifier setting (MDV-HBG). Now, in addition to a Setup algorithm that generates the crs there is a KeyGen algorithm that generates a public key pk along an associated secret key sk. Essentially, we think of crs, pk as together corresponding to the crs in the previous definition, and of sk as the trapdoor. The binding property is essentially the same as before. However, we require that hiding holds even if pk is generated maliciously (and adaptively depending on crs). We show that MDV-HBG allows us to compile NIZKs in the hidden-bits model into reusable MDV-NIZKs. Therefore, we reduce the task of constructing reusable MDV-NIZKs to that of constructing MDV-HBG.

DV-HBG from CDH. We show how to instantiate a designated-verifier DV-HBG based on the computational Diffie-Hellman (CDH) assumption to get our

---

[4] The fact that the prover can adaptively choose com *after* seeing $s_1, \ldots, s_k$ is handled by simply taking a union bound over all possible choices of com.

reusable DV-NIZK from CDH. Our construction relies on the ideas underlying the Cramer-Shoup (1-universal) hash-proof system [CS98, CS02] which can be thought of as an unconditionally secure reusable DV-NIZK for the "equality of two discrete logs" – i.e., given some public group elements g, h we define the language consisting of tuples (g', h') such that $\mathsf{DLOG}_g(g') = \mathsf{DLOG}_h(h')$. In particular, we think of the projection key of the hash-proof system as the CRS of the DV-NIZK, and the hashing key as the associated trapdoor. In the body of our paper, we give our full construction using the specific Cramer-Shoup instantiation, but for the introduction we will treat the Cramer-Shoup reusable DV-NIZK proof system as a black-box.

Our DV-HBG construction works as follows. Let $\mathbb{G}$ be some cyclic group of order p and let g be a generator.

- The Setup algorithm chooses random group elements $h_1, \ldots, h_k$. It also instantiates k copies of the Cramer-Shoup DV-NIZK with respect to the public group elements (g, h_i) respectively. The crs consists of $g, h_1, \ldots, h_k$ together with the k values $\{\mathsf{crs}_i\}_{i \in [k]}$ of the Cramer-Shoup DV-NIZK. The trapdoor $\mathsf{td} = \{\mathsf{td}_i\}_{i \in [k]}$ consists of the k trapdoors for the Cramer-Shoup DV-NIZK.
- The GenBits(crs) algorithm chooses $y \leftarrow \mathbb{Z}_q$ and sets $\mathsf{com} = g^y$. For $i = 1 \ldots, k$, it sets $t_i = h_i^y$, $r_i = \mathsf{hc}(t_i)$, where hc is a hardcore predicate (e.g., Goldreich-Levin [GL89]). Finally it sets $\pi_i = (t_i, \pi_i^{CS})$ where π_i^{CS} is a Cramer-Shoup proof that $\mathsf{DLOG}_g(\mathsf{com}) = \mathsf{DLOG}_{h_i}(t_i)$.
- The Verify algorithm gets r_i and $\pi_i = (t_i, \pi_i^{CS})$ and checks that $r_i = \mathsf{hc}(t_i)$ and that π_i^{CS} is a valid Cramer-Shoup proof using the corresponding trapdoor td_i.

For the statistically binding property we note that given crs, com the values $t_i = h_i^y$ and therefore also the hidden bits $r_i = \mathsf{hc}(t_i)$ are completely determined. The prover cannot lie about t_i and therefore also about r_i by the unconditional reusable security of the Cramer-Shoup proof, and this holds even given oracle access to the Cramer-Shoup verifier. For the computational hiding property we rely on the fact that, given g, h_i, g^y, the CDH assumption ensures that h_i^y is computationally unpredictable and therefore $\mathsf{hc}(h_i^y)$ is indistinguishable from uniform. This holds even given h_j, h_j^y for various random h_j since the distinguisher can sample such values himself by sampling $h_j = g^{x_j}$ and computing $h_j^y = (g^y)^{x_j}$.

MDV-HBG from One-More CDH. Finally, we show how to instantiate our malicious-designated-verifier MDV-HBG based on the one-more CDH assumption to get our reusable MDV-NIZK from one-more CDH. The construction and the security intuition are somewhat involved and so we present them in several stages.

Initial Attempt. As a first attempt, we can try to use the previous construction directly as an MDV-HBG. In particular, we can set the crs to only consist of the uniformly random values $\mathsf{crs} = (h_1, \ldots, h_k)$. The Cramer-Shoup DV-NIZKs then naturally define pk, sk. Here it helps to be concrete about how the Cramer-Shoup

DV-NIZK works. For each i, the Cramer-Shoup proof system defines $\mathsf{pk}_i = h_i^{a_i} g^{b_i}$ and the corresponding $\mathsf{sk}_i = (a_i, b_i)$. The MDV-HBG public keys and secret keys consist of these values $\mathsf{pk} = \{\mathsf{pk}_i\}, \mathsf{sk} = \{\mathsf{sk}_i\}$. Given a commitment $\mathsf{com} = g^y$, recall that the i'th hidden bit is defined by taking a hardcore predicate $r_i = \mathsf{hc}(t_i)$ where $t_i = h_i^y$. The opening to the i'th hidden bit consists of $t_i = h_i^y$ and the Cramer-Shoup proof $\pi_i^{CS} = \mathsf{pk}_i^y$.

Attack on Initial Attempt. Unfortunately, it's clear that the above is not secure as MDV-HBG. For example, if the malicious verifier chooses $\mathsf{pk}_i = h_j$ for $j \neq i$ then, by opening the i'th hidden bit and giving a proof $\pi_i^{CS} = \mathsf{pk}_i^y = h_j^y$, the prover inadvertently also reveals the j'th hidden bit! While the above is easily detectable, the malicious verifier can alternately set $\mathsf{pk}_i = h_j^x$ for a random x and still perform the same attack without being detectable. At the very least, we need to modify our solution to overcome this particular attack.

The Fix. We start with the above "base scheme", which is not secure in the MDV setting, and show how to immunize against the above attack. To do so, we use the "base scheme" to generate ℓ "base hidden values" for some $\ell \gg k$ and then combine them carefully to create the k "actual hidden bits". Recall that the base scheme defines a commitment g^y and the ℓ base hidden values are $t_j = h_j^y$. We can open any base value by giving the opening $\pi_j^{CS} = \mathsf{pk}_j^y$.

Instead of using the base values directly, we define each of the k "actual hidden bits" by combining together a small group of base values and applying a (Goldreich-Levin) hard-core predicate hc. The groups are chosen via a pseudo-random mapping φ which maps each $i \in [k]$ to a small group $\varphi(i) \subseteq [\ell]$. In other words, the i'th actual hidden bit is defined as $r_i = \mathsf{hc}(\{t_j \ : \ j \in \varphi(i)\})$. The mapping φ is chosen by the prover and is a part of com. To open any actual hidden bit $i \in [k]$ the prover opens all of the base hidden values t_j^y and also provides the corresponding Cramer-Shoup proofs pk_j^y for $j \in \varphi(i)$. Note that, since φ is a part of com and we require com to be short, it is important that φ has a short description size and therefore it must be a pseudo-random rather than truly random mapping. For concreteness, we set the number of based hidden values to $\ell = 3k\lambda$ and the group size to $|\varphi(i)| = \lambda$, where λ is the security parameter.

Intuition for the Fix. Intuitively, this prevents the above attacks for the following reason. Assume that the verifier can choose pk maliciously so that the opening of any base value j can inadvertently also reveal some other base value $j' = \psi(j)$, where ψ is some mapping defined implicitly by the choice of pk. Nevertheless, it is likely that each hidden bit i depends on some hidden value $j \in \varphi(i)$ that is not revealed even if we open all the other hidden bits $i' \neq i$. In particular, opening the bits $i' \neq i$ corresponds to giving out the base hidden value j' as well as the inadvertently opened values $\psi(j')$ for each $j' \in \varphi(i')$. But the entire set of revealed values $R = \{j', \psi(j') \ : \ j' \in \varphi(i'), i' \neq i\}$ is of size $|R| \leq 2k\lambda$ and $\varphi(i) \subset [\ell = 3k\lambda]$ appears to be a random and independent subset of size $|\varphi(i)| = \lambda$. Hence it is likely that $\varphi(i)$ contains some value $j \notin R$ which was not revealed. Here we crucially rely on the fact that φ is chosen (pseudo-)randomly by the prover after the verifier chooses pk which defines the mapping ψ.

The One-More CDH Assumption. While the above idea seems to immunize against the particular class of attacks we previously discussed, proving security against general attacks is more challenging. Nevertheless, we manage to do so under the "one-more CDH" assumption. The one-more CDH assumption considers an adversary who is given $g, g^y, h_1, \ldots, h_k$ along with an oracle $O_y(\cdot)$ which takes as input an arbitrary group element f and returns $O_y(f) = f^y$. It says that even if the adversary makes m arbitrary calls to the oracle O_y he cannot predict more than m of the values $\{h_j^y\}$.

Security Under One-More CDH. Our high level proof goes as follows. Assume that a malicious verifier gets to choose $\mathsf{pk} = \{\mathsf{pk}_j\}_{j \in [\ell]}$ maliciously after seeing $\mathsf{crs} = \{h_j\}_{j \in [\ell]}$ and can break hiding. This means that for some $i \in [k]$, if the verifier gets a random com and openings to all the hidden bits *except* for the i'th one, he can distinguish hidden bit i from uniform with non-negligible advantage. Since the i'th hidden bit is defined by taking the Goldreich-Levin hardcore bit of the base hidden values $j \in \varphi(i)$, this means that the verifier can also predict all these values with non-negligible probability. So, if the verifier gets $\varphi, g, g^y, \{h_j^y, \mathsf{pk}_j^y : j \in \varphi(i'), i' \in [k] \setminus \{i\}\}$ then he can predict $\{h_j^y : j \in \varphi(i)\}$. Intuitively, we want to use such a verifier to break one-more CDH.

But in the above scenario, the verifier gets many more values raised to the y power than he is able to output. To get around this, we want to "rewind" the verifier run him on many different choices of φ to get more values $\{h_j^y : j \in \varphi(i)\}$ out of him. But each time we rewind we also need to provide him with the appropriate values $\{h_j^y, \mathsf{pk}_j^y : j \in \varphi(i'), i' \in [k] \setminus \{i\}\}$ so we are again getting fewer powers of y out than we need to put in, which appears to be self-defeating. If φ were truly random, we could get around this by freshly sample $\varphi(i)$ on each rewinding but keep $\varphi(i') : i' \in [k] \setminus i$ fixed – that way we would only need to give out some fixed $2k\lambda$ values $\{h_j^y, \mathsf{pk}_j^y : j \in \varphi(i'), i' \in [k] \setminus \{i\}\}$ but on each rewinding we get some additional fresh values $\{h_j^y : j \in \varphi(i)\}$ out of the verifier, and eventually we get more out than we put in which allows us to break one-more CDH.

Unfortunately φ needs to have a short description, and therefore can only be pseudorandom, in which case it's not clear how to freshly re-sample $\varphi(i)$ while keeping $\varphi(i') : i' \in [k] \setminus i$ fixed. We resolve this issue by using a special form of pseudorandom functions (PRFs) called "somewhere equivocal PRFs" [HJO+16] which essentially allow us to do exactly this while keeping the description of φ short. Furthermore, such somewhere equivocal PRFs were constructed from only one-way functions using the ideas of "distributed point functions" [GI14, BGI15] and therefore don't introduce any additional assumptions.

1.3   Concurrent Works

Concurrently and independently of ours, the works of [CH19] and [KNYY19] present a similar construction of reusable DV-NIZKs from CDH, compiling the hidden-bits NIZK of [FLS99] using the Cramer-Shoup hash-proof system [CS98, CS02, CKS08]. Additionally, they respectively obtain the following results:

- [CH19] gives a construction of NIZKs for all NP assuming LWE, along with a *non-interactive witness intistinguishable* (NIWI) proof for the Bounded Distance Decoding problem.
- [KNYY19] builds pre-processing NIZKs for all NP with *succinct* proofs, namely a pre-processing NIZK from DDH with proofs of size $|C| + \mathsf{poly}(\lambda)$ (where C is a circuit checking the NP relation), and a designated-prover NIZK from (strong) assumptions over pairing-friendly groups, with proof size $|C| + \mathsf{poly}(\lambda)$.

Meanwhile, our work introduces the notion of *malicious designated-verifier* NIZKs (MDV-NIZK), and presents a construction from the One-More CDH assumption.

Organization

Basic definitions and notations are given in Sect. 2. In Sect. 3 we introduce our new notion of Hidden Bits Generator (HBG). In Sect. 4 we show how to use an HBG to construct NIZKs. In Sect. 5 we construct a designated-verifier Hidden Bits Generator assuming CDH. A few extension are mentioned in Sect. 7. In the full version of the paper, we additionally give a construction of a HBG from the CDH assumption over bilinear groups and we construct a HBG from (doubly-enhanced) trapdoor permutations.

2   Preliminaries

We will denote by λ the security parameter. The notation $\mathsf{negl}(\lambda)$ denotes any function f such that $f(\lambda) = \lambda^{-\omega(1)}$, and $\mathsf{poly}(\lambda)$ denotes any function f such that $f(\lambda) = \mathcal{O}(\lambda^c)$ for some $c > 0$.

We define the statistical distance between two random variables X and Y over some domain Ω as: $\mathbf{SD}(X, Y) = \frac{1}{2} \sum_{w \in \Omega} |X(w) - Y(w)|$. We say that two ensembles of random variables $X = \{X_\lambda\}$, $Y = \{Y_\lambda\}$ are *statistically indistinguishable*, denoted $X \overset{s}{\approx} Y$, if $\mathbf{SD}(X_\lambda, Y_\lambda) \leq \mathsf{negl}(\lambda)$.

We say that two ensembles of random variables $X = \{X_\lambda\}$, and $Y = \{Y_\lambda\}$ are *computationally indistinguishable*, denoted $X \overset{c}{\approx} Y$, if, for all (non-uniform) PPT distinguishers Adv, we have $|\Pr[\mathsf{Adv}(X_\lambda) = 1] - \Pr[\mathsf{Adv}(Y_\lambda) = 1]| \leq \mathsf{negl}(\lambda)$.

For a set X, integer k and sequence $x \in X^k$, we denote by x_i the i-th entry in the sequence, for any $i \in [k]$. For a subset $I \subset [k]$, we denote by $x_I = (x_i)_{i \in I}$ the subsequence of x in locations I.

For a probabilistic algorithm $\mathsf{alg}(\cdot)$, we may explicit its internal randomness as follows: $\mathsf{alg}(\,\cdot\,; \mathsf{coins})$.

2.1   The Diffie-Hellman Assumption

A *group generator* $(\mathbb{G}, p, g) \leftarrow \mathsf{GroupGen}(1^\lambda)$ is a PPT algorithm which, on input 1^λ, outputs the description of a cyclic group $\mathbb{G}$ of order p, and a generator g of $\mathbb{G}$. We require that there are efficient algorithms running in time $\mathsf{poly}(\lambda)$ to perform the group operation in $\mathbb{G}$ and to test membership in $\mathbb{G}$. For notational simplicity, we will often shorten such an output $(\mathbb{G}, p, g)$ to $\mathbb{G}$ and assume that g, p are implicit. A *prime-order group generator* additionally ensures that p is prime.

Definition 2.1 (Computational Diffie-Hellman (CDH) assumption). *Let* GroupGen *be a group generator. We say that the* Computational Diffie-Hellman (CDH) assumption *holds relative to* GroupGen *if for all PPT algorithm* $\mathcal{A}$, *we have:*

$$\Pr\left[\mathcal{A}\left(\mathbb{G}, p, g, g^a, g^b\right) = g^{ab} \; : \; (\mathbb{G}, p, g) \leftarrow \mathsf{GroupGen}(1^\lambda), (a, b) \xleftarrow{\$} \mathbb{Z}_p^2\right] \leq \mathsf{negl}(\lambda).$$

Given such a group generator satisfying the CDH assumption, we can consider an associated (randomized) *hard-core bit* $\mathsf{hc} : \mathbb{G} \to \{0, 1\}$ such that for all PPT algorithm $\mathcal{A}$, we have:

$$\Pr\left[\mathcal{A}\left(\mathbb{G}, p, g, g^a, g^b, \tau\right) = \mathsf{hc}(g^{ab} \; ; \; \tau) \; : \; \begin{array}{c} \tau \xleftarrow{\$} \{0, 1\}^{L(\lambda)} \\ (\mathbb{G}, p, g) \leftarrow \mathsf{GroupGen}(1^\lambda) \\ (a, b) \xleftarrow{\$} \mathbb{Z}_p^2 \end{array}\right] \leq 1/2 + \mathsf{negl}(\lambda),$$

where the hard-core bit hc uses $L(\lambda)$ random coins.

Such a hard-core bit can be generically obtained, using the Goldreich-Levin construction [GL89].

2.2 Reusable Designated-Verifier NIZKs

In this section we define the notion of Reusable Designated-Verifier NIZKs (and obtain the standard notion of NIZK as a special case).

Definition 2.2 (Reusable DV-NIZKs). *Let be L an NP language with witness relation R_L. A Reusable Designated-Verifier Non-Interactive Zero-Knowledge (DV-NIZK) Proof for L is a tuple of PPT algorithms* (Setup, $\mathcal{P}, \mathcal{V}$) *where:*

- Setup($1^\lambda, 1^n$): *On input the security parameter λ and statement length n, outputs a common reference string* crs *and a trapdoor* td;
- $\mathcal{P}$(crs, x, w): *On input a common reference string* crs, *a statement x of length n and a witness w, outputs a proof π;*
- $\mathcal{V}$(crs, td, x, π): *On input a common reference string* crs, *a trapdoor* td, *a statement x and a proof π, outputs* accept *or* reject,

such that they satisfy the following properties:

- **Completeness:** *We require that for all $(x, w) \in R_L$, we have:*

$$\Pr\left[\mathcal{V}(\mathsf{crs}, \mathsf{td}, x, \pi) = \mathtt{accept} \; : \; \begin{array}{c} (\mathsf{crs}, \mathsf{td}) \leftarrow \mathsf{Setup}(1^\lambda, 1^{|x|}) \\ \pi \quad \leftarrow \mathcal{P}(\mathsf{crs}, x, w) \end{array}\right] = 1;$$

- **Statistical Soundness:** *Let n and Q be any polynomials, and let $\widetilde{\mathcal{P}}$ be any (computationally unbounded) cheating prover that makes at most $Q(\lambda)$ queries to an oracle $\mathcal{V}(\mathsf{crs},\mathsf{td},\cdot,\cdot)$ which takes as input (x,π), and outputs $\mathcal{V}(\mathsf{crs},\mathsf{td},x,\pi)$). We require that:*

$$\Pr\left[\mathcal{V}(\mathsf{crs},\mathsf{td},x,\pi) = \mathsf{accept} \wedge x \notin L \quad : \quad \begin{array}{l} (\mathsf{crs},\mathsf{td}) \leftarrow \mathsf{Setup}(1^\lambda, 1^{n(\lambda)}) \\ (x,\pi) \leftarrow \widetilde{\mathcal{P}}^{\,\mathcal{V}(\mathsf{crs},\mathsf{td},\cdot,\cdot)}(\mathsf{crs}) \end{array}\right] \leq \mathsf{negl}(\lambda);$$

- **Zero-Knowledge (Selective):** *We require that there exists a PPT simulator Sim such that for any PPT stateful[5] adversary $\mathcal{A}$, the two following distributions are computationally indistinguishable:*

$\mathrm{EXP}^{Real}(1^\lambda):$	$\mathrm{EXP}^{Ideal}(1^\lambda):$
$(x,w) \leftarrow \mathcal{A}(1^\lambda)$	$(x,w) \leftarrow \mathcal{A}(1^\lambda)$
where $(x,w) \in R_L$	*where* $(x,w) \in R_L$
$(\mathsf{crs},\mathsf{td}) \leftarrow \mathsf{Setup}(1^\lambda, 1^{\lvert x \rvert}),\ \pi \leftarrow \mathcal{P}(\mathsf{crs},x,w)$	$(\mathsf{crs},\mathsf{td},\pi) \leftarrow \mathsf{Sim}(1^\lambda, x)$
Output $\mathcal{A}(\mathsf{crs},\mathsf{td},\pi)$	*Output* $\mathcal{A}(\mathsf{crs},\mathsf{td},\pi)$

Our basic definition only considers selective ZK where the statement being proven is chosen ahead of time, prior to seeing the CRS. In Sect. 4.1 we also consider a stronger notion of adaptive ZK.

Our definition of designated-verifier NIZK coincides with that of standard (publicly verifiable) NIZK if the trapdoor td is empty.

Definition 2.3. *A* publicly-verifiable NIZK *is a* reusable designated-verifier NIZK *where the trapdoor td output by Setup is an empty string.*

Remark 2.4 (Bounding the number of queries to the Verify oracle). Notice that for soundness we only allow the unbounded cheating prover to make a *polynomial* number of queries to $\mathcal{V}(\mathsf{crs},\mathsf{td},\cdot,\cdot)$. One would ideally allow the unbounded cheating prover to make *arbitrarily* many queries to $\mathcal{V}$ (matching more closely the publicly-verifiable setting, where a cheating prover can indeed query the verification algorithm on arbitrarily many inputs). It turns out that any DV-NIZK satisfying this stronger notion can be generically turned into a publicly-verifiable one. This is because the cheating prover can query all possible proofs to $\mathcal{V}$ for any $x \notin L$; and therefore soundness can only hold if there are no valid proof of any false statement (with overwhelming probability over the choice of crs), in which case soundness also holds when the prover is given the trapdoor. Therefore this is essentially the best requirement one can hope for as a meaningful notion of reusable DV-NIZKs which is weaker than publicly-verifiable ones.

---

[5] Throughout this paper we follow the convention that whenever a stateful adversary $\mathcal{A}$ is invoked with some inputs it also produces some state which it gets as input on the next invocation.

Remark 2.5 (Single-Theorem vs. Multi-Theorem Zero-Knowledge). The definition of ZK above is often referred to as "single-theorem ZK" since it only requires
zero-knowledge to hold for a *single* statement. However, there is a generic compiler from single-theorem ZK to multi-theorem ZK where zero-knowledge holds
polynomially many statements via the "OR trick" [FLS99]. We note that the
very same transformation directly applies to both the selective and adaptive ZK
setting and also both the publicly-verifiable and the designated-verifier setting.

2.3   NIZKs in the Hidden-Bits Model

We now recall the definition of a NIZK in the hidden-bits model:

Definition 2.6 (NIZK in the Hidden-Bits Model). *Let L be an NP language and n be an integer. A* Non-Interactive Zero-Knowledge Proof in the
Hidden-Bits Model *for L is given by a pair of PPT algorithms $(\mathcal{P}, \mathcal{V})$, and a
polynomial $k(\lambda, n)$, where:*

- *$\mathcal{P}(1^\lambda, r, x, w)$: On input string $r \in \{0,1\}^{k(\lambda,n)}$, a statement x of size $|x| = n$
  and a witness w, output a set of indices $I \subseteq [k]$ and proof π.*
- *$\mathcal{V}(1^\lambda, I, r_I, x, \pi)$: On input a subset $I \subseteq [k]$, a string r_I, a statement x and a
  proof π, outputs* accept *or* reject*,*

such that they satisfy the following properties:

- **Completeness:** *We require that for all $x \in L$ of size $|x| = n$ with witness w
  we have:*

$$
\Pr\left[\mathcal{V}(1^\lambda, I, r_I, x, \pi) = \texttt{accept} \; : \; \begin{array}{c} r \xleftarrow{\$} \{0,1\}^{k(\lambda,n)} \\ (I, \pi) \leftarrow \mathcal{P}(1^\lambda, r, x, w) \end{array} \right] = 1;
$$

- **Soundness:** *We require that for all polynomial $n = n(\lambda)$, and all unbounded
  cheating prover $\widetilde{\mathcal{P}}$, we have:*

$$
\Pr\left[\begin{array}{l} \mathcal{V}(1^\lambda, I, r_I, x, \pi) = \texttt{accept} \\ \wedge\; x \notin L \\ \wedge\; |x| = n \end{array} \; : \; \begin{array}{c} r \xleftarrow{\$} \{0,1\}^{k(\lambda,n)} \\ (x, \pi, I) \leftarrow \widetilde{\mathcal{P}}(1^\lambda, r) \end{array} \right] \leq \mathsf{negl}(\lambda);
$$

- **Zero-Knowledge:** *We require that there exists an efficient simulator* Sim
  *such that for any adversary $\mathcal{A}$ the two following distributions are statistically
  indistinguishable:*

$$
(I, r_I, \pi) \overset{s}{\approx} (I', r'_I, \pi')
$$

*where $(x, w) \leftarrow \mathcal{A}(1^\lambda), r \leftarrow \{0,1\}^{k(\lambda,|x|)}, (I, \pi) \leftarrow \mathcal{P}(1^\lambda, r, x, w), (I', r'_I, \pi') \leftarrow
\mathsf{Sim}(1^\lambda, x).$*

  *When clear from context, we will omit 1^λ as an argument to the algorithms
defined above.*

Remark 2.7 (Amplifying soundness). Let $\ell(\lambda, n)$ be a polynomial. Then, given any NIZK in the hidden-bits model, we can build one with soundness $2^{-\ell(\lambda,n)}$ · negl(λ). This is simply done by running $\ell(\lambda, n)$ copies of the NIZK in parallel, and where the new verification algorithm accepts a proof if and only if all of the executions accept. Note that doing so requires to use $k \cdot \ell(\lambda, n)$ hidden bits instead of k initially.

Theorem 2.8 ([FLS99], **see also** [Gol01, Section 4.10.2])**.** *Every $L \in$ **NP** has a NIZK in the Hidden-Bits Model.*

3 Hidden-Bits Generator

In this section, we define our new notion of Hidden-Bits Generator (HBG). For simplicity, we first define a publicly verifiable version of HBG and then extend the definition to a designated-verifier version (DV-HBG).

Definition 3.1 (Hidden-Bits Generator). *A Hidden-Bits Generator (HBG) is given by a set of PPT algorithms* (Setup, GenBits, Verify)*:*

- Setup($1^\lambda, 1^k$)*: Outputs a common reference string* crs*.*
- GenBits(crs)*: Outputs a triple* $\left(\text{com}, r, \{\pi_i\}_{i \in [k]}\right)$*, where* $r \in \{0, 1\}^k$*.*
- Verify(crs, com, i, r_i, π_i)*: Outputs* accept *or* reject*, where* $i \in [k]$*.*

We require any Hidden-Bits Generator to satisfy the following properties:

Correctness: *We require that for every polynomial $k = k(\lambda)$ and for all $i \in [k]$, we have:*

$$\Pr\left[\text{Verify}(\text{crs}, \text{com}, i, r_i, \pi_i) = \text{accept} \ : \ \begin{matrix} \text{crs} & \leftarrow \text{Setup}(1^\lambda, 1^k) \\ (\text{com}, r, \pi_{[k]}) & \leftarrow \text{GenBits}(\text{crs}) \end{matrix}\right] = 1.$$

Succinct Commitment: *We require that there exists some set $\mathcal{COM}(\lambda)$ and some constant $\delta < 1$ such that $|\mathcal{COM}(\lambda)| \leq 2^{k^\delta \text{poly}(\lambda)}$, and such that for all* crs *output by* Setup($1^\lambda, 1^k$) *and all* com *output by* GenBits(crs) *we have* com $\in \mathcal{COM}(\lambda)$*. Furthermore, we require that for all* com $\notin \mathcal{COM}(\lambda)$*,* Verify(crs, com, $\cdot, \cdot$) *always outputs* reject*.*[6]

Statistical Binding: *There exists an (inefficient) deterministic algorithm* Open(1^k, crs, com) *such that for every polynomial $k = k(\lambda)$, on input 1^k,* crs *and* com*, the algorithms outputs r such that for every (potentially unbounded) cheating prover $\widetilde{\mathcal{P}}$:*

---

[6] The set of commitments $\mathcal{COM}$ should not be thought of as the set of all valid commitments (and indeed it may contain commitments not in the support of GenBits). In particular, the simplest way to satisfy this property is to bound the bit-length of com and have the verifier reject commitments that are too large. Note that additional structural properties about com can be checked by the Verify algorithm.

$$\Pr\left[\begin{array}{ll} r_i^* \neq r_i \\ \wedge \ \mathsf{Verify}(\mathsf{crs}, \mathsf{com}, i, r_i^*, \pi_i) = \mathtt{accept} \end{array} : \begin{array}{ll} \mathsf{crs} & \leftarrow \mathsf{Setup}(1^\lambda, 1^k) \\ (\mathsf{com}, i, r_i^*, \pi_i) & \leftarrow \widetilde{\mathcal{P}}(\mathsf{crs}) \\ r & \leftarrow \mathsf{Open}(1^k, \mathsf{crs}, \mathsf{com}) \end{array}\right] \leq \mathsf{negl}(\lambda).$$

Computationally Hiding: *We require that for all polynomial $k = k(\lambda)$ and $I \subseteq [k]$, the two following distributions are computationally indistinguishable:*

$$\Big(\mathsf{crs}, \mathsf{com}, I, r_I, \pi_I, r_{\bar{I}}\Big)$$
$$\stackrel{c}{\approx}$$
$$\Big(\mathsf{crs}, \mathsf{com}, I, r_I, \pi_I, r'_{\bar{I}}\Big),$$

where $\mathsf{crs} \leftarrow \mathsf{Setup}(1^\lambda, 1^k)$, $(\mathsf{com}, r, \pi_{[k]}) \leftarrow \mathsf{GenBits}(\mathsf{crs})$ *and* $r' \stackrel{\$}{\leftarrow} \{0,1\}^k$.

Designated-Verifier Hidden-Bits Generator. We define the Designated-Verifier version of a Hidden-Bits Generator (DV-HBG) similarly, but with the following differences:

- $\mathsf{Setup}(1^\lambda, 1^k)$: Now outputs $(\mathsf{crs}, \mathsf{td})$, where td is a trapdoor associated to the crs;
- $\mathsf{Verify}(\mathsf{crs}, \mathsf{td}, \mathsf{com}, i, r_i, \pi_i)$ takes the trapdoor td as an additional input, and outputs $\mathtt{accept}$ or $\mathtt{reject}$ as before;
- For *Statistical Binding*, the cheating prover $\widetilde{\mathcal{P}}$ can now make a polynomial number of oracle queries to $\mathsf{Verify}(\mathsf{crs}, \mathsf{td}, \cdots)$. We require that for any such $\widetilde{P}$:

$$\Pr\left[\begin{array}{ll} r_i^* \neq r_i \\ \wedge \ \mathsf{Verify}(\mathsf{crs}, \mathsf{td}, \mathsf{com}, i, r_i^*, \pi_i) = \mathtt{accept} \end{array} : \begin{array}{ll} (\mathsf{crs}, \mathsf{td}) & \leftarrow \mathsf{Setup}(1^\lambda, 1^k) \\ (\mathsf{com}, i, r_i^*, \pi_i) & \leftarrow \widetilde{\mathcal{P}}^{\mathsf{Verify}(\mathsf{crs}, \mathsf{td}, \cdots)}(\mathsf{crs}) \\ r & \leftarrow \mathsf{Open}(1^k, \mathsf{crs}, \mathsf{com}) \end{array}\right] \leq \mathsf{negl}(\lambda).$$

- For *Computational Hiding*, we require that the distributions are indistinguishable given the associated trapdoor td:

$$\Big(\mathsf{crs}, \mathsf{td}, \mathsf{com}, I, r_I, \pi_I, r_{\bar{I}}\Big) \stackrel{c}{\approx} \Big(\mathsf{crs}, \mathsf{td}, \mathsf{com}, I, r_I, \pi_I, r'_{\bar{I}}\Big),$$

where $(\mathsf{crs}, \mathsf{td}) \leftarrow \mathsf{Setup}(1^\lambda, 1^k)$, $(\mathsf{com}, r, \pi_{[k]}) \leftarrow \mathsf{GenBits}(\mathsf{crs})$ and $r' \stackrel{\$}{\leftarrow} \{0,1\}^k$.

4   From Hidden-Bits Generator to NIZKs

We now prove that we can combine any (DV-)HBG with a NIZK in the Hidden-Bits model to get a (Reusable DV-)NIZK in the CRS model. Recall that our basic notion of NIZKs considered selective version of ZK where the statement to be proven is chosen prior to seeing the CRS. In Sect. 4.1 we will then extend our compiler to the adaptive ZK setting.

Theorem 4.1. *Suppose there exists a Hidden-Bits Generator, then there exists a publicly verifiable NIZK. Suppose there exists a designated-verifier Hidden-Bits Generator (DV-HBG), then there exists a reusable designated-verifier NIZK (reusable DV-NIZK).*

For simplicity, we first consider the publicly verifiable version of Theorem 4.1, the minor differences that are needed to extend it to the designated-verifier setting are discussed in the full version of the paper.

Construction. Let L be an NP language and n be an integer. Let $(\mathsf{Setup}^{\mathsf{BG}},$ $\mathsf{GenBits}, \mathsf{Verify})$ be a hidden-bits generator (Definition 3.1), where $|\mathcal{COM}| = |\mathcal{COM}(\lambda)| \leq 2^{k^\delta p(\lambda)}$ for some polynomial p and constant $\delta < 1$. (where k is the number of hidden bits generated).

Given a NIZK in the hidden-bits model for L using $k' = k'(\lambda, n)$ hidden bits (which exists unconditionally by Theorem 2.8), by Remark 2.7, there exists, for all polynomial $q(\lambda, n)$ (which we will set later), a NIZK in the hidden-bits model $(\mathcal{P}^{\mathsf{HB}}, \mathcal{V}^{\mathsf{HB}})$ using $k = k' \cdot q(\lambda, n)$ hidden bits with soundness-error $2^{-q(\lambda,n)} \cdot \mathsf{negl}(\lambda)$.

Consider the following candidate NIZK $(\mathsf{Setup}^{\mathsf{ZK}}, \mathcal{P}, \mathcal{V})$ in the CRS model:

- $\mathsf{Setup}^{\mathsf{ZK}}(1^\lambda, 1^n)$: Compute $\mathsf{crs}^{\mathsf{BG}} \leftarrow \mathsf{Setup}^{\mathsf{BG}}(1^\lambda, 1^k)$, sample $s \xleftarrow{\$} \{0,1\}^k$ and output:
$$\mathsf{crs} = (\mathsf{crs}^{\mathsf{BG}}, s);$$

- $\mathcal{P}(\mathsf{crs}, x, w)$: Compute $(\mathsf{com}, r^{\mathsf{BG}}, \pi_{[k]}) \leftarrow \mathsf{GenBits}(\mathsf{crs}^{\mathsf{BG}})$. Set $r_i = r_i^{\mathsf{BG}} \oplus s_i$ for all $i \in [k]$, and run the hidden-bits prover to get $(I \subseteq [k], \pi^{\mathsf{HB}}) \leftarrow \mathcal{P}^{\mathsf{HB}}(r, x, w)$. Output:
$$\Pi = (I, \pi^{\mathsf{HB}}, \mathsf{com}, r_I, \pi_I).$$

- $\mathcal{V}(\mathsf{crs}, x, \Pi = ((I, \pi^{\mathsf{HB}}, \mathsf{com}, r_I, \pi_I)))$: Compute $r_i^{\mathsf{BG}} = r_i \oplus s_i$ for all $i \in [k]$. Accept if for all $i \in I$, $\mathsf{Verify}(\mathsf{crs}^{\mathsf{BG}}, \mathsf{com}, i, r_i^{\mathsf{BG}}, \pi_i)$ accepts, and if $\mathcal{V}^{\mathsf{HB}}(I, r_I, x, \pi^{\mathsf{HB}})$ also accepts.

We refer the reader to the full version of the paper for a proof that $(\mathsf{Setup}^{\mathsf{ZK}}, \mathcal{P}, \mathcal{V})$ is a NIZK, and how to extend this construction to the designated-verifier setting.

4.1 Adaptive ZK

Our default definition of (reusable designated-verifier) NIZKs considers a *selective* version of the zero-knowledge property, where the statement x is chosen before the CRS. We also consider a stronger *adaptive* zero-knowledge property, where the statement x can depend adaptively on the CRS. Let us begin by defining adaptive ZK.

Definition 4.2 (Adaptive ZK). *A (reusable designated-verifier) NIZK satisfies adaptive Zero-Knowledge (adaptive ZK) if the following holds. We require that there exists a stateful PPT simulator Sim such that for any stateful PPT adversary $\mathcal{A}$ the two following distributions are computationally indistinguishable:*

$$\text{EXP}^{Real}(1^\lambda):$$

$1^n \leftarrow \mathcal{A}(1^\lambda)$
$(\mathsf{crs}, \mathsf{td}) \leftarrow \mathsf{Setup}(1^\lambda, 1^n)$
$(x, w) \leftarrow \mathcal{A}(\mathsf{crs}, \mathsf{td})$
$\qquad$ where $(x, w) \in R_L, |x| = n$
$\pi \leftarrow \mathcal{P}(\mathsf{crs}, x, w)$
$Output\ \mathcal{A}(\pi)$

$$\text{EXP}^{Ideal}(1^\lambda):$$

$1^n \leftarrow \mathcal{A}(1^\lambda)$
$(\mathsf{crs}, \mathsf{td}) \leftarrow \mathsf{Sim}(1^\lambda, 1^n)$
$(x, w) \leftarrow \mathcal{A}(\mathsf{crs}, \mathsf{td})$
$\qquad$ where $(x, w) \in R_L, |x| = n$
$\pi \leftarrow \mathsf{Sim}(x)$
$Output\ \mathcal{A}(\pi)$

The compiler of Theorem 4.1 can be extended to the adaptive setting:

Theorem 4.3. *Suppose there exists a Hidden-Bits Generator, then there exists a publicly verifiable NIZK with adaptive ZK security. Suppose there exists a designated-verifier Hidden-Bits Generator (DV-HBG), then there exists a reusable designated-verifier NIZK (DV-NIZK) with adaptive ZK security.*

We refer to the full version of the paper for the proof of Theorem 4.3.

5   Designated-Verifier Hidden-Bits Generator from CDH

Let $(\mathbb{G}, p, g) \leftarrow \mathsf{GroupGen}(1^\lambda)$ be a prime-order group generator so that $\mathbb{G}$ is a group of prime order p, with a generator g. Let hc be the corresponding Goldreich-Levin [GL89] hard-core bit. Let us define the following hidden-bits generator:

– $\mathsf{Setup}(1^\lambda, 1^k)$: Let $(\mathbb{G}, p, g) \leftarrow \mathsf{GroupGen}(1^\lambda)$. For all $i \in [k]$, pick random $a_i, b_i \overset{\$}{\leftarrow} \mathbb{Z}_p$ and $h_i \overset{\$}{\leftarrow} \mathbb{G}$ and compute:

$$f_i = h_i^{a_i} \cdot g^{b_i}.$$

Sample some random coins γ matching the randomness used by $\mathsf{hc}(\cdot)$. Output:

$$\left(\mathsf{crs} = \big(\mathbb{G}, \{(h_i, f_i)\}_{i \in [k]}, \gamma\big), \ \mathsf{td} = \{(a_i, b_i)\}_{i \in [k]} \right).$$

– $\mathsf{GenBits}(\mathsf{crs})$: Pick a random $y \leftarrow \mathbb{Z}_p$, and compute for all $i \in [k]$: $t_i = h_i^y$ and $u_i = f_i^y$. Output:

$$\mathsf{com} = s = g^y,$$
$$\{r_i = \mathsf{hc}(t_i; \gamma)\}_{i \in [k]},$$
$$\{\pi_i = (u_i, t_i)\}_{i \in [k]}.$$

– $\mathsf{Verify}(\mathsf{crs}, \mathsf{td} = \{(a_i, b_i)\}, \mathsf{com} = s, i, r_i, \pi_i = (u_i, t_i))$: Compute:

$$\rho_i = t_i^{a_i} \cdot s^{b_i},$$

and accept if and only if $\rho_i = u_i$, and $r_i = \mathsf{hc}(t_i; \gamma)$.

Theorem 5.1. *The triple* (Setup, GenBits, Verify) *is a Designated-Verifier Hidden-Bits Generator under CDH.*

We refer to the full version of the paper for a proof of Theorem 5.1.

Combining Theorems 4.3 and 5.1, and Remark 2.5, we obtain the following:

Theorem 5.2 (Reusable DV-NIZK from CDH). *Under the CDH assumption, there exists a reusable DV-NIZK for all NP with statistical soundness, and adaptive, multi-theorem zero-knowledge (Definitions 2.2, 4.2).*

6 Malicious-Designated-Verifier NIZKs

In this section we consider a strengthening of designated-verifier NIZKs to the malicious-designated-verifier setting (MDV-NIZK). In this setting, the trusted setup consists solely of a common random string (CRS). Given the CRS, the (potentially malicious) verifier generates a public key pk along with a secret key sk. The rest of the protocol is otherwise similar to the previous setting: any prover can use the CRS along with the newly generated public key to build non-interactive proofs of (many) NP statements, which can be verified using the corresponding secret key. The main difference is that we require zero-knowledge to hold against malicious verifiers, who can generate arbitrarily malformed public keys pk.

6.1 More Preliminaries

Reusable Malicious-Designated-Verifier NIZK.

Definition 6.1 (Reusable Malicious-Designated-Verifier NIZK (MDV-NIZK)). *Let L be an NP language with witness relation R_L. A Reusable Malicious-Designated-Verifier NIZK (MDV-NIZK) for L is a tuple of PPT algorithms* (Setup, KeyGen, $\mathcal{P}$, $\mathcal{V}$) *where:*

- Setup($1^\lambda, 1^n$): *outputs a common random string* crs;
- KeyGen(crs): *outputs a public key* pk *along with an associated secret key* sk;
- $\mathcal{P}$(crs, pk, x, w): *outputs a proof* π;
- $\mathcal{V}$(crs, sk, pk, x, π): *Outputs* accept *or* reject.

We require those algorithms to satisfy the same completeness and statistical soundness properties as Reusable DV-NIZKs (see Definition 2.2) with direct modifications to match the new syntax above, where now (crs, pk) *together act in place of what was previously just the* crs. *The requirement for zero-knowledge is strengthened to the following:*

Malicious Zero-Knowledge (Adaptive): *We require that there exists a PPT simulator* Sim *such that for any PPT stateful adversary $\mathcal{A}$, the two following distributions are computationally indistinguishable:*

$$\text{EXP}^{Real}(1^\lambda):$$

$1^n \leftarrow \mathcal{A}(1^\lambda)$
$\text{crs} \leftarrow \text{Setup}(1^\lambda, 1^n)$
$(x, w, \text{pk}) \leftarrow \mathcal{A}(\text{crs})$
$\qquad$ where $(x, w) \in R_L, |x| = n$
$\pi \leftarrow \mathcal{P}(\text{crs}, \text{pk}, x, w)$
$Output$ $\mathcal{A}(\pi)$

$$\text{EXP}^{Ideal}(1^\lambda):$$

$1^n \leftarrow \mathcal{A}(1^\lambda)$
$\text{crs} \leftarrow \text{Sim}(1^\lambda, 1^n)$
$(x, w, \text{pk}) \leftarrow \mathcal{A}(\text{crs})$
$\qquad$ where $(x, w) \in R_L, |x| = n$
$\pi \leftarrow \text{Sim}(\text{pk}, x)$
$Output$ $\mathcal{A}(\pi)$

Remark 6.2 (Single-Theorem vs. Multi-Theorem Zero-Knowledge). As in Definition 2.2, the definition above only captures *single-theorem* zero-knowledge. However the same "Or trick" of [FLS99] as in Remark 2.5 allows to generically compile any MDV-NIZK with single-theorem, adaptive (resp. selective) ZK into one satisfying *multi-theorem*, adaptive (resp. selective) ZK.

One-More CDH

We will use in this section a strengthening of the CDH assumption called *One-More CDH*. Intuitively, it states that given a set of challenge elements $\{h_j = g^{b_j}\}$ and the ability to make m queries to an oracle that raises arbitrary elements to some hidden exponent $a \in \mathbb{Z}_p$, it is hard to guess *more* than m of the values $h_j^{a_j} = g^{ab_j}$.

Definition 6.3 (One-More Computational Diffie-Hellman assumption (One-More CDH)). *Let* GroupGen *be a group generator. Let* $\ell = \ell(\lambda)$ *and* $m = m(\lambda)$ *be polynomials. Consider, for any PPT* $\mathcal{A}$, *the following experiment:*

$\text{Exp}^{One\text{-}More\ CDH}(1^\lambda)$

1. $(\mathbb{G}, p, g) \leftarrow \text{GroupGen}(1^\lambda)$

2. $\left(g^a, \{g^{b_i}\}_{i \leq \ell}\right) \overset{\$}{\leftarrow} \mathbb{G}^{1+\ell}$

3. $L \leftarrow \mathcal{A}^{\mathcal{O}_a(\cdot)}(\mathbb{G}, p, g, g^a, \{g^{b_i}\}_{i \leq \ell})$

4. *Output 1 if* $\exists i_1 < \cdots < i_{m+1} \in [\ell]$ *such that* $\forall j \leq m + 1,, g^{a \cdot b_{i_j}} \in L$;
   Otherwise output 0,

where the oracle $\mathcal{O}_a$ *takes as input a group element* $h \in \mathbb{G}$ *and outputs* h^a.
   We say that the One-More CDH *assumption holds relative to* GroupGen[7] *if for all PPT algorithm* $\mathcal{A}$ *making at most* m *queries to* $\mathcal{O}_a$, *we have:*

$$\Pr[\text{Exp}^{One\text{-}More\ CDH}(1^\lambda) = 1] \leq \text{negl}(\lambda).$$

---

[7] Later, we will also use the (mild) additional property that one can *obliviously* sample uniform group elements in $\mathbb{G}$, so that the One-More CDH assumption holds even given the random coins used to sample the group elements in Step 2. (and in particular a and the b_i's should be computationally hidden). Note that most standard groups (such as $\mathbb{Z}_p^*$ or elliptic curves) allow to do so. Looking ahead, if such a property does not hold, the resulting MDV-NIZK (Theorem 6.9) will use a common *reference* string instead.

Remark 6.4 (One-More CDH in Prior Works). A variety of previous works defined assumptions similar to the one above. To our knowledge, the first of this kind was introduced in the context of blind signatures in [Bol03], following the steps of [BNPS03] who first introduced One-More variants of the RSA and Discrete Log assumptions. More recently, another variant was used in the context of Oblivious PRFs (e.g. [JKK14]). The variant of [Bol03] requires the adversary to output one *single* guess for each target index $j \in J$, as opposed to a list of candidates L. As the adversary a-priori cannot test himself whether an element is correct, this makes it more difficult for the adversary to win the game and therefore the assumption of [Bol03] is *weaker* than our version in Definition 6.3. In [JKK14], on the other hand, the adversary is also given oracle access to a procedure that tests whether an element is a correct CDH output associated to some target index, but still has to output a single element for each target index. A direct reduction shows that this assumption is *at least as strong* as our variant: an adversary in the latter can call the oracle of [JKK14] on the whole list L to recover the matching indices.

Somewhere-Equivocable PRFs (SEPRFs)

We recall here the concept of Somewhere-Equivocable pseudorandom function (SEPRF)s, introduced in [HJO+16]. This is a function $\mathsf{PRF}(K, \cdot)$ with two modes of generating a key. There is the standard key generation algorithm which generates a key K honestly. In addition, there is a way to generate a key K' that leaves a "hole" at some particular point x^* but defines the PRF output at all other points; later one can "plug the hole" to any value r by creating a key K^* which agrees with K' on all values other than x^* but on x^* it outputs r. For any x^* and a random r one cannot distinguish between an honestly generated key K and the key K^* created as above. Intuitively, the second mode of key generation ensures that the function $\mathsf{PRF}(K^*, \cdot)$ outputs a truly random and independent value on some specific point x^*.

Definition 6.5 (1-Somewhere-Equivocable   PRFs   (1-SEPRFs) [HJO+16]). *A 1-Somewhere-Equivocable PRF (1-SEPRF) with input size s and output size d is a tuple of PPT algorithms* $(\mathsf{ObvGen}, \mathsf{PRF}, \mathsf{Sim}_1, \mathsf{Sim}_2)$:

- $\mathsf{ObvGen}(1^\lambda)$: *outputs a key K such that $\mathsf{PRF}(K, \cdot)$ maps $\{0,1\}^s$ to $\{0,1\}^d$;*
- $\mathsf{Sim}_1(x^*)$: *on input $x^* \in \{0,1\}^s$, outputs a key K and a state* state*;*
- $\mathsf{Sim}_2(\mathsf{state}, r)$: *on input $r \in \{0,1\}^d$, outputs a key K'.*

such that the following properties hold:

Correctness: *We have that for all $x^* \in \{0,1\}^s$ and $r \in \{0,1\}^d$, if $(K, \mathsf{state}) \xleftarrow{\$}$* $\mathsf{Sim}_1(x^*)$ *and $K' \xleftarrow{\$} \mathsf{Sim}_2(\mathsf{state}, r)$, then:*

$$\mathsf{PRF}(K, x) = \mathsf{PRF}(K', x) \quad \text{if } x \neq x^*$$
$$\mathsf{PRF}(K', x^*) = r.$$

Equivocation security: *For all PPT adversary $\mathcal{A}$ we have:*

$$\left| \Pr \begin{bmatrix} x^* \xleftarrow{\$} \mathcal{A}(1^\lambda) \\ K \xleftarrow{\$} \mathsf{ObvGen}(1^\lambda) \\ \mathcal{A}(K) = 1 \end{bmatrix} - \Pr \begin{bmatrix} x^* \xleftarrow{\$} \mathcal{A}(1^\lambda), r^* \xleftarrow{\$} \{0,1\}^d \\ (K, \mathsf{state}) \leftarrow \mathsf{Sim}_1(x^*) \\ K' \xleftarrow{\$} \mathsf{Sim}_2(\mathsf{state}, r^*) \\ \mathcal{A}(K') = 1 \end{bmatrix} \right| \leq \mathsf{negl}(\lambda).$$

Claim ([HJO+16]). Assuming one-way functions exist, there exist 1-SEPRFs, with key size $\mathcal{O}(s \cdot d \cdot \lambda)$.

6.2   Reusable Malicious-Designated-Verifier HBG (MDV-HBG)

To define a reusable Malicious-Designated-Verifier Hidden-Bits Generator (MDV-HBG), we extend the definition of a DV-HBG in a manner analogous to the difference between DV-NIZKs and MDV-NIZKs. Namely, instead of having a trusted setup that generates a public crs along with a secret key sk for the verifier, we now only have the setup algorithm generate the crs and allow the (potentially malicious) verifier to generate pk, sk on his own via a new KeyGen algorithm. Furthermore, we want to ensure that the generated hidden bits only depend on crs but not on pk; only the openings of the hidden bits can depend on pk.

Definition 6.6 (Reusable Malicious-Designated-Verifier HBG (MDV-HBG)). *A Reusable Malicious-Designated-Verifier HBG is a tuple of PPT algorithms* (Setup, KeyGen, (GenBits.Commit, GenBits.Prove), Verify):

- Setup($1^\lambda, 1^k$): *outputs a common random string* crs.
- KeyGen(crs): *outputs a public key* pk *with an associated secret key* sk.
- GenBits(crs, pk) *is now split into two sub-procedures:*
  - GenBits.Commit(crs): *on input a* crs, *outputs a commitment* com, *some bits* $r \in \{0,1\}^k$ *and a state* state.
  - GenBits.Prove(crs, pk, state): *on input a public key* pk, *a* crs *and a state* state, *produces proofs* $\{\pi_i\}_{i \in k}$.

  It outputs (com, $r, \{\pi_i\}_{i \in [k]}$).
- Verify(crs, sk, com, i, r_i, π_i): *Outputs* accept *or* reject.

We require an MDV-HBG to satisfy the following properties. The first three (correctness, succinctness of the commitments and statistical binding), are direct adaptations of Definition 3.1 to the new syntax:

Correctness: *We require that for every polynomial* $k = k(\lambda)$ *and for all* $i \in [k]$, *we have:*

$$\Pr\left[\mathsf{Verify}(\mathsf{crs}, \mathsf{sk}, \mathsf{com}, i, r_i, \pi_i) = \mathtt{accept} \ : \ \begin{matrix} \mathsf{crs} & \leftarrow \mathsf{Setup}(1^\lambda, 1^k) \\ (\mathsf{pk}, \mathsf{sk}) & \leftarrow \mathsf{KeyGen}(\mathsf{crs}) \\ (\mathsf{com}, r, \pi_{[k]}) & \leftarrow \mathsf{GenBits}(\mathsf{crs}, \mathsf{pk}) \end{matrix}\right] = 1.$$

Succinct Commitment: *We require that there exists some set $\mathcal{COM}(\lambda)$ and some constant $\delta < 1$ such that $|\mathcal{COM}(\lambda)| \leq 2^{k^{\delta}\mathsf{poly}(\lambda)}$, and such that for all crs output by $\mathsf{Setup}(1^{\lambda}, 1^{k})$ and all com output by $\mathsf{GenBits}(\mathsf{crs})$ we have $\mathsf{com} \in \mathcal{COM}(\lambda)$. Furthermore, we require that for all $\mathsf{com} \notin \mathcal{COM}(\lambda)$, $\mathsf{Verify}(\mathsf{crs}, \mathsf{com}, \cdot, \cdot)$ always outputs $\mathtt{reject}$.*

Statistical Binding: *There exists an (inefficient) deterministic algorithm $\mathsf{Open}(1^{k}, \mathsf{crs}, \mathsf{com})$ such that for every polynomial $k = k(\lambda)$, on input 1^{k}, crs and com, the algorithms outputs r such that for every (potentially unbounded) cheating prover $\widetilde{\mathcal{P}}$:*

$$\Pr\left[\begin{array}{c} r_i^* \neq r_i \\ \wedge \;\; \mathsf{Verify}(\mathsf{crs}, \mathsf{sk}, \mathsf{com}, i, r_i^*, \pi_i) = \mathsf{accept} \end{array} : \begin{array}{ll} \mathsf{crs} & \leftarrow \mathsf{Setup}(1^{\lambda}, 1^{k}) \\ (\mathsf{pk}, \mathsf{sk}) & \leftarrow \mathsf{KeyGen}(\mathsf{crs}) \\ (\mathsf{com}, i, r_i^*, \pi_i) & \leftarrow \widetilde{\mathcal{P}}(\mathsf{crs}, \mathsf{pk}) \\ r & \leftarrow \mathsf{Open}(1^{k}, \mathsf{crs}, \mathsf{com}) \end{array} \right] \leq \mathsf{negl}(\lambda).$$

The main conceptual difference with Definition 3.1 comes from the computational hiding property, which now captures security against malicious verifiers:

Computationally Hiding against Malicious Verifiers: *Consider, for an integer k, a bit b, and a stateful PPT adversary $\mathcal{A}$, the following experiment:*

$$\mathsf{Exp}^{Hiding,b}(1^{\lambda}, 1^{k})$$

0. $I \subseteq [k] \leftarrow \mathcal{A}(1^{k})$
1. $\mathsf{crs} \leftarrow \mathsf{Setup}(1^{\lambda}, 1^{k})$
2. $\mathsf{pk} \leftarrow \mathcal{A}(\mathsf{crs})$
3. *Compute* $(\mathsf{com}, r, \{\pi_i\}_{i \in [k]}) \leftarrow \mathsf{GenBits}(\mathsf{crs}, \mathsf{pk})$.

   Set for all $i \notin I :$ $\begin{cases} \rho_i = r_i \; \text{if } b = 0; \\ \rho_i \xleftarrow{\$} \{0,1\} \; \text{otherwise.} \end{cases}$
4. *Output* $: \beta \leftarrow \mathcal{A}(\mathsf{crs}, \mathsf{com}, I, r_I, \pi_I, \{\rho_i\}_{i \notin I})$

We require that for all polynomial $k = k(\lambda)$ and stateful PPT adversary $\mathcal{A}$:

$$\left| \Pr\left[\mathsf{Exp}^{Hiding,0}(1^{\lambda}) = 1 \right] - \Pr\left[\mathsf{Exp}^{Hiding,1}(1^{\lambda}) = 1 \right] \right| \leq \mathsf{negl}(\lambda).$$

6.3   Reusable MDV-NIZK from MDV-HBG

We present here an analogue to Theorem 4.1 in the malicious-verifier setting.

Theorem 6.7. *Suppose there exists a MDV-HBG. Then there exists a reusable MDV-NIZK with adaptive ZK security.*

The proof of Theorem 6.7 is a simple adaptation of the one of Theorem 4.1. We refer the reader to the full version of the paper for more details.

6.4   MDV-HBG from One-More CDH

Notation. Let d, k and ℓ be integers, where ℓ is a power-of-two. Given a function $\varphi : [k] \to [\ell]^d$ and some index $i \in [k]$, we define, for some vector $\mathbf{u}$ of dimension ℓ, the vector:

$$\mathbf{u}_{\varphi(i)} := \left(u_{\varphi(i)_1}, \dots, u_{\varphi(i)_d} \right).$$

In other words, we can think of $\varphi(i)$ as a set of *neighbors* of vertex $i \in [k]$ in the bipartite (multi-)graph $([k], [\ell])$. Furthermore if the vertices $j \in [\ell]$ are labelled with some element u_j, then $u_{\varphi(i)}$ denotes the *list* of labels associated to neighbors of i. Note that vertices in $[k]$ have d neighbors in $[\ell]$ (where there can be multiple occurrences of the same edge). We naturally extend this definition for *sets* of indices: for $I \subseteq [k]$, we define

$$\mathbf{u}_{\varphi(I)} := \left(u_{\varphi(i)_1}, \dots, u_{\varphi(i)_d} \right)_{i \in I}.$$

Let hc be the Goldreich-Levin [GL89] hard-core bit (which, on input a bit-string $x \in \{0,1\}^L$, uses randomness $r \xleftarrow{\$} \{0,1\}^L$ and outputs $\mathsf{hc}(x; r) := (\langle x, r \rangle, r)$).

Construction. Let $(\mathbb{G}, p, g) \leftarrow \mathsf{GroupGen}(1^\lambda)$ be a prime-order group generator so that $\mathbb{G}$ is a group of prime order p, with a generator g. For $\lambda, k \in \mathbb{N}$, let $\ell = \ell(\lambda, k)$ be the least power-of-two greater than $3k\lambda$ (i.e. $\ell = 2^{\lceil \log(3k\lambda) \rceil}$), and let $d = \lambda$. Let $(\mathsf{ObvGen}, \mathsf{PRF}, \mathsf{Sim}_1, \mathsf{Sim}_2)$ be a 1-SEPRF (as defined in Sect. 6.1) where $\mathsf{ObvGen}(1^\lambda)$ outputs keys K such that $\mathsf{PRF}(K, \cdot)$ maps $\{0,1\}^{\lceil \log k \rceil}$ to $\{0,1\}^{d \cdot \log \ell}$ (and in particular maps $[k]$ to $[\ell]$).

Let us define the following hidden-bits generator:

– $\mathsf{Setup}(1^\lambda, 1^k)$: Let $(\mathbb{G}, p, g) \leftarrow \mathsf{GroupGen}(1^\lambda)$. For all $j \in [\ell]$, pick $h_j \xleftarrow{\$} \mathbb{G}$. Output:

$$\mathsf{crs} = (\mathbb{G}, \{h_j\}_{j \in [\ell]}).$$

– $\mathsf{KeyGen}(\mathsf{crs})$: For all $j \in [\ell]$, pick random $a_j, b_j \xleftarrow{\$} \mathbb{Z}_p$, compute:

$$f_j = h_j^{a_j} \cdot g^{b_j},$$

and output:

$$\mathsf{pk} = \{f_j\}_{j \in [\ell]},$$
$$\mathsf{sk} = \{(a_j, b_j)\}_{j \in [\ell]}.$$

– $\mathsf{GenBits}(\mathsf{crs}, \mathsf{pk})$:
  • $\mathsf{GenBits.Commit}(\mathsf{crs})$: Pick a random $y \leftarrow \mathbb{Z}_p$ and set $s = g^y$. Compute for all $j \in [\ell]$: $t_j = h_j^y$. Sample some random coins γ matching the randomness used by $\mathsf{hc}(\cdot)$ taking as input (the bit-representation of) elements in $\mathbb{G}^d$. Sample $K \leftarrow \mathsf{ObvGen}(1^\lambda)$. Parsing the output of $\mathsf{PRF}(K, \cdot)$ as d blocks of $\log \ell$ bits, this defines for all $i \in [k]$:

$$\varphi(i) := (\mathsf{PRF}(K, i)_1, \dots, \mathsf{PRF}(K, i)_d) \in [\ell]^d. \tag{1}$$

Compute for all $i \in [k]$: $r_i = \mathsf{hc}\left((\mathbf{h}^y)_{\varphi(i)} \; ; \; \gamma\right)$, where we recall that by definition $(\mathbf{h}^y)_{\varphi(i)} = \left(h^y_{\mathsf{PRF}(K,i)_1}, \ldots, h^y_{\mathsf{PRF}(K,i)_d}\right)$. Output:

$$\mathsf{com} = (s, \gamma, K),$$
$$\{r_i\}_{i \in [k]},$$
$$\mathsf{state} = (y, K).$$

- GenBits.Prove(crs, pk, state): Parse pk as $\{f_j\}_{j \in [\ell]}$. The key K in state defines a function φ as per Eq. 1. Compute for all $j \in [\ell]$: $t_j = h^y_j$ and $u_j = f^y_j$. Compute for all $i \in [k]$:

$$\pi_i = \{(t_j, u_j)\}_{j \in \varphi(i)}.$$

Output:

$$(\mathsf{com}, r, \{\pi_i\}_{i \in [k]}).$$

– Verify(crs, sk, com, i, r_i, π_i) : Parse $\mathsf{sk} = \{(a_j, b_j)\}_{j \in [\ell]}$, $\mathsf{com} = (s, \gamma, K)$, $\pi_i = \{(t_j, u_j)\}_{j \in \varphi(i)}$. Compute for $j \in \varphi(i)$ (where $\varphi(i)$ is defined as per Eq. 1):

$$\rho_j = t^{a_j}_j \cdot s^{b_j},$$

and accept if and only if $\rho_j = u_j$ for all $j \in \varphi(i)$, and $r_i = \mathsf{hc}\left(\{\mathsf{t}\}_{\varphi(i)} \; ; \; \gamma\right)$.

Theorem 6.8. *Suppose that* (ObvGen, PRF, Sim₁, Sim₂) *is a 1-SEPRF (Definition 6.5). Then, assuming the One-More CDH assumption holds (Definition 6.3,* (Setup, GenBits, Verify) *is a reusable Malicious-Designated-Verifier Hidden-Bits Generator (Definition 6.6).*

We refer the reader to the full version of this paper for a proof of Theorem 6.8.
Combining Claim 6.1, Theorems 6.7 and 6.8, and Remark 6.2, we obtain the following:

Theorem 6.9 (MDV-NIZK from One-More CDH). *Under the One-More CDH assumption (Definition 6.3), there exists a MDV-NIZK for all NP (Definition 6.1) with statistical soundness, and adaptive, multi-theorem zero-knowledge.*

7   Extensions

We informally describe two simple extensions of our construction.

Unbounded Statement Size. In our construction of (reusable DV-)NIZKs, we need to have a bound n on the size of the statements that can be proved and the size of the CRS depends on n. Ideally, we would have a fixed-size CRS which allows us to prove statements of arbitrary size. Indeed, we can achieve this using non-interactive statistically-binding commitments in the CRS model, which exist assuming OWFs [Nao90, Nao91]. Let us fix 3SAT as the NP-complete language. To prove that some 3CNF is satisfiable the prover commits to the satisfying assignments one variable at a time. Then he uses a (reusable DV-)NIZK scheme for each clause separately to show that the 3 relevant committed values satisfy the clause. Note that the size of the statements being proved by the underlying (reusable DV-)NIZK is independent of the size of the actual 3CNF formula. Therefore the above technique bootstraps a (reusable DV-)NIZK for statements of some fixed size which depends only on the security parameter to construct a (reusable DV-)NIZK for statements of arbitrary size.

Proof of Knowledge. While our basic construction is not a proof-of-knowledge it is easy to generically add this property assuming the existence of public-key encryption (PKE). We can add a public-key com of a PKE scheme to the CRS and have the prover encrypt the witness under com and then use the (reusable DV-)NIZK to prove that the ciphertext is an encryption of a valid witness for the statement. The extractor would choose com along with a corresponding decryption key sk and use it to extract the witness.

Acknowledgments. Research supported by NSF grants CNS-1314722, CNS-1413964, CNS-1750795 and the Alfred P. Sloan Research Fellowship. The second author was supported in part by the Israeli Science Foundation (Grant No. 1262/18). We thank Geoffroy Couteau, Dennis Hofheinz, Shuichi Katsumata, Ryo Nishimaki, Shota Yamada, and Takashi Yamakawa for sharing their manuscripts [CH19, KNYY19] and for helpful discussions.

References

[BFM88] Blum, M., Feldman, P., Micali, S.: Non-interactive zero-knowledge and its applications (extended abstract). In: 20th Annual ACM Symposium on Theory of Computing, Chicago, IL, USA, 2–4 May, pp. 103–112. ACM Press (1988)

[BGI15] Boyle, E., Gilboa, N., Ishai, Y.: Function secret sharing. In: Oswald, E., Fischlin, M. (eds.) EUROCRYPT 2015, Part II. LNCS, vol. 9057, pp. 337–367. Springer, Heidelberg (2015). https://doi.org/10.1007/978-3-662-46803-6_12

[BNPS03] Bellare, M., Namprempre, C., Pointcheval, D., Semanko, M.: The one-more-RSA-inversion problems and the security of Chaum's blind signature scheme. J. Cryptol. **16**(3), 185–215 (2003)

[Bol03] Boldyreva, A.: Threshold signatures, multisignatures and blind signatures based on the Gap-Diffie-Hellman-group signature scheme. In: Desmedt, Y.G. (ed.) PKC 2003. LNCS, vol. 2567, pp. 31–46. Springer, Heidelberg (2003). https://doi.org/10.1007/3-540-36288-6_3

[BY93] Bellare, M., Yung, M.: Certifying cryptographic tools: the case of trapdoor permutations. In: Brickell, E.F. (ed.) CRYPTO 1992. LNCS, vol. 740, pp. 442–460. Springer, Heidelberg (1993). https://doi.org/10.1007/3-540-48071-4_31

[CC18]   Chaidos, P., Couteau, G.: Efficient designated-verifier non-interactive zero-knowledge proofs of knowledge. In: Nielsen, J.B., Rijmen, V. (eds.) EUROCRYPT 2018, Part III. LNCS, vol. 10822, pp. 193–221. Springer, Cham (2018). https://doi.org/10.1007/978-3-319-78372-7_7

[CCRR18] Canetti, R., Chen, Y., Reyzin, L., Rothblum, R.D.: Fiat-Shamir and correlation intractability from strong KDM-secure encryption. In: Nielsen, J.B., Rijmen, V. (eds.) EUROCRYPT 2018, Part I. LNCS, vol. 10820, pp. 91–122. Springer, Cham (2018). https://doi.org/10.1007/978-3-319-78381-9_4

[CH19]   Couteau, G., Hofheinz, D.: Towards non-interactive zero-knowledge proofs from CDH and LWE. In: EUROCRYPT (2019)

[CHK03]   Canetti, R., Halevi, S., Katz, J.: A forward-secure public-key encryption scheme. In: Biham, E. (ed.) EUROCRYPT 2003. LNCS, vol. 2656, pp. 255–271. Springer, Heidelberg (2003). https://doi.org/10.1007/3-540-39200-9_16

[CKS08]   Cash, D., Kiltz, E., Shoup, V.: The twin Diffie-Hellman problem and applications. In: Smart, N. (ed.) EUROCRYPT 2008. LNCS, vol. 4965, pp. 127–145. Springer, Heidelberg (2008). https://doi.org/10.1007/978-3-540-78967-3_8

[CL17]   Canetti, R., Lichtenberg, A.: Certifying trapdoor permutations, revisited. IACR Cryptology ePrint Archive 2017/631 (2017)

[CS98]   Cramer, R., Shoup, V.: A practical public key cryptosystem provably secure against adaptive chosen ciphertext attack. In: Krawczyk, H. (ed.) CRYPTO 1998. LNCS, vol. 1462, pp. 13–25. Springer, Heidelberg (1998). https://doi.org/10.1007/BFb0055717

[CS02]   Cramer, R., Shoup, V.: Universal hash proofs and a paradigm for adaptive chosen ciphertext secure public-key encryption. In: Knudsen, L.R. (ed.) EUROCRYPT 2002. LNCS, vol. 2332, pp. 45–64. Springer, Heidelberg (2002). https://doi.org/10.1007/3-540-46035-7_4

[Dam93]   Damgård, I.: Non-interactive circuit based proofs and non-interactive perfect zero-knowledge with preprocessing. In: Rueppel, R.A. (ed.) EUROCRYPT 1992. LNCS, vol. 658, pp. 341–355. Springer, Heidelberg (1993). https://doi.org/10.1007/3-540-47555-9_28

[DDN91]   Dolev, D., Dwork, C., Naor, M.: Non-malleable cryptography (extended abstract). In: 23rd Annual ACM Symposium on Theory of Computing, New Orleans, LA, USA, 6–8 May, pp. 542–552. ACM Press (1991)

[DFN06]   Damgård, I., Fazio, N., Nicolosi, A.: Non-interactive zero-knowledge from homomorphic encryption. In: Halevi, S., Rabin, T. (eds.) TCC 2006. LNCS, vol. 3876, pp. 41–59. Springer, Heidelberg (2006). https://doi.org/10.1007/11681878_3

[DMP88]   De Santis, A., Micali, S., Persiano, G.: Non-interactive zero-knowledge proof systems. In: Pomerance, C. (ed.) CRYPTO 1987. LNCS, vol. 293, pp. 52–72. Springer, Heidelberg (1988). https://doi.org/10.1007/3-540-48184-2_5

[DMP90]   De Santis, A., Micali, S., Persiano, G.: Non-interactive zero-knowledge with preprocessing. In: Goldwasser, S. (ed.) CRYPTO 1988. LNCS, vol. 403, pp. 269–282. Springer, New York (1990). https://doi.org/10.1007/0-387-34799-2_21

[DN00]   Dwork, C., Naor, M.: Zaps and their applications. In: 41st Annual Symposium on Foundations of Computer Science, Redondo Beach, CA, USA, 12–14 November, pp. 283–293. IEEE Computer Society Press (2000)

[FLS99] Feige, U., Lapidot, D., Shamir, A.: Multiple noninteractive zero knowledge proofs under general assumptions. SIAM J. Comput. **29**(1), 1–28 (1999)

[FS87] Fiat, A., Shamir, A.: How to prove yourself: practical solutions to identification and signature problems. In: Odlyzko, A.M. (ed.) CRYPTO 1986. LNCS, vol. 263, pp. 186–194. Springer, Heidelberg (1987). https://doi.org/10.1007/3-540-47721-7_12

[GI14] Gilboa, N., Ishai, Y.: Distributed point functions and their applications. In: Nguyen, P.Q., Oswald, E. (eds.) EUROCRYPT 2014. LNCS, vol. 8441, pp. 640–658. Springer, Heidelberg (2014). https://doi.org/10.1007/978-3-642-55220-5_35

[GL89] Goldreich, O., Levin, L.A.: A hard-core predicate for all one-way functions. In: 21st Annual ACM Symposium on Theory of Computing, Seattle, WA, USA, 15–17 May. ACM Press (1989)

[GMR85] Goldwasser, S., Micali, S., Rackoff, C.: The knowledge complexity of interactive proof-systems (extended abstract). In: 17th Annual ACM Symposium on Theory of Computing, Providence, RI, USA, 6–8 May, pp. 291–304. ACM Press (1985)

[GMR89] Goldwasser, S., Micali, S., Rackoff, C.: The knowledge complexity of interactive proof systems. SIAM J. Comput. **18**(1), 186–208 (1989)

[Gol01] Goldreich, O.: Foundations of Cryptography: Basic Tools, vol. 1. Cambridge University Press, Cambridge (2001)

[Gol04] Goldreich, O.: Foundations of Cryptography: Basic Applications, vol. 2. Cambridge University Press, Cambridge (2004)

[Gol11] Goldreich, O.: Basing non-interactive zero-knowledge on (enhanced) trapdoor permutations: the state of the art. In: Goldreich, O. (ed.) Studies in Complexity and Cryptography. Miscellanea on the Interplay between Randomness and Computation. LNCS, vol. 6650, pp. 406–421. Springer, Heidelberg (2011). https://doi.org/10.1007/978-3-642-22670-0_28

[GOS06] Groth, J., Ostrovsky, R., Sahai, A.: Perfect non-interactive zero knowledge for NP. In: Vaudenay, S. (ed.) EUROCRYPT 2006. LNCS, vol. 4004, pp. 339–358. Springer, Heidelberg (2006). https://doi.org/10.1007/11761679_21

[GR13] Goldreich, O., Rothblum, R.D.: Enhancements of trapdoor permutations. J. Cryptol. **26**(3), 484–512 (2013)

[HJO+16] Hemenway, B., Jafargholi, Z., Ostrovsky, R., Scafuro, A., Wichs, D.: Adaptively secure garbled circuits from one-way functions. In: Robshaw, M., Katz, J. (eds.) CRYPTO 2016, Part III. LNCS, vol. 9816, pp. 149–178. Springer, Heidelberg (2016). https://doi.org/10.1007/978-3-662-53015-3_6

[JKK14] Jarecki, S., Kiayias, A., Krawczyk, H.: Round-optimal password-protected secret sharing and T-PAKE in the password-only model. In: Sarkar, P., Iwata, T. (eds.) ASIACRYPT 2014, Part II. LNCS, vol. 8874, pp. 233–253. Springer, Heidelberg (2014). https://doi.org/10.1007/978-3-662-45608-8_13

[KMO90] Kilian, J., Micali, S., Ostrovsky, R.: Minimum resource zero-knowledge proofs. In: Brassard, G. (ed.) CRYPTO 1989. LNCS, vol. 435, pp. 545–546. Springer, New York (1990). https://doi.org/10.1007/0-387-34805-0_47

[KNYY19] Katsumata, S., Nishimaki, R., Yamada, S., Yamakawa, T.: Designated verifier/prover and preprocessing NIZKs from Diffie-Hellman assumptions. In: EUROCRYPT (2019)

[KW18] Kim, S., Wu, D.J.: Multi-theorem preprocessing NIZKs from lattices. In: Shacham, H., Boldyreva, A. (eds.) CRYPTO 2018, Part II. LNCS, vol. 10992, pp. 733–765. Springer, Cham (2018). https://doi.org/10.1007/978-3-319-96881-0_25

[LS91] Lapidot, D., Shamir, A.: Publicly verifiable non-interactive zero-knowledge proofs. In: Menezes, A.J., Vanstone, S.A. (eds.) CRYPTO 1990. LNCS, vol. 537, pp. 353–365. Springer, Heidelberg (1991). https://doi.org/10.1007/3-540-38424-3_26

[Nao90] Naor, M.: Bit commitment using pseudo-randomness. In: Brassard, G. (ed.) CRYPTO 1989. LNCS, vol. 435, pp. 128–136. Springer, New York (1990). https://doi.org/10.1007/0-387-34805-0_13

[Nao91] Naor, M.: Bit commitment using pseudorandomness. J. Cryptol. 4(2), 151–158 (1991)

[NY90] Naor, M., Yung, M.: Public-key cryptosystems provably secure against chosen ciphertext attacks. In: 22nd Annual ACM Symposium on Theory of Computing, Baltimore, MD, USA, 14–16 May, pp. 427–437. ACM Press (1990)

[PsV06] Pass, R., Shelat, A., Vaikuntanathan, V.: Construction of a non-malleable encryption scheme from any semantically secure one. In: Dwork, C. (ed.) CRYPTO 2006. LNCS, vol. 4117, pp. 271–289. Springer, Heidelberg (2006). https://doi.org/10.1007/11818175_16

[SW14] Sahai, A., Waters, B.: How to use indistinguishability obfuscation: deniable encryption, and more. In: Shmoys, D.B. (ed.) 46th Annual ACM Symposium on Theory of Computing, 31 May–3 June, pp. 475–484. ACM Press, New York (2014)

Designated Verifier/Prover
and Preprocessing NIZKs
from Diffie-Hellman Assumptions

Shuichi Katsumata[1,3(✉)], Ryo Nishimaki[2], Shota Yamada[1],
and Takashi Yamakawa[2]

[1] AIST, Tokyo, Japan
yamada-shota@aist.go.jp
[2] NTT Secure Platform Laboratories, Tokyo, Japan
{ryo.nishimaki.zk,takashi.yamakawa.ga}@hco.ntt.co.jp
[3] The University of Tokyo, Tokyo, Japan
shuichi_katsumata@it.k.u-tokyo.ac.jp

Abstract. In a non-interactive zero-knowledge (NIZK) proof, a prover can non-interactively convince a verifier of a statement without revealing any additional information. Thus far, numerous constructions of NIZKs have been provided in the common reference string (CRS) model (CRS-NIZK) from various assumptions, however, it still remains a long standing open problem to construct them from tools such as pairing-free groups or lattices. Recently, Kim and Wu (CRYPTO'18) made great progress regarding this problem and constructed the first lattice-based NIZK in a relaxed model called NIZKs in the preprocessing model (PP-NIZKs). In this model, there is a trusted statement-independent preprocessing phase where secret information are generated for the prover and verifier. Depending on whether those secret information can be made public, PP-NIZK captures CRS-NIZK, designated-verifier NIZK (DV-NIZK), and designated-prover NIZK (DP-NIZK) as special cases. It was left as an open problem by Kim and Wu whether we can construct such NIZKs from weak paring-free group assumptions such as DDH. As a further matter, all constructions of NIZKs from Diffie-Hellman (DH) type assumptions (regardless of whether it is over a paring-free or paring group) require the proof size to have a multiplicative-overhead $|C| \cdot \mathsf{poly}(\kappa)$, where $|C|$ is the size of the circuit that computes the **NP** relation.

In this work, we make progress of constructing (DV, DP, PP)-NIZKs with varying flavors from DH-type assumptions. Our results are summarized as follows:

- DV-NIZKs for **NP** from the CDH assumption over pairing-free groups. This is the first construction of such NIZKs on pairing-free groups and resolves the open problem posed by Kim and Wu (CRYPTO'18).
- DP-NIZKs for **NP** with short proof size from a DH-type assumption over pairing groups. Here, the proof size has an additive-overhead $|C|+\mathsf{poly}(\kappa)$ rather then an multiplicative-overhead $|C| \cdot \mathsf{poly}(\kappa)$. This

Y. Ishai and V. Rijmen (Eds.): EUROCRYPT 2019, LNCS 11477, pp. 622–651, 2019.
https://doi.org/10.1007/978-3-030-17656-3_22

is the first construction of such NIZKs (including CRS-NIZKs) that does not rely on the LWE assumption, fully-homomorphic encryption, indistinguishability obfuscation, or non-falsifiable assumptions.

– PP-NIZK for **NP** with short proof size from the DDH assumption over pairing-free groups. This is the first PP-NIZK that achieves a short proof size from a weak and static DH-type assumption such as DDH. Similarly to the above DP-NIZK, the proof size is $|C|+\mathsf{poly}(\kappa)$. This too serves as a solution to the open problem posed by Kim and Wu (CRYPTO'18).

Along the way, we construct two new homomorphic authentication (HomAuth) schemes which may be of independent interest.

1   Introduction

1.1   Background

Zero-knowledge (ZK) proof system [57] is an interactive protocol where a prover convinces the validity of a statement to a verifier without providing any additional knowledge. A non-interactive zero-knowledge (NIZK) proof (or argument[1]) [13] is a ZK proof (or argument) where a prover can generate a proof to the validity of a statement *without* interacting with a verifier. Due to the absence of interaction, NIZKs have found tremendous number of applications in cryptography including (but not limited to) chosen-ciphertext secure public key encryption [45,76,84], group/ring signatures [9,37,81], anonymous credentials [35,40], and multi-party computations (MPC) [55]. Furthermore, aside from its practical interests, due to its theoretically appealing nature, studying the types of assumptions which imply NIZKs has also been an active research area for NIZKs [12,47,63,82]. Below, we briefly review the current state of affairs concerning NIZKs.

NIZKs in the CRS model. It is well known that NIZKs for non-trivial languages do not exist in the plain model where there is no trusted setup [56]. Therefore NIZKs for all of **NP** are constructed either in the common reference string (CRS) model [47] or the random oracle model [48,79]. In the former type of NIZK, the prover and the verifier have access to a CRS generated by a trusted third party (hereafter referred to as CRS-NIZK). Thus far, known constructions of CRS-NIZK for **NP** are based on (doubly-enhanced) trapdoor permutation [10,47,53], pairing [63,64], or indistinguishability obfuscation [11,12,86]. Constructing CRS-NIZKs based on other assumptions such as pairing-free groups and lattices remains to be a long standing open problem.

NIZKs in the designated verifier/prover model. As an alternative line of research, NIZKs in a relaxed model have been considered: *designated verifier* NIZKs (DV-NIZKs) and *designated prover* NIZKs (DP-NIZKs). Both notions

---

[1] NIZK arguments are a relaxed notion of NIZK proofs where soundness only holds against computationally bounded adversaries. Throughout the introduction, we simply refer to them as NIZKs.

of NIZKs retain most of the useful security properties of NIZKs with some relaxation. In DV-NIZKs, anybody can generate a proof, but the proof can only be verified by a designated party in possession of a *verification key*. On the other hand, in DP-NIZKs only a designated party in possession of a *proving key* can generate a proof, but the proof can be verified by anybody. Although the two types of NIZKs are relaxation of CRS-NIZKs, they showed to be no easier to construct. There have been a long line of work concerning DV-NIZKs [32–34,42,74,77,91], however, many of these schemes do not satisfy soundness against multiple theorems, which in brief means that soundness does not hold against a cheating prover given unbounded access to a verification oracle (See Sect. 1.4 for more details). Moreover, DV-NIZKs satisfying soundness against multiple theorems [32,34] are built on tools which are already known to imply CRS-NIZKs. It was not until recently that Kim and Wu [71] in a breakthrough result showed how to construct DP-NIZKs supporting **NP** languages from lattices; this is the first NIZKs for all of **NP** in any model that is based on lattice assumptions. They showed a generic construction of DP-NIZKs from homomorphic signatures (HomSig) and instantiated it with the lattice-based HomSig of [60]. However, despite these recent developments, basing the construction of DV-NIZKs or DP-NIZKs for all of **NP** on pairing-free groups still remains unsolved, and Kim and Wu [71] have stated it as an open problem to construct such NIZKs from the decisional Diffie-Hellman (DDH) assumption.

First Contribution. One of our main contributions is solving this open problem and constructing the first DV-NIZKs from the computational Diffie-Hellman (CDH) assumption over *paring-free groups*. As our scheme is DV-NIZKs and not DP-NIZKs, our techniques depart from [71] and follows more closely to the classical techniques of [47]. More details will be provided in Sect. 1.2.

NIZKs with short proof size. An equally important topic for NIZKs is constructing NIZKs with short proof size. Our construction above solves the open problem of constructing DV or DP-NIZKs from paring-free groups, however, the size of proof is rather large. Namely, it is of size $\mathsf{poly}(\kappa, |C|)$, where κ is the security parameter and $|C|$ is the size of circuit computing the **NP** relation $\mathcal{R}$. In particular, the proof size incurs at least a *multiplicative*-overhead of $O(|C|\kappa)$. As far as we know, the only (CRS, DV, DP)-NIZKs for **NP** in the standard model with a short proof size, i.e., a proof with *additive*-overhead $O(|C|) + \mathsf{poly}(\kappa)$ rather than $O(|C|) \cdot \mathsf{poly}(\kappa)$, either requires a knowledge assumption [62], (fully-) homomorphic encryption (FHE) [52], indistinguishability obfuscation (iO) [86], or HomSig with additional compactness properties [71].[2] Notably, we do not know how to construct (CRS, DV, DP)-NIZKs with short proof size from standard assumptions from paring-free groups. In fact, this is the case even if we were to consider paring groups [1,25,63] as none of the aforementioned heavy machineries are implied from such groups. In other words, it is not known whether DH-type assumptions can be used to construct DV or DP-NIZKs with short proof size.

---

[2] In fact, as we show in Table 1, all of these approaches lead to a much more succinct proof size of $|w| + \mathsf{poly}(\kappa)$, where w is the witness.

Second Contribution. Our second contribution is constructing a DP-NIZK *with short proof size* from a DH-type assumption over paring groups by proposing a compact HomSig scheme from a new non-static DH-type assumption (proven to hold in the generic group model) and following the general conversion from HomSig to DP-NIZK by Kim and Wu [71]. More details will be provided in Sect. 1.2.

Our second scheme achieves the first DP-NIZK with short proof size from any DH-type assumptions, however, one caveat is that the assumption is *non-static* and rather strong, and furthermore requires *paring groups*. Therefore, desirably we would like to construct any type of NIZKs with short proof size from weaker and *static* assumptions such as the DDH assumption while only requiring *paring-free groups*. To this end, we consider a further relaxation of NIZKs in the *preprocessing model* (hereafter referred to as PP-NIZK). In this model, there is a trusted preprocessing setup that generates a verification *and* proving key, where only those with the proving (resp. verification) key can generate (resp. verify) proofs. Analogously to the history of DV and DP-NIZKs, even with this added relaxation, PP-NIZKs turned out to be a rather difficult primitive to construct. There have been several works concerning PP-NIZKs [39,41,43,66,70,73], however, all of them were only *bounded-theorem* in the sense that either the soundness or zero-knowledge property hold in a bounded manner. The problem of constructing *unbounded-theorem* PP-NIZKs, which meets the standard criteria of NIZK, was only recently resolved in the aforementioned paper [71], where Kim and Wu showed a generic construction of PP-NIZKs using homomorphic MACs (HomMAC). In particular, depending on whether the signature can be verified publicly (HomSig) or not (Hom-MAC), their generic construction leads to a DP-NIZK or a PP-NIZK, respectively. In fact, it was observed in [71] that using the compact HomMAC proposed by Catalano and Fiore [27] based on the non-static ℓ-computational DH inversion (ℓ-CDHI) assumption [14,21], we can construct PP-NIZKs from a non-static DH-type assumption over paring-free groups. However, they left it as an open problem to construct HomMAC that suffices for PP-NIZKs (with short proof size) from a weaker static assumption such as DDH.

Final Contribution. Our final contribution is constructing a PP-NIZK *with short proof size* from the DDH assumption over *paring-free groups*. We first construct a non-compact HomMAC from the DDH assumption and exploit extra structures in our HomMAC to achieve short proof size when converting it into a PP-NIZK. More details will be provided in Sect. 1.2.

Motivation for studying different types of NIZKs. Although (DV, DP, PP)-NIZKs may be more restricted compared to CRS-NIZKs, they can be useful nonetheless. For example, applications of CRS-NIZKs including group signatures [9,37], anonymous credentials [35,40], electronic cash [36], anonymous authentication [89] may lead to a designated verifier or prover variant by using DV or DP-NIZKs. In some natural scenarios where we do not require public verifiability or require everybody to be able to construct proofs, these alternatives may suffice. Furthermore, as stated in [71], PP-NIZKs can be used instead of CRS-NIZKs to boost semi-honest security to malicious security [55]. Finally, we believe studying different types of NIZKs and understanding which assumptions imply them will provide us with new insights on realizing the long standing open problem of constructing CRS-NIZKs from paring-free groups or lattices.

1.2 Our Results in Detail

As briefly mentioned above, we give new constructions of DV-NIZK, DP-NIZK, and PP-NIZK with different flavors from DH-type assumptions. Our first and third schemes are instantiated on a pairing-free group, and the second scheme requires a pairing group.

1. We construct DV-NIZKs for **NP** from the CDH assumption over pairing-free groups that resists the verifier rejection attack. This is the first construction of such (DV, DP)-NIZK on pairing-free groups and resolves the open problem posed by Kim and Wu [71].
2. We construct DP-NIZKs for **NP** with short proof size from a newly defined non-static (n, m)-computational DH exponent and ratio (CDHER) assumption (proven in the generic group model) over pairing groups. This is the first NIZK in the standard model to achieve a short proof size without assuming the LWE assumption, fully-homomorphic encryption, indistinguishability obfuscation, or non-falsifiable assumptions. The proof size has an additive-overhead $|C| + \mathsf{poly}(\kappa)$ rather then a multiplicative-overhead $|C| \cdot \mathsf{poly}(\kappa)$ where $|C|$ is the size of the circuit that computes the **NP** relation (See Table 1). Moreover, if we make a slight relaxation in the assumption that the **NP** relation is expressed by a "leveled circuit" [20], then the proof size can be made as short as $|w| + |C|/\log \kappa + \mathsf{poly}(\kappa)$ where $|w|$ denotes the witness size. This is the first NIZK (including PP-NIZKs) that achieves *sublinear* proof size in $|C|$. We note that by applying the same technique to the ℓ-CDHI-based construction of PP-NIZK stated in Kim and Wu [71], we can make their proof size sublinear as well, as long as the **NP** relation can be expressed by a leveled circuit.
3. We construct PP-NIZKs for **NP** with short proof size from the DDH assumption over pairing-free groups that are multi-theorem. This is the first PP-NIZK that achieves a short proof size from a weak and static DH-type assumption such as DDH. (In fact, this construction also serves as a solution to the open problem posed by Kim and Wu [71].) Similarly to the above DP-NIZK, the proof size is $|C| + \mathsf{poly}(\kappa)$. Moreover, going through the same technique with additional observations, in case the **NP** relation can be expressed by a leveled circuit, we are able to make the proof size sublinear $|w| + |C|/\log \kappa + \mathsf{poly}(\kappa)$.

Perhaps of an independent interest, along the way to achieve our second result, we propose an HomSig scheme that simultaneously achieves compactness, context-hiding, and online-offline efficiency under the (n, m)-CDHER assumption. This is the first construction of such HomSig schemes on pairing groups.

The comparison table among existing and our NIZK is given in Table 1. We note that we omit schemes that do not support all of **NP**, do not resist the verifier rejection attack, or do not achieve unbounded-theorem soundness or zero-knowledge from the table.

Table 1. Comparison of NIZKs for **NP**.

Reference	Soundness	ZK	Proof size	Model	Assumption
FLS [47]	stat.	comp.	$\mathsf{poly}(\kappa, \|C\|)$	CRS	trapdoor permutation[‡]
Groth [62]	stat.	comp.	$\|C\| \cdot k_{\mathsf{tpm}} \cdot \mathsf{polylog}(\kappa) + \mathsf{poly}(\kappa)$	CRS	trapdoor permutation[‡]
Groth [62]	stat.	comp.	$\|C\| \cdot \mathsf{polylog}(\kappa) + \mathsf{poly}(\kappa)$	CRS	Naccache-Stern PKE
GOS [63]	perf	comp.	$O(\|C\|\kappa)$	CRS	DLIN/SD
GOS [63]	comp.	perf	$O(\|C\|\kappa)$	CRS	DLIN/SD
CHK, DN, Abu [1,25,46]	stat.	comp.	$\mathsf{poly}(\kappa, \|C\|)$	CRS	CDH
Groth [62]	comp.	perf	$O(\kappa)$	CRS	q-PKE and q-CPDH
GGIPSS [52]	stat.	comp.	$\|w\| + \mathsf{poly}(\kappa)$	CRS	FHE and CRS-NIZK
SW [86]	comp.	perf	$O(\kappa)$	CRS	iO+OWF
KW [71]	stat.[*]	comp.	$\|w\| + \mathsf{poly}(\kappa, d)$	DP	LWE
CF+KW [27]+[71]	comp.	comp.	$\|C\| + \mathsf{poly}(\kappa)$	PP	ℓ-CDHI (pairing-free)
Sect. 3	stat.	comp.	$\mathsf{poly}(\kappa, \|C\|)$	DV	CDH (pairing-free)
Sect. 4	comp.	comp.	$\|C\| + \mathsf{poly}(\kappa)$	DP	(n, m)-CDHER
Sect. 4[†]	comp.	comp.	$\|w\| + \|C\|/\log(\kappa) + \mathsf{poly}(\kappa)$	DP	(n, m)-CDHER
Sect. 5	stat.	comp.	$\|C\| + \mathsf{poly}(\kappa)$	PP	DDH (pairing-free)
Sect. 5[†]	stat.	comp.	$\|w\| + \|C\|/\log(\kappa) + \mathsf{poly}(\kappa)$	PP	DDH (pairing-free)

In column "Soundness" (resp."ZK"), perf., stat., and comp. means perfect, statistical, and computational soundness (resp. zero-knowledge), respectively. In column "Proof size", κ is the security parameter, $\|w\|$ is the witness-size, and $\|C\|$ and d are the size and depth of circuit computing the **NP** relation. In column "Assumption", DLIN stands for the decisional linear assumption, SD stands for the subgroup decision assumption, q-PKE stands for the q-power knowledge of exponent assumption, and q-CPDH stands fo the q-computational power Diffie-Hellman assumption.

[*]Though their primary construction only has computational soundness, they sketched a variant that achieves statistical soundness in the latest version [72, Remark 4.10]

[†]Applicable only when C is a leveled circuit.

[‡]If the domain of the permutation is not $\{0,1\}^n$, we further assume they are doubly-enhanced [53].

1.3   Technical Overview

We rely on mainly two approaches to achieve our results. The first approach is an extension of the construction of CRS-NIZKs from trapdoor permutations by Feige, Lapidot, and Shamir [47] (we call it the FLS construction) to the DV setting. The second approach is constructing (DP, PP)-NIZKs using the Kim-Wu conversion [71] from homomorphic authenticators (HomAuth), where HomAuth are shorthand for HomSig and HomMAC. Specifically, we provide new instantiations of context-hiding HomAuth schemes. Our first result is obtained by the first approach, and the second and third results are obtained by the second approach. In the following, we explain these approaches.

Part 1: DV-NIZK from CDH via FLS paradigm. Our DV-NIZK is based on the Feige-Lapidot-Shamir (FLS) paradigm [47], which enables to construct CRS-NIZKs based on trapdoor permutations (TDP). However, we can not directly use the FLS paradigm since we currently do not know how to achieve TDPs from the CDH assumption. In this study, we present a variant of the FLS construction in the DV setting that can be instantiated by the CDH assumption over paring-free groups.

Our starting point is the CRS-NIZK based on the CDH assumption *over pairing groups* [1,25,46]. The idea is to use a function f_ι defined as follows instead of a TDP for the FLS construction: $f_\iota(X, Z) := X$ if (g, X, Y, Z) is a DH tuple and otherwise $\perp$, where $\iota := (g, Y = g^\tau)$. Though f_ι is not a TDP, it is a trapdoor function (TDF) with a structure that is sufficient for implementing the FLS construction. Below, we take a closer look at the construction.

NIZK in the Hidden Bits Model. Before explaining the construction, we recall the notion of NIZK proof systems in the hidden bits model (hereafter referred to as HBM-NIZK) [47]. In [47], HBM-NIZKs is used as a building block for the final CRS-NIZK. In HBM-NIZK, a prover is provided with a randomly generated string $\rho \xleftarrow{\$} \{0, 1\}^\ell$ (referred to as a *hidden random string*) independently from the statement x and witness w for the **NP** language $\mathcal{L}$. Then it generates a proof π_{hbm} along with an index set I indicating the positions in the hidden random string. A verifier given a sub-string $\rho_{|I}$ of the hidden random string ρ on positions corresponding to the index set I along with the statement x and a proof π_{hbm}, either accepts or rejects. Soundness requires that no adversary can generate a valid proof π_{hbm} with an index set I if $x \notin \mathcal{L}$, and the zero-knowledge property requires that a proof provides no additional knowledge to the verifier beyond that $x \in \mathcal{L}$ *If all bits of ρ on positions corresponding to $[\ell] \setminus I$ are hidden* to the verifier. Feige et al. proved that HBM-NIZKs for all of **NP** exist unconditionally.

CRS-NIZK from CDH with pairings. We now describe the CRS-NIZK based on the CDH assumption over pairing groups [1, 25, 46]. We give a direct (high-level) description without using the abstraction by TDFs for clarity.

Setup(1^κ): Output a CRS crs consisting of a group description $(\mathbb{G}, p, g)$ and random group elements $(X_1, ..., X_\ell) \xleftarrow{\$} \mathbb{G}^\ell$ where ℓ is the length of the hidden random string of the underlying HBM-NIZK.

Prove(crs, x, w): The prover samples $\tau \xleftarrow{\$} \mathbb{Z}_p$, computes $Z_i := X_i^\tau$ and lets ρ_i be the hardcore bit of Z_i for all $i \in [\ell]$. Then it uses $\rho := \rho_1 || \cdots || \rho_\ell$ as a hidden random string to generate a proof π_{hbm} along with an index set $I \subset [\ell]$ by the proving algorithm of the underlying HBM-NIZK on (x, w). It outputs a proof $\pi = (\pi_{\text{hbm}}, I, \{Z_i\}_{i \in I}, Y := g^\tau)$.

Verify(crs, x, π): Given a statement x and a proof $\pi = (\pi_{\text{hbm}}, I, \{Z_i\}_{i \in I}, Y := g^\tau)$, the verification algorithm verifies (g, X_i, Y, Z_i) is a DH-tuple for all $i \in I$ by using pairing, and rejects if it is not the case. Then it computes the hardcore bit ρ_i of Z_i for all $i \in I$, and verifies π_{hbm} by the verification algorithm of the underlying HBM-NIZK.

Roughly speaking, soundness and zero-knowledge follow from those of the underlying HBM-NIZK since a hidden random string ρ is somehow "committed" in $(X_1, ..., X_\ell)$ once τ is fixed, and only the sub-string of them corresponding to I is revealed to the verifier.[3] Clearly, the above construction relies on pairing to check if (g, X_i, Y, Z_i) is a DH-tuple during verification. We note that this check is

---

[3] Though a cheating prover can arbitrarily choose $\tau \in \mathbb{Z}_p$, we can negligibly bound its success probability by the union bound if the success probability of a cheating prover of the underlying HBM-NIZK is bounded by $p^{-1} \cdot \text{negl}(\kappa)$.

essential since without it, a cheating prover can arbitrarily choose Z_i for $i \in I$ to control $\rho_{|I}$ to any value, in which case soundness of HBM-NIZK ensures nothing.

Getting rid of pairing. Now, we explain how to get rid of the use of pairing from the above construction in the DV setting. Our main idea is to use the twin-DH technique [26]. Intuitively, the twin-DH technique enables a designated entity to verify whether a tuple $(g, X, Y, Z) \in \mathbb{G}^4$ is a DH-tuple without knowing the discrete logarithm of X or Y and without using pairings, where (g, X) is public, and (Y, Z) may be chosen arbitrarily. More precisely, suppose that an extra element $\widehat{X} := g^\beta / X^\alpha$ is published in addition to (g, X) where $\alpha, \beta \xleftarrow{\$} \mathbb{Z}_p$. Then for $Y = g^\tau$, we may consider $(Z = X^\tau, \widehat{Z} = \widehat{X}^\tau)$ to be a "proof" that (g, X, Y, Z) is a DH-tuple. Namely, a designated verifier who holds α and β can verify the validity of the "proof" by checking if $Z^\alpha \widehat{Z} = Y^\beta$ holds. The main implication of the twin-DH technique is that the above verification is essentially equivalent to checking if $Z = X^\tau$ and $\widehat{Z} = \widehat{X}^\tau$ hold conditioned on the fact that $(Y, Z, \widehat{Z})$ is chosen by a "prover" who does not know $(\alpha.\beta)$.

With this technique in hand, we describe how to modify the above construction to achieve DV-NIZK without pairing: We add extra elements $\widehat{X}_i := g^{\beta_i} / X^{\alpha_i}$ where $\alpha_i, \beta_i \xleftarrow{\$} \mathbb{Z}_p$ for $i \in [\ell]$ in the CRS, give $\{\alpha_i, \beta_i\}_{i \in [\ell]}$ as the verification key to the designated verifier, and add extra elements $\widehat{Z}_i := \widehat{X}_i^\tau$ for $i \in I$ in the proof. Then the verifier can verify that (g, X_i, Y, Z_i) is a DH-tuple by checking if $Z^{\alpha_i} \widehat{Z} = Y^{\beta_i}$ holds *without using pairing*. This enables us to achieve DV-NIZK without pairing.

On adaptive zero-knowledge. Though our main idea is as presented above, the above described construction only achieves *non-adaptive zero-knowledge* which requires an adversary to choose the statement x independently of the CRS. To achieve adaptive zero-knowledge, we need to add some extra structures using the technique of non-committing encryption [24,46]. See Sect. 3 for technical details. We note that the original FLS NIZK proof system is also adaptive zero-knowledge, but it uses specific properties of the underlying HBM-NIZK. Though a similar analysis may also yield alternative construction of DV-NIZKs with adaptive zero-knowledge from the CDH assumption without pairing, we choose the above approach where we do not assume any structure on the underlying HBM-NIZK for a conceptually simpler and modular construction.

Part 2: PP-NIZK via context-hiding HomAuth. Kim and Wu [71] showed a conversion from any context-hiding HomAuth scheme to PP-NIZKs. In particular, they noted that context-hiding HomAuth scheme for $\mathbf{NC}^1$ suffices to instantiate their conversion. In this part, we propose new constructions of context-hiding HomAuth schemes for $\mathbf{NC}^1$, and plug them into their conversion. First, we recall the definition of HomAuth. Roughly speaking, a HomAuth scheme is a digital signature or MAC scheme with a homomorphic property. Namely, given a vector of signatures $\boldsymbol{\sigma}$ for a vector of messages $\mathbf{x}$, anyone can publicly evaluate the signature on a circuit C to generate an evaluated signature σ for a message

$C(\mathbf{x})$. We say that a HomAuth scheme is a HomSig scheme if verification can be done publicly, and is a HomMAC scheme otherwise. As a security requirement of HomAuth scheme, we require that an adversary given $\mathbf{x}$ cannot generate a pair of an evaluated signature σ^* and a circuit C^* such that σ^* is a valid signature for a message $z \neq C^*(\mathbf{x})$ even if the adversary is given access to a verification oracle. In addition, we say that a HomAuth scheme is context-hiding if σ for a message z generated by evaluating a circuit C on a vector of signatures $\boldsymbol{\sigma}$ for $\mathbf{x}$ does not reveal information of $\mathbf{x}$ beyond that $C(\mathbf{x}) = z$.

In this paper, we propose two new constructions of HomAuth schemes for $\mathbf{NC}^1$. The first one is a HomSig scheme based on a new assumption that we call (n, m)- CDHER assumption on a pairing-group. A nice feature of this HomSig scheme is that the size of an evaluated signature is compact (i.e., does not depend on the message vector length or the circuit to evaluate), and has online-offline efficiency. The second one is a HomMAC scheme based on the DDH assumption on a pairing-free group. The function class the second scheme supports is arithmetic circuits over $\mathbb{Z}_p$ of polynomial degree, which is larger than $\mathbf{NC}^1$, and we take advantage of this extra freedom to improve the proof size. We explain these constructions below.

HomSig from CDHER. Here, we informally explain how an attribute-based encryption (ABE) scheme with some special properties can be converted into a HomSig scheme. Our HomSig scheme from the CDHER assumption can be seen as an instantiation of this conversion.

To explain the idea, we first recall the notion of (key-policy) ABE. In an ABE scheme, one can encrypt a message M with respect to some string $\mathbf{x} \in \{0,1\}^\ell$ using some public parameter pp. Furthermore, a secret key is associated with some policy $C : \{0,1\}^\ell \to \{0,1\}$ and the decryption is possible if and only if $C(\mathbf{x}) = 1$. As for security, we require the selective *one-way* security. In a selective one-way security game, an adversary has to declare its target $\mathbf{x}^*$ at the beginning of the game before seeing the public parameter pp. An adversary can further query secret keys for C such that $C(\mathbf{x}^*) = 0$ unbounded polynomially many times throughout the game, and we require that an adversary given an encryption of a *random* message M^* under the string $\mathbf{x}^*$ cannot recover M^*.

We first observe that the security proofs for most selectively secure schemes such as those proposed in [18,59,61,87,92] can be abstracted in the following manner:[4] At the beginning of the game, the reduction algorithm is given a problem instance Ψ of some hard problem (e.g., the bilinear Diffie-Hellman problem). Then, it first runs the adversary to obtain the target $\mathbf{x}^*$. Given Ψ and $\mathbf{x}^*$, the reduction algorithm generates pp along with some simulation trapdoor $\mathsf{td}_{\mathbf{x}^*}$. The reduction algorithm can perfectly simulate the game using $\mathsf{td}_{\mathbf{x}^*}$. Namely, given $\mathsf{td}_{\mathbf{x}^*}$, it can generate correctly distributed secret key sk_C for any C such that $C(\mathbf{x}^*) = 0$. Furthermore, given $\mathsf{td}_{\mathbf{x}^*}$, it can embed the problem instance Ψ into the challenge ciphertext so that it can extract the answer of the hard problem whenever the adversary succeeds in extracting M^*.

---

[4] Actually, these previous works prove the standard indistinguishability security notion rather than one-wayness. However, one-wayness is sufficient for our application.

Our basic idea for constructing HomSig is to use the above reduction algorithm in the real world. To sign on a message $\mathbf{x}$, we generate $\mathsf{td}_{\mathbf{x}}$ and set $\sigma := \mathsf{td}_{\mathbf{x}}$. To evaluate the signature σ on a circuit C such that $C(\mathbf{x}) = 0$, we run the reduction algorithm of the ABE scheme on input $\mathsf{td}_{\mathbf{x}}$ to generate sk_C and set $\sigma := \mathsf{sk}_C$. Here, evaluation of signatures can be done publicly since $\mathsf{td}_{\mathbf{x}}$ is the only secret state required to run the reduction algorithm. A subtle problem with this approach is that we cannot evaluate the signature on a circuit C such that $C(\mathbf{x}) = 1$ since the reduction algorithm does not work for such C. This problem can be easily fixed by defining the scheme so that when evaluating a signature on such C, we generate $\mathsf{sk}_{\neg C}$ instead of sk_C, where $\neg C$ is a circuit that is obtained by flipping the output bit of C by applying the NOT gate. Now, for the signature $\sigma = \mathsf{sk}_C$ to be publicly verifiable, we require it to be possible to efficiently check whether σ is a correctly generated secret key of the ABE given (C, σ). However, this is not such a strong restriction since it is satisfied by many selectively secure ABE schemes such as the ones listed above.

We recall that given $\mathsf{td}_{\mathbf{x}}$, the reduction algorithm can perfectly simulate the selective security game for ABE where $\mathbf{x}$ is the target chosen by the adversary. This in particular implies that sk_C simulated by $\mathsf{td}_{\mathbf{x}}$ follows the same distribution as sk_C generated in the real system which does not use information of $\mathbf{x}$. Then, the context-hiding property of the scheme follows from this fact. Namely, the distribution of $\sigma = \mathsf{sk}_C$ only depends on C and pp, not on $\mathbf{x}$. In other words, σ does not leak any information of $\mathbf{x}$, which meets the requirements of the context-hiding security. Furthermore, the unforgeability of the scheme follows from the one-wayness of the ABE: If the adversary can forge a signature $\sigma = \mathsf{sk}_{C^\star}$ for $C^\star$ such that $C^\star(x) = 1$, then $\mathsf{sk}_{C^\star}$ can be used to decrypt the challenge ciphertext, which contradicts the security of the ABE. We note that the circuit class of the allowed homomorphic evaluation for the resulting HomSig scheme is roughly the same as the circuit class supported by the original ABE scheme.

In order to obtain the aforementioned HomSig scheme for $\mathbf{NC}^1$ with compact signatures, we need a key-policy ABE scheme with constant-size secret keys. Unfortunately, the only construction of ABE scheme [7] which meets the efficiency (i.e., compactness) property we require does not conform to our template that uses the simulation trapdoor $\mathsf{td}_{\mathbf{x}}$. Therefore, we construct a new ABE scheme with the required property which conforms to our template based on the CDHER assumption. The structure of our ABE scheme is inspired by the ciphertext-policy ABE scheme with constant-size ciphertexts (*not* secret keys) due to Agrawal and Chase [3]. To turn their scheme into an ABE scheme with constant-size secret keys, at a high level, we swap the ciphertexts and secret keys of their construction. Since the security of the resulting scheme is not guaranteed by that of the original one, we directly prove its security by adding considerable modification to the previous proof techniques [4,83].

HomMAC from DDH. Here, we explain the construction of HomMAC under the DDH assumption. Our idea is to add the context-hiding property to the non-context-hiding HomMAC proposed by Catalano and Fiore [27] by using functional encryption for inner products (IPFE). First, we recall their non-context-hiding HomMAC, which supports all arithmetic circuits of polynomially bounded degree.[5] The signing/verification key of their construction are $\mathbf{r} \in \mathbb{Z}_p^\ell$ and $s \in \mathbb{Z}_p^*$ where ℓ is the arity of arithmetic circuits it supports, and the evaluation key is a prime p. A signature $\boldsymbol{\sigma} \in \mathbb{Z}_p^\ell$ for a message $\mathbf{x} \in \mathbb{Z}_p^\ell$ is set to be $\boldsymbol{\sigma} := (\mathbf{r} - \mathbf{x})s^{-1}$ mod p.[6] Given an arithmetic circuit f of degree D, a message $\mathbf{x}$, and a signature $\boldsymbol{\sigma}$, the evaluation algorithm computes the coefficients $(c_1, ..., c_D) \in \mathbb{Z}_p^D$ that satisfy $f(\mathbf{r}) = f(\mathbf{x}) + \sum_{j=1}^D c_j s^j$, and sets $\sigma := (c_1, ..., c_D)$ as an evaluated signature. We remark that this can be done by using $\mathbf{x}$, $\boldsymbol{\sigma}$, and p without knowing $(r_1, ..., r_n)$ or s since the signatures satisfy $s\boldsymbol{\sigma} + \mathbf{x} = \mathbf{r}$ mod p. To verify the evaluated signature, the verifier simply checks if the above equation holds by using $\mathbf{r}$ and s included in the verification key. Though the construction is very simple, the scheme satisfies unforgeability even against unbounded-time adversaries. Unfortunately, this construction cannot yet be used for the purpose of PP-NIZKs, since in general it is not context-hiding.

Here, we observe that in the above construction, what a verifier has to know for the verification is only $\sum_{j=1}^D c_j s^j$, and not the entire $(c_1, ..., c_D)$. Moreover, $\sum_{j=1}^D c_j s^j$ does not convey any information on $\mathbf{x}$ beyond $f(\mathbf{x})$ because the term is determined solely by $\mathbf{r}$ and $f(\mathbf{x})$. Therefore if there exists a way to only transfer $\sum_{j=1}^D c_j s^j$ to the verifier, then context-hiding is guaranteed. We remark that a trivial idea of publishing s does not work because it completely breaks the unforgeability. In particular, we want to find a way to let a verifier only know $\sum_{j=1}^D c_j s^j$ without providing s to the evaluator. To solve this problem we rely on IPFE. In an IPFE scheme, both a ciphertext and a secret key are associated with a vector. If we decrypt a ciphertext of a vector $\mathbf{x}$ by a secret key associated with $\mathbf{y}$, then the decryption result is $\langle \mathbf{x}, \mathbf{y} \rangle$, which is an inner product of $\mathbf{x}$ and $\mathbf{y}$. We convert the above non-context-hiding HomMAC to a context-hiding one by using IPFE as follows: In the setup, we additionally generate a public parameter pp and a master secret key msk of IPFE. Then a verifier is provided with a secret key $\mathsf{sk}_{(s,...,s^D)}$ for a vector $(s, ..., s^D)$, and an evaluator is provided with pp. The evaluator sets the evaluated signature to be an encryption ct of $(c_1, ..., c_D)$ instead of $(c_1, , , ., c_D)$ itself. Now, a verifier only learns $\sum_{j=1}^D c_j s^j$ due to the security of IPFE, and thus context-hiding is achieved.

Given the above overview, it may seem that any IPFE scheme suffices for the construction. Moreover, since only one secret key is needed in the construction, it seems that one-key IPFE suffices. Since there are constructions of one-key secure FE even for all circuits based on any PKE scheme [58,85], one may think that we can implement the above construction based on any PKE scheme. However, this is in fact not the case because these FE schemes are malleable. Namely, the standard security notion of FE does not prevent a malicious encryptor from generating an invalid ciphertext. Put differently, the decryption result may be

---

[5] Though the original construction by Catalano and Fiore [27] is based on PRF, we present an information theoretically secure variant of it in a simplified setting where the arity of an arithmetic circuit is bounded.

[6] Though the scheme is not publicly verifiable, we call σ a "signature" for compatibility to HomSig.

controlled. In the context of the above construction, the fact that an evaluator generates a ciphertext ct by the secret key $\mathsf{sk}_{(s,\ldots,s^D)}$ that is decrypted to T does not necessarily mean that it knows $(c_1, \ldots, c_D)$ such that $\sum_{j=1}^{D} c_j s^j = T$. Therefore, although the construction seems to work, we cannot prove unforgeability of the above scheme. To solve this problem, we introduce a notion which we call *extractability* for IPFE. Extractability requires that for any (possibly malformed) ciphertext ct that is decrypted to T with a secret key sk associated with a vector $\mathbf{y}$, we can extract $\mathbf{x}$ such that $\langle \mathbf{x}, \mathbf{y} \rangle = T$ from ct. It is clear that the above problem is resolved if we have an extractable IPFE.

Here, we observe that the IPFE scheme based on the DDH assumption proposed by Agrawal, Libert, and Stehlé [5] satisfies extractability. A subtle problem of their construction is that a decryptor must compute a discrete logarithm for computing a decryption result, and thus the size of the decryption result must be limited to being relatively small. Fortunately, this does not matter in our application since the verification is done by simply checking if a decryption result of IPFE satisfies a certain linear equation which can be performed on the exponent. Concretely, we only need a variant of IPFE that enables a decryptor to learn inner-product on the exponent. Putting all the ideas together, we obtain a context-hiding HomMAC for arithmetic circuits of polynomial degree (which includes $\mathbf{NC}^1$) based on the DDH assumption, which further combined with [71] leads to PP-NIZK proofs based on the DDH assumption. Moreover, we can make the proof size of the PP-NIZK short by incorporating the idea by Katsumata [69]. Namely, the proof size of the resulting PP-NIZK is $|C| + \mathsf{poly}(\kappa)$ where $|C|$ is the size of a circuit that computes a relation to prove. See the full version for details.

PP-NIZK with sublinear proof size. Direct adaptations of the Kim-Wu conversion to compact context-hiding HomAuth for $\mathbf{NC}^1$ yield PP-NIZK with proof sizes $|C| + \mathsf{poly}(\kappa)$. Here, we explain that this can be further reduced to *sublinear size* $|w| + |C|/\log \kappa + \mathsf{poly}(\kappa)$ by making a slight relaxation that a circuit C computing the **NP** relation is expressed as a *leveled circuit* [20]; a circuit whose gates are partitioned into $D + 1$ levels and all incoming wires to a gate of level $i + 1$ come from gates of level i for each $i \in [D]$. To explain this, we first briefly review the Kim-Wu conversion. In their construction, a prover is provided with a secret key K of a symmetric key encryption (SKE) scheme as its proving key, and to prove that (x, w) satisfies $C(x, w) = 1$ for a circuit C, it encrypts w by using K to generate a ciphertext ct, and generates an evaluated signature σ on message "1" under the function $f_{\mathsf{ct},x}$ defined by $f_{\mathsf{ct},x}(K') := C(x, \mathsf{Dec}(K', \mathsf{ct}))$ where Dec is the decryption algorithm of the SKE scheme. A proof consists of ct and σ. A verifier simply verifies that the evaluated signature σ is a valid signature on message "1" under the function $f_{\mathsf{ct},x}$. To implement this construction based on HomAuth for $\mathbf{NC}^1$, we have to express a circuit that computes the **NP** relation in $\mathbf{NC}^1$. This is in general possible by "expanding" the witness to values corresponding to all wires of $C(x, \cdot)$. However, since the size of the expanded witness is as large as the circuit size $|C|$, the proof size of the resulting PP-NIZK is linear in $|C|$. Now, we observe that we actually need not expand the witness to all wires, and we can choose a portion of them based on a similar idea used in [20]. Namely, for a leveled circuit C of depth D, we divide $[D]$ into $\log \kappa$ intervals of length $D/\log \kappa$, and choose "special levels" i in each interval so that the number of gates of level i is the smallest among those in the interval. Then we set an expanded witness to be the original witness appended by values corresponding to all wires *of special levels* of $C(x, \cdot)$. We observe that the consistency of the expanded witness generated in this way still can be verified in $\mathbf{NC}^1$ since successive special levels are at most $2\log \kappa$ apart from each other. Moreover, the size of the expanded witness is at most $|w| + |C|/\log \kappa$ since the number of gates of special levels is at most $|C|/\log \kappa$ by the choice of special levels. Thus, by applying the Kim-Wu conversion with the above expanded witness, we obtain PP-NIZK with proof size $|w| + |C|/\log \kappa + \mathsf{poly}(\kappa)$.

1.4   Other Related Works

Concurrent Works. There are two concurrent and independent works [38,80] that contain similar results to our first result, namely, multi-theorem DV-NIZK from CDH assumption in pairing-free groups. We summarize differences of these results below.

– Couteau and Hofheinz [38] additionally give a construction of (CRS,DV)-NIZK assuming the LWE assumption and a (CRS,DV)-non-interactive witness indistinguishable proof system for bounded distance decoding.
– Quach, Rothblum, and Wichs [80] additionally consider a stronger variant of DV-NIZK called malicious DV-NIZK, and construct it based on a stronger assumption called the one-more CDH assumption in pairing-free groups.
– Constructions of (DP,PP)-NIZKs with compact proofs are unique to this paper.

CRS-NIZK from Lattices. Very recently, Peikert and Shiehian [78] constructed the first CRS-NIZKs for $\mathbf{NP}$ under standard lattice assumptions following the line of researches [22,23,65,68] to instantiate the Fiat-Shamir transform [48] in the standard model.

More discussions on existing (DV, DP, PP)-NIZK. Unlike CRS-NIZKs where proving statements and verifying proofs can be done publicly, in (DV, DP, PP)-NIZKs since we have the notion of secret states, it is not uncommon to have a bound on the number of statements (i.e., theorems) one can prove without compromising soundness or zero-knowledge. In DV-NIZKs, a common issue have been the bound on the number of time the prover can query the verification oracle. Namely, a prover can break the soundness of a DV-NIZK if the verifier uses the same verification key to verify multiple statements. Due to this fact, such DV-NIZKs that require a bound on the number of time a prover can query the verification oracle are called *bounded-theorem*. If the verifier can keep using the same key for multiple statements, then it is called *multi-theorem*. Almost all previous DV-NIZKs for all of **NP** [33,42,74,77,91] suffered from this issue of being bounded-theorem. There are more recent works that avoid the above issue based on a certain type of additively homomorphic encryption [32] or a primitive called oblivious linear-function evaluation [34]. However, instantiating either of these primitives require an assumption that is already known to imply a CRS-NIZK. DP and PP-NIZKs share similar problems, where in this case, zero-knowledge does not hold if the prover uses the same proving key multiple statements. Other than the recent schemes by Kim and Wu [71] and Boyle et al [19], all previous DP or PP-NIZKs [39,41,43,66,70,73] are known to be bounded-theorem. Though it is known that we can convert any bounded theorem NIZK to unbounded theorem NIZK in the CRS setting [47], the conversion heavily relies on the fact that proofs can be generated publicly, and does not seem to work in the PP model. We refer to [71] for more discussions.

Homomorphic authenticators. The notion of homomorphic authenticators (MACs or signatures) originates to Desmedt [44] and was first formalized by Johnson et al. [67]. In the beginning, HomAuth was considered extensively in the context of network coding where the homomorphism were focused on linear functions, yielding a long line of interesting works such as [2,8,15–17,28,29, 31,49,50]. HomAuth for linear functions has also been considered for proofs of retrievability for outsourced storage [6,88]. Boneh and Freeman [16] were the first to consider homomorphism beyond linear functions, showing the first scheme for polynomial function based on lattices. Since then numerous improvements on HomAuth have been made [27,29,51,60]. Gorbunov et al. [60] constructed a HomSig that supports arbitrary circuits with bounded-depth from lattices and Catalano et al. [27] constructed a HomMAC that supports arbitrary arithmetic circuits with bounded-degree from PRFs or DH-type assumptions.

Recently, Tsabary [90] showed a generic conversion of an attribute-based signature (ABS) to HomSig. Using their construction, we may obtain a HomSig with compact signatures starting from an ABS with short signatures. However, the two ABS schemes with short signatures are not a complete fit for the conversion: The scheme by Attrapadung et al. [7] is only selectively-secure and the above conversion is not applicable. The scheme by [75] is constructed on composite-order groups, which is not desirable from the view points of security and efficiency.

Finally, we also mention that our idea of viewing some types of ABE as HomSig seems to be applicable for other ABE schemes such as [61]. This leads to a context-hiding HomSig scheme from the CDH assumption and thus DP-NIZK from the same assumption via the transformation due to Kim and Wu [71]. In addition, we observe that if we start from the ABE for circuits from lattices due to Boneh et al. [18], we recover the existing HomSig scheme by Gorbunov, Vaikuntanathan, and Wichs [60]. While this is not a new result, the observation provides new insights into the connection between them.

2   Preliminaries

We omit basic notations and knowledge on cryptography due to limited space.

2.1   Preprocessing NIZKs

Let $\mathcal{R} \subseteq \{0,1\}^* \times \{0,1\}^*$ be a polynomial time recognizable binary relation. For $(x, w) \in \mathcal{R}$, we call x as the statement and w as the witness. Let $\mathcal{L}$ be the corresponding **NP** language $\mathcal{L} = \{x \mid \exists w \text{ s.t. } (x, w) \in \mathcal{R}\}$. We also write $\mathcal{R}(x, w) \in \{0, 1\}$ as the output of the polynomial time decision algorithm $\mathcal{R}$ on input (x, w), where 0 is for reject and 1 is for accept. Below, we define (adaptive multi-theorem) preprocessing NIZKs for **NP** languages. Some discussions on our presentation of NIZKs are provided below.

Definition 2.1 (NIZK Proofs). *A non-interactive zero-knowledge (NIZK) proof in the preprocessing model Π_{PPNIZK} for the relation $\mathcal{R}$ is defined by the following three polynomial time algorithms:*

$\mathsf{Setup}(1^\kappa) \rightarrow (\mathsf{crs}, k_\mathsf{P}, k_\mathsf{V})$*: The setup algorithm takes as input the security param-eter 1^κ and outputs a common reference string crs, a proving key k_P, and a verification key k_V. This algorithm is executed as the "preprocessing" step.*

$\mathsf{Prove}(\mathsf{crs}, k_\mathsf{P}, x, w) \rightarrow \pi$*: The prover's algorithm takes as input a common refer-ence string crs, a proving key k_P, a statement x, and a witness w and outputs a proof π.*

$\mathsf{Verify}(\mathsf{crs}, k_\mathsf{V}, x, \pi) \rightarrow \top \text{ or } \bot$*: The verifier's algorithm takes as input a common reference string, a verification key k_V, a statement x, and a proof π and outputs $\top$ to indicate acceptance of the proof and $\bot$ otherwise.*

Moreover, an (adaptive multi-theorem) NIZK proof in the preprocessing model Π_{PPNIZK} is required to satisfy the following properties, where the probabilities are taken over the random choice of the algorithms:

Completeness. *For all pairs $(x, w) \in \mathcal{R}$, if we run $(\mathsf{crs}, k_\mathsf{P}, k_\mathsf{V}) \leftarrow \mathsf{Setup}(1^\kappa)$, then we have*

$$\Pr[\pi \leftarrow \mathsf{Prove}(\mathsf{crs}, k_\mathsf{P}, x, w) : \mathsf{Verify}(\mathsf{crs}, k_\mathsf{V}, x, \pi) = \top] = 1.$$

Soundness. *For all (possibly inefficient) adversaries $\mathcal{A}$, if we run $(\mathsf{crs}, k_\mathsf{P}, k_\mathsf{V}) \leftarrow \mathsf{Setup}(1^\kappa)$, then we have*

$$\Pr[(x, \pi) \leftarrow \mathcal{A}^{\mathsf{Verify}(\mathsf{crs}, k_\mathsf{V}, \cdot, \cdot)}(1^\kappa, \mathsf{crs}, k_\mathsf{P}) : x \notin \mathcal{L} \wedge \mathsf{Verify}(\mathsf{crs}, k_\mathsf{V}, x, \pi) = \top] = \mathsf{negl}(\kappa).$$

Here, in case soundness only holds for computationally bounded adversaries $\mathcal{A}$, we say it is a NIZK argument.

(Non-Programmable CRS) Zero-Knowledge. *For all PPT adversaries $\mathcal{A}$, there exists a PPT simulator $\mathcal{S} = (\mathcal{S}_1, \mathcal{S}_2)$ such that if we run $(\mathsf{crs}, k_\mathsf{P}, k_\mathsf{V}) \leftarrow \mathsf{Setup}(1^\kappa)$ and $\tau_\mathsf{V} \leftarrow \mathcal{S}_1(1^\kappa, \mathsf{crs}, k_\mathsf{V})$, then we have*

$$\left| \Pr[\mathcal{A}^{\mathcal{O}_0(\mathsf{crs}, k_\mathsf{P}, \cdot, \cdot)}(1^\kappa, \mathsf{crs}, k_\mathsf{V}) = 1] - \Pr[\mathcal{A}^{\mathcal{O}_1(\mathsf{crs}, k_\mathsf{V}, \tau_\mathsf{V}, \cdot, \cdot)}(1^\kappa, \mathsf{crs}, k_\mathsf{V}) = 1] \right| = \mathsf{negl}(\kappa),$$

where $\mathcal{O}_0(\mathsf{crs}, k_\mathsf{P}, x, w)$ *outputs* $\mathsf{Prove}(\mathsf{crs}, k_\mathsf{P}, x, w)$ *if* $(x, w) \in \mathcal{R}$ *and* $\perp$ *otherwise, and* $\mathcal{O}_1(\mathsf{crs}, k_\mathsf{V}, \tau_\mathsf{V}, x, w)$ *outputs* $\mathcal{S}_2(\mathsf{crs}, k_\mathsf{V}, \tau_\mathsf{V}, x)$ *if* $(x, w) \in \mathcal{R}$ *and* $\perp$ *otherwise.*

Remark 2.1 (Programmable Zero-Knowledge). As also discussed in [72], we can define a slightly weaker variant of zero-knowledge where the simulator is provided the freedom of programming the common reference string crs and verification key k_V.

(Programmable CRS) Zero-Knowledge. For all PPT adversaries $\mathcal{A}$, there exists a PPT simulator $\mathcal{S} = (\mathcal{S}_1, \mathcal{S}_2)$ such that if we run $(\mathsf{crs}, k_\mathsf{P}, k_\mathsf{V}) \leftarrow \mathsf{Setup}(1^\kappa)$ and $(\overline{\mathsf{crs}}, \bar{k}_\mathsf{V}, \bar{\tau}_\mathsf{V}) \leftarrow \mathcal{S}_1(1^\kappa)$, then we have

$$\left| \Pr[\mathcal{A}^{\mathcal{O}_0(\mathsf{crs}, k_\mathsf{P}, \cdot, \cdot)}(1^\kappa, \mathsf{crs}, k_\mathsf{V}) = 1] - \Pr[\mathcal{A}^{\mathcal{O}_1(\overline{\mathsf{crs}}, \bar{k}_\mathsf{V}, \bar{\tau}_\mathsf{V}, \cdot, \cdot)}(1^\kappa, \overline{\mathsf{crs}}, \bar{k}_\mathsf{V}) = 1] \right| = \mathsf{negl}(\kappa),$$

where $\mathcal{O}_0(\mathsf{crs}, k_\mathsf{P}, x, w)$ outputs $\mathsf{Prove}(\mathsf{crs}, k_\mathsf{P}, x, w)$ if $(x, w) \in \mathcal{R}$ and $\perp$ otherwise, and $\mathcal{O}_1(\overline{\mathsf{crs}}, \bar{k}_\mathsf{V}, \bar{\tau}_\mathsf{V}, x, w)$ outputs $\mathcal{S}_2(\overline{\mathsf{crs}}, \bar{k}_\mathsf{V}, \bar{\tau}_\mathsf{V}, x)$ if $(x, w) \in \mathcal{R}$ and $\perp$ otherwise.

This definition captures the zero-knowledge property used in standard NIZKs in the common reference string (CRS) model. In the CRS model, the Setup algorithm outputs a CRS σ used by both the prover and verifier, and the zero-knowledge simulator is allowed to program the CRS σ. Specifically, the proving key and verification key are both set as the CRS σ.

Remark 2.2 (Different types of NIZKs). The definition is general enough to capture many of the existing types of NIZKs. In case $k_\mathsf{P} = k_\mathsf{V} = \perp$, the above definition captures the standard NIZKs in the common reference string (CRS) model, which we refer to as CRS-NIZKs hereafter. Specifically anybody can construct a proof using the public CRS and those proofs are publicly verifiable [47]. On the other hand, in case $k_\mathsf{P} = \perp$ but k_V is required to be kept secret, the above definition captures *designated verifier* NIZKs (DV-NIZKs) [42,77]. Moreover, in case $k_\mathsf{V} = \perp$ but k_P is required to be kept secret, the above definition captures designated prover NIZKs (DP-NIZKs) [71]. Finally, in case both k_P and k_V must be kept secret, it is simply called preprocessing NIZKs (PP-NIZKs) [39].

Remark 2.3 (Bounded and Multi-Theorem NIZK). Unlike CRS-NIZKs where there are nothing to be kept secret, (DV, DP, PP)-NIZKs take more subtle care to construct. Specifically, the latter types of NIZKs may possibly leak secret information when constructing a proof (DP-NIZKs) or verifying a proof (DV-NIZKs). We say the scheme is *bounded-theorem* if the number of statements supported by the scheme to guarantee soundness or zero-knowledge is bounded before setup. Otherwise, we say the scheme is *multi-theorem*. All the NIZKs we construct in this paper are multi-theorem. Finally, we call the scheme *single-theorem* if it only supports one statement.

Remark 2.4 (Adaptive and Non-Adaptive NIZK). One often considers weaker security called *non-adaptive* soundness and zero-knowledge. In non-adaptive soundness, an adversary has to declare the statement x on which he forges a proof before seeing a common reference string. In non-adaptive zero-knowledge, an adversary has to declare a pair of a statement x and its witness w to query the proving oracle before seeing a common reference string. All the NIZKs we construct in this paper satisfy adaptive soundness and zero-knowledge.

NIZKs for Bounded Languages. Throughout this paper, we mainly consider the weaker variant of PP-NIZKs which we call PP-NIZKs *for bounded languages* as was done by Kim and Wu [71]. PP-NIZKs for bounded languages enable one to generate a proof for $(x, w) \in \mathcal{R} \cap (\{0, 1\}^{n(\kappa)} \times \{0, 1\}^{m(\kappa)})$ for a priori bounded polynomials $n(\cdot)$ and $m(\cdot)$. For clarity, we say PP-NIZKs *for unbounded languages* to express PP-NIZKs that do not have the above limitation. As discussed in the full version, we can generically convert any PP-NIZKs for bounded languages to PP-NIZKs for unbounded languages at the cost of making the proof size larger. However, we note that since the conversion makes the proof size larger, the distinction between PP-NIZKs for bounded and unbounded languages are meaningful if we start to consider proof sizes.

3   DV-NIZK from CDH via FLS Transform

In this section, we construct a DV-NIZK from the CDH assumption over pairing-free groups based on the FLS construction [47] for CRS-NIZKs from TDPs. More formally, we prove the following theorem.

Theorem 3.1. *If the CDH assumption holds on a pairing-free group, then there exists an (adaptive multi-theorem) DV-NIZK proof system for all* **NP** *languages.*

The theorem is proven in the following steps:

1. We first construct a variant of DV-NIZK proof system (which we call the *base proof system*) with a special syntax satisfying a relaxed notion of soundness and adaptive *single-theorem* zero-knowledge. We construct it from a NIZK proof system in the hidden-bits model based on the CDH assumption over pairing-free groups. This is done by applying the FLS construction [47] along with the twin-DH technique. A relaxed notion of adaptive zero-knowledge is achieved by using a technique often used in non-committing encryption.
2. We then construct an adaptive *designated-verifier non-interactive witness indistinguishable* (DV-NIWI) proof for all **NP** languages by running many copies of the base proof system in parallel.
3. Finally, we transform our adaptive DV-NIWI proofs into adaptive multi-theorem DV-NIZK proofs by using pseudorandom generators via the transformation of Feige, Lapidot, and Shamir [47] (i.e., the technique of FLS is applicable to the DV-NIZK setting).

3.1   Preliminaries

We introduce the Goldreich-Levin hardcore function $\mathsf{GL}(a; r)$. This is defined by $\mathsf{GL}(a; r) := \langle a, r \rangle := \bigoplus_{j=1}^{u} (a_j \cdot r_j)$ where $a, r \in \{0, 1\}^u$ and σ_j denotes the j-th bit of a string σ. In fact, we use groups in our construction and the input to GL is an element in $\mathbb{G}$. Thus, we interpret a group element $g^{r_i} \in \mathbb{G}$ as a u-bit-string.

Theorem 3.2 (Goldreich-Levin Theorem (adapted)) [54]. *Assuming that the CDH assumption holds, it holds that*

$$\left| \Pr[\mathsf{Expt}_{\mathcal{A}}^{\mathsf{GL\text{-}cdh}}(\kappa, 0) = 1] - \Pr[\mathsf{Expt}_{\mathcal{A}}^{\mathsf{GL\text{-}cdh}}(\kappa, 1) = 1] \right| \leq \mathsf{negl}(\kappa),$$

where the experiment $\mathsf{Expt}_{\mathcal{A}}^{\mathsf{GL\text{-}cdh}}(\kappa, \mathsf{coin})$ *is defined as follows.*

$\underline{\mathsf{Expt}_{\mathcal{A}}^{\mathsf{GL\text{-}cdh}}(\kappa, \mathsf{coin})}$

Samples $(\mathbb{G}, p, g) \xleftarrow{\$} \mathsf{GGen}(1^{\kappa})$, $R \xleftarrow{\$} \{0,1\}^{u}$, *and* $x, y \xleftarrow{\$} \mathbb{Z}_p$.

If $\mathsf{coin} = 1$, *then* $\rho \xleftarrow{\$} \{0,1\}$, *else if* $\mathsf{coin} = 0$, *then* $\rho := \mathsf{GL}(g^{xy}; R)$.

Output $\mathsf{coin}' \leftarrow \mathcal{A}(1^{\kappa}, \mathbb{G}, p, g, g^{x}, g^{y}, R, \rho)$

Next, we introduce a theorem called twin-DH trapdoor test which enables one to check if a tuple (g, X, Y, Z) is a DH-tuple without knowing the discrete logarithm of X or Y by using a special trapdoor.

Theorem 3.3 (Twin-DH Trapdoor Test) [26]. *For any* $(\mathbb{G}, p, g) \leftarrow \mathsf{GGen}(\kappa)$ *and function F, it holds that*

$$\Pr\left[(Z^{\alpha} \widehat{Z} \overset{?}{=} Y^{\beta}) \neq ((Z \overset{?}{=} Y^{x}) \wedge (\widehat{Z} \overset{?}{=} Y^{\hat{x}})) \;\middle|\; \begin{array}{l} X \xleftarrow{\$} \mathbb{G}, \\ \alpha, \beta \xleftarrow{\$} \mathbb{Z}_p, \; \widehat{X} := g^{\beta}/X^{\alpha}, \\ (Y, Z, \widehat{Z}) \leftarrow F((\mathbb{G}, p, g), X, \widehat{X}) \end{array} \right] \leq 1/p,$$

where $X = g^{x}$ *and* $\widehat{X} = g^{\hat{x}}$.

We introduce the notion of witness indistinguishability.

Definition 3.1 (Adaptive WI (in the DV model)). *We say that a proof system Π satisfies adaptive witness indistinguishability if for all PPT adversaries $\mathcal{A}$ that makes arbitrary number of queries (resp. at most 1 query), if we run* $(\mathsf{crs}, k_{\mathsf{V}}) \leftarrow \mathsf{Setup}(1^{\kappa})$, *then we have*

$$\left| \Pr[\mathcal{A}^{\mathcal{O}_0(\mathsf{crs}, \cdot, \cdot, \cdot)}(1^{\kappa}, \mathsf{crs}, k_{\mathsf{V}}) = 1] - \Pr[\mathcal{A}^{\mathcal{O}_1(\mathsf{crs}, \cdot, \cdot, \cdot)}(1^{\kappa}, \mathsf{crs}, k_{\mathsf{V}}) = 1] \right| = \mathsf{negl}(\kappa),$$

where $\mathcal{O}_b(\mathsf{crs}, x, w_0, w_1)$ *outputs* $\mathsf{Prove}(\mathsf{crs}, x, w_b)$ *if* $(x, w_0) \in \mathcal{R} \wedge (x, w_1) \in \mathcal{R}$ *and $\perp$ otherwise.*

Definition 3.2 (Adaptive NIWI). *We say that a proof system Π is adaptive designated-verifier non-interactive witness indistinguishable proof system if Π satisfies completeness, soundness in Definition 2.1 (in the designated-verifier model), and adaptive witness indistinguishability in Definition 3.1.*

We then formally define a NIZK proof in the hidden-bits model, which will be used as a building block in our construction.

Definition 3.3. *A NIZK proof in the hidden-bits model (HBM) for $\mathcal{L}$ is defined by the following two polynomial time algorithms:*

Prove($1^\kappa, x, w, \rho$) $\rightarrow$ (π, I): *The prover's algorithm takes as input the security parameter 1^κ, a statement x, a witness w, and a hidden random string $\rho \in \{0,1\}^{\ell_{hrs}(\kappa)}$, and outputs a proof π and a set of indices $I \subseteq [\ell_{hrs}(\kappa)]$ where $\ell_{hrs}(\cdot)$ is a polynomial of κ.*

Verify($1^\kappa, x, \pi, I, \rho_{|I}$) $\rightarrow$ $\top$ or $\bot$: *The verifier's algorithm takes as input the security parameter, a statement x, a proof π, an index set I, a substring $\rho_{|I} := \{\rho_i\}_{i \in I}$, where ρ_i is the i-th bit of ρ, and outputs $\top$ to indicate acceptance of the proof and $\bot$ otherwise.*

Completeness. *For all $x \in \mathcal{L}$ and w such that $(x, w) \in \mathcal{R}$, we have*

$$\Pr[\rho \xleftarrow{\$} \{0,1\}^{\ell_{hrs}(\kappa)}, (\pi, I) \leftarrow \mathsf{Prove}(1^\kappa, x, w, \rho) : \mathsf{Verify}(1^\kappa, x, \pi, I, \rho_{|I}) = \top] = 1.$$

Soundness. *For all (possibly inefficient) adversaries $\mathcal{A}$, we have*

$$\epsilon_{\mathsf{HBM}} := \Pr[\rho \xleftarrow{\$} \{0,1\}^{\ell_{hrs}(\kappa)}, (x, \pi, I) \leftarrow \mathcal{A}(1^\kappa, \rho) : x \notin \mathcal{L} \wedge \mathsf{Verify}(1^\kappa, x, \pi, I, \rho_{|I}) = \top] = \mathsf{negl}(\kappa).$$

We call ϵ_{HBM} soundness error.

Zero-Knowledge. *There exists a PPT simulator $\mathcal{S}$ such that for all PPT adversaries $\mathcal{A} = (\mathcal{A}_1, \mathcal{A}_2)$, we have*

$$\Big| \Pr[(x, w) \leftarrow \mathcal{A}_1(1^\kappa), \rho \xleftarrow{\$} \{0,1\}^{\ell_{hrs}(\kappa)}, (\pi, I) \leftarrow \mathsf{Prove}(1^\kappa, x, w, \rho) : \mathcal{A}_2(x, \pi, I, \rho_{|I}) = 1]$$
$$- \Pr[(x, w) \leftarrow \mathcal{A}_1(1^\kappa), (\pi, I, \rho_{|I}) \leftarrow \mathcal{S}(1^\kappa, x) : \mathcal{A}_2(x, \pi, I, \rho_{|I}) = 1] \Big| = \mathsf{negl}(\kappa).$$

Theorem 3.4 (NIZK for all NP languages in the HBM [47]). *Unconditionally, there exists NIZK proof systems for all **NP** languages in the HBM with soundness error $\epsilon_{\mathsf{HBM}} \leq 2^{-cn\kappa}$ where $c > 1$ is a constant, n is polynomially related to the size of the circuit computing the **NP** language, κ is the security parameter, and $\ell_{hrs} = \mathsf{poly}(\kappa, n)$.*

3.2 Constructing DV-NIWI

The goal of this subsection is proving the following theorem.

Theorem 3.5. *Assume that the CDH assumption over paring-free group holds, then there exists an adaptive DV-NIWI for all **NP** languages.*

Here, we sketch our high-level construction. First, we present our so-called base proof system bP, and then convert it into an adaptive DV-NIWI proof system. Here, the base proof system bP is *not* a standard DV-NIWI proof system since it has a slightly different syntax. Namely, the proving and verification algorithms of the base proof system take an auxiliary string s as input in addition to (crs, x, w) and (crs, k_V, x, π), respectively. We show that the base proof system satisfies two properties called *relaxed* soundness, which means that an adversary cannot forge a proof *if s is fixed*, and *relaxed* zero-knowledge, which means that a proof can be simulated without a witness *if s is randomly chosen*. Observe that if we were to convert the prover to sample s on its own and include it in the proof, then the syntax fits that of DV-NIWI. However, such a simple conversion of our base proof system bP into a DV-NIWI will not work as the acquired DV-NIWI will not have soundness. Namely, the relaxed soundness of bP does not prevent a cheating prover from forging a proof if he is allowed to choose s himself. To resolve this problem, we use a similar idea used by Dwork and Naor [46]. Our construction of an adaptive DV-NIWI proof system consists of running many copies of the base proof system using a single common auxiliary input s for all copies. Then, when the number of copies is sufficiently large, soundness of the scheme can be proven from the union bound on all possible s. Moreover, since the relaxed zero-knowledge implies witness indistinguishability, and witness indistinguishability is preserved under parallel repetitions, we can prove the witness indistinguishability of our DV-NIWI.

Base proof system. First, we introduce the syntax and security properties of the base proof system bP. Note that bP is merely an intermediate system introduced for a modular exposition and not a standard NIZK proof system.

Definition 3.4 (Syntax of base proof system). *A base proof system* bP *consists of the following three polynomial time algorithms.*

bP.Setup(1^κ) $\to$ (crs, k_V): *The setup algorithm takes as input the security parameter* 1^κ *and outputs a common reference string* crs, *and a verification key* k_V.

bP.Prove(crs, x, w, s) $\to$ π: *The prover's algorithm takes as input a common reference string* crs, *a statement* x, *a witness* w, *and a fixed string* $s \in \{0,1\}^{\ell_{\text{hrs}}(\kappa)}$, *and outputs a proof* π.

bP.Verify(crs, k_V, x, π, s) $\to$ $\top$ *or* $\bot$: *The verifier's algorithm takes as input a common reference string* crs, *a verification key* k_V, *a statement* x, *a proof* π, *and a fixed string* $s \in \{0,1\}^{\ell_{\text{hrs}}(\kappa)}$, *and outputs* $\top$ *to indicate acceptance of the proof and* $\bot$ *otherwise.*

Definition 3.5 (Security of base proof system). *A base proof system is required to satisfy the following three properties.*

Correctness: *For all pairs* $(x, w) \in \mathcal{R}$ *and* $s \in \{0,1\}^{\ell_{\text{hrs}}(\kappa)}$, *if we run* (crs, k_V) $\overset{\$}{\leftarrow}$ bP.Setup(1^κ), *then we have*

$$\Pr[\pi \overset{\$}{\leftarrow} \text{bP.Prove}(\text{crs}, x, w, s) : \text{bP.Verify}(\text{crs}, k_V, x, \pi, s) = \top] = 1$$

Relaxed ϵ-soundness: *For any fixed* $s \in \{0,1\}^{\ell_{\text{hrs}}}$, *it holds that all (possibly inefficient) adversaries* $\mathcal{A}$,

$$\Pr[\text{Expt}_{\mathcal{A}}^{\text{r-snd}}(1^\kappa, s) = \top] < \epsilon,$$

where ϵ *is the soundness error of* bP, *and the experiment* $\text{Expt}_{\mathcal{A}}^{\text{r-snd}}(1^\kappa)$ *is defined as follows.*

$\mathsf{Expt}_{\mathcal{A}}^{\mathsf{r\text{-}snd}}(1^\kappa, s)$

$(\mathsf{crs}, k_\mathsf{V}) \leftarrow \mathsf{bP.Setup}(1^\kappa)$,

$(x^*, \pi^*) \leftarrow \mathcal{A}^{\mathsf{bP.Verify}(\mathsf{crs}, k_\mathsf{V}, \cdot, \cdot, s)}(1^\kappa, \mathsf{crs}, s)$,

If $x^* \notin \mathcal{L} \wedge \mathsf{bP.Verify}(\mathsf{crs}, k_\mathsf{V}, x^*, \pi^*, s) = \top$, *then outputs 1,*

Otherwise, outputs 0.

This is basically the same as the standard soundness except that $\mathcal{A}$ must use a fixed s.

Relaxed zero-knowledge: *There exists a PPT simulation algorithm* $\mathsf{bP}.\mathcal{S} = (\mathsf{bP}.\mathcal{S}_1, \mathsf{bP}.\mathcal{S}_2)$ *that satisfies the following. For all (stateful) PPT adversaries $\mathcal{A}$, we have*

$$\left| \Pr[\mathsf{Expt}_{\mathcal{A}}^{\mathsf{r\text{-}real}}(1^\kappa) = 1] - \Pr[\mathsf{Expt}_{\mathcal{A}, \mathcal{S}}^{\mathsf{r\text{-}sim}}(1^\kappa) = 1] \right| = \mathsf{negl}(\kappa),$$

where experiments $\mathsf{Expt}_{\mathcal{A}}^{\mathsf{r\text{-}real}}$ *and* $\mathsf{Expt}_{\mathcal{A}, \mathcal{S}}^{\mathsf{r\text{-}sim}}$ *are defined as follows.*

$\mathsf{Expt}_{\mathcal{A}}^{\mathsf{r\text{-}real}}$	$\mathsf{Expt}_{\mathcal{A}, \mathcal{S}}^{\mathsf{r\text{-}sim}}$
$(\mathsf{crs}, k_\mathsf{V}) \leftarrow \mathsf{bP.Setup}(1^\kappa)$,	$(\widetilde{\mathsf{crs}}, \widetilde{k}_\mathsf{V}, \widetilde{\tau}_\mathsf{V}) \leftarrow \mathsf{bP}.\mathcal{S}_1(1^\kappa)$,
$(x, w) \leftarrow \mathcal{A}(1^\kappa, \mathsf{crs}, k_\mathsf{V})$,	$(x, w) \leftarrow \mathcal{A}(1^\kappa, \widetilde{\mathsf{crs}}, \widetilde{k}_\mathsf{V})$,
$s \xleftarrow{\$} \{0, 1\}^{\ell_{\mathsf{hrs}}}$,	
If $(x, w) \in \mathcal{R}$, $\pi \leftarrow \mathsf{bP.Prove}(\mathsf{crs}, x, w, s)$,	If $(x, w) \in \mathcal{R}$, $(\pi, s) \leftarrow \mathsf{bP}.\mathcal{S}_2(\widetilde{\mathsf{crs}}, \widetilde{k}_\mathsf{V}, \widetilde{\tau}_\mathsf{V}, x)$,
otherwise $\pi := \bot$,	*otherwise* $\pi := \bot$,
$b' \leftarrow \mathcal{A}(\pi, s)$	$b' \leftarrow \mathcal{A}(\pi, s)$
outputs b'	outputs b'

We present a base proof system $\mathsf{bP} := (\mathsf{bP.Setup}, \mathsf{bP.Prove}, \mathsf{bP.Verify})$ based on a NIZK proof system in the HBM ($\mathsf{HBM.Prove}, \mathsf{HBM.Verify}$) (with hidden-random-string-length $\ell_{\mathsf{hrs}}(\kappa)$) and the CDH assumption. Note that we use the $\mathsf{GGen}(1^\kappa)$ algorithm to generate $(\mathbb{G}, p, g)$ where $2^{2\kappa} \le p$ throughout Sect. 3. Hereafter, we simply write ℓ_{hrs} instead of $\ell_{\mathsf{hrs}}(\kappa)$ for ease of notation.

$\mathsf{bP.Setup}(1^\kappa)$: This algorithm generates the following parameters.

1. Samples $(\mathbb{G}, p, g) \xleftarrow{\$} \mathsf{GGen}(1^\kappa)$.
2. Samples $(\alpha_{i,b}, \beta_{i,b}) \xleftarrow{\$} \mathbb{Z}_p^2$ for all $i \in [\ell_{\mathsf{hrs}}]$ and $b \in \{0, 1\}$ and a common reference string $\overline{\mathsf{crs}} := \{X_{i,b}\}_{i \in [\ell_{\mathsf{hrs}}], b \in \{0,1\}} \xleftarrow{\$} \mathbb{G}^{2\ell_{\mathsf{hrs}}}$ uniformly at random.
3. Sets $\widehat{\mathsf{crs}} := \{\widehat{X}_{i,b}\}_{i \in [\ell_{\mathsf{hrs}}], b \in \{0,1\}} := \{X_{i,b}^{-\alpha_{i,b}} \cdot g^{\beta_{i,b}}\}_{i \in [\ell_{\mathsf{hrs}}], b \in \{0,1\}}$.
4. Samples $R_i \xleftarrow{\$} \{0, 1\}^u$ for all $i \in [\ell_{\mathsf{hrs}}]$ and sets $\overline{R} := \{R_i\}_{i \in [\ell_{\mathsf{hrs}}]}$.
5. Outputs a common reference string $\mathsf{crs} := (\mathbb{G}, p, g) \| \overline{\mathsf{crs}} \| \widehat{\mathsf{crs}} \| \overline{R}$ and a verification key $k_\mathsf{V} := \{(\alpha_{i,b}, \beta_{i,b})\}_{i \in [\ell_{\mathsf{hrs}}], b \in \{0,1\}}$.

We can interpret crs as $(\{X_{i,b}, \widehat{X}_{i,b}, R_i\}_{i \in [\ell_{\mathsf{hrs}}], b \in \{0,1\}}) \in \mathbb{G}^{4\ell_{\mathsf{hrs}}} \times \{0, 1\}^{\ell_{\mathsf{hrs}} u}$, where u is the length of the binary representation of a group element.

$\mathsf{bP.Prove}(\mathsf{crs}, x, w, s)$: This algorithm does the following.

1. Parses $\mathsf{crs} = (\mathbb{G}, p, g) \| \overline{\mathsf{crs}} \| \widehat{\mathsf{crs}} \| \overline{R}$ where $\overline{\mathsf{crs}} = \{X_{i,b}\}_{i \in [\ell_{\mathsf{hrs}}], b \in \{0,1\}}, \widehat{\mathsf{crs}} = \{\widehat{X}_{i,b}\}_{i \in [\ell_{\mathsf{hrs}}], b \in \{0,1\}}, \overline{R} = \{R_i\}_{i \in [\ell_{\mathsf{hrs}}]}$, and $s \in \{0, 1\}^{\ell_{\mathsf{hrs}}}$.
2. Samples $\tau \xleftarrow{\$} \mathbb{Z}_p$.
3. Sets $Z_i := (X_{i,s_i})^\tau$ and $\widehat{Z}_i := (\widehat{X}_{i,s_i})^\tau$ and $\rho_i = \mathsf{GL}(Z_i; R_i)$ for $i \in [\ell_{\mathsf{hrs}}]$.

4. Generates $(\pi_{\mathsf{hbm}}, I) \leftarrow \mathsf{HBM.Prove}(1^\kappa, x, w, \rho)$ where $\rho := \rho_1 \| \cdots \| \rho_{\ell_{\mathsf{hrs}}}$.

5. Outputs a proof $\pi := (\pi_{\mathsf{hbm}}, I, \{(Z_i, \widehat{Z}_i)\}_{i \in I}, g^\tau)$.

$\mathsf{bP.Verify}(\mathsf{crs}, k_V, x, \pi, s)$: This algorithm parses $\pi = (\pi_{\mathsf{hbm}}, I, \{(Z_i, \widehat{Z}_i)\}_{i \in I}, T)$, $k_V := \{(\alpha_{i,b}, \beta_{i,b})\}_{i \in [\ell_{\mathsf{hrs}}], b \in \{0,1\}}$, $\mathsf{crs} = (\mathbb{G}, p, g) \| \overline{\mathsf{crs}} \| \widehat{\mathsf{crs}} \| \overline{R}$ where $\overline{\mathsf{crs}} = \{X_{i,b}\}_{i \in [\ell_{\mathsf{hrs}}], b \in \{0,1\}}$, $\widehat{\mathsf{crs}} = \{\widehat{X}_{i,b}\}_{i \in [\ell_{\mathsf{hrs}}], b \in \{0,1\}}$, $\overline{R} = \{R_i\}_{i \in [\ell_{\mathsf{hrs}}]}$, and $s \in \{0,1\}^{\ell_{\mathsf{hrs}}}$. This algorithm does the following.

- For all $i \in I$,
  1. Verifies that $\mathsf{Test_{TDH}}((\alpha_{i,s_i}, \beta_{i,s_i}), X_{i,s_i}, \widehat{X}_{i,s_i}, T, Z_i, \widehat{Z}_i) = \top$, where $\mathsf{Test_{TDH}}$ is defined in Fig. 1. If any one of the equations does not hold, then outputs $\bot$.
  2. Computes $\rho_i = \mathsf{GL}(Z_i; R_i)$.
- If the proof passes all the tests above, then this algorithm outputs $\mathsf{HBM.Verify}(1^\kappa, x, \pi_{\mathsf{hbm}}, I, \rho_{|I})$.

The trapdoor test $\mathsf{Test_{TDH}}((\alpha, \beta), X, \widehat{X}, Y, Z, \widehat{Z})$

1. Verifies that $Z^\alpha \cdot \widehat{Z} = Y^\beta$. If it holds, then outputs $\top$, else $\bot$.

Fig. 1. The algorithm $\mathsf{Test_{TDH}}((\alpha, \beta), X, \widehat{X}, Y, Z, \widehat{Z})$ verifies that $Z = Y^x$ and $\widehat{Z} = Y^{\widehat{x}}$, that is (g, Y, X, Z) and $(g, Y, \widehat{X}, \widehat{Z})$ where $X = g^x$ and $\widehat{X} = g^{\widehat{x}}$ are DDH-tuples without $(x, \widehat{x})$.

Unlike the idea outlined in the introduction, the CRS in bP consists of a doubled-line of random elements $(X_{i,0}, \widehat{X}_{i,0})$ and $(X_{i,1}, \widehat{X}_{i,1})$ for each $i \in [\ell_{\mathsf{hrs}}]$. These doubled-line of random elements are crucial for achieving *adaptive* zero-knowledge. If we only had a singled-line of random elements as the CRS in the introduction, then we would have the following issue: The only way for the ZK-simulator of bP $\mathcal{S}_{\mathsf{bP}}$ to use the ZK-simulator $\mathcal{S}_{\mathsf{hbm}}$ of the NIZK in the HBM, is to feed $\mathcal{S}_{\mathsf{hbm}}$ the statement x output by the adversary. Now, for the simulated proof π, index set I, and hidden bits $\rho_{|I}$ output by $\mathcal{S}_{\mathsf{hbm}}$ to be useful, we must have $\rho_i = \mathsf{GL}(X_i^\tau; R_i)$ for all $i \in I$ where τ is some element simulated by $\mathcal{S}_{\mathsf{bP}}$. However, due to soundness, if the CRS was only a single-line of random elements $(X_i, \widehat{X}_i)$, then there exists no τ with overwhelming probability such that the above condition holds. Therefore, $\mathcal{S}_{\mathsf{bP}}$ must choose τ and program the singled-line of random elements $(X_i, \widehat{X}_i)$ in the CRS conditioned on $\rho_i = \mathsf{GL}(X_i^\tau; R_i)$ for all $i \in I$ in order to appropriately use $\mathcal{S}_{\mathsf{hbm}}$. However, since ρ_i is only output as the result of feeding $\mathcal{S}_{\mathsf{hbm}}$ with the statement x, $\mathcal{S}_{\mathsf{bP}}$ can only set the CRS *after* it is given the statement x from the adversary. To overcome this problem, we use the technique of non-committing encryption. Namely, we let CRS be a doubled-line of random elements $(X_{i,0}, \widehat{X}_{i,0})$ and $(X_{i,1}, \widehat{X}_{i,1})$. In the real-scheme the fixed string $s \in \{0,1\}^{\ell_{\mathsf{hrs}}}$ dictates which ℓ_{hrs}-random elements $(X_{i,s_i}, \widehat{X}_{i,s_i})_{i \in [\ell_{\mathsf{hrs}}]}$ a prover must use. Then during the adaptive ZK proof, $\mathcal{S}_{\mathsf{bP}}$ will prepare the CRS so that $\{\mathsf{GL}(X_{i,0}^{\tau}; R_i), \mathsf{GL}(X_{i,1}^{\tau}; R_i)\} = \{0,1\}$ *without* seeing the statement x. Then after the adversary outputs the statement x, it runs $\mathcal{S}_{\mathsf{hbm}}$, and samples a string s so that $\rho_i = \mathsf{GL}(X_{i,s_i}^{\tau}; R_i)$ for all $i \in I$.

Security of bP. The following lemmas address the correctness and security of our base proof system. Due to limited space the proof will appear in the full version.

Lemma 3.1 (Correctness). *Our base proof system* bP *satisfies the correctness in Definition 3.5.*

Lemma 3.2 (Relaxed Soundness). *If* HBM *is sound, then* bP *satisfies the relaxed* $(p \cdot \epsilon_{\mathsf{HBM}} + (q_v + 1)/p)$-*soundness defined in Definition 3.5.*

Lemma 3.3 (Relaxed ZK). *If the CDH assumption over pairing-free group holds, then* bP *satisfies the relaxed ZK defined in Definition 3.5.*

Construction of DV-NIWI. Here, we present our adaptive DV-NIWI proof system $\Pi := (\mathsf{Setup}, \mathsf{Prove}, \mathsf{Verify})$ based on the base proof system bP $:=$ (bP.Setup, bP.Prove, bP.Verify) that has relaxed ϵ-soundness for some $\epsilon < 1$. We note that we proved that the base proof system satisfies relaxed $(p \cdot \epsilon_{\mathsf{HBM}} + (q_v + 1)/p)$-soundness in Lemma 3.2, and we can make $(p \cdot \epsilon_{\mathsf{HBM}} + (q_v + 1)/p) < 1$ by choosing a parameter for HBM so that $p \cdot \epsilon_{\mathsf{HBM}}$ is negligible. (This is possible by Theorem 3.4). We set an integer ℓ' so that we have $2^{\ell_{\mathsf{hrs}}} \cdot \epsilon^{\ell'} \leq 2^{-\kappa}$. Then Π is described as follows.

$\mathsf{Setup}(1^{\kappa})$: This algorithm samples $(\mathsf{crs}_j, k_{\mathsf{V}}^{(j)}) \leftarrow$ bP.Setup(1^{κ}) for $j \in [\ell']$. It sets
$\qquad$ crs $:= \mathsf{crs}_1 \| \cdots \| \mathsf{crs}_{\ell'}$ and $k_{\mathsf{V}} := k_{\mathsf{V}}^{(1)} \| \cdots \| k_{\mathsf{V}}^{(\ell')}$, and outputs $(\mathsf{crs}, k_{\mathsf{V}})$.
$\mathsf{Prove}(\mathsf{crs}, x, w) \to \pi$: This algorithm does the following:
$\qquad$ 1. chooses $s \xleftarrow{\$} \{0,1\}^{\ell_{\mathsf{hrs}}}$,
$\qquad$ 2. generates $\pi_j \leftarrow$ bP.Prove$(\mathsf{crs}_j, x, w, s)$ for all $j \in [\ell']$,
$\qquad$ 3. outputs a proof $\pi := (\pi_1, \ldots, \pi_{\ell'}, s)$.
$\mathsf{Verify}(\mathsf{crs}, k_{\mathsf{V}}, x, \pi) \to \top$ or $\bot$: This algorithm parses $\pi = (\pi_1, \ldots, \pi_{\ell'}, s)$. For all $\qquad j \in [\ell']$, it verifies that $\top =$ bP.Verify$(\mathsf{crs}_j, k_{\mathsf{V}}^{(j)}, x, \pi_j, s)$. If the proof passes all the tests, then this algorithm outputs $\top$, otherwise $\bot$.

Our adaptive DV-NIWI proof system Π is complete, sound, and adaptively witness-indistinguishable. The proofs can be found in the full version.

3.3   Transformation from DV-NIWI into Multi-Theorem DV-NIZK

To complete the proof of Theorem 3.1, it remains to show the following theorem.

Theorem 3.6. *If there exists an adaptive DV-NIWI proof systems for all* **NP** *languages and pseudorandom generators, then there exists an adaptive multi-theorem DV-NIZK proof system for all* **NP** *languages.*

We omit the proof since the transformation is essentially the same as that of Feige et al. [47] (from NIWI to multi-theorem NIZK), with the exception that we consider the *designated-verifier* setting.

4   Constructing HomSig from ABE-Simulation Paradigm

We construct a context-hiding HomSig for $\mathbf{NC}^1$ from a new non-static (q-type) assumption on pairing groups that we call the CDHER assumption. Specifically, we first construct a new ABE scheme from the same assumption and then apply the (semi-generic) conversion sketched in Sect. 1.2. We directly give a construction of HomSig instead of constructing it via the new ABE. Using the transformation by Kim and Wu [71], we obtain a DP-NIZK from the same assumption.

Theorem 4.1. *If the CDHER assumption holds on a pairing group, then there exists DP-NIZK for all* **NP** *languages with proof size* $|C| + \mathsf{poly}(\kappa)$, *where* $|C|$ *denotes the size of the circuit that computes the relation being proved.*

As far as we know, this is the first DP-NIZK scheme with short proofs without assuming the LWE assumption, fully-homomorphic encryption, indistinguishability obfuscation, or non-falsifiable assumptions. Furthermore, if the proven **NP** relation can be expressed as a leveled circuit, we can reduce the proof size to $|w| + |C|/\log \kappa + \mathsf{poly}(\kappa)$, where $|w|$ is the length of the witness of the proven relation and a leveled circuit refers to a circuit whose gates can be divided into layers and only gates from the consecutive layers are connected by wires. See the full version for the details.

Besides being a building-block for PP-NIZKs, our HomSig scheme alone may be of an independent interest. In the full version, we extend the scheme to the multi-data setting and demonstrate that it achieves online-offline efficiency. This greatly improves the HomSig scheme with the same properties from the multi-linear map [30] in terms of efficiency and security.

5   HomMAC from Inner Product Functional Encryption

In this section, we give a construction of HomMAC based on a variant of functional encryption for inner-products (IPFE) which we call a *functional encryption for inner-product on exponent* (expIPFE). Namely, we show that an expIPFE scheme that satisfies a property called extractability suffices for constructing statistically unforgeable and computationally context-hiding HomMAC. We also show that the IPFE scheme by Agrawal et al. [5] can be seen as an instantiation of an extractable expIPFE scheme under the DDH assumption. As a result, we obtain a statistically unforgeable and computationally context-hiding HomMAC based on the DDH assumption, which yields statistically sound and computationally (non-programmable CRS) zero-knowledge PP-NIZK based on the DDH assumption (over paring-free groups). Since our HomMAC is not compact, a simple adaptation of their transformation yields PP-NIZK with proof size $O(|C|\kappa) + \mathsf{poly}(\kappa)$. However, by taking advantage of the fact our scheme can deal with arithmetic circuits over $\mathbb{Z}_p$ of polynomial degree, which is larger than $\mathbf{NC}^1$, and incorporating the technique by Katsumata [69], we can reduce the proof size to $|C| + \mathsf{poly}(\kappa)$. See the full version for details. Then we obtain the following theorem.

Theorem 5.1. *If the DDH assumption holds on a pairing free group, then there exists PP-NIZK for all **NP** languages with proof size $|C| + \mathsf{poly}(\kappa)$, where $|C|$ denotes the size of circuit that computes the relation being proved.*

Similarly to the case in Sect. 4, if the proven **NP** relation can be expressed as a leveled circuit, we can further reduce the proof size to $|w| + |C|/\log \kappa + \mathsf{poly}(\kappa)$. See the full version for the details.

Acknowledgement. We would like to thank Geoffroy Couteau for helpful comments on related works and anonymous reviewers of Eurocrypt 2019 for their valuable comments. The first author was partially supported by JST CREST Grant Number JPMJCR1302 and JSPS KAKENHI Grant Number 17J05603. The third author was supported by JST CREST Grant No. JPMJCR1688 and JSPS KAKENHI Grant Number 16K16068.

References

1. Abusalah, H.: Generic instantiations of the hidden bits model for non-interactive zero-knowledge proofs for NP. Master's thesis, RWTH-Aachen University (2013)
2. Agrawal, S., Boneh, D.: Homomorphic MACs: MAC-based integrity for network coding. In: Abdalla, M., Pointcheval, D., Fouque, P.-A., Vergnaud, D. (eds.) ACNS 2009. LNCS, vol. 5536, pp. 292–305. Springer, Heidelberg (2009). https://doi.org/10.1007/978-3-642-01957-9_18
3. Agrawal, S., Chase, M.: A study of pair encodings: predicate encryption in prime order groups. In: Kushilevitz, E., Malkin, T. (eds.) TCC 2016, Part II. LNCS, vol. 9563, pp. 259–288. Springer, Heidelberg (2016). https://doi.org/10.1007/978-3-662-49099-0_10
4. Agrawal, S., Chase, M.: Simplifying design and analysis of complex predicate encryption schemes. In: Coron, J.-S., Nielsen, J.B. (eds.) EUROCRYPT 2017, Part I. LNCS, vol. 10210, pp. 627–656. Springer, Cham (2017). https://doi.org/10.1007/978-3-319-56620-7_22
5. Agrawal, S., Libert, B., Stehlé, D.: Fully secure functional encryption for inner products, from standard assumptions. In: Robshaw, M., Katz, J. (eds.) CRYPTO 2016, Part III. LNCS, vol. 9816, pp. 333–362. Springer, Heidelberg (2016). https://doi.org/10.1007/978-3-662-53015-3_12
6. Ateniese, G., et al.: Provable data possession at untrusted stores. In: Ning, P., De Capitani di Vimercati, S., Syverson, P.F. (eds.) ACM CCS 2007, pp. 598–609. ACM Press, October 2007
7. Attrapadung, N., Hanaoka, G., Yamada, S.: Conversions among several classes of predicate encryption and applications to ABE with various compactness tradeoffs. In: Iwata, T., Cheon, J.H. (eds.) ASIACRYPT 2015, Part I. LNCS, vol. 9452, pp. 575–601. Springer, Heidelberg (2015). https://doi.org/10.1007/978-3-662-48797-6_24
8. Attrapadung, N., Libert, B.: Homomorphic network coding signatures in the standard model. In: Catalano, D., Fazio, N., Gennaro, R., Nicolosi, A. (eds.) PKC 2011. LNCS, vol. 6571, pp. 17–34. Springer, Heidelberg (2011). https://doi.org/10.1007/978-3-642-19379-8_2

9. Bellare, M., Micciancio, D., Warinschi, B.: Foundations of group signatures: formal definitions, simplified requirements, and a construction based on general assumptions. In: Biham, E. (ed.) EUROCRYPT 2003. LNCS, vol. 2656, pp. 614–629. Springer, Heidelberg (2003). https://doi.org/10.1007/3-540-39200-9_38

10. Bellare, M., Yung, M.: Certifying permutations: noninteractive zero-knowledge based on anytrapdoor permutation. J. Cryptol. **9**(3), 149–166 (1996)

11. Bitansky, N., Paneth, O.: ZAPs and non-interactive witness indistinguishability from indistinguishability obfuscation. In: Dodis, Y., Nielsen, J.B. (eds.) TCC 2015, Part II. LNCS, vol. 9015, pp. 401–427. Springer, Heidelberg (2015). https://doi.org/10.1007/978-3-662-46497-7_16

12. Bitansky, N., Paneth, O., Wichs, D.: Perfect structure on the edge of chaos - trapdoor permutations from indistinguishability obfuscation. In: Kushilevitz, E., Malkin, T. (eds.) TCC 2016, Part I. LNCS, vol. 9562, pp. 474–502. Springer, Heidelberg (2016). https://doi.org/10.1007/978-3-662-49096-9_20

13. Blum, M., Feldman, P., Micali, S.: Non-interactive zero-knowledge and its applications (extended abstract). In: 20th ACM STOC, pp. 103–112. ACM Press, May 1988

14. Boneh, D., Boyen, X.: Short signatures without random oracles. In: Cachin, C., Camenisch, J.L. (eds.) EUROCRYPT 2004. LNCS, vol. 3027, pp. 56–73. Springer, Heidelberg (2004). https://doi.org/10.1007/978-3-540-24676-3_4

15. Boneh, D., Freeman, D., Katz, J., Waters, B.: Signing a linear subspace: signature schemes for network coding. In: Jarecki, S., Tsudik, G. (eds.) PKC 2009. LNCS, vol. 5443, pp. 68–87. Springer, Heidelberg (2009). https://doi.org/10.1007/978-3-642-00468-1_5

16. Boneh, D., Freeman, D.M.: Homomorphic signatures for polynomial functions. In: Paterson, K.G. (ed.) EUROCRYPT 2011. LNCS, vol. 6632, pp. 149–168. Springer, Heidelberg (2011). https://doi.org/10.1007/978-3-642-20465-4_10

17. Boneh, D., Freeman, D.M.: Linearly homomorphic signatures over binary fields and new tools for lattice-based signatures. In: Catalano, D., Fazio, N., Gennaro, R., Nicolosi, A. (eds.) PKC 2011. LNCS, vol. 6571, pp. 1–16. Springer, Heidelberg (2011). https://doi.org/10.1007/978-3-642-19379-8_1

18. Boneh, D., et al.: Fully key-homomorphic encryption, arithmetic circuit ABE and compact garbled circuits. In: Nguyen, P.Q., Oswald, E. (eds.) EUROCRYPT 2014. LNCS, vol. 8441, pp. 533–556. Springer, Heidelberg (2014). https://doi.org/10.1007/978-3-642-55220-5_30

19. Boyle, E., Couteau, G., Gilboa, N., Ishai, Y.: Compressing vector OLE. In: Lie, D., Mannan, M., Backes, M., Wang, X. (eds.) ACM CCS 2018, pp. 896–912. ACM Press, October 2018

20. Boyle, E., Gilboa, N., Ishai, Y.: Breaking the circuit size barrier for secure computation under DDH. In: Robshaw, M., Katz, J. (eds.) CRYPTO 2016, Part I. LNCS, vol. 9814, pp. 509–539. Springer, Heidelberg (2016). https://doi.org/10.1007/978-3-662-53018-4_19

21. Camenisch, J., Hohenberger, S., Lysyanskaya, A.: Compact e-cash. In: Cramer, R. (ed.) EUROCRYPT 2005. LNCS, vol. 3494, pp. 302–321. Springer, Heidelberg (2005). https://doi.org/10.1007/11426639_18

22. Canetti, R., et al.: Fiat-Shamir: from practice to theory (2019)

23. Canetti, R., Chen, Y., Reyzin, L., Rothblum, R.D.: Fiat-Shamir and correlation intractability from strong KDM-secure encryption. In: Nielsen, J.B., Rijmen, V. (eds.) EUROCRYPT 2018, Part I. LNCS, vol. 10820, pp. 91–122. Springer, Cham (2018). https://doi.org/10.1007/978-3-319-78381-9_4

24. Canetti, R., Feige, U., Goldreich, O., Naor, M.: Adaptively secure multi-party computation. In: 28th ACM STOC, pp. 639–648. ACM Press, May 1996
25. Canetti, R., Halevi, S., Katz, J.: A forward-secure public-key encryption scheme. J. Cryptol. $20(3)$, 265–294 (2007)
26. Cash, D., Kiltz, E., Shoup, V.: The twin Diffie-Hellman problem and applications. J. Cryptol. $22(4)$, 470–504 (2009)
27. Catalano, D., Fiore, D.: Practical homomorphic message authenticators for arithmetic circuits. J. Cryptol. $31(1)$, 23–59 (2018)
28. Catalano, D., Fiore, D., Nizzardo, L.: Programmable hash functions go private: constructions and applications to (homomorphic) signatures with shorter public keys. In: Gennaro, R., Robshaw, M.J.B. (eds.) CRYPTO 2015, Part II. LNCS, vol. 9216, pp. 254–274. Springer, Heidelberg (2015). https://doi.org/10.1007/978-3-662-48000-7_13
29. Catalano, D., Fiore, D., Warinschi, B.: Efficient network coding signatures in the standard model. In: Fischlin, M., Buchmann, J., Manulis, M. (eds.) PKC 2012. LNCS, vol. 7293, pp. 680–696. Springer, Heidelberg (2012). https://doi.org/10.1007/978-3-642-30057-8_40
30. Catalano, D., Fiore, D., Warinschi, B.: Homomorphic signatures with efficient verification for polynomial functions. In: Garay, J.A., Gennaro, R. (eds.) CRYPTO 2014, Part I. LNCS, vol. 8616, pp. 371–389. Springer, Heidelberg (2014). https://doi.org/10.1007/978-3-662-44371-2_21
31. Catalano, D., Marcedone, A., Puglisi, O.: Authenticating computation on groups: new homomorphic primitives and applications. In: Sarkar, P., Iwata, T. (eds.) ASIACRYPT 2014, Part II. LNCS, vol. 8874, pp. 193–212. Springer, Heidelberg (2014). https://doi.org/10.1007/978-3-662-45608-8_11
32. Chaidos, P., Couteau, G.: Efficient designated-verifier non-interactive zero-knowledge proofs of knowledge. In: Nielsen, J.B., Rijmen, V. (eds.) EUROCRYPT 2018, Part III. LNCS, vol. 10822, pp. 193–221. Springer, Cham (2018). https://doi.org/10.1007/978-3-319-78372-7_7
33. Chaidos, P., Groth, J.: Making sigma-protocols non-interactive without random oracles. In: Katz, J. (ed.) PKC 2015. LNCS, vol. 9020, pp. 650–670. Springer, Heidelberg (2015). https://doi.org/10.1007/978-3-662-46447-2_29
34. Chase, M., et al.: Reusable non-interactive secure computation. IACR Cryptology ePrint Archive, 2018:940 (2018)
35. Chaum, D.: Security without identification: transaction systems to make big brother obsolete. Commun. ACM $28(10)$, 1030–1044 (1985)
36. Chaum, D., Fiat, A., Naor, M.: Untraceable electronic cash. In: Goldwasser, S. (ed.) CRYPTO 1988. LNCS, vol. 403, pp. 319–327. Springer, New York (1990). https://doi.org/10.1007/0-387-34799-2_25
37. Chaum, D., van Heyst, E.: Group signatures. In: Davies, D.W. (ed.) EUROCRYPT 1991. LNCS, vol. 547, pp. 257–265. Springer, Heidelberg (1991). https://doi.org/10.1007/3-540-46416-6_22
38. Couteau, G., Hofheinz, D.: Designated-verifier pseudorandom generators, and their applications. In: Ishai, Y., Rijmen, V. (eds.) EUROCRYPT 2019. LNCS, vol. 11477, pp. 562–592. Springer, Cham (2019)
39. Cramer, R., Damgård, I.: Secret-key zero-knowlegde and non-interactive verifiable exponentiation. In: Naor, M. (ed.) TCC 2004. LNCS, vol. 2951, pp. 223–237. Springer, Heidelberg (2004). https://doi.org/10.1007/978-3-540-24638-1_13
40. Damgård, I.B.: On the randomness of Legendre and Jacobi sequences. In: Goldwasser, S. (ed.) CRYPTO 1988. LNCS, vol. 403, pp. 163–172. Springer, New York (1990). https://doi.org/10.1007/0-387-34799-2_13

41. Damgård, I.: Non-interactive circuit based proofs and non-interactive perfect zero-knowledge with preprocessing. In: Rueppel, R.A. (ed.) EUROCRYPT 1992. LNCS, vol. 658, pp. 341–355. Springer, Heidelberg (1993). https://doi.org/10.1007/3-540-47555-9_28

42. Damgård, I., Fazio, N., Nicolosi, A.: Non-interactive zero-knowledge from homomorphic encryption. In: Halevi, S., Rabin, T. (eds.) TCC 2006. LNCS, vol. 3876, pp. 41–59. Springer, Heidelberg (2006). https://doi.org/10.1007/11681878_3

43. De Santis, A., Micali, S., Persiano, G.: Non-interactive zero-knowledge with preprocessing. In: Goldwasser, S. (ed.) CRYPTO 1988. LNCS, vol. 403, pp. 269–282. Springer, New York (1990). https://doi.org/10.1007/0-387-34799-2_21

44. Desmedt, Y.: Computer security by redefining what a computer is. In: NSPW, pp. 160–166. ACM (1993)

45. Dolev, D., Dwork, C., Naor, M.: Nonmalleable cryptography. SIAM J. Comput. **30**(2), 391–437 (2000)

46. Dwork, C., Naor, M.: Zaps and their applications. SIAM J. Comput. **36**(6), 1513–1543 (2007)

47. Feige, U., Lapidot, D., Shamir, A.: Multiple noninteractive zero knowledge proofs under general assumptions. SIAM J. Comput. **29**(1), 1–28 (1999)

48. Fiat, A., Shamir, A.: How to prove yourself: practical solutions to identification and signature problems. In: Odlyzko, A.M. (ed.) CRYPTO 1986. LNCS, vol. 263, pp. 186–194. Springer, Heidelberg (1987). https://doi.org/10.1007/3-540-47721-7_12

49. Freeman, D.M.: Improved security for linearly homomorphic signatures: a generic framework. In: Fischlin, M., Buchmann, J., Manulis, M. (eds.) PKC 2012. LNCS, vol. 7293, pp. 697–714. Springer, Heidelberg (2012). https://doi.org/10.1007/978-3-642-30057-8_41

50. Gennaro, R., Katz, J., Krawczyk, H., Rabin, T.: Secure network coding over the integers. In: Nguyen, P.Q., Pointcheval, D. (eds.) PKC 2010. LNCS, vol. 6056, pp. 142–160. Springer, Heidelberg (2010). https://doi.org/10.1007/978-3-642-13013-7_9

51. Gennaro, R., Wichs, D.: Fully homomorphic message authenticators. In: Sako, K., Sarkar, P. (eds.) ASIACRYPT 2013, Part II. LNCS, vol. 8270, pp. 301–320. Springer, Heidelberg (2013). https://doi.org/10.1007/978-3-642-42045-0_16

52. Gentry, C., Groth, J., Ishai, Y., Peikert, C., Sahai, A., Smith, A.D.: Using fully homomorphic hybrid encryption to minimize non-interative zero-knowledge proofs. J. Cryptol. **28**(4), 820–843 (2015)

53. Goldreich, O.: Foundations of Cryptography: Volume 2, Basic Applications (2004)

54. Goldreich, O., Levin, L.A.: A hard-core predicate for all one-way functions. In: 21st ACM STOC, pp. 25–32. ACM Press, May 1989

55. Goldreich, O., Micali, S., Wigderson, A.: How to play any mental game or a completeness theorem for protocols with honest majority. In: Aho, A. (ed.) 19th ACM STOC, pp. 218–229. ACM Press, May 1987

56. Goldreich, O., Oren, Y.: Definitions and properties of zero-knowledge proof systems. J. Cryptol. **7**(1), 1–32 (1994)

57. Goldwasser, S., Micali, S., Rackoff, C.: The knowledge complexity of interactive proof systems. SIAM J. Comput. **18**(1), 186–208 (1989)

58. Gorbunov, S., Vaikuntanathan, V., Wee, H.: Functional encryption with bounded collusions via multi-party computation. In: Safavi-Naini, R., Canetti, R. (eds.) CRYPTO 2012. LNCS, vol. 7417, pp. 162–179. Springer, Heidelberg (2012). https://doi.org/10.1007/978-3-642-32009-5_11

59. Gorbunov, S., Vaikuntanathan, V., Wee, H.: Attribute-based encryption for circuits. J. ACM **62**(6), 45:1–45:33 (2015)

60. Gorbunov, S., Vaikuntanathan, V., Wichs, D.: Leveled fully homomorphic signatures from standard lattices. In: Servedio, R.A., Rubinfeld, R. (eds.) 47th ACM STOC, pp. 469–477. ACM Press, June 2015

61. Goyal, V., Pandey, O., Sahai, A., Waters, B.: Attribute-based encryption for fine-grained access control of encrypted data. In: Juels, A., Wright, R.N., De Capitani di Vimercati, S. (eds.) ACM CCS 2006, pp. 89–98. ACM Press, October/November 2006. Available as Cryptology ePrint Archive Report 2006/309

62. Groth, J.: Short pairing-based non-interactive zero-knowledge arguments. In: Abe, M. (ed.) ASIACRYPT 2010. LNCS, vol. 6477, pp. 321–340. Springer, Heidelberg (2010). https://doi.org/10.1007/978-3-642-17373-8_19

63. Groth, J., Ostrovsky, R., Sahai, A.: New techniques for noninteractive zero-knowledge. J. ACM **59**(3), 11:–11:35 (2012)

64. Groth, J., Sahai, A.: Efficient noninteractive proof systems for bilinear groups. SIAM J. Comput. **41**(5), 1193–1232 (2012)

65. Holmgren, J., Lombardi, A.: Cryptographic hashing from strong one-way functions (or: one-way product functions and their applications). In: Thorup, M. (ed.) 59th FOCS, pp. 850–858. IEEE Computer Society Press, October 2018

66. Ishai, Y., Kushilevitz, E., Ostrovsky, R., Sahai, A.: Zero-knowledge proofs from secure multiparty computation. SIAM J. Comput. **39**(3), 1121–1152 (2009)

67. Johnson, R., Molnar, D., Song, D.X., Wagner, D.: Homomorphic signature schemes. In: Preneel, B. (ed.) CT-RSA 2002. LNCS, vol. 2271, pp. 244–262. Springer, Heidelberg (2002). https://doi.org/10.1007/3-540-45760-7_17

68. Kalai, Y.T., Rothblum, G.N., Rothblum, R.D.: From obfuscation to the security of Fiat-Shamir for proofs. In: Katz, J., Shacham, H. (eds.) CRYPTO 2017, Part II. LNCS, vol. 10402, pp. 224–251. Springer, Cham (2017). https://doi.org/10.1007/978-3-319-63715-0_8

69. Katsumata, S.: On the untapped potential of encoding predicates by arithmetic circuits and their applications. In: Takagi, T., Peyrin, T. (eds.) ASIACRYPT 2017, Part III. LNCS, vol. 10626, pp. 95–125. Springer, Cham (2017). https://doi.org/10.1007/978-3-319-70700-6_4

70. Kilian, J., Micali, S., Ostrovsky, R.: Minimum resource zero-knowledge proofs. In: Brassard, G. (ed.) CRYPTO 1989. LNCS, vol. 435, pp. 545–546. Springer, New York (1990). https://doi.org/10.1007/0-387-34805-0_47

71. Kim, S., Wu, D.J.: Multi-theorem preprocessing NIZKs from lattices. In: Shacham, H., Boldyreva, A. (eds.) CRYPTO 2018, Part II. LNCS, vol. 10992, pp. 733–765. Springer, Cham (2018). https://doi.org/10.1007/978-3-319-96881-0_25

72. Kim, S., Wu, D.J.: Multi-theorem preprocessing nizks from lattices. Cryptology ePrint Archive, Report 2018/272 (2018). https://eprint.iacr.org/2018/272.pdf. Version 20180606:204702. Preliminary version appeared in CRYPTO 2018

73. Lapidot, D., Shamir, A.: Publicly verifiable non-interactive zero-knowledge proofs. In: Menezes, A.J., Vanstone, S.A. (eds.) CRYPTO 1990. LNCS, vol. 537, pp. 353–365. Springer, Heidelberg (1991). https://doi.org/10.1007/3-540-38424-3_26

74. Lipmaa, H.: Optimally sound sigma protocols under DCRA. In: Kiayias, A. (ed.) FC 2017. LNCS, vol. 10322, pp. 182–203. Springer, Cham (2017). https://doi.org/10.1007/978-3-319-70972-7_10

75. Nandi, M., Pandit, T.: On the power of pair encodings: frameworks for predicate cryptographic primitives. Cryptology ePrint Archive, Report 2015/955 (2015). http://eprint.iacr.org/2015/955

76. Naor, M., Yung, M.: Public-key cryptosystems provably secure against chosen ciphertext attacks. In: 22nd ACM STOC, pp. 427–437. ACM Press, May 1990

77. Pass, R., Shelat, A., Vaikuntanathan, V.: Construction of a non-malleable encryption scheme from any semantically secure one. In: Dwork, C. (ed.) CRYPTO 2006. LNCS, vol. 4117, pp. 271–289. Springer, Heidelberg (2006). https://doi.org/10.1007/11818175_16

78. Peikert, C., Shiehian, S.: Noninteractive zero knowledge for NP from (plain) learning with errors. IACR Cryptology ePrint Archive, 2019:158 (2019)

79. Pointcheval, D., Stern, J.: Security arguments for digital signatures and blind signatures. J. Cryptol. **13**(3), 361–396 (2000)

80. Quach, W., Rothblum, R., Wichs, D.: Reusable designated-verifier NIZKs for all NP from CDH. In: Ishai, Y., Rijmen, V. (eds.) EUROCRYPT 2019. LNCS, vol. 11477, pp. 593–621. Springer, Cham (2019)

81. Rivest, R.L., Shamir, A., Tauman, Y.: How to leak a secret. In: Boyd, C. (ed.) ASIACRYPT 2001. LNCS, vol. 2248, pp. 552–565. Springer, Heidelberg (2001). https://doi.org/10.1007/3-540-45682-1_32

82. Rothblum, R.D., Sealfon, A., Sotiraki, K.: Towards non-interactive zero-knowledge for NP from LWE. Cryptology ePrint Archive, Report 2018/240 (2018). https://eprint.iacr.org/2018/240

83. Rouselakis, Y., Waters, B.: Practical constructions and new proof methods for large universe attribute-based encryption. In: Sadeghi, A.-R., Gligor, V.D., Yung, M. (eds.) ACM CCS 2013, pp. 463–474. ACM Press, November 2013

84. Sahai, A.: Non-malleable non-interactive zero knowledge and adaptive chosen-ciphertext security. In: 40th FOCS, pp. 543–553. IEEE Computer Society Press, October 1999

85. Sahai, A., Seyalioglu, H.: Worry-free encryption: functional encryption with public keys. In: Al-Shaer, E., Keromytis, A.D., Shmatikov, V. (eds.) ACM CCS 2010, pp. 463–472. ACM Press, October 2010

86. Sahai, A., Waters, B.: How to use indistinguishability obfuscation: deniable encryption, and more. In: Shmoys, D.B. (ed.) 46th ACM STOC, pp. 475–484. ACM Press, May/June 2014

87. Sahai, A., Waters, B.R.: Fuzzy identity-based encryption. In: Cramer, R. (ed.) EUROCRYPT 2005. LNCS, vol. 3494, pp. 457–473. Springer, Heidelberg (2005). https://doi.org/10.1007/11426639_27

88. Shacham, H., Waters, B.: Compact proofs of retrievability. In: Pieprzyk, J. (ed.) ASIACRYPT 2008. LNCS, vol. 5350, pp. 90–107. Springer, Heidelberg (2008). https://doi.org/10.1007/978-3-540-89255-7_7

89. Teranishi, I., Furukawa, J., Sako, K.: k-times anonymous authentication (extended abstract). In: Lee, P.J. (ed.) ASIACRYPT 2004. LNCS, vol. 3329, pp. 308–322. Springer, Heidelberg (2004). https://doi.org/10.1007/978-3-540-30539-2_22

90. Tsabary, R.: An equivalence between attribute-based signatures and homomorphic signatures, and new constructions for both. In: Kalai, Y., Reyzin, L. (eds.) TCC 2017, Part II. LNCS, vol. 10678, pp. 489–518. Springer, Cham (2017). https://doi.org/10.1007/978-3-319-70503-3_16

91. Ventre, C., Visconti, I.: Co-sound zero-knowledge with public keys. In: Preneel, B. (ed.) AFRICACRYPT 2009. LNCS, vol. 5580, pp. 287–304. Springer, Heidelberg (2009). https://doi.org/10.1007/978-3-642-02384-2_18

92. Waters, B.: Ciphertext-policy attribute-based encryption: an expressive, efficient, and provably secure realization. In: Catalano, D., Fazio, N., Gennaro, R., Nicolosi, A. (eds.) PKC 2011. LNCS, vol. 6571, pp. 53–70. Springer, Heidelberg (2011). https://doi.org/10.1007/978-3-642-19379-8_4

Lattice-Based Cryptography

Building an Efficient Lattice Gadget Toolkit: Subgaussian Sampling and More

Nicholas Genise[1]([✉]) [iD], Daniele Micciancio[1] [iD], and Yuriy Polyakov[2] [iD]

[1] University of California, San Diego, La Jolla, USA
{ngenise,daniele}@eng.ucsd.edu
[2] New Jersey Institute of Technology, Newark, USA
polyakov@njit.edu

Abstract. Many advanced lattice cryptography applications require efficient algorithms for inverting the so-called "gadget" matrices, which are used to formally describe a digit decomposition problem that produces an output with specific (statistical) properties. The common gadget inversion problems are the classical (often binary) digit decomposition, subgaussian decomposition, Learning with Errors (LWE) decoding, and discrete Gaussian sampling. In this work, we build and implement an efficient lattice gadget toolkit that provides a general treatment of gadget matrices and algorithms for their inversion/sampling. The main contribution of our work is a set of new gadget matrices and algorithms for efficient subgaussian sampling that have a number of major theoretical and practical advantages over previously known algorithms. Another contribution deals with efficient algorithms for LWE decoding and discrete Gaussian sampling in the Residue Number System (RNS) representation.

We implement the gadget toolkit in PALISADE and evaluate the performance of our algorithms both in terms of runtime and noise growth. We illustrate the improvements due to our algorithms by implementing a concrete complex application, key-policy attribute-based encryption (KP-ABE), which was previously considered impractical for CPU systems (except for a very small number of attributes). Our runtime improvements for the main bottleneck operation based on subgaussian sampling range from 18x (for 2 attributes) to 289x (for 16 attributes; the maximum number supported by a previous implementation). Our results are applicable to a wide range of other advanced applications in lattice cryptography, such as GSW-based homomorphic encryption schemes, leveled fully homomorphic signatures, other forms of ABE, some program obfuscation constructions, and more.

This work was sponsored by the Defense Advanced Research Projects Agency (DARPA) and the Army Research Laboratory (ARL) under Contract Numbers W911NF-15-C-0226 and W911NF-15-C-0233. The views expressed are those of the authors and do not necessarily reflect the official policy or position of the Department of Defense or the U.S. Government.

© International Association for Cryptologic Research 2019
Y. Ishai and V. Rijmen (Eds.): EUROCRYPT 2019, LNCS 11477, pp. 655–684, 2019.
https://doi.org/10.1007/978-3-030-17656-3_23

1   Introduction

Many advanced applications of lattice cryptography require the generation of
a random integer matrix $\mathbf{A} \in \mathbb{Z}_q^{n \times m}$ (with uniform entries modulo q) together
with a *strong trapdoor* (typically a short[1] basis $\mathbf{S}$ for the lattice defined by $\mathbf{A}$
as a parity check matrix). The strong trapdoor is used to efficiently "invert" the
classical Short Integer Solution (SIS) and Learning with Errors (LWE) functions
$f_{\mathbf{A}}(\mathbf{x}) = \mathbf{A}\mathbf{x}$ and $g_{\mathbf{A}}(\mathbf{s}, \mathbf{e}) = \mathbf{s}^t \mathbf{A} + \mathbf{e}^t$ associated to the matrix $\mathbf{A}$. Theoretical
solutions to these trapdoor generation and function inversion problems have
long been known [1,6,7,26,34]. However, the trapdoor constructions of [1,6] and
generic inversion algorithms of [7,26,34] are rather complex, inefficient, and not
suitable for practice.

An important step towards bringing advanced lattice-based cryptographic
applications to practice was taken in [38], where a new notion of trapdoor (and
associated generation algorithm) is proposed. The trapdoor of [38] transforms
the problem of inverting the random functions $f_{\mathbf{A}}, g_{\mathbf{A}}$ to the problem of inverting
the same type of functions $f_{\mathbf{G}}, g_{\mathbf{G}}$, but for a specific, carefully designed "gadget"
matrix $\mathbf{G}$, which admits much simpler and faster inversion algorithms. Gadget
matrices similar to the one of [38] had already been used in a number of previous
works, starting from Ajtai's first construction of "solved instances of the shortest basis problem" [1], and including virtually all works on (fully) homomorphic
encryption schemes based on the LWE problem (e.g., see [12,27]); though different works use the gadget matrix for somehow different purposes. In fact, there
are several inversion problems associated to the gadget matrix $\mathbf{G}$:

- **Digit Decomposition:** This is the problem of expressing an arbitrary vector
  $\mathbf{u} \in \mathbb{Z}_q^n$ as a short vector $\mathbf{x}$ such that $\mathbf{G}\mathbf{x} = \mathbf{u} \pmod{q}$. This is perhaps the
  most basic use of the gadget matrix $\mathbf{G}$, and plays an important role in the
  key-switching and multiplication operations of fully homomorphic encryption
  (FHE) schemes. For example, the binary decomposition gadget matrix $\mathbf{G} =
  [\mathbf{I}, 2\mathbf{I}, \dots, 2^{k-1}\mathbf{I}]$ allows to write any vector with entries in $\mathbb{Z}_q$ as a combination
  $\sum_i 2^i \mathbf{x}_i$ of vectors $\mathbf{x}_i \in \{0,1\}^n$ with the $\{0,1\}$ coefficients corresponding to
  the binary digits of the entries in $\mathbf{u}$.
- **Subgaussian Decomposition:** This is a type of *randomized* digit decompo-
  sition, where a short vector $\mathbf{x}$ satisfying $\mathbf{G}\mathbf{x} = \mathbf{u} \pmod{q}$ is chosen according
  to a distribution with desirable statistical properties. This alternative to the
  standard binary decomposition was suggested in [5] as a method to improve
  the noise growth in homomorphic computations using variants of the GSW
  homomorphic encryption scheme [27], and is potentially applicable to the key-
  switching and homomorphic multiplication operations of many other FHE
  schemes.
- **LWE Decoding:** Given $\mathbf{s}^t \mathbf{G} + \mathbf{e}^t$ for a sufficiently small error vector $\mathbf{e}$, recover
  both $\mathbf{s}$ and $\mathbf{e}$. This is the (deterministic) inversion problem for the standard

---

[1] In the context of lattice cryptography, "short" typically means much smaller than
the modulus q.

(injective) LWE function $g_\mathbf{G}$, which is used, for example, in the decryption algorithms of LWE-based cryptosystems.

- **Discrete Gaussian Sampling:** Produce a sample from a discrete Gaussian distribution over the set of all integer vectors $\mathbf{x}$ such that $f_\mathbf{G}(\mathbf{x}) = \mathbf{u}$. This problem was the main focus of [38], and is used, for example, in hash-and-sign lattice-based signatures and trapdoor delegation for identity-based encryption, among many other applications.

Very efficient gadget inversion algorithms were given in [38], but only for the Discrete Gaussian Sampling and LWE Decoding problems, and in the very special setting where the modulus $q = b^k$ is the power of a small base b. For the case of Discrete Gaussian Sampling, an equally efficient, but more general solution, was recently proposed in [23] for arbitrary modulus q, expanding the range of advanced lattice cryptography applications that admit a reasonably practical implementation. (E.g., see [9,17,19,30,31].) We remark that trapdoor inversion is the most complex operation in many applications of lattice cryptography, and effective solutions to gadget inversion play a critical role in determining the efficiency, quality and other performance characteristics of higher-level algorithms and the final applications.

The main focus of our work is Subgaussian Decomposition, a problem that has received little or no attention so far, and still has the potential to substantially improve the efficiency of many important applications. The importance of subgaussian sampling is easily explained by comparing it to the related problems of Digit Decomposition and Discrete Gaussian Sampling. We recall that lattice-based cryptography directly supports linear homomorphic operations, but ciphertexts are noisy, and their quality degrades when performing homomorphic operations: the noise of a sum $c_0 + c_1$ is the sum of the noise in the original ciphertexts c_0, c_1. More critically, when multiplying a ciphertext by a constant α, the noise scales by a factor α, which can be arbitrarily large. So, one needs to limit linear combinations to use only small coefficients. This is typically done using binary digit decomposition: given encryptions c_i of $2^i m$ (for $i = 0, \ldots, k-1$), one can compute an encryption of αm (for a large $\alpha < 2^k$) by taking a $0-1$ combination $\sum_i \alpha_i c_i$, where $\alpha = \sum_i 2^i \alpha_i$ is the binary representation of α. This way, the resulting noise scales linearly with $k = \log \alpha$, rather than α. Subgaussian decomposition allows to make the resulting noise even smaller: due to cancellations between randomly chosen coefficients, subgaussian decomposition has a "pythagorean additivity" property that makes the noise grow only as $O(\sqrt{k})$. The gain is even more substantial when adding many (say n) ciphertexts, in which case the noise growth is improved by a factor $\sqrt{nk}$. At the other end of the spectrum, pythagorean growth can also be achieved using Discrete Gaussian Sampling, as gaussian distributions are by definition also subgaussians. However, gaussian sampling is considerably more costly than digit decomposition, both in terms of running time, randomness and output quality: even with the improved algorithms of [23,38], Discrete Gaussian Sampling is much more complex than a simple digit decomposition, and it necessarily produces "digits" (i.e., coefficients) that are larger than naive binary decomposition roughly by a factor $\Omega(\sqrt{\log k})$.

The added algorithmic complexity and noise overhead make gaussian sampling unattractive in practice, and, perhaps not surprisingly, all implementations we are aware of use digit decomposition whenever possible.

Subgaussian decomposition has the potential to offer the best of both worlds: pythagorean additivity, but without the $\Omega(\sqrt{\log k})$ noise overhead of a full-blown discrete gaussian sampler. These potential advantages were already outlined in [5], but they were so far considered only of theoretical interest. In fact, none of the subsequent improvements and implementations [15,16,21,40] make use of subgaussian sampling.

Our Contribution. The main contribution of our work is a set of new gadget matrices and algorithms for efficient subgaussian sampling. The improvements are not just theoretical/asymptotical, but very practical, as demonstrated by a concrete complex application: an implementation of a Key-Policy Attribute-Based Encryption (KP-ABE) scheme that speeds up previous implementation efforts by more than one order of magnitude. (See below and Sects. 7 and 8 for details.) On the theoretical side, our algorithms and gadgets result in pythagorean error growth and optimal (essentially linear) time complexity. In practice, the algorithms are easy to implement and have very small hidden constants both in the number of operations they perform and the subgaussian parameters, offering a very attractive alternative to the naive deterministic digit decomposition methods currently used in the implementation of FHE and other related pritimives. Moreover, our gadgets and algorithms have a number of other useful properties that make them even more attractive in practice:

- All our algorithms require very little storage and only a modest (essentially optimal) amount of randomness. In particular, our gadget matrices have a very regular structure, and do not need to be explicitly stored.
- We support an arbitrary modulus q. This is not just of theoretical interest, as fast implementations of lattice cryptography [8,33,35,37] require moduli of special form in order to make use of the Number Theoretic Transform (NTT).
- Our gadgets and algorithms support the "Full RNS" and "double CRT" techniques used to implement lattice cryptography with large modulus without the need for arbitrary-precision arithmetic libraries [8,25,33].

Beside subgaussian decomposition, we also provide very efficient algorithms for LWE Decoding and Discrete Gaussian Sampling that improve previous work [23,38] by supporting arbitrary moduli and Full RNS implementations. (For Discrete Gaussian Sampling, algorithms supporting arbitrary moduli were already provided in [23], but for gadget matrices that do not support Full RNS implementations.)

Taken together, our algorithms provide a complete *lattice gadget toolkit*, offering efficient solutions to the full range of inversion problems encountered in lattice cryptography: Subgaussian Decomposition, LWE Decoding, and Discrete Gaussian Sampling. Our results are not just of theoretical interest, but are also relevant to the implementation and use of advanced lattice cryptography applications.

In order to demonstrate the applicability of our results, we developed an open-source optimized implementation of our algorithms and used it to implement a complex application: a key-policy attribute-based encryption scheme, which utilizes in a single setting most of the new algorithms developed in this paper. Our experimental results strongly support the theoretical analysis, showing both that (with proper algorithms and implementations) the advantages of pythagorean additivity clearly outweigh the modest increase in computational cost over naive bit decomposition, and that the overall impact on the performance of lattice applications can be substantial. Our CPU implementation of homomorphic access policy evaluation for keys and ciphertexts (the most expensive operation in KP-ABE schemes) outperforms the previous CPU implementation (for a comparable level of security) by a factor ranging from 18x to 289x as the number of policy attributes grows from 2 to 16. For higher numbers of attributes, previous CPU implementations were not feasible (only a GPU implementation is known), while we were able to run our implementation within reasonable running times for as many as 128 attributes. Other operations are also faster in our implementation, and memory requirements are also much smaller (by more than a factor of 2x in the simplest case of 2 attributes, and more than one order of magnitude at 16 attributes.) In summary, our results show that using our toolkit gadget inversion is no longer the bottleneck in efficient implementations of lattice cryptography, and it can be profitably used to achieve better performance and scalability both in theory and practice.

While in this paper we focused on the algorithmic core of a general gadget toolkit, and on a specific (but representative) application, our results are applicable to a wide range of other advanced applications in lattice cryptography. These include the use of subgaussian decomposition in GSW-based homomorphic encryption schemes [15,16,21], leveled fully homomorphic signatures [29], other forms of ABE [4], obfuscation of finite automata and branching programs using graph-induced encoding [32], and more.

Techniques. Our efficient lattice gadget toolkit is based on better algorithmic solutions to known problems, but also on a new class of gadget matrices that enable our algorithmic improvements. While gadget matrices of the type used in [38] and our work are quite common in lattice cryptography, they have never been formally defined. In fact, as different applications and algorithms use the gadget matrices in somehow different ways, it was not even clear if one could meaningfully define gadget matrices as abstract mathematical objects, and most of previous works use the term "gadget" informally to identify specific constructions.

The starting point of our investigation is a simple, intuitive definition of gadget matrix, which turns out to be relevant to the solution of all algorithmic problems studied in this work. For any dimension n and modulus q (typically mandated by the application), a gadget of quality β is a matrix $\mathbf{G} \in \mathbb{Z}_q^{n \times w}$ such that any $\mathbf{u} \in \mathbb{Z}_q^n$ can be represented as $\mathbf{Gx} = \mathbf{u}$ for some small integer vector $\mathbf{x} \in \mathbb{Z}^w$ of norm $\|\mathbf{x}\| \leq \beta$. This definition is directly motivated by the Digit Decomposition problem, but as we show in Sect. 3, it is already enough to enable theoretically efficient solutions to all of the algorithmic problems discussed above. (See Theorem 3 and Corollary 1.) The generic solutions obtained from Corollary 1 are not suited for practice, both in terms of algorithmic complexity and output quality. Still, a generic definition of gadget is useful to delimit a design space which extends well beyond the simplest (and perhaps most natural) construction of decomposition gadget $[1, 2, 2^2, \ldots, 2^{k-1}]$ corresponding to the the the standard binary representation of a number as a sequence of bits. Other gadgets used in our work are digit decomposition gadgets $[1, b, b^2, \ldots, b^{k-1}]$ with a larger base $b > 2$, the CRT gadget $[g_1, \ldots, g_k]$ where $g_i = ((q/q_i)^{-1} \bmod q_i) \cdot (q/q_i)$ [11] (for composite moduli with relatively prime factorization $q = \prod_i q_i$) as well as hybrids between the two approaches where each g_i is replaced by an appropriate multiple of a vector of the form $[1, b, b^2, \ldots, b^{k-1}]$.

Our subgaussian decomposition algorithms use ideas and techniques from recent work on discrete gaussian sampling for arbitrary modulus [23]. In particular, we use the **SD** matrix factorization for the lattice defined by **G**, and then perform subgaussian decomposition with respect to matrix **D**, which is sparse and triangular, and admits much faster algorithms. A solution to the original problem is obtained using **S** as a linear transformation. Naturally, the details of our subgaussian decomposition algorithm for **D** are quite different from the algorithms in [23], as that paper solves a different problem (discrete gaussian sampling.) But, as our algorithm can sample an output vector **x** with respect to an arbitrary (not necessarily isotropic) subgaussian distribution, there is no need to apply gaussian correction terms (as done in [23]), and our algorithm is much simpler and more efficient. Our efficient LWE decoding algorithm is analogous. Specifically, the decoding problem can be seen as decoding an input to a lattice basis $q(\mathbf{SD})^{-t} = \mathbf{S}^{-t}(q \cdot \mathbf{D}^{-t})$. Now we can solve the decoding problem by first using $\mathbf{S}^t$ as a linear transformation, then by decoding the transformed input to the lattice generated by $q \cdot \mathbf{D}^{-t}$. This lattice is efficiently decodable since it, too, has a sparse, triangular basis. Our toolkit implementation focuses on providing full ring and RNS support for all gadget algorithms because ring multiplications can be efficiently computed via NTTs and large integer operations can be efficiently performed using native arithmetic in RNS. Full RNS/double CRT constructions based on power-of-two cyclotomic rings provide the best performance for the majority of known lattice cryptography primitives, as illustrated by our experimental results for subgaussian sampling and key-policy attribute-based encryption.

Related Work. The first use of subgaussian decomposition appears in [5] in a theoretical form, not optimized for implementations. While the use of CRT gadgets for digit decomposition in the implementation of FHE schemes [8,33] or even the foundation of Ring LWE [36] is not new, their applicability in the context of Gaussian or subgaussian sampling is, to the best of our knowledge, novel. Our CRT gadget algorithms can be seen as an extension of [11,23,38]. The CRT-like gadget proposed in [32] can be considered as a special case of ours when $b_i = p_i$ (assuming that $q_i = p_i^e$), which implies the gagdet noise width is larger than p_i. In our CRT gadget, b_i's can be chosen independently from

CRT moduli q_i, enabling significantly more efficient implementations in the ring setting. Another related work is a deterministic, balanced digit decomposition for the "double CRT/RNS" gadget [25] in the "LoL" library [18,41] (initially unknown to the authors).

In the ring setting, one method [20] to achieve better-than-generic, $n \log^2 q$, efficiency of "power-of-b" gadget discrete Gaussian sampling is to use the FFO style of discrete Gaussian samplers from [22] and [23, Sect. 4]. This incurs a logarithmic slowdown, $\log n$, in time and space compared to using [23, Sect. 3] on the coefficients independently, which has $n \log q$ time and space efficiency. Further, Sect. 6 of this work extends [23, Sect. 3] to the "double-CRT" setting [25], freeing implementations using discrete Gaussian gadget sampling from multi-precision numbers when the modulus is over 64 bits.

Organization. The rest of the paper is organized as follows. In Sect. 2 we review some preliminary material. In Sect. 3 we present our general definition of gadget matrices. Next, in Sects. 4 and 5 we present our core gadgets and algorithms for subgaussian decomposition and LWE decoding with arbitrary modulus. In Sect. 6 we extend these algorithms to large composite moduli to allow efficient operations in CRT form without the need of multiprecision integer arithmetic. Sections 7 and 8 we present our implementation and experimental results. Note, the full version of the paper [24] contains the missing proofs as well as the description of the generic "subgaussian nearest plane" algorithm in its appendix.

2   Preliminaries

We indicate numbers with lowercase letters, such as $z \in \mathbb{Z}$, vectors as bold lowercase letters, $\mathbf{z} \in \mathbb{Z}^n$, and matrices as uppercase bold letters, $\mathbf{M} \in \mathbb{R}^{n \times n}$. The default norm used is the l_2 norm of a vector unless stated otherwise, though we will often use the max, or l_∞, norm. For a real number r, denote $\lceil r \rfloor$ as the deterministic rounding function to a nearest integer of r. Rounding a real vector is applied analogously, entry-wise. Many computations will be done over the integers modulo q, $\mathbb{Z}_q$. We view $\mathbb{Z}_q$ through its balanced coset representatives in $(-q/2, q/2]$ unless stated otherwise. For a positive integer base b and a non-negative integer $u < b^k$, u's b-ary decomposition is a vector $[u]_b^k = (u_0, \cdots, u_{k-1}) \in \{0, \cdots, b-1\}^k$ and satisfies $\sum_i b^i u_i = u$. When $b = 2$, this is simply u's binary decomposition. Recall the Chinese Remainder Theorem for modular arithmetic. Let q be a positive integer with a prime factorization of $q = p_1^{e_1} \cdots p_l^{e_l} = q_1 \cdots q_l$. Then by the Chinese Remainder Theorem (CRT), we have $\mathbb{Z}_q \cong \mathbb{Z}_{q_1} \times \cdots \times \mathbb{Z}_{q_l}$. The isomorphism $\phi(\cdot)$ is given by $\phi(a) = (a \mod q_1, \cdots, a \mod q_l)$ and its inverse is $\phi^{-1}(a_1, \cdots, a_l) = \sum_i (a_i) q_i^* \hat{q}_i$ where $q_i^* := \frac{q}{q_i}$ and $\hat{q}_i := (q_i^*)^{-1} \mod q_i$.

For a probability distribution χ, we denote $e \leftarrow \chi$ to mean e is sampled from χ. When χ is trivial (often over a number x), we will use $e \leftarrow x$ to be variable assignment as well. We will need the following, known as the *Geršgorin Circle Theorem*.

Theorem 1 (Geršgorin). *Let* **M** *be an* $n \times n$ *matrix with complex entries. For each row* i, *let* r_i *be the sum of its non-diagonal entries' magnitudes:* $r_i = \sum_{j \neq i} |\mathbf{M}(i,j)|$. *Then, the eigenvalues of* **M** *are all in* $\bigcup_i \{z \in \mathbb{C} : |z - M(i,i)| \leq r_i\}$.*

2.1 Subgaussian Random Variables

A random variable X over $\mathbb{R}$ is *subgaussian* [36,43] with parameter $\alpha > 0$ if its (scaled) moment generating function satisfies $\mathbb{E}[\exp(2\pi t X)] \leq \exp(\pi \alpha^2 t^2)$ for all $t \in \mathbb{R}$. Scaling a subgaussian X by any $c \in \mathbb{R}$ to $c \cdot X$ yields a subgaussian random variable with parameter $|c|\alpha$. If X is subgaussian with parameter α, then its tails are dominated by a Gaussian parameterized by α, $\Pr\{|X| \geq t\} \leq 2\exp(-\pi t^2/\alpha^2)$. Any B-bounded centered ($\mathbb{E}[X] = 0$) random variable X is subgaussian with parameter $B\sqrt{2\pi}$. When X is subgaussian with parameter α and Y conditioned on X taking any value is subgaussian with parameter β, $X + Y$ is subgaussian with parameter $\sqrt{\alpha^2 + \beta^2}$. This property is called *Pythagorean additivity*. The proof of the following Lemma is derived by expanding $\mathbb{E}[\exp(2\pi t(X+Y))]$.

Lemma 1. *Let* X, Y *be discrete random variables over* $\mathbb{R}$ *such that* X *is subgaussian with parameter* α *and* Y *conditioned on* X *taking any value is subgaussian with parameter* β. *Then,* $X + Y$ *is subgaussian with parameter* $\sqrt{\alpha^2 + \beta^2}$.*

A random vector **x** over $\mathbb{R}^n$ is *subgaussian* with parameter $\alpha > 0$ if $\langle \mathbf{x}, \mathbf{u} \rangle$ is subgaussian with parameter α for all unit vectors **u**. Using a similar calculation to the above, one can show that if each coefficient of a random vector is subgaussian with parameter α conditioned on the previous coefficients taking any values, then the vector is subgaussian with parameter α. The slightly more general fact below is needed for our algorithms. Its proof is analogous to the proof of Lemma 1.

Lemma 2. *Let* **x** *be a discrete random vector over* $\mathbb{R}^n$ *such that each coordinate* x_i *is subgaussian with parameter* α_i *given the previous coordinates take any values. Then,* **x** *is a subgaussian vector with parameter* $\max_i\{\alpha_i\}$.*

We emphasize this fact, for without it one is left with an unnecessary $\sqrt{n}$ term in the subgaussian parameter of subgaussian vectors. Now, that the sum of independently generated random vectors **x** and **y** subgaussian with parameters α and β is a subgaussian vector with parameter $\sqrt{\alpha^2 + \beta^2}$ immediately follows.

A main algorithm presented in this paper will rely on a linear transformation of a discrete subgaussian vector.

Lemma 3 (Simplified [36, Corollary 2.3]). *Let* **x** *be a subgaussian random vector with parameter* α *and let* **M** *be a linear transformation. Then,* **Mx** *is a subgaussian vector with parameter* $\alpha \lambda_{max}(\mathbf{MM}^T)^{1/2}$ *where* $\lambda_{max}(\cdot)$ *is the largest eigenvalue.*

2.2   Lattices

A lattice is a discrete subgroup of $\mathbb{R}^n$. Equivalently, a lattice Λ can be represented as the set of all integer combinations of a basis $\mathbf{B} = [\mathbf{b}_1, \cdots, \mathbf{b}_k] \in \mathbb{Z}^{n \times k}$, $\Lambda = \{\sum_1^k z_i \mathbf{b}_i : z_i \in \mathbb{Z}\} = \mathcal{L}(\mathbf{B})$. Notice that any permutation of basis vectors is another lattice basis. We only consider full-rank lattices $(k = n)$. A lattice is an integer lattice if it is a sublattice of $\mathbb{Z}^n$. The dual lattice of Λ, denoted as Λ^*, is the set $\Lambda^* = \{\mathbf{z} \in \mathbb{R}^n : \langle \mathbf{z}, \Lambda \rangle \subseteq \mathbb{Z}\}$. Given a basis $\mathbf{B}$ for Λ, its dual basis is $\mathbf{B}^{-t}$ which is also a basis for Λ^*. We will consider direct sums of lattices, $\Lambda = \Lambda_1 \oplus \cdots \oplus \Lambda_l$ and their dual lattices $\Lambda^* = \Lambda_1^* \oplus \cdots \oplus \Lambda_l^*$. The number $\lambda_i(\Lambda)$ is the radius of the smallest ball containing i linearly independent lattice vectors.

Given a basis $\mathbf{B} = [\mathbf{b}_1, \cdots, \mathbf{b}_n]$ for a lattice Λ, its Gram-Schmidt orthogonalization (GSO) is the set of vectors $\widetilde{\mathbf{B}} = [\widetilde{\mathbf{b}}_1, \cdots, \widetilde{\mathbf{b}}_n]$ where $\widetilde{\mathbf{b}}_i$ is the component of $\mathbf{b}_i$ orthogonal to $\text{span}(\mathbf{b}_1, \cdots, \mathbf{b}_{i-1})$. The GSO is not another basis for the lattice in general, but it gives us a tiling of $\mathbb{R}^n$ given by $\mathbb{R}^n = \cup_{\mathbf{x} \in \Lambda}(\mathbf{x} + \mathcal{P}_{1/2}(\widetilde{\mathbf{B}}))$ where $\mathcal{P}_{1/2}(\widetilde{\mathbf{B}}) := \widetilde{\mathbf{B}} \cdot (-1/2, 1/2]^n$. Note that the GSO depends on the order of the vectors given. We define the reverse order GSO analogously. The algorithms presented in this paper will all be instantiations of Babai's greedy decoding algorithm known as the *nearest plane algorithm* [7].

Theorem 2. *There is an algorithm which given* $\mathbf{B}, \widetilde{\mathbf{B}}, \mathbf{t} \in \mathbb{R}^n$ *returns the unique lattice point in* $\mathbf{t} + \mathcal{P}_{1/2}(\mathbf{B}^*)$ *in time* $O(n^2)$ *and memory* $O(n^3)^2$.

Discrete Gaussians. Let $A \subset \mathbb{R}^n$ be a discrete set, and let the (spherical) Gaussian function with width s and center $\mathbf{c} \in \mathbb{R}^n$ be $\rho_{s,\mathbf{c}}(\mathbf{x}) = \exp(-\pi \|\mathbf{x} - \mathbf{c}\|^2/s^2)$. Let $\rho_{s,\mathbf{c}}(A) = \sum_{\mathbf{y} \in A} \rho_{s,\mathbf{c}}(\mathbf{y})$. The smoothing parameter of a lattice [39] for some $\varepsilon > 0$, is denoted as $\eta_\varepsilon(\Lambda)$, and it is defined as the minimum $s > 0$ such that $\rho(s \cdot \Lambda^*) \leq 1 + \epsilon$. When $s = 1$ and $\mathbf{c} = \mathbf{0}$, we denote this as $\rho(\cdot)$. Then, the discrete Gaussian distribution has probability $\rho_{s,\mathbf{c}}(\mathbf{x})/\rho_{s,\mathbf{c}}(A)$ for each $\mathbf{x} \in A$. This distribution is denoted as $D_{A,s,\mathbf{c}}$. Polynomial time discrete Gausisan sampling algorithms for general lattices and their cosets, with width above the GSO length of the input basis (times a small factor, $\omega(\sqrt{\log n})$ or $O(\sqrt{\log n})$), are given in [13,26].

q-ary Lattices. Throughout this paper we will mostly be concerned with *q-ary lattices*. These are full-rank integer lattices with $q \cdot \mathbb{Z}^k$ as a sublattice. Fix an integer $q > 0$ to be used as a modulus and let $m > w > n$. A matrix $\mathbf{A} \in \mathbb{Z}_q^{n \times m}$ is *primitive* if $\mathbf{A}\mathbb{Z}_q^m = \mathbb{Z}_q^n$. Given an $\mathbf{A} \in \mathbb{Z}_q^{n \times m}$, we define the following lattices: $\Lambda_q^\perp(\mathbf{A}) = \{\mathbf{z} \in \mathbb{Z}^m : \mathbf{A}\mathbf{z} = \mathbf{0} \mod q\}$, and $\Lambda_q(\mathbf{A}) = \{\mathbf{v} \in \mathbb{Z}^m : \exists\, \mathbf{s} \in \mathbb{Z}^n, \mathbf{v}^t = \mathbf{s}^t \mathbf{A} \mod q\}$. These lattices satisfy the following duality relation: $\Lambda_q^\perp(\mathbf{A})^* = q \cdot \Lambda_q(\mathbf{A})$. Further, the cosets of $\Lambda_q^\perp(\mathbf{A})$, $\Lambda_{\mathbf{u}}^\perp(\mathbf{A}) := \{\mathbf{z} \in \mathbb{Z}^m : \mathbf{A}\mathbf{z} = \mathbf{u} \mod q\}$, are in bijection with $\mathbb{Z}_q^n$ when $\mathbf{A}$ is primitive. Let $\mathbf{G}$ be an arbitrary, primitive matrix over $\mathbb{Z}_q$. The following sampling problem, defined on the integer cosets of $\Lambda_q^\perp(\mathbf{G})$, is needed for many advanced lattice crypto-schemes.

---

^{2} This assumes the GSO has entries each described in $O(n)$ bits.

Definition 1. *For a primitive* $\mathbf{G} \in \mathbb{Z}_q^{n \times w}$, *the subgaussian decomposition problem with parameter* α *for* $\mathbf{G}$ *is to sample vectors* $\mathbf{x} \in \mathbb{Z}^w$ *subgaussian with parameter* α *such that* $\mathbf{u} = \mathbf{G}\mathbf{x}$ mod q *for arbitrary* $\mathbf{u}$ *given as input.*

Another name for this problem is subgaussian sampling. A generic adaptation of Babai's algorithm (analyzed in the Appendix of the full version [24], called the subgaussian nearest plane algorithm) is used in [5] (AP14) to achieve subgaussian decomposition for a specific $\mathbf{G}$. In general, this generic algorithm runs in time $O(k^2)$, and uses space $O(k^3)$. Another, related problem is the discrete Gaussian sampling problem.

Definition 2. *For a primitive* $\mathbf{G} \in \mathbb{Z}_q^{n \times w}$, *the discrete Gaussian sampling problem with width* s *for* $\mathbf{G}$ *is to sample vectors* $\mathbf{x} \in \mathbb{Z}^w$ *distributed as* $D_{\mathbb{Z}^w, s}$ *conditioned on* $\mathbf{G}\mathbf{x}$ mod $q = \mathbf{u}$ *for arbitrary* $\mathbf{u}$ *given as input.*

Efficient solutions with small s for commonly used $\mathbf{G}$'s are given in [23,38]. Both of the above sampling problems have polynomial time solutions using randomized versions of Babai's algorithm. In addition, we will consider decoding the q-ary code defined by $\mathbf{G}$ for an arbitrary, primitive $\mathbf{G}$.

Definition 3. *For a primitive* $\mathbf{G} \in \mathbb{Z}_q^{n \times w}$, *the LWE decoding problem with tolerance* δ *on* $\mathbf{G}$ *is to return* $\mathbf{s}$ *given* $\mathbf{s}^t\mathbf{G} + \mathbf{e}^t$ mod q *for an error* $\|\mathbf{e}\|_\infty < \delta$.

Specifically, we want to efficiently decode $\mathbf{G}$ while maximizing $\delta \in [0, q/2)$. An efficient LWE decoding algorithm for a specific, commonly used $\mathbf{G}$ ($b = 2$ in the paragraph below) with tolerance $q/4$ is provided in [38].

A $\mathbf{G}$ commonly used in lattice crypto-schemes is defined as follows. Fix an integer $b \in (1, q)$, known as the *base*, and let $k = \lceil \log_b q \rceil$. The block-diagonal gadget matrix is $\mathbf{G} = \mathbf{I}_n \otimes \mathbf{g}^t$ with blocks $\mathbf{g}^t := (1, b, \cdots, b^{k-1})$. A common basis for $\Lambda_q^\perp(\mathbf{g}^t)$ [23] $\mathbf{S}_q$ has a sparse, triangular factorization $\mathbf{S}_q = \mathbf{S}\mathbf{D}$ [23] (restated in Sect. 4.2 in this paper).

3   Gadget Matrices

In order to guide our search for gadget matrices with efficient inversion and sampling algorithms, we give a simple general definition of gadget. The definition is modeled after the properties required by the digit decomposition problem, perhaps the simplest and most natural application of gadgets. But, as we will see, this simple characterization is enough to guarantee (theoretical) solutions to all problems that arise in the application of gadgets in lattice cryptography.

Definition 4. *For any finite additive group* A, *an* A-gadget of size w and quality β *is a vector* $\mathbf{g} \in A^w$ *such that any group element* $u \in A$ *can be written as an integer combination* $u = \sum_i g_i \cdot x_i$ *where* $\mathbf{x} = (x_1, \ldots, x_w)$ *has norm at most* $\|\mathbf{x}\| \leq \beta$.

We are primarily interested in gadgets for $A = \mathbb{Z}_q^n$, in which case the gadget is conveniently represented as a matrix $\mathbf{G} \in \mathbb{Z}_q^{n \times w}$ such that for any $\mathbf{u} \in \mathbb{Z}_q^n$ there is a vector $\mathbf{x} \in \mathbb{Z}^w$ of length $\|\mathbf{x}\| \leq \beta$ such that $\mathbf{Gx} = \mathbf{u} \pmod{q}$. We defined gadgets in terms of abstract groups to emphasize that the dimension n and modulus q should be thought of as part of the problem specification (typically mandated by the target application), while the w and β describe the size and quality of the solution. In particular, for any given n and q, one may consider multiple gadgets achieving different values of w and β. Naturally, smaller w and β are preferable, but as we will see there is a natural tradeoff between these two values, and one may increase β in order to reduce w and vice versa.

Before establishing a formal connection between the above definition and the notion of gadget informally defined in previous work, we make some important observations.

- The matrix $\mathbf{G}$ is necessarily primitive, i.e., $\mathbf{G}\mathbb{Z}_q^w = \mathbb{Z}_q^n$. Moreover, any primitive matrix is a $\mathbb{Z}_q^n$-gadget for a sufficiently large $\beta = \max_{\mathbf{u}} \min\{\|\mathbf{x}\| : \mathbf{Gx} = \mathbf{u} \pmod{q}\}$.
- If $\mathbf{g} \in \mathbb{Z}^k$ is a $\mathbb{Z}_q$-gadget of quality β, then $\mathbf{G} = \mathbf{I} \otimes \mathbf{g}^t \in \mathbb{Z}_q^{n \times w}$ is a $\mathbb{Z}_q^n$-gadget of size $w = kn$ and quality $\sqrt{n}\beta$.
- All definitions and constructions are easily adapted to ideal lattices (as used in the Ring-SIS and Ring-LWE problems) simply by considering "structured gadgets" of the form $\mathbf{G} \otimes [\alpha_1, \ldots, \alpha_n]$ where $[\alpha_1, \ldots, \alpha_n]$ is an appropriate $\mathbb{Z}$-basis of the underlying ring.

Based on the above observations, constructions may focus on the case $n = 1$, i.e., gadget vectors $\mathbf{g} \in \mathbb{Z}_q^w$, and then extend the solution to larger n (and possibly to the ring setting) using general techniques. In fact, this is how larger gadgets are built in all applications we are aware of. However, all the results in this section hold for arbitrary matrices, not necessarily with this tensor structure. So, for the sake of generality, we use matrix notation.

In order to justify our abstract definition of gadget, we show that it guarantees all other properties of gadgets used by lattice cryptography: it maps the gaussian distribution to an almost uniform vector $\mathbf{G}D_{\mathbb{Z},s}^w \approx \mathbb{Z}_q^n$ (as needed by the trapdoor generation algorithm of [38]), and it supports efficient algorithms to invert the LWE function $g_{\mathbf{G}}(\mathbf{x}, \mathbf{e})$, for discrete gaussian sampling on $f_{\mathbf{G}}^{-1}(\mathbf{u})$, and for subgassian decomposition with respect to $\mathbf{G}$. All these properties are proved by bounding the relevant parameters of the lattice $\Lambda_q^{\perp}(\mathbf{G})$ defined by $\mathbf{G}$.

Theorem 3. *For any gadget matrix* $\mathbf{G} \in \mathbb{Z}_q^{n \times w}$ *of quality* β, *the lattice* $L = \Lambda_q^{\perp}(\mathbf{G})$ *has a basis* $\mathbf{S}$ *with orthogonalized length* $\|\tilde{\mathbf{S}}\| \leq 2\beta + \sqrt{w}$, *successive minima* $\lambda_1(L), \ldots, \lambda_w(L) \leq 2\beta + \sqrt{w}$ *and smoothing parameter* $\eta(L) \leq (2\beta + \sqrt{w})\omega(\sqrt{\log n})$.

Note, the proof and theorem easily generalizes to any finite abelian group. Using the bound on the smoothing parameter, and the short (orthogonalized) basis $\mathbf{S} \in \mathbb{Z}^{w \times w}$, we immediately get the following applications. (E.g., for the subgaussian decomposition algorithm see the Appendix of the full version [24].)

Corollary 1. *For any gadget matrix* $\mathbf{G} \in \mathbb{Z}_q^{n \times w}$ *of quality* β *and* $s \geq (2\beta + \sqrt{w})\sqrt{\omega(\log n)}$, *the distribution* $\mathbf{G}D_{\mathbb{Z},s}^w$ *is statistically close to the uniform distribution over* $\mathbb{Z}_q^n$. *Moreover, there are polynomial-time algorithms for the following problems:*

- *Discrete Gaussian Sampling for the function* $f_{\mathbf{G}}(\mathbf{x}) = \mathbf{G}\mathbf{x} \pmod{q}$ *and input distribution* $D_{\mathbb{Z},s}^w$ *with* $s \geq (2\beta + \sqrt{w})\sqrt{\omega(\log n)}$.
- *Subgaussian Decomposition w.r.t* $\mathbf{G}$ *with parameter* $s \geq (2\beta + \sqrt{w}) \cdot \sqrt{2\pi}$.
- *LWE decoding of* $g_{\mathbf{G}}(\mathbf{s}, \mathbf{e})$ *for any* $\mathbf{s} \in \mathbb{Z}_q^n$ *and* $\|\mathbf{e}\|_\infty \leq q/2 \cdot (2\beta + \sqrt{w})$.

We remark that the general solutions provided by this corollary are of theoretical interest, and not suitable for practice. They are provided here only as a general feasibility result, in order to identify classes of good gadget matrices. The rest of the paper is dedicated to showing that by carefully choosing the gadget vector $\mathbf{g}$, one can obtain constructions and algorithms that are not only theoretically efficient, but also easy to implement and extremely fast.

4   Subgaussian Gadget Decomposition

In this section we present our main algorithms for the problem of *subgaussian gadget decomposition*, defined in Sect. 2.2, using the gadget matrix $\mathbf{G} = \mathbf{I}_n \otimes \mathbf{g}^t$. Since this decomposition $\mathbf{G}^{-1}(\mathbf{u}) = (\mathbf{g}^{-1}(u_i))_{i=1}^n$ can be computed one component at a time (even in-parallel!) we restrict our attention to efficiently computing the subgaussian function $\mathbf{g}^{-1} : \mathbb{Z}_q \to \mathbb{Z}^k$ in the one-dimensional case, i.e., for $n = 1$.

The gadgets and algorithms in this section are parametrized by a "base" integer b, which we consider as fixed throughout the section, but can be used to achieve different efficiency/quality trade-offs. We distinguish two cases, depending on whether the modulus is a power $q = b^k$ of the base b, or an arbitrary integer $q < b^k$. In either case, no assumption is made about the factorization of the modulus q. Later, in Sect. 6, we will extend the gadgets and algorithms from this section to provide optimized treatment of large moduli with useful co-prime factorization $q = \prod_i q_i$, where the input $u \in \mathbb{Z}_q$ is given in CRT form $(u \bmod q_1, \ldots, u \bmod q_l)$.

All algorithms in this section use the same gadget $\mathbf{g}^t := (1, b, \cdots, b^{k-1})$, for $k = \lceil \log_b q \rceil$, but with different subgaussian decomposition procedures depending on the whether q is a power of b. Notice that $\mathbf{g}^t$ is a $\mathbb{Z}_q$-gadget of size k and quality $\beta = \sqrt{k}(b/2)$.

The main result of this section is summarized in the following theorem.[3]

Theorem 4. *For any integer base* $b > 1$, *integer modulus* $q > 1$, $k = \lceil \log_b q \rceil$ *and gadget* $\mathbf{g}^t = [1, b, \cdots, b^{k-1}]$, *there is a subgaussian decomposition algorithm* $\mathbf{g}^{-1}$ *as follows:*

---

[3] This theorem is most relevant when q is a relatively small modulus (say $q < 2^{64}$), so that arithmetic operations modulo q can be performed with unit cost. For larger moduli, the theorem will be used as a building block for a more complex algorithm described in Sect. 6 using RNS/CRT representation for the elements of $\mathbb{Z}_q$.

- If $q = b^k$, the algorithm runs in linear $O(k)$ time (and space), uses $\log_2 q$ random bits, and achieves subgaussian parameter at most $(b-1)\sqrt{2\pi}$.
- If $q \neq b^k$, the algorithm runs in linear $O(k)$ time (and space), uses at most $k \log_2 q$ random bits, and achieves subgaussian parameter at most $(b+1)\sqrt{2\pi}$,

Notice how the generic solution obtained by applying Theorem 3 to our gadget $\mathbf{g}$ only implies a polynomial time inversion algorithm with subgaussian parameter $(b+1) \cdot \sqrt{2k\pi}$, and quadratic $O(k^2)$ time complexity (after a cubic time $O(k^3)$ preprocessing). Depending on implementation details, this generic solution would also require the use of high precision floating point numbers[4] and a substantial amount of randomness for high precision sampling. (For completeness, we provide a more detailed analysis of the generic solution in the Appendix of the full version [24].) By contrast, the solution described in Theorem 4 is much more efficient (linear time and space, with no need for preprocessing) and also achieves a smaller subgaussian parameter by a factor of $\sqrt{k}$. Moreover, our specialized algorithms use a relatively small (almost optimal) number of random bits, and can be implemented without the need for high-precision floating-point arithmetic.

A proof of Theorem 4 is given by the algorithms presented and analyzed in the next two subsections for the two separate cases $q = b^k$ and $q < b^k$.

4.1   Power-of-Base Case

Here we consider the subgaussian decomposition problem for the gadget $\mathbf{g} = (1, b, \ldots, b^{k-1})$ when $q = b^k$, and the input is given as a positive coset representative $u \in \{0, 1, \cdots, q-1\}$. Conceptually, our solution to this problem is just a specialized/optimized version of the randomized-rounding variant of Babai's nearest plane algorithm [5,7]. The general algorithm uses the Gram-Schmidt orthogonalization of a basis for the lattice $\Lambda_q^{\perp}(\mathbf{g}^t)$ associated to the gadget $\mathbf{g}$. The optimization is based on the observation (from [38]) that for our gadget $\mathbf{g}$ and modulus $q = b^k$, the lattice $\Lambda_q^{\perp}(\mathbf{g}^t)$ has a very simple basis $\mathbf{S}$, and an even simpler GSO $\tilde{\mathbf{S}}$:

$$\mathbf{S} = \begin{pmatrix} b & & & \\ -1 & \ddots & & \\ & \ddots & b & \\ & & -1 & b \end{pmatrix}, \qquad \tilde{\mathbf{S}} = b \cdot \mathbf{I}.$$

Using this special structure, there is no need to explicitly compute and store the GSO, and the randomized-rounding nearest-plane algorithm can be implemented in linear time and space $O(k)$. The specialized algorithm is best illustrated when $b = 2$, in which case it computes a randomized "bit" decomposition of u as follows:

---

[4] For a general integer basis $\mathbf{B}$, the GSO can have numbers with denominators as large as $\prod_i \|\mathbf{b}_i\|^2$.

Algorithm 1: $\mathbf{g}^{-1}(u)$ for $q = b^k$.

Input: $u \in \{0, 1, \cdots, q-1\}$
Output: subgaussian $\mathbf{x} \in \Lambda_u^\perp(\mathbf{g}^t)$ with parameter $(b-1)\sqrt{2\pi}$

1   Let $\mathbf{x} \leftarrow \mathbf{0}$
2   **for** $i \leftarrow 0, \cdots, k-1$ **do**
3   Let $y \leftarrow u \bmod b \in \{0, \cdots, b-1\}$.
4   **if** $y = 0$ **then**
5        $x_i \leftarrow 0$.
6   **else**
7   with probability y/b, $x_i \leftarrow y - b$, and $x_i \leftarrow y$ otherwise.
8      $u \leftarrow (u - x_i)/b$.
9   **return** $\mathbf{x}$

---

1. For $i = 0, \cdots, k-1$:
   (a) if u is even, then set $x_i \leftarrow 0$,
   (b) if u is odd, then choose $x_i \leftarrow \{-1, +1\}$ uniformly at random
   Update $u \leftarrow (u - x_i)/2$.
2. Return $\mathbf{x} = (x_0, x_1, \cdots, x_{k-1})$.

This is essentially the same as the standard (deterministic) bit decomposition algorithm, except that when the bit is 1, we use a random ± 1 digit. Since ± 1 have the same parity modulo 2, the algorithm works as expected, with the only difference that now each digit is a zero-mean random variable, and the final output is subgaussian with parameter $\sqrt{2\pi}$.

We can modify this algorithm to an arbitrary base b as follows. Let $y := u$ mod $b \in \{0, \cdots, b-1\}$ for an input $u \in \mathbb{Z}_q$. Then, at each step, we pick the coset representative (of u with respect to $\mathbb{Z}_b$) with expectation 0 from the set $\{y-b, y\}$. The resulting algorithm is given in Fig. 1. One can verify that this is the subgaussian nearest plane algorithm (given in the Appendix of the full version [24]) applied to the lattice $\mathcal{L}(\mathbf{S}) = \Lambda_q^\perp(\mathbf{g}^t)$, so the correctness of the algorithm is straightforward. Efficiency is also easily analyzed by inspection. Notice that the algorithm is randomness efficient as it needs only one random number in $\mathbb{Z}_b$ for every interaction, for a total of $k \cdot \log_2(b) = \log_2(q)$ random bits.

We remark that a similar algorithm is analyzed in [4], though with a loose bound on its subgaussian parameter (there is an unnecessary $\sqrt{k}$ factor in their subgaussian analysis). This section's main contribution is how to generalize the algorithm to arbitrary modulus q, as described in the next subsection.

4.2 Arbitrary Modulus, Arbitrary Base

Unfortunately, the (randomized) nearest plane algorithm $\Lambda_q^\perp(\mathbf{g}^t)$ does not specialize well when the modulus q is not a power of b. The reason is that, while we can still use the same gadget $\mathbf{g} = (1, b, \ldots, b^{k-1})$, the corresponding lattice $\Lambda_q^\perp(\mathbf{g}^t)$ has a slightly different basis $\mathbf{S}_q$ whose GSO is not diagonal, and not sparse. Our solution uses a technique developed in [23] for the discrete Gaussian sampling problem. Specifically, we use the fact that $\mathbf{S}_q$ admits a sparse, triangular factorization

$$\mathbf{S}_q = \begin{pmatrix} b & & & q_0 \\ -1 & \ddots & & \vdots \\ & \ddots & b & q_{k-2} \\ & & -1 & q_{k-1} \end{pmatrix} = \begin{pmatrix} b & & & \\ -1 & \ddots & & \\ & \ddots & b & \\ & & -1 & b \end{pmatrix} \begin{pmatrix} 1 & & & d_0 \\ & \ddots & & \vdots \\ & & 1 & d_{k-2} \\ & & & d_{k-1} \end{pmatrix} = \mathbf{SD} \qquad (1)$$

where $(q_0, \cdots, q_{k-1})$ are the (base b) digits of q, and the last column of $\mathbf{D}$ is defined by the simple recurrence $d_i = \frac{d_{i-1}+q_i}{b}$ with initial condition $d_{-1} = 0$. (Note that $b^{i+1}d_i = q \mod b^{i+1} \in \{0, \cdots, b^{i+1} - 1\}$.)

Then, on input $u \in \{0, 1, \cdots, q - 1\}$, we proceed as follows:

1. Compute an arbitrary element $\mathbf{u} \in \mathbb{Z}^k$ of the lattice coset $\Lambda_u^\perp(\mathbf{g}^t)$, for example $\mathbf{u} = (u, 0, \ldots, 0)$.
2. Map $\mathbf{u}$ to $\mathbf{t} = \mathbf{S}^{-1}\mathbf{u}$ by solving a sparse system of linear equations $\mathbf{St} = \mathbf{u} \pmod{q}$.
3. Pick a subgaussian sample from the lattice coset $\mathcal{L}(\mathbf{D}) + \mathbf{t}$.
4. Apply the (sparse) linear transformation $\mathbf{S}$ to the sample, to obtain a subgaussian sample from $\Lambda_u^\perp(\mathbf{g}^t)$.

Here the (randomized) nearest plane algorithm admits a simple and efficient specialization because it is applied to a basis, $\mathbf{D}$, which has a diagonal GSO. The linear transformations $\mathbf{S}^{-1}$ and $\mathbf{S}$ can also be computed in linear time because $\mathbf{S}$ is sparse and triangular. As a result, the algorithm runs in linear time $O(k)$ and does not require any pre-processing. Finally, we get an output with subgaussian parameter $(b+1)\sqrt{2\pi}$ since $\mathbf{S}$ has small spectral norm.

The actual algorithm is given in Algorithm 2. The algorithm directly implements the outline given above, but it is specialized/optimized to avoid the explicit computation of the sparse matrices $\mathbf{S}, \mathbf{D}$, and to use only integer numbers (avoids floating point numbers). Details about the correctness and analysis of the algorithm are provided in the rest of this section.

Lemma 4. *The first loop of Algorithm 2 performs the subgaussian nearest plane algorithm (described generically in the Appendix of the full version, [24]) on the lattice generated by $\mathbf{D}$ around target $\mathbf{t} := -\mathbf{S}^{-1}[u]_b^k$.*

By storing $\mathbf{d} = \mathbf{S}^{-1}[q]_b^k$ in-advance, one can change the code to sample the first $k - 1$ coordinates of $\mathbf{x}$ in-parallel since $\mathcal{L}(\mathbf{d}_0, \cdots, \mathbf{d}_{k-2}) = \mathbb{Z}^{k-1} \oplus \{0\}$.

5   Gadget Decoding

Here we discuss our main algorithm for the problem of *LWE gadget decoding*, defined in Sect. 2.2, on the gadget matrix $\mathbf{G} = \mathbf{I}_n \otimes \mathbf{g}^t$ with entries in $\mathbb{Z}_q$, for an arbitrary modulus q. Given a $\mathbf{v}^t = \mathbf{s}^t\mathbf{G} + \mathbf{e}^t \in \mathbb{Z}_q^{nk}$ as input, we can break the

---

Algorithm 2: $g^{-1}(u)$

Input: $u \in \{0, 1, \cdots, q-1\}$
Output: subgaussian $\mathbf{x} \in \Lambda_u^{\perp}(\mathbf{g}^t)$ with parameter $(b+1)\sqrt{2\pi}$

1   Let $\mathbf{u} \leftarrow [u]_b^k$, $\mathbf{x}, \mathbf{y} \leftarrow \mathbf{0}$
2     $\mathbf{x} \leftarrow \mathbf{0}, \mathbf{q} = [q]_b^k$.
3   set $x_{k-1} \leftarrow 0$ with probability $(q-u)/q$ and $x_{k-1} \leftarrow -1$ otherwise.
4   for $i = k-2, \cdots, 0$ do
5        $u \leftarrow u - u_{i+1}b^{i+1}, q \leftarrow q - q_{i+1}b^{i+1}$.
6   Let $c \leftarrow -(u + x_{k-1}q)$.
7   if $c < 0$ then
8           $p \leftarrow (c + b^{i+1})$, $z \leftarrow -1$.
9   else
10          $p \leftarrow c$, $z \leftarrow 0$.
11   set $x_i \leftarrow z + 1$ with probability p/b^{i+1} and $x_i \leftarrow z$ otherwise.
12   for $i \in \{0, \cdots, k-2\}$ do
13        $y_i \leftarrow b \cdot x_i - x_{i-1} + x_{k-1} \cdot q_i + u_i$.
14     $y_{k-1} \leftarrow -x_{k-2} + x_{k-1} \cdot q_{k-1} + u_{k-1}$.
15   return $\mathbf{y}$.

---

vector into n components of length k, then decode (in-parallel) each component with respect to $\mathbf{g}^t$. Therefore, we focus on decoding $\mathbf{g}^t$ as a gadget for $\mathbb{Z}_q$.

Our algorithm and its respective gadgets are parameterized by an integer "base" b. We consider b as fixed in this section, though varying b for a fixed modulus q yields efficiency/quality trade-offs for these gadgets. Later, in Sect. 6 we present a CRT gadget that can be used to efficiently decode an input given in CRT form.

Let $k = \lceil \log_b q \rceil$ and the gadget be $\mathbf{g}^t = (1, b, \cdots, b^{k-1})$. The vector $\mathbf{g}^t$ is a size k gadget of quality $(b/2)\sqrt{k}$ for $\mathbb{Z}_q$. The results in this section are summarized in the following theorem.

Theorem 5. *For every modulus q, and gadget $\mathbf{g}^t = (1, b, \cdots, b^{k-1})$, there is a time and space $O(k)$ algorithm decoding $\mathbf{g}^t$ with tolerance $q/2(b+1)$.*

A proof of Theorem 5 is given by the algorithm presented in this section. Note, Theorem 3 implies a polynomial time decoding algorithm for $\mathbf{g}^t$ with error tolerance $\|\mathbf{e}\|_\infty \leq q/2\sqrt{k}(b+1)$. Our decoding algorithm is more efficient and has a higher error tolerance by a factor $\sqrt{k}$ than the general gadgets decoding guarantee given by Theorem 3.

An optimized, linear time and space $O(k)$, decoding algorithm is given in [38] for the case $q = b^k$. The reason for this algorithm's efficiency is that the commonly used basis for $\Lambda_{b^k}(\mathbf{g}^t)$ results in a linear time nearest plane algorithm. In more detail, a basis for $\Lambda_{b^k}(\mathbf{g}^t)$ in this case is the triangular matrix $\mathbf{B}_{b^k} = b^k \cdot \mathbf{S}^{-t}$, where $\mathbf{S}$ is the commonly used basis for $\Lambda_{b^k}^{\perp}(\mathbf{g}^t)$ presented in Sect. 2.2, and this basis has a GSO of $(q/b) \cdot \mathbf{I}$.

Algorithm 3: $\text{DECODEG}(\mathbf{v}, b, \mathbf{r}[q]_b^k)$

Input: $\mathbf{v} \in \mathbb{Z}^k$, b, and $\mathbf{q} = [q]_b^k$.
Output: $s \in \mathbb{Z}_q$ where $\mathbf{v} = s\mathbf{g}^t + \mathbf{e}^t$ as long as $\|\mathbf{e}\|_\infty < q/2(b+1)$.

1   **for** $i \leftarrow 0, \cdots, k-2$ **do**
2       $v_i \leftarrow bv_i - v_{i+1}$.
3    $v_{k-1} \leftarrow b \cdot v_{k-1}$.
4   Let $\mathbf{x} \leftarrow \mathbf{0}$ and reg $\leftarrow 0$.
5   **for** $i \leftarrow 0, \cdots, k-2$ **do**
6       $x_i \leftarrow \lceil v_i/q \rfloor$ and reg $\leftarrow$ reg$/b + b^{k-1} \cdot q_i$.
7       $v_{k-1} \leftarrow v_{k-1} + x_i \cdot$ reg.
8    $x_{k-1} \leftarrow \lceil v_{k-1}/b^k \rfloor$.
9   Let $s \leftarrow x_{k-1}$ and reg $\leftarrow 0$.
10   **for** $i \leftarrow k-2, \cdots, 0$ **do**
11   reg $\leftarrow b \cdot$ reg $+ q_{i+1}$.
12      $s \leftarrow s + x_i \cdot$ reg
13   **return** $s \mod q$.

However, the simple decoding idea presented in [38] fails when $q \neq b^k$. Because $\Lambda_q(\mathbf{g}^t)$'s commonly used basis has a dense GSO, Babai's nearest plane algorithm takes time $O(k^2)$ and space $O(k^3)$ when naively applied on $\Lambda_q^\perp(\mathbf{g}^t)$.

Efficient Decoding Algorithm. The intuition for our algorithm is best initially viewed through the case when $q = b^k$. Given an input $\mathbf{v}$, another way to decode the lattice $\Lambda_{b^k}(\mathbf{g}^t)$ is to use $\mathbf{S}^t$ as a linear transformation, decode $\mathbf{S}^t\mathbf{v}$ to the lattice $b^k \cdot \mathbb{Z}^k$ with the nearest plane algorithm, then map the nearest point in $b^k \cdot \mathbb{Z}^k$ back to $\Lambda_{b^k}(\mathbf{g}^t)$. This leads to a slightly stronger condition on the noise vector $\mathbf{e}$ since we now need $\mathbf{S}^t\mathbf{e} \in \mathcal{P}_{1/2}(q \cdot \mathbf{I})$, which is satisfied if $\|\mathbf{e}\|_\infty < q/2(b+1)$. Though there is no need to do this given the algorithm in [38], this is essentially what we will do in the case when $q \neq b^k$.

Overview. The overview of our efficient decoding algorithm for an arbitrary modulus is as follows. First recall the sparse, triangular factorization of $\Lambda_q^\perp(\mathbf{g}^t)$'s commonly used basis given in Sect. 2.2, $\mathbf{S}_q = \mathbf{SD}$. The duality relation for q-ary lattices, $\Lambda_q(\mathbf{g}^t) = q \cdot \Lambda_q^\perp(\mathbf{g}^t)^*$, dictates that a basis for $\Lambda_q(\mathbf{g}^t)$ is $q \cdot \mathbf{S}_q^{-t} = \mathbf{S}^{-t}(q \cdot \mathbf{D}^{-t})$. Luckily, the matrix $\mathbf{D}^{-t}$ is sparse with a diagonal GSO, and $\mathcal{P}_{1/2}(q \cdot \tilde{\mathbf{D}}^{-t}) \supseteq \mathcal{P}_{1/2}(q \cdot \mathbf{I})$ (meaning we can decode as long as $\|\mathbf{e}\|_\infty < q/2(b+1)$). Therefore, we can decode $\mathbf{g}^t$ by the following.

1. Given $\mathbf{v}$, first apply $\mathbf{S}^t$ as a linear transformation.
2. Then, decode the vector $\mathbf{S}^t\mathbf{v}$ to the lattice generated by $q\mathbf{D}^{-t}$ using the nearest plane algorithm.

Both steps can be computed in linear time and space, $O(k)$, given the sparsity of $\mathbf{S}$ and $q\mathbf{D}^{-t}$, and $q\mathbf{D}^{-t}$'s diagonal GSO.

The pseudocode for our algorithm is shown in DECODEG. In short, the algorithm has three components, where each is represented by a loop in the pseudocode. These components are to first compute the linear transformation on the input $\mathbf{v} \leftarrow \mathbf{S}^t\mathbf{v}$, then to run the nearest plane algorithm on the lattice generated by $q \cdot \mathbf{D}^{-t}$, and finally to return s represented as the first entry of the nearest lattice point in $\Lambda_q(\mathbf{g}^t)$ modulo q. The proof of Theorem 5 follows from Lemmas 5 and 6 below.

Lemma 5. *The second loop in DECODEG is an instantiation of Babai's nearest plane algorithm on the lattice $q \cdot \mathbf{D}^{-t}$ given target $\mathbf{S}^t\mathbf{v}$, running in time and space $O(k)$.*

Lemma 6. *The last loop in DECODEG computes $s \mod q$ in time and space $O(k)$.*

6   Gadgets for the CRT Representation

Many applications of lattice gadgets require a large modulus that, for secure and functional sets of parameters, surpasses the native 64-bit integer arithmetic in a modern machine's hardware. One common method to circumvent the use of multi-precision numbers is to pick a modulus of the form $q = \prod q_i$ with each q_i less than 64 bits. Then, one can store an element $u \in \mathbb{Z}_q$ as its *Chinese Remainder* representation (CRT form[5]) as $(u \mod q_1, \cdots, u \mod q_l)$ and perform computations via the Chinese Remainder Theorem, utilizing the ring isomorphism $\mathbb{Z}_q \cong \mathbb{Z}_{q_1} \times \cdots \times \mathbb{Z}_{q_l}$. Simple forms of the gadget matrix (e.g. power of two matrix) are not compatible with this representation because the binary digits of a number cannot be easily recovered from the CRT components without a costly reconstruction phase involving large numbers modulo q.

In this section, we discuss a gadget for the CRT form. As usual, the gadget admits a compact (implicit) representation, and does not need to be computed and stored explicitly. Most importantly, the gadget allows us to use the algorithms in Sects. 4 and 5 in order to perform subgaussian decomposition, discrete Gaussian sampling, and LWE gadget decoding all given input represented in CRT form. This has several theoretical and practical advantages: (1) the algorithms can be directly used by efficient applications that already store their numbers in CRT form, (2) our algorithms can be easily parallelized as they operate on each CRT component independently, (3) all algorithms only require arithmetic on small numbers (at most $\max_i q_i$) even if the modulus $q = \prod_i q_i$ may be very big. (Efficient solutions to Discrete Gaussian Sampling for the individual moduli q_i, as needed by our CRT DGS algorithm, are given in [23,38].) We remark that a balanced, deterministic digit decomposition is provided in [18,41], and an LWE decoding algorithm for a CRT/RNS hybrid gadget for general rings is given in the library's code[6] (without an analysis). Our results are

---

[5] This is also known as the residue number system (RNS) in previous works.
[6] https://github.com/cpeikert/Lol/blob/master/lol/Crypto/Lol/Gadget.hs.

	Algorithm 5: Decoding in CRT form.
Algorithm 4: Sampling in CRT form.	**Input:** $\mathbf{v}^t = s \cdot \mathbf{g}_{CRT} + \mathbf{e}^t \mod q$
Input: $(u_1, \cdots, u_l)$	**Output:** $(s_1, \cdots, s_l)$.
Output: $\mathbf{g}_{CRT}^{-1}(u_1, \cdots, u_l)$.	1 Let $\mathbf{v} = (\mathbf{v}_1, \cdots, \mathbf{v}_l)$ for each
1 **for** $i \in \{1, \cdots, l\}$ **do**	$\mathbf{v}_i \in \mathbb{Z}_q^{k_i}$.
2 $\mathbf{x}_i \leftarrow \mathbf{g}_i^{-1}(u_i)$.	2 **for** $i \in \{1, \cdots, l\}$ **do**
3 **return** $\mathbf{x} = (\mathbf{x}_1, \cdots, \mathbf{x}_l)$.	3 $s_i \leftarrow \text{DECODECRT}(\mathbf{v}_i)$
	4 **return** $(s_1, \cdots, s_l)$.

Fig. 1. Pseudocode for the parallel algorithms given in Theorem 6. We let $\mathbf{g}_i^{-1}(\cdot)$ denote either the subgaussian decomposition algorithm given in Sect. 4 or a discrete Gaussian sampler. The subroutine DECODECRT is a variation of the decoding algorithm given in Sect. 5 and is described in Sect. 6.1.

summarized in the following theorem. We emphasize the analysis below assumes integer operations, including reductions modulo q_i, are done in constant time. This is because our algorithms are best implemented when each q_i is less than 64 bits, avoiding the use of multi-precision numbers.

Theorem 6. *Let q have factorization $q = \prod_{i=1}^{l} q_i$ into coprime factors $\{q_i\}$, $(b_i)_{i=1}^{l}$ be an l-tuple of bases with $b_i < q_i$ for all i, and let $k = \sum k_i$ where $k_i = \lceil \log_{b_i} q_i \rceil$. There exists a gadget, $\mathbf{g}_{CRT}^t$, for $\mathbb{Z}_q$ of size k and quality $\max_i b_i/2$. Further, the gadget satisfies the following properties:*

- *Subgaussian decomposition can be performed in-parallel with l processors, each using time and space $O(k_i)$, consuming less than $k_i \log_2 q_i$ random bits ($(\lceil \log_2(q_i) \rceil$ random bits if $q_i = b_i^{k_i}$)) and with parameter at most $(\max_i(b_i) + 1)\sqrt{2\pi}$.*

- *For any $\epsilon > 0$, discrete Gaussian sampling can be performed in-parallel with l processors, each in time and space $O(k_i)$ with width $s \geq O(b_j^{1.5})\eta_\varepsilon(\mathbb{Z}^{k_j})$ for index j maximizing $\sqrt{2b_j}(b_j + 1) \cdot \eta_\varepsilon(\mathbb{Z}^{k_j})$.*

- *$\mathbf{g}_{CRT}^t$ is decodable in-parallel with l processors in time and space $O(k_i)$ with tolerance $q/(2\max_i(b_i) + 1)$.*

As expected, each processor gets slightly more efficient whenever $q_i = b_i^{k_i}$. The algorithms are represented in Fig. 1.

The CRT Gadget. For each coprime factor q_i, fix the *base-b_i* gadget vector as $\mathbf{g}_i^t := (1, b_i, \cdots, b_i^{k_i-1})$ where $k_i = \lceil \log_{b_i}(q_i) \rceil$. Let $k = \sum_i k_i$, $q_i^* = q/q_i$, and $\hat{q}_i = (q_i^*)^{-1} \mod q_i$. Consider the gadget vector, which we call the *general CRT gadget*, $\mathbf{g}_{CRT}^t = (q_1^* \hat{q}_1 \cdot \mathbf{g}_1^t, \cdots, q_l^* \hat{q}_l \cdot \mathbf{g}_l^t) \mod q \in \mathbb{Z}_q^{1 \times k}$. This is a generalization of the gadgets (or implicit in algorithms) used in [8,11,32,33]. As before, the gadget matrix is the block-diagonal matrix $\mathbf{G} := \mathbf{I}_n \otimes \mathbf{g}_{CRT}^t$. Theorem 6 follows from the fact $\Lambda_q^\perp(\mathbf{g}_{CRT}^t) = \Lambda_{q_1}^\perp(\mathbf{g}_1^t) \oplus \cdots \oplus \Lambda_{q_l}^\perp(\mathbf{g}_l^t)$, Theorem 4, and Proposition 3.1 in [23]. The parallel decoding algorithm is obtained by a slight adaptation

---

Algorithm 4: DECODECRT($\mathbf{v}_i, b_i, \mathbf{t} = [q_i]_{b_i}^{k_i}, q, q_i^*$)

Input: $\mathbf{v}_i \in \mathbb{Z}^{k_i}$, b_i, q_i^*, q, and $\mathbf{t} = [q_i]_{b_i}^{k_i}$.

Output: $s \bmod q_i$ where $\mathbf{v} = s\mathbf{g}^t + \mathbf{e}^t \bmod q$ as long as $\|\mathbf{e}\|_\infty < q/2(b_i + 1)$.

1   **for** $j \leftarrow 0, \cdots, k_i - 1$ **do**

2         $v_j \leftarrow b_j v_j - v_{j+1}$.

3   Let $\mathbf{x} \leftarrow \mathbf{0}$.

4   **for** $j \in \{0, \cdots, k_i - 2\}$ **do**

5         $x_j \leftarrow \lceil v_j/q \rfloor$.

6     $x_{k-1} \leftarrow \lceil (v_{k-1} - \langle \mathbf{c}, \mathbf{x}_0^{k-2} \rangle)/(q_i^* b_i^{k_i}) \rfloor$.

7   Let $s_i \leftarrow x_{k-1}$ and reg $\leftarrow 0$.

8   **for** $j \leftarrow k_i - 2, \cdots, 0$ **do**

9   reg $\leftarrow b \cdot$ reg $+ t_{j+1} \cdot q_i^*$.

10      $s_i \leftarrow s_i + x_j \cdot$ reg.

11   **return** $s_i \bmod q_i$.

---

to DECODEG presented in Sect. 5, and is analyzed in the Sect. 6.1. We prove the direct sum decomposition of $\Lambda_q^\perp(\mathbf{g}_{CRT}^t)$ in the full version of the paper.

6.1 Decoding the CRT Gadget

Here we show how the efficient gadget decoding algorithm from Sect. 5 adapts to the general CRT gadget described in Sect. 6. Recall the decomposition of $\mathbf{g}^t$'s lattice, $\Lambda_q^\perp(\mathbf{g}^t) = \Lambda_{q_1}^\perp(\mathbf{g}_1^t) \oplus \cdots \oplus \Lambda_{q_l}^\perp(\mathbf{g}_l^t) = \mathcal{L}(\mathbf{S}_{q_1}) \oplus \cdots \oplus \mathcal{L}(\mathbf{S}_{q_l})$. The duality relation for q-ary lattices yields $\Lambda_q(\mathbf{g}^t) = q \cdot (\Lambda_q^\perp(\mathbf{g}^t))^* = q \cdot \left(\bigoplus_i \mathcal{L}(\mathbf{S}_{q_i}^{-t} \mathbf{D}_{q_i}^{-t}) \right) = \left(\bigoplus_i \mathcal{L}(\mathbf{S}_{q_i}^{-t} q_i^* \cdot (q_i \cdot \mathbf{D}_{q_i}^{-t})) \right)$.

Now we have a clear way to decode the general CRT gadget. First, break the input into l blocks, $\mathbf{v}^t = s\mathbf{g}^t + \mathbf{e}^t \bmod q = (\mathbf{v}_1^t, \cdots, \mathbf{v}_l^t)$ where $\mathbf{v}_i^t = s \cdot q_i^* \hat{q}_i \mathbf{g}_i^t + \mathbf{e}_i^t \bmod q$. Then, we compute the following. First, transform $\mathbf{v}_i$ to $\mathbf{S}_{q_i}^t \mathbf{v}_i$. Then, decode $\mathbf{S}_{q_i}^t \mathbf{v}_i$ to the lattice $q_i^*(q_i \mathbf{D}_{q_i}^{-t})$. Finally, return $s \bmod q_i$. The pseudocode is given as the algorithm DECODECRT. Another change is that we store the vector $\mathbf{c}$ in memory. Recall, $\mathbf{c}$ has $k - 2$ entries of the form $c_j = -b_i^{k_i-1-j}(q_i \bmod b_i^j)$. Note that the correctness condition of our algorithm is still $\|\mathbf{e}^t\|_\infty < q/2(\max_i(b_i) + 1)$.

Decoding in CRT Form. Here we describe how DECODECRT can decode $\mathbf{v} = s\mathbf{g} + \mathbf{e}$ where the input is given in its CRT representation. The ideas sketched here follow from [33]. The linear transformation $\mathbf{v} \rightarrow \mathbf{S}^t \mathbf{v}$ is easily computed given the CRT form of $\mathbf{v}$. Really, we are only concerned with divisions and integer rounding. In the second loop, note that $x_j \leftarrow \lceil v_j/q \rfloor = \lceil \sum_{o=1}^l [(v \bmod q_o) \cdot (\hat{q}_o/q_o)] \rfloor$. Next we consider the line $x_{k-1} \leftarrow \lceil (v_{k-1} + \langle \mathbf{c}, \mathbf{x}_0^{k-2} \rangle)/(q_i^* b_i^{k_i}) \rfloor$. First, note that $v_{k-1}/(b_i^{k_i} q_i^*) = b_i^{-k_i} \cdot \sum_{o=1}^l (v_{k-1} \bmod q_o) \cdot \hat{q}_o(q_i/q_o)$. This should be a small number in nearly all practical instantiations. Lastly, we note that we return s in CRT form, but we can alter the algorithm to return $s \in (-q/2, q/2]$

via a simple change. The s computed in the last loop is actually $s \cdot q_i^* \hat{q}_i$. So, we can remove the mod q_i in the return statement and sum up the output from the l parallel processors, $\sum_i (s \cdot q_i^* \hat{q}_i) = s \cdot \sum_i (q_i^* \hat{q}_i) = s \cdot 1 \mod q$.

7 Toolkit Implementation and Its Application

7.1 Software Implementation

We implemented most of the algorithms presented in this work in PAL-ISADE [42], a modular open-source lattice cryptography library that includes ring-based implementations of homomorphic encryption, proxy re-encryption, identity-based encryption, attribute-based encryption, and other lattice schemes. More concretely, we added a new lattice gadget toolkit module to PALISADE that implements the following algorithms:

- Subgaussian gadget decomposition (Algorithm 2) for arbitrary moduli and gadget bases.
- Efficient gadget in CRT representation, enabling both trapdoor sampling and subgaussian gadget decomposition in the CRT representation.
- Subgaussian gadget decomposition for cyclotomic rings both in positional and CRT number systems, which wraps around Algorithm 2.

The toolkit module complements/improves the lattice gadget algorithms previously added to PALISADE, such as trapdoor sampling for cyclotomic rings proposed in [23] and implemented in [17,31]. The full lattice gadget capability will be included in the next major public release of PALISADE.

7.2 Optimized Variant of Key-Policy Attribute-Based Encryption

We use the lattice gadget toolkit algorithms to build and implement a full RNS/CRT variant of the short-secret Key-Policy Attribute-Based Encryption (KP-ABE) scheme originally proposed in [10] and implemented for cyclotomic rings in [19]. The KP-ABE scheme is a complex cryptographic primitive that can be used for attribute-based access control applications, as well as a building block for audit log encryption, targeted broadcast encryption, predicate encryption, functional encryption, and some forms of program obfuscation [10,28].

Overview. ABE is a public key cryptography primitive that enables the decryption of a ciphertext by a user only if a specific access policy (defined over ℓ attributes) is satisfied. In the key-policy scenario, a message is encrypted using the attribute values as public keys, and a specific access policy is typically defined afterwards. When the access policy becomes known, a secret key for the policy is generated (using trapdoor sampling in our KP-ABE scheme), and the ciphertexts and public keys are homomorphically evaluated over the policy circuit (using a GSW-type homomorphic multiplication in our KP-ABE scheme).

The short-secret KP-ABE scheme is a tuple of functions, namely Setup, Encrypt, EvalPK, KeyGen, EvalCT, and Decrypt, whose definitions are:

- SETUP($1^\lambda, \ell$) → {MPK, MSK}: Given a security parameter λ and the number of attributes ℓ, a trusted private key generator (PKG) generates a master public key MPK and a master secret key MSK. MPK contains the ABE public parameters while MSK includes the trapdoor that is used by PKG to generate secret keys for access policies.
- ENCRYPT($\mu, \mathbf{x}, $MPK) → $\mathbf{C}$: Using MPK and attribute values $\mathbf{x} \in \{0,1\}^\ell$, sender encrypts the message μ and outputs the ciphertext $\mathbf{C}$.
- EVALPK(MPK$, \mathbf{x}, f$) → PK$_f$: Homomorphically evaluate MPK over a policy (Boolean circuit) $f : \{0,1\}^\ell \to \{0,1\}$ to generate a public key PK$_f$ for the policy f.
- KEYGEN(MSK, MPK, PK$_f$) → SK$_f$: Given MSK, MPK and policy-specific PK$_f$, PKG generates the secret key SK$_f$ corresponding to f. PKG sends SK$_f$ to the receiver that is authorized to decrypt ciphertexts encrypted under f.
- EVALCT($\mathbf{C}, \mathbf{x}, f$) → $\mathbf{C}_f$: Homomorphically evaluate $\mathbf{C}$ over the policy f to generate the ciphertext $\mathbf{C}_f$.
- DECRYPT($\mathbf{C}_f, SK_f$) → $\bar{\mu}$: Given the homomorphically computed ciphertext $\mathbf{C}_f$ and corresponding secret key SK$_f$, find $\bar{\mu}$, which is the same as the original message μ if the receiver has the secret key matching the policy f.

The most computationally expensive operations are EVALPK and EVALCT, which homomorphically evaluate a circuit of depth $\lceil \log_2 \ell \rceil$ using the GSW homomorphic multiplication approach. At each level of a Boolean circuit composed of NAND gates (which are used for benchmark evaluation in [19]), the algorithms compute matrix products $\mathbf{B}_{2i}\mathbf{G}^{-1}(-\mathbf{B}_i)$ and $\left(\mathbf{G}^{-1}(-\mathbf{C}_i)\right)^t \mathbf{C}_{2i}$ for public keys and ciphertexts, respectively. Here, $\mathbf{B}_i \in R_q^{1 \times m}$, $\mathbf{C}_i \in R_q^m$, $R_q = \mathbb{Z}_q[x]/\langle x^n + 1 \rangle$, and $m = \lceil \log_b q \rceil + 2$ (the latter corresponds to the Ring-LWE trapdoor construction). Note that that the gadget $\mathbf{G}$ is extended in this case to m by adding two zero entries to the decomposed digits.

The work [19] presents a CPU implementation of the ring variant of the KP-ABE scheme along with an efficient GPU implementation for policy evaluation and encryption. The CPU implementation was done for a binary gadget base and used the conversion from CRT to the positional number system for digit decomposition both in trapdoor sampling and gadget decomposition. To avoid the linear noise growth $O(nm)$ in gadget decomposition, the authors used a balanced digit decomposition, namely the binary non-adjacent form (NAF), that replaces digits in (0,1) with a zero-centered representation in $(-1,0,1)$. Although this approach allows one to achieve a heuristic growth close to $O(\sqrt{nm})$ in the case of the KP-ABE scheme, the noise properties depend on the randomness of the input, i.e., this approach is deterministic.

The CPU runtimes for policy evaluation and encryption operations in [19] were far from practical (the CPU results only for ℓ up to 8 are presented), and hence the authors developed an efficient GPU implementation for these operations.

For detailed algorithms of the KP-ABE scheme, the reader is referred to [19].

Our Optimizations. We present a full CRT/RNS ring variant of the KP-ABE scheme that leverages the lattice gadget toolkit to significantly (by more than one order of magnitude) speed up the policy evaluation operations. In particular, our implementation includes the following optimizations as compared to [19]:

- The subgaussian gadget decomposition in CRT representation to minimize the noise growth instead of the NAF decomposition with the conversion from CRT representation to positional system. This provides a theoretical guarantee of the square-root noise growth. To achieve the repeatability of randomized decomposition in EVALPK and EVALCT, we use the same seed for the random operations in subgaussian gadget decomposition. The seed is treated as part of the master public key.
- The CRT variant of trapdoor sampling using the gadget decomposition technique discussed in this paper in contrast to the multiprecision digit decomposition in [19].
- The RNS/CRT scaling proposed in [33] for decryption in contrast to the multiprecision scaling.
- Increased gadget base b (both in trapdoor and subgaussian gadget decomposition) instead of the binary base.

Parameter Selection. As the correctness constraint in [19] was derived for the classical binary-base gadget decomposition, we provide here a modified version incorporating the effect of a larger gadget base for the case of subgaussian gadget decomposition:

$$q > 4C_1 s\sigma\sqrt{mn}\left(b\sqrt{mn}\right)^d, \tag{2}$$

where $C_1 = 128$, $s = C \cdot \sigma^2(b+1) \cdot (\sqrt{n\log_b q} + \sqrt{2n} + 4.7)$, $C = 1.8$, $\sigma \approx 4.578$, and $d = \lceil \log_2 \ell \rceil$. Here, C and C_1 are empirical parameters chosen the same way as in [19].

The differences compared to [19] are the b factor in the exponentiation base (as the digits vary between $-b$ and b in subgaussian gadget decomposition) and a $(b+1)$ factor in the expression for s (contributed by Gaussian sampling; see [17,23] for a more detailed discussion of the Gaussian distribution parameter for arbitrary gadget bases).

8 Experimental Results

We ran the experiments in PALISADE version 1.2, which includes NTL version 10.5.0 and GMP version 6.1.2. The evaluation environment was a commodity desktop computer system with an Intel Core i7-3770 CPU with 4 cores rated at 3.40 GHz and 16 GB of memory, running Linux CentOS 7. The compiler was g++ (GCC) 5.3.1.

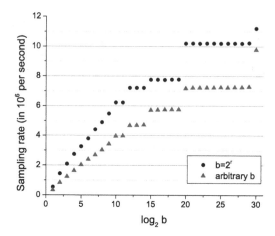

Fig. 2. Runtime baseline of subgaussian sampling rate for native uniformly random integers (w.r.t a 60-bit modulus). When $b = 2^r$, the modulo reduction in digit decomposition is performed by simple bit shifting. When b is arbitrary, the slower hardware modulo operation is used. The plateaus correspond to the same number of digits, i.e., the same value of $\lceil 60/\log_2 b \rceil$.

8.1 Subgaussian Gadget Decomposition

The experiments described in this section were all performed in the single-threaded mode. The goal of these results is to provide the performance baselines for subgaussian gadget decomposition, demonstrate the benefits of the efficient gadget in CRT representation, and illustrate the effect of subgaussian sampling on the noise growth in GSW-type products.

Figure 2 shows the dependence of subgaussian gadget decomposition rate (per decomposed integer) on the gadget base for native (64-bit) integers. The results are shown both for a power-of-two base, which supports fast modulo reduction by bit shifting, and an arbitrary base, which requires a division-based modulo operation on x86 architectures. In our implementation, the native arithmetic is a building block for performing operations in CRT representation for integers that are larger than 60 bits, and, therefore, these results can be used to estimate the runtimes for larger CRT-represented integers. Figure 2 illustrates that the sampling rate increases in a discrete manner as we raise the gadget base because the number of digits is determined by $\lceil 60/\log_2 b \rceil$. The runtime is dominated by the randomized operations (as the difference between a power-of-two-base and arbitrary-base scenarios is relatively small), thus limiting the advantages of choosing the faster power-of-two bases. This suggests that a CRT representation in terms of powers of primes, where the primes are used as the residue bases, might be preferred in some instances (where an efficient implementation of arithmetic over prime powers is available) over power-of-two bases.

Figure 3 illustrates the benefits of using the efficient gadget in CRT representation when working with cyclotomic rings. The conversion from CRT repre-

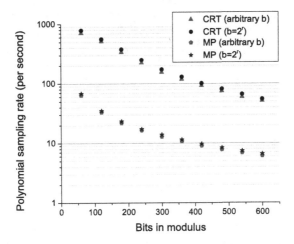

Fig. 3. Comparison of sampling rates for CRT and multiprecision (MP) variants of subgaussian gadget decomposition for ring elements with 4096 coefficients and 60-bit CRT moduli at $r = \lceil \log_2 b \rceil = 20$. The MP variant requires converting from CRT representation to positional number system followed by digit decomposition w.r.t. a large integer.

resentation to the positional system followed by digit decomposition w.r.t a large modulus slows down subgaussian gadget decomposition rate by almost one order of magnitude. We also observe that the difference in performance between a power-of-two base and an arbitrary base is relatively small for both cases.

Figure 4 demonstrates the differences in the noise growth of GSW-type products using the subgaussian and classical binary gadget decomposition methods. For this experiment, we generated an error vector in R^m and iteratively multiplied it by $\mathbf{G}^{-1}(\mathbf{U}_i)$, where $\mathbf{U}_i$ is a vector of uniformly random ring elements in R_q^m at level i. We applied the tree multiplication approach (rather than a sequential evaluation in a right-associative manner, which reduces the noise when dealing with a chained product of fresh encryptions in GSW [5,14]) to emulate the noise growth in evaluating a Boolean policy circuit in the KP-ABE scheme. We considered both the cases when the same $\mathbf{U}$ was used at all levels (correlated ciphertexts) and different $\mathbf{U}_i$ at each level. The results were approximately the same for both scenarios because the classical gadget decomposition matrix is centered at 0.5 (see [19] for a more detailed discussion of the classical gadget decomposition case).

Figure 4 suggests that the noise growth in the subgaussian gadget decomposition case has a square-root dependence on mn ($\beta \approx 0.5$) while the classical gadget decomposition approach results in almost linear noise growth ($\beta \approx 0.9$). Note that the intercept is lower for classical gadget decomposition because the infinity norm of digits is 1 (only 0 or 1 are possible) vs. 2 in the case of subgaussian decomposition (the allowed integer values are in the range from -2 to 2). However, this advantage does not propagate to the second level of the circuit as

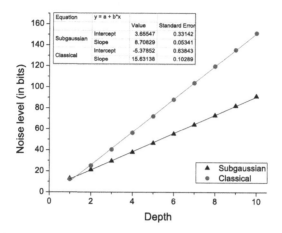

Fig. 4. Noise growth for GSW-type multiplication in the ring-based KP-ABE variant ($k = 180, n = 1024, b = 2$). The base in the exponentiation is $(mn)^{\beta}$, where $m = k+2 = 182$ and β describes the rate of noise growth. The slope of the linear interpolation is $\beta \log_2(mn)$. The values of β are 0.497 and 0.893 for subgaussian and classical gadget decomposition, respectively.

the square-root dependence of subgaussian gadget decomposition already plays a more dominant role here.

8.2   Key-Policy Attribute-Based Encryption

Table 1 shows the performance results for our implementation along with the corresponding results for the implementation in [19]. The first three rows for the results in [19] were obtained using native (64-bit) integer arithmetic and the last row used a multiprecision backend in PALISADE based on NTL/GMP. The experiments were run for 4 threads on a commodity desktop system, i.e., Intel Core i7-3770 CPU with 4 cores at 3.40 GHz and 16 GB of memory running CentOS 7. Both variants were implemented in PALISADE v1.2.

To choose the ring dimension n for both implementations, we ran the LWE security estimator[7] (commit 560525) [3] to find the lowest security levels for the uSVP, decoding, and dual attacks following the standard homomorphic encryption security recommendations [2]. We selected the least value of the number of security bits λ for all 3 attacks on classical computers based on the estimates for the BKZ sieve reduction cost model. All results are presented for at least 128 bits of security.

Table 1 suggests there is a speed-up of 2.1x to 3.2x for key generation, where the lattice trapdoor sampling subroutine is called. The speed-up for encryption is 3.8x to 9.5x, which is mostly attributed to the use of a larger gadget base. The speed-ups for the main bottleneck operations of homomorphic public key and

---

[7] https://bitbucket.org/malb/lwe-estimator.

Table 1. Comparison of performance results for our KP-ABE variant (in bold) vs. the implementation in [19] (in parentheses). $\textsc{EvalCT}^* = \textsc{EvalPK} + \textsc{EvalCT}$ corresponds to the scenario when the policy evaluation of public keys and ciphertexts is done at the same time.

ℓ	k	$\log_2 n$	r	KEYGEN [ms]	ENCRYPT [ms]	EVALCT* [s]	EVALPK [s]	DECRYPT [ms]	RAM [MB]
2	50 (44)	11 (11)	5 (1)	**40** (126)	**7** (33)	**0.023** (0.44)	**0.021** (0.42)	**1.8** (3.0)	**19** (58.5)
4	100 (52)	12 (12)	20 (1)	**64** (143)	**15** (57)	**0.072** (1.76)	**0.064** (1.68)	**3.9** (3.5)	**36.4** (86.3)
8	120 (60)	13 (13)	15 (1)	**151** (317)	**56** (222)	**0.59** (10.8)	**0.53** (10.4)	**8.9** (7.7)	**94.1** (255)
16	180 (70)	13 (13)	20 (1)	**177** (419)	**157** (1,483)	**1.68** (429)	**1.48** (427)	**11.5** (18.1)	**230** (2,867)
32	180	13	15	**206**	**414**	**5.67**	**5.0**	**13.46**	**508**
64	204	13	17	**226**	**1,052**	**13.1**	**11.2**	**16.39**	**1,229**
128	300	14	25	**568**	**6,454**	**98.3**	**85.5**	**45.43**	**7,024**

ciphertext evaluation are in the range from 18x to 289x, which is a combined effect of subgaussian gadget decomposition in CRT and a larger gadget base. The decryption runtimes are comparable, and already fast for both implementations. The memory requirements for our optimized variant are 2.4x to 12.5x smaller.

Note that the performance of the KP-ABE variant in [19] dramatically degrades after switching from the native arithmetic (when $k \leq 60$ bits) to the multiprecision backend (for gadget decomposition), which is observed for $\ell = 16$ in Table 1. This implies the efficient gadgets in CRT representation are critical for supporting deeper Boolean circuits with CPU systems.

We also profiled the contributions of subgaussian gadget decomposition and the number theoretic transforms (NTT) of the digit-decomposed matrix (needed for matrix multiplication) to the runtimes for homomorphic policy evaluation of ciphertexts (EVALCT*). The contribution of subgaussian gadget decomposition was in the range from 15% to 22% w.r.t. the total homomorphic policy evaluation runtime. The contribution of the related NTTs was between 47% and 63%, suggesting that the latter is the main bottleneck of homomorphic circuit evaluation in our KP-ABE variant.

References

1. Ajtai, M.: Generating hard instances of the short basis problem. In: Wiedermann, J., van Emde Boas, P., Nielsen, M. (eds.) ICALP 1999. LNCS, vol. 1644, pp. 1–9. Springer, Heidelberg (1999). https://doi.org/10.1007/3-540-48523-6_1
2. Albrecht, M., et al.: Homomorphic encryption security standard. Technical report, HomomorphicEncryption.org, Cambridge, MA, March 2018
3. Albrecht, M., Scott, S., Player, R.: On the concrete hardness of learning with errors. J. Math. Cryptol. **9**(3), 169–203 (2015)
4. Alperin-Sheriff, J., Apon, D.: Weak is better: tightly secure short signatures from weak PRFs. IACR Cryptology ePrint Archive, 2017:563 (2017)

5. Alperin-Sheriff, J., Peikert, C.: Faster bootstrapping with polynomial error. In: Garay, J.A., Gennaro, R. (eds.) CRYPTO 2014. LNCS, vol. 8616, pp. 297–314. Springer, Heidelberg (2014). https://doi.org/10.1007/978-3-662-44371-2_17
6. Alwen, J., Peikert, C.: Generating shorter bases for hard random lattices. Theory Comput. Syst. **48**(3), 535–553 (2011)
7. Babai, L.: On lovász' lattice reduction and the nearest lattice point problem. Combinatorica **6**(1), 1–13 (1986)
8. Bajard, J.-C., Eynard, J., Hasan, M.A., Zucca, V.: A full RNS variant of FV like somewhat homomorphic encryption schemes. In: Avanzi, R., Heys, H. (eds.) SAC 2016. LNCS, vol. 10532, pp. 423–442. Springer, Cham (2017). https://doi.org/10.1007/978-3-319-69453-5_23
9. Bert, P., Fouque, P.-A., Roux-Langlois, A., Sabt, M.: Practical implementation of ring-SIS/LWE based signature and IBE. In: Lange, T., Steinwandt, R. (eds.) PQCrypto 2018. LNCS, vol. 10786, pp. 271–291. Springer, Cham (2018). https://doi.org/10.1007/978-3-319-79063-3_13
10. Boneh, D., et al.: Fully key-homomorphic encryption, arithmetic circuit ABE and compact garbled circuits. In: Nguyen, P.Q., Oswald, E. (eds.) EUROCRYPT 2014. LNCS, vol. 8441, pp. 533–556. Springer, Heidelberg (2014). https://doi.org/10.1007/978-3-642-55220-5_30
11. Bonnoron, G., Ducas, L., Fillinger, M.: Large FHE gates from tensored homomorphic accumulator. In: Joux, A., Nitaj, A., Rachidi, T. (eds.) AFRICACRYPT 2018. LNCS, vol. 10831, pp. 217–251. Springer, Cham (2018). https://doi.org/10.1007/978-3-319-89339-6_13
12. Brakerski, Z., Gentry, C., Vaikuntanathan, V.: (Leveled) fully homomorphic encryption without bootstrapping. In: Innovations in Theoretical Computer Science - ITCS 2012, pp. 309–325. ACM (2012)
13. Brakerski, Z., Langlois, A., Peikert, C., Regev, O., Stehlé, D.: Classical hardness of learning with errors. In: Symposium on Theory of Computing - STOC 2013, pp. 575–584 (2013)
14. Brakerski, Z., Vaikuntanathan, V.: Lattice-based FHE as secure as PKE. In: Innovations in Theoretical Computer Science - ITCS 2014, pp. 1–12 (2014)
15. Chillotti, I., Gama, N., Georgieva, M., Izabachène, M.: Faster fully homomorphic encryption: bootstrapping in less than 0.1 seconds. In: Cheon, J.H., Takagi, T. (eds.) ASIACRYPT 2016. LNCS, vol. 10031, pp. 3–33. Springer, Heidelberg (2016). https://doi.org/10.1007/978-3-662-53887-6_1
16. Chillotti, I., Gama, N., Georgieva, M., Izabachène, M.: Faster packed homomorphic operations and efficient circuit bootstrapping for TFHE. In: Takagi, T., Peyrin, T. (eds.) ASIACRYPT 2017. LNCS, vol. 10624, pp. 377–408. Springer, Cham (2017). https://doi.org/10.1007/978-3-319-70694-8_14
17. Cousins, D.B., et al.: Implementing conjunction obfuscation under entropic ring LWE. In: Symposium on Security and Privacy - SSP 2018, pp. 354–371 (2018)
18. Crockett, E., Peikert, C.: Λoλ: functional lattice cryptography. In: Weippl, E.R., Katzenbeisser, S., Kruegel, C., Myers, A.C., Halevi, S. (eds.) Proceedings of the 2016 ACM SIGSAC Conference on Computer and Communications Security, Vienna, Austria, 24–28 October 2016, pp. 993–1005. ACM (2016)
19. Dai, W., et al.: Implementation and evaluation of a lattice-based key-policy ABE scheme. IEEE Trans. Inf. Forensics Secur. **13**(5), 1169–1184 (2018)
20. del Pino, R., Lyubashevsky, V., Seiler, G.: Lattice-based group signatures and zero-knowledge proofs of automorphism stability. In: Lie, D., Mannan, M., Backes, M., Wang, X. (eds.) Proceedings of the 2018 ACM SIGSAC Conference on Computer and Communications Security, CCS 2018, pp. 574–591. ACM (2018)

21. Ducas, L., Micciancio, D.: FHEW: bootstrapping homomorphic encryption in less than a second. In: Oswald, E., Fischlin, M. (eds.) EUROCRYPT 2015. LNCS, vol. 9056, pp. 617–640. Springer, Heidelberg (2015). https://doi.org/10.1007/978-3-662-46800-5_24

22. Ducas, L., Prest, T.: Fast fourier orthogonalization. In: Abramov, S.A., Zima, E.V., Gao, X. (eds.) Proceedings of the ACM on International Symposium on Symbolic and Algebraic Computation, ISSAC 2016, pp. 191–198. ACM (2016)

23. Genise, N., Micciancio, D.: Faster Gaussian sampling for trapdoor lattices with arbitrary modulus. In: Nielsen, J.B., Rijmen, V. (eds.) EUROCRYPT 2018. LNCS, vol. 10820, pp. 174–203. Springer, Cham (2018). https://doi.org/10.1007/978-3-319-78381-9_7

24. Genise, N., Micciancio, D., Polyakov, Y.: Building an efficient lattice gadget toolkit: Subgaussian sampling and more. IACR Cryptology ePrint Archive, 2018:946 (2018)

25. Gentry, C., Halevi, S., Smart, N.P.: Homomorphic evaluation of the AES circuit. In: Safavi-Naini, R., Canetti, R. (eds.) CRYPTO 2012. LNCS, vol. 7417, pp. 850–867. Springer, Heidelberg (2012). https://doi.org/10.1007/978-3-642-32009-5_49

26. Gentry, C., Peikert, C., Vaikuntanathan, V.: Trapdoors for hard lattices and new cryptographic constructions. In: Symposium on Theory of Computing - STOC 2008, pp. 197–206 (2008)

27. Gentry, C., Sahai, A., Waters, B.: Homomorphic encryption from learning with errors: conceptually-simpler, asymptotically-faster, attribute-based. In: Canetti, R., Garay, J.A. (eds.) CRYPTO 2013. LNCS, vol. 8042, pp. 75–92. Springer, Heidelberg (2013). https://doi.org/10.1007/978-3-642-40041-4_5

28. Gorbunov, S., Vaikuntanathan, V., Wee, H.: Predicate encryption for circuits from LWE. In: Gennaro, R., Robshaw, M. (eds.) CRYPTO 2015. LNCS, vol. 9216, pp. 503–523. Springer, Heidelberg (2015). https://doi.org/10.1007/978-3-662-48000-7_25

29. Gorbunov, S., Vaikuntanathan, V., Wichs, D.: Leveled fully homomorphic signatures from standard lattices. In: Symposium on Theory of Computing - STOC 2015, pp. 469–477 (2015)

30. Gür, K.D., Polyakov, Y., Rohloff, K., Ryan, G.W., Sajjadpour, H., Savas, E.: Practical applications of improved Gaussian sampling for trapdoor lattices. IACR Cryptology ePrint Archive, 2017:1254 (2017)

31. Gür, K.D., Polyakov, Y., Rohloff, K., Ryan, G.W., Savas, E.: Implementation and evaluation of improved Gaussian sampling for lattice trapdoors. In: Proceedings of the 6th Workshop on Encrypted Computing & Applied Homomorphic Cryptography, WAHC 2018, pp. 61–71 (2018)

32. Halevi, S., Halevi, T., Shoup, V., Stephens-Davidowitz, N.: Implementing BP-obfuscation using graph-induced encoding. In: Computer and Communications Security - CCS 2017, pp. 783–798 (2017)

33. Halevi, S., Polyakov, Y., Shoup, V.: An improved RNS variant of the BFV homomorphic encryption scheme. IACR Cryptology ePrint Archive, 2018:117 (2018)

34. Klein, P.N.: Finding the closest lattice vector when it's unusually close. In: Symposium on Discrete Algorithms - SODA 2000, pp. 937–941 (2000)

35. Lyubashevsky, V., Micciancio, D., Peikert, C., Rosen, A.: SWIFFT: a modest proposal for FFT hashing. In: Nyberg, K. (ed.) FSE 2008. LNCS, vol. 5086, pp. 54–72. Springer, Heidelberg (2008). https://doi.org/10.1007/978-3-540-71039-4_4

36. Lyubashevsky, V., Peikert, C., Regev, O.: A toolkit for ring-LWE cryptography. In: Johansson, T., Nguyen, P.Q. (eds.) EUROCRYPT 2013. LNCS, vol. 7881, pp. 35–54. Springer, Heidelberg (2013). https://doi.org/10.1007/978-3-642-38348-9_3

37. Aguilar-Melchor, C., Barrier, J., Guelton, S., Guinet, A., Killijian, M.-O., Lepoint, T.: NFLLIB: NTT-based fast lattice library. In: Sako, K. (ed.) CT-RSA 2016. LNCS, vol. 9610, pp. 341–356. Springer, Cham (2016). https://doi.org/10.1007/978-3-319-29485-8_20

38. Micciancio, D., Peikert, C.: Trapdoors for lattices: simpler, tighter, faster, smaller. In: Pointcheval, D., Johansson, T. (eds.) EUROCRYPT 2012. LNCS, vol. 7237, pp. 700–718. Springer, Heidelberg (2012). https://doi.org/10.1007/978-3-642-29011-4_41

39. Micciancio, D., Regev, O.: Worst-case to average-case reductions based on Gaussian measures. SIAM J. Comput. **37**(1), 267–302 (2007)

40. Micciancio, D., Sorrell, J.: Ring packing and amortized FHEW bootstrapping. In: Automata, Languages, and Programming - ICALP 2018. LIPIcs, vol. 107, pp. 100:1–100:14 (2018)

41. Peikert, C.: Personal Communication (2018)

42. Polyakov, Y., Rohloff, K., Ryan, G.W.: PALISADE lattice cryptography library. https://git.njit.edu/palisade/PALISADE. Accessed Oct 2018

43. Vershynin, R.: Introduction to the non-asymptotic analysis of random matrices. CoRR, abs/1011.3027 (2010)

Approx-SVP in Ideal Lattices
with Pre-processing

Alice Pellet-Mary, Guillaume Hanrot[(⊠)], and Damien Stehlé

Univ. Lyon, EnsL, UCBL, CNRS, Inria, LIP, 69342 Lyon Cedex 07, France
{alice.pellet__mary,guillaume.hanrot,damien.stehle}@ens-lyon.fr

Abstract. We describe an algorithm to solve the approximate Shortest Vector Problem for lattices corresponding to ideals of the ring of integers of an arbitrary number field K. This algorithm has a pre-processing phase, whose run-time is exponential in $\log|\Delta|$ with Δ the discriminant of K. Importantly, this pre-processing phase depends only on K. The pre-processing phase outputs an "advice", whose bit-size is no more than the run-time of the query phase. Given this advice, the query phase of the algorithm takes as input any ideal I of the ring of integers, and outputs an element of I which is at most $\exp(\widetilde{O}((\log|\Delta|)^{\alpha+1}/n))$ times longer than a shortest non-zero element of I (with respect to the Euclidean norm of its canonical embedding). This query phase runs in time and space $\exp(\widetilde{O}((\log|\Delta|)^{\max(2/3,1-2\alpha)}))$ in the classical setting, and $\exp(\widetilde{O}((\log|\Delta|)^{1-2\alpha}))$ in the quantum setting. The parameter α can be chosen arbitrarily in $[0, 1/2]$. Both correctness and cost analyses rely on heuristic assumptions, whose validity is consistent with experiments.

The algorithm builds upon the algorithms from Cramer *et al.* [EUROCRYPT 2016] and Cramer *et al.* [EUROCRYPT 2017]. It relies on the framework from Buchmann [Séminaire de théorie des nombres 1990], which allows to merge them and to extend their applicability from prime-power cyclotomic fields to all number fields. The cost improvements are obtained by allowing precomputations that depend on the field only.

1 Introduction

The Learning With Errors problem (LWE) introduced by Regev in [Reg05] has proved invaluable towards designing cryptographic primitives. However, as its instance bit-sizes grow at least quadratically with the security parameter to be well-defined, LWE often results in primitives that are not very efficient. In order to improve the efficiency, Stehlé, Steinfeld, Tanaka and Xagawa [SSTX09] introduced the search Ideal-LWE problem which involves polynomials modulo $X^n + 1$ for n a power of two, and Lyubashevsky, Peikert and Regev [LPR10] exhibited the relationship to power-of-two cyclotomic fields, gave a reduction from the latter search problem to a decision variant, and tackled more general rings. This is now referred to as Ring-LWE, and leads to more efficient cryptographic constructions. To support the conjecture that Ring-LWE is computationally intractable, the authors of [SSTX09,LPR10] gave polynomial-time quantum reductions from

© International Association for Cryptologic Research 2019
Y. Ishai and V. Rijmen (Eds.): EUROCRYPT 2019, LNCS 11477, pp. 685–716, 2019.
https://doi.org/10.1007/978-3-030-17656-3_24

the approximate Shortest Vector Problem (approx-SVP) restricted to ideal lattices to Ring-LWE. Approx-SVP consists in finding a non-zero vector of an input lattice, whose norm is within a prescribed factor from the lattice minimum. Ideal lattices are lattices corresponding to ideals of the ring of integers of a number field, for example a power-of-two cyclotomic field in the situation above. When considering a lattice problem for such an ideal, the ideal is implicitly viewed as a lattice via the canonical embedding. A third quantum reduction from approx-SVP for ideal lattices to Ring-LWE was proposed by Peikert, Regev and Stephens-Davidowitz [PRS17]. It has the advantage of working for all number fields.

As is always the case, the value of these reductions highly depends on the intractability of the starting problem, i.e., approx-SVP for ideal lattices: approx-SVP for ideal lattices could even turn out to be computationally easy to solve, hence making these reductions vacuous. We stress that even if this were the case, that would not necessarily mean that there exists an efficient algorithm for Ring-LWE. In this work, we investigate the intractability of ideal approx-SVP for arbitrary number fields.

For arbitrary lattices, the best known trade-off between the run-time and the approximation factor is given by Schnorr's hierarchy of reduction algorithms [Sch87], whose most popular variant is the BKZ algorithm [SE94].

For any real number $\alpha \in [0, 1]$ and any lattice L of dimension n given by an arbitrary basis, it allows one to compute a vector of $L \setminus \{0\}$ which is no more than $2^{\widetilde{O}(n^{\alpha})}$ times longer than a shortest one, in time $2^{\widetilde{O}(n^{1-\alpha})}$ (assuming the bit-size of the input basis in polynomial in n). This trade-off is drawn in blue in Fig. 1.[1] In the case of ideal lattices in a cyclotomic ring of prime-power conductor (i.e., the ring of integers of $\mathbb{Q}(\zeta_m)$ where m is a prime power and ζ_m is a complex primitive m-th root of unity), it has been shown that it is possible to obtain a better trade-off than the BKZ algorithm, in the quantum computation setting. For *principal* ideal lattices, i.e., ideals that can be generated by a single element, the algorithmic blueprint, described in [CGS14,Ber14], consists in first using class group computations to find a generator of the ideal, and then use the so-called log-unit lattice to shorten the latter generator (we note that using the log-unit lattice for this purpose was already suggested in [RBV04]). A quantum polynomial-time algorithm for the first step was provided by Biasse and Song [BS16], building upon the work of [EHKS14]. The second step was carefully analyzed by Cramer, Ducas, Peikert and Regev [CDPR16], resulting in a quantum polynomial-time algorithm for approx-SVP restricted to *principal* ideal lattices, with a $2^{\widetilde{O}(\sqrt{n})}$ approximation factor. (See [HWB17] for a generalization to cyclotomics with degree of the form $p^{\alpha}q^{\beta}$, with p and q prime.) This line of works was extended by Cramer, Ducas and Wesolowski [CDW17] to any (not necessarily principal) ideal lattice of a cyclotomic ring of prime-power conductor. Put together, these results give us the trade-off between approximation factor and run-time drawn in red dashes in Fig. 1. This is better than the BKZ algorithm when the approximation factor is larger than $2^{\widetilde{O}(\sqrt{n})}$.

---

[1] This figure, like all similar ones in this work, is in $(\log_n \log_2)$-scale for both axes.

However, for smaller approximation factors, Schnorr's hierarchy remains the record holder. One could also hope to improve the trade-off for classical computing, by replacing the quantum principal ideal solver of [BS16] by the classical one of Biasse, Espitau, Fouque, Gélin and Kirchner [BEF+17]. However, this classical principal ideal solver runs in sub-exponential time $2^{\widetilde{O}(\sqrt{n})}$, hence combining it with [CDPR16, CDW17] results in a classical approx-SVP algorithm for a $2^{\widetilde{O}(\sqrt{n})}$ approximation factor in time $2^{\widetilde{O}(\sqrt{n})}$. Up to the $\widetilde{O}(\cdot)$ terms, this is exactly the trade-off obtained using Schnorr's hierarchy. Recently, Ducas, Plançon and Wesolowski [DPW19] experimentally analysed the $\widetilde{O}(\cdot)$ term of the $2^{\widetilde{O}(\sqrt{n})}$ approximation factor of the [CDPR16, CDW17] algorithm. This allows them to determine for which dimension n this quantum algorithm outperforms BKZ.

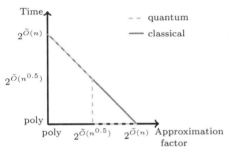

Fig. 1. Prior time/approximation trade-offs for ideal approx-SVP in cyclotomic fields of prime-power conductor. (Color figure online)

Fig. 2. New trade-offs for ideal approx-SVP in the same fields (with a pre-processing of cost $\exp(\widetilde{O}(n))$).

Contributions. We extend the techniques from [CDPR16, CDW17] to all number fields and improve the trade-off above by allowing the algorithm to perform some pre-computations on the number field.

It is a classical fact due to Minkowski [Min67, pp. 261–264] that there exists an absolute constant $c > 1$ such that for all number field K of degree $n \geq 2$ and discriminant Δ, we have $|\Delta| > c^n$. In the sequel, we shall thus state all our upper bounds in terms of $\log |\Delta| \geq \Omega(n)$. Actually, to fix the ideas, one may consider $\log |\Delta| = \widetilde{O}(n)$, which is the case for cyclotomic fields.

Let us consider a number field K of degree n and discriminant Δ. We assume a basis of the ring of integers R of K is given. Our algorithm performs some pre-processing on K, in exponential time $2^{\widetilde{O}(\log |\Delta|)}$. Once this pre-processing phase is completed and for any $\alpha \in [0, 1/2]$, the algorithm can, given any ideal lattice I of R, output a $2^{\widetilde{O}((\log |\Delta|)^{\alpha+1}/n)}$ approximation of a shortest non-zero vector of I in time $2^{\widetilde{O}((\log |\Delta|)^{1-2\alpha})} + T_{\text{c-g}}(K)$. Here $T_{\text{c-g}}(K)$ denotes the time needed to perform class group related computations in K: computing relations between elements of the class group and computing the units of R. Using the results of [BS16, BEF+17, BF14], we can replace $T_{\text{c-g}}(K)$ by $\text{poly}(\log |\Delta|)$ for a quantum computer, and, for a classical computer, by $2^{\widetilde{O}((\log |\Delta|)^{1/2})}$ if K is a cyclotomic field of prime-power conductor and by $2^{\widetilde{O}((\log |\Delta|)^{2/3})}$ for an arbitrary field K. The three algorithms rely on the Generalized Riemann Hypothesis (GRH) and the two sub-exponential algorithms in the classical setting also require additional heuristic assumptions. The correctness and cost analyses of our algorithm rely on these heuristic assumptions, and others. Our contribution is formalized in the theorem below, which is the main result of this article.

Theorem 1.1 (Heuristic, see Theorems 3.4 and 5.1). *Let $\alpha \in [0, 1/2]$ and K be a number field of degree n and discriminant Δ. Assume that a basis of the ring of integers R of K is known. Under some conjectures and heuristics, there exist two algorithms $A_{\text{pre-proc}}$ and A_{query} such that*

- *Algorithm $A_{\text{pre-proc}}$ takes as input the ring R, runs in time $2^{\widetilde{O}(\log |\Delta|)}$ and outputs a hint w of bit-size $2^{\widetilde{O}((\log |\Delta|)^{1-2\alpha})}$;*
- *Algorithm A_{query} takes as inputs any ideal I of R (whose algebraic norm has bit-size bounded by $2^{\text{poly}(\log |\Delta|)}$) and the hint w output by $A_{\text{pre-proc}}$, runs in time $2^{\widetilde{O}((\log |\Delta|)^{1-2\alpha})} + T_{\text{c-g}}(K)$, and outputs an element $x \in I$ such that $0 < \|x\|_2 \le 2^{\widetilde{O}((\log |\Delta|)^{\alpha+1}/n)} \cdot \lambda_1(I)$.*

The hint output by the pre-processing phase has a bit-size that is bounded by the run-time of the query phase. By considering larger hints, the run-time of the query phase could be drastically improved. We give more details below, at the end of the high-level description of the algorithm.

Considering only the query cost, this result is of interest when $\log |\Delta| \le \widetilde{O}(n^{4/3})$ for quantum computations and $\log |\Delta| \le \widetilde{O}(n^{12/11})$ for classical computations. Indeed, in the other cases, the time/quality trade-offs obtained by our algorithm are worse than the ones obtained using Schnorr's hierarchy of algorithms. By letting α vary in $[0, 1/2]$ and considering cyclotomic fields of prime-power conductor, we obtain the trade-offs represented in Fig. 2. For a discussion for more general values of $\log |\Delta|$, we refer to Sect. 5. Going back to cyclotomic fields of prime-power conductor, these new trade-offs improve upon the prior ones, both for quantum and classical computers. Note that in Fig. 2, we only plot the time needed for the query phase of the algorithm, but there is a pre-processing phase of exponential time performed before. Also, the new algorithm is no better than Schnorr's hierarchy in the classical setting when the run-time is sufficiently small. Hence, in Fig. 2, we plotted the trade-offs obtained using Schnorr's hierarchy when they are better than the ones obtained with the new algorithm. The query phase of the new algorithm gives a quantum acceleration for approx-SVP for ideal lattices in cyclotomic fields of prime-power conductor, for all approximation factors $2^{\widetilde{O}(n^\alpha)}$ with $\alpha \in (0, 1)$. This extends [CDW17], which obtained such a quantum acceleration for $\alpha \in [1/2, 1)$. The query phase of the new algorithm also gives a classical acceleration for these fields, but only for $\alpha \in (0, 1/2)$.

Technical overview. Our algorithm is inspired by the algorithms in [CDPR16, CDW17]. Given an ideal I as input, they first allow to find a principal ideal J contained in I (using [CDW17]), and then, they allow to compute a short generator of this ideal J (using [CDPR16]). This short generator is a somehow small element of I. This approach provides a $2^{\widetilde{O}(\sqrt{n})}$ approximation factor for approx-SVP in I. However, it can be shown that we cannot improve this approximation factor using these techniques, even if we increase the run-time of the algorithm. The reason is that, given an arbitrary principal ideal J, it may be that its shortest generator is $2^{\widetilde{O}(\sqrt{n})}$ times longer than its shortest non-zero vector.

We modify the strategy above, as follows. Given any ideal I, we try to find a 'good' principal ideal J contained in I, where we say that a principal ideal is 'good' if its shortest generator is not much larger than its shortest non-zero vector. The precise definition of 'not much larger' will depend on the approximation factor we want to achieve for our approx-SVP instance. Because the Euclidean norm of the shortest non-zero vector of J (broadly) increases with its algebraic norm, we also require that the algebraic norm of J is not much larger than the one of I (note that this was already needed in [CDPR16, CDW17]). To find this 'good' principal ideal J, the main idea of our algorithm is to express the problem as a Closest Vector Problem (CVP) instance in a lattice L depending only on the number field K.

This lattice L is similar to the one appearing in sub-exponential algorithms for computing the class group (see for instance [HM89, Buc88]). More precisely, we first select a set $\mathfrak{B} = \{\mathfrak{p}_1, \ldots, \mathfrak{p}_r\}$ of prime ideals of polynomially bounded algebraic norms, generating the class group. We then compute a generating set of the $\mathfrak{B}$-units, i.e., the set of elements $u \in K$ for which there exists $(e_1, \ldots, e_r) \in \mathbb{Z}^r$ such that $\langle u \rangle = \prod_i \mathfrak{p}_i^{e_i}$. The lattice L is obtained by considering the integer linear combinations of vectors of the form $(\mathrm{Log}\, u, e_1, \ldots, e_r)^T$, where $\langle u \rangle = \prod_i \mathfrak{p}_i^{e_i}$ and Log is the map applying the logarithm function to the canonical embedding, coefficient-wise. This lattice L only depends on the field K and can then be pre-computed and pre-processed.

Given any ideal I, the query phase of our algorithm computes a target vector t from I, and then solves a CVP instance in L with this target vector t. First, we decompose the ideal I in the class group as a product of the ideals of $\mathfrak{B}$. Concretely, we compute $g \in K$ and $(v_1, \ldots, v_r) \in \mathbb{Z}^r$ such that $I = \prod_i \mathfrak{p}_i^{v_i} \cdot \langle g \rangle$. This principal ideal $\langle g \rangle$ is a candidate for our principal ideal J contained in I (assume for the moment that the v_i's are non-positive, so that $\langle g \rangle$ is indeed contained in I). However, as is, we have no guarantee that $\langle g \rangle$ has a short generator. We also have no guarantee that its algebraic norm is not much larger than the one of I (i.e., that the v_i's are small). Hence, our objective is to multiply the principal ideal $\langle g \rangle$ by other principal ideals, until we have a good candidate for J. To do so, we define the vector $t = (-\mathrm{Log}\, g, v_1, \ldots, v_r)^T$. Observe that $\langle g \rangle$ would be a good candidate for J if this vector was short (and with $v_i \leq 0$ for all i). Indeed, this would mean that g is a short generator of $\langle g \rangle$ (because $\mathrm{Log}\, g$ is short), and that $\langle g \rangle = I \cdot \prod_i \mathfrak{p}_i^{-v_i}$ is a small multiple of I (because the $\mathfrak{p}_i$'s have polynomially bounded norms, and the v_i's are small; the non-positivity of the v_i's is used to ensure that the ideal $\prod_i \mathfrak{p}_i^{-v_i}$ is integral). Also, we can see that adding a vector of L to t amounts to multiply the principal ideal $\langle g \rangle$ by another principal ideal (corresponding to the vector of L we are adding). Hence, we can find a good candidate J (together with a short generator of J) by solving a CVP instance in L with target t.

Finally, we need to solve CVP in the lattice L. We do not know any basis for L which would enable us to solve CVP in it efficiently (as opposed to the lattices considered in [CDPR16, CDW17]). However, the lattice L is fixed for a given number field, hence we can pre-process it. For this, we use a CVP with pre-processing (CVPP) algorithm due to Laarhoven [Laa16]. This leads to the time/approximation trade-offs given in Theorem 1.1. In [Laa16], significant effort is spent towards minimizing the constant factors in the exponents. These have recently been improved in [DLW19]. In this work, we neglect these factors for the sake of simplicity, but these would clearly matter in practice.

Laarhoven's CVPP algorithm is such that the bit-size of the output of the pre-processing phase is no larger than the run-time of the query phase[2] (hence, it is also the case for our algorithm). If we do not require this, we could have the following very simple and efficient algorithm for CVPP. First, it computes a short basis B_{sh} of the lattice. Then, it partitions the fundamental parallelepiped associated to B_{sh} into exponentially many small boxes, such that given any point of the parallelepiped, it is easy to determine to which box it belongs. Then, for each of these boxes, the pre-processing algorithm would compute a closest point of the lattice. The output of the pre-processing phase would then be the small basis B_{sh} and the set of all boxes together with their closest lattice point. Finally, given any vector in the real span of the lattice, the query algorithm would reduce it modulo B_{sh} to obtain a vector in the fundamental parallelepiped, and then determine the box of this reduced vector and its associated lattice vector. All this can be done efficiently (assuming we can efficiently access the database) and provides a small factor approximation for CVP, at the expense of a huge database.

Overall, the correctness and cost analyses of our algorithm rely on several heuristic assumptions. Many of them come from previous works [Laa16, BEF+17, BF14] and were already analysed. We introduce three new heuristic assumptions: Heuristics 4, 5 and 6 in Sect. 4. We discuss them by providing some mathematical justifications and some experimental results corroborating them. Concurrently to this work, Stephens-Davidowitz [Ste19] obtained a provable variant of the CVPP trade-offs from [Laa16, DLW19] that we use. Relying on it would allow us to make do with Heuristic 1, which was inherited from [Laa16, DLW19], at the expense of replacing Heuristic 4 by a similar one on the smoothing parameter of the lattice under scope (rather than its covering radius).

Impact. The query phase of the new algorithm can be interpreted as a non-uniform algorithm, as it solves approx-SVP for ideals of K, using a hint

---

[2] Laarhoven also describes a variant of his algorithm in which he uses locality-sensitive hashing to reduce the run-time of the query phase below the bit-size of the advice, but we are not considering this variant here.

depending on K only. As the time needed to compute that hint (i.e., the run-time of $A_{\text{pre-proc}}$) is exponential, the concrete impact is limited. Nevertheless, our result should rather be interpreted as a strong indication that ideal approx-SVP is most likely a weaker problem than approx-SVP for arbitrary lattices: for unstructured lattices, there is no known non-uniform algorithm outperforming Schnorr's hierarchy.

Few cryptographic constructions have their security impacted by faster ideal approx-SVP solvers. A notable example is Gentry's fully homomorphic encryption scheme [Gen09], which was later superseded by faster homomorphic schemes relying on better understood hardness assumptions (see, e.g., [BV11b, BV11a]). Another important example is the Garg, Gentry and Halevi candidate construction of cryptographic multilinear map [GGH13] and its extensions. Because of the large pre-processing time, our algorithm does not provide a concrete attack on those schemes.

More importantly, our result strongly suggests that approx-SVP for ideals of the ring of integers R of a number field K may be weaker than Ring-LWE, for a vast family of number fields. Up to some limited parameter losses, Ring-LWE and approx-SVP for R-modules over K (with ranks ≥ 2) reduce to one another [LS15, AD17]. Therefore, a separation between approx-SVP for ideals and Ring-LWE is essentially the same as a separation between approx-SVP for ideals and approx-SVP for R-modules over K.

Open problems. Throughout the article, we keep track of all the different sub-algorithms that compose our approx-SVP solver. The exact formulation of the total cost of the algorithm, as a function of the costs of the sub-algorithms, is given in Theorem 3.4. When instantiated with the run-times of the currently known algorithms for the different sub-tasks, we obtain the values given in Theorem 1.1. However, the formula with all the sub-algorithms allows to see whether an improvement of the run-time of one of them leads to an improvement of the overall cost of the approx-SVP solver. In particular, for the specific choice of cyclotomic fields of prime-power conductor:

- Improving the approx-CVP solver would lead to an improvement of the slope of the curves in Fig. 2, for approximation factors smaller than $2^{\tilde{O}(\sqrt{n})}$. In a different direction, removing the pre-processing step needed for this approx-CVP solver would remove the pre-processing of the overall approx-SVP algorithm.
- Designing a classical algorithm that performs class group related computations in time less than $2^{\tilde{O}(\sqrt{n})}$ would allow to further extend the (classical) segment of Fig. 2 with slope $-1/2$, until it reaches the cost needed to solve these class group related problems. For example, Biasse described in [Bia17] an algorithm to solve the principal ideal problem in cyclotomic fields of prime-power conductor, with pre-processing. After pre-computations depending on the field only, this algorithm finds a generator of a principal ideal in time less than $2^{\tilde{O}(\sqrt{n})}$ if the ideal has algebraic norm $\leq 2^{\tilde{O}(n^{1.5})}$.

Finally, one could wonder whether it is possible to find significantly faster approx-SVP algorithms for specific families of number fields and/or restricted

families of ideals. For instance, the Bauch *et al.* algorithm from [BBV+17] and
the follow-up algorithm of Biasse and van Vredendaal [BV18] allow to efficiently
solve class group related problems in real multiquadratic number fields in the
classical setting. This means that in these number fields, the classical version of
our algorithm is as efficient as the quantum one (there is no threshold for the
query phase in the classical setting). However, the algorithm still requires an
exponential pre-processing phase for the approx-CVP solver.

Roadmap. In Sect. 2, we recall some necessary background on lattices and number fields. Then, in Sect. 3, we explain how to transform an approx-SVP instance
in any ideal into an approx-CVP instance in some lattice depending only on
the number field. We detail in Sect. 4 some properties of the lattice in which
we want to solve approx-CVP, and we give the trade-offs obtained when using
Laarhoven's algorithm. Finally, in Sect. 5, we instantiate our main theorem with
the best run-times currently known for solving approx-CVPP and class group
related problems.

Supplementary material. The code that was used to perform the experiments
is available on the webpage of the first author.

2   Preliminaries

We let $\mathbb{Z}, \mathbb{Q}, \mathbb{R}$ and $\mathbb{C}$ respectively denote the sets of integer, rational, real and
complex numbers. For a positive real number x, we let $\log x$ denote its binary
logarithm. For two functions $f(n)$ and $g(n)$, we write $f(n) = \widetilde{O}(g(n))$ if there
exists some constant $c > 0$ such that $f(n) = O(g(n) \cdot |\log g(n)|^c)$. We abuse
notations by defining $\widetilde{O}(n^\alpha) = O(n^\alpha \operatorname{poly}(\log n))$ even if $\alpha = 0$ (this will simplify
some statements). For a vector $v \in \mathbb{R}^n$, we let v_i denote the i-th coordinate of v.
We write $\|v\|_1 = \sum_i |v_i|$, $\|v\|_2 = \sqrt{\sum_i v_i^2}$ and $\|v\|_\infty = \max_i |v_i|$. We recall the
following inequalities between these three different norms.

$$\|v\|_\infty \leq \|v\|_2 \leq \|v\|_1, \tag{2.1}$$
$$\|v\|_2 \leq \sqrt{n} \cdot \|v\|_\infty, \quad \|v\|_1 \leq \sqrt{n} \cdot \|v\|_2. \tag{2.2}$$

Note that only the ℓ_2-norm is invariant under orthonormal transformations.

2.1   Lattice Problems

For a lattice L and $i \in \{1, 2, \infty\}$, we let $\lambda_1^{(i)}(L)$ denote the norm of a shortest
non-zero vector of L for the ℓ_i-norm. Similarly, for $k \geq 1$, we let $\lambda_k^{(i)}(L)$ denote
the smallest real number such that there exist k linearly independent vectors
of L whose ℓ_i-norms are no greater than $\lambda_k^{(i)}(L)$. We let $\operatorname{Span}(L)$ denote the
real vector space spanned by the vectors of L. For a point $t \in \operatorname{Span}(L)$, we let
$\operatorname{dist}^{(i)}(t, L) = \inf_{v \in L} \|t - v\|_i$ be the minimal distance between t and any point
of L. We define the covering radius of L as $\mu^{(i)}(L) = \sup_{t \in \operatorname{Span}(L)} \operatorname{dist}^{(i)}(t, L)$.
The determinant (or volume) $\det(L)$ of a full-rank lattice L is the absolute value
of the determinant of any of its bases.

Lemma 2.1 (Minkowski's inequality). *For any full-rank lattice L of dimension n, we have $\lambda_1^{(\infty)}(L) \leq \det(L)^{1/n}$. This implies that $\lambda_1^{(2)}(L) \leq \sqrt{n} \cdot \det(L)^{1/n}$.*

We will consider the following algorithmic problems involving lattices.

Definition 2.2 (Approximate Shortest Vector Problem (approx-SVP)).
Given a lattice L and $i \in \{1, 2, \infty\}$, the approximate Shortest Vector Problem in norm ℓ_i, with approximation factor $\gamma \geq 1$, is to find a vector $v \in L \setminus \{0\}$ such that $\|v\|_i \leq \gamma \cdot \lambda_1^{(i)}(L)$.

Definition 2.3 (Approximate Closest Vector Problem (approx-CVP)).
Given a lattice L, $i \in \{1, 2, \infty\}$ and a target $t \in \mathrm{Span}(L)$, the approximate Closest Vector Problem in norm ℓ_i, with approximation factor $\gamma \geq 1$, is to find a vector $v \in L$ such that $\|v - t\|_i \leq \gamma \cdot \mathrm{dist}^{(i)}(t, L)$.
In this article, we will be essentially interested in a variant of approx-CVP, in which we ask that $\|v - t\|_i \leq \beta$ for some β, independently of $\mathrm{dist}^{(i)}(t, L)$ (i.e., the distance of the found vector is bounded in absolute terms, independently of whether the target is close to the lattice or not). We call this variant approx-CVP'. For $i \in \{1, 2, \infty\}$, we let $T_{\mathrm{CVP}}(i, L, \beta)$ denote the worst-case run-time of the best known algorithm that solves approx-CVP' for the ℓ_i-norm, in the lattice L, with a bound β.

Definition 2.4 (Approx-CVP with Pre-processing (approx-CVPP)).
This problem is the same as approx-CVP, except that the algorithm can perform some pre-processing on the lattice L before it gets the target vector t. Approx-CVPP' is defined analogously. We will then consider the pre-processing time (performed when knowing only L) and the query time (performed once we get the target t). For $i \in \{1, 2, \infty\}$, we let $T_{\mathrm{CVP}}^{\mathrm{pre-proc}}(i, L, \beta)$ (resp. $T_{\mathrm{CVP}}^{\mathrm{query}}(i, L, \beta)$) denote the worst-case run-time of the pre-processing phase (resp. query phase) of the best algorithm that solves approx-CVPP' for the ℓ_i-norm, in the lattice L, with a bound β.

In the following, we will always be interested in the approximate versions of these problems, so we will sometimes omit the 'approx' prefix.

In [Laa16], Laarhoven gives a heuristic algorithm for solving approx-CVPP. The following result is not explicitly stated in [Laa16] (only the two extreme values are given), but the computations can be readily adapted.

Theorem 2.5 ([Laa16, Corollaries 2 and 3]). *Let $\alpha \in [0, 1/2]$. Then, under Heuristic 1 below, there exists an algorithm that takes as pre-processing input an n-dimensional lattice L (given by a basis whose bit-size is polynomial in n) and as query input any vector $t \in \mathrm{Span}(L)$ (with bit-size that is polynomial in n) and outputs a vector $v \in L$ with $\|t - v\|_2 \leq O(n^\alpha) \cdot \mathrm{dist}^{(2)}(t, L)$, with pre-processing time $2^{O(n)}$ and query time $\mathrm{poly}(n) \cdot 2^{O(n^{1-2\alpha})}$ (the memory needed during the query phase is also bounded by $\mathrm{poly}(n) \cdot 2^{O(n^{1-2\alpha})}$).*

The heuristic assumption used in Laarhoven's algorithm states that the lattice L is somehow dense and behaves randomly.

Heuristic 1. There exists a constant $c > 0$ such that the ball of radius $c \cdot \lambda_1^{(2)}(L)$ (in ℓ_2-norm) contains at least 2^n points of L. Moreover, once renormalized, these points 'behave' as uniformly and independently distributed points on the unit sphere.

We can weaken this heuristic assumption by taking $c = \mathrm{poly}(\log n)$, in which case the approximation factor in Laarhoven's algorithm becomes $\tilde{O}(n^\alpha)$ (the pre-processing and query costs remain the same).

We will use this algorithm to heuristically solve approx-CVPP' in Euclidean norm for $\alpha \in [0, 1/2]$, achieving $T_{\mathrm{CVP}}^{\mathrm{pre\text{-}proc}}(2, L, O(n^\alpha) \cdot \mu^{(2)}(L)) = 2^{O(n)}$ and $T_{\mathrm{CVP}}^{\mathrm{query}}(2, L, O(n^\alpha) \cdot \mu^{(2)}(L)) = 2^{\tilde{O}(n^{1-2\alpha})}$.

2.2 Number Fields and Ideals

We let K denote any number field of degree n and R be its ring of integers (i.e., elements of K which are roots of a monic polynomial with integer coefficients). The ring R is a free $\mathbb{Z}$-module of rank n. Let $\sigma_1, \ldots, \sigma_n$ be the n distinct embeddings from K to $\mathbb{C}$ ordered such that for $i \in \{1, \ldots, r_1\}$ we have $\sigma_i : K \to \mathbb{R}$ and for $i \in \{r_1 + 1, \ldots, r_2\}$ we have $\sigma_i = \overline{\sigma_{i+r_2}}$. We have r_1 real embeddings and r_2 pairs of complex conjugate embeddings, with r_1 and r_2 satisfying $r_1 + 2r_2 = n$. We let Δ denote the discriminant of K, i.e., $\Delta = [\det(\sigma_i(b_j))_{i,j}]^2$ for $b_1, \ldots, b_n$ any basis of the $\mathbb{Z}$-module R. Recall from Sect. 1 that Minkowski's bound gives us the following inequality:

$$\log |\Delta| \geq \Omega(n). \tag{2.3}$$

In the following, most of the costs will be expressed in term of $\log |\Delta|$.

We let $R^\times$ denote the group of units of R, that is $R^\times = \{u \in R \mid \exists v \in R, \ uv = 1\}$. Dirichlet's unit theorem states that $R^\times$ is isomorphic to the Cartesian product of a finite cyclic group (formed by the roots of unity contained in K) with the additive group $\mathbb{Z}^{r_1+r_2-1}$.

We associate to an element $x \in K$ the vector $(\sigma_1(x), \ldots, \sigma_{r_1}(x), \mathrm{Re}(\sigma_{r_1+1}(x)), \mathrm{Im}(\sigma_{r_1+1}(x)), \ldots, \mathrm{Re}(\sigma_{r_1+r_2}(x)), \mathrm{Im}(\sigma_{r_1+r_2}(x)))^T \in \mathbb{R}^n$, which we will call the canonical embedding of x. In the following, we will only consider the canonical embedding for the elements of K and R. We will abuse notation by considering that the elements of K and R are real vectors of the above form. Using this representation, the ring R becomes an n-dimensional lattice of $\mathbb{R}^n$. The volume of the lattice R is given by $\det(R) = 2^{-r_2}\sqrt{|\Delta|}$.

A fractional ideal I of K is a subset of K which is stable by addition, and by multiplication with any element of R, and such that $dI \subseteq R$ for some $d \in \mathbb{Z} \setminus \{0\}$. An ideal I is said to be integral if it is contained in R. A non-zero fractional ideal $I \subseteq R$ can be seen as a full-rank lattice in $\mathbb{R}^n$, via the canonical embedding. For an element $g \in K$, we write $\langle g \rangle = gR$, the smallest fractional ideal containing g. Such an ideal is said to be principal. An integral ideal $I \subseteq R$ is said to be prime if the ring R/I is an integral domain. The product of two fractional ideals I and J is defined by $I \cdot J = \{x_1 y_1 + \cdots + x_r y_r \mid r \geq 0, x_1, \ldots, x_r \in I, y_1, \ldots, y_r \in J\}$.

The algebraic norm $\mathcal{N}(I)$ of a non-zero fractional ideal $I \subseteq R$ is the determinant of I when seen as a lattice in $\mathbb{R}^n$ (via the canonical embedding), divided by $\det(R) = 2^{-r_2}\sqrt{|\Delta|}$ (and $\mathcal{N}(\langle 0 \rangle)$ is defined as 0). If I is integral, this is also equal to $|R/I|$. The algebraic norm of a prime ideal is a power of a prime number. For two fractional ideals I and J, the algebraic norm of their product satisfies $\mathcal{N}(I \cdot J) = \mathcal{N}(I) \cdot \mathcal{N}(J)$. The algebraic norm of an element $r \in R$ is defined by $\mathcal{N}(r) = \prod_{i=1}^n \sigma_i(r)$. For any element $r \in R$, we have that $\mathcal{N}(\langle r \rangle) = |\mathcal{N}(r)|$, so in particular $\mathcal{N}(r) \in \mathbb{Z}$.

Let I be a non-zero fractional ideal seen as a lattice. By definition of the norm of I and Minkowski's inequality, we know that $\lambda_1^{(\infty)}(I) \le \mathcal{N}(I)^{1/n} \cdot |\Delta|^{1/(2n)}$. We also have the following lower bound

$$\lambda_1^{(\infty)}(I) \ge \mathcal{N}(I)^{1/n}. \tag{2.4}$$

This lower bound comes from the fact that if $x \in I$ is such that $\|x\|_\infty = \lambda_1^{(\infty)}(I)$, then we have $|\mathcal{N}(x)| = \prod_i |\sigma_i(x)| \ge \mathcal{N}(I)$ (because $\langle x \rangle$ is a sub-lattice of I). This implies that at least one of the $|\sigma_i(x)|$'s is no smaller than $\mathcal{N}(I)^{1/n}$, hence the inequality. When $\log |\Delta| = \widetilde{O}(n)$, these two inequalities imply that $\lambda_1^{(\infty)}(I)$ is essentially $\mathcal{N}(I)^{1/n}$, up to a $2^{\mathrm{poly}(\log n)}$ factor. When $|\Delta|$ increases, so does the gap between the two bounds.

2.3   The Class Group

We let $\mathcal{I}_K$ denote the set of non-zero fractional ideals of K and $\mathcal{P}_K \subseteq \mathcal{I}_K$ denote the subset of non-zero principal fractional ideals. One can prove that for every non-zero fractional ideal I, there is a fractional ideal I^{-1} such that $I \cdot I^{-1} = R$. This gives $\mathcal{I}_K$ a group structure, for which $\mathcal{P}_K$ is a subgroup.

The class group of K is defined as the quotient $Cl_K = \mathcal{I}_K / \mathcal{P}_K$. For any non-zero ideal I of K, we let $[I]$ denote the equivalence class of I in the class group. In particular, we have $\mathcal{P}_K = [R]$. The class group is a finite abelian group and its cardinality h_K is called the class number. We have the following bound:

$$\log h_K = \widetilde{O}(\log |\Delta|). \tag{2.5}$$

This can be derived from the proof of Eq. (2.3), this proof being based on the fact that any class of the class group contains an integral ideal whose norm is bounded as $2^{\widetilde{O}(\log |\Delta|)}$. We also justify it later using Eq. (2.6) (which is significantly stronger).

We know, thanks to a result of Bach [Bac90] that the class group can be generated by ideals of polynomially bounded norms.

Theorem 2.6. (Theorem 4 of [Bac90]). *Under the GRH, the class group of a number field of discriminant Δ is generated by the prime ideals of algebraic norms $\le 12 \log^2 |\Delta|$.*

Moreover, computing all prime ideals of norms $\le 12 \log^2 |\Delta|$ can be done in time polynomial in $\log |\Delta|$. Indeed, these prime ideals can be obtained by factoring all ideals $\langle p \rangle$ where $p \in \mathbb{Z}$ is a prime no greater than $12 \log^2 \Delta$. Further, factoring such an ideal can be done in polynomial time (either using the Kummer-Dedekind theorem if p does not divide the index of $\mathbb{Z}[\theta]/R$, where θ is an algebraic integer such that $K = \mathbb{Q}(\theta)$, or using an algorithm due to Buchmann and Lenstra if p divides $|\mathbb{Z}[\theta]/R|$, see [Coh13, Section 4.8.2] for the former and [Coh13, Section 6.2] for the latter).

We will use the following lemma.

Lemma 2.7. *Let $\mathfrak{B}$ be any finite set of fractional ideals that generates the class group Cl_K. Then we can extract a subset $\mathfrak{B}'$ of $\mathfrak{B}$, of cardinality at most $\log h_K$, which also generates the class group. Moreover, this can be done efficiently if we are given the relations between the elements of $\mathfrak{B}$, in the form of a basis of $\ker(f_{\mathfrak{B}})$ where $f_{\mathfrak{B}} : (e_1, \dots e_r) \in \mathbb{Z}^r \mapsto \prod_i [\mathfrak{p}_i^{e_i}] \in Cl_K$, with $\mathfrak{B} = \{\mathfrak{p}_1, \dots, \mathfrak{p}_r\}$.*

We did not find this exact lemma in previous work, so we give a proof of it for the sake of completeness (even if the technique used to prove it is far from new).

Proof. We know that $\ker(f_{\mathfrak{B}})$ is a lattice of volume h_K contained in $\mathbb{Z}^r$ (it is stable by addition and subtraction, and $|\mathbb{Z}^r / \ker(f_{\mathfrak{B}})| = |Cl_K| = h_K$). Let $R_{\mathfrak{B}} \in \mathbb{Z}^{r \times r}$ be a basis of this lattice, with column vectors. From this basis, we can efficiently compute the Hermite Normal Form (HNF) of the lattice, which we will write $H_{\mathfrak{B}}$. This basis matrix is triangular, and each column corresponds to a relation between the elements of $\mathfrak{B}$ (each row corresponds to an ideal of $\mathfrak{B}$). So we can remove from the set $\mathfrak{B}$ any ideal whose row in $H_{\mathfrak{B}}$ has a 1 on the diagonal. Indeed, if row i has a 1 on the diagonal, this means that we have a relation of the form $[p_i \cdot \prod_{j>i} \mathfrak{p}_j^{e_j}] = [R]$. Hence the ideal class $[\mathfrak{p}_i]$ is in the group generated by $\{[\mathfrak{p}_j]\}_{j>i}$, and so it is not needed to generate the class group. But we know that $\det(H_{\mathfrak{B}}) = \det(\ker(f_{\mathfrak{B}})) = h_K$ is the product of the diagonal elements (which are integers). So we have at most $\log h_K$ ideals with diagonal entries different from 1. Hence, after removing from $\mathfrak{B}$ all ideals whose corresponding row in $H_{\mathfrak{B}}$ has a 1 on the diagonal, we obtain a set $\mathfrak{B}'$ of cardinality at most $\log h_K$ and which still generates the class group. This proof is an efficient algorithm if we are given an initial basis $R_{\mathfrak{B}}$, because we only need to compute an HNF basis, which can be done in time polynomial in the size of the input matrix.   □

Theorem 2.6 states that the class group can be generated by integral ideals of polynomially bounded norms, but this does not give us the existence of many small-norm integral ideals. For instance, if the class group is trivial (i.e., all ideals are principal), then it is generated by $[R]$. More generally, the class group could be generated by a very small number of ideals. In the following, we will need the existence of $\widetilde{\Omega}(\log |\Delta|)$ distinct integral ideals of polynomially bounded norms.

Theorem 2.8 (Theorem 8.7.4 of [BS96]). *Assume the GRH. Let $\pi_K(x)$ be the number of prime integral ideals of K of norm $\leq x$. Then there exists an absolute constant C (independent of K and x) such that*

$$|\pi_K(x) - \mathrm{li}(x)| \leq C \cdot \sqrt{x} \left(n \log x + \log |\Delta| \right),$$

where $\mathrm{li}(x) = \int_2^x \frac{dt}{\ln t} \sim \frac{x}{\ln x}$ (and $\ln$ refers to the natural logarithm).

Instantiating this theorem with $x = (\log|\Delta|)^\kappa$ for some constant $\kappa > 4$, we obtain the following corollary. The bounds in this corollary can be improved, but they suffice for our needs.

Corollary 2.9. *Assume the GRH. Let $\kappa > 4$. For $\log|\Delta|$ sufficiently large, there are $\geq (\log|\Delta|)^{\kappa-2}$ distinct prime integral ideals of norm smaller than $(\log|\Delta|)^\kappa$.*

Proof. We apply Theorem 2.8 with $x = (\log|\Delta|)^\kappa$. As $\mathrm{li}(x) \sim \frac{x}{\ln x}$, we have that $\mathrm{li}(x) \geq (\log|\Delta|)^{\kappa-1}$ holds for $\log|\Delta|$ sufficiently large. Recall that $\log|\Delta| > cn$ for some (explicit) constant c. Hence, the right hand side of the inequality of Theorem 2.8 can be bounded as

$$C \cdot \sqrt{x}\,(n\log x + \log|\Delta|) \leq C(\kappa/c+1) \cdot (\log|\Delta|)^{\kappa/2+1} \cdot \log\log|\Delta|.$$

But, as we chose κ such that $\kappa - 1 > \kappa/2 + 1$, we have, for $\log|\Delta|$ sufficiently large:

$$(\log|\Delta|)^{\kappa-1} - C(\kappa/c+1) \cdot (\log|\Delta|)^{\kappa/2+1} \cdot \log\log|\Delta| \geq (\log|\Delta|)^{\kappa-2},$$

hence proving the corollary. □

We use Theorem 2.6, Corollary 2.9 and Lemma 2.7, to obtain the following.

Corollary 2.10. *Assume the GRH. Then, for $\log|\Delta|$ sufficiently large and for any integer $r \geq \log h_K$, there exists a set $\mathfrak{B} = \{\mathfrak{p}_1, \ldots, \mathfrak{p}_r\}$ of prime integral ideals generating the class group, with $\mathcal{N}(\mathfrak{p}_i) = \mathrm{poly}(\log|\Delta|, r)$ for all i.*

Proof. Combining Theorem 2.6 and Lemma 2.7, we know that there exists a set $\mathfrak{B}$ of cardinality at most r, generating the class group and containing only prime ideals of norms $\leq 12\log^2|\Delta|$. We can then add prime ideals to this set $\mathfrak{B}$, until its cardinality reaches r. Thanks to Corollary 2.9, we know that there are enough prime ideals of norm smaller than $\mathrm{poly}(\log|\Delta|, r)$ (for some fixed poly) to increase the cardinality of $\mathfrak{B}$ up to r. □

2.4 The Log-Unit Lattice

We define $\mathrm{Log}\, x = (\log|\sigma_1(x)|, \ldots, \log|\sigma_n(x)|)^T \in \mathbb{R}^n$, for any $x \in K \setminus \{0\}$. Observe that this is not the usual definition of the logarithmic embedding. The function Log is often defined either as $(\log|\sigma_1(x)|, \ldots, \log|\sigma_{r_1+r_2}(x)|)^T \in \mathbb{R}^{r_1+r_2}$ [Sam13, Section 4.4] or as $(\log|\sigma_1(x)|, \ldots, \log|\sigma_{r_1}(x)|, 2\log|\sigma_{r_1+1}(x)|, 2\log|\sigma_{r_1+r_2}(x)|)^T \in \mathbb{R}^{r_1+r_2}$ [Coh13, Definition 4.9.6]. Indeed, for $i > r_1+r_2$, the $\log|\sigma_i(x)|$'s are redundant because $|\sigma_i(x)| = |\sigma_{i-r_2}(x)|$. However, in our case, it will be more convenient to work with the logarithms of all the embeddings.

Let $E = \{x \in \mathbb{R}^n : x_i = x_{i+r_2}, \forall r_1 < i \leq r_2\}$. We have $\mathrm{Log}(K \setminus \{0\}) \subseteq E$. We let H be the hyperplane of $\mathbb{R}^n$ defined by $H = \{x \in \mathbb{R}^n : \sum_{i=1}^n x_i = 0\}$ and $\mathbf{1}$ be the vector, orthogonal to H, defined as $\mathbf{1} = (1, \ldots, 1)^T \in \mathbb{R}^n$. We write $\pi_H : \mathbb{R}^n \to H$ the orthogonal projection on H, parallel to $\mathbf{1}$. We define $\Lambda = \{\mathrm{Log}\, u, u \in R^\times\}$, which is a lattice of dimension $r_1 + r_2 - 1$ contained in $H \cap E$ (thanks to Dirichlet's unit theorem), called the *log-unit lattice*. We have the following upper bound:

$$\det(\Lambda) \cdot h_K \leq 2^{O(\log|\Delta| + n \log\log|\Delta|)} = 2^{\tilde{O}(\log|\Delta|)}. \tag{2.6}$$

This upper bound comes from the relation between $\det(\Lambda)$, h_K and the residue of the zeta-function ζ_K at $s = 1$ (see [Lou00]). The latter is known considering Λ defined by the logarithmic embedding $(\log|\sigma_1(x)|, \ldots, \log|\sigma_{r_1}(x)|, 2\log|\sigma_{r_1+1}(x)|, 2\log|\sigma_{r_1+r_2}(x)|)^T \in \mathbb{R}^{r_1+r_2}$. However, it can be seen that if one multiplies our lattice Λ by a matrix with blocks of the form $\begin{pmatrix} 1 & 1 \\ -1 & 1 \end{pmatrix}$ (in order to add and subtract $\log|\sigma_i(x)|$ and $\log|\sigma_{i+r_2}(x)|$ for $r_1 < i \leq r_2$), one obtains the log-unit lattice defined by the logarithmic embedding considered in [Lou00]. As multiplying by such a matrix increases the determinant by a factor 2^{r_2}, Inequality (2.6) remains valid in our setup.

This bound, combined with a lower bound on $\det(\Lambda)$ also gives Eq. (2.5). Indeed, using a result of Zimmert [Zim80], we have that $\det(\Lambda) > 0.02 \cdot 2^{-r_2}$ (handling again our unusual definition of Λ).

For any $x \in K$, there exists a unique vector $h \in H \cap E$ and a unique real number a such that $\mathrm{Log}\, x = h + a\mathbf{1}$. In the following, we recall relationships between (h, a) and x. These results are standard (e.g., they are used freely in [CDPR16, Section 6]).

Lemma 2.11. *Let $r \in K$. Then we have $\mathrm{Log}\, r = h + \frac{\log|\mathcal{N}(r)|}{n}\mathbf{1}$, for some $h \in H \cap E$.*

For the sake of completeness, and because we are using an unusual definition of Log, we give a proof of this result below.

Proof. Write $\mathrm{Log}\, r = h + a\mathbf{1}$ for some $h \in H \cap E$ and $a > 0$. First, as $\mathbf{1}$ is orthogonal to H, we have that $\langle \mathbf{1}, \mathrm{Log}\, r \rangle = \langle \mathbf{1}, a\mathbf{1} \rangle = a \cdot n$. But using the definition of $\mathrm{Log}\, r$, we also have that

$$\langle \mathbf{1}, \mathrm{Log}\, r \rangle = \sum_i \log|\sigma_i(r)| = \log|\mathcal{N}(r)|,$$

where we used the fact that $\mathcal{N}(r) = \prod_i \sigma_i(r)$. This completes the proof. $\square$

The following lemma gives a bound on the Euclidean norm of an element $r \in R$ in terms of its decomposition $\mathrm{Log}\, r = h + a\mathbf{1}$.

Lemma 2.12. *For any $r \in K$, if $\mathrm{Log}\, r = h + a\mathbf{1}$ with $h \in H \cap E$ and $a \in \mathbb{R}$, then we have $\|r\|_\infty \leq 2^a \cdot 2^{\|h\|_\infty}$. In particular, this implies that*

$$\|r\|_2 \leq \sqrt{n} \cdot 2^a \cdot 2^{\|h\|_2} = \sqrt{n} \cdot |\mathcal{N}(r)|^{1/n} \cdot 2^{\|h\|_2}.$$

Proof. The second inequality follows from the first one by using Eqs. (2.1) and (2.2) (and Lemma 2.11 for the equality). For the first inequality, recall that by definition of Log, we have that $(\mathrm{Log}\, r)_i = \log|\sigma_i(r)| = h_i + a$ for all i. So, by definition of $\|r\|_\infty = \max_i |\sigma_i(r)|$, we have $\|r\|_\infty = \max_i 2^{h_i+a} \leq 2^a \cdot 2^{\|h\|_\infty}$. $\square$

2.5   Algorithmic Problems Related to Class Group Computations

Let $\mathfrak{B} = \{\mathfrak{p}_1, \ldots, \mathfrak{p}_r\}$ be a set of prime integral ideals generating the class group, obtained for example using Corollary 2.10.

We will be interested in computing the lattice of all the relations between the ideals of $\mathfrak{B}$, i.e., the kernel of the map

$$f_{\mathfrak{B}} : e = (e_1, \ldots, e_r) \in \mathbb{Z}^r \mapsto [\prod_i \mathfrak{p}_i^{e_i}] \in Cl_K.$$

Recall that $\ker(f_{\mathfrak{B}})$ is a full-rank sub-lattice of $\mathbb{Z}^r$ of volume $|\mathbb{Z}^r / \ker(f_{\mathfrak{B}})| = |Cl_K| = h_K$. Let $N_{\mathfrak{B}} = \max_i \mathcal{N}(\mathfrak{p}_i)$. We let $T_{\mathrm{rel}}(N_{\mathfrak{B}}, r)$ denote the time needed to compute a basis of $\ker(f_{\mathfrak{B}})$, together with generators of the corresponding principal ideals, given as input the set $\mathfrak{B}$. We write $T_{\mathrm{decomp}}(N, N_{\mathfrak{B}}, r)$ for the time needed, given $\mathfrak{B}$ and a fractional ideal I of norm $\mathcal{N}(I) = N$, to find a vector $e \in \mathbb{Z}^r$ and an element $g \in K$ such that $I = \prod_i \mathfrak{p}_i^{e_i} \cdot \langle g \rangle$. Note that this decomposition always exists but might not be unique (we only require that $\mathfrak{B}$ generates the class group). Finally, we let $T_{\mathrm{log\text{-}unit}}$ be the time needed to compute a basis of the log-unit lattice of K.

The three problems above are usually solved by computing S-units[3] for a well-chosen set S. This is why, in the following, the same cost bounds hold for the three of them.

In the *quantum* setting, Biasse and Song [BS16] showed that these three problems can be solved in polynomial time for any number field (under GRH). More precisely, they showed that

- $T_{\mathrm{rel}}(N_{\mathfrak{B}}, r) = \mathrm{poly}(\log |\Delta|, r, \log N_{\mathfrak{B}})$;
- $T_{\mathrm{decomp}}(N, N_{\mathfrak{B}}, r) = \mathrm{poly}(\log |\Delta|, \log N_{\mathrm{num}}, \log N_{\mathrm{denom}}, r, \log N_{\mathfrak{B}})$;
- $T_{\mathrm{log\text{-}unit}} = \mathrm{poly}(\log |\Delta|)$;

where N_{num} and N_{denom} refer to the numerator and denominator of N (i.e., $N = N_{\mathrm{num}}/N_{\mathrm{denom}} \in \mathbb{Q}$ with $N_{\mathrm{num}}, N_{\mathrm{denom}}$ in $\mathbb{Z}_{>0}$ and coprime).

In the *classical* setting, these three problems can be solved heuristically in sub-exponential time (under GRH). The first sub-exponential algorithm for all number fields (and which allows n to tend to infinity with $\log |\Delta|$) is due to Biasse and Fieker [BF14]:

- $T_{\mathrm{rel}}(N_{\mathfrak{B}}, r) = \mathrm{poly}(r, \log N_{\mathfrak{B}}) \cdot 2^{\widetilde{O}((\log |\Delta|)^{2/3})}$;
- $T_{\mathrm{decomp}}(N, N_{\mathfrak{B}}, r) = \mathrm{poly}(\log N_{\mathrm{num}}, \log N_{\mathrm{denom}}, r, \log N_{\mathfrak{B}}) \cdot 2^{\widetilde{O}((\log |\Delta|)^{2/3})}$;
- $T_{\mathrm{log\text{-}unit}} = 2^{\widetilde{O}((\log |\Delta|)^{2/3})}$.

Biasse and Fieker actually claim $2^{O((\log |\Delta|)^{2/3+\varepsilon})}$ run-times. Tracing back the source of this ε leads to Biasse's [Bia14, Proposition 3.1]. A careful reading of the proof of the latter shows that the $(\log |\Delta|)^{\varepsilon}$ term is actually a power of $\log \log |\Delta|$,

---

[3] Given a set $S = \{\mathfrak{p}_1, \ldots, \mathfrak{p}_r\}$ of prime integral ideals, the S-units are the elements $\alpha \in K$ such that there exist $e_1, \ldots, e_r \in \mathbb{Z}$ with $\prod_i \mathfrak{p}_i^{e_i} = \langle \alpha \rangle$.

hence, in our notations, it is absorbed by the $\widetilde{O}$ notation. In addition to the GRH, the algorithm of Biasse and Fieker requires two heuristic assumptions, referred to as Heuristic 1 and Heuristic 3 in [BF14]. We recall these two heuristic assumptions below (see [BF14] for more details).

Heuristic 2 ([BF14, **Heuristic 1**]). The probability $P(x, y)$ that an integral ideal of R produced by the Biasse-Fieker [BF14] algorithm, of norm bounded by x, can be factored as a product of prime ideals of norms bounded by y satisfies

$$P(x, y) \geq e^{-(1+o_{x \to \infty}(1)) \cdot u \log u} \quad \text{for} \quad u = \frac{\log x}{\log y}.$$

Heuristic 3 ([BF14, **Heuristic 3**]). Given a set of r elements generating the class group, the algorithm only needs to find $r^{O(1)}$ relations between these elements to generate the full lattice of relations, with probability close to 1.

Smaller cost bounds are known for specific families of number fields. For prime-power cyclotomic fields, the $2^{\widetilde{O}((\log|\Delta|)^{2/3})}$ bounds can be replaced by $2^{\widetilde{O}((\log|\Delta|)^{1/2})}$ [BEF+17]. This algorithm is again heuristic and relies on the same assumptions as [BF14]. For real multiquadratic number fields, efficient classical algorithms allow to solve these three problems [BBV+17, BV18]. Finally, we note that the exponent 2/3 was recently lowered to 3/5 in [Gel17] and can even be decreased further in some cases.

3   From Ideal SVP to CVP in a Fixed Lattice

The main idea of our algorithm is, given an input ideal I, to find a principal ideal $\langle g \rangle \subseteq I$ with a short generator g. This is very similar to [CDW17], where the authors find a $2^{O(\sqrt{n})}$ approximation of a shortest non-zero vector of the ideal I by computing a principal ideal contained in I and then finding a short generator of this principal ideal. The limitation of this approach is that, if we consider any principal ideal contained in I, we cannot hope to find a better approximation than the $2^{O(\sqrt{n})}$ approximation obtained above in the worst case. This is due to the fact that in some principal ideals (including for prime-power cyclotomic fields), the shortest generator can be $2^{O(\sqrt{n})}$ times longer than a shortest non-zero element of the ideal (see [CDPR16]). Instead of looking for any principal ideal contained in I, we consider only those with a 'good' generator (i.e., a generator which is also a very short element of the corresponding principal ideal).

In order to find such an ideal, we merge the two steps of [CDPR16, CDW17] (consisting in first finding a principal multiple of I and then computing a small generator of the principal ideal), by introducing a lattice L that is very similar to the one used for class group computations. This lattice only depends on the number field (and not on the ideal I). We describe it in the next subsection. We then show how to express the problem of finding a principal multiple of I with a small generator as a CVP instance for this fixed lattice.

3.1   Definition of the Lattice L

In this subsection, we define the lattice L which we will use in order to transform our ideal-SVP instance into a CVPP' instance. We also give an algorithm to compute a basis of L and analyze its run-time. The lattice L we are considering is not new. It was already used in previous sub-exponential algorithms computing the class group of a number field [HM89, Buc88, BF14, BEF+17, Gel17]. However, these algorithms usually choose a sub-exponential set of ideals, hence resulting in a lattice L of sub-exponential dimension. Our lattice L will have a dimension which is polynomial in $\log |\Delta|$.

In the following, we fix some integer r such that $\log h_K \le r$ and $r \le \mathrm{poly}(\log |\Delta|)$ (looking forward, the integer r will be related to the dimension of the lattice in which we will solve CVPP', so it would be undesirable to set it too large). Let us also fix a set of prime integral ideals $\mathfrak{B} = \{\mathfrak{p}_1, \ldots, \mathfrak{p}_r\}$ as given by Corollary 2.10. We consider the lattice L of dimension $\nu := r + r_1 + r_2 - 1$, generated by the columns of the following matrix:

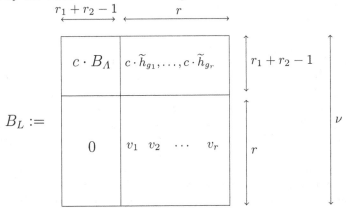

where:

- the scaling parameter $c > 0$ is to be chosen later;
- the matrix $B_\Lambda = (f_{H \cap E}(b_1), \ldots, f_{H \cap E}(b_{r_1+r_2-1}))$ is a basis of $f_{H \cap E}(\Lambda)$, where Λ is the log-unit lattice and $f_{H \cap E} : H \cap E \subset \mathbb{R}^n \to \mathbb{R}^{r_1+r_2-1}$ is an isometry;[4]
- the matrix consisting of the vectors $v_i = (v_{1i}, \ldots, v_{ri})^T$ is a basis of $\ker(f_{\mathfrak{B}})$ (in particular, the ideals $\prod_j \mathfrak{p}_j^{v_{ji}}$ are principal for all i);
- the column vectors $\tilde{h}_{g_i}$ are of the form $f_{H \cap E}(\pi_H(\mathrm{Log}\, g_i))$ for $g_i \in K$ a generator of the fractional principal ideal associated with the relation v_i, i.e., we have $\prod_j \mathfrak{p}_j^{v_{ji}} = \langle g_i \rangle$.

We will explain how to construct L below. This lattice enjoys the following property, which will be used later.

---

[4] As Λ is not full rank in $\mathbb{R}^n$, we change the ambient space such that $f_{H \cap E}(\Lambda)$ becomes full rank in $H \cap E = \mathbb{R}^{r_1+r_2-1}$. Note however that the ℓ_2-norm is preserved by this transformation (this is not the case for the ℓ_1 and ℓ_∞ norms).

Lemma 3.1. *Let w be a vector of L and parse it as $w = (h, v)^T$ with h of dimension $r_1 + r_2 - 1$ and $v = (v_1, \ldots, v_r)$ of dimension r. Then there exists an element $g \in K \setminus \{0\}$ such that $h = c \cdot f_{H \cap E}(\pi_H(\text{Log } g))$ and $\prod_j \mathfrak{p}_j^{v_j} = \langle g \rangle$.*

Proof. We first observe that the result holds for the vectors of the basis B_L. For the r vectors on the right of B_L, this holds by construction. For the $r_1 + r_2 - 1$ vectors on the left, we have that $\prod_j \mathfrak{p}_j^0 = R = \langle u \rangle$ for any unit $u \in R$. So by definition of B_Λ, the property of Lemma 3.1 also holds for the $r_1 + r_2 - 1$ first vectors of B_L.

To complete the proof, it suffices to observe that the property of Lemma 3.1 is preserved by addition (if g_1 corresponds to a vector w_1 and g_2 corresponds to a vector w_2, then $g_1 g_2$ corresponds to the vector $v_1 + v_2$) and by multiplication by -1 (if g corresponds to a vector w, then g^{-1} corresponds to the vector $-w$). All these elements g are invertible as they are obtained by multiplying and inverting non-zero elements of K. □

3.2   Computation of the Lattice L

The lattice L described above only depends on the number field we are working on. A basis of it can then be computed in a pre-processing phase, before the knowledge of the ideal in which we want to find a short non-zero vector. In this subsection, we give an algorithm to compute the lattice L and we show that this algorithm can be performed in time at most exponential in $\log |\Delta|$. As we shall see, this will even be sub-exponential in $\log |\Delta|$. The costly part of the pre-processing phase will be the pre-processing used for the CVPP algorithm.

Algorithm 3.1. Computes a basis B_L as described above

Input: A number field K and an integer $r = \text{poly}(\log |\Delta|)$ such that $\log h_K \leq r$.
Output: The basis B_L described in Section 3.1.
1: Compute the set $\mathfrak{B}'$ of all prime ideals of algebraic norm $\leq 12 \log^2 |\Delta|$.
2: Compute all the relations between the elements of $\mathfrak{B}'$ and the log-unit lattice Λ.
3: Use the relations to extract a set $\mathfrak{B}'' \subseteq \mathfrak{B}'$ generating the class group with $|\mathfrak{B}''| \leq \log h_K$.
4: Compute the set $\mathfrak{P}$ of all prime ideals of norms smaller than some $\text{poly}(\log |\Delta|)$ (choose the bound so that $|\mathfrak{P}| > r$).
5: Create a set $\mathfrak{B}$ by adding to $\mathfrak{B}''$ ideals taken uniformly in $\mathfrak{P}$, until the cardinality of $\mathfrak{B}$ reaches r.
6: Compute a basis of $\ker(f_\mathfrak{B})$ and generators g_i of the fractional principal ideals corresponding to the relations computed.
7: Create the matrix B_L from these r relations, the corresponding g_i and the log-unit lattice Λ computed at Step 2.
8: **return** B_L.

Lemma 3.2. *Assume GRH. Then Algorithm 3.1 outputs a matrix B_L as described above, in time at most*

$$T_{\text{log-unit}} + 2 \cdot T_{\text{rel}}(\text{poly}(\log |\Delta|), \text{poly}(\log |\Delta|)) + \text{poly}(\log |\Delta|).$$

Proof. We analyze the cost of each step of the algorithm, and provide some details for the correctness when needed.

Step 1. We have already seen in Sect. 2 that computing all prime ideals of norm $\leq 12 \log^2 |\Delta|$ can be performed in time polynomial in $\log |\Delta|$. There are $\text{poly}(\log |\Delta|)$ such ideals.

Step 2. Computing all the relations between the elements of $\mathfrak{B}'$ and the log-unit lattice Λ can be performed in time at most $T_{\text{rel}}(\text{poly}(\log |\Delta|), \text{poly}(\log |\Delta|)) + T_{\text{log-unit}}$. The relations between the elements of $\mathfrak{B}'$ are represented as an invertible matrix (whose columns span the kernel of the function $f_{\mathfrak{B}'}$ defined in Sect. 2.5).

Step 3. Extracting a generating set $\mathfrak{B}''$ from $\mathfrak{B}'$ of cardinality at most $\log h_K$ can be done using Lemma 2.7. Because we already have the matrix of relations between the elements of $\mathfrak{B}'$ (and because the size of this matrix is polynomial in $\log |\Delta|$), this can be done in polynomial time (as stated in the lemma).

Step 4. As in Step 1, this can be done in polynomial time, because the bound on the norms of the ideals is polynomial. We obtain a set $\mathfrak{P}$ whose cardinality is polynomial in $\log |\Delta|$.

Step 5. Picking uniform elements in a set of polynomial size can be done efficiently, so this step can be performed in polynomial time (recall that $r = \text{poly}(\log |\Delta|)$). Note that in the previous step, we had that the cardinality of $\mathfrak{B}'$ was at most $\log h_K \leq r$, so we can indeed add ideals to it to reach a set of cardinality r.

Step 6. As in Step 2, computing the kernel of $f_{\mathfrak{B}}$ can be done in time at most $T_{\text{rel}}(\text{poly}(\log |\Delta|), \text{poly}(\log |\Delta|))$. Together with the relations, we also get generators of the corresponding principal ideals.

Step 7. Finally, to compute the matrix B_L, we just need to compute the functions π_H and $f_{H \cap E}$ on the g_i's computed in Step 6. We then put it together with the matrix of relations computed in Step 6 and the log-unit lattice computed in Step 2. This can be done in polynomial time. $\qquad\square$

3.3   From SVP in Ideal Lattices to CVP in L

We now explain how to transform the problem of finding a short non-zero vector in a fractional ideal I of R, into solving a CVP instance with respect to the lattice L described in Sect. 3.1. As explained above, the main idea is to multiply the ideal I by ideals of the set $\mathfrak{B}$, until we obtain a 'good' principal ideal (i.e., with a short generator). In terms of lattice operations in L, our initial lattice I will give us a target vector in the real vector space spanned by L. Multiplying it by ideals of $\mathfrak{B}$ will be the same as adding to the target a vector of the lattice L. Finally, checking whether the resulting ideal is a good principal ideal can be done by checking whether the obtained vector is short. Overall, we are indeed solving a CVP instance in L. We first describe the algorithm, and then prove its correctness and bound its cost.

Algorithm 3.2. Solves ideal SVP using an oracle to solve CVP in L

Input: A non-zero fractional ideal $I \subseteq R$ (given by some basis), the basis B_L defined above and some parameter $\beta = \beta(n) > 0$.
Output: A somehow short non-zero element in I.
1: Compute $v_1, \ldots, v_r \in \mathbb{Z}$ and $g \in K$ such that $I = \prod_j \mathfrak{p}_j^{v_j} \cdot \langle g \rangle$.
2: Let $t = (-c \cdot f_{H \cap E}(h_g), v_1 + \beta, \ldots, v_r + \beta)^T$, where $h_g = \pi_H(\mathrm{Log}\, g)$.
3: Compute $w \in L$ such that $\|t - w\|_\infty \leq \beta$ (see Section 4).
4: Let $g' \in K$ be the element associated to w as in Lemma 3.1.
5: **return** $g \cdot g'$.

Theorem 3.3. *Let us fix $c = n^{1.5}/r$. Let $\beta = \beta(n) > 0$. Then, for any non-zero fractional ideal I of R, Algorithm 3.2 runs in time at most*

$$T_{\mathrm{decomp}}(\mathcal{N}(I), \mathrm{poly}(\log|\Delta|), \mathrm{poly}(\log|\Delta|)) + T_{\mathrm{CVP}}(\infty, L, \beta) + \mathrm{poly}(\log|\Delta|)$$

and outputs a non-zero element $x \in I$ such that $\|x\|_2 \leq 2^{O(\frac{\beta \cdot r \cdot \log\log|\Delta|}{n})} \cdot \mathcal{N}(I)^{1/n}$.

Observe that in the statement of the run-time, the term $T_{\mathrm{CVP}}(\infty, L, \beta)$ will be infinite if β is smaller than $\mu^{(\infty)}(L)$ (no algorithm can find a point of L at distance at most β given any target input). In this case, the run-time of our algorithm might also be infinite (i.e. the algorithm fails).

Proof. **Correctness.** Let us define the fractional ideal $J = \langle g \cdot g' \rangle$. This will be our 'good' principal ideal, i.e., a principal ideal with a small generator, and contained in I. Let us first prove that J is a multiple of I. By Lemma 3.1, we have $w = (c \cdot f_{H \cap E}(\pi_H(\mathrm{Log}\, g')), v'_1, \ldots, v'_r)^T$ with $\langle g' \rangle = \prod_j \mathfrak{p}_j^{v'_j}$. We can then write

$$J = I \cdot \prod_j \mathfrak{p}_j^{-v_j} \cdot \langle g' \rangle \qquad \text{by definition of } g \text{ and the } v_j\text{s}$$

$$= I \cdot \prod_j \mathfrak{p}_j^{-v_j} \cdot \prod_j \mathfrak{p}_j^{v'_j} \qquad \text{by Lemma 3.1}$$

$$= I \cdot \prod_j \mathfrak{p}_j^{v'_j - v_j}.$$

Further, we know that $\|t - w\|_\infty \leq \beta$, and hence we have $v_j \leq v'_j \leq v_j + 2\beta$ for all j. In particular, we have that $v'_j - v_j \geq 0$ and so the ideal $\prod_j \mathfrak{p}_j^{v'_j - v_j}$ is an integral ideal. We conclude that J is contained in I, and in particular $g \cdot g'$ is indeed an element of I. Also, because $g' \neq 0$ (see Lemma 3.1) and $g \neq 0$ (we chose I to be non-zero), then $g \cdot g'$ is a non-zero element of I.

Let us now show that $g \cdot g'$ is short. We will do so by using Lemma 2.12. Let $\mathrm{Log}\, g = h_g + a_g \mathbf{1}$ and $\mathrm{Log}\, g' = h_{g'} + a_{g'} \mathbf{1}$ with h_g and $h_{g'} \in H \cap E$ (note that because $g, g' \in K$, we do not necessarily have $a_g, a_{g'} > 0$). We then have that $\mathrm{Log}(gg') = (h_g + h_{g'}) + (a_g + a_{g'})\mathbf{1}$. By Lemma 2.12, we know that $\|gg'\|_2 \leq \sqrt{n} \cdot |\mathcal{N}(gg')|^{1/n} \cdot 2^{\|h_g + h_{g'}\|_2}$. Therefore, it suffices to bound the two terms $|\mathcal{N}(gg')|^{1/n}$ and $\|h_g + h_{g'}\|_2$.

Let us start by $|\mathcal{N}(gg')|^{1/n}$. By multiplicativity of the algebraic norm, we have that $|\mathcal{N}(gg')|^{1/n} = \mathcal{N}(J)^{1/n} = \mathcal{N}(I)^{1/n} \cdot \prod_j \mathcal{N}(\mathfrak{p}_j)^{\frac{v'_j - v_j}{n}}$. We have chosen the ideals $\mathfrak{p}_j$ with polynomially bounded algebraic norms, and we have seen that $0 \leq v'_j - v_j \leq 2\beta$. Thus, we obtain that $\mathcal{N}(\mathfrak{p}_j)^{\frac{v'_j - v_j}{n}} = 2^{O(\frac{\beta \log \log |\Delta|}{n})}$. By taking the product of the r ideals $\mathfrak{p}_j$, we obtain

$$|\mathcal{N}(gg')|^{1/n} = \mathcal{N}(I)^{1/n} \cdot 2^{O(\frac{\beta \cdot r \cdot \log \log |\Delta|}{n})}.$$

We now consider the term $\|h_g + h_{g'}\|_2$. Recall that $\|w - t\|_\infty \leq \beta$, so in particular, if we consider only the first $r_1 + r_2 - 1$ coefficients of the vectors, we have that $\|c \cdot f_{H \cap E}(h_{g'}) + c \cdot f_{H \cap E}(h_g)\|_\infty \leq \beta$. And if we consider the ℓ_2-norm, we obtain $\|f_{H \cap E}(h_{g'}) + f_{H \cap E}(h_g)\|_2 \leq \sqrt{n}\beta/c$. Using the fact that the ℓ_2-norm is invariant by $f_{H \cap E}$, we conclude that $\|h_{g'} + h_g\|_2 \leq \sqrt{n}\beta/c$.

Finally, combining the two upper bounds above and replacing c by $n^{1.5}/r$, we obtain that

$$\|gg'\|_2 \leq \sqrt{n} \cdot 2^{O(\frac{\beta \cdot r \cdot \log \log |\Delta|}{n})} \cdot \mathcal{N}(I)^{1/n}.$$

Cost. Step 1 can be performed in time $T_{\mathrm{decomp}}(\mathcal{N}(I), \mathrm{poly}(\log |\Delta|), \mathrm{poly}(\log |\Delta|))$. Step 2 can be performed in polynomial time. Step 3 uses a CVP solver and can be done in time $T_{\mathrm{CVP}}(\infty, L, \beta)$. Finally, Step 4 only consists in recovering g' from the vector w, it can be done in polynomial time. Note that for this last step, if we only have the vector w, then we know $\pi_H(\mathrm{Log}\, g')$, but it might not be possible to recover g' from it. On the other hand, the lower part of the vector w also gives us the ideal $\langle g' \rangle$, but then computing g' from it would be costly. In order to perform this step in polynomial time, when creating the matrix B_L we keep in memory the elements g_i corresponding to the different columns. Then, when we obtain w, we only have to write it as a linear combination of the vectors of B_L and we can recover g' as a product of the g_i's. This can also be done in polynomial time. □

Combining Algorithm 3.2 with the pre-processing phase (i.e., computing B_L with Algorithm 3.1 and pre-processing it for approx-CVPP'), we obtain the following theorem.

Theorem 3.4. *Let K be any number field of dimension n and discriminant Δ. Let $\alpha \in [0,1]$, $r = \mathrm{poly}(\log|\Delta|)$ be such that $\log h_K \le r$, and $\nu := r + r_1 + r_2 - 1$. Then, under GRH, there exist two algorithms $A_{pre\text{-}proc}$ and A_{query} such that*

- *Algorithm $A_{pre\text{-}proc}$ takes as inputs the field K and a basis of its ring of integers R, runs in time*

$$T_{\mathrm{CVP}}^{\mathrm{pre\text{-}proc}}(\infty, L, \nu^\alpha) + T_{\mathrm{log\text{-}unit}} + 2 \cdot T_{\mathrm{rel}}(\mathrm{poly}(\log|\Delta|), \mathrm{poly}(\log|\Delta|)) + \mathrm{poly}(\log|\Delta|)$$

  and outputs a hint w of bit-size at most $T_{\mathrm{CVP}}^{\mathrm{query}}(\infty, L, \nu^\alpha)$;
- *Algorithm A_{query} takes as inputs the hint w output by $A_{pre\text{-}proc}$ and any fractional ideal I of R such that the numerator and denominator of $\mathcal{N}(I)$ have bit-sizes bounded by $\mathrm{poly}(\log|\Delta|)$; it runs in time*

$$T_{\mathrm{decomp}}(\mathcal{N}(I), \mathrm{poly}(\log|\Delta|), \mathrm{poly}(\log|\Delta|)) + T_{\mathrm{CVP}}^{\mathrm{query}}(\infty, L, \nu^\alpha) + \mathrm{poly}(\log|\Delta|)$$

  and outputs a non-zero element $x \in I$ such that

$$\|x\|_2 \le 2^{O\left(\frac{\nu^\alpha \cdot r \cdot \log\log|\Delta|}{n}\right)} \cdot \lambda_1^{(2)}(I).$$

The lattice L is as defined in Sect. 3.1 and only depends on the field K. The memory consumption of both algorithms is bounded by their run-times.

Note that we used the fact that $\lambda_1^{(2)}(I) \ge \lambda_1^{(\infty)}(I) \ge \mathcal{N}(I)^{1/n}$ (see Inequality (2.4)) to replace the $\mathcal{N}(I)^{1/n}$ term in Theorem 3.3 by $\lambda_1^{(2)}(I)$.

4   Solving CVP' with Pre-processing

In this section, we describe a possible way of solving approx-CVP' in the lattice L defined previously. Even if our lattice L has some structure, it does not seem easy to solve approx-CVP' in it (not necessarily easier than solving the approx-SVP instance directly for the initial lattice I). However, the lattice L only depends on the field K and not on the ideal I. Hence, in this section, we focus on solving approx-CVP' with pre-processing on the lattice (to which we refer as CVPP'). Combining it with the result of Sect. 3, this will provide an algorithm to solve approx-SVP in ideals, with pre-processing on the field K.

4.1   Properties of the Lattice L

Recall that our lattice L is given by the basis matrix $B_L = \begin{pmatrix} c \cdot B_\Lambda & A_{\mathfrak{B}} \\ 0 & R_{\mathfrak{B}} \end{pmatrix} \in \mathbb{R}^{\nu \times \nu}$, where we let $A_{\mathfrak{B}}$ denote the top-right block of B_L consisting of the vectors $c \cdot \widetilde{h}_{g_i}$, and $R_{\mathfrak{B}}$ be the bottom-right block of B_L containing the relations of the elements of $\mathfrak{B}$. Recall that $R_{\mathfrak{B}}$ is a basis of the kernel of $f_{\mathfrak{B}} : (e_1, \ldots, e_r) \in \mathbb{Z}^r \mapsto [\prod_j \mathfrak{p}_j^{e_j}] \in Cl_K$. Hence we have $\det(R_{\mathfrak{B}}) = |\mathbb{Z}^r / \ker(f_{\mathfrak{B}})| = h_K$.

Equation (2.6) gives that $\det(\Lambda) \cdot h_K \leq 2^{O(\log|\Delta| + n \log\log|\Delta|)}$. Hence, we have that $\det(L) = c^{r_1 + r_2 - 1} \cdot 2^{O(\log|\Delta| + n \log\log|\Delta|)}$. We chose $c = n^{1.5}/r$ in Theorem 3.3. We then obtain the following upper bound on $\det(L)$:

$$\det(L) = \left(\frac{n^{1.5}}{r}\right)^{r_1 + r_2 - 1} \cdot 2^{O(\log|\Delta| + n \log\log|\Delta|)} = 2^{O(\log|\Delta| + n \log\log|\Delta|)}.$$

We still have some freedom for the choice of the parameter r (and hence the dimension $\nu = r + r_1 + r_2 - 1$ of the lattice L), as long as $\log h_K \leq r$. We will choose it sufficiently large to ensure that the root determinant of L is at most constant. On the other hand, the dimension of L should be as small as possible as it impacts the cost of the CVP computations. We fix

$$r = \max(\log h_K, \log|\Delta| + n \log\log|\Delta|).$$

This choice of r satisfies $r \geq \log h_K$ and $\det(L)^{1/\nu} \leq O(1)$. Note that as $\log h_K = \widetilde{O}(\log|\Delta|)$ (see Eq. (2.5)), we have $r \leq \widetilde{O}(\log|\Delta|)$.

In the following, we view the lattice L as random, where the randomness comes from the choice of the set $\mathfrak{B}$ (the initial set $\mathfrak{B}''$ in Algorithm 3.1 is fixed, but then we add to it random prime ideals of polynomially bounded norms to create the set $\mathfrak{B}$). If the created lattice L does not satisfy the conditions we want, we can try another lattice by sampling a new set $\mathfrak{B}$. As we chose r so that $\det(L)^{1/\nu} = O(1)$, we know by Minkowski's inequality that $\lambda_1^{(\infty)}(L) = O(1)$. Then, because L is somehow random, we also expect that all successive minima $\lambda_i^{(\infty)}(L)$ and the covering radius in infinity norm are constant. Hence, we expect to be able to take β as small as $O(1)$ in Algorithm 3.2. We summarize this assumption below.

Heuristic 4. With good probability over the choice of $\mathfrak{B}$, the ℓ_∞-norm covering radius of L satisfies $\mu^{(\infty)}(L) = O(1)$ (and hence $\mu^{(2)}(L) = O(\sqrt{\nu})$).

This heuristic calls for a few comments. First, it is better analyzed as two separate conjectures, one on the log-unit lattice, the other one on the class group lattice. Concerning the latter, assume that the class number is a prime p. Then we can choose $\mathfrak{p}_1$ to be a generator of the class group, and the relation matrix is of the form

$$\begin{pmatrix} p & a_1 & a_2 & \dots & a_r \\ 0 & 1 & 0 & \dots & 0 \\ 0 & 0 & 1 & & 0 \\ \vdots & \vdots & & \ddots & \vdots \\ 0 & 0 & 0 & \dots & 1 \end{pmatrix},$$

where the a_i's in $[0, p-1]$ characterize the elements of the ideal class group. In our setting where each $\mathfrak{p}_i$ (for $i \geq 2$) is picked randomly among small prime ideals, we can thus reasonably assume that the a_i's are uniformly distributed in $[0, p-1]$. Hence, for $(e_i)_{2 \leq i \leq r} \in [-B, B]^{r-1}$ for some constant $B \geq 1$, we can expect that one among the $(2B + 1)^{r-1} = p^c$ (with $c > 1$) fractional ideals

$\prod_{i\geq 2}\mathfrak{p}_i^{e_i}$ is in a class $[\mathfrak{p}_1]^a$ for some $a = O(1)$, which implies that the ℓ_∞-norm covering radius is $O(1)$.

The general case is analogous to this first intuition. Let $\mathfrak{B} = \{\mathfrak{p}_1,\ldots,\mathfrak{p}_s, \mathfrak{p}_{s+1},\ldots,\mathfrak{p}_r\}$ with $\{\mathfrak{p}_1,\ldots,\mathfrak{p}_s\}$ the prime ideals coming from the set $\mathfrak{B}''$ (hence fixed) and $\{\mathfrak{p}_{s+1},\ldots,\mathfrak{p}_r\}$ the ideals uniformly chosen among prime ideals of norm bounded by some polynomial. Because the set $\mathfrak{B}''$ generates the class group, we can find a basis of L of the following form, by taking the HNF matrix for the bottom-right part of B_L.

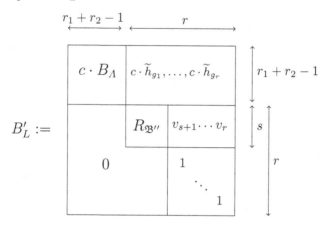

In this matrix, the block matrices B_Λ and $R_{\mathfrak{B}''}$ are fixed, as well as the vectors $\tilde{h}_{g_i}$ for i in $\{1,\ldots,s\}$. However, the vectors v_i and $\tilde{h}_{g_i}$ for $s < i \leq r$ depend on our choices of $\{\mathfrak{p}_{s+1},\ldots,\mathfrak{p}_r\}$. The vectors of $\mathbb{Z}^s/R_{\mathfrak{B}''}$ are in bijection with the elements of the ideal class group (because $\mathfrak{B}''$ generates the class group). So if we assume that the class of a uniform prime ideal of norm polynomially bounded is uniform in the class group, then we would have that the vectors v_i of the matrix above are uniform in $\mathbb{Z}^s/R_{\mathfrak{B}''}$. In a similar way, we will assume that the vectors $\tilde{h}_{g_i}$ are somehow uniform in $\mathbb{R}^{r_1+r_2-1}/\Lambda$ (recall that they correspond to the projection over H of $\mathrm{Log}\,g_i$ for g_i a generator of the principal ideal associated with the lower part of the vector). Let us now explain why, given any target vector $t \in \mathbb{R}^\nu$, we expect to find a vector $v \in L$ at distance $O(1)$ from t. Write $t = (c \cdot \tilde{h}, v, w)^T$ with $\tilde{h}$ of dimension $r_1 + r_2 - 1$, v of dimension s and w of dimension $r - s$. We can assume, without loss of generality, that $|w_i| < 1/2$ for all i (using the last $r - s$ columns of B'_L to reduce t if needed). By taking the subset sums of the last $r - s$ columns of B'_L, we obtain 2^{r-s} vectors of L of the form $t' = (c \cdot \tilde{h}', v', w')^T$, with $w' \in \{0,1\}^{r-s}$. Because we assumed that the v_i and $\tilde{h}_{g_i}$ for $s < i \leq r$ were somehow uniform modulo $R_{\mathfrak{B}''}$ and Λ respectively, we also expect the vectors $\tilde{h}'$ and v' created above to be somehow uniform modulo $R_{\mathfrak{B}''}$ and Λ. Recall that we chose r so that $(\det(c\Lambda) \cdot \det(R_{\mathfrak{B}''}))^{1/r} = O(1)$, hence the volume of $c\Lambda$ and $R_{\mathfrak{B}''}$ satisfies $\det(c\Lambda) \cdot \det(R_{\mathfrak{B}''}) \leq 2^{O(r)}$. We can then assume that we have $2^{r-s} > \det(c\Lambda) \cdot \det(R\mathfrak{B}'')$ (if needed, we can multiply r by a constant factor, which will not change the asymptotics). This means that we expect to find one of the 2^{r-s} vector $t' = (c \cdot \tilde{h}', v', w')^T$ satisfying

$\|(c \cdot \widetilde{h}, v) - (c \cdot \widetilde{h}', v')\|_\infty = O(1)$. And because $|w_i| < 1/2$ and $w_i' \in \{0, 1\}$ we also have $\|t - t'\|_\infty = O(1)$.

We experimentally computed the lattice L for some cyclotomic fields (using Algorithm 3.1). For each lattice L, we then computed an empirical covering radius. To do so, we picked 21 random target vectors t_i in the real span of the lattice. These vectors were sampled from a continuous Gaussian distribution with standard deviation $\sigma = 100$. We then solved the CVP instances in L for these target vectors t_i and let v_i be a closest vector in L (for the ℓ_2-norm). We defined $\widetilde{\mu}^{(2)}(L)$ to be $\max_i \|t_i - v_i\|_2$ and $\widetilde{\mu}^{(\infty)}(L)$ to be $\max_i \|t_i - v_i\|_\infty$.[5] The approximated values of $\mu^{(2)}(L)$ and $\mu^{(\infty)}(L)$ are given in Fig. 3. We observe that, while $\widetilde{\mu}^{(2)}(L)$ increases with the dimension (we expect that it increases as $\sqrt{\nu}$), the approximate covering radius in ℓ_∞-norm $\widetilde{\mu}^{(\infty)}(L)$ seems to remain constant around 1. These experimental results are consistent with Heuristic 4. The code is available as supplementary material.

Conductor of K	Dimension of L	$\widetilde{\mu}^{(2)}(L)$	$\widetilde{\mu}^{(\infty)}(L)$
18	9	1.13	0.755
16	16	1.50	0.899
36	28	1.79	0.795
40	41	2.15	0.893
48	42	2.19	0.840
32	44	2.26	0.794
27	49	2.36	0.901
66	54	2.47	0.989
44	57	2.53	0.815
70	67	2.72	1.03
84	68	2.74	1.27
90	68	2.71	0.814
78	70	2.81	0.882
72	73	2.90	1.00

Fig. 3. Approximate covering radii in ℓ_2 and ℓ_∞ norms for the lattice L, for cyclotomic number fields of different conductors.

4.2   Using Laarhoven's Algorithm

We now consider Laarhoven's algorithm, which solves approx-CVPP in Euclidean norm (we only found algorithms for CVPP with Euclidean norm in the literature, and not for infinity norm). Recall from Sect. 2 that, for a ν-dimensional lattice L, Laarhoven's (heuristic) algorithm gives, for any $\alpha \in [0, 1/2]$:

---

[5] As we solved CVP in L for the ℓ_2-norm, the quantity $\mu^{(\infty)}(L)$ may be over-estimated, but this should not be over-estimated by too much. Further, as we want an upper bound on $\mu^{(\infty)}(L)$, this is not an issue.

$$T_{\text{CVP}}^{\text{pre-proc}}(2, L, O(\nu^\alpha) \cdot \mu^{(2)}(L)) = 2^{O(\nu)},$$
$$T_{\text{CVP}}^{\text{query}}(2, L, O(\nu^\alpha) \cdot \mu^{(2)}(L)) = 2^{\widetilde{O}(\nu^{1-2\alpha})}.$$

As we have assumed (Heuristic 4) that $\mu^{(2)}(L) = O(\sqrt{\nu})$ with good probability over the choice of L, this implies that Laarhoven's algorithm achieves

$$T_{\text{CVP}}^{\text{pre-proc}}(2, L, O(\nu^{1/2+\alpha})) = 2^{O(\nu)} \quad \text{and} \quad T_{\text{CVP}}^{\text{query}}(2, L, O(\nu^{1/2+\alpha})) = 2^{\widetilde{O}(\nu^{1-2\alpha})}.$$

We now have an algorithm that, given any input $t \in \text{Span}(L)$, outputs a vector $v \in L$ such that $\|t - v\|_2 \leq O(\nu^{1/2+\alpha})$, while we would like to have $\|t - v\|_\infty \leq O(\nu^\alpha)$. But when we take a random vector of Euclidean norm bounded by $O(\nu^{1/2+\alpha})$, we expect that with good probability, its coefficients are somehow balanced. Hence we expect its infinity norm to be approximately $\sqrt{\nu}$ times smaller than its Euclidean norm. This is the meaning of the following heuristic assumptions.

First, because we want the output of Laarhoven's algorithm to be somehow random, we argue that we can randomize the input vector of the CVPP algorithm.

Heuristic 5. We assume that in our algorithm, the target vector t given as input to Laarhoven's algorithm behaves like a random vector sampled uniformly in $\text{Span}(L)/L$.

This assumption that t is distributed uniformly in $\text{Span}(L)/L$ may be justified by the fact that, in Algorithm 3.2, we can somehow randomize our target vector t by multiplying our initial fractional ideal I by an integral ideal of small algebraic norm (statistically independent of the $\mathfrak{p}_i$'s chosen for $\mathfrak{B}$).

Heuristic 6. With non-negligible probability over the input target vector t, distributed uniformly in $\text{Span}(L)/L$, the vector v output by Laarhoven's algorithm satisfies $\|t - v\|_\infty \leq \widetilde{O}(\|t - v\|_2/\sqrt{\nu})$.

In order to motivate Heuristic 6, we recall that if a vector is drawn uniformly at random on a sphere, then its ℓ_∞-norm is smaller than its ℓ_2-norm by a factor $O(\log n/\sqrt{n})$, with good probability.

Lemma 4.1. *Let x be sampled uniformly on the unit sphere S^{n-1} in $\mathbb{R}^n$. Then $\Pr(\|x\|_\infty \geq \frac{\sqrt{8}\ln n}{\sqrt{n}}) \leq O(\frac{1}{\sqrt{\ln n}})$.*

Proof. Sampling x uniformly in S^{n-1} is the same as sampling y from a centered spherical (continuous) Gaussian distribution of parameter 1 and then normalizing it by setting $x = \frac{y}{\|y\|_2}$. So we have $\|x\|_\infty = \frac{\|y\|_\infty}{\|y\|_2}$, and it is sufficient to find an upper bound on $\|y\|_\infty$ and a lower bound on $\|y\|_2$. We know that for a centered spherical Gaussian distribution of parameter 1, we have $\Pr(\|y\|_\infty > 2\ln n) = \Pr(\exists i : |y_i| > 2\ln n) \leq \frac{1}{2\sqrt{2\pi\ln n}}$. Moreover, we also have that $\Pr(\|y\|_2 < \sqrt{n/2}) \leq e^{-n/8}$ (see for instante [LM00, Lemma 1]). By the union bound, we finally obtain that $\Pr(\|y\|_\infty/\|y\|_2 > \frac{\sqrt{8}\ln n}{\sqrt{n}}) \leq O(\frac{1}{\sqrt{\ln n}})$. $\square$

Note that the proof also shows that for a continuous Gaussian vector y of dimension n, $\|y\|_\infty/\|y\|_2 = O(\log n/\sqrt{n})$ with good probability. We also have experimental results corroborating Heuristic 6. We implemented our algorithm in Magma, both the generation of the lattice L and the CVP phase using Laarhoven's algorithm (the code is available as supplementary material). We tested our implementation for different cyclotomic fields. The maximum conductor achieved was 90. The maximum dimension of the lattice L that we achieved was 73, for a cyclotomic field of conductor 72. For these cyclotomic fields, we computed the lattice L. Then, we sampled target vectors t in the real span of L, using a Gaussian distribution of parameter $\sigma = 100$, and we ran Laarhoven's CVP algorithm to obtain a vector $v \in L$. We then computed the ratios $\frac{\|t-v\|_2}{\|t-v\|_\infty}$, which we expect to be around $O(\sqrt{\nu}/\log \nu)$. Because we are working in small dimensions, the $\log \nu$ term has a non-negligible impact. So, instead of plotting $\log(\frac{\|t-v\|_2}{\|t-v\|_\infty})$ as a function of $\log \nu$, we compared our ratios with the ones we would have obtained if the vectors were Gaussian vectors. On Fig. 4, the blue dots represent the logarithms of the ratios $\frac{\|t-v\|_2}{\|t-v\|_\infty}$ obtained when choosing a random Gaussian vector t as input of our algorithm. For every fixed conductor, we have several vertically aligned points, because we tried Laarhoven's algorithm for different approximation factors (i.e., different choices of α). The green '+' are obtained by computing $\log(\|x\|_2/\|x\|_\infty)$ for some Gaussian vectors of dimension ν. The red crosses are obtained by taking the median point of a large number of green '+' (not all of them are plotted on the figure).

We observe that the ratios obtained with our algorithm are well aligned with the red crosses. Moreover, even if we have some variance within the blue dots, it is comparable to the variance observed within the green '+'. So Heuristic 6 seems consistent with our empirical experiments (recall that Gaussian vectors provably satisfy Heuristic 6 with good probability).

We conclude that, under Heuristics 4, 5 and 6, and Heuristic 1 present in [Laa16], for any $\alpha \in [0, 1/2]$, Laarhoven's algorithm solves approx-CVPP' with

$$T_{\mathrm{CVP}}^{\mathrm{pre\text{-}proc}}(\infty, L, \nu^\alpha) = 2^{O(\nu)} \text{ and } T_{\mathrm{CVP}}^{\mathrm{query}}(\infty, L, \nu^\alpha) = 2^{\widetilde{O}(\nu^{1-2\alpha})}. \qquad (4.1)$$

5   Summary

We now instantiate Theorem 3.4 with $\nu = \widetilde{O}(\log \Delta)$ and the values given in Sect. 2 and in Eq. (4.1) for $T_{\text{log-unit}}, T_{\text{decomp}}, T_{\text{rel}}, T_{\mathrm{CVP}}^{\mathrm{pre\text{-}proc}}$ and $T_{\mathrm{CVP}}^{\mathrm{query}}$.

Theorem 5.1. Let K be any number field of dimension n and discriminant Δ. Let $\alpha \in [0, 1/2]$. Then, under GRH and Heuristics 1–6, there exist two algorithms $A_{pre\text{-}proc}$ and A_{query} such that

- Algorithm $A_{pre\text{-}proc}$ takes as inputs the field K and a basis of its integer ring R, runs in time $2^{\widetilde{O}(\log|\Delta|)}$ and outputs a hint w of bit-size at most $2^{\widetilde{O}((\log|\Delta|)^{1-2\alpha})}$,

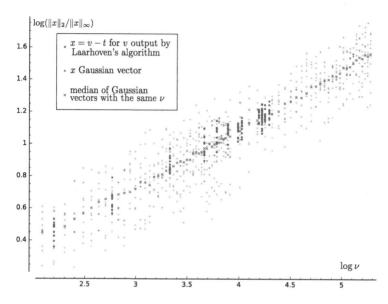

Fig. 4. Comparison of $\log(\|x\|_2/\|x\|_\infty)$ as a function of $\log \nu$ for x a Gaussian vector or $x = t - v$ with t a random target and v the approx-CVP solution output by Laarhoven's algorithm (on our lattice L, in selected cyclotomic fields). (Color figure online)

- *Algorithm A_{query} takes as inputs the hint w output by $A_{pre\text{-}proc}$ and any fractional ideal I of R such that the numerator and denominator of $\mathcal{N}(I)$ have bit-sizes bounded by $\mathrm{poly}(\log|\Delta|)$. It runs in classical time $2^{\widetilde{O}((\log|\Delta|)^{\max(2/3, 1-2\alpha)})}$ or in quantum time $2^{\widetilde{O}((\log|\Delta|)^{1-2\alpha})}$ and outputs an element $x \in I$ such that*
$$0 < \|x\|_2 \le 2^{\widetilde{O}(\frac{(\log|\Delta|)^{\alpha+1}}{n})} \cdot \lambda_1^{(2)}(I).$$

The memory consumption of both algorithms is bounded by their run-times.

In the case where $\log|\Delta| = \widetilde{O}(n)$, we can replace $\log|\Delta|$ by n in all the equations of Theorem 5.1, and we obtain an element x which is a $2^{\widetilde{O}(n^\alpha)}$ approximation of a shortest non-zero vector of I (see Fig. 6). On the other hand, if $\log|\Delta|$ becomes significantly larger than n, then both the run-time and the approximation factor degrade. The cost of the pre-computation phase also becomes larger than $2^{O(n)}$. However, the query phase still improves upon the BKZ algorithm, for some choices of α, as long as $\log|\Delta| = \widetilde{O}(n^{12/11})$ in the classical setting or $\log|\Delta| = \widetilde{O}(n^{4/3})$ in the quantum setting (see Fig. 7). In Figs. 5, 6 and 7, we plot the ratios between time and approximation factor for the BKZ algorithm and the query phase of our algorithm, in the different regimes $\log|\Delta| = \widetilde{O}(n)$ and $\log|\Delta| = \widetilde{O}(n^{1+\varepsilon})$ for some $\varepsilon > 0$.

In the case of prime-power cyclotomic fields, we know that $\log|\Delta| = \widetilde{O}(n)$. Moreover, there is a heuristic algorithm of Biasse *et al.* [BEF+17] satisfying $T_{\mathrm{rel}}, T_{\mathrm{decomp}}, T_{\log\text{-unit}} \approx 2^{\widetilde{O}(n^{1/2})}$. Hence, we obtain the trade-offs shown in Fig. 2 (in the introduction) when applying our algorithm to prime-power cyclotomic

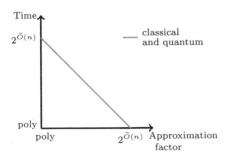

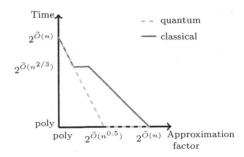

Fig. 5. Prior time/approximation trade-offs for approx-SVP in ideal lattices in any number field of degree n (using the BKZ algorithm).

Fig. 6. New trade-offs for ideal lattices in number fields satisfying $\log|\Delta| = \widetilde{O}(n)$ (with a pre-processing of cost $\exp(\widetilde{O}(n))$).

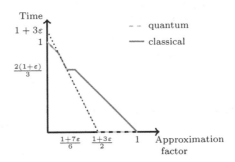

Fig. 7. New trade-offs for ideal lattices in number fields satisfying $\log|\Delta| = \widetilde{O}(n^{1+\varepsilon})$ for some $\varepsilon > 0$ (with a pre-processing of cost $\exp(\widetilde{O}(n^{1+\varepsilon}))$).

fields. Recall that in this special case, we already had an improvement upon the BKZ algorithm in the quantum setting, using the results of [CDPR16] and [CDW17], see Fig. 1.

Acknowledgments. We thank Léo Ducas for his suggestion to use Laarhoven's CVPP algorithm. We thank Oded Regev and Noah Stephens-Davidowitz for illustrating the importance of limiting the witness size by the run-time of the query phase, by pointing out the faster algorithm with exponential-size witness described in the introduction. We also thank Dan Bernstein, Elena Kirshanova and Alexandre Wallet for helpful discussions.

This work was supported in part by BPI-France in the context of the national project RISQ (P141580), by the European Union PROMETHEUS project (Horizon 2020 Research and Innovation Program, grant 780701) and by the ERC Starting Grant ERC-2013-StG-335086-LATTAC.

References

[AD17]   Albrecht, M.R., Deo, A.: Large modulus Ring-LWE $\geq$ Module-LWE. In: Takagi, T., Peyrin, T. (eds.) ASIACRYPT 2017. LNCS, vol. 10624, pp. 267–296. Springer, Cham (2017). https://doi.org/10.1007/978-3-319-70694-8_10

[Bac90]   Bach, E.: Explicit bounds for primality testing and related problems. Math. Comput. **55**(191), 355–380 (1990)

[BBV+17] Bauch, J., Bernstein, D.J., de Valence, H., Lange, T., van Vredendaal, C.: Short generators without quantum computers: the case of multiquadratics. In: Coron, J.-S., Nielsen, J.B. (eds.) EUROCRYPT 2017. LNCS, vol. 10210, pp. 27–59. Springer, Cham (2017). https://doi.org/10.1007/978-3-319-56620-7_2

[BEF+17] Biasse, J.-F., Espitau, T., Fouque, P.-A., Gélin, A., Kirchner, P.: Computing generator in cyclotomic integer rings. In: Coron, J.-S., Nielsen, J.B. (eds.) EUROCRYPT 2017. LNCS, vol. 10210, pp. 60–88. Springer, Cham (2017). https://doi.org/10.1007/978-3-319-56620-7_3

[Ber14]   Bernstein, D.J.: A subfield-logarithm attack against ideal lattices: computational algebraic number theory tackles lattice-based cryptography. The cr.yp.to blog (2014). https://blog.cr.yp.to/20140213-ideal.html

[BF14]   Biasse, J.-F., Fieker, C.: Subexponential class group and unit group computation in large degree number fields. LMS J. Comput. Math. **17**(A), 385–403 (2014)

[Bia14]   Biasse, J.-F.: Subexponential time ideal decomposition in orders of number fields of large degree. Adv. Math. Commun. **8**(4), 407–425 (2014)

[Bia17]   Biasse, J.-F.: Approximate short vectors in ideal lattices of $\mathbb{Q}(\zeta_{p^e})$ with precomputation of $\mathrm{Cl}(\mathcal{O}_K)$. In: Adams, C., Camenisch, J. (eds.) SAC 2017. LNCS, vol. 10719, pp. 374–393. Springer, Cham (2018). https://doi.org/10.1007/978-3-319-72565-9_19

[BS96]   Bach, E., Shallit, J.O.: Algorithmic Number Theory: Efficient Algorithms, vol. 1. MIT Press, Cambridge (1996)

[BS16]   Biasse, J.-F., Song, F.: Efficient quantum algorithms for computing class groups and solving the principal ideal problem in arbitrary degree number fields. In: SODA, pp. 893–902. Society for Industrial and Applied Mathematics (2016)

[Buc88]   Buchmann, J.: A subexponential algorithm for the determination of class groups and regulators of algebraic number fields. Séminaire de théorie des nombres, Paris **1989**(1990), 27–41 (1988)

[BV11a]   Brakerski, Z., Vaikuntanathan, V.: Efficient fully homomorphic encryption from (standard) LWE. In: FOCS 2011, pp. 97–106. IEEE Computer Society (2011)

[BV11b]   Brakerski, Z., Vaikuntanathan, V.: Fully homomorphic encryption from Ring-LWE and security for key dependent messages. In: Rogaway, P. (ed.) CRYPTO 2011. LNCS, vol. 6841, pp. 505–524. Springer, Heidelberg (2011). https://doi.org/10.1007/978-3-642-22792-9_29

[BV18]   Biasse, J.-F., Van Vredendaal, C.: Fast multiquadratic S-unit computation and application to the calculation of class groups. The Open Book Series **2**, 103–118 (2019). https://doi.org/10.2140/obs.2019.2.103

[CDPR16] Cramer, R., Ducas, L., Peikert, C., Regev, O.: Recovering short generators of principal ideals in cyclotomic rings. In: Fischlin, M., Coron, J.-S. (eds.) EUROCRYPT 2016. Part II. LNCS, vol. 9666, pp. 559–585. Springer, Heidelberg (2016). https://doi.org/10.1007/978-3-662-49896-5_20

[CDW17] Cramer, R., Ducas, L., Wesolowski, B.: Short Stickelberger class relations and application to ideal-SVP. In: Coron, J.-S., Nielsen, J.B. (eds.) EUROCRYPT 2017. LNCS, vol. 10210, pp. 324–348. Springer, Cham (2017). https://doi.org/10.1007/978-3-319-56620-7_12

[CGS14] Campbell, P., Groves, M., Shepherd, D.: Soliloquy: a cautionary tale (2014). http://docbox.etsi.org/Workshop/2014/201410_CRYPTO/S07_Systems_and_Attacks/S07_Groves_Annex.pdf

[Coh13] Cohen, H.: A Course in Computational Algebraic Number Theory, vol. 138. Springer, Heidelberg (2013)

[DLW19] Doulgerakis, E., Laarhoven, T., de Weger, B.: Finding closest lattice vectors using approximate Voronoi cells. In: PQCRYPTO. Springer (2019, to appear)

[DPW19] Ducas, L., Plançon, M., Wesolowski, B.: On the shortness of vectors to be found by the Ideal-SVP Quantum Algorithm (2019, to appear)

[EHKS14] Eisenträger, K., Hallgren, S., Kitaev, A., Song, F.: A quantum algorithm for computing the unit group of an arbitrary degree number field. In: Shmoys, D.B. (ed.) 46th Annual ACM Symposium on Theory of Computing, pp. 293–302. ACM Press, May/June 2014

[Gel17] Gelin, A.: Calcul de groupes de classes d'un corps de nombres et applications à la cryptologie. Ph.D. thesis, Paris 6 (2017)

[Gen09] Gentry, C.: Fully homomorphic encryption using ideal lattices. In: Mitzenmacher, M. (ed.) 41st Annual ACM Symposium on Theory of Computing, pp. 169–178. ACM Press, May/June 2009

[GGH13] Garg, S., Gentry, C., Halevi, S.: Candidate multilinear maps from ideal lattices. In: Johansson, T., Nguyen, P.Q. (eds.) EUROCRYPT 2013. LNCS, vol. 7881, pp. 1–17. Springer, Heidelberg (2013). https://doi.org/10.1007/978-3-642-38348-9_1

[HM89] Hafner, J.L., McCurley, K.S.: A rigorous subexponential algorithm for computation of class groups. J. Am. Math. Soc. $\mathbf{2}$(4), 837–850 (1989)

[HWB17] Holzer, P., Wunderer, T., Buchmann, J.A.: Recovering short generators of principal fractional ideals in cyclotomic fields of conductor $p^\alpha q^\beta$. In: Patra, A., Smart, N.P. (eds.) INDOCRYPT 2017. LNCS, vol. 10698, pp. 346–368. Springer, Cham (2017). https://doi.org/10.1007/978-3-319-71667-1_18

[Laa16] Laarhoven, T.: Sieving for closest lattice vectors (with preprocessing). In: Avanzi, R., Heys, H. (eds.) SAC 2016. LNCS, vol. 10532, pp. 523–542. Springer, Cham (2017). https://doi.org/10.1007/978-3-319-69453-5_28

[LM00] Laurent, B., Massart, P.: Adaptive estimation of a quadratic functional by model selection. Ann. Stat. $\mathbf{28}$(5), 1302–1338 (2000)

[Lou00] Louboutin, S.: Explicit bounds for residues of Dedekind zeta functions, values of l-functions at $s = 1$, and relative class numbers. J. Number Theory $\mathbf{85}$(2), 263–282 (2000)

[LPR10] Lyubashevsky, V., Peikert, C., Regev, O.: On ideal lattices and learning with errors over rings. In: Gilbert, H. (ed.) EUROCRYPT 2010. LNCS, vol. 6110, pp. 1–23. Springer, Heidelberg (2010). https://doi.org/10.1007/978-3-642-13190-5_1

[LS15] Langlois, A., Stehlé, D.: Worst-case to average-case reductions for module lattices. Des. Codes Cryptogr. $\mathbf{75}$(3), 565–599 (2015)

[Min67] Minkowski, H.: Gesammelte Abhandlungen. Chelsea, New York (1967)

[PRS17] Peikert, C., Regev, O., Stephens-Davidowitz, N.: Pseudorandomness of Ring-LWE for any ring and modulus. In: STOC 2017, pp. 461–473. ACM (2017)

[RBV04] Rekaya, G., Belfiore, J.-C., Viterbo, E.: A very efficient lattice reduction tool on fast fading channels. In: ISITA (2004)

[Reg05] Regev, O.: On lattices, learning with errors, random linear codes, and cryptography. In: Gabow H.N., Fagin, R. (eds.) 37th Annual ACM Symposium on Theory of Computing, pp. 84–93. ACM Press, May 2005

[Sam13] Samuel, P.: Algebraic Theory of Numbers: Translated from the French by Allan J. Silberger. Courier Corporation, Chelmsford (2013)

[Sch87] Schnorr, C.-P.: A hierarchy of polynomial time lattice basis reduction algorithms. Theoret. Comput. Sci. **53**, 201–224 (1987)

[SE94] Schnorr, C.-P., Euchner, M.: Lattice basis reduction: improved practical algorithms and solving subset sum problems. Math. Program. **66**, 181–199 (1994)

[SSTX09] Stehlé, D., Steinfeld, R., Tanaka, K., Xagawa, K.: Efficient public key encryption based on ideal lattices. In: Matsui, M. (ed.) ASIACRYPT 2009. LNCS, vol. 5912, pp. 617–635. Springer, Heidelberg (2009). https://doi.org/10.1007/978-3-642-10366-7_36

[Ste19] Stephens-Davidowitz, N.: A time-distance trade-off for GDD with preprocessing - instantiating the DLW heuristic (2019). Personal communication

[Zim80] Zimmert, R.: Ideale kleiner Norm in Idealklassen und eine Regulatorabschätzung. Inventiones mathematicae **62**(3), 367–380 (1980)

The General Sieve Kernel
and New Records in Lattice Reduction

Martin R. Albrecht[1], Léo Ducas[2], Gottfried Herold[3], Elena Kirshanova[3], Eamonn W. Postlethwaite[1(✉)], and Marc Stevens[2]

[1] Information Security Group, Royal Holloway, University of London, Egham, UK
eamonn.postlethwaite.2016@rhul.ac.uk
[2] Cryptology Group, CWI, Amsterdam, The Netherlands
[3] ENS Lyon, Laboratoire LIP, Lyon, France

Abstract. We propose the General Sieve Kernel (G6K, pronounced /ʒe.si.ka/), an abstract stateful machine supporting a wide variety of lattice reduction strategies based on sieving algorithms. Using the basic instruction set of this abstract stateful machine, we first give concise formulations of previous sieving strategies from the literature and then propose new ones. We then also give a light variant of BKZ exploiting the features of our abstract stateful machine. This encapsulates several recent suggestions (Ducas at Eurocrypt 2018; Laarhoven and Mariano at PQCrypto 2018) to move beyond treating sieving as a blackbox SVP oracle and to utilise strong lattice reduction as preprocessing for sieving. Furthermore, we propose new tricks to minimise the sieving computation required for a given reduction quality with mechanisms such as recycling vectors between sieves, on-the-fly lifting and flexible insertions akin to Deep LLL and recent variants of Random Sampling Reduction.

Moreover, we provide a highly optimised, multi-threaded and tweakable implementation of this machine which we make open-source. We then illustrate the performance of this implementation of our sieving strategies by applying G6K to various lattice challenges. In particular, our approach allows us to solve previously unsolved instances of the Darmstadt SVP (151, 153, 155) and LWE (e.g. (75, 0.005)) challenges. Our solution for the SVP-151 challenge was found 400 times faster than the time reported for the SVP-150 challenge, the previous record. For exact-SVP, we observe a performance crossover between G6K and FPLLL's state of the art implementation of enumeration at dimension 70.

The research of MA was supported by EPSRC grants EP/P009417/1, EP/S02087X/1 and by the European Union Horizon 2020 Research and Innovation Program Grant 780701; the research of LD was supported by a Veni Innovational Research Grant from NWO under project number 639.021.645 and by the European Union Horizon 2020 Research and Innovation Program Grant 780701; the research of EP was supported by the EPSRC and the UK government as part of the Centre for Doctoral Training in Cyber Security at Royal Holloway, University of London (EP/P009301/1); the research of GH and EK was supported by ERC Starting Grant ERC-2013-StG-335086-LATTAC.

Electronic supplementary material The online version of this chapter (https://doi.org/10.1007/978-3-030-17656-3_25) contains supplementary material, which is available to authorized users.

ⓒ International Association for Cryptologic Research 2019
Y. Ishai and V. Rijmen (Eds.): EUROCRYPT 2019, LNCS 11477, pp. 717–746, 2019.
https://doi.org/10.1007/978-3-030-17656-3_25

1   Introduction

Sieving algorithms have seen remarkable progress over the last few years. Briefly, these algorithms find a shortest vector in a lattice by considering exponentially many lattice vectors and searching for sums and differences that produce shorter vectors. Since the introduction of sieving algorithms in 2001 [AKS01], a long series of works, e.g. [MV10b, BGJ15, HK17], have proposed asymptotically faster variants; the asymptotically fastest of which has a heuristic time complexity of $2^{0.292d+o(d)}$, with d the dimension of the lattice [BDGL16].

Such algorithms for finding short vectors are used in lattice reduction algorithms. These improve the "quality" of a lattice basis (see Sect. 2) and are used in the cryptanalysis of lattice-based cryptography.

On the other hand, lattice reduction libraries such as [dt18a, AWHT16] implement enumeration algorithms, which also find a shortest vector in a lattice. These algorithms perform an exhaustive search over all lattice points within a given target radius by exploiting the properties of projected sublattices. Enumeration has a worst-case time complexity of $d^{\frac{1}{2e}d+o(d)}$ [Kan83, HS07] but requires only polynomial memory.

While, with respect to running time, sieving already compares favourably in relatively low dimensions to simple enumeration[1] (Fincke–Pohst enumeration [FP85] without pruning), the Darmstadt Lattice Challenge Hall of Fame for both approximate SVP [SG10] and LWE [FY15] challenges has been dominated by results obtained using enumeration. Sieving has therefore not, so far, been competitive in practical dimensions when compared to state of the art enumeration with heavy preprocessing [Kan83, MW15] and (extreme) pruning [GNR10] as implemented in e.g. FPLLL/FPyLLL [dt18a, dt18b]. Here, "pruning" means to forego exploring the full search space in favour of focussing on likely candidates. The extreme pruning variant proceeds by further shrinking the search space, and rerandomising the input and restarting the search on failure. In this context "heavy preprocessing" means running strong lattice reduction, such as the BKZ algorithm [Sch87, CN11], which in turn runs enumeration in smaller dimensions, before performing the full enumeration. In short, enumeration currently beats sieving "in practice" despite having asymptotically worse running time. Thus [MW15], relying on the then state of the art, estimated the crossover point between sieving and enumeration for solving the Shortest Vector Problem (SVP) as dimension $d = 146$ (or in the thousands, assuming extreme pruning can be combined with heavy preprocessing without loss of performance).

Contribution. In this work, we report performance records for achieving various lattice reduction tasks using sieving. For exact-SVP, we are able to outperform the pruned enumeration of FPLLL/FPyLLL by dimension 70. For the Darmstadt SVP Challenges (1.05-Hermite-SVP) we solve previously unsolved challenges in dimensions $\{151, 153, 155\}$ (see Fig. 1 and Table 2), and our running

---

[1] For example, the Gauss sieve implemented in FPLLL (`latsieve`) beats its *unpruned* SVP oracle (`fplll -a svp`) in dimension 50.

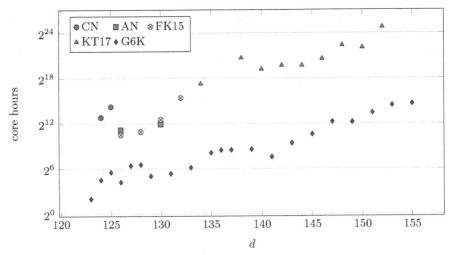

CN: Chen & Nguyen (HoF), BKZ+enum; AN: Aono & Nguyen (HoF), BKZ+enum; FK15: [FK15], RSR; KT17: [TKH18] (Their latest record (SVP-152) from Oct. 2018 is only reported in the HoF. It reports a computation time of 800K CPU-hours. According to personal communications with the authors, this translates to $36 \cdot 800$ K $= 28.8$ Mcore-hours.), RSR; G6K: `WorkOut` with `bgj1`-sieve (this work). "HoF" means data was extracted from the Darmstadt SVP Challenge Hall of Fame [SG10]. Raw data (see Supplementary material).

Fig. 1. New Darmstadt SVP challenge records.

times are at least 400 times smaller than the previous records for comparable instances.

We also solved new instances $(n, \alpha) \in \{(40, 0.005), (50, 0.015), (55, 0.015), (60, 0.01), (65, 0.01), (75, 0.005)\}$ of the Darmstadt LWE challenge (see Table 3). For this, we adapted the strategy of [LN13], which consists of running one large enumeration after a BKZ tour of small enumerations, to G6K. This improves slightly upon the prediction of [ADPS16, AGVW17].

Our sieving performance is enabled by building on, generalising and extending previous works. In particular, the landscape of enumeration and sieving started to change recently with [Duc18a, LM18]. For example, [Duc18a] speculated that the crossover point, for solving SVP, between the SubSieve proposed there and pruned enumeration would be around $d = 90$ if combined with faster sieving than [MV10b]. A key ingredient for this performance gain was the realisation of several "dimensions for free" by utilising heavy preprocessing and Babai lifting (or size reduction) in said free dimensions. This may be viewed as a hybrid of pruned enumeration with sieving, and is enabled by strong lattice reduction preprocessing. In other words, we may consider these improvements as applying lessons learnt from enumeration to sieving algorithms. It is worth recalling here that the fastest enumeration algorithm relies on the input basis being quasi-HKZ reduced [Kan83], but prior to [Duc18a, LM18] sieving was largely oblivious to the quality of the input basis. Furthermore, both [Duc18a, LM18] suggest exploiting the fact that sieving algorithms hold a database of many short vectors, for example by recycling them in future sieving steps. Thus, instead of treating sieving as an SVP oracle outputting a single vector, they implicitly treat it as a stateful machine where the state comprises the current basis and a database of many relatively short vectors.

G6K, an abstract stateful machine. In this work, we embrace and push forward in this direction. After some preliminaries in Sect. 2, we propose the General Sieve Kernel (G6K, pronounced /ʒe.si.ka/) in Sect. 3, an abstract machine for running sieving algorithms, and driving lattice reduction. We define several basic instructions on this stateful machine that not only allow new sieving strategies to be simply expressed and easily prototyped, but also lend themselves to the easy inclusion and extension of previous works. For example, the progressive sieves from [Duc18a, LM18] can be concisely written as

$$\texttt{Reset}_{0,0,0}, \ (\texttt{ER}, \ \texttt{S})^d, \ \texttt{I}_0$$

where S means to sieve, $\texttt{I}_0$ means to insert the shortest vector found into the basis, ER means to increase the sieving dimension and Reset initialises the machine.

Beyond formalising previous techniques, our machine provides new instructions, namely EL, which allows one to increase the sieving dimension "towards the left" (of the basis), and an insertion instruction I which is no longer terminal: it is possible to resieve after an insertion, contrary to [Duc18a]. These instructions increase the range of implementable strategies and we make heavy use of them to achieve the above results.

The General Sieve Kernel also introduces new tricks to further improve efficiency. First, all vectors encountered during the sieve can be lifted "on the fly" (as opposed to only the final set of vectors in [Duc18a]) offering a few extra dimensions for free and thus improved performance. Additionally, G6K keeps insertion candidates for many positions so as to allow a posteriori choices of the most reducing insertion, akin to Deep LLL [SE94] and the latest variants of Random Sampling Reduction (RSR) [TKH18], enabling stronger preprocessing.

Lattice reduction with G6K. Using these instructions, in Sect. 4 we then create reduction strategies for various tasks (SVP, BKZ-like reduction). These strategies encapsulate and extend the contributions and the suggestions made in [Duc18a, LM18], further exploiting the features of G6K. Using the instructions of our abstract stateful machine, our fundamental operation, named the Pump, may be written as

$$\texttt{Reset}_{\kappa,\kappa+\beta,\kappa+\beta}, \ (\texttt{EL}, \ \texttt{S})^{\beta-f}, \ (\texttt{I}, \ \texttt{S}^s)^{\beta-f}.$$

While previous works mostly focus on recursive lattice reduction within sieving, we also explicitly treat and test utilising sieving within the BKZ algorithm. Here, we report both negative and positive results. On the one hand, we report that, at least in our implementation, the elegant idea of a sliding-window sieve for BKZ [LM18] performs poorly and offer a discussion as to why. We also find that the strategy from [Duc18a], consisting of "overshooting" the block size β of BKZ by a small additive factor combined with "jumping" over the same number of indices in a BKZ tour, does not provide a beneficial quality vs. time trade-off. On the other hand, we find that from the second block of a BKZ tour onwards, or always in the progressive BKZ case, cheaper sieving calls (involving less pre-processing) suffice. We also find that opportunistically increasing the number of dimensions for free beyond the optimal values for solving SVP improves the quality vs. time trade-off. Thus, we vindicate the suggestion to move beyond treating sieving merely as an SVP oracle in BKZ.

Implementation. In Sect. 5, we then propose and describe an open-source, tweakable, multi-threaded, low-level optimised implementation of G6K, featuring several sieve variants [MV10b, BGJ15, HK17].[2] Our implementation is carefully optimised to support multiple cores in all time consuming operations, is highly parameterised and makes heavy use of the SimHash test [Cha02, FBB+15, Duc18a]. It combines a C++ kernel with a Python control module. Thus, our higher level algorithms are all implemented in Python for easy experimentation. Our implementation is written with a view towards being extensible and reusable and comes with documentation and tests. We consider hackable and usable software a contribution in its own right.

Performance and Records. Using and tuning our implementation of G6K then allows us to obtain the variety of performance records for solving lattice challenges as described above. We describe our approach in Sect. 6. There, we also describe our experiments for the aforementioned BKZ strategies.

Complementary information on the performance of our implementation is provided in the full version.

Discussion. A natural question is how our results affect the security of lattice-based schemes, especially the NIST PQC candidates. Most candidates have been extremely conservative, and thus we do not expect the classical security claim of any scheme to be directly affected by our results. We note, however, that our results on BKZ substantiate further the prediction made in several analyses of NIST PQC candidates that the cost of the SVP oracle can be somewhat amortised in BKZ [PAA+17, Sec 4.2.6]. Thus, our results provide further evidence that the $8\,d \cdot C_{SVP}$ cost model [ACD+18] is an over-estimate,[3] but they nevertheless do not reach the lower bound given by the "core-hardness" estimates. However, we stress that our work justifies the generally conservative approach and we warn against security estimates based on a state of the art that is still in motion.

On the other hand, the memory consumption of sieving eventually becomes a difficult issue for implementation, and could incur slowdowns due to memory access delays and bandwidth constraints. Though, it is not so clear that these

---

[2] Our implementation is available at https://github.com/fplll/g6k/.

[3] Note that, in addition, this already follows in the enumeration regime from [LN13] which we adapt to the sieving regime in Sect. 6.

difficulties are insurmountable, especially to an attacker having access to custom hardware. For example Kirchner claimed [Kir16] that simple sieving algorithms such as the Nguyen–Vidick sieve are implementable by a circuit with Area = Time = $2^{0.2075n+o(n)}$. Ducas further conjectured [Duc18b] that bgj1 (a simplified version of [BGJ15]) can be implemented with Area = $2^{0.2075n+o(n)}$ and Time = $2^{0.142n+o(n)}$. More concretely, the algorithms that we have implemented mostly consider contiguous streams of data, making the use of disks instead of RAM plausibly not so penalising.

One may also argue that such an area requirement on its own is already unreasonable. Yet, such arguments should also account for what amount of wall-time is considered reasonable. For example, the walltime of a bruteforce search costing 2^{128} CPU-cycles on 2^{64} cores at $4\,\mathrm{GHz}$ runs for 2^{64} cycles = $2^{32}\,\mathrm{s} \approx 134$ years; larger walltimes with fewer cores can arguably be considered irrelevant for practical attacks.

2   Preliminaries

2.1   Notations and Basic Definitions

We start counting at zero. All vectors are denoted by bold lower case letters and are to be read as column vectors. Matrices are denoted by bold capital letters. We write a matrix $\mathbf{B}$ as $\mathbf{B} = (\mathbf{b}_0, \ldots, \mathbf{b}_{n-1})$ where $\mathbf{b}_i$ is the i-th column vector of $\mathbf{B}$. We may also denote $\mathbf{b}_i$ by $\mathbf{B}[i]$ and the j-th entry of $\mathbf{b}_i$ by $\mathbf{B}[i,j]$. If $\mathbf{B} \in \mathbb{R}^{d \times n}$ has full column rank n, the lattice $\mathcal{L}$ generated by the basis $\mathbf{B}$ is denoted by $\mathcal{L}(\mathbf{B}) = \{\mathbf{B}\mathbf{x} \mid \mathbf{x} \in \mathbb{Z}^n\}$. We denote by $(\mathbf{b}_0^*, \ldots, \mathbf{b}_{n-1}^*)$ the Gram–Schmidt orthogonalisation of the matrix $\mathbf{B} = (\mathbf{b}_0, \ldots, \mathbf{b}_{n-1})$. That is, we define

$$\mu_{i,j} = \frac{\langle \mathbf{b}_j^*, \mathbf{b}_i \rangle}{\langle \mathbf{b}_j^*, \mathbf{b}_j^* \rangle} \qquad \text{and} \qquad \mathbf{b}_i^* = \mathbf{b}_i - \sum_{j=0}^{i-1} \mu_{i,j} \cdot \mathbf{b}_j^*.$$

The process of updating $\mathbf{b}_i \leftarrow \mathbf{b}_i - \lfloor \mu_{ij} \rceil \mathbf{b}_j$, for $j \in \{i-1, \ldots, s\}$ with $0 \le s < i$, is known as "size reduction" or "Babai's Nearest Plane" algorithm. We also define $\mathbf{b}_i^\circ = \mathbf{b}_i^* / \langle \mathbf{b}_i^*, \mathbf{b}_i^* \rangle$ and extend this to $\mathbf{B}^\circ$ column wise. For $i \in \{0, \ldots, n-1\}$, we denote the projection orthogonally to the span of $(\mathbf{b}_0, \ldots, \mathbf{b}_{i-1})$ by π_i. For $0 \le \ell < r \le n$, we denote by $\mathbf{B}_{[\ell:r]}$ the local projected basis, $(\pi_\ell(\mathbf{b}_\ell), \ldots, \pi_\ell(\mathbf{b}_{r-1}))$. When the basis is clear from context $\mathcal{L}_{[\ell:r]}$ denotes the lattice generated by $\mathbf{B}_{[\ell:r]}$. We refer to the *left* (resp. the *right*) of a *context* $[\ell : r]$ and by "the context $[\ell : r]$" implicitly refer also to $\mathcal{L}_{[\ell:r]}$ and $\mathbf{B}_{[\ell:r]}$. More generally, we speak of the *left* (resp. the *right*) as a direction to refer to smaller (resp. larger) indices and of contexts becoming larger as $r - l$ grows.

The Euclidean norm of a vector $\mathbf{v}$ is denoted by $|\mathbf{v}|$. The volume of a lattice $\mathcal{L}(\mathbf{B})$ is $\mathrm{Vol}(\mathcal{L}(\mathbf{B})) = \prod_i |\mathbf{b}_i^*|$, an invariant of the lattice. The first minimum of a lattice $\mathcal{L}$ is the length of a shortest non zero vector, denoted by $\lambda_1(\mathcal{L})$. We use the abbreviations $\mathrm{Vol}(\mathbf{B}) = \mathrm{Vol}(\mathcal{L}(\mathbf{B}))$ and $\lambda_1(\mathbf{B}) = \lambda_1(\mathcal{L}(\mathbf{B}))$.

2.2   Sieving, Lattice Reduction and Heuristics

Sieving algorithms build databases of lattice vectors, exponentially sized in the lattice dimension. In the simplest sieves, it is checked whether the sums or differences of any pair of database vectors is shorter than one of the summands or differands. More importantly for G6K as an abstract stateful machine is the property of sieving [NV08, MV10b] that, after sieving in some $\mathcal{L}$, this database contains a constant fraction, which we are able to set, of $\{\mathbf{w} \in \mathcal{L} \colon |\mathbf{w}| \leq R \cdot \mathrm{gh}(\mathcal{L})\}$. Here $\mathrm{gh}(\mathcal{L})$ is the expected length of the shortest vector of a lattice $\mathcal{L}$ (see Definition 2), and R is a small constant determined by the sieve (see Sect. 5.1). It is this information that G6K will leverage when changing context and inserting.

Lattice reduction is the process of taking a basis for a given $\mathcal{L}$ and finding subsequent bases of $\mathcal{L}$ with shorter and closer to orthogonal vectors. Two important notions of reduction are HKZ and BKZ-β reduction. The BKZ algorithm [SE94, CN11] takes as input a lattice basis of $\mathcal{L}$ and a block size β and outputs a BKZ-β reduced basis of $\mathcal{L}$.

Definition 1 (Hermite–Korkine–Zolotarev,   Block-Korkine–Zolotarev). *A size-reduced basis* $\mathbf{B} = (\mathbf{b}_0, \ldots, \mathbf{b}_{d-1})$ *of a lattice* $\mathcal{L}$ *is Hermite–Korkine–Zolotarev (HKZ) reduced if* $|\mathbf{b}_i^*| = \lambda_1(\mathcal{L}_{[i:d]}), \forall i < d$. *It is Block-Korkine–Zolotarev with block size* β *(BKZ-β) reduced if* $|\mathbf{b}_i^*| = \lambda_1(\mathcal{L}_{[i:\min\{i+\beta,d\}]}), \forall i < d$.

Intuitively BKZ reduction requires that a given index in the basis is as short as possible when considering only a local projected sublattice, with the locality parameterised by β. The cost of BKZ increases with β. The LLL algorithm [LLL82] can be thought of as BKZ-2 and is often used as a cheap starting point for lattice reduction. Equally, HKZ reduction can be thought of as BKZ-d and is a strong notion of reduction.

The BKZ algorithm internally calls an SVP oracle in dimension $\leq \beta$, i.e. an algorithm that solves the Shortest Vector Problem (or an approximate variant of it) in dimension β.

The Gaussian heuristic predicts that the number, $|\mathcal{L} \cap \mathcal{B}|$, of lattice points inside a measurable body $\mathcal{B} \subset \mathbb{R}^n$ is approximately $\mathrm{Vol}(\mathcal{B})/\mathrm{Vol}(\mathcal{L})$. Applied to Euclidean n-balls, it leads to the following prediction of $\lambda_1(\mathcal{L})$ for a given $\mathcal{L}$.

Definition 2 (Gaussian Heuristic). *We denote by* $\mathrm{gh}(\mathcal{L})$ *the expected first minimum of a lattice* $\mathcal{L}$ *according to the Gaussian heuristic. For a full rank lattice* $\mathcal{L} \subset \mathbb{R}^d$, *it is given by*

$$\mathrm{gh}(\mathcal{L}) = \sqrt{d/2\pi e} \cdot \mathrm{Vol}(\mathcal{L})^{1/d}. \tag{1}$$

The quality of a basis after lattice reduction can be measured by a quantity called the root Hermite factor.

Definition 3 (Root Hermite Factor). *For a basis* $\mathbf{B}$ *of a d-dimensional lattice, the root Hermite factor is defined as*

$$\delta = \left(|\mathbf{b}_0| / \mathrm{Vol}\,(\mathbf{B})^{1/d} \right)^{1/d}. \tag{2}$$

For BKZ-β, the root Hermite factor is a well behaved quantity. For small block-sizes the root Hermite factor is experimentally calculated [GN08b] and for larger blocksizes [Che13] it follows the asymptotic formula

$$\delta(\beta)^{2(\beta-1)} = (\beta/(2\pi e))(\beta\pi)^{\frac{1}{\beta}}, \tag{3}$$

which tends towards 1. Finally we reproduce the Geometric Series Assumption (GSA) [Sch03] which, given β, heuristically determines the lengths of consecutive Gram–Schmidt basis vectors. It is reasonably accurate for $\beta > 50$ and $\beta \ll d$ [Ngu10, CN11, YD17].

Definition 4 (Geometric Series Assumption). *Let* **B** *be a BKZ-β reduced basis, then the Geometric Series Assumption states that* $|\mathbf{b}_i^*| \approx \delta(\beta)^{-2} |\mathbf{b}_{i-1}^*|$.

3   The General Sieve Kernel

3.1   Design Principles

In this section we propose the General Sieve Kernel (Version 1.0), an abstract machine supporting a wide variety of lattice reduction strategies based on sieving algorithms. It minimises the sieving computation effort for a given reduction quality by:

– offering a mechanism to recycle short vectors from one context to somewhat short vectors in an overlapping context, therefore already starting the sieve closer to completion. This formalises and generalises some of the ideas proposed in [Duc18a, LM18].
– being able to lift vectors to a larger context than the one currently considered. These vectors are considered for insertion at earlier positions. But as an extension to [Duc18a], which only lifted the final database of vectors, G6K is able to lift-and-compare all vectors encountered during the sieve. From this, we expect a few extra dimensions for free.[4]
– deferring the decision of where to insert a short vector until after the search effort. This is contrary to formal definitions of more standard reduction algorithms, e.g. BKZ or Slide [GN08a] reduction, and inspired by Deep LLL and recent RSR variants [TKH18].

The underlying computations per vector are reasonably cheap, typically linear or quadratic in the dimension of the vector currently being considered. The most critical operation, namely the SimHash test [Cha02, FBB+15, Duc18a] may be asymptotically sublinear or even polylogarithmic; in practice it consists of about a dozen x86 non vectorised instructions for vectors of dimension roughly one hundred.

---

[4] Lifting is somewhat more expensive than considering a pair of vectors. We are therefore careful to only lift a fraction of all considered vectors, namely only the considered vectors below a certain length of, say, $\sqrt{1.8} \cdot \mathrm{gh}(\mathcal{L}_{[\ell:r]})$.

3.2   Vectors, Contexts and Insertion

All vectors considered by G6K live in one of the projected lattices $\mathcal{L}_{[\ell:r]}$ of a lattice $\mathcal{L}$. More specifically, they are represented in basis $\mathbf{B}_{[\ell:r]}$ as integral vectors $\mathbf{v} \in \mathbb{Z}^n$ where $n = r - \ell$, i.e. we have $\mathbf{w} = \mathbf{B}_{[\ell:r]} \cdot \mathbf{v}$ for some $\mathbf{w} \in \mathbb{R}^d$. Throughout, we may represent the (projected) lattice vector $\mathbf{w}$ by the vector $\mathbf{v}$. It is convenient, and efficient, to also keep a representation, $\mathbf{v}^\circ \in \mathbb{R}^n$, of $\mathbf{w}$ in the orthonormalised basis $\mathbf{B}_{[\ell:r]}^\circ$. This conversion costs $O(n^2)$.

Below we list the three operations that *extend* or *shrink* a vector to the left or to the right.

- Extend Right (inclusion) er $: \mathcal{L}_{[\ell:r]} \to \mathcal{L}_{[\ell:r+1]}$

$$(v_0, \ldots v_{n-1}) \mapsto (v_0, \ldots v_{n-1}, 0)$$
$$(v_0^\circ, \ldots v_{n-1}^\circ) \mapsto (v_0^\circ, \ldots v_{n-1}^\circ, 0)$$

- Shrink Left (projection) sl $: \mathcal{L}_{[\ell:r]} \to \mathcal{L}_{[\ell+1:r]}$

$$(v_0, \ldots v_{n-1}) \mapsto (v_1, \ldots v_{n-1})$$
$$(v_0^\circ, \ldots v_{n-1}^\circ) \mapsto (v_1^\circ, \ldots v_{n-1}^\circ)$$

- Extend Left (Babai-lift) el $: \mathcal{L}_{[\ell:r]} \to \mathcal{L}_{[\ell-1:r]}$

$$(v_0, \ldots, v_{n-1}) \mapsto (-\lfloor c \rceil, v_0, \ldots, v_{n-1})$$
$$(v_0^\circ, \ldots, v_{n-1}^\circ) \mapsto ((c - \lfloor c \rceil) \cdot |\mathbf{b}_{\ell-1}^*|, v_0^\circ, \ldots v_{n-1}^\circ),$$
$$\text{where } c = \sum_{j=0}^{n-1} \mu_{\ell+j, \ell-1} \cdot v_j.$$

These operations maintain, somewhat, the shortness of vectors. Indeed,

$$|\mathrm{er}(\mathbf{w})| = |\mathbf{w}|, \ |\mathrm{sl}(\mathbf{w})| \approx \sqrt{(r-\ell-1)/(r-\ell)} \cdot |\mathbf{w}|, \ |\mathrm{el}(\mathbf{w})|^2 \le |\mathbf{w}|^2 + \left|\mathbf{b}_{\ell-1}^*\right|^2 / 4.$$

More properly, "shortness" should be considered relative to the Gaussian heuristic of a context, $\mathrm{gh}(\mathcal{L}_{[\ell:r]})$. For BKZ-$\beta$ reduced bases, and growing in accuracy as $r - \ell \to \infty$,

$$\frac{\mathrm{gh}(\mathcal{L}_{[\ell:r]})}{\mathrm{gh}(\mathcal{L}_{[\ell:r+1]})} \text{ and } \frac{\mathrm{gh}(\mathcal{L}_{[\ell:r]})}{\mathrm{gh}(\mathcal{L}_{[\ell+1:r]})} \approx \delta(\beta), \ \frac{\mathrm{gh}(\mathcal{L}_{[\ell:r]})}{\mathrm{gh}(\mathcal{L}_{[\ell-1:r]})} \approx \delta(\beta)^{-1}.$$

We may then calculate an approximate growth factor, relative to the Gaussian heuristics of the contexts, for each of the three operations

$$\frac{|\mathrm{er}(\mathbf{w})| \cdot \mathrm{gh}(\mathcal{L}_{[\ell:r]})}{|\mathbf{w}| \cdot \mathrm{gh}(\mathcal{L}_{[\ell:r+1]})} \approx \delta(\beta), \quad \frac{|\mathrm{sl}(\mathbf{w})| \cdot \mathrm{gh}(\mathcal{L}_{[\ell:r]})}{|\mathbf{w}| \cdot \mathrm{gh}(\mathcal{L}_{[\ell+1:r]})} \approx \sqrt{\frac{r-\ell-1}{r-\ell}} \cdot \delta(\beta),$$

$$\frac{|\mathrm{el}(\mathbf{w})| \cdot \mathrm{gh}(\mathcal{L}_{[\ell:r]})}{|\mathbf{w}| \cdot \mathrm{gh}(\mathcal{L}_{[\ell-1:r]})} \le \delta(\beta)^{-1} \left(1 + \frac{\left|\mathbf{b}_{\ell-1}^*\right|^2}{4 \cdot |\mathbf{w}|^2}\right)^{1/2}.$$

While it would seem natural to also define a Shrink Right operation, we have not found a geometrically meaningful way of doing so. Moreover, we have no algorithmic purpose for it.

Insertion. Performing an insertion (the elementary lattice reduction operation) of a vector is less straightforward. For $i \leq \ell < r$, $n' = r - i$, $n = r - \ell$ an insertion of a vector $\mathbf{w}$ at position i is a local change of basis making $\mathbf{w} = \mathbf{B}_{[i:r]} \cdot \mathbf{v}$ the first vector of the new local projected basis, i.e. applying a unimodular matrix $\mathbf{U} \in \mathbb{Z}^{n' \times n'}$ to $\mathbf{B}_{[i:r]}$ such that $(\mathbf{B}_{[i:r]} \cdot \mathbf{U})[0] = \mathbf{w}$. While doing so, we would like to recycle a database of vectors currently living in the context $[\ell : r]$.

In the case $i = \ell$, this causes no difficulties, and one could apply any change of basis $\mathbf{U}$ to the database. But to exploit dimensions for free, we will typically have $i < \ell$, which is more delicate. If we can ensure that

$$\mathrm{Span}((\mathbf{B} \cdot \mathbf{U})_{[i:\ell+1]}) = \mathrm{Span}(\mathbf{B}_{[i:\ell]} \cup \{\mathbf{w}\}) \tag{4}$$

then one can simply project all the database vectors orthogonally to $\mathbf{w}$, to end up with a database in a new smaller context $[\ell+1 : r]$. If it holds that $\mathbf{v}[j] = \pm 1$ for some $j \in \{\ell, \ldots, r-1\}$ an appropriate matrix $\mathbf{U}$ can be constructed as

$$\mathbf{U} = \left(\mathbf{v} \; \middle| \; \begin{matrix} \mathbf{I}_{j \times j} & \mathbf{0} \\ \mathbf{0} & \mathbf{0} \\ \mathbf{0} & \mathbf{I}_{n'-j-1 \times n'-j-1} \end{matrix} \right). \tag{5}$$

However, it is important that the local projected bases remain somewhat reduced. If not, numerical stability issues may occur. Moreover, the condition that $\mathbf{v}$ contains a ± 1 in the context $[\ell : r]$ is often not satisfied without sufficient reduction. While we must be careful to not alter the vector space inside the sieving context, we can nevertheless perform a full size reduction (upper triangular matrix $\mathbf{T}$ with unit diagonal) on the whole of $\mathbf{B}_{[i:r]}$, as well as two local LLL reductions $\mathbf{U}_\mathrm{L}$ and $\mathbf{U}_\mathrm{R}$ on $\mathbf{B}_{[i:\ell+1]}$ and $\mathbf{B}_{[\ell+1:r]}$.

$$\mathbf{U}' = \mathbf{U} \cdot \mathbf{T} \cdot \begin{pmatrix} \mathbf{U}_\mathrm{L} & \mathbf{0} \\ \mathbf{0} & \mathbf{U}_\mathrm{R} \end{pmatrix}. \tag{6}$$

Note that $\mathrm{Span}((\mathbf{B} \cdot \mathbf{U}')_{[i:\ell+1]}) = \mathrm{Span}((\mathbf{B} \cdot \mathbf{U})_{[i:\ell+1]})$, so that condition (4) is preserved.

3.3   G6K: A Stateful Machine

The General Sieve Kernel is defined by the following internal states and instructions.

State

- A lattice basis $\mathbf{B} \in \mathbb{Z}^{d \times d}$, updated each time an insert is made (Sect. 3.2). Associated with it is its Gram–Schmidt Orthonormalisation basis $\mathbf{B}^\circ$.
- Positions $0 \leq \kappa \leq \ell \leq r \leq d$. We refer to the context $[\ell : r]$ as the *sieving context*, and $[\kappa : r]$ as the *lifting context*. We define $n = r - \ell$ (the sieving dimension).
- A database *db* of N vectors in $\mathcal{L}_{[\ell:r]}$ (preferably short).
- Insertion candidates $\mathbf{c}_\kappa, \ldots, \mathbf{c}_\ell$ where $\mathbf{c}_i \in \mathcal{L}_{[i:r]}$ or $\mathbf{c}_i = \bot$.

Instructions

- Initialisation ($\texttt{Init}_\mathbf{B}$): initialise the machine with a basis $\mathbf{B} \in \mathbb{Z}^{d \times d}$.
- Reset ($\texttt{Reset}_{\kappa,\ell,r}$): empty database, and set (κ, ℓ, r).
- Sieve ($\texttt{S}$): run some chosen sieving algorithm. During execution of the algorithm, well chosen visited vectors are lifted from $\mathcal{L}_{[\ell:r]}$ to $\mathcal{L}_{[\kappa:r]}$ (by iterating el just on these vectors). If such a lift improves (i.e. is shorter than) the best insertion candidate $\mathbf{c}_i$ at position i, then it replaces $\mathbf{c}_i$. We call this optional[5] feature *on-the-fly lifting*.
- Extend Right, Shrink Left, Extend Left ($\texttt{ER}$, $\texttt{SL}$, $\texttt{EL}$): increase or decrease ℓ or r and apply er, sl or el to each vector of the database. All three operations maintain the insertion candidates (except for $\texttt{EL}$ which drops $\mathbf{c}_\ell$).
- Insert ($\texttt{I}$): choose the best insertion candidate $\mathbf{c}_i$ for $\kappa \leq i \leq \ell$, according to a score function, and insert it at position i. The sieving context changes to $[\ell+1 : r]$ and the database is updated as described in Sect. 3.2. If no insertion candidate is deemed suitable, then we simply run $\texttt{SL}$ so as to ensure that the sieving context will end up as expected.[6] When we write $\texttt{I}_i$, we mean that insertion is only considered at position i.
- Grow or Shrink ($\texttt{Resize}_N$): change the database to a given size N. When shrinking, remove the longest vectors from the database. When growing, sample new vectors (using some unspecified sampling algorithm[7]). Typically, we will not explicate the calls to these operations, and assume that calling a sieve includes resizing the database to the appropriate size, for example $N = O(\sqrt{4/3}^n)$ for the 2-sieves of [NV08, MV10b, BGJ15].

Our implementation of this machine offers more functionality, such as the ability to monitor its state and therefore the behaviour of the internal sieve algorithm, and to tune the underlying algorithms.

4   Reduction Algorithms Using G6K

Equipped with this abstract machine, we can now reformulate, improve and generalise strategies for lattice reduction with sieving algorithms. In the following we will assume that the underlying sieve algorithm has a time complexity proportional to C^n, with n the dimension of the SVP instance, and we also define $C' = 1/(1 - 1/C)$. This second constant approximates the multiplicative overhead $\sum_{i=1}^{n} C^i/C^n$ encountered on iterating sieves in dimensions 1 to n. Note that this overhead grows when C decreases. More concretely, depending on the sieve, C can range from $4/3$ down to $\sqrt{3/2}$, giving $C' = 4$ up to $C' \approx 5.45$.

---

[5] The alternative being to only consider the vectors of the final database for lifting.

[6] Note that sl can be viewed as the trivial insertion of the vector $v_\kappa = (1, 0, \dots, 0)$.

[7] When possible we prefer to sample by summing random pairs of vectors from the database.

4.1 The Pump

In this section we propose a sequence of instructions called the Pump. They encompass the progressive sieving strategy proposed in [Duc18a, LM18] as well as the dimensions for free and multi-insertion tricks of [Duc18a]. The original progressive sieving strategy can be written as

$$\texttt{Reset}_{0,0,0}, \ (\texttt{ER, S})^d, \ \texttt{I}_0. \tag{7}$$

Similarly, a SubSieve$_f$ which attempts a partial HKZ reduction using sieving with f dimensions for free can be written as[8]

$$\texttt{SubSieve}_f : \texttt{Reset}_{0,f,f}, \ (\texttt{ER, S})^{d-f}, \ \texttt{I}_0, \texttt{I}_1, \ldots, \texttt{I}_{d-f-1}. \tag{8}$$

We note that due to the newly introduced EL operation, it is also possible to perform the progressive sieving right to left

$$\texttt{Reset}_{0,d,d}, (\texttt{EL, S})^{d-f}, \ \texttt{I}_0, \texttt{I}_1, \ldots, \texttt{I}_{d-f-1}. \tag{9}$$

Perhaps surprisingly, experimentally the left variant of progressive sieving performs substantially better. In combination with certain sieving methods, the right variant even fails completely, this will be discussed in more detail in Sect. 4.5.

To arrive at Pump, note first that G6K maintains insertion candidates at many positions. We can therefore relax the insertion positions of (9) and choose those that appear to be optimal. The choice of insertion position is discussed in Sect. 4.4.

Secondly, due to on-the-fly lifting, we note that the sequence (9) considers many more insertion candidates for the first insertion than for subsequent insertions. Moreover, we noticed that after several insertions, the database contained vectors much longer than recent inserts. By sieving also during the "descent phase", i.e. when inserting and shrinking the sieve context, we remedy this imbalance and expect to obtain a more strongly reduced basis, ideally obtaining an HKZ reduced context.

In summary, we define the parameterised Pump$_{\kappa, f, \beta, s}$ as the following sequence

$$\texttt{Pump}_{\kappa, f, \beta, s} : \texttt{Reset}_{\kappa, \kappa+\beta, \kappa+\beta}, \ \overbrace{(\texttt{EL, S})^{\beta-f}}^{\text{pump-up}}, \ \overbrace{(\texttt{I, S}^s)^{\beta-f}}^{\text{pump-down}}. \tag{10}$$

where $0 \leq \kappa \leq \kappa + \beta \leq d$, $0 \leq f \leq \beta$, and where $s \in \{0, 1\}$ controls whether we sieve during pump-down. One may expect the cost of these extra sieves to be close to a multiplicative factor of 2, but experimentally the factor can reach 3 for certain sieves (e.g. bgj1), as more collisions[9] seem to occur during the descent phase. This feature is mostly useful for weaker reduction tasks such as BKZ, see PumpNJumpBKZTour below.

---

[8] This sequence refers to SubSieve $^+(\mathcal{L}, f)$ with Sieve being progressive [Duc18a].

[9] A collision is when a new vector $\mathbf{v}$ to be inserted in the database equals $\pm \mathbf{v}_2$ for some $\mathbf{v}_2$ already present in the database.

4.2 SVP

To solve the shortest vector problem on the full lattice, starting from an LLL reduced basis $\mathbf{B}$, we proceed as in [Duc18a], that is, we iterate $\texttt{Pump}_{0,f,d,s}$ for decreasing values of f. While only the last $\texttt{Pump}$ delivers the shortest vector, the previous iterations provide a strongly reduced basis (near HKZ reduced), which allows more dimensions for free to be achieved. We expect to obtain further dimensions for free due to on-the-fly lifting.

Similarly, for solving SVP in context $[\kappa : \kappa + \beta]$ (e.g. as a block inside BKZ), we instead make iterative calls to $\texttt{Pump}_{\kappa,f,\beta,s}$.

Note that we can decrease f in larger increments than 1 to balance the cost of the basis reduction effort and the search for the shortest vector itself. Indeed, with increments of 1, the overhead factor C' for $C = \sqrt{3/2}$ is $C' \approx 4.45$. Decreasing f by 2 gives an overhead of $C' = 1/(1 - C^{-2}) = 3$ and by 3 gives $C' = 1/(1 - C^{-3}) \approx 2.19$. Such speed-ups are worth losing 1 or 2 dimensions for free.

We therefore define $\texttt{WorkOut}$ as the following sequence of $\texttt{Pump}$

$$\texttt{WorkOut}_{\kappa,\beta,f,f^+,s} : \texttt{Pump}_{\kappa,\beta-f^+,\beta,s},\ \texttt{Pump}_{\kappa,\beta-2f^+,\beta,s}, \tag{11}$$
$$\texttt{Pump}_{\kappa,\beta-3f^+,\beta,s},\ \cdots\ \texttt{Pump}_{\kappa,f,\beta,s},$$

where f^+ is the increment mentioned above. From experiments on exact-SVP and SVP Challenges, we found it worthwhile to deactivate sieving in the descent phase ($s = 0$), though activating it ($s = 1$) is preferable in other contexts, or to use less memory at a larger time cost. Similarly, for certain tasks (e.g. the SVP Challenges, i.e. 1.05-Hermite-SVP) we found the optimal increment, f^+, to be 2 or 3. This parameter also drives a time-memory trade-off; setting f^+ to 1 saves on memory by allowing for a larger f, but at a noticeable cost in time.

For solving exact-SVP, it is not clear when to stop this process because we are never certain that a vector is indeed the shortest vector of a lattice (except maybe by running a very costly non pruned enumeration). In these cases, one should therefore guess, from experimental data, a good number f of dimensions for free. Note that it is rarely critical to achieve exact-SVP, and lattice reduction algorithms such as BKZ tolerate approximations.

In some cases, such as the Darmstadt SVP Challenge, we do not have to solve exact-SVP, but rather find a vector of a prescribed norm, near the Gaussian heuristic. In this case we do not need to predetermine f and simply iterate the $\texttt{Pump}$ until satisfaction. As a consequence, we also add an extra option to the $\texttt{Pump}$ to allow early aborts when it finds a satisfying candidate $\mathbf{c}_\kappa$. In practice we observe significant savings from this feature, i.e. we observe the $\texttt{Pump}$ aborting before reaching its topmost dimension, or at the beginning of the descent phase.

4.3 BKZ

Having determined the appropriate parameters f, f^+, s for solving SVP-β (made implicit in the following), the naïve implementation of BKZ is given by the following program

$$\text{NaiveTour}_\beta : \text{WorkOut}_{0,\beta}, \quad \text{WorkOut}_{1,\beta+1}, \ \dots$$
$$\text{WorkOut}_{d-\beta,d}, \ \dots, \ \text{WorkOut}_{d-1,d}. \tag{12}$$

Several strategies to amortise the cost of sieving inside BKZ were suggested in [Duc18a, LM18]. These aimed to reduce the cost of a tour of BKZ-β below d (or $d - \beta$) times the cost of SVP in dimension β. Again, these strategies are implementable as a sequence of G6K instructions.

Namely, the sliding-window strategy of [LM18] can be expressed as

$$\text{SlidingWindowTour}_\beta : \text{Reset}_{0,0,0}, \ (\text{ER, S})^\beta, \ (\text{I}_\ell, \ \text{S}, \ \text{ER, S})^{d-\beta}, \ (\text{I}_\ell, \ \text{S})^\beta. \tag{13}$$

It is also possible to combine this strategy with the dimensions for free of [Duc18a]. However, there are two caveats. First, it relies on extend right, which is currently problematic in our implementation of G6K, see Sect. 4.5. Secondly, even if this issue is solved, we remark that inside a BKZ tour it is preferable to run LLL on the full basis periodically. From the sandpile point of view [MV10a, HPS11], not doing so implies that a "bump" accumulates at the right of the reduced blocks, as we try to push the sand to the right. We see no clear strategies to recycle the vectors of a block when calling a full LLL.

Alternatively, [Duc18a] identified two other potential amortisations. First, it is noted that the WorkOut (or even just a Pump) in a block $[\kappa : \kappa + \beta]$ leaves the next block $[\kappa + 1 : \kappa + \beta + 1]$ already quite well reduced. It may therefore not be necessary to do a full WorkOut, but simply run the last Pump of this WorkOut, therefore saving up to a factor of C' in the running time.

The second suggestion of [Duc18a] consists of overshooting the blocksize β, so that a Pump in dimension $\beta' > \beta$ attempts to HKZ reduce a larger block. In particular for parameter j, let $\beta' = \beta + j - 1$ and after a $\text{Pump}_{\kappa,f,\beta'}$ jump by j blocks. This decreases the number of calls to the Pump to d/j and may also slightly improve the quality of the reduction, but increases the cost of the Pump calls by a factor of C^{j-1}. It is argued that such a strategy could give a speed-up factor ranging from 2.2 to 3.6 for a fixed basis reduction quality. In this case we therefore perform the following sequence

$$\text{PumpNJumpTour}_{\beta',f,j} : \text{Pump}_{0,f,\beta'}, \ \text{Pump}_{j,f,\beta'}, \ \text{Pump}_{2j,f,\beta'}, \dots \tag{14}$$

We alter the version above to allow for more opportunism. Since choosing f to almost certainly solve exact-SVP in blocks is costly, we instead embrace the idea of achieving the most basis reduction from a given sieving context. Extending the lift context makes the lift operation more expensive, but gives more insertion candidates, and therefore a new trade-off to be optimised over. Note that while $\text{Pump}_{\kappa',f+\kappa-\kappa',\beta+\kappa-\kappa'}$ for $\kappa' < \kappa$ takes more dimensions for free than $\text{Pump}_{\kappa,f,\beta}$, it still provides the same insertion candidates, $c_\kappa, \dots, c_{\kappa+f}$. It also provides new insertion candidates $c_{\kappa'}, \dots, c_{\kappa-1}$. This is because the sieving contexts do not shrink, and so, provided we take care in the first few blocks, the quality cannot decrease. To achieve this start with Pumps with $f = 0$ and move the sieving context right until the desired f is attained, then continue as before. Set $f' > f$, $\beta = \beta + f' - f$ (i.e. to fix the sieve context sizes), $\beta' = \beta + j - 1$,[10] and perform

$$\text{PumpNJumpTour}_{\beta',f',j} : \text{Pump}_{0,0,\beta'-f'}, \text{Pump}_{0,j,\beta'-f'+j}, \cdots, \text{Pump}_{0,f',\beta'},$$
$$\text{Pump}_{j,f',\beta'}, \text{Pump}_{2j,f',\beta'}, \cdots \tag{15}$$

4.4 Scoring for Inserts

The issue of deciding where in a basis to insert given candidates throughout reduction has already been discussed in [TKH18], in the context of the SVP Challenges. Until the actual shortest vector is found, these insertions have the purpose of improving the basis quality. Inserting at an early position may degrade quality at later positions, because we do not know a priori how inserting c_i will affect $\mathbf{B}_{[\ell:r]}$ for $i \leq \ell < r$. Therefore one must find a good trade-off between making long lasting yet weak improvements at early positions, and strong yet fragile improvements at later positions.

One way to achieve this is to use the scoring proposed in [TKH18], a function over the whole basis which measures the global effect of each potential insert, i.e. checking exactly how inserting c_i affects the $\mathbf{B}_{[\ell:r]}$. We use a simplified variant of this scoring which scores the improvement of each potential insert according to the following local condition

$$\varsigma(i) = \begin{cases} 0, & \text{if } \mathbf{c}_i = \perp \\ \theta^{-i} \cdot |\mathbf{b}_i^*|^2 / |\mathbf{c}_i|^2, & \text{otherwise} \end{cases} \tag{16}$$

for some constant $\theta \geq 1$ and take the maximum over the valid indices. Setting $\theta = 1$ corresponds to always choosing the "most improving" candidate, while setting θ quite large (say 10) corresponds to always inserting at the earliest position.

To optimise θ, we ran $\text{WorkOut}_{0,d,f}$ for $f = 30$ and $d = 110$, measured $\gamma = \text{gh}(\mathcal{L}) / \text{gh}(\mathcal{L}_{[f:d]})$, and chose $\theta = 1.04$ which minimised this quantity γ. We recall [Duc18a] that γ must be below a certain threshold to guarantee the success of exact-SVP in dimension d with f dimensions for free.

The optimal value of θ may differ depending on other parameters, e.g. dimension, approximation factor, and the context, e.g. exact-SVP, 1.05-Hermite-SVP, BKZ, and the question of optimising insertion strategies requires more theoretical and experimental attention. We hope that our open source implementation will ease such future research.

4.5 Issue with Extend Right

As mentioned earlier, our current implementation does not support the ER operation very well. In more detail, the issue is that after running a sieve in the

---

[10] For Fig. 4 we choose yet more opportunism and do not increase β to β'.

context $[\ell : r]$, and applying ER, the vectors in the database are padded with 0 to be defined over the context $[\ell : r + 1]$; geometrically, these vectors remain in the context $[\ell : r]$, and so will all their potential combinations considered by the sieve. While we do add some fresh vectors to increase the database size, the fraction of those fresh vectors in the database is rather small: $1 - \sqrt{3/4} \approx 13\%$. This alone seems to slow down the Gauss sieve when used in right-progressive sieving compared to left-progressive sieving.

The situation is even worse in the faster sieves we implement. Indeed, apart from the reference Gauss sieve, our sieves are not guaranteed to maintain the full-rankness of the database. This is because, for performance purposes, we relax the replacement condition. In the standard Gauss sieve, $\mathbf{x} \pm \mathbf{y}$ may only replace $\mathbf{x}$ or $\mathbf{y}$ if it is shorter. We relax this and allow $\mathbf{x} \pm \mathbf{y}$ to replace the current longest vector $\mathbf{z}$ in the database. Fresh vectors are much longer than the recycled ones, therefore they are quickly replaced by combinations of recycled vectors, effectively meaning there is little representation of the newly introduced basis vector after an ER.

While we tried to implement countermeasures to avoid losing rank, they had a noticeable impact on performance, and were not robust. For this work, we therefore avoid the use of extend right, as reductions based on extend left already perform well. We leave it as an open problem to develop appropriate variants of fast sieve algorithms that avoid this issue.

5   Implementation Details

5.1   Sieving

We implemented several variants of sieving, namely: a Gauss sieve [MV10b], a relaxation of the Nguyen–Vidick sieve [NV08], a restriction of the Becker–Gama–Joux sieve [BGJ15] and a 3-sieve [BLS16, HK17]. All exploit the SimHash speed-up [Cha02, FBB+15, Duc18a].

The first two were mostly implemented for reference and testing purposes, and therefore are not multi-threaded. Nevertheless, we fall back to Gauss sieve in small dimensions for efficiency and robustness; as discussed earlier, Gauss sieve is immune to loss of rank, which we sometimes experienced with other sieves in small dimensions (say, $n < 50$), even when not using extend right.

The termination condition for the sieves follows [Duc18a], namely, they stop when we have obtained a given ratio of the expected number of vectors of norm less than $R \cdot \mathrm{gh}(\mathcal{L}_{[\ell:r]})$. The saturation radius is dictated by the asymptotics of the algorithm at hand, namely, R is such that the sieve uses a database of $N = O(R^n)$ vectors. In particular $R = \sqrt{4/3}$ for all implemented sieves, except for the 3-sieve for which one can choose $R^2 \in [3\sqrt{3}/4, 4/3] \approx [1.299, 1.333]$.

Nguyen–Vidick Sieve (nv) and Gauss Sieve (Gauss). The Nguyen–Vidick sieve finds pairs of vectors $(\mathbf{v}_1, \mathbf{v}_2)$ from the database, whose sum or difference gives a shorter vector, i.e. $|\mathbf{v}_1 \pm \mathbf{v}_2| < \max\{|\mathbf{v}| : \mathbf{v} \in \mathrm{db}\}$. Once such a pair is found, the longest vector from the database gets replaced by $\mathbf{v}_1 \pm \mathbf{v}_2$. The size of the database is a priori fixed to the asymptotic heuristic minimum $2^{0.2075n+o(n)}$ required to find enough such pairs. The running time of the Nguyen–Vidick sieve is quadratic in the database size.

The Gauss sieve algorithm, similar to the Nguyen–Vidick sieve, searches for pairs with a short sum, but the replacement and the order in which we process the database vectors differ. More precisely, the database now is (implicitly) divided into two parts, the so called "list" part and the "queue" part. This separation is encoded in the ordering, with the list part being the first τ vectors. Both parts are kept separately sorted. The list part has the property that the shortness of $\mathbf{v}_1 \pm \mathbf{v}_2$ has been checked for all pairs of vectors $\mathbf{v}_1, \mathbf{v}_2$ in the list. We then only check pairs $(\mathbf{v}_1, \mathbf{v}_2)$, where $\mathbf{v}_1$ comes from the queue part and $\mathbf{v}_2$ from the list part. As opposed to Nguyen–Vidick sieve, once a reduction is found, the longer vector from the pair $(\mathbf{v}_1, \mathbf{v}_2)$ gets replaced by $\mathbf{v}_1 \pm \mathbf{v}_2$, not the longest in the database. In the case where the list vector $\mathbf{v}_2$ gets replaced, the result of the reduction $\mathbf{v}_1 \pm \mathbf{v}_2$ is put into the "queue" part and the search is continued with the same "queue" vector $\mathbf{v}_1$. Otherwise, if the queue vector $\mathbf{v}_1$ was the longest and is replaced, we restart comparing $\mathbf{v}_1$ with all list vectors. A vector is moved from the "queue" to the "list" part once no reduction with the "list" vectors can be found. Asymptotically, the running time and the database size for the Gauss sieve is the same as for the Nguyen–Vidick sieve, but it performs better in practice.

Becker–Gama–Joux Sieve (bgj1). The sieve algorithm from [BGJ15] accelerates the Nguyen–Vidick sieve [NV08] from $2^{0.415n+o(n)}$ down to $2^{0.311n+o(n)}$ by using locality sensitive filters, while keeping the memory consumption to its bare minimum for a 2-sieve, namely $2^{0.2075n+o(n)}$.

This optimal complexity is reached using recursive filtering, however we only implemented a variant of this algorithm with a single level of filtration (hence the name bgj1). We leave it to future work to implement the full algorithm and determine when the second level of filtration becomes interesting.

We briefly describe our simplified version. The algorithm finds reducing pairs in the database by successively filling buckets according to a filtering rule, and doing all pairwise tests inside a bucket. Concretely, it chooses a uniform direction $\mathbf{d} \in \mathbb{R}^n$, $|\mathbf{d}| = 1$, and puts in the bucket all database vectors taking (up to sign) a small angle with $\mathbf{d}$, namely all $\mathbf{v}$ such that $|\langle \mathbf{v}, \mathbf{d} \rangle| > \alpha \cdot |\mathbf{v}|$.

We choose α so that the size of the buckets is about the square root of the size of the database (asymptotically, $\alpha^2 \to 1 - \sqrt{3/4} \approx 0.366^2$). This choice balances the cost of populating the bucket (through testing the filtering condition) and exploring inside the bucket (checking for pairwise reductions). Both cost $O(N) = 2^{0.2075n+o(n)}$; though in practice we found it faster to make the buckets slightly larger, namely around $3.2\sqrt{N}$. Also note that we can apply a SimHash prefiltering before actually computing the inner product $\langle \mathbf{v}, \mathbf{d} \rangle$, but using a larger threshold for the bucketing prefilter than for the reduction prefilter.

Following the heuristic arguments from the literature, and in particular the wedge volume formula [BDGL16, Lemma 2.2], we conclude that this sieve succeeds after about $(2/\sqrt{3} - 1/3)^{-n/2} \approx 2^{0.142n+o(n)}$ buckets, for a total complexity of $2^{0.349n+o(n)}$.

3-sieve (`triple_sieve`). In its original versions [BLS16, HK17], the 3-sieve algorithm aims to reduce memory consumption at the cost of a potential increase in the running time. The 3-sieve algorithm searches not for pairs, but for *triples* of vectors, whose sum gives a shorter vector (hence, the name 3-sieve). Clearly, for a fixed size list of vectors, there are more possible triples than pairs and, therefore, we can start with a shorter list and still find enough reductions. However, a (naïve) search now costs three iterations over the list. To speed-up the naïve search, we can apply filtering techniques similar to the ones used for `bgj1`. In particular, the 3-sieve algorithm with filtering described in [HK17] requires memory $2^{0.1788n+o(n)}$ and runs in time $2^{0.396n+o(n)}$.

For any vector $\mathbf{x}$ from the database, the 3-sieve algorithm of [HK17] filters the database by collecting all vectors $\mathbf{v}$ with a large enough inner product $|\langle \mathbf{x}, \mathbf{v} \rangle|$. For all pairs of these collected vectors $(\mathbf{v}_1, \mathbf{v}_2)$, 3-sieve checks if $|\mathbf{x} \pm \mathbf{v}_1 \pm \mathbf{v}_2|$ gives a short(er) vector. Such an inner product test, as in `bgj1`, helps to identify "promising" vectors which are likely to result in a length reduction. The only subtlety lies in the fact that in order for a triple to give a reduction, the vectors $\mathbf{x}, \mathbf{v}_1, \mathbf{v}_2$ should be far apart, not close to each other as in 2-sieve. We handle this by adjusting the inner product test and choosing the $\pm$ signs appropriately.

The version of the 3-sieve implemented in G6K splits the database into "list" and "queue" parts in the same way as the Gauss sieve above. Further, it combines 2- and 3-sieves. Notice that the filtering process of 3-sieve is basically the same as bucketing in `bgj1`, with a bucket centre defined by a database[11] vector $\mathbf{x}$. When processing the bucket, we check not only whether a pair $(\mathbf{v}_1, \mathbf{v}_2)$ from the bucket gives a shorter vector, but also whether a triple $(\mathbf{x}, \mathbf{v}_1, \mathbf{v}_2)$ may. This additional check has no noticeable impact on performance (we know in which case we potentially are from the signs of the scalar products alone), but has the potential to find more shorter vectors.

As a result, in this combined version of the sieve, we can find more reductions than in 2-sieve if we keep the same database size as for 2-sieve. In such a memory regime, most of the reductions will come from 2-reductions. Setting a smaller database makes the algorithm look for more 3-reductions as 2-reductions become less likely.

As `triple_sieve` finds more reductions than `bgj1` with the same database sizes, we may decrease the size of the database and check how the running time degrades. The results of these experiments are shown in Fig. 2. The leftmost point corresponds to the minimal memory regime for 3-sieve, namely when the database size is set to $2^{0.1788n+o(n)}$, while the rightmost point is for the `bgj1` memory regime, that is the database size is set to $2^{0.2075n+o(n)}$. It turns out that in moderate dimensions (i.e. 80–110), `triple_sieve` performs slightly better if the database size is a bit less than $2^{0.2075n+o(n)}$. Furthermore, these experiments

---

[11] This relies on the fact that we do not use recursive filtering in `bgj1`: the asymptotically optimal choice from [BGJ15] mandates choosing the buckets centres in a structured way, which is not compatible with choosing them as `db` elements.

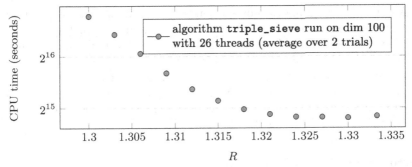

The X-axis is the parameter R such that the database size is set to $3.2 \cdot R^{n/2}$. In particular, the right-most point corresponds to the size of a database set to $3.2 \cdot (4/3)^{n/2}$; for the left-most point this value is set to $3.2 \cdot \left(3\sqrt{3}/4\right)^{n/2}$. Raw data (see Supplementary material).

Fig. 2. Time-memory trade-off for our implementation of the 3-sieve algorithm.

are consistent with theoretical results on the high memory regime for 3-sieve: in [HKL18] it was proven that the running time of 3-sieve quickly drops down if allowed slightly more memory, as Fig. 2 shows.

5.2 The Three Layers: `C++/Cython/Python`

Our implementation consists of three layers.

C++11. The lowest level routines are implemented in C++11. In particular, at this level we define a Siever class which realises G6K for all sieves considered in this work: Gauss, NV, BGJ1 and 3-sieve. The general design is similar to FPLLL where algorithms are objects operating on matrices and Gram–Schmidt objects. In particular, different sieves are realised as methods on the same object (and thus the same database) allowing the caller to pick which sieve to run in a given situation. For example, in small dimensions it is beneficial to run the Gauss sieve and this design decision allows the database to be reused between different sieves. Our C++ layer does not depend on any third party libraries (except pthreads). On the other hand, our C++ layer is relatively low level.

Cython. Cython is a glue language for interfacing between CPython (the C implementation of the Python programming language) and C/C++. We use Cython for this exact purpose. Our Cython layer is relatively thin, mainly making our C++ objects available to the Python layer and translating to and from FPyLLL data structures [dt18b]. The most notable exception is that we implemented the basis change computation of the *insert* instruction I (Eqs. (5) and (6)) in Cython instead of C++. The reason being that we call LLL on the lifting context when inserting (the Cython function `split_lll`) which is realised by calling FPyLLL. That is, while our C++ layer has no external dependencies, the Cython layer depends on FPyLLL.

Python. All our high level algorithms are implemented in (C)Python (2). Our code does not use the functional-style abstractions from Sect. 3, but a more traditional object-oriented approach where methods are called on objects which hold the state. We do provide some syntactic sugar, though, enabling a user to construct new instructions from basic instructions in a function-composition style similar to the notation in Sect. 3. Nevertheless, this simplified abstraction is not able to fully exploit all the features of our implementation, and significant savings may be achieved by using the full expressivity of our library.

5.3   Vector Representation and Data Structures

The data structures of G6K have been designed for high performance sieving operations and we have tried to minimise memory usage where possible. For high performance we retain the following information about each vector $\mathbf{v}$ as an *entry* e in the sieve database db:

- e.x: the vector $\mathbf{v}$ itself as 16-bit integer coordinates in basis $\mathbf{B}_{[\ell:r]}$;
- e.yr: a 32-bit floating point vector to efficiently compute $\langle \mathbf{v}, \mathbf{v}_2 \rangle$, this is a renormalised version of $\mathbf{v}^\circ$;
- e.cv (compressed vector): a 256-bit SimHash of $\mathbf{v}$;
- e.uid (unique identifier): a 64-bit hash of $\mathbf{v}$;
- e.len: the squared length $|\mathbf{v}|$ as a 64-bit floating point number.

The entire database db is stored contiguously in memory, although unordered. This memory is preallocated for the maximum database size within each Pump, to avoid additional memory usage caused by reallocations of the database whenever it grows.

To be able to quickly determine whether a potential new vector is already in the database we additionally maintain a C++ unorderedset (i.e. a hash table) uiddb containing 64-bit hashes uid of all vectors in db.[12] This hash uid $= H(\mathbf{x})$ of $\mathbf{x}$ is simply computed as the inner product of $\mathbf{x}$ with a global random vector in the ring $\mathbb{Z}/2^{64}\mathbb{Z}$, which has the additional benefit that $H(\mathbf{x}_1 \pm \mathbf{x}_2)$ can be computed more efficiently as $H(\mathbf{x}_1) \pm H(\mathbf{x}_2)$. This allows us to cheaply discard collisions without even having to compute $\mathbf{x}_1 \pm \mathbf{x}_2$.

To maintain a sorted database we utilise a compressed database cdb that only stores the 256-bit SimHash, 32-bit floating point length, and the 32-bit db-index of each vector. This requires only 40 bytes per vector and everything is also stored contiguously in memory. It is optimised for traversing the database in order of increasing length and applying the SimHash as a prefilter, since accessing the full entry in db only occurs a fraction of the time.

For the multi-threaded bgj1-sieve, the compressed database cdb is maintained generally sorted in order of increasing length. Initially cdb is sorted, then, during sieving, vectors are replaced one-by-one starting from the back of cdb. It is only resorted when a certain fraction of entries have been replaced. Since we

---

[12] This unorderedset is in fact split into many parts to eliminate most blocking locks during a multi-threaded sieve.

only insert a new vector if its length is below the minimum length of the range of to-be-replaced vectors in cdb, this approach ensures that we always replace the largest vector in db. In the sieve variants that split the database into queue and list ranges, we regularly sort the individual ranges. In our multi-threaded triple_sieve, the vectors removed during a replacement are chosen iteratively from the backs of the two ranges.

Most sieving operations use buckets that are filled based on locality sensitive filters. In bgj1, we use the same datastructure as cdb for the buckets, and thus copy those compressed entries in contiguous memory reserved for that bucket. For triple_sieve, we also store information about the actual scalar product $\langle \mathbf{x}, \mathbf{v} \rangle$ of the bucket elements $\mathbf{v}$ with the bucket centre $\mathbf{x}$ inside the bucket.

5.4   Multi-threading

G6K is able to efficiently use multi-threading for nearly all operations; a detailed efficiency report can be found in the full version. Global per-entry operations such as EL, ER, SL and I-postprocessing are simply distributed over all available threads in the global threadpool.

During multi-threaded sieving we guarantee all write operations to entries in db, cdb and the best lift database to be executed in a thread-safe manner using atomic operations and write locks. (The actual locking strategies differ per implementation.) We always perform all heavy computations before locking and let each thread locally buffer pending writes and execute these writes in batches to avoid bottlenecks in exclusive access of these global resources.

Threads reading entries in db and cdb do not use locking and can thus potentially read partially overwritten entries. While this may result in some wasted computations, no faulty vectors will be inserted in the db: for every new vector we completely recompute its full entry e from e.x including its length and verify it is actually shorter than the length of the to-be-replaced vector before actually replacing it.

Safely resorting cdb during sieving is the most complicated, since threads do not block on reading cdb. Our implementations in G6K resolve this as follows. We let one thread resort cdb and use locking to prevent any insertions (or concurrent resorting) by other threads. We keep the old cdb untouched as a shadow copy for other threads, while computing a new sorted version that we then atomically publish. Afterwards, other threads will then eventually switch to the newer version. Insertions are always performed using cdb and never using a shadow copy, even if e.g. a thread is still using a shadow copy for its main operations, e.g. when building a bucket.

6   New Lattice Reduction Records

The experiments reported in this section are based on bgj1-sieving, except those on BKZ and LWE which are based on triple_sieve, in the high memory regime $(N = \Theta((4/3)^{n/2}))$. The switch occurred when improvements to the latter made

Table 1. Details of the machines used for experiments.

Machine	CPUs	base freq.	cores	threads	HTC*	RAM
L	4xIntel Xeon E7-8860v4	2.2 Ghz	72	72	No	512 GiB
S	2xIntel Xeon Gold 6138	2.0 Ghz	40	80	Yes	256 GiB
C	2xIntel Xeon E5-2650v3	2.3 Ghz	20	40	Yes	256 GiB
A	2xIntel Xeon E5-2690v4	2.6 Ghz	28	56	Yes	256 GiB

* HTC: Hyperthreading Capable.

it faster than the former (especially with pump-down sieve, $s = 1$). While it seemed wasteful to rerun all the experiments, we nevertheless now recommend `triple_sieve` over `bgj1` for optimal performance within our library. The details of the machines used for our various experiments are given in Table 1.

6.1   Exact-SVP

We first report on the efficiency of our implementation of G6K's `WorkOut` ($s = 0$, $f^+ = 1$) when solving exact-SVP. The comparison with pruned enumeration is given in Fig. 3a. While fitted curves are provided, we highlight that they are significantly below asymptotic predictions of $2^{0.349d+o(d)}$ for `bgj1` and thus unreliable for extrapolation.[13] Based on these experiments, we report a crossover with enumeration around dimension 70. Note that we significantly outperform the guesstimates of a crossover at dimension 90 made in [Duc18a].

While our improved speed compared to [Duc18a] is mostly due to having implemented a faster sieving algorithm, the new features of G6K also contribute to this improved efficiency (see the full version for a detailed comparison). In particular the on-the-fly lifting strategy offers a few extra dimensions for free as depicted in Fig. 3b. That is, our new implementation is not only faster but also consumes less memory.

6.2   1.05-Hermite-SVP (a.k.a. Darmstadt SVP Challenges)

The detailed performance of our implementation when solving Darmstadt SVP Challenges is given in Table 2. We also compare the running time of our experiments with prior works in Fig. 1. We warn the reader that the experiments of Table 2 are rather heterogeneous – different machines, different software versions, and different parametrisations were used – and therefore discourage extrapolations. Moreover the design decisions below and the probabilistic nature of the algorithm explain the non monotonic time and space requirements.

The parameters were optimised towards speed by trial and error on many smaller instances ($d \approx 100$). More specifically we ran `WorkOut` with parameters $f = 16 + d/12$, $f^+ = 3$, $s = 1$; choosing $f^+ = 1$ or 2 would cost more time

---

[13] This mismatch with theory can be explained by various kinds of overheads, but mostly by the dimensions for free trick: as $f = \Theta(d/\log d)$ is quasilinear, the slope will only very slowly converge to the asymptotic prediction.

(a) Average time in seconds to solve exact-SVP.

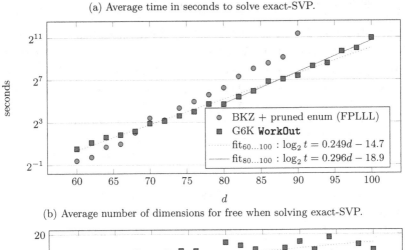

(b) Average number of dimensions for free when solving exact-SVP.

The running time was averaged over 60 trials (12 trials on random bases of 5 different lattices from [SG10]). Each instance was monothreaded, but ran in parallel (20thread/20cores, not hyperthreaded) on machine C. Raw data (see Supplementary material).

Fig. 3. Performance for exact-SVP.

and less memory.[14] The loop was set to exit as soon as a vector of the desired length was found, and if it reached the minimal value of f, it would repeat this largest Pump until success (this repetition rarely happened more than three times). The sieve max dim column reports the actual dimension $d - f_{\text{last}}$ of the last Pump.

6.3 BKZ

To test PumpNJumpTour we compare its quality vs. time performance against BKZ 2.0 [CN11] in FPyLLL and against NaiveTour (see Fig. 4). We generate random q-ary lattice bases, of dimension 180 with 90 equations modulo $q = 2^{30}$. We prereduce the bases using one FPyLLL BKZ tour for each blocksize from 20

---

[14] The number $f = 16 + d/12$ of dimensions for free is only meant to be a local approximation, as we asymptotically expect $f = \Theta(d/\log d)$ even for $O(1)$-approx-SVP [Duc18a].

Table 2. Performance on the Darmstadt SVP challenges.

SVP dim	Norm	Hermite factor	Sieve max dim	Wall time	Total CPU time	Memory usage	Machine
155	3165	1.00803	127	14 d 16 h	1056 d	† 246 GiB	L
153	3192	1.02102	123	11 d 15 h	911 d	† 139 GiB	S
151	3233	1.04411	124	11 d 19 h	457.5 d	† 160 GiB	C
149	3030	0.98506	117	60 h 7 m	4.66 kh	† 59 GiB	S
147	3175	1.03863	118	123 h 29 m	4.79 kh	67.0 GiB	C
145	3175	1.04267	114	39 h 3 m	1496 h	37.7 GiB	C
143	3159	1.04498	110	17 h 23 m	669 h	21.3 GiB	C
141	3138	1.04851	105	4 h 59 m	190 h	10.6 GiB	C
139	3111	1.04303	108	9 h 56 m	380 h	16.2 GiB	C
137	3093	1.04472	107	9 h 26 m	362 h	14.1 GiB	C
136	3090	1.04937	108	9 h 16 m	354 h	16.2 GiB	C
135	3076	1.04968	108	7 h 21 m	277.4 h	16.1 GiB	C
133	3031	1.04133	103	1 h 59 m	71.7 h	8.0 GiB	C
131	2959	1.02362	100	1 h 11 m	41.5 h	5.3 GiB	C
129	2988	1.03813	98	54 m	33.2 h	4.2 GiB	C
128	3006	1.04815	102	2 h 32 m	94.9 h	7.6 GiB	C
127	2972	1.04244	101	2 h 17 m	85.0 h	6.0 GiB	C
126	2980	1.04976	100	31 m	19.2 h	5.6 GiB	C
125	2948	1.04393	99	1 h 18 m	47.6 h	5.2 GiB	C
124	2937	1.04032	98	39 m	23.9 h	4.4 GiB	C
123	2950	1.04994	93	7 m	4.0 h	2.2 GiB	C

†: Not measured, estimate.

to 59 and then report the cumulative time taken by further progressive tours of several BKZ variants.

Contrary to exact-SVP, we find it beneficial for the running time to activate sieving during pump-down for all G6K based BKZ experiments. We further find that `triple_sieve` is noticeably faster than `bgj1`; it seems that the former suffers fewer collisions than the latter when sieving during the pump-down phase.

For all G6K based BKZ experiments we choose the number of dimensions for free following the experimental fit of Fig. 3b, that is $f = 11.5 + 0.075\beta$. We also introduce a parameter $e = f' - f$ to concretise the more opportunistic `PumpNJumpTour` variant discussed at the end of Sect. 4.3.

To measure quality we use an averaged quality measurement, namely, the slope metric of FPyLLL. This slope, ρ, is a least squares fit of the $\log |\mathbf{b}_i^*|^2$. For comparison this metric is preferable to the typical root Hermite factor as it displays much less variance. In the GSA model, the slope ρ relates to the root Hermite factor by $\delta = \exp(-\rho/4)$. We also provide the predictions for

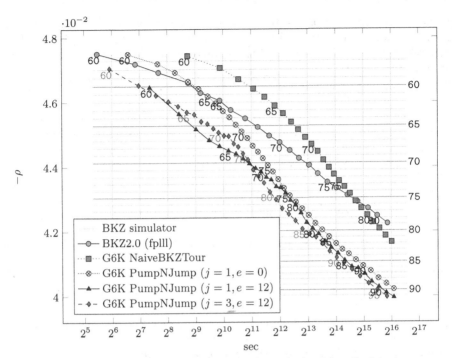

The time and slope are averaged over 8 instances for each algorithm. Each instance was monothreaded, but ran in parallel (40threads/40cores, not hyperthreaded) on machine S. We label the point by β for all multiples of 5. Raw data (see Supplementary material).

Fig. 4. Performance of BKZ-like algorithms.

progressive tours given by the BKZ simulator of [CN11, Wal16]. We note that the simulator is optimistic compared to even the most "textbook" variants, BKZ2.0 and `NaiveTour`, a phenomenon already documented in [YD17, BSW18].

Conclusion. These experiments confirm that it is possible to outperform a naïve application of an SVP-β oracle to obtain a quality equivalent to BKZ-β in less time. Indeed, `PumpNJumpTour`$_{\beta,f,1}$ is about 4 times faster than `NaiveTour`$_{\beta,f}$ for the same reduction quality. Furthermore, the opportunistic variant with $e = 12$ gives even better quality per time, and also only requires a smaller β for the same quality, therefore decreasing memory consumption. These experiments also suggest that jumps $j > 1$ are not beneficial, they require similar running time per quality, but with a larger memory consumption.

6.4 LWE

The Darmstadt LWE challenges [FY15] are labeled by (n, α), where n denotes the dimension of the secret in $\mathbb{Z}_q$, for some q, and α is a noise rate. Concretely the challenges are given as $(\mathbf{A}, \mathbf{b})$ where $\mathbf{As} + \mathbf{e} \equiv \mathbf{b} \mod q$ with $\mathbf{A} \in \mathbb{Z}_q^{m \times n}$ for some m, $\mathbf{s} \in \mathbb{Z}_q^n$, $\mathbf{e} \in \mathbb{Z}^m$ and $\mathbf{b} \in \mathbb{Z}_q^m$. Each entry of $\mathbf{e}$, the error, is sampled independently from the Gaussian distribution over the integers with mean $\mu = 0$ and standard deviation $\sigma = \alpha \cdot q$, while the entries of $\mathbf{A}$ and $\mathbf{s}$ are sampled independently and uniformly from $\mathbb{Z}_q$. Both q and m are constant for a given n, but increase with n.

Our method for solving LWE is via embedding $\mathbf{e}$ into a uSVP instance [Kan87, BG14] but using the success condition originally given in [ADPS16] and experimentally justified in [AGVW17]. We also use the embedding coefficient $t = 1$ following [ADPS16, AGVW17]. We choose the minimal β such that after BKZ-β reduction, $|\pi_{d-\beta}(\mathbf{e})| < \mathrm{gh}(\mathcal{L}_{[d-\beta:d]})$. Therefore $\pi_{d-\beta}(\mathbf{e})$ will be inserted at index $d - \eta$. It is shown in [AGVW17] that size reduction (here lifting) is then enough to lift $\pi_{d-\beta}(\mathbf{e})$ to $\mathbf{e}$. The success condition in [ADPS16, AGVW17] is

$$\sqrt{\beta} \cdot \sigma < \delta(\beta)^{2\beta - d} \cdot \mathrm{Vol}(\mathcal{L})^{1/d}. \tag{17}$$

There is no a priori reason why the β used for BKZ reduction and the dimension of the SVP call (the last full block in some BKZ-β tour) which first finds $\pi_{d-\beta}(\mathbf{e})$, should be equal. For enumeration based algorithms it is customary to run one large enumeration after the smaller enumerations inside BKZ, see [LN13]. To apply this to sieving we alter the above inequality to allow a "decoupling" of these quantities and then balance the expected total time cost.

Let β continue to denote the BKZ block size and η denote the dimension of an SVP call on the lattice $\mathcal{L}_{[d-\eta:d]}$. We obtain the following inequality

$$\sqrt{\eta} \cdot \sigma < \delta(\eta)^{\eta-1} \cdot \delta(\beta)^{\eta-d+1} \cdot \mathrm{Vol}(\mathcal{L})^{1/d}. \tag{18}$$

The left hand side is an approximation of the length $\pi_{d-\eta}(\mathbf{e})$ and the right hand side an approximation of the Gaussian heuristic of $\mathcal{L}_{[d-\eta:d]}$. Indeed

$$\mathrm{gh}(\mathcal{L}_{[d-\eta:d]}) = \sqrt{\eta/2\pi e} \cdot \mathrm{Vol}(\mathcal{L}_{[d-\eta:d]})^{1/\eta} = \sqrt{\eta/2\pi e} \cdot \left(\prod_{i=d-\eta}^{d-1} |\mathbf{b}_i^*| \right)^{1/\eta}, \tag{19}$$

and further $\delta(\eta)^{\eta-1} \sim \sqrt{\eta/2\pi e}$ in increasing η. By combining the GSA and the estimate the root Hermite factor gives for $|\mathbf{b}_0|$, (18) may be derived from (19).

Implemented Strategy and Performance. To solve LWE instances in practice we implemented code which returns triples (β, η, d) that satisfy (18), and choose the number of LWE samples accordingly. We then run `PumpNJumpTour` with $s = 1$, $j = 1$, $e = 12$ and `triple_sieve` as the underlying sieve, and increase β progressively (choosing f as the experimental fit of Fig. 3b). After each tour, we measure the walltime T elapsed since the beginning of the reduction, and predict the maximal dimension ν reachable by pumping up within time T.

Table 3. Performances on Darmstadt LWE challenges.

(n, α)	Estimated(β, η, d)	Successful(β, ν, ν')	CPU time	Wall time	M.
$(65, 0.010)$	$(108, 137, 244)$	$(112, 124, 120)$	2553 h	60 h	A
$(55, 0.015)$	$(106, 135, 219)$	$(110, 125, 103)$	2198 h	34 h 50 m	S
$(40, 0.030)$	$(102, 133, 179)$	$(108, 120, 111)$	1116 h	17 h 43 m	S
$(75, 0.005)$	$(88, 118, 252)$	$(88, 112, 107)^{\ddagger}$	591 h	12 h 26 m	S
$(60, 0.010)$	$(92, 122, 222)$	$(94, 112, 106)^{\dagger}$	579 h	11 h 59 m	S
$(50, 0.015)$	$(87, 118, 194)$	$(81, 111, 95)$	8 h 36 m	1 h 23 m	S

$\dagger$: There was also a failed search after $\beta = 90$, with $\nu = 115$.
$\ddagger$: There was also a failed search after $\beta = 84$, with $\nu = 115$.

We predict whether we expect to find the projected short vector in this pump (ignoring on-the-fly lifting), following the reasoning of [Duc18a]. That is, we check the inequality

$$\sqrt{\nu} \cdot \sigma \leq \sqrt{4/3} \cdot \mathrm{gh}(\mathcal{L}_{[d-\nu:d]}). \tag{20}$$

If this condition is satisfied, we proceed with searching for the LWE solution with this Pump (κ, $f = \nu - \kappa$, $\beta = d - \kappa$, $s = 0$)[15] otherwise, we continue BKZ reduction with larger β. If this search is triggered but fails, we also go back to reducing the basis with progressive BKZ, and reset the timer T. The search may also succeed before reaching the pump dimension ν, in which case we denote by ν' the dimension at which it stops.

Details of the six new Darmstadt LWE records are in Table 3. It should be noted that the CPU-time/walltime ratio can be quite far from e.g. 80, the number of threads on machine S. This is because parallelism only kicks in for sieves in large dimensions (see the full version), while the walltimes of some of the computations were dominated by BKZ tours with medium blocksizes. One could tailor the parameterisation to improve the walltime further, but this would be in vain as we are mostly interested in the more difficult instances, which suffer very little from this issue.

Acknowledgements. We thank Kenny Paterson for discussing a previous version of this draft. We also thank Pierre Karpman for running some of our experiments.

---

[15] One could choose $\kappa = 0$ to be entirely sure not to miss the solution during the lifting phase, but this increases the cost of lifting. Instead, we can choose κ such that $\sqrt{\kappa}\sigma < \mathrm{gh}(\mathcal{L}_{[d-\kappa:d]})$, with a small margin of, say, five dimensions.

References

[ACD+18] Albrecht, M.R., et al.: Estimate all the LWE, NTRU schemes! In: Catalano, D., De Prisco, R. (eds.) SCN 2018. LNCS, vol. 11035, pp. 351–367. Springer, Cham (2018). https://doi.org/10.1007/978-3-319-98113-0_19

[ADPS16] Alkim, E., Ducas, L., Pöppelmann, T., Schwabe, P.: Post-quantum key exchange—a new hope. In: 25th USENIX Security Symposium (USENIX Security 16), Austin, TX. USENIX Association, pp. 327–343 (2016)

[AGVW17] Albrecht, M.R., Göpfert, F., Virdia, F., Wunderer, T.: Revisiting the expected cost of solving uSVP and applications to LWE. In: Takagi, T., Peyrin, T. (eds.) ASIACRYPT 2017. LNCS, vol. 10624, pp. 297–322. Springer, Cham (2017). https://doi.org/10.1007/978-3-319-70694-8_11

[AKS01] Ajtai, M., Kumar, R., Sivakumar, D.: A sieve algorithm for the shortest lattice vector problem. In: 33rd ACM STOC. ACM Press, pp. 601–610, July 2001

[AWHT16] Aono, Y., Wang, Y., Hayashi, T., Takagi, T.: Improved progressive BKZ algorithms and their precise cost estimation by sharp simulator. In: Fischlin, M., Coron, J.-S. (eds.) EUROCRYPT 2016. LNCS, vol. 9665, pp. 789–819. Springer, Heidelberg (2016). https://doi.org/10.1007/978-3-662-49890-3_30

[BDGL16] Becker, A., Ducas, L., Gama, N., Laarhoven, T.: New directions in nearest neighbor searching with applications to lattice sieving. In: Krauthgamer, R. (ed.) 27th SODA. ACM-SIAM, pp. 10–24, January 2016

[BG14] Bai, S., Galbraith, S.D.: Lattice decoding attacks on binary LWE. In: Susilo, W., Mu, Y. (eds.) ACISP 2014. LNCS, vol. 8544, pp. 322–337. Springer, Cham (2014). https://doi.org/10.1007/978-3-319-08344-5_21

[BGJ15] Becker, A., Gama, N., Joux, A.: Speeding-up lattice sieving without increasing the memory, using sub-quadratic nearest neighbor search. Cryptology ePrint Archive, Report 2015/522 (2015). http://eprint.iacr.org/2015/522

[BLS16] Bai, S., Laarhoven, T., Stehle, D.: Tuple lattice sieving. Cryptology ePrint Archive, Report 2016/713 (2016). http://eprint.iacr.org/2016/713

[BSW18] Bai, S., Stehlé, D., Wen, W.: Measuring, simulating and exploiting the head concavity phenomenon in BKZ. In: Peyrin, T., Galbraith, S. (eds.) ASIACRYPT 2018. LNCS, vol. 11272, pp. 369–404. Springer, Cham (2018). https://doi.org/10.1007/978-3-030-03326-2_13

[Cha02] Charikar, M.: Similarity estimation techniques from rounding algorithms. In: 34th ACM STOC. ACM Press, pp. 380–388, May 2002

[Che13] Chen, Y.: Réduction de réseau et sécurité concrète du chiffrement complètement homomorphe. Ph.D. thesis, Thèse de doctorat dirigée par Nguyen, Phong-Quang Informatique Paris 7, p. 1, vol. 133 (2013)

[CN11] Chen, Y., Nguyen, P.Q.: BKZ 2.0: better lattice security estimates. In: Lee, D.H., Wang, X. (eds.) ASIACRYPT 2011. LNCS, vol. 7073, pp. 1–20. Springer, Heidelberg (2011). https://doi.org/10.1007/978-3-642-25385-0_1

[dt18a] The FPLLL Development Team: FPLLL, a lattice reduction library (2018). https://github.com/fplll/fplll

[dt18b] The FPyLLL Development Team: FPyLLL, a lattice reduction library (2018). https://github.com/fplll/fpylll

[Duc18a] Ducas, L.: Shortest vector from lattice sieving: a few dimensions for free. In: Nielsen, J.B., Rijmen, V. (eds.) EUROCRYPT 2018. LNCS, vol. 10820, pp. 125–145. Springer, Cham (2018). https://doi.org/10.1007/978-3-319-78381-9_5

[Duc18b] Ducas, L.: Shortest Vector from Lattice Sieving: a Few Dimensions for Free (talk), April 2018. https://eurocrypt.iacr.org/2018/Slides/Monday/TrackB/01-01.pdf

[FBB+15] Fitzpatrick, R., et al.: Tuning GaussSieve for speed. In: Aranha, D.F., Menezes, A. (eds.) LATINCRYPT 2014. LNCS, vol. 8895, pp. 288–305. Springer, Cham (2015). https://doi.org/10.1007/978-3-319-16295-9_16

[FK15] Fukase, M., Kashiwabara, K.: An accelerated algorithm for solving SVP based on statistical analysis. JIP 23(1), 67–80 (2015)

[FP85] Fincke, U., Pohst, M.: Improved methods for calculating vectors of short length in a lattice, including a complexity analysis. Math. Comput. 44(170), 463–463 (1985)

[FY15] Göpfert, F., Yakkundimath, A.: Darmstadt LWE challenges (2015). https://www.latticechallenge.org/lwe_challenge/challenge.php. Accessed 15 Aug 2018

[GN08a] Gama, N., Nguyen, P.Q.: Finding short lattice vectors within Mordell's inequality. In: Ladner, R.E., Dwork, C. (eds.) 40th ACM STOC. ACM Press, pp. 207–216, May 2008

[GN08b] Gama, N., Nguyen, P.Q.: Predicting lattice reduction. In: Smart, N. (ed.) EUROCRYPT 2008. LNCS, vol. 4965, pp. 31–51. Springer, Heidelberg (2008). https://doi.org/10.1007/978-3-540-78967-3_3

[GNR10] Gama, N., Nguyen, P.Q., Regev, O.: Lattice enumeration using extreme pruning. In: Gilbert, H. (ed.) EUROCRYPT 2010. LNCS, vol. 6110, pp. 257–278. Springer, Heidelberg (2010). https://doi.org/10.1007/978-3-642-13190-5_13

[HK17] Herold, G., Kirshanova, E.: Improved algorithms for the approximate k-list problem in euclidean norm. In: Fehr, S. (ed.) PKC 2017, Part I. LNCS, vol. 10174, pp. 16–40. Springer, Heidelberg (2017). https://doi.org/10.1007/978-3-662-54365-8_2

[HKL18] Herold, G., Kirshanova, E., Laarhoven, T.: Speed-ups and time–memory trade-offs for tuple lattice sieving. In: Abdalla, M., Dahab, R. (eds.) PKC 2018, Part I. LNCS, vol. 10769, pp. 407–436. Springer, Cham (2018). https://doi.org/10.1007/978-3-319-76578-5_14

[HPS11] Hanrot, G., Pujol, X., Stehlé, D.: Analyzing blockwise lattice algorithms using dynamical systems. In: Rogaway, P. (ed.) CRYPTO 2011. LNCS, vol. 6841, pp. 447–464. Springer, Heidelberg (2011). https://doi.org/10.1007/978-3-642-22792-9_25

[HS07] Hanrot, G., Damien, S.: Improved analysis of Kannan's shortest lattice vector algorithm. In: Menezes, A. (ed.) CRYPTO 2007. LNCS, vol. 4622, pp. 170–186. Springer, Heidelberg (2007). https://doi.org/10.1007/978-3-540-74143-5_10

[Kan83] Kannan, R.: Improved algorithms for integer programming and related lattice problems. In: 15th ACM STOC. ACM Press, pp. 193–206, April 1983

[Kan87] Kannan, R.: Minkowski's convex body theorem and integer programming. Math. Oper. Res. 12(3), 415–440 (1987)

[Kir16]   Kirchner, P.: Re: sieving vs. enumeration, May 2016. https://groups.
          google.com/forum/#!msg/cryptanalytic-algorithms/BoSRL0uHIjM/
          wAkZQlwRAgAJ

[LLL82]   Lenstra, A.K., Lenstra, H.W., Lovasz, L.: Factoring polynomials with ratio-
          nal coefficients. Mathematische Annalen **261**(4), 515–534 (1982)

[LM18]   Laarhoven, T., Mariano, A.: Progressive lattice sieving. In: Lange, T.,
          Steinwandt, R. (eds.) PQCrypto 2018. LNCS, vol. 10786, pp. 292–311.
          Springer, Cham (2018). https://doi.org/10.1007/978-3-319-79063-3_14

[LN13]   Liu, M., Nguyen, P.Q.: Solving BDD by enumeration: an update. In: Daw-
          son, E. (ed.) CT-RSA 2013. LNCS, vol. 7779, pp. 293–309. Springer, Hei-
          delberg (2013). https://doi.org/10.1007/978-3-642-36095-4_19

[MV10a]   Madritsch, M., Vallée, B.: Modelling the LLL algorithm by sandpiles. In:
          López-Ortiz, A. (ed.) LATIN 2010. LNCS, vol. 6034, pp. 267–281. Springer,
          Heidelberg (2010). https://doi.org/10.1007/978-3-642-12200-2_25

[MV10b]   Micciancio, D., Voulgaris, P.: Faster exponential time algorithms for the
          shortest vector problem. In: Charika, M. (ed.) 21st SODA. ACM-SIAM,
          pp. 1468–1480, January 2010

[MW15]   Micciancio, D., Walter, M.: Fast lattice point enumeration with minimal
          overhead. In: Indyk, P. (ed.) 26th SODA. ACM-SIAM, pp. 276–294, Jan-
          uary 2015

[Ngu10]   Nguyen, P.Q.: Hermités constant and lattice algorithms. In: Nguyen, P.,
          Valle, B. (eds.) The LLL Algorithm. Information Security and Cryptogra-
          phy, pp. 19–69. Springer, Heidelberg (2010). https://doi.org/10.1007/978-
          3-642-02295-1_2

[NV08]   Nguyen, P.Q., Vidick, T.: Sieve algorithms for the shortest vector problem
          are practical. J. Math. Cryptol. **2**(2), 181–207 (2008)

[PAA+17] Poppelmann, T., et al.: Newhope, Technical report, National Institute
          of Standards and Technology (2017). https://csrc.nist.gov/projects/post-
          quantum-cryptography/round-1-submissions

[Sch87]   Schnorr, C.-P.: A hierarchy of polynomial time lattice basis reduction algo-
          rithms. Theor. Comput. Sci. **53**, 201–224 (1987)

[Sch03]   Schnorr, C.P.: Lattice reduction by random sampling and birthday meth-
          ods. In: Alt, H., Habib, M. (eds.) STACS 2003. LNCS, vol. 2607, pp. 145–
          156. Springer, Heidelberg (2003). https://doi.org/10.1007/3-540-36494-
          3_14

[SE94]   Schnorr, C.P., Euchner, M.: Lattice basis reduction: improved practical
          algorithms and solving subset sum problems. Math. Program. **66**(1), 181–
          199 (1994)

[SG10]   Schneider, M., Gama, N.: Darmstadt SVP Challenges (2010). https://
          www.latticechallenge.org/svp-challenge/index.php. Accessed 17 Aug 2018

[TKH18]   Teruya, T., Kashiwabara, K., Hanaoka, G.: Fast lattice basis reduction
          suitable for massive parallelization and its application to the shortest vector
          problem. In: Abdalla, M., Dahab, R. (eds.) PKC 2018, Part I. LNCS, vol.
          10769, pp. 437–460. Springer, Cham (2018). https://doi.org/10.1007/978-
          3-319-76578-5_15

[Wal16]   Walter, M.: Sage implementation of Chen and Nguyen's BKZ simulator
          (2016). http://pub.ist.ac.at/~mwalter/src/sim_bkz.sage

[YD17]   Yu, Y., Ducas, L.: Second order statistical behavior of LLL and BKZ. In:
          Adams, C., Camenisch, J. (eds.) SAC 2017. LNCS, vol. 10719, pp. 3–22.
          Springer, Cham (2018). https://doi.org/10.1007/978-3-319-72565-9_1

Misuse Attacks on Post-quantum Cryptosystems

Ciprian Băetu$^{(\boxtimes)}$, F. Betül Durak, Loïs Huguenin-Dumittan,
Abdullah Talayhan, and Serge Vaudenay

EPFL, 1015 Lausanne, Switzerland
b.cip11@gmail.com
http://lasec.epfl.ch

Abstract. Many post-quantum cryptosystems which have been proposed in the National Institute of Standards and Technology (NIST) standardization process follow the same meta-algorithm, but in different algebras or different encoding methods. They usually propose two constructions, one being weaker and the other requiring a random oracle. We focus on the weak version of nine submissions to NIST. Submitters claim no security when the secret key is used several times. In this paper, we analyze how easy it is to run a key recovery under multiple key reuse. We mount a classical key recovery under plaintext checking attacks (i.e., with a plaintext checking oracle saying if a given ciphertext decrypts well to a given plaintext) and a quantum key recovery under chosen ciphertext attacks. In the latter case, we assume quantum access to the decryption oracle.

1 Introduction

By anticipating that quantum computers will eventually compute discrete logarithms and integer factorization [22], all public key cryptosystems which are being used today will break down. There is an urgent need to replace them. For this purpose, the US National Institute of Standards and Technology (NIST) initiated a standardization process for post-quantum algorithms. The call for proposals expired in 2017 resulting with many submissions. Among these, only a very few types of algorithms are proposed such as lattice-based, code-based, hash-based, or isogeny-based. One of the most promising algorithms is lattice-based.

Many of the lattice-based cryptosystems are inspired by the Regev cryptosystem [20] such as the Lyubashevsky-Peikert-Regev cryptosystem [18]. We can easily extract a common pattern in proposals which followed them. All proposed constructions require never to reuse the secret key because of possible attacks. In other words, the proposed cryptosystems are ephemeral in the sense that the secret key is meant to be used only once. The approach of designers is to start from this ephemeral construction and then to transform it into a strongly secure (i.e. with reusable keys) key encapsulation mechanism (KEM) by using

© International Association for Cryptologic Research 2019
Y. Ishai and V. Rijmen (Eds.): EUROCRYPT 2019, LNCS 11477, pp. 747–776, 2019.
https://doi.org/10.1007/978-3-030-17656-3_26

the Fujisaki-Okamoto transformation or one of its variants [12,13,15,25]. What transformations share in common is that they imply a computation overhead. Concretely, after normal decryption, a re-encryption is made to check if the ciphertext was correctly formed. This re-encryption does not seem so useful to non-experts. Additionally, these transformations need a random oracle which has no practical existence.

When there is a cryptosystem which is practical but comes with a warning and an extension which looks only motivated by academic people, we believe that users will eventually try to use the weakly secure cryptosystems and pay little attention to the warning, or even misunderstand the strengthened version, just because the threat is not clear. For this reason, we should understand what are the risks under misuse of keys.

Another observation was made by Lepoint [17]. He checked that several implementations of the strongly secure KEM have side channels leaking the result of the decryption under the weak cryptosystem. It comes from a mis-implementation of the Fujisaki-Okamoto (FO) transform. In the FO transformation, the decryption is done, then the verification checks that the ciphertext is well-formed. The result is released only if the test passes. However, Lepoint has shown that side channels in implementations were leaking the result in any case, no matter whether the ciphertext was well-formed or not.

In 2015, the NSA [16] reported some concerns about recurring problems with key leakage in key agreement protocols. They suggested to explicitly check that ciphertexts are well-formed by using the FO transform. This recommendation was followed by designers, as mentioned above.

In 2016, Fluhrer [11] published an attack based on the key reuse. In his attack, an adversary encrypts a message by deviating a bit from the protocol. Then, he sends the ciphertext for decryption and checks if the decryption matches what he expected. After a few trials, the adversary recovers the secret key. The attack applies to all protocols using a special signaling (a.k.a. error-reconciliation) function. In 2017, Ding et al. [10] expanded this attack to a class of key agreement protocols based on ring-LWE with signaling. Our goal is to apply the Fluhrer attack model to more protocols and to minimize the number of key reuse to recover the key and to be able to assess how weak those protocols are under key reuse.

At CT-RSA'2019, Bauer et al. [5] presented an attack on the weak version of NewHope-CPA-PKE [1]. For $n = 1\,024$, they recover the secret with high probability using 2^{14} queries to a *key mismatch oracle*, what we herein call a *plaintext checking oracle* (PCA).

Our contribution. In this paper, we first define the meta-structure of constructions with ephemeral keys. Then, we formalize the noise learning problem which is required to break these constructions. We identify optimal bounds in terms of the number of oracle calls to solve this problem. Then, we mount a classical key recovery under plaintext checking attack (KR-PCA), which is also the model of Fluhrer attacks [11]. This model makes sense when an adversary can play with a server with a modified ciphertext and check if it still decrypts to the same plaintext as before. Compared to Fluhrer [11], we apply it to different classes of protocols, we optimize the number of oracle calls, and we identify the link with the noise learning problem. We give optimal lower bounds for the number of oracle calls to solve this problem. We finally propose a quantum variant of the attack in an "imaginary" model where the adversary has quantum access to a decryption oracle. This is a key recovery under chosen ciphertext attack (KR-CCA). Our quantum attack is based on the GKZ algorithm [14] to solve LWE from a superposition of inputs. We also adapt the AJOP attack [2] based on the Bernstein-Vazirani algorithm [7] to solve the LPN problem from a super-position of inputs.[1] The AJOP-based attack has better performances but is more restrictive. It only works for cryptosystems of a special form and it also assumes a quantum decryption oracle working with addition in a special group instead of the XOR. Our result shows that a single use of the key leads to a full or partial key recovery with a probability of success proving the attacks are a big threat.

Table 1. Attacks on post-quantum cryptosystems. For two types of attacks (classical KR-PCA and quantum KR-CCA), we report the number of oracle calls as O, the probability of success as P, the number of collected linear equations in $\mathbf{Z}_q$ as E, and the number of unknowns in $\mathbf{Z}_q$ as U. We also indicate for information the expected total number $T = \frac{OU}{PE}$ of oracle calls obtained to recover the full key with probability 1 by iterating the attacks.

		classical KR-PCA attack				GKZ-based quantum KR-CCA attack				AJOP-based quantum KR-CCA attack			
	U	O	P	E	(T)	O	P	E	(T)	O	P	E	(T)
EMBLEM128	2^{10}	2^9	1	2^5	(2^{14})	2	2^{-16}	2^{10}	(2^{17})	1	1	2^{10}	(1)
R.EMBLEM128	2^9	2^{13}	1	2^9	(2^{13})	2	2^{-24}	2^9	(2^{25})	1	2^{-1}	2^9	(2)
Frodo-640	2^{12}	2^{10}	1	2^6	(2^{16})	2	2^{-13}	2^9	(2^{17})	1	2^{-2}	2^{12}	(2^2)
KINDI256	2^{10}	2^{12}	1	2^8	(2^{14})	2	2^{-14}	2^{10}	(2^{15})	1	2^{-1}	2^{10}	(2)
Lepton Light I	2^{13}	2^{13}	1	2^{12}	(2^{14})	-	-	-	-	-	-	-	-
LIMA227-2p	2^{10}	2^{14}	1	2^{10}	(2^{14})	2	2^{-17}	2^{10}	(2^{18})	1	2^{-1}	2^{10}	(2)
LIMA152-sp	2^{10}	2^{15}	1	2^{10}	(2^{15})	2	2^{-24}	2^{10}	(2^{25})	1	2^{-1}	2^{10}	(2)
Lizard536	2^{17}	-	-	-	-	2	2^{-9}	2^9	(2^{18})	1	2^{-1}	2^9	(2^9)
RLizard536	2^{10}	-	-	-	-	2	2^{-8}	2^{10}	(2^9)	1	2^{-1}	2^{10}	(2)
LOTUS128	2^{16}	2^{11}	1	2^7	(2^{20})	2	2^{-13}	2^9	(2^{21})	1	2^{-1}	2^9	(2^8)
NewHope512	2^9	-	-	-	-	2	2^{-28}	2^9	(2^{29})	1	2^{-1}	2^9	(2)
TitaniumStd128	2^{11}	2^{12}	1	2^8	(2^{15})	2	2^{-16}	2^{10}	(2^{18})	1	2^{-1}	2^{10}	(2^2)

We report our results for both types of attacks in Table 1. *For information only*, the table indicates the total number of oracle calls we need, by iterating, to recover the full key with probability 1. We should, however, take this mea-surement with care. This is because running a single instance of the attack may

---

[1] The AJOP attack was released after we submitted this paper. For completeness, we include its adaptation here.

be enough to decrease the security of the cryptosystem and to recover the secret by other means.

2   A Meta-**PKC** Construction

We define a cryptosystem as follows.

Definition 1 (Public-key cryptosystem). *A* public-key cryptosystem (PKC) *with security parameter λ consists of four algorithms:*

- PKC.setup$(1^\lambda; \text{coinS}) \overset{\$}{\to} \text{pp}$
- PKC.gen$(\text{pp}; \text{coinA}) \overset{\$}{\to} (\text{sk}, \text{pk})$
- PKC.enc$(\text{pp}, \text{pk}, \text{pt}; \text{coinB}) \overset{\$}{\to} \text{ct}$
- PKC.dec$(\text{pp}, \text{sk}, \text{ct}) \to \text{pt}'$

Correctness *implies that for any* pt, *by running all these algorithms, we obtain* pt $=$ pt$'$ *with probability* $1 - \text{negl}(\lambda)$, *over the random selection of the coins.*

In this paper, we consider two types of security notions.

Definition 2 (KR-PCA and KR-CCA). *We use the key recovery game with oracle $\mathcal{O}$ of Fig. 1. We consider two types of oracles:* PCO *and* DecO *which are defined on the figure. The KR-PCA game uses $\mathcal{O} = $ PCO. The KR-CCA game uses $\mathcal{O} = $ DecO.*

PCO(ct, pt) is a plaintext checking oracle which receives ct and pt, runs the decryption and only returns one bit saying if it decrypts to pt. KR-PCA is an adaptive key recovery attack. Security against KR-PCA is implied by IND-CCA security. KR-PCA attacks are *not* in the IND-CPA security framework. Hence, a PKC could be IND-CPA secure but still vulnerable to a KR-PCA attack.

Game $\text{KR}_{\mathcal{A}}^{\mathcal{O}}(\lambda)$:
1: pick coinS, coinA
2: setup$(1^\lambda; \text{coinS}) \to \text{pp}$
3: gen$(\text{pp}; \text{coinA}) \to (\text{sk}, \text{pk})$
4: $\mathcal{A}^{\mathcal{O}(\cdot)}(\text{pp}, \text{pk}) \to \text{sk}'$
5: **return** $1_{\text{sk}=\text{sk}'}$

Oracle PCO(ct, pt):
1: dec$(\text{pp}, \text{sk}, \text{ct}) \to \text{pt}'$
2: **return** $1_{\text{pt}'=\text{pt}}$

Oracle DecO(ct):
3: **return** dec$(\text{pp}, \text{sk}, \text{ct})$

Fig. 1. KR-PCA and KR-CCA games.

The PCA model makes sense in several cases. For instance, in the client-server protocol where the encryption is used to transport a symmetric key to start secure messaging, an adversary can try to encrypt a symmetric key by deviating from the protocol. He generates malformed ciphertexts which may decrypt to the chosen symmetric key or not. By sending the malformed ciphertext to the server, the adversary can easily see if secure messaging with the server is possible, hence simulate a PCO oracle. Clearly, it is devastating that such an attack would lead to a key recovery.

We define the following algebra. We consider six *additive Abelian groups* S_{sk}, S_A, S_B, S_t, S_U, and S_V and four *bilinear mappings* which are all denoted with $\times$. The four bilinear mappings have domains $S_A \times S_{sk} \to S_B$, $S_U \times S_{sk} \to S_V$, $S_t \times S_A \to S_U$, and $S_t \times S_B \to S_V$. We assume associativity in the sense that

$$(t \times A) \times sk = t \times (A \times sk)$$

for all $t \in S_t$, $A \in S_A$, and $sk \in S_{sk}$. Hence, multiplication works as in the diagram on Fig. 2.

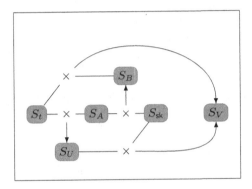

Fig. 2. Algebra: the four bilinear functions on the six spaces. For instance, one element of S_t multiplied by one element of S_A gives one element of S_U.

We also assume that there is a norm $\| \cdot \|$ on S_{sk}, S_B, S_t, S_U, and S_V (i.e. all spaces except S_A for which we need no norm). By definition, the norm is positive and satisfies the triangular inequality. We assume that we can upper bound $\|x \times y\|$ in terms of $\|x\|$ and $\|y\|$ for the four bilinear functions.

Finally, we assume two functions encode : $\mathcal{M} \to S_V$ and decode : $S_V \to \mathcal{M}$ such that encode is injective. The image set $C = \text{encode}(\mathcal{M})$ is called a *code*. Elements of the code are *codewords*. The packing radius of C is denoted as ρ_- and the covering radius of C is denoted as ρ_+. Around every $W \in S_V$, balls of radius ρ_- contain no more than one codeword and balls of radius ρ_+ contain at least one codeword:

$$\forall V, V' \in C \quad \forall W \in S_V \quad (\|V - W\| \leq \rho_- \wedge \|V' - W\| \leq \rho_-) \Longrightarrow V = V'$$
$$\forall W \in S_V \quad \exists V \in C \quad \|V - W\| \leq \rho_+$$

Additionally,

$$\forall W \in S_V \quad \text{decode}(W) = \arg\min_{\text{pt}} \|W - \text{encode}(\text{pt})\|$$

in the sense that encode(decode(W)) is *one* closest codeword to W (which may be ambiguous if there is no unique closest codeword). This implies

$$\forall W \in S_V \quad \forall \mathsf{pt} \quad \|W - \mathsf{encode}(\mathsf{pt})\| \le \rho_- \implies \mathsf{decode}(W) = \mathsf{pt}$$
$$\forall W \in S_V \quad \|W - \mathsf{encode}(\mathsf{decode}(W))\| \le \rho_+$$

An element δ such that $\|\delta\| \le \rho_-$ will be called *sparse*. In what follows, we use small letters to designate sparse elements.

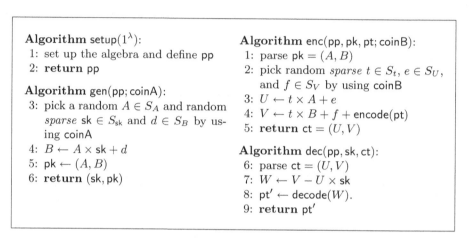

Fig. 3. The meta-cryptosystem defined on the algebra.

We define a PKC as on Fig. 3 in which the choice of the algebra, norm, encoding/decoding, and the probability distributions are left free. Thanks to bilinearity and associativity, we have $W = \delta + \mathsf{encode}(\mathsf{pt})$ with

$$\delta = t \times d + f - e \times \mathsf{sk} \tag{1}$$

This value δ will be called the *noise*. By controlling the size of t, d, f, e, sk (with their respective probability distribution), we can make sure that the noise δ is sparse. Hence, $\mathsf{decode}(W) = \mathsf{pt}$.

In what follows, for an element $X \in \mathbf{Z}_q$, $|X|$ is the absolute value of $X \in \mathbf{Z}_q$ when represented modulo q such that $-\frac{q}{2} \le X \le \frac{q}{2}$. As an example of norm over $\mathbf{Z}_q^n$, we can consider the L_∞ norm $\|X\| = \max(|X_1|, \ldots, |X_n|)$, or the L_1 norm $\|X\| = |X_1| + \cdots + |X_n|$, or a combination of both $\|X\| = \max_i \sum_j |X_{i,j}|$.

As an example of encode/decode with the L_∞ norm, we could consider $\mathcal{M} = \{0, 1, \ldots, z-1\}^n$ (z is the alphabet size) and $\mathsf{encode}(\mathsf{pt}) = L \cdot \mathsf{pt}$ in $S_V = \mathbf{Z}_q^n$, for a positive integer L such that $L > 2\rho_-$ and $zL \le q$. We define $\mathsf{decode}(X) = \lfloor \frac{X}{L} \rceil$ component-wise when the coordinates of X are taken modulo q.

We only give three examples below. More are provided in Appendix A.

Example 3 (Frodo). FrodoPKE [8] works with $S_{sk} = S_B = \mathbf{Z}_q^{n\bar{n}}$, $S_A = \mathbf{Z}_q^{n^2}$, $S_t = S_U = \mathbf{Z}_q^{\bar{m}n}$, and $S_V = \mathbf{Z}_q^{\bar{m}\bar{n}}$ with the L_∞ norm and $\mathcal{M} = \{0,1\}^{\ell\bar{m}\bar{n}}$. It uses $q = 2^{15}$, $\ell = 2$, $\bar{m} = \bar{n} = 8$, and $n = 640$ (for the Frodo-640 parameters). The bilinear mappings are matrix multiplications. I.e., elements of $S_{sk} = S_B$ are $n \times \bar{n}$ matrices, elements of S_A are $n \times n$ matrices, elements of $S_t = S_U$ are $\bar{m} \times n$ matrices, and elements of S_V are $\bar{m} \times \bar{n}$ matrices. Frodo uses the L_∞ norm (when considering elements as vectors). Encoding ℓ bits per $\mathbf{Z}_q$ elements is done by taking them as an integer and multiplying them by $L = q2^{-\ell}$. Decoding takes the ℓ most significant bits. So, each $\mathbf{Z}_q$ element is at a distance up to $\rho_+ = \rho_- = q2^{-\ell-1}$ to a codeword. Elements of t, d, f, e, sk, are sampled in $\{-11, \ldots, +11\}$ with Gaussian-looking distribution.

Example 4 (NewHope). NewHope-CPA-PKE [1] defines $S_{sk} = S_A = S_B = S_t = S_U = S_V = \mathbf{Z}_q^n$. Elements are considered as polynomials in variable X modulo $X^n + 1$. Bilinear mappings are simply multiplications of polynomials in this structure. Message bits are encoded by multiplication to $L = q/2$ and represented twice in NewHope512-CPA-PKE. Namely, if $Y = \mathsf{encode}(\mathsf{pt})$, the bit pt_i of the message appears at position i and $i + 256$ of Y by $Y_i = Y_{i+256} = \mathsf{pt}_i \frac{q}{2}$. How decoding works is also important. Namely, if $Y \in S_V$, the algorithm decodes $b = (\mathsf{decode}(Y))_i$ to the value b minimizing $|Y_i - b\frac{q}{2}| + |Y_{i+256} - b\frac{q}{2}|$, i.e. the L_1 distance. For this reason, it uses an L_1 norm on pairs of components at position i and $i + 256$ and the L_∞ norm over all i. Namely,

$$\|Y\| = \max_i(|Y_i| + |Y_{i+256}|)$$

Hence, $\rho_- = \frac{q}{2}$ and $\rho_+ = \frac{q}{2}$. For t, d, f, e, sk, sparse elements of $\mathbf{Z}_q$ are sampled by taking the difference of the Hamming weight of two uniformly distributed random bytes. Hence, they are in $\{-8, \ldots, +8\}$. NewHope deviates a bit from our meta-construction in the sense that encryption replaces V by its compression $\bar{V} = \lceil \frac{p}{q}V \rfloor$ and decryption replaces $\bar{V}$ by $V' = \lceil \frac{q}{p}\bar{V} \rfloor$. This adds an error bounded by $\frac{q}{p}$ in δ. In NewHope512-CPA-PKE, the parameters are $n = 512$, $q = 12\,289$, and $p = 8$.

Example 5 (Lepton). Lepton.CPA [26] defines $S_{sk} = S_A = S_B = S_t = S_U = \mathbf{Z}_2^n$ and $S_V = \mathbf{Z}_2^\ell$ with the Hamming weight norm. All bilinear functions are the $\mathrm{GF}(2^n)$ multiplications represented with the trinomial $X^n + X^m + 1$, except for $S_t \times S_B \to S_V$ which does the multiplication then truncate to ℓ bits. For t, d, f, e, and sk, of weight bounded by k, it is proven that δ has a weight bounded by $4k^2 + k + 2m - 2$. Encoding uses a BCH code and a repetition code to correct at least this amount of error. The Light I version of Lepton.CPA uses $n = 8\,100$, $k = 16$, $m = 9$, and $\ell = 4\,572$. The BCH code is a $[508, 256]$ binary code which can decode up to 30 errors. After BCH-encoding, it uses a repetition code with 9 repetitions. By design, $\|\delta\| \leq 1\,056$.

3   The Noise Learning Problem

In this section, we will introduce a noise learning problem with examples for different metrics. Solving it is the heart of the KR-PCA attack that we describe in the next section. We prove lower bounds for the number of oracle calls to solve this problem.

3.1 Definition

We consider a sampling algorithm S generating a sparse δ (which we call *noise*), a threshold ρ, and an algorithm $\mathsf{learn}^{\mathsf{BOO}(\cdot)}$ which makes r queries to a *bounded offset oracle* BOO. We will later give a few instances of the learning problem. We define $\mathsf{Adv}(\mathsf{LEARN}_S^\rho(\lambda))$ as the probability that the game on Fig. 4 gives 1.

Game $\mathsf{LEARN}_S^\rho(\lambda)$:
1: pick coinS
2: $\mathsf{setup}(1^\lambda; \mathsf{coinS}) \to \mathsf{pp}$
3: pick the noise δ using the sampling algorithm S
4: $\mathsf{learn}^{\mathsf{BOO}(\cdot)}(\mathsf{pp}) \to \delta'$
5: **return** $1_{\delta = \delta'}$

Oracle $\mathsf{BOO}(x)$:
1: **return** $1_{\|\delta + x\| \le \rho}$

Fig. 4. The noise learning game with threshold ρ.

In all cases, the idea of the learning algorithm is to start with $x = 0$ and to gradually increase $\|x\|$ (i.e. reduce the sparsity of $\delta + x$) until the decoding is no longer the same. This is a hill-climbing method. Then, we can explore the top of the hill with small modifications of this critical x and analyze the decoding algorithm to deduce the value of δ.

3.2 Lower Bounds for the Noise Learning Problem

Theorem 6. *Given a probability distribution for δ, we assume that an algorithm* learn *has an advantage of 1 in the* **LEARN** *game, using r queries. We have*

$$E(r) \ge H(\delta)$$

where H is the Shannon entropy.

Proof. Since we do not consider the running time, we assume without loss of generality that learn uses the random coins minimizing $E(r)$. Hence, we consider it as deterministic. We let $C(\delta)$ be the sequence of answers from the oracle BOO when δ is sampled. When running learn alone with oracle answers simulated by the sequence $b_1, \ldots, b_r$, we let $D(b_1 \cdots b_r) = \delta'$ be the output from learn. We let r_δ be the length of $C(\delta)$. Due to the hypothesis that the advantage is 1, we have

$D(C(\delta)) = \delta$ for all δ. Thanks to this property, the Kraft Inequality says that $\sum_\delta 2^{-r_\delta} \le 1$. We have

$$E(r_\delta) - H(\delta) = \sum_\delta \Pr[\delta] \log_2(2^{r_\delta} \Pr[\delta])$$

$$\ge \frac{1}{\ln 2} \sum_\delta \Pr[\delta] \left(1 - \frac{1}{2^{r_\delta} \Pr[\delta]}\right)$$

$$= \frac{1}{\ln 2}\left(1 - \sum_\delta 2^{-r_\delta}\right)$$

$$\ge 0$$

by using $\ln x \ge 1 - \frac{1}{x}$. $\qquad\square$

Theorem 7. *Given a probability distribution for δ, we assume that an algorithm* learn *has an advantage of p in the* LEARN *game, using r queries. We have*

$$E(r|\text{success}) \ge H_\infty(\delta) + \log_2 p$$

where H_∞ is the min-entropy and success *is the event that* LEARN *returns 1.*

Proof. As in Theorem 6, we assume without loss of generality that learn is deterministic. Let S be the distribution of δ and we define $\bar{S} = [S|\delta' = \delta]$, i.e. the distribution S conditioned to $\delta' = \delta$. The advantage with δ of distribution $\bar{S}$ is 1. Due to the previous result, we have that $E_{\bar{S}}(r) \ge H(\bar{S})$. We have

$$H(\bar{S}) = -\sum_{\delta=\delta'} \frac{\Pr[\delta]}{p} \log_2 \frac{\Pr[\delta]}{p} \ge \sum_{\delta=\delta'} \frac{\Pr[\delta]}{p}(H_\infty(\delta) + \log_2 p) = H_\infty(\delta) + \log_2 p$$

and

$$E_{\bar{S}}(r) = E_S(r|\delta = \delta') = E_S(r|\text{success})$$

therefore, $E(r|\text{success}) \ge H_\infty(\delta) + \log_2 p$. $\qquad\square$

3.3 Example: Learning a Small Integer

We consider the learning problem over $\mathbf{Z}_q$. We assume that ρ and p are defined parameters. If $\rho = \lfloor \frac{q}{2} \rfloor$, the BOO oracle always answers 1 and the problem is unsolvable. Hence, we assume that $\rho < \lfloor \frac{q}{2} \rfloor$. We design a learning algorithm with parameter p by a cut-and-choose algorithm as shown on Fig. 5. We first assume that ρ is known.

Theorem 8. *Assume that $\rho < \lfloor \frac{q}{2} \rfloor$. The algorithm on Fig. 5 succeeds with probability at least p. The number r of oracle calls satisfies*

$$E(r) \le 1.41 + 2.41 \times (H(\delta) + \log_2 p) + r_1^{\max}$$

with $r_1^{\max} = \frac{2\rho}{\lfloor \frac{q}{2} \rfloor - \rho}$ if $\rho > \frac{q}{6} - \frac{1}{3}$ and $r_1^{\max} = 0$ otherwise. When δ is uniform in $[-\rho, +\rho]$, we have

$$E(r) \le H(\delta) + \log_2 p + 1 + r_1^{\max}$$

Algorithm learn(pp):
1: define q from pp
2: set $a = -\rho$ ▷ we have $\mathsf{BOO}(-a - \rho) = 1$
3: set $b = \rho$
4: **if** $\rho > \frac{q}{6} - \frac{1}{3}$ **then**
5:    $\ell = \lfloor \frac{q}{2} \rfloor - \rho$ ▷ $-\rho \le a \le \delta \le b$
6:    **while** $\max_{\delta \in [a,b]} \Pr[\delta] < p \cdot \Pr[\delta \in [a,b]]$ and $b - a \ge \ell$ **do**
7:       $b' = a + \ell - 1$
8:       **if** $\mathsf{BOO}(-b' - 1 - \rho) = 1$ **then** $a \leftarrow b' + 1$ **else** $b \leftarrow b'$
9:    **end while**
10: **end if**   ▷ we have $\mathsf{BOO}(-b - 1 - \rho) = 0$ and $b - a \le \ell - 1$
11: **while** $\max_{\delta \in [a,b]} \Pr[\delta] < p \cdot \Pr[\delta \in [a,b]]$ **do**   ▷ we have $a \le \delta \le b$
12:    cut $[a, b] = [a, c] \cup [c + 1, b]$ which minimizes

$$\max \left(\Pr[\delta \in [a, c]], \Pr[\delta \in [c + 1, b]] \right)$$

▷ the median of $(\delta | [a \le \delta \le b])$ is either c or $c + 1$
13:    **if** $\mathsf{BOO}(-c - 1 - \rho) = 1$ **then** $a \leftarrow c + 1$ **else** $b \leftarrow c$
14: **end while**
15: pick $\delta = \arg\max_{\delta \in [a,b]} \Pr[\delta]$
16: **return** δ

Fig. 5. Learning an integer with probability p.

For $\rho \approx \frac{q}{4}$, $r_1^{\max}$ is typically 2. If we neglect this overhead, we deduce that the learning algorithm is optimal up to a factor of 2.41 in general. It is further optimal for a uniform distribution.

Proof. Let $\ell = \lfloor \frac{q}{2} \rfloor - \rho$ and $b' = a + \ell - 1$. If $-\rho \le a \le \delta \le b$, we have $\delta - b' - 1 - \rho = \delta - a - \ell - \rho \ge -\ell - \rho = -\lfloor \frac{q}{2} \rfloor$ and $\delta - b' - 1 - \rho = \delta - a - \ell - \rho \le \delta - \ell \le \rho$ thus $\|\delta - b' - 1 - \rho\| \le \rho$ is equivalent to $\delta - b' - 1 - \rho \ge -\rho$. This means that $\mathsf{BOO}(-b' - 1 - \rho) = 1$ is equivalent to $\delta \ge b' + 1$. This shows that the condition $-\rho \le a \le \delta \le b$ is preserved in the loop in Step 6–9.

The loop in Step 6–9 terminates as soon as soon as $\mathsf{BOO}(-b' - 1 - \rho) = 0$. The previous argument shows that the loop terminates when $\delta \le b'$. Hence, the number r_1 of iterations of this loop is bounded by $r_1 \le \frac{2\rho}{\ell}$ which is $r_1^{\max}$.

The purpose of both loops is to find an interval $[a, b]$ containing δ such that $\max_{\delta \in [a,b]} \Pr[\delta] < p \cdot \Pr[\delta \in [a, b]]$. The loop in Step 6–9 either directly finds this interval or finds one containing δ such that $b - a \le \ell - 1$. If $\rho \le \frac{q}{6} - \frac{1}{3}$, this loop is skipped but we already have $b - a = 2\rho \le \ell - 1$.

We can repeat the previous argument: if δ and c belong to $[a, b]$, $b - a \le \ell - 1$, and $b - a \le 2\rho$, then $\delta - c - 1 - \rho \ge \delta - a - \ell - \rho \ge -\ell - \rho = -\lfloor \frac{q}{2} \rfloor$ and $\delta - c - 1 - \rho \le b - a - \rho \le \rho$ thus $\|\delta - c - 1 - \rho\| \le \rho$ is equivalent to $\delta - c - 1 - \rho \ge -\rho$. This means that $\mathsf{BOO}(-c - 1 - \rho) = 1$ is equivalent to $\delta \ge c + 1$. Hence, we prove by induction that at every step of the loop in Step 11–14, δ belongs to $[a, b]$, $b - a \le \ell - 1$, and $b - a \le 2\rho$.

When the algorithm terminates, it returns δ in the interval with maximal likelihood. The loop enforces that $\Pr[\delta | a \leq \delta \leq b] \geq p$ for all δ in the interval. Hence, the probability that the best δ in the interval is correct is at least p.

Let r_2 be the number of iterations. Let a_1, b_i, c_i be the values of a, b, c in the ith iteration. In the ith iteration, we let $p_i = \Pr[a_i \leq \delta \leq b_i]$, $x_i \in \{c_i, c_i + 1\}$ be the median point in the $[a_i, b_i]$ interval, $m_i = \Pr[\delta = x_i]$, and we set $d_i = 1_{x_i \in [a_{i+1}, b_{i+1}]}$. We have $a_0 = -\rho$, $b_0 = \rho$, $p_0 = 1$.

When cutting $[a_i, b_i]$ in two intervals $[a_i, c_i]$ and $[c_i + 1, b_i]$, we let x_i in the most probable of these two intervals. Hence, $p_{i+1} \leq \frac{p_i + d_i m_i}{2}$.

Another property of the x_i is that when $d_i = 1$, it becomes one border of the next interval. Hence, if it is taken in a further iteration as a median point, it can only be the final iteration reducing the interval to the point x_i, which is the r_2th iteration. We deduce that the sequence of x_i is non-repeating until the r_2th iteration. One consequence is that $m_i + m_{i+1} + \cdots + m_{r_2-1} \leq p_i$ for any $i < r_2$.

Let $S_i = d_i m_i + d_{i+1} m_{i+1} + \cdots + d_{r_2-1} m_{r_2-1}$. We have $S_i \leq p_i$. We also have

$$p_{i+1} \leq \frac{p_i + S_i - S_{i+1}}{2}$$

for all $i < r_2$. By induction, we prove

$$p_i \leq 2^{-i} + 2^{-i-1} S_0 + \sum_{j=0}^{i-1} 2^{-i+j-1} S_j - \frac{1}{2} S_i$$

for all $i < r_2$. We deduce

$$2^i p_i \leq \frac{3}{2} + \frac{1}{2} \sum_{j=0}^{i-1} 2^j p_j$$

and hence $2^i p_i \leq \left(\frac{3}{2}\right)^{i+1}$ by induction. Therefore, $p_i \leq \frac{3}{2} \left(\frac{3}{4}\right)^i$ for $i < r_2$.

For $i \geq \frac{\log\left(\frac{2\Pr[\delta]}{3p}\right)}{\log \frac{3}{4}}$, we have $\Pr[\delta] \geq \frac{3}{2} \left(\frac{3}{4}\right)^i p \geq p_i p$. So, any δ with probability of occurrence $\Pr[\delta]$ makes $r_2 \leq \frac{\log\left(\frac{2\Pr[\delta]}{3p}\right)}{\log \frac{3}{4}}$ iterations. Hence,

$$E(r_2) \leq \frac{\log \frac{2}{3}}{\log \frac{3}{4}} + \frac{1}{\log_2 \frac{3}{4}} E\left(\log_2 \frac{\Pr[\delta]}{p}\right)$$

$$= \frac{\log \frac{2}{3}}{\log \frac{3}{4}} - \frac{1}{\log_2 \frac{3}{4}} (H(\delta) + \log_2 p)$$

$$\leq 1.41 + 2.41(H(\delta) + \log_2 p)$$

We treat the uniform case similarly. We let $\ell_i = b_i - a_i + 1$. We have $\ell_0 = 2\rho + 1$. We have $\ell_{i+1} \leq \frac{\ell_i + 1}{2}$. Hence, $\ell_i \leq 2^{1-i}\rho + 1$. For $i \geq \log_2(2\rho) + \log_2 p$, we have $\ell_i p \leq 1 + p$. The ℓ_i sequence is strictly decreasing until $i = r_2$. Hence, if $i < r_2$, then $\ell_{i+1} \leq \ell_i - 1$ so $\ell_{i+1}p \leq 1$ which implies $i + 1 = r_2$. Therefore $r_2 \leq i + 1$. Hence,

$$E(r_2) \leq \log_2(2\rho) + \log_2 p + 1 \leq H(\delta) + \log_2 p + 1$$

The number of oracle calls is $r = r_1 + r_2$.   □

In $\mathbf{Z}_q^n$ with the L_∞ norm, we can run this algorithm for each coordinate and use $r \leq 1.41 + 2.41 \times n \log_2(2\rho + 1)$ queries to learn δ (using that the entropy of a single component is bounded by $\log_2(2\rho + 1)$).

Learning an integer with unknown threshold. The previous algorithm does not work if the threshold ρ is unknown. However, we could learn the offset x such that $\delta + x$ reaches this unknown threshold, either $+\rho$ or $-\rho$, by using an upper bound ρ' on this threshold. We write the algorithm on Fig. 6 for the uniform distribution and $\rho' \leq \frac{q}{6} - \frac{1}{3}$, for simplicity. (For $\rho' > \frac{q}{6} - \frac{1}{3}$, we apply the same strategy as previously to first find a small interval and run the algorithm on this interval.) The algorithm first learns $\rho - \delta$ on Step 8. Similarly, it learns $-\rho - \delta$ on Step 15. Finally, it deduces both δ and ρ.

Algorithm learn(pp):
1: define q from pp
2: set $a = 0$ ▷ we have $\mathsf{BOO}(a) = 1$
3: set $b = 2\rho' + 1$ ▷ we have $\mathsf{BOO}(b) = 0$
4: **while** $b > a + 1$ **do**   ▷ we have $a \leq \rho - \delta < b$
5:      set $c = \lfloor \frac{a+b}{2} \rceil$
6:      **if** $\mathsf{BOO}(c) = 1$ **then** $a \leftarrow c$ **else** $b \leftarrow c$
7: **end while**
8: $x \leftarrow a$ ▷ $x = \rho - \delta$
9: set $a = -2\rho' - 1$ ▷ we have $\mathsf{BOO}(a) = 0$
10: set $b = 0$ ▷ we have $\mathsf{BOO}(b) = 1$
11: **while** $b > a + 1$ **do**   ▷ we have $a < -\rho - \delta \leq b$
12:     set $c = \lfloor \frac{a+b}{2} \rceil$
13:     **if** $\mathsf{BOO}(c) = 0$ **then** $a \leftarrow c$ **else** $b \leftarrow c$
14: **end while**
15: $y \leftarrow b$ ▷ $y = -\rho - \delta$
16: **return** $-\frac{x+y}{2}$ ▷ we deduce $\rho = \frac{x-y}{2}$ as well

---

Fig. 6. Learning an integer with a bound $\rho' \leq \frac{q}{6} - \frac{1}{3}$ on the unknown threshold ρ.

3.4   Example: Learning a Vector Which Is L_1-Small

When δ is a vector with components in $\mathbf{Z}_q$ and the norm is the L_1 norm, we can learn δ easily when $\rho \leq \frac{q}{6} - \frac{1}{3}$. The idea is first to learn δ_1 and $\rho - |\delta_2| - \cdots - |\delta_n|$.

Then we can iterate and learn every δ_i. We can easily see that the number of oracle calls is roughly $(2n - 1)\log_2 \rho$.

This is more complicated when $\rho > \frac{q}{6} - \frac{1}{3}$. Below, we study the $\rho = \frac{q}{2}$ case in dimension $n = 2$ which is significant for NewHope [1].

Lemma 9. *We consider the* LEARN *game over* $\mathbf{Z}_q^2$ *with the* L_1 *norm and* $\rho = \frac{q}{2}$. *Given* $c \in [-\rho, \rho]$, *we have* $\mathsf{BOO}(c - \rho, 0) = 1 \iff |\delta_1 + c| \geq |\delta_2|$.

Similarly, $\mathsf{BOO}(0, c - \rho) = 1 \iff |\delta_2 + c| \geq |\delta_1|$.

Proof. If $\delta_1 + c - \rho \geq -\frac{q}{2}$, we observe that $\delta_1 + c - \rho \leq \frac{q}{2}$ since $c \leq \rho$, so $\|\delta + (c - \rho, 0)\| = |\delta_1 + c - \rho| + |\delta_2|$. Hence, $\mathsf{BOO}(c - \rho, 0) = 1$ is equivalent to $|\delta_2| - \rho \leq \delta_1 + c - \rho \leq -|\delta_2| + \rho$. The second inequality is always true because $\|\delta\| \leq \rho$ and $c \leq \rho$. $\mathsf{BOO}(c - \rho, 0) = 1$ is therefore equivalent to $|\delta_2| - \rho \leq \delta_1 + c - \rho$, thus to $|\delta_2| \leq \delta_1 + c$. Furthermore, $\delta_1 + c - \rho \geq -\frac{q}{2}$ implies that $\delta_1 + c \geq 0$ (since $\rho = \frac{q}{2}$). Hence, $\mathsf{BOO}(c - \rho, 0) = 1$ is equivalent to $|\delta_2| \leq |\delta_1 + c|$.

If $\delta_1 + c - \rho \leq -\frac{q}{2}$, we observe that $\delta_1 + c - \rho \geq -\frac{3q}{2}$, so $\|\delta + (c - \rho, 0)\| = |\delta_1 + c - \rho + q| + |\delta_2|$. Hence, $\mathsf{BOO}(c - \rho, 0) = 1$ is equivalent to $|\delta_2| - \rho \leq \delta_1 + c - \rho + q \leq -|\delta_2| + \rho$. The first inequality is always granted, hence $\mathsf{BOO}(c - \rho, 0) = 1$ is equivalent to $\delta_1 + c \leq -|\delta_2|$. Furthermore, $\delta_1 + c - \rho \leq -\frac{q}{2}$ implies that $\delta_1 + c \leq 0$, therefore $\mathsf{BOO}(c - \rho, 0) = 1$ is equivalent to $-|\delta_1 + c| \leq -|\delta_2|$ as well.

The equivalence between $\mathsf{BOO}(0, c - \rho) = 1$ and $|\delta_2 + c| \geq |\delta_1|$ is obtained by swapping the two coordinates of the BOO oracle and of δ. $\qquad\square$

Theorem 10. *We consider the* LEARN *game over* $\mathbf{Z}_q^2$ *with the* L_1 *norm and* $\rho = \frac{q}{2}$. *There exists a learning algorithm making up to* $3 \log_2 \rho + 2$ *oracle calls.*

Proof. Using Lemma 9 with $c = 0$, we can first learn if $|\delta_1| \leq |\delta_2|$ by making an oracle call $\mathsf{BOO}(0, -\rho)$. We assume without loss of generality that $|\delta_1| < |\delta_2|$. (If not, we flip the pair of entries in the BOO oracle and learn a flipped δ; the $|\delta_1| = |\delta_2|$ case will solve by itself.) Hence, the inequality $|\delta_1 + c| \geq |\delta_2|$ is not satisfied for $c = 0$ but it may be for $c = \rho$ or $c = -\rho$, or even both. By two cut-and-choose algorithms, we find a threshold c making it an equality in both intervals $[-\rho, 0]$ and $[0, \rho]$, if there exists one. This requires up to $2 \log_2 \rho$ oracle calls. If there are two, the smallest one has the same sign as δ_1. If there is one, it has the same sign as δ_1. The $|\delta_1| = |\delta_2|$ case gives $c = 0$ and any oracle query with $c \neq 0$ gives the sign of δ_1. In all cases, we learn the sign of δ_1 and the smallest value c such that $|\delta_1 + c| = |\delta_2|$. We can also learn if δ_1 is null and isolate this case as we already learned $\delta_1 = 0$ and $|\delta_2| = |c|$. For $\delta_1 \neq 0$, we continue with Lemma 9 again. We have $\mathsf{BOO}(0, -|c| - 1 - \rho) = 1 \iff |\delta_2 - |c| - 1| \geq |\delta_1|$, which is equivalent to $\delta_2 < 0$. Hence, we learn the sign of δ_2 too with one oracle call. So far, with $2 \log_2 \rho + 2$ oracle calls, we found the sign of δ_1 and δ_2, and a smallest c such that $|\delta_1 + c| = |\delta_2| = \frac{1}{2}\|\delta + (c, 0)\|$.

Let $\varepsilon_i \in \{-1, +1\}$ such that $\varepsilon_i \delta_i \geq 0$ for $i = 1, 2$. Let $d \in [0, \rho]$. We consider oracle calls of form $\mathsf{BOO}(c + \varepsilon_1 \frac{d}{2}, \varepsilon_2 \frac{d}{2})$. We have $|\delta_1 + c| = \frac{\|\delta + (c,0)\|}{2} \leq \frac{\rho}{2}$ so $|\delta_1 + c + \varepsilon_1 \frac{d}{2}| \leq \rho$ and similarly, $|\delta_2 + \varepsilon_2 \frac{d}{2}| \leq \rho$. Hence,

$$\left\| \delta + \left(c + \varepsilon_1 \frac{d}{2}, \varepsilon_2 \frac{d}{2} \right) \right\| = \left| \delta_1 + c + \varepsilon_1 \frac{d}{2} \right| + \left| \delta_2 + \varepsilon_2 \frac{d}{2} \right| = \|\delta\| + d$$

By cut-and-choose, we learn $d \in [0, \rho]$ such that $\|\delta\| + d = \rho$. We deduce $\|\delta\|$ then δ completely.   □

3.5   Example: Learning a String of Small Hamming Weight

As another example, consider that δ lives in $\{0, 1\}^n$ and that $\|x\|$ is the Hamming weight of x. We consider $\oplus$, the bitwise XOR, as an addition. We assume $\rho \leq \frac{n}{2}$.

Algorithm learn(pp):
1: define n from pp
2: set $x = (0, 0, \ldots, 0)$ ▷ we have $\mathsf{BOO}(x) = 1$
3: set $y = (1, 1, \ldots, 1)$ ▷ we have $\mathsf{BOO}(x \oplus y) = 0$
4: **while** $\|y\| > 1$ **do**   ▷ during the loop, $\mathsf{BOO}(x) = 1$ and $\mathsf{BOO}(x \oplus y) = 0$
5:     split at random $y = u \oplus v$ with $u \wedge v = 0$, $\|u\| = \left\lfloor \frac{\|y\|}{2} \right\rfloor$, and $\|v\| = \left\lceil \frac{\|y\|}{2} \right\rceil$,
6:     **if** $\mathsf{BOO}(x \oplus u) = 1$ **then**
7:         $x \leftarrow x \oplus u$
8:         $y \leftarrow v$
9:     **else**
10:         $y \leftarrow u$
11:     **end if**
12: **end while**
13: $z \leftarrow x$ ▷ we know that $\|\delta \oplus x\| = \rho$
14: set i_{done} such that $y_{i_{\mathsf{done}}} = 1$
15: **for** $i = 1$ to n except $i \neq i_{\mathsf{done}}$ **do**
16:     set y to x with the i-th bit flipped
17:     **if** $\mathsf{BOO}(y) = 1$ **then** in z, flip the i-th bit
18: **end for**
19: **return** z

Fig. 7. Learning a bitstring.

We can have a learning algorithm by a cut-and-choose algorithm.

Theorem 11. *The algorithm on Fig. 7 succeeds with probability* 1. *The number of iterations r satisfies*

$$r \leq n + \log_2 n$$

The analysis is similar as before.

When δ is uniformly distributed among the strings of Hamming weight at most ρ, the entropy is

$$H(\delta) = \log_2 \left(\sum_{w=0}^{\rho} \binom{n}{w} \right) \leq nH\left(\frac{\rho}{n}\right)$$

where H is the binary entropy function.

For $\rho = \frac{n}{2}$, we obtain $H(\delta) \leq n$. The number of queries is asymptotically equivalent to n and it is nearly optimal.

For $\rho = \frac{n}{4}$, we obtain $H(\delta) \leq 0.82n$, meaning that the algorithm is not optimal.

4 Key-Recovery with a Plaintext Checking Oracle

With the meta-PKC construction, there is a fixed sk and the adversary can play with an oracle $\mathsf{PCO}(U, V, \mathsf{pt})$ checking if the noise is sparse, i.e. if $\mathsf{encode}(\mathsf{pt})$ is the closest codeword to $V - U \times \mathsf{sk}$. One strategy consists of learning the noise $\delta = t \times d + f - e \times \mathsf{sk}$ by playing with PCO. Actually, by defining $\mathcal{O}(x) = \mathsf{PCO}(U, V + x, \mathsf{pt})$, we simulate a bounded offset oracle to run the LEARN game.

Learning δ gives a linear equation with unknown d and sk. We can eliminate d using the known relation $B = A \times \mathsf{sk} + d$. With a few equations (depending on the selected algebra), we deduce sk.

More precisely, if we can learn δ sampled by $\mathcal{S}$ which generates δ like $\delta = t \times d + f - e \times \mathsf{sk}$, we assume that for some k (typically, $k = 1$ if all sets are equal), there exists an algorithm such that given k equations of form $(t_i \times A + e_i) \times \mathsf{sk} = t_i \times B + f_i - \delta_i$ with sk unknown, it solves sk. We use the algorithm on Fig. 8.

We obtain the following result.

Theorem 12. *Let t, d, f, e, sk follow the random distributions of the cryptosystem. We define*

$$\delta = t \times d + f - e \times \mathsf{sk}$$

*With probability 1, the algorithm of Fig. 8 gives at every iteration of the **for** loop, one linear equation over S_V with unknown $\mathsf{sk} \in S_{\mathsf{sk}}$, using the same number of oracle calls as learn with the distribution of δ.*

In the case that $S_{\mathsf{sk}} = \mathbf{Z}_q^{n_{\mathsf{sk}}}$, $S_A = \mathbf{Z}_q^{n_A}$, $S_B = \mathbf{Z}_q^{n_B}$, $S_t = \mathbf{Z}_q^{n_t}$, $S_U = \mathbf{Z}_q^{n_U}$, $S_V = \mathbf{Z}_q^{n_V}$, when the considered norm is the L_∞ norm, this means that we obtain n_V linear equations over $\mathbf{Z}_q$ with n_{sk} unknowns. We use $H(\delta)$ oracle calls and we can approximate $H(\delta) \approx n_V \log_2(2\rho_+)$. (This can only over-estimate the entropy.)

One difficulty is to know what happens when the distance of $V - U \times \mathsf{sk}$ to the code is between ρ_- and ρ_+. If $\rho_- = \rho_+$, there is no problem. If decryption aborts when the number of errors exceeds ρ_-, then ρ_- becomes de facto the threshold to be used in the LEARN game. Otherwise, the learn algorithm must be refined.

Example 13 (Frodo continued). The Frodo-640 parameters are $n_{\mathsf{sk}} = n_B = n\bar{n} = 640 \times 8$ and $n_V = \bar{m}\bar{n} = 8 \times 8$. Since $\rho_+ = \rho_- = q2^{-\ell-1} = 2^{12}$, we need about 2^{10} oracle calls to recover 2^6 equations in 2^{12} unknowns. (By iterating, we have a full key recovery using 2^{16} oracle calls.)

Algorithm $\mathcal{A}^{\mathsf{PCO}(\cdot)}(\mathsf{pp},\mathsf{pk})$:
1: parse $\mathsf{pk} = (A, B)$
2: **for** $i = 1$ to k **do**
3:     pick pt_i at random
4:     run $\mathsf{enc}(\mathsf{pp},\mathsf{pk},\mathsf{pt}_i) \to \mathsf{ct}_i$ ▷ this defines t_i, e_i, f_i
5:     parse $\mathsf{ct}_i = (U_i, V_i)$
6:     run $\mathsf{learn}^{\mathcal{O}(\cdot)}(\mathsf{pp}) \to \delta_i$ ▷ deduce $(t_i \times A + e_i) \times \mathsf{sk} = t_i \times B + f_i - \delta_i$
7: **end for**
8: solve $\begin{pmatrix} t_1 \times A + e_1 \\ \vdots \\ t_k \times A + e_k \end{pmatrix} (\mathsf{sk}) = \begin{pmatrix} t_1 \times B + f_1 - \delta_1 \\ \vdots \\ t_k \times B + f_k - \delta_k \end{pmatrix}$
9: **return** sk

Oracle $\mathcal{O}(x)$:
10: **return** $\mathsf{PCO}(U_i, V_i + x, \mathsf{pt}_i)$

Fig. 8. KR-PCA attack based on learning.

Example 14 (NewHope continued). NewHope512 defines $n_{\mathsf{sk}} = n_B = n_V = n = 512$, $\rho_- = \frac{q}{4}$, $\rho_+ = \frac{q}{2}$, and $p = 8$. Decoding in NewHope is based on the cumulative distance. We should rather apply the algorithm in Theorem 10 for the L_1 norm in dimension two to learn each pair of components of δ. However, the use of compression of V into $\bar{V}$ in NewHope makes the attack learning an approximation of δ instead of δ completely. Essentially, we learn $\lceil \frac{p}{q}\delta \rfloor$ (i.e., the $\log_2 p$ most significant bits of each of the n_V components of δ). However, we already know that δ is sparse, so we learn zero bits only. The attack does not work but we can adopt another strategy of well selecting $(U, \bar{V})$ to progressively learn the bits of sk. This was done by Bauer et al. [5].

Example 15 (Lepton continued). With Lepton.CPA Light I, δ is a string of length $\ell = 4572$ of Hamming weight bounded by 1056. Encoding consists of $d = 9$ repetitions of a BCH code which can correct up to 30 errors. One problem is that decoding fails if decoding the repetition code results in more than $t = 30$ errors in BCH decoding. Nevertheless, we can adapt the attack based on the learn algorithm of Fig. 7. In learn, instead of considering a string of ℓ bits, we consider a string of $\frac{\ell}{d}$ packets of d bits in which the packet 1 represents the packet of d bits set to 1, and flipping a packet means xoring it to the packet 1. Thus, we learn which packets of δ have an error using up to $\frac{\ell}{d} + \log_2 \frac{\ell}{d}$ queries. Then, for each packet, we modify δ to have exactly $t - 1$ incorrect other packets and we apply the learn algorithm of Fig. 7 at the bit level. For each packet, we need up to $d + \log_2 d$ oracle calls. In total, the number of oracle calls is bounded by $\ell + \frac{\ell}{d}\log_2 d + \frac{\ell}{d} + \log_2 \frac{\ell}{d} \approx 2^{13}$. This gives 4572 equations in 8100 unknowns. So we use $k = 2$ and recover the entire sk using 2^{14} oracle calls.

5   Quantum Key-Recovery with a Decryption Oracle

5.1   GKZ-Based Attack

We build a KR-CCA attack which is inspired by the quantum LWE solving algorithm from Grilo, Kerenidis, and Zijlstra (GKZ) [14]. This algorithm works with a quantum superposition of LWE entries. In this attack, we consider an adversary with quantum access to a decryption oracle. More precisely, we assume that the oracle makes the following mapping:

$$|\mathsf{ct} \ x \ Z\rangle \mapsto |\mathsf{ct} \ (x \oplus \mathsf{Dec}(\mathsf{sk}, \mathsf{ct})) \ Z\rangle$$

Instead of calling this oracle on a chosen ct, we will call it on a superposition of ct states.

We denote $\omega_q = e^{\frac{2i\pi}{q}}$, a qth primitive root of unity.

We consider our meta PKC construction in which all groups are powers of $\mathbf{Z}_q$ for simplicity: $S_{\mathsf{sk}} = \mathbf{Z}_q^{n_{\mathsf{sk}}}$, $S_A = \mathbf{Z}_q^{n_A}$, $S_B = \mathbf{Z}_q^{n_B}$, $S_t = \mathbf{Z}_q^{n_t}$, $S_U = \mathbf{Z}_q^{n_U}$, $S_V = \mathbf{Z}_q^{n_V}$.

We let $I \subseteq \{1, \ldots, n_V\}$ be a set of indices i.

We split the quantum state into several registers:

- one register $U \in S_U$;
- one register $V \in S_V$;
- a plaintext in $\mathcal{M}$;
- one register $Z \in \mathbf{Z}_q^I$.

We assume that we have an operator L mapping

$$|U \ V \ \mathsf{pt} \ Z\rangle \mapsto |U \ V \ \mathsf{pt} \ (Z \oplus (V - \mathsf{encode}(\mathsf{pt}))_I)\rangle$$

where V_I denotes the restriction of the vector V on indices in I. This means that we compute the ith coordinates of $V - \mathsf{encode}(\mathsf{pt})$ and XOR it to the working register Z.

We run the algorithm on Fig. 9. Steps 6–7 of our attack are equivalent to the GKZ algorithm [14]. As usual quantum algorithms, the algorithm is deterministic until we perform a measurement (in Step 7). What follows the measurement in Step 7 is done by a classical computer.

We define $W_U = V - U \times \mathsf{sk}$, $\mathsf{pt}_U = \mathsf{decode}(W_U)$, $Z_U = (V - \mathsf{encode}(\mathsf{pt}_U))_I$, $\delta_U = W_U - \mathsf{encode}(\mathsf{pt}_U)$. Due to the property of the encoding/decoding algorithms,[2] we have $\|\delta_U\| \leq \rho_+$. We define

$$\psi_U = (\delta_U)_I = (W_U - \mathsf{encode}(\mathsf{decode}(W_U)))_I \tag{2}$$

Hence, $Z_U = (U \times \mathsf{sk})_I + \psi_U$ with ψ_U small. Hence, the state after Step 4 resembles a quantum superposition of LWE entries, but with a spurious register pt. This spurious register has a dramatic impact on the probability of later measurements, as it will be shown below. It was noticed by Ambainis, Magnin,

---

[2] We recall that we assume that decoding is defined over the entire S_V space.

Input: $I \subseteq \{1, \ldots, n_V\}$ and $V \in S_V$
Decryption oracle: $|U\ V\ x\ Z\rangle \mapsto |U\ V\ x \oplus \mathsf{Dec}(\mathsf{sk}, (U, V))\ Z\rangle$
1: set the quantum state to $|0\ V\ 0\ 0\rangle$
2: make a quantum Fourier transform on the first register and obtain $q^{-\frac{n_U}{2}} \sum_U |U\ V\ 0\ 0\rangle$
3: make a decryption oracle call and obtain $q^{-\frac{n_U}{2}} \sum_U |U\ V\ \mathsf{pt}_U\ 0\rangle$
4: apply the L operator and obtain $q^{-\frac{n_U}{2}} \sum_U |U\ V\ \mathsf{pt}_U\ Z_U\rangle$
5: make a decryption oracle call again and obtain $q^{-\frac{n_U}{2}} \sum_U |U\ V\ 0\ Z_U\rangle$
6: make a quantum Fourier transform on the U and Z registers and obtain

$$q^{-n_U - \frac{\#I}{2}} \sum_{U, \alpha, \beta} \omega_q^{(\alpha \cdot U) + (\beta \cdot Z_U)} |\alpha\ V\ 0\ \beta\rangle$$

7: measure the two active registers to get (α, β)
8: solve the linear system $\alpha_j + (\beta \cdot (e_j \times \mathsf{sk})_I) = 0$, $j = 1, \ldots, n_U$, where e_j is the U vector with coordinates set to 0 except the jth bit which is set to 1
9: **if** not solvable **then** abort
10: set s to the solution
11: **if** s not sparse **then** abort
12: **return** s

Fig. 9. GKZ-based key recovery with quantum access to a decryption oracle.

Roetteler, and Roland [3] that getting rid of it is hard in general. Fortunately, we can call the decryption oracle again to clear pt completely. Step 5 is doing it. This is why we need a double query to the decryption oracle.

We call a pair $(\alpha, \beta) \in S_U \times \mathbf{Z}_q^I$ *good* if $\beta \neq 0$ and α satisfies the property $\alpha_j + (\beta \cdot (e_j \times \mathsf{sk})_I) = 0$ for $j = 1, \ldots, n_U$. For all $\beta \neq 0$, we have a unique α such that (α, β) is good. When this property is satisfied, since $U = \sum_j U_j e_j$, we have $(U \times \mathsf{sk})_I = \sum_j U_j (e_j \times \mathsf{sk})_I$, thus

$$
\begin{aligned}
(\alpha \cdot U) + (\beta \cdot Z_U) &= (\alpha \cdot U) + (\beta \cdot (U \times \mathsf{sk})_I) + (\beta \cdot \psi_U) \\
&= (\beta \cdot \psi_U) + \sum_j U_j ((\alpha \cdot e_j) + (\beta \cdot (e_j \times \mathsf{sk})_I)) \\
&= (\beta \cdot \psi_U)
\end{aligned}
$$

We compute the probability p_g to measure a good (α, β) pair. We have

$$
\begin{aligned}
p_g &= \sum_{(\alpha, \beta)\ \text{good}} \left| q^{-n_U - \frac{\#I}{2}} \sum_U \omega_q^{(\alpha \cdot U) + (\beta \cdot Z_U)} \right|^2 \\
&= \sum_\beta \frac{1}{q^{2n_U + \#I}} \left| \sum_U \omega_q^{\beta \cdot \psi_U} \right|^2 - q^{-\#I}
\end{aligned}
$$

where the $q^{-\#I}$ term cancels the $\beta = 0$ term in the sum. By using $|z|^2 = z \times \bar{z}$ over complex numbers, we have

$$
\begin{aligned}
p_g &= \sum_{\beta} \frac{1}{q^{2n_U + \#I}} \left(\sum_{U} \omega_q^{\beta \cdot \psi_U} \right) \left(\sum_{U'} \omega_q^{-\beta \cdot \psi_{U'}} \right) - q^{-\#I} \\
&= \sum_{\beta} \frac{1}{q^{2n_U + \#I}} \sum_{U,U'} \omega_q^{\beta \cdot (\psi_U - \psi_{U'})} - q^{-\#I} \\
&= \frac{1}{q^{2n_U}} \sum_{U,U'} \frac{1}{q^{\#I}} \sum_{\beta} \omega_q^{\beta \cdot (\psi_U - \psi_{U'})} - q^{-\#I} \\
&= \frac{1}{q^{2n_U}} \sum_{U,U'} 1_{\psi_U = \psi_{U'}} - q^{-\#I} \\
&= \Pr[\psi_U = \psi_{U'}] - q^{-\#I} \\
&= 2^{-H_2(\psi_U)} - q^{-\#I}
\end{aligned}
$$

where H_2 is the collision entropy (or Rényi of degree 2).

Without the second decryption call, with the same method, we would have obtained $p_g = 2^{-H_2(\psi_U, \mathsf{pt}_U)} - q^{-\#I} 2^{-H_2(\mathsf{pt}_U)}$ which is too small. This shows the importance of this second call, just to clear one register, although it works against intuition when we are used to classical computing.

Our analysis is based on the decode function being defined on the entire domain S_V. However, instances of our construction may rely on a partially defined decode algorithm. In that case, we may have $\mathsf{Dec}(\mathsf{sk}, \mathsf{ct}) = \bot$. By convention, we set $\mathsf{encode}(\bot) = 0$. The same analysis gives

$$
\begin{aligned}
p_g &= \Pr[\psi_U = \psi_{U'}] - q^{-\#I} \\
&\approx \Pr[\mathsf{pt}_U \neq \bot]^2 \left(\Pr[\psi_U = \psi_{U'} | \mathsf{pt}_U \neq \bot, \mathsf{pt}_{U'} \neq \bot] - q^{-\#I} \right)
\end{aligned}
$$

This may be too small if $\Pr[\mathsf{pt}_U \neq \bot]$ is small. This is the case with Lepton.

One crucial thing is that the distribution of ψ_U comes from (2), where U is uniform, sk comes from the key generation algorithm, and V is fixed. In particular, ψ_U does not follow the normal distribution defined by (1) from the encryption/decryption process which would have a lower $H_2(\psi_U)$. If we had a way to sample ψ from (1) without any spurious register, we would have a better attack.

What we obtain is the following result:

Theorem 16. *Let $I \subseteq \{1, \ldots, n_V\}$ be a set of indices, let $V \in S_V$ be arbitrarily fixed, let $\mathsf{sk} \in S_{\mathsf{sk}}$ following the random distribution of the cryptosystem, and let $U \in S_U$ be uniformly distributed. We define*

$$
\psi_U = (V - U \times \mathsf{sk} - \mathsf{encode}(\mathsf{decode}(V - U \times \mathsf{sk})))_I
$$

The algorithm of Fig. 9 gives a pair (α, β) at Step 7 such that $\beta \neq 0$ and

$$
\forall j \in \{1, \ldots, n_U\} \qquad \alpha_j + (\beta \cdot (e_j \times \mathsf{sk})_I) = 0
$$

with probability $2^{-H_2(\psi_U)} - q^{-\#I}$ *and using* 2 *oracle calls. This is a set of* n_U *linear equations with* $n_{\sf sk}$ *unknowns over* $\mathbf{Z}_q$.

For $\#I = 1$ and ψ_U uniform in $[-\rho_+, +\rho_+]$, we can assume that $H_2(\psi_U) = \log_2(2\rho_+ + 1)$. We obtain that the success probability is roughly $\frac{1}{2\rho_+} - \frac{1}{q}$. Clearly, it requires $\rho_+ < \frac{q}{2}$.

Note that when this fails (with probability $1-p_g$) and $n_{\sf sk} \leq n_U$, we can easily filter out those cases because we can eliminate the cases when the equations have no solution or when the solution is not sparse. Hence, either we recover part of $\sf sk$, or we abort. Therefore, we can iterate this attack to recover (at least some part of) $\sf sk$ with a better probability.

For $n_{\sf sk} = n_U$, we iterate $1/p_g$ times on average until one sparse equation is found.

For $n_{\sf sk} > n_U$, we should, in general, treat the problem on a case-by-case basis by studying the structure of the equations we obtain but there are some general methods we can apply when $n_{\sf sk}/n_U$ is small. We could indeed iterate $\frac{n_{\sf sk}}{p_g n_U}$ times to be sure to get enough sets of n_U equations to recover the $n_{\sf sk}$ unknowns. To recover them, we can try all the $\left(\frac{n_{\sf sk}}{p_g n_U}\right)^{\frac{n_{\sf sk}}{n_U}}$ combinations of $\frac{n_{\sf sk}}{n_U}$ sets. Each combination gives $n_{\sf sk}$ equations. We can solve each combination until one sparse $\sf sk$ is found.

However, for each j, the equation $\alpha_j + (\beta \cdot (e_j \times {\sf sk})_I) = 0$ typically depends on a fixed subset of coordinates of $\sf sk$ and we should better apply those methods for each of these subset separately.

We assume $n_{\sf sk} \leq n_U$. If there is a single coordinate i of $U \times {\sf sk}$ which depends on all coordinates of $\sf sk$, by using $I = \{i\}$, we recover the entire $\sf sk$ with probability p or abort. Hence, iterating p_g^{-1} times fully recover $\sf sk$ with $2p_g^{-1}$ decryption calls.

It is not always possible to find a coordinate i which depends on the entire $\sf sk$. For instance, in the case of Frodo, $U \times {\sf sk}$ is a matrix multiplication so each coordinate i depends on one column of $\sf sk$ only.

Interestingly, $H_2(\psi_U) \leq H(\psi_U)$ so the number of queries when iterating is lower bounded by $2^{1+H(\psi_U)}$. This is a big difference with the previous classical attack, which recovers one $\mathbf{Z}_q$ element within only $H(\psi_U)$ queries. However, it is hard to compare an attack finding a piece of the key using two oracle calls and with probability p to an attack finding linear equations using many oracle calls and succeeding with probability 1. For applications where the number of key reuse is strictly limited, the former is more devastating.

Example 17 (Frodo continued). The Frodo-640 parameters are $n_{\sf sk} = n_U$ but we need to recover each of the $\bar{n} = 8$ columns separately if we take $\#I = 1$. We approximate $2^{-H_2(\psi_U)} \approx \frac{1}{2\rho_+}$. Since $\rho_+ = q2^{-\ell-1} = 2^{12}$ and $q = 2^{15}$, by using two oracle calls, we recover one column of 640 values with probability 2^{-13}. (By iterating, we need 2^{17} oracle calls to fully recover $\sf sk$.) We can recover two columns at the same time with $\#I = 2$ in two oracle calls, but with probability 2^{-26}.

Example 18 (NewHope continued). With NewHope512, ψ_U on a single component has no information (this is due to $\rho_+ = \frac{q}{2}$ which does not characterize ψ_U compared to any other $\mathbf{Z}_q$ element). Indeed, we obtain $p = 0$ with $\#I = 1$. We can use $\#I = 2$ with $I = \{i, i + 256\}$. This uses two positions encoding the same bit. We can see that ψ_U encodes two values which are either both smaller than $\frac{q}{4}$ (with probability $\frac{1}{2}$) or with exactly one being smaller than $\frac{q}{4}$. Hence, two ψ_U pairs collide with probability $\frac{1}{4}\left(\frac{2}{q}\right)^2 + \frac{1}{8}\left(\frac{2}{q}\right)^2 = \frac{3}{2q^2}$. It gives $p_g \approx \frac{1}{2q^2}$. This means that with two oracle calls we recover the full secret with probability 2^{-28}. (By iterating, we need $4q^2$ decryption calls, i.e. 2^{29}.) Compressing V in NewHope does not modify our attack as V is constant.

Example 19 (Lepton continued). Lepton.CPA Light I considers $q = 2$. With $\#I = 1$, we obtain $p = 0$. The encoding function in Lepton is obtained by first applying a BCH encoding, then using a repetition code. We can focus on the repetition code and take two repeating bits in the set I with $\#I = 2$. The distribution of ψ_U from (2) should be of form $\Pr[\psi_U = 00] = \frac{1}{2}$, $\Pr[\psi_U = 01] = \Pr[\psi_U = 10] = \frac{1}{4}$. Hence, $\Pr[\psi_U = \psi_{U'}] = \frac{3}{8}$ and we obtain $p = \frac{1}{8}$. Hence, we could recover sk with probability $\frac{1}{8}$.[3] Unfortunately, decode is partially defined, due to the BCH code, and we have $\Pr[\mathsf{pt}_U \neq \bot] \approx 2^{-93}$ which is too low. That is why the attack does not work for Lepton.

5.2 AJOP-Based Attack

We let a and b be two integers. We partition $[a, a + q - 1]$ into c intervals $I_k = [a + kb, a + kb + b - 1]$ for $k = 0, \ldots, c - 2$ and $I_{c-1} = [a + (c - 1)b, a + q - 1]$, with $b = \lceil \frac{q}{c} \rceil$. We define $\mathsf{RF}(x) = k$ such that $x \in I_k$ modulo q. Hence, we can consider RF as a function from $\mathbf{Z}_q$ to $\mathbf{Z}_c$.

Lemma 20. *Given $f \in \mathbf{Z}_q^n$ such that $\{u \cdot f; u \in \mathbf{Z}_q^n\} = \mathbf{Z}_q$ (we call such f regular),[4] we define*

$$p_{q,c} = q^{-2n} \left| \sum_{u \in \mathbf{Z}_q^n} \omega_c^{-\mathsf{RF}(v - u \cdot f)} \omega_q^{u \cdot (-f)} \right|^2$$

We write $\varepsilon = b - \frac{q}{c}$. For $c = \mathcal{O}(1)$ and $q \to +\infty$, we have

$$p_{q,c} \geq q^{-2} \left(\left| \frac{\sin \frac{\pi c \varepsilon}{q}}{\sin \frac{\pi \varepsilon}{q}} \times \frac{\sin \frac{\pi b}{q}}{\sin \frac{\pi}{q}} \right| - \left| \frac{\sin \frac{\pi c \varepsilon}{q}}{\sin \frac{\pi}{q}} \right| \right)^2 = \frac{c^2}{\pi^2} \sin^2 \frac{\pi b}{q} - o\left(\frac{1}{q}\right)$$

If c divides q, we have $p_{q,c} \geq \frac{c^2}{\pi^2} \sin^2 \frac{\pi}{c}$.

---

[3] In this computation, we took the worst case for ambiguous decoding (e.g. when both 01 and 10 decode to 00). If now 01 decode to 00 and 10 decode to 11, the distribution of ψ_U becomes $\Pr[\psi_U = 00] = \Pr[\psi_U = 01] = \frac{1}{2}$ and we obtain $p = \frac{1}{4}$.

[4] For q prime, every nonzero f is regular. For $q = 2^n$, every f with at least one odd component is regular.

Proof. Due to the fact that $u \mapsto v - u \cdot f$ being balanced, we have

$$p_{q,c} = q^{-2} \left| \sum_{x \in \mathbf{Z}_q} \omega_c^{-\mathsf{RF}(x)} \omega_q^{x-v} \right|^2 = q^{-2} \left| \sum_{x \in \mathbf{Z}_q} \omega_c^{-\mathsf{RF}(x)} \omega_q^x \right|^2 = q^{-2} \left| \sum_{k \in \mathbf{Z}_c} \sum_{x \in I_k} \omega_c^{-k} \omega_q^x \right|^2$$

Hence,

$$p_{q,c} \geq q^{-2} \left(\left| \sum_{k \in \mathbf{Z}_c} \sum_{x=a}^{a+b-1} \omega_c^{-k} \omega_q^{x+kb} \right| - \left| \sum_{x=a+q-(c-1)b}^{a+b-1} \omega_c^{-c+1} \omega_q^{x+(c-1)b} \right| \right)^2$$

$$= q^{-2} \left(\left| \sum_{k \in \mathbf{Z}_c} \sum_{x=0}^{b-1} \omega_c^{-k} \omega_q^{x+kb} \right| - \left| \sum_{x=b-c\varepsilon}^{b-1} \omega_q^x \right| \right)^2$$

$$= q^{-2} \left(\left| \sum_{k \in \mathbf{Z}_c} \sum_{x=0}^{b-1} \omega_q^{k\varepsilon} \omega_q^x \right| - \left| \sum_{x=b-c\varepsilon}^{b-1} \omega_q^x \right| \right)^2$$

$$= q^{-2} \left(\left| \frac{\sin \frac{\pi c \varepsilon}{q}}{\sin \frac{\pi \varepsilon}{q}} \times \frac{\sin \frac{\pi b}{q}}{\sin \frac{\pi}{q}} \right| - \left| \frac{\sin \frac{\pi c \varepsilon}{q}}{\sin \frac{\pi}{q}} \right| \right)^2$$

by using $\omega_q = e^{\frac{2i\pi}{q}}$ and $\omega_c = e^{\frac{2i\pi}{c}}$. We have $0 \leq \varepsilon < 1$, $c = \mathcal{O}(1)$. When $q \to +\infty$, this bound tends towards $\frac{c^2}{\pi^2} \sin^2 \frac{\pi b}{q}$ with a $o(\frac{1}{q})$ difference. $\qquad\square$

In Fig. 10, we adapt the AJOP algorithm to make a KR-CCA attack using a single query to a quantum oracle making

$$|U \ \ V \ \ z\rangle \mapsto |U \ \ V \ \ z + \mathsf{Dec}(\mathsf{sk}, U, V)\rangle$$

where the addition is in $\mathbf{Z}_c^{n_V}$. It works assuming a special form of the cryptosystem. Compared to our GKZ-based attack, this is more restrictive but it uses a single oracle call and has a better success probability.

Theorem 21. *We consider meta-PKC constructions of the following form. We assume that $\mathcal{M} = \mathbf{Z}_c^{n_V}$, $\mathsf{decode}(W)_j = \mathsf{RF}(W_j)$, and that we can write $U = (U_1, \ldots, U_m) \in \mathbf{Z}_q^{n_U^1} \times \cdots \mathbf{Z}_q^{n_U^m}$, $(U \times \mathsf{sk})_j = U_{g(j)} \cdot f_j(\mathsf{sk})$ for some functions g and f_j, for $j = 1, \ldots, n_V$. Given a subset J over which g is injective, the algorithm on Fig. 10 recovers all $f_j(\mathsf{sk})$ for $j \in J$ with probability $p \geq p_{q,c}^{\#J}$ with $p_{q,c}$ defined in Lemma 20, when they are regular.*

Proof. We compute the probability $\Pr[\alpha]$:

$$\Pr[\alpha] = q^{-2n_U}c^{-n_V}\left\|\sum_{U,z}\left(\prod_{j\in J}\omega_c^{z_j-(\mathsf{pt}_U)_j}\right)\omega_q^{U\cdot\alpha}|\alpha \ V \ z\rangle\right\|^2$$

$$= q^{-2n_U}c^{-n_V}\sum_z\left|\sum_U\left(\prod_{j\in J}\omega_c^{z_j-(\mathsf{pt}_U)_j}\right)\omega_q^{U\cdot\alpha}\right|^2$$

$$= q^{-2n_U}\left|\sum_U\left(\prod_{j\in J}\omega_c^{-(\mathsf{pt}_U)_j}\right)\omega_q^{U\cdot\alpha}\right|^2$$

$$= q^{-2n_U}\left|\sum_U\left(\prod_{j\in J}\omega_c^{-\mathsf{RF}(V_j-U_{g(j)}\cdot f_j(\mathsf{sk}))}\right)\omega_q^{U\cdot\alpha}\right|^2$$

$$= q^{-2n_U}\left|\sum_U\prod_{j\in J}\left(\omega_c^{-\mathsf{RF}(V_j-U_{g(j)}\cdot f_j(\mathsf{sk}))}\omega_q^{U_{g(j)}\cdot\alpha_{g(j)}}\right)\prod_{j\notin g(J)}\omega_q^{U_j\cdot\alpha_j}\right|^2$$

For α such that $\alpha_{g(j)} = -f_j(\mathsf{sk})$ for all $j \in J$ and $\alpha_j = 0$ for $j \notin g(J)$, we have $\Pr[\alpha] = p_{q,c}^{\#J}$ when the $f_j(\mathsf{sk})$ are regular. $\qquad\square$

Example 22 (Frodo continued). Frodo has $c = 2^\ell$. We regroup U by rows, i.e., $U_{g(j)}$ is the $g(j)$th row of U and $f_j(\mathsf{sk})$ is the $g(j)$th column of sk. We have $\bar{m}$ columns in sk. We recover $\#J$ columns with probability at least $p_{q,c}^{\#J}$. The Frodo-640 parameters are $q = 2^{15}$, $\ell = 2$, and $\bar{m} = 8$. This gives $\varepsilon = 0$ and $p_{q,c} \geq 81\%$. We can be greedy with $\#J = \bar{m}$ and fully recover sk with probability greater than 18% which we approximate as 2^{-2} in the table.

Example 23 (NewHope continued). To adapt the attack to NewHope, we observe

$$\mathsf{decode}(W)_j = \mathsf{RF}(|W_j| + |W_{j+256}|) = 1_{|W_j|+|W_{j+256}|\leq\frac{q}{2}}$$

and we cut an interval of $2q$ values into $c = 2$ intervals. We use $\#J = 1$. For simplicity, we use $J = \{0\}$ so $j = 0$. We modify the algorithm by sampling $\bar{V}_0$ and $\bar{V}_{256}$ and letting all other components of $\bar{V}$ constant, and by making the Fourier transform on the $\bar{V}_0$ and $\bar{V}_{256}$ registers as well. We obtain that we measure the final state

$$q^{-n_U}p^{-2}\sum_{\substack{\alpha,U\in S_U\\ \bar{V}_0,\bar{V}_{256},\beta_1,\beta_2\in\mathbf{Z}_p}}c^{-\frac{n_V}{2}}\sum_{z\in\mathbf{Z}_c^{n_V}}(-1)^{z_0-(\mathsf{pt}_U)_0}\omega_q^{U\cdot\alpha}\omega_p^{\bar{V}\cdot\beta}|\alpha \ \beta \ z\rangle$$

where $\bar{V}\cdot\beta$ means $\bar{V}_0\beta_1 + \bar{V}_{256}\beta_2$. By similar computation as before, we obtain

$$\Pr[\alpha,\beta] = q^{-2n_U}p^{-4}\left|\sum_{U,\bar{V}_0,\bar{V}_{256}}(-1)^{(\mathsf{pt}_U)_0}\omega_q^{U\cdot\alpha}\omega_p^{\bar{V}\cdot\beta}\right|^2$$

Input: $J \subseteq \{1, \ldots, n_V\}$ and $V \in S_V$
Decryption oracle: $|U \ V \ z\rangle \mapsto |U \ V \ z + \mathsf{Dec}(\mathsf{sk}, (U, V))\rangle$
1: prepare the state $|0 \ V \ (1_{j \in J})_{j=1, \ldots, n_V}\rangle$ in $\mathbf{Z}_q^{n_U} \times \mathbf{Z}_q^{n_V} \times \mathbf{Z}_c^{n_V}$
2: make a quantum Fourier transform on all registers except V and obtain

$$q^{-\frac{n_U}{2}} \sum_{U \in S_U} c^{-\frac{n_V}{2}} \sum_{z' \in \mathbf{Z}_c^{n_V}} \left(\prod_{j \in J} \omega_c^{z'_j} \right) |U \ V \ z'\rangle$$

3: apply the decryption oracle and get (by writing $z = z' + \mathsf{pt}_U$)

$$q^{-\frac{n_U}{2}} \sum_{U \in S_U} c^{-\frac{n_V}{2}} \sum_{z \in \mathbf{Z}_c^{n_V}} \left(\prod_{j \in J} \omega_c^{z_j - (\mathsf{pt}_U)_j} \right) |U \ V \ z\rangle$$

4: make a quantum Fourier transform on the first register and obtain

$$q^{-n_U} \sum_{\alpha, U \in S_U} c^{-\frac{n_V}{2}} \sum_{z \in \mathbf{Z}_c^{n_V}} \left(\prod_{j \in J} \omega_c^{z_j - (\mathsf{pt}_U)_j} \right) \omega_q^{U \cdot \alpha} |\alpha \ V \ z\rangle$$

5: measure the first register and obtain α with some probability $\Pr[\alpha]$

Fig. 10. AJOP-based key recovery with quantum access to a decryption oracle.

Let s be such that $U \cdot s = (U \times \mathsf{sk})_0$. Clearly, there is a one-to-one mapping between sk and s. The probability that $\alpha = -s$, $\beta_1 = 1$, $\beta_2 = 0$ is

$$\Pr[-s, 1, 0] = q^{-2n_U} p^{-4} \left| \sum_{U, \bar{V}_0, \bar{V}_{256}} (-1)^{(\mathsf{pt}_U)_0} \omega_q^{-(U \times \mathsf{sk})_0} \omega_p^{\bar{V}_0} \right|^2$$

$$= q^{-4} p^{-4} \left| \sum_{u, v \in \mathbf{Z}_q, \bar{V}_0, \bar{V}_{256} \in \mathbf{Z}_p} \left(1 - 2 \cdot 1_{|u| + |v| \leq \frac{q}{2}} \right) \omega_q^{u - V'_0} \omega_p^{\bar{V}_0} \right|^2$$

$$= 4q^{-4} p^{-2} \left| \sum_{\substack{u, v \in \mathbf{Z}_q \\ |u| + |v| \leq \frac{q}{2}}} \omega_q^u \sum_{\bar{V}_0 \in \mathbf{Z}_p} \omega_q^{-V'_0} \omega_p^{\bar{V}_0} \right|^2$$

with $V'_j = \lceil \frac{q}{p} \bar{V}_j \rceil$, $u = V'_0 - (U \times \mathsf{sk})_0$, and $v = V'_{256} - (U \times \mathsf{sk})_{256}$. We can compute experimentally this sum as $\Pr[-s, 1, 0] \approx 16.4\%$. We can also compute literally for $p = q$ (i.e., we ignore compression). The terms of the inner sum are

1. We have

$$
\Pr[-s, 1, 0] = 4q^{-4} \left| 2 \sum_{v=0}^{\frac{q-1}{2}} \sum_{u=-\frac{q-1}{2}+v}^{\frac{q-1}{2}-v} \omega_q^u - \sum_{u=-\frac{q-1}{2}}^{\frac{q-1}{2}} \omega_q^u \right|^2
$$

$$
= 4q^{-4} \left| 2 \sum_{v=0}^{\frac{q-1}{2}} \frac{\omega_q^{\frac{q}{2}-v} - \omega_q^{-\frac{q}{2}+v}}{\omega_q^{\frac{1}{2}} - \omega_q^{-\frac{1}{2}}} - \frac{\omega_q^{\frac{q}{2}} - \omega_q^{-\frac{q}{2}}}{\omega_q^{\frac{1}{2}} - \omega_q^{-\frac{1}{2}}} \right|^2
$$

$$
= 4q^{-4} \left| 2 \frac{\left(\omega_q^{\frac{q+1}{4}} - \omega_q^{-\frac{q+1}{4}}\right)^2}{\left(\omega_q^{\frac{1}{2}} - \omega_q^{-\frac{1}{2}}\right)^2} - \frac{\omega_q^{\frac{q}{2}} - \omega_q^{-\frac{q}{2}}}{\omega_q^{\frac{1}{2}} - \omega_q^{-\frac{1}{2}}} \right|^2
$$

Since $\omega_q^{\frac{q}{2}} = -1$ and $\omega_q^{\frac{q}{4}} = i$, we obtain

$$
\Pr[-s, 1, 0] = 16q^{-4} \left| \frac{\omega_q^{\frac{1}{4}} + \omega_q^{-\frac{1}{4}}}{\omega_q^{\frac{1}{2}} - \omega_q^{-\frac{1}{2}}} \right|^4 = \left| \frac{2}{q\left(\omega_q^{\frac{1}{4}} - \omega_q^{-\frac{1}{4}}\right)} \right|^4 = \frac{1}{q^4 \sin^4 \frac{\pi}{2q}} \sim \left(\frac{2}{\pi}\right)^4
$$

which is also 16.4%. We also have $\Pr[s, -1, 0] = \Pr[-s, 1, 0]$. Similarly, let s' be such that $U \cdot s' = (U \times \mathsf{sk})_{256}$. We have $\Pr[\mp s', 0, \pm 1] = \Pr[s, 1, 0]$ and all four cases reveal sk. Hence, we recover sk with probability about 66%. The compression of V in NewHope does not modify the attack.

6   Conclusion

We have shown how to make efficient key recovery attacks against the weak version of many post-quantum cryptosystems under classical PCA mode or quantum CCA mode, even with one or two CCA queries.

When trying to adapt this attack strategy to various algorithms, we observed that a binary code followed by an encoding in the high bits in $\mathbf{Z}_q$ makes the attack more difficult. However, the repetition code helps the attacker quite a lot.

Our attacks do not work for some algorithms. For instance, KINDI does not encode the plaintext but rather a seed (our attack only works because decryption outputs the seed for some reason). Compressing V does not harm quantum attacks but makes our classical attack impossible. Compressing U sometimes makes the quantum attack harder (this is the case in Kyber [6] but not in Lizard [9]). Lepton does not decode on the entire domain. We let as future work to investigate if attacks are still possible in those cases. Attacking other NIST applications is also left as an open problem.

A   Post-quantum Cryptosystems

We list here several algorithms for which we could adapt our attacks. The algorithms are available from

For the KR-PCA attack, we estimate to $(\log_q \#S_V) \log_2(2\rho_+ +1)$ the number of oracle calls. For the GKZ-based attack, the probability of success is estimated to $\frac{1}{(2\rho_+)^{\#I}} - \frac{1}{q^{\#I}}$. For the AJOP-based attack, the probability of success is $p_{q,c}^{\#J}$.

EMBLEM. EMBLEM-CPA [21] works with $S_A = \mathbb{Z}_q^{m \times n}$, $S_{sk} = \mathbb{Z}_q^{n \times k}$, $S_B = \mathbb{Z}_q^{m \times k}$, $S_t = \mathbb{Z}_q^{v \times m}$, $S_U = \mathbb{Z}_q^{v \times n}$ and $S_V = \mathbb{Z}_q^{v \times k}$. The bilinear mappings are matrix multiplications. The message space is $\{0,1\}^\ell$ and a message is encoded by t-bit chunks. Each block of t-bits is padded with a 1 bit and 0 bits to match a length of $\log_2(q)$ bits. Then, all $\frac{\ell}{t}$ blocks are arranged in a $v \times k$ matrix. Thus, for a message pt, each $\log_2(q)$-bits element of the matrix $M = \mathsf{encode(pt)}$ is $\mathsf{pt}_{i,j} \|1\| 00 \ldots 0$, where $\mathsf{pt}_{i,j}$ is a t-bit block of the original message. Decoding takes the t most significant bits of each element and concatenate them to obtain the original message. Therefore, we have $\rho_- = \rho_+ = q2^{-t-1}$. Components of sk, t are sampled in $[-B, B]$ uniformly at random and components of d, e, f are sampled from the discrete Gaussian distribution on $\mathbb{Z}$ with standard deviation σ. This is similar to Frodo. Hence, we have nk unknowns and each δ gives vk equations. The GKZ-based attack with $\#I = 1$ recovers one column of n unknowns. The AJOP-based attack uses $\varepsilon = 0$, $c = 2^t$, and $\#J = k$. For 128-bit security, the following parameters are used: $m = 1\,003$, $n = 770$, $\ell = 256$, $q = 2^{24}$, $\sigma = 25$, $t = 8$, $B = 1$, v and k can be tuned such that $v \times k \times t = \ell = 256$, typically $v = 32, k = 1$. We compute $p_{q,c} \approx 1$.

R.EMBLEM-CPA is a variant of EMBLEM where the variables are considered as polynomials in X modulo $X^n + 1$ with coefficients in $\mathbb{Z}_q$. It has $S_{sk} = S_A = S_B = S_t = S_U = \mathbb{Z}_q^n$ and $S_V = \mathbb{Z}_q^{\ell/t}$ with L_∞ norm. The bilinear mappings are polynomial multiplications. A message $m \in \{0,1\}^\ell$ is encoded as in EMBLEM-CPA, except that now the $\frac{\ell}{t}$ encoded blocks are polynomial coefficients and not matrix entries. As before, we have $\rho_- = \rho_+ = q2^{-t-1}$. There is a small subtlety at encryption and decryption: since $\mathsf{encode}(m) \in \mathbb{Z}_q^{\ell/t}$, we compute $V = \mathsf{trunc}(t \times B + f, \ell/t) + \mathsf{encode}(m)$ and $W = V - \mathsf{trunc}(U \times sk, \ell/t)$, where $\mathsf{trunc}(x, l)$ takes only the first ℓ components of a vector x. Coefficients of sk, t are sampled in $[-B, B]$ uniformly at random and coefficients of d, e, f are sampled from a discrete Gaussian distribution on $\mathbb{Z}$ with standard deviation σ. For 128-bit security, the following parameters are proposed: $n = 463$, $\ell = 256$, $q = 2^{25}$, $\sigma = 25$, $t = 1$, $B = 1$. We have n unknowns and each δ give them all. The number of oracle calls is about $n(\log_2 q - t)$ in the classical attack. The probability of success in the quantum attack is $\frac{2^t}{q}$ for the GKZ-based one, and $p_{q,c}$ for the AJOP-based one.

KINDI. KINDI-CPA [4] works with the ring $\mathcal{R}_q = \mathbb{Z}_q[X]/(X^n + 1)$. It has $S_A = \mathcal{R}_q^{\ell^2}$, $S_{sk} = S_B = S_t = S_U = \mathcal{R}_q^\ell$ and $S_V = \mathcal{R}_q$. The norm is L_∞. The bilinear mappings are matrix multiplications and scalar product when the elements are vectors, where elements are considered as polynomials in $\mathcal{R}_q$. The public key B is compressed by dropping the k least significant bits of all coefficients.

The encoding of a message pt is more complex than in other LWE schemes. A random polynomial s_1 with binary coefficients is uniformly sampled from $\mathcal{R}_2$. This polynomial is used as a seed for a PRNG function (Shake) that returns a one time-pad $\bar{u}$ and the value t. The message is encrypted into $u = \bar{u} \oplus$ pt by one-time pad and encoded in a value $e \in \mathcal{R}_q^\ell$ and $f \in \mathcal{R}_q$. The ciphertexts are computed as $(U, V) = (t \times A + e, t \times B + f + \mathsf{encode}(s_1))$ where $\mathsf{encode}(s_1) = L \cdot s_1$ with $L = \frac{q}{2}$. Then, the decryption $V - U \times \mathsf{sk}$ recovers s_1 thus t then e and f, then u. The value of s_1 also gives $\bar{u}$ which decrypts u into pt. We have $\rho_- = \rho_+ = \frac{q}{4}$. Elements of A are sampled uniformly at random from $\mathcal{R}_q$, elements of sk, d, t and the one-time pad are sampled uniformly at random from $\mathcal{R}_q$ where the coefficients of the polynomials are in $[-p, p)$ and e, f are derived from the message xored with the one-time pad. For KINDI256-CPA, the parameters used are $n = 256, \ell = 3, p = 4, k = 2, q = 2^{14}$. In our KR-PCA attack, we have to be aware that tampering V results in having junk decryption in the last bits, so we must assume that the PCO oracle ignores those last bits. Adapting the quantum attacks may not be possible because they need s_1 and we cannot recover s_1 from pt. Surprisingly, the decryption in KINDI kindly returns s_1 in addition to the plaintext. So, the quantum attacks work well, with $\varepsilon = 0$, $c = 2$, $\#I = \#J = 1$.

LIMA. LIMA-CPA [23] has $S_{\mathsf{sk}} = S_A = S_B = S_t = S_U = S_V = \mathbb{Z}_q^n$ with the L_∞ norm. Elements are considered as polynomials in $\mathbb{Z}_q[X]/\langle g \rangle$. LIMA-CPA comes in two variants, namely LIMA-2p and LIMA-sp. In LIMA-2p, the polynomial g is $X^n + 1$ with $q \equiv 1 \mod 2n$ and in LIMA-sp, g is a trinomial of degree $n = p - 1$ and p is a safe prime (i.e. $p = 2q + 1$ for a prime q). Each bit of a message is encoded into a 0 or $q/2$. Therefore, we have $\rho_- = \rho_+ = \frac{q}{4}$. The sparse elements sk, d, t, e, f are sampled in $\{-B, \ldots, B\}$ from an approximation of a centered discrete Gaussian distribution of standard deviation $\sigma = \sqrt{(B+1)/2}$. A subtlety is that a pair (t, e) is accepted only if for $y_i = t_i + e_i$, it has

$$\left| \sum_{i=0}^{n-1} y_i \right| \leq 11 \times \sqrt{2 \times n} \times \sigma$$

for LIMA-2p and

$$\left| \sum_{i=0}^{k} y_i + \sum_{i=1}^{n-1} y_i + \sum_{i=k+2}^{n-1} y_i \right| \leq 11 \times \sqrt{4 \times n} \times \sigma$$

for LIMA-sp and any $k \in \{0, \ldots, n-1\}$. For a classical 227-bit security LIMA-2p-CPA, the parameters used are $B = 19, n = 1\,024, q = 133\,121$. For a classical 152-bit security, LIMA-sp-CPA uses $B = 19, n = 1\,018$ and $q = 12\,521\,473$. The quantum attacks work with $c = 2$, $\#I = \#J = 1$, and $p_{q,c} = 41\%$.

Lizard. Lizard-CPA [9] has $S_A = \mathbb{Z}_q^{m \times n}$, $S_{\mathsf{sk}} = \{-1, 0, 1\}^{n \times \ell}$, $S_B = \mathbb{Z}_q^{m \times \ell}$, $S_t = \{-1, 0, 1\}^m$, $S_U = \mathbb{Z}_p^n$, and $S_V = \mathbb{Z}_p^\ell$. The norm is L_∞. Bilinear mappings are matrix multiplications in these structures. Each bit of a message is encoded into 0 or $q/2$ but U, V are scaled by a p/q factor, then pt $\in \{0, 1\}^\ell$ is encoded into 0 or $p/2$. Therefore, we have $\rho_- = \rho_+ = p/4$. Actually, encryption is based on the LWR problem, hence with deterministic e and f. Decryption has form $\mathsf{Dec}(\mathsf{sk}, U, V) = \lceil \frac{2}{p}(V - U \times \mathsf{sk}) \rfloor$, which fits the quantum attacks. Elements of sk are sampled from the distribution $\Pr[x = 1] = \Pr[x = -1] = \gamma/2$, $\Pr[x = 0] = 1 - \gamma$, elements of d are sampled in $\mathbb{Z}_q$ from a discrete Gaussian distribution of parameter $\sigma = \alpha q$, t is sampled uniformly at random in $\{x \in \{-1,0,1\}^m : \mathsf{HW}(x) = h\}$, where $\mathsf{HW}(x)$ counts the number of non-zero elements of x, and e, f are zero. Proposed parameters are $n = 544$, $m = 840$, $q = 1\,024$, $p = 256$, $\ell = 256$, $\gamma = \frac{1}{2}$, $\alpha = \frac{1}{171}$, and $h = 128$. The quantum attacks work with $\varepsilon = 0$, $c = 2$, $\#I = \#J = 1$.

RLizard-CPA is a variant of Lizard which works with rings. It has $S_A = S_B = \mathbb{Z}_q^n$, $S_U = S_V = \mathbb{Z}_p^n$, and $S_{\mathsf{sk}} = S_t = \{-1,0,1\}^n$. Elements are considered as polynomials in these structures and bilinear mappings are polynomial multiplications in the corresponding ring. Messages are encoded similarly as in Lizard-CPA. Elements sk, t are sampled uniformly at random in $\{x \in \{-1,0,1\}^m : \mathsf{HW}(x) = h\}$ with $h = h_{\mathsf{sk}}$ and $h = h_t$, respectively. Coefficients of d are sampled according to a discrete Gaussian distribution of parameter σ in $\mathbb{Z}_q$. Proposed parameters are $n = 1\,024$, $q = 1\,024$, $p = 256$, $\alpha = \frac{1}{154}$ and $h_{\mathsf{sk}} = h_t = 128$.

LOTUS. LOTUS-PKE-CPA [19] is the same as Lindner-Peikert scheme. We have $S_A = \mathbb{Z}_q^{n \times n}$, $S_{\mathsf{sk}} = S_B = \mathbb{Z}_q^{n \times \ell}$, $S_t = S_U = \mathbb{Z}_q^n$, and $S_v = \mathbb{Z}_q^\ell$ with the L_∞ norm. Each bit of a message is multiplied by $\lfloor \frac{q}{2} \rfloor$. Elements of sk, d, t, e, f are sampled from a centered discrete Gaussian distribution of standard deviation σ. Therefore, we have $\rho_+ = \rho_- = \lfloor \frac{q}{4} \rfloor$. For LOTUS128-CPA, we have $n = 576$, $q = 8\,192$, $\ell = 128$, $\sigma = 3$. For key recovery, we have $n \times \ell$ unknowns and ℓ equations for each sample δ_i, hence we need n samples. The quantum attacks work with $\varepsilon = 0$, $c = 2$, $\#I = \#J = 1$.

Titanium. Let $\mathcal{R}_{q,n}$ be the set of polynomials in X with degree less than n and coefficients in $\mathbb{Z}_q$. Titanium has $S_A = \mathcal{R}_{q,n}^m$, $S_{\mathsf{sk}} = \mathcal{R}_{q,n+d+k-1}$, $S_B = \mathcal{R}_{q,d+k}^m$, $S_t = \mathcal{R}_{q,k+1}^m$, $S_U = \mathcal{R}_{q,n+k}$ and $S_V = \mathcal{R}_{q,d}$ with the L_∞ norm. The bilinear mappings use the middle product $\odot$ defined as follows: Let $a \in \mathcal{R}_{q,d_a}$ and $b \in \mathcal{R}_{q,d_b}$ s.t. $d_a + d_b - 1 = d + 2k$ for some integers d_a, d_b, d, k. The middle product $\odot_d : \mathcal{R}_{q,d_a} \times \mathcal{R}_{q,d_b} \to \mathcal{R}_{q,d}$ is the map

$$a \odot_d b = \left\lfloor \frac{(a \times b) \mod X^{k+d}}{X^k} \right\rfloor$$

i.e. we take the d terms of $a \times b$ of degree $k, k+1, \ldots, k+d-1$ and divide by X^k. Titanium extends it to vector multiplication as the dot product with $\odot_d$ for component multiplications and to polynomial-vector multiplication as the component-wise middle product with the polynomial. All bilinear mappings are middle products as described above, except for the $S_t \times S_A \to S_U$, which is the dot product with polynomial multiplication in $\mathbb{Z}_q[X]$. A message pt is encoded as a polynomial in $\mathcal{R}_{2,d}$ with each coefficient scaled by $\lfloor \frac{q}{p} \rfloor$. Therefore, we have $\rho_- = \rho_+ = \lfloor \frac{q}{p} \rfloor / 2$. The secret key sk is sampled uniformly at random in S_t and d is sampled by taking the difference of the Hamming weight of two uniformly distributed η-bits values, this approximates a discrete Gaussian distribution. For t, $N_t = (k+1) \times m$ coefficients need to be sampled in $\mathbb{Z}_q$. In order to tune the variance, N_1 of them are sampled uniformly in $\{-B_1/2, \ldots, B_1/2\} \setminus \{0\}$ and $N_t - N_1$ of them are sampled uniformly in $\{-B_2/2, \ldots, B_2/2\} \setminus \{0\}$. The elements e, f are null. For TitaniumStd128-CPA [24] with NIST security level I, the parameters are $n = 1\,024$, $k = 511$, $d = 256$, $m = 9$, $q = 86\,017$, $p = 2$, $\eta = 4$, $N_1 = 3\,816$, $B_1 = 2^6$, $B_2 = 2^7$. The quantum attacks work with $c = p$ and $\#I = \#J = 1$.

References

1. Alkim, E., Ducas, L., Pöppelmann, T., Schwabe, P.: Post-quantum Key Exchange - A New Hope. https://eprint.iacr.org/2015/1092
2. Alagic, G., Jeffery, S., Ozols, M., Poremba, A.: On Quantum Chosen Ciphertext Attacks and Learning with Errors. https://eprint.iacr.org/2018/1185
3. Ambainis, A., Magnin, L., Roetteler, M., Roland, J.: Symmetry-Assisted Adversaries for Quantum State Generation. CoRR, vol. abs/1012.2112 (2010). https://arxiv.org/pdf/1012.2112.pdf
4. El Bansarkhani, R.: Kindi. http://kindi-kem.de/
5. Bauer, A., Gilbert, H., Renault, G., Rossi, M.: Assessment of the key-reuse resience of NewHope. In: Matsui, M. (ed.) CT-RSA 2019. LNCS, vol. 11405, pp. 272–292. Springer, Cham (2019). https://doi.org/10.1007/978-3-030-12612-4_14. https://eprint.iacr.org/2019/075
6. Bos, J., et al.: CRYSTALS-Kyber: a CCA-secure module-lattice-based KEM. In: IEEE European Symposium on Security and Privacy EuroS&P'2018, London, UK, pp. 353–367. IEEE (2018) https://eprint.iacr.org/2017/634
7. Bernstein, E., Vazirani, U.: Quantum complexity theory. SIAM J. Comput. **26**(5), 1411–1473 (1997)
8. Bos, J.W., et al.: Frodo: take off the ring! Practical, quantum-secure key exchange from LWE. In: 23rd ACM Conference on Computer and Communications Security, Vienna, Austria, pp. 1006–1018. ACM Press (2016). https://eprint.iacr.org/2016/659
9. Cheon, J.H., Kim, D., Lee, J., Song, Y.: Lizard: cut off the tail! A practical post-quantum public-key encryption from LWE and LWR. In: Catalano, D., De Prisco, R. (eds.) SCN 2018. LNCS, vol. 11035, pp. 160–177. Springer, Cham (2018). https://doi.org/10.1007/978-3-319-98113-0_9. https://eprint.iacr.org/2016/1126
10. Ding, J., Alsayigh, S., Saraswathy, R.V., Fluhrer, S., Lin, X.: Leakage of signal function with reuse keys in RLWE key exchange. In: IEEE International Conference on Communications ICC 2017, Paris, France, pp. 1–6. IEEE (2017)
11. Fluhrer, S.: Cryptanalysis of Ring-LWE Based Key Exchange with Key Share Reuse. https://eprint.iacr.org/2016/085
12. Fujisaki, E., Okamoto, T.: Secure integration of asymmetric and symmetric encryption schemes. In: Wiener, M. (ed.) CRYPTO 1999. LNCS, vol. 1666, pp. 537–554. Springer, Heidelberg (1999). https://doi.org/10.1007/3-540-48405-1_34
13. Fujisaki, E., Okamoto, T.: J. Cryptol. **26**, 80–101 (2013)
14. Grilo, A.B., Kerenidis, I., Zijlstra, T.: Learning with Errors is Easy with Quantum Samples. CoRR, vol. abs/1702.08255 (2017). https://arxiv.org/pdf/1702.08255.pdf

15. Hofheinz, D., Hövelmanns, K., Kiltz, E.: A modular analysis of the Fujisaki-Okamoto transformation. In: Kalai, Y., Reyzin, L. (eds.) TCC 2017. LNCS, vol. 10677, pp. 341–371. Springer, Cham (2017). https://doi.org/10.1007/978-3-319-70500-2_12. https://eprint.iacr.org/2017/604

16. Kirkwood, D., Lackey, B.C., McVey, J., Motley, M., Solinas, J.A., Tuller, D.: Failure is not an option: standardization issues for post-quantum key agreement. Presented at the NIST Workshop on Cybersecurity in a Post-Quantum World (2015). https://www.nist.gov/news-events/events/2015/04/workshop-cybersecurity-post-quantum-world. https://csrc.nist.gov/csrc/media/events/workshop-on-cybersecurity-in-a-post-quantum-world/documents/presentations/session7-motley-mark.pdf

17. Lepoint, T.: Algorithmic of LWE-Based Submissions to NIST Post-Quantum Standardization Effort. Presented at the Post-Scryptum Spring School (2018). https://postscryptum.lip6.fr/. https://postscryptum.lip6.fr/tancrede.pdf

18. Lyubashevsky, V., Peikert, C., Regev, O.: On ideal lattices and learning with errors over rings. In: Gilbert, H. (ed.) EUROCRYPT 2010. LNCS, vol. 6110, pp. 1–23. Springer, Heidelberg (2010). https://doi.org/10.1007/978-3-642-13190-5_1

19. Phong, L.T., Hayashi, T., Aono, Y., Moriai, S.: LOTUS. https://www2.nict.go.jp/security/lotus/index.html

20. Regev, O.: On lattices, learning with errors, random linear codes, and cryptography. J. ACM **56**(6), 34 (2009)

21. Seo, M., Park, J.H., Lee, D.H., Kim, S., Lee, S.-J.: Emblem. https://pqc-emblem.org

22. Shor, P.W.: Algorithms for quantum computation: discrete logarithms and factoring. In: Proceedings of the 35th IEEE Symposium on Foundations of Computer Science, Santa Fe, New Mexico, USA, pp. 124–134. IEEE (1994)

23. Smart, N.P., et al.: Lima 1.1: a PQC Encryption Scheme. https://lima-pq.github.io

24. Steinfeld, R., Sakzad, A., Zhao, R.K.: Titanium. http://users.monash.edu.au/~rste/Titanium.html

25. Targhi, E.E., Unruh, D.: Post-quantum security of the Fujisaki-Okamoto and OAEP transforms. In: Hirt, M., Smith, A. (eds.) TCC 2016. LNCS, vol. 9986, pp. 192–216. Springer, Heidelberg (2016). https://doi.org/10.1007/978-3-662-53644-5_8. https://eprint.iacr.org/2015/1210

26. Yu, Y., Zhang, J.: Lepton: Key Encapsulation Mechanisms from a Variant of Learning Parity with Noise. NIST Round 1 submission to Post-Quantum Cryptography (2017). https://csrc.nist.gov/projects/post-quantum-cryptography/round-1-submissions

Author Index

Abusalah, Hamza II-277
Agarwal, Navneet II-381
Aggarwal, Divesh I-531, II-442
Agrawal, Shweta I-191
Aharonov, Dorit III-219
Alamati, Navid II-55
Albrecht, Martin R. II-717
Alwen, Joël I-129
Anand, Sanat II-381
Ananth, Prabhanjan II-532
Applebaum, Benny II-504, III-441
Aragon, Nicolas III-728
Asharov, Gilad II-214
Attrapadung, Nuttapong I-34
Aviram, Nimrod II-117

Backes, Michael III-281
Badrinarayanan, Saikrishna I-593
Băetu, Ciprian II-747
Ball, Marshall I-501
Barak, Boaz I-226
Bar-On, Achiya I-313
Bartusek, James III-636
Beimel, Amos III-441
Ben-Sasson, Eli I-103
Bernstein, Daniel J. II-409
Bitansky, Nir III-667
Blazy, Olivier III-728
Boyle, Elette II-3
Brakerski, Zvika II-504, III-219, III-619

Canteaut, Anne III-585
Chen, Hao II-34
Cheu, Albert I-375
Chiesa, Alessandro I-103
Chillotti, Ilaria II-34
Choudhuri, Arka Rai II-351, II-532
Chung, Kai-Min II-442, III-219
Coladangelo, Andrea III-247
Coretti, Sandro I-129
Couteau, Geoffroy II-473, II-562

Dachman-Soled, Dana I-501
De Feo, Luca III-759

Dinur, Itai I-343, III-699
Dodis, Yevgeniy I-129
Döttling, Nico I-531, II-292, III-281
Ducas, Léo II-717
Dunkelman, Orr I-313
Durak, F. Betül II-747
Dutta, Avijit I-437
Dziembowski, Stefan I-625

Eckey, Lisa I-625

Farràs, Oriol III-441
Faust, Sebastian I-625
Fehr, Serge III-472
Fisch, Ben II-324
Fuchsbauer, Georg I-657

Gaborit, Philippe III-728
Galbraith, Steven D. III-759
Ganesh, Chaya I-690
Garg, Sanjam III-33
Gay, Romain III-33
Gellert, Kai II-117
Genise, Nicholas II-655
Ghosh, Satrajit III-154
Goel, Aarushi II-532
Goyal, Vipul I-562, II-351
Green, Ayal III-219
Grilo, Alex B. III-247
Guan, Jiaxin III-500

Haitner, Iftach III-667
Hajiabadi, Mohammad III-33
Hamlin, Ariel II-244
Hanrot, Guillaume II-685
Hanzlik, Lucjan III-281
Hauck, Eduard III-345
Hauteville, Adrien III-728
Herold, Gottfried II-717
Hesse, Julia I-625
Hoang, Viet Tung II-85
Hofheinz, Dennis II-562
Hong, Cheng III-97
Hopkins, Samuel B. I-226

Hostáková, Kristina I-625
Hubert Chan, T.-H.   I-720, II-214
Huguenin-Dumittan, Loïs   II-747

Jaeger, Joseph I-467
Jager, Tibor II-117
Jain, Aayush I-226, I-251
Jain, Abhishek II-351, II-532
Jeffery, Stacey III-247
Jost, Daniel I-159

Kales, Daniel I-343
Kamara, Seny II-183
Kamath, Chethan II-277
Katsumata, Shuichi II-622, III-312
Katz, Jonathan III-97
Keller, Nathan I-313
Kiltz, Eike III-345
Kirshanova, Elena II-717
Klein, Karen II-277
Klooß, Michael I-68
Kluczniak, Kamil III-281
Kohl, Lisa II-3
Kolesnikov, Vladimir III-97
Kölsch, Lukas I-285
Komargodski, Ilan III-667
Kothari, Pravesh I-226
Kowalczyk, Lucas I-3
Kulkarni, Mukul I-501

Lai, Ching-Yi III-219
Lai, Russell W. F.   II-292
Lallemand, Virginie III-585
Lange, Tanja II-409
Leander, Gregor III-585
Lehmann, Anja I-68
Lepoint, Tancrède III-636
Leurent, Gaëtan III-527
Li, Ting III-556
Lin, Han-Hsuan II-442
Lin, Huijia I-251, I-501
Liu, Qipeng III-189
Loss, Julian III-345
Lu, Wen-jie III-97
Lyubashevsky, Vadim III-619

Ma, Fermi III-636
Malavolta, Giulio II-292
Malkin, Tal I-501

Martindale, Chloe II-409
Matt, Christian I-251
Maurer, Ueli I-159
Micciancio, Daniele II-655, III-64
Miller, David II-85
Moataz, Tarik II-183
Montgomery, Hart II-55
Mularczyk, Marta I-159

Nadler, Niv III-699
Nandi, Mridul I-437
Nayak, Kartik II-214
Neumann, Patrick III-585
Nielsen, Jesper Buus I-531
Nilges, Tobias III-154
Nir, Oded III-441
Nishimaki, Ryo II-622

Obremski, Maciej I-531
Orlandi, Claudio I-690
Orrù, Michele I-657
Ostrovsky, Rafail II-244

Panny, Lorenz II-409
Pass, Rafael I-720, II-214
Patranabis, Sikhar II-55
Pellet-Mary, Alice II-685
Persiano, Giuseppe I-404
Peter, Naty III-441
Peyrin, Thomas III-527
Pietrzak, Krzysztof II-277
Pinkas, Benny III-122
Polyakov, Yuriy II-655
Postlethwaite, Eamonn W.   II-717
Prabhakaran, Manoj II-381
Promitzer, Angela I-343
Purwanto, Erick I-531

Quach, Willy II-593

Ramacher, Sebastian I-343
Rechberger, Christian I-343
Ren, Ling II-214
Riabzev, Michael I-103
Rothblum, Ron D.   II-593
Roy, Arnab II-55
Rupp, Andy I-68

Sahai, Amit I-226, I-251
Sattath, Or III-219

Schneider, Jonas   III-281
Schneider, Thomas   III-122
Scholl, Peter   II-3
Seurin, Yannick   I-657
Shi, Elaine   I-720, II-214
Shumow, Dan   II-151
Smith, Adam   I-375
Song, Yifan   I-562
Song, Yongsoo   II-34
Spooner, Nicholas   I-103
Srinivasan, Akshayaram   I-593
Stehlé, Damien   II-685
Stevens, Marc   II-717
Sun, Yao   III-556

Talayhan, Abdullah   II-747
Talnikar, Suprita   I-437
Tessaro, Stefano   I-467
Tkachenko, Oleksandr   III-122
Trieu, Ni   II-85
Tsabary, Rotem   II-504
Tschudi, Daniel   I-690

Ullman, Jonathan   I-375

Vaikuntanathan, Vinod   III-619
Vaudenay, Serge   II-747

Vidick, Thomas   II-442, III-247
Virza, Madars   I-103

Walter, Michael   II-277
Wang, Xiao   III-97
Ward, Nicholas P.   I-103
Wee, Hoeteck   I-3
Weiss, Mor   II-244
Weizman, Ariel   I-313
Wesolowski, Benjamin   III-379
Wichs, Daniel   II-244, II-593, III-619
Wiemer, Friedrich   III-585
Woodage, Joanne   II-151

Yamada, Shota   II-622, III-312
Yamakawa, Takashi   II-622
Yanai, Avishay   III-122
Yeo, Kevin   I-404
Yogev, Eylon   III-667
Yuan, Chen   III-472

Zeber, David   I-375
Zémor, Gilles   III-728
Zhandary, Mark   III-500
Zhandry, Mark   III-3, III-189, III-408, III-636
Zhilyaev, Maxim   I-375

Printed in the United States
By Bookmasters